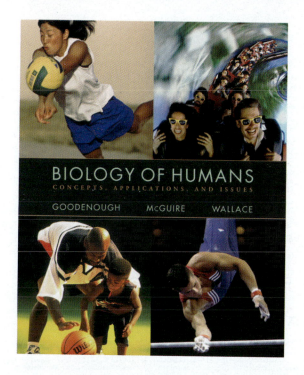

BIOLOGY OF HUMANS: CONCEPTS, APPLICATIONS, AND ISSUES is designed to motivate students to learn basic biological concepts. Knowing that students have a deep interest in their own bodies, the authors develop these concepts with human and personal applications keenly in mind. In addition, they continually focus on how human biology relates to health, social, and environmental issues to motivate a broader understanding of the subject.

Students are interested in learning, especially when topics are relevant to their lives. *Biology of Humans* sets forth three major goals: (1) Help students understand basic biological concepts; (2) Apply basic concepts in interesting ways that will benefit readers; (3) Equip students with reasoning skills developed in the context of biology that can be applied to choices in other areas of their lives. To achieve these goals the following features are deployed throughout the book.

WRITING AND CONTEXT

- Accessible writing that smoothly integrates just the right amount of detail with intriguing applications keeps students engaged
- Topic heads are written as declarative sentences to give students an overview and framework for review and understanding
- Superb chapter summaries guide students to understand major ideas
- Select chapters end with an *exploring further . . .* box that makes connections between two or more chapters

APPLICATIONS

- Six *Special Topic* Chapters highlight important areas that warrant detailed explanation and exploration, but remain optional to instructors
- Interesting applications are woven into the text at every opportunity
- Contemporary, important issues are addressed in special boxes entitled *Health Issues, Social Issues,* and *Environmental Issues*
- Each chapter ends with questions focused on *Applying the Concepts*

REASONING SKILLS

- Periodic interludes entitled *stop and think* and *what would you do?* ask students to reflect and reason as well as to consider bioethical issues
- Most boxes end with provocative questions that invite students to ponder broader aspects of an issue
- *Applying the Concepts* questions invite critical thinking

GRAPHICS

- The illustration program anticipates students' questions and guides them through concepts and processes
- Visually compelling art seizes students' interest
- "Overview" figures, as in the digestive system, recur to give students context for detailed illustrations
- Imaginative combinations of illustations and photomicrographs connect the abstract to the real

"The explanation of complex topics and principles was, I thought, masterfully done."

—JOHN RINEHART
EASTERN OREGON UNIVERSITY

"I think the writing style is perfect for the students I teach. It is a friendly, informative style. I think it makes the concepts and processes quite clear."

—SUSAN BEAR
ST. JOHN'S UNIVERSITY

"It is as if Goodenough is explaining the concepts to a group of friends, always including the interesting issues associated with each of the science topics. I know from 23 years of experience that that approach will be very much appreciated by the students."

—PAUL GANNON
EL PASO COMMUNITY COLLEGE

When a blood vessel is cut, a series of reactions stops blood flow. These reactions are similar to those you might initiate if the garden hose you are using springs a leak. Your initial response might be to squeeze the hose, in hopes of stopping water flow. Likewise, the body's immediate response to blood vessel injury is for the vessel to constrict (squeeze shut). This constriction of the injured vessel is caused by direct injury to the muscles lining the blood vessel and by chemicals released from the injured tissue.

The next response is to plug the hole (Figure 11–9). Your thumb might do the job on your garden hose; platelets form the plug that seals the leak in a vessel. The **platelet plug** is formed when platelets cling to cables of collagen, a protein fiber, in the exposed blood vessel surface. When the platelets attach to collagen, they become activated and change in several ways: They swell, form many cellular extensions, and stick together. Platelets produce a chemical called thromboxane that makes platelets stick to one another. Thromboxane also attracts other platelets to the wound. Aspirin prevents the formation of thromboxane and, therefore, inhibits clot formation, explaining how a daily dose of aspirin may prevent the formation of blood clots that can block blood vessels nourishing heart tissue, resulting in the death of heart cells (a heart attack). It also explains how aspirin can cause excessive bleeding.

"The writing style is fit to our students: simple, direct, and easy to understand. I like it. It is very important to have a writing style that can motivate our students to read their text."

—LEE. H. LEE
MONTCLAIR STATE UNIVERSITY

NATURAL KILLER CELLS A third type of white blood cell, called natural killer (NK) cells, roams the body in search of abnormal cells, which are then quickly killed. In a sense, NK cells function as the body's police walking a beat. They are not seeking a specific villain. Instead, they respond to any suspicious character, which, in this case, is a cell whose cell membrane has been altered by the addition of proteins that are unfamiliar to the NK cell. The prime targets of NK cells are cancerous cells and cells infected with viruses. Cancerous cells routinely form but are quickly destroyed by NK cells and prevented from spreading (Figure 13.5).

"The first person/story-telling style invites the reader to continue reading in the book. It provides a very friendly atmosphere for learning biology."

—CLARE HAYS
METROPOLITAN STATE UNIVERSITY

The coordination of sensory input and motor output by the cerebellum involves two important processes: comparison and prediction. During every move you make, the cerebellum continuously compares the actual position of each part of the body with where it *ought* to be at that moment (with regard to the intended movement) and makes the necessary corrections. Try to touch the tips of your two index fingers together above your head. You probably missed on the first attempt. However, the cerebellum makes the necessary corrections, and you will likely succeed on the next attempt. At the same time, the cerebellum calculates future positions of a body part during a movement. Then, just before that part reaches the intended position, the cerebellum sends messages to stop the movement at a specific point. Therefore, when you scratch an itch on your cheek, the hand stops before slapping your face!

"The writing is not overly technical or difficult. It is easy to read and follow. None of the sections are overly long. Care has been taken to illustrate the more difficult/abstract concepts in more easily understood terms."

—MARY JO KENYON
NORTH DAKOTA STATE UNIVERSITY

"There is a casual, almost conversational style in many places. This is great, it is more welcoming to the student for whom science is not their life's work."

—DONALD GLASSMAN
DES MOINES AREA COMMUNITY COLLEGE

"The writing style is engaging, and I believe my students would benefit from using a text like this one. It makes human biology approachable (the terms are clearly defined, concepts clearly presented, with an excellent summary at the end of the chapter)."

—MICHAEL P. FRANKLIN
CALIFORNIA STATE UNIVERSITY, BAKERSFIELD

1. **An adjustable entrance to the respiratory system.** The larynx provides an opening to the respiratory system that can be adjusted; that is, it can be opened to allow air to pass into the lungs and closed to prevent other matter, such as food, from entering the lungs. Because the esophagus (the tube leading to the stomach) is behind the larynx, material must pass over the opening to the respiratory system to reach the digestive system. If solid material such as food enters the respiratory system, it could lodge in one of the tubes conducting air to the lungs and prevent air flow. Fluid entering the lungs is equally dangerous because it can cover the respiratory surfaces, decreasing the area of gas exchange. Normally, foreign material is prevented from entering the lower respiratory system during swallowing because the larynx rises and causes a flap of cartilage called the **epiglottis** to move downward and form a lid over the glottis. In this way, the epiglottis covers the **glottis**, the opening in the larynx through which air passes, and prevents food from entering the lungs. You can feel the movement of the larynx if you put your fingers on your Adam's apple while swallowing. Because of this movement, you cannot breathe and swallow at the same time. (Try it!)

Clear and Engaging Writing Motivates Students

15a

SPECIAL TOPIC

Nutrition and Weight Control

The Food Guide Pyramid Helps Us Plan a Healthy Diet

Nutrients Provide Energy or Have a Structural or Functional Role in the Body
- Lipids include fats, oils, and cholesterol
- Carbohydrates include sugars, starches, and dietary fiber
- Proteins are chains of amino acids
- Vitamins are needed in small amounts to promote and regulate the body's chemical reactions
- Minerals play structural and functional roles in the body
- Water is essential and needed in large amounts

Body Energy Balance Depends on Calories Gained in Food and Calories Used

Obesity Is Body Weight 20 Percent or More above the Body Weight Standard

Dietary Guidelines for Americans Promote Health

Successful Weight-Loss Programs Usually Involve Reducing Calorie Intake, Increasing Calorie Use, and Changing Behavior

Anorexia Nervosa and Bulimia Are Eating Disorders That Create Calorie Deficits

6 SPECIAL TOPIC CHAPTERS focus on topics of high student interest.

exploring further ...

In Chapter 7, we learned that neurons communic using chemicals called neurotransmitters. Neurotr into receptors on the membrane of the receiving channels. Depending on the type of channel oper is either excited or inhibited. Different neurotrans ent behavioral systems. Norepinephrine and sero well-being. Dopamine helps regulate emotions.

In this chapter, we learned that different parts o specialized for different functions. The limbic syst "pleasure center." The sympathetic nervous syst emergency situations. To bring about its effects sympathetic nervous system releases the neurotr

Next, in Chapter 8A, Special Topic: Drugs and th psychoactive drugs—those that affect one's men psychoactive drugs bring about their effects by i effects of specific neurotransmitters. In this way, specific regions of the brain and alter one's ment effects on the nervous system, drugs can affect o also consider the health risks associated with dru

THESE BOXES connect a chapter to preceding and following chapters.

Shortness of breath, the main symptom of emphysema, has several causes. Two causes are the decreased surface area for gas exchange and increased dead air space. As the disease progresses, gas exchange becomes even more difficult because the alveolar walls thicken with fibrous connective tissue. The oxygen that does make it to the alveoli has difficulty crossing the connective tissue to enter the blood. Thus, a person with emphysema constantly gasps for air.

APPLICATIONS are woven into the text more than any other textbook.

SOCIAL ISSUE

Building Muscle, Fair and Square? Ana

We have always appreciated strength, probably because it's associated with a "can-do" ability. Of course, such appreciation has led us not only to strength itself but also to the display of strength, that is, bodybuilding. Unfortunately, some people are always looking for shortcuts.

One of the biggest dangers to young

ENVIRONMENTAL ISSUE

Noise Pollution

It is difficult to escape the din of modern life—noise from airports, city streets, loud appliances, stereos. Noise poll...

...your hearing ...ssive noise is ...ne-third of all ... on the hair ...

damaging hair cells. More commonly, however, hearing loss results from prolonged exposure to volumes over 55 dB. Hearing can be damaged by exposure to noise loud enough to make it difficult to converse with someone. Some damage probably occurred if sounds seem muffled after you leave the noisy area. Your ears can endure sound at 90 dB for about 8 hours. For every 5 dB above that, it takes half as long for damage to begin. Thus, soun...

HEALTH ISSUE

Meningitis: Bacterial and West Nile

Meningitis is an inflammation of the meninges, the protective coverings of the brain and spinal cord. Many types of bacteria and certain viruses can cause it. Diagnosis is usually done by studying the cerebrospinal fluid. If bacteria are the cause, the person is treated with antibiotics. The treatment for viral meningitis includes med...

symptoms may develop within hours or take several days. Not all symptoms may be present. bacteria may spread to the blood, causing blood poisoning (septicemia), which may cause a skin rash that can start anywhere on the body. The rash initially looks like a cluster of tiny pinpricks. However, if they are not treated with antibiotics, the spots will join and look like fresh bruises. The bacteria can also cause gangrene, whic...

ISSUES BOXES are designed to capture student attention and to get them to connect science to broader realms.

APPLYING THE CONCEPTS

1. Joe's kidneys are failing, and he needs a kidney transplant. Luckily, he has an identical twin who wishes to donate one of his kidneys. Why would a kidney from his identical twin be better than one from another family member or a complete stranger? Explain your answer with reference to the plasma membrane of Joe's cells.

2. Yo...
th...
qu...

3. W...
tes...

EACH CHAPTER ENDS WITH QUESTIONS
that ask students to apply concepts.

Rich in Applications and Connections

Art development editors and authors evaluated each illustration to determine how it could best be rendered to teach the concept or process at hand. A color palette was selected for its visual impact as well as its ability to provide contrast, and ultimately to project well as PowerPoint slides or acetates. Colors consistently code for recurring elements in the art program. Photomicrographs complement figures where appropriate.

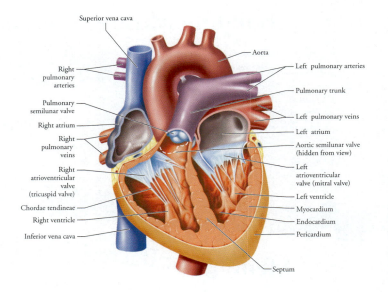

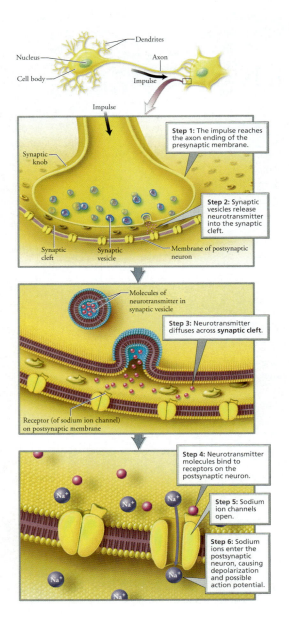

"The artwork provided is probably the best I have seen….they provide a clear, crisp view with a very striking 3-D effect. The colors are subtle, but provide a nice impression. Wonderfully done."

—WILLIAM C. KLEINELP
MIDDLESEX COMMUNITY COLLEGE

"The art pieces are excellent."

—STEVE STOCKING
SAN JOAQUIN DELTA COLLEGE

"The images are nice and sharp, with good attention to detail, which makes them visually pleasing and easy to understand."

—MARY JO KENYON
NORTH DAKOTA STATE UNIVERSITY

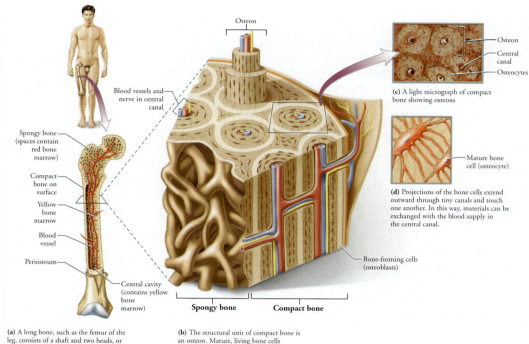

Osteon

Blood vessels and nerve in central canal

Spongy bone (spaces contain red bone marrow)

Compact bone on surface

Yellow bone marrow

Blood vessel

Periosteum

Central cavity (contains yellow bone marrow)

Spongy bone

Compact bone

Osteon
Central canal
Osteocytes

(c) A light micrograph of compact bone showing osteons

Mature bone cell (osteocyte)

(d) Projections of the bone cells extend outward through tiny canals and touch one another. In this way, materials can be exchanged with the blood supply in the central canal.

Bone-forming cells (osteoblasts)

(a) A long bone, such as the femur of the leg, consists of a shaft and two heads, or enlarged ends. Compact bone is located on the outer surface of the bone. Spongy bone is found in the heads.

(b) The structural unit of compact bone is an osteon. Mature, living bone cells (osteocytes) are found in small spaces within the hard matrix.

Art That Teaches

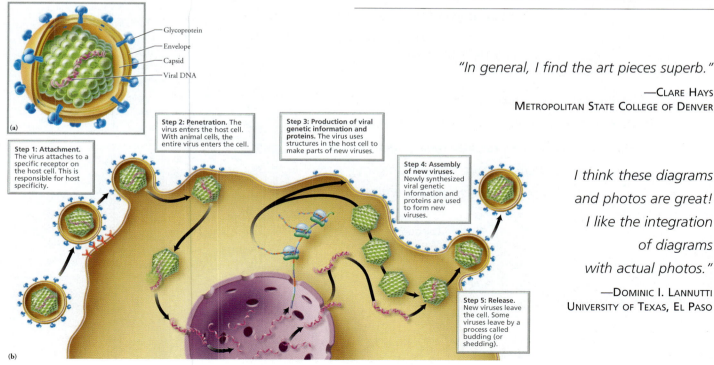

Glycoprotein
Envelope
Capsid
Viral DNA

(a)

Step 1: Attachment. The virus attaches to a specific receptor on the host cell. This is responsible for host specificity.

Step 2: Penetration. The virus enters the host cell. With animal cells, the entire virus enters the cell.

Step 3: Production of viral genetic information and proteins. The virus uses structures in the host cell to make parts of new viruses.

Step 4: Assembly of new viruses. Newly synthesized viral genetic information and proteins are used to form new viruses.

Step 5: Release. New viruses leave the cell. Some viruses leave by a process called budding (or shedding).

(b)

stop and think

Women who have just given birth are often encouraged to nurse their babies as soon as possible after delivery. How might an infant's suckling promote completion of, and recovery from, the birth process? Consider that the placenta (afterbirth) must still be expelled after the birth of the baby and that the uterus must return to an approximation of its prepregnancy form.

what would you do?

Imagine that you and your spouse desperately want children. However, both of you are carriers Tay-Sachs disease and thus could pass the gene to your children. The options for any child that you might have are as follows: Your child may not have the gene for Tay-Sachs and may be healthy, may have the disease and die in early childhood, or may be a carrier as you are. Your parents urge adoption over having your own children. Your spouse prefers not to adopt but to use prenatal screening to check if your fetus has the disease. What would you do?

"I like the approach of Goodenough et al—presenting, then applying the concepts; prodding students to "Stop and Think;" and emphasizing that the choices they make today have far-reaching effects that go beyond themselves."

—Nancy Pencoe
State University of West Georgia

"I really like the presentation style of this text. It has numerous side-bars and thought questions that will provoke students into really digging in to the material and, more importantly, its relevance to their own lives."

—John Rinehart
Eastern Oregon University

but instead be used by employers as well as by life and health insurers. As an employer, if you had information about the genetic makeup of prospective employees, would you choose to invest time and money in training a person who carried an allele that increased the risk of cancer, heart disease, Alzheimer's disease, or alcoholism? As an insurer, would you knowingly cover a carrier?

The results of gene testing can have both positive and negative consequences for those being tested and for their families: Who should decide whether screening should be done, for which genes, on whom, and in which communities? A flippant answer might be, "There oughta be a law!" Should we leave ethical issues to judges and legislators? Should moral matters be decided by society or clergy? Or should they be personal decisions?

If gene testing is done, who should be told the results? If the affected person is an infant, should the parents always be told the results, even if the condition is poorly understood? How do we balance helping such children with the possibility of stigmatizing them?

We live in a world of limited resources. Thus, as soon as it is decided who should be tested, we must decide who should pay the bill. Both testing and treatment are expensive. Should testing be done only when treatment or preventive measures are available? How much say should the agent that pays for the procedure have in who is tested and who receives medical treatment?

There are no easy answers. It is time for each of us to think about the issues raised by these and similar questions. ✦

Critical Thinking and Reasoning Skills

OneKey

OneKey offers the best teaching and learning resources all in one place. Conveniently organized by textbook chapter, these compiled resources help you save time and help your students reinforce and apply what they have learned in class. The student media program in OneKey includes Web Tutorials, Practice Tests, Electronic Flash Cards and Vocabulary Quizzes, and more.

OneKey is available in Blackboard, WebCT, and CourseCompass formats to serve all of your course management needs.

Companion Web Site: Student Web Tutorials

www.prenhall.com/goodenough

Web Tutorials, available on the student Companion Web Site or in OneKey, review concepts from the text in an engaging tutorial format. 83 Tutorials address important concepts through hundreds of animations, interactive activities, and review quizzes that enable students to master the most challenging topics. Icons have been placed throughout the book to indicate that relevant Web Tutorials are available.

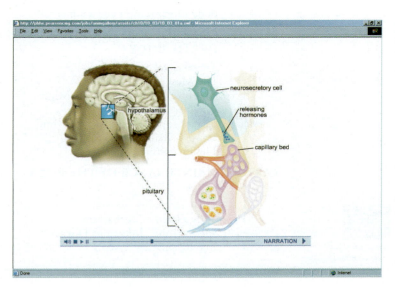

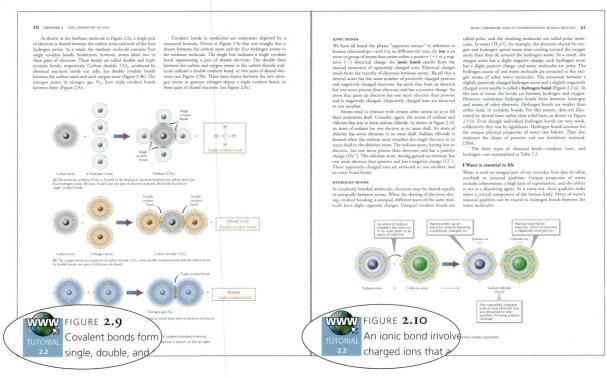

WWW FIGURE 2.9
TUTORIAL 2.2
Covalent bonds form single, double, and

WWW FIGURE 2.10
TUTORIAL 2.2
An ionic bond involve charged ions that a

NEXT-GENERATION MEDIA

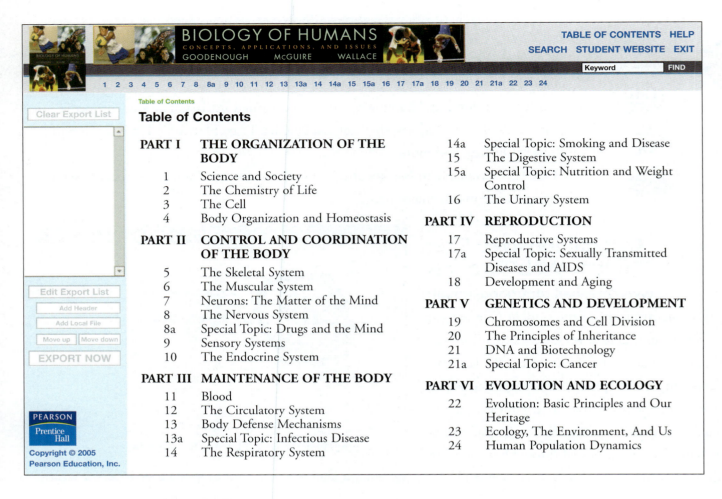

Clear Export List

Edit Export List
Add Header
Add Local File
Move up   Move down
EXPORT NOW

PEARSON
Prentice Hall

Copyright © 2005
Pearson Education, Inc.

For Instructors

The new Instructor Resource Center on CD provides instructors with one comprehensive, easy-to-use, electronic resource that will complement any teaching style. Everything you need to prepare and present a lecture is included on one CD. All of these resources are also available on-line for you to access through OneKey.

- Create Presentations
- Customize Instructor Resources
- Create Tests
- Access Student Resources All In One Place

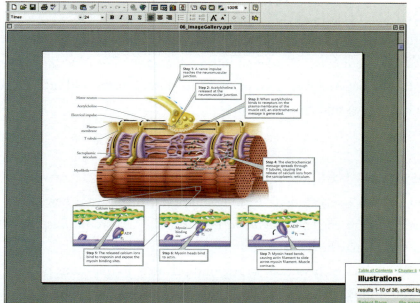

PowerPoint® Slides

Ready-made PowerPoint slides with embedded illustrations, lecture outlines, and animations are available for each chapter. All illustrations are available in both labeled and unlabeled formats.

Image Gallery

A complete image gallery with labeled and unlabeled illustrations gives you flexibility in creating custom, high-resolution presentations.

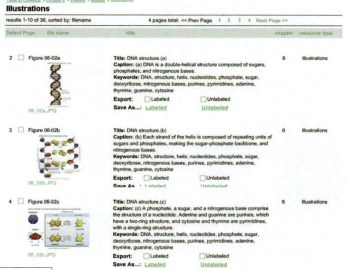

Instructor's Manual and Test Item File

Electronic versions of the Instructor's Manual and Test Item File allow for complete customization.

All student resources are also available on the CD for presentation purposes.

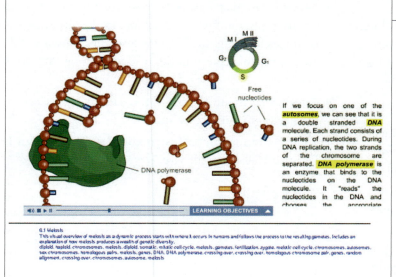

Research Navigator

Give students access to the most current information available via EBSCO's Content Select Academic Journal Database, *The New York Times* Search by Subject Archive, "Best of the Web" Link Library and information on the latest news and current events.

Classroom Response System

Do you need to assess student learning quickly? Do you want new and effective means to engage students in Lecture? The Classroom Response System (CRS) may be just the key.

CRS enables each student to respond privately to questions posed, giving you immediate feedback that will provide valuable insight into student learning. Multiple-choice questions will be available electronically for use in several of the most popular CRS systems.

Instructors 1st

For qualified adopters, Prentice Hall is proud to introduce **Instructors 1st**— the first integrated service committed to meeting your customization, support, and training needs for your course.

For your customization needs:

- A Prentice Hall Technology Consultant and a Custom Editor from our sister company, Pearson Custom Publishing, to work with you to ensure your course materials are tailored to meet your specific needs by customizing and integrating any print or multimedia products, within our capabilities.

For your support needs:

- One dedicated 800 number and e-mail address for trained technical support for faculty registered in **Instructors 1st:** 887-547-4538, M–F 8:00–5:00 CST and 1st@prenhall.com.

- A Technology Consultant to follow up with a designated contact at your school to ensure you are getting the support you need.

- A service agreement from your Prentice Hall publishing representative outlining our customization, service, and training commitment in writing.

For your training needs:

- Access to the **Instructors 1st** resource Website, designed by faculty using the Web for teaching, for faculty who want resources for using the Web for teaching. This site will also build academic communities of faculty using Prentice Hall materials.

- Customized training, from Web-based to in-services to meet the needs of your faculty and staff.

For more information on **Instructors 1st,** contact your local Prentice Hall representative.

Supplements

INSTRUCTOR RESOURCE CENTER ON CD

(0-13-046162-8)

- Ready-made PowerPoint slides for each chapter with labeled and unlabeled art and a brief chapter outline
- All student tutorials and animations
- Image gallery with labeled and unlabeled art
- Instructor's Resource Manual in an editable format
- Instructor's Test Item File in an editable format

TRANSPARENCY PACK

(0-13-046150-4)

- Includes 400 illustrations from the text on full-color acetates.

INSTRUCTOR'S GUIDE

(0-13-046157-1)

By Jennifer Warner of the University of North Carolina at Charlotte and Katherine Rasmussen of the University of South Dakota.

This groundbreaking resource provides tips for teaching non-majors by making the material relevant, interesting, and interactive. Each chapter includes:

- Lecture Activity suggestions that include objectives, procedures, and assessment suggestions
- Objectives that identify goals for students and instructors
- Answer key to all end-of-chapter questions

TEST ITEM FILE

(0-13-046158-X)

Over 1500 test questions were carefully created by Dominic Lannutti at El Paso Community College and reviewed by a panel of educators.

TEST GEN-EQ

(0-13-046159-8)

This text-specific testing program is networkable for administering tests. It also allows you to edit existing test items or add your own questions to create a nearly unlimited number of quizzes and tests.

LABORATORY MANUAL FOR BIOLOGY OF HUMANS

(0-13-144960-5)

By Mimi Bres and Arnold Weisshaar, Prince George's Community College.

- Emphsis throughout on linking concepts to thinking and problem solving skills.
- Twenty laboratory activities written in an informal style which is easily accessible and interesting to students with no previous background in Biology.
- Focused on collaborative, small group activities that encourage student interactions and maximize laboratory resources.
- Designed to provide an understanding of the basic principles of human anatomy and physiology, genetics and evolutionary change, ecology, and the impact of human actions on the environment.
- Designed to run smoothly in large and small laboratory sections, each exercise is suitable for completion in two- or three-hour laboratory periods and can be divided to accommodate 90-minute sessions. The equipment and supplies needed are cost effective and easily accessible.

STUDY GUIDE

(0-13-143784-4)

By Claudia Douglass of Central Michigan University.

This guide consists of learning objectives, key concepts, study tips, chapter summaries, critical-thinking questions, short-answer questions, labeling exercises, and fill-in-the-blank questions. A multiple choice "Practice Test" is included at the end of each chapter to help students assess their understanding.

All the Support
You Need

BIOLOGY OF HUMANS

BIOLOGY OF HUMANS

CONCEPTS, APPLICATIONS, AND ISSUES

JUDITH GOODENOUGH
University of Massachusetts, Amherst

BETTY McGUIRE
Smith College

ROBERT A. WALLACE

PEARSON

Prentice Hall

Upper Saddle River, NJ 07458

Library of Congress Cataloguing-in-Publication Data

Goodenough, Judith
   Biology of humans: concepts, applications, and issues / Judith Goodenough,
 Betty McGuire, Robert A. Wallace Pearson Prentice Hall. – 1st ed.
   p.cm

 Includes index.
   ISBN 0-13-046020-6

 1. Human biology. I. McGuire, Betty. II. Wallace, Robert A. III. Title.

QP34.5.G658 2005

612—dc22   2004017675

Executive Editor: Gary Carlson
Editor in Chief, Science: John Challice
Project Manager: Crissy Dudonis
Editor in Chief, Development: Carol Trueheart
Development Editor: Karen Nein
Production Editor: Debra A. Wechsler
Executive Managing Editor: Kathleen Schiaparelli
Assistant Managing Editor: Beth Sweeten
Managing Editor, Media: Nicole M. Jackson
Vice President of Production and Manufacturing: David W. Riccardi
Media Editor: Travis Moses-Westphal
Marketing Manager: Andrew Gilfillan
Manufacturing Buyer: Alan Fischer
Manufacturing Manager: Trudy Pisciotti
Director of Creative Services: Paul Belfanti
Creative Director: Carole Anson
Art Director: Kenny Beck
Interior Design: Anne DeMarinis
Cover Design: Anne DeMarinis / Kenny Beck

Managing Editor, Audio and Visual Assets: Patricia Burns
AV Production Manager: Ronda Whitson
AV Production Editor: Jessica Einsig
Art Development: Jay McElroy
Assistant Manager, Formatting: Allyson Graesser
Formatters: Beth Gschwind / Vicki Croghan / William Johnson/
   Julita Nazario
Director, Image Resource Center: Melinda Reo
Manager, Rights and Permissions: Zina Arabia
Interior Image Specialist: Beth Brenzel
Cover Image Specialist: Karen Sanatar
Image Permission Coordinator: Joanne Dippel
Photo Researcher: Jerry Marshall
Editorial Assistant: Susan Zeigler
Cover Images: Volleyball player: Michael Kevin Daly/ Corbis/
   Bettmann; roller coaster: Robert Landau/Corbis/Bettmann;
   father and son; Jim Erickson/Corbis/Bettmann; gymnast:
   Brian Snyder/Corbis/Bettmann

©2005 Pearson Education, Inc.
Pearson Prentice Hall
Pearson Education, Inc.
Upper Saddle River, NJ 07458

Pearson Prentice Hall™ is a trademark of Pearson Education, Inc.

Printed in the United States of America
10 9 8 7 6 5 4 3 2 1

ISBN 0-13-046020-6

Pearson Education LTD., *London*
Pearson Education Australia PTY, Limited, *Sydney*
Pearson Education Singapore, Pte. Ltd.
Pearson Education North Asia Ltd., *Hong Kong*
Pearson Education Canada, Ltd., *Toronto*
Pearson Educación de Mexico, S.A. de C.V.
Pearson Education—Japan, *Tokyo*
Pearson Education Malaysia, Pte. Ltd.

Special-Interest Essays

ENVIRONMENTAL ISSUE ESSAYS

HEALTH ISSUE ESSAYS

SOCIAL ISSUE ESSAYS

Brief Contents

Contents

THE ORGANIZATION OF THE BODY · PART I

Science and Society   1

The Chemistry of Life   15

The Cell   40

PART II CONTROL AND COORDINATION OF THE BODY

5 The Skeletal System 89

6 The Muscular System 104

Neurons: The Matter of the Mind 118

The Nervous System 133

8a

SPECIAL TOPIC
Drugs and the Mind 152

9

Sensory Systems 164

The Endocrine System   187

MAINTENANCE OF THE BODY   **PART III**

Blood   213

12 The Circulatory System   229

13 Body Defense Mechanisms   255

13a

14

SPECIAL TOPIC

Smoking and Disease   312

The Digestive System   321

SPECIAL TOPIC

Nutrition and Weight Control   340

The Urinary System   356

REPRODUCTION | **PART IV**

Reproductive Systems   378

17a

SPECIAL TOPIC

Sexually Transmitted Diseases and AIDS   402

18

Development and Aging   415

Chromosomes and Cell Division   438

The Principles of Inheritance   459

21

DNA and Biotechnology   475

21a

SPECIAL TOPIC

Cancer   499

Evolution: Basic Principles and Our Heritage 511   22

Ecology, The Environment, And Us 533   23

24 Human Population Dynamics 552

To Stephen, my husband,
best friend, personal hero,
and the funniest person I know

To Aimee and Heather, my daughters,
who fill me with love, joy, and amazement

To Betty Levrat, my mother,
an excellent role model
and endless source of support and encouragement

To "The Group," friends for more than thirty years,
who help me hold it all together

To "The Secret Ingredient," still the best band I know.
Thanks for the laughs and the memories.
—J. G.

To Willy, Kate, and Owen Bemis,
and to Dora and James McGuire
—B. M.

About the Authors

Judith Goodenough

Judith received her B.S. from Wagner College (Staten Island, NY) and her doctorate from New York University. She has 30 years of teaching experience at the University of Massachusetts, Amherst, specializing in introductory level courses. The insights into student concerns and problems gained from 25 years of teaching Human Biology and 18 years of team-teaching The Biology of Social Issues have helped shape this book. In 1986, Judith was honored with a "Distinguished Teaching Award." In addition to teaching, she coordinates the introductory biology laboratories at UMass. Judith has written articles in peer-reviewed journals, contributed chapters to several introductory biology texts, and written numerous laboratory manuals. With the author team of McGuire and Wallace, she wrote *Perspectives on Animal Behavior*.

Judith Goodenough

Betty McGuire

Betty McGuire

Betty received her B.S. in Biology from Pennsylvania State University, where she also played varsity basketball. She went on to receive an M.S. and Ph.D. in Zoology from the University of Massachusetts at Amherst, and then spent two happy years as a postdoctoral researcher at the University of Illinois, Champaign—Urbana. Her field and laboratory research emphasize the social behavior and reproduction of small mammals. She has published numerous research papers, co-authored the text *Perspectives on Animal Behavior*, and is an Associate Editor for *Mammalian Species*, a publication of the American Society of Mammalogists. Betty has taught Human Biology, Introductory Biology, Vertebrate Biology, and Animal Behavior at Smith College.

Robert A. Wallace

The late Robert Wallace received a B.A. in Fine Art and Biology from Harding University, an M.A. in Muscle Histochemistry from Vanderbilt University, and a Ph.D. in Behavioral Ecology from the University of Texas at Austin. He subsequently taught at a number of colleges and universities in the United States and Europe, including the Richard Bland College of William and Mary, the University of Maryland—Overseas Division, Duke University, and the University of Florida. He is the author of seven previously published biology textbooks, including *Biology: The World of Life*, and two mass-market science books, *The Genesis Factor* and *How They Do It*, as well as numerous scientific articles on a variety of subjects. Robert was also a Fellow of the Explorers Club of New York and the Royal Geographical Society of London. He received the Orellana Medal from the government of Ecuador in recognition of his work with the medicinal plants of vanishing tribes in that country.

Preface

Humans are, by nature, curious, and this book is intended to stimulate the curiosity of nonscience students toward gaining an appreciation for the intricacy of human life and our place in the ecosphere. Once piqued, curiosity must rely upon conceptual substance for understanding. We inform students by building a conceptual framework that allows them to better understand their everyday experiences with their bodies and with the world around them. Connections between biological concepts and social issues and the application of these concepts to familiar experiences will support you in your classroom instruction and discussion by helping students see the importance—and excitement—of science in their lives.

The first goal of this textbook is to present the important concepts of human anatomy, physiology, development, genetics, evolution, and ecology. Then, after thoroughly explaining the basic concepts, we apply them in ways that will both interest and benefit the student. For example, a discussion of Alzheimer's disease, depression, and Parkinson's disease follows an explanation of neurotransmitters. When the content is relevant, it gives students a reason to want to learn the information. The chapters on organ systems explain how a healthy system functions, how that system might malfunction, measures to avoid a malfunction, and what current medicine can offer when systems are compromised or fail. Topics that students are likely to encounter in the media on an almost daily basis—smoking, contraception, STDs, cancer, bioterrorism, antibiotic-resistant bacteria—will help students make connections between real-life and classroom activities. Connections between concepts and environmental issues will help students develop a global perspective about environmental issues.

This text answers some very practical questions, including: What type of exercise benefits the heart? How does someone cope with insomnia? How does one protect against unwanted pregnancy and prevent the spread of sexually transmitted diseases? Each of us enters this world with a most intricate machine—our body—but we do not come equipped with an owner's manual. In a sense, this book can be the student's owner's manual. Understanding the information in it and applying it to our own lifestyles and health choices can help each of us live longer, happier, and more productive lives.

The second goal is to help students develop reasoning skills, so they can use the information in situations they face every day in life. Woven throughout each chapter are "stop and think" questions that ask students to apply information to a new situation. When a topic or issue is controversial, the discussion presents both sides of the argument, together with the supporting evidence. Scattered throughout each chapter are "what would you do?" questions that ask the student's opinion or challenge the student to take a stand on a particular issue as well as to identify the criteria used in reaching that decision. These questions foster the practice of thinking through issues, examining the information available, and making decisions based on that information. Additional critical-thinking questions reside on the Companion Web Site.

The third goal is to help students understand how the choices they make can affect the quality of life for themselves, society, and the planet. The material learned in the textbook or during lecture often bears on social or environmental issues that are important to us all. This text will help you, as the instructor, heighten students' awareness of their impact on the biosphere and prepare students to be responsible citizens and voters. Society is currently grappling with many pressing biological issues—the cloning of human cells, stem cell research, genetically modified foods, gene therapy, organ transplants, defining death, dealing with bioterrorism, and preventing and treating HIV infections, among others—and students need the tools to understand these issues and make informed decisions about them.

To reach these three goals, the text engages the student with applications that will interest them personally and discussions of critical issues facing society. The writing makes the information easily accessible. Each chapter begins with an outline of the major section headings to provide a preview and a framework upon which concepts are developed. These section headings are presented as complete sentences that clearly state the main concept of each section. Illustrated tables offer a way to organize and summarize information to help students see the big picture as well as the details.

The visual program stimulates learning with simple, beautiful illustrations that are supported and enhanced by effective pedagogy. Vibrant, three-dimensional figures show appropriate depth and detail and are clearly accessible and understandable for your students. The illustrations—including molecules and human tissues and organs—are visually consistent in form and structure throughout the text. "Voice balloons" draw the student's attention to a particularly important process or teaching point in many of the figures. Key figures pull concepts together to present the big picture. For example, the chapters on organ systems include figures that show both the anatomical structure and the function of the components of each system. These figures, such as Figure 14.2 of the respiratory system, help students understand the important relationship between structure and function. Micrographs paired with illustrations help students more easily interpret the micrograph by comparing it with the coupled illustration. Reference figures help students locate a particular structure within the body. Figure 4.2, which illustrates types of connective tissue, is an example that pairs micrographs to illustrations and also shows the student where each type of connective tissue is found. Many illustrations provide surrounding context for the particular structure being examined. For example, Figure 8.3, a sagittal section through the brain, includes the surrounding facial features and an illustration of a head to indicate orientation and perspective. Flowcharts visually walk students through a process so they can follow the discussion step by step as it moves through a sequence of explanations. Difficult concepts are presented using step-by-step figures, with a brief explanation of each step. By breaking a difficult concept into smaller components, such figures help the student understand each step and see how the steps fit together. For example, Figure 9.13 guides the student through the sequence of events involved in hearing, from sound waves hitting the eardrum to the brain's interpretation of neural information from the ear.

Finally, color is used in the visual program to effectively organize information. Where appropriate, color delineates the steps in a process. For example, in Figure 13.15, subtle differences in background shading distinguish different steps in the immune response. In Figure 12.13, a depiction of the electrical activity of the heart, color indicates the progress of a process.

How the Book Is Organized

The text begins with a discussion of the chemistry of life and then moves to cells, tissues, organs, organ systems, and finally to populations and ecosystems. Rarely is it possible to cover all the topics in a human biology text in one semester. Instructors must make difficult decisions about what to include and the order of presentation, and there are many excellent ways of presenting the material. For this reason, the chapters in this text do not depend heavily on material covered in earlier chapters. The independence of chapters allows you to tailor the use of this text to your course. Cross-references are given to direct students to relevant discussions in other chapters.

Features of the Book

CHAPTER OUTLINES

Each chapter begins with an outline that provides a framework upon which the student can organize the information presented. An outline identifies the important concepts and serves as a map of the relationships among these concepts.

SPECIAL TOPIC CHAPTERS

The Special Topic chapters (8a, Drugs and the Mind; 13a, Infectious Disease; 14a, Smoking and Disease; 15a, Nutrition and Weight Control; 17a, Sexually Transmitted Diseases and AIDS; and 21a, Cancer) expand "pure biology" to cover issues that are likely to be of personal interest and therefore motivating to students. The topics address personal health issues and are more thoroughly developed than they could be in an essay. Even if you do not assign these special topics for students to read, we hope that the issues are so pertinent that they will read or at least refer to these chapters as guides to healthier lifestyles.

"STOP AND THINK" QUESTIONS

"Stop and think" questions are scattered throughout each chapter and are intended to engage students in the learning process and to promote active learning. They invite the student to pause, think about the concept explained, and apply that information to a new situation. They provide periodic checks for the student to determine whether he or she understands the basic concepts.

"WHAT WOULD YOU DO?" QUESTIONS

"What would you do?" questions are also scattered throughout each chapter and raise ethical questions about issues that society faces today. These help the student see the relevance of information learned in a biology classroom to real-life problems or decisions that society must make, including fluoridation of water, routine screening for prostate cancer, the use of animal organs to save human lives, the export of pesticides to developing countries, and the means of slowing the growth of human populations. There is no right answer to any of these questions. They simply point out to the student that there are broad implications to many of the topics discussed. You may choose to use these questions to begin a classroom presentation, to stimulate discussions, or as questions to leave students thinking about a topic outside the classroom.

ESSAYS

The essays apply information and focus on health, social, or environmental issues. Instructors might use these essays to engage students in the classroom or to provide them with information needed for informed in-class discussions. **Health Issue** essays deal primarily with personal health topics. They provide current information on certain health problems that they, their family, or their friends might encounter. Some of the topics discussed in Health Issue essays are acne, osteoporosis, treatments for the common cold, and heartburn. These essays will help students better understand what their physicians may be telling them. The **Social Issue** essays explore some of the ethical or social issues related to the topics under consideration. The topics of Social Issue essays include anabolic steroids, gene testing, stem cells, and cloning. Finally, the **Environmental Issue** essays deal with the ways in which human activities alter the environment or the ways in which the environment influences human health or well-being. Among the topics discussed in Environmental Issues essays are acid rain, asbestos, noise pollution, and global warming.

Pedagogical Features

When students are studying outside the classroom, this text will help them understand the concepts presented during lecture. Some of the features are designed to reinforce details and others to reinforce concepts that were presented in class.

Headings and Summaries   The headings are presented as sentences, which state the main point of the sections that follow. Thus, students see the big picture and focus on the explanations that follow. The Reviewing the Concepts summary sections are organized using the main headings of the chapter. Relevant page numbers are included to guide a student from the summary of a topic back to the full text discussion of that topic.

Key Terms and Glossary   This text minimizes technical language because it is intended for students who are not science majors.

Nonetheless, some terms must be used even though they may not be familiar to students. The important terms are boldfaced throughout the chapter and are listed as key terms at the end of the chapter. This list also provides chapter page numbers, indicating where each term is defined. A glossary at the end of the book includes all key terms used in the book.

Questions   The questions posed at the end of each chapter have a variety of formats. Some are simply for content review; others require more critical thinking to apply the information to new situations. Review questions that require a written answer are followed by the page number of the relevant discussion. This practice will engage students in the learning process by encouraging them to review and understand the relevant material instead of memorizing the answer. Answers to the multiple-choice and fill-in questions are provided in an appendix. Hints for answering the Applying the Concepts questions are also included in an appendix. These hints help students identify the information needed to answer each question; instead of providing a quick answer, the hints guide a student's thinking process.

Icons   Icons in the book direct the student to Web tutorials on the Web Site for this text. The Web tutorials cover processes and difficult concepts with animations, interactive exercises, and a quiz to help students assess their understanding of the topic after viewing the animation or completing the exercise.

Relevant Web Sites   The front inside text cover lists 160 relevant and useful Web Sites for students who want to explore a topic in more detail. Because the text discusses many health issues that students or their families and friends may face, the list of Web Sites includes resources that can provide support for people with various health problems, including cancer, drug and alcohol abuse, and smoking.

Supplements

A package of supplemental material has been carefully designed for the student and instructor to further facilitate learning and teaching. It includes:

FOR INSTRUCTORS

Instructor Resource Center on CD   This is an all-inclusive resource bank for instructors. Instructors will be able to access textbook art, PowerPoint presentations, animations, lecture outlines, and assessment content. The CD also provides instructors with the content needed to build an online course in the WebCT, Blackboard, or CourseCompass course management systems.

Transparencies   Includes approximately 400 illustrations from the text on full-color acetates.

Instructors' Manual   This manual contains material that will assist instructors in the preparation of lectures, including lecture outlines and hints, suggestions for classroom activities, answers to the end-of-chapter questions, and suggestions for assignments.

Test Item File   This file includes over 1000 questions instructors can use to prepare exams. Question types include multiple choice, true/false, short answer/fill-in-the-blank, matching, and labeling.

TestGen EQ Computerized Testing Software   In addition to the printed Test Item File, the test questions are also available as part of the TestGen EQ Testing Software, a text-specific testing program that can be networked for administering tests. It also allows instructors to view and edit existing test questions or add their own questions to create a nearly unlimited number of quizzes and tests.

FOR THE STUDENT:

Companion Web Site   The Companion Web Site **WWW.PREN-HALL.COM/GOODENOUGH** includes many resources to help students master the course. *Web Tutorials* consisting of animations, interactive exercises, and quizzes help students master difficult concepts. Web Tutorial icons have been included throughout the book. In addition, the Companion Web Site includes *Practice Tests, Vocabulary Review* quizzes with Flashcards, *Critical-Thinking Questions,* and extensive study aids including *Learning Objectives, Key Concepts,* a detailed *Chapter Summary,* and *Chapter Study Tips.*

Student Study Guide (printed study aid)   By Claudia Douglass, Central Michigan University   The Student Study Guide is designed to help students master the topics and concepts covered in the textbook. The study guide includes different types of review questions including multiple choice, fill-in-the-blank, critical thinking, matching, and labeling. It also includes objectives, summaries, key terms, and student study tips.

Laboratory Manual for Biology of Humans   by Mimi Bres and Arnold Weisshaar, Prince George's Community College   The subject matter and approach to learning of this lab manual make it very appropriate to accompany *Biology of Humans*. It is designed to provide an understanding of the basic principles of human anatomy and physiology, genetics and evolutionary change, ecology, and the impact of human actions on the environment.

Acknowledgments

The authors are only part of the team that produces a book. We thank a few of the dedicated people who helped with this book.

Gary Carlson, our executive editor, launched this text with energy and enthusiasm. He included us in decisions and supported us throughout the book's development. When we had problems, he was eager to help. His unflagging faith in our project is greatly appreciated. He has been involved at almost every level, and the result is clear evidence of his effectiveness. Carol Trueheart, editor in chief of development, helped shape our vision of the project. She and Gary were instrumental in bringing that vision to reality. They watched over our efforts and helped us at every turn.

Our developmental editor, Karen Nein, read numerous drafts of our chapters, each time with a fresh eye and incredible attention to detail. She standardized the format of chapters, improved presentation of material in the text and figures, and helped refine our vision. Her critical judgment is reflected throughout the book.

Sue Zeigler, editorial assistant, found us outstanding reviewers and delivered the reviews to us without delay.

Margo Quinto copyedited the book. Her detailed, accurate, and consistent touch was the best we have experienced. The text was refined even more by the skills of proofreaders Alison Lorber, William Bassham, and L. Leann Kanda.

Our production editor, Debra Wechsler, brought together manuscript, art, and photos with skillful diligence. Her "can do" attitude eased the pressures of book production. She kept a keen eye on schedules and helped us prioritize. She is professional and a pleasure to work with.

The art program was designed and developed by two medical illustrators: Jay McElroy from Prentice Hall with the help of Jack Haley from Imagineering, Inc. Their art development has created an accurate and pedagogically sound art program. We especially enjoyed working with Jay McElroy on art development. He was always responsive to questions and suggestions and somehow managed to keep a wonderful demeanor throughout this complex and demanding project. Imagineering Media Services skillfully rendered the art. Photo Researcher Jerry Marshall persistently worked to find the best and most original photographs. Kenny Beck and Anne DeMarinis created an elegant design that highlights all the pedagogical elements. Formatting manager, Allyson Graesser, skillfully implemented the design.

Project manager, Crissy Dudonis, coordinated production of the ancillary materials that accompany our text. She brought together media, lecture aids, and study aids to create a supplemental package that is beneficial to instructors and students. Development of media supplements was ably directed by Travis Moses-Westphal.

We owe a great deal to our late friend and colleague, Bob Wallace. We are thankful for all he taught us. His wife, Jayne Wallace, provided information about medicines in the rain forest and has continued to support our efforts.

Smith College student Lauren Dutton compiled the list of Web sites for our front-end pages as part of a Special Studies project during her last semester at Smith. Her commitment to health education is reflected in the care that she took in selecting well-established Web sites to provide accurate, current information.

University of Massachusetts students Jessica Connolly, Ericka Swierzowski, and Beth Flynn read many chapters at different stages of development. As they read, they identified explanations that needed clarification. They also assisted in the preparation of the glossary.

FROM JUDITH GOODENOUGH

I thank my family and friends who supported and encouraged me at every stage of this project. My husband, Steve, was a cheerleader and convinced me that I would complete this project. Without his witty quips, I would have lost my sanity. He reassures me that I'll always be his first wife. My daughters, Aimee and Heather, inspire me and continually remind me that the people you love should always come first. The willingness of my mother, Betty Levrat, to help in any way, allowed me to focus on writing.

Margaret Tillson cheered me up and helped me cope when things got tense. Margaret's willingness to pick up the slack at UMass allowed me to take vacation days to work at home during the semester. Lee Estrin was patient and understanding when I postponed our usual "I need a break" weekends. Lee is one of "the group," dear friends who have always provided moral support and advice.

FROM BETTY MCGUIRE

I thank my husband, Willy Bemis, for support and encouragement throughout this project, and my children, Kate and Owen, for (usually) waiting patiently for me to complete just one more sentence or paragraph. Dora, Jim, and Cathy McGuire were endlessly encouraging and understanding, as they have been throughout my life. Lowell Getz, my friend and research colleague, waited with good humor as one after another of our papers took a back seat to a book chapter.

We must thank the dozens of reviewers who so carefully read chapters. They were immensely helpful in shaping the book and in catching errors. Of course, any errors that remain are our responsibility, and we hope you will inform us if you find any.

REVIEWERS

John Aliff, *Georgia Perimeter College*
Susan Bear, *Saint Joseph's University*
Virgina Bliss, *Framingham State University*
Robert Boyd, *Auburn University*
Christine Curran, *University of Cincinnati*
Michael A. Davis, *Central Connecticut State University*
Jeff W. Dawson, *University of Cambridge*
Claudia Douglass, *Central Michigan University*
Lucinda H. Elliott, *Shippensburg University*
Fledia Ellis, *Gulf Coast Community College*
Beth Erviti, *Greenfield Community College*
Jane Finan, *Lewis-Clark State College*
Deb Firak, *McHenry County College*
Michael P. Franklin, *California State University, Bakersfield*
Paul Gannon, *El Paso Community College*
Donald Glassman, *Des Moines Area Community College*
Sheldon R. Gordon, *Oakland University*
Margarit Gray, *Anderson College*
Esther Hager, *Passaic County Community College*
P.C. Hirsch, *East LA College*
Donald B. Hoagland, *Westfield State College*
Clare Hoffman Hays, *Metropolitan State College of Denver*
Nina Holland, *University of California, Berkeley*
John Hoover, *Millersville University*
Robert J. Huskey, *University of Virginia*
R. Kent Johnson, *San Jose City College*
Carol Kasper, *MacMurray College*
Mary Jo Kenyon, *North Dakota State University*
Sherrene D. Kevan, *Wilfrid Laurier University*
William Kleinelp, *Middlesex Community College*
Dominic Lannutti, *El Paso Community College*
Lee Lee, *Montclair State University*
Maureen Leupold, *Gennessee Community College*
James Loats, *Black Hawk College*
Patricia Matthews, *Grand Valley State University*
Ann Maxwell, *Angelo State University*
Linda McNally, *Davidson College*

David P. S. Mork, *St. Cloud State University*

James Murphy, *Monroe Community College*

Nancy L. Pencoe, *State University of West Georgia*

Bernice Richards, *Northern Essex Community College*

John Rinehart, *Eastern Oregon University*

Matthew Schmidt, *Empire State College and Stony Brook University*

Mark A. Schneegurt, *Wichita State University*

Harold F. Sears, *University of South Carolina, Union*

Richard Shippee, *Vincennes University*

Jeff Simpson, *Metropolitan State College of Denver*

Paulette Snyder, *Metropolitan State College of Denver*

Karen Stine, *Ashland University*

Steven Stocking, *San Joaquin Delta College*

Chad Thompson, *SUNY, Westchester Community College*

Janis Thompson, *Lorain County Community College*

Michael Troyan, *Pennsylvania State University*

Ann Vernon, *St. Charles Community College*

Mary A. Vetter, *Luther College, University of Regina*

Susan Wadkowski, *Lakeland Community College*

Jennifer Warner, *University of North Carolina—Charlotte*

Rosemary Waters, *Fresno City College*

Lura Williamson, *University of New Orleans*

Brian D. Wisenden, *Minnesota State University, Moorhead*

Sabrina D. Woodford, *University of Cambridge*

Scott Zimmerman, *University of Wisconsin, Stout*

ACCURACY REVIEWERS

William Bassham

L. Leann Kanda

MEDIA REVIEWERS

Christine Curran, *University of Cincinnati*

Michael Davis, *Central Connecticut State University*

Claudia Douglass, *Central Michigan University*

Lucinda Elliott, *Shippensburg University*

Deb Firac, *McHenry County College*

Eric Gillock, *Fort Hays State University*

Margarit Gray, *Anderson College*

Clare Hoffman Hays, *Metropolitan State College of Denver*

Donna Huffman, *Calhoun Community College*

George Jacob, *Xavier University*

William Kleinelp, *Middlesex County College*

Mary Katherine Lockwood, *University of New Hampshire*

James Mulrooney, *Central Connecticut State University*

Paulette Snyder, *Metropolitan State College of Denver*

Mary Vetter, *University of Regina*

Susan Wadkowski, *Lakeland Community College*

Jennifer Warner, *University of North Carolina—Charlotte*

Cheryl Watson, *Central Connecticut State University*

Arnold Weisshaar, *Prince Georges Community College*

Brian Wisenden, *Minnesota State University, Moorhead*

Science and Society

All Living Things Share Basic Characteristics

Living Organisms Share an Evolutionary History

Life Has Many Levels of Organization

The Scientific Method Helps to Gather Information and Reach Conclusions

- Inductive and deductive reasoning help solve problems
- Clinical trials follow strict guidelines

Critical Thinking Helps Us Make Informed Decisions

ENVIRONMENTAL ISSUE Medicinal Plants and the Shrinking Rain Forest

FIGURE **I.I**
The rosy periwinkle (*Catharanthus roseus*) is a source of two anticancer drugs.

The dugout canoe, with four on board, turns from the muddy river and pushes onto the slippery bank. A tribesman cheerfully steps off into mud up to his knees and, laughing at his friend pushing with a pole from the back of the canoe, pushes and heaves and moves the boat as far as possible onto the bank and ties it to a tree.

You carefully make your way forward, then stop and reach back for a plastic-wrapped bundle that your friend offers. You pass it to the tribesman. He carries it to a grassy area and then comes back for another until all the equipment is ashore. The tribesman helps you and your friend to the grassy area, knowing that North Americans are not too good at this sort of thing.

Once ashore, you peer through the tangled, viny vegetation into the dark interior of the forest. An insect whines. Already you are soaked with sweat. Still, this is what you came for—to explore this part of the Amazon rain forest, searching for medicinal plants. You know that two chemicals from the rosy periwinkle of Madagascar are already being used to treat leukemia and Hodgkin's disease (Figure 1.1). You also know that the Amazon rain forest, home to millions of species of plants, is being destroyed rapidly. You want to learn its secrets while you still can.

To learn those secrets, you need to talk to the native people about the plants they use for healing. You have spent months preparing—learning about the culture of the native people, cataloging the information already available on local plants and their uses, learning how to collect and preserve the plants, and gathering tools and supplies. A physician helped prepare simple descriptions and photographs of diseases, which will be helpful when you discuss diseases with local healers. If healers recognize a disease with similar symptoms, they will show you the plants they use to treat those ailments. In laboratories at home, chemists and biologists will study the plants you have collected to determine their medicinal value. ■

All Living Things Share Basic Characteristics

As you make your way deeper into the rain forest, the bewildering variety of unfamiliar plants, the brilliant colors of the flowers, and the strange songs of unseen birds make you keenly aware of the diversity of life around you. In fact, that diversity is equaled in very few places, such as a few tropical coastal areas and perhaps some coral reefs. As you continue exploring the forest, how do you know if something unfamiliar to you is alive? In some cases, the question is easy to answer. Although the leaves around you have different shapes and sizes, you know they are all leaves. And a tree is a tree. But what about the gray matter on that tree trunk? How can you tell if it is alive?

Defining life would seem to be easy, but it is not. In fact, no single definition would suit all life scientists. For example, if we say you can tell something is alive if it reproduces, someone is likely to note that oil droplets "reproduce" when they join and grow so big they fall apart and form smaller droplets. If we say you can tell something is alive if it grows, what should we conclude about crystals? They grow, but they are not alive. And so it goes. It seems there is no single defining feature of life.

No single definition covers all life, so we find that instead of defining life, we can only characterize it. That is, we can only list the traits associated with life. Most biologists agree that, in general, the following statements characterize life.

1. **Living things contain nucleic acids, proteins, carbohydrates, and lipids.** The same set of slightly more than 100 elements is present in various combinations in everything on the earth—living or nonliving. However, living things can combine certain elements to create molecules that are found in all living organisms: nucleic acids, proteins, carbohydrates, and lipids. The nucleic acid DNA (deoxyribonucleic acid) is especially important because the molecules can make copies of themselves, an ability that enables organisms to reproduce (Figure 1.2a). The molecules of life are discussed in Chapter 2.

2. **Living things are composed of cells.** Cells are the smallest units of life. Some organisms have only a single cell; others, such as humans, are composed of trillions of cells (Figure 1.2b). All cells come from preexisting cells. Because cells can divide to form new cells, reproduction, growth, and repair are possible. Cells are discussed in Chapter 3.

3. **Living things reproduce.** Living things have ways of generating new individuals that carry some of the genetic material of the parents. Some organisms, such as bacteria, reproduce simply by making new and virtually exact copies of themselves. Other organisms, including humans, reproduce by combining genetic material with another individual. (Figure 1.2c). Reproduction is discussed in Chapter 17, and cell division in Chapter 19.

4. **Living things use energy and raw materials. Metabolism** refers to the sum total of all chemical reactions that occur within the cells of living things. Through metabolic activities, organisms take in energy and transform it to do many kinds of work. Metabolism maintains life and allows organisms to grow (Figure 1.2d). Chemical reactions involved in the transformation of energy are discussed in Chapter 2.

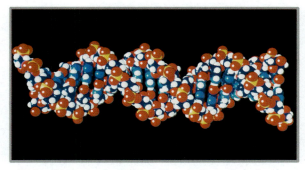

a. Living things contain nucleic acids, proteins, carbohydrates, and lipids.
This is a computer-generated model of the nucleic acid DNA, which carries genetic information.

d. Living things use energy and raw materials.
This mother orangutan is sharing bananas with her offspring. This food will provide energy and raw materials for the maintenance, repair, and growth of their bodies.

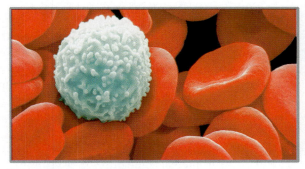

b. Living things are composed of cells.
These are red blood cells (disks) and a white blood cell (sphere).

e. Living things respond.
This chameleon sees and catches its prey.

c. Living things reproduce.
All organisms reproduce their own kind.

f. Living things maintain homeostasis.
This child cools himself, a behavioral response to maintain a constant body temperature.

g. Populations of living things evolve and have adaptive traits.
The orchid is adapted to live perched on the branches of trees. It uses the other plant for support so that it can be in sunlight and produce its own food by photosynthesis.

FIGURE **1.2**
Characteristics of life

5. **Living things respond.** A boxer tries to avoid the blow of an opponent, and a chameleon catches its prey (Figure 1.2e). For a living thing to respond, it must first detect a stimulus and then have a way to react. As you will learn, your sensory organs detect stimuli; your nervous system processes sensory input; and your skeletal and muscular systems allow you to respond. The skeletal and muscular systems are discussed in Chapters 5 and 6. The nervous system is discussed in Chapter 8, and sensory organs in Chapter 9.

6. **Living things maintain homeostasis. Homeostasis** is the ability to maintain a relatively constant internal environment. We generally find that life can exist only within certain limits and that living things tend to behave in ways that will keep their body systems functioning within those limits. For example, if you become too cold, you shiver (a metabolic response). Shivering produces heat that warms your body. Alternatively, if you become too hot, your sweat glands will be activated. The blood capillaries in the skin will open up to bring more blood to the surface of the body to cool the body down. In addition, you may seek ways of cooling yourself (a behavioral response) (Figure 1.2f). We will discuss homeostasis in Chapter 4 and as we present each system of the body.

7. **Populations of living things evolve and have adaptive traits.** Members of a population bearing beneficial genetic traits will survive and reproduce better than members of the population that lack these traits. As a result of this process, called natural selection, each of the amazing organisms you see around you has **adaptive traits**—that is, traits that help it survive and reproduce in its natural environment. Plants in the rain forest have many adaptations that allow them to thrive in that environment. Most plants have shallow root systems because the topsoil is thin and nutrients are near the surface. As a result, tall trees develop supports like cathedral buttresses to hold them up, while vines climb over both roots and trees to reach the light. Many plants grow, not in the ground, but high above it in the canopy. These plants, called epiphytes, include orchids and are plants that are rooted on the surfaces of other plants. (Figure 1–2g). Most rain forest leaves have a pointed "drip tip" to help them shed water. Many animals also have adaptations—the ability to fly or climb, for example—that enable them to reach the plants for food. Evolution is discussed in Chapter 22.

stop and think

Scientists have sent a mechanical rover to explore Mars. They have found evidence that there was once water on Mars, which means that it had the potential to support life. If material from Mars were brought back to the earth, which characteristics of life could you look for to determine whether the material was once alive?

Living Organisms Share an Evolutionary History

At least 10 million species of organisms live on the earth. Living things are classified, or organized, in a way that shows evolutionary relationships among them. Scientists assume that shared characteristics in organisms indicate a common ancestry. Therefore, the organisms that are most similar are grouped together. One scheme of classification recognizes six kingdoms, as shown in Figure 1.3. Members of the kingdoms Bacteria and Archaea are prokaryotes; that is, they are single-celled organisms that lack a nucleus or other internal compartments. Organisms in other kingdoms have eukaryotic cells, which contain a nucleus and complex internal compartments called organelles. A third kingdom, Protista, consists primarily, but not exclusively, of single-celled organisms. Kingdom Protista includes algae, slime molds, and the protozoans. The fungi (such as molds and mushrooms) are placed in kingdom Fungi. The plant and animal kingdoms are well known.

Within each kingdom, organisms are categorized into large groups whose members share distinctive characteristics that set those organisms apart from others of the kingdom. Therefore, the kingdoms are divided into smaller groups called phyla (singular, phylum). Members of a phylum are more closely related to one another than to members of other phyla. Each phylum is subdivided into classes to better organize the array of organisms and to show

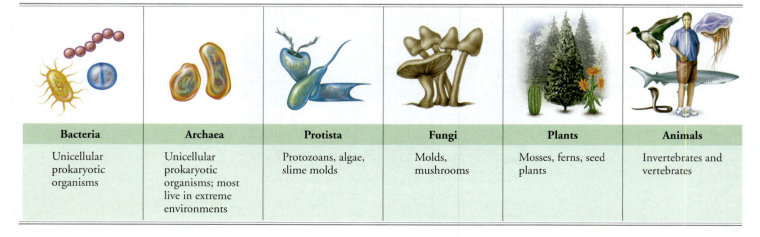

Bacteria	Archaea	Protista	Fungi	Plants	Animals
Unicellular prokaryotic organisms	Unicellular prokaryotic organisms; most live in extreme environments	Protozoans, algae, slime molds	Molds, mushrooms	Mosses, ferns, seed plants	Invertebrates and vertebrates

FIGURE **1.3**
The six kingdoms of life

Kingdom	Animalia (animals)	Multicellular; heterotrophic (ingest food); lack cell walls; have nervous and muscle tissue; most reproduce sexually.
Phylum	Chordata (chordates)	Possess a notochord (a dorsal, hollow nerve cord); pharyngeal gill slits; tail that extends beyond the anus.
Subphylum	Vertebrata (vertebrates)	Possess bone and a backbone; at least two semicircular canals in inner ear; paired fins or limbs.
Class	Mammalia (mammals)	Have mammary glands; hair; maintain constant body temperature; three middle ear bones; mature red blood cells lack nucleus.
Order	Primates	Have rotating shoulder joints; grasping feet and hands; fingernails (instead of claws); forward-looking eyes; well-developed cerebrum (the part of the brain most directly related to intelligence); reliance on learned behavior; small litter size.
Family	Hominidae (human family)	Walk upright on two legs; have well-developed brain; language; culture.
Genus	*Homo*	Walk upright on two legs; continued development of brain; language; culture.
Species	*Homo sapiens*	Walk upright on two legs; continued development of brain; language; culture.

FIGURE **1.4**
The classification of humans

finer relationships. Classes are divided into orders, orders into families, families into genera (singular, genus), and genera into species.

Humans are chordates (phylum) and more specifically vertebrates (subphylum). Humans are mammals (class). Two characteristics that make us mammals are that we have hair and that we feed our young milk produced by mammary glands. However, we are rather unusual mammals that share a suite of characteristics with lemurs, monkeys, and apes. These features include forward-looking eyes and a well-developed brain (cerebral cortex). Humans, monkeys, and apes also have opposable thumbs (a thumb that can touch the fingertips of the other four fingers). Together with lemurs, monkeys, and apes we are primates (order). Even smaller details, such as tooth structure and skeletal characteristics serve to separate the primates into separate families and smaller groupings (Figure 1.4). Humans are hominids (family) and their scientific name is *Homo sapiens* (genus followed by species).

What sets humans apart as species *Homo sapiens?* Human characteristics include large brain size relative to body size, a two-legged gait, and an opposable thumb. Probably nothing distinguishes humans more than culture. *Culture* may be regarded as a set of social influences that produce an integrated pattern of knowledge, belief,

and behavior (Figure 1.5). Other animals, of course, have social interactions, as we see in everything from competition and cooperation to territoriality, hierarchies, and mating behavior. But in other species, social influences are not so pronounced.

FIGURE **1.5**
Social interactions are an important thread in the fabric of human life.

Keep in mind, however, that there is no single human culture. As you follow your guides through the rain forest you will eventually collect your plants and then return to their village. You will quickly learn that their culture is different from yours. As you travel on to stay with other groups, you will learn that there are many different cultures in the rain forest. People in the same environment do not adapt to it in the same way. If you try to describe your culture to them, your description may elicit anything from astonishment to howls of laughter. Even so, you notice that some things, especially love of children and the need to belong to a group, seem the same almost everywhere. In every way, culture powerfully defines the human condition.

Life Has Many Levels of Organization

As we study human biology, we will see that life can be organized on many levels (Figure 1.6). Cells, the smallest unit of life, are composed of molecules. A multicellular organism may have tissues, groups of similar cells that perform specific functions. Organs consist of different types of tissue that work together for a specific function. Two or more organs working together to perform specific functions form an organ system. Humans have 11 organ systems, as we will see in Chapter 4.

Life can also be organized at levels above the individual. A **population** is all the individuals of the *same species* (individuals that can interbreed) living in a distinct geographic area. A population might be the yellow-bellied marmots living in a meadow or the four-eyed butterfly fish living in a coral reef. A **community** is all living species that can potentially interact in a particular geographic area. Communities can exist on land or in the water. For example, a community might be *all the species* that live and interact in an alpine meadow or in a pond.

Ecosystems are defined areas in which certain living organisms interact with their physical environment. The size of the defined area comprising an ecosystem varies with the interest of the person studying it. An ecosystem can be as large as the whole earth, a particular forest, or even a single rotting log within a forest. Regardless of its size, an ecosystem is relatively self-contained.

The **biosphere** is that part of the earth in which life is found; it encompasses all of the earth's living organisms. In essence, the biosphere is where light, minerals, water, and gases interplay in a thin veil over the earth to produce environments that permit life to exist. The biosphere extends only about 11 kilometers (7 miles) above sea level and the same distance below, to the deepest trenches of the sea. If the earth were the size of a basketball, the biosphere would have the depth of about one coat of paint. In this thin layer over one small planet, we find all of the *known* life in the entire universe.

The Scientific Method Helps to Gather Information and Reach Conclusions

Humans are an irrepressibly curious species. Our intelligence has been coupled with an inquisitiveness that leads us in innumerable directions, each according to his or her own tendencies, background, and abilities. Because we are curious and intelligent, we often ask questions about things we observe. Sometimes we may try to answer those questions by doing experiments. So, we will look at how scientific investigation proceeds and how you might apply these steps in your own investigations.

There is no such thing as *the* scientific method—not if it refers to a formalized ritual for performing experiments. Instead, the scientific method involves certain logic in an approach to gathering information and reaching conclusions. It often begins with an observation that raises a question. Next, an educated guess attempts to answer the question, and that guess must be testable. Generally, the tentative explanation will lead to a prediction. If the prediction holds true when it is tested, it will support the explanation. If the original guess is not supported, an alternative guess is generated and tested.

1. **Make careful observations and ask a question about the observations.** If, for example, you were walking in a rain forest and came across a patch of plants growing alone, with no other plants growing around them, you would record and describe that observation. In the rain forest, this situation is true for lemon ant plants (Figure 1.7). Growing in simple-looking, knee-high patches, it harbors small ants considered delicious by the local people. Considering the density of plants that you see in nearby areas of the forest floor, this observation of relatively isolated groups of plants strikes you as odd. You want to learn more about this plant. You determine the area of the patch, the approximate number of plants in it, and the size of the cleared area around the plant. You also look for another patch, make the same measurements, and determine if there is some correlation between the size of the patch and the area of the clearing.

2. **Develop a testable hypothesis (educated guess) about the answer to the question about the observations.** Once you have your measurements and find that the larger the patch, the larger the cleared area, the observation raises the question, "Why don't other plants grow near the lemon ant plant?" You start to formulate ideas about why. Your first guess is that the plants in the patch are producing a chemical that inhibits the growth of other plants. This statement is your **hypothesis**—a testable explanation for your observation. A hypothesis that cannot be tested is worthless because it cannot be supported or refuted. However, a hypothesis that is shown to be incorrect can still be useful in developing alternative hypotheses.

3. **Make a prediction and test it with a controlled experiment.** Now you must plan an experiment to test your hypothesis. If your hypothesis is correct, you might predict that the chemical alone, without the intact plant, would inhibit growth of other plants. You select one lemon ant plant and liquefy its tissues in alcohol, using a blender. You now have an extract—a mixture of all the chemicals in the plant that are soluble in alcohol. This extract can be used to treat plant seeds to determine its effects on growth.

You must design your experiment so that there can be only one explanation for the results. So you run a **controlled experiment**, an experiment in which the subjects (in this case, plant seeds) are randomly divided into two groups, usually called

Molecule	The chemical components of cells	
Cell	The smallest unit of life	
Tissue	A group of similar cells that perform the same function	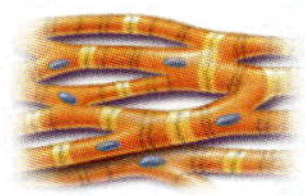
Organ	A structure with two or more tissues working together to perform a function	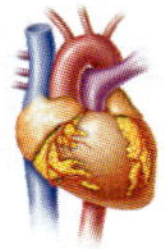
Organ systems	At least two organs working together to perform a function	
Individual	A single organism	
Population	All individuals of the same species in an area	
Community	All the species in an ecosystem that can interact	
Ecosystem	A community and its physical environment	
Biosphere	The part of the earth that supports life	

FIGURE **1.6**
Levels of organization of life

"control" and "experimental" (Figure 1.8a). Both groups are treated in the same way except for *one* special factor, the **variable**, whose effect you want to determine. In this case, the variable is the plant's chemicals. You want to know if those chemicals inhibit the growth of other plants, so you prepare two identical plots of ground and seed them both with the same variety of plants. You then spread the plant extract over a specified area of the experimental plot and leave the control plot untreated. Sure enough, no new plants grow in the area that you treated with the alcohol extract, but plants grow well in the untreated areas.

FIGURE I.7
Patches of lemon ant plants have little or no growth around them.

Is your hypothesis supported? Yes, but not completely. Is it possible that the alcohol, not the plant's chemicals in the extract, inhibited plant growth? In the first experiment, you had two variables: the alcohol and the extract itself. So you experiment again. This time, however, you have three plots, as identical to one another as possible. You till the soil, plant the seeds, and water each plot. Next, you treat one plot with the alcohol extract; you treat another plot with alcohol only; and you leave the third plot untreated (Figure 1.8b). Then you wait. Sure enough, plant growth is retarded only in the area treated with the extract. Plants grow in the plot treated only with alcohol and in the untreated plot. Your experimental results indicate that the chemicals in the plant tissue are inhibiting the growth of other plants.

4. **Draw a conclusion based on the results of the experiment.** Next, you arrive at a conclusion. Your conclusion, in this case, may be that the lemon ant plant produces a chemical that inhibits the growth of other plants.

   Note, however, that although your results support your hypothesis, they do not *prove* your hypothesis. If the results supported the hypothesis, you would make additional predictions based on your hypothesis and test them. Although hypothesis can be proven wrong when the results do not support it, results that are consistent with a hypothesis do not prove that it is correct. There may be other

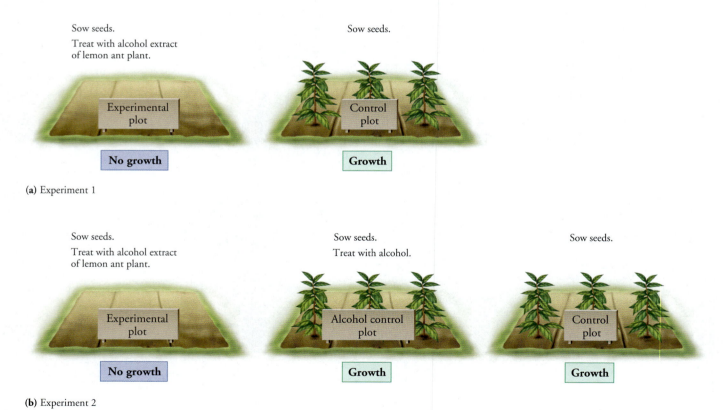

FIGURE I.8
A controlled experiment is set up in duplicate. The two groups should be exactly alike except for the variable. In both experiments, the plant seeds are the subjects. (a) In experiment 1, the variable is the extract of the lemon ant plant. (b) In experiment 2, a second control is added to the design to determine whether the alcohol used to prepare the extract was responsible for the inhibition of growth.

hypotheses that would make the same prediction, and there may be other explanations for the results. For instance, suppose the lemon ants that live on the plant deposit a chemical on the surface of the host plant that washes off with rain and inhibits plant growth in the surrounding area. The ant-produced chemical might also be in your extract along with the plant's chemicals. Thus, experiments often lead to revised hypotheses and new experiments (Figure 1.9).

What would you do if the plant extract had not inhibited the growth of other plants? This result would mean that your hypothesis needed more thought. So, you would have to reconsider the situation and try to find another explanation. Remember, this explanation is a new hypothesis and will call for new experiments (with new controls) and new conclusions. Maybe the inhibitory chemical is not soluble in alcohol. Maybe the chemical is a gas. Maybe a chemical does not cause the inhibition of growth. Essentially, you must now look for new variables. You might want to go back to the forest and note whether the plant is always found growing under another kind of plant. Are the two plants interacting? The lemon ant plant is host to lemon ants. Is the inhibitory effect due to the ants? Maybe the lemon ant plant was not *inhibiting* the growth of other plants. Maybe it was better at obtaining something necessary for the growth of other plants, such as water or a specific nutrient. How could you test these ideas?

Another important part of scientific inquiry is that experiments be repeated and yield similar results. Other scientists should be able to follow your procedure and obtain similar results, not just in that part of the rain forest but any place where the lemon ant plant is found. It can be very difficult to identify all the factors that might affect the outcome of the experiment. In this example, rainfall, the amount of light, and soil characteristics might influence the results. Each time the experiment is repeated, additional factors may be identified as important.

As we have seen, the formation of a hypothesis represents one level of the scientific process. In some cases, related hypotheses that have been confirmed repeatedly fit together and form a **theory**—a broad-ranging explanation for some aspect of the universe. Because of its breadth, a theory cannot be tested by a single experiment. Instead, a theory is the result of many observations, hypotheses, and experiments. A theory leads to additional predictions and experiments to continue testing it. Among the few hypotheses that have been tested thoroughly enough to be considered theories are the cell theory of life (all cells come from preexisting cells) and the theory of evolution by natural selection.

▊ Inductive and deductive reasoning help solve problems

As you looked at the nearly bare ground around the lemon ant plants and wondered what was causing it, your mind was positioning itself to attack the problem, probably without your knowing it. You answered the question of why other plants did not grow near the lemon ant plants using two types of reasoning: inductive and deductive.

Inductive reasoning involves the accumulation of facts through observation until the sheer weight of the evidence allows some logical general statement about nature. In the case of the cleared area around the plant, you collected as much information as possible about this plant, others like it, the soil, the leaves, and anything else that might have a bearing. When as much information as possible was gathered, you used the specific facts to draw a general conclusion, such as "some plants interfere with the growth of other plants." You used inductive reasoning to develop a testable hypothesis.

Deductive reasoning involves making some general statement and then drawing, or deducing, conclusions from it. The statement is usually in the form of an "if-then" premise. *If* this plant is making something that interferes with the growth of other

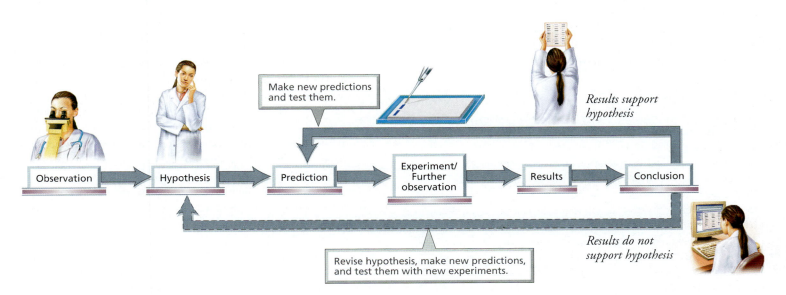

Make new predictions and test them.

Results support hypothesis

Observation → Hypothesis → Prediction → Experiment/ Further observation → Results → Conclusion

Revise hypothesis, make new predictions, and test them with new experiments.

Results do not support hypothesis

FIGURE **1.9**

Scientific process involves observation, creating testable hypotheses, experimentation, drawing conclusions, revising hypotheses, and designing new experiments.

plants, *then* there should be a cleared area around this plant; *then* this plant should have a way of producing some chemical; *then* that chemical should be detectable; *then* the plant should have a delivery system to transport the chemical to the soil, and so on. You used deductive reasoning to make predictions that would support your hypothesis and to decide whether the results supported or refuted your hypothesis.

▌Clinical trials follow strict guidelines

When you began this journey, your goal was to find medicinal plants (see the Environmental Issue essay, *Medicinal Plants and the Shrinking Rain Forest*). The guides were helpful, introducing you to several native healers. You collected many species of plants identified by the healers as being medicinal and made extensive notes on how they are used and the ailments for which they are beneficial. Now, you will preserve the plants for the trip home, where extracts of the plants will undergo extensive testing before they reach the consumer.

Before testing a new drug or treatment on humans, you would want to know that it is not harmful (Table 1.1). Usually the drug is tested first on animals such as laboratory rodents. Most medical advances, including vaccinations, chemotherapy, new surgical techniques, and organ transplants, began with animal studies. Strict rules safeguard the care and use of animals in research and testing.

If no ill effects are discovered in animals receiving the drug, studies on humans, called *clinical trials*, may begin. In all phases of testing, the studies are done on people who volunteer. In Phase I, the drug is screened for safety on fewer than 100 healthy people. At this stage, researchers hope to learn whether they can safely give the drug to humans. If the drug is safe, researchers will need to know the maximum safe dose.

If found to be safe, the drug is tested further. In Phase II, a few hundred people with the target disease are given the drug to see if it works for its intended purpose. If it does, final questions can be addressed in Phase III: How does the new drug compare with alternative treatments? What are its side effects? Thousands of participants are involved in Phase III of testing. The Food and Drug Administration (FDA) approves only those drugs or treatments that have passed all phases of testing.

what would you do?

The job of the FDA is to ensure the safety and effectiveness of new drugs and treatments. It must balance the patients' desires for access to new ways for treating illnesses that are sometimes life-threatening against the government's desire to protect patients from unsafe treatments. The drug approval process is painstakingly slow. Do you think that the FDA should bypass certain steps of the approval process to make new drugs available to critically ill patients who may not be able to wait? If you do, what criteria should be used to decide the degree of illness that would warrant treatment with a drug that was not yet approved? Who should be held responsible if early access to a drug of unknown safety causes serious side effects or the premature death of a patient? The FDA? The drug company? The patient? If you were ill and there were a drug for your illness in clinical trials, would you participate in those trials? What factors would influence your decision?

TABLE **1.1**

TESTS PERFORMED ON A NEW DRUG BEFORE IT IS APPROVED BY THE FOOD AND DRUG ADMINISTRATION (FDA)
Tests on laboratory animals
Is the drug safe for use on animals?
Clinical trials
Phase I   Is the drug safe for humans?
Phase II   Does the drug work for its intended purpose?
Phase III   How does the new drug compare with other available treatments?

Clinical studies must be properly designed. Recall that a good experiment requires both an experimental group and a control group. The experimental group would receive the drug under consideration. The control group would receive a **placebo**, a drug that appears to be identical to the drug being tested but has no known effect on the condition being studied. Study participants are randomly assigned to either the control group or the experimental group and do not know whether they are receiving the treatment or a placebo. When neither the researcher nor the study participants know which people are receiving treatment and which are receiving the placebo, the study is described as being *double blind*. It is important that participants not know whether they are receiving the placebo or the drug because their expectations about the effects the drug could bias the results. Researchers should not know which people are in the experimental or control groups because their expectations or desire for a particular result could affect the interpretation of the data.

Finally, it is extremely important that study participants give their **informed consent**. An informed consent document lists all the possible harmful effects of the drug or treatment and must be signed before a person can take part in the study. Examples of experiments on humans in which informed consent was not obtained or was obtained under questionable circumstances are described in the Social Issue essay in Chapter 2, *The Ethics of Radiation Research on Humans*.

Human health can also be studied without clinical trials. How might we investigate the effects of air pollution on asthma, a condition in which airway constriction causes breathing problems? We might address such a question by looking at patterns that occur within populations—an *epidemiological* study. We look for a correlation between the variable of interest (air pollution) and its suspected effects (worsening of asthma). Suppose that your hypothesis is that air pollution aggravates asthma. You might predict, then, that the number of people admitted to hospitals for asthma-related problems would increase with increased levels of air pollution.

Several studies have shown that the number of hospital admissions for asthma and other respiratory problems do increase with the level of air pollution. In one study, the residents of Brisbane City, Australia, were followed for 8 years. During this time, the levels of ozone, nitrogen dioxide, sulfur dioxide, and particulate matter—all components of air pollution—were measured every 30 minutes at nearby government sites. The number of admissions to hospital

ENVIRONMENTAL ISSUE

Medicinal Plants and the Shrinking Rain Forest

Many cultures have known about the healing power of plants for centuries.

Historically, knowledge of the healing powers of plants was gained by trial and error and passed through generations by word of mouth. For example, many cultures have long known that tea made from willow bark relieves pain and reduces fever. Scientists later learned that willow bark contains salicylic acid, the active ingredient in aspirin. Digitalis, a heart medication, was discovered when a patient with an untreatable heart condition drank an herbal medicine provided by a local gypsy. His health improved! The active ingredient in the potion was purple foxglove, which, like willow bark, is frequently mentioned in ancient texts as a healing herb.

Medicinal plants are still important today. More than 25% of the prescription medicines sold in the United States contain chemicals that came from plants. Many more healing chemicals were first discovered in plants known to have medicinal power by native people and later synthesized in laboratories (Table 1.A).

Although medicinal plants can be found almost everywhere, a particularly rich source is the tropical rain forest. The American Cancer Society has identified approximately 3000 plants that have anticancer properties, and 70% of them are found in the rain forest. As a result, there are now more than 100 drug companies funding projects to learn about medicinal plants from native people living in tropical rain forests.

The problem is that most plants that have proved to be medically useful are found in the tropics, regions where the human population is growing rapidly. Forests are being cut to create living space and to foster economic development. For example, the rosy periwinkle, the source of two cancer drugs, is found in Madagascar, a region where humans have destroyed 90% of the vegetation. It has been estimated that roughly nine-tenths of the original tropical rain forest will have been destroyed by 2030. With 155,000 of the known 250,000 plants existing in tropical rain forests and less than 2% of the known species having been tested for their medicinal value, we have no way of knowing what potential new medicines are being destroyed. 🌿

TABLE **1.A SELECTED MEDICINAL PLANTS AND THEIR USES**

SOURCE PLANT	COMPOUND	MEDICINAL USE
Willow bark and meadowsweet	Salicylic acid (active ingredient in aspirin)	Pain relief, fever reduction
Foxglove	Digitalin	Treatment of heart conditions (increase intensity of heart contraction, slows heart rate)
Quinine tree	Quinine	Malaria preventive
Opium poppy	Morphine	Pain relief
Eucalyptus tree	Menthol	Decongestant
Rauvolfia root	Reserpine	Treatment of high blood pressure
Belladonna	Atropine	Dilates pupils for eye exams
Pacific yew	Taxol	Treatment of ovarian and breast cancer
Rosy periwinkle	Vincristine	Treatment of Hodgkin's disease (a type of cancer)
	Vinblastine	Treatment of leukemia

emergency rooms for respiratory symptoms was also recorded. When the levels of ozone and particulate matter increased, so did the number of hospital admissions for asthma.

Conversely, we would also expect that the rate of asthma attacks would go down as air pollution levels decreased. The 1996 Olympic Games in Atlanta, Georgia, provided an opportunity to test this idea. To reduce traffic congestion in the city during the games, Georgia officials provided 24-hour public transportation, increased park-and-ride services, closed the city to private automobiles, and promoted alternative work schedules and telecommuting. As a result, the level of air pollution was lower during the games than it was during the 4 weeks preceding and the 4 weeks following the games. Consistent with our prediction, the frequency of asthma attacks in children during the interval of lower air pollution dropped by 11% to 44%, depending on the hospital being monitored. Although these results are certainly consistent with our predictions, they do not prove our hypothesis.

An alternative explanation might exist for the decrease in asthma attacks during the Olympic Games. Molds can also trigger asthma attacks. The Olympics took place during the time of year when mold counts are usually low. Thus, two variables—air pollution and mold—could have changed over the course of the study. In a scientific study, the second variable, in this case mold counts, is called a **confounding variable**. When there are two variables, we cannot say for sure which one caused the effect. So now we must ask if the number of asthma attacks among children dropped because the mold counts were lower. Probably not. The levels of mold did not change between the time of the Olympic Games and the weeks preceding and following the games. Furthermore, the daily mold count and the number of asthma attacks were not correlated at any time during the study in the Atlanta area.

So we see that new medications and new understandings of how things in our environment affect our health come only after careful and detailed study.

Critical Thinking Helps Us Make Informed Decisions

It is likely that few of you perform controlled experiments in your daily lives. However, each of us must evaluate scientific claims. They come to us in many ways—in advertisements, through the news, and in friendly conversation. We often must make decisions on the basis of these claims. How can we decide whether such claims are true? Critical-thinking skills can help us analyze the information and make informed decisions.

The key to becoming a critical thinker is to ask questions. The following list is not exhaustive, but it may help guide your thinking process.

1. **Is the information consistent with information gathered from other sources?** The best way to answer this question is to gather as much information as possible. Do not passively accept the "fact" as true. Do some research. If the information is controversial, listen to the debate among the "experts."

2. **How reliable is the source of the information?** Investigate the source of the information to determine whether that person or group has the training to know whether the "fact" is correct. Is there any reason to think that the information presented may be biased? Who stands to gain if you believe the claim? For example, the Food and Drug Administration is probably a more reliable source of information on the effectiveness of a drug than is the drug company marketing the drug.

3. **Was the information obtained through proper scientific procedures?** Information gathered through controlled experiments is more reliable than anecdotal evidence, which is unverified. For example, your friend might tell you that his muscles have gotten larger since he started using some special exercise equipment. But you cannot be sure unless measurements were taken before and after he began using the equipment. Even if your friend can prove his muscles are bigger, there is no guarantee that exercising with this equipment will build *your* muscles.

4. **Were the results of the experiment interpreted correctly?** Consider, for example, the headline advertising capsules containing fish oils: "Fish Oils Increase Longevity." It would be easy (but incorrect) to conclude from this headline that you will live longer if you take fish oil supplements. This headline is based on an experiment in which *dietary* fish oil was altered in *rats*. Rats fed a diet high in fish oils lived longer than did rats on a diet low in fish oil. The claim that taking fish oil supplements will increase longevity is not a valid conclusion based on the experiment. First, the study was done on rats, not on humans. Second, the amount of dietary fish oil, not the amount of fish oil from capsules that supplement dietary fish oil, was the variable in this experiment. Supplements of fish oil may not have the same effect as dietary fish oils. It could be that taking fish oil supplements would boost the amount of fish oil in your body to unhealthy levels.

5. **Are there other possible explanations for the results?** Suppose you learn that the fish oil headline is based on a study that showed that people who eat fish at least three times a week live longer than those who eat fish less frequently. In this case, the data indicate that there is a correlation between fish in the diet and length of life. However, a correlation between two factors does not prove that one *caused* the other. There may be other lifestyle habits that differ between these groups. For instance, people who eat fish may exercise more frequently or have less stressful jobs or live in areas with less pollution. One of these other factors may be responsible for the increase in longevity.

You will be forced to make many decisions in your life. Some of them will affect society. Should we eliminate genetically modified food? Should stem cell research be permitted? Should we vaccinate the general public against smallpox? Should companies polluting the atmosphere be taxed? We will raise these and similar questions in this text. The media will raise others.

Although you may never be a lawmaker who makes decisions on these issues, you are a voter who can help choose the lawmakers who will decide. Scientists can provide some "facts" that may be useful as we all struggle to answer the complex questions facing society, but they cannot provide answers. Each of us must stay informed and exercise critical-thinking skills to make decisions about not only these issues but also the new ones that inevitably will arise as our knowledge grows.

REVIEWING THE CONCEPTS

All Living Things Share Basic Characteristics (pp. 2–4)

1. Life cannot be defined, only characterized.
2. Living things contain nucleic acids, proteins, carbohydrates, and lipids; are made of cells; reproduce, metabolize, respond, and maintain homeostasis. Populations of living things evolve.

WEB TUTORIAL 1.1 Signs of Life

Living Organisms Share an Evolutionary History (pp. 4–6)

3. Living organisms are classified in a manner that shows the evolutionary relationships among them. One classification system recognizes six kingdoms: Bacteria, Archaea, Protista, Fungi, Plants, and Animals.
4. Humans are classified as animals (kingdom), chordates (phylum), vertebrates (subphylum), mammals (class), primates (order), *Homo* (genus), *Homo sapiens* (species).
5. Humans share traits with other species, but they also have many features not shared by most other animals, among the most important of which is culture.

Life Has Many Levels of Organization (p. 6)

6. Human biology can be studied at different levels. Within an individual, increasing levels of study are molecules, cells, tissues, organs, organ systems, and individual. Above the individual, increasing levels of study are populations, communities, ecosystems, and the biosphere.

The Scientific Method Helps to Gather Information and Reach Conclusions (pp. 6–12)

7. The scientific method is a sequence of activities that involves observation, forming a question and a hypothesis (a testable explanation for the observation), experimentation (performed with controls), and drawing a conclusion, which may lead to further experimentation.

8. As evidence in support of a hypothesis mounts, it may become so impressive that the hypothesis becomes a theory, a well-supported explanation of nature.

WEB TUTORIAL 1.2 Scientific Method

9. With inductive reasoning, the conclusion is a general statement summarizing a large number of specific observations. Deductive reasoning progresses from the general to the specific. It uses "if-then" logic.
10. Experiments on humans and other animals follow strict rules. The drug approval process usually begins with studies on animals. If no ill effects in animals are observed, the drug is then tested on humans. The drug trials determine whether the drug is safe for humans (Phase I), whether it works for its intended purpose (Phase II), and finally whether it is more effective than existing treatments (Phase III).
11. The experimental design in human experiments often includes an experimental group that receives the treatment and a control group that receives a placebo, which is a drug that has no known effect on the condition being treated. In a double blind experiment, neither the study participant nor the researcher knows which participants are receiving the real treatment. Study participants must sign an informed consent, indicating that they were made aware of the possible harmful consequences of the treatment.
12. In epidemiological studies, we look for a correlation between the variable of interest and its suspected effects by examining patterns within populations.

Critical Thinking Helps Us Make Informed Decisions (p. 12)

13. Critical thinking involves asking questions and evaluating evidence to draw conclusions and make informed decisions.

KEY TERMS

adaptive trait *p. 4*
biosphere *p. 6*
community *p. 6*
confounding variable *p. 12*

controlled experiment *p. 6*
deductive reasoning *p. 9*
ecosystem *p. 6*
homeostasis *p. 4*

hypothesis *p. 6*
inductive reasoning *p. 9*
informed consent *p. 10*
metabolism *p. 2*

population *p. 6*
placebo *p. 10*
theory *p. 9*
variable *p. 7*

THINKING ABOUT THE CONCEPTS

1. List seven features that characterize life. *pp. 2–4*
2. How are humans classified? *pp. 4–6*
3. Distinguish among a population, a community, an ecosystem, and the biosphere. *p. 6*
4. What is a hypothesis? How does it differ from a theory? *pp. 6,9*
5. Define a controlled experiment. *pp. 6–7*
6. Differentiate between inductive and deductive reasoning. *pp. 9–10*
7. What is a placebo? *p. 10*
8. What is meant by a double blind experiment? *p. 10*
9. Describe the procedure in an epidemiological study. *p. 10*

10. A theory is
    a. a testable explanation for an observation.
    b. a conclusion based on the results of an experiment.
    c. a broad-ranging explanation for natural events that has been extensively tested over time.
    d. the factor that is altered in a controlled experiment.
11. The condition in which the physical and chemical conditions inside the body are maintained within tolerable ranges is called _____.
12. A(n) _____ is a testable explanation for an observation.
13. A trait that increases the chance that an organism will survive and reproduce in its natural environment is described as being _____.

APPLYING THE CONCEPTS

1. One of the natives you met in the rain forest told you that one of the plants you collected makes it easier to breathe during episodes when breathing becomes difficult. You suspect that it might be a good treatment for asthma, a condition in which constriction of airways causes breathing problems. You are able to isolate a component of this plant as a drug. Tests on animals show that it is effective. Tests on humans show that the drug can be given safely to humans. Design an experiment to test the hypothesis that this drug eases breathing during an asthma attack.

2. Find an article or advertisement that makes a scientific claim. Use your critical-thinking skills to evaluate the claim.

The Chemistry of Life

Basic Chemistry Aids in Understanding Human Biology

- Atoms contain protons, neutrons, and electrons
- Elements combine to form compounds
- Chemical bonds form between the atoms of a compound
- Water is essential to life

Carbohydrates, Lipids, Proteins, and Nucleic Acids Are the Major Molecules of Life

- Carbohydrates supply energy to cells
- Lipids store energy and form cell membranes
- Proteins provide structure and speed up chemical reactions
- Nucleic acids include DNA and RNA

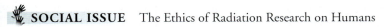 **SOCIAL ISSUE**   The Ethics of Radiation Research on Humans

 **ENVIRONMENTAL ISSUE**   What Is Happening to the Rain?

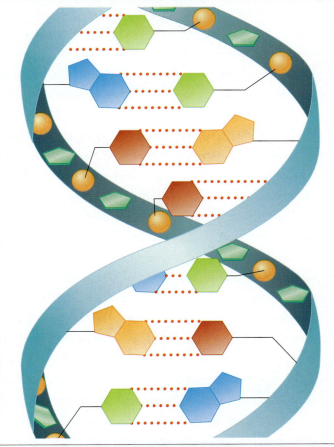

Kathryn and her mother did their best to maintain optimistic attitudes, but Thursday afternoon was distressing. Since her mother had been diagnosed with non-Hodgkin's lymphoma, Kathryn had spent every morning at the Cancer Center's chemotherapy unit, waiting to drive them back to the hotel, care for her mother, then return the next day for another round of treatment. Each afternoon Kathryn would try to cheer her mother with encouraging stories she'd read about cancer survivors, and then they would head to a nearby park or to the shops. But this afternoon was different. Kathryn saw that her mother's hair was thinning and her skin was darkening. Neither of them could eat their lunch. Kathryn felt her tears rising as she feared for her mother's survival. Although the doctors assured her that the chemotherapy could save her mother's life, at this point it seemed only to be making her sicker. What Kathryn's mother was experiencing was an inevitable consequence of the effects of chemical treatments on the body's own natural chemistry.

Chemotherapy directly attacks chemical components in cells—primarily nucleic acids and proteins. Because cancer cells are somewhat more active than most cells—dividing uncontrollably—they are slightly more sensitive to chemotherapy than most other cells. Other cells in our bodies also divide rapidly, such as those in hair follicles and the intestinal lining. These too suffer the assaults of chemotherapy. As Kathryn's mother was discovering, it's a fine line between destroying normal cells and destroying the lymphoma cells. Although she stands a good chance of conquering her lymphoma, Kathryn's mother will have to experience the temporary debilitating effects of her treatment.

In this chapter, we learn some basic principles of chemistry and lay the foundation for what follows. We begin with atoms and end with the major molecules of life. We explore how these major molecules form the foundations of our bodies and minds and give us the energy for life. ■

Basic Chemistry Aids in Understanding Human Biology

The world around you is made of many physical substances such as the grass or concrete on which you walk, the water you drink, the air you breathe, and even this book that you are reading. All of these substances that make up our world are called matter. In basic terms, **matter** is anything that takes up space and has mass. All forms of matter are made up of tiny particles called atoms.

▌Atoms contain protons, neutrons, and electrons

Atoms are units of matter that cannot be broken down by chemical means. Atoms, the most basic units of matter, are composed of even smaller subatomic particles that include protons, neutrons, and elec-

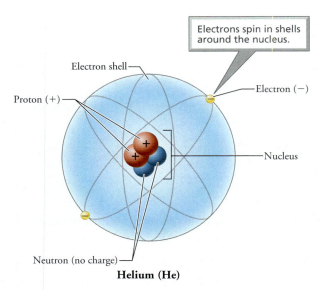

Electrons spin in shells around the nucleus.

Electron shell

Proton (+)

Electron (−)

Nucleus

Neutron (no charge)

Helium (He)

FIGURE **2.1**

An atom of helium, showing protons and neutrons (in the nucleus) and electrons

trons. These subatomic particles are characterized by their location within the atom, their electrical charge, and their mass. **Mass** is the measure of how much matter is contained within an object.

Each atom consists of two regions: the nucleus at the center and a surrounding "cloud" of electrons. As you can see in Figure 2.1, the nucleus contains protons and neutrons. Electrons spin rapidly around the nucleus in orbits called shells. The shell closest to the nucleus can hold up to 2 electrons. The next shell out can hold up to 8 electrons. Atoms with more than 10 electrons have additional shells. Neutrons, as their name implies, have no electrical charge; they are neutral. Protons, in contrast, have a positive charge. As a result, the nucleus of an atom has a positive charge. Electrons have a negative charge and generally balance, or neutralize, the positive charge of the nucleus. Put another way, because most atoms have the same number of positively charged protons and negatively charged electrons, they have no net charge and thus are "neutral."

Protons have essentially the same mass as neutrons. Their mass is measured in special units called atomic mass units, or amu. Electrons are much smaller than either neutrons or protons and have almost no mass. Table 2.1 summarizes the basic characteristics of protons, neutrons, and electrons.

TABLE **2.1**

SUBATOMIC PARTICLES			
PARTICLE	**LOCATION**	**CHARGE**	**MASS**
Proton	Nucleus	1 positive unit	1 atomic mass unit
Neutron	Nucleus	None	1 atomic mass unit
Electron	Outside the nucleus	1 negative unit	Negligible

ELEMENTS

An **element** is a "pure" form of matter that cannot be broken down into simpler substances by ordinary chemical means. You are probably familiar with many elements such as gold, silver, iron, and oxygen. The earth is made up of a little over 100 elements, 92 of which occur naturally. Only about 20 elements are found in the human body, which consists mostly of carbon, oxygen, hydrogen, and nitrogen.

All the atoms that make up an element have the same number of protons in the nucleus. All the iron atoms that make up an iron bar, for example, have the same number of protons. The same is true for the element gold; all of its atoms contain the same number of protons in the nucleus. The number of protons in the atom's nucleus is called the **atomic number**.

The periodic table lists the elements and their characteristics. Figure 2.2 depicts a simplified periodic table. Note that each element has a name and a one- or two-letter symbol. The symbol for the element carbon, for example, is C, and that for chlorine is Cl. Each element has an atomic number and an atomic weight. Recall that the atomic number is simply the number of protons in the nucleus of an atom. Carbon has an atomic number of 6, which means it has 6 protons in its nucleus. Recall that each proton and neutron has an approximate mass of 1 atomic mass unit, or amu. The mass of an electron is so small that it is usually considered zero. Because electrons have negligible mass, and protons and neutrons each have an atomic mass of 1, the atomic weight for any element equals the number of protons plus the number of neutrons. Oxygen has an atomic weight of 16, indicating that it has 8 protons and 8 neutrons in its nucleus.

ISOTOPES AND RADIOISOTOPES

Although all the atoms of a particular element contain the same number of protons, they can have different numbers of neutrons. Such differences result in atoms of the same element having slightly different atomic weights. Atoms that have the same number of protons but differ in the number of neutrons are called **isotopes**. Over 300 isotopes occur naturally on the earth. The element carbon, for example, has three isotopes. All carbon atoms have 6 protons in the nucleus. Although most carbon atoms have 6 neutrons, some have 7 or 8. The isotopes of carbon thus have atomic weights of 12, 13, and 14, respectively, depending on the number of neutrons in the nucleus. These isotopes are written ^{12}C (the most common form in nature), ^{13}C, and ^{14}C.

Some isotopes are unstable and emit either particles or energy in the form of radiation. Such isotopes are called **radioisotopes**. About 60 naturally occurring isotopes emit radiation. Many more radioisotopes have been made in laboratories.

Radiation can be very dangerous and can damage the body directly or indirectly. Direct damage to a person receiving radiation may include a low white blood cell count, development of some cancers, and damage to organs and glands (Figure 2.3). In other cases, radiation may not produce any noticeable injury to the individual receiving the radiation. Nonetheless, radiation may alter the hereditary material carried in the cells of the person's reproductive system, possibly causing defects in the individual's offspring.

In stark contrast to the harmful effects of radiation are its medical uses. Medical professionals use small doses of radiation to generate visual images of internal body parts. These images may be used to diagnose irregularities in the body's structure or function. Radioactive iodine, for example, is often used to identify disorders of the thyroid gland. Located in the neck, the thyroid gland normally accumulates the element iodine, which it uses to regulate growth and metabolism. Small doses of iodine-131 (^{131}I), a radioactive isotope of iodine, can be given to a patient suspected of having metabolic problems. The radioactive iodine is taken up by the patient's thyroid gland. Medical instruments can then detect the radiation and translate its pattern into an image of the thyroid gland, as shown in Figure 2.4. This image can then be used in the diagnosis of disorders. The small amount of radioactive iodine used does not damage the thyroid gland or surrounding structures.

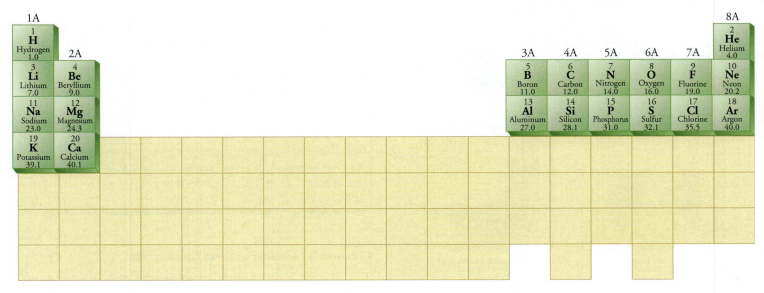

FIGURE **2.2**

A simplified version of the periodic table, showing elements with atomic numbers from 1 through 20

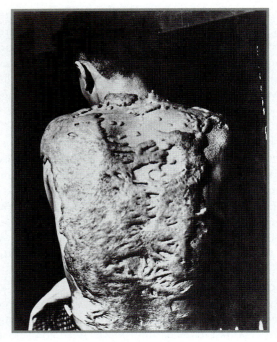

FIGURE **2.3**
An example of direct damage caused by radiation. This individual survived the atomic explosion at Nagasaki, Japan, on August 9, 1945. Damage to the reproductive cells of people exposed to radiation has proved more difficult to observe than such direct damage. Children of bomb survivors are still being monitored.

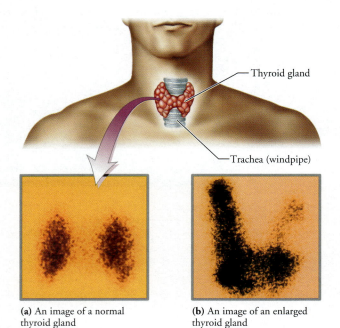

(a) An image of a normal thyroid gland

(b) An image of an enlarged thyroid gland

FIGURE **2.4**
Radioactive iodine can be used to generate images of the thyroid gland. Such images may be used to diagnose metabolic disorders.

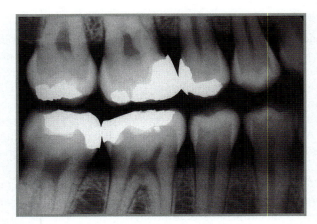

FIGURE **2.5**
A dental x-ray

However, larger doses can be used to kill thyroid cells when the gland is enlarged and overactive. A more familiar medical use of small doses of radiation is the x-ray (Figure 2.5).

Radiation also can be used to kill cancer cells. Cancer cells divide more rapidly and have higher rates of metabolic activity than do most normal cells. For these reasons, cancer cells are more susceptible to the destructive effects of radiation. During radiation therapy, beams of gamma rays are directed at a tumor to kill cancer cells. Such beams usually cause relatively minor damage to surrounding healthy tissue because efforts are made to carefully shield such tissue. Because radiation can be used either to kill or to heal, research into its effects and uses often raises ethical concerns. These concerns are particularly apparent when the subjects of the experiments are human (see the Social Issue essay, *The Ethics of Radiation Research on Humans*).

what would you do?

When food is irradiated, it is exposed to beams of electrons, X rays, or gamma rays emitted by radioactive forms of certain elements. The food does not become radioactive as a result. Irradiating food has many advantages, including (1) ridding food of harmful microorganisms (pathogens), insect pests, and parasites; (2) delaying spoilage and increasing shelf life; and (3) sterilizing the material in which the food is packaged. Supporters of food irradiation note that test animals that have been fed irradiated food have shown no adverse effects. Opponents, however, point to the environmental risks of building and operating food irradiation plants. Opponents also note that carefully controlled experiments are still needed to show that irradiated food is safe for people of all ages and nutritional states. Several foods, including wheat flour, poultry, and white potatoes, have been approved for irradiation in the United States. These irradiated foods have a distinctive logo (Figure 2.6) on their packaging. Would you eat irradiated food? Explain your answer.

▌Elements combine to form compounds

Two or more elements may combine with each other to form a **compound**. The characteristics of compounds are usually different from those of their elements. Consider, for example, what happens when you join the element sodium (Na) with the element chlorine

FIGURE **2.6**
Logo for irradiated foods. This logo and words such as "Treated with radiation" should appear on food that has been irradiated.

(a) The element sodium is a solid metal.

(b) Elemental sodium reacts explosively with water.

(c) The element chlorine is a yellow gas.

(d) When the elements sodium and chlorine join, they form table salt, a compound quite different from its elements.

FIGURE **2.7**
The characteristics of compounds are usually different from those of their elements.

(Cl). Sodium is a silvery metal that explodes when it comes into contact with water. Chlorine is a deadly yellow gas. When the two elements join, the result is a crystalline solid called sodium chloride (NaCl)—plain table salt (Figure 2.7).

▌ Chemical bonds form between the atoms of a compound

The atoms in a compound are held together by chemical bonds. There are three types of chemical bond: covalent, ionic, and hydrogen. In covalent and ionic bonds, the atoms behave as if they are trying to complete their electron shells. Recall that these shells are the specific areas within the atom in which electrons are most likely to be found. As shown in Figure 2.8, shells can be drawn as circles around the atomic nucleus. As you can see in the figure, the innermost shell contains 2 electrons. The hydrogen atom has only 1 electron and only one shell. The next shell can hold up to 8 electrons. Atoms with more than 10 electrons have additional shells. The number of electrons in the outermost shell of an atom determines the type of chemical bond that forms between atoms.

COVALENT BONDS

A **covalent bond** forms when two or more atoms share the electrons in their outer shells. Consider the compound methane (CH_4). Methane is formed by the sharing of electrons between one atom of carbon and four atoms of hydrogen. Notice in Figure 2.9a that the outer shell of a carbon atom contains only four electrons, even though it can hold eight. Also note that hydrogen atoms only have one electron, although the first shell can hold two electrons. A carbon atom can fill its outer shell by joining with four atoms of hydrogen, each with only one electron in its outer shell. By forming a covalent bond with the carbon atom, each hydrogen atom fills its outer shell. We see, then, that the covalent bonds between the carbon atom and hydrogen atoms of methane result in filled outer shells for all five atoms involved.

A **molecule** is a chemical structure held together by covalent bonds. Molecules are described by a formula that contains the symbols for all of the elements included in that molecule. If more than one atom of a given element is present in the molecule, subscripts are used to denote the precise number of atoms. For example, sucrose (table sugar) has the molecular formula $C_{12}H_{22}O_{11}$. So, one molecule of sucrose contains 12 atoms of carbon, 22 atoms of hydrogen, and 11 atoms of oxygen. Numbers placed in front of the molecular formula indicate more than one molecule. For example, three molecules of sucrose are described by the formula $3C_{12}H_{22}O_{11}$.

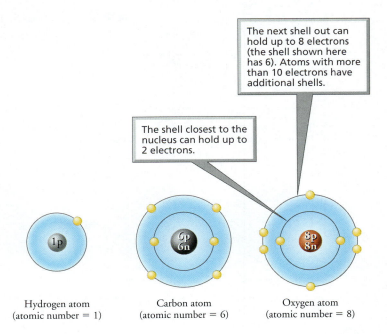

The next shell out can hold up to 8 electrons (the shell shown here has 6). Atoms with more than 10 electrons have additional shells.

The shell closest to the nucleus can hold up to 2 electrons.

Hydrogen atom (atomic number = 1)

Carbon atom (atomic number = 6)

Oxygen atom (atomic number = 8)

FIGURE **2.8**
Atoms of hydrogen, carbon, and oxygen. Each of the concentric circles around the nucleus represents a shell occupied by electrons.

As shown in the methane molecule in Figure 2.9a, a single pair of electrons is shared between the carbon atom and each of the four hydrogen atoms. As a result, the methane molecule contains four single covalent bonds. Sometimes, however, atoms share two or three pairs of electrons. These bonds are called double and triple covalent bonds, respectively. Carbon dioxide, CO_2, produced by chemical reactions inside our cells, has double covalent bonds between the carbon atom and each oxygen atom (Figure 2.9b). The nitrogen atoms in nitrogen gas, N_2, have triple covalent bonds between them (Figure 2.9c).

Covalent bonds in molecules are sometimes depicted by a structural formula. Notice in Figure 2.9a that one straight line is drawn between the carbon atom and the four hydrogen atoms in the methane molecule. The single line indicates a single covalent bond representing a pair of shared electrons. The double lines between the carbon and oxygen atoms in the carbon dioxide molecule indicate a double covalent bond, or two pairs of shared electrons (see Figure 2.9b). Three lines drawn between the two nitrogen atoms in gaseous nitrogen depict a triple covalent bond, or three pairs of shared electrons (see Figure 2.9c).

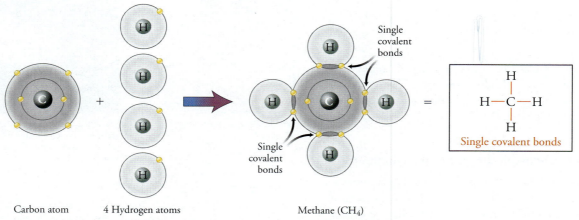

Carbon atom   4 Hydrogen atoms   Methane (CH_4)

(a) The molecule methane (CH_4) is formed by the sharing of electrons between one carbon atom and four hydrogen atoms. Because, in each case one pair of electrons is shared, the bonds formed are single covalent bonds.

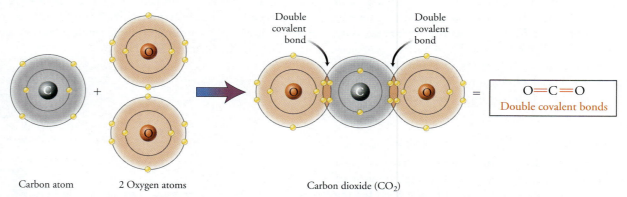

Carbon atom   2 Oxygen atoms   Carbon dioxide (CO_2)

(b) The oxygen atoms in a molecule of carbon dioxide (CO_2) form double covalent bonds with the carbon atom. In double bonds, two pairs of electrons are shared.

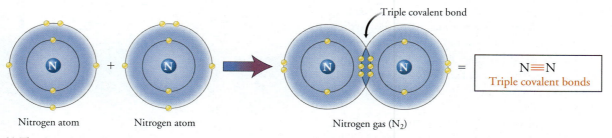

Nitrogen atom   Nitrogen atom   Nitrogen gas (N_2)

(c) The nitrogen atoms in nitrogen gas (N_2) form a triple covalent bond in which three pairs of electrons are shared.

WWW FIGURE **2.9**

TUTORIAL 2.2

Covalent bonds form when electrons are shared between atoms. Shown here are examples of covalent bonding involving single, double, and triple covalent bonds. For each example, the structural formula is shown on the far right.

IONIC BONDS

We have all heard the phrase "opposites attract" in reference to human relationships—and it is no different for ions. An **ion** is an atom or group of atoms that carries either a positive (+) or a negative (−) electrical charge. An **ionic bond** results from the mutual attraction of oppositely charged ions. Electrical charges result from the transfer of electrons between atoms. Recall that a neutral atom has the same number of positively charged protons and negatively charged electrons. An atom that loses an electron has one more proton than electrons and has a positive charge. An atom that gains an electron has one more electron than protons and is negatively charged. Oppositely charged ions are attracted to one another.

Atoms tend to interact with certain other atoms so as to fill their outermost shell. Consider, again, the atoms of sodium and chlorine that join to form sodium chloride. As shown in Figure 2.10, an atom of sodium has one electron in its outer shell. An atom of chlorine has seven electrons in its outer shell. Sodium chloride is formed when the sodium atom transfers the single electron in its outer shell to the chlorine atom. The sodium atom, having lost an electron, has one more proton than electrons and has a positive charge (Na^+). The chlorine atom, having gained an electron, has one more electron than protons and has a negative charge (Cl^-). These oppositely charged ions are attracted to one another, and an ionic bond forms.

HYDROGEN BONDS

In covalently bonded molecules, electrons may be shared equally or unequally between atoms. When the sharing of electrons during covalent bonding is unequal, different parts of the same molecule have slight opposite charges. Unequal covalent bonds are called polar, and the resulting molecules are called polar molecules. In water (H_2O), for example, the electrons shared by oxygen and hydrogen spend more time circling around the oxygen atom than they do around the hydrogen atom. As a result, the oxygen atom has a slight negative charge; each hydrogen atom has a slight positive charge; and water molecules are polar. The hydrogen atoms of one water molecule are attracted to the oxygen atoms of other water molecules. The attraction between a slightly positively charged hydrogen atom and a slightly negatively charged atom nearby is called a **hydrogen bond** (Figure 2.11a). In the case of water, the bonds are between hydrogen and oxygen. However, sometimes hydrogen bonds form between hydrogen and atoms of other elements. Hydrogen bonds are weaker than either ionic or covalent bonds. For this reason, they are illustrated by dotted lines rather than solid lines, as shown in Figure 2.11b. Even though individual hydrogen bonds are very weak, collectively they can be significant. Hydrogen bonds account for the unique physical properties of water (see below). They also maintain the shape of proteins and our hereditary material, DNA.

The three types of chemical bond—covalent, ionic, and hydrogen—are summarized in Table 2.2.

❚ Water is essential to life

Water is such an integral part of our everyday lives that we often overlook its unusual qualities. Unique properties of water include cohesiveness, a high heat of vaporization, and the ability to act as a dissolving agent. As it turns out, these qualities make water a critical component of the human body. Many of water's unusual qualities can be traced to hydrogen bonds between the water molecules.

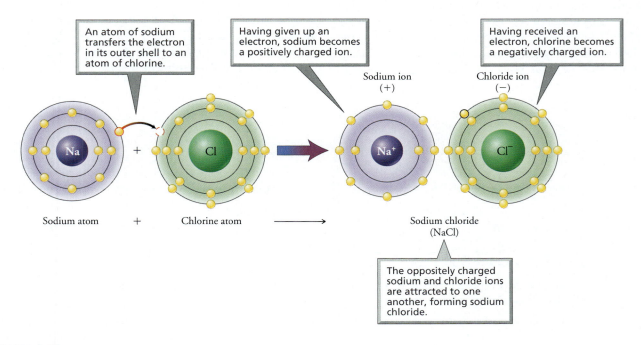

www FIGURE **2.10**

An ionic bond involves the transfer of electrons between atoms. Such a transfer creates oppositely charged ions that are attracted to one another.

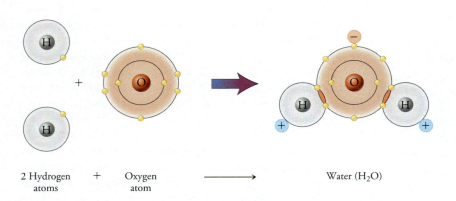

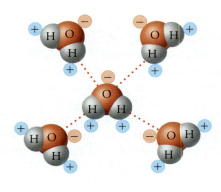

2 Hydrogen + Oxygen ⟶ Water (H₂O)
atoms atom

(a) Water is formed when an oxygen atom covalently bonds (shares electrons) with two hydrogen atoms. As a result of unequal sharing of electrons, oxygen carries a slight negative charge and the hydrogen atoms a slight positive charge.

(b) The hydrogen atoms from one water molecule are attracted to the oxygen atoms of other water molecules. This relatively weak attraction (shown by dotted lines) is called a hydrogen bond.

 FIGURE **2.11**
The hydrogen bonds of water

PROPERTIES OF WATER

Life depends on the properties of water. As we have seen, water is a polar molecule; the polarity results in hydrogen bonding. Hydrogen bonds cause water molecules to stick together. When water is a liquid, the hydrogen bonds are very fragile—frequently forming, breaking, and re-forming. Let's consider some of the ways that the polarity and hydrogen bonding of water are related to the properties of water that make it so vital to life.

Because of the polarity of its molecules, water interacts with many substances. Water is an excellent solvent, easily dissolving both polar and charged substances. Thus, water serves as the body's main transport medium. Nutrients, gases, and wastes carried by the vessels of the circulatory system dissolve in the water of blood. Metabolic wastes are excreted from the body in urine, another watery medium.

Water also helps prevent dramatic changes in body temperature. Its role in keeping a relatively constant body temperature comes from its high heat capacity, which simply means that a great deal of heat is required to raise its temperature. Hydrogen bonds hold multiple water molecules together, so a large amount of heat is required to break these bonds. About 67% of the human body is water. To put this figure into perspective, if a person weighs 68 kilograms (150 pounds), water makes up about 45 kilograms (100 pounds) of the body weight. Because humans, as well as other organisms, are made up largely of water, they are well suited to resist changes in body temperature and to keep a relatively stable internal environment. Water in blood also helps redistribute heat within our bodies. Our fingers don't usually freeze on a frigid day because heat is carried to them by blood from muscles where the heat is generated.

TABLE **2.2**

CHEMICAL BONDS			
TYPE	**BASIS FOR ATTRACTION**	**STRENGTH**	**EXAMPLE**
Covalent	Sharing of electrons between atoms; the sharing between atoms may be equal or unequal	Strongest	CH₄ (methane)
Ionic	Transfer of electrons between atoms creates oppositely charged ions that are attracted to one another	Strong	NaCl (table salt)
Hydrogen	Attraction between a hydrogen atom with a slight positive charge and another atom (often oxygen) with a slight negative charge	Weak	Between a hydrogen atom on one water molecule and an oxygen atom on another water molecule

The Ethics of Radiation Research on Humans

In the United States between about 1945 and 1971, a number of people were exposed to radiation as part of research into its effects and uses (Figure 2.A). These people included terminally ill hospital patients, mentally delayed school children, prisoners, and scientists. Did the research subjects or their families understand the risks of these experiments? Did anyone warn them? In other words, did the participants, or someone speaking for them, give their informed consent?

In one set of radiation experiments, the military's security concerns took precedence over fully informing study participants. Between 1945 and 1947, 18 terminally ill volunteers received injections of a substance called "product." The true identity of "product" was not disclosed to the subjects. The substance was plutonium, an element used to make one type of atomic bomb. The rationale for the study was that many workers might have been exposed to plutonium during construction of the bomb and many more might be exposed at nuclear facilities in the years ahead, yet little was known about the effects of plutonium on humans. The military needed to know about such effects. Plutonium is metabolized differently by different species. So experiments on the effects of plutonium on other animals could not be applied to humans. Also, although several thousand workers at atomic facilities were probably exposed to plutonium, the precise levels of exposure were unknown. The exposures, after all, had been accidental. The researchers believed it necessary to inject known quantities of the element into human subjects. They would then monitor the movement of plutonium through the body. In this way, they could determine whether plutonium was rapidly excreted by humans or held in their tissues for years.

The scientists believed the research was necessary. Indeed, they anticipated potentially valuable results. Perhaps, for example, the results would help set safety standards for plutonium exposure. Nevertheless, failure to fully inform the volunteers in the plutonium experiments raised ethical questions. Is it ethical to use terminally ill volunteers in potentially harmful experiments? Were terminally ill patients chosen because the researchers could avoid the issue of long-term harm? Were terminally ill patients convenient subjects because any plutonium remaining in the body could be measured at a not-too-distant autopsy? Could the need for an autopsy interfere with the doctor-patient relationship by making a quick death desirable? And what if some of the patients had been misdiagnosed and were not close to death after all? Because of such questions, terminally ill people are generally no longer used in research, except when the research involves an experimental treatment for their illness.

continued ⟶

ETHICS IN RADIOACTIVITY EXPERIMENTS ON HUMANS

In some of the radiation studies cited in newspapers in late 1993 and in an earlier congressional report, known as the Markey Report, subjects did not freely consent to the experiments. In other studies it is doubtful whether informed consent was obtained. But in some of the studies informed consent was truly given. Here are examples from each category.

Date	Experiment
Possible Infliction of Harm or No Informed Consent	
1945–47	18 supposedly terminal patients were injected with high doses of plutonium to learn whether the body absorbed it.
1946–47	6 hospital patients were injected with uranium salts to determine the dose that produced injury to the kidneys.
1963–70	64 prison inmates had their testicles exposed to x-rays to relate radiation damage to sperm production.
1963–71	67 prison inmates had their testicles exposed to x-rays to measure radiation damage to sperm production.
Questionable Consent	
1946	17 mentally delayed teenagers at the Fernald School in Waltham, Massachusetts, ate meals with trace amounts of radioactive iron to learn about iron absorption in the body.
1953–57	11 comatose brain cancer patients were injected with uranium to learn whether it is absorbed by brain tumors.
1954–56	32 mentally delayed teenagers at the Fernald School drank milk with trace amounts of radioactive calcium to learn whether oatmeal impeded its absorption by the body.
Informed Consent	
1945	10 researchers and workers at Clinton Laboratory, in Oak Ridge, Tennessee, voluntarily exposed patches of their skin to radioactive phosphorus.
1951	14 researchers at Hanford Nuclear Reservation voluntarily exposed patches of their skin to gaseous tritium.
1963	54 hospital patients volunteered to take trace amounts of radioactive lanthanum to measure effects on the large intestine.
1965	Trace doses of radioactive technetium were given to 8 healthy volunteers to determine its utility as a medical diagnostic tool.

FIGURE 2.**A**

A sample of radiation experiments performed on people

SOCIAL ISSUE (CONTINUED)

What about prisoners? Is it appropriate to use them? Prisoners served as subjects in experiments on the effects of high doses of radiation on sperm production. The idea for this research grew out of a 1962 accident at a nuclear facility. Three male workers at the facility were exposed to high doses of radiation. Officials found they could tell the men nothing about the possible impact of such exposure on their ability to father children. Prisoners were attractive research subjects because their confinement made follow-up easy. Because they volunteered to participate in the study, the prisoners sometimes received letters of cooperation in their files. Some received benefits such as improved food or housing. Such "enticements" raise the issue of whether the prisoners truly consented or were somehow coerced into participating in the research. Today, many people believe that prisoners should not be used as research subjects in medical experiments. This belief centers on the ease with which prisoners can be pressured to "volunteer" for experiments.

The principle of informed consent was legally established in the United States during the 1930s. Informed consent received international attention during the mid to late 1940s. At this time, Nazi doctors were on trial for torturing, in the name of science, people being held in concentration camps. However, for years afterward, various research groups in the United States violated the principle. In a number of other radiation studies with human subjects, the consent was questionable. In still others, informed consent was properly obtained. (Extreme examples of the latter include studies in which researchers conducted the experiments on themselves.)

In 1974, the U.S. Department of Health, Education, and Welfare established regulations for all research with human subjects. However, it was not until 1991 that these regulations were fully operational in all government agencies. Today, researchers working with human subjects must obtain the informed consent of all participants in their studies. Nevertheless, new ethical questions continue to arise. Should physicians provide AIDS patients with the experimental drugs they sometimes demand? Such patients may be more than willing to participate in research trials. However, the issue of informed consent may be clouded by lack of adequate information on the risks of the experiment. Without the experiment, of course, researchers might never discover new treatments. How can we get the data and assess the risks without conducting the experiments? Is informed consent always necessary? Should doctors be able to try out new methods to bring people out of a coma—people who cannot consent? Where should the lines be drawn?

Water also helps prevent the body from overheating. Water has a high heat of vaporization, which means that a great deal of heat is required to make water evaporate (that is, change from a liquid to a gas). Water molecules that evaporate from a surface carry away a lot of heat, cooling the surface. We rely on the evaporation of water in sweat to cool the body surface and prevent overheating (Figure 2.12).

FIGURE 2.12
The evaporation of water in sweat cools the surface of this runner's body. Water has a high heat of vaporization, so when water molecules in sweat evaporate, they carry away a lot of heat.

stop and think

Sharp increases in body temperature can cause heat stroke, a condition that may damage the brain. Explain why heat stroke is more likely to occur on a hot humid day than on an equally hot dry day.

ACIDS AND BASES

When a water molecule dissociates, or breaks up, it forms a positively charged hydrogen ion (H^+) and a negatively charged hydroxide ion (OH^-):

$$\underset{\text{Water}}{H{-}O{-}H} \leftrightarrow \underset{\text{Hydrogen ion}}{H^+} + \underset{\text{Hydroxide ion}}{OH^-}$$

Dissociation is not easily accomplished. As a result, water molecules are much more common in the human body than are H^+ and OH^-. In fact, the amount of H^+ in the body must be precisely regulated. Even slight changes in the concentration of H^+ can be disastrous, disrupting chemical reactions within cells.

Acids and bases are defined by what happens when each is added to water. An **acid** is anything that releases hydrogen ions (H^+) when placed in water. A **base** is anything that produces hydroxide ions (OH^-) when placed in water. Hydrochloric acid (HCl), for example, dissociates in water to produce hydrogen ions (H^+) and chloride ions (Cl^-). Because HCl increases the concentration of H^+ in solution, it is classified as an acid. Sodium hydroxide ($NaOH$), on the other hand, dissociates in water to produce sodium ions (Na^+) and hydroxide ions (OH^-). Because NaOH increases the concentration of OH^- in solution, it is classified as a

base. The OH^- produced when NaOH dissociates reacts with H^+ to form water molecules and thus reduces the concentration of H^+ in solution. Therefore, acids increase the concentration of H^+ in solution, and bases decrease the concentration of H^+ in solution.

THE pH SCALE

We often want to know more than simply whether a substance is an acid or a base. For example, how strong an acid is HCl? Asked in a different way, how acidic is the solution created when HCl is added to water? Questions such as these can be answered using the **pH scale** (Figure 2.13). This scale ranges from 0 to 14, with a pH of 7 being neutral, a pH of less than 7 being acidic, and a pH of greater than 7, basic. But what exactly is pH? Usually, the amount of H^+ in a solution is very small. For example, the concentration of H^+ in a solution with a pH of 6 is 1×10^{-6} (or 0.000001) moles per liter (a mole, here, is not a small animal, but a unit of measurement that indicates a specific number of atoms, molecules, or ions). Similarly, the concentration of H^+ in a solution with a pH of 5 is 1×10^{-5} (or 0.00001) moles per liter. Technically, pH is the negative logarithm of the concentration of H^+ in a solution. According to the pH scale, the lower the pH, the greater the acidity, or concentration of H^+, in a solution. Each lower pH unit has 10 times the amount of H^+ than the preceding unit. So a solution with a pH of 5 is 10 times more acidic than a solution with a pH of 6. And a solution with a pH of 4 is 100 times more acidic than one with a pH of 6. Some of the characteristics of acids and bases, including their values on the pH scale, are summarized in Table 2.3.

BUFFERS

Typically, biological systems function within a narrow range of pH values. Substances called **buffers** prevent dramatic changes in pH. Buffers remove excess H^+ from solution when concentrations increase. Buffers add H^+ when concentrations decrease. For example, an important buffering system that keeps the pH of blood at about 7.4 is the carbonic acid (H_2CO_3)–bicarbonate (HCO_3^-) system. When carbon dioxide is added to water it forms carbonic acid, which dissociates into hydrogen ions and bicarbonate ions:

$$CO_2 + H_2O \leftrightarrow H_2CO_3 \leftrightarrow H^+ + HCO_3^-$$

| Carbon dioxide | Water | Carbonic acid | Hydrogen ion | Bicarbonate ion |

The buffering action of carbonic acid and bicarbonate results from the fact that when levels of H^+ decrease in the blood, carbonic acid dissociates, adding H^+ to solution. When levels of H^+ increase in the blood, the H^+ combines with bicarbonate and is removed from solution. Such action is essential because even slight changes in the pH of blood—say, a drop from 7.4 to 7.0 or an increase to 7.8—can cause death in a few minutes. The critical link between pH and life is also illustrated by the impact of acid rain on our environment and health (see the Environmental Issue essay, *What Is Happening to the Rain?*).

In the human body, almost all biochemical reactions occur around pH 7 and are maintained at that level by powerful buffering systems. An important exception occurs in the stomach, where pH values from about 1 to 3 are found. Hydrochloric acid (HCl), secreted by cells that line the stomach, promotes the breakdown of proteins. Normal digestion requires an acid stomach, and the stomach has several mechanisms

to protect it from the acid. However, sometimes stomach acid backs up into the esophagus, and "heartburn" is the uncomfortable result. Taking an antacid can treat the discomfort of heartburn. Antacids consist of weak bases that temporarily relieve the pain of stomach acid in the esophagus by neutralizing some of the hydrochloric acid.

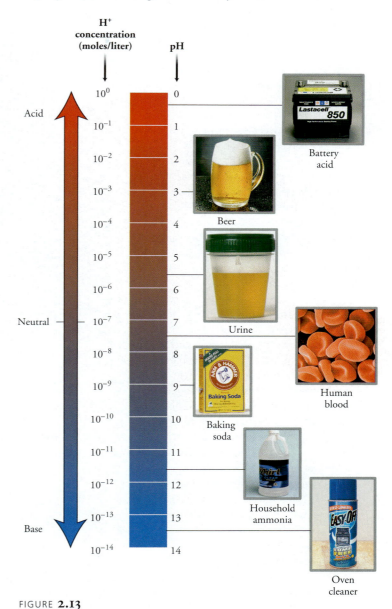

FIGURE **2.13**

The pH scale and the pH of some body fluids and commonly encountered substances

TABLE **2.3**

ACIDS AND BASES COMPARED		
CHARACTERISTIC	**ACID**	**BASE**
Dissociation in Water	Releases H^+	Releases OH^-
pH	Less than 7	Greater than 7
Example	HCl (hydrochloric acid)	NaOH (sodium hydroxide)

ENVIRONMENTAL ISSUE

What Is Happening to the Rain?

Statues that have withstood not only the critics but also the ravages of time are part of our vast heritage. Some have stood for centuries gazing over city streets and sidewalks at generations of passersby. But now the statues have had to be moved, and many have been virtually destroyed (Figure 2.B). Rain has dissolved away the features—rain that has in recent years become a solution of acid.

Acid rain, usually defined as rain with a pH lower than 5.6—the pH of natural precipitation—is caused largely by the burning of fossil fuels in cars, factories, and power plants. The sulfur dioxide and nitrogen oxides produced by these activities react with water in the atmosphere to form sulfuric acid (H_2SO_4) and nitric acid (HNO_3). These acids fall to the earth as rain, snow, or fog, with pH values sometimes as low as 1.5.

The effects of acid rain on the environment have been devastating (Figure 2.C). On land, acid rain has been linked to the decline of forests. Trees, for example, become stressed and more susceptible to disease when nutrient uptake is disrupted by the increased acidity of soils. In aquatic environments, acid rain has been linked to the decrease and sometimes total elimination of populations of fish and amphibians. Embryos of spotted salamanders, for example, develop abnormally under somewhat acidic conditions and die when pH values are less than about 5. Acid rain also causes mercury and aluminum to leach from soils and rocks. These harmful elements eventually make their way into lakes and streams, where they contaminate aquatic life.

Acid rain is also harmful to human health. The pollutants that cause acid rain form fine particles of sulfate and nitrate. These particles are easily inhaled into our respiratory system. Once inside, they cause irritation and respiratory illnesses such as asthma and bronchitis.

Because most acid rain is caused by human activity (a small percentage of the pollutants that cause acid rain are released into the atmosphere during volcanic eruptions and other natural processes), it is within our power to reduce, if not eliminate, the problem. Many of the devastating effects of acid rain could be reversed by tighter controls on industrial emissions. The installation of antipollution devices in our cars also would be helpful. And, of course, we could opt for walking or biking instead of driving and could carpool with others when driving is necessary. 🌲

(a) A figure, known as "The Vampire," on a cathedral in Paris around 1920

(b) The same figure, about 1990

FIGURE 2.**B**

Acid rain is caused by the release into the atmosphere of nitrogen and sulfur oxides during the burning of fossil fuels. Such rain is slowly eating away our works of art. The damage to this French statue took place over a period of about 70 years.

(b) In some areas, acid rain has reduced or eliminated populations of aquatic organisms. Acidic conditions have been shown, for example, to disrupt the development of spotted salamanders or to kill the embryos outright.

(a) Acid rain has destroyed parts of our forests.

FIGURE 2.**C**

Effects of acid rain

Carbohydrates, Lipids, Proteins, and Nucleic Acids Are the Major Molecules of Life

Most of the molecules we have discussed so far have been rather small and simple. Many of the molecules of life, however, are enormous by comparison and have complex architecture. Some proteins, for example, are made up of thousands of atoms linked together in a chain that repeatedly coils and folds upon itself. The giant molecules of life are **macromolecules**.

Macromolecules that consist of many small repeating molecular subunits linked in a chain are called **polymers**. The small molecular subunits that form the building blocks of the polymer are called **monomers**. We might think of a polymer as a pearl necklace, with each pearl being a monomer. As we shall see, a protein is a polymer, or chain of amino acids linked together. And glycogen, the storage form of carbohydrates in animals, is a polymer of glucose molecules.

Polymers form through **dehydration synthesis**. In this process, monomers are linked through the removal of a water molecule: one monomer donates OH, and the other donates H. With the removal of water, the two monomers covalently bond to one another. By the same token, polymers can be broken apart by **hydrolysis**, the addition of water across the covalent bonds. The H from the water molecule attaches to one monomer and the OH to the adjoining monomer, thus breaking the covalent bond between the two. Hydrolysis plays a critical role in digestion. Most foods consist of polymers too large to pass from our digestive tract into the bloodstream and on to our cells. Thus, the polymers are hydrolyzed into their component monomers. The monomers can then be absorbed into the bloodstream for transport throughout the body. Dehydration synthesis and hydrolysis are summarized in Figure 2.14.

▌Carbohydrates supply energy to cells

Carbohydrates provide fuel (energy) for the human body. Carbohydrates are made up entirely of carbon, hydrogen, and oxygen. Each carbohydrate molecule has twice as many hydrogen atoms as oxygen atoms. Most carbohydrates are commonly called sugars and starches. These carbohydrates can be classified by size into mono-saccharides, disaccharides, and polysaccharides. Some common carbohydrates are described in Table 2.4.

MONOSACCHARIDES

Monosaccharides are the smallest molecular units of carbohydrates. They are also called simple sugars. Monosaccharides contain from three to seven carbon atoms. In fact, monosaccharides can be classified by the number of carbon atoms they contain. A sugar that contains five carbons is pentose; one with six carbons is hexose; and so on. Monosaccharides that contain the same number of carbons can, however, differ in structure. Glucose, galactose, and fructose, for example, are six-carbon sugars with the molecular formula $C_6H_{12}O_6$. Each, however, has a slightly different structure, as you can see in Figure 2.15.

DISACCHARIDES

Disaccharides are double sugars that form when two monosaccharides covalently bond to each other. The disaccharide sucrose (table sugar) is formed when the monosaccharides glucose and fructose join (Figure 2.16). Two glucose molecules joined together form the disaccharide maltose, an important ingredient of beer. A final example of a disaccharide is lactose, the principal carbohydrate of milk and milk products. Lactose is formed by the joining of glucose and galactose. Infants and children usually produce adequate amounts of lactase, the enzyme needed to break lactose back into glucose and galactose for digestion and use by the body. (Enzymes, discussed later in this chapter, are usually proteins; they function to speed up chemical reactions). However, many adults do not produce sufficient lactase. For lactase-deficient adults, consumption of milk and milk products can lead to diarrhea, cramps, and bloating. These symptoms are caused by undigested lactose passing into the large intestine, where the disaccharide serves as an energy source for resident bacteria. The bacteria, in turn, produce gas and lactic acid that irritate the bowels. The milk industry has responded to the problem of lactase deficiency by producing lactose-reduced milk. Tablets and caplets that contain the enzyme lactase can also be taken before consuming dairy products.

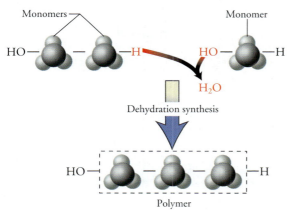

(a) Polymers are formed by dehydration synthesis, the removal of a water molecule, and the joining of two monomers.

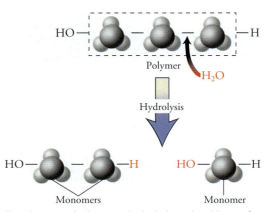

(b) Polymers are broken apart by hydrolysis, the addition of a water molecule, and the breaking of bonds between monomers.

WWW FIGURE **2.14**
TUTORIAL 2.4 Formation and breaking apart of polymers

POLYSACCHARIDES

Polysaccharides are complex carbohydrates that are formed when monosaccharides (most commonly glucose) join to form long chains. Most polysaccharides store energy or provide structure. In plants, the storage polysaccharide is starch; in animals it is glycogen. Humans store glycogen mainly in the cells of the liver and muscles. Glycogen can be broken down to release energy-laden glucose molecules.

Cellulose is a structural polysaccharide found in the cell walls of plants (Figure 2.17). Humans lack the enzymes necessary to digest cellulose, and, as a result, it passes unchanged through our digestive tract. Nonetheless, cellulose is an important form of dietary fiber that helps fecal matter move through the large intestines. Including fiber in our diet may reduce the incidence of colon cancer.

▌ Lipids store energy and form cell membranes

Lipids are compounds, such as fats, that do not dissolve in water. This characteristic of lipids is due to the fact that they are nonpolar (having no electrical charges), and water is polar. Because of this difference, water shows no attraction for lipids and vice versa. So, water and lipids do not mix. Three types of lipids that are important to human health are triglycerides, phospholipids, and steroids.

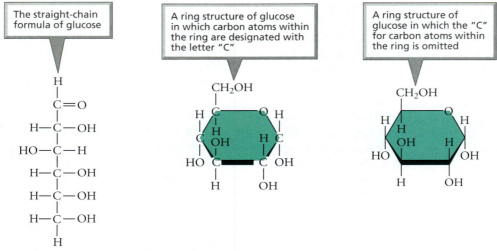

(a) Three representations of the monosaccharide glucose ($C_6H_{12}O_6$)

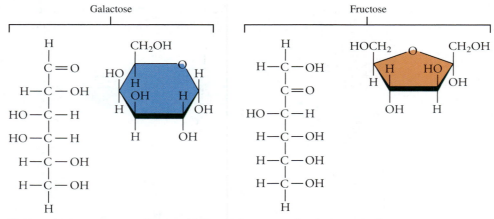

(b) Two other hexose sugars—galactose and fructose—also are examples of monosaccharides. Galactose and fructose have the same molecular formula as glucose ($C_6H_{12}O_6$), but slightly different structures.

WWW FIGURE **2.15**

TUTORIAL 2.4

Monosaccharides are simple sugars. Most monosaccharides have a backbone of 3 to 6 carbon atoms. Many of these carbon atoms have hydrogen (H) and a hydroxyl group (OH) attached to them. In the fluid within our cells, the carbon backbone usually forms a ringlike structure.

TABLE **2.4**

COMMON CARBOHYDRATES			
CARBOHYDRATE	MOLECULAR FORMULA	SOURCE	COMPONENT MONOSACCHARIDES
Monosaccharides			
Glucose	$C_6H_{12}O_6$	Blood, fruit, honey	
Fructose	$C_6H_{12}O_6$	Fruit, honey	
Galactose	$C_6H_{12}O_6$	From hydrolysis of lactose (milk sugar)	
Disaccharides			
Sucrose	$C_{12}H_{22}O_{11}$	Sugar cane, maple syrup	Glucose, fructose
Maltose	$C_{12}H_{22}O_{11}$	From hydrolysis of starch; ingredient in beer	Glucose
Lactose	$C_{12}H_{22}O_{11}$	Component of milk	Glucose, galactose
Polysaccharides			
Starch	*	Potatoes, corn, some grains	Glucose
Glycogen	*	Stored in muscle and liver cells	Glucose
Cellulose	*	Cell walls of plants	Glucose
Chitin	*	Outer coverings of insects, crustaceans	Glucose

* These complex carbohydrates consist of chains containing hundreds of glucose molecules joined to each other in long strings.

TRIGLYCERIDES

We know triglycerides as fats and oils. **Triglycerides** are made of one molecule of glycerol and three fatty acids. Fatty acids are carbon and hydrogen bonded together in chains that have the acidic group COOH at one end. The fatty acids bond to glycerol through dehydration synthesis (Figure 2.18a). Triglycerides are described as saturated, unsaturated, or polyunsaturated. These descriptions are based on the number of double bonds between the carbon atoms of the fatty acids (Figure 2.18b). Saturated fatty acids have no double bonds linking the carbon atoms. These fatty acids are "saturated with hydrogen" because the carbon atoms are bonded to as many hydrogen atoms as possible. Fatty acids with one or more double bonds are described as unsaturated. More specifically, fatty acids that contain one double bond within their carbon skeleton are monounsaturated. Fatty acids containing two or more double bonds are polyunsaturated. Basically, these fatty acids are "not saturated with hydrogen," because they could bond to more hydrogen atoms if the double bonds between their carbon atoms were broken.

Fats and oils are rich sources of energy that provide about twice as much energy per gram as carbohydrates or proteins. The high energy density of fat makes it an ideal way for the body to store energy for the long term. Our bulk would be much greater, indeed, if over the long run, we stored excess energy as carbohydrates or proteins, given their relatively low energy yield compared with fat. (The carbohydrate glycogen, described earlier,

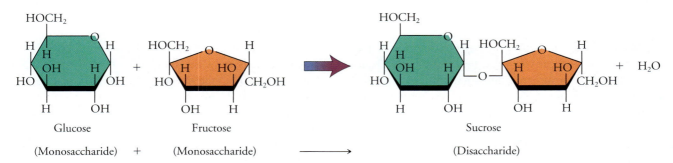

Glucose   Fructose   Sucrose

(Monosaccharide) + (Monosaccharide) ⟶ (Disaccharide)

FIGURE **2.16**

Disaccharides are formed by joining two monosaccharides. Here, a molecule of glucose and one of fructose combine to form sucrose.

Micrograph of cellulose fibrils in plant cell wall

Cell walls

Plant cells

FIGURE **2.17**
Cellulose is a structural polysaccharide found in the cell walls of plants.

functions in short-term rather than long-term energy storage. Glycogen gets us from one meal to the next rather than from one day to the next.)

Let's consider the functions of fat in the body. Fat is critical in long-term energy storage. Excess triglycerides, carbohydrates, and proteins from the foods we consume are converted into small globules of fat that are deposited in the cells of adipose tissue. There, the fat remains until broken down by cells to release the energy needed to keep vital processes going, even if we skip a meal or two. In addition to its role in long-term energy storage, fat serves a protective function in the body. Thin layers of fat surround major organs such as the kidneys. These fat layers cushion the organs against physical shock from falls or blows. Fat also serves as insulation and is necessary for the absorption and transport of lipid-soluble vitamins from the intestines to the cells that use them. Despite the importance of fats and oils to human health, in excess they can be very dangerous, particularly to our circulatory system.

PHOSPHOLIPIDS

Phospholipids are made of a molecule of glycerol bonded to two fatty acids and a negatively charged phosphate group. Linked to the phosphate group are other small molecules (called a variable group), which are usually polar. This chemical composition results in phospholipids having two regions with very different characteristics. As you will notice in Figure 2.19a, one region, made up of the fatty acids, is nonpolar and forms the hydrophobic, or "water-fearing," tail. The other region is polar

and forms the hydrophilic, or "water-loving," head. The tails, being hydrophobic, are excluded from water. The heads, being hydrophilic, interact with water. The separation of phospholipids into hydrophilic heads and hydrophobic tails is important in the structure of plasma (cell) membranes. At the surface of cells, phospholipids are arranged in a double layer, called a bilayer (Figure 2.19b). In this arrangement, the hydrophilic heads point outward, contacting the watery solutions inside and outside the cell. The hydrophobic tails, on the other hand, point inward and help hold the membrane together.

STEROIDS

A **steroid** consists of four carbon rings attached to molecules that vary from one steroid to the next. Cholesterol, one of the most familiar steroids, is a component of the plasma membrane and is also the foundation from which steroid hormones, such as estrogen and testosterone, are made. Cholesterol, estrogen, and testosterone are shown in Figure 2.20. A high level of cholesterol in the blood is considered a risk factor for heart disease.

▌ Proteins provide structure and speed up chemical reactions

A **protein** is made of one or more chains of amino acids that are held together by peptide bonds. Thousands upon thousands of different kinds of proteins are found in the human body. Protein structure, as you will see, can be quite elaborate, consisting of chains that twist, turn, and fold upon themselves. Some proteins provide structural support (for example, collagen, a major component of bone); others function in transport (as do hemoglobin molecules, which carry oxygen in the blood). Proteins are also involved in movement (such as actin and myosin filaments, which move muscles) and regulation of chemical reactions (as when enzymes speed up reactions). Despite diversity in structure and function, all proteins are polymers made from a set of only 20 monomers important to human life—the amino acids.

AMINO ACIDS

Amino acids are the building blocks of proteins. They consist of a central carbon atom bound to a hydrogen atom (H), an amino group (NH_2), an acidic carboxyl group (COOH), and a side chain designated by the letter R (Figure 2.21). Amino acids differ from each other in their side chains. Some amino acids, called nonessential amino acids, can be synthesized by our bodies. Other amino acids, called essential amino acids, cannot be synthesized by our bodies and must be obtained from the foods we eat.

The amino acids that form proteins are linked through dehydration synthesis. A peptide bond links the carboxyl group (COOH) of one amino acid to the amino group (NH_2) of the adjacent amino acid, as shown in Figure 2.22. Chains containing only a few amino acids are called **peptides**. (Dipeptides contain two amino acids, tripeptides contain three amino acids, and so on.) Chains containing 10 or more amino acids are called **polypeptides**. The term *protein* is used for polypeptides with at least 50 amino acids.

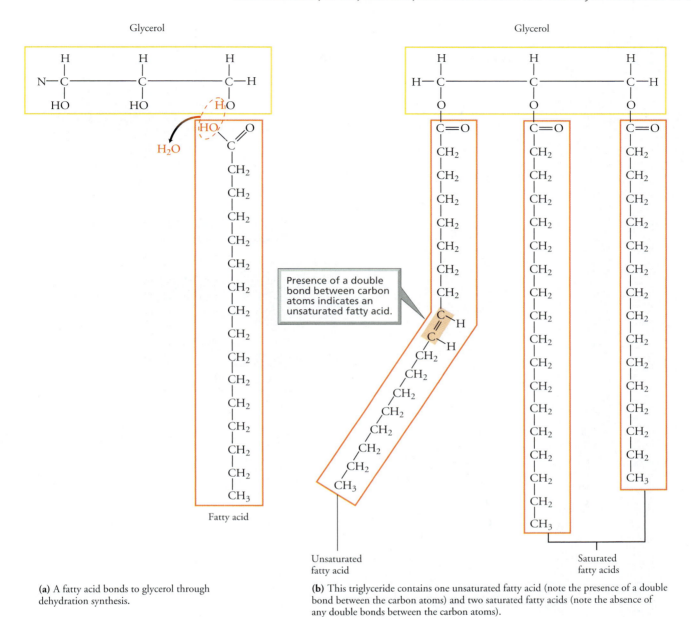

(a) A fatty acid bonds to glycerol through dehydration synthesis.

(b) This triglyceride contains one unsaturated fatty acid (note the presence of a double bond between the carbon atoms) and two saturated fatty acids (note the absence of any double bonds between the carbon atoms).

FIGURE **2.18**

Triglycerides are composed of a molecule of glycerol joined to three fatty acids.

PROTEIN STRUCTURE

Proteins have four distinct levels of structure: primary, secondary, tertiary, and quaternary. The **primary structure** of a protein is the precise sequence of amino acids. This sequence, determined by the genes, dictates a protein's structure and function. Even slight changes in the primary structure can alter the shape of a protein and its ability to function. The disease sickle-cell anemia is an inherited blood disorder. This disease primarily affects populations in central Africa as well as about 1 in 500 African Americans. Sickle-cell anemia results from the substitution of one amino acid for another during synthesis of the protein hemoglobin, which carries oxygen in our red blood cells. This single substitution in a molecule that contains hundreds of amino acids creates a misshapen hemoglobin molecule that then alters the shape of red blood cells. Instead of being flattened disks, the red blood cells are sickle shaped. These oddly shaped red blood cells can clog the tiny vessels of the brain and heart, resulting in death.

The **secondary structure** of proteins results when the chain of amino acids bends and folds, forming coils, spirals, and pleated sheets. These shapes form as a result of hydrogen bonding between different parts of the polypeptide chain. All three types of secondary

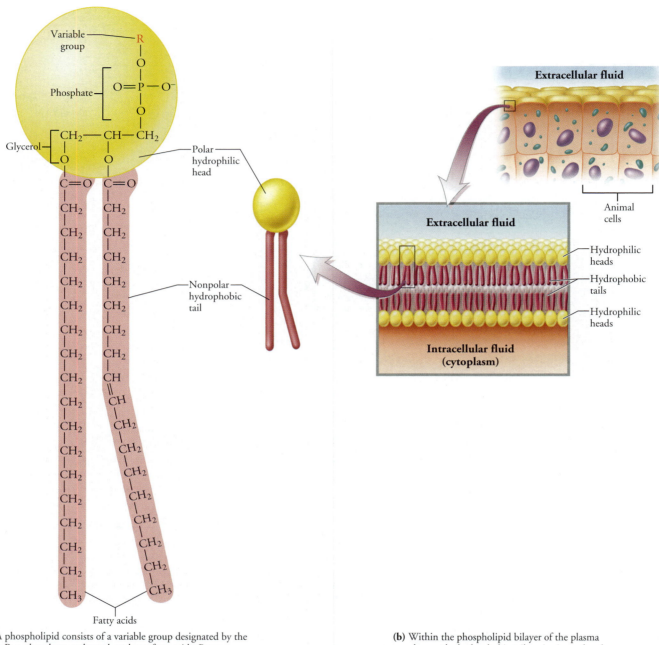

(a) A phospholipid consists of a variable group designated by the letter R, a phosphate, a glycerol, and two fatty acids. Because the variable group is often polar and the fatty acids nonpolar, phospholipids have a polar hydrophilic (water-loving) head and a nonpolar hydrophobic (water-fearing) tail.

(b) Within the phospholipid bilayer of the plasma membrane, the hydrophobic tails point inward and help hold the membrane together. The outward-pointing hydrophilic heads mix with the watery environments inside and outside the cell.

 FIGURE **2.19**

Structure of a phospholipid. Phospholipids are important components of the plasma membrane, the outermost boundary of a cell that separates the internal and external watery environments.

structure may be found at different locations within a given protein. What relevance does the secondary structure of proteins have to human health? Misfolding of a protein normally found on the surface of nerve cells can transform the protein into an infectious agent known as a prion. Prions have been implicated in a number of diseases, including Creutzfeldt-Jakob disease in humans and "mad-cow disease" (see Chapter 13a).

The **tertiary structure** is the overall three-dimensional shape of the protein. Hydrogen, ionic, and covalent bonds between different side chains may all contribute to tertiary structure. Changes in the environment of a protein, such as increased heat or changes in pH, can cause the protein to unravel and lose its three-dimensional shape. This process is called **denaturation**. Change in the shape of a protein results in loss of function.

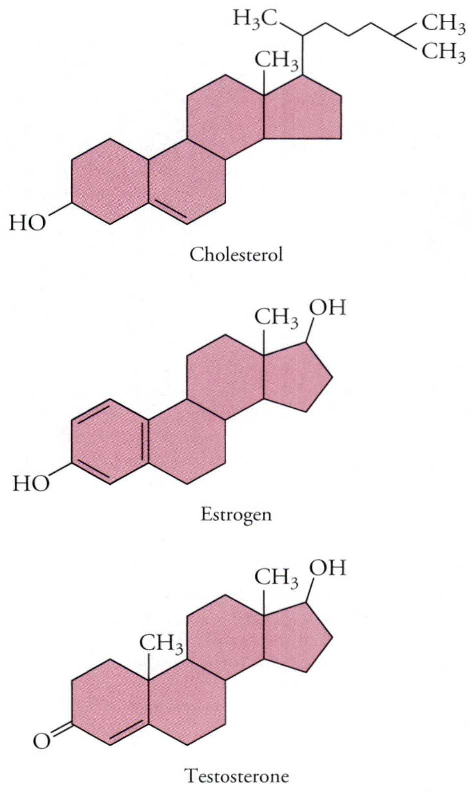

Cholesterol

Estrogen

Testosterone

FIGURE 2.20

The steroid cholesterol is a component of cell membranes. Cholesterol is also the substance from which steroid hormones such as estrogen and testosterone are made. All steroids have a structure consisting of four carbon rings. Steroids differ in the groups attached to these rings.

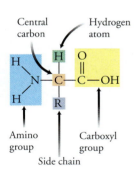

FIGURE 2.21

Structure of an amino acid. Amino acids differ from one another in the type of R group (side chain) they contain.

Finally, some proteins consist of two or more polypeptide chains, each chain forming a subunit. **Quaternary structure** results from the mutually attractive forces between two or more subunits. The forces that hold the subunits together are largely the result of oppositely charged side chains. The four levels of protein structure are summarized in Figure 2.23.

ENZYMES

Life is possible because of enzymes. Without them, most chemical reactions within our cells occur far too slowly to sustain life. **Enzymes** are substances—almost always proteins, but sometimes RNA molecules—that speed up chemical reactions without being consumed in the process. Typically, reactions with enzymes proceed 10,000 to 1,000,000 times faster than the same reactions without enzymes.

The basic process by which an enzyme speeds up a chemical reaction can be summarized by the following equation:

$$E \;+\; S \;\rightarrow\; ES \;\rightarrow\; E \;+\; P$$

Enzyme Substrate Enzyme-substrate complex Enzyme Product

The particular substance that an enzyme works on is called its substrate. For example, the enzyme sucrase speeds up the reaction in which sucrose is broken down into glucose and fructose. In this reaction, sucrose is the substrate, and glucose and fructose are products. Similarly, the enzyme maltase speeds up the breakdown of maltose (the substrate) into molecules of glucose (the product). From these examples you can see that an enzyme's name may resemble the name of its substrate. The examples we have given concern decomposition reactions in which a substance is broken down into its component parts. Enzymes also play a role in synthesis reactions.

During reactions such as those described above, the substrate binds to the enzyme at a specific location on the enzyme known as the **active site**. While bound to one another, the enzyme and substrate are known as an **enzyme-substrate complex**. This binding orients the substrate molecules so they can react. The substrate is converted to products that leave the active site. The enzyme then binds to another substrate molecule. The entire process occurs very rapidly. In fact, one estimate suggests that within one second a typical enzyme converts about 1000 molecules of substrate into product. Figure 2.24 summarizes the steps involved in an enzymatic reaction.

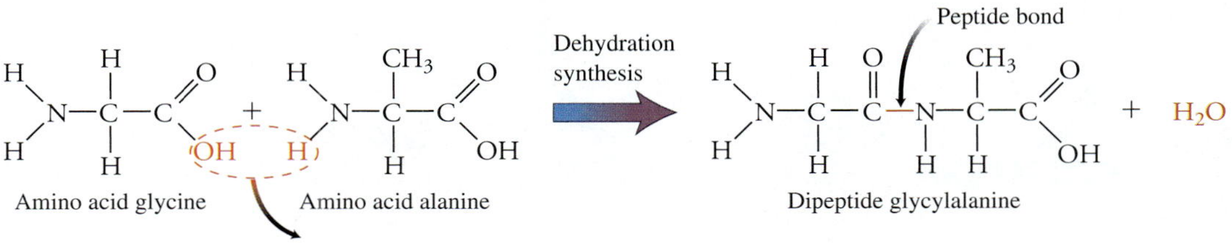

Amino acid glycine   Amino acid alanine

Dipeptide glycylalanine

$+ \; H_2O$

FIGURE 2.22

Formation of a peptide bond between two amino acids through dehydration synthesis. The carboxyl group (COOH) of one amino acid bonds to the amino group (NH_2) of the adjacent amino acid, releasing water.

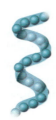

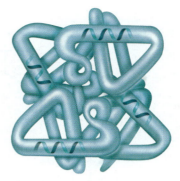

(a) Primary structure is the specific sequence of amino acids. Each amino acid is depicted here as a bead within the polypeptide chain.

(b) Secondary structure, such as the helix shown here, results from the bending and coiling of the chain of amino acids.

(c) Tertiary structure is the three-dimensional shape of proteins.

(d) Some proteins have two or more polypeptide chains, each chain forming a subunit. Quaternary structure results from the attractive forces between two or more subunits.

FIGURE **2.23**
Levels of protein structure

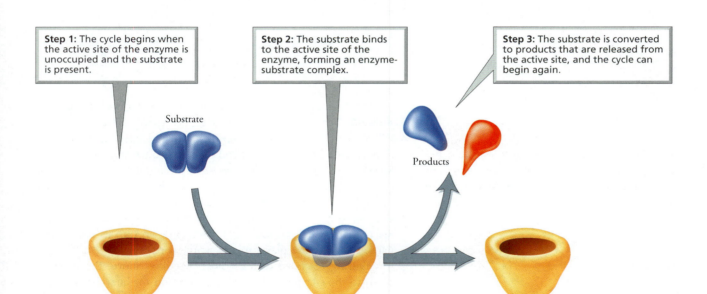

Step 1: The cycle begins when the active site of the enzyme is unoccupied and the substrate is present.

Step 2: The substrate binds to the active site of the enzyme, forming an enzyme-substrate complex.

Step 3: The substrate is converted to products that are released from the active site, and the cycle can begin again.

Substrate

Products

Enzyme

Enzyme-substrate complex

Enzyme

(a) A decomposition reaction involving an enzyme

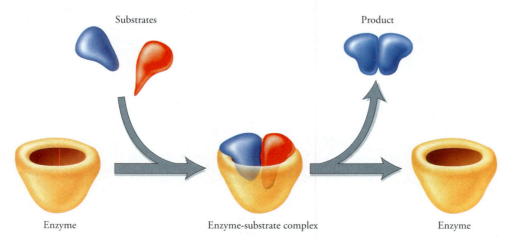

Substrates

Product

Enzyme

Enzyme-substrate complex

Enzyme

(b) A synthesis reaction involving an enzyme

FIGURE **2.24**
The working cycle of an enzyme

Enzymes are very specific, with each one capable of binding to only one or at most a few substrates. Maltase, for example, will act only on maltose and not on sucrose, a structurally similar compound. This specificity is due to the unique shape of each enzyme's active site. Basically, the shape of an enzyme's active site corresponds to the shape of its substrate; the two fit together like pieces of a jigsaw puzzle.

Sometimes enzymes need **cofactors**, nonprotein substances that help enzymes convert substrate to product. Some cofactors permanently reside at the active site of the enzyme. Other cofactors bind to the active site at the same time as the substrate. Some cofactors are inorganic substances such as zinc or iron. Other cofactors are organic substances such as vitamins. Organic cofactors are called **coenzymes**.

Nucleic acids include DNA and RNA

In our discussion of protein structure we mentioned that genes determine primary structure, the sequence of amino acids. Genes, our units of inheritance, are parts of a long polymer called deoxyribonucleic acid, or DNA. DNA is one of two members of the class of macromolecules known as the nucleic acids.

NUCLEOTIDES

The nucleic acid deoxyribonucleic acid (DNA) forms the genes and holds a pattern that determines the sequence of amino acids during protein synthesis. Ribonucleic acid (RNA) is a similar molecule and also plays a role in protein synthesis. Both DNA and RNA are polymers of smaller units called nucleotides. **Nucleotides**, the building blocks of nucleic acids, join through dehydration synthesis to form long chainlike molecules. Every nucleotide monomer consists of a five-carbon (pentose) sugar bonded to one of five nitrogenous bases and at least one phosphate group (Figure 2.25). The five nitrogenous bases are adenine, guanine, cytosine, thymine, and uracil. Some of the bases have a single ring made of carbon and nitrogen atoms (the bases cytosine, thymine, and uracil). Other nitrogenous bases have two such rings (the bases adenine and guanine). The sequence of bases in DNA and RNA determines the sequence of amino acids in a protein.

DNA AND RNA

There are several key differences in the structures of RNA and DNA, as summarized in Table 2.5. RNA is a single strand of nucleotides. The five-carbon sugar in RNA is ribose. The nitrogenous bases in RNA are cytosine, adenine, guanine, and uracil (Figure 2.26). In contrast, DNA is a double-stranded chain. The two strands of DNA, held together by hydrogen bonds formed between the nitrogenous bases, twist around one another to form a double helix. The five-carbon sugar in DNA is deoxyribose. The nitrogenous bases in DNA are adenine, thymine, cytosine, and guanine (Figure 2.27).

ATP

At this moment within your cells, the nucleotide adenosine triphosphate (ATP) is losing a phosphate. As the phosphate is lost, energy stored in holding the phosphate is released. Your

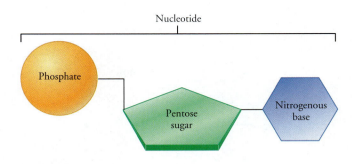

Nucleotide

FIGURE 2.25

Structure of a nucleotide. Nucleotides consist of a five-carbon (pentose) sugar bonded to a phosphate molecule and one of five nitrogenous bases. Nucleotides are the building blocks of nucleic acids.

cells trap that energy and use it to perform work. It is because of this activity that you are able to sit there so attentively, reading this book. ATP consists of the sugar ribose, the base adenine, and three phosphate groups. ATP is formed during an energy-requiring reaction. In this reaction, an inorganic phosphate covalently bonds to adenosine diphosphate (ADP). The energy absorbed during the reaction is stored in the phosphate bond of the ATP molecule. The phosphate bonds of ATP molecules are very unstable. When cells require energy, the phosphate bond is broken, releasing energy. ADP and inorganic phosphate are left (Figure 2.28). The energy released by the splitting of ATP is then available for other chemical reactions occurring at the same time in the cell.

ATP is often described as the energy currency of cells. All energy from the breakdown of molecules such as glucose must be channeled through ATP before it can be used by the body.

TABLE **2.5**

STRUCTURAL DIFFERENCES BETWEEN RNA AND DNA		
CHARACTERISTIC	**RNA**	**DNA**
Sugar	Ribose	Deoxyribose
Bases	Adenine, guanine, cytosine, uracil	Adenine, guanine, cytosine, thymine
Number of strands	One	Two; twisted to form double helix

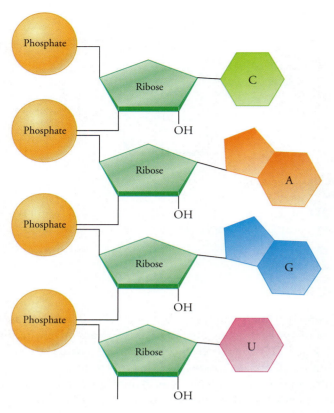

FIGURE 2.26

RNA is a single-stranded nucleic acid. RNA is formed by the linking together of nucleotides composed of the sugar ribose, a phosphate group, and the nitrogenous bases cytosine (C), adenine (A), guanine (G), and uracil (U).

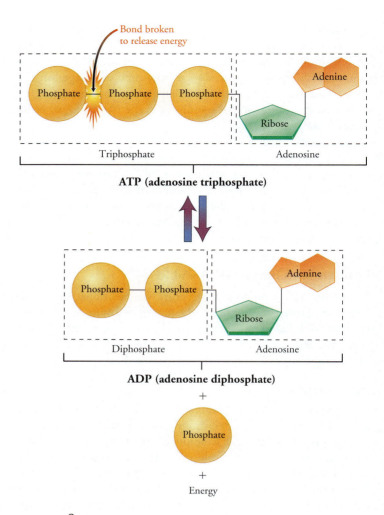

ATP (adenosine triphosphate)

ADP (adenosine diphosphate)
+
Phosphate
+
Energy

FIGURE 2.28

Structure and function of adenosine triphosphate (ATP). This nucleotide consists of the sugar ribose, the base adenine, and three phosphate groups. The phosphate bonds of ATP are unstable. When cells need energy, the last phosphate bond is broken, yielding adenosine diphosphate (ADP), a phosphate molecule, and energy.

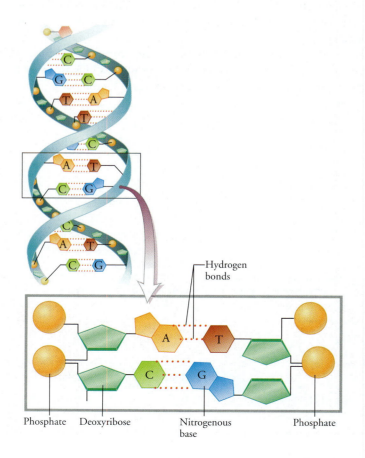

Phosphate   Deoxyribose   Nitrogenous base   Phosphate

FIGURE 2.27

DNA is a nucleic acid containing two chains of nucleotides twisted around one another to form a double helix. The two chains are held together by hydrogen bonds between the nitrogenous bases. Note that each nucleotide of DNA contains the pentose sugar deoxyribose, a phosphate group, and one of the following four nitrogenous bases: adenine (A), thymine (T), cytosine (C), and guanine (G).

REVIEWING THE CONCEPTS

Basic Chemistry Aids in Understanding Human Biology (pp. 16–26)

1. Atoms consist of subatomic particles called protons, neutrons, and electrons. Protons have a positive charge. Neutrons have no charge. Both of these subatomic particles are found in the nucleus and have atomic masses equal to one. Electrons have a negative charge, weigh almost nothing, and orbit around the nucleus in shells.

WEB TUTORIAL 2.1 Atoms, Ions and Bonding

2. Each element is made of atoms, all of which contain the same number of protons. Elements cannot be broken down by ordinary chemical means. The atomic number of an element is the number of protons in one of its atoms. The atomic weight of an element equals the number of protons plus the number of neutrons.

3. Isotopes are atoms that have the same number of protons but different numbers of neutrons. Some isotopes, called radioisotopes, emit radiation. Radiation can cause illness or death to the person exposed to it. Radiation also can be used to diagnose and treat certain illnesses.

4. Elements combine with each other to form compounds. The characteristics of compounds are usually different from those of the elements that form them.

WEB TUTORIAL 2.2 Chemical Bonding

5. When atoms come together to form compounds, chemical bonds form between them. Covalent bonds form when atoms share electrons. Ionic bonds form when electrons are transferred, rather than shared, between atoms. Atoms that have lost or gained electrons have an electric charge and are called ions. The attraction between oppositely charged ions is an ionic bond. Sometimes the sharing of electrons during covalent bonding is unequal. Unequal sharing of electrons results in polar covalent bonds and forms polar molecules. Hydrogen bonds are weak attractive forces between the atoms of these polar molecules.

WEB TUTORIAL 2.2 Chemical Bonding

6. Water is an important component of the human body, owing, in part, to its unusual characteristics. Water is an excellent solvent because of the polarity of its molecules and the hydrogen bonding between them. These properties also give water its high heat capacity, making possible the maintenance of a constant internal body temperature. The high heat of vaporization of water prevents overheating through perspiration and facilitates cooling of the body surface.

WEB TUTORIAL 2.3 Chemistry and Water

7. When added to water, an acid increases the number of hydrogen ions (H^+), and a base increases the number of hydroxide ions (OH^-). The OH^- produced by the dissociation of a base reacts with H^+ to form water molecules. Acids increase the concentration of H^+ in solution. Bases decrease the concentration of H^+ in solution.

8. The strengths of acids and bases can be measured on the pH scale. This scale ranges from 0 to 14, with a pH of 7 being neutral, a pH of less than 7 being acidic, and a pH of greater than 7 being basic. Biological systems usually function within a narrow pH range. Buffers prevent dramatic changes in pH. Buffers remove excess H^+ from solution when concentrations increase and add H^+ when concentrations decrease.

WEB TUTORIAL 2.4 Monomers and Polymers

Carbohydrates, Lipids, Proteins, and Nucleic Acids Are the Major Molecules of Life (pp. 27–36)

9. A polymer is a large molecule made of many smaller molecules, called monomers. The monomers are linked together in a chain. Polymers form through dehydration synthesis (removal of a water molecule) and are broken apart by hydrolysis (addition of a water molecule).

10. Carbohydrates are sugars and starches. These substances provide fuel for the human body. Monosaccharides (simple sugars) such as glucose, galactose, and fructose are the smallest molecular units of carbohydrates. Disaccharides (double sugars) are formed when two monosaccharides join through dehydration synthesis. Polysaccharides (many sugars) are complex carbohydrates. Polysaccharides are formed when large numbers of monosaccharides (usually glucose) join through dehydration synthesis. Examples of polysaccharides include starch and glycogen, two energy-storage carbohydrates. Cellulose, another polysaccharide, is a structural carbohydrate.

WEB TUTORIAL 2.4 Monomers and Polymers

11. Lipids, such as triglycerides, phospholipids, and steroids, are nonpolar molecules that do not dissolve in water. Triglycerides (fats and oils) are made of glycerol and three fatty acids. Fat functions in long-term energy storage, insulation, and protection of internal organs from physical shock. Phospholipids are made of glycerol bonded to two fatty acids and a phosphate group; other small molecules are attached to the phosphate. Phospholipids are important components of plasma membranes. Steroids consist of four carbon rings attached to functional groups. Cholesterol, one of the most familiar steroids, is a component of the plasma membrane and also serves as the foundation for steroid hormones.

WEB TUTORIAL 2.5 Carbohydrates, Lipids, Proteins, and Nucleotides

12. Although proteins have diverse structures and functions, they are all polymers made from a set of about 20 amino acids. Amino acids consist of a central carbon bound to a hydrogen atom, an amino group, an acidic carboxyl group, and a side chain (R). Amino acids, linked together through dehydration synthesis, form chains. These chains are called peptides when they contain only a few amino acids and polypeptides when they contain 10 or more amino acids. Proteins are chains of 50 or more amino acids.

13. There are four levels of protein structure. Primary structure is the specific sequence of amino acids in a protein. Secondary structure results from the bending and folding of the amino acid chain into coils, spirals, or pleated sheets. Tertiary structure is the three-dimensional shape of the proteins. Some proteins consist of two or more polypeptide chains, each chain forming a subunit. Attractive forces between the subunits of these proteins hold the subunits together and produce quaternary structure.

14. Enzymes are proteins, or sometimes RNA molecules, that speed up chemical reactions without being consumed in the process. During a chemical reaction, the substrate binds to the enzyme at a specific location on the enzyme called the active site. While bound to one another, the enzyme and substrate are known as the enzyme-substrate complex. At this time, the substrate is converted to products that then leave the active site.

15. Deoxyribonucleic acid (DNA) and ribonucleic acid (RNA) are polymers of nucleotides. A nucleotide consists of a five-carbon sugar bonded to one of five nitrogenous bases and a phosphate group. DNA, a double-

stranded molecule, forms the genes and holds a pattern that determines the sequence of amino acids during protein synthesis. The sugar in DNA is deoxyribose. The nitrogenous bases in DNA are adenine, guanine, cytosine, and thymine. RNA is a single-stranded molecule that also plays a major role in protein synthesis. The sugar in RNA is ribose, and the nitrogenous bases are adenine, guanine, cytosine, and uracil.

WEB TUTORIAL 2.5 Carbohydrates, Lipids, Proteins, and Nucleotides

16. Adenosine triphosphate (ATP) is the energy currency of cells. ATP is a nucleotide made of the sugar ribose, the base adenine, and three phosphate groups. When cells require energy, one of the unstable phosphate bonds is broken and energy is released.

KEY TERMS

acid *p. 24*	covalent bond *p. 19*	lipid *p. 28*	polypeptide *p. 30*
active site *p. 33*	dehydration synthesis *p. 27*	macromolecule *p. 27*	polysaccharide *p. 28*
amino acid *p. 30*	denaturation *p. 32*	mass *p. 16*	primary structure *p. 31*
atom *p. 16*	disaccharide *p. 27*	matter *p. 16*	protein *p. 30*
atomic number *p. 17*	element *p. 17*	molecule *p. 19*	quaternary structure *p. 33*
base *p. 24*	enzyme *p. 33*	monomer *p. 27*	radioisotope *p. 17*
buffer *p. 25*	enzyme-substrate complex *p. 33*	monosaccharide *p. 27*	secondary structure *p. 31*
carbohydrate *p. 27*	hydrogen bond *p. 21*	nucleotide *p. 35*	steroid *p. 30*
cellulose *p. 28*	hydrolysis *p. 27*	peptide *p. 30*	tertiary structure *p. 32*
coenzyme *p. 35*	ion *p. 21*	pH scale *p. 25*	triglyceride *p. 29*
cofactor *p. 35*	ionic bond *p. 21*	phospholipid *p. 30*	
compound *p. 18*	isotope *p. 17*	polymer *p. 27*	

THINKING ABOUT THE CONCEPTS

1. What is an atom? *p. 16*
2. Compare protons, neutrons, and electrons with respect to their charge, mass, and location within an atom. *p. 16*
3. What is an isotope? *p. 17*
4. Describe the damage to the body caused by radiation. *p. 17*
5. How is radiation used to diagnose or cure illness? *pp. 17–18*
6. Describe ionic, covalent, and hydrogen bonds. Give an example of each. *pp. 19–21*
7. What characteristics of water make it a critical component of the body? *pp. 21–24*
8. Define acids and bases according to what happens when they are added to water. *pp. 24–25*
9. What is the pH scale? *p. 25*
10. How are polymers formed and broken? Give three examples of polymers and their component monomers. *pp. 27–28*
11. Name two important energy-storage polysaccharides and one important structural polysaccharide. *p. 28*
12. Describe the structure of a phospholipid. What important roles do they play in the human body? *p. 30*
13. What are saturated, monounsaturated, and polyunsaturated triglycerides? What are the functions of triglycerides in the body? *pp. 29–30*
14. Describe the four levels of protein structure. *pp. 31–33*
15. Compare the structures of RNA and DNA. *pp. 35–36*
16. Describe the structure and function of ATP. *pp. 35–36*
17. Compare and contrast the ways in which proteins and nucleic acids are used in the body. *pp. 30, 35*
18. Hydrogen bonds
    a. are stronger than either ionic or covalent bonds.
    b. form between a slightly positively charged hydrogen atom and a slightly negatively charged atom nearby.
    c. maintain the shape of proteins and DNA.
    d. b and c

19. Water
    a. is a nonpolar molecule and therefore an excellent solvent.
    b. has a high heat capacity and therefore helps maintain a constant body temperature.
    c. has a low heat of vaporization and therefore helps prevent overheating of the body.
    d. makes up about 25% of the human body.
20. Acids
    a. release hydrogen ions when placed in water.
    b. produce hydroxide ions when placed in water.
    c. have a pH greater than 7.
    d. prevent normal digestive activities in the stomach.
21. Carbohydrates
    a. consist of chains of amino acids.
    b. supply our cells with energy.
    c. contain glycerol.
    d. function as enzymes.
22. Triglycerides
    a. have one molecule of glycerol and three fatty acids.
    b. are poor sources of energy.
    c. are saturated when there are two or more double bonds linking carbon atoms.
    d. are major components of plasma membranes.
23. Enzymes
    a. speed up chemical reactions and are consumed in the process.
    b. are always proteins.
    c. are usually nonspecific and therefore capable of binding to many different substrates.
    d. have locations, known as active sites, to which the substrate binds.
24. Changes in temperature or pH can cause a protein to lose its three-dimensional shape and become nonfunctional. This process is called _____.

25. DNA and RNA are polymers of smaller units called _____.
26. The _____ structure of a protein is the precise sequence of amino acids.
27. _____ are arranged in a double layer (bilayer) that forms the plasma membrane of cells.
28. In plants, the storage polysaccharide is _____, and in animals it is _____.
29. The nucleotide _____ is the energy currency of cells.

30. Buffers prevent dramatic changes in _____.
31. _____ bonds form when outer shell electrons are shared between atoms.
32. _____ is the process through which monomers are linked together through the removal of a water molecule to form polymers. _____ is the process that breaks polymers apart by the addition of water.

APPLYING THE CONCEPTS

1. A friend eyes your lunch and begins to lecture you on the perils of high-fat foods. You decide to acknowledge the health risks of eating a high-fat diet but also to point out to her the important roles that lipids play in your body. What will you say?
2. Bill claims that eating fruits and vegetables is overrated because humans lack the enzyme needed to digest cellulose and thus cellulose passes unchanged through our digestive tract. Is Bill correct that we gain nothing from eating food that contains cellulose?

3. For years, your mother has been advocating drinking several large glasses of water each day. Having just read this chapter, you now understand the critical roles that water plays in your body. You vow to explain during your next phone call home what you have learned. What will you say to your mother? Also, why do you think dehydration might be dangerous?

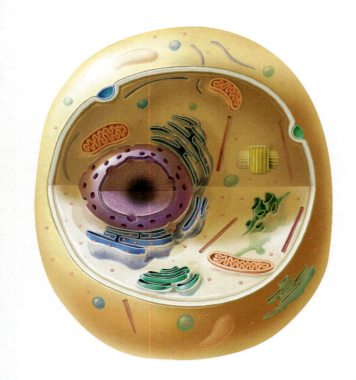

The Cell

Our Cells Are Highly Structured

- ■ Eukaryotic cells are structurally more complex than prokaryotic cells
- ■ Cells are very small
- ■ The plasma membrane has diverse functions
- ■ Organelles are specialized compartments within cells
- ■ The cytoskeleton provides support and movement

Our Cells Use Cellular Respiration and Fermentation to Generate ATP

- ■ Cellular respiration requires oxygen
- ■ Fermentation does not require oxygen

 ENVIRONMENTAL ISSUE   Asbestos: The Deadly Miracle Material

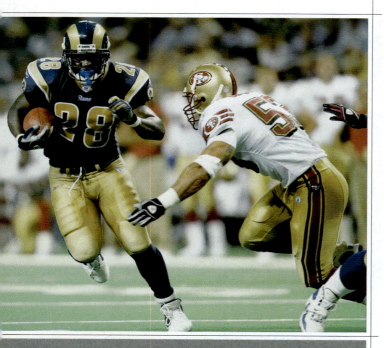

Morinville quarterback Joe Norton found receiver Charlie Mendoza in the right corner of the end zone with 15 seconds remaining in the fourth quarter. Norton, playing with a strained ankle in the second half, passed for 25 yards on the winning drive, taking advantage of Mendoza's legendary speed and agility. Mendoza rushed for 90 yards and two touchdowns, sealing his place as MVP (Most Valuable Player) in the championship game. Rob Douglas launched a stunning 47-yard field goal with one minute remaining in the third quarter and kept Beaumont at bay with a series of punts and kickoffs into the end zone and inside the 5-yard line.

Coach Anderson was beside himself with praise. "The offensive line deserves tremendous credit," he said. "With their blocking and Norton's and Mendoza's performances no one could have beat us today, even though the score was close. And we couldn't have done it without superb defense. We held Beaumont at key times in the second half and the two sacks on Tony Jones sealed it. You fellas deserve immense credit. It's a historic win."

As any coach knows, a team cannot function without the coordinated efforts of players in a wide range of specialized positions. Likewise, the bodies of each team member cannot contribute to performance in a game without the coordinated actions of a vast number of cell types. Our bodies contain over 10 trillion cells that specialize in more than 100 different functions. To throw a pass, muscle cells must contract in response to electrical signals from the brain. To react instantly to a possible tackle, cells of the adrenal glands release hormones such as norepinephrine that trigger near-instantaneous actions and reactions. Kidney cells must continuously remove wastes from the burning of cellular fuels. Intestinal and liver cells rush nutrients to be burned and converted to energy by all other cell types. Neurons coordinate the senses and intellect of each player.

In this chapter we discover how cells work by examining the structures that all specialized cells have in common. We begin with the plasma membrane, the outer boundary of a cell, and work inward. We examine the various organelles, small compartments within cells that have specialized functions. Finally, we explore the ways in which cells obtain the energy they need to do their work and run the body. ■

Our Cells Are Highly Structured

All cells come in one of two basic types—eukaryotic cells and prokaryotic cells—as shown in Figure 3.1. This distinction is based on whether a cell has internal compartments that are enclosed by membranes. These membrane-bound compartments are called **organelles**, or "little organs." **Eukaryotic cells** have membrane-bound organelles. **Prokaryotic cells** do not.

Eukaryotic cells are structurally more complex than prokaryotic cells

Eukaryotic cells are found in plants, animals, and all other organisms except bacteria and archaea. All the cells that make up your body are eukaryotic. Eukaryotic cells have extensive internal membranes. Membranes also enclose some of the organelles within eukaryotic cells (Figure 3.1). In addition, eukaryotic cells have DNA enclosed within a well-defined, membrane-bound region called the nucleus.

Prokaryotic cells are unique to bacteria and another microscopic group of organisms called archaea. You are probably already familiar with bacteria, some of which inhabit your body. Archaea may be less familiar. They include species that inhabit extreme environments such as the Great Salt Lake or the hot sulfur springs of Yellowstone National Park. Prokaryotic cells are much simpler and typically much smaller than eukaryotic cells. Most prokaryotic cells are surrounded by a rigid cell wall. As shown in Figure 3.1, the DNA of prokaryotic cells is circular and is not set apart from the rest of the cell in a nucleus. Table 3.1 summarizes the major differences between eukaryotic and prokaryotic cells.

Cells are very small

Both eukaryotic and prokaryotic cells are quite small, so you need a microscope to see them. The small size of cells can be explained by the critical events that occur at their surfaces and by a physical relationship known as the surface area-to-volume ratio (Figure 3.2). Nutrients enter a cell and wastes leave a cell at its surface. The **surface area-to-volume ratio** dictates that increases in the volume of a cell occur faster than increases in surface area. Thus, a large cell would struggle to survive because of problems moving substances across its inadequate surface. Simply put, small cells are more efficient than large cells and, therefore, better able to survive.

The plasma membrane has diverse functions

We begin our examination of the cell at its outer surface—the **plasma membrane**. This remarkably thin outer boundary of the cell controls the movement of substances both into and out of the cell. A cell's interior is critically balanced. Substances are not permitted to move randomly into and out of it. Both prokaryotic and eukaryotic cells have a plasma membrane. Eukaryotic cells also contain several internal membranes that divide the cell into many compartments. Each compartment contains its own assortment of enzymes and is specialized for particular functions. In general, the principles described for the plasma membrane also apply to the membranes inside the cell.

STRUCTURE OF THE PLASMA MEMBRANE

The plasma membrane is made of lipids, proteins, and carbohydrates. Recall from Chapter 2 that phospholipids are major components of the plasma membrane. These molecules have hydrophilic (water-loving) heads and hydrophobic (water-fearing) tails. The phospholipid molecules form a double layer, called the lipid bilayer, at the surface of the cell, as shown in Figure 3.3. The hydrophilic heads of the outer layer point out-

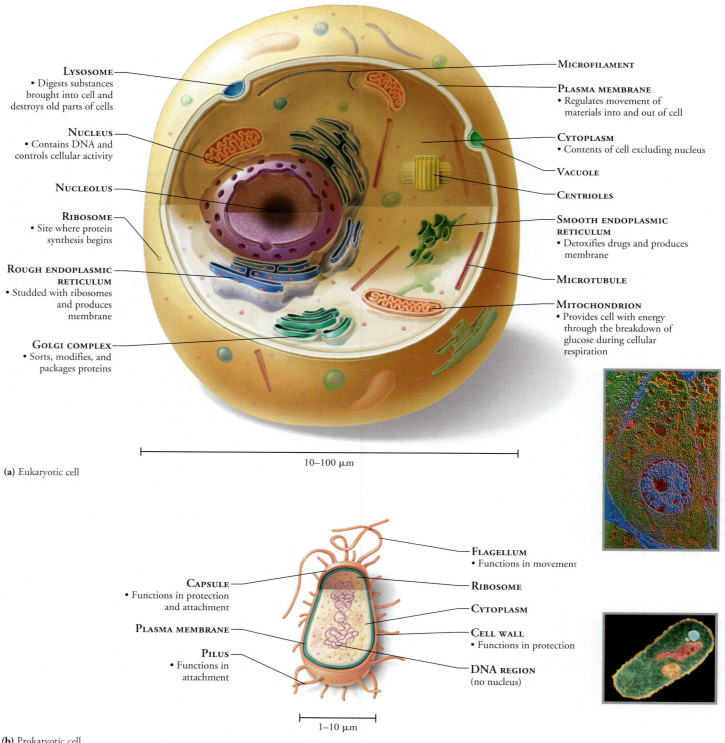

LYSOSOME
• Digests substances brought into cell and destroys old parts of cells

NUCLEUS
• Contains DNA and controls cellular activity

NUCLEOLUS

RIBOSOME
• Site where protein synthesis begins

ROUGH ENDOPLASMIC RETICULUM
• Studded with ribosomes and produces membrane

GOLGI COMPLEX
• Sorts, modifies, and packages proteins

MICROFILAMENT

PLASMA MEMBRANE
• Regulates movement of materials into and out of cell

CYTOPLASM
• Contents of cell excluding nucleus

VACUOLE

CENTRIOLES

SMOOTH ENDOPLASMIC RETICULUM
• Detoxifies drugs and produces membrane

MICROTUBULE

MITOCHONDRION
• Provides cell with energy through the breakdown of glucose during cellular respiration

10–100 μm

(a) Eukaryotic cell

CAPSULE
• Functions in protection and attachment

PLASMA MEMBRANE

PILUS
• Functions in attachment

FLAGELLUM
• Functions in movement

RIBOSOME

CYTOPLASM

CELL WALL
• Functions in protection

DNA REGION
(no nucleus)

1–10 μm

(b) Prokaryotic cell

FIGURE 3.1
Eukaryotic cells, such as (a) the generalized animal cell shown here, have internal membrane-bound organelles. Prokaryotic cells, such as (b) a bacterium, lack internal membrane-bound organelles.

ward and contact extracellular fluid. **Extracellular fluid** (also known as interstitial fluid) is the watery solution outside the cell. The hydrophilic heads of the inner layer point inward and contact cytoplasm. **Cytoplasm** is the jelly-like solution inside

the cell. The cytoplasm includes all contents of the cell between the plasma membrane and the nucleus. Within the lipid bilayer, the hydrophobic tails point toward each other and hold the plasma membrane together.

TABLE **3.1**

COMPARISONS OF EUKARYOTIC AND PROKARYOTIC CELLS		
FEATURE	**EUKARYOTIC CELLS**	**PROKARYOTIC CELLS**
Organisms	Plants, animals, fungi, protists	Bacteria, archaea
Size	10–100 μm across	1–10 μm across
Membrane-bound organelles	Present	Absent
DNA form	Coiled, linear strands	Circular
DNA location	Nucleus	Cytoplasm
Internal membranes	Many	Rare
Cytoskeleton	Present	Absent

Interspersed in the phospholipid bilayer are proteins, as seen in Figure 3.3. Some proteins are embedded in the membrane. Some of these proteins completely span the bilayer; others do not. Still other proteins are simply attached to the inner or outer surface of the membrane. These surface proteins are often attached to an exposed portion of an embedded protein. Molecules of cholesterol are scattered throughout the lipid bilayer.

As you can see in Figure 3.3, carbohydrates attach only to the outer surface of the plasma membrane. Some of these carbohydrates are attached to lipids. These substances are called glycolipids. Most of these carbohydrates, however, are attached to proteins, forming glycoproteins. Glycoproteins often function in cell recognition.

The structure of the plasma membrane is often described as a **fluid mosaic**. The proteins interspersed throughout the lipid molecules give the membrane its mosaic quality. Many proteins are able to move sideways, to some degree, in the bilayer of phospholipids, giving the membrane its fluid quality.

FUNCTIONS OF THE PLASMA MEMBRANE

The plasma membrane performs several functions for the cell. First, by forming the boundary between a cell's internal and external environment, the plasma membrane maintains the cell's structural integrity. Second, the plasma membrane regulates the movement of substances into and out of the cell. Although movement of materials across the membrane is extensive, it is regulated. The plasma membrane permits some substances to move across and denies access to others. For this reason, the membrane is often described as being **selectively permeable**. You will read more about the transport of materials across the plasma membrane later in this chapter.

The plasma membrane also functions in cell-cell recognition. Cells distinguish one type of cell from another by recognizing molecules on the surface of the plasma membrane. Often the recognition molecules are glycoproteins. Membrane carbohydrates differ from one species to another and among individuals of the same species. Even different cell types within an individual have different membrane carbohydrates. This variation allows the body to recognize foreign invaders such as bacteria. Your body would recognize such invaders because the bacteria lack the surface molecules of your own cells. Bacteria, in turn,

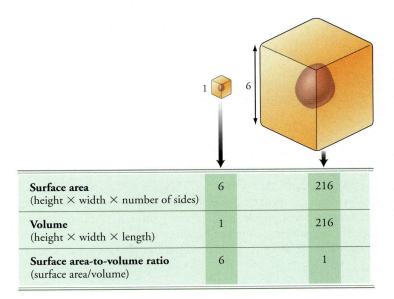

Surface area (height × width × number of sides)	6	216
Volume (height × width × length)	1	216
Surface area-to-volume ratio (surface area/volume)	6	1

FIGURE **3.2**

Most cells are very small. The small size can be explained because the ratio of surface area to volume decreases as cell size increases.

"read" the surface molecules of your cells. They use this information to preferentially settle on certain cells in your body.

Cell-cell recognition is the basis for rejection of tissue grafts and organ transplants. Each individual has a characteristic glycoprotein "fingerprint" on the surface of his or her cells. Hence, the transplanted cells are viewed as foreign (unless they come from an identical twin) and will tend to be rejected by the recipient's immune system.

Another important function of the plasma membrane is communication between cells. For example, hormones secreted by one group of cells may bind to specific proteins, called **receptors**, in the plasma membranes of other cells. These receptors, in turn, relay the message to proteins inside the cell. Once inside the cell, the message is transmitted to other nearby molecules. Through a series of chemical reactions, the message ultimately initiates a response by the recipient cell, perhaps causing it to release another hormone or glucose.

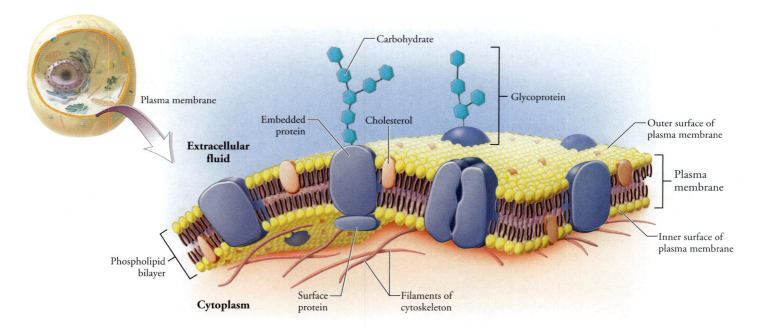

FIGURE **3.3**

The structure of the plasma membrane of a cell according to the fluid mosaic model

The plasma membrane plays an important role in sticking cells together. **Cell adhesion molecules** (CAMs, for short) poke through the plasma membranes of most cells (Figure 3.4). CAMs help hold cells together, especially when tissues and organs form during embryonic development (tissues are groups of cells that work together to perform a common function; organs are structures made of two or more different tissues). CAMs are also involved in many processes in the adult that involve cellular movement and interaction. For example, CAMs play a role in healing wounds. The cells of blood vessels near a wound appear to increase their production of CAMs. These sticky molecules promote healing by snagging repair cells, such as white blood cells, as they move by.

The functions of the plasma membrane are summarized in Table 3.2.

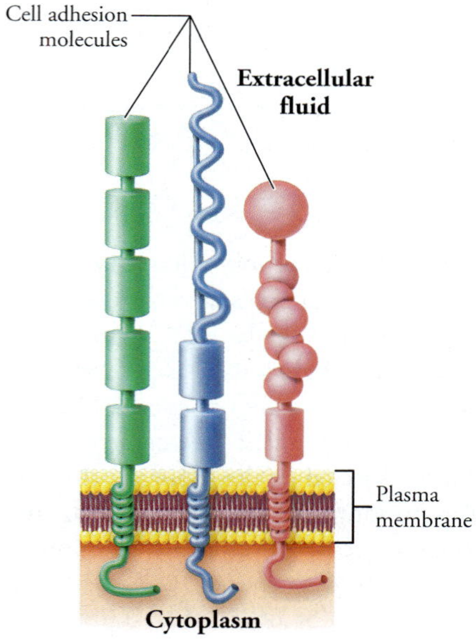

stop and think

Cancer is characterized by the uncontrolled proliferation of cells. Metastasis occurs when cancerous cells separate from a primary tumor. These cells move throughout the body, creating new tumors. How might cell adhesion molecules be involved in metastasis? Would you expect them to be reduced or increased at the time of metastasis?

FIGURE **3.4**

Cell adhesion molecules (CAMs). These uniquely shaped molecules hold cells together to form tissues and organs and are involved in cellular interaction and movement.

MOVEMENT ACROSS THE PLASMA MEMBRANE

Recall that an important function of the plasma membrane is the control of the movement of substances into and out of the cell. Materials cross the plasma membrane in several ways. Perhaps the most straightforward method is simple diffusion.

SIMPLE DIFFUSION The random movement of a substance from a region of higher concentration to a region of lower concentration is **simple diffusion**. Once the concentration of the substance is the same on both sides of the membrane, movement of the substance does not simply stop. Instead, the substance continues to move randomly back and forth across the membrane. The rate of movement in each direction, however, is the same. Substances such as carbon dioxide and oxygen diffuse through the lipid bilayer (Figure 3.5).

TABLE **3.2**

FUNCTIONS OF THE PLASMA MEMBRANE
Maintenance of structural integrity of the cell
Regulation of movement of substances into and out of the cell
Recognition between cells
Communication between cells
Sticking cells together to form tissues and organs

FACILITATED DIFFUSION   Water-soluble substances cannot move through the bilayer by simple diffusion. Some of these molecules are "carried" across the plasma membrane by attaching to specific proteins within the plasma membrane. Some of these proteins transport molecules from one side of the membrane to the other. Other proteins form channels through which the molecules can move. **Facilitated diffusion** is the movement of a substance from a region of higher concentration to a region of lower concentration with the aid of a membrane protein. Molecules of glucose, for example, enter fat cells by facilitated diffusion involving a carrier protein (Figure 3.6).

OSMOSIS   Osmosis is a special type of diffusion. **Osmosis** involves the movement of water across the plasma membrane or

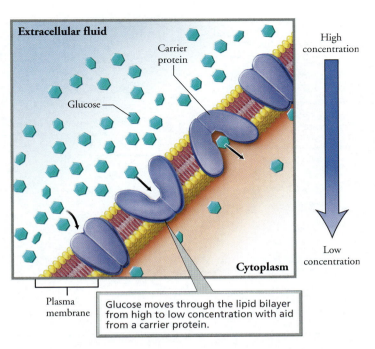

FIGURE **3.6**
Facilitated diffusion is the movement of a substance from a region of high concentration to a region of lower concentration with the aid of a membrane protein that acts as a channel or a carrier protein.

any other selectively permeable membrane from a region of high water concentration to a region of low water concentration. Consider what happens when a substance such as table sugar (the solute) is dissolved in water (the solvent) in a membranous bag through which water, but not sugar, can move. Keep in mind that when solute concentration is low, water concentration is high, and when solute concentration is high, water concentration is low. If the membranous bag is placed into a **hypertonic solution** in which the concentration of sugar is higher outside the bag than inside, more water moves out of the bag than in, causing the bag to shrivel (Figure 3.7a). If, on the other hand, the bag is placed into an **isotonic solution**, one with the same concentration of sugar as inside the bag, there is no net movement of water in either direction, and the bag maintains its normal shape (Figure 3.7b). When the bag is placed into a **hypotonic solution** in which the concentration of sugar is lower outside the bag than inside, more water moves into the bag than out, causing the bag to swell and possibly burst (Figure 3.7c).

The bag in our example can be replaced by a red blood cell, as shown in Figure 3.7. The solution can be replaced by the surrounding blood plasma. Note the changes in the shape of red blood cells in response to different levels of solute concentration in the plasma.

ACTIVE TRANSPORT   Another mechanism of membrane transport is **active transport**—the movement of molecules across the plasma membrane usually against a concentration gradient with the aid of a carrier protein and energy supplied by the cell. Adenosine triphosphate (ATP) provides the energy for most active transport (Figure 3.8). So far in our discussion, we have described substances moving from regions of higher

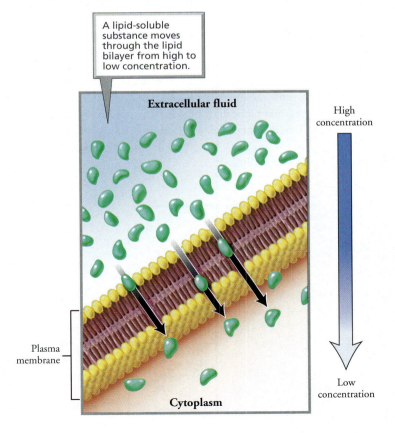

FIGURE **3.5**
Diffusion is the random movement of a substance from a region of high concentration to a region of lower concentration.

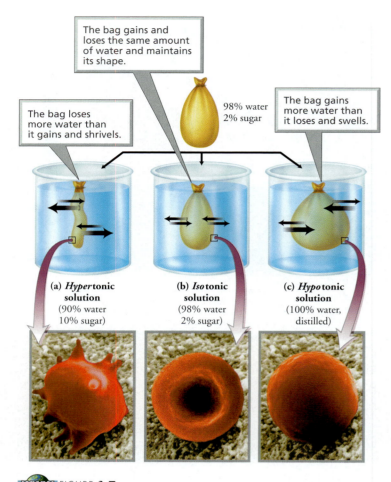

The bag gains and loses the same amount of water and maintains its shape.

98% water 2% sugar

The bag loses more water than it gains and shrivels.

The bag gains more water than it loses and swells.

(a) *Hypertonic* solution (90% water 10% sugar)

(b) *Isotonic* solution (98% water 2% sugar)

(c) *Hypotonic* solution (100% water, distilled)

FIGURE **3.7**

TUTORIAL 3.2

Osmosis is a special type of diffusion in which water moves across a selectively permeable membrane. The drawings at the top of the figure show what happens when a membranous bag through which water but not sugar can move is placed in solutions that are (a) hypertonic, (b) isotonic, or (c) hypotonic. The width of the arrows corresponds to the amount of water moving into and out of the bag. The photographs show what happens to red blood cells when placed in similar solutions. Red blood cells are normally shaped like flattened disks as in (b).

concentration to regions of lower concentration. In active transport, substances usually move from regions of lower concentration to higher concentration. This type of movement is described as "against the concentration gradient" and occurs when cells need to concentrate certain substances. For example, the cells in our bodies contain higher concentrations of potassium ions (K^+) and lower concentrations of sodium ions (Na^+) than their surroundings. Through active transport, the plasma membrane helps maintain these conditions. The membrane pumps potassium ions into the cell and sodium ions out of the cell. In this example, both potassium and sodium are moving from regions of lower concentration to regions of higher concentration. Carrier proteins and energy (ATP) are needed for the journey.

ENDOCYTOSIS Most small molecules cross the plasma membrane by way of diffusion, facilitated diffusion, or active transport. Large molecules, single-celled organisms such as bacteria, and fluid containing dissolved substances enter cells through endocytosis. During **endocytosis**, a region of the plasma membrane surrounds the substance to be ingested and then pinches off from the rest of the membrane. The membrane encloses the substance in a saclike structure called a **vesicle**. The vesicle is released into the cell. Two types of endocytosis are phagocytosis ("cell eating") and pinocytosis ("cell drinking"). **Phagocytosis** occurs when cells engulf large particles or bacteria (Figure 3.9a). **Pinocytosis** occurs when cells engulf droplets of fluid (Figure 3.9b). In pinocytosis, all of the dissolved substances within the droplet are brought into the cell.

EXOCYTOSIS The process by which large molecules leave cells is **exocytosis** (Figure 3.10). Some cells that produce hormones package their products in membrane-bound vesicles. These vesicles move within the cell toward the plasma membrane. When the vesicle reaches the plasma membrane, the membrane of the vesicle fuses with the plasma membrane and the vesicle releases the hormone outside the cell. Similarly, nerve cells release chemicals by exocytosis.

Table 3.3 summarizes the ways in which substances move across the plasma membrane.

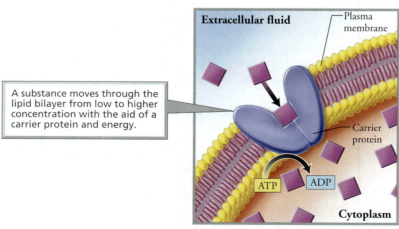

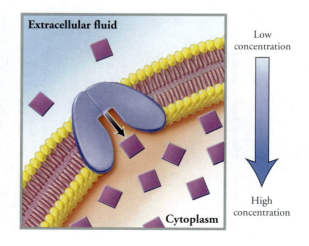

Extracellular fluid

Plasma membrane

A substance moves through the lipid bilayer from low to higher concentration with the aid of a carrier protein and energy.

Carrier protein

ATP ADP

Cytoplasm

Extracellular fluid

Low concentration

High concentration

Cytoplasm

FIGURE **3.8**

TUTORIAL 3.3

Active transport is the movement of molecules across the plasma membrane from an area of low concentration to one of higher concentration with help from a carrier protein and energy usually in the form of ATP.

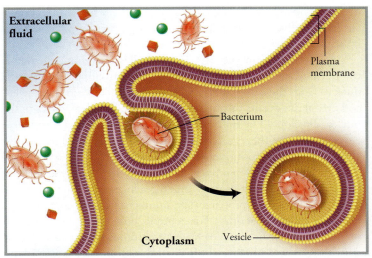

(a) Phagocytosis ("cell eating") occurs when cells engulf bacteria or other large particles.

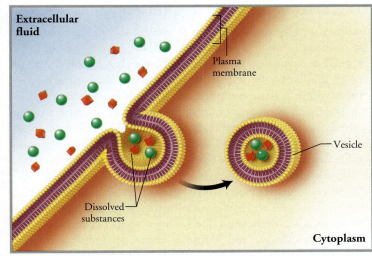

(b) Pinocytosis ("cell drinking") occurs when cells engulf droplets of extracellular fluid and the dissolved substances therein.

WWW TUTORIAL 3.4 FIGURE **3.9**

Endocytosis—phagocytosis or pinocytosis—occurs when a localized region of the plasma membrane surrounds a bacterium, large molecule, or fluid containing dissolved substances and then pinches off to form a vesicle that moves into the cell.

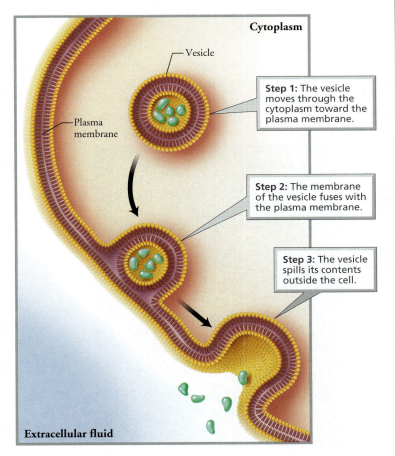

Step 1: The vesicle moves through the cytoplasm toward the plasma membrane.

Step 2: The membrane of the vesicle fuses with the plasma membrane.

Step 3: The vesicle spills its contents outside the cell.

WWW TUTORIAL 3.4 FIGURE **3.10**

Cells package large molecules in membrane-bound vesicles, which then leave by exocytosis.

stop and think

People with kidney disorders often undergo dialysis. During this process, the patient's blood is passed through a long coiled tube that is submerged in a tank filled with a fluid called dialyzing fluid. The tubing is made of porous cellophane that allows small molecules to diffuse out of the blood into the surrounding fluid. Urea is a waste product that must be removed from the blood. Glucose molecules, on the other hand, should remain in the blood. Using the concepts of diffusion and concentration gradients, how would you set the concentrations of urea and glucose in the dialyzing fluid to achieve these goals?

TABLE **3.3**

MECHANISMS OF TRANSPORT ACROSS THE PLASMA MEMBRANE

MECHANISM	DESCRIPTION
Simple diffusion	Random movement from region of high to low concentration
Facilitated diffusion	Movement from region of high to low concentration with the aid of a carrier or channel protein
Osmosis	Movement of water from region of high water concentration (low solute concentration) to low water concentration (high solute concentration)
Active transport	Movement often from region of low to high concentration with the aid of a carrier protein and energy usually from ATP
Endocytosis	Materials are engulfed by plasma membrane and drawn into cell in a vesicle
Exocytosis	Membrane-bound vesicle from inside the cell fuses with the plasma membrane and spills contents outside of cell

Organelles are specialized compartments within cells

You have seen the many roles that the plasma membrane plays in the life of the cell. Inside the eukaryotic cell, the primary role of membranes is to delineate compartments. Specific chemical processes critical to the life of the cell occur within these compartments. Like offices within a large company, these compartments have specific functions such as production or shipping. With compartments, segregated combinations of molecules can carry out specific tasks. These compartments are the organelles that you read about at the beginning of the chapter. Organelles are bounded by membranes and distributed in the jelly-like cytoplasm within the cell (refer again to Figure 3.1).

Several organelles are involved in the manufacture of cell products, such as lipids and proteins. Some organelles give directions for the manufacture of these products. Other organelles are involved in making, modifying, and shipping these products. Still other organelles break down substances for cells or process energy.

NUCLEUS

The **nucleus** of the cell contains almost all the genetic information of the cell (Figure 3.11). The DNA within the nucleus influences cellular structure and function because it contains a code for the production of proteins. All our cells contain the same genetic information. Thus, the character of a particular cell, whether it is a muscle cell or a liver cell, is determined in large part by the directions it receives from its nucleus.

A double membrane called the **nuclear envelope** surrounds the nucleus, as shown in Figure 3.11. The envelope separates the nucleus from the cytoplasm. Communication between the nucleus and cytoplasm occurs through openings called **nuclear pores** in the nuclear envelope. The traffic of selected materials across the nuclear envelope allows the cytoplasm to influence the nucleus. It also allows the nucleus to influence the cytoplasm.

The genetic information within the nucleus is in the form of chromosomes. **Chromosomes** are threadlike structures made of DNA and associated proteins. The number of chromosomes varies from one species to another. For example, humans have 46 chromosomes (23 pairs), house mice have 40 chromosomes, and domestic dogs have 78 chromosomes. Individual chromosomes are visible with a light microscope only during cell division when they shorten and condense (Figure 3.12a). At all other times, the chromosomes are extended and not readily visible. In this dispersed state, the genetic material is called **chromatin** (Figure 3.12b). The chromatin and the environment within the nucleus constitute the **nucleoplasm**. We will discuss chromosomes and cell division in Chapter 19.

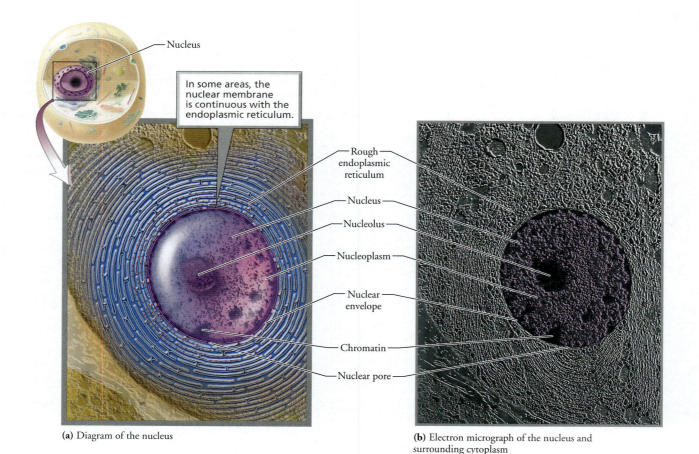

Nucleus

In some areas, the nuclear membrane is continuous with the endoplasmic reticulum.

Rough endoplasmic reticulum

Nucleus

Nucleolus

Nucleoplasm

Nuclear envelope

Chromatin

Nuclear pore

(a) Diagram of the nucleus

(b) Electron micrograph of the nucleus and surrounding cytoplasm

FIGURE **3.11**
The nucleus contains almost all the genetic information of a cell.

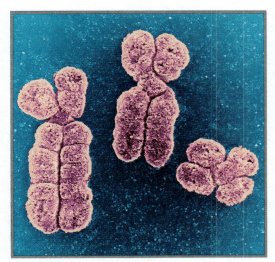

(a) Individual chromosomes are visible during cell division when they shorten and condense.

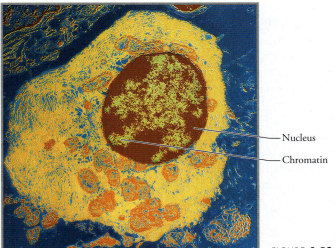

(b) At all other times, the chromosomes are extended and not readily visible. In this dispersed state, the genetic material is called chromatin.

Nucleus

Chromatin

FIGURE **3.12**
Chromosomes are composed of DNA and associated proteins.

The **nucleolus** is a specialized region within the nucleus (see Figure 3.11). It forms and disassembles during the course of the cell cycle (see Chapter 19). The nucleolus is not surrounded by a membrane. It is simply a region of DNA involved in the production of a type of RNA called ribosomal RNA (rRNA). Ribosomal RNA is a component of **ribosomes**, which are sites where protein synthesis begins. A ribosome consists of two subunits, each composed of rRNA and protein. These subunits leave the nucleus and enter the cyto-plasm, where they come into contact with messenger RNA (mRNA) and join to become a functional ribosome. Messenger RNA is a molecule that contains a copy of the information encoded in the DNA for making proteins (Figure 3.13). The ribosomes read the instructions from mRNA and translate the information into a chain of amino acids. The resulting chains of amino acids will fold and may undergo additional modifications before becoming a functional protein.

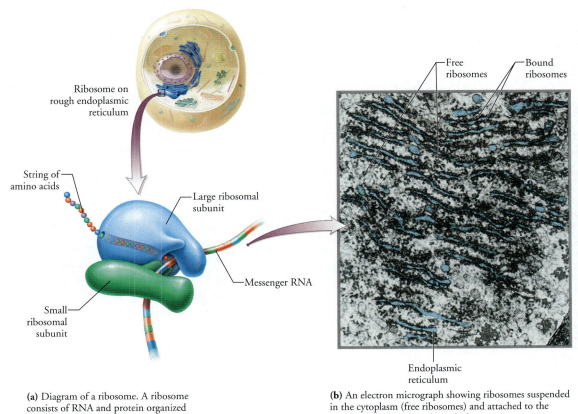

Ribosome on rough endoplasmic reticulum

String of amino acids

Large ribosomal subunit

Messenger RNA

Small ribosomal subunit

Free ribosomes

Bound ribosomes

Endoplasmic reticulum

(a) Diagram of a ribosome. A ribosome consists of RNA and protein organized into two subunits, one large and one small.

(b) An electron micrograph showing ribosomes suspended in the cytoplasm (free ribosomes) and attached to the endoplasmic reticulum (bound ribosomes).

FIGURE **3.13**
Ribosomes are sites where protein synthesis begins.

ENDOPLASMIC RETICULUM

The **endoplasmic reticulum (ER)** is part of an extensive network of channels that is connected to the plasma membrane, nuclear envelope, and some organelles (Figure 3.14). In some regions, the ER is studded with ribosomes and is called **rough endoplasmic reticulum (RER)**. The amino acid chains made by the attached ribosomes are threaded through the RER's membrane to its internal space. Here, the chains are processed and modified. RER enzymes attach carbohydrates to many of the amino acid chains, forming glycoproteins. The products are then enclosed in vesicles formed from the RER membrane. The vesicles are transferred to the Golgi complex for further processing and packaging.

Another function of rough endoplasmic reticulum is membrane production. Because the RER membrane is continually used to form vesicles for shipping, it must constantly be replenished. Where does the new material come from? Recall that proteins and phospholipids are major components of cell membranes. The protein components of the RER membrane come from the RER itself. The lipid components come mostly from the smooth ER.

Smooth endoplasmic reticulum (SER) lacks ribosomes. As mentioned, one function of SER is the production of phospholipids, which become part of cell membranes. The SER (particularly in liver cells) detoxifies alcohol and other drugs such as phenobarbital. Typically, enzymes of SER chemically modify drugs, making them more water soluble and easier to eliminate from the body.

GOLGI COMPLEX

Resembling a stack of dinner plates, the **Golgi complex** is the protein processing and packaging center of the cell (Figure 3.15). Protein-filled vesicles from the rough endoplasmic reticulum (RER) arrive at the "receiving side" of the Golgi complex. The vesicles fuse with the membrane of the Golgi complex and empty their contents to the inside. The Golgi complex then chemically modifies many of the proteins as they move, by way of vesicles, from one membranous disk in the stack to the next. After such processing is complete, the Golgi complex sorts the proteins, much as a postal worker sorts letters. The Golgi complex then sends the proteins to their specific destinations. Some of the proteins emerging from the "shipping side" are packaged in vesicles and sent to the plasma membrane, where the proteins may be exported from the cell or incorporated into the plasma membrane. Other proteins are packaged in lysosomes.

LYSOSOMES

How does the cell digest substances and rid itself of worn-out parts? If it simply released digestive enzymes, it would destroy itself. Instead, intracellular digestion occurs mainly within lysosomes. **Lysosomes** are roughly spherical organelles that are surrounded by a single membrane. They are packed with about 40 different digestive enzymes. The enzymes and membranes of lysosomes are made by the rough endoplasmic reticulum and then sent to the Golgi complex for further treatment and packaging. Eventually, enzyme-filled lysosomes bud from the Golgi complex and begin their diverse roles in digestion within the cell. The digestion of food in our gastrointestinal tract is a different phenomenon; it occurs outside of cells and is described in Chapter 15.

Lysosomes break down foreign invaders such as bacteria that enter the cell. Lysosomes also break down macromolecules. Consider, for example, what happens when a cell engulfs a particle of food. You can follow this process in Figure 3.16. During the process of phagocytosis (Step 1), a vesicle encircles the food, forming a food vacuole. A lysosome released from the Golgi complex then fuses with the food vacuole (Step 2) and releases its digestive enzymes, which break down the food into smaller molecules. These molecules diffuse out of the vesicle into the cytoplasm, where they can be used by the cell (Step 3). Indigestible residues may be expelled from the cell by exocytosis (Step 4), or they may be stored in vesicles (residual bodies) (Step 5).

Lysosomes are present in most cells. They are particularly numerous in cells that specialize in phagocytosis. Certain white blood cells ingest bacteria that have invaded the body. These bacteria are usually destroyed once lysosomes fuse with the phagocytic vacuoles that contain them. Certain bacteria, such as those that cause leprosy and tuberculosis, have developed ways to evade destruction once engulfed by white blood cells. These bacteria have substances on their outer surfaces that prevent lysosomes from fus-

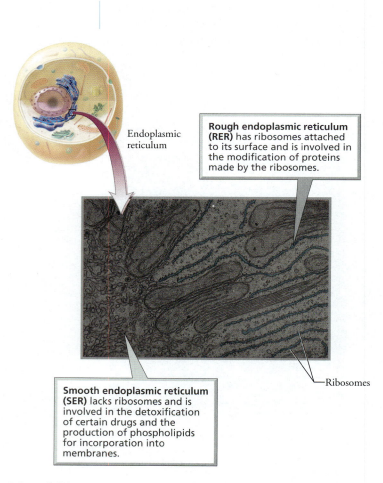

Endoplasmic reticulum

Rough endoplasmic reticulum (RER) has ribosomes attached to its surface and is involved in the modification of proteins made by the ribosomes.

Smooth endoplasmic reticulum (SER) lacks ribosomes and is involved in the detoxification of certain drugs and the production of phospholipids for incorporation into membranes.

Ribosomes

FIGURE **3.14**

The endoplasmic reticulum (ER), shown in this electron micrograph, consists of two continuous regions: rough ER and smooth ER.

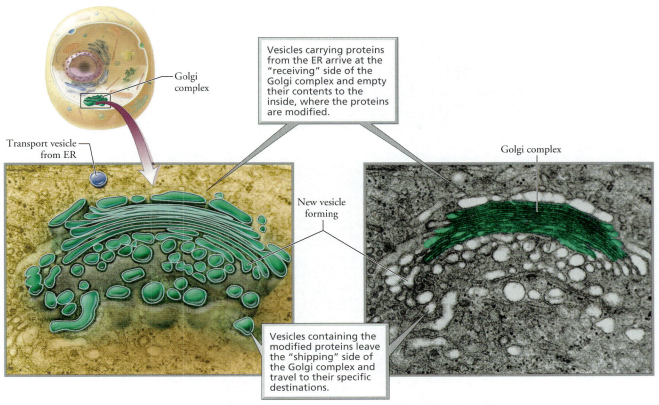

Vesicles carrying proteins from the ER arrive at the "receiving" side of the Golgi complex and empty their contents to the inside, where the proteins are modified.

Golgi complex

Transport vesicle from ER

New vesicle forming

Golgi complex

Vesicles containing the modified proteins leave the "shipping" side of the Golgi complex and travel to their specific destinations.

(a) Diagram of the Golgi complex. This organelle serves as the site for protein processing and packaging within the cell.

(b) Electron micrograph showing the Golgi complex and its associated vesicles

FIGURE **3.15**
The Golgi complex

ing with the vacuole that contains them. So, instead of being destroyed by lysosomal enzymes, these bacteria thrive and reproduce inside the cell—eventually killing the cell.

Lysosomes also destroy obsolete parts of the cell itself. Old organelles and macromolecules are broken down (see Figure 3.16). The resultant materials can be used by the cell to make new structures. An organelle called the mitochondrion, for example, lasts only about 10 days in a typical liver cell before being destroyed by lysosomes. Such "housecleaning" ensures proper functioning of the cell. The cleaning also promotes the recycling of essential materials.

We have seen the important roles played by lysosomal enzymes in the digestion of macromolecules and organelles. You should not be surprised to learn, then, that the absence of a single lysosomal enzyme can have devastating consequences. Lysosomal storage diseases result from the buildup of the molecules that would normally be broken down by the missing enzyme. Such molecules accumulate in the lysosomes and cause them to swell. Ultimately, the accumulating molecules interfere with cell function. Lysosomal storage diseases are inherited and progress with age. We will focus on Tay-Sachs disease.

Tay-Sachs disease is more common in Jewish children of eastern European ancestry than non-Jewish children. It is caused by the absence of the lysosomal enzyme hexosaminidase (Hex A). This enzyme breaks down lipids in nerve cells. When Hex A is missing, the lysosomes swell with undigested lipids (Figure 3.17). Infants with Tay-Sachs disease appear normal at birth. They begin to deteriorate by about 6 months of age as abnormal amounts of lipid continually accumulate in the nervous system. Typically, motor skills that had developed, such as crawling and grasping, progressively decline. Senses, such as sight and hearing, also deteriorate. Recurrent seizures are common in 2-year-olds. By the age of 4 or 5, Tay-Sachs causes paralysis and death. At present there is no cure for this disease. However, identification of the missing enzyme has led to the development of a blood test to detect individuals that carry the gene for Tay-Sachs. Called carriers, these individuals do not have the disease but could pass the gene to their offspring.

what would you do?

Imagine that you and your spouse desperately want children. However, both of you are carriers of Tay-Sachs disease and thus could pass the gene to your children. The options for any child that you might have are as follows: Your child may not have the gene for Tay-Sachs and may be healthy, may have the disease and die in early childhood, or may be a carrier as you are. Your parents urge adoption over having your own children. Your spouse prefers not to adopt but to use prenatal screening to check if your fetus has the disease. What would you do?

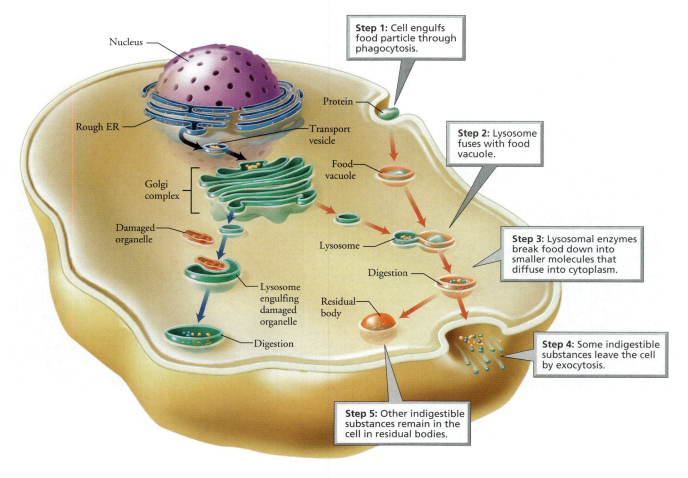

Step 1: Cell engulfs food particle through phagocytosis.

Step 2: Lysosome fuses with food vacuole.

Step 3: Lysosomal enzymes break food down into smaller molecules that diffuse into cytoplasm.

Step 4: Some indigestible substances leave the cell by exocytosis.

Step 5: Other indigestible substances remain in the cell in residual bodies.

Nucleus

Rough ER

Transport vesicle

Golgi complex

Damaged organelle

Lysosome engulfing damaged organelle

Digestion

Protein

Food vacuole

Lysosome

Digestion

Residual body

FIGURE **3.16**

Lysosome formation and function in cellular digestion. Lysosomes, released from the Golgi complex, digest substances imported from outside the cell (see pathway on right). Lysosomes also digest obsolete parts of the cell itself (see pathway on left).

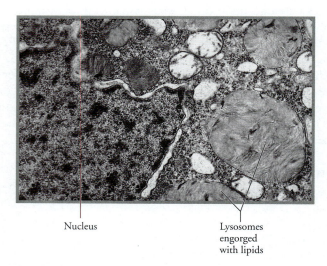

Nucleus

Lysosomes engorged with lipids

FIGURE **3.17**

An electron micrograph of a nerve cell from the brain of a person with Tay-Sachs disease. Note the lysosomes engorged with undigested lipids in the nerve cell shown here.

So far we have considered a lysosomal disease caused by a genetic error leading to a missing enzyme. Some environmental factors also interfere with lysosomes. Such interference can cause disease. In the Environmental Issue essay, *Asbestos: The Deadly Miracle Material,* we describe the impact of asbestos on lysosomes and health.

MITOCHONDRIA

Most cellular activities require energy. Energy is needed, for example, to transport certain substances across the plasma membrane. Energy is also needed to fuel many of the chemical reactions that occur in the cytoplasm. Specialized cells such as muscle cells and nerve cells also require energy to perform their particular activities. The energy needed by cells is provided by mitochondria (singular, mitochondrion). **Mitochondria** are organelles within which most of cellular respiration occurs. Cellular respiration is a three-phase process in which oxygen and an organic fuel such as glucose are consumed and energy in the form of ATP is released. The first phase takes place in the cytoplasm. The

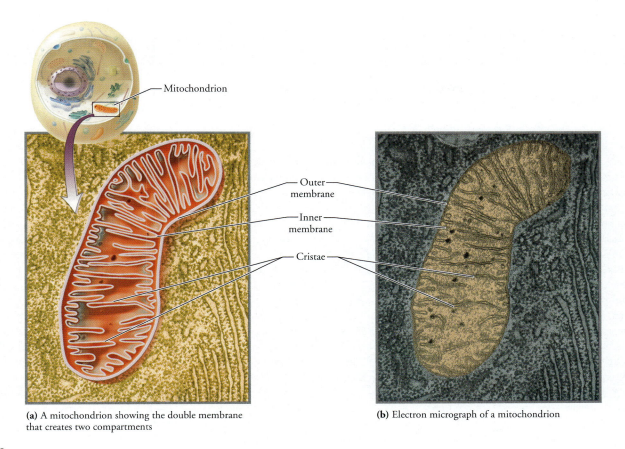

(a) A mitochondrion showing the double membrane that creates two compartments

(b) Electron micrograph of a mitochondrion

FIGURE **3.18**
Mitochondria are sites of energy conversion in the cell.

remaining two phases occur in the mitochondria. The process of cellular respiration will be described later in this chapter.

The number of mitochondria varies considerably from cell to cell and is roughly correlated with a cell's demand for energy. It is safe to say that most cells contain several hundred to thousands of mitochondria. Like the nucleus, but unlike other organelles, mitochondria are bounded by a double membrane (Figure 3.18). The inner and outer membranes create two separate compartments within a mitochondrion. These areas serve as sites for some of the reactions in cellular respiration. The infoldings of the inner membrane of a mitochondrion are called **cristae**. The cristae are the site of the last phase of cellular respiration. Finally, mitochondria contain ribosomes and a small percentage of a cell's total DNA (the rest being found in the nucleus).

Table 3.4 summarizes the functions of organelles.

▌The cytoskeleton provides support and movement

Inside the cell is a complex network of fibers called the **cytoskeleton**. Three types of fibers make up the cytoskeleton. **Microtubules** are the thickest of the three fibers; **microfilaments** are the thinnest. **Intermediate filaments** are a diverse group of fibers with diameters that fall in between those of microtubules and microfilaments. Microtubules and microfilaments are often disassembled and reassembled. In contrast, intermediate filaments tend to be more permanent.

MICROTUBULES

Microtubules are straight, hollow rods made of the protein tubulin. Near the nucleus is a microtubule-organizing center within which is a pair of centrioles. Each **centriole** is composed of nine sets of

TABLE **3.4**

SUMMARY OF MAJOR ORGANELLES AND THEIR FUNCTIONS	
ORGANELLE	**FUNCTION**
Nucleus	Contains almost all the genetic information and influences cellular structure and function
Rough endoplasmic reticulum (RER)	Studded with ribosomes (sites where the synthesis of proteins begins); produces membrane
Smooth endoplasmic reticulum (SER)	Detoxifies drugs; produces membrane
Golgi complex	Sorts, modifies, and packages products of RER
Lysosomes	Digest substances imported from outside the cell; destroy old or defective cell parts or cells
Mitochondria	Provide cell with energy through the breakdown of glucose during cellular respiration

ENVIRONMENTAL ISSUE

Asbestos: The Deadly Miracle Material

During the first three-quarters of the twentieth century, over 30 million tons of asbestos were used in the United States. About half a million people are estimated to have died as a result. What is asbestos? Why do we use it? And why do we sometimes die from it?

Asbestos is a fibrous mineral containing silica (Figure 3.A). It comes in several forms and is strong, flexible, and resistant to heat and corrosion. Because of these properties, asbestos has proved extremely valuable in construction. It has been used, for example, as an insulator on ceilings and pipes and has been sprayed on the walls of schools to make them more soundproof and fireproof.

The very same properties of asbestos that make it an ideal building material—its fibrous nature and durability—also can make it deadly. For example, when fibers of asbestos insulation are dislodged, small, light particles float in the air. These asbestos particles can be inhaled into the lungs. There is some evidence that the different forms of asbestos differ in the time they persist in lung tissue. Nevertheless, particles of at least one form appear resistant to degradation. These particles remain in the lungs for life (Figure 3.B).

Asbestosis is the most common disease caused by exposure to asbestos. It results from the dangerous interaction between asbestos and lysosomes and usually occurs in people who have worked either directly with asbestos or in buildings with exposed insulation. Asbestosis often takes 20 or more years to develop. Apparently, cells responsible for cleaning the respiratory passages engulf small particles of asbestos inhaled into the lungs. Lysosomes inside the cleaning cells then fuse with the vacuoles containing the asbestos particles. Lysosomal enzymes cannot break down the asbestos particles. Instead, the particles destabilize the membranes of the lysosomes, causing massive release of enzymes. The lysosomal enzymes destroy the cells of the respiratory tract. The result is irreversible scarring of lung tissue. Such scarring, detectable on a chest x-ray, eventually interferes with the exchange of gases in the lungs. People with asbestos-damaged lungs experience chronic coughing and shortness of breath. These symptoms become more severe over time, even without further exposure to asbestos. People with severe asbestosis may die from impaired respiratory function.

At present, there is no effective treatment for asbestosis. The focus, therefore, has been on prevention. Since the late 1970s, many governments have begun to regulate the use and removal of asbestos. In the United States, the use of asbestos for insulation and fireproofing or for new purposes is banned. However, some asbestos-containing products have not been banned; thus, it remains the responsibility of consumers to inquire whether asbestos is present in products they wish to purchase. In addition, asbestos is still present in many buildings constructed before the bans on its use in insulation or fireproofing went into effect. This point was made painfully clear when asbestos was detected in the debris and dust at the site where the World Trade Centers collapsed on September 11, 2001. Asbestos had been used as insulation in parts of the towers.

What can be done about the asbestos already present in our schools and workplaces? It is generally recommended and often required that exposed asbestos be removed, enclosed by other building materials, or covered with a sealant. Experts should determine which method is best for dealing with asbestos in a particular setting. Experts also should be involved in the sealing or removal. This step is critical because the greatest risk of asbestos exposure occurs when it is improperly removed. Such removal

---

triplet (three) microtubules arranged in a ring (Figure 3.19). Some microtubules provide support to the cell. Other microtubules near the plasma membrane maintain cell shape. Microtubules also form tracks along which organelles or vesicles travel (Figure 3.20). For example, secretory vesicles budded from the Golgi complex make their way to the plasma membrane by moving along a microtubule track. Finally, microtubules play a role in the separation of chromosomes during cell division. In cell division, the nucleus and cytoplasm of a cell split into two daughter cells (Chapter 19). Cancer is characterized by uncontrolled cell division and is often treated with chemotherapy. Some chemotherapeutic drugs halt cell division either by dismantling microtubules or by preventing their assembly. You will learn more about cancer in Chapter 21a.

Microtubules are responsible for the structure and movement of two types of cell extensions called cilia (singular, cilium) and flagella (singular, flagellum). **Cilia** are numerous, short hairlike extensions from the cell. Cilia resemble oars and move with a back-and-forth motion. Cilia are found, for example, on the surfaces of cells lining the respiratory tract (Figure 3.21), where they sweep debris

trapped in mucus away from the lungs. Smoking destroys these cilia and hampers cleaning of respiratory surfaces. A **flagellum** resembles a whip and moves in an undulating manner. Flagella are much longer than cilia. Most cells with flagella have a single flagellum. Each human sperm is propelled by a single flagellum (Figure 3.22).

Cilia and flagella differ in length, number per cell, and pattern of movement. Nevertheless, they have a similar arrangement of microtubules at their core. This arrangement consists of nine pairs of microtubules arranged in a ring. In the center of the ring are two single microtubules. This arrangement is called a 9 + 2 pattern and is shown in Figure 3.23. The microtubules of a cilium or flagellum are anchored to the cell by a **basal body**. Like the centrioles discussed above, basal bodies contain nine triplets of microtubules arranged in a ring (Figure 3.24).

Immotile cilia syndrome is a genetic disorder found in about 1 person out of every 20,000. The syndrome involves severe loss of function in cilia and flagella throughout the body. People with immotile cilia syndrome experience chronic respiratory problems, and males may be sterile because of immotile sperm. Loss of ciliary can release large numbers of particles into the air. People at high risk of being exposed to asbestos—plumbers, electricians, insulation workers, and carpenters, to name a few—must be informed of the health risks of their jobs. Workers at risk of exposure to asbestos should insist upon frequent testing of the air in their workplace. Such monitoring occurred at the World Trade Center site and surrounding neighborhoods. Workers at Ground Zero were instructed to wear respirators and other protective equipment (Figure 3.C). The results from air testing at the World Trade Center site and surrounding blocks have suggested to most experts that the risk of developing asbestos-related disease is quite small. The conclusion of low disease risk is based on the short-term, limited exposure experienced by most people at Ground Zero. (Long-term workers at the site should be protected from asbestos and other contaminants by their protective equipment.) We will know in 20 or 30 years if the experts are correct. 🌲

FIGURE 3.**A**
The deadly miracle material asbestos. The qualities of asbestos that make it a valuable material for construction—strength, durability, and flexibility—also make it hazardous to our health. Particles of asbestos released into the air may be inhaled into the lungs, where they cause asbestosis (scarring of lung tissue).

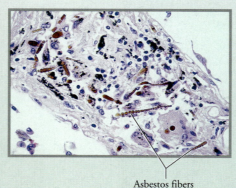

Asbestos fibers

FIGURE 3.**B**
Asbestos fibers in lung tissue. The destruction of lung tissue characteristic of asbestosis results from the release of lysosomal enzymes inside the cells responsible for cleaning the respiratory tract.

FIGURE 3.**C**
A worker at Ground Zero, New York City. Protective equipment, such as a respirator, can safeguard those exposed to asbestos and other contaminants in the workplace.

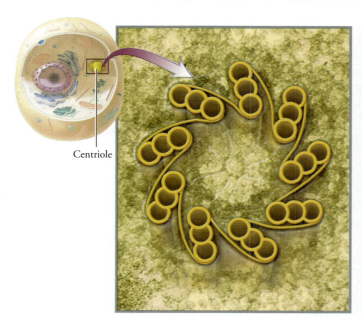

Centriole

(a) Diagram of a centriole. Each centriole is composed of nine sets of triplet microtubules arranged in a ring.

(b) Electron micrograph showing the microtubules of a centriole

FIGURE **3.19**
Centrioles may help organize microtubules.

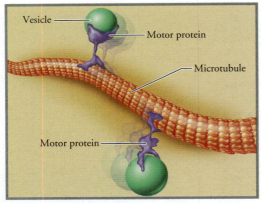

(a) Vesicles carried by proteins travel along a microtubule.

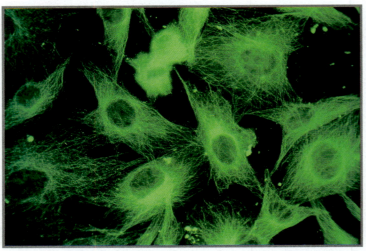

(b) An electron micrograph showing the microtubule component of the cytoskeleton for several cells. The round area in the center of each cell is the nucleus.

FIGURE **3.20**
Microtubules serve as tracks along which organelles or vesicles move.

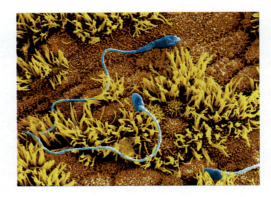

FIGURE **3.22**
Sperm cells in a fallopian tube. Each sperm cell has a single whiplike flagellum. The undulating motion of the flagellum propels the sperm cell. Microtubules are responsible for movement of flagella.

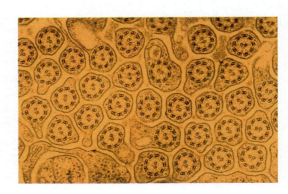

FIGURE **3.23**
Structure of cilia. Several cilia in cross section showing the 9 + 2 arrangement of microtubules. Flagella (not shown) have a similar arrangement of microtubules.

FIGURE **3.21**
Cilia on cells lining the respiratory tract. The cilia move in a back-and forth motion and sweep debris trapped in mucus away from the lungs. Microtubules are responsible for the movement of cilia.

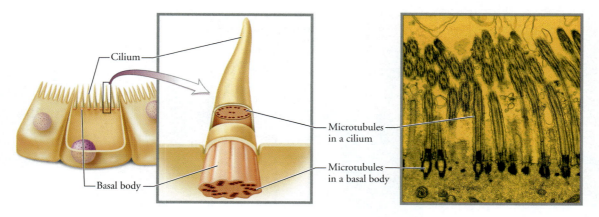

FIGURE **3.24**

Basal bodies. Cilia and flagella are anchored to cells by basal bodies. Basal bodies contain an arrangement of microtubules identical to that found in centrioles (nine sets of triplet microtubules).

and flagellar function appears to be caused by the absence of arm-like structures that normally link the nine pairs of microtubules in the 9 + 2 pattern (Figure 3.25).

MICROFILAMENTS

Microfilaments are solid rods made of the protein actin. These fibers are best known for their role in muscle contraction when they slide past thicker filaments made of the protein myosin. Microfilaments provide cells with support. They are also responsible for localized movements of the plasma membrane such as when extensions of the cell, called **pseudopodia** (*pseudo*, false; *pod*, foot) produce cell movement along a surface (Figure 3.26). This type of motion is called **amoeboid movement**. It is characteristic of some white blood cells that creep through infected tissue devouring bacteria and debris. Finally, microfilaments play a role in cell division when a band of them contracts and pinches the cell in two.

INTERMEDIATE FILAMENTS

Intermediate filaments are a diverse group of ropelike fibers. Their protein composition varies from one type of cell to another. Intermediate filaments help to maintain cell shape. They also anchor certain organelles in place.

Table 3.5 summarizes the structures and functions of microtubules, microfilaments, and intermediate filaments.

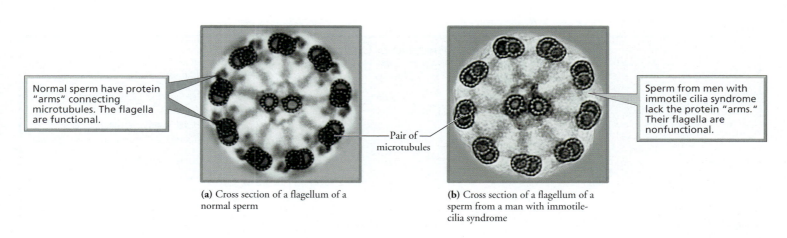

Normal sperm have protein "arms" connecting microtubules. The flagella are functional.

Sperm from men with immotile cilia syndrome lack the protein "arms." Their flagella are nonfunctional.

Pair of microtubules

(a) Cross section of a flagellum of a normal sperm

(b) Cross section of a flagellum of a sperm from a man with immotile-cilia syndrome

FIGURE **3.25**

Loss of function of cilia and flagella

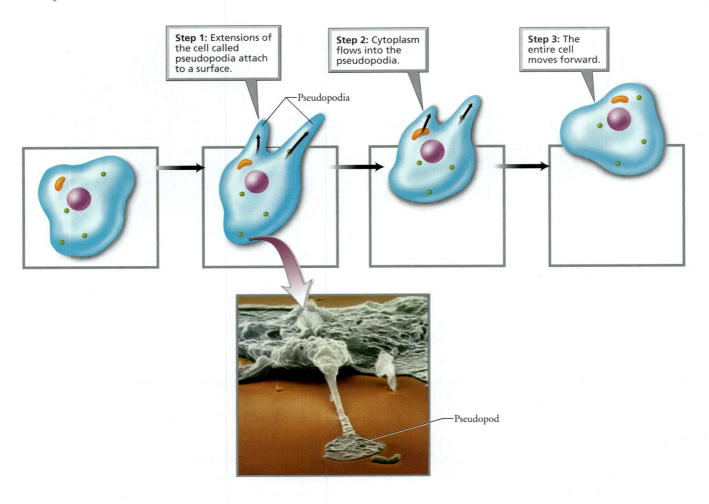

Step 1: Extensions of the cell called pseudopodia attach to a surface.

Step 2: Cytoplasm flows into the pseudopodia.

Step 3: The entire cell moves forward.

Pseudopodia

Pseudopod

FIGURE **3.26**
Amoeboid movement. Microfilaments are responsible for amoeboid movement.

TABLE **3.5**

STRUCTURES AND FUNCTIONS OF CYTOSKELETAL ELEMENTS			
ELEMENT	**STRUCTURE**	**PROTEIN**	**FUNCTIONS**
Microtubules	Hollow rods	Tubulin	• Cell movement (cilia, flagella) • Support • Tracks for organelles and vesicles • Chromosome movements in cell division
Microfilaments	Solid rods	Actin	• Muscle contraction • Support • Cell movement (pseudopodia) • Pinch cell in two in cell division
Intermediate filaments	Ropelike fibers	Different proteins; depends on cell type	• Maintain cell shape • Support • Anchor organelles

Our Cells Use Cellular Respiration and Fermentation to Generate ATP

Living requires work, and work requires energy. Logic tells us, therefore, that living requires energy. So, where do we get our energy? The short answer is that we get our energy from the food we eat. Our digestive system breaks down complex food molecules, such as carbohydrates, proteins, and fats, into their simpler components (see Chapter 15). These simpler molecules—such as amino acids and simple sugars—are then absorbed into the bloodstream and carried to our cells, where they are converted into fuel for use by the cells. This conversion requires the breakdown of these simple molecules and the release of energy once stored in their chemical bonds. Some of this energy is used to make ATP, the energy-rich molecule that our cells can use to do their work. The rest is given off as heat. Our focus here is on how the simple sugar glucose is converted into fuel within our cells. We will trace the two pathways used by cells—cellular respiration and fermentation—to break apart glucose molecules for energy. Cellular respiration requires oxygen; fermentation does not.

The transfer of electrons from one atom or molecule to another within our cells is a key feature of the chemical reactions that capture energy from fuel. **Oxidation** is the loss of electrons from one atom or molecule to another. **Reduction** is the gain of electrons by one atom or molecule from another. During the conversion of glucose into useful energy, it is oxidized in a series of small steps involving several intermediate compounds.

Cellular respiration requires oxygen

Cellular respiration is the oxygen-requiring pathway by which glucose is broken down by cells. The complete breakdown of glucose yields carbon dioxide, water, and energy. The pathway is an elaborate series of chemical reactions that involves the passage of energy from one molecule to another. Why is cellular respiration so critical to cells? Basically, because our cells cannot directly use the energy stored in the bonds of glucose. Instead, this energy must be released during the breakdown of glucose and then used by cells to form chemical bonds in ATP. Recall from Chapter 2 that ATP (adenosine triphosphate) molecules contain the chemical energy that cells can use. ATP is formed from ADP (adenosine diphosphate) and inorganic phosphate (Pi) in a process that requires energy. The complete breakdown of glucose during cellular respiration is shown by the following equation:

$$C_6H_{12}O_6 + 6O_2 \Rightarrow 6CO_2 + 6H_2O + energy$$

Glucose   Oxygen   Carbon dioxide   Water

Cellular respiration has three major phases: (1) glycolysis, (2) the citric acid cycle, and (3) the electron transport chain. All three phases occur at the same time within cells. Glycolysis takes place in the cytoplasm of the cell. The citric acid cycle and the electron transport chain take place in mitochondria. You will see that glycolysis and the citric acid cycle involve a series of reactions in which the products from one reaction become the substrates (raw materials) for the next reaction.

GLYCOLYSIS

Glycolysis (*glyco*, sugar; *lysis*, splitting) involves the splitting of glucose, a six-carbon sugar, into two three-carbon sugars. These three-carbon sugars are then oxidized to form two molecules of pyruvate (Figure 3.27). This splitting occurs in several steps. Each step involves

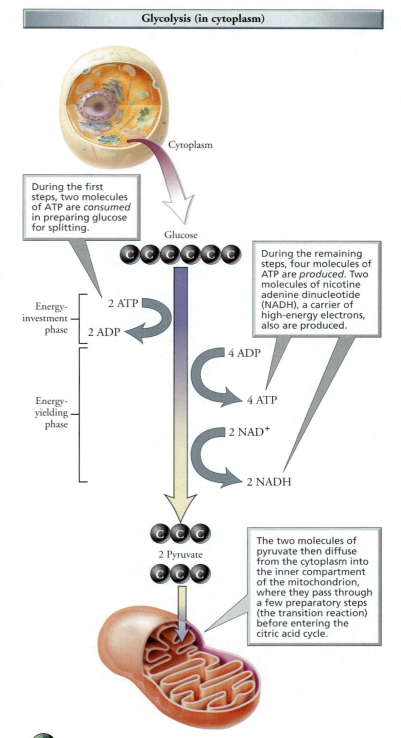

During the first steps, two molecules of ATP are *consumed* in preparing glucose for splitting.

During the remaining steps, four molecules of ATP are *produced*. Two molecules of nicotine adenine dinucleotide (NADH), a carrier of high-energy electrons, also are produced.

Energy-investment phase — 2 ATP / 2 ADP

Energy-yielding phase — 4 ADP / 4 ATP / 2 NAD$^+$ / 2 NADH

2 Pyruvate

The two molecules of pyruvate then diffuse from the cytoplasm into the inner compartment of the mitochondrion, where they pass through a few preparatory steps (the transition reaction) before entering the citric acid cycle.

 FIGURE **3.27**
Glycolysis is a several-step sequence of reactions in the cytoplasm. Glucose, a six-carbon sugar, is split into two three-carbon molecules of pyruvate.

a different, specific enzyme. During the first steps, two molecules of ATP are consumed (energy is needed to prepare glucose for splitting). During the remaining steps, four molecules of ATP are produced, for a net gain of two ATP. Glycolysis also produces two molecules of nicotine adenine dinucleotide (NADH), which are generated when electrons are donated to the coenzyme NAD^+. Glycolysis does not require oxygen and releases only a small amount of the chemical energy stored in glucose. Most of the energy remains in the two molecules of pyruvate. These molecules move from the cytoplasm into the inner compartment of the mitochondria, where they pass through a few preparatory steps (the transition reaction) and then enter the next major phase of cellular respiration—the citric acid cycle.

CITRIC ACID CYCLE

Once inside the inner compartment of the mitochondrion, pyruvate reacts with a substance called coenzyme A (CoA). This reaction is called the transition reaction. The **transition reaction** results in the removal of one carbon (in the form of carbon dioxide, CO_2) from each pyruvate (Figure 3.28). The resulting two-carbon molecule, called an acetyl group, then binds to CoA forming acetyl CoA. A molecule of NADH is also produced from each pyruvate.

Still in the inner compartment of the mitochondrion, acetyl CoA enters a cyclic series of chemical reactions known as the **citric acid cycle**. The cycle is named after the first product (citric acid, or citrate) formed along its route (Figure 3.29). The cycle is sometimes called the Krebs cycle, after the scientist Hans Krebs, who described many of the reactions. We will not consider each of the chemical reactions in this eight-step series. We will simply say that the citric acid cycle completes the oxidation of glucose and yields two molecules of ATP (one from each acetyl CoA that enters the cycle) and several molecules of NADH and $FADH_2$ (flavin adenine dinucleotide). NADH and $FADH_2$ are carriers of high-energy electrons. These molecules, along with those produced during glycolysis and the transition reaction, enter the electron transport chain, the final phase of cellular respiration. The citric acid cycle also produces CO_2 as waste.

ELECTRON TRANSPORT CHAIN

During the final phase of cellular respiration, the molecules of NADH and $FADH_2$ produced by earlier phases pass their electrons to a series of carrier proteins. These carrier proteins are embedded in the inner membrane of the mitochondrion. The proteins are known as the **electron transport chain** (Figure 3.30). (Recall that the inner membrane of the mitochondrion is highly folded, providing space for thousands of copies of the carrier proteins.) During the transfer of electrons from one molecule to the next, energy is released. This energy is then used to make ATP. Eventually, the electrons are passed to oxygen, the final electron acceptor. Oxygen has a critical role in cellular respiration. In the absence of oxygen, electrons accumulate in the carrier molecules, halting the citric acid cycle and cellular respiration. When oxygen is present, however, electron transfer and the citric acid cycle continue. Upon accepting electrons, oxygen then combines with two hydrogen ions to form water. The electron transport chain produces 32 molecules of ATP per molecule of glucose.

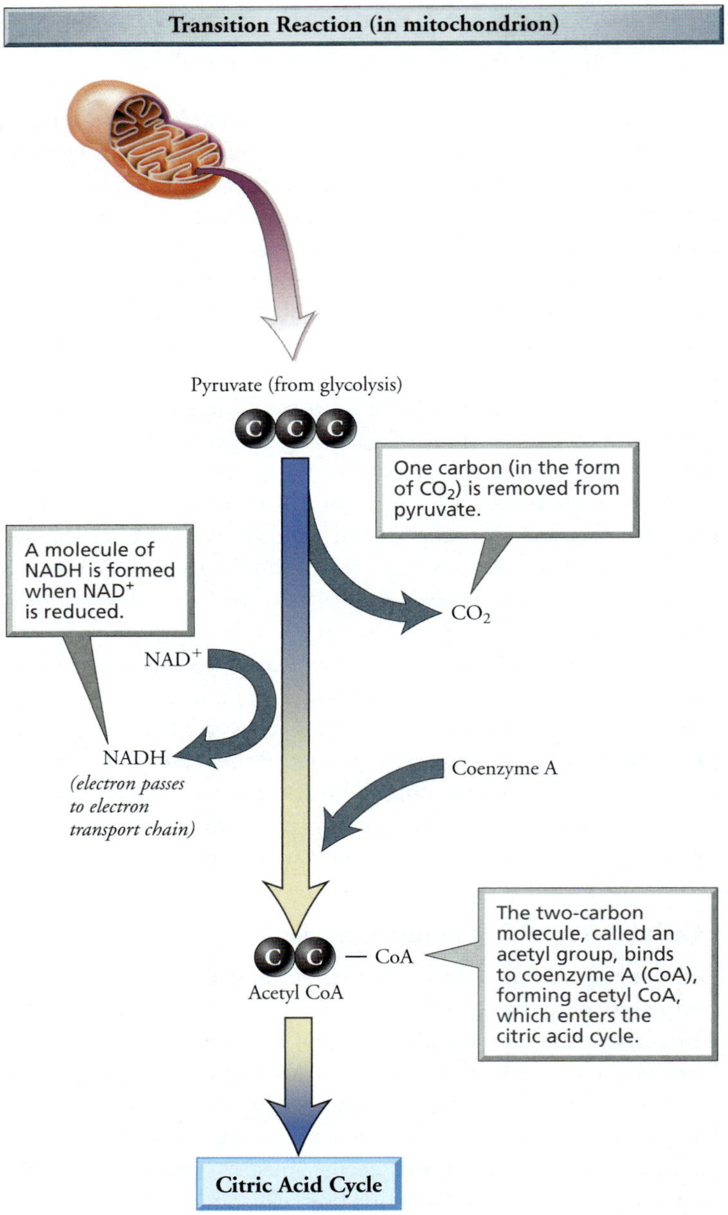

FIGURE **3.28**

The transition reaction takes place inside the mitochondrion and is the link between glycolysis and the citric acid cycle.

Some deadly poisons, such as cyanide, are known to inhibit cellular respiration by blocking the flow of electrons along the electron transport chain. Symptoms of cyanide poisoning in humans include gasping, convulsions, coma, and death in cases of significant exposure. Cyanide, you might be surprised to learn, is present in several plants—most notably in the roots of the bitter cassava, a tropical plant from which tapioca is made. Heating the roots of the plant destroys the concentrated cyanide within. After heating, starch is extracted from the roots and used to make tapioca. So you can eat your pudding with near total peace of mind!

Citric Acid Cycle (in mitochondrion)

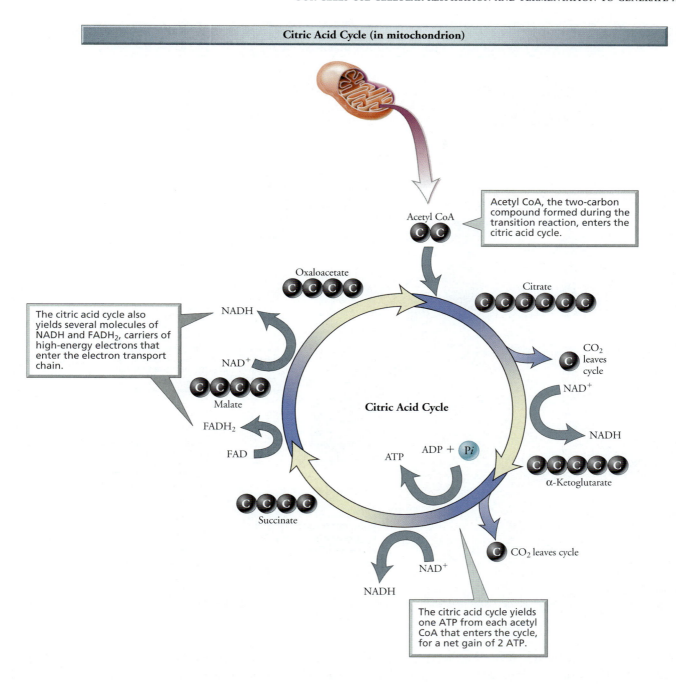

Acetyl CoA, the two-carbon compound formed during the transition reaction, enters the citric acid cycle.

Acetyl CoA

Oxaloacetate

The citric acid cycle also yields several molecules of NADH and FADH$_2$, carriers of high-energy electrons that enter the electron transport chain.

NADH

NAD$^+$

Malate

FADH$_2$

FAD

Succinate

Citric Acid Cycle

ATP   ADP + Pi

NADH

NAD$^+$

Citrate

CO_2 leaves cycle

NAD$^+$

NADH

α-Ketoglutarate

C CO_2 leaves cycle

The citric acid cycle yields one ATP from each acetyl CoA that enters the cycle, for a net gain of 2 ATP.

FIGURE **3.29**

The citric acid cycle is a cyclic series of eight chemical reactions that occurs inside the mitochondrion and yields 2 molecules of ATP and several molecules of NADH and FADH$_2$ per molecule of glucose.

The process of cellular respiration is summarized in Figure 3.31. Basic descriptions of each phase can be found in Table 3.6. All together, cellular respiration produces 36 molecules of ATP per molecule of glucose: 2 ATP from glycolysis, 2 ATP from the citric acid cycle, and 32 ATP from the electron transport chain.

▍Fermentation does not require oxygen

You have seen that cellular respiration depends on oxygen as the final electron acceptor in the electron transport chain. Without oxygen, the transport chain comes to a halt, stopping the citric acid

cycle and cellular respiration. The question arises, then, is there a way for cells to harvest energy when molecules of oxygen are scarce? The answer is yes, and the pathway is fermentation.

Fermentation is the breakdown of glucose without oxygen. It begins with glycolysis, which as you recall, occurs in the cytoplasm and does not require oxygen. From one molecule of glucose, glycolysis produces two molecules each of pyruvate, the electron carrier NADH, and ATP. The chemical reactions of fermentation that follow glycolysis also take place in the cytoplasm. These reactions involve a transfer of electrons from NADH to pyruvate or a

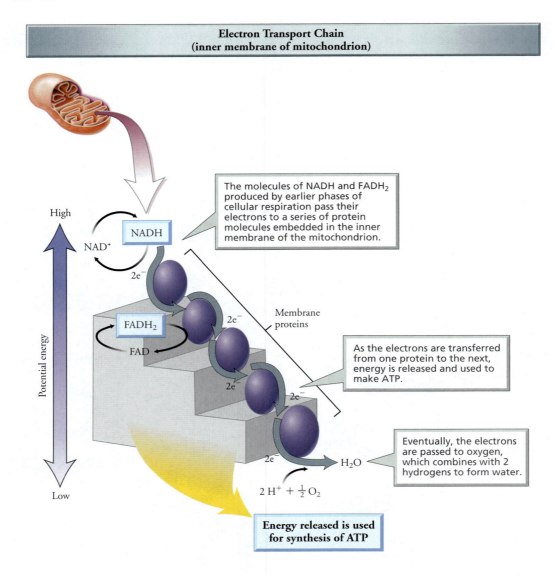

Electron Transport Chain
(inner membrane of mitochondrion)

High

NAD$^+$

NADH

The molecules of NADH and FADH$_2$ produced by earlier phases of cellular respiration pass their electrons to a series of protein molecules embedded in the inner membrane of the mitochondrion.

2e$^-$

Potential energy

FADH$_2$

FAD

2e$^-$

Membrane proteins

2e$^-$

2e$^-$

As the electrons are transferred from one protein to the next, energy is released and used to make ATP.

2e$^-$

H$_2$O

Eventually, the electrons are passed to oxygen, which combines with 2 hydrogens to form water.

Low

$2\,H^+ + \frac{1}{2}O_2$

Energy released is used
for synthesis of ATP

FIGURE **3.30**

The electron transport chain is the final phase of cellular respiration. This phase yields 32 ATP per molecule of glucose.

derivative of pyruvate. This transfer of electrons is critical because it regenerates NAD$^+$. NAD$^+$ is essential for the production of ATP through glycolysis. In fermentation, the final electron acceptor is pyruvate or one of its derivatives. Recall that in cellular respiration, oxygen is the final electron acceptor in the electron transport chain. Fermentation is a very inefficient way for cells to harvest energy. It nets only 2 molecules of ATP compared with the 36 molecules of ATP produced by cellular respiration.

There are several types of fermentation that differ in the waste products formed from pyruvate. We will consider lactic acid fermentation and alcohol fermentation.

LACTIC ACID FERMENTATION

During **lactic acid fermentation**, NADH passes electrons directly to pyruvate. The waste product produced is lactate or lactic acid (Figure 3.32a). During strenuous exercise, the oxygen supply in our muscle cells is low. Under these conditions, the cells increase lactic acid fermentation to ensure the continued production of ATP. The muscle pain we often experience after intense exercise is caused, in part, by the accumulation of lactic acid produced by fermentation. In time, the soreness disappears as the lactic acid moves out of the muscle cells and into the bloodstream, where it is carried to the liver.

ALCOHOL FERMENTATION

Unlike lactic acid fermentation, **alcohol fermentation** is a two-step process (Figure 3.32b). First, carbon dioxide (CO$_2$) is released from pyruvate, leaving a two-carbon derivative of pyruvate called acetaldehyde. In the second step, NADH passes electrons to acetaldehyde, generating ethyl alcohol or ethanol. In alcohol fermentation, electrons are not passed directly to pyruvate. Instead, the electrons are passed to acetaldehyde, a derivative of pyruvate.

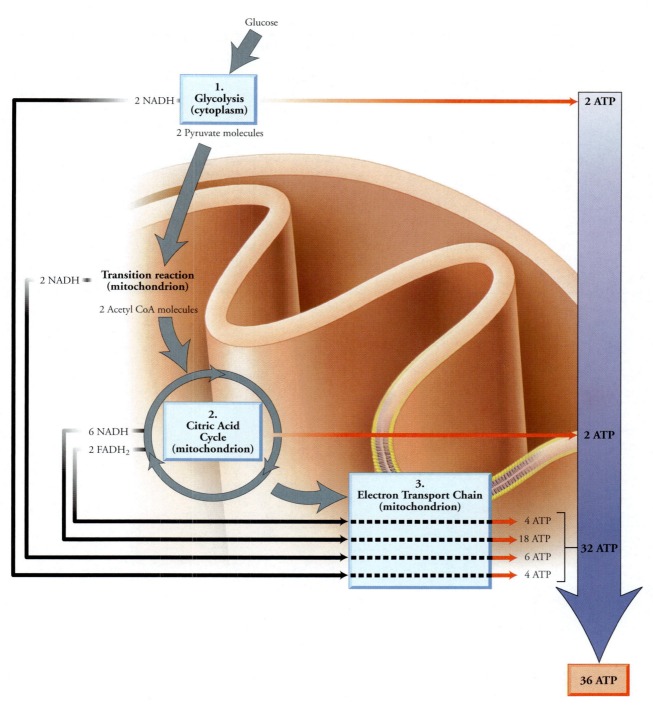

FIGURE **3.31**

Summary of the phases of cellular respiration. Cellular respiration produces 36 ATP per molecule of glucose (2 ATP from glycolysis, 2 ATP from the citric acid cycle, and 32 ATP from the electron transport chain).

Yeast, a single-celled fungus, switches to alcohol fermentation when oxygen becomes scarce in its environment. We often add yeast cells to bread dough. Within the low-oxygen environment of dough, yeast uses alcohol fermentation. The CO_2 produced during the first step of alcohol fermentation causes the bread to rise. The ethanol produced as a waste product dissipates during baking. Brewers do not see the ethanol as a waste, but as a product (beer, wine, or liquor) to be bottled and sold.

TABLE 3.6

THE PHASES OF CELLULAR RESPIRATION

PHASE	LOCATION	DESCRIPTION	MAIN PRODUCTS
Glycolysis	Cytoplasm	Several-step process by which glucose is split into 2 pyruvate	2 pyruvate 2 ATP 2 NADH
Transition reaction and	Mitochondria	One CO_2 is removed from each pyruvate; the resulting molecules bind to CoA, forming 2 acetyl CoA	2 acetyl CoA 2 NADH
Citric acid cycle	Mitochondria	Cyclic series of eight chemical reactions by which acetyl CoA is broken down	2 ATP 2 $FADH_2$ 6 NADH
Electron transport chain	Mitochondria	Electrons from NADH and $FADH_2$ are passed from one protein to the next, releasing energy for ATP synthesis	32 ATP H_2O

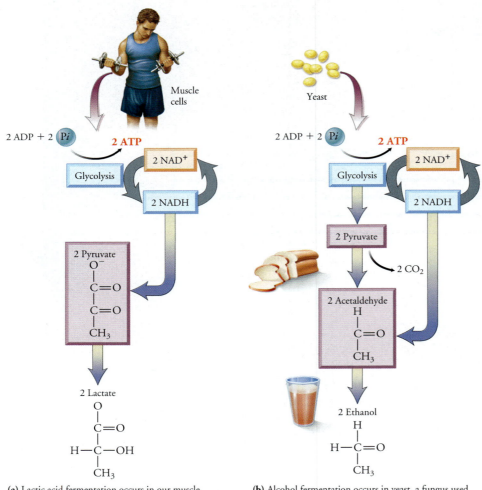

(a) Lactic acid fermentation occurs in our muscle cells during strenuous exercise. It consists of glycolysis plus chemical reactions in which NADH passes electrons directly to pyruvate, forming the waste product lactate (lactic acid).

(b) Alcohol fermentation occurs in yeast, a fungus used in making bread and beer. It consists of glycolysis plus a two-step chemical process. In the first step, CO_2 is released from pyruvate, leaving a two-carbon derivative of pyruvate called acetaldehyde. In the second step, NADH passes electrons to acetaldehyde, generating ethanol (ethyl alcohol).

FIGURE 3.32

Two types of fermentation. Fermentation occurs in the cytoplasm, does not require oxygen, and yields only 2 molecules of ATP per molecule of glucose.

REVIEWING THE CONCEPTS

Our Cells Are Highly Structured (pp. 41–58)

1. There are two main types of cell. Prokaryotic cells, unique to bacteria and archaea, lack membrane-bound organelles. Eukaryotic cells, found in all other organisms, have membrane-bound organelles.

WEB TUTORIAL 3.1 Cells

2. Because increases in cell volume occur faster than increases in cell surface area, a large cell faces problems associated with inadequate surface area. For that reason, most cells are very small.

3. The plasma membrane is made of phospholipids arranged in a bilayer. Proteins and molecules of cholesterol are interspersed throughout the bilayer. Carbohydrates attach only to the outer surface of the membrane. The structure of the plasma membrane is often described as a fluid mosaic.

4. The plasma membrane maintains the integrity of the cell, regulates movement of substances into and out of the cell, functions in cell-cell recognition, promotes communication between cells, and sticks cells together to form tissues and organs.

5. Substances cross the plasma membrane in several ways. Some substances cross by random movement from high to low concentration (simple diffusion); others need help from a carrier or channel protein (facilitated diffusion). Water also moves across the plasma membrane from high to low concentration (osmosis). Substances being concentrated by cells move across the plasma membrane from low to high concentration with the help of ATP and a carrier protein (active transport). Finally, cells may engulf outside materials by surrounding the material with plasma membrane (endocytosis), or they may release substances to the external environment when internal vesicles fuse with the plasma membrane and spill their contents to the outside (exocytosis).

WEB TUTORIAL 3.2 Plasma Membranes, Osmosis, Diffusion
WEB TUTORIAL 3.3 Passive and Active Transport
WEB TUTORIAL 3.4 Exocytosis and Endocytosis

6. Within the eukaryotic cell, membranes delineate compartments within which specific chemical processes occur. Organelles are small, specialized components suspended in the cytoplasm of cells. The nucleus contains almost all the genetic information and thus holds the code for the cell's structure and many of its functions. The nucleolus, a specialized region within the nucleus, makes ribosomal RNA. Ribosomes are involved in protein synthesis. Endoplasmic reticulum (ER) functions in membrane production and may be studded with ribosomes (rough endoplasmic reticulum, RER) or free of ribosomes (smooth endoplasmic reticulum, SER). The Golgi complex sorts, modifies, and packages products of the RER. The amino acid chains made at the ribosomes attached to RER are threaded into the RER, where they are processed and then sent to the Golgi complex for further processing. Lysosomes digest substances imported from outside the cell or break down old or defective components within cells. Mitochondria process energy for cells.

7. The cytoplasm is filled with a complex network of fibers called the cytoskeleton. Three types of fibers make up the cytoskeleton: microtubules, microfilaments, and intermediate filaments. Microtubules are hollow rods of tubulin that function in cell movement (cilia and flagella), support, and the movement within cells of chromosomes, organelles, and vesicles. Microfilaments are solid rods of actin that function in muscle contraction, cell support, movement (pseudopodia), and cell division. Intermediate filaments are ropelike fibers made of diverse proteins; these filaments maintain cell shape, anchor organelles, and provide support.

Our Cells Use Cellular Respiration and Fermentation to Generate ATP (pp. 59–64)

8. Cells require energy to work. We get this energy from the food we eat. Carbohydrates, fats, and proteins that we consume are each broken down by our digestive system into smaller units such as simple sugars, fatty acids, and amino acids. These simpler substances are absorbed into the bloodstream and carried to our cells. Within our cells, they are converted to fuel for use by our cells. Basically, the energy stored in their chemical bonds is released. Some of this energy is stored in the chemical bonds of ATP. Some is given off as heat.

9. Cells use two pathways—cellular respiration and fermentation—to break down the carbohydrate glucose. Cellular respiration requires oxygen and yields 36 molecules of ATP per molecule of glucose. Fermentation does not require oxygen and yields only 2 molecules of ATP per molecule of glucose.

WEB TUTORIAL 3.4 Cellular Respiration

10. Cellular respiration has three main phases: glycolysis, the citric acid cycle, and the electron transport chain. These phases occur at the same time within cells. Glycolysis occurs in the cytoplasm and splits glucose into pyruvate. Glycolysis also produces NADH (a carrier of high-energy electrons) and results in a net gain of 2 ATP. The molecules of pyruvate move from the cytoplasm into the inner compartment of the mitochondrion, where they pass through a few preparatory steps (the transition reaction) before entering the citric acid cycle. The citric acid cycle is an eight-step cycle that completes the breakdown of glucose into carbon dioxide. The citric acid cycle also yields 2 ATP and carriers of high-energy electrons (NADH and FADH$_2$). During the electron transport chain, the carriers of high-energy electrons produced during glycolysis and the citric acid cycle pass their electrons to a series of proteins embedded in the inner membrane of the mitochondrion; oxygen is the final electron acceptor. Energy released during the transfer of electrons yields 32 ATP.

11. Fermentation is the breakdown of glucose without oxygen. It occurs in the cytoplasm and begins with glycolysis, the splitting of glucose into pyruvate. The remaining chemical reactions involve the transfer of electrons from NADH to pyruvate or a derivative of pyruvate. When compared with cellular respiration, fermentation is an inefficient way for cells to harvest energy.

KEY TERMS

active transport *p. 45*
alcohol fermentation *p. 62*
amoeboid movement *p. 57*
basal body *p. 54*
cell adhesion molecule (CAM) *p. 44*
cellular respiration *p. 59*
centriole *p. 53*

chromatin *p. 48*
chromosome *p. 48*
cilia *p. 54*
citric acid cycle *p. 60*
cristae *p. 53*
cytoplasm *p. 42*
cytoskeleton *p. 53*
electron transport chain *p. 60*

endocytosis *p. 46*
endoplasmic reticulum (ER) *p. 50*
eukaryotic cell *p. 41*
exocytosis *p. 46*
extracellular fluid *p. 42*
facilitated diffusion *p. 45*
fermentation *p. 61*
flagellum *p. 54*

fluid mosaic *p. 43*
glycolysis *p. 59*
Golgi complex *p. 50*
hypertonic solution *p. 45*
hypotonic solution *p. 45*
intermediate filament *p. 53*
isotonic solution *p. 45*
lactic acid fermentation *p. 62*

THINKING ABOUT THE CONCEPTS

1. How do prokaryotic and eukaryotic cells differ?   p. 41
2. Describe the structure of the plasma membrane.   pp. 41–43
3. List five functions of the plasma membrane.   pp. 43–44
4. What is the difference between simple and facilitated diffusion? Give examples of each.   pp. 44–45
5. Describe endocytosis and exocytosis.   pp. 46–47
6. What is contained within the nucleus of a cell?   p. 48
7. List three functions of lysosomes.   pp. 50–52
8. What are lysosomal storage diseases?   pp. 51–52
9. Name the three types of fibers that make up the cytoskeleton. Describe their structures and functions.   pp. 53–56, 58
10. What is the basic function of cellular respiration?   p. 59
11. How do cellular respiration and fermentation differ with respect to the number of ATP molecules produced and the requirement for oxygen?   pp. 61–62
12. Prokaryotic cells
    a. have internal membrane-bound organelles.
    b. are usually larger than eukaryotic cells.
    c. lack internal membrane-enclosed organelles.
    d. have linear strands of DNA within a nucleus.
13. The plasma membrane
    a. is selectively permeable.
    b. contains lipids that function in cell-cell recognition.
    c. has cell adhesion molecules that prevent cells from sticking together.
    d. is made of nucleic acids.
14. Facilitated diffusion is
    a. the random movement of a substance from a region of higher concentration to a region of lower concentration.
    b. the movement of water across the plasma membrane.
    c. the movement of molecules across the plasma membrane against a concentration gradient with the aid of a carrier protein and energy supplied by the cell.
    d. the movement of a substance from a region of higher concentration to a region of lower concentration with the aid of a membrane protein.

15. Almost all the genetic information of a cell is found in the
    a. endoplasmic reticulum.
    b. Golgi complex.
    c. nucleus.
    d. mitochondria.
16. Ribosomes
    a. are found on smooth endoplasmic reticulum.
    b. translate information from mRNA into a chain of amino acids.
    c. process and modify proteins.
    d. break down foreign invaders and old organelles.
17. Mitochondria
    a. process energy for cells.
    b. lack ribosomes and DNA.
    c. are bounded by a single membrane.
    d. function in cell digestion.
18. Microtubules
    a. are found in eukaryotic cilia and flagella.
    b. are made of the protein actin.
    c. are responsible for amoeboid movement.
    d. pinch a cell in two during cell division.
19. Glycolysis
    a. occurs in the mitochondria.
    b. requires oxygen.
    c. begins the oxidation of glucose within the cytoplasm of our cells.
    d. nets 32 molecules of ATP per molecule of glucose.
20. _____ is the jelly-like solution within a cell that contains everything between the nucleus and the plasma membrane.
21. _____ is called "cell drinking."
22. The _____ is a specialized region within the nucleus that is involved in the production of rRNA.
23. _____ is the final electron acceptor in the electron transport chain.
24. Our muscle cells may switch from cellular respiration to _____ when oxygen is low.

APPLYING THE CONCEPTS

1. Joe's kidneys are failing, and he needs a kidney transplant. Luckily, he has an identical twin who wishes to donate one of his kidneys. Why would a kidney from his identical twin be better than one from another family member or a complete stranger? Explain your answer with reference to the plasma membrane of Joe's cells.

2. Your friend wonders aloud why beer has "natural carbonation." Now that you have learned about alcohol fermentation, you can answer his question. What would you say?
3. Would you expect to find more mitochondria in muscle cells or bone cells? Explain your answer.

4

Body Organization and Homeostasis

The Organization of the Human Body Increases in Complexity from Cells to Organ Systems

- Groups of similar cells form tissues
- Many tissues are held together by specialized junctions
- Tissues form organs that often work together in organ systems
- Most organs are housed in body cavities that are lined with membranes

The Skin is an Organ System

- Skin protects, regulates body temperature, and excretes
- The skin has two layers
- Skin color is determined by pigment and blood flow
- Hair, nails, and glands are skin derivatives
- Skin functions with other organ systems

Homeostasis Maintains Relative Internal Constancy

- Homeostasis is maintained by negative feedback mechanisms
- The hypothalamus regulates body temperature

HEALTH ISSUE   Fun in the Sun?

HEALTH ISSUE   Acne: The Misery, the Myths, and the Medications

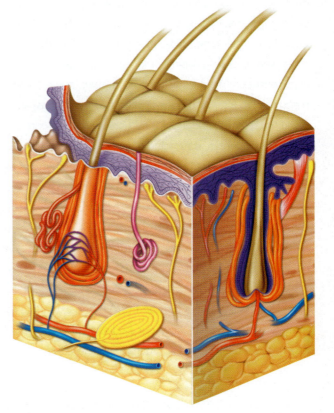

Shattered glass and chunks of metal lay across Interstate 5. The force of the impact was sufficient to throw the driver horizontally down the pavement for over 35 feet. He lay still in the center of the road. His passenger, inside the car wearing her seat belt, was conscious but not moving. Motorists pulled over, called for help, and ran toward the two young victims.

Luckily, both the driver and his passenger were breathing, and only the driver was bleeding noticeably. However, they were in a serious life-threatening situation. The trauma of the crash had sent them into shock. Their blood pressures were dropping; their body temperatures were dropping rapidly, and the passenger was gradually losing consciousness. Their bodies were shutting down, with inadequate blood flow to internal organs such as the heart, lungs, and brain. If the bleeding, heat loss, and low blood pressure went untreated for more than a few minutes, they could die. Help arrived in time.

The overwhelming effects of shock on an otherwise healthy body illustrate how our bodies depend on a carefully maintained equilibrium and how internal or external trauma can trigger a rapid shutdown of all body systems. How does the body normally organize its structures and interact with the environment to maintain a stable, constant internal equilibrium that allows organs and systems to function?

In this chapter, we take a close look at our bodies as integrated systems, beginning with cells and tissues and extending to organ systems. We then discuss how our bodies maintain relatively constant internal conditions, at every organizational level. ■

The Organization of the Human Body Increases in Complexity from Cells to Organ Systems

Think for a moment about the multitude of functions taking place in your body at this moment. Your heart is beating. You are breathing. You are reading, so some nerve cells are active and your eyes are moving. Your body can carry out these functions because its cells are specialized to perform specific tasks. But cell specialization is not enough. Specialized cells are organized into tissues, organs, and organ systems.

Groups of similar cells form tissues

A **tissue** is a group of cells that work together to serve a common function. Human tissues come in four primary types: epithelial tissue, connective tissue, muscle tissue, and nervous tissue. **Epithelial tissue** covers body surfaces, lines body cavities and organs, and forms glands. **Connective tissue** serves as a storage site for fat, plays an important role in immunity, and provides the body and its organs with protection and support. **Muscle tissue** is responsible for movement of the organism and of substances through the organism. **Nervous tissue** coordinates body activities that result from initiation and transmission of nerve impulses from one part

of the body to another. As you read the chapter, you will learn more about each of these types of tissue.

EPITHELIAL TISSUE

When you look in the mirror, you see epithelial tissue, also called epithelium. Your skin and the lining of your mouth are examples of epithelial tissue. Epithelial tissue is made up of closely packed cells that form a continuous layer, or sheet, that lines the body's inner and outer surfaces. There are two basic types of epithelial tissue. **Membranous epithelium** forms the outer layer of the skin, covers some internal organs, and forms the inner lining of blood vessels and several body systems (digestive, urinary, reproductive, and respiratory). Depending on its location, it protects, secretes, and absorbs. **Glandular epithelium** is specialized to secrete a product.

Epithelium can also be classified according to the shape of the cells and the number of cell layers (Figure 4.1). Epithelial cells are identified by three basic shapes. The shapes of the cells are suited to their functions.

1. **Squamous epithelium** is made up of flattened, or scale-like, cells. These cells form linings, in the blood vessels or lungs, for instance. Their flattened shape allows oxygen and carbon dioxide to diffuse easily across the lining. In blood vessels, the smooth surface of the lining also reduces friction. The many layers of squamous epithelial cells in the outer layer of skin provide protection from the elements.

2. **Cuboidal epithelium** consists of cube-shaped cells that are specialized for secretion and absorption. These cells are found in many glands. They secrete different substances—including enzymes, buffers, and hormones—depending on their specialization. Cuboidal epithelium is also found in the lining of kidney tubules.

3. **Columnar epithelium** is made up of tall, rectangular, column-shaped cells. Like cuboidal epithelium, columnar epithelium is specialized for secretion and absorption. Columnar cells lining the small intestine are specialized for absorption. The surface area for absorption is greatly increased by numerous small fingerlike folds on the exposed surfaces.

With respect to number of layers, epithelial tissues may be either **simple epithelial tissue**—with only a single layer of cells—or **stratified epithelial tissue**—with cells stacked in several layers. All three epithelial cell shapes—squamous, cuboidal, and columnar—can be either simple or stratified. Table 4.1 summarizes the types of epithelial tissue.

Glandular epithelium forms the secretory parts of glands. These glands form during embryonic development as membranous epithelium grows inward, forming a duct. Glands that maintain their connection to the surface epithelium by way of ducts are called exocrine glands. **Exocrine glands** secrete their products via ducts onto body surfaces or into body cavities or organs. Exocrine glands include the salivary glands of the mouth and the oil and sweat glands of the skin. Sometimes glandular epithelial cells lose their connection to the surface epithelium and form endocrine glands. **Endocrine glands** lack ducts and secrete their products (hormones) into the spaces just outside the cells. Ultimately, hormones diffuse into the bloodstream and are carried throughout the body.

Simple epithelium

Stratified epithelium

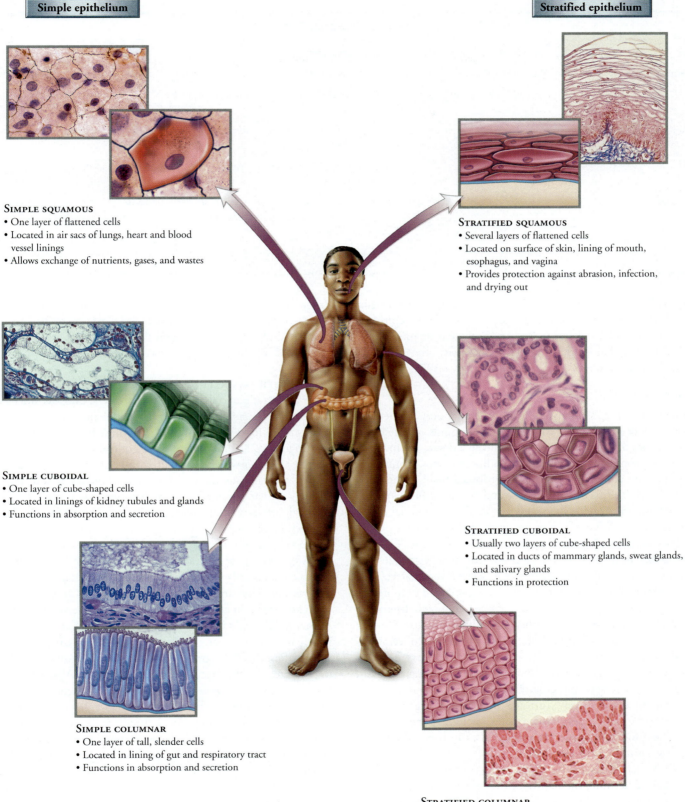

SIMPLE SQUAMOUS
- One layer of flattened cells
- Located in air sacs of lungs, heart and blood vessel linings
- Allows exchange of nutrients, gases, and wastes

SIMPLE CUBOIDAL
- One layer of cube-shaped cells
- Located in linings of kidney tubules and glands
- Functions in absorption and secretion

SIMPLE COLUMNAR
- One layer of tall, slender cells
- Located in lining of gut and respiratory tract
- Functions in absorption and secretion

STRATIFIED SQUAMOUS
- Several layers of flattened cells
- Located on surface of skin, lining of mouth, esophagus, and vagina
- Provides protection against abrasion, infection, and drying out

STRATIFIED CUBOIDAL
- Usually two layers of cube-shaped cells
- Located in ducts of mammary glands, sweat glands, and salivary glands
- Functions in protection

STRATIFIED COLUMNAR
- Several layers of tall, slender cells
- Rare, located in urethra (tube through which urine leaves the body)
- Functions in protection and secretion

FIGURE **4.1**
Types of membranous epithelial tissue. Epithelial tissue is named for the shape of the cell and for the number of cell layers.

TABLE **4.1**

TYPES OF EPITHELIAL TISSUE			
SHAPE	**NUMBER OF LAYERS**	**EXAMPLE LOCATIONS**	**FUNCTIONS**
Squamous (flat, scale-like cells)	Simple (single layer)	Lining of heart and blood vessels, air sacs of lungs	Allows passage of materials by diffusion
	Stratified (more than one layer)	Linings of mouth, esophagus, and vagina; outer layer of skin	Protects underlying areas
Cuboidal (cube-shaped cells)	Simple	Kidney tubules, secretory portion of glands and their ducts	Secretion; absorption
	Stratified	Ducts of sweat glands, mammary glands, and salivary glands	Protects underlying areas
Columnar	Simple	Most of digestive tract (stomach to anus), air tubes of lungs (bronchi), excretory ducts of some glands, uterus	Absorbs; secretes mucus, enzymes, and other substances
	Stratified	Rare; urethra, junction of esophagus and stomach	Protects underlying areas, secretes

Epithelial tissue is supported by a layer beneath the epithelial tissue called the **basement membrane,** which binds the epithelial cells to underlying connective tissue. The basement membrane is a noncellular layer that helps epithelial tissue resist stretching and forms its boundary.

CONNECTIVE TISSUE

Connective tissue has many forms and functions. Sometimes described as the body's glue, connective tissue binds together and supports tissues of the body. Certain connective tissues are involved in transport (blood) and energy storage (adipose tissue). Connective tissue is the most abundant and widely distributed tissue in the body.

All connective tissues contain cells and an extracellular matrix, which consists of protein fibers and a noncellular material called ground substance. Whereas all other types of tissue consist primarily of cells, connective tissue is made up mostly of matrix. The cells are distributed in the matrix like pieces of fruit in a gelatin dessert. Connective tissue cells and nearby cells secrete the matrix. The matrix contains three types of protein fibers in varying proportions depending on the type of connective tissue. **Collagen fibers** are strong and ropelike and can withstand pulling because of their great tensile strength. **Elastic fibers** have a randomly coiled structure and can stretch and recoil like a spring. Elastic fibers are common in areas where great elasticity is needed, including the skin, lungs, and blood vessels. **Reticular fibers** are thin strands of collagen[1] that branch extensively, forming interconnecting networks. These networks help support soft tissues, including the liver and spleen. The ground substance may be solid as in bone, fluid as in blood, or gelatinous as in cartilage.

The characteristics of the matrix of connective tissue, rather than its cells, determine the nature of the tissue. Thus, connective tissue can form soft packing around an organ (loose connective tissue), form ropes of incredible strength that bind bones together (ligaments), be fluid (blood), bear weight (bone), or provide flexible support (cartilage). The many types of connective tissue can be grouped into two categories: connective tissue proper (loose and dense connective tissue) and specialized connective tissue (cartilage, bone, and blood). Table 4.2 and Figure 4.2 summarize the characteristics of the different types of connective tissue.

CONNECTIVE TISSUE PROPER Loose and dense connective tissues differ in the ratio of cells to extracellular fibers. **Loose connective tissue** contains many cells, and the fibers of the matrix are fewer in number and more loosely woven than those in dense connective tissue (Figure 4.2). One type of loose connective tissue, **areolar connective tissue**, is widely distributed and contains many cells embedded in a gelatinous matrix. Because the matrix is soft and easily shaped, areolar connective tissue functions as a universal packing material between other tissues. It is found, for example, between muscles, permitting one muscle to move freely over the other. It also anchors the skin to underlying tissues and organs.

The second type of loose connective tissue is **adipose tissue,** which contains cells that are specialized for fat storage. Dietary carbohydrates, proteins, and fats in excess of the body's needs are converted to fat and stored in fat cells. Thus adipose tissue serves as a storehouse of energy. We could not live for more than a few days without food if it were not for stored fat. Because fat is a poor heat conductor, it serves as insulation that helps prevent the loss of body heat, especially when it is located in the layer of areolar connective tissue located just beneath the skin. Adipose tissue is also found around certain organs, including the kidneys, where it functions as a shock absorber.

Other sites of fat deposits, particularly those in the abdomen, hips, and thighs, affect our outward appearance. Fat cells become larger as they take up fat, making the region of the deposit plumper. As fat is released from the cells, they deflate like a collapsing balloon. The person may then appear thinner. However, the fat cells are not killed; they are simply smaller. So, the fat can easily—too easily—be regained in the same areas. The number of fat cells is determined during infancy and childhood, and mature fat cells cannot divide. In children, excess nutrients may increase the number of fat cells

[1]Reticular fibers have the same subunits as collagen fibers, but they are put together in a slightly different way.

formed. Some nutritionists believe that a child who overeats may be setting the stage for obesity later in life.

Dense connective tissue contains many tightly woven fibers and is found in ligaments (structures that join bone to bone), tendons (structures that join muscle to bone), and the dermis (layer of skin below the epidermis).

In both loose and dense connective tissue, the protein fibers are produced by **fibroblasts**, cells of the connective tissue that also repair tears in body tissues. For example, when the skin is cut, fibroblasts move to the area of the wound and produce collagen fibers that help close the wound and provide a surface upon which the outer layer of skin can grow and cover the damage.

SPECIALIZED CONNECTIVE TISSUE Specialized connective tissue, as shown in Figure 4.2, comes in three types: cartilage, bone, and blood. **Cartilage** is tough, but flexible. It serves as cushions between certain bones and helps maintain the structure of certain body parts, including the ears. In cartilage, the cells (chondrocytes) sit within spaces in the matrix called lacunae. The matrix is a firm gel that contains protein fibers for strength and a somewhat gelatinous ground substance responsible for the tissue's resilience. Cartilage lacks blood vessels and nerves, so nutrients reach cartilage cells by diffusion from nearby capillaries. Because this process is fairly slow, cartilage heals more slowly than bone, a tissue with a rich blood supply.

The human body has three types of cartilage:

- **Hyaline cartilage** is the most abundant of the three types. It contains numerous cartilage cells in a matrix of collagen fibers and a bluish white, gel-like ground substance. Known commonly as gristle, hyaline cartilage is found at the ends of long bones (look carefully at your next drumstick), where it allows one bone to easily slide over another. It also forms part of the nose, ribs, larynx, and trachea. Hyaline cartilage provides support and flexibility.

- **Elastic cartilage** is more flexible than hyaline cartilage because of the large number of wavy elastic fibers in its matrix. Elastic cartilage is found in the external ear, where it provides strength and elasticity.

- **Fibrocartilage** contains fewer cells than either hyaline or elastic cartilage. Like hyaline cartilage, the cells of fibrocartilage sit in a matrix that contains collagen fibers. Fibrocartilage forms the outer part of the shock-absorbing disks between the vertebrae of the spine. If the protective fibrocartilage ring of an intervertebral disk tears, the inner spongy tissue bulges outward and presses on nerves. This condition, known as a slipped or herniated disk, can cause great pain.

Bone, together with cartilage and joints, makes up the skeletal system. To many people's surprise, bone is an active, ever-changing tissue with a good blood supply that promotes prompt healing. Bone has many functions: protection and support for internal structures; movement by working with muscles; storage of lipids (in yellow marrow), calcium, and phosphorus; and production of blood cells (in red marrow). The matrix secreted by bone cells is hardened by calcium, enabling bones to provide rigid support. Collagen fibers also strengthen bone.

Bone tissue is classified as compact or spongy, depending on its underlying organization. **Compact bone** consists of structural units called osteons. The bone cells (osteocytes) in each osteon are organized in concentric rings of matrix surrounding a large central canal, which contains nerves and blood vessels (Figure 4.2). Tiny canals connect the bone cells to central canals and provide routes by which nutrients can reach the cells and wastes can leave. Some of the spaces and cavities within bone contain yellow marrow, a storage site for lipids. Others contain red marrow, a production site for blood cells. **Spongy bone** contains small, bony struts and plates and is found in areas where bone is stressed from many directions. The struts are oriented along lines of stress. You will read more about the structure and function of bones in Chapter 5.

Blood is a specialized connective tissue that consists of a liquid matrix, called **plasma**, in which **formed elements** (cells and

TABLE **4.2**

TYPES OF CONNECTIVE TISSUE		
TYPE	**EXAMPLE LOCATIONS**	**FUNCTIONS**
Connective tissue proper		
Loose, areolar	Between muscles, surrounding glands, wrapping small blood vessels and nerves	Wraps and cushions organs
Loose, adipose (fat)	Under skin, around kidneys and heart	Stores energy, insulates, cushions organs
Dense	Tendons; ligaments	Attaches bone to bone (ligaments) or bone to muscle (tendons)
Specialized connective tissue		
Cartilage (semisolid)	Nose (tip); rings in respiratory air tubules; external ear	Provides flexible support, cushions
Bone (solid)	Skeleton	Provides support and protection (by enclosing), and levers for muscles to act on
Blood (fluid)	Within blood vessels	Transports oxygen and carbon dioxide, nutrients, hormones, and wastes; helps fight infections

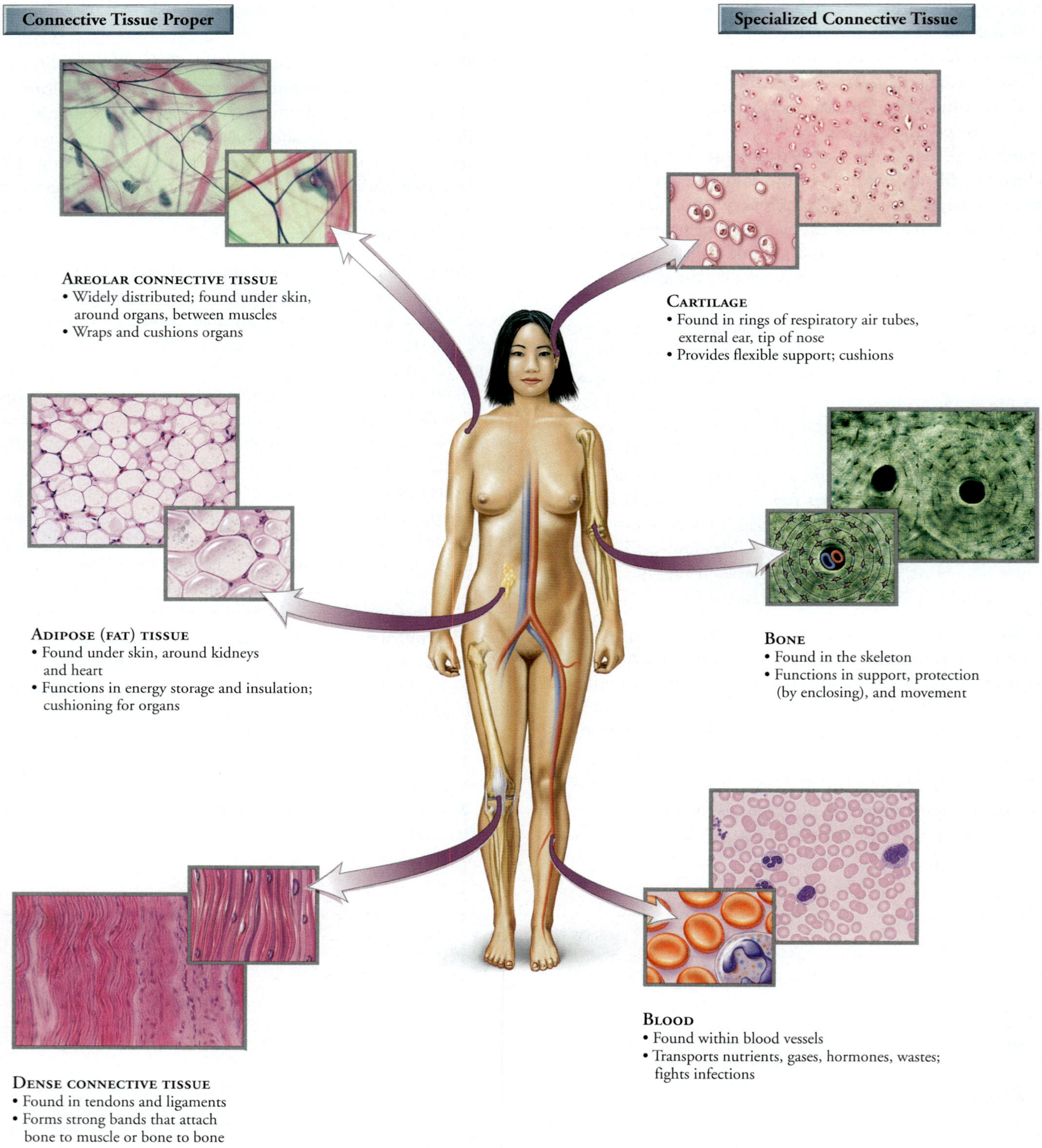

Connective Tissue Proper

Specialized Connective Tissue

AREOLAR CONNECTIVE TISSUE
• Widely distributed; found under skin, around organs, between muscles
• Wraps and cushions organs

CARTILAGE
• Found in rings of respiratory air tubes, external ear, tip of nose
• Provides flexible support; cushions

ADIPOSE (FAT) TISSUE
• Found under skin, around kidneys and heart
• Functions in energy storage and insulation; cushioning for organs

BONE
• Found in the skeleton
• Functions in support, protection (by enclosing), and movement

DENSE CONNECTIVE TISSUE
• Found in tendons and ligaments
• Forms strong bands that attach bone to muscle or bone to bone

BLOOD
• Found within blood vessels
• Transports nutrients, gases, hormones, wastes; fights infections

FIGURE **4.2**

Types of connective tissue

platelets) are suspended (Figure 4.2). The "fibers" in blood are soluble proteins. These fibers become visible only when blood clots. An important function of blood is to transport materials, many of which are dissolved in plasma. Plasma is mostly water plus dissolved substances such as oxygen, carbon dioxide, nutrients, and hormones. One of the formed elements, the red blood cells, transport oxygen to cells and some carbon dioxide away from cells. The other two formed elements are white blood cells, which help fight infection, and platelets, which help with clotting. Both white blood cells and platelets help protect the body. You will read more about blood in Chapter 11.

MUSCLE TISSUE

Muscle tissue is composed of muscle cells (fibers) that contract when stimulated. As shown in Figure 4.3, there are three types of muscle tissue: skeletal, cardiac, and smooth. Their characteristics are summarized in Table 4.3. You will read more about muscle tissue in Chapters 6 and 12.

- **Skeletal muscle** tissue is so named because it is usually attached to bones. When skeletal muscle tissue contracts, it usually moves a part of the body. Because skeletal muscle is under conscious control, it is described as voluntary muscle. Skeletal muscle cells are long cylinder-shaped cells, each of which contains several nuclei. Skeletal muscle cells have striations, which are alternating light and dark bands that are visible under a light microscope. The striations are caused by the orderly arrangement of the contractile proteins actin and myosin.

- **Cardiac muscle** tissue is found only in the walls of the heart, where its contractions are responsible for pumping blood to the rest of the body. Cardiac muscle cells resemble branching cylinders and have striations and typically only one nucleus. The cells of cardiac muscle are attached to one another by special junctions at the plasma membranes of the cells that strengthen cardiac tissue and promote rapid conduction of impulses throughout the heart.

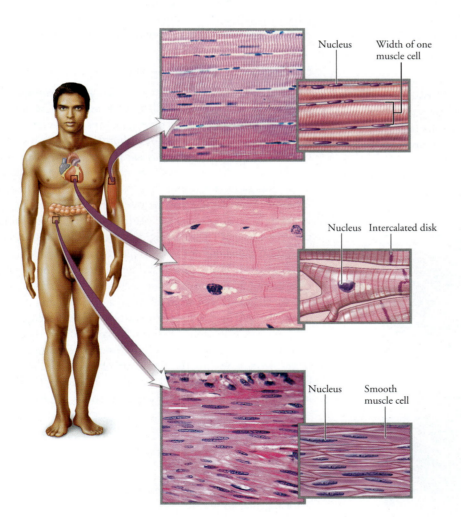

Nucleus   Width of one muscle cell

SKELETAL MUSCLE
- Long cylindrical striated cells with many nuclei
- Voluntary contraction
- Most are found attached to the skeleton
- Responsible for voluntary movement

Nucleus   Intercalated disk

CARDIAC MUSCLE
- Branching striated cells, one nucleus
- Involuntary contraction
- Found in wall of heart
- Pumps blood through the body

Nucleus   Smooth muscle cell

SMOOTH MUSCLE
- Cells tapered at each end, one nucleus
- Involuntary contraction
- Found in walls of hollow internal organs, such as the intestines, and tubes, such as blood vessels
- Contractions in digestive system move food along
- When arranged in circle, controls diameter of tube

FIGURE **4.3**
Types of muscle tissue

TABLE **4.3**

TYPES OF MUSCLE TISSUE			
TYPE	**DESCRIPTION**	**EXAMPLE LOCATIONS**	**FUNCTIONS**
Skeletal	Long, cylindrical cells; multiple nuclei per cell; obvious striations	Muscles attached to bones	Provides voluntary movement
Cardiac	Branching striated cells; one nucleus; specialized junctions between cells	Wall of heart	Contracts and propels blood through the circulatory system
Smooth	Cells taper at each end; single nucleus; arranged in sheets; no striations	Walls of digestive system, blood vessels, and tubules of urinary system	Propels substances or objects through internal passageways

- **Smooth muscle** tissue is involuntary and is found in the walls of blood vessels and airways, where its contraction reduces the flow of blood or air. Smooth muscle is also found in the walls of organs such as the stomach, intestines, and bladder, where it aids in mixing and propelling food through the digestive tract and in eliminating wastes. The cells of smooth muscle tissue taper at each end, contain a single nucleus, and lack striations. You will read more about muscle tissue in Chapters 6 and 12.

NERVOUS TISSUE

The final major type of tissue is nervous tissue. Nervous tissue makes up the brain, spinal cord, and nerves. The nervous system consists of two general cell types, neurons and accessory cells called neuroglia (Figure 4.4). **Neurons** convert stimuli (light, for example) into nerve impulses, which they conduct to other neurons, muscle cells, or glands. Although neurons come in many shapes and sizes, most have three parts: the cell body, dendrites, and an axon. The cell body houses the nucleus and most organelles. Dendrites are highly

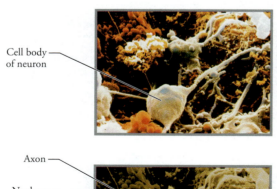

Cell body of neuron

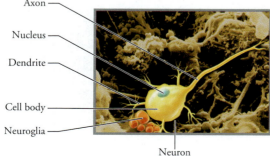

Axon

Nucleus

Dendrite

Cell body

Neuroglia

Neuron

FIGURE **4.4**
Neuroons and neuroglia

branched processes that provide a large surface area for the reception of signals from other neurons. A neuron generally has one axon, a long process that usually conducts impulses away from the cell body. **Neuroglia** (or more simply, glial cells) support, insulate, and protect neurons. They do not generate or conduct nerve impulses, but they do increase the rate at which impulses are conducted by neurons and provide neurons with nutrients from nearby blood vessels. You will read more about nervous tissue in Chapters 7 and 8.

Many tissues are held together by specialized junctions

Many tissues, especially epithelial tissue, have membrane specializations that attach adjacent cells to each other. There are three kinds of junctions between cells: tight junctions, adhesion junctions, and gap junctions. The structure of each type of junction suits the function of the tissue. In *tight junctions* (Figure 4.5a), the membranes of neighboring cells are actually attached, forming a leak-proof seal. Tight junctions are found in the lining of the urinary tract and intestine. Here, secure seals between cells prevent urine or digestive juices from flowing across the epithelium through the minute spaces between adjacent cells. Less rigid than tight junctions, *adhesion junctions* (also called desmosomes) (Figure 4.5b) resemble rivets holding adjacent tissue layers together. The plasma membranes of adjacent cells do not actually touch. Instead they are held together by intercellular filaments that are attached to a thickening in the membrane. Thus, the cells are held together but can still slide slightly relative to one another. Adhesion junctions are common in tissues that must withstand stretching, such as the skin and heart muscle. *Gap junctions* (Figure 4.5c) link the cytoplasm of adjacent cells through small holes, allowing physical and electrical continuity between cells. In heart and smooth muscle cells, gap junctions help synchronize electrical activity and contraction.

Tissues form organs that often work together in organ systems

An **organ** is a structure composed of two or more different tissues that work together to perform a specific function. Organs usually do not function as independent units but instead work together as an **organ system**—a group of organs with a common function. For example, organs such as the trachea, bronchi, and lungs constitute the respiratory system. The common function of these organs is to bring oxygen into

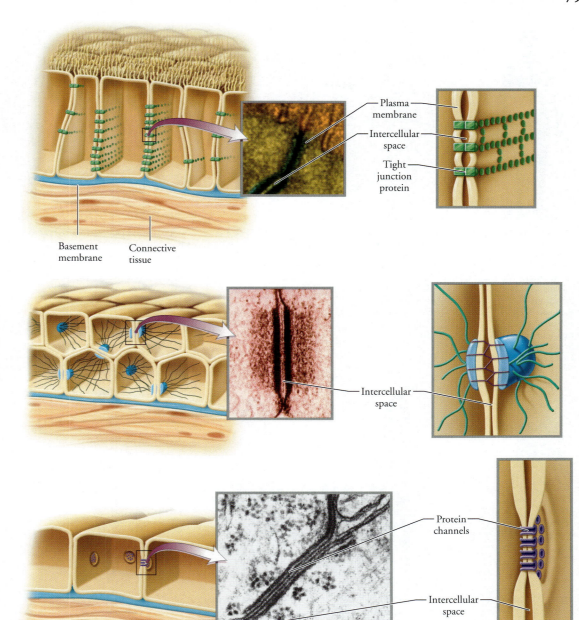

(a) Tight junction
• Creates an impermeable junction that prevents the exchange of materials between cells
• Found between epithelial cells of the digestive tract, where they prevent digestive enzymes and microorganisms from entering the blood

Basement membrane   Connective tissue

Plasma membrane

Intercellular space

Tight junction protein

(b) Adhesion junction
• Holds cells together despite stretching
• Found in tissues that are often stretched, such as the skin and the opening of the uterus

Intercellular space

(c) Gap junction
• Allows cells to communicate by allowing small molecules and ions to pass from cell to cell
• Found in epithelia in which the movement of ions coordinates functions, such as the beating of cilia; found in excitable tissue such as heart and smooth muscle

Protein channels

Intercellular space

FIGURE **4.5**
Specialized cell junctions

the body and remove carbon dioxide from the body. Figure 4.6 shows the 11 major organ systems of the human body and their functions.

Most organs are housed in body cavities that are lined with membranes

Most of the organs are suspended in internal body cavities (Figure 4.7). These cavities have two important functions. First, they help protect the vital organs from damage due to the bumps and thumps that occur as we walk and jump. Second, they allow organs to slide past one another and change shape. Sliding and changing shape are important when the lungs fill with air, the stomach fills with food, the urinary bladder fills with urine, or we bend or stretch.

There are two main body cavities—the anterior and posterior cavities—each of which is further subdivided. The anterior cavity is divided into the *thoracic* (chest) *cavity* and the *abdominal cavity*. The thoracic cavity is subdivided into the pleural cavities, which house the lungs, and the pericardial cavity, which holds the heart. The abdominal cavity contains the digestive system, the urinary system, and the reproductive system. A muscle sheet called the diaphragm separates the thoracic and

INTEGUMENTARY SYSTEM
- Protects underlying tissues from abrasion and dehydration
- Provides cutaneous sensation
- Regulates body temperature
- Immune function
- Synthesizes vitamin D
- Excretion

SKELETAL SYSTEM
- Attachment for muscles
- Encloses and protects organs
- Stores calcium and phosphorus
- Produces blood cells
- Stores fat

MUSCULAR SYSTEM
- Moves body and maintains posture
- Internal transport of fluids
- Generation of heat

NERVOUS SYSTEM
- Regulates and integrates body functions via neurons

ENDOCRINE SYSTEM
- Regulates and integrates body functions via hormones

CIRCULATORY SYSTEM
- Transports nutrients, respiratory gases, wastes, and heat
- Transports cells and antibodies for immune response
- Transports hormones
- Regulates pH through buffers

LYMPHATIC SYSTEM
- Returns tissue fluids to bloodstream
- Protects against infection and disease

RESPIRATORY SYSTEM
- Exchanges respiratory gases with the environment

DIGESTIVE SYSTEM
- Physical and chemical breakdown of food
- Absorbs, processes, stores and controls release of digestive products

URINARY SYSTEM
- Maintains constant internal environment through the excretion of nitrogenous waste

REPRODUCTIVE SYSTEM
- Produces and secretes hormones
- Produces and releases egg and sperm cells and accessory secretions
- Forms placental attachment with fetus (females only)

FIGURE **4.6**

The major organ systems of the human body

abdominal cavities. The posterior cavity is subdivided into the *cranial cavity*, which encloses the brain, and the *spinal cavity*, which houses the spinal cord.

Body cavities and surfaces of organs are covered with membranes—sheets of epithelium supported by connective tissue. Membranes form physical barriers that protect underlying tissues. The body has four types of membrane.

1. **Mucous membranes** line passageways that open to the exterior of the body, including those of the respiratory, digestive, reproductive, and urinary systems. Some mucous membranes, including the mucous membrane of the small intestine, are specialized for absorption. Others, those of the respiratory system for instance, secrete mucus. Here, mucus traps bacteria and viruses that could cause illness.

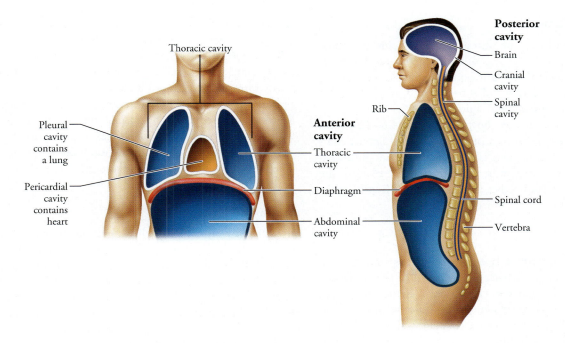

FIGURE **4.7**
Body cavities. The internal organs are suspended in body cavities that protect the organs and allow organs to slide past one another as the body moves.

2. **Serous membranes** line the thoracic and abdominal cavities and the organs within them. They secrete a fluid that lubricates the organs within the cavities.

3. **Synovial membranes** line the cavities of freely movable joints, including the knee. These membranes secrete a fluid that lubricates the joint, easing movement.

4. **Cutaneous membrane,** or skin, covers the outside of the body. Unlike other membranes, it is thick, relatively waterproof, and dry.

We will consider the structure and function of the skin in the next section. Details about the structure and functions of the other organ systems are presented in subsequent chapters.

The Skin is an Organ System

We have all been told that "beauty is only skin deep," but our skin does much more than just make us attractive. Our skin is the part of our body that directly encounters the environment. In doing so, the skin acts as a shield against many types of environmental dangers. We will see that this is only one of the many important functions of skin.

▍Skin protects, regulates body temperature, and excretes

An integument is an outer covering, so the skin is sometimes called the **integumentary system**. Our skin is indeed an outer covering, but it is much more than that. Few of us realize the many services performed by the skin and its derivatives—hair, nails, sweat glands, and oil glands.

One of the main functions of our skin is protection. The skin is essentially a physical barrier between the contents of the body and the outside world and, as a result, it protects the tissues it covers from invasion by foreign bacteria and particles and ultraviolet radiation as well as from physical and chemical insult. In addition, cells in the skin called macrophages help protect the body from infection.

The skin has many other functions. Because the outermost layer of skin cells contains the water-resistant protein keratin, the skin plays a vital role in preventing excessive water loss from underlying tissues. In addition, the skin plays a role in the regulation of body temperature. Although we perspire almost constantly, our sweat glands dramatically increase their output of perspiration during times of strenuous exercise or high environmental temperatures. Evaporation of perspiration from the surface of the skin helps rid the body of excess heat. Later in the chapter we will see how changes in the flow of the blood to the skin help maintain body temperature. Perspiration contains water, salts, and wastes such as urea. Thus, the skin functions in excretion as well.

Modified cholesterol molecules are located in the outer layer of skin and are converted to vitamin D when exposed to ultraviolet radiation. Vitamin D then travels by way of the bloodstream to the liver and kidneys, where it is chemically modified to assume its role in stimulating the absorption of calcium and phosphorus from the food we eat.

The skin also contains components of the nervous system that detect temperature, touch, pressure, and pain stimuli. By receiving such stimuli, receptors in our skin help keep us informed about the conditions around us. With these functions in mind, we will now study the structure of skin and its derivatives.

The skin has two layers

On most parts of your body, the skin is less than 5 mm (less than a quarter of an inch) thick, yet it is one of your largest organs. It represents about one-twelfth of your body weight. As we will see, its surface, an area of 1.5 to 2 m² (1.8 to 2.4 yd²), is our first line of defense against environmental assaults.

The skin has two layers, as shown in Figure 4.8. The thin, outer layer, the **epidermis**, forms a protective barrier against environmental hazards. The inner layer is the **dermis**; it contains blood vessels, nerves, sweat and oil glands, and hair follicles. Beneath the skin is a layer of loose connective tissue called the **hypodermis**, which anchors the skin.

THE EPIDERMIS

The epidermis is the outermost layer of skin and is composed of epithelial cells. The epidermis itself is composed of several layers. The outermost layer, the part you can touch, is made up of dead skin cells. Thus, when we look at someone, most of what we see on the surface is dead. These dead cells are constantly shed at a rate of about 30,000 to 40,000 each minute. In fact, much of the dust in any room consists of the dead cells shed from the skin's surface. When you go swimming or soak in the bathtub for a long time, the dead cells in the outer layer of skin absorb water and swell, causing the skin to wrinkle. The wrinkling is particularly noticeable in regions of skin with the thickest layer of dead cells, such as the palms of the hands and soles of the feet.

The skin does not get thinner as the dead cells are shed because they are replaced from below. The deepest layer of the epidermis contains rapidly dividing cells. As new cells are produced in this layer, older cells are pushed toward the skin surface. On their way to the surface, the cells flatten and die as a result of their increasing distance from the dermis, their supplier of nutrients. Along this death route, keratin gradually replaces the cytoplasmic contents of the cells. Keratin is a tough fibrous protein that gives the epidermis its protective properties. About two weeks to a month pass from the time a new cell is formed to the time it is lost from the surface of the skin.

Certain viruses, called papilloma viruses, sometimes cause the newly forming skin cells to increase their already high rate of multiplication, and small, raised masses known as warts are the result. Warts may be removed by various means, including cryosurgery with liquid nitrogen (essentially freezing the tissue to death), surgical removal, and chemical destruction with nonprescription drops. The best news, however, is that after a period of time, most warts go away without any treatment.

THE DERMIS

In most areas of the body, the dermis, as shown in Figure 4.8, is a much thicker layer than the epidermis. The dermis lies just beneath the epidermis and consists primarily of connective tissue. In addition, it contains blood vessels, hair follicles, oil glands, the ducts of sweat glands, sensory structures, and nerve endings. Another difference is that, unlike the epidermis, the dermis does not wear away. This durability explains why tattoos—designs created when tiny droplets of ink are injected into the dermal layer—are permanent (Figure 4.9). The only way to remove a tattoo is through surgical means, including the use of lasers or "shaving" (abrading) of the skin. Because the dermis is laced with nerves and sensory receptors, getting a tattoo hurts.

Blood vessels are present in the dermis but not in the epidermis. Nutrients reach the epidermis by moving out of dermal blood vessels and diffusing through tissue fluid into the layer above. Such tissue fluid is probably quite familiar to you. Trauma to the skin, such as that caused by a burn or an ill-fitting shoe rubbing against your heel, causes this fluid to accumulate in areas between the epidermis and dermis, and the accumulated fluid separates the layers and forms blisters.

stop and think

Burns—tissue damage caused by exposure to heat, radiation, electric shock, or chemicals—can be classified according to the depth to which the tissue damage penetrates. First-degree burns are those that are confined to the upper layers of the epidermis. A first-degree burn causes reddening and slight swelling of the skin. In second-degree burns, damage extends through the epidermis into the upper regions of the dermis. Blistering, pain, and swelling occur. Third-degree burns extend through the epidermis and dermis and into underlying tissues. Severe burns, particularly those covering large portions of the body, are life threatening. Given your knowledge of skin functions, what would you predict the immediate medical concerns to be when third-degree burns are suspected?

The lower layer of the dermis consists of dense connective tissue with collagen and elastic fibers. This combination of fibers allows the skin to stretch and then return to its original shape. Unfortunately, the skin's elasticity has its limits. Pregnancy or a substantial weight gain may exceed the ability of skin to stretch and return to its earlier size, and stretch marks, tears in the dermis, result. The resilience of our skin also decreases as we age. The most pronounced effects begin in the late forties. Around this time, collagen fibers begin to stiffen and decrease in number, and elastic fibers thicken and lose their elasticity.

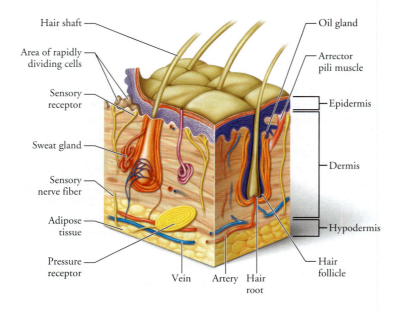

Hair shaft
Area of rapidly dividing cells
Sensory receptor
Sweat gland
Sensory nerve fiber
Adipose tissue
Pressure receptor
Vein   Artery   Hair root
Oil gland
Arrector pili muscle
Epidermis
Dermis
Hypodermis
Hair follicle

FIGURE **4.8**

Structures of the skin and underlying hypodermis

FIGURE **4.9**

Tattoos—designs created when droplets of ink are injected into the dermis—are essentially permanent because the dermis is not shed, as is the epidermis.

These changes, combined with reductions in moisture and the amount of fat in the hypodermis, produce wrinkles and sagging skin.

Certain wrinkles, such as frown lines, are caused by the contraction of facial muscles. A controversial and popular treatment for these wrinkles is the injection of Botox®, the toxin from the bacterium that causes botulism. When Botox® is injected into facial muscles, they become temporarily paralyzed. The skin smooths out over the next two to three days because the muscle contractions that form the wrinkles cannot occur. The muscles regain the ability to contract over the next several months, and the Botox® injection needs to be repeated (Figure 4.10).

THE HYPODERMIS

Just below the epidermis and dermis is a layer of loose connective tissue called the hypodermis. The hypodermis anchors the skin to underlying tissues and organs. Although usually not considered part of the skin, the hypodermis shares some of the skin's functions, including cushioning blows and helping to prevent extreme changes in body temperature. It contains about half of the body's fat stores. Interestingly (and often alarmingly), the distribution of these fat stores changes as we mature. In infants and toddlers, this layer of subcutaneous fat—often called baby fat—covers the entire body. As we mature, however, some fat stores are redistributed. In women, fat often accumulates in the breasts, hips, and thighs. Both sexes, however, must contend with the tendency of fat cells to accumulate in the abdominal hypodermis, where they help sculpt the all-too-familiar pot belly, or in the sides of the lower back, where they form "love handles."

Liposuction, a procedure for vacuuming fat from the hypodermis, is the most commonly performed cosmetic surgery in North America. The physician makes a small incision in the skin above the area of unwanted fat and inserts a fine tube into the fat. The tube is moved back and forth to loosen the fat cells, which are then sucked through the tube into a container. Liposuction is not a way of losing a lot of weight because only a small amount of fat—not more than a few pounds—can be removed. However, it is a way of

sculpting the body and removing bulges. Furthermore, because the fat cells are removed, fat does not usually return to those areas. Although liposuction is generally a safe procedure, it is not risk free. Patients have died from the procedure. Blood clots can form and travel to the lungs, causing death. People who are considering liposuction should choose their doctor carefully.

what would you do?

There are many social issues concerning liposuction. The procedure is becoming increasingly popular, even among teenagers. What does this trend say about the value we place on physical beauty? Should there be age restrictions on liposuction? Health insurers do not cover elective procedures, so physicians can set their own fees. It is alleged that some physicians who perform liposuction take unnecessary risks to increase profit. If you decided to have liposuction, what criteria would you use to choose a physician?

▍ Skin color is determined by pigment and blood flow

Two interacting factors produce skin color: (1) the quantity and distribution of pigment and (2) blood flow. The pigments are produced by **melanocytes**, cells with spiderlike extensions that are found at the base of the epidermis. The pigment produced, called **melanin**, comes in two forms: a yellow-to-red form and the more common black-to-brown form. Melanin is released from the melanocytes and is taken up by surrounding epidermal cells. In this way, melanin, produced only in melanocytes, colors the entire epidermis.

All people have about the same number of melanocytes. Thus, differences in skin color are due to differences in the form and amount of melanin produced and the way it is dispersed. A person's genetic

FIGURE **4.10**

Botox® injections before and after

makeup determines whether the yellowish red or the brown form of melanin is produced. When the yellowish red form of melanin is produced, the person generally has blond hair, blue eyes, and a light complexion. These people often have freckles, which are simply localized concentrations of pigment. When the brown form of melanin is produced, the person has skin that is yellow, brown, black, or white. The actual skin color depends on the amount of pigment produced and the size and dispersion of pigment granules. Albinism is an inherited condition in which an individual's melanocytes are incapable of producing melanin. It occurs in about 1 in 10,000 people and is not specific to a particular race. Such individuals, known as albinos, lack pigment in their eyes, hair, and skin (Figure 4.11a). The absence of pigment in the eyes and skin leads to a pinkish color because the underlying blood supply shows through. Vitiligo is a condition in which melanocytes disappear either partially or completely from certain areas of the skin, leaving patches of white skin in their wake (Figure 4.11b).

Another pigment that influences skin color is carotene. This orange yellow pigment accumulates in the dermis and in cells of the outermost layer of the epidermis. Carotene is found in foods such as carrots, apricots, and oranges. It can lend an orange color to the skin of persons who overindulge in carotene-rich fruits and vegetables. Finally, hemoglobin, the pigment that carries oxygen in blood, is responsible for the pinkish color of Caucasian skin.

Circulation can influence skin color. When well-oxygenated blood flows through vessels in the dermis, the skin has a pinkish or reddish tint that is most easily seen in light-skinned people. During times of intense embarrassment, this color may heighten as blood flow to the skin, particularly in the areas of the face and neck, increases. This response, known as blushing, is impossible to stop. Intense emotions may also cause color to disappear from the skin. A sudden fright, for example, may cause a rapid drop in blood supply to the skin, causing the person to become extremely pale. Skin color may also change in response to changing levels of oxygen in the blood. In contrast to well-oxygenated blood, which is bright red, poorly oxygenated blood is a much deeper red and gives the skin a bluish appearance. This condition occurs in response to extremely cold temperatures (poor oxygenation is why your lips appear blue after emerging from a cold dip in the ocean) or as a result of disorders of the circulatory or respiratory system. When you do not get enough sleep, the amount of oxygen in the blood may be slightly lower than usual, and the low oxygen level causes the blood to become a darker, purplish color. In some people, the darker color shows through the thin skin under the eyes as dark circles.

Some people change their skin color by sitting in the sun. Melanocytes respond to the ultraviolet (UV) radiation of sunlight by increasing the production of melanin. In this protective response, melanin absorbs some of the radiation before it reaches the lower layers of the epidermis and dermis. Tanning is the buildup of melanin in the skin in response to UV exposure. Although some ultraviolet radiation is useful for the production of vitamin D, too much can be harmful. See the Health Issue essay, *Fun in the Sun*?

❚ Hair, nails, and glands are skin derivatives

Many seemingly diverse structures are derived from the epidermis—hair, nails, oil glands, sweat glands, and teeth. Next, we consider the first four of these epidermal derivatives in terms of their structure, functions, and roles in everyday life. (Teeth are discussed in Chapter 15.)

HAIR

Hair usually covers almost all of our body, except for certain areas including the lips, palms of the hands, and soles of the feet. Although hair is not essential for human life, its presence or absence on certain parts of the body can have a strong psychological impact. Indeed, the magnitude of the emotional response of some individuals to the growth or loss of hair seems particularly surprising when we learn that hair is essentially nothing more than dead cells filled with keratin and packed into a column.

We may wonder, then, what functions does hair serve? An important function is protection. Hair on the scalp protects the head from UV radiation. Hair in the nostrils and external ear canals keeps particles and bugs from entering these structures. Our eyes are protected by eyebrows and eyelashes. Hair also has a sensory role. Receptors associated with hair follicles are sensitive to touch.

(a)

(b)

FIGURE **4.11**

Conditions involving melanocytes. (a) Albinism, a genetic condition in which an individual's melanocytes cannot produce melanin, occurs throughout the world. Here, an Angolan mother holds her albino child. (b) Vitiligo is a condition in which melanocytes disappear from areas of the skin, leaving white patches.

HEALTH ISSUE

Fun in the Sun?

It is a beautiful sunny day at the shore, and the beach is packed with people. Some jump in the surf; some make sand castles along the water's edge; and others sit propped up in a chair or stretched out on a blanket or towel on the sand. It is hard for us to equate this scene with disfigurement and death. Nonetheless, we should make the connection, and make it soon, because skin cancers are increasing at an alarming rate. At present, about one in six Americans will get skin cancer, and the incidence continues to rise annually. Here, we will first consider how the skin responds to sunlight and then describe what skin cancers are, how they form, and what can be done to treat and prevent them.

The ultraviolet (UV) radiation of sunlight causes the melanocytes of the skin to increase their production of the pigment melanin. Melanin is taken up by surrounding epidermal cells, where it functions to absorb the UV before the radiation can travel any deeper into tissues and damage the genetic information of the cell. A tan is the buildup of melanin in the skin in response to sunlight. Unfortunately, this somewhat protective buildup is not instantaneous. Indeed, sunburn—damage to skin accompanied by reddening and peeling—often follows the first day at the beach. Too much sun destroys the cells at the surface of the skin. The redness of sunburned skin results from the dilation of blood vessels in the dermis, a response prompted by the sun-damaged epidermal cells. Sunburned skin peels because the large-

scale destruction of skin cells caused by the overexposure to the sun's rays provokes an increased rate of production of new cells. These new cells push the burned cells off the skin surface, sometimes in small strands, and in severe cases, in large sheets.

Painful though sunburn may be, its discomfort cannot compare to that of some skin cancers. In skin cancer, the genetic material of skin cells is altered by UV radiation. As a result, the cells grow and divide uncontrollably, forming a tumor. Three types of skin cancer are caused by overexposure to the sun: (1) basal cell carcinoma, (2) squamous cell carcinoma, and (3) melanoma (Figure 4.A). Basal cell carcinoma arises in the layer of rapidly dividing cells in the epidermis. It is the most common type of skin cancer. Squamous cell carcinoma arises in the newly formed skin cells as they flatten and move toward the skin surface and is the second most common form of skin cancer. Melanoma is the least common and most dangerous type of skin cancer. It arises in melanocytes, the pigment-producing cells of the skin. Unlike basal or squamous cell carcinomas, melanomas, when left untreated, often metastasize (spread rapidly) throughout the body. The cancer cells first infiltrate the lymph nodes and later the vital organs. The survival rate in persons whose melanoma is found before it has metastasized is about 90%. This rate drops to about 14%, however, if the cancerous cells have spread throughout the body. Melanomas can be caught at an early stage by carefully examining your skin while keeping the ABCD mnemonic of the American Cancer Society in mind:

A stands for asymmetry. Most melanomas are irregular in shape.

B stands for border. Melanomas often have a diffuse, unclear border.

C stands for color. Melanomas usually have a mottled appearance and contain colors such as brown, black, red, white, and blue.

D stands for diameter. Growths with a diameter of more than 5 mm (about 0.2 in.) are threatening.

All three types of skin cancer are more common in fair-skinned than in dark-skinned people. Skin cancer particularly affects persons who are middle-aged or elderly, live in the tropics or at high altitude, and spend a large amount of time outdoors. People whose immune system is suppressed also appear more susceptible to skin cancer. In addition, episodes of severe sunburn during childhood or adolescence seem to predispose people to developing melanomas years, even decades, later. Whatever your skin color, lifestyle, or history of exposure to the sun, an unusual growth on your skin warrants a visit to a dermatologist.

Most skin cancers are treated either by cutting out the tumor and nearby tissue or by radiation therapy. In some cases, cancerous cells may be destroyed through laser surgery, cryosurgery (freezing), or electrodesiccation (exposing the tumor to an electric current that causes the cells to dry out and die). Many skin cancers occur on the face, and the disfigurement caused by the removal of large chunks of the nose or ears may necessitate reconstructive surgery. The unattractive results of too much exposure to the sun do not, however, stop here. Excessive exposure to

continued →

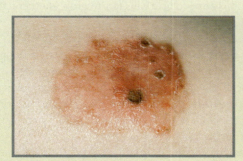
(a) Basal cell carcinoma

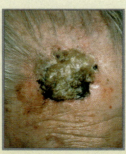
(b) Squamous cell carcinoma

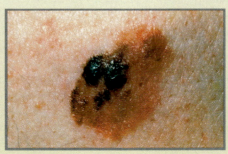
(c) Melanoma

FIGURE 4.**A**
Three skin cancers

UV radiation also causes the skin to wrinkle, sag, and take on a leathery appearance.

What can be done to prevent skin cancer? The best way to avoid getting skin cancer is to avoid prolonged exposure to the sun. However, if you must be out in the sun, apply a sunscreen and avoid the midday hours between 10:00 A.M. and 3:00 P.M, when the ultraviolet rays are the strongest. Sunscreens that have a sun protection factor (SPF) of at least 15 are best. Apply sunscreen about 45 minutes before going out into the sun, and reapply after swimming or perspiring. Sunscreens, it is important to realize, are not foolproof. Most sunscreens block the higher-energy portion of the sun's UV radiation, known as UV-B, while providing only limited protection against the lower-energy portion, called UV-A.

Whereas UV-B causes skin to burn, recent research suggests that exposure to UV-A weakens the body's immune system, possibly impairing its ability to fight melanoma. Ironically, by providing protection from sunburn, sunscreens have had the potentially devastating effect of enabling people to spend more time in the sun, possibly increasing their risk of developing melanoma.

Sunblocks such as zinc oxide, the white ointment used by lifeguards and others who spend long hours in the sun, differ from sunscreens in that they totally deflect ultraviolet rays. For this reason, sunblocks are usually applied to particularly sensitive or already burned areas of the face such as the nose and lips.

In addition to using chemical protection, it is helpful to wear a wide-brim hat while out in the

sun. The hat will reduce the rays that hit you by 70%. Tightly woven clothing also offers protection from UV radiation. In fact, without such protective clothing, it is hard to avoid the sun's rays. You should know that ultraviolet rays can pass through clouds, can penetrate water up to about 3 feet, and reflect off surfaces such as the sand and patio decks to reach you under your umbrella. A final preventive measure is to avoid tanning salons. For many years, tanning salons have claimed to use "safe" wavelengths of UV radiation because they did not use skin-reddening UV-B. But these "safe" wavelengths are actually UV-A. Given the apparent link between UV-A and increased risk of melanoma, the potential danger of these "safe" wavelengths is now obvious. ♂♀

A hair consists of a shaft and a root (see Figure 4.8). The shaft projects above the surface of the skin, and the root extends below the surface into the dermis or hypodermis, where it is embedded in a structure called the hair follicle. Nerve endings surround the follicle and are sensitive to touch. This sensitivity explains why we are aware of even slight movements of the hair shaft. (Try to move just one hair without feeling it.) Each hair is also supplied with an oil gland that opens onto the follicle and supplies the hair with an oily secretion that makes it soft and pliant. In the dermis, a tiny smooth muscle called the arrector pili is attached to the hair follicle. This muscle can pull on the follicle, causing the hair to stand up. Contraction of the muscle is usually associated with fear as well as with cold. The tiny mound of flesh that forms at the base of the erect hair is sometimes called a goose bump.

A strand of hair is made up of dead, keratinized cells. The outermost layer of a hair strand, the cuticle, consists of a single layer of thin, flat cells that are scale-like in appearance and packed with keratin. Exposed to the elements and abrasion, the cuticle sometimes wears off at the tip of hair shafts, causing the underlying cortex to frizz, forming dreaded split ends. The cuticle may also be damaged or worn away by exposure to chlorine in swimming pools. To avoid this damage, many swimmers wear a bathing cap or wash their hair soon after emerging from the water.

Within the cuticle is a thick layer of pigment-containing cells that gives hair its color. Hair color is genetically determined and is based largely on the type and amount of melanin it contains. A large amount of brown melanin makes hair dark brown or black; lesser amounts yield light brown or blond hair. As we age, the production of melanin by hair follicles slows down. Eventually, no melanin is produced, and new hairs are gray or white.

For at least 3000 years, people have been changing their hair color, often to restore color to graying hair. Hair dyes may be temporary or permanent. Temporary dyes contain large molecules that

coat only the surface of the hair and wash off with shampooing. Permanent dyes are made up of smaller molecules that move through the cuticle to the middle layer of the hair, where they fill in air spaces. Once inside the hair, they cannot be washed away. In contrast to the few minutes to an hour that dyeing usually takes, natural changes in hair color occur quite gradually. Contrary to what some horror stories would have us believe, it is impossible for hair to turn white overnight, unless, of course, some bleaching agent is used.

Hair grows as cells deep within the follicle divide and push older cells upward toward the skin surface; as the older cells emerge, the hair on your scalp gets longer, growing about 12.5 to 15 cm (5 to 6 in.) a year. Along their upward route, these cells die, flatten, fill with keratin, and bind together to form the hair. Thus the part of a hair extending above the skin surface is simply a column of dead cells. A follicle on the scalp typically produces new cells for a period of 2 to 6 years and then stops, entering a resting phase that will last a few months. At the end of this time, the root of the hair detaches from the base of the follicle and the hair falls out, often aided by brushing or combing. A few months later, this same follicle will start to grow another hair.

NAILS

Nails protect the sensitive tips of fingers and toes. Although the nail itself is dead and lacks sensory receptors, it is embedded in sensitive tissue. Even the slightest pressure on a nail that touches an object usually is detected. In this way, nails may serve as sensory "antennas." Nails can help you manipulate objects, as when you undo the knot in your shoelace.

Similar to hair, nails are modified skin tissue hardened by the protein keratin. Nails differ from hair, however, in that they grow continuously. Compared with the rate at which hair grows, the pace of nail growth is quite slow. Whereas a hair may grow 12.5 to 15 cm (5 to 6 in.) a year, fingernails and toenails typically grow only about 3.75 and 1.25 cm (1.5 and .05 in.) per year, respectively.

GLANDS

Three types of glands—oil, sweat, and wax—are found in the skin. Although all three types develop from epidermal cells, they differ in their locations, structures, and functions.

Oil (sebaceous) **glands** are found virtually all over the body except on the palms of the hands and soles of the feet. They secrete sebum, an oily substance made of fats, cholesterol, proteins, and salts. The secretory part of these glands is located in the dermis, as shown in Figure 4.8. In some instances, oil glands open directly onto the surface of the skin, but in most cases they open onto hair follicles. Sebum lubricates hair and skin, waterproofs skin, and contains substances that inhibit growth of certain bacteria. Sometimes, however, the duct of an oil gland becomes blocked, allowing sebum to accumulate. Next, bacteria invade the gland and hair follicle. The resulting condition is called acne. See the Health Issue essay, *Acne: The Misery, the Myths, and the Medications.*

As their name implies, **sweat glands** produce sweat, which is largely water plus some salts, lactic acid, vitamin C, and metabolic wastes such as urea. Although some wastes are eliminated through sweating, the principal function of sweat is to help in the regulation of body temperature through the evaporation of sweat from the skin surface.

There are two types of sweat glands. One type releases its sweat directly onto the skin surface. These glands function throughout life and are most common in the skin of the forehead, palms of the hands, and soles of the feet. The other type, found mostly in the armpits, pubic area, and the pigmented region around the nipples, begins functioning at puberty and discharges its secretion onto hair follicles. This secretion contains the same basic components of true sweat plus some fatty substances and proteins. Although the secretions are initially odorless, they may take on a musky odor as a result of the bacteria that thrive on them on the skin surface. Antiperspirants and hygiene sprays are designed either to inhibit such secretions or to mask their odor.

Wax glands are modified sweat glands found in the tissue of the external ear canal. The secretion of wax glands, called ear wax, is a mixture of substances secreted by wax and oil glands. In combination with tiny hairs in the external ear canal, ear wax functions to prevent foreign material from reaching the eardrum. Some people overproduce ear wax, and it becomes impacted. Because impacted ear wax may block sound waves from reaching the eardrum, it should be removed either by a medical professional using a blunt instrument or by periodic irrigation with a syringe.

Mammary glands are modified sweat glands. Although technically part of the integumentary system, we consider mammary glands in our discussion of the female reproductive system in Chapter 17.

Skin functions with other organ systems

Many of the skin's functions link it to other organ systems. For example, vitamin D synthesized in the skin is essential for the digestive system's absorption of calcium and phosphorus from the food we eat. Calcium and phosphorus, in turn, are important to bone growth and maintenance. Calcium also plays a role in muscle contraction and nerve functioning. Here, then, is an example of links among four organ systems—integumentary, digestive, skeletal, and muscular.

The skin also works closely with the nervous and endocrine systems. Recall, for example, that there are receptors in the skin sensitive to pressure, touch, temperature, and pain. The nervous system also influences many aspects of the skin's role in the regulation of body temperature, including the activity of sweat glands and regulation of blood flow. Finally, hormones influence the activity of oil glands and certain sweat glands as well as the distribution and growth of hair. Thus, the integumentary, endocrine, and reproductive systems are linked. The close working relationships among organ systems will become especially apparent in the next section on homeostasis.

Homeostasis Maintains Relative Internal Constancy

Although conditions outside the body sometimes vary dramatically, our bodies maintain a relatively constant internal environment. The pH of blood, for example, is about 7.4, the concentration of glucose in the blood is about 0.9 mg/ml, and body temperature usually hovers around 36.2°C (98.2°F). This internal constancy is called **homeostasis.** It occurs at all levels of body organization, from cells to organ systems. The word *homeostasis* means "to stay the same." As external conditions vary, the body's internal processes must also shift to counteract the change and keep internal conditions relatively constant.

Homeostasis is maintained by negative feedback mechanisms

Homeostasis is maintained primarily through **negative feedback mechanisms**—mechanisms in which the outcome of an event affects (feeds back on) the system, slowing or reversing the initial event. Homeostatic mechanisms do not maintain absolute constancy. Instead, they dampen fluctuations around a set point. Thus, homeostasis is a dynamic rather than a static state. Homeostatic mechanisms have three components. First, a receptor detects change in the internal or external environment. Second, a control center such as the brain integrates information coming in from receptors and selects a response. Third, an effector such as a muscle or gland carries out the response.

Consider how a negative feedback mechanism controls the temperature in your home during the frigid winter months. The thermostat, which serves as a temperature-sensing receptor and the control center, is set to the desired temperature. If the temperature inside your home falls below the desired set point, the thermostat turns on the heating system. When the internal temperature climbs to the set point, the thermostat turns off the heating system. Thus, the temperature fluctuates around the set point. Now, let's apply these principles to see how homeostatic mechanisms regulate body temperature.

The hypothalamus regulates body temperature

The body's thermostat and temperature control center is located in a region of the brain called the hypothalamus. Its set point usually averages 36.2°C (98.2°F), although it differs slightly from one person to the next. Body temperature must not vary too far from this mark, because even slight changes in the core temperature have dramatic effects on the body's metabolism. Body temperature is sensed at the surface by skin receptors and deep inside the body by receptors that sense the temperature of the blood. The brain receives input from both types of receptor. If body temperature is below the set point, the brain initiates mechanisms that increase heat production and conserve heat. When body temperature is above the set point, the brain initiates mechanisms that promote heat loss (Figure 4.12).

How do our bodies maintain a relatively constant internal temperature? Let's begin by considering what happens when we find ourselves in an environment where the temperature is above our set point, say 38°C (100.4°F). Thermoreceptors in the skin detect heat and activate nerve cells that send a message to the brain. The brain then sends nerve impulses to the sweat glands to increase their secretions. As perspiration evaporates, the surface of the skin

cools, lowering body temperature. When body temperature drops below 37°C (98.6°F), signals are no longer sent from the brain to the sweat glands. In this homeostatic system, thermoreceptors in the skin are the receptors, the brain is the control center, and the sweat glands are the effectors. This system is a negative feedback mechanism because the effect (cooling of the skin) feeds back on the system and inhibits further lowering of body temperature.

Strenuous exercise can increase body temperature. In this situation, heat sensors in the brain respond to increases in the temperature of the blood and initiate responses. Sweat gland activity is increased, and blood vessels in the dermis are dilated (widened). The latter response releases more heat to the surrounding air and gives the skin the flushed appearance we get during strenuous

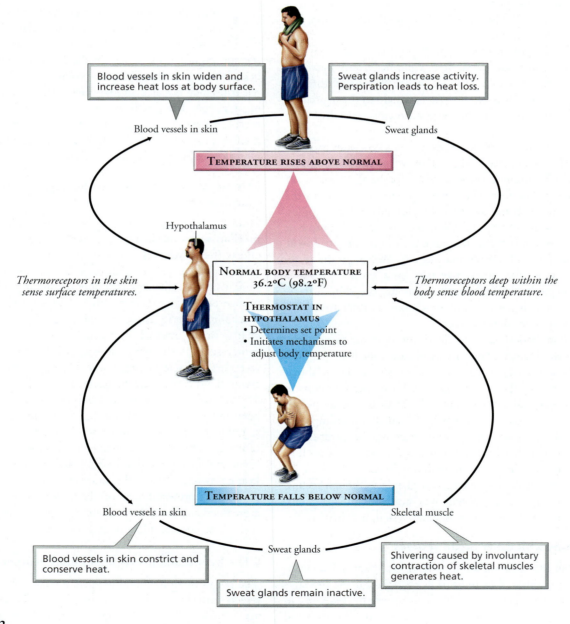

Blood vessels in skin widen and increase heat loss at body surface.

Sweat glands increase activity. Perspiration leads to heat loss.

Blood vessels in skin

Sweat glands

TEMPERATURE RISES ABOVE NORMAL

Hypothalamus

Thermoreceptors in the skin sense surface temperatures.

NORMAL BODY TEMPERATURE 36.2°C (98.2°F)

Thermoreceptors deep within the body sense blood temperature.

THERMOSTAT IN HYPOTHALAMUS
• Determines set point
• Initiates mechanisms to adjust body temperature

TEMPERATURE FALLS BELOW NORMAL

Blood vessels in skin

Skeletal muscle

Sweat glands

Blood vessels in skin constrict and conserve heat.

Sweat glands remain inactive.

Shivering caused by involuntary contraction of skeletal muscles generates heat.

WWW FIGURE **4.12**
TUTORIAL 4.1 A homeostatic mechanism for regulating body temperature

Acne: The Misery, the Myths, and the Medications

Acne and adolescence go hand in hand. In fact, about four out of five teenagers have acne, a skin condition that will probably dog them and distress them well into their twenties and possibly beyond. What is acne? What causes this most common of skin disorders? Is the cause really the chocolate, potato chips, and sodas that often make up the teenage diet? Why do males usually experience more severe cases of acne than do females? Can acne be treated?

Simple acne is a condition that affects hair follicles associated with oil glands. When adolescence begins, oil glands increase in size and step up their production of the oil, sebum. These changes are prompted by increasing levels of "male" hormones called androgens in the blood of both males and females. These hormones are secreted by the testes, ovaries, and adrenal glands. The changes they induce in the activity and structure of oil glands set the stage for development of acne. It should come as no surprise, then, that acne occurs on areas of the body where oil glands are largest and most numerous. The face, chest, upper back, and shoulders are hot spots for acne. Because most females have lower levels of androgens than do males, their acne is typically less severe. Flareups in acne around the time of menstruation are common, however, and may be linked to prog-

esterone, a hormone secreted after ovulation. Unfortunately, the precise details of the link between progesterone and acne are unknown.

Acne is basically the inflammation that results when sebum and dead cells clog the duct where the oil gland opens onto the hair follicle (Figure 4.B). A follicle obstructed by sebum and cells is called a whitehead. Sometimes the sebum in plugged follicles oxidizes and mixes with the skin pigment melanin, forming a blackhead. Melanin, not dirt or bacteria, lends the dark color to these blemishes. The next stage of acne is pimple formation. A pimple begins with the formation of a red raised bump, often with a white dot of pus at the center. The bump occurs when obstructed follicles rupture and spew their contents into the surrounding epidermis. Rupture of an obstructed follicle may occur naturally by the general buildup of sebum and cells or may be induced by squeezing the area. The sebum, dead cells, and bacteria that thrive on such things then cause a small infection, called a pimple or pustule, that will usually heal within a week or two without leaving a scar. In severe cases of acne, the rupture of plugged follicles produces large cysts called boils. Boils extend into the dermis and leave scars when they heal.

Eating nuts, chocolate, pizza, potato chips, or any of the other "staples" of the teenage

diet does not cause acne. Some of the afflicted, however, find that they have to avoid certain foods. Also, acne is not caused by poor hygiene. Follicles plug from below, so dirt or oil on the skin surface is not responsible for causing acne. (Most doctors do, however, recommend washing the face two or three times a day with hot water to help open plugged follicles.) Two factors that do appear to contribute to acne are heredity and stress. Individuals are more likely to have acne if their parents had acne. Acne often flares up during times of stress, presumably as a result of stress-induced changes in the levels of hormones.

Treatments for acne fall into two main categories: topical (those applied to the skin) and oral (those taken by mouth). Medicines applied to the skin usually require several weeks of treatment before improvements are noticeable. A 5% solution of benzoyl peroxide is a useful topical preparation that is marketed under several names, some of which are sold over the counter (such as Oxy-5® or OxyClear®). Benzoyl peroxide, when applied once or twice a day as a cream after washing the area affected by acne, is a powerful antibacterial agent that kills bacteria living in the follicles. Another option is Tretinoin (retinoic acid, Retin-A®), available only by prescription. This topical medication is a derivative of vitamin A that helps

continued ⟶

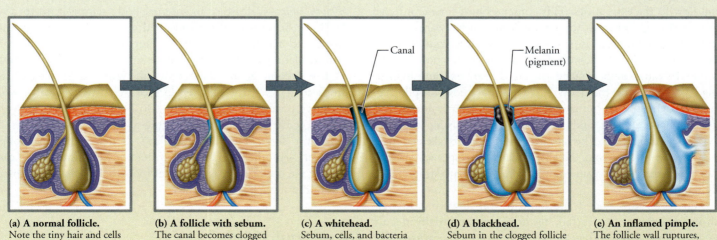

(a) A normal follicle.
Note the tiny hair and cells of the oil gland.

(b) A follicle with sebum.
The canal becomes clogged with sebum, dead cells, and bacteria.

(c) A whitehead.
Sebum, cells, and bacteria accumulate in the follicle.

(d) A blackhead.
Sebum in the clogged follicle oxidizes and mixes with melanin.

(e) An inflamed pimple.
The follicle wall ruptures, releasing the contents of a whitehead or blackhead into the surrounding epidermis.

FIGURE 4.**B**
The stages of acne

thin the epidermis, reduce the stickiness of dead cells, and increase the rate at which such cells are sloughed off. These actions help follicles remain clear and push out existing whiteheads and blackheads.

Severe cases of acne often call for oral, or systemic, medicines, such as tetracycline or minocycline, that work by inhibiting bacteria inhabiting the follicle. After an initial large dose, a low dose is given over a long period of time. Few health problems are caused by long-term use of tetracycline or minocycline, although some women experience a yeast infection of the vagina. Isotretinoin (Accutane®), another derivative of vitamin A available only with a prescription, is probably the most effective treatment for severe acne. This drug works by poisoning the oil glands,

causing them to shrink and reduce their output of sebum. Some of this shrinkage is permanent. As a result, Accutane® differs from all other treatments in that it suppresses acne even after the treatment has stopped. Most patients take Accutane® for 4 months. About 30% may need a second 4-month course of the drug. Although highly effective, Accutane® should be used with extreme caution, particularly by women of child-bearing age. Increased birth defects have been found in children born to women using Accutane®. More general side effects of Accutane® result from its overall drying of the body, leading to such things as chapped lips, flaking skin, nosebleeds, and hair loss. Finally, estrogen in the form of birth control pills is sometimes prescribed for women with severe acne. Like

Accutane®, estrogen seems to work by shrinking oil glands (although not permanently as Accutane® does) and reducing sebum production.

There are also ways to reduce the scarring that sometimes results from severe acne. Once the acne has subsided, doctors may recommend dermabrasion (skin planing), in which the scarred area is frozen and the top layer is sanded off with a rapidly rotating brush. Dermabrasion usually is restricted to the face, where it helps to remove shallow pitted scars. Another way to remove scars is by grafting a small amount of unscarred skin from behind the ear onto the scarred area of the face. A final alternative is to use chemical peeling agents to make a depressed scar more level with the surrounding skin surface. ♀♂

exercise. In addition, the arrector pili muscles relax so that damp cooling hair lies close to the skin. There may also be behavioral responses such as seeking shade or removing your sweatshirt.

What happens when the body temperature drops below the set point? Subtle drops in body temperature are detected largely by thermoreceptors in the skin, which send a message to the brain. The brain then sends nerve impulses to sweat glands, ordering a decrease in their activity. Impulses are also sent to vessels in the dermis, telling them to constrict. Constriction of these vessels reduces blood flow to the extremities. This conserves heat for the internal organs, giving credence to the saying "cold hands, warm heart." Another response to decreasing body temperature is contraction of arrector pili muscles. As described previously, contraction of these tiny muscles causes hairs to stand on end, thereby trapping an insulating layer of air near the body. This response, known as piloerection, is less effective in humans than in more heavily furred animals. The body also responds to cooling by increasing metabolic activity to generate heat. The repeated contraction of skeletal muscles is known as shivering. Finally, behavioral responses such as folding your arms across your chest may help combat a drop in core body temperature.

Sometimes the body's mechanisms for lowering higher-than-normal temperatures fail, resulting in potentially deadly *hyperthermia*—abnormally elevated body temperature. For example, some marathon runners and people immersed in a hot tub with water of too high temperature (say up around 114°F) have died as a result of elevated core temperatures. In both situations, efficient sweating and evaporation of perspiration from

the body surface were prevented by high humidity (in the case of the runners) or the surrounding water (in the case of the hot tubbers). The condition, commonly called heat stroke, is marked by confusion and dizziness. If the core temperature reaches about 42°C (107°F), the heartbeat becomes irregular, oxygen levels in the blood drop, the liver ceases to function, and unconsciousness and death soon follow. Few people can survive core temperatures of 43°C (110°F).

What happens if the body's temperature drops too far? *Hypothermia*—a decrease in internal body temperature to 35°C (95°F) or below—results. Continued drops in body temperature disrupt normal functioning of the nervous system and temperature-regulating mechanisms. People suffering from hypothermia usually become giddy and confused. When their temperature drops to 33°C (91.4°F), they lose consciousness. Finally, when body temperature drops to 30°C (86°F), blood vessels are completely constricted and temperature-regulating mechanisms are fully shut down. Death soon follows. If detected soon enough, however, hypothermia can be treated. The most severe cases require the use of dialysis machines to artificially warm the blood and pump it back into the body.

stop and think

Frostbite is damage to tissues exposed to cold temperatures. Given what you know about the body's response to cold temperature, why are fingers and toes particularly susceptible to frostbite?

REVIEWING THE CONCEPTS

The Organization of the Human Body Increases in Complexity from Cells to Organ Systems (pp. 68–77)

1. Tissues are groups of cells that work together to perform a common function. There are four main types of tissue in the human body: epithelial (covers body surfaces, forms glands, and lines internal cavities and organs), connective (acts as storage site for fat, plays role in immunity, and provides protection and support), muscle (provides movement), and nervous tissue (coordinates body activities through initiation and transmission of nerve impulses).

2. Epithelial tissue comes in two types, membranous epithelium (covers organs and forms the outer layer of the skin and the inner lining of blood vessels and several body systems) and glandular epithelium (forms secretory portions of exocrine and endocrine glands).

3. All connective tissues contain cells and an extracellular matrix composed of protein fibers and ground substance. The various types of connective tissue are connective tissue proper (loose and dense connective tissue) and specialized connective tissue (cartilage, bone, and blood). Many connective tissues function in binding and supporting structures within the body; others function in transport (blood) and fat storage (adipose tissue).

4. Loose connective tissues, such as areolar and adipose tissue, have more cells and fewer, more loosely woven protein fibers in their matrix than do dense connective tissues. Dense connective tissues are found in ligaments (structures that join bone to bone), tendons (structures that join muscle to bone), and the dermis (the layer of skin below the epidermis).

5. Cartilage consists of cartilage cells (chondrocytes) embedded in a matrix that contains protein fibers for strength and a gelatinous ground substance for resilience. We have three types of cartilage in our bodies: hyaline cartilage (found, for example, at the ends of long bones), elastic cartilage (found, for example, in the external ear), and fibrocartilage (found, for example, in intervertebral disks).

6. Bone can be classified as compact or spongy. Compact bone has an underlying organization in which bone cells (osteocytes) are organized in concentric rings of matrix composed of collagen fibers and mineral salts. A large central canal contains blood vessels. Nutrients reach the osteocytes by moving from these canals through smaller canals. Spongy bone is less dense than compact bone and is made of an irregular network of collagen fibers covered by a calcium matrix.

7. Blood consists of formed elements (red blood cells, white blood cells, and platelets) suspended in a liquid matrix (plasma). The protein fibers are normally dissolved in the plasma and play a role in blood clotting. Red blood cells transport oxygen and carbon dioxide; white blood cells aid in fighting infections; and platelets function in blood clotting.

8. Muscle tissue is composed of muscle fibers that contract when stimulated, generating force. There are three types of muscle tissue: skeletal, cardiac, and smooth. Skeletal muscle tissue is usually found attached to bone, is voluntary, and has cross-striations that are visible under a microscope and several nuclei in each cell. Cardiac muscle tissue is found in the walls of the heart, is involuntary, and has cross-striations and usually only one nucleus in each cell. Smooth muscle tissue is found in the walls of blood vessels, airways, and organs. It is involuntary and lacks striations. A smooth muscle cell tapers at each end and has a single nucleus.

9. Nervous tissue consists of neurons. The nervous system also contains accessory cells called neuroglia. Whereas neurons convert stimuli into nerve impulses that they conduct to glands, muscles, or other neurons, neuroglia increase the rate at which impulses are conducted by neurons and provide neurons with nutrients from nearby blood vessels.

10. Three types of specialized junctions hold tissues together. Tight junctions prevent the exchange of materials between cells. Adhesion junctions link cells by intercellular filaments attached to thickenings in the plasma membrane. Gap junctions have small pores that allow physical and chemical communication between cells.

11. An organ is a structure that is composed of two or more different tissues and has a specialized function.

12. Organs that participate in a common function are collectively called an organ system. The 11 major organ systems of the human body are: integumentary (skin), skeletal, muscular, nervous, endocrine, circulatory, lymphatic, respiratory, digestive, excretory, and reproductive.

13. Internal organs are located in body cavities. There are two main body cavities. The posterior cavity is subdivided into the cranial cavity, in which the brain is located, and the spinal cavity, in which the spinal cord is located. The anterior cavity is subdivided into the thoracic (chest) cavity and the abdominal cavity. The thoracic cavity is further divided into the pleural cavities, which contain the lungs, and the pericardial cavity, which contains the heart. Membranes line body cavities and spaces within organs.

The Skin Is an Organ System (pp. 77–83)

14. The integumentary system includes the skin and its derivatives such as hair, nails, and sweat and oil glands. The functions of these structures include protection of underlying tissues from abrasion and dehydration; excretion in the form of sweat; regulation of body temperature; synthesis of vitamin D; detection of stimuli associated with touch, temperature, and pain; and body defense mechanisms.

15. The skin has two layers. The outermost layer of the skin is composed of epithelial cells and is called the epidermis. Just below the epidermis is the dermis, a much thicker layer composed of connective tissue and containing nerves and blood vessels. Below the dermis is the hypodermis, a layer of loose connective tissue that anchors the skin to underlying tissues.

16. The epidermis is a renewing barrier. Cells produced in the deepest layer are pushed toward the skin surface. During this journey, the cells flatten and die as a result of their increasing distance from the blood supply of the dermis and the replacement of their cytoplasmic contents with keratin. These cells are the most numerous cells of the epidermis and are eventually shed at the skin surface.

17. The dermis does not wear away. It contains blood vessels, nerves, and glands.

18. Skin color is determined, in part, by the quantity and distribution of pigments, primarily melanin. Melanin, released by melanocytes at the base of the epidermis, is taken up by neighboring cells on their way to the surface and thus colors the entire epidermis.

19. Circulation influences skin color through changes in either the amount of blood flow to certain regions of the body or the oxygen content of the blood.

20. Hair is a derivative of skin. The primary function of hair is protection. A hair consists of a shaft and a root. The shaft projects above the skin surface, and the root extends below the surface into the dermis or hypodermis, where it is embedded in a follicle. New hair cells, produced deep within the follicle, push older cells upward toward the skin surface. On their way to the surface, these cells die. The part of a hair extending above the skin surface is thus a column of dead cells.

21. Nails are modified skin tissue hardened by keratin. Nails provide protection for the tips of our fingers and toes and help us grasp and manipulate small objects.

22. Oil, sweat, and wax glands are derivatives of the skin. Oil glands open either directly onto the surface of the skin or more typically onto a hair follicle. Sebum, the oily substance secreted by these glands, lubricates the skin and hair, prevents desiccation, and inhibits the growth

of certain bacteria. There are two types of sweat glands. One type functions throughout life and aids in regulation of body temperature through its secretion of sweat. The other type begins functioning at puberty and discharges a thicker secretion composed of sweat, fatty substances, and proteins. Wax glands are modified sweat glands that secrete wax in the external ear canal.

23. The integumentary system does not function alone but instead works closely with other organ systems.

Homeostasis Maintains Relative Internal Constancy (pp. 83–86)

24. Homeostasis is the relative internal constancy that occurs at all levels of body organization. It is a dynamic state (small fluctuations occur around a set point) maintained primarily through negative feedback mechanisms.

25. Thermoreceptors in the skin detect heat and activate nerve cells that send a message to the brain. In response, the brain sends impulses to the sweat glands to produce more perspiration. As perspiration evaporates, the body surface cools and body temperature decreases. This is an example of a negative feedback mechanism because the "product" of the system (cooling of the skin) feeds back into the system and inhibits further lowering of body temperature. Homeostatic mechanisms consist of receptors (here, thermoreceptors in the skin), a control center (here, the brain), and effectors (here, the sweat glands).

WEB TUTORIAL 4.1 Homeostasis

KEY TERMS

adipose tissue   p. 70
areolar connective tissue   p. 70
basement membrane   p. 70
blood   p. 71
bone   p. 71
cardiac muscle   p. 73
cartilage   p. 71
collagen fiber   p. 70
columnar epithelium   p. 68
compact bone   p. 71
connective tissue   p. 68
cuboidal epithelium   p. 68
cutaneous membrane   p. 77
dense connective tissue   p. 71

dermis   p. 78
elastic cartilage   p. 71
elastic fibers   p. 70
endocrine gland   p. 68
epidermis   p. 78
epithelial tissue   p. 68
exocrine gland   p. 68
fibroblast   p. 71
fibrocartilage   p. 71
formed element   p. 71
glandular epithelium   p. 68
homeostasis   p. 83
hyaline cartilage   p. 71
hypodermis   p. 78

integumentary system   p. 77
loose connective tissue   p. 70
melanin   p. 79
melanocyte   p. 79
membranous epithelium   p. 68
mucous membrane   p. 76
muscle tissue   p. 68
negative feedback mechanism   p. 83
nervous tissue   p. 68
neuroglia   p. 74
neuron   p. 74
oil gland   p. 83
organ   p. 74
organ system   p. 74

plasma   p. 71
reticular fiber   p. 70
serous membrane   p. 77
simple epithelial   p. 68
skeletal muscle   p. 73
smooth muscle tissue   p. 74
spongy bone   p. 71
squamous epithelium   p. 68
stratified epithelial tissue   p. 68
sweat gland   p. 83
synovial membrane   p. 77
tissue   p. 68

THINKING ABOUT THE CONCEPTS

1. What are the four types of tissue found in the human body? (*p. 68*)
2. Compare the organization of epithelial and connective tissues. How are differences in the matrix of different types of connective tissue related to their functions? (*pp. 68–73*)
3. Why does bone heal more rapidly than cartilage? (*p. 71*)
4. What type of tissue is blood? (*pp. 71–73*)
5. Compare and contrast skeletal, cardiac, and smooth muscle with respect to structure and function. (*pp. 73–74*)
6. What types of cells are found in nervous tissue? What are their functions? (*p. 74*)
7. List the functions of the integumentary system. (*p. 77*)
8. Describe the roles of pigments and blood flow in determining skin color. (*pp. 79–80*)
9. Describe how hair grows. (*p. 82*)
10. What are oil glands? What role do these glands play in acne? (*pp. 83, 85*)
11. Describe the functions of the sweat glands. (*p. 83*)
12. Give an example of an interaction between the integumentary system and another organ system. (*p. 83*)
13. What is homeostasis? (*p. 83*)

14. Describe the body's homeostatic mechanisms for raising and lowering core temperature. (*pp. 83–84*)
15. The four basic tissue types in the body are
    a. simple, cuboidal, squamous, columnar.
    b. neural, epithelial, muscle, connective.
    c. blood, nerves, bone, cartilage.
    d. fat, cartilage, muscle, neural.
16. The lining of the intestine is composed primarily of _____ cells.
    a. epithelial
    b. muscle
    c. connective tissue
    d. nerve
17. Cells that form pads that cushion the vertebrae are
    a. bone.
    b. muscle.
    c. epithelium.
    d. cartilage.
18. The mineral that makes bone hard is _____.
19. The maintenance of body processes within a relatively constant range is called _____.

APPLYING THE CONCEPTS

1. Joan is a young woman who was caught in a fire and suffered third-degree burns. What are some of the potentially life-threatening problems she might experience?

2. When you exercise, your breathing rate, blood pressure, and heart rate increase. Is this a violation of homeostasis? Explain.

5

The Skeletal System

Bones Function in Support, Movement, Protection, Storage of Minerals and Fat, and Blood Cell Production

Bones Have a Hard Outer Layer of Compact Bone Surrounding Spongy Bone

Bone Is Living Tissue
- Most of the skeleton begins as a cartilage model
- Hormones regulate bone growth

Bone Fractures Are Healed by Fibroblasts and Osteoblasts

Bones Are Continuously Remodeled

The Human Skeleton Has Two Parts
- The axial skeleton is composed of the skull, vertebral column, sternum, and rib cage
- The appendicular skeleton is composed of the pectoral girdle, the pelvic girdle, and the limbs

Joints Are Junctures between Bones
- Synovial joints permit flexibility

HEALTH ISSUE Osteoporosis: Fragility and Aging

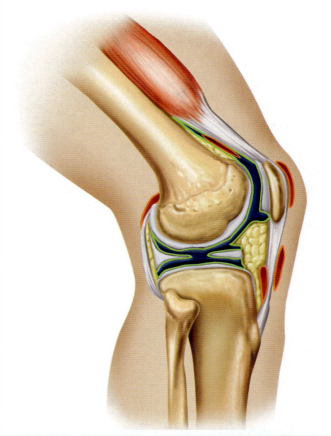

At age 4 Jamal wished he could fly. He swooped around the backyard with a small blanket for a cape, flapping his arms to get the lift that would propel him upward. Birds flew so easily, why couldn't he? He soon figured out that he would have a better chance if he started with a little elevation, so he stood on the second step of the ladder to his backyard slide. Starting from the step seemed to work a bit better than starting from the ground. He was in the air for a microsecond longer. He decided to go up two more steps and again starting from a higher point seemed to work, except that his outstretched hands stung a little when he hit the ground. He decided to go to the sixth stair of the slide and again dive forward, cape extended. In less than a second he hit the ground again, hands first, and immediately felt the pain. He screamed loudly, and in a matter of seconds his mom was there to pick him up. A trip to the emergency room and an x-ray indicated that Jamal had sprained his wrist. The doctor gave him a lollipop, told him that he was lucky he hadn't broken the bones in his wrist, and never, never to jump off of ladders again.

The fact that Jamal cannot fly and the fact that he could have broken his wrist in his fall are both related to the design, structure, and alignment of bones and muscles in humans. In this chapter, we will examine the structure of bone and see why it is able to support our bodies against gravity, holding us upright. We'll see that bone is a dynamic, living tissue and consider how bones grow. We will also see and a few reasons why some of us are taller than others. ■

Bones Function in Support, Movement, Protection, Storage of Minerals and Fat, and Blood Cell Production

The **skeleton** is a framework of bones and cartilage that functions in movement and the protection of internal organs. Specifically, the skeleton has the following functions:

1. **Support.** It provides a rigid framework that supports soft tissues.
2. **Movement.** It provides places of attachment for muscles.
3. **Protection.** It encloses internal organs, such as the heart and lungs, which are within the chest cavity, and the brain, which lies within the skull.
4. **Storage of minerals.** The bones store minerals, particularly calcium and phosphorus, that can be released to the rest of the body when needed.
5. **Storage of fat.** It stores energy-rich fat in yellow bone marrow (the soft tissue within some bones). The fat can be metabolized and the energy released when needed.
6. **Blood cell production.** It produces blood cells in the red marrow of certain bones.

Bones Have a Hard Outer Layer of Compact Bone Surrounding Spongy Bone

The 206 bones that form the internal scaffolding and girders of the human body come in a variety of shapes and sizes. They are generally classified on the basis of their shape—long, short, flat, or irregular. Bones usually contain some degree of both compact and spongy bone; the exact proportion depends on the shape and size of the bone.

Compact bone is very dense bone with only a few internal spaces (Figure 5.1). Compact bone is what you see when you look at the outside of any bone. For instance, compact bone forms most of the shaft of long bones, such as those of the arms and legs. Compact bone is covered by a glovelike **periosteum**, the membrane covering that nourishes bone. The periosteum contains blood vessels and nerves as well as cells involved in bone growth and repair. When a bone is bruised or fractured, most of the pain results from injury to the periosteum.

Spongy bone is formed from a latticework of thin layers of bone with open areas between. These layers form internal struts that brace the bone from within. Spongy bone is largely found in small, flat bones, such as most of the bones of the skull, and in both the head (enlarged end) and near the ends of the shaft of long bones. In adults, the small spaces of some spongy bones (including the ribs, pelvis, backbone, skull, and long-bone ends) are filled with **red marrow**, where blood cells form. The cavity in the shaft of adult long bones is filled with **yellow marrow**, a fatty tissue for energy storage.

Bone Is Living Tissue

Compact bone is highly organized living tissue. The structural unit of compact bone is microscopic and is called an **osteon**, which you can see in Figure 5.1. Each osteon consists of mature bone cells, called **osteocytes** (*osteo*, bone; *cyte*, cell), which are arranged in concentric rings around a central canal. Each osteocyte lies within a tiny cavity (lacuna) in the hardened matrix. Tiny canals connect with nearby lacunae and eventually with the central canal. In this way, oxygen and nutrients can pass from the blood vessels of the central canal to the osteocytes, and wastes can be carried away.

Bone is, indeed, a living tissue, but its characteristics come from its nonliving component—the matrix. Secreted by the bone cells, the matrix makes bone both hard and resilient. Bone's hardness and rigidity come from mineral salts, primarily calcium and phosphorus. Woven throughout the matrix are strands of the strong elastic protein collagen. Without the calcium and phosphorus salts, bone would be rubbery and flexible like a garden hose. Without collagen, bone would be brittle and would crumble like chalk. Sometimes bones do bend, causing bowlegs, in disorders such as rickets, in which the amount of calcium salts in the bones is greatly reduced (Figure 5.2).

stop and think

Strontium-90 is a radioactive material that enters the atmosphere after atomic explosions. This material can end up in human bodies via milk from cows that grazed on contaminated grass. Strontium-90 can replace calcium in bone and then kill nearby cells or alter their genetic information. Explain why exposure to strontium-90 can lead not just to bone cancer but also to disruption of blood cell formation.

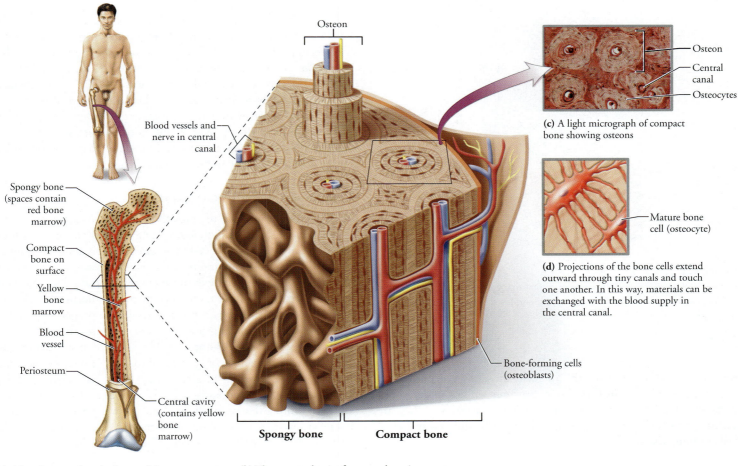

(a) A long bone, such as the femur of the leg, consists of a shaft and two heads, or enlarged ends. Compact bone is located on the outer surface of the bone. Spongy bone is found in the heads.

(b) The structural unit of compact bone is an osteon. Mature, living bone cells (osteocytes) are found in small spaces within the hard matrix.

(c) A light micrograph of compact bone showing osteons

(d) Projections of the bone cells extend outward through tiny canals and touch one another. In this way, materials can be exchanged with the blood supply in the central canal.

FIGURE **5.1**
The structure of bone

▌ Most of the skeleton begins as a cartilage model

During embryonic development, most of the skeleton is first formed of cartilage, a firm yet flexible connective tissue (Figure 5.3). Mature bone cells cannot divide because they are enclosed in a solid matrix. However, cartilage cells are able to divide and form new cartilage cells. Thus, the cartilage model can grow and expand as the fetus grows rapidly. Beginning in the third month of development, the cartilage begins to be replaced by bone.

During embryonic development, the formation of a long bone, such as the bones of the arms or legs, begins with the formation of a collar of bone around the shaft of the cartilaginous model (Figure 5.4). Bone-forming cells, called **osteoblasts** (*osteo*, bone; *blast*, beginning or bud), form the collar. The osteoblasts form from cartilage cells that are transformed as blood vessels grow around the cartilage. The bony collar supports the shaft as the cartilage within it breaks down, forming the marrow cavity. Once a cavity is produced within the shaft, osteoblasts migrate into the space and form spongy bone. Osteoblasts secrete the collagen (as well as other

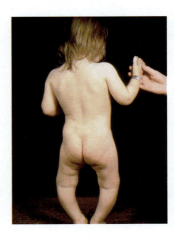

FIGURE **5.2**
The legs of this child are bent because of rickets, a condition caused by insufficient vitamin D, which is needed for proper absorption of calcium from the digestive system. Calcium salts make the matrix of bone very hard. Collagen cells in the matrix add strength. In rickets, the bones become soft and somewhat pliant.

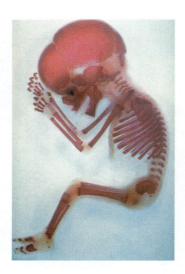

FIGURE **5.3**
During fetal development, the skeleton is first made of cartilage. The cartilaginous model of this 16-week-old human fetus will be gradually replaced by bone.

organic materials) of the matrix, and calcium salts are deposited on the collagen. These cells form bone, but they cannot undergo cell division. Once they form the matrix around themselves, they are called osteocytes, which are mature bone cells and the principal cells in bone tissue.

At about the time of birth, bone growth centers form in the ends of long bones, and spongy bone begins to fill these regions. Two regions of cartilage will remain. One is a cap of cartilage that covers the surfaces that rub against other bones in a joint. The second is a plate of cartilage, called the epiphyseal plate, that separates the end of the bone from the shaft. The epiphyseal plate is commonly called the growth plate. Cartilage cells within the epiphyseal plate divide, forcing the end of the bone farther away from the shaft. As bone replaces the newly formed cartilage in the region closest to the shaft, the bone becomes longer.

Hormones regulate bone growth

Parents proudly measure the growth of a child, inch by inch, on a growth chart. During childhood, bone growth is powerfully stimulated by growth hormone (GH), released by the anterior pituitary gland. Growth hormone prompts the liver to release growth factors that produce a surge of growth in the epiphyseal plate. The activity of growth hormone is, in turn, modified by thyroid hormones. Thyroid hormones ensure that the skeleton grows with the proper proportions.

At puberty, a growth spurt often occurs. At this time, the legs of pants can seem to shrink almost weekly. These dramatic changes are orchestrated by the increasing levels of male or female sex hormone (testosterone and estrogen, respectively) that accompany puberty. Initially the sex hormones stimulate the cartilage cells of the epiphyseal plates into a frenzy of cell division. But we generally stop growing taller toward the end of our teenage years (age 18 in females and 21 in males) because of later changes initiated by the sex hormones.

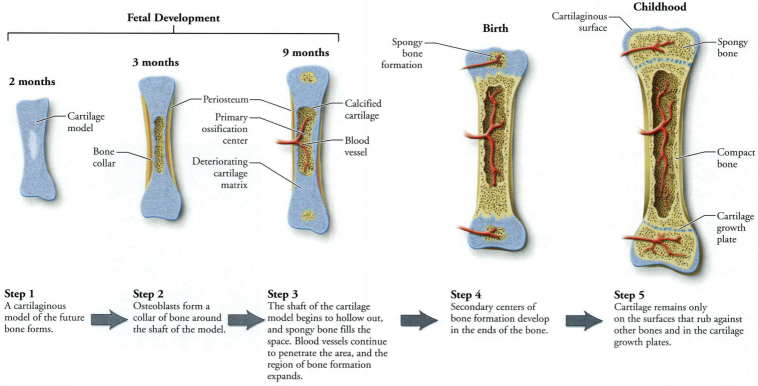

Step 1
A cartilaginous model of the future bone forms.

Step 2
Osteoblasts form a collar of bone around the shaft of the model.

Step 3
The shaft of the cartilage model begins to hollow out, and spongy bone fills the space. Blood vessels continue to penetrate the area, and the region of bone formation expands.

Step 4
Secondary centers of bone formation develop in the ends of the bone.

Step 5
Cartilage remains only on the surfaces that rub against other bones and in the cartilage growth plates.

FIGURE **5.4**
Steps of bone formation in long and short bones of an embryo

The cartilage cells in the epiphyseal plates gradually divide less frequently. The plates become thinner as cartilage is replaced by bone. Finally, the bone of the ends fuses with that of the shaft.

stop and think

We have considered the way the long bones grow and the mechanism by which growth hormone causes the growth of long bones. Why would it be ineffective for a short, middle-aged person to be treated with growth hormone to stimulate growth?

Bone Fractures Are Healed by Fibroblasts and Osteoblasts

In spite of their great strength, bones occasionally break. Fortunately, bone tissue can heal. The first thing that happens with a break is bleeding (even with a simple fracture) followed by clot (hematoma) formation at the break (Figure 5.5). Within a few days, fibroblasts—cells of a type of connective tissue—grow inward from the periosteum and invade the clot. The fibroblasts secrete collagen fibers that form a mass called a callus, which links the ends of the broken bone. Some of the fibroblasts then transform into cartilage-producing cells that secrete cartilage in the callus.

Next, osteoblasts from the periosteum invade the callus and begin to transform the cartilage into new bone material. As this transformation happens, the callus becomes larger than the bone itself and protrudes from it. In time, however, the extra material will be broken down, and the bone will return to normal size.

Bones Are Continuously Remodeled

After the initial growth process is completed, the life-long process of bone deposition and absorption is called remodeling. As we have seen, bone is deposited by osteoblasts. Another kind of bone cell, called an **osteoclast**, breaks down bone, and the minerals are reabsorbed by the body. The process of remodeling is continuous and occurs at different rates in different parts of bones. For example, the end of the femur (thighbone) closest to the knee is completely replaced every 5 to 6 months, but the shaft of the same bone is replaced more slowly.

An important factor in determining the rate and extent of bone remodeling is the degree of stress to which a bone region is exposed. Bone forms in response to stress and is destroyed when it is not stressed. Weight-bearing exercise, such as walking or jogging, thickens the layer of compact bone, leading to stronger bones. Bones that are frequently used may actually change shape. Because of continual practice, the knuckles of pianists and the big toes of ballet dancers can enlarge. On the other hand, bones that are not stressed lose mass. After a few weeks without stress, a bone could lose nearly a third of its mass. For instance, this loss might occur in an astronaut in the weightless environment of space flight or right here on the earth in a person using crutches with a leg in a cast.

Bone that is broken down faster than it is built becomes weak and fragile. **Osteoporosis**, a condition in which there is a progressive loss in bone density is discussed in the Health Issues essay, *Osteoporosis: Fragility and Aging.*

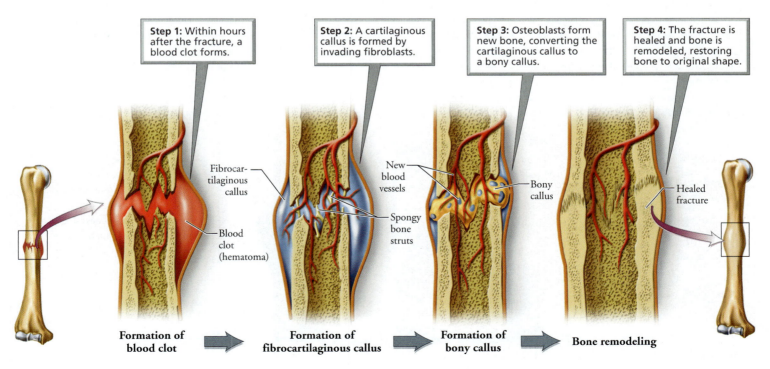

Step 1: Within hours after the fracture, a blood clot forms.

Step 2: A cartilaginous callus is formed by invading fibroblasts.

Step 3: Osteoblasts form new bone, converting the cartilaginous callus to a bony callus.

Step 4: The fracture is healed and bone is remodeled, restoring bone to original shape.

Fibrocartilaginous callus

Blood clot (hematoma)

New blood vessels

Spongy bone struts

Bony callus

Healed fracture

Formation of blood clot → **Formation of fibrocartilaginous callus** → **Formation of bony callus** → **Bone remodeling**

FIGURE 5.5
The progress of healing in a bone

HEALTH ISSUE

Osteoporosis: Fragility and Aging

Osteoporosis is a decrease in bone density that occurs when the destruction of bone outpaces the formation of new bone during ongoing bone remodeling. This destruction causes bones to become thin, brittle, and susceptible to fracture. One of the greatest problems with osteoporosis is the constant threat of broken bones. We frequently hear of elderly people falling and breaking a hip. A fall, blow, or lifting action that would not bruise a person with healthy bones could easily cause a bone to fracture in a person with severe osteoporosis. Osteoporosis affects about 20 million Americans, most of them elderly white women.

As we have seen, bone remodeling occurs throughout life. Until we reach about age 35, bone is formed faster than it is broken down. Thus, bones are strongest and densest during our mid-thirties. Peak bone density is influenced by a number of factors, including sex, race, nutrition, exercise, and overall health. In men, bone mass is generally 30% higher than in women. The bones of African Americans are generally 10% denser than those of Caucasians and Asians. Within each group, bone density is influenced by diet, especially the dietary levels of calcium and vitamin D, which are essential for forming hard bones, and by the history of weight-bearing exercise, such as walking. Recall that such exercise stresses bones and promotes bone formation.

Each of us begins to lose bone density after our mid-thirties, but the degree to which bones will become dangerously weakened depends largely on how dense the bones were at their peak. Osteoporosis is sometimes very apparent. The afflicted person becomes hunched and stooped and, as vertebrae lose mass, may become even shorter (Figure 5.A). It is sometimes startling to see that a middle-aged or elderly person whom you haven't seen in a while has actually gotten shorter!

Women are at greater risk for developing osteoporosis than are men. This difference is not just because they have less bone mass at peak but also because the rate of bone loss is accelerated for several years after menopause. Menopause, the time in a woman's life when she no longer produces mature eggs or menstruates, marks a severe reduction in the production of the female hormone estrogen. Estrogen stimulates bone formation. Without the high levels of estrogen, bone formation slows. However, the destruction of bone continues, and the imbalance results in the reduction of bone mass. Estrogen is also important in the absorption of calcium from the intestines, and calcium is needed to make bones hard.

It is noteworthy that young women who undergo athletic training extensive enough to stop menstruation are also placing their bones at risk. It is not uncommon for intense physical exercise to disrupt menstrual cycles because of alterations in hormone levels. Researchers have shown that, in such cases, bone density can also be affected. In one study, 20% of the young female athletes who failed to ovulate during even one cycle out of the whole year suffered as much as a 4% loss in bone density.

A number of factors other than estrogen levels have been implicated in the onset of osteoporosis. For example, size is important. Short people are at greater risk, perhaps because they generally start with less bone mass. People with a good supply of body fat are less at risk because fat can be converted to estrogen. Heavy drinkers are at higher risk, because alcohol interferes with estrogen function. Smoking is also bad for bones, because it can reduce estrogen levels. People who do not take in enough calcium, or who cannot absorb it because of insufficient vitamin D, have thinner bones because calcium is necessary for bone growth. Calcium-rich foods include milk products as well as broccoli, spinach, shrimp, and soybean products. Certain drugs, such as caffeine (a diuretic), tetracycline, and cortisone can promote osteoporosis. Finally, sedentary people have thinner bones than do active people. Exercise places stress on the bones, and the bones respond by becoming thicker. Thus, weight-bearing exercise is recommended.

We can expect osteoporosis and other afflictions associated with aging to become more common in the population as people tend to live longer. Thus, it would seem prudent for each of us to start an early program of prevention by eating a diet rich in calcium and vitamin D and by engaging in regular weight-bearing exercise, such as walking or jogging.

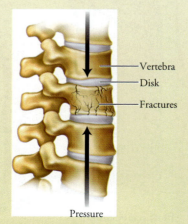

(a) Normal bending can place pressure on vertebrae that can cause small fractures.

Vertebra
Disk
Fractures
Pressure

(b) Osteoporosis can be seen in this colored x-ray image of vertebrae as the pink regions in the bones.

(c) A loss of height and a stooped posture result as the weakened bones of the vertebrae become compressed.

FIGURE 5.A
Osteoporosis is a loss of bone density that occurs when bone destruction outpaces bone deposition during the continuous process of bone remodeling. Bones become brittle and are easily fractured.

The Human Skeleton Has Two Parts

As can be seen in Figure 5.6, the bones of the human body are arranged into the axial skeleton and the appendicular skeleton. The **axial skeleton** (shown in the darker color in the figure) includes the skull, the vertebral column (backbone), and the bones of the chest region (sternum and rib cage). The **appendicular skeleton** (shown in the lighter color) includes the pectoral girdle (shoulders), the pelvic girdle (pelvis), and the limbs (arms and legs).

The axial skeleton is composed of the skull, vertebral column, sternum, and rib cage

We will focus on only the major bones of the 80 bones that make up the axial skeleton. We will start with the skull. The skull is divided into two areas—the bones of the cranium and those of the face (Figure 5.7). The cranium houses the brain and the structures of hearing. The facial bones give us our facial features and house certain sensory systems of the head, including vision, smell, and taste. In general, the axial skeleton protects and supports internal organs.

THE CRANIUM

The cranium is formed from eight (or sometimes more) flattened bones. The single **frontal bone** forms the forehead and the front of the brain case. The top and two sides of the skull are formed by the two **parietal bones**, which join at the midline. At the back of the head lies the **occipital bone**, which surrounds the **foramen magnum**, the opening through which the spinal cord passes.

Before and shortly after birth, the bones of the cranium are held together by membranous areas called the **fontanels**, often referred to as the soft spots (Figure 5.8). During birth, the fontanels allow the skull to be compressed, easing the passage of the head through the birth canal. The fontanels accommodate the rapid enlargement of the brain during fetal growth and infancy. The fontanels will be replaced by bone by the age of 2.

On either side of the cranium are the **temporal bones**, part of which form what we think of as our temples. The **sphenoid bone**, with its bow-tie shape, forms the floor of the cranium. The **ethmoid bone**, the smallest bone in the cranium, separates the cranial cavity from the nasal cavity. The olfactory nerves from the nasal cavity that are responsible for our sense of smell communicate with the brain through tiny holes in the ethmoid.

THE FACIAL BONES

The face is composed of 14 bones (Figure 5.7). They support several sensory structures and serve as attachments for most muscles of the face.

Nasal bones form the bridge of the nose. They are paired and fused at the midline. Inside the nose, a partition called the **nasal septum** (composed of the vomer and part of the ethmoid bone) divides the left and right chambers.

Cheekbones are formed largely from the paired **zygomatic bones**. Flattened areas of these bones form part of the bottom of the eye sockets. The zygomatic bone also gives rise to an extension that joins with one from the temporal bone to form the zygomatic arch (the "cheekbone" itself).

The smallest facial bones are the two **lacrimal bones**, located at the corners of the eyes near the nose. A duct passes through each bone and drains tears from the eyes into the nasal chambers; this connection explains why our nose runs when we cry.

The jaw is formed by two pairs of bones. The upper jaw is composed of two **maxillae**, fused at the midline, directly below the nasal septum. Most other facial bones are joined to them, illustrating the importance of the upper jaw to facial structure. The maxillae also form part of the hard palate, the roof of the mouth. Behind the palate lie two **palatine bones** that form the rest of the roof. When the maxillae fail to join as the face develops, a cleft palate results. In this case, the mouth cavity and the nasal cavity are not fully separated. This condition is easily corrected surgically.

The lower jaw is called the **mandible**. It, too, is formed from two bones connected at the midline. The mandible is connected to the skull at the temporal bone, forming a hinge called the temporomandibular joint. (Joints are places where bones meet and will be discussed later in this chapter.) This joint allows the mouth to open and close. Emotional stress causes some people to clench or grind their teeth, sometimes unconsciously. This action can cause physical stress on the temporomandibular joint, causing headaches, toothaches, or even earaches. The condition is known as temporomandibular joint (TMJ) syndrome.

THE VERTEBRAL COLUMN

The **vertebral column**, more familiarly known as the backbone, is a series of bones, descending from the cranium, through which the spinal cord passes (Figure 5.9). Each of these bones is called a **vertebra**. The vertebrae (plural of vertebra) are divided according to where they lie along the length of the vertebral column. There are:

- 7 **cervical** (neck) **vertebrae**
- 12 **thoracic** (chest) **vertebrae**
- 5 **lumbar** (lower back) **vertebrae**
- 1 **sacrum** (formed by the fusion of five vertebrae)
- 1 **coccyx** (or tailbone)

Scoliosis, which means "twisted disease," is an abnormal curvature of the spine to the left or right. The most common form of scoliosis has no known cause and affects over 1.5 million adolescents, primarily females. It usually begins and progresses through the adolescent growth spurt. Treatment, if needed, is intended to straighten the spine with a brace or by surgery.

The added strength of the fused sacral vertebrae is necessary because the sacrum joins the pelvic girdle. Great stress is placed on the sacrum owing to the weight of the vertebral column and the powerful movements of the leg. The coccyx, on the other hand, may be fused for precisely the opposite reason. It serves no function, being regarded as a vestigial tail—an evolutionary relic—whose bones may have joined as a side effect of growth without movement. (The upper part of the coccyx, however, is supplied with nerves and can cause extreme pain if somehow broken.)

Between the vertebrae of the cervical, thoracic, and lumbar regions lie **intervertebral disks**, pads of fibrocartilage that help cushion the bones of the vertebral column. Because of their smooth, lubricated surfaces, these disks help in the movement of the column. As the

Axial skeleton

BONES OF THE SKULL

Cranial bones
• Surround and protect the brain and sensory organs

Facial bones
• Form the underlying structure of the face; support teeth

BONES OF THE RIB CAGE
• Protect internal organs, assist breathing

Sternum (breastbone)

Ribs (12 pairs)

VERTEBRAL COLUMN (BACKBONE)

Vertebrae (26 bones)
• Surround and protect the spinal cord; support for upper part of body; provide attachment for muscles

Intervertebral disks
• Cartilaginous pads between vertebrae; absorb shock; permit flexibility

Appendicular skeleton

BONES OF THE PECTORAL GIRDLE AND UPPER APPENDAGES
• Bones with extensive muscle attachments, arranged for great freedom of movement

Clavicle (collarbone)

Scapula (shoulder blade)

Humerus (upper arm bone)

Radius (forearm bone)

Ulna (forearm bone)

Carpals (wrist bones)

Metacarpals (palm bones)

Phalanges (thumb, finger bones)

BONES OF THE PELVIC GIRDLE AND LOWER APPENDAGES

Pelvic girdle (six fused bones)
• Protects internal pelvic organs; supports weight of vertebral column

Femur (thighbone)
• Largest, strongest bone in the body; important in movement of body and upright posture

Patella (kneecap)
• Protects knee joint

Tibia (lower leg bone)
• Support for upper body

Fibula (lower leg bone)
• Support for upper body

Tarsals (ankle bones)

Metatarsals (sole bones)

Phalanges (toe bones)

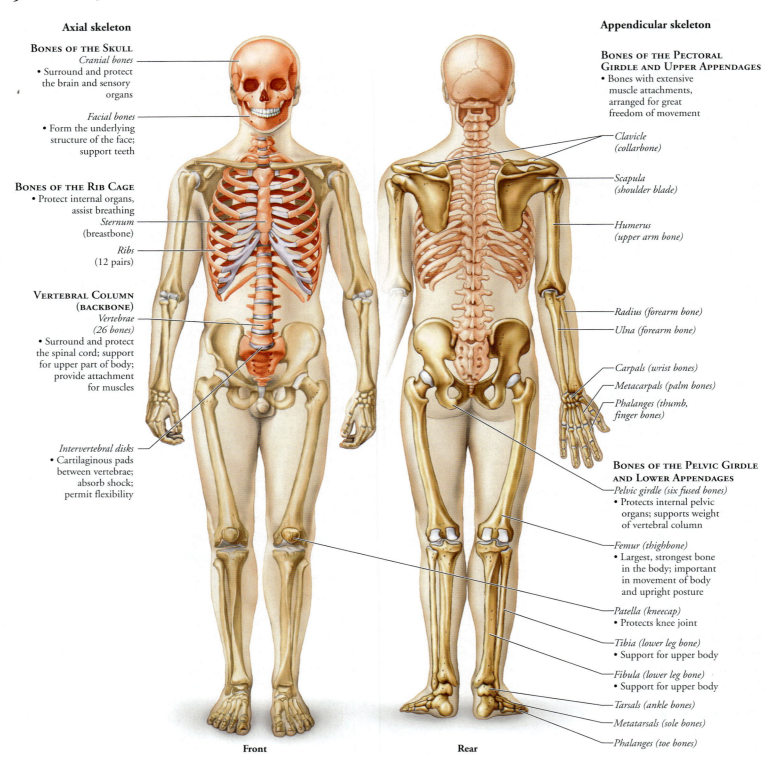

Front　　　**Rear**

FIGURE **5.6**

The major bones of the human body. The orange bones constitute the axial skeleton, and the light brown bones, the appendicular skeleton. Cartilage appears as gray areas of the skeleton.

intervertebral disks become compressed over the years, the person may become shorter. This compression, alone or coupled with the effects of osteoporosis, can have a pronounced effect on height as a person ages.

Other problems can also arise from abnormalities in these disks. For example, if too much pressure is applied to the vertebral column, the disks can bulge from between the vertebrae. If the disks bulge inward, pressing against the spinal cord itself, they can interfere with muscle control and perception of incoming stimuli. If the bulging disk presses outward against a spinal nerve that branches from the spinal cord, great pain can result. The sciatic

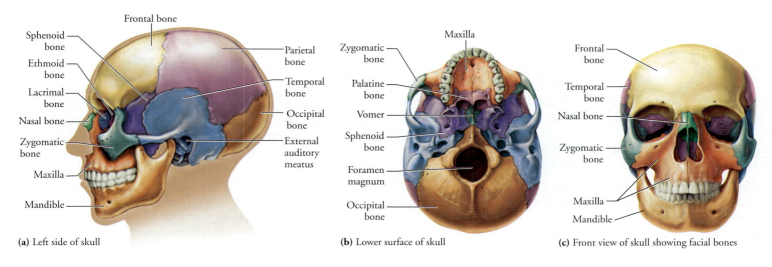

Sphenoid bone
Ethmoid bone
Lacrimal bone
Nasal bone
Zygomatic bone
Maxilla
Mandible
Frontal bone
Parietal bone
Temporal bone
Occipital bone
External auditory meatus

(a) Left side of skull

Zygomatic bone
Palatine bone
Vomer
Sphenoid bone
Foramen magnum
Occipital bone
Maxilla

(b) Lower surface of skull

Frontal bone
Temporal bone
Nasal bone
Zygomatic bone
Maxilla
Mandible

(c) Front view of skull showing facial bones

FIGURE **5.7**
The major bones of the skull and face

FIGURE **5.8**
The bones of the skull of a human newborn are not fused but are instead held together by fibrous connective tissue. These "soft spots" allow the skull bones to move during the birth process, easing the passage of the skull through the birth canal. By the age of 2, the soft spots will be replaced by bone.

nerve, a large nerve that extends down the back of the leg, is one of the most frequently affected nerves. Sciatica, the resulting inflammation, can be cripplingly painful.

A particularly mysterious human ailment is lower back pain. It accounts for more missed workdays than any medical problem besides colds. The source of the pain is notoriously difficult to pinpoint. A person with a slipped disk may or may not have pain, and a person whose back looks perfectly normal may complain of terrible pain. Part of the problem seems to be that the pain may originate from muscle rather than bone, and muscles do not show up on x-rays. One source of back problems, however, may be weak abdominal muscles that cannot counteract the pull of the powerful back muscles, a situation that can cause the vertebrae to misalign.

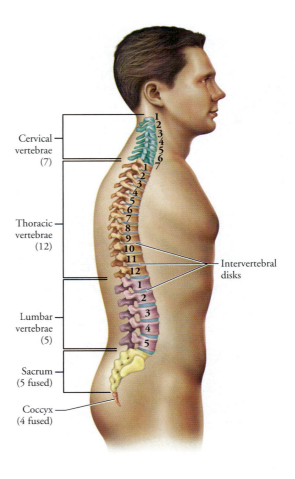

Cervical vertebrae (7)
Thoracic vertebrae (12)
Lumbar vertebrae (5)
Sacrum (5 fused)
Coccyx (4 fused)
Intervertebral disks

FIGURE **5.9**
A side view of the vertebral column

THE RIB CAGE

Twelve pairs of ribs attach at the back to the thoracic vertebrae (Figure 5.10). At the front, the upper 10 pairs of ribs attach directly or indirectly to the breastbone (sternum) by cartilage. Their flexibility permits the ribs to take some blows without breaking and to move during breathing. The last two pairs of ribs do not attach to the sternum and are called floating ribs.

The appendicular skeleton is composed of the pectoral girdle, the pelvic girdle, and the limbs

The appendicular skeleton is composed of the pelvic and pectoral girdles and the attached limbs. A girdle is a skeletal structure that supports the arms (the **pectoral girdle**) or the legs (the **pelvic girdle**). The pectoral girdle, then, connects the arms to the rib cage, and the pelvic girdle connects the legs to the vertebral column. In general, the appendicular skeleton allows you to change position and location in the environment.

THE PECTORAL GIRDLE

The pectoral girdle is composed of the **scapulae** (shoulder blades) and the **clavicles** (collarbones) (Figure 5.11). Each clavicle makes a relatively rigid connection between a scapula and the sternum. The clavicles are more curved in males than in females (one way to tell the sex of a skeleton). One corner of the roughly triangular scapula bears a socket into which fits one end of the **humerus**, the upper arm bone. The humerus joins at the elbow with the **radius** and **ulna**, the bones of the lower arm, or forearm. The ulna extends past the junction with the humerus to form the elbow. The radius and ulna meet the eight **carpals** at the wrist, and these join with the five **metacarpals**, which are the bones of the hand. The hand gives rise to three **phalanges** in each finger and two in each thumb.

There is a narrow opening, or tunnel, through the carpal bones that form the wrist. Through this so-called carpal tunnel passes a nerve that controls sensations in the fingers and some muscles in the hand. **Tendons**, bands of connective tissue that attach muscles to bones, also pass through the carpal tunnel. Repeated motion in the hand or wrist can cause the tendons to become inflamed and press against the nerve. As a result, there may be numbness or tingling in the affected hand and pain that may affect the wrist, hand, and fingers. This condition, known as carpal tunnel syndrome, is becoming increasingly common as increasing numbers of people operate the keyboard of a computer or play video games. Barbers, cab drivers, and pianists have

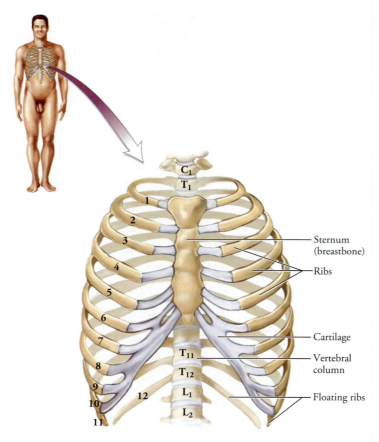

FIGURE **5.10**
The bones of the rib cage

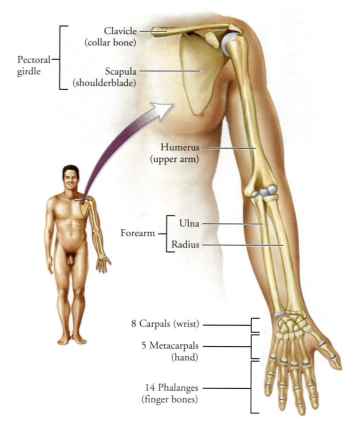

FIGURE **5.11**
The pectoral girdle and arm

experienced the same problem. The increased incidence of carpal tunnel syndrome has caused the computer industry to redesign their keyboards and games. Employers have responded by providing adjustable work stations and rest and exercise periods.

what would you do?

There has been a rash of lawsuits against computer manufacturers and video-game designers by people who have developed carpal tunnel syndrome as a result of endless hours spent using keyboards and video-game controllers. If you developed carpal tunnel syndrome from using a keyboard or video game, would you sue the manufacturer? What criteria would you use to decide?

THE PELVIC GIRDLE

The pelvic girdle is much more rigid than the pectoral girdle, as you know if you have ever tried to shrug your hips. The pelvic girdle consists of two **pelvic bones**, which are attached to the sacrum (Figure 5.12). These bones arch around to the front, where they join to a cartilage disk at the **pubic symphysis**. The male and female hips are easily distinguishable because the opening in the female is wider to facilitate childbirth.

The **femur**, or thighbone, rotates within a socket in the pelvis. At the knee, the femur joins the **tibia** (the shinbone). The **fibula** is a smaller bone that runs down the side of the lower leg. The junction where the tibia joins the femur is covered by a **patella** (kneecap). At the ankle, the lower leg bones meet the **tarsals**, or ankle bones. These are connected to the **metatarsals**, or foot bones. The metatarsals, in turn, are connected to the phalanges, the toe bones.

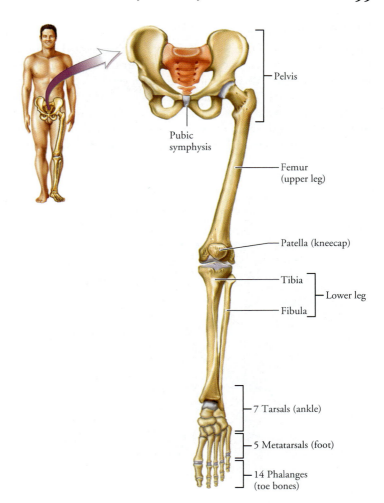

FIGURE **5.12**
The pelvic girdle and leg

Joints Are Junctures between Bones

Joints are the places where bones meet. They can be classified according to their structure (fibrous, cartilaginous, or synovial) or by the degree of movement they permit. Some joints allow no movement; others permit slight movement; and still others are freely movable.

Fibrous joints are held together by fibrous connective tissue. They have no joint cavity, and most do not permit movement. For example, the joints between the skull bones of an adult, called sutures, are immovable because the bones are interlocked and held together tightly by fibrous connective tissue (Figure 5.13). These joints can actually move ever so slightly. This ability is fortunate, because the movements serve as a shock absorber when you hit your head.

Cartilage, which is rather rigid, holds bones together in *cartilaginous joints*. Some cartilaginous joints are immovable, and others allow slight movement. We find such joints between vertebrae, in the attachment of ribs to the sternum, and in the pubic symphysis, the joint between the two pelvic bones. In a pregnant woman, hormones loosen the cartilage of the pubic symphysis, allowing the pelvis to widen to ease childbirth.

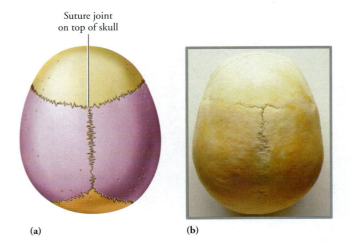

(a) (b)

FIGURE **5.13**
The bones of the skull are joined by suture joints. (a) Top view of the skull. (b) Photograph of suture joints in an adult skull.

▌Synovial joints permit flexibility

Most of the joints of the body are freely movable, **synovial joints**. Because of these joints, muscles can maneuver the body into thousands of positions. All synovial joints share certain common features (Figure 5.14). The surfaces of the joints that move past one another have a thin layer of hyaline cartilage. The cartilage reduces friction, allowing the bones to slide over one another without grating and grinding. Synovial joints are surrounded by a two-layered

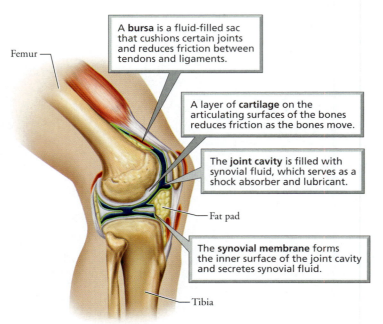

A **bursa** is a fluid-filled sac that cushions certain joints and reduces friction between tendons and ligaments.

A layer of **cartilage** on the articulating surfaces of the bones reduces friction as the bones move.

The **joint cavity** is filled with synovial fluid, which serves as a shock absorber and lubricant.

Femur

Fat pad

The **synovial membrane** forms the inner surface of the joint cavity and secretes synovial fluid.

Tibia

(a) Synovial joints, such as the knee shown here, permit a great range of movement.

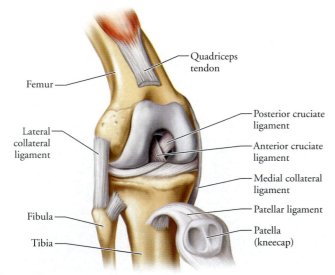

Quadriceps tendon

Femur

Posterior cruciate ligament

Lateral collateral ligament

Anterior cruciate ligament

Medial collateral ligament

Fibula

Patellar ligament

Tibia

Patella (kneecap)

(b) Ligaments hold bones together, support the joint, and direct the movement of the bones.

FIGURE **5.14**
The knee is an example of a synovial joint.

capsule. The inner layer of the capsule secretes viscous, clear fluid (synovial fluid) into the space between the two bones, called the synovial cavity. The synovial fluid lubricates and cushions the joint. The outer layer of the capsule is continuous with the covering membranes of the bones forming the joint. The entire synovial joint is reinforced with ligaments. **Ligaments** are strong straps of connective tissue that hold the bones together, support the joint, and direct the movement of the bones.

Although all synovial joints share these features, there are different types of synovial joints, each permitting a different type and range of motion. **Hinge joints**, such as the knee and elbow, are so named because, like a hinge on a door, they permit motion in only one plane. A **ball-and-socket joint** allows movement in all planes. The shoulder and hip are examples of ball-and-socket joints; the ball at the head of one bone fits into a socket on another bone. Notice that you can swing your arm around in a complete circle. Some of the ways that body parts can move at synovial joints are shown in Figure 5.15.

DAMAGE TO JOINTS

Damage to a ligament is called a **sprain**. Stretching a ligament may cause slight sprains, but a torn ligament results in swelling and enough pain to inhibit movement. Like tendons, ligaments have few blood vessels and heal slowly. They are covered with a concentration of pain receptors that are very sensitive to stretching and swelling. Thus, the injury may not be as severe as the pain would suggest. As with most musculoskeletal swelling, first treatment may involve reducing the swelling with ice.

A common knee injury among athletes, such as gymnasts and those in sports such football or soccer, is a tear in the anterior cruciate ligament (ACL). Why is this ligament so vulnerable? When the knee is bent, the ACL acts as a restraining wire that restricts front-to-back twisting movement between the thighbone (femur) and the shinbone (tibia). The ACL is stretched when there is an external blow to the bent knee, as might occur during a tackle or when muscular force is applied to the knee by the athlete as might occur during landings in a tumbling run. If the force applied to the two bones is greater than the strength of the ligament, the ACL can tear.

In regions of the body where movements might cause friction between moving parts, such as might occur around synovial joints, the body has its own "ball bearings"—fluid-filled sacs called **bursae** (singular, *bursa*; meaning pouch or purse). Bursae surround and cushion certain joints. They resemble synovial sacs in that they are lined with synovial membranes. However, bursae are found in places where skin rubs over bone as the joint moves and between tendons and bones, muscles and bones, and ligaments and bones.

Repeated pressure on a bursa or injury to a nearby joint can cause the bursa to become inflamed and swell with excess fluid, a condition called bursitis (*itis* refers to an inflammation). Bursitis is characterized by intense pain that becomes worse when the joint is moved and cannot be relieved by resting in any position. Nonetheless, bursitis is not serious and usually subsides on its own within a week or two. In severe cases, a physician may drain some of the excess fluid to remove the pressure.

Flexion
Motion that *decreases* the angle
between the bones of the joint,
bringing the bones closer together

Extension
Motion that *increases* the
angle between the bones of
the joint

Adduction
Movement of a body part
toward the body midline

Abduction
Movement of a body part
away from the body midline

Rotation
Movement of a body part
around it own axis

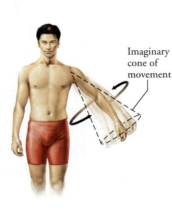

Imaginary
cone of
movement

Circumduction
Movement of a body part in a
wide circle so that the motion
describes a cone

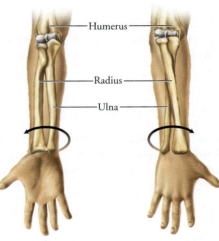

Humerus

Radius

Ulna

Supination
Rotation of the forearm
so that the palm faces up

Pronation
Rotation of the forearm
so that the palm faces
down

FIGURE **5.15**
Types of movement at synovial joints

ARTHRITIS

Arthritis always involves joint inflammation. Nevertheless, there are more than 100 kinds of arthritis, some far more serious than others. Osteoarthritis is a degeneration of the surfaces of a joint, caused by wear and tear. Over time, any joint surface that undergoes friction is bound to wear down. Finally, the slippery cartilage at the ends of the bones forming a joint begins to disintegrate until the bones themselves come into contact and grind against each other, causing intense pain and stiffness. Osteoarthritis is most likely to occur in the weight-bearing joints, such as the hip, knee, and spine, and occasionally in the finger joints or the wrist.

There is a new generation of drugs,[1] such as Celebrex® and Vioxx®, used to treat arthritis. They work the same way as the older medications, such as ibuprofen (Motrin® or Advil®) or naproxen (Aleve®). They all inhibit prostaglandins, hormone-like substances that trigger inflammation. Although the newer drugs are probably no more effective than the older ones, they are gentler to the digestive system.

Prevention is always better than treatment. One tip is to control your weight. This strategy will take a load off your knees and hips, reducing the risk of arthritis in those joints. Another tip is to exercise. Lifting weights will strengthen the muscles around your

---

[1]This new class of drugs is referred to as cox-2 inhibitors.

joints, which will help support them. Stretching will give you a greater range of movement.

Another, far more threatening form of arthritis (and, unfortunately, one of the most common) is rheumatoid arthritis (discussed further in Chapter 13). Rheumatoid arthritis is marked by an inflammation of the synovial membrane. As the synovial membrane becomes inflamed, excess synovial fluid accumulates in the joint, causing swelling, pain, and stiffness. Eventually, the constant irritation can lead to the destruction of the cartilage. The cartilage may then be replaced by fibrous connective tissue that further impedes the movement of the joint.

Rheumatoid arthritis differs from other types of arthritis in that it is apparently an autoimmune disease. That is, it is caused by an immune response of the body toward its own synovial membranes; the body attempts to reject its own tissue just as it would some invasive foreign matter. Rheumatoid arthritis can vary in its severity, but it is a permanent condition. It normally affects the joints of the fingers, wrist, knees, neck, ankles, and hips. Sometimes the only effective treatment is to replace the joint with an artificial one (Figure 5.16).

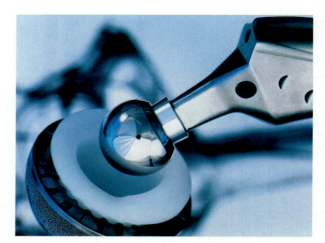

FIGURE **5.16**
Replacement prosthetic of the hip

REVIEWING THE CONCEPTS

Bones Function in Support, Movement, Protection, Storage of Minerals and Fat, and Blood Cell Production (p. 90)

1. The skeleton is a framework of bone and cartilage that supports and protects the internal organs and allows for movement. It also serves as a storage site for minerals (calcium and phosphorus) and fat. Blood cells are produced in the red marrow of certain bones.

Bones Have a Hard Outer Layer of Compact Bone Surrounding Spongy Bone (p. 90)

2. Compact bone is dense bone found on the outside of all bones. It is covered by a membrane, the periosteum, that nourishes the bone cells. Spongy bone is a latticework of bony struts found in flat bones and near the ends of long bones.

3. The shaft of a long bone has two knobby ends, or heads. The shaft of a long bone has a cavity filled with fatty yellow marrow. The spongy bone of certain bones in adults is filled with red marrow.

Bone is Living Tissue (pp. 90–93)

4. The structural unit of bone, called an osteon, consists of a central canal with bone cells (osteocytes) arranged around it in concentric circles. Cellular extensions of osteocytes touch one another through tiny canals, allowing for the exchange of materials between cells and the central canal. Bone matrix is hardened by calcium salts and strengthened by strands of collagen.

5. In an embryo, most of the skeleton first forms in cartilage. The cartilage is gradually replaced by bone. The process begins when osteoblasts form a bony collar around the shaft of the bone. Cartilage within the shaft then begins to break down and is replaced by bone. The cartilage in the heads (epiphyses) is then replaced by bone. Epiphyseal plates of cartilage remain and allow bones to grow in length until the person reaches ages 18 to 21.

6. Bone growth is stimulated by growth hormone from the anterior pituitary gland. Sex hormones (estrogen and testosterone) initially stimulate growth, but later they cause the epiphyseal plates to disappear, ceasing growth.

WEB TUTORIAL 5.1 Bone Growth

Bone Fractures Are Healed by Fibroblasts and Osteoblasts (p. 93)

7. Repair of a bone fracture involves the formation of a blood clot and then a cartilaginous callus that links the broken ends of the bone. The cartilage is gradually replaced by bone.

WEB TUTORIAL 5.2 Bone Repair

Bones Are Continuously Remodeled (pp. 93–94)

8. Bone is a living tissue that is constantly being remodeled. Osteoblasts deposit new bone, and osteoclasts break down bone. Stress is an important factor in bone remodeling because bone forms when it is stressed and is destroyed when it is not stressed.

The Human Skeleton Has Two Parts (pp. 95–99)

9. The skeleton can be divided into the axial skeleton and the appendicular skeleton.

10. The axial skeleton protects and supports internal organs.

11. The appendicular skeleton allows you to change position and location.

Joints Are Junctures between Bones (pp. 99–102)

12. Joints, the places where bones meet, can be classified according to the degree of movement they permit. Some joints, such as the sutures between the skull bones, allow no movement.

13. Synovial joints are freely movable. Synovial joints have cartilage on the adjoining bone surfaces. They are surrounded by a synovial cavity filled with synovial fluid and are held together by ligaments.

14. Arthritis is inflammation of a joint. Osteoarthritis occurs when the surface of a joint degenerates because of use. Rheumatoid arthritis is an autoimmune disease.

KEY TERMS

appendicular skeleton *p. 95*	intervertebral disk *p. 95*	osteoporosis *p. 93*	sprain *p. 100*
arthritis *p. 101*	joint *p. 99*	pectoral girdle *p. 98*	synovial joint *p. 100*
axial skeleton *p. 95*	ligament *p. 100*	pelvic girdle *p. 98*	tendon *p. 98*
ball-and-socket joint *p. 100*	osteoblast *p. 91*	periosteum *p. 90*	yellow marrow *p. 90*
bursa *p. 100*	osteoclast *p. 93*	red marrow *p. 90*	
compact bone *p. 90*	osteocyte *p. 90*	skeleton *p. 90*	
hinge joint *p. 100*	osteon *p. 90*	spongy bone *p. 90*	

THINKING ABOUT THE CONCEPTS

1. List six functions of the skeleton. (*p. 90*)
2. Compare compact and spongy bone. (*p. 90*)
3. Describe the structure of a long bone. Where are the yellow and red marrow found? (*p. 90*)
4. Diagram an osteon, including the osteocytes and central canal. (*p. 90–91*))
5. Describe the formation of bone in a fetus. Explain how bone growth continues after birth. (*pp. 91–92*)
6. Explain how a bone heals after it has been fractured. (*p. 93*)
7. What is bone remodeling? Explain the role of osteoblasts and osteoclasts. How does stress affect remodeling? (*p. 93*)
8. Describe the axial and appendicular parts of the skeleton. (*pp. 95–99*)
9. Describe a synovial joint. (*pp. 99–100*)
10. The functional units of compact bone are called
    a. myofilaments.
    c. striations.
    b. osteons.
    d. osteocytes.
11. Which of the following persons would be expected to have the *densest* bones?
    a. an elderly woman whose favorite pastime is knitting
    b. a 16-year-old female who swims regularly
    c. a 33-year-old male African American who loves to drink milk and plays basketball regularly
    d. a 50-year-old-male workaholic with a desk job
12. The skull, vertebral column, sternum, and rib cage constitute the _____ skeleton.
13. Bone is hardened by the mineral _____.
14. The cells that form bone are called _____.
15. A _____ is a tear in a ligament that may cause pain after the injury.

APPLYING THE CONCEPTS

1. Edith is a 64-year-old woman who has been confined to a wheelchair since she was 35 because she was in a car accident and her legs are paralyzed. Her leg bones are very thin and weak. Why?
2. Makiri is a 19-year-old woman who is pregnant with her first child. Her diet consists primarily of junk food. During the second month of her pregnancy, she trips and breaks her arm in the fall. Tests show that her bones have become weak owing to the loss of calcium. What has caused Makiri's bone weakness?

6

The Muscular System

The Muscular System Produces Movement and Maintains Posture

Most Skeletal Muscles Work in Pairs

Sarcomeres Are the Contractile Units of Muscle
- Skeletal muscle contracts when actin filaments slide across myosin filaments
- Regulatory proteins and calcium ions control contraction
- Nerves stimulate muscle contraction

The Strength of Muscle Contraction Depends on the Number of Motor Units Involved

The Strength of Contraction Increases If a Muscle Is Stimulated Before It Has Relaxed

ATP for Muscle Contraction Comes from Many Sources

Slow-Twitch and Fast-Twitch Muscle Cells Differ in Contraction Speed and Duration

Aerobic Exercise Increases Endurance; Resistance Exercise Builds Muscle

 **SOCIAL ISSUE** Building Muscle, Fair and Square? Anabolic Steroid Abuse

Enjoying his walk on the mountain trail, Michael felt exhilarated by the fresh breeze, the scent of pine needles, and the warmth of the sun. He had been walking for about an hour, his mind finally freed of worries: balancing the demands of his job, 16 credit hours of courses at the community college, mid-terms, and the stress of a recent breakup with Heather. He hadn't had a decent night's sleep for what seemed like weeks. This walk, he thought to himself, was the first real exercise he had had since the Heather debacle. The only problem at the moment was that he was getting increasingly thirsty, because he had forgotten his water bottle. Better head for home. No sooner had he changed direction then he felt a sudden pain in his left calf. He couldn't walk. It felt as if the whole muscle had locked up in a severe cramp, a painful, involuntary muscle contraction. He knew that he had to stretch the muscles in his leg gradually to work out the cramp. What he didn't realize was that the cramp was caused by common factors: fatigue (from insomnia), lack of regular exercise, and dehydration. Fortunately, in a matter of minutes Michael had relieved the pain by stretching the calf muscles, and after a brief rest was able to walk home.

In this chapter, we will see why muscle contraction rarely results in cramps. We will consider muscle structure and the mechanism of contraction. Then, we will see how nerves control muscle contraction and energy sources that fuel contraction. ■

The Muscular System Produces Movement and Maintains Posture

There are three kinds of muscles—skeletal, cardiac, and smooth. They have distinct qualities and functions, but they all have four traits in common.

1. **Muscles are excitable.** They respond to stimuli.
2. **Muscles are contractile.** They have the ability to shorten.
3. **Muscles are extensible.** They have the ability to stretch.
4. **Muscles are elastic.** They can return to their original length after being shortened or lengthened.

In this chapter we will concentrate on skeletal muscles, the ones we usually think of when we hear the word *muscles*. Smooth and cardiac muscles are discussed in Chapters 4 and 12, respectively. Unlike cardiac and smooth muscle, skeletal muscle is under voluntary control. We can contract it when we want to.

Most Skeletal Muscles Work in Pairs

The body has more than 600 skeletal muscles, some of which are shown in Figure 6.1. Muscles allow you to smile with pleasure and scowl in anger. They allow you to move, some of us more gracefully than others. And they allow you to maintain posture, keeping you erect despite the pull of gravity.

Most major muscles involved in the general movement of the body are attached to bones. Indeed, each end of the muscle is attached to a bone by a **tendon**, which is a band of connective tissue. One end of the muscle, the **origin**, is attached to the bone that remains relatively stationary during a movement. The other end, the **insertion**, is attached to the bone that moves.

Most muscles are arranged in pairs or groups that cooperate in movement. Muscles that work together to cause movement in the same direction are called *synergistic muscles*. Most muscles, however, are arranged in pairs so that the actions of the members of the pair are opposite to one another; that is, muscles are usually arranged in **antagonistic pairs** (Figure 6.2). As any muscle contracts and pulls on a bone, a muscle with the opposite action must relax. An example of such cooperation is provided by the biceps, located on the top of the upper arm (the muscles people like to show off), and the triceps on the back of the arm (the ones involved in push-ups). Whereas the contraction of the biceps causes the arm to flex and bend at the elbow, the contraction of the triceps causes the arm to extend and straighten at the elbow.

Contraction of a muscle pulls on the tendon that connects that muscle to a bone. Excessive stress on a tendon can cause it to become inflamed, a condition called *tendinitis*. In some cases, calcium deposits form and cause pain in the joint.

Unfortunately, the causes of tendinitis are not well understood. It is probably largely due to overuse, misuse (as when lifting improperly), and age. Sedentary people who move relatively little are likely to overuse a joint that is not accustomed to use. People who often use the muscles and joints suffer less from overuse. Because tendons are poorly supplied with blood vessels, they heal slowly. One of the most effective treatments is rest. The general rule is, if it hurts, do not use it.

Sarcomeres Are the Contractile Units of Muscle

Now we'll look at the brawny bulk that gives shape to the body, gradually delving deep into the fine structure forming the mechanism responsible for muscle contraction. Skeletal muscle cells can be enormously long as cells go, up to several centimeters. Indeed, muscle cells in a thigh muscle may be 30 cm (1 ft.) in length.

When viewed under the microscope, skeletal muscle cells have pronounced bands that look like stripes, so this type of muscle is also called **striated** (striped) **muscle** (Figure 6.3). The striations are formed from the orderly arrangement of many elongated **myofibrils** within the muscle cell. Each myofibril contains groups of long **myofilaments**, which are composed of two protein filaments: the thicker **myosin filaments** and a greater number of the thinner **actin filaments**. Myofilaments make up about 80% of the cell volume.

Each muscle cell is divided into contractile units called **sarcomeres**. The ends of each sarcomere are marked by dark bands of protein called Z lines. One end of each actin filament is attached to a Z line. The actin and myosin filaments within a sarcomere partially overlap. When the muscle contracts, the degree of overlap increases.

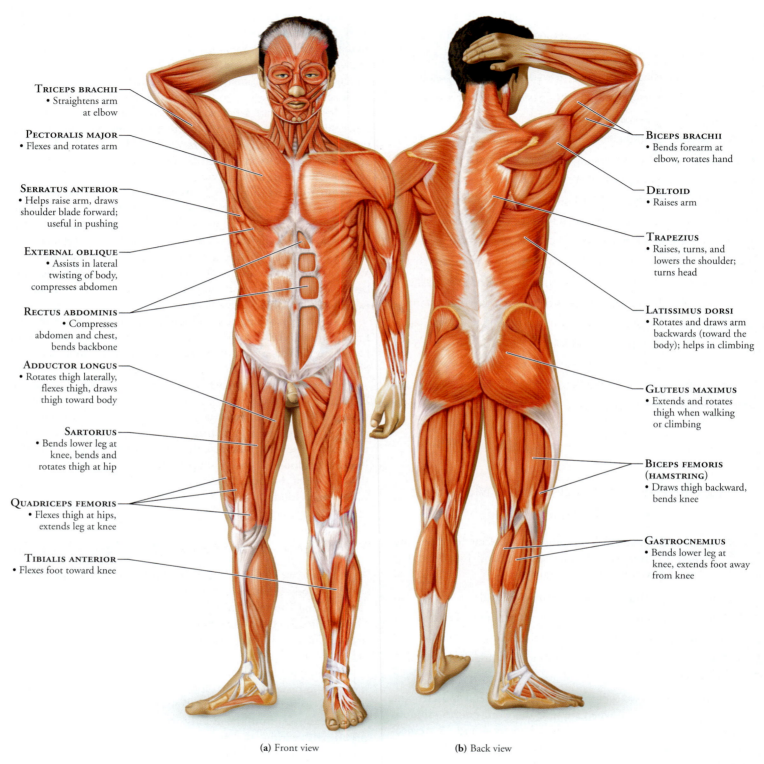

TRICEPS BRACHII
• Straightens arm
  at elbow

PECTORALIS MAJOR
• Flexes and rotates arm

SERRATUS ANTERIOR
• Helps raise arm, draws
  shoulder blade forward;
  useful in pushing

EXTERNAL OBLIQUE
• Assists in lateral
  twisting of body,
  compresses abdomen

RECTUS ABDOMINIS
• Compresses
  abdomen and chest,
  bends backbone

ADDUCTOR LONGUS
• Rotates thigh laterally,
  flexes thigh, draws
  thigh toward body

SARTORIUS
• Bends lower leg at
  knee, bends and
  rotates thigh at hip

QUADRICEPS FEMORIS
• Flexes thigh at hips,
  extends leg at knee

TIBIALIS ANTERIOR
• Flexes foot toward knee

BICEPS BRACHII
• Bends forearm at
  elbow, rotates hand

DELTOID
• Raises arm

TRAPEZIUS
• Raises, turns, and
  lowers the shoulder;
  turns head

LATISSIMUS DORSI
• Rotates and draws arm
  backwards (toward the
  body); helps in climbing

GLUTEUS MAXIMUS
• Extends and rotates
  thigh when walking
  or climbing

BICEPS FEMORIS
(HAMSTRING)
• Draws thigh backward,
  bends knee

GASTROCNEMIUS
• Bends lower leg at
  knee, extends foot away
  from knee

(a) Front view

(b) Back view

FIGURE **6.1**
Some major muscles of the body

Skeletal muscle contracts when actin filaments slide across myosin filaments

According to the sliding filament model of muscle contraction, a muscle contracts when actin filaments slide along the myosin filaments. This movement increases the degree of overlap between actin and myosin filaments and shortens the sarcomere. When many sarcomeres shorten, the entire muscle contracts (Figure 6.4).

The question arises, then, how do the actin filaments move? To answer this question, we must consider the structure of the thin and thick myofilaments in more detail. Spherical actin molecules

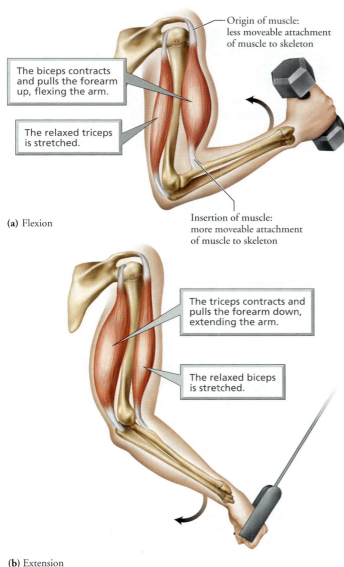

The biceps contracts and pulls the forearm up, flexing the arm.

Origin of muscle: less moveable attachment of muscle to skeleton

The relaxed triceps is stretched.

Insertion of muscle: more moveable attachment of muscle to skeleton

(a) Flexion

The triceps contracts and pulls the forearm down, extending the arm.

The relaxed biceps is stretched.

(b) Extension

FIGURE **6.2**

The antagonistic action of the triceps and biceps muscles during flexion and extension, showing origins and insertions of the muscles

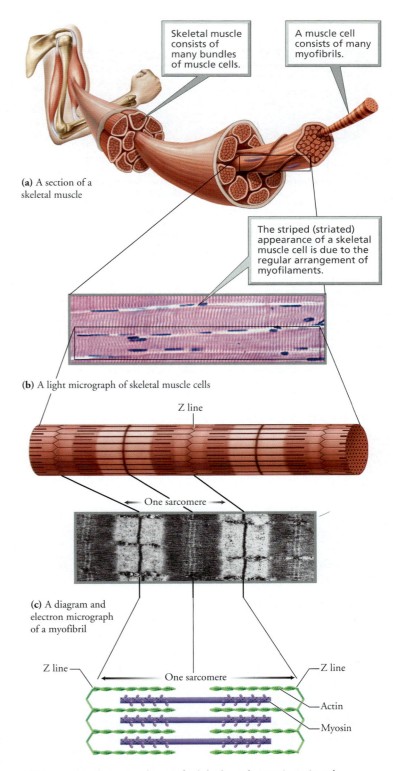

Skeletal muscle consists of many bundles of muscle cells.

A muscle cell consists of many myofibrils.

(a) A section of a skeletal muscle

The striped (striated) appearance of a skeletal muscle cell is due to the regular arrangement of myofilaments.

(b) A light micrograph of skeletal muscle cells

Z line

One sarcomere

(c) A diagram and electron micrograph of a myofibril

Z line

One sarcomere

Z line

Actin

Myosin

(d) A sarcomere, the contractile unit of a skeletal muscle, contains actin and myosin myofilaments.

 FIGURE **6.3**

The structure of a skeletal muscle

make up almost all of a thin myofilament. Many actin molecules join to form a chain that resembles a string of beads. A thin myofilament consists of two strands of actin twisted together to form a helix. The thick filaments are composed of the protein myosin. Each myosin molecule is shaped like a golf club but with two heads side by side on the same shaft. In a myosin filament, the "shafts" of several hundred myosin molecules adhere to one another and the "heads" protrude from each end of the bundle in a spiral pattern.

The club-shaped ends of the myosin molecule, the so-called *myosin heads*, are essential to the movement of actin filaments and, therefore, to muscle contraction. Myosin has several features important to muscle contraction. First, a myosin head is attached to the shaft with a hingelike connection that allows the head to pivot back and forth relative to the shaft. Second, myosin heads can bind both actin and the energy-laden molecule ATP. Third, myosin heads have an enzyme that can split ATP and release its stored energy to power contraction.

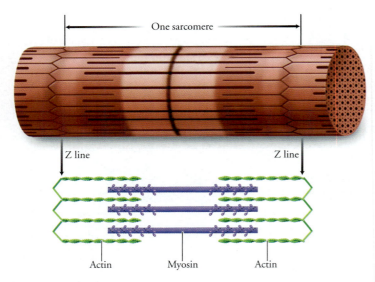

One sarcomere

Z line   Z line

Actin   Myosin   Actin

(a) Sarcomere relaxed

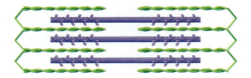

(b) Sarcomere contracted

FIGURE **6.4**

Each myofibril is packed with actin filaments and myosin filaments. When a muscle contracts, the actin filaments slide past myosin filaments. Movements of the heads of myosin filaments pull the actin filaments inward, toward the center of the sarcomere.

Muscle contraction involves cyclic interactions between myosin and actin (Figure 6.5).

- **Resting sarcomere.** At the start of each cycle of contraction, the myosin heads have already split a molecule of ATP to ADP and inorganic phosphate P_i. The energy released from splitting ATP is stored and activates the myosin heads. Activation causes the myosin heads to swivel in a way that straightens them out toward the borders of the sarcomere. This step is analogous to cocking a pistol.

- **Cross-bridge attachment.** The myosin heads attach themselves to the nearest actin filament. When the myosin head is bound to an actin molecule, it forms a bridge between the thick and thin filaments. For this reason, myosin heads are also called **cross-bridges**.

- **Pivoting of myosin head.** Binding to actin causes the energy stored in the activated myosin heads to be released, which, causes the heads to bend forcefully back to their original bent positions. Because actin filaments are bound to the myosin heads, actin filaments are pulled toward the midline of the sarcomere. This so-called power stroke is analogous to pulling the trigger on a pistol. Then the ADP and inorganic phosphate pop off the myosin head.

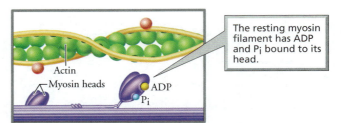

Actin
Myosin heads   ADP
                P_i

Resting sarcomere

The resting myosin filament has ADP and P_i bound to its head.

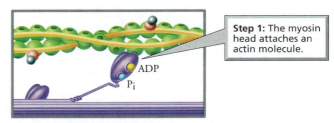

ADP
P_i

Cross-bridge attachment

Step 1: The myosin head attaches an actin molecule.

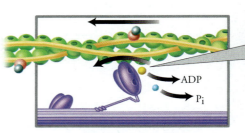

ADP
P_i

Pivoting of myosin head

Step 2: The myosin head bends, causing the actin filament to slide across the myosin filament. The ADP and P_i are released from the myosin head.

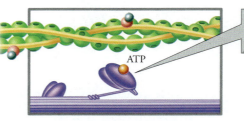

ATP

Cross-bridge detachment

Step 3: Free ATP binds to myosin, causing it to release actin.

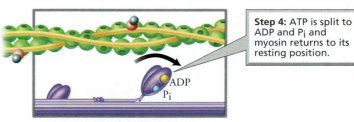

ADP
P_i

Myosin reactivation

Step 4: ATP is split to ADP and P_i and myosin returns to its resting position.

FIGURE **6.5**

The sliding filament model of muscle contraction

TUTORIAL 6.2

- **Cross-bridge detachment.** New ATP molecules then bind to the myosin heads, causing the myosin heads to disengage from the actin.
- **Myosin reactivation.** Myosin heads split the ATP and store the energy. Then the contraction cycle begins again. This cycle of events is repeated hundreds of times each second.

stop and think

It takes an ATP molecule to break the cross-bridges so that a new one may be formed. Thus, without ATP, cross-bridges cannot be broken and the muscle becomes stiff. When a person dies, ATP is no longer formed. How does this fact explain the stiffening of muscles, known as rigor mortis, that begins about 3 to 4 hours after death?

▌ Regulatory proteins and calcium ions control contraction

The proteins that regulate muscle contraction are **troponin**, and **tropomyosin** and together they form the troponin-tropomyosin complex. During muscle *relaxation*, the actin-myosin binding sites—where the myosin heads attach to the actin filaments—are covered by the troponin-tropomyosin complex. Contraction occurs when calcium ions enter the sarcomere and bind to troponin, causing it to change shape. This change causes tropomyosin to shift position, thereby exposing the myosin binding sites on actin. Actin is then able to form new cross-bridges with myosin (Figure 6.6).

Where do calcium ions come from? Calcium ions are stored in the **sarcoplasmic reticulum**, an elaborate form of smooth endoplasmic reticulum found in muscle cells. Like the sleeve of a loose lace shirt, the sarcoplasmic reticulum surrounds each myofibril in a muscle cell. (Recall that a myofibril is a bundle of actin and myosin.) Membrane pumps powered by the energy of ATP move calcium ions from the cytoplasm into the sarcoplasmic reticulum for storage. The pumps keep the calcium isolated there until it is needed for contraction. Then it is released. Also scattered through the cell are a number of **transverse tubules** (T tubules), which are tiny, cylindrical inpocketings of the muscle cell's plasma membrane. The T-tubules carry signals from motor neurons deep into the muscle cell to virtually every sarcomere.

▌ Nerves stimulate muscle contraction

So, what controls the availability of calcium ions? The stimulus that ultimately leads to the release of calcium ions, and therefore to muscle contraction, is a nerve impulse from a motor neuron (discussed in Chapter 7). A motor neuron branches when it is close to the muscle. The junction between the tip of each motor neuron branch and a skeletal muscle cell tip of each branch is called a **neuromuscular junction** (Figure 6.7). When a nerve impulse reaches a neuromuscular junction, it causes the release of the chemical acetylcholine from small packets in the tip of the motor neuron. Acetylcholine then diffuses across a small gap onto the surface of the muscle cell, where it binds to special receptors on the muscle cell membrane. Acetylcholine

causes changes in the permeability of the muscle cell membrane, creating an electrochemical message similar to a nerve impulse. This message travels along the plasma membrane, into the T-tubules and then to the sarcoplasmic reticulum, where calcium ions are stored. The message causes channels in the sarcoplasmic reticulum to open, releasing calcium ions. The calcium ions then combine with troponin, the myosin binding sites on actin are exposed, and the muscle contracts. All this happens every time you absent-mindedly scratch your head, and it happens a lot faster than it takes to describe it.

When the impulse stops, the events are reversed. Membrane pumps quickly clear the sarcomere of calcium ions and the troponin-tropomyosin complexes block the binding sites, resulting in the muscle relaxing. The contraction of other muscles stretches the sarcomere to its original length.

The Strength of Muscle Contraction Depends on the Number of Motor Units Involved

Skeletal muscles are stimulated to contract by motor neurons (nerve cells). A motor neuron that brings an impulse from the brain telling a muscle cell to contract does, in fact, stimulate a number of muscle cells in that area. A motor neuron and all the muscle cells it stimulates is called a **motor unit** (Figure 6.8). All the

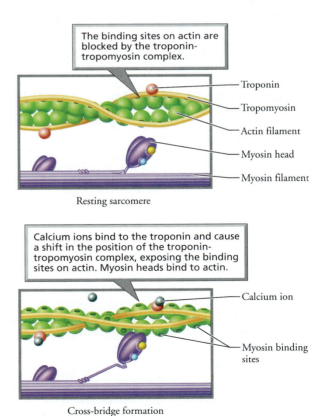

The binding sites on actin are blocked by the troponin-tropomyosin complex.

Troponin
Tropomyosin
Actin filament
Myosin head
Myosin filament

Resting sarcomere

Calcium ions bind to the troponin and cause a shift in the position of the troponin-tropomyosin complex, exposing the binding sites on actin. Myosin heads bind to actin.

Calcium ion
Myosin binding sites

Cross-bridge formation

FIGURE **6.6**
The availability of calcium ions controls muscle contraction.

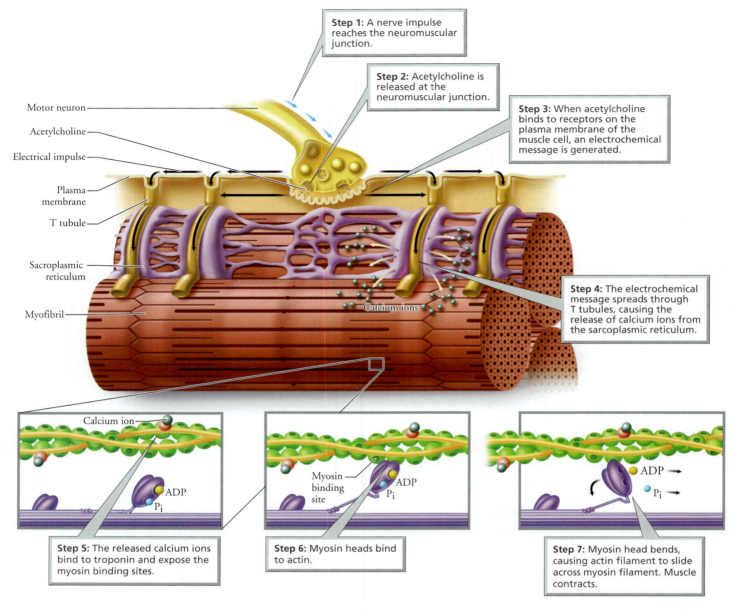

Step 1: A nerve impulse reaches the neuromuscular junction.

Step 2: Acetylcholine is released at the neuromuscular junction.

Step 3: When acetylcholine binds to receptors on the plasma membrane of the muscle cell, an electrochemical message is generated.

Motor neuron

Acetylcholine

Electrical impulse

Plasma membrane

T tubule

Sacroplasmic reticulum

Myofibril

Calcium ions

Step 4: The electrochemical message spreads through T tubules, causing the release of calcium ions from the sarcoplasmic reticulum.

Calcium ion

ADP

P_i

Step 5: The released calcium ions bind to troponin and expose the myosin binding sites.

Myosin binding site

ADP

P_i

Step 6: Myosin heads bind to actin.

ADP

P_i

Step 7: Myosin head bends, causing actin filament to slide across myosin filament. Muscle contracts.

FIGURE **6.7**

The connection between a motor neuron and a muscle cell is called a neuromuscular junction.

muscle cells in a given motor unit contract together. On average, there are 150 muscle cells in a motor unit, but this number is quite variable. Muscles responsible for finely controlled movements, such as those of the fingers or eyes, have a small number of muscle cells in a motor unit. In contrast, muscles with less precise movements, such as the hip or calf, have many muscle cells in a motor unit. Whereas a motor unit in tiny eye muscles may have only three muscle cells, a motor unit in a calf muscle may have thousands of muscle cells.

The nervous system increases the strength of muscle contraction by increasing the number of motor units involved, a

process called *recruitment*. The muscle cells of a given motor unit are generally spread throughout the muscle. Thus, if a single motor unit is stimulated, the entire muscle contracts weakly. Although the same muscles are involved in lifting a table and lifting a fork, the number of motor units summoned in those muscles is greater when lifting the table.

Our movements are generally smooth and graceful, rather than jerky, because the nervous system carefully choreographs the involvement of motor units. Movement is smooth because not all motor units are active simultaneously. In addition, different motor units are active for different amounts of time.

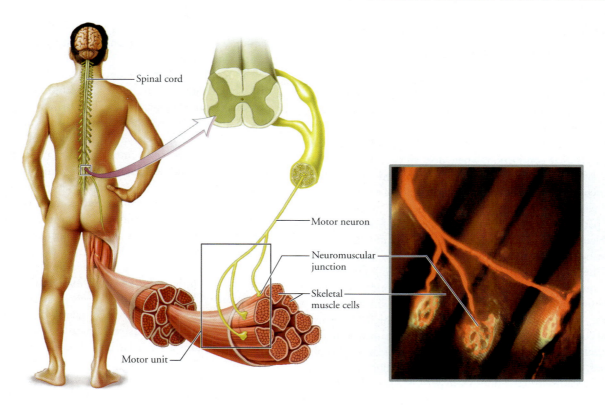

FIGURE **6.8**

A motor unit includes a motor neuron and the muscle cells it stimulates.

The Strength of Contraction Increases If a Muscle Is Stimulated Before It Has Relaxed

If a neuron is artificially stimulated briefly in the laboratory, all the muscle cells it innervates will contract, causing a **muscle twitch** (Figure 6.9a). The interval between the reception of the stimulus and the time when contraction begins is called the latent period. The contraction phase is quite rapid and is followed by a slower relaxation phase as the muscle returns to its resting state.

If a second stimulus is given before the muscle is fully relaxed, the second twitch will be stronger than the first. Because the second contraction is added to the first, the phenomenon is described as **summation** (Figure 6.9b). Summation occurs because an increasing number of muscle cells are stimulated to contract.

When stimuli arrive even more frequently, the muscle twitches are added one onto the next, and the contraction becomes increasingly stronger. If the stimuli occur so frequently that there is no time for relaxation before the next stimulus arrives, the muscle goes into a constant, powerful contraction called **tetanus** (Figure 6.9c). Tetanus cannot continue indefinitely. Eventually, the muscle will become unable to produce sufficient ATP to fuel contraction, and lactic acid will accumulate. As a result, the muscle will become unable to contract despite continued stimulation, a condition called *fatigue*.

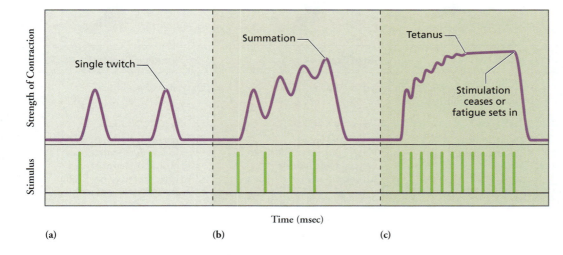

FIGURE **6.9**

Muscle contraction shown graphically: (a) muscle twitch, (b) summation, (c) tetanus

ATP for Muscle Contraction Comes From Many Sources

When a single muscle cell is contracting, it can require as much as 600 trillion ATP molecules per second simply to form and break the cross-bridges responsible for contraction. Even small muscles contain thousands of muscle cells.

A fundamental question presents itself: If ATP is the only source of energy for muscle contraction, where does all the ATP come from? There are a number of sources typically employed in a particular sequence, depending on the duration and intensity of exercise: (1) ATP in muscle cells, (2) creatine phosphate, (3) anaerobic (without oxygen) metabolic pathways (fermentation), and (4) aerobic (with oxygen) respiration (Figure 6.10). (Cell respiration is discussed in Chapter 2.)

A resting muscle stores some ATP, but this reserve is used quickly. When you start to exercise vigorously, the ATP reserves in the active muscles are depleted within about 6 seconds. You are able to exercise for another 30 seconds or so because when the muscles were resting, energy was transferred to another high-energy compound, called creatine phosphate, that is stored in muscle tissue. A resting muscle contains about six times as much creatine phosphate than stored ATP. Creatine phosphate has a high-energy bond between the creatine and the phosphate parts of the molecule. When needed, creatine phosphate releases its stored energy to convert ADP to ATP, which, in turn, can be used to power muscle contraction.

Activities such as diving, weight lifting, and sprinting, which require a short burst of intense activity, are powered by ATP and creatine phosphate reserves.

Once the supply of creatine phosphate is diminished, ATP must be generated from either anaerobic metabolic pathways or aerobic respiratory pathways. The primary fuel for either pathway is glucose. The glucose that fuels muscle contraction comes mainly from glycogen, a large chain of glucose molecules. Indeed, a muscle cell contains quite a bit of glycogen, about 1.5% of its total weight. Muscles depend on these glycogen stores to fuel contraction. Endurance sports may deplete glycogen reserves. The accompanying feeling of overwhelming fatigue is known to runners as "hitting the wall," to cyclists as "bonking," and to boxers as becoming "arm weary."

When an active muscle cell runs short of ATP and creatine phosphate, enzymes convert glycogen to glucose. The circulatory system can supply enough oxygen for aerobic respiratory pathways to produce ATP to power *low* levels of activity, even if continued for a prolonged time. However, anaerobic pathways produce ATP two and one-half times faster than do aerobic pathways. Therefore, during strenuous activity lasting 30 or 40 seconds, anaerobic pathways supply the ATP to fuel muscle contraction. Indeed, we find that burstlike activities such as tennis or soccer rely nearly completely on anaerobic pathways of ATP production. In anaerobic respiratory pathways, the pyruvic acid produced in glycolysis is converted to lactic acid.

$$\text{Creatine phosphate} \xrightarrow{\quad\text{ADP}\quad\text{ATP}\quad} \text{Creatine}$$

$$\text{Glycogen} \rightarrow \text{Glucose} \xrightarrow[\text{No oxygen required}]{\text{ADP} \smile \text{ATP}} \text{Pyruvic acid} \rightarrow \text{Lactic acid}$$

6 seconds	25 seconds	10 minutes	End of exercise	After prolonged exercise
ATP stored in muscles	ATP formed from creatine phosphate and ADP	ATP generated from glycogen stored in muscles and broken down to form glucose		Oxygen debt paid back
		Oxygen limited • Glucose oxidized to lactic acid	Oxygen present • Heart beats faster to deliver oxygen more quickly • Myoglobin releases oxygen	Breathe heavily to deliver oxygen • Lactic acid used to produce ATP • Creatine phosphate restored • Oxygen restored to myoglobin • Glycogen reserves restored

FIGURE **6.10**
Energy sources for muscle contraction

During more prolonged muscular activity, the body gradually switches back to aerobic pathways for producing ATP. Aerobic pathways require oxygen, which can come from either of two sources. One oxygen source is the oxygen bound to hemoglobin in the blood supply. As activity continues, the heart rate increases, and blood is pumped more quickly. At the same time, blood is shunted to the neediest tissues. Another oxygen source is myoglobin, an oxygen-binding pigment within muscle cells. Aerobic pathways produce more than 90% of the ATP required for intense activity lasting more than 10 minutes. Such pathways also produce nearly 100% of the ATP that powers a truly prolonged intense activity, such as running a marathon.

$$\text{Glycogen} \rightarrow \text{Glucose} \xrightarrow[\text{Oxygen required}]{\text{ADP} \smile \text{ATP}} \text{Carbon dioxide} + \text{Water}$$

After prolonged exercise, a person continues to breathe heavily for several minutes. The extra oxygen relieves the **oxygen debt** that was caused because the muscles used more ATP than was provided by aerobic metabolism. Lactic acid is converted back to pyruvic acid and then oxidized to carbon dioxide and water through aerobic respiratory pathways. In addition, the oxygen that was released by myoglobin is replaced, and glycogen and creatine phosphate reserves are restored.

stop and think

Some athletes take dietary creatine supplements to improve their performance. Creatine does seem to boost performance in sports that require short bursts of energy but not in those that require endurance. Explain why this difference might occur. Creatine does not increase muscle mass, yet it can enhance performance in sprint sports. How is this effect possible?

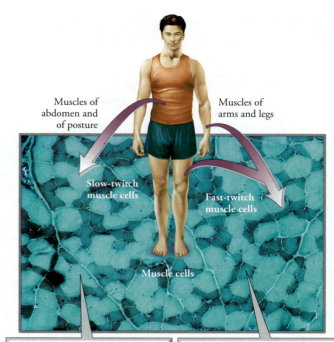

Muscles of abdomen and of posture

Muscles of arms and legs

Slow-twitch muscle cells

Fast-twitch muscle cells

Muscle cells

Slow-twitch muscle cells:	Fast-twitch muscle cells:

• Designed for endurance
• Contract slowly
• Strong, sustained contractions
• Steady supply of energy
  – Many mitochondria (structures for aerobic production of ATP)
  – Many capillaries
  – Packed with the oxygen-binding pigment myoglobin
	• Designed for rapid, powerful response
	• Contract rapidly
	• Short, powerful contraction because there is more actin and myosin than in slow-twitch cell
	• Depend more heavily on anaerobic metabolic pathways to generate ATP, so fatigue rapidly

FIGURE **6.11**
Slow- and fast-twitch muscle cells

Slow-Twitch and Fast-Twitch Muscle Cells Differ in Contraction Speed and Duration

There are two general types of muscle cells: slow-twitch and fast-twitch (Figure 6.11). **Slow-twitch cells** contract slowly when stimulated but with enormous endurance. These cells are dark and reddish because they are packed with the oxygen-binding pigment myoglobin and because they are richly supplied with capillaries. Slow-twitch cells also contain abundant mitochondria, the cellular structures essential for aerobic production of ATP. As a result, slow-twitch cells are specialized to deliver prolonged, strong contractions.

In contrast, **fast-twitch cells** contract rapidly and powerfully, with far less endurance. Fast-twitch cells have a form of an enzyme that can split ATP bound to myosin more quickly than the same enzyme in slow-twitch cells. Because they can make and break cross-bridges more quickly, they can contract more rapidly than slow-twitch cells. In addition, compared with their slow-twitch cousins, fast-twitch cells have a wider diameter because they are packed with more actin and myosin. This feature allows them to contract powerfully. However, fast-twitch cells, rich in glycogen deposits, depend more heavily on anaerobic means of producing ATP. As a result, fast-twitch cells tire more quickly than slow-twitch cells.

The two kinds of cells are distributed unequally throughout any human body. The abdominal muscles do not need to contract rapidly, but they need to be able to contract steadily to hold our paunch in at the beach and to balance the powerful, slow-twitch back muscles, enabling us to stand upright. Fast-twitch cells, on the other hand, are more common in the legs and arms, structures that must sometimes move quickly.

Human athletes show some of these biochemical cellular differences as well. Whereas the muscles of endurance athletes, such as marathoners, are made up of about 80% slow-twitch cells, those of sprinters are about 60% fast-twitch cells. By the way, if you are a "fast-twitch person," you can build a certain level of endurance, but you will be limited to a degree because such differences are genetic. Endurance runners, on the other hand, dread those times when, at the end of a long race, some well-trained competitor, loaded with fast-twitch cells, sprints past them at the finish line.

Aerobic Exercise Increases Endurance; Resistance Exercise Builds Muscle

In adolescents and adults, exercise can have great influence on the further development of muscle. Different kinds of exercise, though, can produce different results. For example, *aerobic exercise*, such as walking, jogging, or swimming (exercises in which

SOCIAL ISSUE

Building Muscle, Fair and Square? Anabolic Steroid Abuse

We have always appreciated strength, perhaps because it's associated with a "can-do" ability. Of course, such appreciation has led us not only to strength itself but also to the display of strength, that is, bodybuilding. Unfortunately, some people are always looking for shortcuts.

One of the biggest dangers to young people in sports is the use of anabolic (building) steroids. Anabolic steroids are synthetic hormones that mimic the male sex hormone testosterone. Anabolic steroids, taken in large amounts, stimulate the body to build muscle and to increase strength dramatically (Figure 6.A). They accomplish these improvements by stimulating protein formation in muscle cells and by reducing the amount of rest needed between workouts.

Although anabolic steroids are available only by prescription, about 6.5% of adoles-

cent boys and 1.9% of adolescent girls obtain and regularly use the drugs without a prescription. Commonly called 'roids, juice, or slop, steroids are swallowed as a pill or injected. Steroid use in the United States is a $500 million business and growing. Because most steroids are obtained illegally through the black market or from foreign countries, the purity and quality of the drugs are questionable at best.

We might ask, then, why steroid use is so attractive. One reason seems to be that steroids provide an easy route to the body image so prized by society. Nearly one-third of high school males who use steroids admit that they use the drugs to acquire a muscular, well-built look. Steroid use seems to be particularly widespread among high school senior boys. Perhaps they are unsure of themselves as they prepare to leave the nest and enter the "real world."

Among athletes, however, the reason for steroid use is clear: It gives them a distinct competitive edge, and everyone loves a winner. Almost all international athletic committees prohibit steroid use. Indeed, more than a few medals have had to be returned when the cheating was discovered. *Cheating* is the right word because steroids give the user an unfair advantage.

Anabolic steroids have more than 70 possible unwanted side effects that can range in severity from acne to liver cancer. Among the risks are injuries that result from one intended effect of the drug—increased muscle strength—without an accompanying increase in the strength of tendons and ligaments. Such injuries may take a long time to heal. But most commonly and most seriously affected by steroid abuse are the liver, the cardiovascular system, and the reproductive system. Effects on the cardiovascular system,

---

enough oxygen is delivered to the muscles to keep them going for long periods), will foster the development of new capillaries that service the muscles and more mitochondria to facilitate energy usage. Aerobic exercise also increases muscle coordination, improves digestive tract movement, and increases the strength of the skeleton by placing force on the bones. Cardiovascular and respiratory system improvements help muscles to function more efficiently, and the heart is enlarged so that each stroke pumps more blood.

Aerobic exercises, however, generally do not increase muscle size. Muscular development, the kind that helps you look good on the beach, comes mostly from *resistance exercise*, such as you get from lifting heavy weights. To build muscle mass, one must force muscles to exert more than 75% of their maximum force. These exercises can be very brief because only three sets of six contractions each are needed to build bulk and increase strength. The added size apparently comes from an increase in the diameter of existing muscle cells. Some researchers, however, suggest that heavy exercise splits or tears the muscle cells and that each of these parts then regrows to a larger size.

To build and firm muscles, people head to the gyms to use exercise equipment, such as Nautilus®, or to lift free weights (Figure 6.12). Exercise machines automatically adjust the resist-

FIGURE **6.12**

Muscles get larger when they are repeatedly forced to exert more than 75% of their maximum force.

including heart attacks and strokes, may not show up for years. However, the effects on the reproductive system are more immediate. The testicles of male steroid users often become smaller. Males also may become sterile and impotent (unable to achieve an erection). These effects on the reproductive system occur because the anabolic steroids inhibit natural testosterone production. Female steroid abusers develop irreversible masculine traits, such as a deeper voice, growth of body hair, loss of scalp hair, smaller breasts, and an enlarged clitoris (during embryological development, the clitoris develops from the same structure that develops into the penis in a male).

Steroid use can have psychological effects too. For instance, it may promote aggression and a feeling of invincibility, often referred to as "'roid rage." Users may also develop severe depression. And steroids are addictive.

Some athletes avoid the legal issue of steroid use by taking a legal drug called androstenedione that is available without prescription. The liver converts this drug to testosterone, a muscle-building steroid. Although there are no laws prohibiting the use of androstenedione, the National Football League, the National Collegiate Athletic Association, and the International Olympic Committee have banned its use among their athletes. The long-term effects are largely unknown, but androstenedione increases blood cholesterol levels, thereby increasing the risk of heart disease. It is also thought to increase one's risk of prostate cancer.

The person who abuses steroids is the one immediately responsible for the behavior. Nevertheless, one cannot help but wonder how great an influence the ideals of a society that idolizes winners may have been in shaping the behavior. ✌

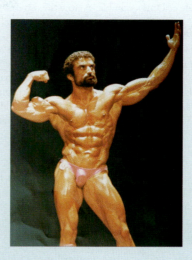

FIGURE **6.A**
Anabolic steroids are often abused as an easy way to build muscle and strength.

---

ance that muscles encounter during an exercise. Muscles are weaker at some parts of their range of motion than at others. A cam (built on a "snail shape") in the exercise machine makes the lifting easier where the muscle is weaker and more difficult in the range the muscle can handle. "Free weights"—barbells and dumbbells—are much harder to control and more dangerous to use than an exercise machine.

what would you do?

The International Olympics Committee and the World Anti-Doping Agency are concerned about the future possibility of creating genetically modified athletes. Gene therapy techniques (see Chapter 21) might be developed to insert genes into atheletes that could improve athletic performance. For example, researchers exploring muscle deterioration have developed gene therapy that repairs and rebuilds muscles in old mice. Clinical trials will soon begin on humans who have muscle-wasting disorders. Expert warn that athletes may someday use similar gene therapy to build more muscle, faster, and with less effort. What would you do? If you were a competitive athlete, would you use gene therapy to repair damaged muscle tissue? Would you use gene therapy to build muscle tissue to enhance performance? Do you think that it should be illegal for athletes to use gene therapy to repair and rebuild muscle?

One problem with building muscle is that you have to keep at it. If you train just in the spring to look good at the beach, you have an uphill battle, because all the mass of muscle you built last year began to disappear only 2 weeks after the training sessions stopped.

A frequent question is, are men naturally stronger than women? In general, yes. Women have about 35% muscle mass, and men have 42%. One reason is that the level of the male hormone testosterone, which builds muscles, is generally higher in men than in women (see the Social Issue essay, *Building Muscle, Fair and Square?*). Also, men are generally larger than women and so have more muscle mass to begin with. However, the individual muscle cells of men and women have the same strength.

REVIEWING THE CONCEPTS

The Muscular System Produces Movement and Maintains Posture (p. 105)

1. There are three types of muscle: skeletal, smooth, and cardiac. All types of muscle cells are excitable, contractile, extensible, and elastic.

Most Skeletal Muscles Work in Pairs (p. 105)

2. Many of the skeletal muscles of the body are arranged in antagonistic pairs so that the actions of the members of the pair are opposite to one another. For instance, one member of a pair usually causes flexion (bending at the joint) and the other, extension (straightening at the joint).

3. Muscles are attached to bones by tendons. The muscle origin is the end attached to the bone that is more stationary during a movement, and the insertion is the end attached to the bone that moves.

Sarcomeres Are the Contractile Units of Muscle (pp. 105–109)

4. An entire muscle is wrapped in a connective tissue sheath. A muscle cell is packed with myofibrils, which are composed of myofilaments (the contractile proteins actin and myosin).

WEB TUTORIAL 6.1 Muscle Structure
WEB TUTORIAL 6.2 Muscle Function

5. A muscle contracts when a myosin head binds to actin, swivels, and pulls actin toward the midline of the cell, causing the filaments to slide past one another and increasing their degree of overlap.

6. Contraction is controlled by the availability of calcium ions. Calcium ions interact with two proteins on the actin filament—troponin and tropomyosin—that in turn determine whether myosin can bind to actin. Calcium ions are stored in the sarcoplasmic reticulum and released when a motor nerve sends an impulse.

7. Motor nerves contact muscle cells at neuromuscular junctions. When the nerve impulse reaches a neuromuscular junction, acetylcholine is released and causes a change in the membrane permeability of the muscle cell that, in turn, causes calcium ions to be released from the sarcoplasmic reticulum.

The Strength of Muscle Contraction Depends on the Number of Motor Units Involved (pp. 109–110)

8. A motor neuron and all the muscle cells it stimulates are collectively called a motor unit.

The Strength of Contraction Increases If a Muscle Is Stimulated Before It Has Relaxed (p. 111)

9. The response of a muscle cell to a single brief stimulus is called a twitch. If a second stimulus arrives before the muscle has relaxed, the second contraction builds upon the first. This phenomenon is known as summation. Frequent stimuli cause a sustained contraction, called tetanus.

ATP for Muscle Contraction Comes from Many Sources (pp. 112–113)

10. Each time a cross-bridge between myosin and actin forms and is broken, two ATP molecules are used. The sources of ATP are (1) stored ATP, (2) creatine phosphate, (3) anaerobic metabolic pathways (fermentation), and (4) aerobic respiration.

Slow-Twitch and Fast-Twitch Muscle Cells Differ in Contraction Speed And Duration (p. 113)

11. Slow-twitch muscle cells contract slowly but with enormous endurance. Fast-twitch muscle cells contract rapidly and powerfully but with less endurance.

Aerobic Exercise Increases Endurance; Resistance Exercise Builds Muscle (pp. 113–115)

12. Muscle development is a result of resistance exercise. When muscles are forced to exert more than 75% of their maximal force, myofilaments are added to existing muscle cells and the diameter of the cells is increased.

KEY TERMS

actin filament *p. 105*	muscle twitch *p. 111*	oxygen debt *p. 113*	tendon *p. 105*
antagonistic pair *p. 105*	myofibril *p. 105*	sarcomere *p. 105*	tetanus *p. 111*
cross-bridge *p. 108*	myofilament *p. 105*	sarcoplasmic reticulum *p. 109*	transverse tubule *p. 109*
fast-twitch cell *p. 113*	myosin filament *p. 105*	slow-twitch cell *p. 113*	tropomyosin *p. 109*
insertion *p. 105*	neuromuscular junction *p. 109*	striated muscle *p. 105*	troponin *p. 109*
motor unit *p. 110*	origin *p. 105*	summation *p. 111*	

THINKING ABOUT THE CONCEPTS

1. Why are most skeletal muscles arranged in antagonistic pairs? Give an example that illustrates the roles of each member of an antagonistic pair of muscles. *p. 105*

2. Describe a skeletal muscle, including definitions of muscle cells, myofibrils, and myofilaments. *p. 105*

3. What is the sliding filament model of muscle contraction? *pp. 106–107*

4. What causes actin to move during muscle contraction? *pp. 108–109*

5. Explain the roles of troponin, tropomyosin, and calcium ions in regulating muscle contraction. *p. 109*

6. Explain how the events that occur when a motor nerve impulse reaches a neuromuscular junction and calcium ions are released from the sarcoplasmic reticulum lead to muscle contraction. *pp. 109–110*

7. Define a motor unit. Explain how motor units vary depending on the degree of control a person has over a particular muscle. *pp. 109–110*

8. Define muscle twitch, summation, and tetanus. *p. 111*

9. List the sources of ATP for muscle contraction, and explain when each source is typically called upon. *pp. 112–113*

10. Differentiate between slow- and fast-twitch muscle cells. *p. 113*

11. What type of exercise can build muscles? What accounts for the increase in muscle size? *pp. 113–115*

12. A single motor neuron and all the muscle cells it stimulates is called a
    a. sarcoplasmic reticulum.
    b. neuromuscular junction.
    c. motor unit.
    d. wave summation.

13. In a muscle, energy is stored in the form of
    a. creatine phosphate.
    b. ADP.
    c. myosin.
    d. glucose.

14. A muscle cell contracts when _____ filaments and _____ filaments slide past one another.

15. The contractile unit of a muscle is called a/an _____.

APPLYING THE CONCEPTS

1. Hakeem is a 22-year-old who ate some of his Aunt Sophie's canned tomatoes for dinner last night. This morning, his speech is slurred and he is having trouble standing or walking, so his brother rushes him to the emergency room. The doctor tells Hakeem that he probably has botulism, a type of food poisoning. The toxin produced by this bacterium prevents the release of acetylcholine at the neuromuscular junction. Explain how Hakeem's symptoms are related to the effects of the toxin.

2. The muscle cells of people with muscular dystrophy lack a protein, called dystrophin, that helps maintain the sarcoplasmic reticulum of muscle cells. As a result, the sarcoplasmic reticulum allows calcium ions to leak out. Why would this condition affect muscle function?

3. Chickens are ground-dwelling birds. They run to escape from predators, and they can fly only short distances. How do these activities explain the distribution of "white" and "dark" meat on a chicken? (Meat is skeletal muscle.)

7

Neurons: The Matter of the Mind

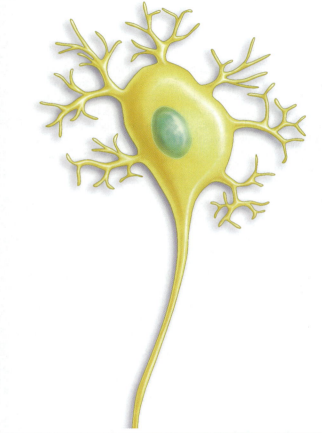

The Nervous System Has Two Main Parts

Neurons and Neuroglial Cells Are the Cells of the Nervous System
- Neuroglial cells support, protect, insulate, and nurture neurons
- Neurons can be sensory, motor, or associative

Neurons Have Dendrites, a Cell Body, and an Axon
- Axons and dendrites are bundled together to form nerves
- The myelin sheath increases the rate of conduction and helps in repair

The Nerve Impulse Is a Bioelectrical Signal
- Ions move passively through ion channels
- The sodium-potassium pump uses ATP to transport sodium ions out and potassium ions in
- The inside of a resting neuron has a negative charge relative to the outside
- An action potential is a reversal and restoration of the charge difference across the membrane
- The sodium-potassium pump restores the original distribution of ions
- Action potentials are all-or-none events
- A neuron cannot fire during the refractory period

Synaptic Transmission Is Communication between Neurons
- Synaptic transmission involves the release of neurotransmitter and the opening of ion channels
- Synapses provide interaction points in the nervous system
- The neurotransmitter is quickly removed from the synapse
- Different neurotransmitters play different roles

 **ENVIRONMENTAL ISSUE**　Environmental Toxins and the Nervous System

Competitive chess was Amelia's passion. At age 12 she had won every junior chess tournament in her state. It was said that she could remember every chess game she had ever played and that she had categorized all of them by opening, general strategy, and closing moves. Her IQ was extremely high, and she used it to its capacity in competitive chess. Today she is part of a chess club exhibition playing 15 games simultaneously with a 5-minute time limit for each game. She walks from board to board, looks briefly at the chess pieces, makes her move and quickly presses the timer button on the clock. As she is moving from board to board, she is thinking about a difficult match she played 2 weeks ago and retracing the moves. An hour passes, and, after winning all 15 games and giving a brief interview to a television reporter, Amelia walks to her parents' car to return home to finish her calculus assignment. Although Amelia is gifted almost beyond measure, her brain functions with the same biological materials we all possess.

In this chapter, we will explore the universe of the mind, beginning with the cells from which the brain and nervous system are built. We will discover how these cells work and how they communicate with one another. In Chapter 8, The Nervous System, we will explore the parts and functions of the brain and spinal cord. ■

The Nervous System Has Two Main Parts

The *nervous system* integrates and coordinates all the body's varied activities. As you will learn in greater detail in Chapter 8, the nervous system is made up of two primary parts: (1) the central nervous system, which is made up of the brain and spinal cord, and (2) the peripheral system, which is made up of all the nervous tissue in the body outside the brain and spinal cord.

Neurons and Neuroglial Cells Are the Cells of the Nervous System

The nervous system has two types of specialized cells. **Neurons** (nerve cells) are excitable cells that generate and transmit messages. **Neuroglial cells** (also called simply glial cells) are supporting cells that outnumber the neurons by about 10 to 1.

Neuroglial cells support, protect, insulate, and nurture neurons

The nervous system has several types of glial cells, each with different jobs to do. Some glial cells provide structural support for the neurons of the brain and spinal cord. Glial cells also provide a steady supply of chemicals called nerve growth factors that stimulate nerve growth. Without nerve growth factors, neurons die. Other glial cells form insulating sheaths around the long projections (axons) extending from certain neurons. This sheath, called the myelin sheath, has several important roles and will be discussed in more detail later in this chapter.

Unlike neurons, glial cells are able to reproduce. In fact, when neurons die, glial cells multiply, filling up the vacant space. A consequence of this trait is that the uncontrolled division of glial cells forms most brain tumors.

Neurons can be sensory, motor, or associative

The basic unit of the nervous system is the neuron, or nerve cell (Figure 7.1). Neurons are responsible for an amazing variety of functions, which can be grouped into three general categories.

- **Sensory** (or afferent) **neurons** conduct information *toward* the brain and spinal cord from the sensory receptors, which are structures specialized to gather information about the conditions within and around our bodies.

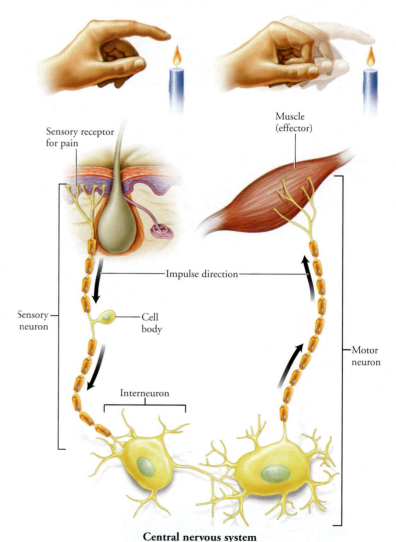

FIGURE **7.1**
The nervous system consists of sensory neurons, interneurons, and motor neurons. Trace the pathway of the impulse from a sensory receptor that detects a change in the external or internal environment along a sensory neuron to an interneuron. In this diagram, only one sensory neuron is shown. An interneuron usually receives input from many sensory neurons. The interneuron integrates the information from the sensory neurons and stimulates a motor neuron. The motor neuron then conducts the information to a muscle or a gland (an effector).

- **Motor** (or efferent) **neurons** carry information *away from* the brain and spinal cord to an **effector**—either a muscle or a gland.

- Association neurons, commonly called **interneurons**, are located between sensory and motor neurons. They are found only within the brain and spinal cord, where they integrate or interpret the sensory signals, thereby "deciding" on the appropriate response. Interneurons are by far the most numerous nerve cells in the body. They account for over 99% of the body's neurons.

Neurons Have Dendrites, a Cell Body, and an Axon

Neurons are specialized for communicating with other cells, and their shape reflects their functions (Figure 7.2). **Dendrites** are numerous short, branching projections from the neuron that create a huge surface for receiving signals from other cells. The information then travels toward an enlarged central region called a cell body, which has all the normal organelles, including a nucleus, for maintaining a cell. When a neuron responds to an incoming signal, it transmits its message along the **axon**, a single long extension from the cell body. The message is sent along the axon to either another neuron or an effector, such as a muscle or a gland. In some cases, the axon allows the neuron to communicate over long distances. The end of the axon has many branches specialized to release a chemical, called a neurotransmitter, that alters the activity of the effector. In summary, the dendrites and cell body are typically the *receiving* portions of the neuron; the axon is the *sending* portion of the cell.

To appreciate the dimensions of a neuron, we will pretend for a moment that there is such a thing as a "typical neuron." We will choose as our example a motor neuron, one that carries a message from the spinal cord to a muscle. Imagine an enlarged cell body of a typical neuron to be about the size of a tennis ball. The axon of this neuron would be about 1 mile (1.6 km) long but only about one half-inch (1.3 cm) in diameter. The dendrites, the shorter but more numerous projections of the neuron, would fill an average-sized living room.

▌Axons and dendrites are bundled together to form nerves

A **nerve** consists of parallel axons, dendrites, or both from many neurons that are bundled together and covered with tough connective

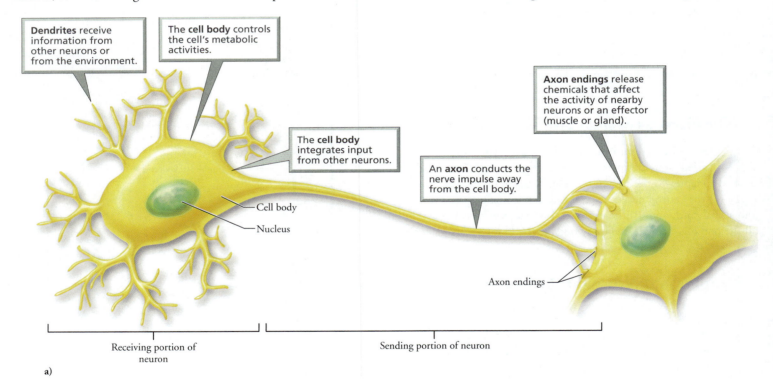

Dendrites receive information from other neurons or from the environment.

The **cell body** controls the cell's metabolic activities.

The **cell body** integrates input from other neurons.

Axon endings release chemicals that affect the activity of nearby neurons or an effector (muscle or gland).

An **axon** conducts the nerve impulse away from the cell body.

Cell body

Nucleus

Axon endings

Receiving portion of neuron

Sending portion of neuron

a)

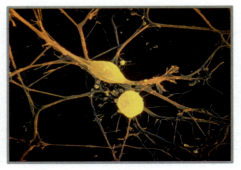

b)

FIGURE **7.2**

The structure of a neuron

tissue. Depending on the type of neurons forming the nerve, it can be classified as a sensory, motor, or mixed nerve (a nerve made up of both sensory and motor neurons).

The myelin sheath increases the rate of conduction and helps in repair

Most of the axons outside the brain and spinal cord and some of those within are enclosed in an insulating layer called the **myelin sheath** (Figure 7.3). The myelin sheath is composed of wrappings of the plasma membrane of glial cells. Outside of the brain and spinal cord, **Schwann cells** form the myelin sheath. A Schwann cell wraps around the axon many times, forming a spiral of membrane that looks somewhat like a jellyroll. Thus, the myelin sheath serves as a kind of living electrical tape that insulates individual axons, preventing messages from short-circuiting between neurons. It also allows the signal to move along the axon faster. The myelin sheath is kept alive by the nucleus and cytoplasm of the Schwann cell, which are squeezed to the periphery as the sheath forms.

A single Schwann cell encloses only a small region, about 1 mm, of an axon. Along the longest axons in your body, those that conduct impulses from the spinal cord to your toes for instance, there may be as many as 500 Schwann cells. The regions along an axon between adjacent Schwann cells are

exposed to the extracellular environment. These unsheathed regions are called *nodes of Ranvier*. This arrangement is very important to the speed at which a neuron transmits messages. With the myelin sheath in place, a nerve impulse "jumps" successively from one node of Ranvier to the next. This type of transmission, called **saltatory conduction** (*saltare*, to jump), is up to 100 times faster than signal conduction would be on an unmyelinated axon of the same diameter. For this reason, the axons that are involved in conduction of signals over long distances are typically myelinated.

It is easy to see how this "jumping" mode of transmission increases the speed at which the message travels. We frequently see a similar effect during a basketball game. With seconds left in the game, dribbling the ball the length of the court would take too much time. Passing the ball through a series of players is faster. Likewise, an impulse passed from one node to the next, as occurs in myelinated nerves, moves faster than one traveling along the length of the axon, as occurs in unmyelinated nerves.

The myelin sheath also plays a role in helping to repair a neuron in a cut or crushed nerve. When a neuron in the peripheral nervous system is cut, the part of the axon that has been separated from the cell body can no longer receive life-sustaining materials from the cell body. The axon begins to degenerate within a few

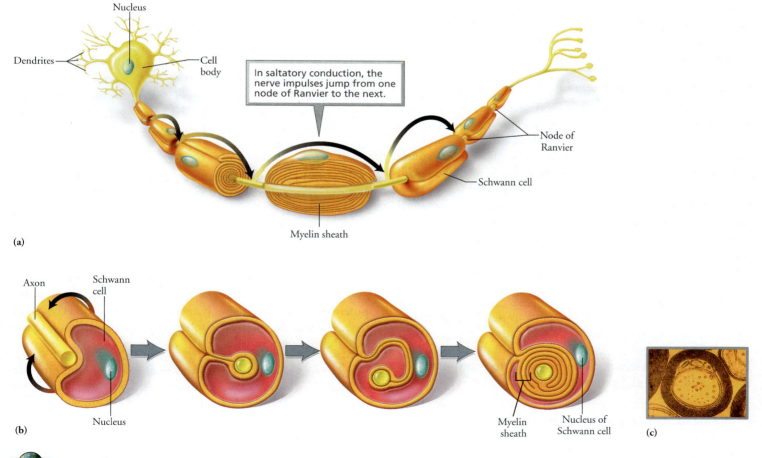

Nucleus

Dendrites

Cell body

In saltatory conduction, the nerve impulses jump from one node of Ranvier to the next.

Node of Ranvier

Schwann cell

Myelin sheath

(a)

Axon

Schwann cell

Nucleus

Myelin sheath

Nucleus of Schwann cell

(b)

(c)

WWW FIGURE **7.3**

TUTORIAL 7.2

The myelin sheath. (a) An axon protected by a myelin sheath. Note the Schwann cells that form the myelin sheath and the nodes of Ranvier—areas of exposed axon that allow for saltatory conduction. (b) The myelin sheath forms from multiple wrappings of Schwann cells. (c) An electron micrograph of the cut end of a myelinated axon.

minutes of the injury, and Schwann cells begin to clear all the axon fragments from the myelin sheath. Then, the axon stump begins "sprouting" from the cell body and grows toward the proper cell, guided by a tube of myelin produced by Schwann cells.

MULTIPLE SCLEROSIS

The importance of the myelin sheath becomes dramatically clear in people with multiple sclerosis (MS). This disease involves the progressive destruction of myelin sheaths within the brain and spinal cord. The damaged regions of myelin then become hardened scars called scleroses (hence the name of the disease). The scleroses interfere with the transmission of nerve impulses and allow short-circuiting between normally unconnected conduction paths, which delays or completely blocks the signals from one brain region to another. Depending on the part of the nervous system affected, the result can be paralysis or the loss of sensation, including loss of vision.

The Nerve Impulse is a Bioelectrical Signal

Neurons are specialized for communication. A nerve's message, which is called a nerve impulse, or action potential, is a bioelectrical signal caused by sodium ions (Na^+) and potassium ions (K^+) crossing the neuron's membrane to enter and leave the cell. We will consider the action potential again shortly to see the causes and effects of these ion movements. But first we will consider how ions cross a membrane.

■ Ions move passively through ion channels

The key to the informational signals of a neuron is its membrane. As are most living membranes, the plasma membrane of a neuron is selectively permeable; it allows some substances through but not others. The membrane contains many pores, called **ion channels**, that provide the only means for ions to cross the membrane without using cellular energy (Figure 7.4). Each ion channel is specific for one or a very few types of ions. For example, sodium channels allow the passage of only sodium ions. Potassium channels allow only potassium ions to pass through the membrane. Thus, ion channels function as molecular sieves, selecting which ions can cross the membrane. Some of the channels are normally open, whereas a "gate" regulates the opening of other channels. A change in the shape of the protein constituting the gate either opens the channel—allowing ions to pass through—or closes it—preventing ions from crossing the membrane.

■ The sodium-potassium pump uses ATP to transport sodium ions out and potassium ions in

Sodium and potassium ions are also moved actively across the membrane by **sodium-potassium pumps**, which are special proteins in the cell membrane. These pumps use cellular energy in the form of ATP to pump ions against their concentration gradients. Each pump ejects three sodium ions (Na^+) from within the cell while bringing in two potassium ions (K^+). So, more positive charges leave the neuron than enter it.

■ The inside of a resting neuron has a negative charge relative to the outside

It will be easier to understand how and why the ions move during an action potential if we first consider a neuron that is not transmitting an action potential—that is, a neuron in its resting state. As we will see, however, *resting* is hardly the word to describe what is going on at this stage. The membrane of a resting neuron maintains a charge difference across its surface such that the inside surface is more negative than the outside. This charge difference across the membrane, called the **resting potential**, results from the unequal distribution of ions across the membrane.

In a resting neuron, sodium and potassium ions are unequally distributed across the plasma membrane. There are roughly 10 times more sodium ions outside the membrane than inside. Furthermore, there are about 30 times more potassium ions inside than outside. Potassium ions tend to leak out because they are more concentrated inside the axon. (Recall from Chapter 3 that substances tend to move from an area of higher concentration to one of lower concentration.) To a lesser extent, sodium ions leak in. However, sodium-potassium pumps maintain the resting potential by pumping out sodium ions while moving potassium ions back in.

The result of these ion distributions is that the inner surface of a resting neuron's membrane is typically about 70 millivolts (mV) more negative (-70 mV) than the outer surface. This voltage is about 5% that of a size AA flashlight battery.

Although the sodium-potassium pumps of neurons consume a lot of energy to maintain the resting potential, the energy is not wasted. The resting potential allows the neuron to be ready to respond more quickly than it could if the membrane were electrically neutral in its resting state. This situation is somewhat analogous to charging a car's battery so that the car will start as soon as the key is turned.

■ An action potential is a reversal and restoration of the charge difference across the membrane

What happens when a neuron is stimulated, as might occur when another neuron releases a chemical called a neurotransmitter? Shortly, we will consider in more detail some of the excitatory signals—those that might generate a nerve impulse. Here, we will summarize that part of the story and consider the effect of an excitatory signal reaching a neuron's membrane. In brief, the **action potential**, or nerve impulse, is a sudden reversal in the charge difference across the membrane followed by its restoration. Let's see why this sequence occurs.

1. **The loss of the charge difference across the membrane (depolarization) occurs as sodium ions (Na^+) enter the axon.** An excitatory stimulus causes the gates on sodium channels to open. Sodium ions then enter the neuron, and their positive charge begins to reduce the negative charge within. The reduction of the charge difference across the membrane is called *depolarization*. The action potential begins when membrane depolarization reaches a certain value called the **threshold**. When the threshold is reached, the gates on sodium channels open.

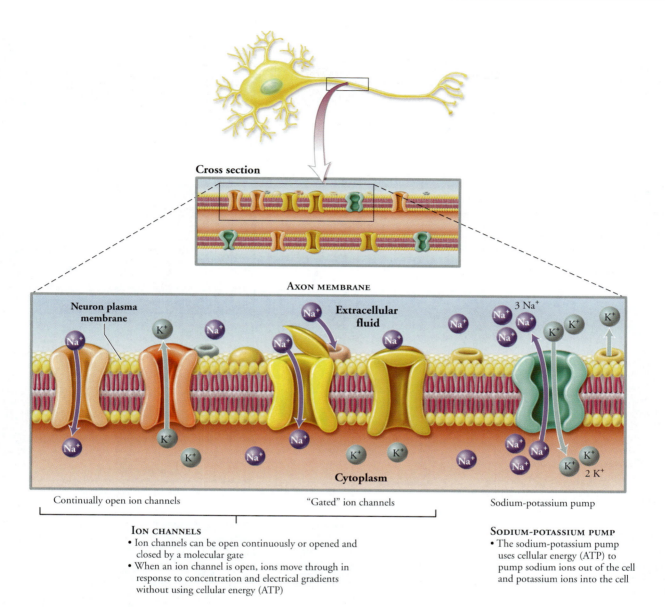

FIGURE **7.4**

The plasma membrane of a neuron contains two types of pathways through which ions can pass.

Enough sodium ions enter through the open gates to create a net positive charge in that region, as shown in Figure 7.5. Sodium ions continue to enter for a brief instant, driven by their concentration gradient. However, the gates on these sodium channels automatically snap shut (or inactivate) within about half a millisecond, quickly halting the inward rush of sodium ions.

2. **The return of the membrane potential to close to its resting value (repolarization) occurs as potassium (K^+) ions leave the axon.** About halfway through the action potential, the gates on potassium channels open. Potassium ions now leave the cell. The exodus of potassium ions with their positive charge causes the interior of the neuron to become negative once again relative to the outside. The outward flow of potassium ions returns the membrane potential to close to its resting value. The restoration of the charge difference across the membrane is called *repolarization*.

The action potential, then, is a reversal of the charge difference across the membrane caused by the inward flow of sodium ions, followed immediately by the restoration of the charge difference caused by the outward flow of potassium ions. These changes occur in a "wave" that moves along the axon away from the cell body, as will be discussed shortly.

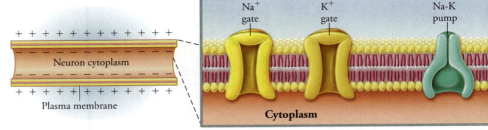

RESTING NEURON
Plasma membrane is charged, with the inside negative relative to the outside.

Neuron cytoplasm

Plasma membrane

Na⁺ gate   K⁺ gate   Na-K pump

Cytoplasm

ACTION POTENTIAL
The charge difference across the membrane reverses and then is restored.

Step 1: The loss of the charge difference across the membrane (depolarization) occurs as sodium ions (Na⁺) enter the axon. The inside of the membrane becomes positively charged.

Excitatory stimulus
↓
Sodium channels open
↓
Sodium ions (Na⁺) enter the neuron

Step 2: The return of the membrane potential to near its resting value (repolarization) occurs as potassium (K⁺) ions leave the axon.

Potassium channels open
↓
Potassium ions (K⁺) leave the neuron

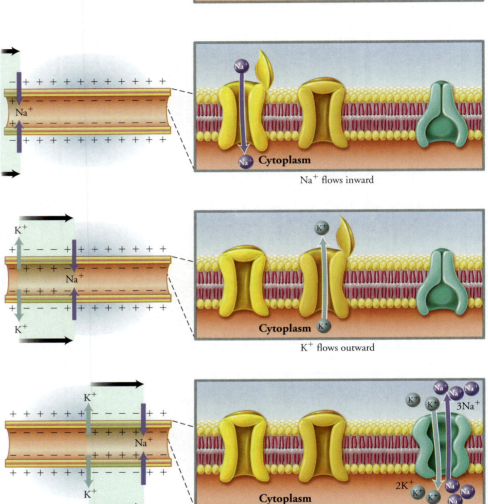

Na⁺

Na⁺   **Cytoplasm**

Na⁺ flows inward

K⁺

Cytoplasm   K⁺

K⁺ flows outward

CHARGE RESTORATION
The sodium-potassium pump restores the original distribution of ions.

K⁺   Na⁺   3Na⁺

Cytoplasm   2K⁺

Na-K pump restores the original ion distribution

FIGURE 7.5
The resting state and the propagation of an action potential along an axon

The sodium-potassium pump restores the original distribution of ions

Notice that, at the end of an action potential, the charge distribution across the membrane is returned to the resting potential. However, there are slightly more sodium ions and slightly fewer potassium ions inside the cell than before. This imbalance is corrected as the sodium-potassium pump restores the original

ion distribution. The action of the sodium-potassium pump is slow. Therefore, it does not contribute directly to the events of the action potential.

Action potentials are all-or-none events

The action potential is described as a *wave* of changes in the charge across the neuron's plasma membrane because the events do not occur simultaneously along the entire length of the axon. Instead, as sodium ions enter in one region of the membrane and the charge inside the membrane becomes less negative in that region of the cell, the change in charge causes the opening of the sodium channel gates in an adjacent region. Then, as sodium ions enter that region, the gates in the next adjacent region are opened. As a result, the changes in charge across the membrane spread along the axon. Therefore, once started, action potentials do not diminish, just as the last domino in a falling row falls with the same energy as the first. And, the intensity of the nerve impulse does not vary with the strength of the stimulus that triggered it. If an action potential occurs at all, it is always of the same intensity as any other action potential. This "all-or-none" nerve cell conduction is similar to the firing of a gun in that the force of the bullet is not changed by how hard you pull the trigger.

▌A neuron cannot fire during the refractory period

Immediately after an action potential, the neuron cannot be stimulated again for a brief instant, called the **refractory period**. During the refractory period, the sodium channels are closed and cannot be opened by depolarization. Consequently, a second action potential cannot be generated. A prolonged stimulus that is above threshold will cause a series of discrete nerve impulses, not a bigger impulse. Up to a point, then, the frequency of impulses increases with increasing strength of the stimulus.

stop and think

A bloom of dinoflagellates, single-celled marine algae, causes "red tides." These organisms contain a chemical called saxitoxin (STX). Extremely small concentrations of STX prevent sodium channels from opening. Clams, scallops, and mussels consume the dinoflagellates. Because the shellfish are insensitive to the toxin, STX accumulates in their tissues. What effect would you expect STX to have on nerve transmission in humans who accidentally consume tainted shellfish?

Synaptic Transmission Is Communication Between Neurons

When a nerve impulse reaches the end of an axon, in almost all cases the message must be relayed to the adjacent cell across a small gap. Although the gap is just billionths of a meter wide, the impulse cannot cross it. Communication with the adjacent cell requires a change in the nature of the message from a bioelectrical signal to a chemical signal. The action potential in the first cell causes the release of a chemical from the tips of its axon. That chemical, called a **neurotransmitter**, diffuses across the gap, usually in less than half a millisecond, and conveys a message to the adjacent cell.

The junction between a neuron and another cell is called a **synapse**. The structure of a synapse between two neurons is shown in Figure 7.6. The gap between the cells is called the *synaptic cleft*.

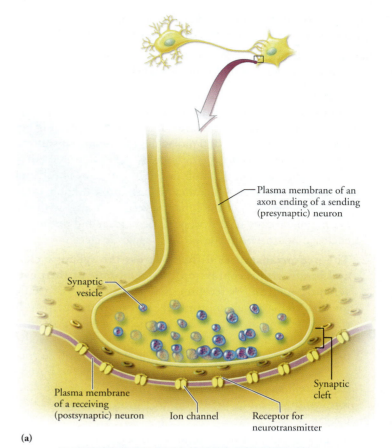

(a)

(b)

FIGURE **7.6**

Structure of a synapse. (a) The axon ending of the presynaptic neuron is separated from the dendrite or cell body of the postsynaptic neuron by a small gap called a synaptic cleft. Within the axon ending (synaptic knob) are small sacs, called synaptic vesicles, filled with neurotransmitter molecules. (b) An electron micrograph of a synapse.

Recall that the axon branches near the end of its length. Each branch ends with a small bulblike swelling called a *synaptic knob*. The neuron sending the message is the *presynaptic neuron* (before the synapse). The neuron receiving the message is the *postsynaptic neuron* (after the synapse).

Synaptic transmission involves the release of neurotransmitter and the opening of ion channels

Now let's consider the events that occur in the synapse as the message is sent from one neuron to the next (Figure 7.7).

1. **Packets of neurotransmitter are released from synaptic knobs of the presynaptic neuron.** Within the synaptic knobs of the presynaptic neuron are tiny sacs containing between 10,000 and 100,000 molecules of a neurotransmitter. These sacs are called *synaptic vesicles*. When the nerve impulse reaches a synaptic knob, it causes the gates of calcium ion channels to open. Calcium ions rush into the knob, causing the membranes of the synaptic vesicles to fuse with the plasma membrane of the presynaptic neuron and dump the enclosed neurotransmitter into the synaptic cleft.

2. **The neurotransmitter diffuses across the synaptic cleft and binds with receptors on the membrane of the postsynaptic neuron.** A receptor is a protein that is specific for a particular neurotransmitter. It recognizes the neurotransmitter much as a lock "recognizes" a key. A neurotransmitter can affect only a cell that has receptors specific for it. Thus, only certain neurons can be affected by a specific neurotransmitter. The more receptors on the postsynaptic cell, the more vigorous its response. When the neurotransmitter binds with the receptor, the interaction triggers a response.

3. **When a neurotransmitter binds to its receptor, an ion channel is opened, and the opening of the channel can excite or inhibit the postsynaptic neuron.** The effect a neurotransmitter has on the cell receiving the message depends on the type of ion channel opened. It is really the receptor that determines which ion channels will open and what the effect of a given neurotransmitter will be.

An *excitatory synapse* is one in which the binding of the neurotransmitter to the receptor opens sodium channels and increases the likelihood that an action potential will begin in the postsynaptic cell. The result is a temporary local change in the charge difference across the postsynaptic membrane, called a **graded potential** because it can vary in strength. If enough receptor sites bind with neurotransmitter to cause depolarization to threshold value, an action potential is generated in the postsynaptic cell.

We see, then, that the localized graded depolarization of the postsynaptic membrane acts like pressure on the trigger of a gun. Slight pressure on the trigger will not cause the gun to fire. However, when the pressure on the trigger reaches a critical threshold level, each bullet will be discharged from the gun with the same force as any other bullet.

In contrast, an *inhibitory synapse* is one in which the binding of the neurotransmitter decreases the likelihood that an action potential will be generated in the postsynaptic neuron. The postsynaptic cell is inhibited because its resting potential becomes more negative than usual. As a result, it will require more neurotransmitter with an excitatory effect than usual to reach threshold.

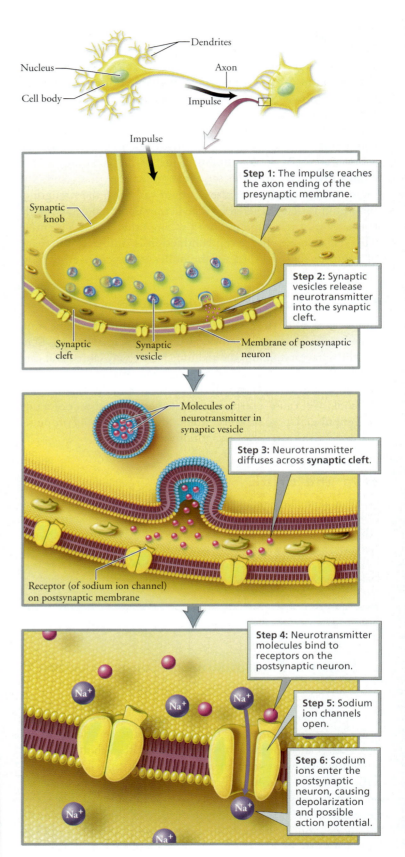

Step 1: The impulse reaches the axon ending of the presynaptic membrane.

Step 2: Synaptic vesicles release neurotransmitter into the synaptic cleft.

Step 3: Neurotransmitter diffuses across **synaptic cleft**.

Step 4: Neurotransmitter molecules bind to receptors on the postsynaptic neuron.

Step 5: Sodium ion channels open.

Step 6: Sodium ions enter the postsynaptic neuron, causing depolarization and possible action potential.

FIGURE **7.7**

Transmission across an excitatory synapse

TUTORIAL 7.4

▌Synapses provide interaction points in the nervous system

Postsynaptic cells integrate excitatory and inhibitory synaptic input. The balance of input determines whether an action potential will be generated. A single nerve impulse in a single synaptic knob will not generate an impulse in the postsynaptic cell. It generally requires at least 50 nerve impulses arriving almost simultaneously before the postsynaptic cell will fire. A neuron may have as many as 10,000 synapses with other neurons (Figure 7.8). Some of these synapses will have excitatory effects on the postsynaptic membrane. Others will have inhibitory effects. The summation of excitatory and inhibitory effects will determine whether the cell will generate an action potential. The integration of synaptic input provides an opportunity to modify the response of the receiving cell. This integration results in finer control over responses, just as having both an accelerator and a brake gives you finer control over the movement of a car.

▌The neurotransmitter is quickly removed from the synapse

The effects of neurotransmitters are temporary because they are removed from the synapse. If they were not, they would continue to excite or inhibit the postsynaptic membrane indefinitely. Depending on the neurotransmitter, disposal may be accomplished in one of two ways. First, enzymes can deactivate a neurotransmitter. For example, the enzyme acetylcholinesterase removes the neurotransmitter **acetylcholine** from the synapse. Second, the neurotransmitter may be actively pumped back into the axon tip. See the Environmental Issue essay, *Environmental Toxins and the Nervous System*.

stop and think

The neurotransmitter acetylcholine triggers contraction of voluntary muscles. Myasthenia gravis is an autoimmune disease in which the body's defense mechanisms attack the acetylcholine receptor at the junction between nerve and muscle. As a result, a person with myasthenia gravis is weak. Repeated movements become feeble quite rapidly because the amount of acetylcholine released with each nerve impulse decreases after neurons have fired a few times in rapid succession. The low number of acetylcholine receptors in people with myasthenia gravis makes them extremely sensitive to even the slightest decline in the availability of acetylcholine. Why would you expect drugs that inhibit acetylcholinesterase to be helpful in treating myasthenia gravis?

▌Different neurotransmitters play different roles

TUTORIAL 7.5

What's your mood today? Are you on top of the world or down in the dumps? Are you anxious about something—about everything? The neurotransmitters that nerve cells use to communicate with one another are largely responsible for our moods. Indeed, the communication among the neurons makes possible a plethora of emotions. But an imbalance in neurotransmitters can cause mental disorders. So the messages nerve cells send one another are very important.

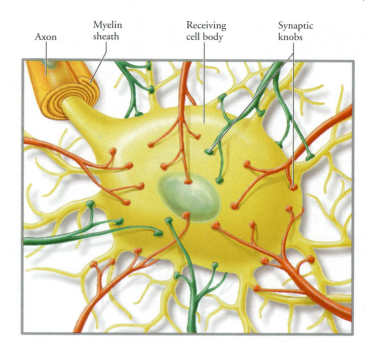

FIGURE **7.8**

Synapses are interaction points within the nervous system. A neuron may have as many as 10,000 synapses. Some synapses have an excitatory effect on the membrane of the postsynaptic neuron and increase the likelihood that the neuron will fire. Other synapses have an inhibitory effect and reduce the likelihood that the postsynaptic neuron will fire. The net effect of all synapses will determine whether an action potential is generated in the postsynaptic neuron. (The shape of the synaptic knobs shown in this electron micrograph is distorted as a result of the preparation process.)

About 50 substances are involved in communication between neurons within the brain. Why are there so many? One reason seems to be that different neurotransmitters are involved with different behavioral systems. Norepinephrine, for instance, is important in the

ENVIRONMENTAL ISSUE

Environmental Toxins and the Nervous System

Every day, many of us are exposed to factors that can harm our nervous system. The exposure may occur because of our job, our lifestyle, the foods we eat, or the drugs we take. Because we are born with all the neurons we will ever have, the death of neurons can lead to irreversible loss of function. Some damage may take years to become apparent, but other effects are immediate.

Pesticides are chemicals that are sprayed on plants to kill organisms that might damage the plant, such as insects. Organophosphate insecticides, such as Malathion®, Parathion®, and Diazinon®, are poisons that kill insects by excessively boosting activity at certain synapses in the insect's body. These insecticides work by inhibiting acetylcholinesterase, the enzyme that breaks down the neurotransmitter acetylcholine. As a result, acetylcholine accumulates in the synapses and has a continuous effect.

These insecticides work the same way in the human body. Insecticides accidentally poison approximately 500,000 people around the world, primarily farm workers, each year; 5000 to 14,000 die as a result. One symptom of pesticide poisoning is muscle spasms. These continuous involuntary contractions occur because acetylcholine is the transmitter that triggers contraction of voluntary muscles, such as those in the arms or legs.

Mercury, a naturally occurring heavy metal, damages the nervous system. In adults, symptoms of mercury poisoning include numbness in the arms and legs, sensory disturbance, lack of coordination, weakness and tremor, and slowed

---

regulation of mood, in the pleasure system of the brain, and in arousal. Norepinephrine is thought to produce an energizing "good" feeling. It is also thought to be essential in hunger, thirst, and sex drive. Serotonin is thought to promote a generalized feeling of well-being. Dopamine helps regulate emotions. It is also used in pathways that control complex movements.

A change in the level of a neurotransmitter affects the behaviors controlled by neurons that communicate using that neurotransmitter. Neurotransmitter levels may change as a result of taking certain drugs (see Chapter 8a). Changes in neurotransmitter levels also cause certain diseases.

ALZHEIMER'S DISEASE

Alzheimer's disease is progressive and results in loss of memory, particularly for recent events, followed by sometimes severe personality changes (Figure 7.9). A formerly good-natured person may become moody and irritable. In Alzheimer's disease, a tremendous number of neurons in the parts of the brain important in memory and intellectual functioning (hippocampus and cerebral cortex, see Chapter 8) are lost. Some of the neurons in these regions communicate using the neurotransmitter acetylcholine. In a person with Alzheimer's disease, the level of acetylcholine drops as much as 90%, possibly causing the loss of memory and mental capacity. In addition, the brain of a person with Alzheimer's disease is pocked with clusters of proteins, some between the neurons (amyloid plaques) and others within the neurons (neurofibrillary tangles). The amyloid plaques and neurofibrillary tangles are the prime suspects for the cause of the death of acetylcholine-producing neurons.

FIGURE **7.9**
Former President Ronald Reagan died in June 2004 at the age of 93. He had Alzheimer's disease since at least 1994. His wife Nancy Reagan described Alzheimer's disease as "the long good-bye" because the illness robs people of their memories.

If the loss of acetylcholine is responsible for some of the symptoms of Alzheimer's, it seems reasonable to treat Alzheimer's disease with drugs that will raise or at least maintain acetylcholine levels. Although such drugs (Aricept®, Exelon®, and Reminyl®) are available and they do improve the memory and intellectual ability of some people with Alzheimer's, they are far from being a complete answer to the problem. These drugs do not help everyone, and any improvement is rapidly lost when the person stops taking the drugs.

and slurred speech. Continued exposure can cause paralysis, convulsions, and death. If a pregnant woman is exposed to mercury, the metal can cause abnormalities in the development of the nervous system of the fetus. Nursing mothers can pass mercury to their babies in breast milk.

People can be exposed to mercury from a variety of sources. During the nineteenth century, mercury was used in the hat-making industry. The effects of mercury caused symptoms of mental derangement in hat makers, giving rise to the saying "mad as a hatter." Today, mercury is still used in a variety of industrial processes.

Wastewater from these industrial plants can enter natural bodies of water, carrying mercury with it. Mercury can also enter the air in the emissions of coal-fired power plants or from sites where municipal or medical waste is burned. From the air, mercury can be deposited on land or in the water.

Bacteria in the water convert the mercury to methyl mercury. Other organisms eat the bacteria, and other organisms, in turn, consume them. As a result, mercury accumulates in the bodies of fish and other organisms (see Chapter 23). Today, exposure most commonly results from eating mercury-contaminated

food, especially fish. The Food and Drug Administration (FDA) advises pregnant women and women of child-bearing age not to eat any shark, swordfish, king mackerel, or tilefish because of their high mercury content and to limit consumption of all fish to two servings a week. In 2004, the FDA added that these women should also limit their consumption of albacore (white) tuna to only one serving per week. The FDA suggests that consumers should eat up to two servings per week of fish and shellfish that are lower in mercury, such as salmon, pollock, catfish, canned light tuna, and shrimp. ✻

DEPRESSION

An insufficient amount of the neurotransmitter serotonin, as well as dopamine and norepinephrine, is thought to play a role in depression and suicidal behavior. We all get depressed at times. However, more than 19 million Americans experience depression that lasts for weeks, months, or years and interferes with their ability to function in daily life. Depression is common and can affect anyone regardless of age, sex, race, or income. Signs of depression are a loss of interest and pleasure in the activities and hobbies that were previously pleasurable; anxiety; sleep problems; decreased energy; and feelings of sadness, hopelessness, worthlessness, and guilt. Depression takes the joy out of life and complicates certain medical conditions such as heart disease, cancer, diabetes, epilepsy, and osteoporosis.

Many depressed people commit suicide, which is the third-ranking cause of death for teenagers. Studies comparing the brains of people who committed suicide with those of people who died of other causes reveal structural and chemical differences. The brains of people who took their own lives indicate a problem producing and using serotonin.

Depression can be treated successfully. Unfortunately, few of the millions of people suffering from depression recognize the symptoms and seek help. Antidepressant drugs affect the functioning of the neurotransmitters responsible for the problem: norepinephrine and serotonin. Older medications affect both neurotransmitters simultaneously. Newer medications, including Prozac®, Zoloft®, and Paxil®, specifically affect serotonin functioning. These drugs increase the level of serotonin in the synapse by reducing its rate of removal.

what would you do?

Few people would question the wisdom of helping someone who is depressed or suicidal to live a normal life. There is, however, a spectrum of severity for each personality disorder—from a nail-biting habit to obsessive-compulsive behavior, for example. Researchers are even looking for the biological roots of personality traits, such as shyness or impulsiveness. They believe that the levels of key neurotransmitters also affect these traits. If neurotransmitter levels do affect these traits, we may someday be able to design our own personalities. Should minor personality problems be treated with drugs? Should a personality "flaw" be treated? What do you think?

PARKINSON'S DISEASE

Parkinson's disease is a progressive disorder that results from the death of dopamine-producing neurons that lie in the heart of the brain's movement control center (Figure 7.10). A person with Parkinson's disease moves slowly, usually with a shuffling gait and a hunched posture. In addition, involuntary muscle contractions may interfere with intended movements. These contractions cause tremors (involuntary rhythmic shaking) of the hands or head when the muscles alternately contract and relax. Involuntary muscle contractions may also cause muscle rigidity when some muscles contract continuously. This muscle rigidity may cause someone with Parkinson's disease to suddenly "freeze" in the middle of a movement.

As the dopamine-producing neurons in the brain's movement control center die, dopamine levels begin to fall. Initially, the symptoms are subtle and are written off as part of the aging process. By the time symptoms of Parkinson's disease are apparent, 80% of the neurons in this small area of the brain have already died.

Attempts to treat Parkinson's disease have focused on replacing dopamine or helping the brain get by with the remaining dopamine. Unfortunately, treatment is not as easy as swallowing a few dopamine pills, because dopamine is prevented from reaching the brain by the blood-brain barrier that shields the brain from many substances (see Chapter 8). Instead, patients are given other substances that can reach the brain. The most common and effective treatment combines two drugs. The first is L-dopa, an amino acid that the brain converts to dopamine. The second is carbidopa, which prevents dopamine formation outside of the brain, where it causes undesirable side effects. Unfortunately, because carbidopa cannot stop the steady loss of dopamine-producing neurons, it loses effectiveness as the disease progresses. Some patients are treated with drugs that inhibit the enzyme that breaks down dopamine, resulting in enhanced levels of dopamine.

A controversial experimental treatment for Parkinson's disease, the transplantation of stem cells, shows some promise. Unlike most cells, stem cells are capable of differentiating into virtually any type of tissue. Although stem cells can be collected from certain adult tissues, the stem cells of an embryo are more versatile. However, the use of human embryonic stem cells has generated intense political debate (see Chapter 21).

FIGURE **7.10**

Actor Michael J. Fox and former heavyweight champion Mohammad Ali have Parkinson's disease, a progressive debilitating disease characterized by slowed movements, tremors, and rigidity. The symptoms are caused by an insufficient amount of the neurotransmitter dopamine, which is brought on by the death of dopamine-making nerve cells in a movement control center of the brain. The Michael J. Fox Foundation for Parkinson's Research (www.michaeljfox.org) funds research on the early diagnosis and treatment of Parkinson's disease.

REVIEWING THE CONCEPTS

The Nervous System Has Two Main Parts (pp. 119)

1. The nervous system integrates and coordinates the body's activities. It is made up of the central nervous system (the brain and spinal cord) and the peripheral nervous system (nervous tissue outside the brain and spinal cord).

Neurons and Neuroglial Cells Are the Cells of the Nervous System (pp. 119–120)

2. The nervous system has two types of specialized cells: neurons (nerve cells) and neuroglial cells.
3. Neuroglial cells outnumber neurons and have several important functions. They help to hold neurons together and provide structural support, supply nerve growth factors, and form myelin sheaths around certain axons.
4. There are three general categories of neurons. Sensory (or afferent) neurons conduct information from the sensory receptors toward the central nervous system. Motor (or efferent) neurons conduct information away from the central nervous system to an effector. Association neurons (interneurons) are positioned between sensory and motor neurons and are located in the central nervous system.

WEB TUTORIAL 7.1 Reflex Arc

Neurons Have Dendrites, a Cell Body, and an Axon (pp. 120–122)

5. Neurons are specialized for communicating with other cells. A "typical" neuron has a cell body that houses the organelles that maintain the cell. Numerous branching fibers called dendrites conduct messages *toward* the cell body. A single long axon conducts impulses *away* from the cell body.
6. Many axons are enclosed in an insulating layer called the myelin sheath. The myelin sheath forms from multiple wrappings of the plasma membrane of Schwann cells. The myelin sheath greatly increases the rate at which impulses are conducted along an axon. The sheath also plays a role in the regeneration of cut axons in the peripheral nervous system.

WEB TUTORIAL 7.2 Myelinated Neurons and Saltatory Conduction

The Nerve Impulse Is a Bioelectrical Signal (pp. 122–125)

7. The message conducted by a neuron, called a nerve impulse or an action potential, is caused by sodium ions (Na^+) and potassium

ions (K$^+$) crossing the neuron's plasma membrane to enter and leave the cell.

8. Ion channels are small pores through the plasma membrane through which ions move. Ion channels are usually specific to one or a few types of ions. Ions move through ion channels without the use of cellular energy.

9. The sodium-potassium pump uses cellular energy in the form of ATP to pump sodium ions out of the cell and potassium ions into the cell against their concentration gradients.

10. In the resting state, a neuron has an electrical potential difference, called the resting potential, across its plasma membrane. The resting potential is generated by the unequal distribution of ions across the membrane. Sodium ions are in greater concentration outside the neuron than inside. Potassium ions are in greater concentration inside than outside. The membrane also holds many negatively charged proteins inside the cell. The resting potential makes the neuron more negative inside than outside. The resting neuron's membrane is not very permeable to sodium ions. If some leak in, they are pumped out by the sodium-potassium pump.

11. The action potential begins when a region of the membrane suddenly becomes permeable to sodium ions. If enough sodium ions enter to reach threshold, an action potential begins. The gates on sodium channels open, and many of these ions then enter the cell, making the interior of the cell in that region of the membrane temporarily positive (depolarization). Potassium ions then leave the interior, making the inside once again more negative than the outside (repolarization). The change in the distribution of ions sweeps along the axon as a wave of depolarization and repolarization called an action potential.

12. At the end of an action potential, the sodium-potassium pump moves sodium ions out of the neuron and potassium ions into the neuron, restoring the original ion distribution.

13. The action potential acts as a wave of changes along the neuron's plasma membrane. Sodium ions enter one region of the membrane, making the charge inside the membrane less negative. The change in charge causes the opening of the sodium channel gates in an adjacent region of the membrane. Once initiated, an action potential sweeps to the end of the axon without diminishing in strength.

14. The refractory period is a brief instant immediately following an action potential in which the neuron cannot be stimulated.

WEB TUTORIAL 7.3 Action Potential

Synaptic Transmission Is Communication Between Neurons (pp. 125–130)

15. The point where one neuron meets another one is called a synapse. The neuron sending the message (the presynaptic neuron) and the neuron receiving the message (the postsynaptic neuron) are separated by a small gap called the synaptic cleft.

16. The arrival of a nerve impulse at the axon ending of the presynaptic neuron causes calcium ions to enter the cell. These ions cause synaptic vesicles that store neurotransmitters to fuse with the plasma membrane of the presynaptic neuron and release their contents into the synaptic cleft. The neurotransmitter then diffuses across the gap and binds to receptors on the membrane of the postsynaptic neuron.

17. If the synapse is excitatory, sodium ions enter the cell and increase the likelihood that the postsynaptic neuron will generate a nerve impulse. If the synapse is inhibitory, the charge difference across the membrane of the receiving neuron is increased, making it less likely that the postsynaptic neuron will generate a nerve impulse.

18. Postsynaptic cells integrate excitatory and inhibitory input from many cells. If threshold is reached, an action potential is generated in the postsynaptic cell.

19. The neurotransmitter is quickly removed from the synapse either by enzymatic breakdown or by being transported back into the presynaptic neuron.

20. Many neurotransmitters are used in the brain. Different neurotransmitters are involved with different behavioral systems. Disturbances in brain chemistry affect mood and behavior. Alzheimer's disease, which is characterized by a progressive loss of memory, is associated with a loss of acetylcholine in certain parts of the brain. Reduction in the functioning of dopamine, norepinephrine, and serotonin can cause depression and suicidal behavior. Low levels of dopamine in another brain region cause Parkinson's disease, which is characterized by slow movements, tremors, and muscle rigidity.

WEB TUTORIAL 7.4 Synapses

WEB TUTORIAL 7.5 Neurotransmitters Impact the Brain

KEY TERMS

acetylcholine *p. 127*	interneuron *p. 120*	neuron *p. 119*	sensory neuron *p. 119*
action potential *p. 122*	ion channel *p. 122*	neurotransmitter *p. 125*	sodium-potassium pump *p. 122*
axon *p. 120*	motor neuron *p. 120*	refractory period *p. 125*	synapse *p. 125*
dendrite *p. 120*	myelin sheath *p. 121*	resting potential *p. 122*	threshold *p. 122*
effector *p. 120*	nerve *p. 120*	saltatory conduction *p. 121*	
graded potential *p. 126*	neuroglial cell *p. 119*	Schwann cell *p. 121*	

THINKING ABOUT THE CONCEPTS

1. List the three types of neurons and give their general functions. *pp. 119–120*
2. Draw a typical neuron and label the following: cell body, nucleus, dendrites, and axon. *p. 120*
3. What are the functions of neuroglial cells? *pp. 119, 120–121*
4. Explain how a myelin sheath is formed. What are the functions of the myelin sheath? *p. 121*

5. Describe the distribution of sodium ions and potassium ions during a neuron's resting state. Explain how the movements of these ions affect the charge difference across the membrane. *p. 122*
6. Why is the resting potential important? *p. 122*
7. What happens to sodium ions at the beginning of an action potential? *pp. 122–123*

8. Describe the events that bring about the restoration of the charge difference across the membrane (repolarization). *pp. 123–124*
9. How is the original ion distribution of sodium and potassium ions restored? *p. 124*
10. What is the refractory period? *p. 125*
11. Draw a synapse between two neurons. Label the following: presynaptic neuron, postsynaptic neuron, synaptic cleft, synaptic vesicles, neurotransmitter molecules, and receptors. *p. 125*
12. How do the events at an excitatory synapse differ from those at an inhibitory synapse? *p. 126*
13. How is the action of a neurotransmitter terminated? *p. 127*
14. Oubain is a drug that causes the axon to lose its membrane potential. In other words, there is no separation of charges across the membrane. The most likely effect that this loss of potential would have on the neuron is
    a. to lower the threshold and make it easier to generate action potentials.
    b. to slow the rate at which action potentials move along the axon.
    c. that action potentials could go only halfway down the axon.
    d. that action potentials could not be generated.
15. Choose the *incorrect* statement.
    a. Neurotransmitters diffuse across the myelin sheath.
    b. An inhibitory neurotransmitter makes it less likely that an action potential will be generated in the presynaptic (before the synapse) neuron.
    c. Neurotransmitters are stored in synaptic vesicles.
    d. Most interneurons are found in the central nervous system.
16. In botulism, a type of food poisoning, the poison produced by bacteria in the spoiled food prevents the person's synaptic vesicles from fusing with the neuron's membrane. You would expect that this effect would
    a. cause excessive destruction of neurotransmitter in the synaptic cleft.
    b. destroy myelin.
    c. prevent the message of the presynaptic cell from reaching the postsynaptic cell.
    d. cause neurotransmitter to clog the synaptic cleft.
17. The synaptic cleft
    a. is a chemical that allows two neurons to communicate with one another.
    b. is a gap between two neurons.
    c. is a gap between two Schwann cells forming the myelin sheath.
    d. allows saltatory conduction.
18. The _____ is an insulating layer formed by Schwann cells that increases the rate of an action potential.
19. The _____ uses cellular energy to move three sodium ions out of the axon and two potassium ions into the axon.
20. Parkinson's disease is caused by the loss of neurons that produce the neurotransmitter _____.

APPLYING THE CONCEPTS

1. Amber has taken the sedative Valium®, a drug that has a shape similar to that of molecules of the neurotransmitter GABA. When Valium binds to the GABA receptors, it causes the same effects as the binding of GABA itself. GABA is an inhibitory neurotransmitter. On the basis of this information, explain why Valium® has a calming effect on the nervous system.
2. A nerve gas (diisopropyl fluorophosphate) used during war blocks the action of acetylcholinesterase, the enzyme that breaks down acetylcholine in the synapse. What effects would you expect this gas to have at the synapses that use acetylcholine?
3. Mohammed has a kidney condition that raises the level of potassium ions in the extracellular fluid that bathes cells. What effect would you expect this condition to have on his ability to generate nerve impulses (action potentials)?
4. Kerry is experiencing a weakness in her legs. A physician tells her that she has Guillain-Barré syndrome, which is a progressive but reversible condition in which the myelin sheath is lost from parts of the nervous system. Explain why the loss of myelin would cause weakness in Kerry's legs.

The Nervous System

The Nervous System Consists of the Central and Peripheral Nervous Systems

Bone, Membranes, and Cerebrospinal Fluid Protect the CNS

The Brain Is the Central Command Center
- The cerebrum is the conscious part of the brain
- The thalamus allows messages to pass to the cerebral cortex
- The hypothalamus is essential to homeostasis
- The cerebellum is an area of sensory-motor coordination
- The medulla oblongata controls many of life's basic processes
- The pons connects the brain and spinal cord
- The limbic system is involved in emotions and memory
- The reticular activating system filters sensory input

The Spinal Cord Transmits Messages to and from the Brain and Is a Reflex Center

The Peripheral Nervous System Consists of the Somatic and Autonomic Nervous Systems
- The somatic nervous system controls conscious functions
- The autonomic nervous system controls internal organs

Disorders of the Nervous System Vary in Health Significance
- Headaches have several possible causes
- Stroke results when the brain is deprived of blood
- Coma results in lack of response to all sensory input
- Spinal cord injury results in impaired function below the site of injury

HEALTH ISSUE   Meningitis: Bacterial and West Nile Virus

HEALTH ISSUE   To Sleep, Perchance to Dream

John's fingers trembled as he rang Tegan's doorbell. How would their first date go? Would he know the right things to say? The right things to do? His mind was racing. In the theater, he could hardly smell or taste the hot buttery popcorn or feel the cold ice in his cola, as it spilled onto his lap. And then, there was the film. His pulse raced as aliens launched their missiles, fired multibarreled guns, and wielded chainsaws in combat scenes of mass destruction. He dodged in his seat as smashed cars and shattered glass seemed to fly off the screen directly at him. He laughed at the crisp one-liners. On the way home, Tegan smiled and thanked John for a lovely evening. She had really enjoyed the witty, well-paced script, she said, and the skillful choreography of destruction. Could they see another film next week, he asked nervously? "Of course," she said, "perhaps the new Brad Pitt film at the Cineplex downtown." Awkwardly, they kissed goodnight.

Although they were unlikely aware of it, John and Tegan's night at the movies was sponsored by their brains and nervous systems. Fear, long-term planning, language, automatic visceral responses to sounds, sights, tastes, pain, and smell—all are functions of the body, mind, and emotions, and all depend on the actions of billions of neurons in the peripheral and central nervous systems. How can this array of cellular switches, tangled neurons, and brain structures control such a wide range of human responses?

In this chapter, we explore the architecture of the nervous system. We also learn how the parts of the nervous system work together as a coordinated whole. We will also discover some disorders of the brain and spinal cord that affect every part of the human body and mind. ■

The Nervous System Consists of the Central and Peripheral Nervous Systems

If you were to view the nervous system apart from the rest of the body, you would see a dense area of neural tissue in the head and a cord of neural tissue that extends down the middle of the back (Figure 8.1). These parts are the brain and spinal cord, and they constitute the **central nervous system (CNS)**, which integrates and coordinates all voluntary and involuntary nervous functions. Extending from the brain and spinal cord, are many communication "cables"—the nerves that carry messages to and from the CNS. The nerves branch extensively, forming a vast network. There are also small clusters of nerve cell bodies, called **ganglia** (singular, *ganglion*). The nerves and ganglia are located outside of the CNS and make up the **peripheral nervous system (PNS)**. The PNS keeps the central nervous system in continuous contact with almost every part of the body.

The peripheral nervous system can be further subdivided on the basis of function into the somatic nervous system and the autonomic nervous system. The **somatic nervous system** consists of nerves that carry information to and from the CNS, resulting in sensations and voluntary movement. The **autonomic nervous system**, on the other hand, governs the involuntary, unconscious activities that maintain a relatively stable internal environment. The autonomic nervous system has two parts that generally cause opposite effects on the muscles or glands they control. The **sympathetic nervous system** is in charge during stressful or emergency conditions. In contrast, the **parasympathetic nervous system** adjusts bodily function so that energy is conserved during nonstressful times.

Although we have referred to various parts of the nervous system, remember that the parts function as a single unit, as can be seen in the following scenario. Imagine for a moment that you are meditating in the park; your eyes are closed, and you are resting. While you are relaxing, the parasympathetic nervous system is ensuring that your life-sustaining bodily activities continue. Then someone grasps your hand. Sensory receptors in the skin (part of the somatic nervous system) respond to the pressure and warmth of the hand. They send messages over sensory nerves to the spinal cord. Neurons within the spinal cord relay the messages to the brain. The brain integrates incoming sensory information, "deciding" on an appropriate response. The brain may then generate messages that cause your eyes to open. If the sight of the person holding your hand generates strong emotion, the sympathetic nervous system may cause your heart to beat faster and perhaps even your breathing rate to increase.

Bone, Membranes, and Cerebrospinal Fluid Protect the CNS

The brain and spinal cord are made up of many closely packed neurons. Neurons are very fragile, and most cannot divide and produce new cells. Therefore, a neuron that is damaged or dies cannot be replaced. The brain and spinal cord are protected by bony cases (the skull and vertebral column), membranes (the meninges), and a fluid cushion (cerebrospinal fluid).

The **meninges** are three protective connective tissue coverings of the brain and spinal cord (Figure 8.2). The outermost layer, the dura mater, is tough and leathery. Beneath the dura mater is the arachnoid, which is anchored to the next lower layer of meninges by thin threadlike extensions that resemble a spider's web (hence the name of the layer). The innermost layer is the pia mater, which is molded around the brain.

A variety of bacteria and viruses can cause inflammation of the meninges, a condition called *meningitis*. Regardless of the cause of infection, meningitis is a very serious condition because the infection can spread to the underlying nervous tissue, causing encephalitis (an inflammation of the brain). An outbreak of bacterial meningitis causes alarm on a college campus. The West Nile virus, which is becoming a problem throughout the United States, also causes a sometimes fatal form of meningitis. These conditions are discussed in the Health Issue essay, *Meningitis: Bacterial and West Nile Virus.*

The **cerebrospinal fluid** fills the space between the arachnoid and the pia mater as well as the internal cavities of the brain (ventricles) and spinal cord (central canal) (Figure 8.2). This fluid is

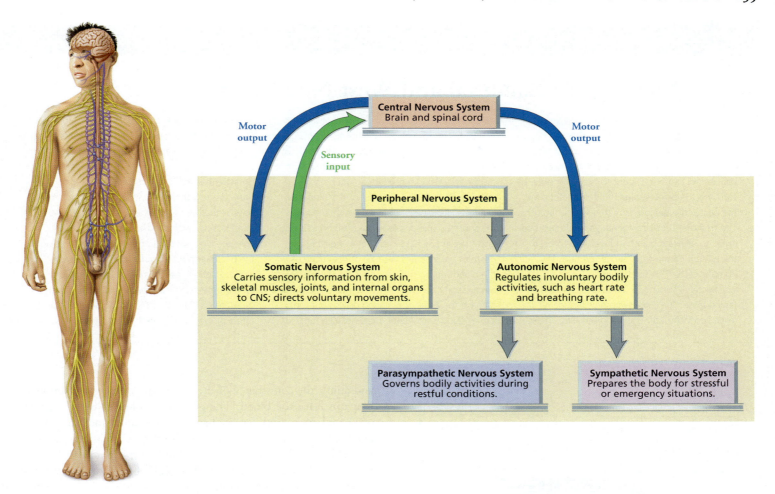

FIGURE **8.1**

The nervous system. Although the human nervous system functions as an integrated whole, it is convenient to talk about its functional parts. The brain and spinal cord (shown in pink), form the central nervous system (CNS), which integrates incoming information and directs appropriate responses. Nerves carry information to and from the central nervous system. These nerves, together with collections of nerve cell bodies called ganglia, constitute the peripheral nervous system (PNS). The PNS may be divided into the somatic and the autonomic nervous systems. The somatic nervous system (shown in yellow) carries sensory information from the skin, skeletal muscles, and joints to the CNS and also directs our voluntary movements, such as those of the arms and legs. The autonomic nervous system (shown in blue and purple) serves as an "automatic pilot" that regulates our involuntary bodily activities, such as heartbeat and breathing rate. The autonomic nervous system is composed of the parasympathetic nervous system, which governs activities during restful conditions, and the sympathetic nervous system, which prepares the body to face stressful or emergency conditions.

formed in the ventricles of the brain and circulates from the ventricles through the central canal of the spinal cord. Eventually, cerebrospinal fluid is reabsorbed into the blood.

Cerebrospinal fluid has several important functions:

- **Shock absorption.** Just as an air bag protects the driver of a car by preventing impact with the steering wheel, the cerebrospinal fluid protects the brain by cushioning its impact with the skull during blows or other head trauma.

- **Support.** Because the brain floats in the cerebrospinal fluid, it is not crushed under its own weight.

- **Nourishment.** The cerebrospinal fluid delivers nutrients and chemical messengers and removes waste products.

The central nervous system is also protected by the **blood-brain barrier,** a mechanism that selects the substances permitted to enter the cerebrospinal fluid from the blood. This barrier forms because the cells of the capillary walls that supply blood to the brain and spinal cord are held together much more tightly than are cells in capillaries in the rest of the body. These tight junctions force substances in the blood to pass through the cells of the capillaries instead of between the cells. Thus, the membranes of the capillary cells filter

HEALTH ISSUE

Meningitis: Bacterial and West Nile Virus

Meningitis is an inflammation of the meninges, the protective coverings of the brain and spinal cord.

Many types of bacteria and certain viruses can cause it. Diagnosis is usually done by studying the cerebrospinal fluid. If bacteria are the cause, the person is treated with antibiotics. The treatment for viral meningitis includes medicines to alleviate pain and fever and to keep the patient comfortable while the body's immune system fights the virus.

Bacterial meningitis is a serious, sometimes fatal disease. The initial symptoms include a high fever, headache, stiff neck, and vomiting. These symptoms may develop within hours or take several days. Not all symptoms may be present. The bacteria may spread to the blood, causing blood poisoning (septicemia), which may cause a skin rash that can start anywhere on the body. The rash initially looks like a cluster of tiny pinpricks. However, if they are not treated with antibiotics, the spots will join and look like fresh bruises. The bacteria can also cause gangrene, which is the death of body tissues because of the failure of the blood supply. The gangrene may require the amputation of arms or legs. The bacteria can also damage the kidneys, requiring dialysis or even a kidney transplant. About 1 case in 10 is fatal.

Freshmen college students living in dormitories are at increased risk of getting bacterial meningitis. Part of the reason is the means by which the bacteria are spread. Many people carry the bacteria that can cause meningitis in their throat without having any symptoms of illness. Nonetheless, coughing, sneezing, or intimate kissing can spread the bacteria. Presumably, the close living quarters make it easier for the bacteria to be spread. Upperclassmen are less susceptible, perhaps because they have built up immune defenses against the bacteria. Vaccines are available that protect against several forms of meningitis. The protection lasts at least

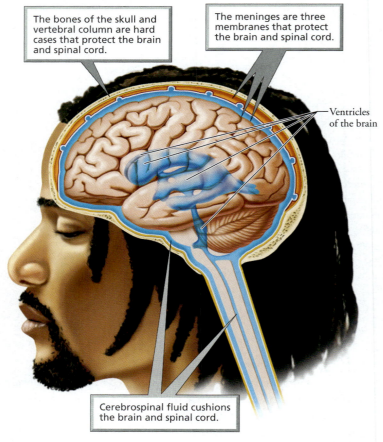

The bones of the skull and vertebral column are hard cases that protect the brain and spinal cord.

The meninges are three membranes that protect the brain and spinal cord.

Ventricles of the brain

Cerebrospinal fluid cushions the brain and spinal cord.

FIGURE **8.2**

The central nervous system is protected by the meninges, the cerebrospinal fluid, and the bones of the skull and vertebral column.

and adjust the consistency of the filtrate by selecting the substances that can leave the blood. The plasma membranes of the capillary walls are largely lipid. So, lipid-soluble substances, including oxygen and carbon dioxide, can pass through easily. Certain drugs, including caffeine and alcohol, are lipid soluble, explaining why they can have a rapid effect on the brain. The blood-brain barrier often frustrates physicians by keeping potentially life-saving, infection-fighting, or tumor-suppressing drugs away from brain tissue.

The Brain Is the Central Command Center

In a sense, your brain is more "you" than is any other part of your body, because it holds your emotions and the keys to your personality. Yet, if you were to look at your brain, you probably would deny any relationship to it. The brain is the consistency of soft cheese and weighs less than 1600 g (3 lb), which is probably less than 3% of your body weight. But your brain is the essential "you." It is the origin of your secret thoughts and desires; it remembers your most embarrassing moment; and it regulates your other body systems so that they function harmoniously while you concentrate on other activities. So, let's look at the brain to appreciate better how its many circuits are organized to accomplish these amazing feats. Selected areas of the brain are shown in Figure 8.3 and are described in the discussion that follows.

▌ The cerebrum is the conscious part of the brain

The **cerebrum** is the largest and most prominent part of the brain. It is, quite literally, your "thinking cap." Accounting for 83% of the total brain weight, the cerebrum gives you most of your human characteristics.

3 years, so a freshman who is vaccinated should not require a booster during the college years. Some colleges are now requiring that incoming freshmen be vaccinated against meningitis.

The West Nile virus, which is transmitted by infected mosquitoes, can also cause meningitis and encephalitis (brain inflammation). The first reported cases of West Nile virus in North America were in New York City in 1999. Since then, the disease has spread to nearly every state, and the spread doesn't show signs of slowing. The virus can infect certain vertebrates, including humans, horses, birds, and occasionally dogs and cats. Testing mosquitoes and dead birds, especially crows and starlings, for the presence of the virus is one way to track its spread. Most

people under the age of 50 have few or no symptoms when they are infected with the West Nile virus. Because symptoms are similar to those of the flu—fever, headache, and muscle and joint pain—many people who have become infected are unaware of it. However, older people have weaker immune systems. If they become infected, they are more likely to develop meningitis or encephalitis, illnesses that can cause brain damage, paralysis, or death.

You can protect yourself from West Nile virus by avoiding wet and humid areas that harbor mosquitoes. If you must enter areas where mosquitoes are likely to be, wear light-colored clothing that covers your body and use insect repellent.

Human-to-human transfer of the West Nile virus is possible but less likely. Transfusion of blood contaminated with the virus has caused illness in about two dozen people. Transmission of West Nile virus in contaminated blood is now preventable, because there is now a test that can be used to screen blood donations for the virus. During the summer of 2003, West Nile virus was found in the blood of 600 donors, and their blood was kept out of blood banks. There have also been a few instances in which the virus has spread from a pregnant woman to her fetus and to an infant through breast feeding. 👫

The many ridges and grooves on the surface of the cerebrum make it appear wrinkled. Some furrows are deeper than others. The deepest indentation is in the center and runs from front to back. This groove, called the longitudinal fissure, separates the cerebrum into two hemispheres (Figure 8.4). Each hemisphere receives sensory information from and directs the movements of the opposite side of the body. Furthermore, as we will see, the hemispheres process information in slightly different ways and are, therefore, specialized for slightly different mental functions.

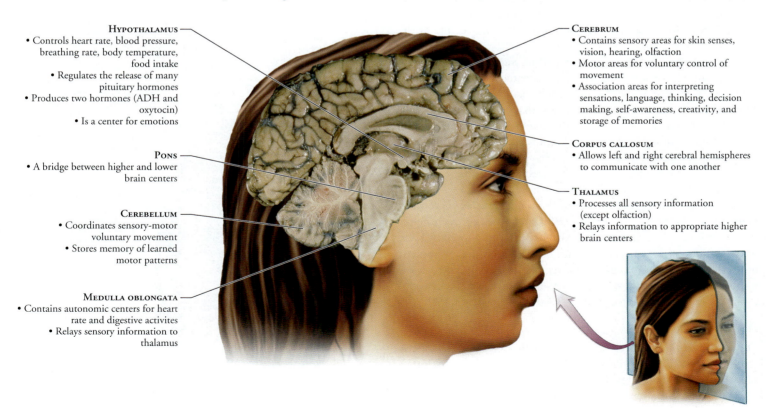

HYPOTHALAMUS
- Controls heart rate, blood pressure, breathing rate, body temperature, food intake
- Regulates the release of many pituitary hormones
- Produces two hormones (ADH and oxytocin)
- Is a center for emotions

PONS
- A bridge between higher and lower brain centers

CEREBELLUM
- Coordinates sensory-motor voluntary movement
- Stores memory of learned motor patterns

MEDULLA OBLONGATA
- Contains autonomic centers for heart rate and digestive activites
- Relays sensory information to thalamus

CEREBRUM
- Contains sensory areas for skin senses, vision, hearing, olfaction
- Motor areas for voluntary control of movement
- Association areas for interpreting sensations, language, thinking, decision making, self-awareness, creativity, and storage of memories

CORPUS CALLOSUM
- Allows left and right cerebral hemispheres to communicate with one another

THALAMUS
- Processes all sensory information (except olfaction)
- Relays information to appropriate higher brain centers

FIGURE **8.3**

A section through the brain from front to back (sagittal section), indicating the functions of selected structures

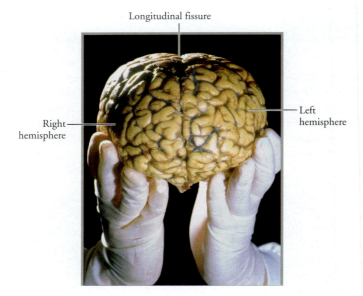

Right hemisphere

Left hemisphere

FIGURE **8.4**

A photograph of the human brain from the front, showing the left and right cerebral hemispheres

The thin outer layer of each hemisphere is called the **cerebral cortex** (Figure 8.5). (*Cortex* means bark or rind.) The cerebral cortex consists of billions of neuroglial cells, nerve cell bodies, and unmyelinated axons and is described as **gray matter**. Although the cerebral cortex is only 1 to 2 mm (about $\frac{1}{8}$ in.) thick, it is highly folded. These folds, or convolutions, triple the surface area of the cortex.

Beneath the cortex is the cerebral **white matter**, which appears white because it consists primarily of myelinated axons. Recall from Chapter 7 that myelin sheaths increase the rate of conduction along axons and are, therefore, found on axons that conduct information over long distances. The axons of the cerebral white matter allow various regions of the brain to communicate with one another and with the spinal cord. A very important band of white matter, called the *corpus callosum*, connects the two cerebral hemispheres so they can communicate with one another.

Other grooves on the surface of the brain serve as anatomical landmarks, indicating the boundaries of four lobes on each hemisphere: frontal, parietal, temporal, and occipital (Figure 8.6). Each of these lobes has its own specializations. Although the assignment of a specific function to a particular region of the cerebral cortex is imprecise, it is generally agreed that there are three types of functional areas: sensory, motor, and association.

SENSORY AREAS

Our awareness of sensations depends on the sensory areas of the cerebral cortex. The various sensory receptors send information to sensory areas of the cortex, each sense to a different region. If you stood on a street corner watching a parade go by, you would hear the band play because information from your ears would be sent to the auditory area in the temporal lobe. You would see the flag wave because information from your eyes would be sent to the

visual area in the occipital lobe. When you catch a whiff of popcorn, information is sent from the olfactory (smell) receptors in your nose to the olfactory area in the temporal lobe of the cortex. As you eat that popcorn, you know it is too salty because information from the taste receptors is sent to gustatory areas in the parietal lobe.

You know that you are standing in the hot sun and that your belt is too tight because information from receptors in the skin regarding touch, pain, and temperature and in the joints and skeletal muscles is sent to the **primary somatosensory area**. This region forms a band in the parietal lobes that stretches over the cortex from the region of one ear to the other (Figure 8.7). Sensations from different parts of the body are represented in different regions of the primary somatosensory area (of the hemisphere on the opposite side of the body). The greater the degree of sensitivity, the greater the area of cortex devoted to that body part. Thus, if areas of equal size are considered, your most sensitive body parts, such as the tongue, hands, face, and genitals, have more of the cortex devoted to them than do less sensitive areas, such as the forearm.

MOTOR AREAS

If you decide to join the parade, the **primary motor area** (Figure 8.7) of the cerebral cortex will send messages to your skeletal muscles. This motor area controls voluntary movement. It forms a band in the frontal lobe, just anterior to the primary

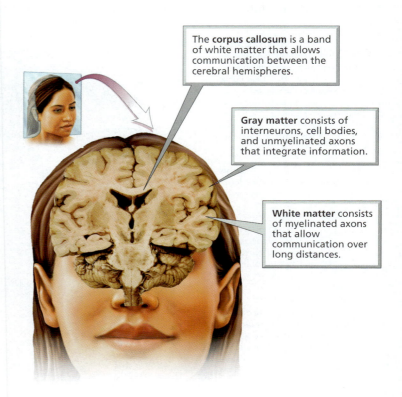

The **corpus callosum** is a band of white matter that allows communication between the cerebral hemispheres.

Gray matter consists of interneurons, cell bodies, and unmyelinated axons that integrate information.

White matter consists of myelinated axons that allow communication over long distances.

FIGURE **8.5**

A cross section through the brain, showing the gray matter, white matter, and selected deeper structures

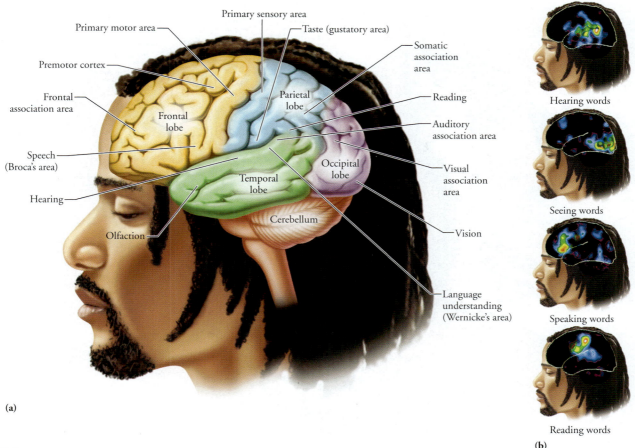

FIGURE **8.6**

The cerebral cortex. (a) The cerebral cortex has four lobes. Some of the functions associated with each lobe are indicated. (b) These PET scans of the brain show regions of increased blood flow during different tasks. Increased blood flow indicates regions that are active while hearing words, seeing words, speaking words, and reading words. Notice the relation between active regions of the cerebral cortex during these tasks and the cortical areas for various language skills shown in part (a).

somatosensory area. The motor area is meticulously arranged in a manner similar to the somatosensory area. Each point on the surface corresponds to the movement of a different part of the body. The parts of the body over which we have finer control, such as the tongue and fingers, have greater representations on the motor cortex than do regions with less dexterity, such as the trunk of the body.

Just in front of the motor cortex is the *premotor cortex*. It coordinates learned motor skills that are patterned or repetitious, such as typing or playing a musical instrument. The premotor cortex coordinates the movement of several muscle groups at the same time. When a pattern of movement is repeated many times, the proper pattern of stimulation is stored in the premotor cortex. For example, as a guitar player practices playing a particular song many times, the pattern of stimulation needed to play that song is stored in the premotor cortex. Then, each time the song is played, the premotor cortex will stimulate the primary cortex in the pattern needed to play that song without thinking about where on the strings fingers should be placed.

ASSOCIATION AREAS

The association areas of the cerebral cortex communicate with the sensory and motor areas to analyze and act on sensory input. Next to each primary sensory area is an association area that analyzes input. As the sensory association areas communicate with one another and with other parts of the brain, you begin to recognize what you are sensing. Each sensory association area communicates with the general interpretation area. This area intergrates the input from sensory association areas with stored sensory memories and assigns meaning to the experience. For example, on a dark night, your eyes may detect a small moving object. If the object then rubs against your legs and purrs, you may recognize it as the neighbor's friendly cat. However, if it turns away from you and raises its tail, you will recognize it as a skunk.

Information is then sent to the most complicated of all association areas, the **prefrontal cortex** (the most anterior part of the frontal lobe), which will decide how you should respond. In making the decision, the prefrontal cortex will predict the consequences of possible responses and judge which response will be best for you

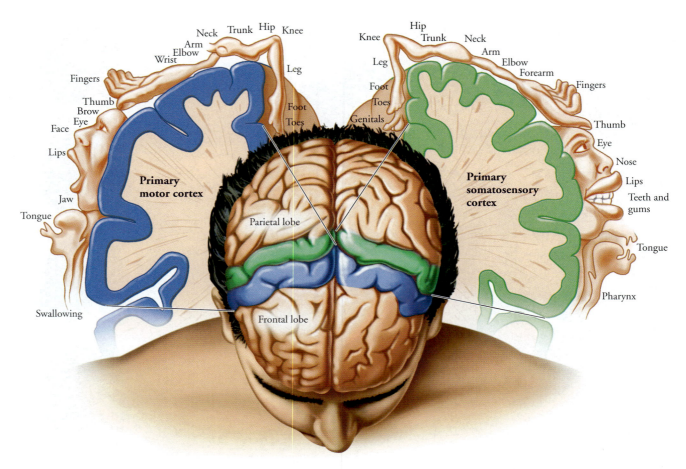

FIGURE **8.7**

The primary motor and the primary somatosensory regions of the cerebral cortex are meticulously arranged, with each area corresponding to a particular part of the body. The general arrangement is similar in the two regions. However, the amount of sensory cortex devoted to each body part increases with the degree of sensitivity of that part. The largest portions of the cerebral motor cortex are devoted to the body parts over which we have the greatest degree of motor control.

in that situation. The prefrontal cortex is also important in reasoning, long-term planning, producing abstract ideas, judgment, complex learning, intellect, and personality.

In our everyday lives, the two cerebral hemispheres gather essentially the same sensory information about the world around us. However, that input often is processed in slightly different ways by each hemisphere. The left hemisphere generally processes information analytically. It dissects the situation to understand the component parts and considers one aspect at a time. It is best in verbal and mathematical skills. Language abilities are generally restricted to one hemisphere. Language is centered in the left hemisphere of most people. The right hemisphere, on the other hand, puts the parts together to understand the whole. The right hemisphere excels at generating visual images and mental images of sound, touch, and smell to compare relationships. As a result, the right hemisphere is better at space and pattern perception, identifying objects on the basis of shape, and recognizing faces. Perhaps this emphasis is why the right hemisphere is generally better at music and art.

You get the benefit of both ways of thinking, however, because the two hemispheres communicate with one another through the corpus callosum (see Figure 8.3). Therefore, when you find your new puppy with one of your favorite shoes in his mouth, the right cerebral hemisphere may be the first to recognize him, but the left hemisphere allows you to respond, "Bad dog!"

The thalamus allows messages to pass to the cerebral cortex

The cerebral hemispheres sit comfortably on the thalamus (see Figure 8.3). The **thalamus** is often described as the gateway to the cerebral cortex because all messages to the cerebral cortex must first pass through the thalamus. The thalamus is important in sensory experience, motor activity, stimulation of the cerebral cortex, and memory. Sensory input from every sense except smell and from all parts of the body is delivered to the thalamus. The thalamus then sorts the information by function and relays it to the appropriate regions of the brain for processing. Some regions of the thalamus do more than just relay information to the cortex; they integrate information. At the thalamic level of processing, you have a general idea whether the sensation is pleasant or unpleasant. If you step on a tack, for instance, you may experience pain by the time the messages reach the thalamus, but you will not know where it hurts until the message is directed to the cerebral cortex.

The hypothalamus is essential to homeostasis

Below the thalamus is the **hypothalamus** (*hypo,* "under"), a small region of the brain that is essential to homeostasis—the maintenance of a stable environment within the body (discussed in Chapter 4). The hypothalamus, shown in Figure 8.3, influences blood pressure, heart rate, digestive activity, breathing rate, and many other vital physiological processes. Because the hypothalamus receives input from the cerebral cortex, it can make your heart beat faster when you just see or think of something exciting or dangerous—a rattlesnake about to strike, for instance. In addition, the hypothalamus acts as the body's "thermostat" and adjusts body temperature to its set point. By regulating food intake, hunger, and thirst, the hypothalamus helps maintain the body's constant internal environment. The hypothalamus also coordinates the activities of the nervous system and the endocrine (hormonal) system, as we will see in Chapter 10.

As part of the limbic system (discussed shortly), the hypothalamus is also part of the circuitry for emotions. Specific regions of the hypothalamus play a role in the sex drive and are involved with the perception of pain, pleasure, fear, and anger.

Another area of the hypothalamus (the suprachiasmatic nucleus) serves as a "master biological clock" that synchronizes all the other biological clocks within the body and sets the timing for all our biological rhythms. Most of your physiological processes, as well as many of your physical abilities and even your mental sharpness, vary predictably throughout the day. And, day after day, these variations repeat themselves with such regularity that they are called biological rhythms. Homeostasis does not describe a completely constant internal state; it is more than the regulation of the body's functioning within certain limits. Instead, homeostasis is often rhythmic, so physiological processes fluctuate in a predictable manner.

The cerebellum is an area of sensory-motor coordination

The **cerebellum** (see Figure 8.3) is the part of the brain responsible for sensory-motor coordination. It acts as an automatic pilot that produces smooth, well-timed voluntary movements and controls equilibrium and posture. The cerebellum receives sensory information regarding the position of joints and the degree of tension in muscles and tendons throughout the body. It integrates this information with input from the eyes and the equilibrium receptors in the ears. As a result, the cerebellum can determine the body's position, as well as where it is going, at any given instant.

The coordination of sensory input and motor output by the cerebellum involves two important processes: comparison and prediction. During every move you make, the cerebellum continuously compares the actual position of each part of the body with where it *ought* to be at that moment (with regard to the intended movement) and makes the necessary corrections. Try to touch the tips of your two index fingers together above your head. You probably missed on the first attempt. However, the cerebellum makes the necessary corrections, and you will likely succeed on the next attempt. At the same time, the cerebellum calculates future positions of a body part during a movement. Then, just before that part reaches the intended position, the cerebellum sends messages to stop the movement at a specific point. Therefore, when you scratch an itch on your cheek, the hand stops before slapping your face!

The medulla oblongata controls many of life's basic processes

The **medulla oblongata** is often simply called the medulla. This marvelous inch of nervous tissue contains reflex centers for some of life's most vital physiological functions—including the pace of the basic breathing rhythm, the force and rate of heart contraction, and blood pressure. The medulla (see Figure 8.3) connects the spinal cord to the rest of the brain. Therefore, all sensory information going to the upper regions of the brain and all motor messages leaving the brain are carried by nerve tracts running through the medulla.

stop and think

Why would a brain tumor that destroyed the functioning of nerve cells in the medulla lead to death more quickly than a tumor of the same size on the cerebral cortex?

The pons connects the brain and spinal cord

The **pons**, which means "bridge," connects lower portions of the CNS with higher brain structures. More specifically, it connects the spinal cord and cerebellum with the cerebrum, thalamus, and hypothalamus. In addition, the pons has a discrete region that assists the medulla in regulating respiration.

The limbic system is involved in emotions and memory

The **limbic system** is a collective term for several structures involved in emotions and memory (Figure 8.8). It is called a system because it includes parts of several brain regions and the neural pathways that connect them. The structures in the limbic system are grouped on the basis of function rather than anatomy.

The limbic system is our emotional brain. It allows us to experience countless emotions including rage, pain, fear, sorrow, joy, and sexual pleasure. Emotions are important because they motivate behavior that will increase the chance of survival. Fear, for example, may have evolved to focus the mind on the threatening things in the environment.

Connections between the cerebrum and the limbic system allow us to have *feelings* about *thoughts*. These connections also keep us from responding to emotions, such as rage, in ways that would be unwise. The limbic system includes the hypothalamus. It is also connected to lower brain centers, such as the medulla, that control the activity of our internal organs. Therefore, we also have "gut" responses to emotions.

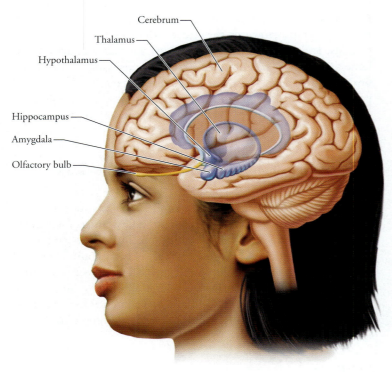

FIGURE **8.8**

The limbic system is our emotional brain. It allows us to feel fear, sorrow, joy, and sexual arousal. These emotions lead to behaviors that increase the chance of survival. The diagram shows the limbic system as a three-dimensional structure within the brain viewed from the left side.

Certain structures involved with our sense of smell (olfaction) are part of the limbic system. As a result, we often have emotional responses to odors. For example, we usually consider the odor of baking cinnamon rolls as pleasant and the odor of skunk as repulsive.

The **hippocampus** is a part of the limbic system that is essential to converting short-term memory to long-term memory. **Short-term memory** holds a small amount of information for a few seconds or minutes, as when you look up a phone number and remember it long enough to place the call. **Long-term memory** stores large amounts of information for hours, days, or years. The amygdala links memories to emotional states.

▌The reticular activating system filters sensory input

The **reticular activating system (RAS)** is a very extensive network of neurons that runs through the medulla and projects to the cerebral cortex. The RAS functions as a net, or filter, for sensory input. Our brain is constantly flooded with a tremendous amount of sensory information, about 100 million impulses each second, most of them trivial. The RAS filters out the repetitive familiar stimuli—the sound of street traffic, paper rustling, the coughing of the person next to you, or the pressure of clothing. However, infrequent or important stimuli pass through the RAS to the cerebral cortex and, therefore, reach our consciousness. Consequently, if you are very tired, you might

fall asleep with the television on but wake up when someone whispers your name.

In addition, the RAS is an activating center. Unless inhibited by other brain regions, the RAS activates the cerebral cortex, keeping it alert and "awake." Consciousness occurs only while the RAS stimulates the cerebral cortex. When sleep centers in other regions of the brain inhibit activity in the RAS, sleep results. In essence, then, the cerebrum "sleeps" whenever it is not stimulated by the RAS. Sensory input to the RAS stimulates the cerebral cortex and raises consciousness levels, explaining why it is usually easier to sleep in a dark quiet room than in an airport terminal. Conscious activity in the cerebral cortex can also stimulate the RAS, which will, in turn, stimulate the cerebral cortex. Therefore, thinking about a problem may keep you awake all night.

stop and think

When a boxer is hit very hard in the jaw, his head and, therefore, his medulla and RAS are twisted sharply. Why might this twisting result in a knockout in which the boxer loses consciousness?

The Spinal Cord Transmits Messages to and from the Brain and Is a Reflex Center

The other major component of the central nervous system besides the brain is the spinal cord. The **spinal cord** is a tube of neural tissue that is continuous with the medulla at the base of the brain and extends about 45 cm (17 in.) to just below the last rib. For most of its length, the spinal cord is about the diameter of your little finger. The spinal cord is slightly thicker in two regions, just below the neck and at the end of the cord, because of the large group of nerves connecting these regions of the cord with the arms and legs. Running the length of the spinal cord is the central canal, which is filled with cerebrospinal fluid.

The spinal cord is encased in and protected by the stacked bones of the vertebral column (Figure 8.9). Pairs of spinal nerves (part of the peripheral nervous system) arise from the spinal cord and exit through the openings between the vertebrae to serve a specific part of the body. Each member of the pair of spinal nerves serves either the right or the left side of the body. A disk of cartilage that serves as a cushion separates the vertebrae.

The spinal cord has two functions: (1) to transmit messages to and from the brain and (2) to serve as a reflex center. The transmission of messages is performed primarily by white matter, which is found in the outer regions of the spinal cord. Within the white matter are myelinated nerves grouped into tracts. Ascending tracts carry sensory information up to the brain. Descending tracts carry motor information from the brain to a nerve leaving the spinal cord.

The second function of the spinal cord is to serve as a reflex center. Spinal reflexes are essentially "decisions" made by the spinal cord. A reflex action is an automatic response to a stimulus. Reflexes are prewired in a circuit of neurons, called a **reflex arc.**

Back view   **View from left side**   **Top view**

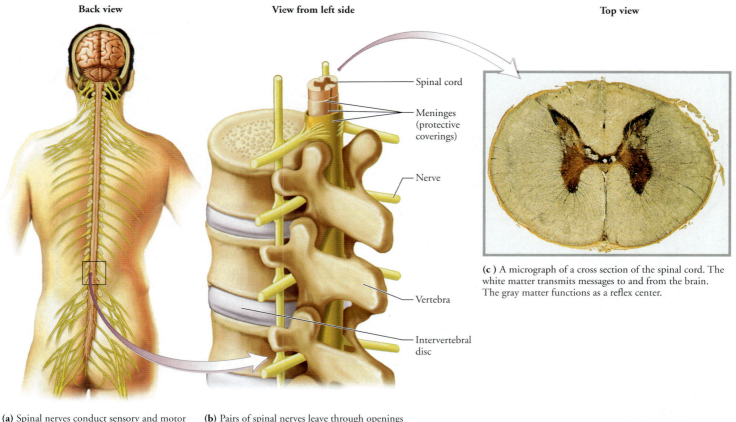

Spinal cord

Meninges (protective coverings)

Nerve

Vertebra

Intervertebral disc

(c) A micrograph of a cross section of the spinal cord. The white matter transmits messages to and from the brain. The gray matter functions as a reflex center.

(a) Spinal nerves conduct sensory and motor information between the central nervous system and a specific region of the body.

(b) Pairs of spinal nerves leave through openings between the vertebrae.

FIGURE **8.9**

The spinal cord is a column of neural tissue that is protected by the bones of the vertebral column.

The circuit consists of a receptor, a sensory neuron (which brings information from the receptors toward the central nervous system), usually at least one interneuron, a motor neuron (which brings information from the central nervous system toward an effector), and an effector (Figure 8.10). Shaped somewhat like a butterfly in the central region of the spinal cord, the gray matter houses the interneurons and the cell bodies of motor neurons involved in reflexes.

Spinal reflexes are beneficial when a speedy reaction is important to a person's safety. Consider, for example, the withdrawal reflex. When you step on a piece of broken glass, impulses speed toward the spinal cord over sensory nerves. Within the gray matter of the spinal cord, the sensory neuron synapses with an interneuron. The interneuron, in turn, synapses with a motor neuron that sends a message to the appropriate muscle to contract and lift your foot off the glass.

While the spinal reflexes were removing the foot from the glass, pain messages from the cut foot were sent to the brain through ascending tracts in the spinal cord. However, it takes longer to get a message to the brain than it does to get one to the spinal cord because the distance and number of synapses involved are greater. Therefore, by the time pain messages reach the brain, you have already withdrawn your foot. Nonetheless, once the sensory information reaches the conscious brain, decisions can be made about how to care for the wound.

The Peripheral Nervous System Consists of the Somatic and Autonomic Nervous Systems

The nerves and the ganglia (collections of neurons and their associated nerve fibers) of the peripheral nervous system (PNS) carry information between the central nervous system (CNS) and the rest of the body. The peripheral nervous system consists of spinal nerves and cranial nerves.

The body has 31 pairs of **spinal nerves**, each of which originates in the spinal cord and services a specific region of the body (Figure 8.11a). All spinal nerves carry both sensory and motor fibers. The fibers from sensory neurons are grouped. The resulting bundle is called the *dorsal root* of the spinal nerve, because it enters the spinal cord from the dorsal, or posterior, side. The cell bodies of the sensory neurons are located in a ganglion in the dorsal root. The axons of motor neurons leave the ventral (front side) of the spinal cord in a bundle called the *ventral root*. The cell bodies of

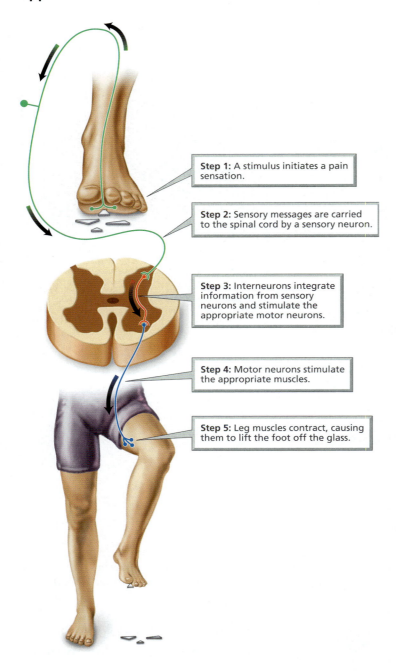

Step 1: A stimulus initiates a pain sensation.

Step 2: Sensory messages are carried to the spinal cord by a sensory neuron.

Step 3: Interneurons integrate information from sensory neurons and stimulate the appropriate motor neurons.

Step 4: Motor neurons stimulate the appropriate muscles.

Step 5: Leg muscles contract, causing them to lift the foot off the glass.

FIGURE **8.10**

A reflex arc consists of a sensory receptor, a sensory neuron, usually at least one interneuron, a motor neuron, and an effector.

motor neurons are located in the gray matter of the spinal cord. The dorsal and ventral roots join to form a single spinal nerve, which passes through the opening between the vertebrae.

The 12 pairs of **cranial nerves** (Figure 8.11b) arise from the brain and service the structures of the head and certain body parts such as the heart and diaphragm. Some cranial nerves carry only sensory fibers, others carry only motor fibers, and others carry both types of fiber.

The somatic nervous system controls conscious functions

The peripheral nervous system is subdivided into the somatic nervous system and the autonomic nervous system. The somatic nervous system carries sensory messages that tell us about the world around us and within us, and it controls movement. Sensory messages carried by somatic nerves result in sensations, including light, sound, and touch. The somatic nervous system also controls our voluntary movements, allowing us to smile, stomp a foot, sing a lullaby, or frown as we sign a check.

The autonomic nervous system controls internal organs

The autonomic nervous system automatically adjusts the functioning of our body organs so that an internal stability (homeostasis) is maintained and the body is able to meet the demands of the world around it. The somatic nervous system sends information about conditions within the body to the autonomic nervous system. Then the autonomic nervous system makes the appropriate adjustments. Its activities alter digestive activity, open or close blood vessels to shunt blood to areas that need it most, and alter heart rate and breathing rate.

Recall that the autonomic nervous system consists of two branches: the sympathetic and the parasympathetic nervous systems. The sympathetic nervous system gears the body to face an emergency or stressful situation, such as fear, rage, or vigorous exercise. Thus, the sympathetic nervous system prepares one for fight or flight. In contrast, the parasympathetic nervous system adjusts body function so that energy is conserved during relaxation.

Both the parasympathetic and the sympathetic nervous systems send nerve fibers to innervate most, but not all, internal organs (Figure 8.12). When both systems innervate an organ, they have opposite effects on its function. If one system stimulates, the other system inhibits. The antagonistic effects are brought about by different neurotransmitters. Whereas sympathetic neurons release mostly norepinephrine at their target organs, parasympathetic neurons secrete acetylcholine at their target organs.

The sympathetic nervous system acts as a whole, bringing about all its effects at once. This action is possible because its neurons are connected through a chain of ganglia. A unified response is exactly what is needed in an emergency. To meet the threat, the sympathetic nervous system increases breathing rate, heart rate, and blood pressure. It also orchestrates the responses needed to increase the amount of glucose and oxygen delivered to the cells. The adrenal glands are stimulated to release two hormones, epinephrine and norepinephrine, into the bloodstream. These hormones back up and prolong the effects of sympathetic stimulation. In a crisis, digesting the previous meal is hardly a priority, and digestive activity is inhibited.

The effects of the parasympathetic nervous system occur more independently of one another. After the emergency, organ systems return to a relaxed state at their own pace, based on the anatomy of the system. Organs can respond to the parasympathetic nervous system independently because the ganglia containing the parasympathetic neurons that stimulate organs are not connected in a chain, as they are in the sympathetic nervous system. Instead, the ganglia of the parasympathetic nervous system are located near the organs.

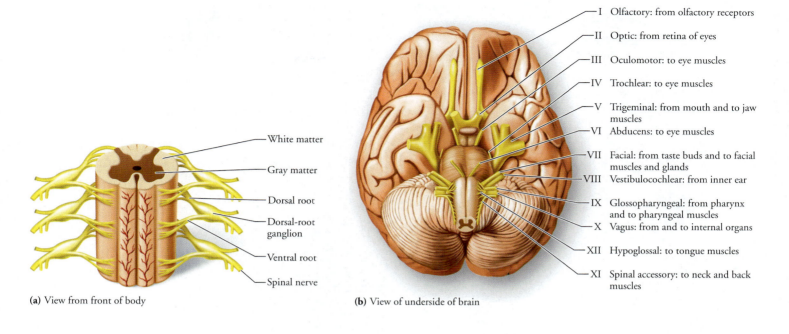

I Olfactory: from olfactory receptors
II Optic: from retina of eyes
III Oculomotor: to eye muscles
IV Trochlear: to eye muscles
V Trigeminal: from mouth and to jaw muscles
VI Abducens: to eye muscles
VII Facial: from taste buds and to facial muscles and glands
VIII Vestibulocochlear: from inner ear
IX Glossopharyngeal: from pharynx and to pharyngeal muscles
X Vagus: from and to internal organs
XII Hypoglossal: to tongue muscles
XI Spinal accessory: to neck and back muscles

White matter
Gray matter
Dorsal root
Dorsal-root ganglion
Ventral root
Spinal nerve

(a) View from front of body

(b) View of underside of brain

FIGURE **8.11**

(a) Spinal and (b) cranial nerves. The 12 pairs of cranial nerves can be seen in this view of the underside of the brain. Most cranial nerves service structures within the head, but some, such as the vagus, service the lower body organs. The descriptions indicate whether the neuron carries sensory information (toward the brain) or motor information (away from the brain).

Disorders of the Nervous System Vary in Health Significance

Disorders of the nervous system vary tremendously in severity and impact on the body. Some disorders, such as a headache, can be mild and often are more of a nuisance than a health problem. Insufficient sleep can sometimes cause more problems that expected (see the Health Issue essay, *To Sleep, Perchance to Dream*). Other disorders, such as stroke, coma, and spinal cord injury, can have devastating effects on a person's well-being.

▌ Headaches have several possible causes

Excessive exercise may make your muscles hurt. However, thinking too much cannot cause a headache. The brain has no pain receptors, so a headache is not a brainache. Headaches often present a no-win situation; they can be caused by stress or by relaxation, by hunger or by eating the wrong food, or by too much or too little sleep. The most common type of headache is a tension headache, affecting some 60% to 80% of people who suffer from frequent headaches. The pain of a tension headache is usually a dull, steady ache that is often felt as a tight band around the head, as if your head were in a vise. Migraine headaches are usually confined to one side of the head, often centered behind one eye. A migraine headache typically causes a throbbing pain that increases with each beat of the heart. It is sometimes called a "sick headache" because it often causes nausea and vomiting. Some migraine sufferers experience an aura, a group of sensory symptoms, just before an attack. The aura may be visual disturbances (a blind spot, zigzag lines, flashing lights), auditory hallucinations, or numbness. Migraines are set off by an imbalance in the brain's chemistry. The level of one of the brain's chemical neurotransmitters, serotonin, is low. With too little serotonin, pain messages flood the brain. Cluster headaches are headaches that tend to occur in groups, occurring two or three times a day for days or weeks. More common in men than in women, cluster headaches cause severe pain that lasts for a few minutes to a few hours, often awakening a person from a sound sleep.

what would you do?

There is an experimental pain treatment for severe headaches. A tiny electrode is implanted in the skin and placed near the nerve responsible for the pain. The device is powered by a battery that is implanted near the collarbone. Electric pulses are continually delivered through the electrode, blocking the pain signals in the nerve and stopping the pain. If you suffered from severely painful headaches would you opt for this treatment? What criteria would you use to decide?

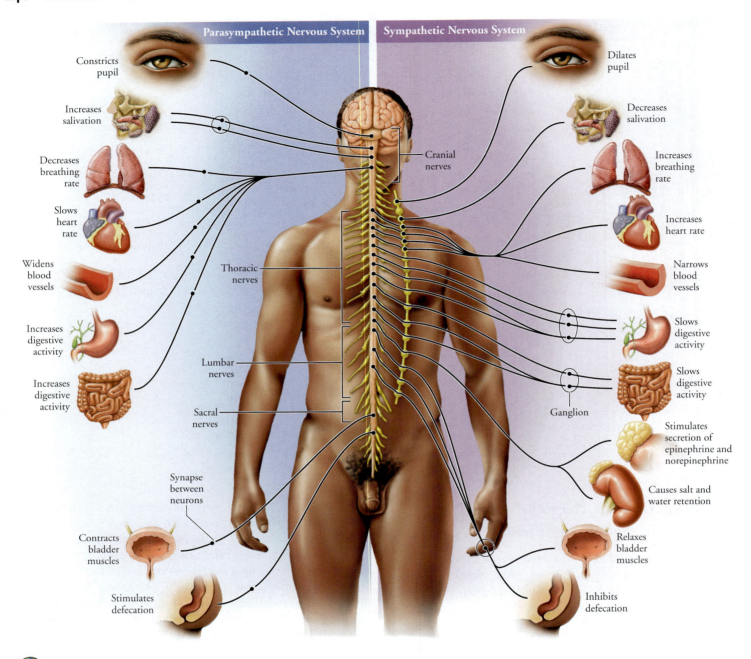

FIGURE **8.12**

The structure and function of the autonomic nervous system. Most organs are innervated by fibers from both the sympathetic and the parasympathetic nervous systems. When this dual innervation occurs, the two branches of the autonomic nervous system have opposite effects on the level of activity of that organ. A chain of ganglia links the sympathetic nervous system. It usually acts as a unit, with all its effects occurring together. The ganglia of the parasympathetic nervous system are near the organ serviced. Its effects are more localized.

Stroke results when the brain is deprived of blood

A stroke, also called a cerebrovascular accident, is the death of nerve cells in a region of the brain. The extent and location of the mental or physical impairment caused by a stroke depend on the region of the brain involved. If the left side of the brain is affected, the person may lose sensations in or the ability to move parts of the right side of his or her body. Because the language centers are usually in the left hemisphere, the person may have difficulty speaking. When the stroke dam-

ages the right rear of the brain, some people show what is called the neglect syndrome and behave as if the left side of everything, even their own bodies, does not exist. The person may comb only the hair on the right side of the head or eat only the food on the right side of the plate.

Neurons have a high demand for both oxygen and glucose. Therefore, when the blood supply to a portion of the brain is shut off, the affected neurons begin to die within minutes. Common causes of strokes include blood clots that block a vessel, hemorrhage from the

To Sleep, Perchance to Dream

Not all sleep is the same. Indeed, during a typical night's sleep, each hour and a half you will cycle through five stages of sleep. If the natural sleep pattern is disturbed, you may feel as though you have had a bad night's sleep, even if you have "slept" for 8 hours. Stage 1 sleep occurs while you "drift off to sleep." During this period of transition from wakefulness to sleep, you become less aware of your surroundings. You slowly tune out lights and low level noises (Figure 8.A). When you are exhausted, Stage 1 sleep can take over regardless of how much you struggle to stay awake. Within the next few minutes, Stage 2 sleep usually occurs. Muscle tension is lower than it is during wakefulness and breathing and heart rate also decrease. More than half the night's sleeping time is spent in Stage 2 sleep. Soon afterward, you sink even deeper into sleep—Stages 3 and 4. After about 20 minutes of Stage 4 sleep, you usually switch back to Stage 3 followed by Stage 2. Then, you will begin an interval of Stage 5 sleep, called paradoxical sleep, so named because the type of brain waves that occur are also seen during alert mental activity. Therefore, it seems that the brain is quite active at this time, but it is more difficult to awaken a person in paradoxical sleep than it is at any other stage. Paradoxical sleep is also called rapid-eye-movement (REM) sleep because your eyes move rapidly behind closed eyelids during this stage. At this time, your heart and breathing rate can be quite variable. During REM sleep, almost every skeletal muscle in your body except for those of eyes and ears, is virtually paralyzed. This paralysis

may keep us from hurting ourselves or others.

Adults repeatedly cycle through these stages all night (Figure 8.A). The first bout of REM sleep usually occurs about 90 minutes after falling asleep and lasts about 10 minutes. The REM period of each successive cycle is a little longer. By morning, the interval of REM sleep may be about an hour long. So, the second half of a night's sleep is not identical to the first half. There is much more REM in the second half. For this reason, several short "cat naps" do not provide the same quality of sleep as a single, longer period of sleep, even if the total number of hours of sleep is identical.

Rapid-eye-movement (REM) sleep is significant because most dreams that have a storylike progression of events occur at this time. About 80% of the people who are awakened during REM sleep and asked what was going on say they were dreaming. Only about 20% report that they were dreaming if awakened during other stages of sleep. However, the dreams that occur outside of REM sleep are more likely to be a thought, an image, or an emotion than a story. Indeed, most nightmares occur during Stage 3 and 4 sleep.

How much sleep does one need to remain healthy? There is no set answer to this question. You need less sleep as you get older. A newborn sleeps 18 hours a day; young adults generally sleep 7 to 8 hours; and elderly people need only $4\frac{1}{2}$ to $6\frac{1}{2}$ hours. Sleep requirements also vary among individuals. A few people need as little as 3 hours of sleep a night. Others do not feel their best unless they get 10 hours or more. So, you need as much sleep as it takes to

feel well and function efficiently the next day—and no more.

Millions of Americans cheat on sleep, however. Students pull "all-nighters" studying for exams. Some students must also fit jobs into their study schedules. Parents try to juggle jobs and family, and sometimes school as well. Workers cope with long shifts and long commutes. And when does *fun* fit into the schedule? People with such busy schedules may not have enough time for a good night's sleep.

Without enough sleep, people usually can manage to get through the day doing simple things—walking, seeing, hearing—but they cannot think clearly. They cannot make appropriate judgments, and their attention span is short. Thus, sleepiness interferes with the ability to learn. Sleepy students often sit through class in a daze, experiencing it like a dream. Some nod off.

Sleep deprivation can also be hazardous to health. Drowsiness is a major cause of industrial accidents and traffic fatalities. In a national poll, 20% of American drivers reported having fallen asleep at the wheel within the past year. Driving on Friday night is a greater risk than it is on Monday night, because so many drivers have been sleep deprived all week.

Ironically, because of the pressures of life, sleep does not come easily. Tens of millions of Americans will lie awake tonight, some of them for hours, suffering from insomnia. Insomnia occurs in different patterns—difficulty falling asleep, waking up during the night, or waking up earlier than desired. If you get insomnia, don't worry. Most people have insomnia at some time. Occasional bouts of poor sleep will

continued ⟶

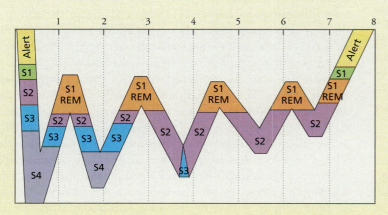

FIGURE **8.A** During a good night's sleep a person generally cycles through the five stages of sleep every 90 minutes.

HEALTH ISSUE (CONTINUED)

not harm your health. Furthermore, scientific studies have shown that people who complain about an inability to sleep are poor judges of the amount of sleep they actually get. About one-third of those who think they are lying awake are actually asleep. Some even dream that they are awake and trying to fall asleep. Most insomniacs overestimate the amount of time it takes to fall asleep and underestimate the duration of sleep. Most elderly people report problems with sleep. For some of them, there is a physical cause, such as chronic pain. However, many elderly people have not adjusted to their reduced need for sleep and worry when their customary sleep schedule changes, even though they are getting all the sleep they need.

What should you do if you occasionally have trouble sleeping? Here are some suggestions.

1. Establish a regular bedtime and a regular waking time. Do not sleep late, even on weekends. Do not nap.

2. Do not lie in bed if you cannot sleep. The harder you try to fall asleep, the more aroused you become and the less likely you are to be able to fall asleep. Learn to associate the bedroom with sleep by getting out of bed if you have not fallen asleep in 10 to 15 minutes and doing something else until you are sleepy again.

3. Relax before bedtime: read a book, watch television, take a warm shower—whatever helps you unwind.

4. Do not use caffeinated beverages, such as coffee, tea, and cola, for 6 hours before bedtime. Chocolate also contains caffeine, so a cup of hot cocoa at bedtime is not a good idea.

5. Establish a regular pattern of exercise. Mild exercise promotes sleep and reduces stress. However, do not exercise too close to bedtime.

6. Avoid drinking alcohol for 2 hours before bedtime. A small amount of alcohol makes

you drowsy, but alcohol also disrupts sleep patterns. It does not increase the duration of sleep, and it decreases REM sleep.

7. Avoid using sleeping pills, even those sold over the counter in the drugstore, if possible. At best, they decrease the amount of time it takes to fall asleep by only 10 to 20 minutes and lengthen the night's sleep by only 20 to 40 minutes. Similar to alcohol, sleeping pills decrease REM sleep, and the REM sleep that occurs is not normal. Sleeping pills generally lose effectiveness within 2 weeks, so there is a tendency to increase the dosage. The effects of sleeping pills may linger into the next day, causing difficulty with coordination or memory. Furthermore, they have side effects, some of which are dangerous. The antihistamines in over-the-counter sleeping pills can provoke or worsen episodes of asthma, urinary retention, or glaucoma (increased pressure within the eye that can cause blindness). ♀♂

---

rupture of a blood vessel in one of the protective layers surrounding the brain (the pia mater), or the formation of fatty deposits that block a vessel. High blood pressure, heart disease, diabetes, smoking, obesity, and excessive alcohol intake increase the risk of stroke.

Coma results in lack of response to all sensory input

Although a comatose person seems to be asleep—eyes closed and no recognizable speech—a coma is not deep sleep. A person in a coma is totally unresponsive to all sensory input and cannot be awakened. Although the cerebral cortex is most directly responsible for consciousness, damage to the cerebrum is rarely the cause of coma. Instead, coma is caused by trauma to neurons in regions of the brain responsible for stimulating the cerebrum, particularly those in the reticular activating system or thalamus. Coma can be caused by mechanical shock—as might be caused by a blow to the head—tumors, infections, drug overdose, or failure of the liver or kidney.

Spinal cord injury results in impaired function below the site of injury

The spinal cord contains pathways that allow the brain to communicate with the rest of the body. Therefore, damage to the spinal cord can impair sensation and motor control below the site of injury. The extent and location of the injury will determine how

long these symptoms persist, as well as the degree of permanent damage. Depending on which tracts are damaged, injury results in the loss of sensation, paralysis, or both. If the cord is completely severed, there is a complete loss of sensation and voluntary movement below the level of the cut.

Restoring the ability to function to people with spinal cord injuries is an active area of research. Some researchers are trying to reestablish neural connections by stimulating nerve growth through treatments with nerve growth factors. Others are exploring the possibility of treatments using embryonic stem cells (discussed in Chapter 21). Stem cells retain the ability to develop into nerve cells. Mice with spinal cord injuries that were treated with embryonic stem cells recovered some ability to move. Another approach to restoring the ability to move is to use computers to electronically stimulate specific muscles and muscle groups. The stimulation is delivered through wires that are either implanted under the skin or woven into the fabric of tight-fitting clothing. A small computer, which is usually worn at the wrist, directs the stimulation to the appropriate muscles. This technology has helped some people with a spinal cord injury to walk again. It has also helped people, including actor Christopher Reeve, to breathe without a respirator for periods of time (Figure 8.13). In this case, electrodes stimulate the diaphragm, a muscle important in breathing.

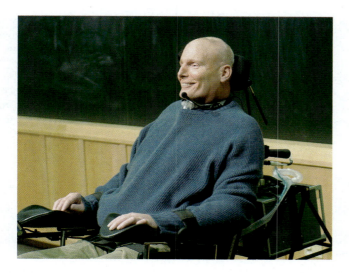

FIGURE **8.13**

When he fell from a horse, actor Christopher Reeve damaged his spinal cord in a region close to where it connects to the brain, cutting off communication between the brain and the spinal cord below the injury. The actor, who once played Superman, seems to be living up to the reputation of that character. He can now breathe for hours without a respirator. He has normal sensation over 65% to 75% of his body. He can move some fingers. In a swimming pool, he can move most joints, take a step, and push off the wall. Reeve is the first person to regain so much function so long after such a severe injury. Activity-based therapy, including activity stimulated by electrical stimulation, is credited with helping Reeve's body make the most of the remaining healthy spinal nerve cells.

exploring further ...

In Chapter 7, we learned that neurons communicate with one another using chemicals called neurotransmitters. Neurotransmitter molecules fit into receptors on the membrane of the receiving neuron and open ion channels. Depending on the type of channel opened, the receiving neuron is either excited or inhibited. Different neurotransmitters play roles in different behavioral systems. Norepinephrine and serotonin create a feeling of well-being. Dopamine helps regulate emotions.

In this chapter, we learned that different parts of the nervous system are specialized for different functions. The limbic system of the brain is a "pleasure center." The sympathetic nervous system prepares the body for emergency situations. To bring about its effects on muscles or glands, the sympathetic nervous system releases the neurotransmitter epinephrine.

Next, in Chapter 8A, Special Topic: Drugs and the Mind, we will consider psychoactive drugs—those that affect one's mental state. We will see that psychoactive drugs bring about their effects by increasing or decreasing the effects of specific neurotransmitters. In this way, neurotransmitters affect specific regions of the brain and alter one's mental state. In addition to effects on the nervous system, drugs can affect other organs. So, we will also consider the health risks associated with drug use.

REVIEWING THE CONCEPTS

The Nervous System Consists of the Central and Peripheral Nervous Systems (p. 134)

1. The nervous system is divided into the central nervous system (the brain and spinal cord) and the peripheral nervous system (all the neural tissue outside the central nervous system). Ganglia are clusters of nerve cell bodies located outside the central nervous system. The peripheral nervous system can be further subdivided into the somatic nervous system and the autonomic nervous system.

Bones, Membranes, and Cerebrospinal Fluid Protect the CNS (pp. 134–136)

2. The brain and the spinal cord are protected by the bony cases of the skull and vertebral column, membranes (the meninges), and a fluid cushion (cerebrospinal fluid).

3. The meninges are three protective layers of connective tissue that cover the brain and spinal cord. Bacteria and viruses can cause inflammation of the meninges, resulting in the condition called meningitis.

4. The cerebrospinal fluid, located between the arachnoid and pia mater, serves as a shock absorber for the brain, supports the brain, and provides nourishment to the brain.

5. The blood-brain barrier is a mechanism that selects the substances permitted to enter the cerebrospinal fluid from the blood. This barrier prevents potentially damaging substances from entering the brain or spinal cord.

The Brain Is the Central Command Center (pp. 136–142)

6. The brain serves as the body's central command center, coordinating and regulating the body's other systems.

7. The cerebrum is the thinking, conscious part of the brain. It consists of two hemispheres. Each hemisphere receives sensory impressions from and directs the movements of the opposite side of the body. The cerebrum has an outer layer of gray matter called the cerebral cortex and an underlying layer of white matter consisting of myelinated nerve tracts that allow communication between various regions of the brain.

8. The cerebral cortex has three types of functional areas: sensory, motor, and association. Our awareness of sensation depends on the sensory areas of the cerebral cortex. Motor areas of the brain allow for the movement of different parts of the body. Association areas communicate with the sensory and motor areas to analyze and act on sensory input.

9. The thalamus is an important relay station for all sensory experience except smell. It also plays a role in motor activity, stimulation of the cerebral cortex, and memory.

10. The hypothalamus is essential in maintaining a stable environment within the body. It regulates many vital physiological functions, such as blood pressure, heart rate, breathing rate, digestion, and body temperature. The hypothalamus also coordinates the activities of the nervous and endocrine systems through its connection to the pituitary gland. As part of the limbic system, the hypothalamus is a center for emotions. It also serves as a "master biological clock."

11. The primary function of the cerebellum is sensory-motor coordination. The cerebellum integrates information from the motor cortex and sensory pathways to produce smooth movements. The cerebellum also stores memories of learned motor skills.

12. The medulla oblongata regulates breathing, heart rate, and blood pressure. It also serves as a pathway for all sensory messages to higher brain centers and motor messages leaving the brain.

13. The pons connects lower portions of the central nervous system with higher brain structures. It essentially connects the spinal cord and cerebellum to the cerebrum, thalamus, and hypothalamus.

14. The limbic system, which includes several brain structures, is largely responsible for emotions. The hippocampus, which is part of the limbic system, is essential to converting short-term memory to long-term memory.

15. The reticular activating system is a complex network of neurons that filters sensory input and keeps the cerebral cortex in an alert state.

The Spinal Cord Transmits Messages to and from the Brain and Is a Reflex Center (pp. 142–143)

16. The spinal cord is a tube of nerve tissue extending from the medulla to approximately the bottom of the rib cage. The spinal cord has two functions: to conduct messages between the brain and the body and to serve as a reflex center.

The Peripheral Nervous System Consists of the Somatic and Autonomic Nervous Systems (pp. 143–144, 146)

17. The peripheral nervous system consists of 31 pairs of spinal nerves, each of which originates in the spinal cord and services a specific region of the body, and 12 pairs of cranial nerves, each of which arises from the brain and services the structures of the head and certain body parts such as the heart and diaphragm.

18. The peripheral nervous system is divided into the somatic nervous system, which governs conscious sensations and voluntary movements, and the autonomic nervous system, which is concerned with our unconscious, involuntary internal activities.

19. The autonomic nervous system can be divided into the sympathetic and parasympathetic nervous systems, two branches with antagonistic actions. The sympathetic nervous system gears the body to face stressful or emergency situations. The parasympathetic nervous system adjusts body functioning so that energy is conserved during restful times.

WEB TUTORIAL 8.1 Autonomic Nerve Pathways

WEB TUTORIAL 8.2 Sympathetic and Parasympathetic Pathways

Disorders of the Nervous System Vary in Health Significance (pp. 145–149)

20. Headaches can range from relatively mild (tension headaches) to severe (migraine and cluster headaches). They usually are not life threatening.

21. A stroke is caused by the death of nerve cells due to an interruption of blood flow. The effects depend on the region of the brain affected.

22. A coma, a condition in which a person is totally unresponsive to sensory input, can be caused by a blow to the head, tumors, infections, drugs, or failure of the liver or kidney.

23. Because the spinal cord contains the pathways of communication between the brain and the rest of the body, damage to the spinal cord impairs functioning below the site of injury.

KEY TERMS

autonomic nervous system   *p. 134*
blood-brain barrier   *p. 135*
central nervous system (CNS)   *p. 134*
cerebellum   *p. 141*
cerebral cortex   *p. 138*
cerebrospinal fluid   *p. 134*
cerebrum   *p. 136*
cranial nerve   *p. 144*
ganglion   *p. 134*

gray matter   *p. 138*
hippocampus   *p. 142*
hypothalamus   *p. 141*
limbic system   *p. 141*
long-term memory   *p. 142*
medulla oblongata   *p. 141*
meninges   *p. 134*
parasympathetic nervous system   *p. 134*

peripheral nervous system (PNS)   *p. 134*
pons   *p. 141*
prefrontal cortex   *p. 139*
primary motor area   *p. 138*
primary somatosensory area   *p. 138*
reflex arc   *p. 142*
reticular activating system (RAS)   *p. 142*

short-term memory   *p. 142*
somatic nervous system   *p. 134*
spinal cord   *p. 142*
spinal nerve   *p. 143*
sympathetic nervous system   *p. 134*
thalamus   *p. 140*
white matter   *p. 138*

THINKING ABOUT THE CONCEPTS

1. Distinguish between the central nervous system and the peripheral nervous system. List the components of each.   *p. 134*

2. Describe three ways the brain and spinal cord are protected.   *pp. 134–136*

3. What are the functions of cerebrospinal fluid?   *pp. 134–135*

4. What is gray matter? What is white matter?   *p. 138*

5. Describe the three types of functional areas of the cerebral cortex.   *pp. 138–140*

6. In what way are the organization of the primary somatosensory area and that of the primary motor area of the cerebral cortex similar? In what way do these areas differ?   *pp. 138–140*

7. Describe the specializations of the left and right cerebral hemispheres.   *p. 140*

8. List five functions of the hypothalamus.   *p. 141*

9. What is the function of the cerebellum?   *p. 141*

10. Which functional system of the brain is responsible for emotions?   *p. 141*

11. Describe the two functions of the reticular activating system.   *p. 142*

12. List the two functions of the spinal cord and relate each function to the structure of the spinal cord.   *pp. 142–143*

13. You are cooking dinner and carelessly touch the hot burner on the stove. You remove your hand before you are even aware of the pain. Using the anatomy of a spinal reflex arc, explain how you could react before you were aware of the pain.   *pp. 142–143*

14. What are the two divisions of the peripheral nervous system? What type of response does each control?   *p. 144*

15. Compare and contrast the functions of the sympathetic and parasympathetic nervous systems. *pp. 144, 146*
16. List some effects of sympathetic stimulation, and explain how these prepare the body for an emergency. *p. 144*
17. You are watching a football game with your friends. A wide receiver makes an incredible catch and then runs 20 yards, skillfully dodging defensive players to make a touchdown. Your friend Joe says, "Amazing! How *does* he do that?" The receiver's outstanding sensory-motor coordination is largely due to the actions of his
    a. cerebellum.
    b. medulla.
    c. reticular activating system.
    d. hypothalamus.
18. As you sit here studying, you are unlikely to be aware of the pressure of your clothes against your body and the rustling of paper as other students turn pages. The part of the brain that "decides" that these are unimportant stimuli is the
    a. reticular activating system.
    b. cerebellum.
    c. hypothalamus.
    d. medulla.

19. Belinda was riding a bicycle without a helmet and was struck by a car. She hit the back of her head very hard in the fall. The physician is quite concerned because the medulla is located at the base of the skull. She explains to the parents that injury to the medulla could result in
    a. the loss of coordination so that the child may never regain the motor skills needed to ride a bicycle.
    b. the loss of speech.
    c. amnesia (the loss of all memory).
    d. death because many life-support systems are controlled here.
20. The neural center that regulates body temperature is the _____.
21. The region of the brain that regulates basic physiological processes such as breathing and heart rate is the _____.
22. The branch of the nervous system that prepares the body to respond to emergency situations is the _____.
23. The brain region responsible for intelligence and thinking is the _____.

APPLYING THE CONCEPTS

1. Your Aunt Rosa had a stroke; that is, some of the neurons in her brain died or were injured when a blood clot or hemorrhage reduced the blood supply to them. She can understand what you say to her, but she cannot speak to answer you. She has trouble moving her right arm. What region of the brain was affected by the stroke? How do you know?
2. Joe and Henry were both in car accidents, and both suffered spinal cord damage. Joe's injury was in the lower back, and Henry's was in the neck region. The degree of injury to the spinal cord is similar in both Joe and Henry. Would the resulting problems be equal in severity? Explain. Would either one or both require a respirator to breathe for him? Why? Describe some of the difficulties that you might expect Joe and Henry to have.
3. When you have a cold, you might take a decongestant. Some decongestants contain pseudoephedrine, which mimics the effects of the sympathetic nervous system. What side effects might you expect? Would you expect this medication to make you drowsy?
4. When you have dental work done, the dentist often administers a local anesthetic in the gums near the region that requires drilling. You are usually advised not to eat anything until the anesthetic wears off. This advice is given out of concern for your tongue, not your teeth. Why?

8a

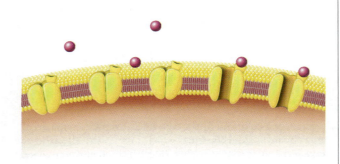

It seemed like the worst headache of his life. At 7:15 AM Francine had just walked into the kitchen and opened a stuck cabinet door. It sounded like a freight train grinding to a halt. Ken moaned something about keeping the noise down. He wanted a cup of coffee so badly but he couldn't bear walking near the coffee maker. Just looking at it kicked his salivary glands into high gear and heightened his extreme need for caffeine at this moment. He was determined to quit, however. His anxiety at mid-day had been getting so great that he had to do something and his physician had suggested cutting back or getting off coffee altogether. He decided to go cold turkey. If only he could get through the next few days, he thought, he'd make it.

In this chapter, we will consider how drugs affect the nervous system and alter our state of mind. We will see why the use of some drugs, caffeine for example, leads to a need to continue its use. Then we will take a look at some of the more common mind-altering drugs. ■

Psychoactive Drugs Alter Communication Among Neurons

A drug that alters one's mood or emotional state often is described as a psychoactive drug. The mind-altering effects of these drugs result from their ability to alter communication between nerve cells. As you learned in Chapter 7, neurons communicate with one another using chemicals called neurotransmitters. Neurotransmitters are released by one neuron, diffuse across a small gap, and bind to specific receptors on another neuron, triggering changes in the activity of the second neuron. Under normal conditions, the action of the neurotransmitter is stopped almost immediately, because the neurotransmitter is broken down by enzymes, reabsorbed into the cell that released it, or simply diffuses away.

A psychoactive drug may alter this communication between neurons in any of several ways (Figure 8a.1). It may stimulate the release of a neurotransmitter, thereby enhancing the response of the receiving neuron. Alternatively, the drug may inhibit the release of a neurotransmitter and dampen the response of the receiving neuron. A drug may also increase and prolong the effect of a neurotransmitter by delaying its removal from the synapse. If the drug is chemically similar to the normal neurotransmitter, it may bind to the receptor and affect the activity of the receiving neuron in the same manner as the neurotransmitter. Finally, a drug may bind to the receptor and prevent the neurotransmitter from acting at all.

One of the problems with using psychoactive drugs is that the user may develop some level of dependence on the drug. So before we consider the drugs themselves, let's consider the matter of dependence.

Drug Dependence Causes Continued Drug Use

It is difficult to provide a precise definition of drug dependence. A loose interpretation might be, "It is what causes a person to continue using a drug."

Tolerance is a progressive decrease in the effectiveness of a drug. As tolerance to a particular drug develops, a person must take larger or more frequent doses of the drug to produce the same effect. Tolerance develops partly because the body steps up its production of enzymes that break down the drug and partly because of changes in the nerve cells that make them less responsive to the drug. When the normal level of neurotransmitter is no longer sufficient to adequately stimulate the receiving nerve cells, the person cannot function normally without the drug. At this point, we describe the person as being physically dependent on the drug.

Cross-tolerance occurs when tolerance to one drug results in a lessened response to another, usually similar drug. If one abuses codeine, for instance, tolerance develops for codeine as well as for other drugs that have similar effects on the nervous system, such as morphine and heroin.

Another reason that certain drugs cause users to continue using them is that the drugs stimulate the "pleasure" centers in the limbic system of the brain (see Chapter 8). An animal with electrodes implanted in the pleasure center will quickly learn to press a lever to stimulate this brain region. If permitted, it will self-stimulate repeatedly, sometimes hundreds of times an hour, until it is exhausted. When the experiment is changed so that a small dose of a drug is released into the blood of an animal when it presses a lever, the animal will learn to press the lever to self-administer the drug—if the drug stimulates the pleasure center. Drugs that stimulate the pleasure center include cocaine, amphetamine, morphine, and nicotine. (Nicotine is discussed further in Chapter 14a.)

Several issues should be considered when deciding whether to use a drug. One is safety—both short term and long term. Safety issues include health risks associated with the use of the drug and the degree to which using the drug leads to tolerance and dependence. Pregnant women have an additional issue to consider: the possible effects of the drug on the growing fetus. Most drugs can move from the mother's blood to the fetus's blood. In this case, the drug affects the fetus directly and causes effects similar to those experienced by the mother. Other drugs can harm the fetus by constricting blood vessels in the umbilical cord, shutting down the lifeline through which oxygen and nutrients reach the fetus.

As we consider some of the psychoactive drugs, we will discuss their habit-forming potential as well as possible health risks associated with their use.

Alcohol Depresses the Central Nervous System

In every alcoholic drink, the alcohol is ethanol. Ethanol is produced as a by-product of fermentation when yeast cells break down sugar to release energy for their own use. The taste of the beverage is determined by the source of the sugar, which comes from the fruit or vegetable that is fermented. For example, grapes are fermented in wine, juniper berries in gin, barley in beer, and malt in scotch.

The effects of alcohol on a person's behavior depend on the blood alcohol level, which is measured as the number of grams of alcohol in 100 milliliters of blood. In other words, 1 g of alcohol in 100 ml of blood is a blood alcohol level of 1%. The blood alcohol level in turn depends on several factors, including how much

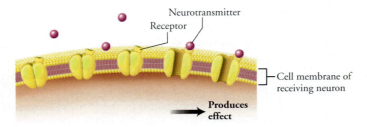

(a) The natural sequence of events: molecules of neurotransmitter released by one neuron diffuse across a gap and fit into receptors on the membrane of a receiving neuron, causing a response.

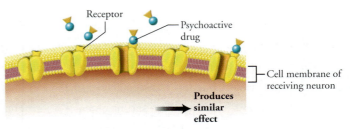

(d) A psychoactive drug may fit into the receptors for a neurotransmitter, causing a similar response by the receiving neuron.

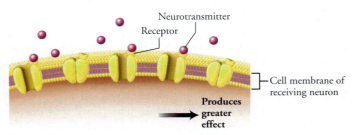

(b) A psychoactive drug may increase the number of neurotransmitter molecules released, increasing the response of the receiving neuron.

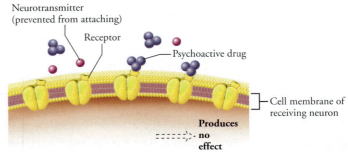

(e) A psychoactive drug may fit into the receptors for a neurotransmitter and prevent the neurotransmitter from entering the receptor, blocking the response by the receiving neuron.

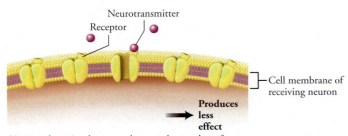

(c) A psychoactive drug may decrease the number of neurotransmitter molecules released, decreasing the response of the receiving neuron.

FIGURE **8a.1**

Psychoactive drugs alter one's mental state by affecting communication between neurons.

alcohol is consumed; how rapidly it is consumed (over what period of time); and its rate of absorption, distribution, and metabolism.

A "drink" can mean different things to different people. For some it is a can of beer, for others a glass of wine, and for others scotch on the rocks. Whatever the drink, the effect on one's body will be determined by the *amount of ethanol* in the beverage, which can vary tremendously (Table 8a.1). Natural fermentation produces beer and wine. Natural fermentation cannot yield more than 15% alcohol, because the alcohol kills the yeast cells producing it. Liquor (distilled spirits) is produced by distillation, a process that concentrates the alcohol. The alcohol content of distilled spirits is measured as "proof." One degree of proof equals 0.5% alcohol. Most distilled spirits (vodka, gin, scotch, whiskey, rum, brandy, and cognacs) are 80 proof, which means they contain 40% alcohol. But some brands are 90 proof (45% alcohol) or even 100 proof (50%). Clearly, a 3-ounce[1] martini, which contains only distilled spirits and therefore about 1.2 ounces of alcohol, is more intoxicating than an 8-oz mug of beer, which contains about 0.4 oz of alcohol.

Because of the variability in alcohol content among beverages, there is a need to standardize the meaning of a drink. Generally, the standard drink contains 0.5 oz of pure ethanol. A standard drink is one bottle or can of beer, a glass of wine, or a jigger of 80 proof distilled spirits, such as gin, whisky, vodka, or scotch (Figure 8a.2).

[1]In this chapter, the term *ounce* refers to a fluid ounce, which is equal to about 29 ml.

TABLE **8a.1**

THE ALCOHOLIC CONTENT OF SELECTED BEVERAGES	
ALCOHOLIC BEVERAGE	**PERCENT ALCOHOL (ETHANOL)**
Light beer	4.5
Most beer	~5
Flavored malt beverages (coolers)	5
Ice beer	5.5–5.9
Dark beer (stout, porter, bock)	6–7
Malt liquor	8
French and German wines	8.5–10
American wine (most)	12–14
Sherry and port	18–21
Distilled liquor (vodka, gin, scotch, whiskey, rum, brandy, cognac)	Most 40% (80 proof); some 45% (90 proof) or 50% (100 proof)

The rate of alcohol absorption depends on its concentration

The intoxicating effects of alcohol begin when it is absorbed from the digestive system into the blood and is delivered to the brain. As a rule, the rate of absorption of alcohol depends on the concentration of alcohol: the higher the concentration, the faster the rate of absorption. So wine, beer, or distilled spirits diluted with a mixer will be absorbed more slowly than pure liquor. The choice of mixer also influences the rate of absorption. Carbonated beverages speed the rate of absorption because of the pressure of the gas bubbles.

Although very few substances are absorbed across the walls of the stomach, about 20% of the alcohol consumed is absorbed there. The remaining alcohol is absorbed through the intestines.

Because alcohol can be absorbed from the stomach, one begins to feel the effects of a drink quickly, usually within about 15 minutes. The presence of food in the stomach slows alcohol absorption because it dilutes the alcohol, covers some of the stomach membranes through which alcohol would be absorbed, and slows the rate at which the alcohol passes into the intestines.

Alcohol is distributed to all body tissues

Ethanol is a small molecule that is soluble in both fat and water, so it is distributed to all body tissues. Therefore, overall body size affects one's blood alcohol level and the degree of intoxication. After consuming equal amounts of alcohol, a large person would have a lower blood alcohol level and, therefore, be less intoxicated than would a small, slender person (Figure 8a.3).

The rate of elimination of alcohol from the body cannot be increased

Ninety-five percent of the alcohol that enters the body is metabolized (broken down) before it is eliminated. Most of that metabolism occurs in the liver, which converts alcohol to carbon dioxide and water. The rate of metabolism is slow, about one-third of an ounce of pure ethanol per hour. Unlike most other drugs, the rate of metabolism does not increase with the concentration of alcohol.

In practical terms, it takes slightly more than an hour for the liver to break down the alcohol contained in one standard drink—a can of beer or a glass of wine. Because alcohol cannot be stored in the body, it continues to circulate in the bloodstream until it is metabolized. Therefore, if more alcohol is consumed in an hour than is metabolized, both the blood alcohol level and the degree of intoxication increase (Figure 8a.4).

There is no way to increase the rate of alcohol metabolism by the liver and, therefore, no way to sober up quickly. A cup of coffee may slightly counter the drowsiness caused by alcohol, but it does

12 oz Most beer	5 oz Wine (12% alcohol)	1.5 oz Distilled spirits (80 proof gin, whisky, vodka, scotch)	8 oz Mixer Carbonated beverage	6 oz Mixer Fruit juice
1 bottle or can	1 glass	1 jigger	1 glass	1 glass
140–150 calories	110–200 calories	~100 calories	~70–120 calories	35–105 calories

FIGURE **8a.2**

A standard drink contains 0.5 ounce of alcohol. Different types of alcoholic beverages vary in their alcohol content. The volume of beverage in a standard drink varies with its alcoholic content.

EFFECTS OF DRINKING ON DRIVING

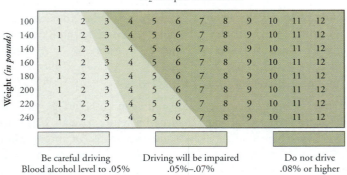

Drinks (2-hr period)
1½ oz liquor or 12 oz beer

Be careful driving	Driving will be impaired	Do not drive
Blood alcohol level to .05%	.05%–.07%	.08% or higher

FIGURE **8a.3**

Alcohol consumption can impair driving. The blood level of alcohol depends on both the number of drinks consumed and body size. A small person has a higher blood alcohol level than a large person after consuming the same amount of alcohol. The blood level of alcohol determines the effect on the nervous system.

PERCENT OF BLOOD ALCOHOL CONCENTRATION IN AN AVERAGE MAN AT HOURLY INTERVALS AFTER DRINKING 1, 2, 4, OR 6 OUNCES OF 80-PROOF SPIRITS

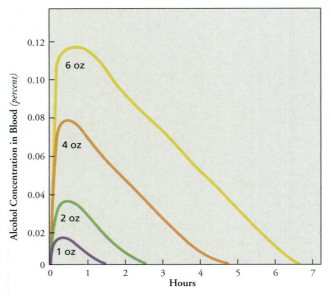

FIGURE **8a.4**

Alcohol remains in the blood and has mind-altering effects until it is broken down in the liver. The liver breaks down alcohol at a slow constant rate of about one-third ounce of pure ethanol an hour. Therefore, it takes slightly more than an hour to metabolize the alcohol in one standard drink. It is impossible to become sober by increasing the rate at which alcohol is eliminated from the body. Source: Reproduced with permission from *The Encyclopedia Britannica*, © 2004 by Encyclopedia Britannica, Inc.

not reduce the level of intoxication. Furthermore, because muscles do not metabolize alcohol, exercise does not help. So walking around the block will not make a person sober, nor will a cold shower.

If a man and a woman consume the same amount of alcohol, the woman will usually feel the effects before the man does. Part of the explanation for this difference is based on anatomy. Women are usually smaller than men. In addition, on average, a woman's body contains a higher percentage of fat than does a man's. Alcohol dissolves in fat more slowly than in water. As a result, the alcohol is diluted more slowly in a woman's body, and the effects are prolonged. But there's more to the story; the difference also has to do with differences in metabolism. Although the liver is the primary site for alcohol metabolism, some alcohol is broken down by an enzyme found in the stomach lining. Alcohol that is metabolized in the stomach never enters the bloodstream and cannot cause intoxication. Women have less of this enzyme in their stomach linings than do men. Consequently, a woman absorbs about 30% more of the alcohol in a drink than a man does. When weight differences between an average man and woman are also taken into account, 2 oz of liquor can have approximately the same effect on a woman as 4 oz would have on a man.

A small amount of alcohol, about 5%, is eliminated from the body unchanged through the lungs or in the urine. Alcohol eliminated from the lungs is the basis of the breathalyzer test that may be administered by law enforcement officers who want an on-the-spot sobriety check.

▌ Alcohol has many health-related effects

Approximately 70% of Americans over the legal drinking age consume alcohol. Nearly 10% of those who drink do so in excess or have lost some degree of control over their habit. One-third of the teenagers in America consume alcohol at least once a week.

Alarmingly, in 2002 more than a quarter of high school seniors reported that they had consumed five or more drinks in a row within the preceding two weeks. Because alcohol has a negative effect on virtually every organ of the body, people do not have to be alcoholics for alcohol to impair their health.

THE NERVOUS SYSTEM

Many people believe that alcohol is a stimulant, but it is a depressant. In other words, it slows down the activity of all the neurons of the brain, beginning with the higher cortical, or "thinking," centers. Alcohol often is mistakenly thought to be a stimulant because it depresses the inhibitory neurons first, allowing the excitatory ones to take over. As alcohol removes the "brakes" from the brain, normal restraints on behavior may be lost. Release from inhibitory controls also tends to reduce anxiety and often creates a sense of well-being. However, discrimination, control of fine movements, memory, and concentration are gradually lost as well.

The brain centers that are involved in balance and coordination are affected next, causing a staggering gait. Numbed nerve cells send slower messages, resulting in slower reflexes. The brain regions responsible for consciousness are eventually inhibited, causing a person to pass out. Still higher concentrations of alcohol can cause coma and death from respiratory failure.

Finally, alcohol kills nerve cells. Recall from Chapter 7 that nerve cells cannot be replaced. As the nerve cells die, the brain actually becomes smaller. The frontal lobes of the cerebral cortex, where judgment, thought, and reasoning are centered, are the first to shrink. The loss of intellectual ability may be the first complication of chronic alcohol use.

NUTRITION

Alcohol is high in calories but has little nutritional value. It is not just the calories in alcohol that make a person fat. Part of the problem is that the body "prefers" to utilize alcohol over other nutrients. When alcohol is present in the body, it is metabolized for energy before fats are used. Unused fat is then stored in places such as thighs, hips, and bellies. (The term *beer belly* is well deserved!) Alcohol also robs the body of other nutrients. It decreases the absorption of certain vitamins, including folate, thiamine, B_{12}, and B_6. In addition, alcohol causes the kidneys to pump important substances such as potassium ions, zinc, calcium, magnesium, and folate out of the body.

THE LIVER

Excessive alcohol consumption damages the liver, an organ that performs many vital functions in the body. Severe damage to the liver is, therefore, a serious threat to life. Liver damage occurs because alcohol metabolism preempts fat metabolism in the liver, causing fats to accumulate in the liver. Four or five drinks daily for several weeks are enough to begin fat accumulation in liver cells. At this early stage, however, the liver cells are not yet harmed, and, with abstinence, they can be restored to normal. With continued drinking, accumulating fat causes liver cells to enlarge, sometimes so much so that the cells rupture or grow into cysts that replace normal cells. The fat reduces blood flow through the liver, causing inflammation known as alcoholic hepatitis. Signs of alcoholic hepatitis include fever and tenderness in the upper abdominal region. Gradually, fibrous scar tissue may form, a condition known as cirrhosis, which further impedes blood flow and impairs liver functioning. Cirrhosis can lead to intestinal bleeding, kidney failure, fluid accumulation, and eventually death, if drinking continues. Indeed, cirrhosis of the liver, the ninth leading cause of death in the United States, is most often caused by alcohol abuse.

CANCER

A person who drinks heavily is at least twice as likely to develop cancer of the mouth, tongue, or esophagus than is a nondrinker. Evidence also exists that the risk of cancer from both drinking and smoking cigarettes is greater than the sum of the risks caused by either drinking or smoking alone.

HEART AND BLOOD VESSELS

Here's the good news. *Moderate* amounts of alcohol can be good for the heart. Teetotalers are more likely to suffer heart attacks than are persons who drink moderately, say a drink a day. The frequency of alcohol consumption is also important. Men who consume three or four drinks a week have a lower risk of heart attack than do nondrinkers. One reason may be that the relaxing effect of alcohol helps to relieve stress. But alcohol also seems to raise the levels of the "good" form of a cholesterol-carrying particle—HDL—in the blood. This form of cholesterol reduces the likelihood that fats in the blood will be deposited in the walls of blood vessels, clog the vessels, and reduce the blood supply to vital organs such as the heart or brain. In addition, moderate amounts of alcohol reduce the likelihood that blood clots will form when they should not. Such clots can block a blood vessel nourishing the heart muscle, thereby causing a heart attack. Thus, persons who imbibe moderately, generally live longer than nondrinkers.

When alcohol is consumed in more than moderate quantities, it damages the heart and blood vessels. It weakens the heart muscle itself, reducing the heart's ability to pump blood. It also promotes the deposit of fat in the blood vessels, making the heart work harder to pump blood through them. And it raises blood pressure. Although a drink a day has little impact on blood pressure, consuming larger quantities may have a substantial effect. Together, these effects—damage to heart muscle, blood vessels clogged with fatty materials, and high blood pressure—can enlarge the heart to twice its normal size (Figure 8a.5).

ACCIDENTS

Accidents—at home or on the road—are more likely to happen when a person has been drinking. According to a 2002 study by the Task Force on College Drinking, 1400 college students die every year in alcohol-related accidents. That number is roughly half the number of casualties in the World Trade Center tragedy that occurred on September 11, 2001.

Driving while intoxicated causes more than 20,000 deaths a year in the United States, nearly half of the nation's traffic fatalities. A blood alcohol level of only 0.04% to 0.05%, which can result from only two or three drinks, decreases peripheral vision, decreases the light sensitivity of the eye by 30% (equivalent to wearing sunglasses at night), slows recovery from headlight glare, and reduces reaction time by as much as 25%. At the same time, alcohol impairs the ability to concentrate and make judgments of the distance and speed of objects. If the driver is lucky, these impairments may not have dire consequences. However, considering that a single mile of city driving requires a driver to make roughly 300 split-second decisions, such impairments can make the difference between life and death.

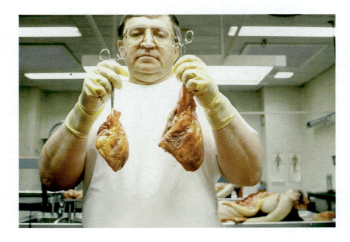

FIGURE **8a.5**

The alcoholic's heart on the right is nearly twice the normal size.

EFFECTS ON FETAL DEVELOPMENT

Alcohol moves freely from a drinking mother's blood to the fetus. Indeed, the blood alcohol level of the fetus is the same as the drinking mother's. If a pregnant woman consumes even one or two drinks a day, she increases the chance that her baby will be of lower than normal birth weight. Low birth weight, in turn, is associated with many complications soon after birth. Heavy drinking during pregnancy increases the risk of miscarriage and stillbirth. After Down syndrome and spina bifida, alcohol is the third leading cause of birth defects associated with mental retardation. Of the three leading causes, alcohol use is the only cause that is preventable.

In the 1970s, researchers began to notice a pattern of growth abnormalities and birth defects common among children of women who drank heavily during pregnancy. This collection of effects has since been named the fetal alcohol syndrome (FAS), although all characteristics are not always present in any one infant. Fetal alcohol syndrome is associated with mental retardation, growth deficiency, and characteristic facial features (Figure 8a.6).

The effects of FAS may be caused by chronic alcohol abuse throughout pregnancy or by binge drinking at critical times during fetal development. The fetus is particularly sensitive to chemicals, such as alcohol, during the first 3 months of pregnancy, when most organ systems are forming. The best advice for pregnant women is to avoid alcohol consumption entirely, because safe levels are not known.

ALCOHOLISM

Although alcohol is a legal drug, alcoholism is America's number one drug problem. The problem with identifying alcoholism is that there is no such thing as a typical alcoholic. Alcoholics can be young or old, rich or poor, and of any race, economic status, or profession.

Different alcoholics have different drinking patterns. Some binge and some chronically overindulge. But what they all share is a loss of control over their drinking. When an alcoholic takes the first sip of alcohol, he or she cannot predict how much or how long drinking will continue.

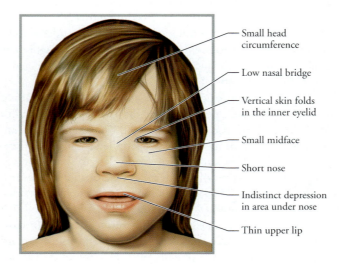

- Small head circumference
- Low nasal bridge
- Vertical skin folds in the inner eyelid
- Small midface
- Short nose
- Indistinct depression in area under nose
- Thin upper lip

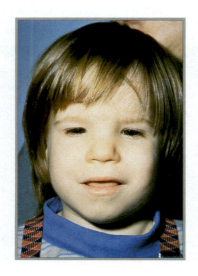

Marijuana's Psychoactive Ingredient Is THC

Marijuana is the most widely used illegal drug in the United States today. It consists of the leaves, flowers, and stems of the Indian hemp plant, *Cannabis sativa*. The principal psychoactive ingredient (the component that brings about marijuana's mind-altering effects) is delta-9-tetrahydrocannabinol, or THC.

The effects of marijuana depend on the concentration of THC in the preparation and the amount consumed. The average THC content of marijuana smoked in the United States has increased since the 1980s from about 0.4% to as much as 12%. In small to moderate doses, THC produces feelings of well-being and euphoria. In large doses, THC can cause hallucinations and paranoia. Anxiety may even reach panic proportions at very high doses.

Marijuana is not addicting in the sense that it produces severe, unpleasant withdrawal symptoms. Nonetheless a withdrawal syndrome has been identified. It includes symptoms such as restlessness, irritability, mild agitation, insomnia, nausea, and cramping. Withdrawal symptoms are not common, and if they do occur, they are usually relatively mild and short lived.

▌Marijuana binds to THC receptors in the brain

Researchers are just beginning to understand how marijuana brings about its effects. When THC binds to certain receptors on nerve cells[2], it triggers a cascade of events that ultimately lead to the "high" that the user experiences. (Figure 8a.7). Once the receptors were discovered, researchers quickly set about to find the natural substance that binds to them, reasoning that the receptors did not evolve millions of years ago just in case someone decided to smoke marijuana. Researchers then sorted through thousands of chemicals in pulverized

FIGURE **8a.6**
Characteristic facial features of children with fetal alcohol syndrome.

[2]There is a slightly different type of THC receptor found primarily on cells of the immune system.

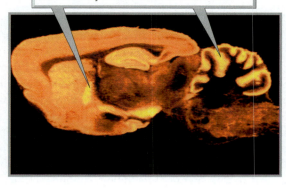

Marijuana receptors are most highly concentrated in regions involved in thinking and memory and in motor control areas.

FIGURE **8a.7**

The map of marijuana receptors shows that they are most highly concentrated in brain regions where THC is thought to be active. The areas with receptors appear yellow in this photograph showing a section of a rat's brain.

pig brains to find one that would bind to THC receptors. The first natural substance that binds the THC receptors to be identified was a newly discovered chemical messenger, named *anandamide,* after the Sanskrit word meaning "internal bliss" (*ananda*). Anandamide functions as the brain's own THC. The normal jobs of anandamide probably include the regulation of mood, memory, pain, appetite, movement, and other activities. In addition, THC seems to stimulate the release of dopamine in the reward pathways of the brain.

Long-term marijuana use has many effects on the body

The media have popularized many claims that long-term use of marijuana carries with it a plethora of health hazards. Is there evidence for these claims? Let's consider what is known about the risks of marijuana use.

RESPIRATORY SYSTEM

Marijuana usually is smoked. It should not be surprising, then, that the most clearly harmful effects of marijuana are on the respiratory system. However, the damage, it seems, is done by the residual materials in the smoke and not by the THC itself. It is true that people who smoke marijuana on a daily basis smoke fewer joints (marijuana cigarettes) than cigarette smokers smoke cigarettes. However, compared with a regular cigarette, a joint has 50% more tar, which contains cancer-causing chemicals. Also, marijuana smoke is usually inhaled deeply and held within the lungs. As a result, three times as much tar is deposited in the airways, and five times as much carbon monoxide is inhaled. Carbon monoxide prevents red blood cells from carrying needed oxygen to the cells of the body.

The end result of prolonged smoking of marijuana is marked trauma to the respiratory system. Similar to cigarette smoke, marijuana smoke inflames air passages and reduces breathing capacity. People who smoke marijuana, but not tobacco, are more likely to report symptoms such as coughing, excess phlegm, and wheezing than are people who do not smoke at all.

At this time, there is no clear evidence that marijuana causes cancer in humans. However, because lung cancer usually takes decades to develop, it is still too soon to say. It has now been 30 years since the widespread use of marijuana began among young people, so we may soon see an association between marijuana and cancer, if one exists. We do know, though, that the tissues of the lungs and airways of a marijuana smoker show the same progression of changes that are observed in cigarette smokers who develop lung cancer. Furthermore, some researchers have suggested that a combination of tobacco and marijuana smoke may be more likely to lead to cancer than either one alone.

THE HEART

Smoking marijuana makes the heart beat faster, sometimes to double its normal rate. In some people, marijuana also increases blood pressure. Either change increases the heart's workload. These changes could pose a threat to people with preexisting cardiovascular problems, such as high blood pressure or atherosclerosis (fatty deposits in the arteries). Although the risk is small, marijuana increases one's risk of immediate heart attack.

REPRODUCTION

Some studies have shown that THC can interfere with reproductive functions in both males and females. At least some of these disturbances may result from the structural similarity between THC and the female hormone estrogen. Males who smoke marijuana often have lower levels of the male sex hormone testosterone and produce fewer sperm than do nonusers. If a man is having difficulty becoming a father and smokes marijuana, quitting might boost fertility. Testosterone and sperm levels return to normal when THC is cleared from the body.

The effects of marijuana on the female reproductive system are not clear. Although we know that THC interferes with ovulation in female monkeys, we know less about what it does to human females. Some doctors, however, report menstrual problems and reproductive irregularities in women who smoke marijuana.

Marijuana impairs driving ability

Marijuana has a detrimental effect on driving skills and performance. In particular, it impairs the accuracy in steering a car, known as tracking ability. The decline in tracking ability has been clearly demonstrated in experiments on restricted driving courses. For example, volunteers found it very difficult to follow a figure-eight loop of road after smoking marijuana.

Does marijuana use lead to use of other drugs?

Through the years, it has been suggested that marijuana use leads to the use of harder drugs, such as heroin or cocaine. Such a relationship could be expected since a person who uses one illegal drug is more likely to be interested in using others than is someone who has never used any drug. It is argued that buying marijuana also might introduce a user to dealers who could supply other drugs. So it should not be surprising to learn that marijuana use often precedes the use of other, more dangerous drugs. However, this connection does not show that the use of one drug *causes* the use of another. For example, whereas most people who smoke marijuana have previously consumed alcohol, most people who drink alcohol never smoke marijuana. Likewise, most people who smoke marijuana never develop an interest in other drugs.

Stimulants Excite the Central Nervous System

Stimulants are drugs that excite the central nervous system (see Chapter 8). We'll now take a closer look at three stimulants: cocaine, amphetamines, and caffeine.

Cocaine augments the neurotransmitters dopamine and norepinephrine

Cocaine is a drug that is extracted from the leaves of the coca plant (*Erythroxylon coca*), which grows naturally in the mountainous regions of South America. When the cocaine powder is inhaled into the nasal cavity ("snorted"), it reaches the brain within a few seconds and produces an effect almost as intense as when it is injected. Smoking the drug is an even more effective delivery route. Forms of cocaine that can be smoked—specifically, freebase and crack—are obtained by further extraction and purification of the cocaine. Crack is an extremely potent form of cocaine used by two-thirds of the cocaine addicts (several hundred thousand) in the United States.

Cocaine brings about a rush of intense pleasure (euphoria), a sense of self-confidence and power, clarity of thought, and increased physical vigor. It does so by increasing the levels of two "feel good" neurotransmitters: dopamine and norepinephrine. The euphoria is caused primarily by cocaine's effect on dopamine, a neurotransmitter used by nerve cells in the pleasure centers of the brain. Normally, dopamine is almost immediately reabsorbed into the nerve cell that released it, and its effect on the next nerve cell ceases. Cocaine, however, interferes with the reuptake of dopamine, thus increasing and prolonging dopamine's effect. As a result, cocaine loads the brain's pleasure centers and stimulates them. Cocaine also increases the effects of another neurotransmitter, norepinephrine. Norepinephrine brings about the effects of the sympathetic nervous system, the part of the nervous system that prepares the body to face an emergency. Thus, cocaine also triggers the bodily responses that would equip the body to face stress: increased heart rate and blood pressure, narrowing of certain blood vessels, dilation of pupils, a rise in body temperature, and a reduction of appetite. Cocaine, then, makes the user feel alert, energetic, and confident.

The effects of cocaine are short lived and are followed by a "crash." A cocaine high may last from 2 to 90 minutes, depending on how the cocaine enters the body. When the high wears off, it is generally followed by a period of deep depression, anxiety, and extreme fatigue. To relieve these uncomfortable feelings, the user often craves more cocaine. The higher the high, the lower the crash and, therefore, the more intense the craving for cocaine. For this reason, crack, which is estimated to be about 75% pure compared with the 10% to 35% purity of street cocaine, causes higher highs and lower crashes. As a result, crack is extremely addicting.

CARDIOVASCULAR RISKS

Cocaine may cause heart attack or stroke, conditions that occur when an interruption of blood flow deprives heart cells or brain neurons of oxygen and nutrients. One way that cocaine blocks blood flow is by constricting arteries. It can cause spasms in the arteries supplying the heart, for instance, causing a heart attack. Blood flow to the heart can also be disturbed by irregularities in heartbeat. By disabling the nerves that regulate heartbeat, cocaine can cause disturbed heartbeat rhythms that result in chest pain and heart palpitations (that uncomfortable feeling of being aware of your heart beating). Cocaine use may even cause the heart to stop beating. In addition, cocaine increases blood pressure, which may cause a blood vessel to burst.

RESPIRATORY RISKS

Cocaine can cause respiratory failure that can lead to death. Stimulation of the central nervous system is always followed by depression. Thus, as the stimulatory effects of cocaine wear off, the respiratory centers in the brain that are responsible for breathing become depressed, or inhibited. Breathing may become shallow and slow and may stop completely.

Damage to the respiratory system itself caused by cocaine use depends largely on how the drug is taken. If it is snorted, cocaine causes damage to the nerves, lining, and blood vessels of the nose. It can dry out the nose's delicate mucous membranes until they crack and bleed almost continuously. Symptoms of a sinus infection—a perpetually runny nose and a dull headache spanning the bridge of the nose—are common. On a more startling level, the partition between the two nasal cavities may even disintegrate. Because cocaine is a painkiller, considerable damage to the nose may occur before it is discovered.

Smoking crack, on the other hand, damages the lungs and airways. The chronic irritation leads to bronchitis and may also cause lung damage from reduced oxygen flow into the lungs or blood flow through the lungs.

Amphetamines augment the neurotransmitters dopamine and norepinephrine

Amphetamines are synthetically produced stimulants that closely resemble two of the brain's natural neurotransmitters: dopamine and norepinephrine. There are many forms of amphetamine, including dextroamphetamine and methamphetamine, also known as crystal meth. Amphetamine is sold as the prescription drug Desoxyn®. Ritalin®, another prescription drug, is used to treat attention deficit hyperactivity disorder (ADHD). But methamphetamine is illegally sold as a powder that can be injected, snorted, swallowed, or smoked. All are stimulants of the central nervous system. Similar to cocaine, amphetamines make the user feel good—exhilarated, energetic, talkative, and confident. They suppress appetite and the need for sleep. Amphetamines are active for a longer period of time than is cocaine—hours as opposed to minutes.

Amphetamines can be swallowed in a pill or injected intravenously. A crystalline form of methamphetamine, called ice, is smoked to produce effects similar to those of crack cocaine. Methamphetamine can produce hazardous effects, including blood vessel spasm, blood clot formation, insufficient blood flow to the heart, and accumulation of fluid in the lungs.

Amphetamines bring about their physical and psychological effects by causing the release of certain neurotransmitters, such as norepinephrine and especially dopamine. Norepinephrine is used for communication between certain neurons within the brain. It is also the neurotransmitter that brings about the effects of the sympathetic nervous system, which prepares the systems of the body to face an emergency. Thus, amphetamines elevate blood pressure, increase heart rate, and open up airways in the lungs. High intravenous doses administered during a spree of amphetamine use over several days may cause a paranoid mania, similar to that associated with schizophrenia. The personality changes are caused by several factors, including excessive stimulation of norepinephrine and dopamine neurotransmission as well as chronic lack of sleep.

Tolerance to amphetamines develops (accompanied by simultaneous tolerance to cocaine) along with both physical and psychological dependence. Withdrawal symptoms include extreme fatigue, depression, and increased appetite. Also, the pleasurable feelings caused by amphetamine use can lead to a compulsion to overuse the drug.

Amphetamines cross the placenta and can affect the developing fetus. Women who use amphetamines during pregnancy are more likely to give birth prematurely and to have a low-birth-weight baby than are nonusers.

Caffeine blocks the effect of the inhibitory chemical adenosine

Coffee, tea, and cola drinks are known for their arousing effects. The sense of wakefulness and increased energy come primarily from caffeine, although other substances also may contribute. The caffeine content of selected beverages and foods is shown in Table 8a.2. Caffeine occurs naturally in more than 60 plants, but Americans overwhelmingly prefer theirs from coffee, derived from the seeds of the coffee plant (*Coffea* spp.). Indeed, Americans drink more than half of all the coffee produced in the world.

Caffeine acts by inhibiting adenosine activity. Adenosine is a neuromodulator, a chemical released by a neuron that affects the activity of other, nearby neurons. The effects of a neuromodulator are more prolonged than those of a neurotransmitter. Adenosine depresses the nervous system by preventing the neurons from releasing their neurotransmitters, including norepinephrine and dopamine. When caffeine is present in the adenosine receptors, adenosine cannot act and slow the activity of other neurons. Thus, caffeine stimulates the central nervous system.

Because the neurons of the cerebral cortex (the thinking part of the brain) are so sensitive to caffeine, it serves as a "pick-me-up." With caffeine, as with many other stimulants, dosage is important. A low dose, such as the amount obtained from one or two cups of coffee, increases wakefulness and mental alertness, resulting in faster and clearer thought.

However, more than one or two cups can cause "coffee nerves," which make one jumpy and restless. Very high doses (12 or more cups a day) can cause agitation, anxiety, tremors, rapid breathing, and irregular heartbeats. Death from caffeine is not likely, however. A lethal dose is about 10 g, which is equivalent to 80 to 100 cups of coffee a day.

Caffeine can increase the risk of heart attack. First, it increases heart rate. (Take your pulse before and after a cup or two of coffee.) Second, it increases blood pressure. So people with heart problems are advised to avoid products with caffeine.

Caffeine affects other muscles, too. It relaxes the muscles in the digestive tract and in the walls of the airways in the respiratory system. The increased glucose supply spurred by caffeine's effect on metabolism makes voluntary muscles, such as those in the arms and legs, tire less easily.

Caffeine increases a pregnant woman's risk of miscarriage, especially during the first 3 months of pregnancy. One to three cups of coffee a day can increase the risk of miscarriage 30%. Five cups a day can double the risk of miscarriage. Because the pregnancy is not usually confirmed until the fetus is 8 to 10 weeks old, it would be wise for pregnant women and those who would like to become pregnant to avoid caffeine.

Although most people can drink moderate amounts of caffeine—up to 250 mg a day, the equivalent of two cups of coffee—they should be aware that caffeine is addicting, even if only a cup of coffee a day is consumed. Withdrawal symptoms include headache (sometimes severe), depression, anxiety, and, occasionally, flulike symptoms, all of which lead to irritability.

TABLE **8a.2**

CAFFEINE CONTENT IN SELECTED BEVERAGES AND FOODS	
ITEM	CAFFEINE CONTENT (mg)
Coffee (5-oz cup)	50–150
Tea, major U.S. brands (5-oz cup)	20–90
Tea, imported brands (5-oz cup)	25–100
Cola soft drink (12 oz)	35–55
Dark chocolate (semisweet) (1 oz)	15–30
Milk chocolate (1 oz)	1–10

Hallucinogenic Drugs Alter Sensory Perception

The hallucinogenic drugs are grouped together because they have similar effects, despite their diverse chemical structures. These effects include visual, auditory, or other distortions of sensation as well as vivid, unusual changes in thought and emotions.

The psychedelic drugs include some natural and some synthetic forms. There are about six natural psychedelic drugs, the best known of which are mescaline, which comes from peyote cactus, and psilocybin, which is found in certain mushrooms. There are also synthetic forms, including LSD (lysergic acid diethylamide) and MDMA,[3] which is popularly known as ecstasy.

These drugs are thought to act by augmenting the action of the neurotransmitters serotonin, norepinephrine, or acetylcholine. Some of the best-known psychedelic drugs, including LSD, psilocybin, DMT (dimethyltryptamine), and bufotenin, bind to the serotonin receptors in the brain, thereby mimicking the natural effects of serotonin. Ecstasy binds to serotonin receptors and promotes the release of serotonin and dopamine. Mescaline, on the other hand, is structurally similar to norepinephrine.

The normal physiological reactions to psychedelic drugs are not especially harmful, but the distortions of reality that are created by the drug experience may lead to some actions that are quite dangerous. *Bad trip* is the term most often used to describe an unpleasant reaction to a psychedelic drug. The symptoms may include paranoia, panic, depression, and confusion. The trip usually ends within 24 hours, when drug levels decrease. It is nearly impossible to determine who will have a bad trip, but the likelihood is greater if the person using the drug has preexisting emotional problems or is uncomfortable with the company or setting.

Brief recurrences of sensory or emotional alterations produced by the drug are called flashbacks. Almost any aspect of the drug experience may be relived during a flashback. Flashbacks are usually brief and merely annoying, but they can be frightening.

Tolerance for psychedelic drugs develops quickly. Furthermore, cross-tolerance is the rule; that is, a person who has become tolerant of one psychedelic drug will be tolerant to others as well. Craving and withdrawal reactions are unknown, and laboratory animals offered LSD will not take it voluntarily.

Ecstasy deserves additional consideration because its use is growing rapidly. It is sometimes called the "love drug" or "hug drug" because users often say that they feel at peace with themselves and at ease with others. Similar to any stimulant, ecstasy causes increases in heart rate, blood pressure, and body temperature—sometimes to dangerous levels. Ecstasy increases energy levels, allowing people to dance all night (Figure 8a.8). Thus, it can cause dehydration and trigger heat stroke, which occasionally causes death. Deaths are rare, however. The release of serotonin from neurons causes a euphoric high. The temporary depletion of serotonin can cause depression and anxiety in the days following ecstasy use. Ecstasy can also cause nausea, vomiting, and dizziness. Another danger is that ecstasy pills may contain drugs other than MDMA. Among the drugs commonly found are the cough suppressant dextromethorphan, caffeine, ephedrine, and pseudoephedrine. Even the poison strychnine has been found.

Phencyclidine, or PCP, also known as "angel dust," and ketamine, also known as "Special K," are classified as psychedelic anesthetics. That is, they cause hallucinations and alter physical sensations. Both are used as pet tranquilizers. Unlike other psychedelic drugs, they do not work by raising serotonin levels.

Ketamine is gaining popularity as a "club drug." It can be snorted as a powder or popped in pills. Its effects range from a dreamy feeling to hallucinations to a deep, paralyzing out-of-body experience. High doses can cause seizures and coma.

Sedatives Depress the Central Nervous System

Sedatives are drugs that depress the central nervous system. They include barbiturates and benzodiazepines, such as Valium®. Similar to alcohol, the depressant drugs affect inhibitory neurons first. As a result, low doses first produce a relief from anxiety and a mild euphoria. As in so many cases involving drugs, however, the effects are dose dependent. In the case of sedatives, higher doses also inhibit excitatory neurons, and sleep follows.

The effects of different depressants are additive. So if you drink alcohol and also take a sedative, the depressant effects will be greatly intensified. Combining depressant drugs can easily lead to overdose. The respiratory system can, in fact, become so depressed that breathing stops. Administering a central nervous system stimulant, amphetamine, for example, cannot reverse an overdose of depressants. Although the drugs may have antagonistic effects, each brings about its effects by binding to different receptors or by affecting different regions of the brain. So taking a stimulant and a depressant at the same time is like simultaneously stepping on the brake and the accelerator of a car. Slamming the accelerator does not remove the brake. To stop an overdose, it would be necessary to remove the drug from its receptors. A stimulant may temporarily arouse the person, but when it wears off, the resulting rebound depression added to the residual depression caused by the sedative could be fatal.

Continued use of sedatives leads to tolerance as well as both psychological and physical dependence. Cross-tolerance also develops, so tolerance to one sedative also reduces the response to other

FIGURE **8a.8**
Ecstasy increases energy levels and allows people to dance all night.

sedatives. Tolerance develops because the liver produces more of the enzymes that break down the drug and, to some extent, because the nerve cells become less sensitive to its effects. As tolerance develops, overdose becomes a greater risk. Increased doses are required for the drug's sedating effect, but the neurons in the breathing center are still sensitive to a lethal dose.

A dangerous depressant that has been gaining popularity lately is GHB (gamma-hydroxybutrate). It is a clear liquid that is mixed with another beverage. This fast-acting sedative has the reputation as a "date rape" drug because men have used it to make women vulnerable to sexual assault. The most common negative effects of GHB are dizziness, nausea, vomiting, confusion, and sometimes blackouts. Overdoses can be fatal because GHB inhibits the respiratory system.

The Opiates Reduce Pain

The opiates are natural or synthetic drugs that affect the body in ways similar to morphine, the major pain-relieving agent in opium. The opiates and related drugs have two different faces. On one hand, they have a high potential for abuse because tolerance and physical dependence occur. On the other hand, they are medically important because they alleviate severe pain.

Morphine and codeine, which come from the opium poppy, were among the first opiates used. Heroin is a synthetic derivative of morphine that is more than twice as effective. Heroin is usually injected intravenously. There are now forms that can be smoked or inhaled into the nasal cavity. Regardless of how it enters the body, heroin reaches the brain very quickly, producing a feeling that can be described only in ecstatic or sexual terms. The rush of euphoria produced by heroin has made it the drug preferred by opiate abusers and spurs its continued use.

The opiates, including heroin, exert their effects by binding the receptors for the body's endogenous (natural, internally produced) opiates: endorphins, enkephalins, and dynomorphins. These are the neurotransmitters involved in the perception of pain and fear, among other things.

Heroin and other commonly abused opiates have effects similar to those of morphine: euphoria, pain suppression, and reduction of anxiety. They also slow the breathing rate. An overdose may cause the user to fall into a coma and stop breathing. Overdose is always a potential problem because heroin is bought on the street for illegal use. The buyer has no idea of its strength. Street supplies are often diluted with sugar. If a heroin addict who is accustomed to a diluted drug injects heroin that is much more potent than he or she thinks and is used to, death due to overdose often occurs. Extremely constricted pupils in an unconscious person is a sign of heroin overdose.

Surprisingly, habitual use of heroin has few directly toxic effects. Many of the problems arise because heroin addicts suffer from a general disregard of good health practices. Other problems are associated with the injections themselves. Frequent intravenous injection of any drug is associated with certain ailments. Because of the constant puncturing, veins can become inflamed. In addition, shared needles may spread disease-causing organisms, including the viruses that cause AIDS and hepatitis, and the bacterium that causes syphilis.

Sensory Systems

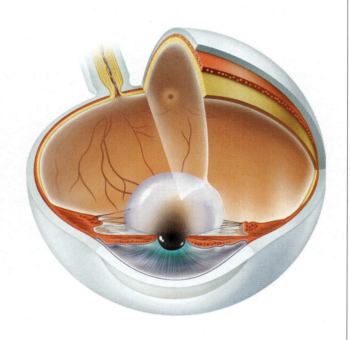

Sensory Receptors Generate Electrochemical Messages in Response to Stimuli

Receptors Are Specialized to Respond to Different Types of Stimuli

Receptors for the General Senses Are Widespread Throughout the Body
- Touch and pressure are detected by receptors in the skin
- Temperature is detected by cold and heat receptors
- Body and limb position are detected by muscle spindles and Golgi tendon organs
- Pain is caused by any sufficiently strong stimulus

Vision Depends on the Eye
- The eye focuses light on the light-sensitive retina
- The cornea and lens focus the image on the retina
- Light changes the shape of pigment molecules, which results in neural messages
- Color vision depends on cones

Hearing Depends on the Ear
- The ear collects and amplifies sound waves and converts them to neural messages
- Variations in the movements of the basilar membrane determine loudness and pitch
- Hearing loss can be conductive or sensorineural
- Ear infections can occur in the ear canal or in the middle ear

Balance Depends on the Vestibular Apparatus of the Inner Ear

Smell and Taste Are the Chemical Senses

HEALTH ISSUE   Correcting Vision Problems

ENVIRONMENTAL ISSUE   Noise Pollution

It's the ninth inning, and the local Triple A baseball team is ahead by one run. There are two outs, and runners are on second and third, the catcher having missed a wild pitch that rolled all the way to the backstop, signaling the first- and second-base runners to race full speed and slide into second and third. It's now a full count: 3 balls and 2 strikes on the batter. The next pitch will decide the game. Jim can barely hear himself shout as he and hundreds of other fans jeer the batter, hoping he'll strike out. Danielle tugs anxiously on Jim's shirt, fearing the worst. The pitcher winds up and delivers the pitch. It's a low inside strike. The game is over. They've won! Seeing the joy on the field and hearing the triumphant shouts of the home crowd, Jim and Danielle are positively thrilled to be at the baseball game. They love the smell of hotdogs and garlic fries, the sound of the crowd roaring, cheering, and booing. They like to feel the warmth of the sun on a Friday afternoon. And they have to taste everything the vendors wave in front of them. On this afternoon, their senses are directing their every pleasure.

Our senses help us perceive the world around us. In this chapter, we will explore the body's general senses (such as touch, pressure, vibration, temperature, and pain) and our special senses (vision, hearing, balance, smell, and taste). We will take a closer look at vision and hearing and examine the intricate structures of the eye and ear. We also will explore the relationship between smell and taste. ■

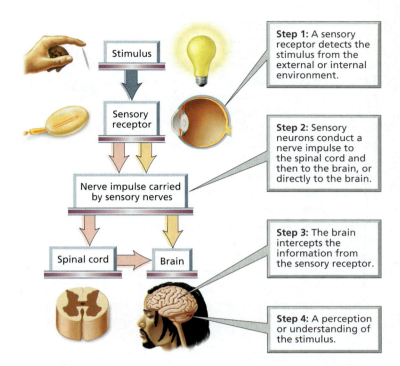

Step 1: A sensory receptor detects the stimulus from the external or internal environment.

Step 2: Sensory neurons conduct a nerve impulse to the spinal cord and then to the brain, or directly to the brain.

Step 3: The brain intercepts the information from the sensory receptor.

Step 4: A perception or understanding of the stimulus.

FIGURE **9.1**
An overview of the steps involved in sensation

Sensory Receptors Generate Electrochemical Messages in Response to Stimuli

Information about the external and internal world comes to us in different forms of energy, including light, sound, chemicals, and pressure. **Sensory receptors** are structures that are specialized to respond to stimuli, or changes, in their environment by generating electrochemical messages caused by ions crossing the plasma membrane of the neuron.. If the stimulus is strong enough, these messages eventually become nerve impulses. The nerve impulses are then conducted to the brain, where they are interpreted to form our perceptions of the world around us (Figure 9.1).

Sensory receptors share several characteristics. For one thing, they are selective. That is, each type responds best to one form of energy. Photoreceptors, for instance, respond to light. The response of a sensory receptor, called a **receptor potential**, is an electrochemical message (a change in the degree of charge difference across a membrane) that varies in magnitude with the strength of the stimulus. As a result, sensory receptors can produce graded signals that vary in intensity. For instance, the louder the sound, the larger the receptor potential—up to a point. When the receptor potential reaches threshold level, an action potential (nerve impulse) is generated. Most types of sensory receptors gradually stop responding when they are continuously stimulated, a phenomenon called **sensory adaptation**. As receptors adapt, we become less aware of the stimulus. For example, the musty smell of an antique store may be obvious to a person who just walked in, but it goes unnoticed by the salesclerk working there. Some receptors, such as those for pressure and touch, adapt quickly. For this reason,

we quickly become unaware of the feeling of our clothing against our skin. Other receptors adapt slowly or not at all. The receptors in muscles and joints that report on the position of body parts, for instance, never adapt. Their continuous input is essential for coordinated movement and balance.

Receptors Are Specialized to Respond to Different Types of Stimuli

Receptors are classified according to the stimulus to which they respond. Several classes of receptors are traditionally recognized:

1. **Mechanoreceptors** are responsible for the sensations we describe as touch, pressure, hearing, and equilibrium. They respond to distortions in the receptor itself or in nearby cells. In addition, mechanoreceptors detect changes in blood pressure and tell us the position of our body.

2. **Thermoreceptors** detect changes in temperature.

3. **Photoreceptors** detect changes in light intensity.

4. **Chemoreceptors** respond to chemicals. We describe the input from the chemoreceptors of the mouth as taste (gustation) and those from the nose as smell (olfaction). Other chemoreceptors monitor levels of chemicals, such as carbon dioxide, oxygen, and glucose, in body fluids.

5. **Pain receptors** respond to very strong stimuli, which usually result from physical or chemical damage to tissues. Instead of being classified in a category of their own, pain receptors are sometimes classified as chemoreceptors, because they often respond to chemicals liberated by damaged tissue, and occasion-

ally as mechanoreceptors, because they are stimulated by physical changes, such as swelling, in the damaged tissue.

Receptors may also be classified by their location. **Exteroceptors** are located near the surface of the body and respond to changes in the environment. **Interoceptors** are inside the body and monitor conditions there. Although we often are unaware of the activity of interoceptors, they play a vital role in maintaining homeostasis. They are an important part of the feedback loops that regulate blood pressure, blood chemistry, and breathing rate. Interoceptors may also cause us to feel pain, hunger, or thirst, thereby prompting us to take appropriate action.

The **general senses**—touch, pressure, vibration, temperature, a sense of body and limb position, and pain—arise from receptors in the skin, muscles, joints, bones, and internal organs. The **special senses** are smell, taste, vision, hearing, and the sense of balance, or equilibrium. These senses usually come to mind when we think of the senses, largely because we are so dependent on them as we try to perceive the nature of our world. The receptors of the special senses are located in the head, usually within specific structures.

Receptors for the General Senses Are Widespread Throughout the Body

The receptors for general senses are widespread throughout the body. Some monitor conditions within the body; others provide information about the world around us. The receptors rely on free nerve endings or encapsulated endings. Free nerve endings are branched tips of dendrites of sensory neurons. An encapsulated ending is one in which a connective tissue capsule encloses the tips of the dendrites of a sensory neuron.

▌Touch and pressure are detected by receptors in the skin

During the seventh week of pregnancy, long before the eyes and ears have formed, a sense of touch is already functioning in the embryo. Right from the start, infants explore their world using the sense of touch. Throughout life, touch is a way of learning about the world and of communicating with one another.

Light touch, such as might occur when the cat brushes past your legs, depends on several types of receptors (Figure 9.2). Wrapped around the base of the fine hairs of the skin are free nerve endings, which detect any bending of the hairs. *Merkel disks*, free nerve endings that end on special cells, also sense light touch. They are found on both the hairy and the hairless parts of the skin. Free nerve endings and Merkel disks tell us that something has touched the skin. *Meissner's corpuscles* tell us exactly where we have been touched. These encapsulated nerve endings are common on the hairless, very sensitive areas of skin, such as the lips, nipples, and fingertips.

The sensation of pressure generally lasts longer than does touch, and pressure is felt over a larger area. *Pacinian corpuscles*, which consist of onion-like layers of tissue surrounding a nerve ending, are scattered in the deeper layers of skin and the underlying tissue. They respond only when the pressure is first applied and, therefore, are important in sensing vibration. *Ruffini corpuscles* are encapsulated endings that respond to continuous pressure.

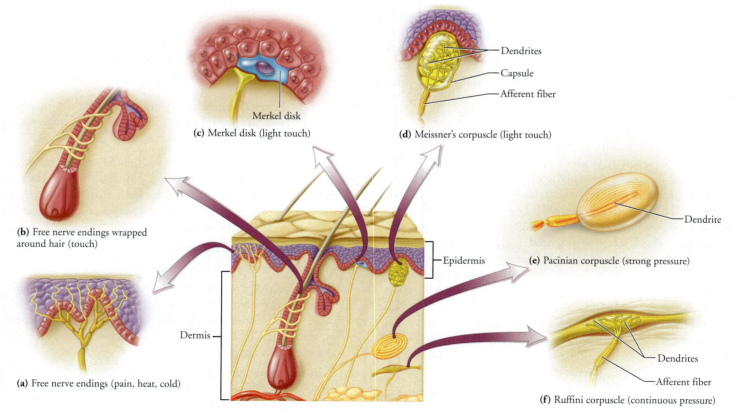

(c) Merkel disk (light touch)

(d) Meissner's corpuscle (light touch) — Dendrites, Capsule, Afferent fiber

(b) Free nerve endings wrapped around hair (touch)

(e) Pacinian corpuscle (strong pressure) — Dendrite

Epidermis

Dermis

(a) Free nerve endings (pain, heat, cold)

(f) Ruffini corpuscle (continuous pressure) — Dendrites, Afferent fiber

FIGURE **9.2**

General sense receptors of the skin allow us to sense touch, pressure, temperature, and pain.

Temperature is detected by cold and heat receptors

Thermoreceptors respond to changes in temperature. They are widely distributed throughout the body and are especially common around the lips and mouth. In humans, thermoreceptors may simply be two types of specialized free nerve endings. The so-called cold receptors are most active from 10° to 20°C (50° to 68°F). The heat receptors are active as the temperature climbs to 25°C (77°F) but become inactive at temperatures above 45°C (113°F). (Temperatures higher than this activate pain receptors.) You may have noticed that the sensation of hot or cold fades rapidly. For example, because the thermoreceptors adapt rapidly, the water in a hot tub may feel scalding at first, but it very quickly feels comfortably warm.

Body and limb position are detected by muscle spindles and Golgi tendon organs

Whether you are at rest or in motion, the brain "knows" the location of all your body parts. It continuously scans the signals from muscles and joints to check body alignment and ensure coordination. *Muscle spindles* are specialized muscle fibers with sensory nerve endings wrapped around them that report to the brain whenever a muscle is stretched. *Golgi tendon organs* measure the degree of muscle tension. They are highly branched nerve fibers located in tendons, the connective tissue bands that connect muscles to bones. The information from muscle spindles and Golgi tendon organs is used along with information from the inner ear (as we will see later in this chapter) to coordinate our movements.

Pain is caused by any sufficiently strong stimulus

The receptors for pain are free nerve endings that are found in almost every tissue of the body. Any stimulus strong enough to damage tissues, including heat, cold, touch, and pressure, will cause pain. Tissue damage activates enzymes that form a chemical called bradykinin. Bradykinin then triggers the release of inflammatory chemicals, such as histamine and prostaglandins, that begin the healing process. Together, bradykinin, histamine, and prostaglandins alert free nerve endings of the injury. The stimulated sensory neurons then carry the message to the brain, where it is interpreted as pain. Aspirin reduces pain because it interferes with the production of prostaglandins.

Many of our internal organs also have pain receptors. However, pain originating in an internal organ is sometimes experienced as pain in an uninjured region of the skin (Figure 9.3). This phenomenon is called **referred pain**. For example, the pain of a heart attack is often experienced as pain in the left arm. This pain probably occurs because sensory neurons from the internal organ and a particular region of the skin communicate with the same neurons in the spinal cord. The brain interprets the input as coming from the skin.

Our perceptions of pain involve both the sensation of pain and our emotional response to that pain. Pain is an important mechanism that warns the body and protects it from injury. For example, pain usually prevents a person with a broken leg from causing additional damage by moving the limb. Nonetheless, few of us appreciate the value of pain while we are experiencing it. Also, pain that persists long after the warning is needed can be debilitating.

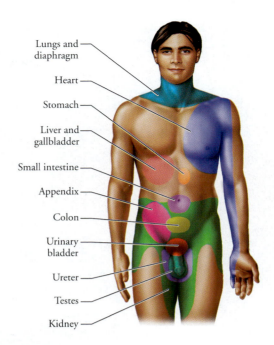

FIGURE **9.3**
Referred pain. Pain from certain internal organs is sensed as originating in particular uninjured regions of the skin.

Vision Depends on the Eye

We, as humans, are very visual creatures. We may not see detail as well as an eagle or movement as well as an insect, but we see far better than most other mammals.

The eye focuses light on the light-sensitive retina

Each of your eyeballs is an irregular sphere about 25 mm (1 in.) in diameter. As shown in Figure 9.4, the eyeball consists of three layers. Table 9.1 summarizes the structures of the eye and their functions. The outermost layer is a tough, fibrous covering with two obviously different regions: the sclera and the cornea. The **sclera**, often called the white of the eye, protects and shapes the eyeball and serves as an attachment site for the muscles that move the eye. In the front and center of the eye, the transparent **cornea** bulges slightly outward and provides the window through which light enters the eye. As light passes through the curved surface of the cornea, it is bent toward the light-sensitive surface at the rear of the eye. Unlike most tissues in the body, the cornea lacks blood vessels. Although the cornea has no blood supply, it does have pain receptors. Indeed, a tear in the cornea is extremely painful.

The middle layer of the eye has three distinct regions—the choroid, the ciliary body, and the iris. The **choroid** is a layer that contains many blood vessels that supply nutrients and oxygen to

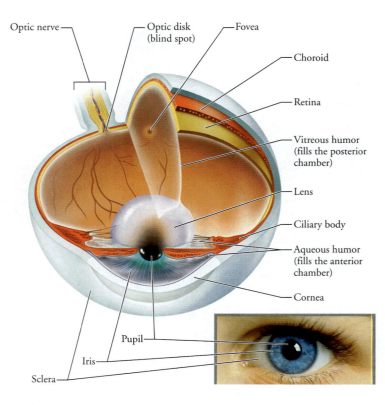

the tissues of the eye. The choroid layer also contains the brown pigment melanin, which absorbs light after it strikes the light-sensitive layer. This absorption of light helps prevent the reflection of light within the eye, resulting in sharper vision.

The **ciliary body** is a ring of tissue, primarily muscle, that encircles the lens, holding it in place and controlling its shape. As we will see, the shape of the lens is important in focusing light on the light-sensitive layer of the eye.

The **iris**, the colored portion of the eye that can be seen through the cornea, regulates the amount of light that enters the eye. It is shaped like a flat doughnut. The doughnut hole is the opening through the center of the iris, called the **pupil**, through which light enters the eye.

The iris contains smooth muscle fibers that automatically adjust the size of the pupil to admit the appropriate amount of light to the eye. The pupil becomes larger (dilates) in dim light and becomes smaller (constricts) in bright light. In fact, its size can vary

FIGURE **9.4**
Structure of the human eye. Light enters the eye through the transparent cornea and then passes through the pupil. The lens focuses light on the light-sensitive retina, which contains the rods and cones.

TABLE **9.1**

STRUCTURES OF THE EYE AND THEIR FUNCTIONS		
STRUCTURE	**DESCRIPTION**	**FUNCTION**
Outer layer		
Sclera	Outer layer of the eye	Protects the eyeball
Cornea	Transparent dome of tissue forming the outer layer at the front of the eye	Refracts light, focusing it on the retina
Middle layer		
Choroid	Pigmented layer containing blood vessels	Absorbs stray light; delivers nutrients and oxygen to tissues of eye
Ciliary body	Encircles lens; contains the ciliary muscles	Controls shape of lens; secretes aqueous humor
Iris	Colored part of the eye	Regulates the amount of light entering the eye through the pupil
Pupil	Opening at the center of the iris	Opening for incoming light
Inner layer		
Retina	Layer of tissue that contains the photoreceptors (rods and cones); also contains bipolar and ganglion cells involved in retinal processing	Receives light and generates neural messages
Rods	Photoreceptor	Responsible for black and white vision and vision in dim light
Cones	Photoreceptor	Responsible for color vision and visual acuity
Fovea	Small pit in the retina that has a high concentration of cones	Provides detailed color vision
Other structures of the eye		
Lens	Transparent, semispherical body of tissue behind the iris and pupil	Fine focusing of light onto retina
Aqueous humor	Clear fluid found between the cornea and the lens	Refracts light and helps maintain shape of the eyeball
Vitreous humor	Gelatinous substance found within the chamber behind the lens	Refracts light and helps maintain shape of the eyeball
Optic nerve	Group of axons from the eye to the brain	Transmits impulses from the retina to the brain

16-fold. Pupil size is also affected by emotions. The pupils dilate when you are frightened or when you are very interested in something. They constrict when you are bored. Candlelight creates a romantic setting partly because its dimness dilates the pupils, making the lovers appear more attentive and interested.

The innermost layer of the eye is the **retina**, which contains almost a quarter-billion photoreceptors, the structures that respond to light by generating electrical signals. The retina contains two types of photoreceptors: rods and cones. The structure and function of rods and cones are discussed in more detail later in the chapter. The **rods** are more numerous than the cones and are responsible for our black and white vision. Rods are exceedingly sensitive to light and are capable of responding to light measuring one 10-billionth of a watt—the equivalent of a match burning 50 miles (80.5 km) away on a clear, pitch-dark night! (Of course, the ideal conditions necessary to actually *see* a burning match at such a distance are impossible to attain.) The rods allow us to see in dimly lit rooms and in pale moonlight. By detecting changes in light intensity across the visual field, rods contribute to the perception of movements. The **cones** are responsible for color vision. Unlike the rods, the cones produce sharp images. The cones are most concentrated in a small region in the center of the retina called the **fovea**. Only the fovea contains a sufficient number of cones, about 150,000/mm^2, to produce detailed color vision. Therefore, when we want to see the fine details of something, the image must be focused on the fovea. However, the fovea is only the size of the head of a pin. At any given moment, therefore, only about a thousandth of our visual field is in sharp focus. Eye movements bring different parts of the visual field to the fovea.

stop and think

There are no rods in the fovea. Rods are more concentrated at the periphery of the retina. How does this arrangement explain why you can see an object better at night if you do not look directly at it?

The processing of electrical signals from the rods and cones begins before the signals leave the retina. The messages are first sent to bipolar cells, which begin to process the information, and then to ganglion cells (Figure 9.5). Between these two cell layers are *horizontal cells* and *amacrine cells*, which send signals laterally across the retina. Together, these cells convert the input from the retina into patterns, such as edges and spots. The bundled axons of the ganglion cells form the **optic nerve**, which carries the processed message from the eye to the brain, where the message is interpreted. The region where the optic nerve leaves the retina has no photoreceptors. As a result, we cannot see an image that strikes this area. This area is, therefore, called the **blind spot**. You are usually not aware of your blind spot because the missing parts are automatically "filled in" as the visual information is processed by the brain.

stop and think

Most bipolar cells receive input from several rods but from only one cone in the fovea. How might this difference explain why rods permit us to see at lower light intensities than do cones? How might it explain why the sharpest images are formed when the object is focused on the fovea?

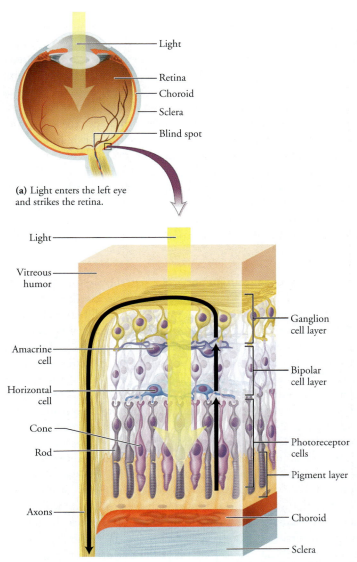

(a) Light enters the left eye and strikes the retina.

(b) When light is focused on the retina, it passes through the ganglion cell layer and bipolar cell layer before reaching the rods and cones. In response to light, the rods and cones generate electrical signals that are sent to bipolar cells and then to ganglion cells. These cells begin the processing of visual information.

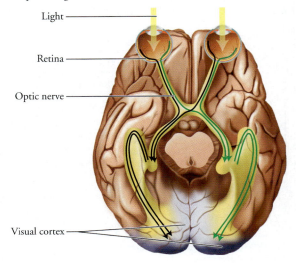

FIGURE **9.5**

Neural pathways of the retina convert light to nerve signals.

In addition to the three layers, the eyeball can also be divided into two fluid-filled cavities (see Figure 9.4). The posterior cavity, located at the back of the eye between the lens and the retina, is filled with a jellylike fluid called **vitreous humor**. This fluid helps keep the eyeball from collapsing and holds the thin retina against the wall of the eye. The anterior cavity, located at the front of the eye between the cornea and the lens, is filled with a fluid called **aqueous humor**. This clear fluid supplies nutrients and oxygen to the cornea and lens and carries away their metabolic wastes. In addition, the aqueous humor creates pressure within the eye, helping to maintain the shape of the eyeball. Unlike the vitreous humor, which is produced during embryonic development and is never replaced, aqueous humor is replaced about every 90 minutes. It is continuously produced from the capillaries of the ciliary body, circulates through the anterior cavity, and is drained into the blood through a network of channels that surround the eye.

GLAUCOMA

If the drainage of aqueous humor is blocked, the pressure within the eye may increase to dangerous levels. This condition, called *glaucoma*, is the second most common cause of blindness (after cataracts, which are discussed shortly). The accumulating aqueous humor pushes the lens partially into the posterior cavity of the eye, increasing the pressure there, and compressing the retina and optic nerve. This pressure, in turn, collapses the tiny blood vessels that nourish the rods and cones and the fibers of the optic nerve. Deprived of nutrients and oxygen, the photoreceptors and nerve fibers begin to die and vision fades. Unfortunately, glaucoma is painless and progressive. Late signs include blurred vision, headaches, and seeing halos around objects. Because many people do not realize they have a problem until some vision has been lost, eye specialists recommend that everyone over 40 years of age be tested for glaucoma every year. Early glaucoma is treated with eye drops and medications that either reduce the rate of formation of aqueous humor or enhance its drainage. If these treatments fail to reduce the pressure in the eye, an artificial drainage channel can be created surgically.

▌ The cornea and lens focus the image on the retina

Sharp, clear vision requires that light rays be focused on the retina; focusing is accomplished by bending the light rays that enter the eye.

Light rays are bent, or refracted, when the speed at which they are traveling changes, as occurs when they enter a medium of a different density. It is this bending of light rays that focuses the image on the retina. Because of the way that the curved cornea bends the light rays, the image created on the retina is upside down and backward.

The most important structures in focusing the image are the cornea and the lens. Because the cornea is the only transparent structure of the eye exposed directly to air, more than 75% of the bending of light that enters the eye occurs here. But the cornea has a fixed shape, so it always bends light to the same degree. The cornea cannot make the adjustments needed to focus on objects at varying distances.

The lens, however, is elastic, and its shape can be changed to focus on both near and distant objects by a process called **accommodation**. To understand how the shape of the lens changes, picture it as an underinflated round balloon. If you pulled on the sides of the balloon, it would flatten. When you released the tension, the balloon would assume its usual, rounder shape. When focusing on a nearby object, the lens becomes rounder and thicker, thereby increasing the degree of bending in the light ray (Figure 9.6).

In the eye, changes in the tension on the lens and therefore changes in lens shape are controlled by the ciliary muscle. The lens is attached to the ciliary muscle by ligaments. Because the ciliary muscle is circular, its diameter becomes smaller when it contracts, much like pursed lips. This contraction relaxes the tension on the ligaments, and the lens is free to assume the rounded shape needed to focus on nearby objects. Relaxation of the ciliary muscle causes the lens's diameter to increase; this increase in diameter, in turn, increases the tension on the ligaments and the lens. Consequently, the lens flattens and focuses light from more distant objects on the retina. As we age, the lens becomes less elastic and does not round up to focus on nearby objects as easily, explaining why we hold the newspaper farther away as we become older.

CATARACTS

A **cataract** is a lens that has become cloudy or opaque, usually because of aging. Cataracts are the most common eye problem that affects men and women who are older than 50. Typically, the lens takes on a yellowish hue that blocks light on its way to the retina. In the beginning, cataracts may cause a person to see the world

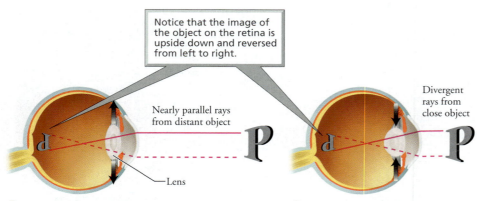

Notice that the image of the object on the retina is upside down and reversed from left to right.

Nearly parallel rays from distant object

Lens

Divergent rays from close object

DISTANT OBJECT – FLATTENED LENS:
• Ciliary muscles relax
• Ligaments to lens stretch
• Lens flattens

CLOSE OBJECT – ROUNDED LENS:
• Ciliary muscles contract
• Ligaments to lens loosen
• Lens becomes rounded

FIGURE **9.6**

Accommodation. The lens changes shape to view objects at varying distances.

through a haze that can limit activities and cause automobile accidents, early retirement, and falls that can result in hip fractures. As the lens becomes increasingly opaque, the fog thickens. Indeed, cataracts are the leading cause of blindness worldwide. When cataracts make it impossible for a person to perform everyday tasks, the clouded lens can be surgically removed and replaced with an artificial lens. Cataract surgery has a success rate of more than 90%.

Most cataracts form as a natural part of the aging process, but they also can be a consequence of drug reactions, injuries, or diabetes mellitus, a condition in which the pancreas does not secrete enough of the hormone insulin. People who smoke 20 or more cigarettes a day are twice as likely as a nonsmoker to develop cataracts. Exposure to the ultraviolet B portion of sunlight, the same part of sunlight that causes sunburn and an increased risk of skin cancer, is also thought to be a contributing factor in the formation of cataracts. Adults and children should wear sunglasses with specially treated lenses that absorb nearly all of the ultraviolet rays. Regular tinted glasses reduce the amount of visible light reaching the eyes but do nothing to reduce the amount of ultraviolet (UV) light entering the eye. In fact, untreated sunglasses may do more harm than good, because they reduce the intensity of light entering the eye enough that people do not squint in bright sunlight, thus allowing more UV light in. So, it is worth the extra money for sunglasses that will filter out UV light.

In addition to changes in lens shape, sharp vision requires coordinated eye movements to keep the image focused on the retina's fovea. Both eyes focus on the same group of objects. When looking at a distant object, both eyes are directed at the same angle. However, as we move closer to the object, our eyes must be directed toward the midline if they are to remain fixed on the object, a movement called **convergence**. Muscles attached to the eyeball, the extrinsic eye muscles, move the eye. Convergence is important because it keeps the image focused on the fovea. The nearer the object, the greater the degree of convergence required.

When you have been reading or doing other close work for a long time, the constant contraction of both the ciliary muscles needed for focusing over short distances and of the eye muscles causing convergence can cause eyestrain. Eyestrain can be relieved by gazing into the distance for a while, allowing these muscles to relax. Table 9.2 summarizes the processes necessary for sharp vision.

FOCUSING PROBLEMS

The three most common visual problems—farsightedness, nearsightedness, and astigmatism—are focusing problems. Therefore, normal vision can be restored with corrective lenses (Table 9.3). In *farsightedness,* distant objects are seen more clearly than nearby ones because the eyeball is too short or the lens is too thin, causing the closer image to be focused behind the retina (Figure 9.7). Although distant objects can be seen clearly, the lens cannot become round enough to bend the light sufficiently to focus on nearby objects. Corrective lenses that are thicker in the middle than at the edges (convex) cause the light rays to converge a bit before they enter the eye. The lens can then focus the image on the retina. Other solutions to vision problems are discussed in the Health Issue essay, *Correcting Vision Problems.*

About 25% of the American population is nearsighted; that is, they can see nearby objects more clearly than those far away. *Nearsightedness* (myopia) occurs when the eyeball is elongated or when the lens is too thick. This condition causes the image to focus in front of the retina. Nearsighted people see nearby objects clearly because the lens becomes round enough to focus the image

TABLE **9.2**

PROCESSES IMPORTANT TO SHARP VISION		
PROCESS	**STRUCTURE(S) INVOLVED**	**RESULT**
Refraction	Cornea	Produces most bending of light
Accommodation	Lens	Changes shape to further bend light and focus image on retina
Convergence	Extrinsic eye muscles	Directs eyes so that image falls on the fovea

TABLE **9.3**

FOCUSING PROBLEMS			
PROBLEM	**DESCRIPTION**	**CAUSE**	**CORRECTION**
Farsightedness	See distant objects more clearly than nearby objects	Eyeball too short or lens too thin; lens cannot become round enough	Convex lens; increases corneal curvature
Nearsightedness	See nearby objects more clearly than distant objects	Eyeball too long or lens too thick; lens cannot flatten enough	Concave lens; decreases corneal curvature
Astigmatism	Visual image is distorted	Irregularities in curvature of cornea or lens	Lenses that correct for the asymmetrical bending of light

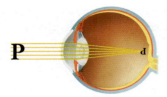

(a) Normal eye
- Close and distant object seen clearly
- Image focuses on retina

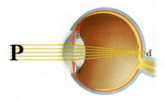

(b) Farsightedness
- Distant objects seen clearly
- Close objects out of focus
- Short eyeball causes image to focus behind retina

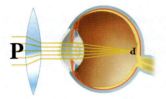

Convex lens
- Causes light rays to converge so that the image focuses on the retina

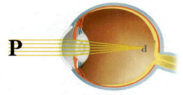

(c) Nearsightedness (myopia)
- Close objects seen clearly
- Distant objects out of focus
- Long eyeball causes image to focus in front of retina

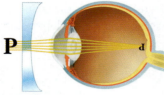

Concave lens
- Causes light rays to diverge so that the image focuses on the retina

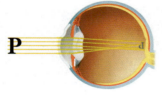

(d) Astigmatism
- Image blurred
- Irregular curvature of cornea or lens causes light rays to focus unevenly

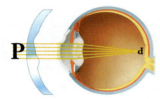

Uneven lens
- Focuses entire image on retina

FIGURE **9.7**

Focusing problems such as farsightedness, nearsightedness, and astigmatism are caused when the image of an object is not focused on the retina. These vision problems can be corrected with the use of specific lenses.

on the retina. However, the lens simply cannot flatten enough to bring the focused image to the retina, so distant objects appear blurred. Lenses that are thinner in the middle than at the edges (concave) can correct nearsightedness. These lenses cause the light rays to diverge slightly before entering the eye.

Although genetics undoubtedly plays a role in the development of nearsightedness, frequent close work, such as reading or working at a computer is also a cause. When you do a lot of close work, the frequent contractions of the ciliary muscles that change the shape of the lens increase the pressure within the eye. This pressure can cause the eye to stretch and elongate, causing nearsightedness. Eye specialists recommend that you look up from the page or away from the computer screen at frequent intervals—particularly if nearsightedness runs in your family.

Irregularities in the curvature of the cornea or lens will cause distortion of the image because they cause the light rays to converge unevenly. This condition is called *astigmatism*. Vision can be restored to normal by corrective lenses that compensate for the asymmetrical bending of light rays.

Light changes the shape of pigment molecules, which results in neural messages

We said earlier that humans are very visual creatures, but it may surprise you to know that the eye contains 70% of all the sensory receptors in the human body. The function of all photoreceptors, the rods and cones, is to respond to light with neural messages that are sent to the brain, where they are translated into images of our surroundings (Figure 9.8).

Our world is, in fact, so visual that few of us ever stop to wonder how vision works. So let's take a moment to consider the process, as shown in Figure 9.9. First, light waves in the visible spectrum (that is, the wavelengths that our eyes can detect, 380 to 750 nanometers) strike an object. At least some wavelengths are reflected off the object to the retina of the eye. The light is absorbed by pigment molecules in the photoreceptors, causing a chemical change in the molecules. This change, in turn, changes the permeability of the plasma membrane of the photoreceptor, resulting in electrochemical changes in the receptor (the receptor potential mentioned earlier) that initiate a neural message when they reach a certain strength. The optic nerve carries nerve signals from the retina to the back of the brain, where the world outside takes shape. In a sense, then, seeing is an illusion, because there are no pictures inside our heads.

The light-absorbing portion of the pigment molecules in all photoreceptors is a compound called **retinal**, which is bound to a protein called an **opsin**. There are four types of opsins and thus four types of photoreceptors: rods and three types of cones. Retinal preferentially absorbs different wavelengths (colors) of the visible spectrum, depending on the type of opsin to which it is bound.

The rods, you may recall, allow us to see in dim light, but only in shades of gray. The pigment in rods, called **rhodopsin**, is packaged in membrane-bound disks, which are stacked like coins in the outer segment of the rod. When light strikes a rod, the retinal portion of rhodopsin changes shape and splits from the protein portion, which is called scotopsin. The splitting of rhodopsin triggers events that reduce the permeability of the plasma membrane to sodium ions, eventually leading to changes in the activity of bipolar cells and ganglion cells. In the dark, rhodopsin is resynthesized. It is difficult to see when you first walk into a dark room from a brightly lit area because the rhodopsin has been split. As rhodopsin is resynthesized, vision becomes sharper.

HEALTH ISSUE

Correcting Vision Problems

Although most people wear glasses to correct vision problems, almost 20 million Americans use contact lenses instead. The contact lens sits on the cornea over a layer of tears. The outer surface of the lens is the corrective surface, and the inner surface fits snugly on the cornea. There are two types of contact lenses. Rigid gas-permeable lenses are made of slightly flexible plastic. These lenses are tough and hard wearing, but some people find that they irritate the eyes. Soft lenses are made of a more malleable plastic. Although soft lenses are gentler on the eyes, they become scratched and damaged more easily.

Contact lenses must be cleaned and disinfected regularly. Infections, caused by bacteria, fungi, or *Acanthamoeba* (a type of amoeba), can develop behind the lens. Tears leave deposits of protein, fats, and calcium that help infectious organisms adhere to the contact lens. At best, these deposits are difficult to remove. The more infrequently the lenses are cleaned, the more difficult it becomes to remove the deposits, and the more likely infection becomes. Rubbing the lenses with special lens cleaners helps remove these deposits. The no-rub contact lens solutions contain chemicals that break down the protein deposits so that rinsing alone will remove them.

Extended-wear contact lenses of both types have been developed, some of which can be left in place for a month or more. Initially, eye spe-cialists thought that less frequent placing and removal of lenses would reduce trauma to the cornea. A serious safety concern with contact lenses that are left in overnight or for extended periods of time is ulcerative keratitis, a condition in which the cells of the cornea may be rubbed away by the contact lens, sometimes leading to infection and scarring. If not promptly treated, ulcerative keratitis can lead to blindness. The problem with leaving contact lenses in overnight seems to be a reduced oxygen supply to the cornea, which causes it to swell.

Many people who are tired of depending on glasses or contact lenses have opted to undergo laser eye surgery. This procedure is popularly known as LASIK, which stands for laser-assisted *in situ* keratomileusis. LASIK permanently changes the shape of the cornea. The procedure involves cutting a flap in the cornea, using pulses from a computer-controlled laser to reshape the middle layer of the cornea, and then replacing the flap. This treatment is used to correct for nearsighted-ness, farsightedness, and astigmatism. People who need different lenses for reading and distance vision may still require glasses for reading. How-ever, it is possible to correct one eye for close-up work and the other for distance. Some people find it difficult to adjust to having the focus of their eyes set differently, so this method should be tried with contact lenses before permanently changing the shape of one's corneas.

Other procedures can be used to reshape the cornea. LASEK (laser-assisted subepithelial ker-atomileusis) differs from LASIK in that the surgeon creates a flap in the epithelium only instead of in the entire cornea, eliminating some of the compli-cations due to the deeper corneal flaps. LASEK is used mostly for people who are poor candidates for LASIK because their corneas are thin or flat. In another procedure, photorefractive keratectomy (PRK), a surgeon uses short bursts of a laser beam to shave a microscopic layer of cells off the corneal surface and flatten it. A computer calcu-lates and controls the laser exposure.

The shape of the cornea can also be altered using a surgical procedure that does not involve lasers. A surgeon can place corneal ring segments, two tiny crescent-shaped pieces of plastic, in the cornea to flatten it. This brief procedure, lasting only about 15 minutes, reduces nearsightedness. Although the ring segments are intended to be permanent, they can be removed if necessary.

Special hard contact lenses are designed so that they can reshape the cornea. This proce-dure, called orthokeratology, produces long-last-ing but not permanent, changes in the cornea. At first, these special contact lenses are worn for 8 hours a day. After the cornea has been reshaped for clearest vision, the lenses can be worn a few hours every few days. If their use is discontinued, the cornea gradually returns to its natural shape. ♟♟

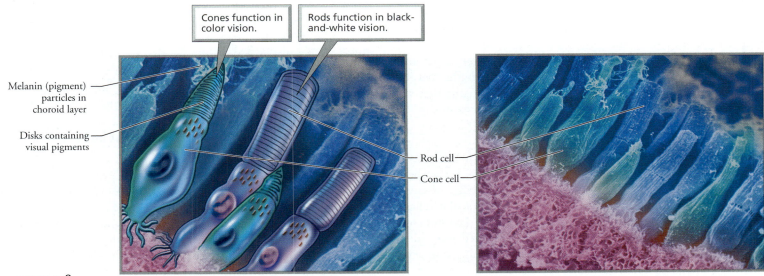

Cones function in color vision.

Rods function in black-and-white vision.

Melanin (pigment) particles in choroid layer

Disks containing visual pigments

Rod cell

Cone cell

FIGURE 9.8

The rods and cones, photoreceptors in the retina, are named for their shapes. Their function is to generate neural messages in response to light. The outer portion of each photoreceptor is packed with membrane-bound disks that contain pigment molecules.

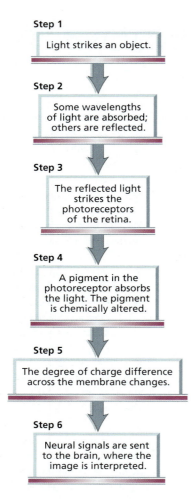

Step 1
Light strikes an object.

Step 2
Some wavelengths of light are absorbed; others are reflected.

Step 3
The reflected light strikes the photoreceptors of the retina.

Step 4
A pigment in the photoreceptor absorbs the light. The pigment is chemically altered.

Step 5
The degree of charge difference across the membrane changes.

Step 6
Neural signals are sent to the brain, where the image is interpreted.

FIGURE **9.9**
An overview of the visual process

▌Color vision depends on cones

What makes you see red? Actually, red, or any other color, is determined by wavelengths of light. White light, such as that emitted by an ordinary light bulb or the sun, consists of all wavelengths. (Rainbows are created when the various wavelengths of white light are separated as they pass through tiny droplets of water in the air.) When light strikes an object, some of the wavelengths may be absorbed and others reflected. We see the wavelengths that are reflected. Thus, an apple looks red because it reflects mostly red light and absorbs most of the other wavelengths.

We see color because we have three types of cones, which are called blue, green, and red. The cones are named for the wavelengths they absorb best, not for their color. When light is absorbed, the cone is stimulated. Each type of cone absorbs a range of wavelengths, and the ranges overlap quite a bit. As a result, colored light stimulates each type of cone to a different extent. For instance, when we look at a bowl of fruit, light reflected by a red apple stimulates red cones, light reflected by a ripe yellow banana stimulates both red and green cones, and light reflected from blueberries stimulates both blue and green cones. The brain then interprets color according to how strongly each type of cone is stimulated.

COLOR BLINDNESS

Color blindness is a condition in which certain colors cannot be distinguished. Most people who are colorblind see some colors, but they tend to confuse certain colors with others. A lack, or a reduced number, of one of the types of cones causes the confusion. A person who lacks red cones sees deep reds as black. In contrast, a person who lacks green cones sees deep reds but cannot distinguish among reds, oranges, and yellows. Absence of blue cones is extremely rare.

People who are colorblind generally function normally in everyday life. They compensate for the inability to distinguish certain colors by using other cues, such as intensity, shape, or position. Red and green, the universal traffic colors for stop and go, are the most commonly confused colors. However, it is usually possible to pick out the brightest of the lights in the traffic signal, and all traffic signals are arranged the same way—red on top and green on the bottom. School children are regularly tested for color blindness so that both students and teachers are aware of the condition. Color blindness may cause difficulty with classwork, such as reading maps or graphs that are presented in color. If the teacher is aware of the condition, steps can be taken to avoid frustration and confusion for the student faced with such tasks.

Hearing Depends on the Ear

From a social perspective, hearing is perhaps the most important of our senses because speech plays such an important role in the communication that binds society. But hearing can also make us aware of things in the external world that we cannot see—the approach of friend or foe on a pitch-black night, for instance. Hearing can also help us better understand events that we can see, as when a mother determines by the sound of the cry whether her baby is hungry or in pain. And hearing can enrich the quality of our lives, as when we listen to music or hear waves crashing on the shore.

In every instance, what we hear are sound waves produced by vibration. Vibrating objects, such as guitar strings, the surface of the stereo speaker, or vocal cords, move rapidly back and forth. Each vibration pushes repeatedly against the surrounding air, creating sound waves, as shown in Figure 9.10. The loudness of sound is determined by the amplitude of the sound wave (the difference between the height of the peaks relative to the troughs). The pitch is determined by its frequency (the number of cycles per second). The more cycles, the higher the pitch. Sound waves travel through the air (or water) to reach our ears.

▌The ear collects and amplifies sound waves and converts them to neural messages

The ear has three parts—the outer ear, the middle ear, and the inner ear—as shown in Figure 9.11 and summarized in Table 9.4. The **outer ear** functions as a receiver. The part of the outer ear visible on the outside of the head consists of a fleshy flap, called the **pinna**. Shaped like a funnel, the pinna gathers the sound and channels it into the **external auditory canal,** the canal that leads from the pinna to the eardrum. The pinna also accentuates the frequencies of the most important speech sounds, making speech easier to pick out from background noise. Because of their shape and placement at the

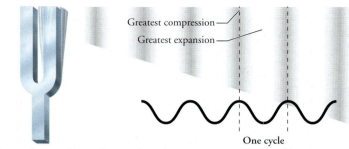

(a) Sound is caused by a vibrating object that causes pressure waves in the air (or water).

Greatest compression
Greatest expansion
One cycle

Low amplitude
Higher amplitude
(b) The amplitude (height) of the wave determines the loudness of the sound.

Low frequency
Higher frequency
(c) The frequency (cycles per second) of the waves determines the pitch of the sound.

sides of the head, the pinnas also help us determine the direction from which a sound comes.

The sheet of tissue that separates the outer ear from the middle ear is the eardrum, or **tympanic membrane**. It is as thin as a sheet of paper and as taut as the head of a tambourine. When sound waves strike the eardrum, it vibrates at the same frequency as the sound wave and transfers the vibrations to the middle ear.

stop and think

As we age, the tissues of the eardrum thicken and become less flexible. How do these changes partially explain why sensitivity to high-frequency sounds is usually lost as we grow older?

The **middle ear** serves as an amplifier. It consists of an air-filled cavity within the temporal bone of the skull that is spanned by the three smallest bones of the body: the **malleus** (hammer), **incus** (anvil), and **stapes** (stirrup). Together, these bones function as a system of

FIGURE **9.10**

A sound wave and the effects of amplitude (height) and frequency (cycles per second)

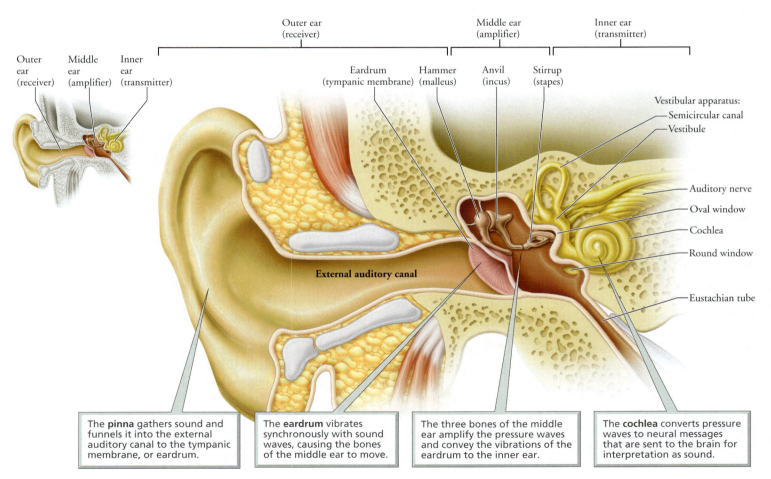

Outer ear (receiver)
Middle ear (amplifier)
Inner ear (transmitter)

Outer ear (receiver)
Middle ear (amplifier)
Inner ear (transmitter)

Eardrum (tympanic membrane)
Hammer (malleus)
Anvil (incus)
Stirrup (stapes)

Vestibular apparatus:
Semicircular canal
Vestibule

Auditory nerve
Oval window
Cochlea
Round window

Eustachian tube

External auditory canal

The **pinna** gathers sound and funnels it into the external auditory canal to the tympanic membrane, or eardrum.

The **eardrum** vibrates synchronously with sound waves, causing the bones of the middle ear to move.

The three bones of the middle ear amplify the pressure waves and convey the vibrations of the eardrum to the inner ear.

The **cochlea** converts pressure waves to neural messages that are sent to the brain for interpretation as sound.

WWW FIGURE **9.11**
TUTORIAL 9.2
Structure of the ear. The human ear has three parts: the outer ear (the receiver), the middle ear (the amplifier), and the inner ear (the transmitter).

TABLE 9.4

STRUCTURES OF THE EAR AND THEIR FUNCTIONS		
STRUCTURE	**DESCRIPTION**	**FUNCTION**
Outer ear		
Pinna	Fleshy, funnel-shaped part of the ear protruding from the side of the head	Collects and directs sound waves
External auditory canal	Canal between pinna and tympanic membrane	Directs sound to the middle ear
Middle ear		
Eardrum (tympanic membrane)	Membrane spanning the end of the external auditory canal	Vibrates in response to sound waves
Hammer (malleus), anvil (incus), and stirrup (stapes)	Three tiny bones of the middle ear	Amplify the vibrations of the tympanic membrane and transmit vibrations to inner ear
Eustachian tube	A tube that connects the middle ear with the throat	Allows equalization of pressure in middle ear with external air pressure
Inner ear		
Cochlea	Fluid-filled, bony, snail-shaped chamber	Houses organ of Corti and has openings called oval window and round window
Organ of Corti	Contains hair cells	The organ of hearing
Oval window	Membrane between the middle and inner ear that the stapes presses against	Transmits the movements of the stapes to the fluid in the inner ear
Round window	Membrane at the end of the lower canal in cochlea	Relieves pressure created by the movements of the oval window
Vestibular apparatus	Fluid-filled chambers and canals	Monitors position and movement of the head
Vestibule (utricle and saccule)	Two fluid-filled chambers	Maintain static equilibrium (body and head stationary, information on position of head)
Semicircular canals	Three fluid-filled chambers oriented at right angles to one another	Maintain dynamic equilibrium (body or head moving)

levers to convey the airborne sound waves from the eardrum to the **oval window**, a sheet of tissue that forms the threshold of the inner ear. The malleus is attached to the inner surface of the eardrum. Therefore, the vibrations of the eardrum in response to sound cause the malleus to rock back and forth. The rocking of the malleus, in turn, causes the incus and then the stapes to move. The base of the stapes fits into the oval window. When the stapes moves, it pushes against the oval window, conveying the vibrations of the eardrum to the inner ear. When very loud sounds strike the ear, two small muscles in the middle ear contract and restrain the movements of these bones to prevent irreparable damage to the delicate receptor cells of the inner ear.

The force of vibrations of the eardrum is amplified 22 times in the middle ear. The magnification of force is necessary to transfer the vibrations to the fluid of the inner ear. The amplification occurs both because of the lever system arrangement of the bones of the middle ear and because of the size difference between the eardrum and the oval window. This size difference concentrates the pressure against the oval window.

If the air pressure is not about equal on both sides of the eardrum, it cannot vibrate freely. The reason is simple. Unequal pressure makes the eardrum bulge inward or outward, holding it in place and causing discomfort, pain, and difficulty hearing. For instance, when a person ascends quickly to a high altitude, the atmospheric pressure is lower than the pressure in the middle ear, and the eardrum bulges outward. Unequal pressure usually is alleviated by the **Eustachian tube**, which connects the middle ear cavity with the upper region of the throat. Most of the time, the Eustachian tube is closed and flattened. However, swallowing or yawning opens it briefly, allowing the air pressure to equalize with the pressure in the middle ear cavity. The sensation of pressure suddenly equalizing often is described as ear "popping." It is likely to occur at times when the pressure on the external eardrum changes, as when riding up or down in an elevator, taking off or landing in an airplane, or scuba diving.

The **inner ear** is a transmitter that generates neural messages in response to pressure waves caused by sound waves, which are sent to the brain for interpretation. The inner ear contains two sensory organs only one of which, the **cochlea**, is concerned with hearing. The other sensory organ, the vestibular apparatus, is concerned with sensations of body position and movement and will be discussed in the next section.

The cochlea is the true seat of hearing, so we should consider some of the details of its structure. About the size of a pea, the cochlea is a bony tube about 35 mm (1.4 in.) long and is coiled about two and a half times, somewhat like the shell of a snail (as you can see in Figure 9.12). (*Cochlea* is from Latin for "snail.") The wider end of the tube, where the snail's head would be, has two membrane-covered openings. The upper opening is the oval window into which the stapes fits. The lower opening, called the **round window**, serves to relieve the pressure created by the movements of the oval window.

The internal structure of the cochlea is easier to understand if we imagine the cochlea uncoiled so that it forms a long straight tube, as shown in. We would then see that two membranes divide the interior of the cochlea into three longitudinal compartments, each filled with fluid. Like the finger on a glove, the central compartment (the cochlear duct) ends blindly and does not extend completely to the end of the cochlea. As a result, the upper and lower compartments (the vestibular canal and the tympanic canal) are connected at the end of the tube nearest to the tip of the coil. The basilar membrane is the floor of the central compartment. The **organ of Corti**, the portion of the cochlea most directly responsible for the sense of hearing, is supported on the basilar membrane (Figure 9.12). The receptors for hearing are called **hair cells**. Each hair cell has about 100 "hairs," slender projections from its upper surface. The 25,000 hair cells are arranged on the basilar mem-

brane in rows that resemble miniature picket fences. The hair cells stimulate the nerve cells that carry nerve impulses to the brain. The roof of the organ of Corti is formed by the **tectorial membrane**, which projects over and is in contact with the hair cells.

When the stapes moves to and fro against the oval window, it sets up corresponding movements in the fluid of the inner ear (see Figure 9.13). The resulting pressure waves are transmitted to the fluid of the upper compartment and then to the lower compartment because they are continuous. The movements of the fluid cause the basilar membrane to swing up and down; this swinging, in turn, causes the processes (projections) on the hair cells to be pressed against the tectorial membrane. The bending of the hairs ultimately results in nerve impulses in the auditory nerve.

Variations in the movements of the basilar membrane determine loudness and pitch

The louder the sound, the greater the pressure changes in the fluid of the inner ear and, therefore, the stronger the bending of the basilar membrane. Because hair cells have different thresholds of stimulation, more vigorous vibrations in the basilar membrane stimulate more hair cells. The brain interprets the increased number of impulses as louder sound.

How do we determine the pitch of a sound? Very low-pitched sound, below 150 Hz (Hz, hertz = cycles per second), is interpreted by the brain by the frequency of impulses in the auditory

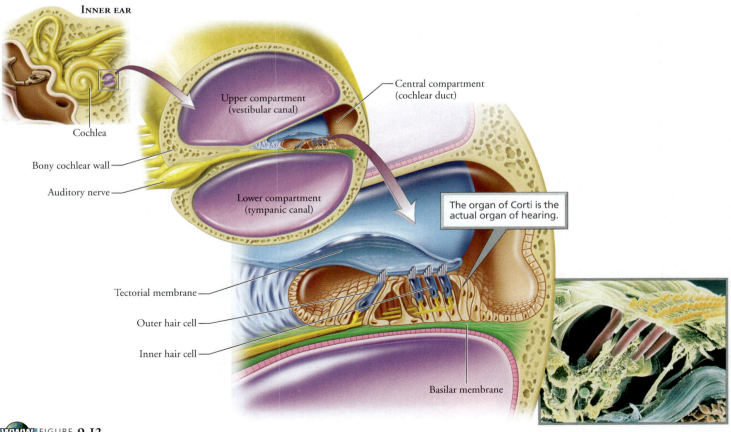

INNER EAR

Cochlea

Bony cochlear wall

Auditory nerve

Upper compartment (vestibular canal)

Lower compartment (tympanic canal)

Central compartment (cochlear duct)

The organ of Corti is the actual organ of hearing.

Tectorial membrane

Outer hair cell

Inner hair cell

Basilar membrane

FIGURE **9.12**

TUTORIAL 9.3

The cochlea houses the organ of Corti. The organ of Corti rests on the basilar membrane. The hair cells on the basilar membrane are the receptors for hearing.

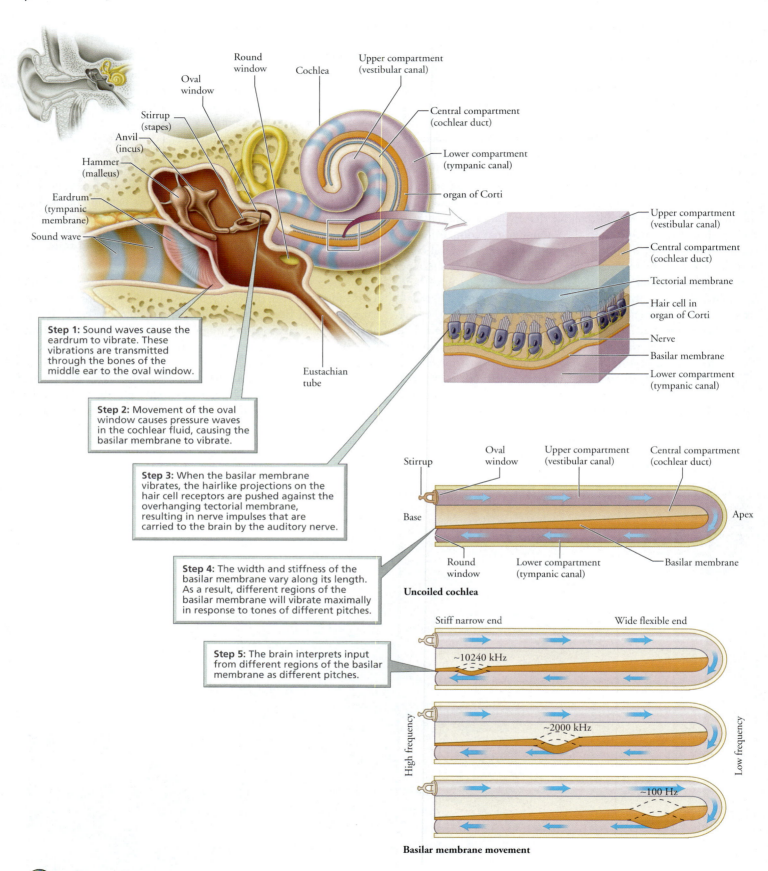

Step 1: Sound waves cause the eardrum to vibrate. These vibrations are transmitted through the bones of the middle ear to the oval window.

Step 2: Movement of the oval window causes pressure waves in the cochlear fluid, causing the basilar membrane to vibrate.

Step 3: When the basilar membrane vibrates, the hairlike projections on the hair cell receptors are pushed against the overhanging tectorial membrane, resulting in nerve impulses that are carried to the brain by the auditory nerve.

Step 4: The width and stiffness of the basilar membrane vary along its length. As a result, different regions of the basilar membrane will vibrate maximally in response to tones of different pitches.

Step 5: The brain interprets input from different regions of the basilar membrane as different pitches.

Uncoiled cochlea

Stiff narrow end   Wide flexible end

~10240 kHz

~2000 kHz

~100 Hz

High frequency   Low frequency

Basilar membrane movement

FIGURE **9.13**

The sequence of events from sound vibration to a nerve impulse begins when sound enters the external auditory canal.

TUTORIAL 9.4

nerve. The brain's interpretation of the pitch of sounds above 150 Hz is based on the region of the basilar membrane that is stimulated by the sound. In a sense, the cochlea is like a spiral piano keyboard. The basilar membrane varies in width and flexibility along its length. Near the oval window, the basilar membrane is narrow and stiff. Like the shorter strings on a piano, this region vibrates maximally in response to high-frequency sound. At the tip of the cochlea, the basilar membrane is wider and floppier. Low-frequency sounds cause maximum vibrations in this region of the basilar membrane. Thus, sounds of different pitches activate hair cells at different places along the basilar membrane. The brain then interprets input from different regions of the basilar membrane as sounds of different pitch.

▎ Hearing loss can be conductive or sensorineural

An estimated 28 million Americans have some degree of hearing loss, and 2 million of them are completely deaf. Hearing loss that is severe enough to interfere with social and job-related communication is among the most common chronic neural impairments in the United States.

There are two types of hearing loss: conductive loss and sensorineural loss. Conductive loss results when airborne sounds are not conducted through the external auditory canal to the eardrum or over the bones of the middle ear to the inner ear. An obstruction anywhere along this route can prevent sound messages from stimulating the nerves in the inner ear. The external auditory canal can become clogged with wax or other foreign matter. In this case, the insertion of cotton-tip swabs usually aggravates the situation by further clogging the canal and should, therefore, be avoided. Other possible causes for conductive hearing loss are thickening of the eardrum, which might occur with chronic infection, or perforation of the eardrum, which might occur as a result of trauma. Excess fluid in the middle ear, which often occurs with middle ear infections, can also cause conductive hearing loss. A common age-related cause of conduction deafness is otosclerosis, a condition in which the bones of the middle ear become fused and can no longer transmit sound. In this case, the only way that sound can be transmitted to the inner ear is through the bones of the skull, which is much less efficient. Otosclerosis is routinely corrected by surgery.

Sensorineural deafness is caused by damage to the hair cells in the inner ear or to the nerve supply of the inner ear. Sensorineural deafness is commonly caused by the gradual loss of hair cells throughout life. Some of the loss is simply due to aging. After age 20, we lose about 1 Hz of our total perceptive range of 20,000 Hz every day. However, much of the damage to hair cells is caused by exposure to loud noise (see the Environmental Issue essay, *Noise Pollution*). Neural damage can also be caused by infections, including mumps, rubella (German measles), syphilis, and meningitis. Drugs such as those used for treating tuberculosis and certain cancers can also cause neural damage.

One of the first signs of sensorineural damage is the inability to hear high-frequency sounds. Because consonants, which are needed to decipher most words, include higher frequencies than do vowels, speech becomes difficult to understand. So beware if you find yourself asking, "Could you repeat that?"

The most effective way of combating hearing loss is the use of a hearing aid. The basic job of a hearing aid is to amplify sound. One type of hearing aid presents amplified sound to the eardrum. Another type uses a vibrator or stimulator that is placed in the bone of the skull behind the ear. Sound is then conducted through the bones of the skull to the inner ear.

Hearing aids often are unable to help profoundly deaf people, but cochlear implants can pierce the silence for some of them. A cochlear implant is an array of electrodes that transforms sound into electrical signals that are delivered to the nerve cells near the cochlea. The cochlear implant is surgically inserted into the cochlea through the round window. A cochlear implant is not a miracle cure. Only about 20% of implant recipients can hear well enough to understand most spoken sentences and to use the telephone easily. Another 20% of those with cochlear implants are not helped at all. For most recipients, the benefit is somewhere between those two extremes. They are able to hear enough sound to make lip reading easier.

what would you do?

Many deaf people consider themselves to be a cultural group, not a medical anomaly. Some people argue, for instance, that communication is the real problem faced by the deaf and that the solution to the problem should, therefore, be a social one, not a medical one. If medical technology is available, should we use it to "fix" conditions such as deafness? If you had a child who was born deaf, would you want a cochlear implant for that child? Why or why not? (Assume that the child is not yet able to make that decision for himself or herself.)

▎ Ear infections can occur in the ear canal or in the middle ear

An external ear infection is one in the ear canal that leads to the eardrum. Swimmer's ear, the most common infection, is precipitated by water becoming trapped in the canal and creating an environment favorable for the growth of bacteria. The first symptom of swimmer's ear is itching, which is usually followed by pain that can become intense and constant. Chewing food or touching the earlobe sharpens the pain. The usual treatment consists of antibiotic ear drops, heat, and pain medication. To prevent swimmer's ear, always be sure that water drains from the ear canal after swimming or bathing, by shaking the head if need be.

Middle ear infections usually result from infections of the nose and throat that work their way through the Eustachian tubes connecting the throat and middle ear. At least half of all children get an ear infection at some time; some children get four or five ear infections a year. Middle ear infections are more common in children than in adults because the Eustachian tubes become tilted downward during growth. In children, the tubes are nearly horizontal, making it easier for infectious organisms to travel through them than through the tilted, curved tubes of adults. The signs of a middle ear infection are a stabbing earache, impaired hearing, and a feeling of fullness in the ear, often accompanied by a fever. A middle ear infection usually is treated with an antibiotic.

ENVIRONMENTAL ISSUE

Noise Pollution

It is difficult to escape the din of modern life—noise from airports, city streets, loud appliances, stereos. Noise pollution threatens your hearing and your health. Exposure to excessive noise is to blame for the hearing loss of one-third of all hearing-impaired people.

Loud noise damages the hairs on the hair cells of the inner ear. When the hairs are exposed to too much noise, they become worn down, lose the ability to move, and can become fused together (Figure 9.A). Unfortunately, there is no way to undo the damage; you cannot get spare parts for your ears.

The loudness of noise is measured in decibels (dB). The decibel scale is logarithmic. An increase of 10 dB generally makes a given sound twice as loud. The decibel ratings and effects of some familiar sounds are given in Table 9.A. Most people judge sounds over 60 dB to be intrusive, over 80 dB to be annoying, and over 100 dB to be extremely bothersome. The federal Occupational Safety and Health Administration (OSHA) has set 85 dB as the sound safety limit for 8 hours of exposure. The threshold for physical pain is 140 dB.

The louder a sound, the shorter the exposure time necessary to damage the ear. Even a single, explosively loud sound is capable of damaging hair cells. More commonly, however, hearing loss results from prolonged exposure to volumes over 55 dB. Hearing can be damaged by exposure to noise loud enough to make it difficult to converse with someone. Some damage probably occurred if sounds seem muffled after you leave the noisy area. Your ears can endure sound at 90 dB for about 8 hours. For every 5 dB above that, it takes half as long for damage to begin. Thus, sound at 95 dB will damage your ears in only 4 hours. At 110 dB, the average rock concert or stereo headset at full blast can damage your ears in as little as 30 minutes.

Hearing loss is expected in the elderly, but a surprising number of young people also have impaired hearing. The culprit is most likely noise—probably in the form of music.

If you have ever walked away from a noisy area, such as a concert or a construction site, and your ears were ringing or everything sounded as if you were under water, you have experienced what is called a temporary threshold shift. It is a sign that some of the hair cells in the inner ear were damaged.

Noise has other harmful effects. The stress responses to daytime noise can carry over into the night, causing sleep disturbances that can make you groggy, tense, and forgetful the next day. Noise triggers stress responses that may increase heart rate and blood pressure. Even common kitchen appliances such as blenders or mixers produce sound levels that cause pupils to dilate, the mouth to become dry, muscles to tense, and the digestive system to slow down. People become irritable when exposed to noise for prolonged periods, and their irritability may aggravate tension within a family. Many an argument has flared up when a person who has spent the day in a noisy workplace comes home to find the television or stereo blaring or children crying instead of the peace and quiet that was craved.

How can you protect yourself from bad vibes? Don't listen to loud music. Keep the tunes low enough that you can still hear other sounds. No one else should be able to hear the music you are listening to with earphones. When you cannot avoid loud noise, as when you are mowing the lawn, vacuuming, or attending a rock concert, wear earplugs to protect your hearing. You can find earplugs at most drugstores, sporting goods stores, and music stores. ≢

Balance Depends on the Vestibular Apparatus of the Inner Ear

The **vestibular apparatus**, a fluid-filled maze of chambers and canals within the inner ear, is responsible for monitoring the position and movement of the head. The receptors in the vestibular apparatus are hair cells similar to those in the cochlea. When the hairs on these cells are bent, as will occur during head movements or changes in velocity, the hair cells send messages to the brain. The brain uses this input to maintain balance.

The two components of the vestibular apparatus are the semicircular canals and the vestibule. The **semicircular canals** are three canals in each ear that contain sensory receptors. The canals help us stay balanced as we move. Because they monitor precisely any sudden movements of the head, including those caused by acceleration and deceleration, the canals are important in our equilibrium when the body or head is moving (dynamic equilibrium) (Figure 9.14). At the base of each semicular canal is an enlarged region called the **ampulla**. Within the ampulla is a tuft of hair cells. The hairlike projections from these cells are embedded in a pointed "cap" of stiff, pliable, gelatinous material called the **cupula**. When you move your head, fluid in the canal lags, causing the cupula to bend the hair cells and stimulate them.

The three semicircular canals in each ear are oriented at right angles to one another. One canal is parallel to the side of the head, another to the plane of the face, and the third to the horizon. Head movements cause the fluid in the canals to move in the opposite direction of head movement in the same way that sudden acceleration slams passengers against the car seat. Whereas nodding your head to say yes will cause the fluid in the canals parallel to the sides of the head to swirl, shaking it from side to side to say no will cause fluid in the canals parallel to the horizon to move. Tilting your head to look under the bed will cause fluid in the canals that are parallel to your face to move back and forth. Everyday movements are rarely in a single plane, but even the most complex movement can be analyzed in terms of motion in three planes. Indeed, every movement

TABLE **9.A**

EFFECTS OF NOISE POLLUTION		
SOUND SOURCE	LOUDNESS (dB)	EFFECT FROM PROLONGED EXPOSURE
Jet plane at takeoff	150	Eardrum rupture
Deck of aircraft carrier	140	Very painful; traumatic injury
Rock-and-roll band (at maximum volume)	130	Irreversible damage
Jet plane at 152 m (500 ft)	110	Loss of hearing
Subway, lawn mower	100	
Electric blender	90	Annoying
Washing machine, freight train at 15 m (50 ft)	80	
Traffic noise	70	Intrusive
Normal conversation	65	
Chirping bird	60	
Quiet neighborhood (daytime)	50	
Soft background music	40	Quiet
Library	30	
Whisper	20	Very quiet
Breathing, rustling leaves	10	
	0	Threshold of hearing

FIGURE **9.A**

The hair cells of the inner ear can be permanently damaged by loud noise. The micrograph on the left shows a portion of the organ of Corti from a normal guinea pig, showing the characteristic three rows of outer hair cells and a single inner row. The micrograph on the right shows the organ of Corti from a guinea pig that had been exposed to loud (120 dB) sound. Note that scars have replaced some injured cells. On most of the hair cells that have survived, the hairs are no longer in an orderly pattern.

of your head causes fluid movement in at least one of the semicircular canals. This movement stimulates the hair cells, and they send messages to the brain that are interpreted as head movement in the opposite direction as the bending of hair cells. Within a few seconds after the movement of the body stops, the movement of fluid in the semicircular canals also ceases. Therefore, the hair cells are no longer stimulated when we travel at a constant velocity and, without visual cues, we cannot tell that we are moving.

stop and think

When you stop moving, there is a time lag before the fluid in the semicircular canals stops swirling. How does this time lag account for the sensation of dizziness after you have been twirling around for a while?

The **vestibule**, the other part of the vestibular apparatus, is important in static equilibrium. In other words, it helps us keep our balance when we are not moving. The vestibule consists of the utricle and the saccule, two fluid-filled cavities. Thanks to the utricle and the saccule, you know which end is up, literally (Figure 9.14). These cavities tell the brain the position of the head with respect to gravity when the body is not moving. They also respond to acceleration and deceleration but not to rotational changes as the semicircular canals do. Both the utricle and the saccule contain hair cells overlain with a gelatinous material in which small granules of calcium carbonate, a chalklike substance, are embedded. These granules, called **otoliths**, make the gelatin heavier than the surrounding material and, therefore, make it slide over the hair cells whenever the head is moved. The movement of the gelatin stimulates the hair cells, which send messages to the brain, and the brain interprets the messages to determine the position of the head relative to gravity.

The utricle and saccule sense different types of movement. The utricle senses the forward tilting of the head, as well as forward motion, because its hair cells are on the floor of the chamber, oriented vertically when the head is upright. In contrast, the hair cells

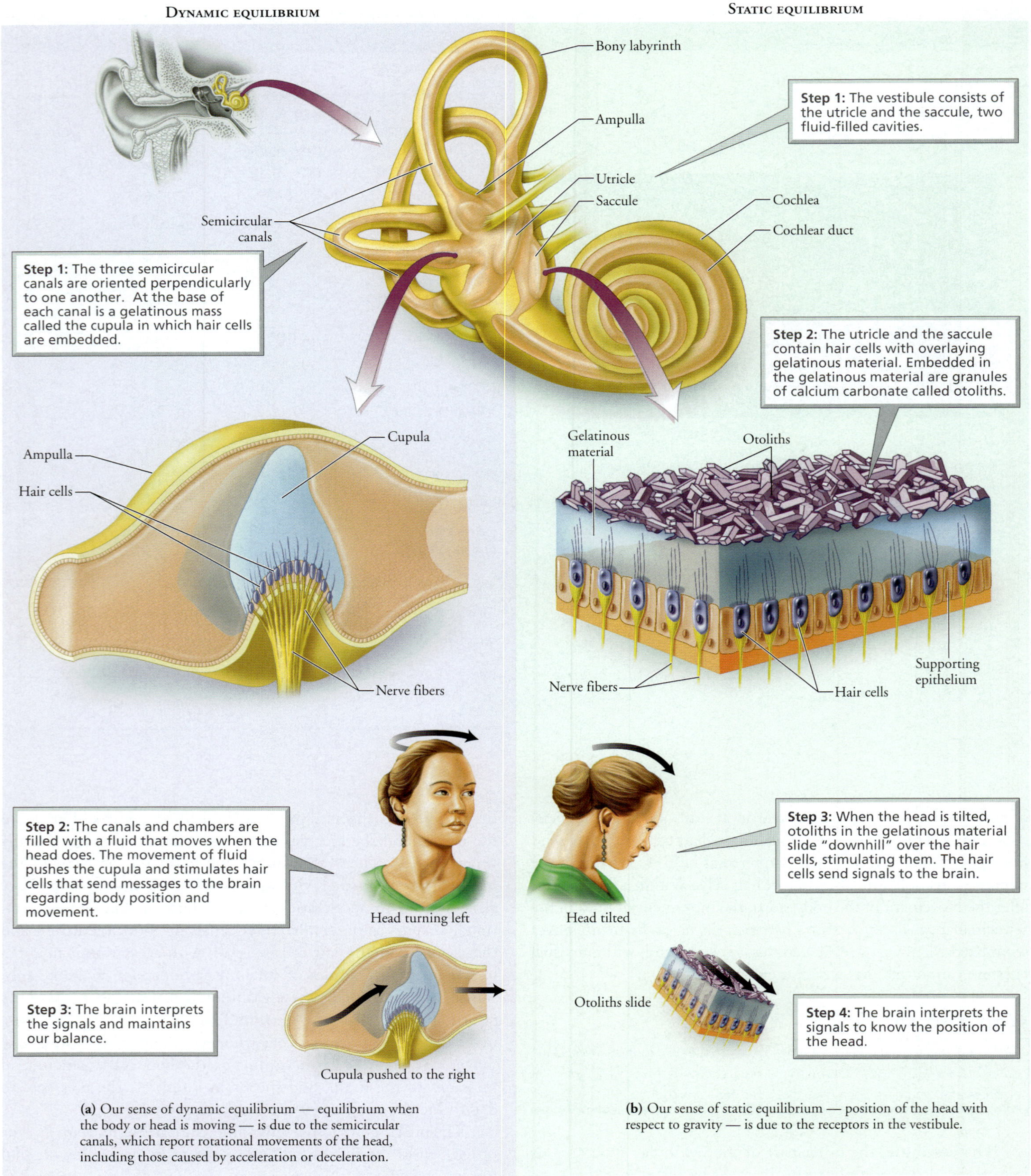

DYNAMIC EQUILIBRIUM

STATIC EQUILIBRIUM

Bony labyrinth

Ampulla

Utricle

Saccule

Semicircular canals

Cochlea

Cochlear duct

Step 1: The vestibule consists of the utricle and the saccule, two fluid-filled cavities.

Step 1: The three semicircular canals are oriented perpendicularly to one another. At the base of each canal is a gelatinous mass called the cupula in which hair cells are embedded.

Step 2: The utricle and the saccule contain hair cells with overlaying gelatinous material. Embedded in the gelatinous material are granules of calcium carbonate called otoliths.

Cupula

Gelatinous material

Otoliths

Ampulla

Hair cells

Nerve fibers

Nerve fibers

Hair cells

Supporting epithelium

Step 2: The canals and chambers are filled with a fluid that moves when the head does. The movement of fluid pushes the cupula and stimulates hair cells that send messages to the brain regarding body position and movement.

Head turning left

Head tilted

Step 3: When the head is tilted, otoliths in the gelatinous material slide "downhill" over the hair cells, stimulating them. The hair cells send signals to the brain.

Step 3: The brain interprets the signals and maintains our balance.

Otoliths slide

Cupula pushed to the right

Step 4: The brain interprets the signals to know the position of the head.

(a) Our sense of dynamic equilibrium — equilibrium when the body or head is moving — is due to the semicircular canals, which report rotational movements of the head, including those caused by acceleration or deceleration.

(b) Our sense of static equilibrium — position of the head with respect to gravity — is due to the receptors in the vestibule.

FIGURE **9.14**
Dynamic and static equilibrium

of the saccule are on the wall of the chamber, oriented horizontally when the head is upright. They respond when you move vertically, as when you jump up and down.

Motion sickness—that dreadful feeling of dizziness and nausea that sometimes causes vomiting—is thought to be caused by a mismatch of sensory input from the vestibular apparatus and the eyes. For example, many people get carsick from reading in the car. When looking down at a book, your eyes tell your brain that your body is stationary. However, as the car changes speed, turns, and hits bumps in the road, the vestibular system detects motion. Seasickness results when the vestibular apparatus tells the brain your head is rocking back and forth, but the deck under your feet looks level. The brain is somehow confused by the conflicting information, and the result is motion sickness. Staring at the horizon so you can see that you are moving can sometimes relieve the feeling. Over-the-counter drugs to prevent motion sickness, such as Dramamine®, work by inhibiting the messages from the vestibular apparatus.

Smell and Taste Are the Chemical Senses

Smell is perhaps the least appreciated of our senses. You could easily get along without it, but life would not be as interesting. For one thing, about 80% of what we usually think of as the flavor of a food is really due to our sense of smell. This relationship explains why food often tastes bland when your nose is congested.

You have about 5 million olfactory (smell) receptors, located, not in the nostrils, but in a small patch of tissue the size of a postage stamp in the roof of each nasal cavity (Figure 9.15). The receptors are neurons with long cilia (cell hairs) that project outward from the cells that line the nasal cavity and are covered by a coat of mucus, which keeps them moist and is a solvent for odorous molecules. The hairs move, gently swirling the mucus. Olfactory receptors, by the way, are one of the few neurons known to be replaced during life—about every 60 days.

An odor is detected when odor molecules are carried into the nasal cavity in air, dissolve in the mucus, and bind to receptors on the hairs of the olfactory receptors, thereby stimulating the receptor. If threshold is reached, the message is carried to the two olfactory bulbs of the brain. The olfactory bulbs process the information from the olfactory receptors and pass it on to the limbic system and to the cerebral cortex (see Chapter 8), where it is interpreted. Interestingly, the limbic system is a center for emotions and memory. Thus, we rarely have neutral responses to odors. In general, the smell of freshly baked bread is pleasing, but the smell of a skunk is repulsive. The perfume industry makes a fortune from the association between scents and sexuality. Odors can also trigger a flood of long-forgotten memories. If your first kiss was near a blooming lilac bush, for instance, a simple sniff of lilac may take you back in time and place.

We have about 1000 types of olfactory receptors, with which we can distinguish about 10,000 odors. Each receptor responds to several odors. Thus, the brain relies on input from more than one type of receptor to identify an odor.

There are four primary tastes: sweet, salty, sour, and bitter. These four tastes are enough to answer the important question

about food or drink: Should we swallow it or spit it out? We are prompted to swallow if it tastes sweet, because sweetness implies a rich source of calories and thus energy. Salty tastes also prompt swallowing, because salts will replace those lost in perspiration. Sourness is tough to call. Sour is the taste of unripe fruit that would have more food value later on. As fruit ripens, starches break down into sugars that create a sweet taste and mask the sourness. So it is often better to reject sour fruits and wait for them to ripen. However, some sour fruits, such as oranges, lemons, and tomatoes, are rich sources of vitamin C, an essential vitamin. Bitter is easy. We reject foods that taste bitter. Bitterness usually indicates that food is poisonous or spoiled.

We have about 10,000 **taste buds**, the structures responsible for our sense of taste. Most taste buds are on the tongue, but some are scattered on the inner surface of the cheeks, on the roof of the mouth, and in the throat. The taste buds on the tongue are located along the sides of the papillae, those small bumps that give the tongue a slightly rough feeling. Each papilla contains 100 to 200 taste buds. The cells of a taste bud are completely replaced about every 10 days, so you need not worry about losing your

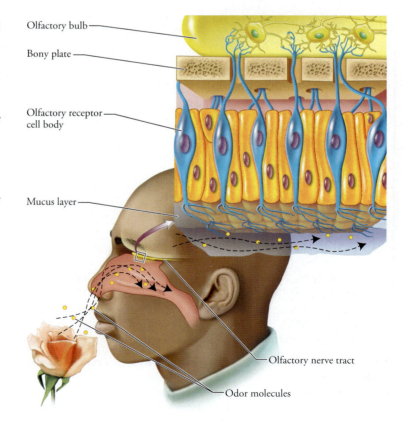

Olfactory bulb
Bony plate
Olfactory receptor cell body
Mucus layer
Olfactory nerve tract
Odor molecules

FIGURE **9.15**

Our sense of smell resides in a small patch of tissue in the roof of each nasal cavity. Within each cavity are some 5 million olfactory receptors.

sense of taste for life when you burn your mouth eating hot pizza, as almost always happens.

A taste bud is the interface between chemicals dissolved in saliva and the sensory neurons that will convey information to the brain. Each taste bud is a lemon-shaped structure containing about 40 modified epithelial cells (Figure 9.16). Some of these cells are the taste cells that respond to chemicals; others are supporting cells. The taste cells have long microvilli, called *taste hairs*, that project into a pore at the tip of the taste bud. These taste hairs bear the receptors for certain chemicals found in food. Dissolved in water, the food molecules enter the pore and stimulate the taste hairs.

Although taste cells are not neurons, they generate electrical signals, which are then sent to the dendrites of sensory nerve cells wrapped around the taste cell.

Each taste bud responds to all four basic tastes, but it is usually more sensitive to one or two of them than to the others. Because of the way the taste buds are distributed, different regions of the tongue have *slightly* different sensitivity to these tastes. In general, the tip of the tongue is most sensitive to sweet tastes, the back to bitter, and the sides to sour. Sensitivity to salty tastes is fairly evenly distributed on the tongue.

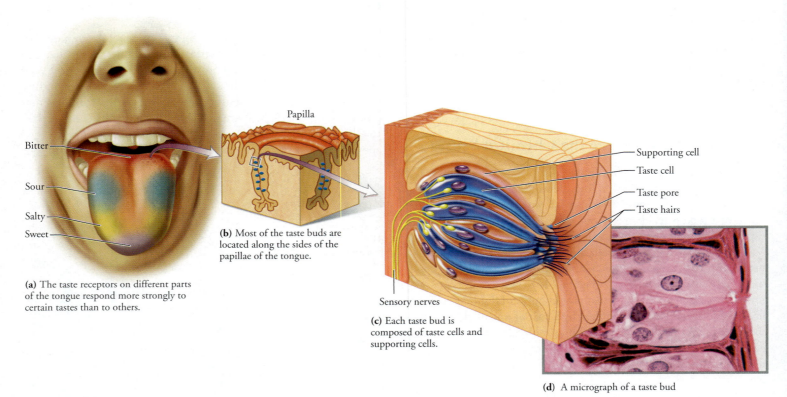

Papilla

Bitter
Sour
Salty
Sweet

(a) The taste receptors on different parts of the tongue respond more strongly to certain tastes than to others.

(b) Most of the taste buds are located along the sides of the papillae of the tongue.

Sensory nerves

(c) Each taste bud is composed of taste cells and supporting cells.

Supporting cell
Taste cell
Taste pore
Taste hairs

(d) A micrograph of a taste bud

FIGURE **9.16**
Taste buds

REVIEWING THE CONCEPTS

Sensory Receptors Generate Electrochemical Messages in Response to Stimuli (pp. 165)

1. Changes in external and internal environments stimulate sensory receptors, which in turn generate electrochemical messages that eventually are converted to nerve impulses that are conducted to the brain.

Receptors Are Specialized to Respond to Different Types of Stimuli (pp. 165–166)

2. There are five types of sensory receptors. Mechanoreceptors are responsible for touch, pressure, hearing, equilibrium, blood pressure, and body position. Thermoreceptors detect changes in temperature. Photoreceptors detect light. Chemoreceptors monitor chemical levels within the body and are responsible for smell and taste. Pain receptors respond to strong stimuli caused by physical or chemical damage to tissues.

3. Exteroceptors are located near the body surface and respond to envi-

ronmental changes. Interoceptors are located inside the body and monitor internal conditions.

Receptors for the General Senses Are Widespread Throughout the Body (pp. 166–167)

4. The general senses are touch, pressure, temperature, sense of body and limb position, and pain.

5. A light touch to the skin is detected by free nerve endings in the skin and receptors called Merkel disks. Meissner's corpuscles are encapsulated nerve endings that tell us exactly where we have been touched. Pacinian corpuscles respond to pressure.

6. Thermoreceptors, which are widely distributed throughout the body, respond to changes in temperature.

7. Muscle spindles, specialized muscle fibers with sensory nerve endings wrapped around them, and Golgi tendon organs, highly branched

nerve fibers in tendons, help the brain monitor the location of all parts of the body.

8. Pain receptors, located in almost every tissue of the body, respond to any strong stimulus that results from physical or chemical damage to tissues.

Vision Depends on the Eye (pp. 167–174)

9. The eye has three layers. The sclera and the cornea make up the outer layer. The choroid, ciliary body, and iris make up the middle layer. The retina, which contains the photoreceptors (rods and cones), is the innermost layer.

10. The cornea and the lens work together to focus images on the retina. The lens can accommodate (change its shape) to focus on near or distant objects.

11. A cataract is a lens that has become cloudy or opaque, usually because of aging or injury, drug reaction, or disease.

12. The three most common focusing problems are farsightedness (difficulty in seeing nearby objects), nearsightedness (difficulty in seeing distant objects), and astigmatism (irregularities in the cornea that cause distortion of the image). All can be corrected with the use of lenses.

13. When pigment molecules in the rods or cones (photoreceptors) absorb light, a chemical change occurs in the pigment molecule. The resulting change in the permeability of the membrane of the photoreceptor sets up a neural message that is carried by the optic nerve to the brain.

14. Color vision depends on the cones. The three types of cones are named for the color of light they absorb—green, red, and blue—and allow us to see colors. Color blindness, a condition in which certain colors cannot be distinguished, is caused by the lack of a sufficient number of one of these types of cones.

WEB TUTORIAL 9.1 Vertebrate Vision

Hearing Depends on the Ear (pp. 174–179)

15. Sound is produced by the vibrations generated by a vibrating object, which travel through the air to reach our ears.

16. The ear is made up of three regions: the outer ear, the middle ear, and the inner ear. The outer ear, which functions as a receiver, consists of the pinna and the external auditory canal. The middle ear, which serves as an amplifier, consists of the tympanic membrane (eardrum), three small bones (malleus, incus, and stapes), and the Eustachian tube. The inner ear is a transmitter and consists of the cochlea and vestibular apparatus.

17. Hearing is the sensing of sound waves caused by vibration. Sound enters the outer ear and vibrates the eardrum (tympanic membrane). These vibrations move the malleus, which moves the incus and the stapes. The stapes conveys these vibrations to the inner ear by means of the oval window.

18. Pressure on either side of the eardrum is regulated by the Eustachian tube.

19. The cochlea is a coiled tube enclosed in bone about 35 mm (1.4 in.) long. It is separated into three longitudinal tubes. The middle tube contains the organ of Corti, which is lined with hair cells and is the portion of the cochlea most responsible for hearing.

20. A loud sound causes greater pressure in the fluid of the inner than does a soft sound and causes greater bending of the basilar membrane, stimulating more hair cells.

21. The basilar membrane changes width along the length of the cochlea. Sounds of different pitches activate hair cells at different places along the basilar membrane.

22. The two types of hearing loss are conductive and sensorineural. Conductive loss occurs when sounds are prevented from reaching the inner ear. Sensorineural hearing loss occurs when there is damage to the hair cells or nerve supply of the inner ear. Means of combating hearing loss include hearing aids (amplifiers) and cochlear implants.

23. Ear infections can occur in the ear canal leading to the middle ear or in the middle ear (caused when bacteria from the nose and throat enter the middle ear through the Eustachian tubes). Chronic infections can lead to the buildup of fluid in the middle ear and eventually to hearing loss.

WEB TUTORIAL 9.2 Structure of the Ear

WEB TUTORIAL 9.3 Inner Ear: Anterior View

WEB TUTORIAL 9.4 Hair Cell

Balance Depends on the Vestibular Apparatus of the Inner Ear (pp. 180–183)

24. Balance is controlled by the vestibular apparatus, which consists of the semicircular canals and the vestibule. The semicircular canals monitor sudden movements of the head. The vestibule is made of two components, the saccule and the utricle, which tell the brain the position of the head with respect to gravity.

Smell and Taste Are the Chemical Senses (pp. 183–184)

25. Smell receptors are located in the nasal cavity. They are lined with cilia and are coated by mucus. Odorous molecules dissolve in the mucus and bind to the receptors' hairs, stimulating the receptors. Information is passed to the olfactory bulbs and then to the limbic system and the cerebral cortex.

26. Taste buds are responsible for our sense of taste. They are on the tongue, inside the cheeks, on the roof of the mouth, and in the throat. They sense the four basic tastes of sweet, salty, sour, and bitter.

KEY TERMS

accommodation *p. 170*	external auditory canal *p. 174*	organ of Corti *p. 177*	round window *p. 177*
ampulla *p. 180*	exteroceptor *p. 166*	otolith *p. 181*	sclera *p. 167*
aqueous humor *p. 170*	fovea *p. 169*	outer ear *p. 174*	semicircular canal *p. 180*
blind spot *p. 169*	general sense *p. 166*	oval window *p. 176*	sensory adaptation *p. 165*
cataract *p. 170*	hair cell *p. 177*	pain receptor *p. 165*	sensory receptor *p. 165*
chemoreceptor *p. 165*	incus *p. 175*	photoreceptor *p. 165*	special sense *p. 166*
choroid *p. 167*	inner ear *p. 176*	pinna *p. 174*	stapes *p. 175*
ciliary body *p. 168*	interoceptor *p. 166*	pupil *p. 168*	taste bud *p. 183*
cochlea *p. 176*	iris *p. 168*	receptor potential *p. 165*	tectorial membrane *p. 177*
cone *p. 169*	malleus *p. 175*	referred pain *p. 167*	thermoreceptor *p. 165*
convergence *p. 171*	mechanoreceptor *p. 165*	retina *p. 169*	tympanic membrane *p. 175*
cornea *p. 167*	middle ear *p. 175*	retinal *p. 172*	vestibular apparatus *p. 180*
cupula *p. 180*	opsin *p. 172*	rhodopsin *p. 172*	vestibule *p. 181*
Eustachian tube *p. 176*	optic nerve *p. 169*	rod *p. 169*	vitreous humor *p. 170*

THINKING ABOUT THE CONCEPTS

1. What is a receptor potential? *p. 165*
2. Define sensory adaptation and give two examples. *p. 165*
3. What are the five classes of receptors? Which would give rise to the general senses? Which would give rise to special senses? *pp. 165–166*
4. When stimulated, what do Merkel disks tell us? *p. 166*
5. What are the functions of the two distinct regions of the outer layer of the eye? *p. 167*
6. How is light focused on the retina? *p. 170*
7. How is light converted to a neural message? *pp. 172–174*
8. How are sound waves produced? How are loudness and pitch determined? *pp. 175–176*
9. What is the visible part of the outer ear, and how do its shape and placement help in the reception of sound? Give two examples. *p. 174*
10. What is the function of the eardrum (tympanic membrane)? *p. 175*
11. Why is it necessary for the force of the vibrations to be amplified in the middle ear? How is amplification accomplished? *pp. 175–176*
12. What are the effects of unequal air pressure on either side of the eardrum? How does the Eustachian tube help regulate air pressure? *p. 176*
13. What are the two structures for equilibrium in the inner ear? Explain the structure of each and how it allows them to sense body position and motion. *pp. 176, 180*
14. How does the basilar membrane respond to pitch? *pp. 177–179*
15. What are the two types of hearing loss? Explain how they differ. *p. 179*
16. What causes motion sickness? How can it be "cured"? *p. 183*
17. Where are the olfactory receptors located? Explain their structure as it is related to their function. *p. 183*
18. What are the four primary tastes? *p. 183*
19. What are the structures responsible for taste? Where are they found? *pp. 183–184*
20. Describe the structure of a taste bud. *p. 184*

21. Pacinian corpuscles sense
    a. pressure.
    b. light touch.
    c. warmth.
    d. pain.
22. The organ of Corti is important in sensing
    a. body movement.
    b. sound.
    c. light.
    d. degree of muscle contraction.
23. The greatest concentration of cones is found in the
    a. sclera.
    b. lens.
    c. ciliary body.
    d. fovea.
24. Noah is a 4-year-old boy who has a middle ear infection. The cause of the infection could be
    a. bacteria that spread to the ear from a sore throat.
    b. excessive wax buildup in the ear canal.
    c. water that was trapped in the ear after bathing.
    d. noise pollution.
25. The blind spot of the eye is
    a. located in the cornea.
    b. the region where the optic nerve leaves the eye.
    c. the region of the eye where rods outnumber cones.
    d. the region where the ciliary muscle attaches.
26. The receptors responsible for color vision are the _____.
27. The _____ of the eye changes shape to focus.
28. The snail-shaped structure concerned with hearing is the _____.
29. The semicircular canals sense _____.

APPLYING THE CONCEPTS

1. Molly needs glasses for driving but not for reading. What is the name for her visual problem? What type of corrective lens would she need?
2. Sarah has a viral infection that is causing vertigo (dizziness). In which part of the ear is the infection?
3. Manuel is a guitar player in a rock band. He complains that he is having difficulty hearing high-pitched sounds. Explain the relationship between his hearing problem and his occupation.

The Endocrine System

The Endocrine System Communicates via Chemical Messages

- Endocrine glands and organs with endocrine tissue release hormones
- Lipid-soluble and water-soluble hormones interact differently with target cells
- Feedback mechanisms regulate the secretion of hormones
- Nervous control sometimes overrides feedback mechanisms

Hormones Influence Growth, Development, Metabolism, and Behavior

- Pituitary hormones often prompt other glands to release hormones
- Thyroid hormones regulate metabolism and decrease blood calcium
- Parathyroid hormone increases blood calcium
- The adrenal glands secrete stress hormones
- Hormones of the pancreas regulate blood glucose
- Hormones of the thymus gland promote maturation of white blood cells
- The pineal gland secretes melatonin

Prostaglandins Act Locally

HEALTH ISSUE   Is It Hot in Here, or Is It Me? Hormone Replacement Therapy and Menopause

HEALTH ISSUE   Hormones and Our Response to Stress

HEALTH ISSUE   Melatonin: Miracle Supplement or Potent Drug Misused by Millions?

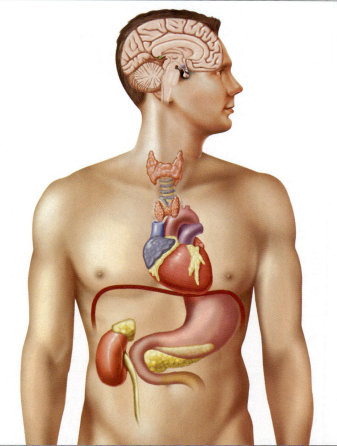

Basketball was Kazuo's passion. He loved afternoon practices, even the drills that most of his teammates detested. At 5 feet 4 inches, he realized that he had to make up in speed, skill, and desire what he lacked in size; most of the guards he played against were six feet or taller. It was only through an injury to the starting guard that he had been able to join the starting lineup the last six games, so he had to make the most of this opportunity to play and impress the coach. So far he had done well, averaging eight points and eight assists per game, but tonight he had to do even better. He would be matched against Conklin, the 6 foot 3 inch all-conference point guard for the Tigers. It was a grueling game. With six seconds left to play and the score tied, and after being beaten all night by Conklin's use of his height advantage to score over him, Kazuo suddenly reached around Conklin's back and knocked the ball away. He felt a huge surge of energy as he dove to the ball, got back on his feet, and raced the rest of the way down the court to the basket. With Conklin in mid-air above him and almost rim-high, Kazuo drove his pivot foot into the floor, sprung to the left, switched the ball to his left hand, twisted his hips for torque, and deftly spun the ball off the backboard and into the hoop. The buzzer rang just as Kazuo hit the floor, suddenly exhausted and exhilarated at the same time.

Fortunately for Kazuo, his late-in-the-game surge of energy came in the form of epinephrine, a hormone released from his adrenal gland. In this chapter, we will consider the endocrine glands and the hormones they secrete. We also will focus on what those hormones do. We will see that organs communicate with each other and with other tissues by hormones. We will learn that once hormones are produced, they enter the bloodstream and are carried along to their destinations, where they initiate long-term changes in our bodies, such as growth and development, and more short-term changes, such as Kazuo's winning surge of energy. In the end, we will see that proper functioning of the body depends on this system of internal communication. The endocrine system complements our more rapid system of internal communication, the nervous system (Chapters 7 and 8). Indeed, because the nervous and endocrine systems share the common function of regulating and coordinating the activities of all body systems, some consider them to be one system—the neuroendocrine system. ■

The Endocrine System Communicates Via Chemical Messages

Our bodies contain two types of glands, exocrine glands and endocrine glands (Figure 10.1). **Exocrine glands** secrete their products into ducts that empty onto the body surface, into the spaces within organs, or into a body cavity. Oil (sebaceous) glands, for example, are exocrine glands that secrete oil into ducts that open onto the surface of the skin. Salivary glands, another familiar example of exocrine glands, secrete saliva into ducts that open into the mouth. **Endocrine glands** are made of secretory cells that release their products, called hormones, into the fluid just outside the cells. Endocrine glands do not secrete their products into ducts. Instead, the hormones move from cells of the gland to the fluid just outside the cells, where they diffuse directly into the bloodstream. Once in the bloodstream, hormones circulate to virtually all the cells of the body, though they typically affect particular cells, as we will see.

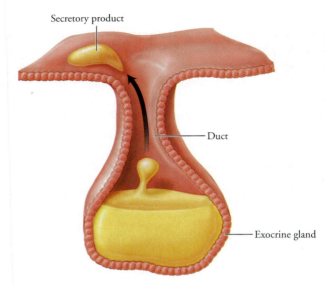

(a) Exocrine glands secrete their products into ducts that open onto the surface of the body, into the spaces within organs, or into cavities within the body.

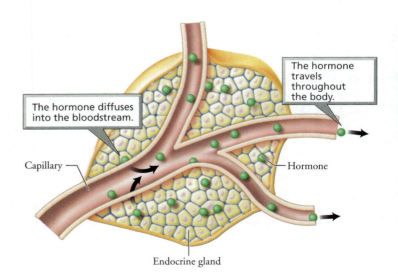

(b) Endocrine glands release their products, called hormones, into the fluid just outside cells. The hormones then diffuse into the bloodstream to be transported throughout the body.

FIGURE 10.1
Exocrine and endocrine glands

Endocrine glands and organs with endocrine tissue release hormones

The endocrine system consists of endocrine glands and organs that contain some endocrine tissue; these latter organs have other functions besides hormone secretion (Figure 10.2). The major endocrine glands are the pituitary gland, thyroid gland, parathyroid glands, adrenal glands, and pineal gland. Organs with some endocrine tissue include the hypothalamus, thymus, pancreas, ovaries, testes, heart, and placenta. Organs of the digestive and urinary systems, such as the stomach, small intestine, and the kidneys, also have endocrine tissue. Our discussion here will focus on the major endocrine glands. We also will cover three organs with endocrine tissue: hypothalamus, thymus, and pancreas. The remaining organs with endocrine tissue will be discussed in the chapters concerned with their other functions. For example, the kidneys are discussed in Chapter 16 on the urinary system, and the ovaries and testes are discussed in Chapter 17 on the reproductive system. Table 10.1 summarizes the endocrine glands and organs with endocrine tissue and the hormones they release.

Lipid-soluble and water-soluble hormones interact differently with target cells

Hormones are the chemical messengers of the endocrine system. They are released in very small amounts by cells of endocrine glands and some organs and enter the bloodstream to travel throughout the body. Although hormones contact virtually all cells, most affect only a particular type of cell, called a **target cell**. Target cells have **receptors**, protein molecules either located in the cell's cytoplasm or embedded in its plasma membrane. The receptors recognize and bind to specific hormones. Once a hormone binds to its specific receptor, it begins to exert its effects on the cell. These effects include alteration of either protein synthesis, secretory activity, energy utilization, or properties of the plasma membrane. Cells other than target cells lack the correct receptors and are unaffected by the hormone.

The mechanisms by which hormones influence target cells depend on the chemical makeup of the hormone. Hormones are classified as being either lipid soluble or water soluble. Lipid-soluble hormones include steroid hormones. **Steroid hormones** are a group of closely related hormones derived from cholesterol. The ovaries, testes, and adrenal glands are the main organs that secrete steroid hormones. Lipid-soluble hormones move easily through the plasma membrane because it is a lipid bilayer. Once inside the target cell, a steroid hormone combines with receptor molecules in the cytoplasm. The hormone-receptor complex then moves into the nucleus of the cell, where it attaches to DNA and activates certain genes. Ultimately, such activation leads to the synthesis of specific proteins by the target cell. (The precise steps involved in protein synthesis are described in Chapter 21). These proteins may include enzymes that stimulate or inhibit particu-

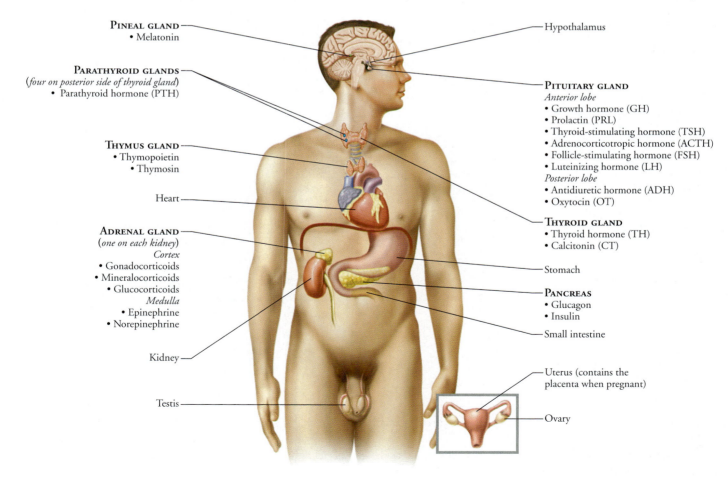

PINEAL GLAND
• Melatonin

PARATHYROID GLANDS
(*four on posterior side of thyroid gland*)
• Parathyroid hormone (PTH)

THYMUS GLAND
• Thymopoietin
• Thymosin

Heart

ADRENAL GLAND
(*one on each kidney*)
Cortex
• Gonadocorticoids
• Mineralocorticoids
• Glucocorticoids
Medulla
• Epinephrine
• Norepinephrine

Kidney

Testis

Hypothalamus

PITUITARY GLAND
Anterior lobe
• Growth hormone (GH)
• Prolactin (PRL)
• Thyroid-stimulating hormone (TSH)
• Adrenocorticotropic hormone (ACTH)
• Follicle-stimulating hormone (FSH)
• Luteinizing hormone (LH)
Posterior lobe
• Antidiuretic hormone (ADH)
• Oxytocin (OT)

THYROID GLAND
• Thyroid hormone (TH)
• Calcitonin (CT)

Stomach

PANCREAS
• Glucagon
• Insulin

Small intestine

Uterus (contains the placenta when pregnant)

Ovary

FIGURE **IO.2**

The endocrine system. The endocrine system is made up of endocrine glands and organs that contain some endocrine tissue.

TABLE 10.1

SOME ENDOCRINE GLANDS AND THEIR HORMONES		
GLAND OR ORGAN	**HORMONE**	**FUNCTION**
Anterior lobe of pituitary	Growth hormone (GH)	Stimulates growth, particularly of muscle, bone, and cartilage Stimulates breakdown of fat
	Prolactin (PRL)	Stimulates breasts to produce milk
	Thyroid-stimulating hormone (TSH)	Stimulates synthesis and release of hormones from thyroid gland
	Adrenocorticotropic hormone (ACTH)	Stimulates synthesis and release of glucocorticoid hormones from adrenal glands
	Follicle-stimulating hormone (FSH)	Stimulates gamete production in males and females Stimulates secretion of estrogen by the ovaries
	Luteinizing hormone (LH)	Causes ovulation and stimulates ovaries to secrete estrogen and progesterone Stimulates cells of testes to develop and secrete testosterone
Posterior lobe of pituitary	Antidiuretic hormone (ADH)	Promotes water reabsorption by the kidneys
	Oxytocin (OT)	Stimulates milk ejection from the breasts Stimulates uterine contractions during childbirth
Thyroid	Thyroid hormone (TH)	Regulates metabolism and heat production Promotes normal development and functioning of nervous, muscular, skeletal, and reproductive systems
	Calcitonin (CT)	Decreases blood levels of calcium (stimulates absorption of calcium by bone)
Parathyroid	Parathyroid hormone (PTH)	Increases blood levels of calcium (stimulates breakdown of bone and rate at which calcium is removed from urine and absorbed from gastrointestinal tract)
Adrenal cortex	Gonadocorticoids (androgens, estrogens)	Amounts secreted by adults are so low that effects are probably insignificant
	Mineralocorticoids (aldosterone)	Increase sodium reabsorption by kidneys Increase potassium excretion by kidneys
	Glucocorticoids (cortisol, corticosterone, cortisone)	Stimulate glucose synthesis Inhibit the inflammatory response
Adrenal medulla	Epinephrine	Fight-or-flight response Response to stress
	Norepinephrine	Fight-or-flight response Response to stress
Pancreas	Glucagon	Increases blood glucose level (prompts liver to increase conversion of glycogen to glucose and formation of glucose from fatty and amino acids)
	Insulin	Decreases blood glucose level (stimulates transport of glucose into cells, inhibits breakdown of glycogen to glucose, prevents conversion of fatty and amino acids to glucose)
Thymus	Thymopoietin, thymosin	Promote maturation of white blood cells
Pineal	Melatonin	May influence daily rhythms, fertility, and aging

lar metabolic pathways. Figure 10.3 summarizes the mode of action of steroid hormones.

Water-soluble hormones include molecules made of amino acids. Recall that amino acids are the building blocks of proteins. Some water-soluble hormones are only slightly modified amino acids and thus are relatively small molecules. Other water-soluble hormones con-sist of chains of amino acids that form larger molecules called peptides or proteins. Water-soluble hormones cannot pass through the lipid bilayer of the plasma membrane and therefore cannot enter target cells. Thus, they exert their effects indirectly by binding to receptors on the surface of the target cell. The hormone (considered the first messenger) binds to a receptor on the plasma membrane. This binding activates a

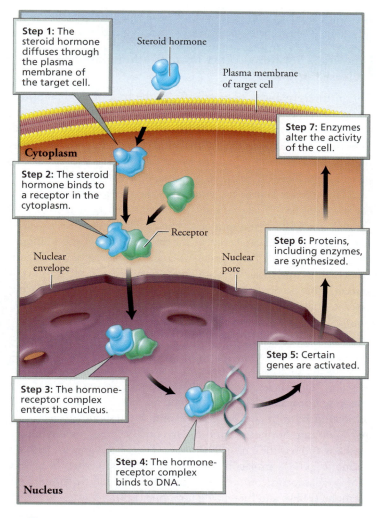

Step 1: The steroid hormone diffuses through the plasma membrane of the target cell.

Steroid hormone

Plasma membrane of target cell

Cytoplasm

Step 7: Enzymes alter the activity of the cell.

Step 2: The steroid hormone binds to a receptor in the cytoplasm.

Receptor

Step 6: Proteins, including enzymes, are synthesized.

Nuclear envelope

Nuclear pore

Step 5: Certain genes are activated.

Step 3: The hormone-receptor complex enters the nucleus.

Step 4: The hormone-receptor complex binds to DNA.

Nucleus

FIGURE **10.3**

Mode of action of steroid hormones

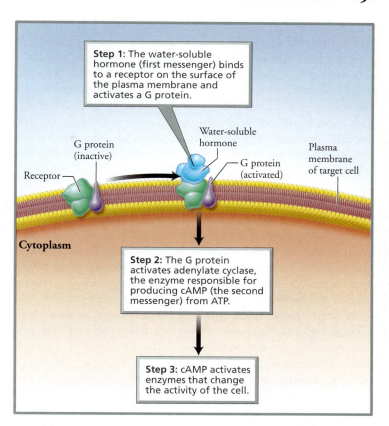

Step 1: The water-soluble hormone (first messenger) binds to a receptor on the surface of the plasma membrane and activates a G protein.

Water-soluble hormone

G protein (inactive)

Receptor

G protein (activated)

Plasma membrane of target cell

Cytoplasm

Step 2: The G protein activates adenylate cyclase, the enzyme responsible for producing cAMP (the second messenger) from ATP.

Step 3: cAMP activates enzymes that change the activity of the cell.

FIGURE **10.4**

Mode of action of some water-soluble hormones: the second messenger system of cAMP. Because water-soluble hormones cannot cross the plasma membrane, they indirectly affect the activities of target cells.

molecule (considered the second messenger) in the cytoplasm. **Second messengers** are molecules within the cell that influence the activity of enzymes and ultimately the activity of the cell to produce the effect of the hormone. Intermediaries, called G proteins, are attached to the receptor in the plasma membrane, where they often link first and second messengers. A system involving the second messenger cyclic adenosine monophosphate (cAMP) is illustrated in Figure 10.4.

Feedback mechanisms regulate the secretion of hormones

We now know how hormones work at the cellular level. Let's turn our attention to the factors that stimulate and regulate the release of hormones from endocrine glands. Stimuli that prompt endocrine glands to manufacture and release hormones are other hormones, signals from the nervous system, and changes in the levels of certain ions or nutrients in the blood.

Recall from Chapter 4 that body homeostasis is a relatively constant internal environment. Such constancy most often is achieved through **negative feedback mechanisms**, homeostatic mechanisms

in which the outcome of a process feeds back on the system, shutting down the process. Negative feedback mechanisms regulate the secretion of some hormones. Typically, a gland releases a hormone, and then rising blood levels of that hormone inhibit its further release. The feedback is described as negative because it acts to inhibit further release of the hormone. Some endocrine glands are sensitive to the particular condition they regulate rather than to the level of the hormone they produce. For example, the pancreas secretes the hormone insulin in response to high levels of glucose in the blood. Insulin prompts the liver to store glucose, and thus the blood level of glucose declines. The pancreas senses the low glucose in the blood and stops secreting insulin. The level of glucose in the blood also is regulated by glucagon, another hormone secreted by the pancreas. Whereas insulin lowers the glucose level in the blood, glucagon raises the level of glucose in the blood. In other words, insulin and glucagon have opposite, or antagonistic, effects. The various mechanisms that regulate glucose in the blood are discussed in more detail later in the chapter and are summarized in Figure 10.20. When feedback mechanisms are functioning properly, blood levels of most hormones fluctuate within a fairly narrow range. Hormonal disorders may result when negative feedback mechanisms fail. Under these conditions, too much or too little hormone is produced.

Secretion of hormones is sometimes regulated by **positive feedback mechanisms**, mechanisms in which the outcome of a process feeds back on the system, further stimulating the process. For example, during childbirth, the pituitary gland releases the hormone oxytocin (OT), which stimulates the uterus to contract. Uterine contractions then stimulate further release of oxytocin. The oxytocin stimulates even more contractions (see Figure 10.10). The feedback is described as positive because it acts to stimulate, rather than to inhibit, the release of oxytocin. Eventually, some change breaks the positive feedback cycle. In the case of childbirth, expulsion of the baby and the placenta terminates the feedback cycle. When we discuss the various glands and their hormones in the sections that follow, we also will describe the feedback mechanisms involved in their regulation.

▍Nervous control sometimes overrides feedback mechanisms

Sometimes the nervous system overrides the controls of the endocrine system. During times of severe stress, for example, the nervous system overrides the mechanism controlling the release of the hormone insulin. Recall that the level of insulin in the blood is controlled by a negative feedback system. Because of this negative feedback system, levels of glucose in the blood are normally maintained within a relatively narrow range. Under conditions of severe stress, such as those associated with strong emotional reactions, heavy bleeding, or starvation, the nervous system permits levels of blood glucose to rise much higher than normal. This adjustment gives cells access to the large amounts of energy needed to cope with the stressful situation.

Hormones Influence Growth, Development, Metabolism, and Behavior

Now we'll take a look at the endocrine glands. For each gland, we will consider its location and general structure, the hormones it secretes, and the effects of these hormones. We

also will consider disorders associated with the gland and hormone secretion. We begin with the pituitary gland, a gland the size of a pea.

▍Pituitary hormones often prompt other glands to release hormones

The **pituitary gland** is suspended from the base of the brain, just above the roof of the mouth, by a short stalk (Figure 10.5). The gland consists of two lobes, the anterior lobe and the posterior lobe. These lobes differ in size and in their relationship with the **hypothalamus**—the area at the base of the brain to which the pituitary is connected by its stalk. The lobes of the pituitary secrete different hormones.

The anterior lobe of the pituitary is the larger of the two lobes. A network of tiny capillaries runs from the base of the hypothalamus through the stalk of the pituitary. The capillaries connect to veins that lead into more capillaries in the anterior lobe of the pituitary gland (Figure 10.6). Any system whereby a capillary bed drains to veins that, in turn, drain into another capillary bed is called a **portal system.** This particular portal system is called the **pituitary portal system.** It allows hormones of the hypothalamus to control the secretion of hormones from the anterior lobe of the pituitary. Neurosecretory cells, nerve cells that synthesize and secrete hormones in the hypothalamus, release hormones into the pituitary portal system that travel to the anterior lobe, where they stimulate or inhibit hormone secretion. Substances that stimulate hormone secretion are called **releasing hormones**. Those that inhibit hormone secretion are called **inhibiting hormones**. The anterior pituitary responds to releasing and inhibiting hormones from the hypothalamus by modifying its synthesis and secretion of six hormones. These hormones are growth hormone (GH), prolactin (PRL), thyroid-stimulating hormone (TSH), adrenocorticotropic hormone (ACTH), follicle-stimulating hormone (FSH), and luteinizing hormone (LH).

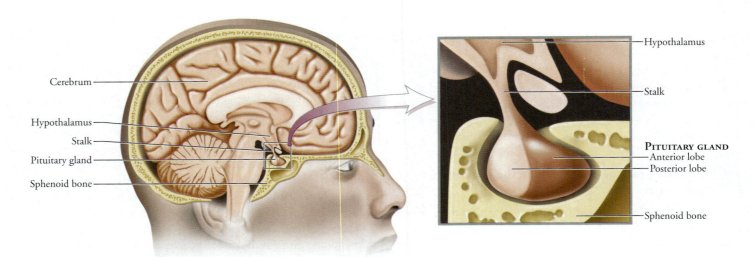

(a) Side view of the pituitary gland

(b) Close-up of the pituitary gland showing how it is attached to the hypothalamus by a short stalk

FIGURE **10.5**

Location and structure of the pituitary gland

The posterior lobe of the pituitary is very small, just larger than the head of a pin. It consists of neural tissue that releases hormones. Recall that there is a circulatory connection—the pituitary portal system—between the hypothalamus and the anterior lobe. In contrast, there is a neural connection between the hypothalamus and the posterior lobe. As shown in Figure 10.6, neurosecretory cells from the hypothalamus project directly into the posterior lobe. These nerve cells release oxytocin (OT) and antidiuretic hormone (ADH).

THE ANTERIOR LOBE

The anterior lobe of the pituitary produces and secretes six major hormones. We will begin with **growth hormone (GH)**, the primary function of which is to stimulate growth through increases in cell size and rates of cell division. Cells of bone, muscle, and cartilage are most susceptible to GH, but cells of other tissues are affected as well. The effects of GH on body growth are thought to occur indirectly, through intermediary hormones, called somatomedins, produced by the liver. The presence of GH stimulates the cells of the liver to release somatomedins, which, in turn, increase protein synthesis in target cells, thereby enhancing cell growth and division. Growth hormone also plays a role in glucose conservation by making fats more available as a source of fuel.

Two hormones of the hypothalamus regulate the synthesis and release of GH. Growth hormone-releasing hormone (GHRH) stimulates the release of GH. Growth hormone-inhibiting hormone (GHIH) inhibits the release of GH. Levels of GH are normally maintained within an appropriate range. However, excesses or deficiencies of the hormone can produce dramatic effects on growth. Abnormally high production of GH in childhood, when the bones are still capable of growing in length, results in giantism. **Giantism** is a condition characterized by rapid growth and eventual attainment of heights up to

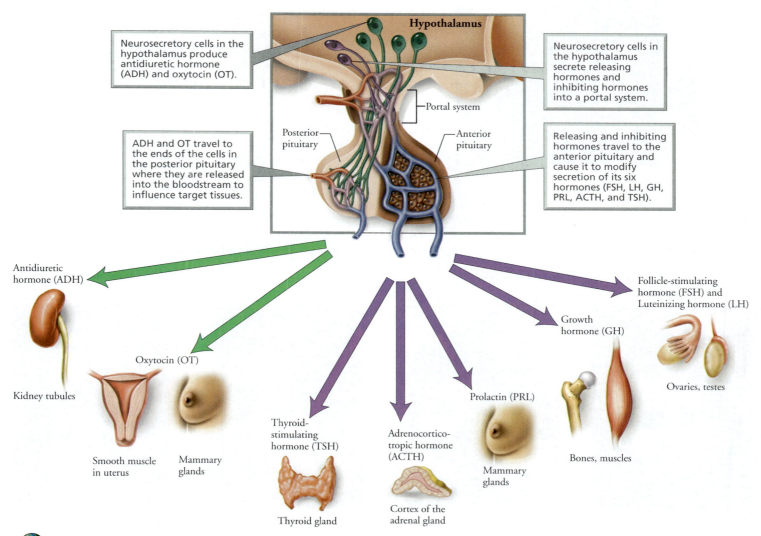

Neurosecretory cells in the hypothalamus produce antidiuretic hormone (ADH) and oxytocin (OT).

Neurosecretory cells in the hypothalamus secrete releasing hormones and inhibiting hormones into a portal system.

ADH and OT travel to the ends of the cells in the posterior pituitary where they are released into the bloodstream to influence target tissues.

Releasing and inhibiting hormones travel to the anterior pituitary and cause it to modify secretion of its six hormones (FSH, LH, GH, PRL, ACTH, and TSH).

Hypothalamus

Portal system

Posterior pituitary

Anterior pituitary

Antidiuretic hormone (ADH)

Kidney tubules

Oxytocin (OT)

Smooth muscle in uterus

Mammary glands

Thyroid-stimulating hormone (TSH)

Thyroid gland

Adrenocorticotropic hormone (ACTH)

Cortex of the adrenal gland

Prolactin (PRL)

Mammary glands

Growth hormone (GH)

Bones, muscles

Follicle-stimulating hormone (FSH) and Luteinizing hormone (LH)

Ovaries, testes

FIGURE 10.6

TUTORIAL 10.3

The two lobes of the pituitary gland, their connections with the hypothalamus, and the hormones they secrete

8 or 9 feet (Figure 10.7). Increased production of GH in adulthood, when the bones can thicken but not lengthen, causes **acromegaly**. Literally meaning "enlarged extremities," acromegaly is characterized by enlargement of the tongue and a gradual thickening of the bones of the hands, feet, and face (Figure 10.8). Giantism and acromegaly usually are caused by a tumor of the anterior pituitary. Both conditions are associated with decreased life expectancy. Tumors can be treated with surgery, radiation, or drugs that reduce GH secretion and tumor

size. Insufficient production of GH in childhood results in **pituitary dwarfism**. Typically, pituitary dwarfs are sterile and attain a maximum height of about 4 feet (Figure 10.9). Administering growth hormone in childhood can treat pituitary dwarfism but not other forms of dwarfism.

In the past, the use of growth hormone for the treatment of medical conditions was extremely limited because GH was scarce, given that the hormone had to be extracted from the pituitary glands of cadavers. Because of its scarcity, GH was used only to treat medical conditions such as pituitary dwarfism. Beginning in the late 1970s, however, GH could be made in the laboratory. With the greater availability of GH came research on its uses in the treatment of aging in adults and below-average height in children. Many aspects of normal aging, such as thinning of the skin, decreased muscle mass, and increased body fat, appear to be reversed by the administration of GH. For children who are of below-average height, administration of GH appears to add a few inches of adult height.

FIGURE **10.7**
Robert Wadlow, a pituitary giant, was born in 1918 at a normal size but developed a pituitary tumor as a young child. The tumor caused increased production of growth hormone. Robert never stopped growing until his death at 22 years of age, by which time he had reached a height of 8 feet 11 inches.

what would you do?

A recent review of studies in which children of below-average height were administered growth hormone (GH) concluded that such therapy could yield an additional 2 inches of adult height. Opponents of using GH for the treatment of below-average height in children believe it is wrong to give a powerful and potentially harmful hormone to children who are basically healthy. Rather than administering GH to such children, opponents suggest working instead to increase societal acceptance of persons who are short. Opponents also cite the cost of GH therapy. For example, each additional inch of adult height gained costs about $35,000. Finally, those opposed to GH therapy point to side effects, such as swelling of tissues and increased blood pressure. Do you think that GH should be used to "treat" below-average height in children? What would you do if you were the parent of a healthy child who was destined to be very short? What if your child were a pituitary dwarf; would you approve the use of GH then?

(a)

(b)

(c)

(d)

FIGURE **10.8**
Acromegaly. Excess secretion of growth hormone in adulthood, when the bones can thicken but not lengthen, causes acromegaly, a gradual thickening of the bones of the hands, feet, and face. The disorder was not apparent in this female at ages (a) 9 or (b) 16, but it became apparent by ages (c) 33 and (d) 52.

FIGURE **IO.9**
Pituitary dwarfism is caused by insufficient growth hormone in childhood.

Prolactin (PRL), another hormone secreted by the anterior lobe of the pituitary gland, is produced mainly during lactation, the period in which a female produces and ejects milk from the mammary glands. Prolactin stimulates the mammary glands to produce milk. Oxytocin, a hormone secreted from the posterior pituitary, causes the ducts of the mammary glands to eject milk. Prolactin interferes with female sex hormones, explaining why most mothers fail to have regular menstrual cycles and fail to conceive while nursing their newborn. (Lactation should not, however, be relied upon as a method for birth control, because the suppression of female hormones and ovulation lessens as mothers breast-feed their infants less frequently. Thus, a mother who breast-feeds infrequently may ovulate, and the released egg could be fertilized.)

Growth of a pituitary tumor may cause excess secretion of PRL. In females, excess PRL may cause infertility and the production of milk when birth has not occurred. In males, PRL appears to be involved in the production of mature sperm in the testis, but its precise role is not yet clear. Nevertheless, production of too much PRL, as might occur with a pituitary tumor, can cause sterility and impotence in men. Some hormones from the hypothalamus stimulate and others inhibit production and secretion of PRL.

The remaining hormones produced by the anterior lobe of the pituitary gland influence other endocrine glands. Hormones that influence another endocrine gland are called **tropic hormones**. Two such hormones secreted by the anterior lobe of the pituitary are thyroid-stimulating hormone and adrenocorticotropic hormone. **Thyroid-stimulating hormone (TSH)** acts on the thyroid gland in the neck to stimulate synthesis and release of thyroid hormones. A releasing hormone from the hypothalamus controls the release of TSH. **Adrenocorticotropic hormone (ACTH)** controls the synthesis and secretion of glucocorticoid hormones from the outer portion (cortex) of the adrenal glands (see Figure 10.2). A releasing hormone from the hypothalamus controls release of ACTH.

Two other tropic hormones secreted by the anterior lobe of the pituitary gland influence the gonads (ovaries in the female and testes in the male) and are therefore considered gonadotropins.

Follicle-stimulating hormone (FSH) promotes development of egg cells and secretion of the hormone estrogen from the ovaries in females. In males, FSH promotes production of sperm. **Luteinizing hormone (LH)** causes ovulation, the release of a future egg cell by the ovary in females. Luteinizing hormone also stimulates the ovaries to secrete estrogen and progesterone. These two hormones prepare the uterus for implantation of a fertilized ovum and the breasts for production of milk. In males, LH stimulates cells within the testes to develop and secrete the hormone testosterone. These particular cells within the testes are called interstitial cells (Chapter 17), and thus LH in males is sometimes called interstitial cell–stimulating hormone (ICSH). The same releasing hormone from the hypothalamus controls FSH and LH.

The hormones produced by the anterior lobe of the pituitary are summarized in Table 10.1.

THE POSTERIOR LOBE

The posterior pituitary does not produce any hormones. However, neurons of the hypothalamus manufacture **antidiuretic hormone (ADH)** and **oxytocin (OT)**. These hormones travel down the nerve cells into the posterior pituitary, where they are stored and released.

The main function of ADH is to conserve body water by decreasing urine output. ADH accomplishes this task by causing the kidneys to remove water from the fluid destined to become urine. The water is then returned to the blood. Alcohol temporarily inhibits secretion of ADH, causing increased urination following alcohol consumption. The increased output of urine causes dehydration and the resultant headache and dry mouth typical of many hangovers. Antidiuretic hormone is also called **vasopressin**. This name comes from its role in constricting blood vessels and raising blood pressure, particularly during times of severe blood loss. A deficiency of ADH may result from damage to either the posterior pituitary or the area of the hypothalamus responsible for the hormone's manufacture. Such a deficiency results in **diabetes insipidus**, a condition characterized by excessive urine production and resultant dehydration. Mild cases may not require treatment. Severe cases may cause extreme fluid loss; death through dehydration can result. Treatment usually involves administration of synthetic ADH in a nasal spray. Diabetes insipidus (*diabetes*, overflow; *insipidus*, tasteless) should not be confused with **diabetes mellitus** (*mel*, honey). The latter is a condition in which large amounts of glucose are lost in the urine as a result of an insulin deficiency. Both conditions, however, are characterized by increased production of urine. (We will look at diabetes mellitus again when we discuss the hormones of the pancreas.) Look again at what their names mean. Can you guess how physicians in the past distinguished between these two forms of diabetes?

Oxytocin (OT) is the second hormone produced in the hypothalamus and released by the posterior pituitary. The name oxytocin (*oxy*, quick; *tokos*, childbirth) reveals one of its two main functions, stimulating the uterine contractions of childbirth (Figure 10.10). During pregnancy, the cells of the uterus become increasingly sensitive to oxytocin. This increased sensitivity may result from an increase late in gestation in the abundance of oxytocin receptors on smooth muscle cells of the uterus. Eventually, the uterus begins to contract in response to OT. During labor, the

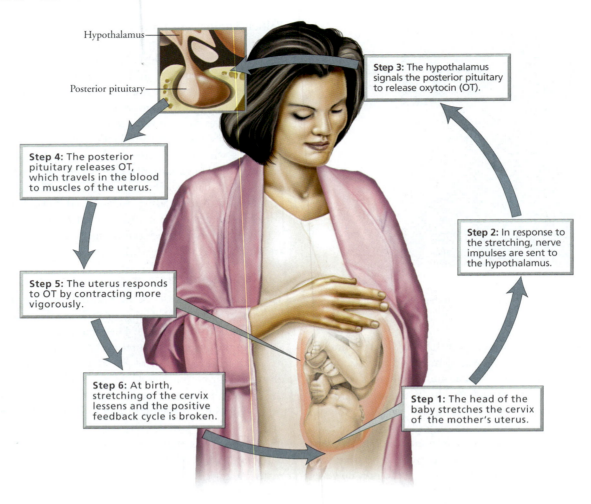

Hypothalamus

Posterior pituitary

Step 3: The hypothalamus signals the posterior pituitary to release oxytocin (OT).

Step 4: The posterior pituitary releases OT, which travels in the blood to muscles of the uterus.

Step 2: In response to the stretching, nerve impulses are sent to the hypothalamus.

Step 5: The uterus responds to OT by contracting more vigorously.

Step 6: At birth, stretching of the cervix lessens and the positive feedback cycle is broken.

Step 1: The head of the baby stretches the cervix of the mother's uterus.

FIGURE 10.10

The steps by which oxytocin stimulates uterine contractions during childbirth

baby's head begins to stretch the narrow portion of the uterus called the cervix. Nerve impulses are sent to the hypothalamus. The hypothalamus signals the posterior pituitary to release additional OT. The released OT travels in the blood to the uterus, where it stimulates ever-stronger contractions of the uterus. Finally, the baby is pushed down the birth canal and out into waiting hands. As described earlier, the control of OT during the induction of labor is an example of a positive feedback mechanism. Oxytocin stimulates uterine contractions that stimulate further release of OT. Once the baby is born, stretching of the cervix lessens, and the positive feedback cycle is broken. Pitocin is a synthetic form of OT used to induce labor.

The second major function of oxytocin is to stimulate milk ejection (letdown) from the mammary glands. Milk ejection occurs in response to stimuli from an infant (Figure 10.11). Recall that prolactin secreted by the anterior pituitary stimulates the mammary glands to produce, but not to eject, milk. Once a baby begins to suck at the nipple, touch receptors in the nipple send nerve impulses to the hypothalamus. The hypothalamus prompts the posterior pituitary to release OT. The OT travels to the mammary glands, where it stimulates cells

to eject milk. Men also secrete OT, and there is some evidence that this hormone facilitates the transport of sperm in the male reproductive tract. Oxytocin also may be involved in male sexual behavior.

The hormones released by the posterior lobe of the pituitary are summarized in Table 10.1.

stop and think

Women who have just given birth are often encouraged to nurse their babies as soon as possible after delivery. How might an infant's suckling promote completion of, and recovery from, the birth process? Consider that the placenta (afterbirth) must still be expelled after the birth of the baby and that the uterus must return to an approximation of its prepregnancy form.

Thyroid hormones regulate metabolism and decrease blood calcium

The **thyroid gland** is a shield-shaped structure in the front of the neck, as shown in Figure 10.12a. The characteristic deep red color of the thyroid gland stems from its prodigious blood supply.

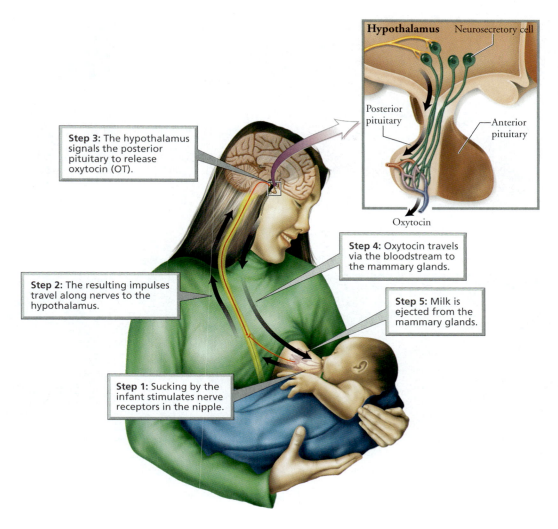

Step 3: The hypothalamus signals the posterior pituitary to release oxytocin (OT).

Step 2: The resulting impulses travel along nerves to the hypothalamus.

Step 1: Sucking by the infant stimulates nerve receptors in the nipple.

Step 4: Oxytocin travels via the bloodstream to the mammary glands.

Step 5: Milk is ejected from the mammary glands.

Hypothalamus Neurosecretory cell

Posterior pituitary

Anterior pituitary

Oxytocin

FIGURE **10.11**

The steps by which oxytocin stimulates milk ejection (letdown) from the mammary glands

Within the thyroid are small spherical chambers called follicles (Figure 10.12b, c). Cells line the walls of the follicles and produce thyroglobulin, the substance from which **thyroid hormone (TH)** is made. (Thyroid hormone is the name given to two very similar hormones: thyroxine, also called T4 because it has four iodine atoms, and triiodothyronine, also called T3 because it has three iodine atoms.) Other endocrine cells in the thyroid, called parafollicular cells, secrete the hormone calcitonin (Figure 10.12c).

Nearly all body cells have receptors for thyroid hormone. Not surprisingly, the hormone has broad effects. Thyroid hormone regulates the body's metabolic rate and production of heat. It also maintains blood pressure and promotes normal development and functioning of several organ systems. Thyroid hormone affects cellular metabolism by stimulating protein synthesis, the breakdown of lipids, and the use of glucose for production of ATP. The pituitary gland and hypothalamus control the release of thyroid hormone. Falling levels of thyroid hormone in the blood prompt the hypothalamus to secrete a releasing hormone. This releasing hormone

stimulates the anterior pituitary to release thyroid-stimulating hormone (TSH), which, in turn, causes the thyroid to release more TH.

Undersecretion of thyroid hormone during fetal development or infancy causes cretinism. **Cretinism** is characterized by dwarfism, mental retardation, and slowed sexual development (Figure 10.13). If a pregnant woman produces sufficient thyroid hormone, many of the symptoms of cretinism do not appear until after birth. It is at this time that the infant relies solely on its own malfunctioning thyroid gland to supply the needed hormones. Most infants are tested for proper thyroid function shortly after birth. Giving oral doses of thyroid hormone can prevent cretinism. Undersecretion of thyroid hormone in adulthood causes **myxedema**, a condition in which fluid accumulates in facial tissues. Other symptoms of undersecretion of thyroid hormone include decreased alertness, body temperature, and heart rate. Oral administration of thyroid hormone can prevent these symptoms.

Oversecretion of thyroid hormone causes **Graves' disease**, an autoimmune disorder in which a person's own immune system

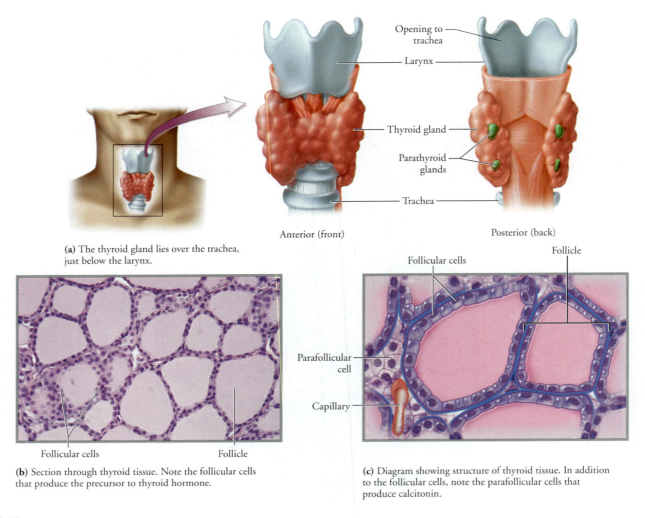

(a) The thyroid gland lies over the trachea, just below the larynx.

(b) Section through thyroid tissue. Note the follicular cells that produce the precursor to thyroid hormone.

(c) Diagram showing structure of thyroid tissue. In addition to the follicular cells, note the parafollicular cells that produce calcitonin.

FIGURE **10.12**
Location and structure of the thyroid gland

FIGURE **10.13**
Cretinism, a disorder of the thyroid gland. This condition is characterized by mental retardation, dwarfism, and delayed sexual development. Cretinism is caused by undersecretion of thyroid hormone during fetal life or infancy.

produces Y-shaped proteins called antibodies that mimic the action of thyroid-stimulating hormone (TSH). The antibodies stimulate the thyroid gland, causing it to enlarge and overproduce its hormones. The symptoms of Graves' disease include increased metabolic rate and heart rate. Sweating, nervousness, and weight loss also occur in people with this disease. Many patients with Graves' disease also have **exophthalmos**, protruding eyes caused by the accumulation of fluid (Figure 10.14). Treatment of Graves' disease may involve the use of drugs to block synthesis of thyroid hormones. Alternatively, thyroid tissue may be reduced through surgery or the administration of radioactive iodine. The thyroid gland accumulates iodine; thus, ingestion of radioactive iodine selectively destroys thyroid tissue.

Iodine is needed for production of thyroid hormone. A diet deficient in iodine can produce a **simple goiter**, an enlarged thyroid gland (Figure 10.15). When intake of iodine is inadequate, the level of TH is low, and the low level triggers secretion of TSH. Thyroid-stimulating hormone stimulates the thyroid gland to increase production of thyroglobulin. (Recall that this

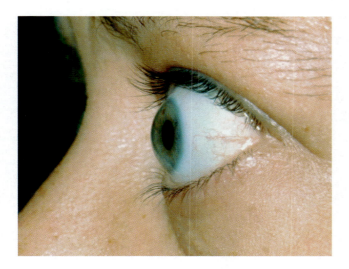

FIGURE **10.14**
Exophthalmos, a disorder of the thyroid gland. Oversecretion of thyroid hormone leads to an accumulation of fluid behind the eyes, causing the eyes to bulge out. Exophthalmos often is associated with Graves' disease.

FIGURE **10.15**
Simple goiter. In response to an iodine-deficient diet, the thyroid gland enlarges, causing a goiter.

is the substance from which TH is made.) The lack of iodine prevents formation of TH from the accumulating thyroglobulin. Responding to the continued low levels of TH, the pituitary continues to release increasing amounts of TSH, which cause the thyroid to enlarge in a futile effort to filter more iodine from the blood. Goiters were quite common, especially in parts of the Midwestern United States (deemed the Goiter Belt), where iodine-poor soil and little access to iodine-rich shellfish led to diets deficient in iodine. The incidence of goiter in the United States dramatically decreased once iodine was added to most table salt beginning in the 1920s. Simple goiter can be treated by iodine supplements or administration of TH.

Calcitonin (CT), another hormone produced by the thyroid gland, helps regulate the concentration of calcium in the blood. Calcium is important to the proper functioning of neurons and muscle cells, and thus its levels must be regulated. When the level of calcium in the blood is high, calcitonin stimulates the absorption of calcium by bone and inhibits the breakdown of bone, thereby lowering the level of calcium in the blood. Calcitonin also lowers blood calcium by stimulating an initial increase in the excretion of calcium in the urine. When the level of calcium in the blood is low, the parathyroid glands, which we discuss next, are prompted to release their hormone.

Parathyroid hormone increases blood calcium

The **parathyroid glands** are four small, round masses at the back of the thyroid gland (Figure 10.12a). These glands secrete **parathyroid hormone (PTH)**, also called **parathormone**. As mentioned above, calcitonin from the thyroid gland lowers the level of calcium in the blood. There is, however, another side to the balancing act of maintaining proper blood calcium levels. This other side involves PTH,

which increases levels of calcium in the blood. Low levels of calcium in the blood stimulate the parathyroid glands to secrete PTH, which causes calcium to move from bone and urine into the blood. Parathyroid hormone exerts its effects by stimulating (1) bone-destroying cells called osteoclasts that release calcium from bone into the blood, (2) the removal of calcium from the urine and its return to the blood, and (3) the rate at which calcium is absorbed into the blood from the gastrointestinal tract. Parathyroid hormone also inhibits bone-forming cells called osteoblasts and thereby reduces the rate at which calcium is deposited in bone.

The feedback system by which CT and PTH regulate levels of calcium in the blood is summarized in Figure 10.16.

Surgery on the neck or thyroid gland may damage the parathyroid glands. The resultant decrease in PTH causes decreased blood calcium. The decreased calcium produces nervousness and muscle spasms. In severe cases, death may result from spasms of the larynx and paralysis of the respiratory system. Parathyroid hormone is difficult to purify. Thus, deficiencies are usually not treated by administering the hormone. Instead, calcium is given either in tablet form or through increased dietary intake. A tumor of the parathyroid gland can cause excess secretion of PTH. Oversecretion of PTH pulls calcium from bone tissue, causing increased blood calcium and weakened bones. High levels of calcium in the blood may lead to kidney stones, calcium deposits in other soft tissue, and decreased activity of the nervous system.

Hormones released by the thyroid and parathyroid glands are summarized in Table 10.1.

The adrenal glands secrete stress hormones

The body's two **adrenal glands**, each about the size of an almond, are located at the tops of the kidneys (*ad*, upon; *renal*, kidney).

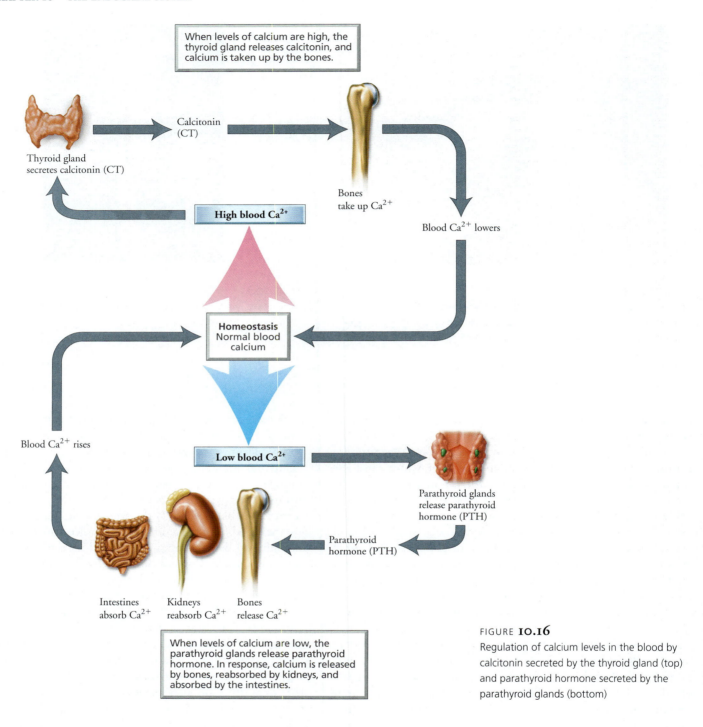

When levels of calcium are high, the thyroid gland releases calcitonin, and calcium is taken up by the bones.

Thyroid gland secretes calcitonin (CT)

Calcitonin (CT)

Bones take up Ca^{2+}

Blood Ca^{2+} lowers

High blood Ca^{2+}

Homeostasis Normal blood calcium

Blood Ca^{2+} rises

Low blood Ca^{2+}

Parathyroid glands release parathyroid hormone (PTH)

Parathyroid hormone (PTH)

Intestines absorb Ca^{2+}

Kidneys reabsorb Ca^{2+}

Bones release Ca^{2+}

When levels of calcium are low, the parathyroid glands release parathyroid hormone. In response, calcium is released by bones, reabsorbed by kidneys, and absorbed by the intestines.

FIGURE **10.16**

Regulation of calcium levels in the blood by calcitonin secreted by the thyroid gland (top) and parathyroid hormone secreted by the parathyroid glands (bottom)

Each gland has two distinct regions (Figure 10.17). The outer region, the **adrenal cortex**, secretes more than 20 different steroid hormones. These hormones fall into three groups, the gonadocorticoids, mineralocorticoids, and glucocorticoids. The inner region, called the **adrenal medulla**, secretes epinephrine and norepinephrine.

The adrenal cortex secretes **gonadocorticoids**, male and female sex hormones known as **androgens** and **estrogens,** respectively. In males, the adrenal cortex secretes both androgens and estrogens. The same is true of the adrenal cortex in

females. In normal adult males, androgen secretion by the testes far surpasses that by the adrenal cortex. Thus, the effects of adrenal androgens in adult males are probably insignificant. In females, the ovaries and placenta also produce estrogen. In menopause, however, the ovaries decrease secretion of estrogen and eventually stop secreting it. The gonadocorticoids from the adrenal cortex may somewhat alleviate the effects of decreased ovarian estrogen. One option for menopausal women is hormone replacement therapy. The advantages and disadvantages of hormone replacement in menopause are considered in the

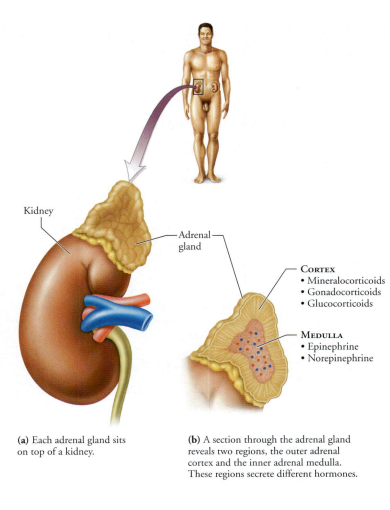

Kidney

Adrenal gland

CORTEX
• Mineralocorticoids
• Gonadocorticoids
• Glucocorticoids

MEDULLA
• Epinephrine
• Norepinephrine

(a) Each adrenal gland sits on top of a kidney.

(b) A section through the adrenal gland reveals two regions, the outer adrenal cortex and the inner adrenal medulla. These regions secrete different hormones.

FIGURE **10.17**
Location and structure of an adrenal gland

that the pituitary gland secretes adrenocorticotropic hormone (ACTH), which stimulates the cortex of the adrenal glands to secrete its hormones. Thus, Addison's disease can also be caused by inadequate secretion of ACTH by the pituitary. Addison's disease can be treated with tablets containing the missing hormones.

stop and think

High blood pressure can signal abnormal aldosterone secretion. Would high blood pressure be associated with the undersecretion or oversecretion of aldosterone?

The **glucocorticoids** are hormones secreted by the adrenal cortex that affect glucose homeostasis. Glucocorticoids act on the liver to promote the conversion of fat and protein to intermediate substances that are ultimately converted into glucose. The glucocorticoids also act on adipose tissue to prompt the breakdown of fats to fatty acids that are released into the bloodstream, where they are available for use by the body's cells. Glucocorticoids also conserve glucose by inhibiting the uptake of glucose by muscle and fat tissue.

Glucocorticoids inhibit the inflammatory response. These hormones slow the movement of white blood cells to the site of injury. Glucocorticoids also inhibit the activity of white blood cells that are already at the site by reducing the likelihood that other cells will release chemicals to promote inflammation. The downside of glucocorticoids is that they inhibit wound healing. This fact explains why steroid creams containing glucocorticoids are applied only to the surface of the skin to treat superficial rashes. These creams should not be applied to open wounds. Some examples of glucocorticoids are cortisol, corticosterone, and cortisone.

Cushing's syndrome results from prolonged exposure to high levels of cortisol. Body fat is redistributed and fluid accumulates in the face (Figure 10.18). Additional symptoms include fatigue, high

Health Issue essay, *Is It Hot in Here, or Is It Me? Hormone Replacement Therapy and Menopause.*

The **mineralocorticoids** are hormones secreted by the adrenal cortex that affect mineral homeostasis and water balance. The primary mineralocorticoid is **aldosterone**, a hormone that acts on cells of the kidneys to increase reabsorption of sodium ions (Na^+) into the blood. This reabsorption prevents depletion of Na^+ and increases water retention. Aldosterone also acts on kidney cells to promote the excretion of potassium ions (K^+) in urine. **Addison's disease** is a disorder caused by the undersecretion of cortisol and aldosterone. This disease appears to be an autoimmune disorder in which the body's own immune system perceives cells of the adrenal cortex as foreign and destroys them. Destruction of cells of the adrenal cortex results in inadequate production of their hormones, which in turn, causes weight loss, fatigue, electrolyte imbalance, poor appetite, and poor resistance to stress. A peculiar bronzing of the skin also is associated with Addison's disease. Recall

(a) Patient diagnosed with Cushing's syndrome

(b) Same patient after treatment

FIGURE **10.18**
Cushing's syndrome. Prolonged exposure to cortisol causes fluid to accumulate in the face. Most often, Cushing's syndrome is caused by the administration of cortisol for allergies or inflammation.

Is It Hot in Here, or Is It Me?
Hormone Replacement Therapy and Menopause

Many endocrine disorders are characterized by little or no secretion of certain hormones. Cretinism, for example, is caused by the undersecretion of thyroid hormone during fetal life or infancy. In other cases, hormone secretion declines as part of the normal aging process. Estrogen, a hormone produced primarily by the ovaries, is secreted in decreasing amounts beginning when women are about age 30. This decrease eventually leads into menopause, the termination of ovulation and menstruation. Menopause usually occurs when women are between 45 and 55 years of age. Undersecretion of hormones, whether caused by malfunctioning glands or normal aging, often is

treated with hormone replacement therapy (HRT). Hormone replacement therapy involves replacing the deficient hormone through means such as injections, pills, patches, or creams.

Hormone replacement has had many successes, but it is not without controversy. Controversy is particularly intense when HRT is used to treat normal declines in hormone levels. As mentioned above, levels of ovarian hormones decline as women age; menopause is the ultimate result. Some women face a host of uncomfortable symptoms in conjunction with the eventual end of ovulation and menstruation. Such symptoms include hot flashes, night sweats, vaginal dryness, stress incontinence (involuntary loss of

small amounts of urine when laughing, coughing, or sneezing), heightened mood swings, and memory loss. Symptoms may last only a few months or a few years. Declining estrogen levels also have been linked to increased risk of heart disease, stroke, and osteoporosis (increased bone loss leading to increased risk of fracture; see Chapter 5). Given all these effects, it is not surprising that for many years doctors administered ovarian hormones to millions of women at or past menopause. In most cases, doctors prescribed estrogen and progestin (a form of progesterone). Estrogen alone, estrogen replacement therapy (ERT), was given if a woman's uterus had been surgically removed (this surgical procedure

blood pressure, and elevated glucose levels. A tumor on either the adrenal cortex or the anterior pituitary may cause the oversecretion of cortisol that leads to Cushing's syndrome. (Recall that the anterior pituitary secretes adrenocorticotropic hormone, ACTH, which stimulates the release of hormones from the adrenal cortex.) Tumors are treated with radiation, drugs, or surgery. Cushing's syndrome also may result from giving glucocorticoid hormones to treat asthma, lupus, and rheumatoid arthritis. Treatment in medically induced cases of Cushing's syndrome typically entails a gradual reduction of the glucocorticoid dose. Ideally, the dose is reduced to the lowest level necessary to control the existing disorder.

The adrenal medulla produces **epinephrine** (adrenaline) and **norepinephrine** (noradrenaline). These hormones help us respond to stress and are critical in the **fight-or-flight response**. Imagine, for example, that you are walking home alone late at night and a stranger suddenly steps out in front of you from the bushes. Impulses received by your hypothalamus are sent by neurons to your adrenal medulla. These impulses cause cells in your adrenal medulla to increase output of epinephrine and norepinephrine. In response to these hormones, your heart rate, respiratory rate, and blood glucose levels rise. Blood vessels associated with the digestive tract constrict because digestion currently is not of prime importance. Vessels associated with skeletal and cardiac muscles dilate, allowing more blood, glucose, and oxygen to reach them. These substances also reach your brain in greater amounts, leading to the increased mental alertness essential to fleeing or fighting. We explore further how our bodies react to stress in the Health Issue essay, *Hormones and Our Response to Stress.*

The hormones secreted by the adrenal glands are summarized in Table 10.1.

Hormones of the pancreas regulate blood glucose

The **pancreas** is located in the abdomen just behind the stomach; it contains both endocrine and exocrine cells (Figure 10.19). The exocrine cells secrete digestive enzymes into ducts that empty into the small intestine. The role of the pancreas in digestion will be discussed in Chapter 15. The endocrine cells occur in small clusters called **pancreatic islets** (or islets of Langerhans). These clusters contain two major types of hormone-producing cells. One type produces the hormone glucagon; the other produces the hormone insulin.

Glucagon increases glucose in the blood. Glucagon accomplishes this increase by prompting cells of the liver to increase conversion of glycogen (the storage polysaccharide in animals) to glucose (a simple sugar, or monosaccharide). Glucagon also stimulates formation of glucose from lactic acid and amino acids. The liver releases the resultant glucose molecules into the bloodstream, causing a rise in blood sugar level. Amino acids in the blood are taken up by the liver to form new glucose molecules. Thus, a secondary effect of glucagon is a lowering of amino acid levels in the blood. Declining levels of blood glucose mainly stimulate glucagon secretion. Secretion also is stimulated by increasing levels of amino acids in the blood, as might occur after eating a large steak.

In contrast to glucagon, **insulin** decreases glucose in the blood. Insulin lowers blood sugar in several ways. First, insulin stimulates transport of glucose into muscle cells, white blood cells, and connective tissue cells. Second, insulin inhibits the breakdown of is known as a hysterectomy). Both HRT and ERT have some benefits. Both therapies, when used for a few years, relieve hot flashes and night sweats. However, HRT and ERT must be used for more than 10 years to achieve the benefit of reduced risk of osteoporosis. Bone loss resumes once therapy is stopped.

In recent years, research has revealed a dark side to HRT and ERT. In 2002, the U.S. government formally listed all forms of estrogen used in replacement therapies as "known human carcinogens." Estrogen replacement therapy seems to increase a woman's risk of developing endometrial cancer (cancer of the lining of the uterus). This discovery led doctors to use lower doses of estrogen and to combine it with progestin, which protects the uterine lining against cancer. There also is the possibility that HRT and ERT, particularly when used for many years, increase the risk of breast cancer. More than 50 studies since the late 1970s have addressed this question. Some studies have found increased risk; others have not. Finally, hormone replacement therapy and estrogen replacement therapy also may increase the risk of blood clots.

The future of HRT and ERT has recently grown darker still. Early studies suggested that HRT and ERT reduced the risk of heart disease and stroke and slowed the progression of Alzheimer's disease. These claims are now being questioned in a report called the *International Position Paper on Women's Health and Menopause*, published in 2002. The early claims of health benefits came from observational studies in which women themselves decided whether to undergo hormone therapy. More recent studies have used randomized control trials in which patients are assigned at random to taking the hormone or a placebo. These new, more rigorous studies have failed to substantiate the health benefits of HRT and ERT regarding heart disease, stroke, and Alzheimer's. The health benefits found in the earlier observational studies may simply reflect a flawed experimental design. Perhaps the women who elected to take hormones were healthier and more concerned with their health to begin with than were the women who elected not to take hormones.

In view of the risks of HRT and ERT, many physicians are now urging women to consider the many other ways available to stave off heart disease, stroke, and osteoporosis. These physicians urge women to avoid smoking, to eat a diet rich in calcium and low in fat, and to exercise regularly. Nonhormonal drugs may be better and safer for preventing bone fractures and lowering cholesterol and blood pressure. Some alternative practitioners advise getting estrogen from dietary sources such as yams and soybeans. These foods are certainly weaker sources than prescription estrogen and may reduce the risks.

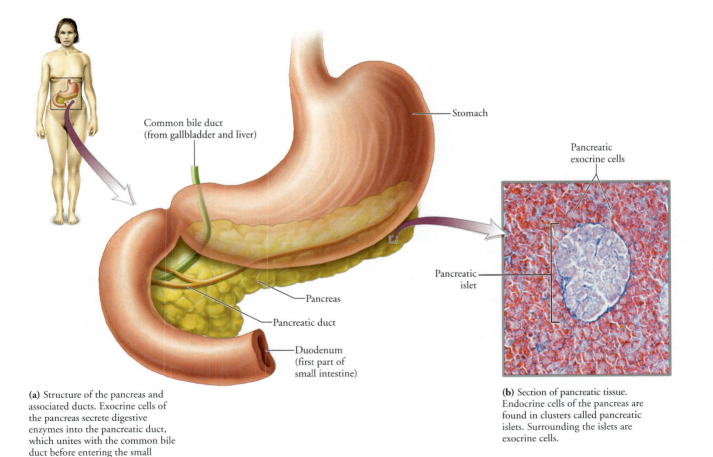

(a) Structure of the pancreas and associated ducts. Exocrine cells of the pancreas secrete digestive enzymes into the pancreatic duct, which unites with the common bile duct before entering the small intestine.

(b) Section of pancreatic tissue. Endocrine cells of the pancreas are found in clusters called pancreatic islets. Surrounding the islets are exocrine cells.

Common bile duct (from gallbladder and liver)

Stomach

Pancreatic exocrine cells

Pancreatic islet

Pancreas

Pancreatic duct

Duodenum (first part of small intestine)

FIGURE **10.19**

Location and structure of the pancreas

Hormones and Our Response to Stress

Stress can be broadly defined as mental or physical tension. Rarely does a day go by that we are not subjected to stress. Awaiting the start of an exam, some personal performance, or an interview can be stressful. Our bodies usually can deal with everyday stresses and maintain the relative constancy of our internal environment. Sometimes, however, stress is extreme in its intensity and duration, and our coping mechanisms prove inadequate. At such times, stress triggers the hypothalamus to initiate the general adaptation syndrome (GAS), a series of physiological adjustments made by our bodies in response to extreme stress.

The general adaptation syndrome consists of three phases: alarm, resistance, and exhaustion. The alarm phase is also known as the fight-or-flight response. Recall that the fight-or-flight response is initiated by epinephrine from the adrenal medulla. The response immediately funnels huge amounts of glucose and oxygen to the organs most critical in responding to danger.

Sometimes the adjustments of the alarm phase are sufficient to overcome stress. At other times, stress is so intense and long lasting that the individual enters the resistance phase. Changes wrought by the resistance phase are more long term than are those of the alarm phase. Also, rather than being stimulated by nerve impulses from the hypothalamus, the resistance phase is initiated by releasing hormones from the hypothalamus. The releasing hormones stimulate the anterior pituitary to secrete several hormones. Some of these hormones stimulate other glands to secrete their hormones. Glucocorticoids from the adrenal cortex are the main hormones of the resistance phase. Two primary effects of glucocorticoids are to mobilize the body's protein and fat reserves and to conserve glucose for use by cells of the nervous system. The hormones of the resistance phase also result in conservation of body fluids.

The resistance phase is sustained by the body's fat reserves. This phase may last for weeks or months, but it cannot go on indefinitely. Sooner or later, lipid reserves are exhausted and structural proteins must be broken down to meet energy demands. Eventually, organs are unable to meet the heavy demands of the resistance phase, and they begin to fail. This is the exhaustion phase. Without immediate attention, death may result from collapse of one or more organ systems. The phases of the general adaptation syndrome are summarized in Figure 10.A.

Stress can have dramatic effects on our health, especially when it is prolonged and uncontrollable. It is now known that stress depresses wound healing, increases our susceptibility to infections, and leads to disorders such as hypertension, irritable bowel syndrome, and asthma. Some studies have shown that stress puts people at greater risk for developing chronic diseases. Overall, prolonged stress appears to shorten the life span.

Given the connection between stress and health, it is important to reduce some of the stress in our lives. It also is important to develop ways to cope with unavoidable stress. A first step in reducing stress is to be realistic in assessing the levels of stress associated with life events, such as starting college or getting married. Because not everyone is affected in the same way by the same life event, it is important to consider what you find stressful and then take positive steps to reduce your exposure to stress.

There are several strategies for coping with stress. One strategy is to develop things in your life that make you feel good and optimistic and help you to function more effectively. Many people achieve such support from developing close relationships with family or friends and from a job or hobbies that they enjoy. People often are better able to cope with stress with such support systems in place. Coping skills also may be improved through counseling or self-help groups. Other commonly used means to alleviate the effects of stress include relaxation techniques and regular exercise. A more specialized option is biofeedback. Biofeedback is a procedure used to help a person recognize the symptoms of stress

---

glycogen to glucose. Finally, insulin prevents conversion of amino and fatty acids to glucose. As a result of these actions, insulin promotes protein synthesis, fat storage, and the use of glucose for energy. Rising blood levels of glucose, amino acids, or fatty acids trigger insulin release. Once cells begin to take up these substances, their levels in the blood fall, and insulin secretion is inhibited. Figure 10.20 summarizes the regulation of glucose in the blood by insulin and glucagon.

Insulin has dramatic effects on our health. More than 120 million people worldwide, over 15 million of them Americans, suffer from diabetes mellitus. Diabetes mellitus is a group of metabolic disorders characterized by an abnormally high level of glucose in the blood. There are several types of diabetes mellitus. **Type 1 diabetes mellitus** used to be known as insulin-dependent diabetes or Type I diabetes. It also was called juvenile-onset diabetes, because it usually develops in people younger than 25 years of age. Type 1 diabetes is an autoimmune disorder that represents about 5% to 10% of all diagnosed cases of diabetes. A person's own immune system attacks the cells of the pancreas responsible for insulin production. Symptoms include nausea, vomiting, thirst, and excessive urine production. Treatment involves daily, and sometimes multiple, injections of insulin. Exercise and careful monitoring of diet and blood glucose levels also are essential. Ingestion of insulin is not yet useful, because insulin is a protein hormone that can be broken down in the digestive tract. However, work is under way to package insulin in resistant microcapsules. There is also the possibility that insulin could be administered as a nasal spray. Other research has focused on whether insulin-producing cells can be transferred into the pancreas of a diabetic.

Type 2 diabetes mellitus was also known as non-insulin-dependent diabetes and Type II diabetes. It was formerly called adult-onset diabetes because it usually develops after age 40, although it recently has begun showing up in younger people. Type 2 diabetes accounts for between 90% and 95% of diabetes cases.

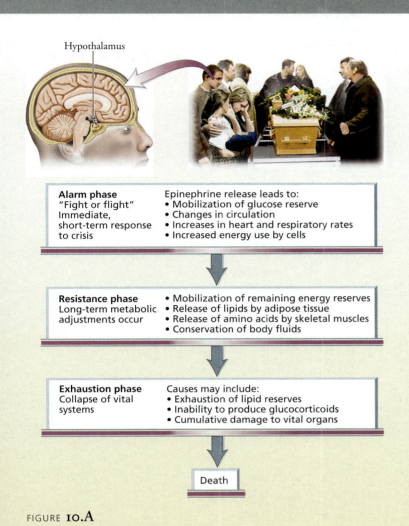

Hypothalamus

| Alarm phase
"Fight or flight"
Immediate,
short-term response
to crisis | Epinephrine release leads to:
• Mobilization of glucose reserve
• Changes in circulation
• Increases in heart and respiratory rates
• Increased energy use by cells |

| Resistance phase
Long-term metabolic
adjustments occur | • Mobilization of remaining energy reserves
• Release of lipids by adipose tissue
• Release of amino acids by skeletal muscles
• Conservation of body fluids |

| Exhaustion phase
Collapse of vital
systems | Causes may include:
• Exhaustion of lipid reserves
• Inability to produce glucocorticoids
• Cumulative damage to vital organs |

Death

FIGURE **10.A**
The general adaptation syndrome

and learn how to control them. During a biofeedback session, a health care professional connects a patient to a machine that monitors physiological indicators of stress, such as heart rate or muscle tension (Figure 10.B). The health care worker then discusses a stressful situation with the patient. The machine gives off signals when the patient begins to show signs of stress. Increased tension in muscles might prompt a clicking sound. The patient may be able to decrease muscle tension, through deep breathing and relaxation. This decrease can be monitored by decreases in the frequency of clicks. Eventually, patients are able to recognize and cope with signs of stress without the help of the machine. 👫

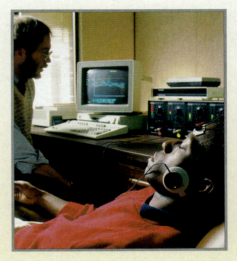

FIGURE **10.B**
Biofeedback, monitoring the body's physiological response to stress, is one way in which people can learn to recognize the symptoms of stress and how to cope with them.

This form is characterized by decreased sensitivity to insulin. The lower sensitivity may result from a decrease in insulin receptors on target cells. Many people with Type 2 diabetes are overweight. Treatment typically involves diet, exercise, and weight loss. Oral medications that increase insulin sensitivity or production also may be given. About 40% of people with Type 2 diabetes need insulin injections. This was one reason for dropping the name non-insulin-dependent diabetes.

There are two other forms of diabetes. When diabetes mellitus develops in women during gestation, it is called gestational diabetes. This condition occurs in 2% to 5% of all pregnancies. It usually begins in the second half of pregnancy and resolves after delivery of the baby. The placenta is the organ that supplies the growing fetus with nutrients and oxygen and carries away wastes and carbon dioxide (Chapter 18). The placenta also produces hormones, some of which may block the effects of insulin. Normally, such blocking effects are overcome by the pancreas's producing more insulin.

Sometimes, however, the production of insulin is insufficient, and gestational diabetes results. This form of diabetes typically resolves after delivery because the placenta is expelled a few minutes after the baby. Treatment plans for gestational diabetes typically include testing blood glucose levels, possibly taking insulin, eating a healthy diet, and engaging in regular physical activity. "Other specific types" of diabetes represent 1% to 2% of all diagnosed cases. These types include insulin deficiencies resulting from damage to the pancreas from disease, infection, or drugs.

Diabetes has serious complications. Diabetics are at increased risk for blindness, kidney disease, heart disease, high blood pressure, and atherosclerosis (buildup of fatty deposits in the arteries). Many diabetics suffer from gum disease and damage to their nervous system. The latter may include loss of sensation and impotence. Poor circulation and problems with nerves in the lower legs may make amputation of the lower limbs necessary. Indeed, more than half of lower limb amputations in the United States involve diabetics. Given these

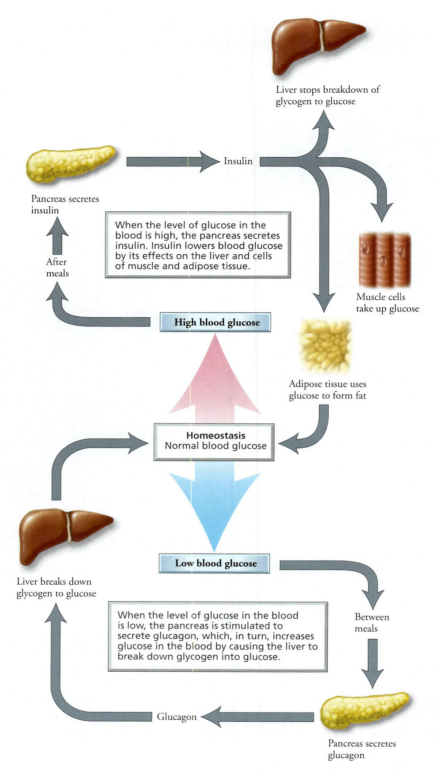

Liver stops breakdown of
glycogen to glucose

Pancreas secretes
insulin

Insulin

When the level of glucose in the
blood is high, the pancreas secretes
insulin. Insulin lowers blood glucose
by its effects on the liver and cells
of muscle and adipose tissue.

After
meals

Muscle cells
take up glucose

High blood glucose

Adipose tissue uses
glucose to form fat

Homeostasis
Normal blood glucose

Liver breaks down
glycogen to glucose

Low blood glucose

When the level of glucose in the blood
is low, the pancreas is stimulated to
secrete glucagon, which, in turn, increases
glucose in the blood by causing the liver to
break down glycogen into glucose.

Between
meals

Glucagon

Pancreas secretes
glucagon

FIGURE **IO.2O**
The regulation of glucose level in
the blood by insulin secreted by
the pancreas (top) and glucagon
secreted by the pancreas (bottom)

serious complications, it is important that diabetes be diagnosed and treated early. It also is essential to prevent diabetes when possible. Today, the fasting plasma glucose test is recommended for diagnosing Type 1 and Type 2 diabetes mellitus. The risk factors for diabetes include obesity, high blood pressure, parent or sibling with diabetes, and a history of gestational diabetes. People without symptoms and risk factors should be tested at age 45 and every 3 years thereafter. People without symptoms but with one or more risk factors should be tested before age 45 and more frequently than once every 3 years.

Too much insulin sometimes results from a tumor of the pancreas. More typically, however, a diabetic person mistakenly injects too much insulin. The result is depressed levels of glucose in the blood. Initial symptoms of low blood glucose include anxiety, sweating, hunger, weakness, and disorientation. Because brain cells fail to function properly when starved of glucose, the next results may be convulsions and unconsciousness. The progressive events associated with severe depletion of blood glucose are known as **insulin shock**. Insulin shock can prove fatal unless blood sugar levels are raised.

The hormones secreted by the pancreas are summarized in Table 10.1.

Hormones of the thymus gland promote maturation of white blood cells

The **thymus gland** lies just behind the breastbone, on top of the heart (see Figure 10.2). It is more prominent in infants and children than in adults because it decreases in size as we age. The thymus secretes hormones such as **thymopoietin** and **thymosin**. These hormones are involved in the maturation of white blood cells called T lymphocytes. T lymphocytes, also known as T cells, are involved in the body's defense mechanisms. The thymus and T lymphocytes play an important role in immunity and will be discussed in greater detail in Chapter 13.

The hormones secreted by the thymus gland are summarized in Table 10.1.

The pineal gland secretes melatonin

The **pineal gland** lies at the center of the brain (Figure 10.21). About the size of a pea, this tiny gland contains secretory cells that produce the hormone **melatonin**. Levels of circulating melatonin are greater at night than during daylight hours. This pattern can be explained by the observation that the pineal gland receives input from visual pathways. Neurons of the retina are stimulated by light entering the eye, and impulses are sent to the hypothalamus. These messages eventually reach the pineal gland, where they inhibit secretion of melatonin.

Research in the past few decades has suggested diverse roles for melatonin. Melatonin may, as it does in some nonhuman animals, inhibit production of the pigment melanin by melanocytes of the skin. Melatonin may influence daily rhythms. Sleep and, for some people, seasonal changes in depression appear to be influenced by melatonin. Melatonin may inhibit fertility hormones and slow the aging process. The links between melatonin and sleep, fertility, and aging have led to wild, over-the-counter purchasing of melatonin. This situation has been described as "melatonin mania." We examine the promise and possible risks of taking melatonin in the Health Issue essay, *Melatonin: Miracle Supplement or Potent Drug Misused by Millions?*

One disorder associated with too much melatonin is **seasonal affective disorder (SAD)**. This form of depression is associated with winter, when short day length results in overproduction of melatonin. (Recall that melatonin is secreted in the absence of light.) Too much melatonin causes symptoms such as lethargy, long periods of sleep, low spirits, and a craving for carbohydrates. The symptoms usually appear around October and end about April. Three quarters of persons who suffer from SAD are female. Treatment of SAD often involves repeated exposure to very bright light for about an hour each day. The intense light inhibits melatonin production.

The potential effects of melatonin are summarized in Table 10.1.

Prostaglandins Act Locally

Now that we have described the endocrine glands and their functions, let's consider a class of chemical messengers that do not travel to distant sites within the body. These messengers exert their effects locally. They act either on the secreting cells themselves or on nearby tissues. Prostaglandins are one example of these so-called local signaling molecules.

Prostaglandins are lipid molecules continually released by the plasma membranes of most cells. Different types of cells secrete different prostaglandins. At least 16 different prostaglandin molecules function within the human body. Prostaglandins are not considered true hormones because they produce their effects locally rather than at distant sites. Nevertheless, prostaglandins have remarkably diverse effects. Prostaglandins play roles in the prevention of blood clotting, regulation of body temperature, opening of airways to the lungs, and the body's inflammatory response. Prostaglandins also affect the reproductive system. Menstrual cramps are thought to be caused by prostaglandins released by cells of the uterine lining. These prostaglandins act on the smooth muscle of the uterus, causing muscle contractions and cramping. Drugs such as aspirin inhibit the synthesis of prostaglandins and thus may lessen the discomfort of menstrual cramps. Prostaglandins also are found in semen, the fluid discharged from the penis at ejaculation. Once in the female reproductive tract, prostaglandins in semen cause the smooth muscles of the uterus to contract. This contraction may help sperm continue their journey farther into the female reproductive tract.

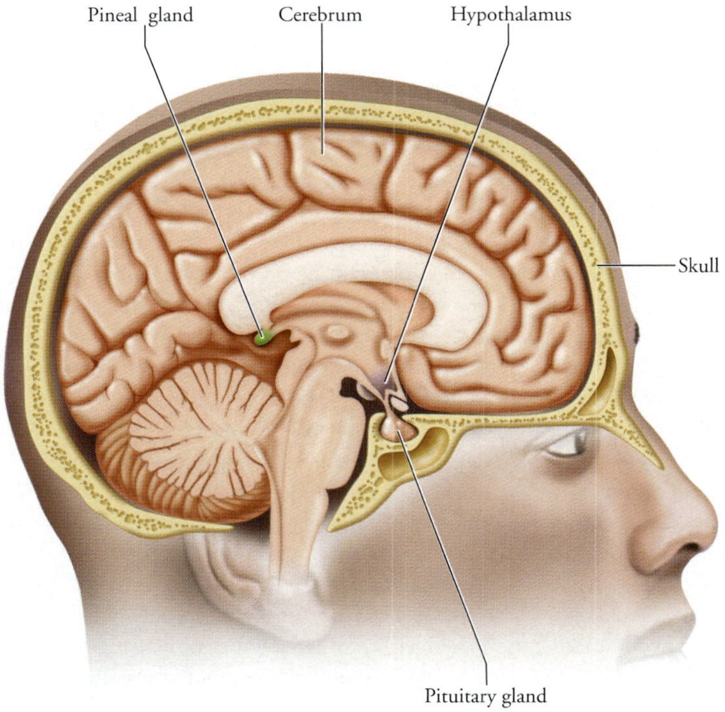

FIGURE **10.21**

The pineal gland. Located at the center of the brain, the pineal gland secretes the hormone melatonin.

HEALTH ISSUE

Melatonin: Miracle Supplement or Potent Drug Misused by Millions?

Melatonin, a hormone secreted by the pineal gland, has long been billed as a wonder drug. There are claims that melatonin can cure insomnia, prevent pregnancy, boost immunity, and slow the aging process. All that—and melatonin is apparently nontoxic, even at extremely high doses. No wonder health food stores are struggling to keep their shelves stocked. What are the bases for these claims regarding the miracles of melatonin? Should this potent hormone be sold as an over-the-counter dietary supplement to self-prescribing consumers?

For people who toss and turn and watch the clock on the nightstand hoping for a few hours of shut-eye, melatonin can bring sleep. Research, beginning in the 1980s and continuing today, has shown that remarkably low doses of melatonin can induce sleep. Although most agree that melatonin is an effective sleep aid, scientists debate the precise mechanism by which melatonin produces its soporific effect. Some believe melatonin induces sleep by acting on the body's biological clock. Others speculate that the mechanism is independent of the biological clock. Some complaints about melatonin also have surfaced; it seems that nightmares, disrupted sleep, and next-day grogginess occurred in some instances when high doses of melatonin were taken.

Melatonin's prospects for contraceptive use in humans stem from its observed role in regulating fertility in other mammals. In the Syrian hamster, for example, melatonin functions in birth control. Syrian hamsters are seasonal breeders, being infertile during winter and sexually active in the spring (Figure 10.C). This cycle appears to be regulated by the pineal gland through seasonal changes in photoperiod (day–night cycle). Beginning in late September, when daylight hours start to wane, the reduced light reaching the pineal gland stimulates greater production of melatonin. Increased melatonin, in turn, causes ovarian

function to go on hold in females. In male hamsters, high levels of melatonin cause a decrease in both testes size and testosterone output. Winter is a sexually quiet time for Syrian hamsters. It is not until early spring that their reproductive systems spontaneously awaken and breeding begins.

A point in favor of melatonin's possible use in birth control in humans is its apparent lack of toxicity. Some hormones currently being used for contraception appear, over the long term, to have carcinogenic effects. Synthetic forms of estrogen and progesterone, for example, are used in the "combination" birth control pill. Recall that estrogen stimulates development of breast tissue during each reproductive cycle. Recent data suggest a link between risk of breast cancer and a woman's cumulative lifetime exposure to estrogen. These data, together with melatonin's documented effects on reducing fertility in other animals, have sparked research devoted to replacing the estrogen in combination birth control pills with melatonin. The contraceptive success rate of melatonin-progesterone pills appears similar to that of estrogen-progesterone pills.

Melatonin also seems to boost immunity and to slow aging. Recall that the thymus gland, responsible for the production of infection-fighting white blood cells called T lymphocytes, is largest in infants and young children and shrinks as we age. Old mice also show shrinkage of the thymus. Such shrinkage can be reversed in geriatric mice by injections of melatonin. Indeed, the immune systems of injected animals showed signs of revitalization. Similarly dramatic results were obtained in aging research when the pineal glands of 10 old mice were switched with the pineal glands of 10 young mice. The young mice with the "old" pineals aged rapidly and died in what would normally be middle age. The old mice with the "young" pineals lived about 30% longer than untreated mice. These results led

researchers to describe the pineal as the "aging clock," speculating that melatonin is the pineal's way of translating its timekeeping into changes in the body. Such findings also raised the possibility that melatonin supplements might compensate for aging pineal glands (the amount of melatonin secreted by the pineal declines with age).

Melatonin also may slow aging through its effective scavenging of free radicals, molecular fragments that contain an unpaired electron. Free radicals have been implicated in many chronic diseases of old age, such as cancer and heart disease. Substances, such as melatonin, that destroy free radicals are called antioxidants. Free radicals are normally generated by some cells of the body; their role is to destroy bacteria and old cells. Environmental agents such as drugs, toxins, and radiation also generate free radicals. What evidence is there to support melatonin as an effective antioxidant? To begin with, white blood cells incubated with melatonin sustained 70% less damage from radiation than did untreated cells. Also, in rats, injections of melatonin before and after exposure to a toxic herbicide protected these animals from the liver and lung damage found in similarly exposed animals that did not receive melatonin treatment.

Melatonin certainly seems to be the answer for many of life's ills. But should consumers be gobbling down melatonin pills and lozenges purchased at health food stores? At present, melatonin is being sold as a dietary supplement and thus is not subject to intense scrutiny by the Food and Drug Administration (FDA). No one knows the possible long-term effects of high doses of melatonin. Another concern is that production of the hormone by companies not subjected to regulation raises questions about the strength and purity of the final marketed product. And questions remain about the benefits of melatonin. Health claims of the miracles of melatonin have been piling up at such a fast rate that the FDA has barely had time to evaluate whether such claims are real. As a result, some people in the medical profession caution all consumers to wait until melatonin is subjected to further scrutiny and tighter regulation. Others argue, however, that melatonin is probably safe for most people but that certain groups should avoid it until further information is available. Specifically, avoidance of melatonin is urged for pregnant or nursing women (the effects of high doses on fetuses or infants are unknown), children (who normally produce melatonin in large amounts), and women trying to conceive (given melatonin's apparent effectiveness as a contraceptive agent). Avoidance of melatonin is also indicated for people with autoimmune disorders or cancers of the immune system (melatonin stimulates the immune system and thus could worsen these conditions). Finally, people taking other medications should consult their physician before taking melatonin. Such consultation should help people avoid dangerous drug interactions.

(a) A Syrian hamster

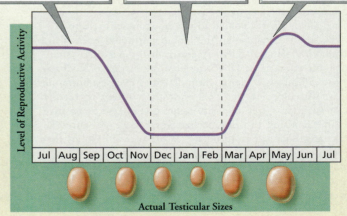

In September, the level of reproductive activity begins to decline. This decline is due to increased production of melatonin by the pineal gland in response to shortening periods of daylight.

The reproductive systems of female and male hamsters shut down for the winter (note, for example, the decrease in testis size shown at the bottom of the figure). This shutdown occurs in response to high levels of melatonin.

In early spring, the reproductive systems spontaneously activate, and breeding begins.

(b) The reproductive cycle of the Syrian hamster

FIGURE **10.C**
Melatonin influences reproduction in hamsters.

REVIEWING THE CONCEPTS

The Endocrine System Communicates via Chemical Messages (pp. 188–192)

1. Exocrine glands secrete their products into ducts that open onto the body surface, into spaces within organs, or into a body cavity. In contrast, endocrine glands lack ducts and release their products (hormones) into the spaces just outside cells. The hormones then diffuse into the bloodstream. Endocrine glands and organs that contain some endocrine tissue constitute the endocrine system.

2. Hormones, the chemical messengers of the endocrine system, contact virtually all cells within the body. However, hormones affect only target cells, those cells with receptors that recognize and bind specific hormones.

3. Steroid hormones are lipid soluble and derived from cholesterol. They are secreted primarily by the ovaries, testes, and adrenal glands. Steroids move easily through the plasma membrane of target cells into the cytoplasm. Once in the cytoplasm, they combine with a receptor molecule, forming a hormone-receptor complex, which moves into the nucleus of the cell, where it directs synthesis of specific proteins, including enzymes that stimulate or inhibit particular metabolic pathways.

4. Water-soluble hormones include amino acid derivatives, peptides, and proteins. These hormones cannot pass through the lipid bilayer of the plasma membrane. Thus, they exert their effects indirectly by activating second messenger systems. The hormone, considered the first messenger, binds to a receptor on the plasma membrane. The binding activates a molecule in the cytoplasm, considered the second messenger, that carries the hormone's message inside the cell, where it changes the activity of enzymes and chemical reactions.

5. Endocrine glands are stimulated to manufacture and release hormones by chemical changes in the blood, hormones released by other endocrine glands, and messages from the nervous system. Hormone secretion is usually regulated by negative feedback mechanisms. Typically, a gland is stimulated to release a hormone and then rising blood levels of that hormone inhibit its further release. Sometimes, secretion of hormones is regulated by positive feedback mechanisms. In these mechanisms, release of a hormone causes a change that acts to stimulate (rather than to inhibit) further production and release of that hormone.

6. Under certain conditions, the nervous system may override the controls of the endocrine system. During severe stress, the nervous system permits levels of glucose in the blood to rise much higher than normal. Cells can then use the extra energy to cope with the stress.

WEB TUTORIAL 10.1 Modes of Action of Hormones

WEB TUTORIAL 10.2 Hormonal Feedback Loops

Hormones Influence Growth, Development, Metabolism, and Behavior (pp. 192–207)

7. The pituitary gland is suspended from the hypothalamus by a short stalk. The gland has two parts, an anterior lobe and a posterior lobe. The anterior lobe is influenced by the hypothalamus through a circulatory connection known as the pituitary portal system. Nerve cells in the hypothalamus release hormones into the portal system, and these hormones travel to the anterior lobe, where they stimulate or inhibit release of hormones. The anterior pituitary releases six hormones (growth hormone, GH; prolactin, PRL; thyroid-stimulating hormone, TSH; adrenocorticotropic hormone, ACTH; follicle-stimulating hormone, FSH; and luteinizing hormone, LH). Four of the six hormones are tropic hormones that influence other endocrine glands (TSH, ACTH, FSH, LH). In contrast to the circulatory connection between the hypothalamus and the anterior lobe of the pituitary gland, the connection between the hypothalamus and the posterior lobe is neural. Neurosecretory cells from the hypothalamus extend down into the posterior lobe, where they store and release oxytocin (OT) and antidiuretic hormone (ADH).

8. The thyroid gland lies at the front of the neck. This gland produces thyroid hormone (TH) and calcitonin (CT). Thyroid hormone has broad effects, including regulating metabolic rate, heat production, and blood pressure. Calcitonin maintains low levels of calcium in the bloodstream. It does so by promoting the absorption of calcium by bone and by inhibiting the breakdown of bone.

9. The parathyroid glands are four small masses of tissue at the back of the thyroid gland. The parathyroids secrete parathyroid hormone (PTH, or parathormone), an antagonist to CT. As such, PTH is responsible for raising blood levels of calcium by stimulating the movement of calcium from bone and urine to the blood.

10. Each of two adrenal glands sits perched on top of a kidney and has two regions. The outer region is the adrenal cortex, and the inner region is the adrenal medulla. The adrenal cortex secretes several hormones that fall into three groups, the gonadocorticoids, mineralocorticoids, and glucocorticoids. The adrenal medulla, on the other hand, produces epinephrine (adrenaline) and norepinephrine (noradrenaline). These two hormones initiate the fight-or-flight response.

11. The pancreas secretes the hormones glucagon (increases glucose in the blood) and insulin (decreases glucose in the blood). Type 1 diabetes mellitus is an autoimmune disorder in which a person's own immune system attacks the insulin-producing cells of the pancreas, causing insulin deficiency. This disorder usually develops in people younger than 25 years old and is treated by daily insulin injections. Type 2 diabetes mellitus is characterized by a decreased sensitivity to insulin that usually develops after about 40 years of age. Type 2 diabetes is much more common than Type 1, and is usually treated through diet, exercise, and weight loss. Sometimes, insulin injections are also needed to treat Type 2 diabetes.

12. The thymus gland lies on top of the heart and plays an important role in immunity. This gland secretes hormones that influence the maturation of certain white blood cells called T lymphocytes.

13. The pineal gland lies at the center of the brain and secretes the hormone melatonin. Melatonin appears to be responsible for establishing biological rhythms, triggering sleep, inhibiting fertility, and slowing aging.

WEB TUTORIAL 10.3 Hypothalamic Control of the Pituitary

Prostaglandins Act Locally (p. 207)

14. Prostaglandins are lipid molecules released by the plasma membranes of many cells. These molecules exert their effects locally by affecting nearby cells or tissues. Prostaglandins are known to influence physiological processes, including regulation of body temperature, blood clotting, and contraction of smooth muscle.

KEY TERMS

acromegaly *p. 194*
Addison's disease   *p. 201*
adrenal cortex   *p. 200*
adrenal glands   *p. 199*
adrenal medulla   *p. 200*
adrenocorticotropic hormone
    (ACTH) *p. 195*
aldosterone   *p. 201*
androgen   *p. 200*
antidiuretic hormone (ADH)
    p. 195
calcitonin (CT)   *p. 199*
cretinism   *p. 197*
Cushing's syndrome   *p. 201*
diabetes insipidus   *p. 195*
diabetes mellitus   *p. 195*
endocrine gland   *p. 188*
epinephrine   *p. 202*
estrogen   *p. 200*
exocrine gland   *p. 188*

exophthalmos   *p. 198*
fight-or-flight response   *p. 202*
follicle-stimulating hormone (FSH)
    p. 195
giantism   *p. 193*
glucagon   *p. 202*
glucocorticoids   *p. 201*
gonadocorticoids   *p. 200*
Graves' disease   *p. 197*
growth hormone (GH)   *p. 193*
hormone   *p. 189*
hypothalamus   *p. 192*
inhibiting hormone   *p. 192*
insulin   *p. 202*
insulin shock   *p. 206*
luteinizing hormone (LH)   *p. 195*
melatonin   *p. 207*
mineralocorticoids   *p. 201*
myxedema   *p. 197*

negative feedback mechanism
    p. 191
norepinephrine   *p. 202*
oxytocin (OT)   *p. 195*
pancreas   *p. 202*
pancreatic islets   *p. 202*
parathormone   *p. 199*
parathyroid glands   *p. 199*
parathyroid hormone (PTH)
    p. 199
pineal gland   *p. 207*
pituitary dwarfism   *p. 194*
pituitary gland   *p. 192*
pituitary portal system   *p. 192*
portal system   *p. 192*
positive feedback mechanism
    p. 191
prolactin (PRL)   *p. 195*
prostaglandin   *p. 207*

receptor   *p. 189*
releasing hormone   *p. 192*
seasonal affective disorder (SAD)
    p. 207
second messenger   *p. 191*
simple goiter   *p. 198*
steroid hormone   *p. 189*
target cell   *p. 189*
thymopoietin   *p. 207*
thymosin   *p. 207*
thymus gland   *p. 207*
thyroid gland   *p. 196*
thyroid hormone (TH)   *p. 197*
thyroid-stimulating hormone (TSH)
    p. 195
tropic hormone   *p. 195*
Type 1 diabetes mellitus   *p. 204*
Type 2 diabetes mellitus   *p. 204*
vasopressin   *p. 195*

THINKING ABOUT THE CONCEPTS

1. How do endocrine glands differ from exocrine glands? Give examples of each. *pp. 188-189*
2. Given that hormones contact virtually all cells in the body, why are only certain cells affected by a particular hormone? *p. 189*
3. How do lipid-soluble and water-soluble hormones differ in their mechanisms of action? *pp. 189-191*
4. Compare negative and positive feedback mechanisms with regard to regulation of hormone secretion. Provide an example of each. *pp. 191-192*
5. How do the anterior and posterior lobes of the pituitary gland differ in size and relationship with the hypothalamus? *pp. 192-193*
6. List the hormones secreted by the anterior lobe of the pituitary and their functions. *pp. 192-195*
7. List the hormones released by the posterior lobe of the pituitary and their functions. *pp. 195-196*
8. What are the effects of thyroid hormone? *pp. 196-199*
9. Describe the feedback system by which calcitonin and parathyroid hormone regulate levels of calcium in the blood. *pp. 199-200*
10. What are the major functions of the glucocorticoids, mineralocorticoids, and gonadocorticoids secreted by the adrenal cortex? *pp. 200-202*
11. What is the fight-or-flight response? Which hormones are critical in initiating this response? *p. 202*
12. What hormones are secreted by the pancreas? What are their functions? *pp. 202-204*
13. Explain the differences between Type 1 and Type 2 diabetes mellitus. *pp. 204-206*
14. What is the basic function of hormones secreted by the thymus gland? *p. 207*
15. What roles might melatonin play in the body? *p. 207*
16. What are prostaglandins? How do prostaglandins differ from true hormones? *p. 207*
17. Which of the following does *not* characterize the anterior lobe of the pituitary gland?
    a. releases oxytocin and antidiuretic hormone
    b. circulatory connection to the hypothalamus
    c. larger of the two lobes
    d. secretes growth hormone and prolactin
18. A diet deficient in iodine may produce
    a. cretinism.
    b. Graves' disease.
    c. Cushing's syndrome.
    d. goiter.
19. Which of the following does *not* characterize the adrenal medulla?
    a. inner region of the adrenal gland
    b. secretes epinephrine and norepinephrine
    c. secretes glucocorticoids
    d. secretes hormones involved in fight-or-flight response
20. Type 1 diabetes
    a. is more common than Type 2 diabetes.
    b. usually develops after age 40.
    c. is an autoimmune disorder.
    d. is characterized by decreased sensitivity to insulin due to decreased insulin receptors on target cells.
21. Overproduction of melatonin by the pineal gland may cause
    a. seasonal affective disorder.
    b. diabetes insipidus.
    c. acromegaly.
    d. Addison's disease.
22. _____ hormones combine with receptor molecules in the cytoplasm of target cells, whereas _____ hormones bind to receptors on the surface of target cells and activate second messengers.
23. Oversecretion of growth hormone in childhood causes _____. Oversecretion in adulthood causes _____.
24. The hormone _____ lowers blood levels of calcium, whereas the hormone _____ increases blood levels of calcium.
25. In males, androgens are produced by the testes and the _____.
26. The hormone _____ lowers glucose in the blood, whereas the hormone _____ increases glucose in the blood.

APPLYING THE CONCEPTS

1. Mary has an itchy rash on the surface of her skin, and Rick has cut his finger on glass. Would either person benefit from applying a steroid cream containing cortisone? Why? Why not?

2. Matt is a thin 20-year-old. He is weak, disoriented, and sweating profusely. His friend brings him to the emergency room of the local hospital and explains that Matt is diabetic. Which type of diabetes does Matt likely have? What might explain his current condition? What might be done to help him?

3. It is winter in Massachusetts and Theresa has felt "down" and lethargic since the fall. She has trouble getting out of bed in the morning, and, once up, she craves carbohydrates. What might explain Theresa's symptoms? What might alleviate them, and why?

Blood

Blood Functions in Transportation, Protection, and Regulation

Blood Consists of Plasma and Formed Elements
- Plasma is the liquid portion of blood
- Stem cells give rise to the formed elements
- Platelets are cell fragments essential to blood clotting
- White blood cells help defend the body against disease
- Red blood cells transport oxygen
- Disorders of red and white blood cells have different effects depending on the type of cell affected

Blood Types Are Determined by Antigens on the Surface of Red Blood Cells

Blood Clotting Occurs in a Regulated Sequence of Events

 SOCIAL ISSUE   Blood Stem Cells

 HEALTH ISSUE   Transfusions and Artificial Blood

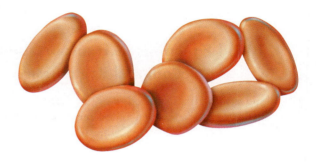

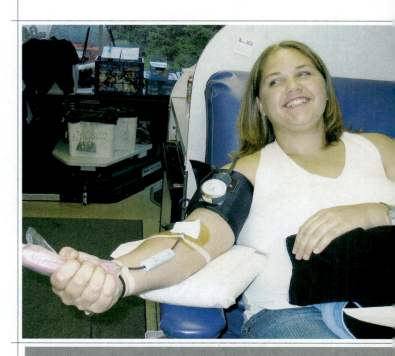

It is 4:00 A.M. The sound of a fired gun causes neighbors to awake with alarm. A few brave souls venture forth from their houses; another calls 911. They see a young man lying next to the sidewalk in a pool of blood. He has a wound in his chest and is moaning softly. They're relieved. He's alive. Ten minutes later, an ambulance arrives, and two rescue workers begin work to stop the bleeding. They recognize that blood is a critical life-sustaining fluid, and this young man will die if the uncontrolled bleeding continues for even minutes longer. What is so special about blood? Why does this young man's life depend on the emergency team's ability to stop the bleeding?

We'll answer some of these questions in this chapter as we consider the functions and composition of blood. In addition, we will see how blood type is determined and why it is so important. Finally, we will consider how blood clots to prevent blood loss from a wound. ■

carries away the wastes that cells produce. But blood does more than passively move its precious cargoes around the body. Its white blood cells help protect us against disease-causing organisms, and its clotting mechanisms help protect us from blood loss when a vessel is damaged. In addition, buffers in the blood help regulate the acid-base balance of body fluids. Blood also helps regulate body temperature by absorbing heat produced in metabolically active regions and distributing it to cooler regions and to the skin, where the heat is dissipated. We see, then, that the diverse functions of blood can be grouped into three categories: transportation, protection, and regulation.

Blood Functions in Transportation, Protection, and Regulation

Blood is sometimes referred to as the river of life. The title is apt because, like a river, blood serves as the body's transportation system. It carries vital materials to the cells and

Blood Consists of Plasma and Formed Elements

Blood *is* thicker than water. Blood is not only thicker; it is also denser. An important reason for the different properties of blood and water is that blood contains cells suspended in its watery fluid. In fact, a single drop of blood contains more than 250 million blood cells. You may recall from Chapter 4 that blood is classified as a connective tissue because it contains cellular elements suspended in a matrix. The liquid matrix is called plasma, and the cellular elements are collectively called the formed elements, as shown in Figure 11.1.

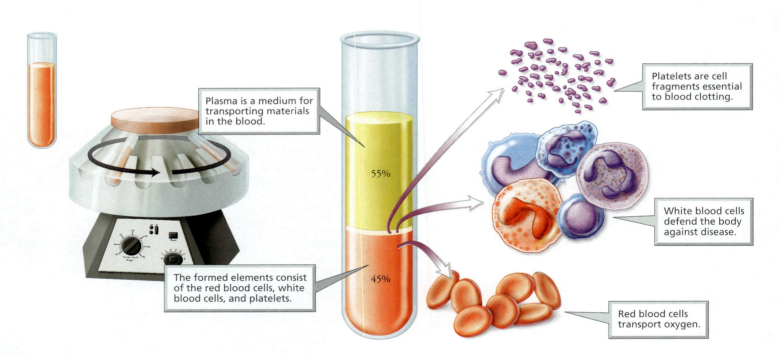

Plasma is a medium for transporting materials in the blood.

The formed elements consist of the red blood cells, white blood cells, and platelets.

55%

45%

Platelets are cell fragments essential to blood clotting.

White blood cells defend the body against disease.

Red blood cells transport oxygen.

FIGURE **11.1**

Whole blood consists of a straw-colored liquid, called plasma, in which cellular elements, called formed elements, are suspended. Blood can be separated into its major components when it is placed in a test tube with a substance that prevents coagulation and then spun in a centrifuge. The uppermost layer consists of plasma. The formed elements are found in two layers below the plasma. Just below the plasma is a thin layer consisting of platelets and white blood cells (leukocytes). The red blood cells (erythrocytes) are packed at the bottom of the test tube.

Plasma is the liquid portion of blood

Plasma is a straw-colored liquid that makes up about 55% of blood. It serves as the medium for transporting materials within the blood. Plasma consists primarily of water, but 7% to 8% of it consists of dissolved substances. Almost every substance that is transported within the blood is dissolved in the plasma. Thus, the plasma contains nutrients (such as simple sugars, amino acids, lipids, and vitamins), ions (such as sodium, potassium, and chloride), dissolved gases (including carbon dioxide, nitrogen, and a small amount of oxygen), and every hormone. In addition to transporting materials to the cells, the plasma carries away cellular wastes. For example, urea from protein breakdown and uric acid from nucleic acid breakdown are carried to the kidneys, where they can be removed from the body.

Most of the dissolved substances (solutes) in the blood are *plasma proteins*. The plasma proteins help balance water flow between the blood and the cells. You may recall from Chapter 3 that water moves by osmosis across biological membranes from an area of lesser solute concentration to an area of greater solute concentration. Without the plasma proteins, water would be drawn out of the blood by the proteins in cells. As a result, fluid would accumulate in the tissues, causing swelling.

Most of the 50 or so types of plasma proteins fall into one of three general categories: albumins, globulins, and clotting proteins. The albumins make up more than half of the plasma proteins. They are most important in the blood's water-balancing ability. The globulins are a group of proteins with a variety of functions. Some globulins transport lipids, including fats and cholesterol, as well as fat-soluble vitamins. Other globulins are antibodies, which provide protection against many diseases. The third category of plasma protein includes clotting proteins, such as fibrinogen.

Stem cells give rise to the formed elements

The plasma is the vehicle for transporting many substances around the body. However, the **formed elements**—platelets, white blood cells, and red blood cells—perform some of the key functions of the blood. The descriptions and functions of the formed elements of the blood are summarized in Table 11.1.

Red bone marrow, a framework of connective tissue that fills the cavities within many bones, is the birthplace and nursery for the formed elements. This framework supports fat cells and undifferentiated cells, called **stem cells**, that divide and give rise to all formed elements (Figure 11.2). In infants and children, red marrow fills most bone cavities. With age, however, the marrow of most long bones fills with fat. In this way, red marrow becomes yellow marrow and serves as a reserve of energy that can be used if body fat becomes depleted. In adults, then, red bone marrow is located primarily in the flat bones of the skull and pelvis, the sternum (breastbone), the ribs, and the heads of the humerus (bone of the upper arm) and of the femur (thighbone). However, after a severe and sustained blood loss, yellow marrow can be converted back to red marrow, increasing the rate of blood cell production.

Platelets are cell fragments essential to blood clotting

Platelets, sometimes called thrombocytes (*thromb-o*, clot; *cyte*, cell) are essential to blood clotting. Platelets are actually fragments of larger precursor cells called megakaryocytes. Platelets are formed in the red bone marrow when these precursor cells break apart. The fragments are released into the blood at the astounding rate of about 200 billion a day. Platelets mature during the course of a week and then circulate in the blood for about 10 to 12 days before dying. Platelets contain several substances important in stopping the loss of blood through damaged blood vessels. This vital function of platelets will be considered later in this chapter.

White blood cells help defend the body against disease

White blood cells (WBCs), or **leukocytes** (*leuk-*, white; *-cyte*, cell), perform certain mundane housekeeping duties—such as removing wastes, toxins, or damaged, abnormal cells—and serve as warriors in the body's fight against disease (see Chapter 13). Although leukocytes represent less than 1% of whole blood, we simply could not live without them. We would succumb to the microbes that surround us. Because the number of WBCs increases when the body responds to microbes, white blood cell counts are often used as an index of infection.

White blood cells are also produced in the red bone marrow[1]. They are nucleated cells that circulate in the bloodstream. However, white blood cells are not confined to the bloodstream. By squeezing between the cells that form the walls of blood vessels, white blood cells can leave the circulatory system and move to a site of infection, tissue damage, or inflammation (Figure 11.3). Having slipped out of the capillary into the fluid bathing the cells, white blood cells roam through the tissue spaces. White blood cells gather in areas of tissue damage or infection because they are attracted by chemicals released by invading microbes or by damaged cells. Certain types of white blood cells may then engulf the "offender" in a process called **phagocytosis** (*phago-*, to eat; *-cyt-*, cell; *-sis*, process of) (see Chapter 3). Here we will only briefly consider the white blood cells; we will discuss them and their many tactics for defending our bodies in more detail in Chapter 13.

TYPES OF WHITE BLOOD CELLS

The five types of white blood cells can be classified into one of two groups—the granulocytes and the agranulocytes—based on cytoplasmic differences. **Granulocytes** have granules in their cytoplasm. The granules are sacs containing chemicals that are used as weapons to destroy invading pathogens, especially bacteria. The **agranulocytes** lack cytoplasmic granules or have very small granules.

GRANULOCYTES: NEUTROPHILS, EOSINOPHILS, AND BASOPHILS
Depending on the color of the granules after they have been stained

[1]One type, the lymphocytes, also may be produced in the lymphoid tissues, such as lymph nodes.

TABLE **11.1**

THE FORMED ELEMENTS OF BLOOD

TYPE OF FORMED ELEMENT	DESCRIPTION	NO. OF CELLS/MM^3	LIFE SPAN	CELL FUNCTION
Platelets 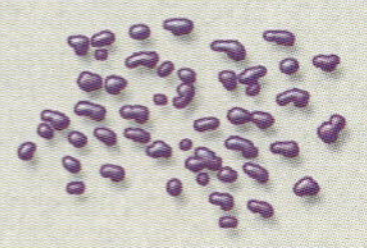	Fragments of a megakaryocyte; small, purple-stained granules in cytoplasm	250,000–500,000	5–10 days	Play role in blood clotting

Leukocytes (white blood cells; WBCs)

Granulocytes

Neutrophils 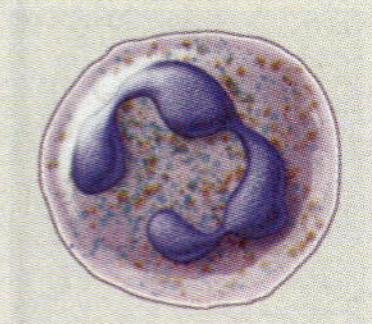	Multilobed nucleus, clear-staining cytoplasm, inconspicuous granules	3000–7000	6–72 hours	Consume bacteria by phagocytosis
Eosinophils 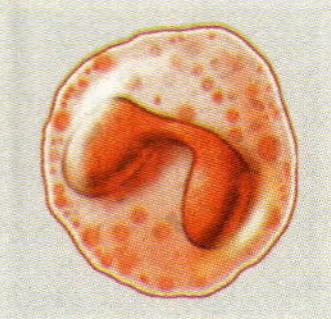	Large granules in cytoplasm, pink-staining cytoplasm, bilobed nucleus	100–400	8–12 days	Consume antibody-antigen complex by phagocytosis; attack parasitic worms
Basophils 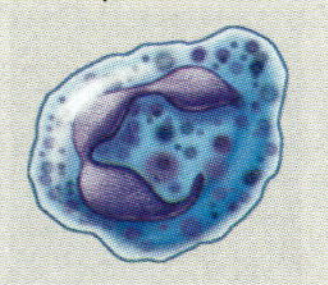	Large, purple cytoplasmic granules; bilobed nucleus	20–50	3–72 hours	Release histamine

Agranulocytes

Monocytes 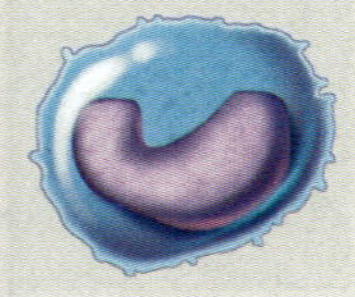	Gray-blue cytoplasm with no granules; U-shaped nucleus	100–700	Several months	Give rise to macrophages, which consume bacteria, dead cells, and cell parts by phagocytosis
Lymphocytes 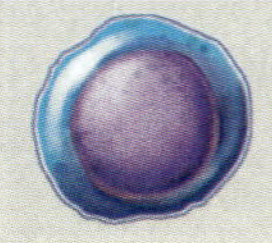	Round nucleus that almost fills the cell	1500–3000	Many years	Attack damaged or diseased cells or produce antibodies
Erythrocytes (red blood cell; RBCs) 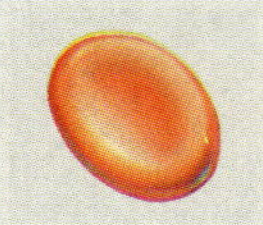	Biconcave disk, no nucleus	4–6 million	About 120 days	Transport oxygen and carbon dioxide

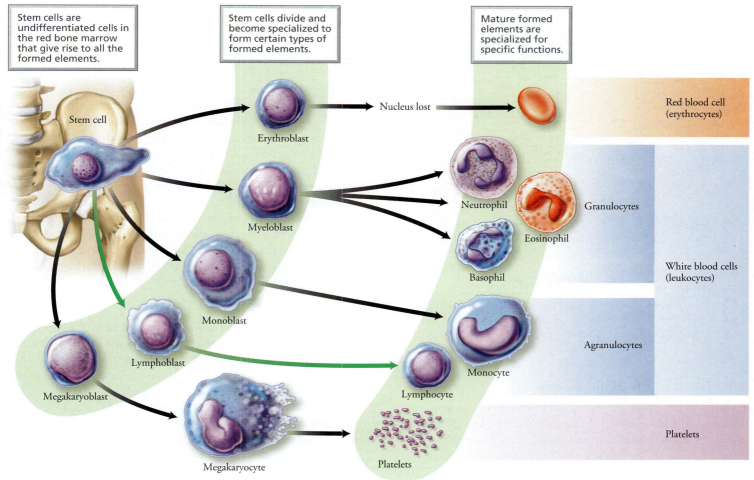

Stem cells are undifferentiated cells in the red bone marrow that give rise to all the formed elements.

Stem cells divide and become specialized to form certain types of formed elements.

Mature formed elements are specialized for specific functions.

Stem cell

Erythroblast

Nucleus lost

Red blood cell (erythrocytes)

Myeloblast

Neutrophil

Eosinophil

Granulocytes

Basophil

Monoblast

White blood cells (leukocytes)

Lymphoblast

Monocyte

Agranulocytes

Megakaryoblast

Lymphocyte

Megakaryocyte

Platelets

Platelets

FIGURE **11.2**

All formed elements originate in the red bone marrow from an undifferentiated cell called a stem cell. Stem cells divide and give rise to two lines of cells (shown in black and green) which, through additional cell division and differentiation, give rise to all types of blood cells.

for microscopic study, the granulocytes are classified as neutrophils, eosinophils, or basophils.

- **Neutrophils**, the most abundant of all white blood cells, are the blood cell soldiers on the front lines. Arriving at the site of infection before the other types of white blood cells, neutrophils immediately begin to engulf microbes by phagocytosis, thus curbing the spread of the infection. After engulfing a dozen or so bacteria, a neutrophil dies. But, even in death, it helps the body's defense by releasing chemicals that attract more neutrophils to the scene. Dead neutrophils, along with bacteria and cellular debris, make up pus, the yellowish liquid we usually associate with infection.

- **Eosinophils** are important in the body's defense against parasitic worms, such as tapeworms and hookworms. They also lessen the severity of allergies by phagocytizing antibody-antigen complexes and inactivating inflammatory chemicals.

- **Basophils** release histamine, a chemical that attracts other white blood cells to the site of infection and causes blood vessels to dilate (widen), thereby increasing blood flow to the affected area. They also play a role in some allergic reactions.

AGRANULOCYTES: MONOCYTES AND LYMPHOCYTES The agranulocytes lack visible granules in the cytoplasm. Agranulocytes are classified as monocytes or lymphocytes.

- **Monocytes**, the largest of all formed elements, leave the bloodstream and enter various tissues, where they develop into macrophages. Macrophages are phagocytic cells that engulf invading microbes, dead cells, and cellular debris.

- **Lymphocytes** are classified into two types: B lymphocytes and T lymphocytes. The *B lymphocytes* give rise to plasma cells, which, in turn, produce antibodies. Antibodies are proteins that recognize specific molecules (antigens) on the surface of microbes that have invaded the body. Antibodies bind with antigens and help prevent the microbe or other foreign cell from harming the body. There are several types of *T lymphocytes*, specialized white blood cells that play roles in the body's defense mechanisms. Some T lymphocytes kill cells not recognized as coming from the body, including bacteria and body cells that are infected with a virus or are cancerous. Other *T lymphocytes* help activate the immune system. Still others suppress the immune system. We will discuss lymphocytes further in Chapter 13.

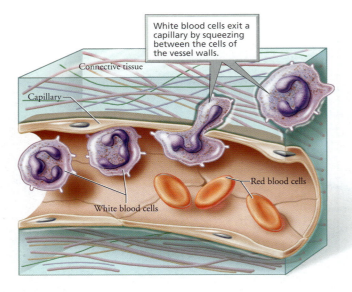

White blood cells exit a capillary by squeezing between the cells of the vessel walls.

Connective tissue

Capillary

Red blood cells

White blood cells

FIGURE **11.3**
White blood cells can squeeze between the cells that form the wall of a capillary. They then enter the fluid surrounding body cells and, attracted by chemicals released by microbes or damaged cells, gather at the site of infection or injury.

▍Red blood cells transport oxygen

Red blood cells (RBCs), also called **erythrocytes**, pick up oxygen in the lungs and ferry it to all the cells of the body. Red blood cells also carry about 23% of the total carbon dioxide, a metabolic waste product. They are by far the most numerous cells in the blood. Indeed, they number 4–6 million/mm^3 of blood and constitute approximately 45% of the total blood volume.

The shape of red blood cells, as shown in Figure 11.4, is marvelously suited to their function of picking up and transporting oxygen. Red blood cells are quite small, and each is shaped like a biconcave disk. That is, it is a flattened cell indented on each side. The biconcave shape maximizes the surface area of the cell. Because of the greater surface area, oxygen can enter the red blood cell more rapidly than if the disk were flat. Speed is important because a red blood cell spends only a second or two in the capillaries of the lungs (the smallest blood vessels, those in which gas exchange occurs). A red blood cell is also unusually flexible, allowing it to squeeze through capillaries with a diameter much smaller than red blood cells. Each red blood cell is packed with **hemoglobin**, the oxygen-binding pigment. As a red blood cell matures in the red bone marrow, it loses its nucleus and most organelles. Thus, it is scarcely more than a sac of hemoglobin molecules. With roughly a third of its weight contributed by hemoglobin, each red blood cell can transport about a billion molecules of oxygen. That precious load of oxygen is not used by the red blood cells because they lack mitochondria and, therefore, can produce ATP only by anaerobic mechanisms (Chapter 3).

Each red blood cell is packed with approximately 280 million molecules of hemoglobin. Once inside a red blood cell, oxygen binds to the hemoglobin molecules. As can be seen in Figure 11.5, each hemoglobin molecule is made up of four subunits. Each subunit consists of a polypeptide chain (the globin) and a heme group. The heme group includes an iron ion that actually binds to the oxygen. Therefore, each hemoglobin molecule can carry up to four molecules of oxygen. The compound formed when hemoglobin binds with oxygen is called, logically enough, *oxyhemoglobin*.

As wonderfully adapted as the hemoglobin molecule is for carrying oxygen, it has more than 200 times the affinity for carbon monoxide, a product of the incomplete combustion of any carbon-containing fuel, than it does for oxygen. In other words, if concentrations of carbon monoxide and oxygen were identical in inhaled air, for every one molecule of hemoglobin that picked up an oxygen molecule, 200 molecules of hemoglobin would bind to carbon monoxide. The primary source of carbon monoxide is automobile exhaust, but it can also come from indoor sources, including improperly vented heaters and leaky chimneys. The reason that carbon monoxide can be deadly is that it binds to the oxygen-binding site on hemoglobin and prevents blood from carrying life-giving oxygen molecules to the cells. As a result, the cells cannot carry out cellular respiration, and the person can die. Carbon monoxide is a particularly insidious poison because it is odorless and tasteless.

LIFE CYCLE OF RED BLOOD CELLS

The birth process of a red blood cell, which takes about 6 days to complete, involves many changes in the cell's activities and structure. First, the very immature cell becomes a factory for hemoglobin molecules. After the cell is packed with hemoglobin, its nucleus is pushed out. A structural metamorphosis then occurs, culminating in a cell with the biconcave shape typical of a red blood cell. These mature red blood cells then leave the bone marrow and enter the bloodstream. In this way, red marrow produces roughly 2 million red blood cells a second, for a cumulative total of more than half a ton in your lifetime.

A red blood cell lives for only about 120 days. During that time, it travels through approximately 100 km (62 mi) of blood vessels, being bent, bumped, and squeezed repeatedly. Its life span is probably limited by the lack of a nucleus to maintain the cell and direct needed repairs. Without a nucleus, for instance, protein syn-

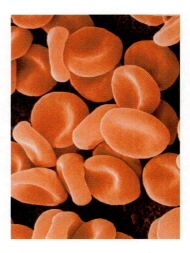

FIGURE **11.4**
Red blood cells serve to ferry oxygen from the lungs to the needy tissues. Each red blood cell is a small biconcave (indented on both sides) disk. This design maximizes the surface area for gas exchange. Lacking a nucleus and other organelles, a red blood cell is essentially a bag packed with the oxygen-binding pigment hemoglobin.

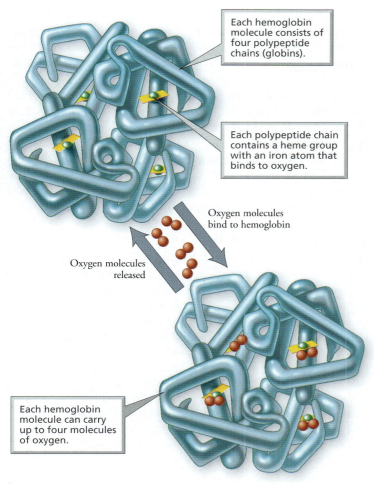

Each hemoglobin molecule consists of four polypeptide chains (globins).

Each polypeptide chain contains a heme group with an iron atom that binds to oxygen.

Oxygen molecules bind to hemoglobin

Oxygen molecules released

Each hemoglobin molecule can carry up to four molecules of oxygen.

FIGURE **11.5**

The structure of hemoglobin, the pigment in red blood cells that transports oxygen from the lungs to the cells

thesis needed to replace key enzymes cannot take place, so the cell becomes increasingly rigid and fragile.

The liver and spleen are the "graveyards" for red blood cells, where worn-out red blood cells are removed from circulation. The old, inflexible red blood cells tend to become stuck in the tiny circulatory channels of these organs. Macrophages (large phagocytic cells that are formed from monocytes) then engulf and destroy the dying red blood cells. The liver degrades the hemoglobin released from destroyed red blood cells to its protein (globin) component and heme. The protein is digested to amino acids, which can be used to make other proteins. The iron from the heme is salvaged and sent to the red marrow for recycling.

The remainder of heme is degraded to a yellow pigment, called bilirubin, which is excreted by the liver in bile. Bile is released into the small intestine, where it assists in the digestion of fats. It is carried along the digestive system to the large intestine with undigested food and becomes a component of feces. The color of feces is partly due to bilirubin that has been broken down by intestinal bacteria.

Products formed by the chemical breakdown of heme also create the yellowish tinge in a bruise that is healing. A bruise, or black-and-blue mark, results when tiny blood vessels or capillaries are ruptured and blood leaks into the surrounding tissue. As the tissues use up the oxygen, the blood becomes darker and, viewed through the overlying tissue, looks black or blue. Gradually, the red blood cells degenerate, releasing hemoglobin. The breakdown products of hemoglobin then make the bruise appear yellowish.

stop and think

Hepatitis is an inflammation of the liver that can be caused by certain viruses or exposure to certain drugs. Hepatitis impairs the liver's ability to handle bilirubin properly. A symptom of hepatitis is jaundice, a condition in which the skin develops a yellow tone. Explain why hepatitis causes jaundice.

Red blood cell production is regulated according to the needs of the body, especially the need for oxygen (Figure 11.6). Most of the time, red blood cell production matches red blood cell destruction. There are circumstances, blood loss for instance, that trigger a homeostatic mechanism that speeds up the rate of red blood cell production. This mechanism is initiated when the oxygen supply to the body's cells drops. Certain cells in the kidney sense the reduced oxygen, and they respond by producing the hormone **erythropoietin**. Erythropoietin then travels to the red marrow, where it steps up both the division rate of stem cells and the maturation rate of immature red blood cells. When maximally stimulated by erythropoietin, the red marrow can increase red blood cell production tenfold—to 20 million cells per second! The resulting increase in red blood cell numbers should soon be adequate to meet the oxygen needs of body cells.

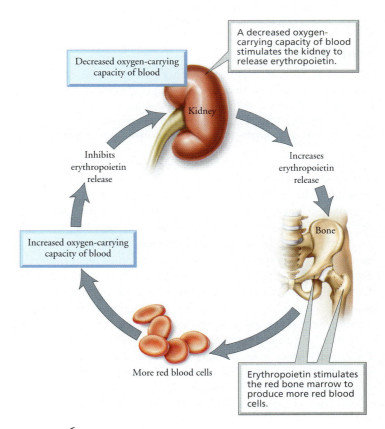

Decreased oxygen-carrying capacity of blood

A decreased oxygen-carrying capacity of blood stimulates the kidney to release erythropoietin.

Kidney

Inhibits erythropoietin release

Increases erythropoietin release

Increased oxygen-carrying capacity of blood

Bone

More red blood cells

Erythropoietin stimulates the red bone marrow to produce more red blood cells.

FIGURE **11.6**

The production of red blood cells is regulated by the oxygen levels in the blood.

Disorders of red and white blood cells have different effects depending on the type of cell affected

Disorders of red and white blood cells have many different causes. The problems associated with each disorder will depend on the type of blood cells affected, because red and white blood cells have different functions.

DISORDERS OF RED BLOOD CELLS

Anemia, a condition in which the blood's ability to carry oxygen is reduced, can result from too little hemoglobin, too few red blood cells, or both. The symptoms of anemia include fatigue, headaches, dizziness, paleness, and breathlessness. In addition, an anemic person's heart often beats faster to help compensate for the blood's decreased ability to carry oxygen. The accelerated pumping can cause heart palpitations—the uncomfortable awareness of one's own heartbeat. Although anemia is not usually life threatening, it does increase susceptibility to illness, and it stresses the heart and lungs. It can also affect the quality of life because lack of energy and low levels of productive activity often go hand in hand.

Worldwide, the most common cause of anemia is an insufficiency of iron in the body, which leads to an inadequate hemoglobin production. *Iron-deficiency anemia* can be caused by a diet that contains too little iron, by an inability to absorb iron from the digestive system, or from blood loss, such as might occur because of menstrual flow or peptic ulcers. Treatment for iron-deficiency anemia involves dealing with the cause of the depletion of iron in the body and restoring iron levels to normal by eating foods that are rich in iron—such as meat, leafy green vegetables, and fortified cereals—or by taking pills that contain iron.

stop and think

Between 25 and 65 ml of blood is lost in each menstrual cycle. Why would this blood loss make women in their reproductive years more likely than men of the same age to suffer from iron-deficiency anemia (a condition in which an insufficiency of iron causes reduced hemoglobin synthesis)?

Blood loss will obviously lower red blood cell counts, but so will any condition that results in the destruction of red blood cells exceeding the production of red blood cells. For example, in hemolytic anemias, red blood cells are ruptured because of infections, defects in the membranes of red blood cells, transfusion of mismatched blood, or hemoglobin abnormalities. *Sickle-cell anemia* is an example of a hemolytic anemia caused by abnormal hemoglobins. In sickle-cell anemia, the abnormal hemoglobin (hemoglobin S) causes the red blood cells to become deformed to a crescent (or sickle) shape when the oxygen content is low. The distorted cells are fragile and rupture easily. They also clog small blood vessels and promote clot formation. Because these events prevent oxygen-laden blood from reaching the tissues, they can cause episodes of extreme pain. Interestingly, although sickle-cell anemia is an inherited condition, the *fetal*[2] hemoglobin does not cause sickling, even in persons who eventually show symptoms of sickle-cell anemia in childhood. An anti-

cancer drug that is also used to treat sickle-cell anemia, hydroxyurea, is thought to work by stimulating the production of fetal hemoglobin in the affected child or adult. Hydroxyurea reduces both the frequency of these painful episodes and the need for blood transfusions in people with sickle-cell anemia.

Red blood cell numbers also drop when the production of red blood cells is halted or impaired, as occurs in *pernicious anemia*. The production of red blood cells depends on a supply of vitamin B_{12}. Absorption of this vitamin from the small intestine depends on a chemical called intrinsic factor, which is produced by the stomach lining. People with pernicious anemia do not produce intrinsic factor and are, therefore, unable to absorb vitamin B_{12}. They are treated with injections of B_{12}.

what would you do?

Blood doping is a practice that boosts red blood cell numbers and, therefore, the blood's ability to carry oxygen. It involves withdrawing and storing some blood—a process that triggers stepped-up production of red blood cells by the body—and then reinfusing the stored blood. Some athletes who compete in aerobic events, such as swimming or cycling, have used blood doping to increase their endurance and speed. Although blood doping is considered unethical and is prohibited in Olympic competition, it is very difficult to detect. It could be detected if the athlete had an exceptionally high number of red blood cells or if the presence of a hormone that stimulates red blood cell production (erythropoietin) were found in the urine. Tests can detect only about 50% of the athletes who practice blood doping. Considering that roughly half of the athletes who practice blood doping will escape detection even if they are tested, do you think that those who are caught should be punished?

DISORDERS OF WHITE BLOOD CELLS

Infectious mononucleosis, or simply mono, is a viral disease of the lymphocytes caused by the Epstein-Barr virus. The infection is highly contagious and causes an increase in the number of lymphocytes with an atypical appearance. Because mono often is spread from person to person by oral contact, it is sometimes called the kissing disease. However, mono can be spread just as easily by sharing eating utensils or drinking glasses. Mono is most common among teenagers and young adults, particularly those away from home at school. It often strikes at stressful times, such as during final exams, when resistance is low.

The initial symptoms of mono are similar to those of influenza: fever, chills, headache, sore throat, and an overwhelming awareness of illness. Within a few days, the glands in the neck, armpits, and groin become painfully swollen. Mono must simply run its course. The major symptoms generally subside within a few weeks, but fatigue may linger much longer.

Leukemia is a cancer of the white blood cells that causes uncontrolled multiplication of white blood cells, causing the number of white blood cells to increase. The cancerous cells are all descendants of a single abnormal cell. Although there are large numbers of white blood cells, they remain unspecialized and are, therefore, unable to defend the body against infectious organisms. However, they divide more rapidly and live longer than do normal cells. The abnormal cells "take over" the bone marrow, preventing

[2]During embryonic growth in the womb, red blood cells contain a slightly different form of hemoglobin than that found in adults.

the development of normal blood cells, including red blood cells, white blood cells, and platelets.

Symptoms of leukemia generally are caused either by the insufficient number of normal blood components or by the invasion of organs by abnormal white blood cells. The increased number of white blood cells crowds out the other formed elements. Insufficient numbers of platelets cause gum bleeding and frequent bruising. Anemia, which causes chronic fatigue, breathlessness, and pallor, results from reduced levels of red blood cells. Because their white blood cells do not function properly, leukemia patients may suffer from repeated respiratory or throat infections, herpes, or skin infections. Also, bone tenderness may be experienced because the immature white blood cells pack the red marrow. Headaches, another symptom of leukemia, may be caused by anemia or by the effects of abnormal white blood cells in the brain.

The treatment of leukemia usually involves radiation therapy and chemotherapy to kill the rapidly dividing cells. In addition, transfusions of red blood cells and platelets may be given to alleviate anemia and prevent excessive bleeding.

Today, children with acute leukemia often are cured with bone marrow transplants. A person, most often a family member, whose tissue type closely matches that of the patient must be located and agree to serve as a donor. After a donor has been found, the bone marrow in the leukemia patient must be destroyed by irradiation and

drugs. This drastic treatment leaves the patient completely vulnerable to any type of infection. The donor is then anesthetized, and some bone marrow is removed from his or her pelvic bones. Next, the donor's bone marrow is given to the patient intravenously, just like a blood transfusion. The marrow cells find their way to the patient's marrow and begin to grow. Soon, the donated bone marrow cells begin to produce healthy, new white blood cells, as well as other formed elements. As white blood cell numbers reach normal levels, usually within 2 months, the patient regains resistance to infection. Treatment with stem cells from umbilical cord blood seems to be as effective as bone marrow transplant in helping some children go into remission, as discussed in the Social Issue essay, *Blood Stem Cells*.

Blood Types Are Determined by Antigens on the Surface of Red Blood Cells

Blood types are determined by the presence of certain genetically determined glycoproteins (proteins attached to carbohydrates) on the surface of red blood cells. Glycoprotein molecules act as antigens (molecules that are recognized as foreign) if they are transfused into a person whose red blood cells lack this protein. Blood types are determined by and named after the specific antigens found on the surface of the red blood cells (Table 11.2).

TABLE **11.2**

TRANSFUSION RELATIONSHIPS AMONG BLOOD TYPES				
BLOOD TYPE	**ANTIGENS ON RED BLOOD CELLS**	**ANTIBODIES IN PLASMA**	**BLOOD TYPES THAT CAN BE RECEIVED IN TRANSFUSIONS**	**INCIDENCE OF BLOOD TYPE IN U.S.**
A	A	Anti-B	A, O	Caucasian, 40% African American, 27% Asian, 28% Native American, 8%
B	B	Anti-A	B, O	Caucasian, 10% African American, 20% Asian, 27% Native American, 1%
AB (universal recipient)	A and B	None	A, B, AB, O	Caucasian, 5% African American, 4% Asian, 5% Native American, 0%
O (universal donor)	None	Anti-A, anti-B	O	Caucasian, 45% African American, 49% Asian, 40% Native American, 91%

Blood Stem Cells

If blood is considered to be the river of life, then stem cells are the springs that feed that river. Stem cells give rise to red blood cells, white blood cells, and platelets. Indeed, they produce about 260 billion new cells (about an ounce of blood) every day.

Because stem cells are able to produce an endless supply of blood cells, they hold a promise for treating a host of blood conditions. For instance, inherited disorders such as sickle-cell anemia and beta-thalassemia, in which abnormal forms of hemoglobin are produced, could become curable. Some patients with sickle-cell anemia and others with beta-thalassemia have already been treated successfully with stem cell therapy. Stem cells that produce normal hemoglobin can be transplanted into a person with the inherited anemia. Then normal red blood cells can begin replacing abnormal ones.

Stem cells could also be used to regenerate important immune system cells that form the foundation of the body's defense system. This situation would open the door for new cancer therapies, especially for leukemia. Research is under way for developing ways to use stem cells to treat certain autoimmune diseases; that is, diseases in which the body's immune system mistakenly attacks a part of the body (discussed in Chapter13). Two autoimmune diseases that have been treated in this way are rheumatoid arthritis, an inflammation of the joints, and diabetes Type 1, a condition in which too little of the hormone insulin is produced. For instance, diabetes Type 1 develops when the body's own T lymphocytes attack the islet cells in the pancreas, the cells that produce insulin.

Although researchers had no difficulty dreaming up life-saving uses for stem cells, the trick was to find them. Fewer than one in every

---

Although there are at least 30 common varieties of antigens, those of the ABO and Rh blood groups cause serious, often fatal, complications if improperly transfused. We will, therefore, consider these blood types more carefully.

When asked about your blood type, you will probably respond by indicating one of the types in the ABO series: A, B, AB, or O. Red blood cells with only the antigen A on their surface are type A. When only the B antigen is on the red blood cell surface, the blood is type B. Blood with both A and B antigens on the red blood cell surface is designated type AB. When neither A nor B antigens are present, the blood is type O.

Normally, a person has antibodies in the plasma against those antigens that are not on his or her own red blood cells. Thus, individuals with type A blood have antibodies against the B antigen (anti-B antibodies), and those with type B blood have antibodies against A (anti-A antibodies). Because individuals with type AB blood have both antigens on their red blood cells, they have neither antibody. Those with type O blood have neither antigen, so they have both anti-A and anti-B antibodies in their plasma.

Blood type is determined by mixing a drop of blood with a solution containing anti-A antibodies and another drop of blood with a solution containing anti-B antibodies. If clumping occurs, it means the antigen is present (Figure 11.7).

When a person is given a blood transfusion with donor blood containing foreign antigens, the antibodies in the recipient's blood will cause the donor's cells to clump, or **agglutinate**. This clumping of the donor's cells is damaging, perhaps even fatal. The clumped cells get stuck in small blood vessels and block blood flow to body cells. Or they may break open, releasing their cargo of hemoglobin. The hemoglobin clogs the filtering system in the kidneys, causing death.

It is important, therefore, to be sure that the blood types of the donor and recipient are compatible, which means that the recipient's blood does not contain antibodies to antigens on the red blood cells of the donor. The plasma of the donor's blood may contain antibodies against antigens on the recipient's red blood cells, but these will be diluted as they enter the recipient's circulation. Therefore, the donor's antibodies are not a major problem.

To determine whether a blood transfusion will be safe, then, we must consider the antigens on the donor's cells and the antibodies in the recipient's blood. For example, if a person with blood type A is given a transfusion of blood type B or of type AB, the naturally occurring anti-B antibodies in the recipient's blood will cause the red blood cells of the donor to clump because they have the B antigen. The transfusion relationships among blood types in the ABO series are shown in Table 11.2. Type O blood is sometimes called the universal donor because its red blood cells have neither anti-A nor anti-B antigens. Therefore, type O blood will not clump if there are anti-A or anti-B antibodies in the recipient's plasma. Type AB blood is called the universal recipient because its plasma lacks antibodies to both A and B antigens. It will not cause clumping of donated blood that has A or B antigens (see the Health Issue essay, *Transfusions and Artificial Blood*).

The A and B antigens are not the only important antigens found on the surface of red blood cells. The Rh factor, which is actually a series of antigens, is also important. The name Rh comes from the beginning of the genus name, *Rhesus,* of the monkey in which the antigen was first discovered. People who have any of the Rh antigens on their red blood cells are considered Rh-positive (Rh^+). When Rh antigens are missing from the red blood cell surface, the individual is considered Rh-negative (Rh^-).

The Rh factor differs from the case with the ABO system. An Rh-negative person will not form anti-Rh antibodies unless he or she has been exposed to the Rh antigen. For this reason, an Rh-negative individual should be given only Rh-negative blood in a transfusion. If he or she is mistakenly given Rh-positive blood, it will stimulate the production of anti-Rh antibodies. A transfusion reaction will not occur after the first such transfusion because it takes thousand marrow cells is a stem cell. Nonetheless, techniques have been developed to identify and isolate the stem cells from the marrow cells.

Blood from the umbilical cord also is a good source of stem cells. Stem cells do not settle into the bone marrow until a few days after birth. During fetal development, stem cells circulate in the bloodstream and, therefore, travel through vessels in the umbilical cord when circulating to and from the placenta. Because the umbilical cord blood is rich in stem cells, it is possible to collect blood from the umbilical cord at birth and know that it contains some stem cells.

The use of umbilical cord blood as a source of stem cells avoids the ethical (and legal) problems of obtaining stem cells from embryos. Although stem cells from embryos are more plentiful and more potent than those from umbilical cord blood, the embryo must be destroyed to obtain the stem cells. Furthermore, the federal government currently funds only stem cell research involving existing lines of embryonic stem cells. We will revisit the issue of using embryo stem cells in Chapter 21.

Stem cells from umbilical cord blood offer exciting promises of medical benefits, most of which have yet to be realized. Nonetheless, there are already umbilical cord blood banks that store cord blood from newborn babies. In some cases, parents pay a set-up fee and an annual service fee to store their baby's cord blood, just in case it is needed some day. Some banks collect and store thousands of samples of cord blood in the hopes of having a tissue match for most individuals in the population. Researchers have developed ways to amplify umbilical cord blood stem cells by growing them in the laboratory.

The existence of umbilical cord blood banks raises some social issues. Should stored blood be reserved only for the possible future need of the donor, or should it be made available for anyone in need whose tissue type matches? Should cord blood banks be privately owned or funded by the government? ✍

---

time for the body to start making anti-Rh antibodies. After a second transfusion of Rh-positive blood, the antibodies in the recipient's plasma will react with the antigens on the red blood cells of the donated blood. This reaction may lead to the death of the patient.

The Rh factor may also be of medical importance in pregnancies in which the mother is Rh-negative and the fetus is Rh-positive, a situation that can occur if the father is Rh-positive (see Chapter 20) (Figure 11.8). Ordinarily, the maternal and fetal blood supplies do not mix during pregnancy. However, some mixing may occur during delivery when some blood vessels are damaged. If the baby's red blood cells, which bear Rh antigens, accidentally pass into the bloodstream of the mother, she will produce anti-Rh antibodies. There are usually no ill effects associated with the first introduction of the Rh antigen. However, if antibodies are present in the maternal blood from a previous pregnancy with an Rh-positive child or from a transfusion of Rh-positive blood, the anti-Rh antibodies may pass into the blood of the fetus. This transfer can occur because anti-Rh antibodies, unlike red blood cells, can cross the placenta (a structure

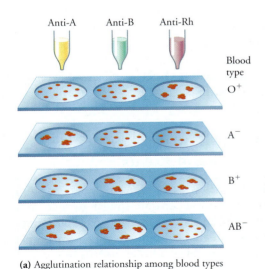

(a) Agglutination relationship among blood types

(b) Micrographs showing no agglutination and agglutination

(c) Diagrams of antigens and antibodies with no agglutination and with agglutination

FIGURE **11.7**

Blood is typed by mixing it with serum known to contain antibodies specific for a certain antigen. If blood containing that antigen is mixed with the serum, the blood will agglutinate (clump). Thus, one drop of blood is mixed with serum containing anti-A, and another drop with serum containing anti-B. A third drop of blood is mixed with serum containing antibodies to Rh. Agglutination in response to an antibody reveals the presence of the antigen.

HEALTH ISSUE

Transfusions and Artificial Blood

The need for blood for transfusion often exceeds the amount supplied by donations. Although more people are donating blood than ever before, the number of people needing transfusions is growing rapidly. Our society is aging, and there are increasing numbers of elective surgeries and medical treatments that require transfusions. Certain willing donors are banned from donating blood. For example, concern over the spread of the human version of mad cow disease (Creutzfeldt-Jakob disease) led to a 1999 Food and Drug Administration ruling preventing blood donations from people who lived in the United Kingdom for more than 6 months between 1980 and 1996. This concern is coupled with others about the safety of the blood supply. Of particular concern is the possibility of the transmission of a virus that causes hepatitis or HIV, the virus that causes AIDS, through tainted blood. All of these concerns have led to some changes in thinking about transfusions.

Today, before elective surgery, patients often have their own blood withdrawn and stored so that it can be reinfused, if necessary, during surgery. Although these transfusions have become quite common because of the patients' concern about disease transmission, they can be very expensive for hospitals and health insurance companies. Why? The most important reason is that the patient does not always need all the blood that was donated ahead of time. The unneeded blood is usually discarded, but the cost of collecting and storing the blood remains. Furthermore, the safety of blood collected from volunteers has increased greatly in recent years.

If a person is given a transfusion of an incompatible blood type, it almost always causes death. Although the risk of receiving blood of the wrong type is low, mistakes are sometimes made, especially in situations in which the medical personnel are hurried, such as in emergency rooms or intensive care units. For this reason, some emergency rooms stock only type O blood, the universal donor. The problem with this plan is that it can lead to a shortage of type O blood and wasted units of blood types A and B.

An Rh$^+$ male and an Rh$^-$ female have a 50% chance of having an Rh+ baby.

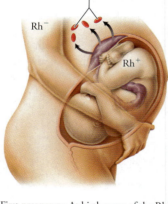

First pregnancy: At birth some of the Rh$^-$ blood of the fetus may enter the mother's circulation.

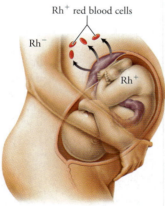

The mother forms anti-Rh antibodies over the next few months.

Second pregnancy with an Rh$^+$ fetus: Anti-Rh antibodies may pass into the fetus's blood, causing its blood cells to burst.

that forms during pregnancy that allows the exchange of selected substances between the maternal and fetal circulatory systems). These anti-Rh antibodies may destroy the fetus's red blood cells. As a result, the child may be stillborn or very anemic at birth. This condition is called *hemolytic disease of the newborn.*

The incidence of hemolytic disease of the newborn has decreased in recent years because of the development of a means of destroying any Rh-positive fetal cells in the maternal blood supply before they can stimulate the mother's cells to produce her own anti-Rh antibodies. The Rh-positive cells are killed by injecting RhoGAM, a serum containing antibodies against the Rh antigens, before she gives birth or shortly afterward. Rh antigens are thus prevented from being "set" in the memory of the mother's immune system. The injected antibodies disappear after a few months. Therefore, no antibodies linger to affect the fetus in a subsequent pregnancy.

Blood Clotting Occurs in a Regulated Sequence of Events

When a blood vessel is cut, a series of reactions stops blood flow. These reactions are similar to those you might initiate if the garden hose you are using springs a leak. Your initial response might be to squeeze the hose, in hopes of stopping water flow. Likewise, the body's immediate response to

FIGURE **11.8**

Rh incompatibility can result when an Rh-negative (Rh$^-$) woman is pregnant with an Rh-positive (Rh$^+$) baby, if the woman has been previously exposed to Rh$^+$ blood.

Whole blood—your own or someone else's—has some disadvantages when used for transfusions. It must be typed and matched. It must be screened for a host of diseases, among them syphilis, HIV-1, HIV-2, human T-cell lymphotrophic virus type 1, and both hepatitis B and C. At best, the shelf life of whole blood is slightly over a month. These drawbacks to the use of whole blood have spurred attempts to produce synthetic blood components.

When a person has lost blood, it is often important to boost the oxygen-carrying ability of the remaining blood. For this purpose, blood substitutes that contain cell-free hemoglobin are being developed. Most of the hemoglobin substitutes are made from hemoglobin from outdated human blood, but at least one is made from cow's blood. Unfortunately, molecules of free hemoglobin cannot be infused into the body. If they are, the four protein chains of the hemoglobin molecules fall apart and clog the kidney's filtering system. The problem, then, is to keep hemoglobin as an intact molecule. One approach to solving this problem is to chemically rejoin the hemoglobin molecules. A second approach is to stabilize the molecules by linking them into larger complexes.

Blood substitutes containing cell-free hemoglobin have yet another problem. They may increase a patient's chances of developing bacterial infections, which may develop because the cell-free hemoglobin provides an easy iron source for reproducing bacteria.

With other blood substitutes, oxygen is not carried by hemoglobin in any form but rather by oxygenating agents such as perfluorochemical (PFC) emulsions, compounds that can carry roughly 70% of the oxygen that could be transported by whole blood. One such blood substitute that is in the late stages of development, Oxygent®, is intended for use during surgery and circulates for about 2 days before it is naturally eliminated.

Artificial blood will certainly be a milestone in medical history—one that will someday save thousands of lives a year. However, its intended use is for the emergency needs of trauma victims and patients undergoing surgery. The artificial blood will sustain patients until they can replenish their own blood or receive donated blood. Blood drives will not become a thing of the past. 👫

blood vessel injury is for the vessel to constrict (squeeze shut). This constriction of the injured vessel is caused by direct injury to the muscles lining the blood vessel and by chemicals released from the injured tissue.

The next response is to plug the hole (Figure 11.9). Your thumb might do the job on your garden hose; platelets form the plug that seals the leak in a vessel. The **platelet plug** is formed when platelets cling to cables of collagen, a protein fiber, in the exposed blood vessel surface. When the platelets attach to collagen, they become activated and change in several ways: They swell, form many cellular extensions, and stick together. Platelets produce a chemical called thromboxane that makes platelets stick to one another. Thromboxane also attracts other platelets to the wound. Aspirin prevents the formation of thromboxane and, therefore, inhibits clot formation, explaining how a daily dose of aspirin may prevent the formation of blood clots that can block blood vessels nourishing heart tissue, resulting in the death of heart cells (a heart attack). It also explains how aspirin can cause excessive bleeding.

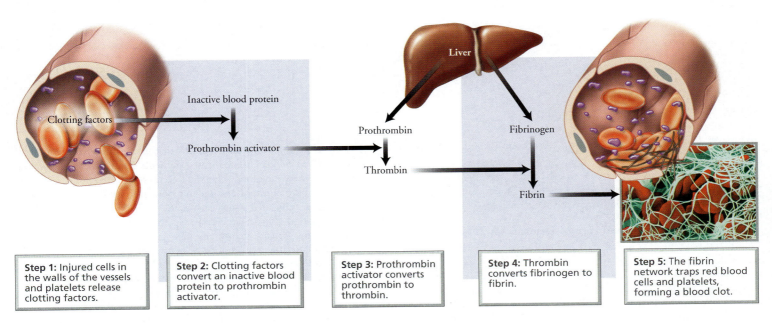

Liver

Clotting factors

Inactive blood protein

Prothrombin activator

Prothrombin

Thrombin

Fibrinogen

Fibrin

Step 1: Injured cells in the walls of the vessels and platelets release clotting factors.

Step 2: Clotting factors convert an inactive blood protein to prothrombin activator.

Step 3: Prothrombin activator converts prothrombin to thrombin.

Step 4: Thrombin converts fibrinogen to fibrin.

Step 5: The fibrin network traps red blood cells and platelets, forming a blood clot.

FIGURE **11.9**
Selected steps in the blood-clotting process

The next step in stopping blood loss through a damaged blood vessel is the formation of the clot itself. There are more than 30 steps in the process of clot formation, but here we will simplify the process. Clot formation begins when clotting factors are released from injured tissue and from platelets. At the site of the wound, the clotting factors convert an inactive blood protein to **prothrombin activator**,[3] which converts **prothrombin**, a plasma protein produced by the liver, to an active form, **thrombin**. Thrombin then causes a remarkable change in another plasma protein produced by the liver, **fibrinogen**. The altered fibrinogen forms long strands of **fibrin**, which form a web that traps blood cells and forms the clot. The clot is a barrier that prevents further blood loss through the wounded vessel.

stop and think

Thromboplastin, a chemical important in the initiation of clot formation, is released from both damaged tissue and activated platelets. How does this fact explain why a scrape, which causes a great deal of tissue damage, generally stops bleeding more quickly than a clean cut, such as a paper cut or one that might occur with a razor blade?

If even one of the many factors needed for clotting is lacking, the process can be slowed or completely blocked. Vitamin K is needed for the liver to synthesize prothrombin and three other clotting factors. Thus, without vitamin K, clotting does not occur. We have two sources of vitamin K. One is the diet. Vitamin K is found in leafy green vegetables, tomatoes, and vegetable oils. A second source is from bacteria living in our intestines. These bacteria manufacture vitamin K, some of which we absorb for our own use. Antibiotic treatment for serious bacterial infections can kill gastrointestinal bacteria and lead to a vitamin K deficiency in as few as 2 days. Vitamin K is used rapidly by body tissues, so both sources are needed for proper blood clotting.

Hemophilia is an inherited condition in which the affected person bleeds excessively owing to a faulty gene needed to produce a

---

[3]Prothrombin activator is an enzyme called prothrombinase.

clotting factor. The different forms of hemophilia are due to faulty genes that fail to produce one of the many clotting factors needed for blood clotting. Because of the way hemophilia is inherited, the condition usually occurs in males (see Chapter 20). Symptoms appear when the affected child first becomes active. Crawling, for instance, causes bruises on the elbows and knees, and cuts tend to bleed longer than usual. Excessive internal bleeding can damage nerves or, when it occurs in joints, permanently cripple the hemophiliac.

Treatment for hemophilia involves restoring the missing clotting factor. Bleeding episodes can be controlled with repeated transfusions of fresh plasma or with injections of concentrated clotting factor. The plasma transfused to treat hemophilia is combined from the plasma donations of many people to concentrate the clotting factor. Each injection of concentrated clotting factor is prepared from 2000 to 5000 individual donations. Today, one of the clotting factors (factor VIII) is being manufactured using recombinant DNA technology.

The failure of blood to clot can shorten the life of a person with hemophilia, but in people without hemophilia the formation of blood clots when they are not needed can have more immediate health consequences because they can disrupt blood flow. A blood clot in an unbroken blood vessel that stays in place is called a *thrombus*. Clots can also drift through the circulatory system until they become lodged in a narrow vessel. When the tiny vessels that nourish the heart or brain become clogged with a clot, the consequences can be severe—disability or even death. A blood clot that drifts through the circulatory system is called an *embolus*.

After a wound has healed, clots are normally dissolved by an enzyme called **plasmin**, which is formed from an inactive protein, **plasminogen**. Plasmin dissolves clots by digesting the fibrin strands that form the framework of the clot.

stop and think

Heparin is a drug that inactivates thrombin. It is sometimes administered to patients when it is desirable to inhibit the clotting response. How would heparin act to achieve these ends?

REVIEWING THE CONCEPTS

Blood Functions in Transportation, Protection, and Regulation (p. 214)

1. Blood transports vital material to cells and wastes away from cells. White blood cells defend against disease. Blood clotting prevents excessive blood loss. Blood also helps regulate body temperature.

Blood Consists of Plasma and Formed Elements (pp. 214–221)

2. Blood is a type of connective tissue that contains formed elements in a liquid (plasma) matrix.

3. Plasma is the liquid portion of the blood. It contains the substances transported by the blood, including nutrients, ions, dissolved gases, hormones, and waste products. Plasma proteins, which constitute most of the substances dissolved in the plasma, aid in the balance of water flow between blood and cells. Plasma proteins are grouped into one of three categories: albumins, globulins, or clotting proteins (fibrinogen).

4. Stem cells are undifferentiated cells that divide and give rise to all of the formed elements. The formed elements of the blood are platelets, leukocytes (white blood cells), and erythrocytes (red blood cells).

5. Platelets play an important role in blood clotting. They are fragments of a larger cell called a megakaryocyte.

6. White blood cells (leukocytes) help the body fight off disease and help remove wastes, toxins, and damaged cells.

7. There are five types of leukocytes: neutrophils, eosinophils, basophils, monocytes, and lymphocytes. Neutrophils are the most abundant type and immediately phagocytize foreign microbes. Eosinophils help the body defend against parasitic worms and play a role in allergic reactions. Basophils can increase the flow of blood by releasing histamines. Monocytes are the largest leukocytes and, after maturing into macrophages, they actively fight chronic infections. Lymphocytes are involved in body defense responses.

8. Red blood cells (erythrocytes) are flexible cells packed with hemoglobin, an oxygen-binding pigment. Originating in the red bone marrow, erythrocytes have a life span of about 120 days. Red blood cell production is controlled by erythropoietin, a hormone produced by the kidney in response to low oxygen. Worn and dead red blood cells are removed from circulation and broken down in the liver and spleen.

9. Anemia is a reduction in the blood's ability to carry oxygen. There are several forms of anemia, including iron-deficiency anemia, hemolytic anemias, such as sickle-cell anemia, and pernicious anemia, caused by a lack of the intrinsic factor produced by the stomach lining that is needed to absorb the vitamin B_{12} necessary for red blood cell production.

10. Infectious mononucleosis is a highly contagious viral disease of lymphocytes. Leukemia is a cancer of the white blood cells. There are high numbers of white blood cells, but they do not function properly to defend against infectious agents.

Blood Types Are Determined by Antigens on the Surface of Red Blood Cells (pp. 221–224)

11. Blood types are determined by the presence of certain antigens (glycoproteins) on the surface of red blood cells: the ABO and Rh groups. The plasma contains antibodies against A and B antigens if the antigens are not present on the red blood cells. Antibodies to Rh antigens are formed only after exposure to Rh-positive blood. If blood containing foreign antigens is introduced during a transfusion, a reaction between the antibodies and antigens can cause agglutination and clogging in the recipient's bloodstream, which can lead to death.

12. The Rh incompatibility becomes important to an Rh-negative woman who has been previously exposed to Rh-positive blood and who is pregnant with an Rh-positive fetus. Antibodies to the Rh antigens can cross the placenta and destroy the red blood cells of the fetus, a condition called hemolytic disease of the newborn.

WEB TUTORIAL 11.1 Blood Types and Surface Glycoproteins

Blood Clotting Occurs in a Regulated Sequence of Events (pp. 224–226)

13. The prevention of blood loss involves three mechanisms: blood vessel spasm, platelet plug formation, and clotting. Clotting (coagulation) is initiated when platelets and damaged tissue release clotting factors that lead to the production of a chemical called prothrombin activator, which converts the blood protein prothrombin to thrombin. Thrombin then converts the blood protein fibrinogen to fibrin. Strands of fibrin form a mesh that traps red blood cells and forms the clot. Hemophiliacs are missing an important clotting factor and, therefore, bleed excessively.

KEY TERMS

agglutinate *p. 222*	fibrinogen *p. 226*	monocyte *p. 217*	prothrombin *p. 226*
agranulocyte *p. 215*	formed element *p. 215*	neutrophil *p. 217*	prothrombin activator *p. 226*
anemia *p. 220*	granulocyte *p. 215*	phagocytosis *p. 215*	red blood cell (RBC) *p. 218*
basophil *p. 217*	hemoglobin *p. 218*	plasma *p. 215*	stem cell *p. 215*
eosinophil *p. 217*	hemophilia *p. 226*	plasmin *p. 226*	thrombin *p. 226*
erythrocyte *p. 218*	leukemia *p. 220*	plasminogen *p. 226*	white blood cell (WBC) *p. 215*
erythropoietin *p. 219*	leukocyte *p. 215*	platelet *p. 215*	
fibrin *p. 226*	lymphocyte *p. 217*	platelet plug *p. 225*	

THINKING ABOUT THE CONCEPTS

1. What is plasma? What are its functions? *p. 215*
2. What are the three categories of plasma proteins? *p. 215*
3. List the three types of formed elements and describe the function of each. *pp. 215–218*
4. Compare the size, structure, and numbers of leukocytes with the size, structure, and numbers of erythrocytes. *p. 216*
5. Leukocytes function in defending the body against foreign invaders. List the five types of white blood cells and describe the role each plays in body defense. *p. 217*
6. Describe the characteristics of red blood cells that make them specialized for delivering oxygen to the tissues. *p. 218*
7. Describe the structure of hemoglobin. *pp. 218–219*
8. Where are red blood cells produced? How is the production of red blood cells controlled? *pp. 218–219*
9. Describe what happens to worn or damaged red blood cells and their hemoglobin. *p. 219*
10. Why would pernicious anemia be treated with regular injections of vitamin B_{12} instead of by dietary supplements of this vitamin? *p. 220*
11. Bleeding, fatigue, and repeated infections are symptoms of leukemia, a cancer that causes white blood cell numbers to increase. Explain how the dramatic increase in white blood cells is related to the symptoms of leukemia. *pp. 220–221*
12. How are blood types determined? What happens if a person is given a blood transfusion with blood of an incompatible type? *pp. 221–222*
13. What blood type(s) can a person with type B blood receive? Explain. *p. 222*
14. What is hemolytic disease of the newborn? What causes it? *pp. 222–224*

15. After a blood vessel is cut, what mechanisms prevent blood loss? *pp. 224–226*
16. Describe the steps involved in blood clotting. *p. 226*
17. What is the difference between the blood clotting that occurs after an injury and the agglutination that occurs after a mismatched transfusion?
18. Type B blood contains which antibodies?
    a. A
    b. B
    c. both A and B
    d. neither A nor B
19. An important function of white blood cells is
    a. blood clotting.
    b. transportation of oxygen.
    c. fighting infection.
    d. maintaining blood pressure.
20. If erythropoietin levels were elevated, the most likely cause would be
    a. infection.
    b. anemia.
    c. leukemia.
    d. a kidney problem.
21. The primary function of red blood cells is to _____.
22. Hemolytic disease of the newborn may result if the mother is Rh _____ and the fetus is Rh _____.
23. _____ is a protein in red blood cells that transports oxygen.
24. The protein that forms a net that traps red blood cells and platelets and forms blood clots is _____.

APPLYING THE CONCEPTS

1. Erin has leukemia, and her white blood cell count is elevated. White blood cells are important in fighting disease. Why does Erin have an elevated risk of infection?

2. John is a runner from New York City who is on his college track team. The championship meet is in Denver, which is called the Mile-High City because of its elevation above sea level. The oxygen concentration in the air we breathe decreases as elevation increases. Because this meet is so important, John plans to arrive in Denver several weeks before the meet. Why? Would you expect John's red blood cell count to be higher or lower than normal when he returns home to New York City, which is at sea level?

3. Janie is a 25-year-old woman who has uterine polyps that cause heavy vaginal bleeding. She complains that she tires easily with physical activity and always feels fatigued. The doctor orders that a blood test be done to determine what percentage of the blood is red blood cells. Why? Would you expect the percentage to be higher or lower than normal? The doctor suggests that she take iron supplements. Why?

4. Raul is in a car accident and is taken to the emergency room. He has type AB blood. Which blood types can he receive?

5. Elizabeth has type Rh-negative blood. She is pregnant for the second time. What information would the doctor want to know about the first and second fetuses to know whether to expect a problem with this pregnancy?

6. Sarala was given an antibiotic that caused her platelet count to fall. What symptoms would be expected?

The Circulatory System

The Cardiovascular System Consists of the Blood Vessels and the Heart

The Blood Vessels Conduct Blood in Continuous Loops
- Arteries carry blood away from the heart
- Capillaries are sites of exchange with body cells
- Veins return blood to the heart

The Heart Is a Muscular Pump
- The heart functions as two separate pumps
- Blood flows through the heart in two circuits
- Coronary circulation serves the heart muscle
- The cardiac cycle is the sequence of heart muscle contraction and relaxation
- The rhythmic contraction of the heart is due to its internal conduction system
- An electrocardiogram is a recording of the electrical activities of the heart
- Blood pressure is the force blood exerts against blood vessel walls

Cardiovascular Disease Is a Major Killer in the United States
- High blood pressure can kill without producing symptoms
- Atherosclerosis is a buildup of lipids in the artery walls
- Coronary artery disease is atherosclerosis in the coronary arteries

Heart Attack Is the Death of Heart Muscle

The Lymphatic System Functions in the Circulatory and Immune Systems

HEALTH ISSUE   The Cardiovascular Benefits of Exercise

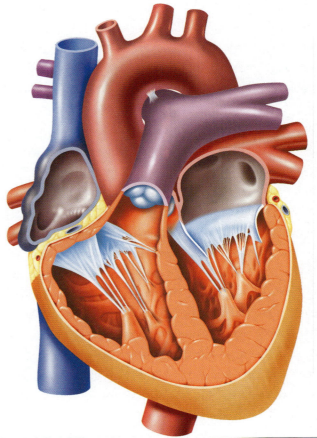

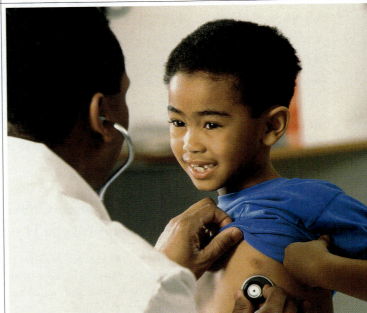

It was past midnight, and Susan, a 38-year-old air traffic controller, was again suffering from angina, severe chest pains. It was a completely frustrating puzzle to her that she would be having these symptoms night after night and occasionally during the day for the past week. She had normal blood pressure; she had never smoked; she was in a normal weight range; and she exercised strenuously five days a week. For comfort, Susan snuggled her cat, Harold, and tried to put herself into a meditative state both to reduce the pain and to calm her anxieties. After several hours and only intermittent periods of little or no pain, Susan called the emergency room and talked to the physician on duty. He recommended that she have someone drive her to the hospital or, better yet, call 911 and come by ambulance to the emergency room. At that hour of the night, she fortunately did not have to wait for treatment. After giving Susan supplemental oxygen and performing a number of blood tests, the doctor called in a heart specialist. The specialist performed more tests, including a coronary angiogram (a procedure to examine blood flow through blood vessels), and informed Susan that an important coronary artery was almost completely blocked.

He explained that this was a serious condition that was causing the frequent angina, because the blocked artery was not allowing enough oxygen to supply the heart. Therefore, he recommended a coronary angioplasty, a procedure using a balloon to stretch open the blockage in the heart artery, allowing adequate blood supply and oxygen to reach the heart muscle. Susan agreed to have the angioplasty done. During her several-month period of cardiac rehabilitation and follow-up appointments, Susan was given a diet and exercise plan that would help reduce the future possibility of heart problems.

In this chapter, we will learn more about angina and see why Susan was right to be concerned about her chest pains. We will learn about the structure and the function of the heart and grow to appreciate even more its marvelous design. An understanding of how the heart functions might help us to properly care for it and maintain its health. Finally, we will examine the link between the **cardiovascular system**—the heart and blood vessels—and the network of vessels that are part of the lymphatic system. Together, the cardiovascular system and the lymphatic vessels constitute the **circulatory system**. ■

The Cardiovascular System Consists of the Blood Vessels and the Heart

The heart is, indeed, important to life, but without the other parts of the cardiovascular system—the blood and blood vessels—it would be useless. The **heart** is a muscular pump that contracts rhythmically, providing the force that drives blood through a system of tubules called blood vessels. The blood delivers a contin-

uous supply of oxygen and nutrients to the cells of the body and washes away metabolic waste products before they poison the cells.

Why is the cardiovascular system so critical to survival? It is the body's transportation network, similar in some ways to the highways within a country. The cardiovascular system provides a means for distributing vital chemicals from one part of the body to another quickly enough to sustain life. Our bodies are too large and complex for diffusion alone to distribute materials. The cardiovascular system is more than a system of pipelines, however. It responds to the body's changing needs. Heart rate and the diameter of certain blood vessels are continuously adjusted.

The Blood Vessels Conduct Blood in Continuous Loops

Once every minute, or about 1440 times each day, the blood circulates through a life-sustaining loop of blood vessels (Figure 12.1). This loop is extremely long. If all the vessels in an adult's body were placed end to end, they would stretch about 100,000 km (60,000 mi), long enough to circle the earth's equator more than twice!

The blood vessels do not form a single long tube. Instead, they are arranged in branching networks. With each circuit through the body, blood is carried away from the heart in an *artery*, which branches and becomes narrower, giving rise to *arterioles*. An arteriole leads to a network of microscopic vessels called *capillaries* that allow the exchange of materials between the blood and body cells. The capillaries eventually merge to form *venules*, which eventually join to form larger tubes called *veins*. The venules and veins return the blood to the heart. We see, then, that the path of blood is

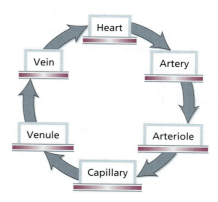

All blood vessels share some common features, but each type has its own traits. Each type of vessel is marvelously adapted for its specific function (Figure 12.2). The hollow interior of a blood vessel, through which the blood flows, is called the *lumen*. The inner lining of a vessel that comes into contact with the blood flowing through the lumen is composed of flattened, tight-fitting cells. This lining, called the endothelium, forms a smooth surface that minimizes friction and allows the blood to flow over it easily. The lumen and endothelium are characteristic of all blood vessels.

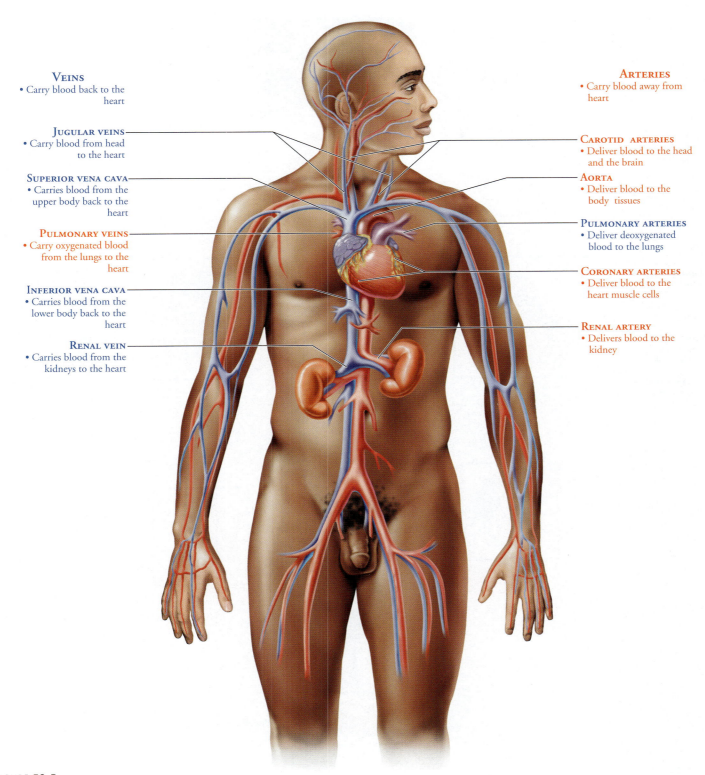

VEINS
- Carry blood back to the heart

JUGULAR VEINS
- Carry blood from head to the heart

SUPERIOR VENA CAVA
- Carries blood from the upper body back to the heart

PULMONARY VEINS
- Carry oxygenated blood from the lungs to the heart

INFERIOR VENA CAVA
- Carries blood from the lower body back to the heart

RENAL VEIN
- Carries blood from the kidneys to the heart

ARTERIES
- Carry blood away from heart

CAROTID ARTERIES
- Deliver blood to the head and the brain

AORTA
- Deliver blood to the body tissues

PULMONARY ARTERIES
- Deliver deoxygenated blood to the lungs

CORONARY ARTERIES
- Deliver blood to the heart muscle cells

RENAL ARTERY
- Delivers blood to the kidney

FIGURE **12.1**

A diagrammatic view of the principal arteries and veins. Red indicates blood that is high in oxygen. Blue indicates blood that is low in oxygen.

▌Arteries carry blood away from the heart

Arteries are muscular tubes that transport blood *away* from the heart, delivering it rapidly to the body tissues. The innermost layer of an arterial wall, as shown in Figure 12.2, is the endothelium. Immedi-

ately outside the endothelium is the middle layer of the arterial wall, which contains elastic fibers and circular layers of smooth muscle. The elastic fibers allow an artery to stretch and then return to its original shape. The smooth muscle provides the artery with the ability to con-

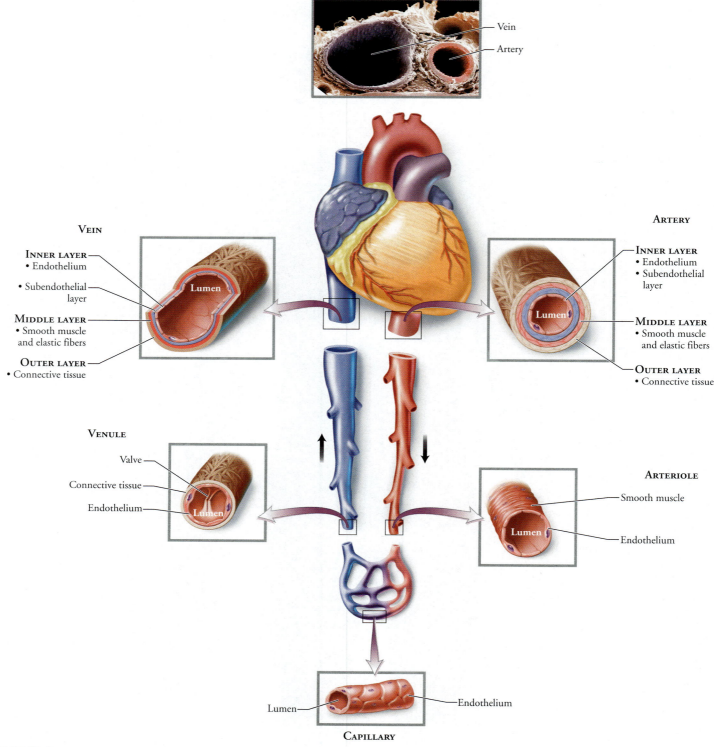

FIGURE 12.2
The structure of blood vessels

tract. The outer layer of an arterial wall is a sheath of connective tissue that contains elastic fibers and collagen. This layer adds strength to the arterial wall and anchors the artery to surrounding tissue.

The elastic fibers have two important functions: (1) They help the artery tolerate the pressure shock caused by blood surging into the artery when the heart contracts; and (2) they help maintain a relatively even pressure within the artery, despite large changes in the volume of blood within the artery. Consider, for instance, what happens when the heart contracts and sends blood into the **aorta,** the body's main artery. Each beat of the heart causes 70 ml (about $\frac{1}{4}$ cup) of blood to pound against the wall of the aorta like a tidal wave. A rigid pipe could not withstand the repeated pressure surges, but the elastic walls of the arteries stretch with each wave of blood and recoil when the surge has moved along, resulting in the intermittent waves being dampened into a continuous stream.

The alternate expansion and recoiling of arteries create a pressure wave, called a **pulse**, that moves along the arteries with each heartbeat. You can feel the pulse by slightly compressing with the fingers any artery that lies near the body's surface, such as the one at the wrist or the one under the angle of the jaw.

As previously mentioned, the middle layer of an artery wall also contains smooth muscle that enables the artery to contract. When the muscle contracts and the diameter of the lumen becomes narrower, the process is called **vasoconstriction**. On the other hand, when the smooth muscle relaxes and the arterial lumen increases in diameter, the process is called **vasodilation**. Vasoconstriction *reduces* blood flow through the artery. Conversely, vasodilation *increases* blood flow through the artery. The smooth muscle is best developed in small to medium-sized arteries. These arteries serve to regulate the distribution of blood, adjusting flow to suit the needs of the body. Vasoconstriction is also important when an artery is cut. Because arterial blood is under great pressure, much blood can be lost in a short time through the wound. In this situation, the artery may constrict, helping to minimize blood loss.

The wall of an artery sometimes becomes weakened by disease, inflammation, injury, or a defect existing at birth. When the arterial wall becomes weakened, the pressure of the blood flowing through the weakened area may cause the artery to swell outward like a balloon, forming an *aneurysm*. Most aneurysms do not cause symptoms, but the condition can be threatening just the same. The primary risk is that the aneurysm will burst, causing blood loss. As a result, the tissues serviced by that vessel will be deprived of oxygen and nutrients, a situation that can be fatal. Even if the aneurysm does not rupture, it can cause the formation of life-threatening blood clots. A clot can break free of the site of formation and float through the circulatory system until it lodges in a small vessel, where it can block blood flow and cause tissue death beyond that point. In some cases, an aneurysm can be repaired surgically.

The smallest arteries, called **arterioles**, are barely visible to the unaided eye. Their walls have the same three layers found in arteries, but the middle layer is primarily smooth muscle with only a few elastic fibers.

Arterioles have two extremely important regulatory roles. First, they are the prime controllers of **blood pressure**, which is the pressure of blood against the vessel walls (discussed later in this chapter). Second, they serve as gatekeepers to a network of capillaries. A capillary network can be open or closed, depending on whether the smooth muscle of the arteriole leading to it allows blood through. In this way, metabolically active cells receive more blood than metabolically inactive ones. Minute by minute, arterioles respond to input from hormones, the nervous system, and local conditions, constantly modifying blood pressure and flow to meet the body's changing needs.

Capillaries are sites of exchange with body cells

Capillaries are microscopic blood vessels that connect arterioles and venules. The capillaries are well suited to their primary function: the exchange of materials between the blood and the body cells, as shown in Figure 12.3. Because capillaries are only

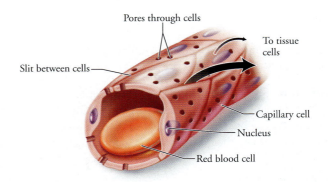

(a) Substances are exchanged between the blood and tissue fluid across the plasma membrane of the capillary.

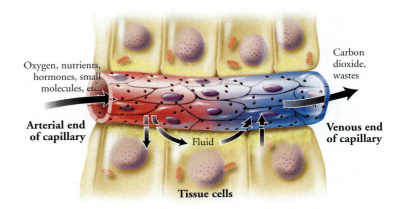

(b) At the arterial end of a capillary, blood pressure forces fluid out of the capillary to the fluid surrounding tissue cells. At the venous end, fluid is drawn back into the capillary by osmotic pressure.

(c) Capillaries are so narrow that red blood cells must travel through them in single file.

FIGURE **12.3**
Capillaries are the sites where materials are exchanged between the blood and the body cells.

one cell layer thick, substances move easily between the blood and the fluid surrounding the cells. The plasma membrane of the capillary's endothelial cells is an effective selective barrier that determines which substances can cross. Some substances that cross the capillary walls do not pass *through* the endothelial cells. Instead, these substances filter through small slits *between* adjacent endothelial cells. The slits between the cells are just large enough for some fluids and small dissolved molecules to pass through.

Materials can move between the capillaries and the body cells in a number of ways:

1. **Diffusion.** Simple diffusion is the most important way that capillary exchange occurs. Gases such as oxygen and carbon dioxide, hormones, glucose, amino acids, and other small molecules simply move down their concentration gradient directly through a capillary cell or the small gaps between the capillary cells.

2. **Pressure-driven movements.** The movement of many materials gets a boost from the flow of water into and out of the capillary (see Figure 12.3). Two competing forces—blood pressure and osmotic pressure—determine the direction of water movement. Blood pressure is the force of blood against the vessel that contains it. You may recall from Chapter 3 that osmosis is the movement of water across a semipermeable membrane in response to differing concentrations of solutes (dissolved particles). Osmotic pressure, then, indicates the force of that water movement. These competing pressures affect the flow of fluid across capillaries. First, the fluid is pushed out of the capillary at its beginning by the high blood pressure there. As blood flows through the capillary bed, blood pressure falls. At the same time, osmotic pressure increases because the blood becomes more concentrated as proteins are left behind when water is pushed out. At the end of the bed, water is drawn back into the capillary because the osmotic pressure there exceeds blood pressure. Substances that are small enough to fit through the gaps between capillary cells can simply "go with the flow" of water into or out of the capillary.

3. **Endocytosis or exocytosis.** A few materials are actively ferried across the capillary wall in tiny vesicles by endocytosis or exocytosis. In these processes, a pocket forms on the membrane of a capillary cell. Fluid and suspended materials are drawn into that pocket. The pocket then breaks off, forming a vesicle that crosses the cytoplasm and unloads its contents on the other side.

Taken together, capillaries provide a tremendous surface area for the rapid exchange of materials between body and blood. In fact, an adult's body contains about 10 billion capillaries with a combined length of over 8000 km (5000 mi). Capillaries are arranged in highly branched networks. The networks bring capillaries very close to nearly every cell. There are so many capillaries that few cells are more than 0.125 mm (0.005 in.) from a capillary. Your fingernails provide windows that allow you to appreciate the vast capillary network in the body. You may have noticed that the tissue beneath a fingernail normally has a pink tinge. The color results from blood flowing through numerous capillaries there. Gentle pressure on the nail causes the tissue to turn white as the blood is pushed from those capillaries.

Although a single capillary is scarcely wide enough for a single red blood cell to squeeze through it, there are so many capillaries that their *combined* cross-sectional area is enormous, much greater than that of the arteries or veins. Because of the large cross-sectional area of the capillaries, the blood flows much more slowly through them than through the arteries or veins. The slower rate of flow in the capillaries provides more time for the exchange of materials (Figure 12.4).

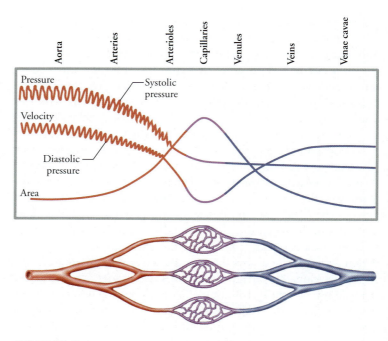

FIGURE **12.4**

The capillaries are so numerous that their total cross-sectional area is much greater than that of arteries or veins. Thus, blood pressure drops and blood flows more slowly as it passes through a capillary bed. The slower rate of flow allows time for the exchange of materials between the blood and the tissues.

Blood flow through capillaries can also be adjusted to the body's metabolic needs. The adjustments are made possible by the design of capillary networks. The network of capillaries servicing a particular area is called a **capillary bed** (Figure 12.5). The number of capillaries in a bed generally ranges from 10 to 100, depending on the type of tissue. A ring of smooth muscle called a *precapillary sphincter* surrounds the capillary at the arterial side of the bed and regulates blood flow into it. Contraction of the precapillary sphincter squeezes the capillary shut.

The precapillary sphincters act as valves that open and close capillary beds. For instance, while you are resting on the beach after finishing a picnic lunch, the capillary beds servicing the digestive organs will be open, and nutrients from the food will be absorbed. If you decide to go for a swim, the capillary beds of the digestive organs will close down, and those in the skeletal muscles will open. This redirection of blood flow can lead to abdominal cramps, which could have unfortunate consequences—especially if you are in deep water.

Veins return blood to the heart

Capillaries merge, forming the smallest of veins, called **venules.** Venules then join and form veins. **Veins** are blood vessels that return the blood to the heart.

Although veins share some structural features with arteries, there are some important differences. The walls of veins have the same three layers found in arterial walls, but the middle layer in veins has little smooth muscle (Figure 12.2). As a result, the walls of veins are thinner and more collapsible than those of arteries.

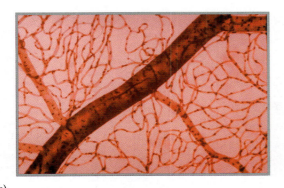

(a)

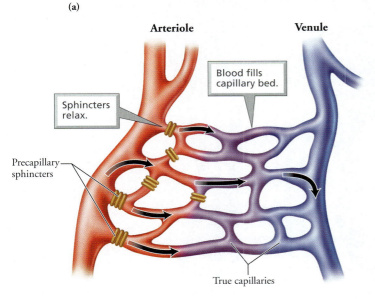

Arteriole

Venule

Blood fills capillary bed.

Sphincters relax.

Precapillary sphincters

True capillaries

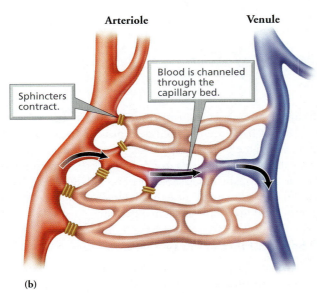

Arteriole

Venule

Blood is channeled through the capillary bed.

Sphincters contract.

(b)

FIGURE **12.5**

(a) A capillary bed is a network of capillaries. (b) The entrance to each capillary (the arterial side) is guarded by a ring of muscle called a precapillary sphincter. Blood flow through a capillary bed is regulated according to the body's metabolic needs.

Blood pressure in veins is generally low, so there is no danger that the thin walls will burst. The lumens of veins are larger than those of arteries of equal size. Together, the thin walls and large lumens allow veins to hold a large volume of blood. Veins serve as blood reservoirs, holding up to 65% of the body's total blood supply.

The same amount of blood that is pumped out of the heart must be conducted back to the heart, but it must be moved along the veins without assistance from the high pressure generated by the heart's contractions. How is this movement possible? Two design features of veins help. First, the relatively large lumens of veins reduce resistance to blood flow. But this feature alone would not be sufficient to move blood against the force of gravity—from the foot back to the heart, for instance. Second, veins often contain valves that act as one-way turnstiles that allow blood to move toward the heart but prevent it from flowing backward. Venous valves are pockets of connective tissue that emerge from the lining of the vein, as shown in Figure 12.6a.

Still, we have not identified a force that could move blood back to the heart. In the head and neck, of course, that force is gravity. But what happens in the lower parts of the body? The force is produced by simple muscle contraction. Virtually every time a skeletal muscle contracts, it squeezes nearby veins. This pressure pushes blood past the valves toward the heart. The mechanism propelling the blood is not unlike the one that causes toothpaste to squirt out of the uncapped end of the tube regardless of where the tube is squeezed. When skeletal muscles relax, any blood that moves backward fills the valves. As the valves fill with blood, they extend into the lumen of the vein, closing the vein and preventing blood from reversing direction (Figure 12.6b). Thus, the skeletal muscles are always squeezing the veins and driving blood toward the heart.

A simple experiment can demonstrate the effectiveness of venous valves. Allow your hand to hang by your side until the veins on the back of your hand become distended. Place two fingertips from the other hand at the end of one of the distended veins nearest to the knuckles. Then, leaving one fingertip pressed on the end of the vein, move the other toward the wrist, pressing firmly and squeezing the blood from the vein. Lift the fingertip near the knuckle and notice that blood immediately fills the vein. Repeat the procedure, but this time lift the fingertip near the wrist. You will see the vein remain flattened, because the valves prevent the backward flow of blood.

stop and think

If an artery is cut, blood is lost in rapid spurts. In contrast, blood loss through a cut vein has an even flow. What accounts for these differences?

VARICOSE VEINS

Varicose veins are veins that have become distended because blood is prevented from flowing freely, so it accumulates, or "pools," in the vein. As you may recall, the walls of veins are relatively thin, so they are easily stretched, especially if the vein is near the body surface, where there is little surrounding tissue to support it. Stretching is also more likely to occur in veins that must support a great column of blood, such as those in the legs.

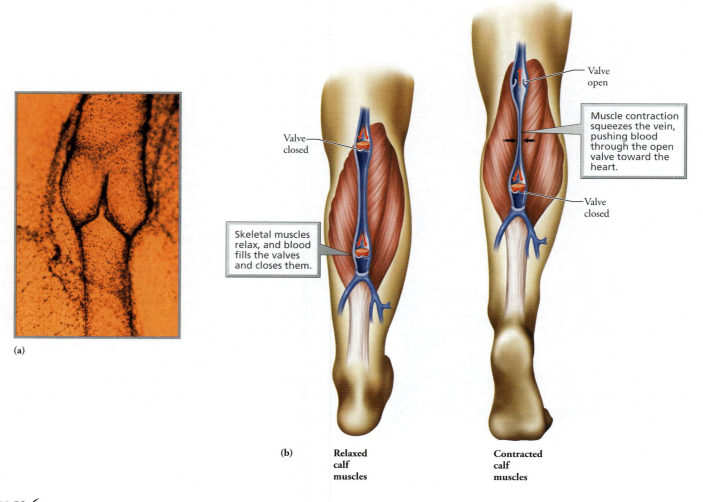

FIGURE **12.6**

(a) A micrograph of a vein showing a valve. (b) Pocketlike valves on the lining of veins assist the return of blood to the heart against gravity by preventing the backflow of blood.

Once a section of a vein has been stretched, extra blood can accumulate there and place stress on the valve below. Eventually, that valve can become stretched and ineffective, so the blood the valve once held falls back to the next valve. As a result, varicose veins may appear as bulging veins or as bluish purple, spiderlike lines.

stop and think

Why would sitting with your legs crossed favor the formation of varicose veins?

The Heart Is a Muscular Pump

The heart is an incredible, muscular pump that generates the force needed to circulate blood. It beats about 72 times a minute every hour of every day. To appreciate the work done by the heart, form a fist and alternately clench and relax it 70 times a minute. How many minutes does it take before the muscles of your hand are too tired to continue? In contrast, the heart does not fatigue. It beats over 100,000 times each day, which adds up to about 2 billion beats over a lifetime. The pumping done by the heart is equally remarkable. It pumps slightly less than 5 liters (10 pints) of blood a minute through its chambers, which adds up to over 9400 liters (2500 gallons) per day.

▌The heart functions as two separate pumps

The structure of the heart is key to its ability to function as a pump. The **myocardium**, which is most of the bulk of the heart, is composed of cardiac muscle tissue. The myocardium's ability to contract is responsible for the heart's incredible pumping action. The **endocardium** is a thin layer that lines the cavities of the heart. By reducing friction, the smooth surface of the endocardium lessens the resistance to blood flow through the heart. The **pericardium** is a fibrous sac that holds the heart in the center of the chest (thoracic) cavity without hampering its movements, even when the heart is contracting vigorously.

The heart appears to be a single structure, but the right and left halves function as two separate pumps (Figure 12.7). As we will see shortly, the right side of the heart pumps blood to the lungs, where it picks up oxygen. The left side pushes the blood to the body cells. The two pumps are physically separated by a partition called a septum. Each pump, or side of the heart, consists of two chambers: an upper chamber, called an **atrium** (plu-

ral, atria), and a lower chamber, called a **ventricle**. The two atria function as receiving chambers for the blood returning to the heart. The two ventricles function as the main pumps of the heart. The contraction of the ventricles forces blood out of the heart under great pressure. When we think about the work of the heart, we are, in fact, thinking about the work of the ventricles. It should not be surprising, then, that the ventricles are

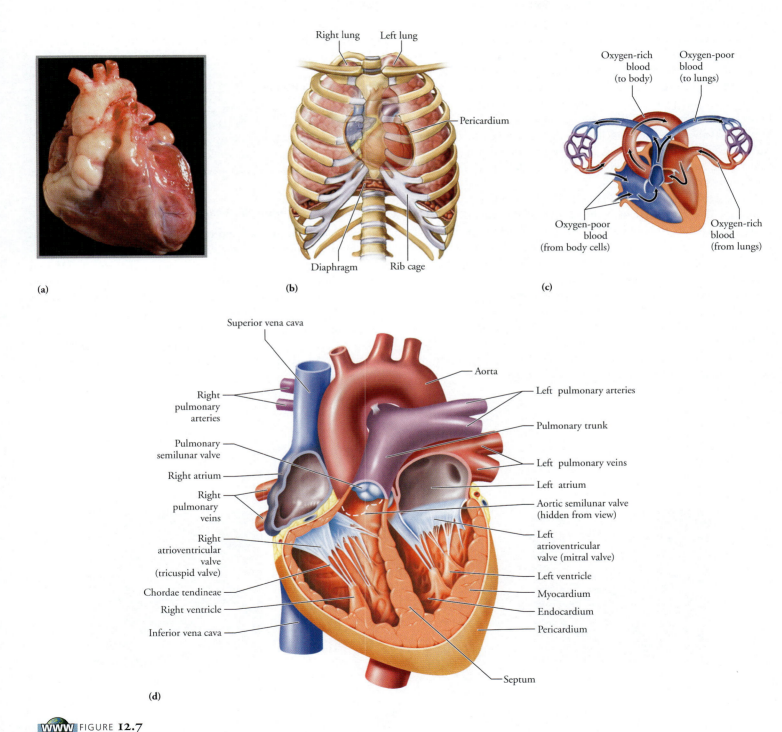

FIGURE 12.7

(a) The human heart. (b) The heart is located in the chest cavity. (c) Blood flows through the heart from the atria to the ventricles. (d) This diagram of a human heart shows the four chambers, the major vessels connecting to the heart, and the two pairs of heart valves.

much larger chambers than the atria and have thicker, more muscular walls.

Two pairs of valves ensure that the blood flows in only one direction through the heart. The first pair, the **atrioventricular (AV) valves**, is located between each atrium and ventricle, as shown in Figure 12.8. The AV valves are connective tissue flaps, called cusps, anchored to the wall of the ventricle by strings of connective tissue called the **chordae tendineae**—the heartstrings. These strings prevent the AV valves from flapping back into the atria under the pressure developed when the ventricles contract. The AV valve on the right side of the heart has three flaps and is called the tricuspid valve. The AV valve on the left side of the heart has two flaps and is called the bicuspid, or mitral, valve.

The second pair of valves, the **semilunar valves**, is located between each ventricle and its connecting artery. The cusps of the semilunar valves are small pockets of tissue attached to the inner wall of their respective arteries. These valves fill with blood, preventing the backflow of blood into the ventricles from the aorta or pulmonary artery.

The familiar sounds of the heart, which are often described as "lub-dup," are associated with the closing of the valves. The first heart sound ("lub") is produced when the AV valves snap shut as the ventricles begin to contract. The higher pitched second heart sound ("dup") represents the closure of the semilunar valves and the beginning of ventricular relaxation.

Heart murmurs, which are heart sounds other than "lub-dup," are created by turbulent blood flow. Although heart murmurs are sometimes heard in normal, healthy people, they can indicate a heart problem. For instance, malfunctioning valves often disturb blood flow through the heart, causing the swishing or gurgling sounds of heart murmurs. Several conditions can cause valves to malfunction.

In some cases, thickening of the valves narrows the opening and impedes blood flow. In other cases, the valves do not close properly and, therefore, allow the backflow of blood. In either case, the heart is strained because it must work harder to move the blood.

Defective valves can be replaced with artificial valves. Replacement heart valves can be made of animal tissues, such as a pig's aortic valve, or from synthetic materials (Figure 12.9). One popular model is designed as a ball-and-cage. Blood can flow in one direction by forcing the ball away from the ring into the cage. Blood flow in the opposite direction is prevented because blood forces the ball firmly into the ring.

stop and think

Abnormally short (or long) chordae tendineae of the mitral (bicuspid) valve can cause a condition known as mitral valve prolapse, in which the mitral valves do not close properly. Why would mitral valve prolapse cause heart murmurs?

▌Blood flows through the heart in two circuits

You may recall that the left and right sides of the heart function as two separate pumps. The two sides of the heart work simultaneously, but each pump circulates the blood through a different route, as shown in Figure 12.10. The right side of the heart pumps blood through the **pulmonary circuit**, which transports blood to and from the lungs. The left side of the heart pumps blood through the **systemic circuit**, which transports blood to and from body tissues. This arrangement prevents oxygenated blood (blood rich in oxygen) from mixing with deoxygenated blood (blood that is low in oxygen).

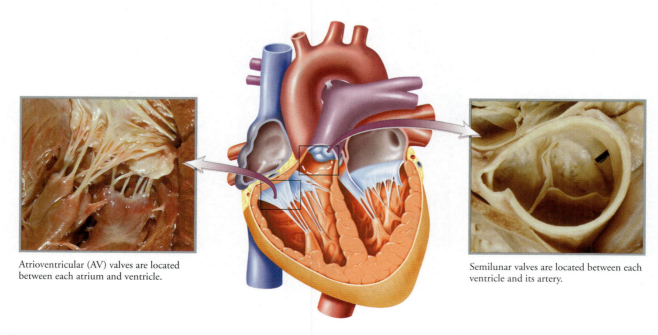

Atrioventricular (AV) valves are located between each atrium and ventricle.

Semilunar valves are located between each ventricle and its artery.

FIGURE **12.8**

The valves of the heart keep blood flowing in one direction.

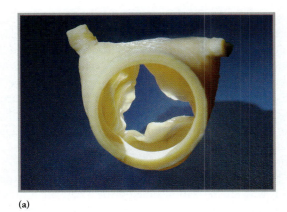

(a)

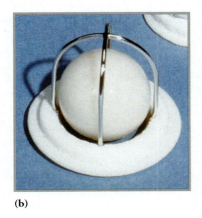

(b)

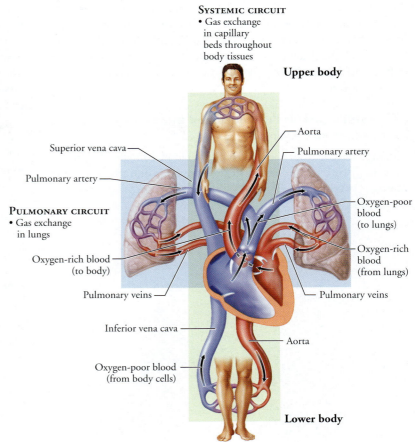

SYSTEMIC CIRCUIT
• Gas exchange in capillary beds throughout body tissues

Upper body

Superior vena cava

Pulmonary artery

PULMONARY CIRCUIT
• Gas exchange in lungs

Oxygen-rich blood (to body)

Pulmonary veins

Inferior vena cava

Oxygen-poor blood (from body cells)

Lower body

Aorta

Pulmonary artery

Oxygen-poor blood (to lungs)

Oxygen-rich blood (from lungs)

Pulmonary veins

Aorta

FIGURE **12.9**
Defective heart valves can be replaced by artificial valves made either from animal tissues or from synthetic materials. Here you see (a) a pig's aortic semilunar valve sewn onto a frame and (b) the ball-and-cage model.

FIGURE **12.10**
Circuits of blood flow. The right side of the heart pumps blood through the pulmonary circuit, which carries blood to and from the lungs. The left side of the heart pumps blood through the systemic circuit, which conducts blood to and from the body tissues.

The pulmonary circuit begins in the right atrium, as veins return oxygen-poor blood from the systemic circuit. (You can trace the flow of blood through the heart in Figure 12.10 as you read the following description.) The blood then moves from the right atrium to the right ventricle. Contraction of the right ventricle pumps poorly oxygenated blood to the lungs through the pulmonary trunk (main pulmonary artery), which divides to form the left and right **pulmonary arteries**. In the lungs, oxygen diffuses into the blood, and carbon dioxide diffuses out. The oxygen-rich blood is delivered to the left atrium through four **pulmonary veins**, two from each lung. (Note that the pulmonary circulation is an *exception* to the general rule that arteries carry oxygen-rich blood, and veins carry oxygen-poor blood. Exactly the opposite is true of vessels in the pulmonary circulation.) The pathway of blood pumped through the pulmonary circuit by the right side of the heart is

Right atrium → AV valve (tricuspid) → Right ventricle →

Pulmonary semilunar valve → Pulmonary trunk →

Pulmonary arteries → Lungs → Pulmonary veins → Left atrium

The systemic circuit begins when oxygen-rich blood enters the left atrium (see Figure 12.10). Blood then flows to the left ventricle.

When the left ventricle contracts, oxygenated blood is pushed through the largest artery in the body, the aorta. The aorta arches over the top of the heart and gives rise to the smaller arteries that eventually feed the capillary beds of the body tissues. The venous system collects the oxygen-depleted blood and eventually culminates in veins that return the blood to the right atrium. These veins are the *superior vena cava*, which delivers blood from regions above the heart, and the *inferior vena cava*, which returns blood from regions below the heart. Thus, the pathway of blood through the systemic circuit pumped by the left side of the heart is

Left atrium → AV (bicuspid or mitral) valve → Left ventricle →

Aortic semilunar valve → Aorta → Body tissues →

Inferior vena cava or superior vena cava → Right atrium

■ Coronary circulation serves the heart muscle

Cells of the heart muscle themselves are not nourished by blood flowing through the heart's chambers. Instead, an extensive network of vessels, known as the **coronary circulation**, services the tissues of the heart. The first two arteries that branch off the aorta are the coronary arteries (Figure 12.11). These arteries give rise to numerous branches, ensuring that the heart receives a rich supply of oxygen and nutrients. After passing through the capillary beds that nourish the heart tissue, blood enters cardiac veins, which join to form the *coronary sinus*. Blood flows from the coronary sinus to the right atrium.

The cardiac cycle is the sequence of heart muscle contraction and relaxation

Although the two sides of the heart pump blood through different circuits, they work in tandem. First, the two atria contract. Next, both ventricles contract, squeezing blood into their respective arteries.

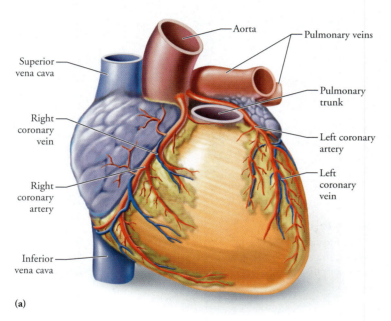

(a)

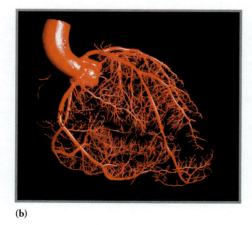

(b)

FIGURE **12.11**

Coronary circulation. (a) The coronary vessels deliver a rich supply of oxygen and nutrients to the heart muscle cells and remove the metabolic wastes. (b) This cast of the coronary blood vessels reveals the complexity of the coronary circuit.

We see, then, that a "heartbeat" is not a single event. Each beat involves contraction, which is called **systole**, as well as relaxation, which is called **diastole**. All the events associated with the flow of blood through the heart chambers during a single heartbeat are collectively called the **cardiac cycle**, as illustrated in Figure 12.12. First, all chambers relax and blood passes through the atria and enters the ventricles. When the ventricles are about 70% filled, the atria contract and push their contents into the ventricles. Then the atria relax and the ventricles begin their contraction phase. Upon completion of this contraction, the whole heart again relaxes. If we were to add the contraction time of the heart during a day and compare it with the relaxation time during a day, the heart's workday might turn out to be equivalent to yours. In 24 hours, the heart spends a total of about 8 hours working (contracting) and 16 hours relaxing. However, unlike your workday, the heart's day is divided into repeating cycles of work and relaxation.

The rhythmic contraction of the heart is due to its internal conduction system

If a human heart is removed, as in a transplant operation, and placed in a dish, it will continue to beat, keeping a lonely and useless rhythm until its tissues die. In fact, if a few cardiac muscle cells are grown in the laboratory, they too will beat on their own, each twitch a reminder of the critical role the intact organ plays. Clearly, then, the heart muscle does not require outside stimulation to beat. Instead, the tendency is intrinsic, within the heart muscle itself.

Another remarkable observation has been made of heart muscle cells grown in a laboratory dish. Although isolated heart cells twitch independently of others, if two cells should touch, they will begin beating in unison. This, too, is part of their basic nature and has a lot to do with the type of connections between cells. The cell membranes of adjacent cardiac muscle cells weave together extensively, forming specialized boundaries called *intercalated disks*. Cell junctions in the intercalated disks mechanically and electrically couple the connected cardiac muscle cells. Adjacent cells are held together so tightly that they do not rip apart during contraction, and the pull of contraction is transmitted from one cell to the next. At the same time, the junctions permit electrical communication between adjacent cells, allowing the electrical events responsible for contraction to spread rapidly over the heart by passing from cell to cell. So the cells contract in a coordinated manner because of the heart's construction. However, this intrinsic tendency to contract still needs some outside control to contract at the proper rate.

The tempo of the heartbeat is set by a cluster of specialized cardiac muscle cells, called the **sinoatrial (SA) node,** located in the right atrium near the junction of the superior vena cava (Figure 12.13). Because the SA node sends out impulses that initiate each heartbeat, it is often referred to as the **pacemaker**. About 70 to 80 times a minute, the SA node sends out an electrical signal that spreads through the muscle cells of the atria, causing them to contract. The signal reaches another cluster of specialized muscle cells called the **atrioventricular (AV) node,** located in the partition between the two atria, and stimulates it.

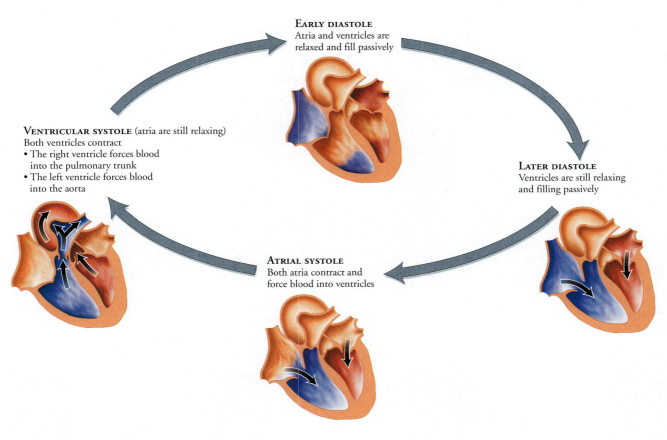

EARLY DIASTOLE
Atria and ventricles are
relaxed and fill passively

LATER DIASTOLE
Ventricles are still relaxing
and filling passively

ATRIAL SYSTOLE
Both atria contract and
force blood into ventricles

VENTRICULAR SYSTOLE (atria are still relaxing)
Both ventricles contract
• The right ventricle forces blood
  into the pulmonary trunk
• The left ventricle forces blood
  into the aorta

WWW
TUTORIAL
12.3
FIGURE **12.12**

The cardiac cycle—the sequence of heart muscle relaxation and contraction. The atria contract together and the ventricles contract together. Red indicates blood high in oxygen. Blue indicates blood low in oxygen.

The AV node then relays the stimulus by means of a bundle of specialized muscle fibers, called the **atrioventricular bundle,** that runs along the wall between the ventricles. The bundle forks into right and left branches and then divides into many other specialized cardiac muscle cells, called **Purkinje fibers,** which penetrate the walls of the ventricles. Thus, the electrical signal is conducted to the ventricles through the atrioventricular bundle and quickly fans out through the ventricle walls, carried by the Purkinje fibers. The rapid spread of the impulse through the ventricles ensures that they contract smoothly.

When the heart's conduction system is faulty, cells can begin to contract independently. Such cellular independence can result in rapid, irregular contractions of the ventricles, called *ventricular fibrillation,* which renders the ventricles useless as pumps and stops circulation. The brain no longer receives the blood it needs to function. Death occurs unless an effective heartbeat is restored quickly. Ventricular fibrillation can be stopped by delivering a strong electric shock to the heart, causing the entire heart to contract at once; in many cases, the SA node will once again begin to function normally. Although costly, an implantable defibrillator (a device that electrically shocks the heart so that all its cells contract at once, allowing the natural pacemaker to restart the heart) can provide a life-saving "jump start" to the heart as needed and immediately.

Problems with the conduction system of the heart can sometimes be treated with an artificial pacemaker, a small device that monitors the heart rate and rhythm and responds to abnormalities if they occur. For instance, if the heart rate becomes too slow, the pacemaker will send electrical signals to the heart through an electrode. Today, many types of pacemakers can vary the heart rate to suit the body's needs. The pacemaker can be worn on a belt, if it will be needed for only a short time, such as during the temporary disturbance of the heart's rhythm that can follow a heart attack. However, if the need for a pacemaker will be permanent, it can be inserted surgically under the skin on the chest wall.

REGULATION OF HEART RATE

The pace or rhythm of the heartbeat changes constantly in response to activity or excitement. The autonomic nervous system and certain hormones make the necessary adjustments so that heart rate suits the body's needs. During times of stress, the sympathetic nervous system increases the rate and force of heart contractions. At this time, the adrenal medulla produces epinephrine. This hormone can then prolong the effects of the sympathetic nervous system. In contrast, when restful conditions prevail, the parasympathetic nervous system dampens heart activity, bringing it in line with the body's more modest metabolic needs.

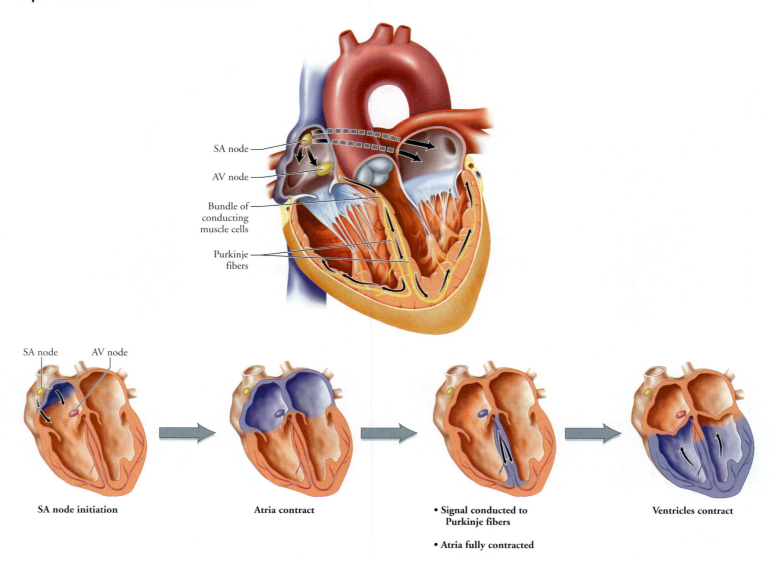

SA node

AV node

Bundle of conducting muscle cells

Purkinje fibers

SA node   AV node

SA node initiation

Atria contract

• Signal conducted to Purkinje fibers

• Atria fully contracted

Ventricles contract

FIGURE **12.13**

The conduction system of the heart consists of specialized cardiac muscle cells that speed electrical signals through the heart. The sinoatrial (SA) node serves as the heart's internal pacemaker that determines the heart rate. Electrical signals from the SA node spread through the walls of the atria, causing them to contract. The signals then stimulate the atrioventricular (AV) node, which in turn sends the signals along the atrioventricular bundle to its forks and finally to the many Purkinje fibers that penetrate the ventricular walls. The Purkinje fibers distribute the signals to the walls of the ventricles, causing them to contract. Electrical activity is indicated by the blue color in the wall of the heart.

what would you do?

There are thousands more people in need of a heart transplant than there are available donor hearts. An artificial heart, AbioCor®, is now approved by the Food and Drug Administration. To qualify as a recipient, a person must have a 70% chance of dying within the next 30 days. Some of the recipients have died within a day of receiving the artificial heart. Most lived several months. As of mid-2004, the longest-living recipient survived for 17 months. If you were in need of a transplant, would you volunteer to participate in the testing of an artificial heart? What criteria would you use to decide?

An electrocardiogram is a recording of the electrical activities of the heart

The electrical events that spread through the heart with each heart-beat actually travel throughout the body because body fluids are good conductors. Electrodes on the body surface can detect these electrical events. The electrical changes associated with each heart-beat cause deflections (movements) in the tracing made by the recording device. An **electrocardiogram** (an **ECG** or **EKG**) is a recording of the electrical activities of the heart.

A typical ECG consists of three distinguishable deflection waves, as shown in Figure 12.14. The first wave, called the P wave, these waves are remarkably consistent in a healthy heart, abnormal patterns can indicate heart problems.

Blood pressure is the force blood exerts against blood vessel walls

We hear a lot about blood pressure, usually when someone frets about how high it is or brags about how low it is. We have already referred to blood pressure, but what exactly is it? Blood pressure is the force exerted by the blood against the walls of the blood vessels. When the ventricles contract, they push blood into the arteries under great pressure. This pressure is the driving force that moves blood through the body, and the same pressure pushes outward against vessel walls. So blood pressure must be great enough to circulate the blood but not so great that it stresses the heart and blood vessels.

Blood pressure in the arteries varies predictably during each heartbeat. It is highest during the contraction of the ventricles (systole), when blood is being forced into the arteries. The **systolic pressure**, the highest pressure in the artery during each heartbeat, of a typical, healthy adult is less than 120 mm of mercury (mm Hg).[1] Blood pressure is lowest when the ventricles are relaxing (diastole). In a healthy adult, the lowest pressure, or **diastolic pressure** is less than 80 mm Hg. A person's blood pressure is usually expressed as two values—the systolic followed by the diastolic. For instance, normal adult blood pressure is said to be less than 120/80. (Do you know what your blood pressure is?)

Blood pressure is measured with a device called a *sphygmomanometer*,[2] which consists of an inflatable cuff that wraps around the upper arm attached to a device that can measure the pressure within the cuff. The idea is to measure blood pressure by applying a counter pressure within the cuff by pumping air into it. As the air is slowly released from the cuff, a technician uses a stethoscope to listen for blood flow through the brachial artery, which runs along the inner surface of the arm (Figure 12.15).

Cardiovascular Disease is a Major Killer in the United States

Cardiovascular disease is the single biggest killer of men and women in the United States. It affects slightly more men than women because, until menopause, women gain some natural protection from the female hormone estrogen. Ironically, although slightly fewer women than men have heart attacks, women who do have a heart attack are twice as likely to die within the following weeks than are men. Because heart attacks are commonly thought of as a male problem, women and their physicians often fail to recognize the symptoms, a failure that delays treatments that could be lifesaving. So let's examine common problems with the heart and blood vessels with an eye to how we can reduce our risks of developing them. See the Health Issue essay, *The Cardiovascular Benefits of Exercise.*

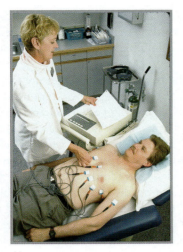

(a)

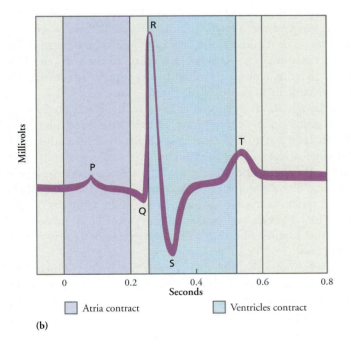

(b)

FIGURE **12.14**

(a) A person having an ECG recorded. (b) The electrical activity that accompanies each heartbeat can be visualized in an electrocardiogram (ECG). The P wave is generated as the electrical signals from the SA node spread across the atria and cause them to contract. The QRS wave represents the spread of the signal through the ventricles and ventricular contraction. The T wave occurs as the ventricles recover and return to the electrical state that preceded contraction.

accompanies the spread of the electrical signal over the atria and the atrial contraction that follows. The next wave, the QRS wave, reflects the spread of the electrical signal over the ventricles and ventricular contraction. The third wave, the T wave, represents the return of the ventricles to the electrical state that preceded contraction (ventricular repolarization). Because the pattern and timing of

---

[1]Pressure is measured as the height to which that pressure could push a column of mercury (Hg).

[2]*Sphygmomanometer* is pronounced sfig-mo-mah-*nom*-eter.

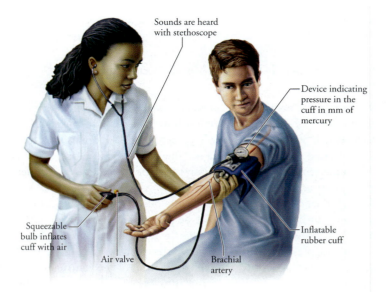

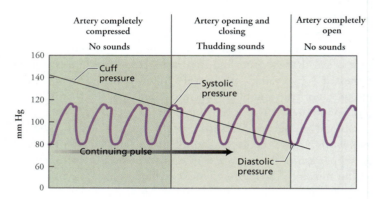

FIGURE **12.15**

Blood pressure is measured with a sphygmomanometer, which consists of an inflatable cuff and a means of measuring the pressure within the cuff. The cuff is placed around the upper arm and inflated so that it compresses the brachial artery. The pressure in the cuff is slowly released. As the cuff pressure is released, blood is able to spurt through the constricted artery only at the points of highest blood pressure. The pressure when "tapping" sounds are first heard is the systolic pressure, the blood pressure when the heart is contracting. As the pressure in the cuff drops, a point is reached when the sounds disappear. The blood is now flowing continuously through the brachial artery. The pressure in the cuff at this point is the diastolic pressure, the blood pressure when the heart is relaxing.

High blood pressure can kill without producing symptoms

High blood pressure, or **hypertension**, plagues over 58 million Americans. It accounts for more outpatient prescriptions than any other disease in America. Hypertension is often called the silent killer. It is *silent* because it does not produce any telltale symptoms. It is a *killer* because it can cause fatal problems, usually involving the heart, brain, blood vessels, or kidneys. Hypertension damages the heart in a number of ways but primarily by causing the heart to work harder to keep the blood moving. In response, the heart muscle thickens, and the heart enlarges; the enlarged heart works less

efficiently and has difficulty keeping up with the body's needs. At the same time, the increased workload increases the heart's need for oxygen and nutrients. If these cannot be delivered rapidly enough, a heart attack can result.

In addition to harming the heart, high blood pressure can also damage the blood vessels by causing clogging or rupture. High blood pressure promotes the development of **atherosclerosis**, a condition in which fatty deposits form within the walls of arteries and, in some cases, obstruct blood flow. Atherosclerosis (*ather-o*, yellow, fatty deposit; *sclerosis*, a hardening) is a leading cause of heart attacks (death of heart muscle cells) and strokes (death of brain cells). Furthermore, the added stress placed on the arteries when the blood pressure is high can cause an artery to rupture, leading to blood loss, which can be fatal. The blood vessels of the brain generally have less supporting tissue than other blood vessels do. Thus, these vessels are particularly likely to burst with high blood pressure. The kidneys are also vulnerable to the effects of hypertension. The damage begins as the walls of the arterioles that supply the filtering system of the kidney thicken and the lumens become narrowed, thereby reducing blood flow. The reduced blood flow can cause kidney damage. The kidneys make matters worse by responding to the reduced blood flow by secreting renin, a chemical that leads to further increases in blood pressure in an ever-escalating cycle.

Although about 90% of the cases of hypertension have no *known* cause, many contributing factors have been identified. Between 10% and 40% of cases have a genetic basis. In other cases, the kidneys have an impaired ability to handle sodium, which results in fluid retention; fluid retention increases blood pressure by increasing blood volume. In yet other people, the sympathetic nervous system reacts too strongly to stressful conditions by narrowing blood vessels and increasing heart rate. Thus, more blood per minute is pumped through vessels that provide a greater resistance to flow.

At what point should blood pressure be considered high? This question is one of the medical profession's most vexing problems. Most physicians would agree that a blood pressure of 160/90 is high and should be treated. But uncertainty clouds the treatment issue when the person's diastolic pressure is between 80 and 89. Although drug treatment may help in borderline cases, it usually must be continued for life. A diagnosis of high blood pressure may influence other aspects of a person's life, such as life insurance premiums. So, sometimes, only lifestyle changes are recommended for mild cases of hypertension (140–159/90–99).

A high upper (systolic) value usually suggests that the person's arteries have become hardened by mineral deposits and are no longer able to dampen the high pressure of each heartbeat. This condition is known as *arteriosclerosis*. The lower (diastolic) value is generally considered more important because it indicates the pressure when the heart is relaxing.

When the diagnosis of hypertension is clear, many drugs can be prescribed, each of which combats a specific mechanism that contributes to high blood pressure. Some drugs, the diuretics for instance, decrease blood volume by increasing the excretion of sodium and fluids. Other drugs cause the blood vessels to dilate (become wider), which decreases peripheral resistance by making blood flow easier.

A number of changes in lifestyle are recommended to treat or prevent hypertension.

1. **Control weight.** A close relationship exists between being overweight and high blood pressure. Thus, maintaining normal body weight can help control blood pressure. Many overweight people with high blood pressure benefit from shedding just a few extra pounds. The best way to lose weight is to eat a moderate, balanced diet, reduce fat intake, and increase physical activity.

2. **Exercise regularly.** Aerobic exercise, such as brisk walking, jogging, swimming, or cycling, performed for at least 20 minutes three times a week, helps lower blood pressure and keeps it low.

3. **Do not smoke.** Cigarette smoke contains nicotine, a drug that increases heart rate and constricts blood vessels; both of those effects increase blood pressure.

4. **Limit dietary salt.** Some people with hypertension can lower their blood pressure by lowering the amount of salt in their diet. Salt can affect fluid retention and, therefore, blood volume.

5. **Limit alcohol intake.** Alcohol consumption can elevate blood pressure. It can also interact with medications prescribed to treat hypertension.

▍Atherosclerosis is a buildup of lipids in the artery walls

Atherosclerosis, as mentioned earlier, is a buildup of fatty deposits in the walls of arteries that is fueled by an inflammatory response. In some cases, the deposits narrow the artery (Figure 12.16). Such narrowing causes problems because it reduces blood flow through the vessel, choking off the vital supply of oxygen and nutrients to the tissues served by that vessel. But, contrary to the beliefs held just a few years ago, atherosclerosis is more than just a plumbing problem—a clog in a passive pipeline. Inflammatory chemicals in the artery wall cause weakening of the wall over the deposit, allowing the deposit to rupture. This rupture then triggers the formation of a blood clot. Either way, blood flow through the vessel is restricted. The starved cells lying downstream may die, an event that can be fatal, especially if the affected cells are those of the heart (a heart attack) or of the brain (a stroke).

The inflammation process underlying atherosclerosis is the same one that wards off infection when you scrape your knee. However, in this case it can have dire consequences. Infection or irritants, including chemicals in cigarette smoke or cholesterol deposits in the artery lining, may initiate the inflammation. Low-density lipoproteins (LDLs), the so-called bad form of cholesterol, leave the blood and collect amid the cells of the artery lining. This accumulation is most likely to occur when the LDL concentration in the blood is high. It may be easier for LDLs to collect when the lining of the artery is damaged slightly, as might occur because of excessively rapid or turbulent blood flow caused by high blood pressure. Within the arterial wall, the LDLs undergo oxidation, a chemical process similar to the one that causes pipes to rust. These chemical changes in the LDLs cause the cells of the artery lining to enlist the body's defense responses: inflammatory chemicals and defense cells. Certain defense cells secrete a growth factor that attracts smooth muscle cells from the middle layer of the artery and causes them to proliferate. Other cells engulf lipids, such as cholesterol, enlarge, and form fatty streaks on the artery lining. As these cells continue to scavenge lipids, the fatty streak enlarges and forms plaque, a bumpy, fatty layer in the artery wall. The plaque can bulge into the artery channel, blocking blood flow, or it can expand outward in the artery wall. In either event, the plaque is covered with a fibrous cap that keeps pieces from breaking away. The fat-filled cells secrete inflammatory substances that weaken the cap. A small break in that protective cap can allow the plaque to rupture. At the same time, these

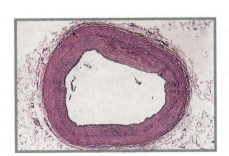

(a) A normal artery

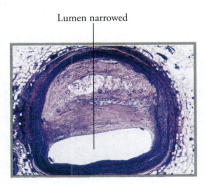

Lumen narrowed

(b) An artery partially obstructed with plaque

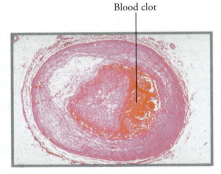

Blood clot

(c) An artery completely obstructed with plaque and a blood clot

FIGURE **12.16**

Atherosclerosis is a low-level inflammatory response in the wall of an artery associated with the formation of fat-filled plaques. The arterial wall thickens because of the proliferation of smooth muscle and the accumulation of lipids, primarily cholesterol. Together, the smooth muscle and cholesterol form plaque, which is typically covered with a protein cap. Plaque can obstruct blood flow through the artery, thus depriving the cells that would be fed life-sustaining blood by the artery. Plaque can also rupture, causing a blood clot to form. The clot may then completely clog the vessel and cause the death of tissue downstream.

The Cardiovascular Benefits of Exercise

What would you say if you were offered something that would reduce your risk of heart attack, stroke, diabetes, and cancer while controlling your weight, strengthening your bones, relieving anxiety and tension, and improving your memory? "Impossible!" you might say. "What's the catch?" There is none. This seemingly magical elixir of life is regular aerobic exercise, which uses the same large muscle groups rhythmically and continuously and elevates heart rate and breathing rate for at least 15 to 20 minutes.

Although exercise has many beneficial effects throughout the body, here we will con-sider only the benefits to the cardiovascular sys-tem. Exercise benefits the heart in several ways: It makes the heart a more efficient pump, thus reducing its workload. A well-exercised heart beats more slowly than that of a sedentary per-son—during both exercise and rest. For instance, the heart rate of a well-trained athlete is about 40 to 60 beats per minute, compared with the average 72 beats per minute. The lower heart rate gives the heart more time to rest between beats. At the same time, however, the well-exercised heart pumps more blood with each beat.

Exercise also increases the oxygen supply to the heart muscle by widening the coronary arteries, thereby increasing blood flow to the heart. Also, because the capillary beds within the heart muscle become more extensive with regular exercise, oxygen and nutrients can be delivered to the heart cells and wastes can be removed more quickly.

Furthermore, exercise helps to ensure con-tinuous blood flow to the heart. One way it accomplishes this benefit is by increasing the body's ability to dissolve blood clots that can lead to heart attacks or strokes. Exercise stimu-lates the release of a natural enzyme, called tis-sue plasminogen activator (tPA), that prevents blood clotting. The effect of tPA is immediate and lasts for as long as $1\frac{1}{2}$ hours after you stop

cells secrete chemicals that promote blood clotting. The clot (thrombus) can limit or completely halt blood flow through the ves-sel. Because plaques that rupture do not necessarily bulge into the artery and cause symptoms of atherosclerosis, heart attacks often occur in patients without previous symptoms.

stop and think

C-reactive protein (CRP) is an inflammatory chemical released by injured cells in the artery lining. Why might CRP prove to be a better predictor of atherosclerosis than is blood cholesterol level?

Any artery can be affected by atherosclerosis. In some people, it occurs mainly in the vessels of the legs. The carotid arteries that deliver blood to the brain may also be clogged with plaque. In this case, the risk is that the reduced flow will cause brain cells to die, resulting in a stroke. For some reason, however, the most vulnera-ble arteries are those of the heart. Therefore, we will consider this problem of the heart in more detail.

Coronary artery disease is atherosclerosis in the coronary arteries

Coronary artery disease (CAD) is a condition in which fatty deposits associated with atherosclerosis form within coronary arter-ies, those that nourish the heart muscle. Coronary artery disease is the underlying cause of the vast majority of heart attacks.

The problem of supplying blood to the heart muscle is one of supply and demand. The harder the heart works, the more oxygen it needs to continue pumping. If the blood flow through the coro-nary arteries cannot meet the heart's demand, warning symptoms, such as chest pain or irregularities in heartbeat, may occur. A tem-porary reduction in blood supply due to obstructed blood flow, called ischemia, causes reversible damage to heart muscle.

A temporary shortage of oxygen to the heart is accompanied by angina pectoris—chest pain, usually experienced in the center of the chest or slightly to the left. The name *angina* comes from the Latin word *angere*, meaning "to strangle." The name is apt, since the pain of angina is often described as suffocating, viselike, or choking. Typically, the pain begins during physical exertion or emotional stress, when the demands on the heart are increased and the blood flow to the heart muscle can no longer meet the needs. The pain stops after a period of rest. This type of angina is called stable angina. However, blood flow can also be shut down by spasms (intense contractions) of the smooth muscle in the walls of the coronary arteries. The spasms also produce chest pain. In this case, the pain is called variant angina, because it occurs during rest rather than during exertion.

Angina serves as a warning that part of the heart is receiv-ing insufficient blood through the coronary arteries, but it does not cause permanent damage to the heart. The warning should be taken seriously, however, because each year up to 15% of those people who have angina die suddenly from a heart attack. The lack of oxygen stimulates white blood cells called neu-trophils to release a flood of harmful substances. These sub-stances can trigger ventricular fibrillation, a condition in which the ventricles beat extremely rapidly but ineffectively. As a result, the brain no longer gets oxygen, and the victim slips into unconsciousness. Death follows if a normal heartbeat is not quickly restored by cardiopulmonary resuscitation (CPR) or by a defibrillator.

Although coronary artery disease is usually diagnosed from the symptoms of angina and a physical examination, another procedure allows a physician to spot areas in the coronary arteries that have

exercising. Besides this, exercise stimulates the development of collateral circulation, that is, additional blood vessels that provide alternative pathways for blood flow. As a result, blood flows continuously through the heart, even if one vessel becomes blocked.

At the same time, there are blood changes that allow more oxygen to be delivered to the cells. The amount of hemoglobin, the oxygen-binding protein in red blood cells, increases. In addition, the blood volume and the numbers of red blood cells increase.

The blood vessels also profit from regular exercise. Exercise lowers the risk of coronary artery disease (CAD) by lowering blood pressure. Exercise also lowers blood pressure by keeping arteries elastic and relieving nervous tension.

Besides lowering blood pressure, exercise lowers the risk of coronary artery disease by shifting the balance of lipids in the blood. High-density lipoproteins (HDLs), which are the "good" form of cholesterol-carrying particles that remove cholesterol from the arterial walls, increase with exercise. Low-density lipoproteins (LDLs), the "bad" form of cholesterol-carrying particles that deposit cholesterol in arterial walls, decrease.

Not all exercise promotes cardiovascular fitness. Cardiovascular conditioning requires that the exercise be aerobic, which means that it promotes the use of oxygen and can be sustained for at least 2 minutes without causing shortness of breath. Thus, to reap cardiovascular benefits, you must exercise hard enough and

long enough. How hard? The exercise must be vigorous enough to elevate your heart rate to the target zone. The heart rate target zone is between 70% and 85% of your maximal attainable heart rate, which can be determined by subtracting your age in years from 222 beats per minute. How long? The exercise must continue for at least 20 minutes and be performed at least 3 days a week, with no more than 2 days between sessions.

Doing *something* active on a regular basis can improve the quality of life. Moderate activity helps you feel better—emotionally and physically. This difference has been likened to traveling first class instead of coach. 👫

become narrowed by atherosclerosis. In this procedure, which is called coronary angiography, a contrast dye that is made visible by x-rays is released in the heart so that the coronary vessels can be seen on film. A catheter (a slender, flexible tube) is inserted into an artery in the arm or leg and then threaded through the blood vessels until it reaches the heart. The dye is then squirted into the openings of the coronary arteries. The movement of the dye through the arteries is then recorded in a series of high-speed x-rays.

Coronary artery disease can be treated with medicines or with surgery. Among the medicines commonly used are those that dilate (widen) blood vessels, such as nitroglycerin. Certain other drugs

specifically dilate the coronary arteries. Wider blood vessels make it easier for the heart to pump blood through the circuit. When the coronary arteries dilate, more blood is delivered to the heart muscle. Also used are drugs that dampen the heart's response to stimulation from the sympathetic nervous system, thus decreasing its need for oxygen.

Three surgical operations used to treat coronary artery disease are balloon angioplasty, atherectomy, and coronary artery bypass. *Angioplasty* involves widening the channel of an artery that is narrowed by soft, fatty plaque by inflating a tough, plastic balloon inside the artery (Figure 12.17). The tiny, deflated balloon is

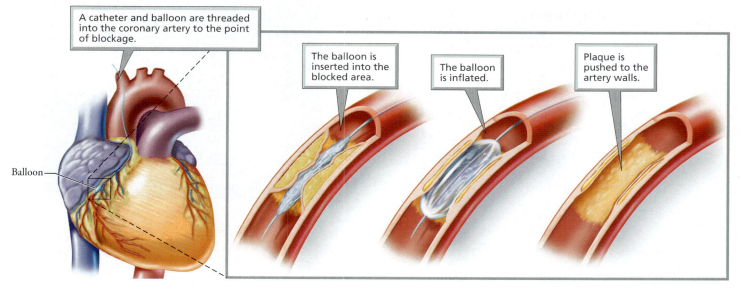

A catheter and balloon are threaded into the coronary artery to the point of blockage.

The balloon is inserted into the blocked area.

The balloon is inflated.

Plaque is pushed to the artery walls.

Balloon

FIGURE **12.17**

Balloon angioplasty opens a partially blocked artery.

attached to the end of a long tube, called a catheter. It is inserted through an artery in the arm or leg and then is pushed to the blocked spot in a coronary artery; x-rays are used to follow its progress. The balloon is inflated under pressure, stretching the artery and pressing the soft plaque against the wall. The artery is immediately widened.

Balloon angioplasty is a mixed bag of costs and benefits. The success rate is over 90%, and death rate during the procedure is low—only about 1%. Unfortunately, in about 20% of the patients, the artery narrows again within a few months, and additional treatment is required.

A more direct approach to widening the blocked artery is removing the deposits responsible for blockage, a procedure called *atherectomy*. Again, a catheter is guided to the site of blockage, and one of several surgical instruments is moved through the catheter. With one type of instrument, plaque is scraped away from the artery wall. Another type attacks the plaque with laser beams. The plaque fragments are then sucked into the catheter so that they do not block smaller vessels downstream.

After angioplasty or atherectomy, physicians commonly insert a metal-mesh tube called a stent into the treated arteries. The stent prevents the arteries from collapsing and prevents loose pieces of plaque from being swept into the bloodstream. Although stents boost the percentage of arteries that stay open, they sometimes trigger an inflammatory response or become clogged again.

An alternative surgical procedure is a *coronary bypass*, in which a segment of a leg vein is removed and grafted so that it provides an alternate pathway between the aorta and a coronary artery—past the point of obstruction (Figure 12.18). Blood flow then "bypasses" the clogged region of the artery. A coronary bypass usually is performed if the blockage in a coronary artery is located in a critical spot or if there are simply too many bottlenecks to correct using balloon angioplasty. Although coronary bypass surgery provides great relief from angina, it does not seem to lengthen a person's life.

stop and think

In a coronary bypass operation, why is it important for the surgeon to suture the vein in place in the correct orientation? What would happen if the piece of vein were sutured in backward?

Heart Attack Is the Death of Heart Muscle

In a heart attack, technically known as a *myocardial infarction*, a part of the heart muscle dies because of an insufficient blood supply. (*Myocardial* refers to heart muscle; *infarct* refers to dead tissue.) Heart muscle cells begin to die if they are cut off from their essential blood supply for more than 2 hours. Depending on the extent of damage, the effects of a heart attack can spread quickly throughout the body: The brain receives insufficient oxygen; the lungs fill with fluid; and the kidneys fail. Within a short time, white blood cells swarm in to remove the damaged tissue. Then, over the next 8 weeks or so, scar tissue replaces the dead cardiac muscle (Figure 12.19). Because scar tissue cannot contract, part of the heart permanently loses its pumping ability.

Heart attacks have several possible causes. The most common type of heart attack is a coronary thrombosis, which means that it is caused by a blood clot blocking a coronary artery. This blockage generally does not happen unless the artery already contains

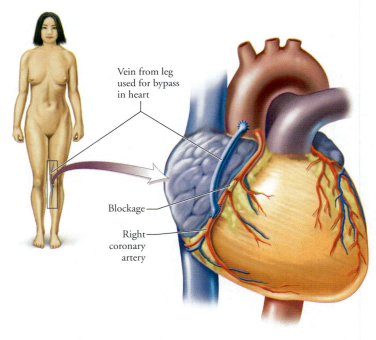

FIGURE **12.18**

In coronary bypass surgery, a section of a leg vein is removed. One end of the vein is attached to the heart's main artery, the aorta, and the other to a coronary artery, bypassing the obstructed region. The grafted vein provides a pathway through which blood can reach the previously deprived region of heart muscle.

Vein from leg used for bypass in heart

Blockage

Right coronary artery

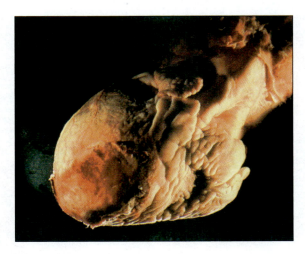

FIGURE **12.19**

The ravages of a prior heart attack are visible as scar tissue at the bottom of this lifeless heart. Scar tissue replaced cardiac muscle when the blood supply to the heart muscle was shut down. Scar tissue cannot contract. Therefore, that part of the life-sustaining pump becomes ineffective. The most common cause of a heart attack is a coronary thrombosis, in which a blood clot blocks blood flow through a coronary artery.

plaques of atherosclerosis, as occurs in coronary artery disease. In some instances, the blood clot is formed elsewhere in the body but is swept along in the bloodstream and lodges in a coronary artery. In still other instances, the blockage may be temporary, caused by a constriction of a coronary artery, called a coronary artery spasm.

Chest pain is a common symptom of heart attack, especially in men. In some cases, the pain is a severe, crushing pain that begins in the center of the chest and often spreads down the inside of one or both arms (most commonly the left one), as well as up to the neck and shoulders. Although the pain is usually severe enough to cause the victim to stop whatever he or she is doing, it is not always so overwhelming that it is recognized as a heart attack. A heart attack may also cause nausea and dizziness, which can prompt the victim to write off the symptoms as an upset stomach. Oddly, persons who experience severe pain may be the lucky ones, because they are more likely to realize that the cause is a heart attack and seek immediate help. Doubt about the cause of the symptoms and denial of a heart attack often go hand in hand. Denial is unfortunate because it often delays treatment, and treatment within the first few hours after a heart attack can make the difference between life and death. Unfortunately, the average time between the start of a heart attack and seeking medical help is 2 hours. Roughly 25% of heart attack victims die within that period, usually because the heart develops an abnormal rhythm.

If a large enough section of heart muscle is damaged by a heart attack, the heart may not be able to continue pumping blood to the lungs and the rest of the body at an adequate rate. When the heart becomes an inefficient pump, the condition is known as *heart failure*. The symptoms of heart failure include shortness of breath, fatigue, weakness, and fluid accumulation in the lungs or limbs. Although the heart can never be restored to its former health, the symptoms of heart failure can be treated with drugs. For instance, digitalis can increase the strength of heart contractions, and diuretics reduce fluid accumulation, thus lessening the heart's workload. Drugs that are vasodilators relax constricted arteries, thereby reducing resistance to blood flow and blood pressure. Together, these drugs help a weakened heart pump more efficiently.

Sometimes, drugs and surgery cannot halt progressive heart failure. In this case, heart transplant surgery may provide hope for some patients, but generally only for those younger than 60 years of age. First, a donor heart that provides an acceptable tissue match must be found. During transplant surgery, a heart-lung machine takes over the circulation while the weakened heart is removed, and the new heart is sewn in place. The patient is then treated with drugs that lessen the chances of rejection. The one year survival rate is 86% and 70% of transplant recipients live at least 5 years.

The Lymphatic System Functions in the Circulatory and Immune Systems

The **lymphatic system** consists of a fluid, called **lymph**, that is identical to interstitial fluid (the fluid that bathes all the cells of the body), **lymphatic vessels** through which the lymph flows, and various **lymphoid tissues** and organs scattered throughout the body (Figure 12.20).

The functions of the lymphatic system are as diverse as they are essential to life.

1. **Return excess interstitial fluid to the bloodstream.** The lymphatic system maintains blood volume by returning excess interstitial fluid to the bloodstream. Only 85% to 90% of the fluid that leaves the blood capillaries and bathes the body tissues as interstitial fluid is reabsorbed by the capillaries. The rest of that fluid, which amounts to 2 to 3 liters (slightly more than 2 to 3 quarts) a day, is absorbed by the lymphatic system and returned to the circulatory system. This job is important. If the surplus interstitial fluid were not drained, it would cause the tissue to swell; the volume of blood would drop to potentially fatal levels; and the blood would become too viscous (thick) for the heart to pump.

   A dramatic example of the importance of returning fluid to the blood is provided by elephantiasis, a condition in which parasitic worms block lymphatic vessels (Figure 12.21). The blockage can cause a substantial buildup of fluid in the affected body region, followed by the growth of connective tissue. Elephantiasis is so named because it results in massive swelling and the darkening and thickening of the skin in the affected region, making the region resemble the skin of an elephant. Elephantiasis is a tropical disease, transmitted by mosquitoes, that affects over 400 million people.

2. **Transport products of fat digestion from the small intestine to the bloodstream.**

3. **Help defend against disease-causing organisms.** The lymphatic system produces lymphocytes, white blood cells that defend the body against specific disease-causing organisms, including bacteria, viruses, parasites, and abnormal cells. In this way, the system helps protect against disease and cancer.

Let's now consider the structures of the lymphatic system in more detail to shed some light on how it carries out such varied jobs. The structure of the lymphatic vessels is crucial to their ability to absorb the excess interstitial fluid. The extra fluid enters microscopic tubules, called *lymphatic capillaries*, which form a branching network that penetrates between the cells and the capillaries in almost every tissue of the body (except teeth, bones, bone marrow, and the central nervous system) (Figure 12.22). The lymphatic capillaries differ from the blood capillaries in two ways. First, lymphatic capillaries end blindly, like the fingers of a glove. In essence, they serve as drainage tubes. Fluid enters the "fingertips" easily, and lymph moves through the system in only one direction. Second, lymphatic capillaries are much more permeable than blood capillaries, a feature that is key to their ability to absorb the digestive products of fats as well as excess interstitial fluid. The permeability is due largely to the unique structure of lymphatic capillaries. The endothelial cells that form the wall of lymph capillaries loosely overlap one another, and each is anchored to surrounding tissue by fine filaments. As a result, the endothelial cells function as one-way valves that respond to the pressure of accumulating interstitial fluid. When fluid builds up in the tissue surrounding a capillary, the anchoring filaments pull on the edges of the endothelial cells, causing the flaps to separate and form gaps in the wall of the

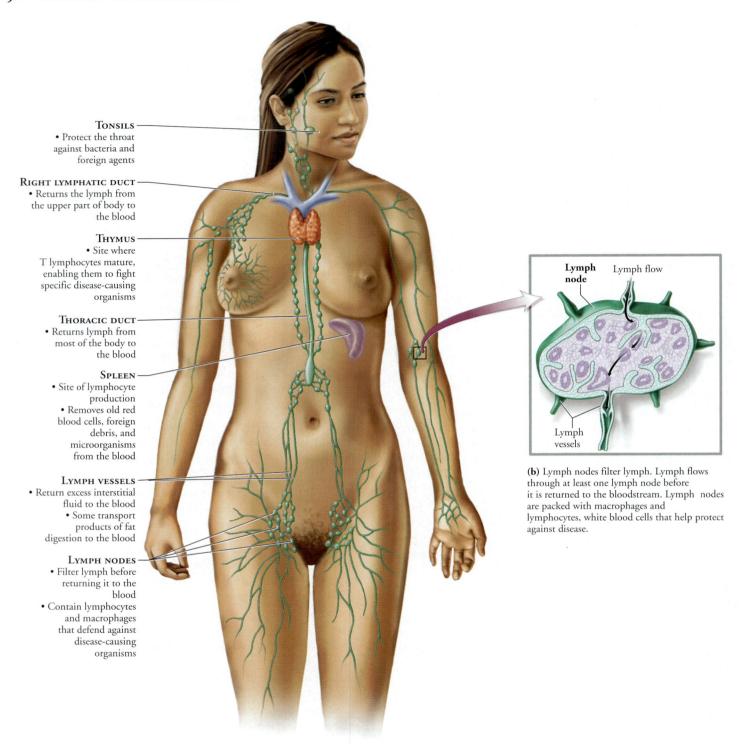

Tonsils
• Protect the throat against bacteria and foreign agents

Right lymphatic duct
• Returns the lymph from the upper part of body to the blood

Thymus
• Site where T lymphocytes mature, enabling them to fight specific disease-causing organisms

Thoracic duct
• Returns lymph from most of the body to the blood

Spleen
• Site of lymphocyte production
• Removes old red blood cells, foreign debris, and microorganisms from the blood

Lymph vessels
• Return excess interstitial fluid to the blood
• Some transport products of fat digestion to the blood

Lymph nodes
• Filter lymph before returning it to the blood
• Contain lymphocytes and macrophages that defend against disease-causing organisms

Lymph node Lymph flow

Lymph vessels

(b) Lymph nodes filter lymph. Lymph flows through at least one lymph node before it is returned to the bloodstream. Lymph nodes are packed with macrophages and lymphocytes, white blood cells that help protect against disease.

(a) The lymphatic system returns the fluid to the bloodstream that previously left the capillaries to bathe the cells, protects against disease-causing organisms, and transports products of fat digestion from the small intestine to the bloodstream.

FIGURE **12.20**

The lymphatic system is a system of lymphatic vessels containing a clear fluid, called lymph, and various lymphatic tissues and organs located throughout the body.

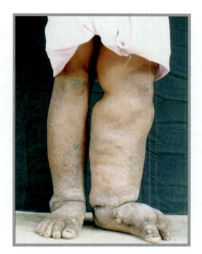

FIGURE **12.21**

The leg of a person with elephantiasis. In this condition, parasitic worms plug lymphatic vessels and prevent the return of fluid from the tissues to the circulatory system.

lymphatic capillary. Interstitial fluid then enters the capillary. (After the interstitial fluid enters a lymphatic vessel, it is called lymph.) When the pressure of interstitial fluid drops below that of the lymphatic capillary, the flap is forced shut and the lymph cannot leave the capillary.

The lymphatic capillaries drain into larger lymphatic vessels, which become progressively larger tubes with thicker walls. Lymph is eventually returned to the circulatory system through one of two large ducts that join with the large veins at the base of the neck.

With no pump to drive it, lymph flows slowly through the lymphatic vessels, driven by the same forces that move blood through the veins. That is, the contractions of nearby skeletal muscles compress the lymphatic vessels, pushing the lymph along. One-way valves similar to those in veins prevent backflow. Pressure changes in the thorax (chest cavity) that accompany breathing also help pull the lymph upward from the lower body. Gravity assists the flow from the upper body.

The lymphatic vessels are studded with **lymph nodes**, which cleanse the lymph as it slowly filters through. The lymph nodes contain macrophages and lymphocytes, white blood cells that play an essential role in the body's defense system. Macrophages engulf bacteria, cancer cells, and other debris, clearing them from the lymph. Lymphocytes serve as the surveillance squad of the immune system. They are continuously on the lookout for specific disease-causing invaders, as we will see in Chapter 13. Swollen and painful lymph nodes indicate infection.

Besides the lymph nodes, there are several other *lymphoid organs*. Among these are the *tonsils*, which form a ring around the entrance to the throat, where they help protect against disease organisms that are inhaled or swallowed. The *thymus gland*, located

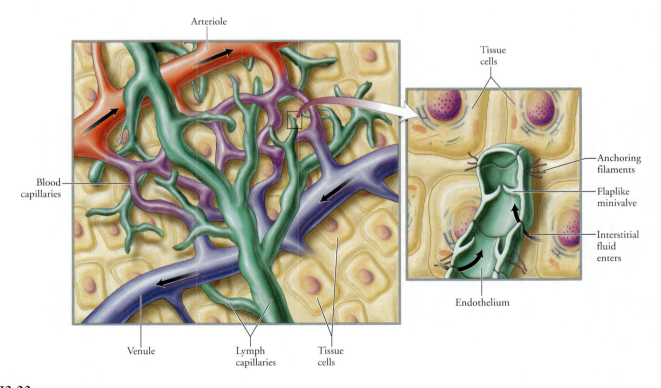

FIGURE **12.22**

The lymphatic capillaries are microscopic, blind-ended tubules through which surplus interstitial fluid enters the lymphatic system to be returned to the bloodstream. They weave between the cells and the blood capillaries. The endothelial cells that form the walls of lymphatic capillaries function as one-way valves that open in response to pressure from accumulating tissue fluid.

in the chest, is another lymphoid organ. It plays its part during early childhood by processing T lymphocytes, thus enabling them to function against specific disease-causing organisms. On the left side of the abdominal region is the largest lymphoid organ, the *spleen*. In addition to serving as a birthplace for lymphocytes, the spleen clears the blood of old and damaged red blood cells and platelets. Finally, isolated clusters of lymph nodules along the small intestine, known as *Peyer's patches*, keep bacteria from breaching the intestinal wall.

stop and think

Cancer cells often break loose from their original site, a process called metastasis. The cancer cells have easy access to the highly permeable lymphatic capillaries. The lymphatic vessels then provide a route for the cancer cells to spread to nearly every part of the body. Explain why the lymph nodes are often studied to determine whether the cancer has spread. Why are the lymph nodes near the original cancer site often removed?

REVIEWING THE CONCEPTS

The Cardiovascular System Consists of the Blood Vessels and the Heart (p. 230)

1. The heart serves as a pump that pushes blood through blood vessels.

WEB TUTORIAL 12.1 Cardiovascular System Overview

The Blood Vessels Conduct Blood in Continuous Loops (pp. 230–236)

2. Blood circulates through a branching network of blood vessels that forms a long loop. The path of blood through this loop is from the heart to arteries to arterioles to capillaries to venules to veins and then back to the heart.

3. Arteries are elastic, muscular tubes that carry blood away from the heart. They withstand the high pressure of blood as it is pumped from the heart by stretching and then returning to their original shape. These changes help maintain a relatively even blood pressure within the arteries even with large changes in the volume of blood within them.

4. The pressure change along an artery as it expands and recoils is called a pulse. Contraction of the smooth muscle in arterial walls, which is called vasoconstriction, narrows the space through which blood flows and reduces blood flow through the artery. Vasodilation occurs when the smooth muscle in arterial walls relaxes, increasing blood flow through the vessel.

5. Arteries branch to form narrower tubules called arterioles. Arterioles are important in the regulation of blood pressure, and they regulate blood flow through capillary beds.

6. The exchange of materials between the blood and tissues takes place across the thin walls of capillaries. Capillaries are arranged in highly branched networks that provide a tremendous surface area for the exchange of materials. Each network of capillaries is called a capillary bed. A ring of muscle called a precapillary sphincter determines whether blood flows into a capillary bed or is channeled through it.

7. Capillaries merge to form venules, and these, in turn, merge to form veins, which conduct blood back to the heart. Because veins have thin walls and large lumens, they function as blood reservoirs. Blood is returned to the heart against gravity as nearby skeletal muscles contract and push blood along within the veins. Valves within the veins prevent blood from flowing backward while the skeletal muscles are relaxing. Pressure differences generated by breathing also help draw blood toward the heart from the lower torso.

The Heart Is a Muscular Pump (pp. 236–243)

8. Every minute, the heart beats about 72 times and moves slightly less than 5 liters (10 pints) of blood through its chambers.

9. Most of the wall of the heart is composed of cardiac muscle and is called the myocardium. The heart is enclosed in a double-layered sac called the pericardium, which allows the heart to beat while still con-

fining it near the midline of the thoracic cavity. The endocardium is a thin inner lining.

10. The right and the left halves of the heart function as two separate pumps. Each side consists of two chambers: an upper chamber called the atrium and a lower, thick-walled chamber called the ventricle. The smaller, thin-walled atria function primarily as receiving chambers that accept blood returning to the heart and pump it a short distance to the ventricles. When the ventricles contract, they push blood through the arteries, toward the body cells.

11. Blood circulates in one direction through the body because of the action of two pairs of valves. The atrioventricular (AV) valves are located between each atrium and ventricle. The semilunar valves are located between each ventricle and its connecting artery. The valves function by responding passively to pressure differences on either side. The heart sounds, "lub-dup," are caused by blood turbulence associated with the closing of the heart valves. Defective valves can produce abnormal heart sounds, called heart murmurs.

12. The right side of the heart pumps blood to the lungs through a loop of vessels called the pulmonary circuit. The left side of the heart pumps blood to all parts of the body except the lungs through a loop of vessels called the systemic circuit.

13. The heart has its own network of vessels, called the coronary circuit, that services the heart tissue itself.

14. Each heartbeat consists of contraction (systole) and relaxation (diastole). The atria contract together, and then the ventricles contract in unison. The events associated with each heartbeat are collectively called the cardiac cycle.

15. The rhythmic contraction of the heart is due to its internal conduction system. Contraction spreads throughout the heart because junctions called intercalated disks connect cardiac muscle cells. A cluster of specialized cardiac muscle cells, called the sinoatrial (SA) node, sets the tempo of the heartbeat and is, therefore, called the pacemaker. The electrical signal to contract begins in the SA node, which is located in the right atrium near the superior vena cava, and spreads over the atria, causing them to contract. When the electrical signal reaches another cluster of specialized muscle cells, called the atrioventricular (AV) node, the stimulus is quickly relayed along the atrioventricular bundle that runs through the wall between the two ventricles and then fans out into the ventricular walls through the Purkinje fibers.

16. The rate of heartbeat is regulated by the autonomic nervous system to suit the body's level of activity. Whereas the sympathetic branch of the autonomic nervous system speeds up heart rate during stressful times, the parasympathetic nervous system slows heart rate during restful times.

17. An electrocardiogram (ECG or EKG) is a recording of the electrical events associated with each heartbeat. Each heartbeat produces three distinguishable deflections in the ECG.

18. Blood pressure, the force created by the heart to drive the blood around the body, is measured as the force of blood against the walls of blood vessels. The blood pressure in an artery peaks when a ventricle contracts. This is called the systolic pressure. In contrast, the lowest blood pressure of each cardiac cycle, the diastolic pressure, is caused by the elastic recoil of arteries while the heart is relaxing between contractions. Blood pressure is measured with a device called a sphygmomanometer.

WEB TUTORIAL 12.2 The Heart

WEB TUTORIAL 12.3 Cardiac Cycle

Cardiovascular Disease Is a Major Killer in the United States (pp. 243–248)

19. Cardiovascular diseases are those that affect the heart and blood vessels.
20. Hypertension (high blood pressure) is called a silent killer because it does not always produce symptoms and can have fatal consequences, involving the heart, brain, blood vessels, or kidneys. The heart enlarges, but it works less efficiently. High blood pressure damages blood vessels by promoting the formation of fatty deposits in arteries that block blood flow. High blood pressure may cause vessels to rupture. The kidneys can be damaged when the arteries leading to their filtering system are narrowed. Most cases of hypertension have no known cause.
21. Atherosclerosis is an inflammatory response associated with a buildup of fatty deposits within artery walls. The fatty material is covered with a fibrous cap and is called plaque. The plaque can bulge into the artery channel, restricting blood flow. Plaques that do not bulge into the artery may still be vulnerable to rupture, triggering blood clot formation that can block blood flow.
22. When the vessel affected by atherosclerosis is a coronary artery that supplies blood to heart tissues, the condition is known as coronary artery disease (CAD). A temporary reduction in blood supplied to the heart (ischemia) is accompanied by chest pain, called angina pectoris. Angina is a warning of a heart problem, but it does not cause permanent damage to the heart.

Heart Attack Is the Death of Heart Muscle (pp. 248–249)

23. A region of heart muscle dies during a heart attack (myocardial infarction). In a surviving patient, the dead cells are gradually replaced by scar tissue. The most common cause of a heart attack is a coronary thrombosis, in which a blood clot blocks a coronary artery. Coronary artery disease (CAD) makes a coronary thrombosis more likely. Another cause of a heart attack is a coronary artery spasm. Symptoms of a heart attack include severe chest pain and, often, nausea. Drugs that break up blood clots can often restore circulation to the heart muscle and save the person's life.
24. In heart failure, the heart becomes an inefficient pump. Symptoms include shortness of breath, fatigue, and fluid accumulation in the lungs and limbs. Heart failure is commonly treated with drugs that strengthen the heartbeat, reduce fluid retention, and widen blood vessels.

The Lymphatic System Functions in the Circulatory and Immune Systems (pp. 249–252)

25. The lymphatic system consists of lymph, lymphatic vessels, lymphoid tissue, and lymphoid organs.
26. Three vital functions of the lymphatic system are to return interstitial fluid to the bloodstream, to transport products of fat digestion from the digestive system to the bloodstream, and to defend the body against disease-causing organisms or abnormal cells.
27. Tissue fluid enters lymphatic capillaries—microscopic tubules that end blindly and are more permeable than blood capillaries. The fluid, then called lymph, is moved along larger lymphatic vessels by contraction of nearby skeletal muscles. The lymphatic vessels have valves to prevent the backflow of lymph.
28. Lymph nodes filter lymph and contain cells that actively defend against disease-causing organisms.
29. Lymphoid organs include the lymph nodes, tonsils, thymus gland, spleen, and Peyer's patches of the small intestine.

KEY TERMS

aorta *p. 232*	cardiovascular system *p. 230*	lymph *p. 249*	Purkinje fiber *p. 241*
arteriole *p. 233*	chordae tendineae *p. 238*	lymphatic system *p. 249*	semilunar valve *p. 238*
artery *p. 231*	circulatory system *p. 230*	lymphatic vessel *p. 249*	sinoatrial (SA) node *p. 240*
atherosclerosis *p. 244*	coronary artery disease (CAD) *p. 246*	lymph node *p. 251*	systemic circuit *p. 238*
atrioventricular bundle *p. 241*	coronary circulation *p. 239*	lymphoid tissue *p. 249*	systole *p. 240*
atrioventricular (AV) node *p. 240*	diastole *p. 240*	myocardium *p. 236*	systolic pressure *p. 243*
atrioventricular (AV) valve *p. 238*	diastolic pressure *p. 243*	pacemaker *p. 240*	vasoconstriction *p. 233*
atrium *p. 237*	electrocardiogram (ECG, EKG) *p. 242*	pericardium *p. 236*	vasodilation *p. 233*
blood pressure *p. 233*		pulmonary artery *p. 239*	vein *p. 234*
capillary *p. 233*	endocardium *p. 236*	pulmonary circuit *p. 238*	ventricle *p. 237*
capillary bed *p. 234*	heart *p. 230*	pulmonary vein *p. 239*	venule *p. 234*
cardiac cycle *p. 240*	hypertension *p. 244*	pulse *p. 233*	

THINKING ABOUT THE CONCEPTS

1. Trace the flow of blood from the heart and back to it by naming, in order, the general types of vessels through which the blood flows. *p. 230*
2. What is a pulse? *p. 233*
3. Explain how vasoconstriction and vasodilation regulate blood flow through arteries. *p. 233*
4. What are two important functions of arterioles? *p. 233*
5. What determines whether blood flows through any particular capillary bed? *pp. 234, 235*
6. List three mechanisms responsible for the exchange of materials across capillary walls. *p. 234*
7. Compare the structure of arteries, capillaries, and veins. Explain how the structure is suited to the function of each type of vessel. *pp. 230-235*

8. Explain how blood is returned to the heart from the lower torso against the force of gravity. *p. 235*
9. Describe the structure of the heart. Explain how it functions as two separate pumps. *pp. 236-238*
10. Describe the structure of the heart valves. Explain how they function. *p. 238*
11. Trace the path of blood from the left ventricle to the left atrium, naming each major vessel associated with the heart and the heart chambers in the correct sequence. *pp. 238-239*
12. Describe the cardiac cycle. *p. 240*
13. Explain how clusters or bundles of specialized cardiac muscle cells coordinate the contraction associated with each heartbeat. *pp. 240-241*
14. What is hypertension? In what ways does it damage the body? *p. 244*
15. What is atherosclerosis? Why is it harmful? How does it develop? *pp. 245-246*
16. Explain the relationships among coronary artery disease, angina pectoris, and heart attack. *pp. 246-248*
17. What is a heart attack? What are the common causes? *p. 248*
18. List three important functions of the lymphatic system. *p. 249*
19. Compare the structure of lymphatic capillaries with the structure of blood capillaries. How does the structure of lymphatic capillaries allow them to absorb tissue fluid? *pp. 249-251*
10. What is the function of lymph nodes? *pp. 251-252*
21. The semilunar valves prevent the backflow of blood from the
    a. arteries to the ventricles.
    b. veins to the atria.
    c. ventricles to the atria.
    d. arteries to the atria

22. If you cut all the nerves to the heart but kept the heart alive
    a. the heart would stop beating.
    b. the heart would continue beating.
    c. only systole would occur.
    d. only diastole would occur.

For questions 24 and 25, imagine that you have been miniaturized and are riding through the circulatory system using a red blood cell as a life raft.

23. You are in the big toe traveling toward the heart. The last vessel you pass through before entering the heart is the
    a. aorta.
    b. inferior vena cava.
    c. coronary artery.
    d. pulmonary vein.
24. You are nearly deafened by the first heart sound, which is caused by
    a. the opening of the atrioventricular valves.
    b. the closing of the atrioventricular valves.
    c. the opening of the semilunar valves.
    d. the closing of the semilunar valves.
25. The _____ is a cluster of specialized heart muscle cells that determines the heart rate by initiating each cardiac cycle.
26. A weakened area on an artery wall that can balloon outward is called _____.
27. Oxygen and nutrients move through the walls of _____ to reach the body cells.

APPLYING THE CONCEPTS

1. Yumei is a 75-year-old woman who broke her hip and had difficulty walking for several weeks. She passed the time watching television. After her hip healed, she suddenly had a heart attack while food shopping. The doctor said that the heart attack was caused by a blood clot in a leg vein that was swept away with the flow of blood. Explain the connection between these events.
2. Amelia is a friend of yours. When you call her to ask about dinner plans, she tells you that she is feeling under the weather. She aches all over and has swollen "glands" in her neck. You explain that these are not really glands, because they do not secrete anything. What are they? Why are they swollen?
3. Nicotine, a substance in cigarette smoke, causes vasoconstriction (narrowing of certain blood vessels) and increases heart rate. Explain how these effects lead to high blood pressure. Explain why cigarette smokers are more likely to die of cardiovascular diseases than are non-smokers.

Body Defense Mechanisms

The Body's Defense System Targets Pathogens and Cancerous Cells

The Body Has Three Lines of Defense
- The first line of defense is physical and chemical barriers that prevent entry by pathogens
- The second line of defense includes defensive cells and proteins, inflammation, and fever
- The third line of defense, the immune system, has specific targets and memory

The Immune System Mounts Antibody-Mediated Responses and Cell-Mediated Responses
- Macrophages present antigens to helper T cells to trigger an immune response
- Helper T cells activate B cells and T cells to destroy the specific antigen
- B cells mount an antibody-mediated immune response against antigens free in the blood or bound to a cell surface
- Cytotoxic T cells mount a cell-mediated immune response to destroy antigen-bearing cells
- Suppressor T cells turn off the immune response
- Immunological memory allows for a more rapid response on subsequent exposure to the antigen

Immunity Can Be Active or Passive

Monoclonal Antibodies Are Used in Research, Clinical Diagnosis, and Disease Treatment

The Immune System Can Cause Problems
- Autoimmune disorders occur when the immune system attacks the body's own cells
- Allergies are immune responses to harmless substances

HEALTH ISSUE Rejection of Organ Transplants

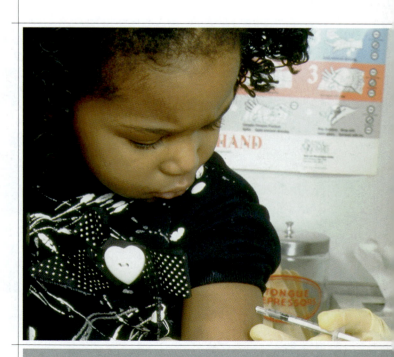

Sarah was eating dinner with her family when she started to feel slightly nauseous. "Feeling like this on top of my headache is all I need," she thought. She told her mother that she didn't feel well and Mom reached over to feel her forehead. "Oh dear," she said, "You have a fever. Let's get you into bed." On the way up the stairs, Sarah felt weak, as if her muscles were protesting even this slight exertion. She briefly wondered if she might have the flu. The next day her symptoms increased. A rash appeared on her arm and it was clearly time for a visit to the doctor. Fortunately, the family physician was well aware that Colorado had been experiencing a rapid escalation in cases of West Nile virus that summer. Sarah's doctor ordered diagnostic tests, which determined that Sarah had indeed contracted the virus. The virus entered Sarah's body by way of a mosquito bite from a mosquito that had previously fed on the blood of an infected bird. Sarah's symptoms disappeared within a couple of days, as is the case with most West Nile virus patients who exhibit symptoms. (Most do not show symptoms.) In the worst cases, the brain becomes inflamed and the result may be death, particularly in people over 60 years of age.

In this chapter, we will study how the body reacts to an invasion such as the West Nile virus and many other organisms that it perceives as threats, even those that may live on us or in us (Figure 13.1). We will see that there are three lines of defense. The first line of defense is physical and chemical barriers that protect the body's surface and openings. If a microbe enters the body, it encounters the second line of defense, which consists of defensive cells and proteins and the inflammatory response. The third line of defense is the immune response, which recognizes, destroys, and remembers the particular microbe encountered. We will also learn that long-lasting resistance to a microbe can be acquired actively by becoming ill or by being immunized. Finally, we will consider some potential problems caused by the immune system. ■

FIGURE **13.1**

Millions of bacteria live on the surface of your skin. Most of these bacteria are harmless. Because they are so well entrenched, they often can outcompete dangerous microbes and, in this way, help defend you from infection.

The Body's Defense System Targets Pathogens and Cancerous Cells

Your body generally defends you against anything that is not recognized as being part of or belonging in your body. Common targets of your defense system include organisms that cause disease or infection, as well as body cells that have turned cancerous.

The bacteria, viruses, protozoans, fungi, parasitic worms, and prions (infectious proteins) that cause disease are called **pathogens** (discussed in Chapter 13A). Not all representatives of these groups cause disease. Some bacteria are beneficial. They flavor our cheese, help rid the planet of corpses through decomposition, and live in or on our own bodies, where they keep potentially harmful bacterial in check.

Cancerous cells also threaten our well-being. A cancer cell was once a normal body cell, but because of changes in its genes, it can no longer regulate its cell division. These renegade cells can multiply until they take over the body, upsetting its balance, choking its pathways, and ultimately causing great pain and sometimes death.

The Body Has Three Lines of Defense

The body's "strategy" for defending against foreign organisms, cells, or molecules has three phases.

1. Keep the foreign cells or molecules out of the body.
2. Attack *any* foreign cell or molecule that enters the body.
3. Destroy the *specific* type of foreign cell or molecule that enters the body.

The body has three lines of defense, each accomplishing one of the phases in its defensive strategy. The first line of defense—chemical and physical surface barriers—is always prepared. The second line of defense is internal cellular and chemical defenses that become active if the surface barriers are penetrated. The first and second lines of defense involve nonspecific mechanisms that are effective against *any* foreign organisms or substances. The third line of defense is the immune system (discussed shortly), which destroys *specific* targets, usually disease-causing organisms, and remembers those targets to mount a quick response if that target enters the body again. Thus, the immune system is a specific mechanism of defense. The three lines of defense against pathogens are summarized in Figure 13.2.

The first line of defense is physical and chemical barriers that prevent entry by pathogens

The skin and mucous membranes that form the first line of defense provide physical barriers that help keep foreign substances from entering the body, and they produce several protective chemicals.

PHYSICAL BARRIERS: SKIN AND MUCOUS MEMBRANES

SKIN   Like a suit of armor, unbroken skin shields the body from the hostile world by providing an effective barrier to foreign substances (Figure 13.3). A layer of dead cells forms the tough outer layer of skin. These cells are filled with the fibrous protein keratin, which makes the skin nearly impenetrable, waterproof, and resistant to the disruptive toxins (poisons) and

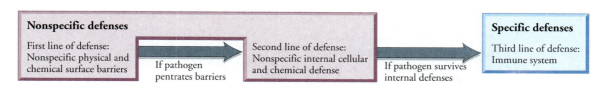

Nonspecific defenses		Specific defenses
First line of defense: Nonspecific physical and chemical surface barriers	**Second line of defense:** Nonspecific internal cellular and chemical defense	**Third line of defense:** Immune system

If pathogen pentrates barriers → If pathogen survives internal defenses →

FIGURE 13.2

The body's three lines of defense against pathogens

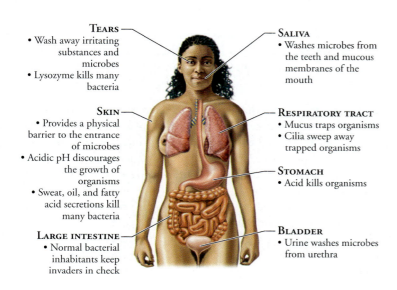

TEARS
• Wash away irritating substances and microbes
• Lysozyme kills many bacteria

SKIN
• Provides a physical barrier to the entrance of microbes
• Acidic pH discourages the growth of organisms
• Sweat, oil, and fatty acid secretions kill many bacteria

LARGE INTESTINE
• Normal bacterial inhabitants keep invaders in check

SALIVA
• Washes microbes from the teeth and mucous membranes of the mouth

RESPIRATORY TRACT
• Mucus traps organisms
• Cilia sweep away trapped organisms

STOMACH
• Acid kills organisms

BLADDER
• Urine washes microbes from urethra

FIGURE 13.3

The body's first line of defense is physical and chemical barriers that serve as nonspecific defenses that protect against *any* threats to our well-being. Collectively, they prevent the threat from entering the body, confine it to a local region, kill it, remove it, or slow its growth.

enzymes of most would-be invaders. Some of the strength of this barrier results from the tight connections binding the cells together. Furthermore, the dead cells are continuously shed and are replaced at the rate of about a million cells every 40 minutes. As dead cells flake off, they take with them any microbes that have somehow managed to latch on.

MUCOUS MEMBRANES   The inner surfaces of the body are guarded by mucous membranes that line the respiratory, digestive, urinary, and reproductive systems. Mucous membranes are more vulnerable than skin, so most organisms that enter the body do so by penetrating mucous membranes.

CHEMICAL BARRIERS

In addition to providing a tough, physical barrier, the skin provides some chemical protection against invaders. Sweat and oil produced by glands in the skin wash away microbes, and their acidity slows bacterial growth. In addition, the oils contain chemicals that kill bacteria.

The mucous membranes lining the digestive and respiratory passages produce sticky mucus that traps many microbes. The lining of the stomach produces hydrochloric acid and protein-digesting enzymes that destroy pathogens. Furthermore, the acidity of urine slows bacterial growth, and urine periodically flushes microbes from the lower urinary tract. Saliva and tears contain an enzyme called **lysozyme** that kills bacteria by disrupting their cell walls.

The second line of defense includes defensive cells and proteins, inflammation, and fever

The second line of defense involves nonspecific internal defenses against any pathogen that breaks through the physical and chemical barriers and enters the body. These defenses, which include defensive cells and proteins, inflammation, and fever, are summarized in Table 13.1.

TABLE **13.1**

THE SECOND LINE OF DEFENSE—NONSPECIFIC INTERNAL DEFENSES		
DEFENSE	**EXAMPLE**	**FUNCTION**
Defensive cells	Phagocytic cells such as neutrophils and macrophages	Engulf invading organisms
	Eosinophils	Kill parasites
	Natural killer cells	Kill many invading organisms and cancer cells
Defensive proteins	Interferons	Slow the spread of viruses in the body
	Complement system	Stimulates histamine release; promotes phagocytosis; kills bacteria; enhances inflammation
Inflammation	Widening of blood vessels and increased capillary permeability lead to redness, heat, swelling, and pain	Brings in defensive cells and speeds healing
Fever	Abnormally high body temperature	Slows the growth of bacteria; speeds up body defenses

DEFENSIVE CELLS

PHAGOCYTES Specialized "scavenger" cells called **phagocytes** (*phage*, to eat; *cyte*, cell) engulf pathogens, damaged tissue, or dead cells by the process of phagocytosis. In this way, phagocytes serve as the front-line soldiers in the body's defense system and as janitors that clean up debris. As a phagocyte engulfs a foreign particle, cytoplasmic extensions flow from the phagocytic cell, bind to the particle, and pull it inside the cell, as shown in Figure 13.4. Once the particle is inside the cell, it is enclosed within a membrane-bound vesicle and quickly destroyed by digestive enzymes.

The body has several types of phagocytes. Neutrophils are one type of phagocyte. *Neutrophils* arrive at the site of attack before the other types of white blood cells and immediately begin to consume the pathogens, primarily bacteria, by phagocytosis. Other white blood cells (monocytes) leave the vessels of the circulatory system and enter the tissue fluids, where they develop into large **macrophages** (*macro*, big; *phage*, to eat). Macrophages have hearty and less discriminating appetites than neutrophils and attack and consume virtually anything that is not recognized as belonging in the body—including viruses, bacteria, and damaged tissue.

EOSINOPHILS A second type of white blood cell, *eosinophils*, attack pathogens that are too large to be consumed by phagocytosis, such as parasitic worms. Eosinophils get close to the parasite and discharge destructive enzymes that destroy the organism.

NATURAL KILLER CELLS A third type of white blood cell, called **natural killer (NK) cells**, roams the body in search of abnormal cells, which are then quickly killed. In a sense, NK cells function as the body's police walking a beat. They are not seeking a specific villain. Instead, they respond to any suspicious character, which, in this case, is a cell whose cell membrane has been altered by the addition of proteins that are unfamiliar to the NK cell. The prime targets of NK cells are cancerous cells and cells infected with viruses. Cancerous cells routinely form but are quickly destroyed by NK cells and prevented from spreading (Figure 13.5).

When an NK cell touches a cell with an abnormal surface, the NK cell immediately attaches to its target and delivers a "kiss of death," which comes in the form of proteins that create numerous pores in the target cell. The pores make the target cell "leaky." The cell can no longer maintain a constant internal environment and eventually bursts.

DEFENSIVE PROTEINS

The second line of defense also includes defensive proteins. We will discuss two types of defensive proteins: interferons, which slow viral reproduction, and the complement system, which assists other defensive mechanisms.

INTERFERON A cell that has been infected with a virus can do little to help itself. But cells infected with a virus can help cells that are not yet infected. Before certain virally infected cells die, they secrete small proteins called **interferons** that act to slow the spread of viruses already in the body. As the name implies, interferons interfere with viral activity.

Interferons mount a two-pronged attack. First, they help rid the body of virus-infected cells by attracting macrophages and natural killer cells that destroy the infected cells immediately. Second, interferons protect cells that are not yet infected with the virus. When released, an interferon diffuses to neighboring cells and stimulates the cells to produce proteins that prevent viruses from making copies of themselves in those cells (Figure 13.6). Because viruses cause disease by replicating inside body cells, preventing replication curbs the disease. Interferon helps protect uninfected cells from *all* strains of virus, not just the one responsible for the initial infection.

When interferons and their broad-spectrum antiviral activity were first discovered, researchers hoped that interferons could be used to treat viral diseases. Researchers also hoped that interferons could cure cancer because they stimulate the activity of natural killer cells that

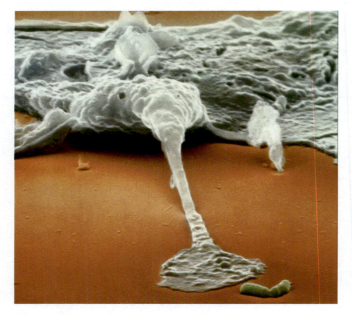

FIGURE **13.4**
A macrophage ingesting bacteria (the rod-shaped structure). The bacteria will be pulled inside the cell within a membrane-bound vesicle and quickly killed.

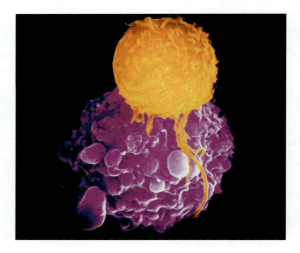

FIGURE **13.5**
A natural killer cell (shown in orange) attacking a cancer cell (shown in purple). NK cells patrol the body, bumping and touching other cells as they do. When NK cells contact a cell with an altered cell surface, such as a cancer cell or a virus-infected cell, a series of events is immediately initiated. The NK cell attaches to the target cell and releases proteins that create pores in the target cell, making the membrane leaky and causing the cell to burst.

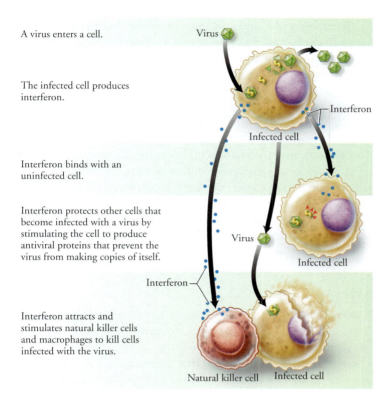

A virus enters a cell.

The infected cell produces interferon.

Interferon binds with an uninfected cell.

Interferon protects other cells that become infected with a virus by stimulating the cell to produce antiviral proteins that prevent the virus from making copies of itself.

Interferon attracts and stimulates natural killer cells and macrophages to kill cells infected with the virus.

FIGURE **13.6**

The mechanism by which interferon protects against viruses

attack cancer cells. Although interferon has not turned out to be the wonder drug people had hoped for, it has been shown to be effective against certain cancers and viral infections. For instance, interferon is often successful in combating a rare form of leukemia (hairy cell leukemia) and Kaposi's sarcoma, a form of cancer that often occurs in people with AIDS. Interferon has also been approved for treating the hepatitis C virus, which can cause cirrhosis of the liver and liver cancer; the human papilloma virus, which causes genital warts; and the herpes virus, which causes genital herpes. In laboratory tests, interferon is active against the virus that causes SARS (severe acute respiratory syndrome), and researchers hope it will be an effective treatment for the deadly respiratory infection. One form of interferon slows the progression of multiple sclerosis, a debilitating disease of the nervous system. Also, tests are now under way to determine whether interferon can slow the progression of an HIV infection before AIDS develops.

THE COMPLEMENT SYSTEM The **complement system**, or simply *complement*, is a group of at least 20 proteins whose activities enhance, or "complement," the body's other defense mechanisms. These proteins enhance, or complement, both nonspecific and specific defense mechanisms. The effects of complement include the following:

- **Destruction of pathogen** Complement can act *directly*, by punching holes in the target cell's membrane (Figure 13.7). The cell is no longer able to maintain a constant internal environment. Water enters the cell, causing it to burst, just as we saw when the NK cells secreted proteins that made the target cell's membrane leaky.

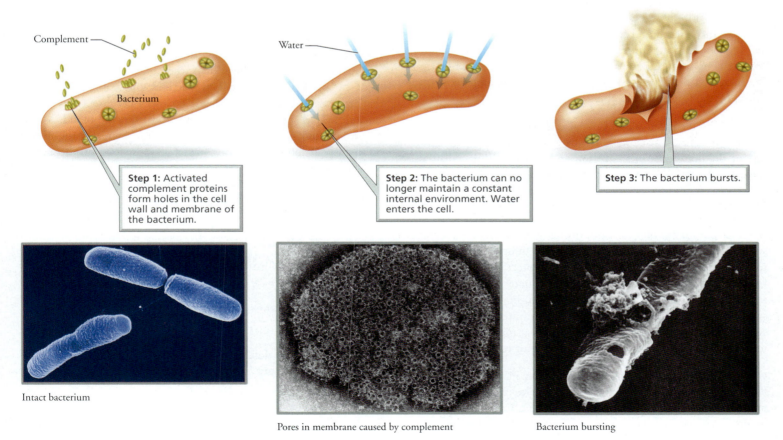

Step 1: Activated complement proteins form holes in the cell wall and membrane of the bacterium.

Step 2: The bacterium can no longer maintain a constant internal environment. Water enters the cell.

Step 3: The bacterium bursts.

Intact bacterium

Pores in membrane caused by complement

Bacterium bursting

FIGURE **13.7**

Complement can destroy a pathogen by acting directly.

- **Enhancement of phagocytosis** Complement enhances phagocytosis in two ways. Complement proteins attract macrophages and neutrophils to the site of infection to remove the foreign cells. One of the complement proteins binds to the surface of the microbe, making it easier for macrophages and neutrophils to "get a grip" on the intruder and devour it.

- **Stimulation of inflammation** (discussed shortly) Complement also causes blood vessels to become wider and more permeable. These changes allow increased blood flow to the area and increased access for white blood cells.

INFLAMMATION

When body tissues are injured or damaged, a series of events called the **inflammatory response**, or reaction, occurs. This response destroys invaders and helps repair and restore damaged tissue. We are all familiar with the four cardinal signs of inflammation that occur at the site of a wound—redness, heat (or warmth), swelling, and pain. These signs announce that certain cells and chemicals have combined efforts to contain infection, clean up the damaged area, and heal the wound. Let's consider the causes of the cardinal signs and how they are related to the benefits of inflammation (Figure 13.8).

REDNESS Redness is caused by increased blood flow to the damaged area that occurs because blood vessels dilate (become wider). The dilation is caused by **histamine**, which is released by **mast cells** (small mobile connective tissue cells) and **basophils** (a type of white blood cell).

The increased blood flow to the site of injury delivers phagocytes, blood-clotting proteins, and defensive proteins including complement and antibodies (discussed shortly). At the same time, the increased blood flow washes away dead cells and toxins produced by the invading microbes.

HEAT The increased blood flow also elevates the temperature in the area of injury. The elevated temperature increases the metabolic rate of the body cells in the region and speeds healing. Heat also increases the activities of phagocytic cells and other defensive cells.

SWELLING The injured area swells because histamine also makes capillaries more permeable, or leakier, than usual. Fluid seeps into the tissues from the bloodstream, bringing with it many beneficial substances. Blood-clotting factors enter the injured area and begin to wall off the region, thereby helping to protect surrounding areas from injury and preventing the loss of blood. The seepage also increases the oxygen and nutrient supply to the cells. In addition, swelling can help healing directly. If the injured area is a joint, swelling can hamper movement, an effect that might seem to be an inconvenience, but it permits the injured joint to rest and recover.

PAIN There are several causes for the pain in an inflamed area. For example, the excessive fluid that has leaked into the tissue presses on cells and contributes to the sensation. Some soreness might be caused by bacterial toxins. Injured cells also release pain-causing chemicals, such as prostaglandins. Pain usually causes a person to protect the area and prevent additional injury.

Soon after the inflammatory response begins, phagocytes swarm to the injured site, attracted by chemicals released when tissue is damaged. First on the scene are the neutrophils. Within minutes, the neutrophils squeeze through capillary walls into the fluid around cells and begin engulfing pathogens, toxins, and dead body cells. A neutrophil can engulf about two dozen bacteria before being killed by the poisons produced by the engulfed bacteria and by its own powerful chemicals produced to destroy those bacteria.

Not long after the neutrophils begin the battle, reinforcements arrive—additional neutrophils and macrophages that continue the body's counterattack over the long haul. Macrophages are also important in cleaning debris, such as dead body cells, from the damaged area. As the infection continues, dead cells, including microbes, body tissue cells, and phagocytes, may begin to ooze from the wound as pus.

FEVER

A *fever* is an abnormally high body temperature (Figure 13.9). Fevers are caused by *pyrogens* (*pyro*, fire; *gen*, producer), chemicals that set the "thermostat" in the brain (the hypothalamus) to a higher set point. Bacteria release toxins that sometimes act as pyrogens. It is interesting to note, however, that the body produces its own pyrogens as part of its defensive strategy. Regardless of the source, pyrogens stimulate certain brain cells to secrete prostaglandins. The prostaglandins, in turn, raise the set point of the body's thermostat. Physiological responses, such as shivering, are then initiated to raise body temperature (discussed in Chapter 4). Thus, we have the "chills" while the fever is rising. When the fever breaks, the set point is lowered, and physiological responses such as perspiring lower the body temperature until it reaches the new set point.

A mild or moderate fever helps the body fight bacterial infections by slowing the growth of bacteria and stimulating body defense responses. Bacterial growth is slowed because a mild fever causes the liver and spleen to remove iron from the blood. Many bacteria require iron to reproduce. Fever also increases the metabolic rate of body cells; the higher rate speeds up defensive responses and repair processes. On the other hand, a very high fever (over 105°F, 40.6°C) is dangerous. It can inactivate enzymes needed for biochemical reactions within body cells.

stop and think

Aspirin inhibits the synthesis and release of prostaglandins. Explain how this action contributes to aspirin's pain-killing effect. Why might it be wise to avoid taking aspirin to reduce a fever?

The third line of defense, the immune system, has specific targets and memory

When the body's first and second lines of defense fail, the body's specific defenses respond and target the particular pathogen or foreign molecule that has entered the body. The third line of defense, the **immune system,** provides the specific responses. The organs of the lymphatic system (see Chapter 12) are important components

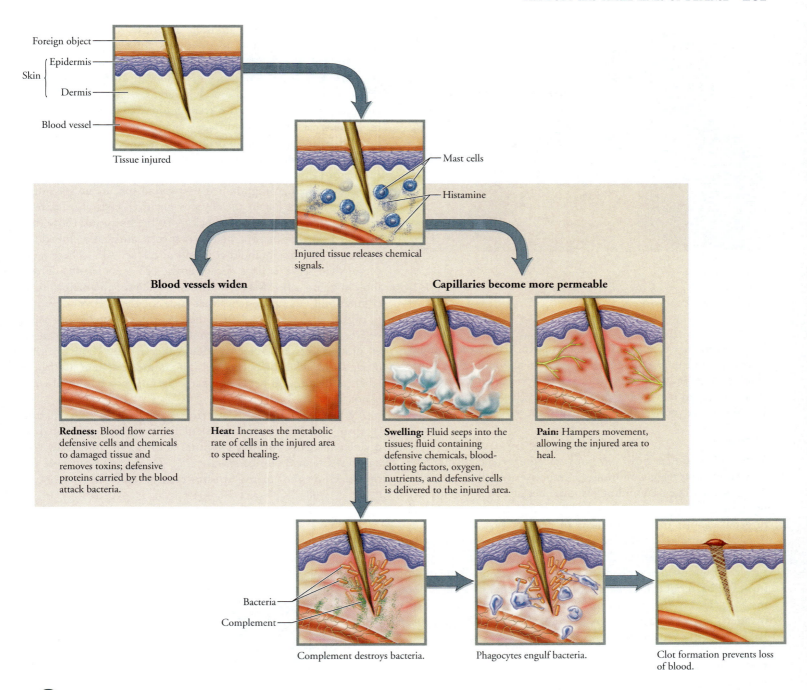

Foreign object
Skin { Epidermis
Dermis
Blood vessel

Tissue injured

Mast cells
Histamine

Injured tissue releases chemical signals.

Blood vessels widen

Redness: Blood flow carries defensive cells and chemicals to damaged tissue and removes toxins; defensive proteins carried by the blood attack bacteria.

Heat: Increases the metabolic rate of cells in the injured area to speed healing.

Capillaries become more permeable

Swelling: Fluid seeps into the tissues; fluid containing defensive chemicals, blood-clotting factors, oxygen, nutrients, and defensive cells is delivered to the injured area.

Pain: Hampers movement, allowing the injured area to heal.

Bacteria
Complement

Complement destroys bacteria.

Phagocytes engulf bacteria.

Clot formation prevents loss of blood.

FIGURE **13.8**
TUTORIAL 13.1

The inflammatory response is a general response to tissue injury or the invasion of foreign microbes. It serves to defend against pathogens and to clear the injured area of pathogens and dead body cells, allowing repair and healing to occur. The four cardinal signs of inflammation are redness, heat, swelling, and pain.

of the immune system. However, the immune system is not an organ system in an anatomical sense. Instead, the immune system is defined by its *function*: the recognition and destruction of specific pathogens or foreign molecules. The body's specific defenses working together are called an *immune response*.

The immune system has two characteristics: (1) it is specific, and (2) it has memory. Let's look at an example. You may have noticed that after you recover from a certain disease, say chickenpox, you are not likely to get it again. You might get measles or the flu afterward, but you will not get chickenpox. You are *immune* to chickenpox, meaning that you have long-lasting resistance specific to the chickenpox virus.

From this simple observation, you can understand several important characteristics of the immune system. First, it is specific for one particular pathogen, in this case, the virus that causes chickenpox. The immune system recognizes a pathogen as a foreign substance (not belonging in the body) and then acts to immobilize, neutralize, or destroy it. You know the immune system was effective

FIGURE **13.9**
Although a fever might make us feel uncomfortable, it can help the body fight disease.

because you recovered from the chickenpox. Second, the immune system has memory. If you are exposed to the virus that causes chickenpox 20 years after you recovered from the original illness, the immune system remembers the virus and attacks it so quickly and vigorously that you will not become ill again.

ANTIGENS

You might wonder how the body recognizes specific targets. Substances that trigger an immune response are called **antigens**. Antigens provoke an immune response because they are not recognized as belonging in the body. The immune response is then directed against the specific type of antigen that provoked it. Typically, antigens are large molecules, such as proteins, polysaccharides, or nucleic acids. Antigens are often found on the surface of an invader, embedded in the plasma membrane of an invading bacterial cell or part of the protein coat of a virus, for instance. However, pieces of invaders and chemicals secreted by invaders, such as bacterial toxins, can also serve as antigens. Each antigen is recognized by its characteristic shape.

LYMPHOCYTES

Certain white blood cells, called lymphocytes, are responsible for both the specificity and the memory of the immune response. There are two principal types of lymphocytes. They are the **B lymphocytes**, or more simply *B cells*, and the **T lymphocytes**, or *T cells*. Both types form in the bone marrow, but they mature in different organs of the body. It is thought that B cells mature in the bone marrow. (The B stands for bursa, a pouch of lymphatic tissue in a chicken's digestive system, where B cells were first identified. However, the B is now a useful reminder that B cells mature in the bone marrow.) The T cells, on the other hand, mature in the thymus gland, which overlies the heart. (The T stands for thymus gland.)

As lymphocytes mature, they are programmed to recognize only one particular type of antigen. This recognition is the basis

of the specificity of the immune response. Each lymphocyte develops receptors—molecules with a unique shape—on its surface. Indeed, thousands of *identical* receptor molecules pepper the surface of each lymphocyte. When an antigen fits into a lymphocyte's receptors, the body's defenses target that particular antigen. Because of the tremendous diversity of receptor molecules, each type on a different lymphocyte, a few of the billions of lymphocytes in your body are programmed to respond to each of the thousands of different antigens that you will be exposed to in your lifetime.

Besides being programmed to attack a specific antigen during their maturation in the thymus gland, T cells also learn to distinguish between cells that belong in the body and those that do not. Each cell in your body has special molecules embedded in the plasma membrane that label the cell as "self." These molecules then serve as flags declaring the cell as a "friend." These molecules are called **MHC markers**, named for the major histocompatibility complex genes that code for them. Any substance or organism that lacks a "self" label is considered to be "nonself," or "foe." The "self" labels on your cells are different from those of another person (except an identical twin) and those of other organisms, including pathogens. The immune system uses these labels to distinguish between what is part of your body and what is not (Figure 13.10).

When an antigen is detected, B cells and T cells bearing receptors specific for that particular invader are stimulated to divide repeatedly, forming two lines of cells. One line of descendant cells is made up of *effector cells*, which are responsible for the attack on the enemy. Effector cells generally live for only a

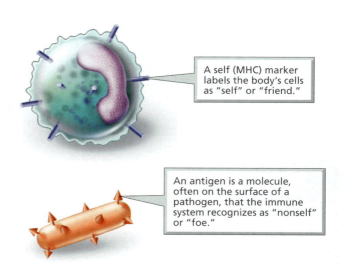

A self (MHC) marker labels the body's cells as "self" or "friend."

An antigen is a molecule, often on the surface of a pathogen, that the immune system recognizes as "nonself" or "foe."

FIGURE **13.10**
All the nucleated cells in your body have molecular MHC markers on their surface that label them as "self." Foreign substances, including potential disease-causing organisms, have molecules on their surfaces that are not recognized as belonging in the body. Foreign molecules that are capable of triggering an immune response are called antigens.

few days. Thus, after the invader has been eliminated from the body, the number of effector cells declines. The other line of descendant cells is composed of *memory cells*, long-lived cells that "remember" that particular invader and mount a rapid, intense response to it if it should ever appear again. The quick response of memory cells is the mechanism that prevents you from getting the same illness twice.

Next, we'll see how these pieces fit together to form your body's highly effective immune response. Table 13.2 summarizes the functions of the cells involved in the immune response, and Table 13.3 summarizes the steps in the immune response. You can refer to these tables as you read the following descriptions of the immune response.

The Immune System Mounts Antibody-Mediated Responses and Cell-Mediated Responses

There are some striking similarities between the body's immune defenses and a nation's military defense system. The body has scouts, called macrophages, that roam the body looking for an invader. If an invader is found, the scout will alert the commander-in-chief of the military forces and provide an exact description of the villain. The scout must, of course, provide the appropriate password so that the commander knows it is not a spy providing misinformation. The immune system's commander-in-chief is a subset of T cells, called helper T cells. When properly alerted, helper T cells call out the body's defensive forces.

A nation's military often has several branches, the Army and the Navy, for instance. Primed to respond to slightly different forms of enemy invasion, each branch is armed with certain types of weapons. When activated, the military is usually looking for a specific threat, say little green people with purple hair. The Navy may be called into action if the enemy is encountered at sea, whereas the Army will come to the defense if the enemy is on land.

The body also has two types of specific defense. The two recognize and destroy the same antigens but do so in different ways.

- **Antibody-mediated immune responses** defend primarily against antigens that are free in body fluids, including toxins or extracellular pathogens such as bacteria or free viruses. The warriors of this branch of immune defense are the effector B cells (plasma cells) and their weapons are Y-shaped proteins called **antibodies**, which neutralize and remove potential threats from the body. When called into action, a B cell transforms into a type of effector cell called a plasma cell, whose job it is to secrete into the body fluids many copies of a specific antibody programmed to recognize and bind to the antigen posing the threat. Antibodies help eliminate the antigen from the body.

- **Cell-mediated immune responses** involve living cells and protect against cellular threats, including body cells that have become infected with viruses or other pathogens and cancer cells. The lymphocytes responsible for cell-mediated immune responses are a type of T cell, the cytotoxic T cell (discussed shortly). Once activated, cytotoxic T cells quickly destroy the infected or cancerous cells by causing them to burst.

TABLE **13.2**

CELLS INVOLVED IN THE IMMUNE RESPONSE	
CELL	**FUNCTIONS**
Macrophage	**An antigen-presenting cell**
	• Engulfs and digests antigens
	• Places a piece of consumed antigen on its plasma membrane
	• Presents the antigen to a helper T cell
	• Activates the helper T cell
T Cells	
Helper T cell	**The "on" switch for both lines of immune response**
	• After activation by macrophage, it divides, forming effector helper T cells and memory helper T cells
	• Helper T cells activate B cells and T cells
Cytotoxic T cell (effector T cell)	**Responsible for cell-mediated immune responses**
	• When activated by helper T cell, it divides to form effector cytotoxic T cells and memory cytotoxic T cells
	• Destroys cellular targets, such as virus-infected body cells, bacteria, fungi, parasites, and cancer cells
Suppressor T cell	**The "off" switch for immune responses**
	Suppresses the activity of B cells and T cells after the foreign cell or molecule has been successfully destroyed
B Cells	**Involved in antibody-mediated responses**
	When activated by helper T cell, it divides to form plasma cells and memory cells
Plasma cell	**Effector in antibody-mediated response**
	Secretes antibodies specific to the invader
Memory cells	**Responsible for memory of immune system**
	• Generated by B cells or any type of T cell during an immune response
	• Enable quick and efficient response on subsequent exposures of the antigen
	• May live for years

TABLE **13.3**

STEPS IN THE IMMUNE RESPONSE	
1. **Threat**	Foreign cell or molecule enters the body
2. **Detection**	Macrophage detects foreign cell or molecule and engulfs it
3. **Alert**	• Macrophage puts antigen from the pathogen on its surface and finds the helper T cell with correct receptors for that antigen • Macrophage presents antigen to the helper T cell • Macrophage alerts the helper T cell that there is an invader that "looks like" the antigen • Macrophage activates the helper T cell
4. **Alarm**	Helper T cell activates both lines of defense to fight that specific antigen
5. **Build specific defense (clonal selection)**	• Antibody-mediated defense—B cells are activated and divide to form plasma cells that secrete antibodies specific to the antigen • Cell-mediated defense—T cells divide to form cytotoxic T cells that attack cells with the specific antigen
6. **Defense**	• Antibody-mediated defense—antibodies specific to antigen eliminate the antigen • Cell-mediated defense—cytotoxic T cells cause cells with the antigen to burst
7. **Continued surveillance**	Memory cells formed when helper T cells, cytotoxic T cells, and B cells were activated remain to provide swift response if the antigen is detected again
8. **Withdrawal of forces**	Once the antigen has been destroyed, suppressor T cells shut down the immune response to that antigen

Macrophages present antigens to helper T cells to trigger an immune response

Recall that macrophages are phagocytic cells that roam the body, engulfing any foreign material or organisms they may encounter. Within the macrophage, the engulfed material is digested into smaller pieces. The macrophage then moves some of these pieces to its own surface, where they are bound to the MHC "self" markers on its membrane. These self markers identify the macrophage as a "friend." However, the pieces of antigen that the macrophage places on its surface function as a kind of "wanted poster" that tells lym-

phocytes that there is an invader and reveals what the invader looks like. The displayed antigens trigger the immune response. Although a few other kinds of cells can display antigens,[1] the macrophage is an important type of **antigen-presenting cell (APC)**.

The macrophage then presents the antigen to a **helper T cell**,[2] the kind of T cell that serves as the main switch for the entire immune response. The macrophage cannot alert just *any* helper T cell. It must alert a helper T cell bearing receptors that recognize the specific antigen being presented. These specific helper T cells constitute only a tiny fraction of the entire T cell population. Finding the right helper T cell is like looking for a needle in a haystack. The macrophage wanders through the body until it literally bumps into an appropriate helper T cell. The encounter most likely occurs in a lymph node, a bean-shaped structure that contains huge numbers of lymphocytes of all kinds (discussed in Chapter 12).

Helper T cells activate B cells and T cells to destroy the specific antigen

When the antigen-presenting macrophage meets the appropriate helper T cell and binds to it, the macrophage secretes a chemical called *interleukin 1* that activates the helper T cell (Figure 13.11). Within hours, an activated helper T cell begins to secrete its own chemical messages in the form of *interleukin 2*. The helper T cell's message calls into active duty the appropriate B cells and T cells with the ability to bind to the particular antigen that triggered the response.

Before the appropriate B and T cells can be activated to mount a full-fledged defense, they must be sensitized. In the case of T cells, sensitization is accomplished by another cell, usually a macrophage, but they can be sensitized by any infected body cell that presents the antigen bound to the cell's self (MHC) marker. Sensitization of a T cell requires simultaneous recognition of self and nonself. B cells, on the other hand, can bind to antigens that are free in body fluids. This binding sensitizes a B cell and makes it more responsive to activation from helper T cells. (Sensitization can be likened to starting the engine of a car. Activation by interleukin 2 is analogous to putting the car in gear.)

When the appropriate "virgin"[3] B cells or T cells are activated, they begin to divide repeatedly. The result is a clone (a population of genetically identical cells) that is specialized to protect against that particular antigen.

We see, then, that the body produces highly specialized armies of B cells and T cells designed to eliminate a specific antigen from the body. The process by which this production occurs, called **clonal selection**, underlies the entire immune response. Each lymphocyte is equipped to recognize a specific antigen.

---

[1]B cells and dendritic cells (cells with long extensions found in lymph nodes) are also antigen-presenting cells.

[2]A helper T cell is also known as a *T4 cell* or a *CD4 cell,* after the receptors on its surface.

[3]A "virgin" cell is one that has been preprogrammed to respond to a particular antigen but has not been previously activated to respond.

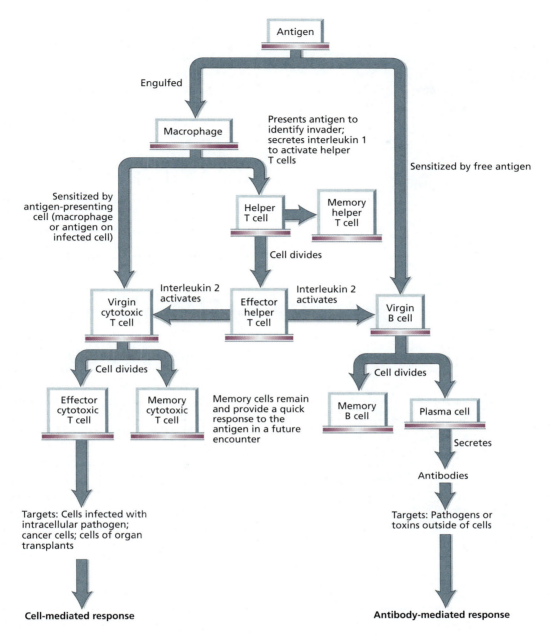

FIGURE **13.11**

An overview of the immune response. When an antigen-presenting macrophage encounters the appropriate helper T cell, it binds to it. The binding causes the macrophage to secrete a chemical called interleukin 1, which activates a helper T cell. The activated helper T cell then locates an appropriate B cell or T cell and secretes interleukin 2, a chemical that activates the B cells and T cells to begin a full-fledged defense against the invader.

Only a few lymphocytes are able to recognize each of millions of different antigens. By binding to the receptors on the lymphocyte surface, an antigen *selects* a lymphocyte that was preprogrammed during its maturation with receptors that recognize that particular antigen. That lymphocyte is then stimulated to divide and produces a clone of millions of identical cells programmed to recognize that antigen (Figure 13.12).

The following analogy may be helpful in understanding clonal selection. Consider a small bakery with only sample cookies on display. A customer chooses a particular cookie, and places an order for many cookies of that type. The cookies are then prepared especially for that person. The sample cookies do not take a lot of space, so a wide selection can be available. The baker does not waste energy making unnecessary cookies. Your

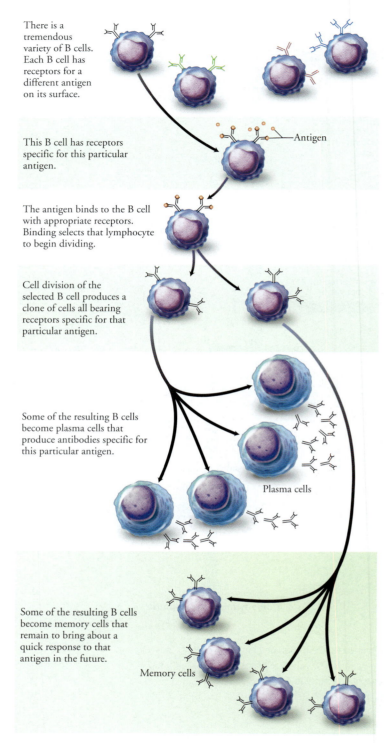

There is a tremendous variety of B cells. Each B cell has receptors for a different antigen on its surface.

This B cell has receptors specific for this particular antigen.

Antigen

The antigen binds to the B cell with appropriate receptors. Binding selects that lymphocyte to begin dividing.

Cell division of the selected B cell produces a clone of cells all bearing receptors specific for that particular antigen.

Some of the resulting B cells become plasma cells that produce antibodies specific for this particular antigen.

Plasma cells

Some of the resulting B cells become memory cells that remain to bring about a quick response to that antigen in the future.

Memory cells

FIGURE **I3.I2**

Clonal selection is the process by which the specificity of the immune response develops. This figure shows clonal selection of B cells, but a similar process occurs with T cells. As a B cell or T cell matures, it develops receptors capable of recognizing and binding to one specific antigen. When an antigen enters the body, it binds to a B cell or T cell that bears the appropriate receptor, and the reaction causes the cell to proliferate. In this way, millions of exact copies (a clone) of cells specialized to recognize that particular antigen are created. There are two types of cells in the clone: effector cells (in the case of B cells, these are plasma cells) and memory cells.

body prepares samples of many kinds of lymphocytes. When an antigen selects the appropriate lymphocyte, the body produces many additional copies of the lymphocyte chosen by that particular antigen.

We have already mentioned that there are two types of cells in such a clone: memory cells and effector cells. Before discussing the role of memory cells, let's look more closely at exactly *how* the effector cells protect us.

stop and think

A primary target of HIV, the human immunodeficiency virus that leads to AIDS, is the helper T cell. Why does the virus's preference for the helper T cell impair the immune system more than if another type of lymphocyte were targeted?

B cells mount an antibody-mediated immune response against antigens free in the blood or bound to a cell surface

The effector cells produced when activated B cells divide are called **plasma cells**, and they secrete antibodies into the bloodstream (Figure 13.13). *Antibodies* are Y-shaped proteins that recognize a specific antigen by the shape of the molecule. Each antibody is specific for one particular antigen. The specificity results from the shape of the proteins that form the tips of the Y (Figure 13.14). Because of their shapes, the antibody and antigen fit together like a lock and a key.

Antibodies can bind only to antigens that are free in body fluids or on the surface of a cell. Their main targets are extracellular microbes, including bacteria, fungi, and protozoans. Many bacteria produce harmful chemicals called toxins that can make us ill. Antibodies can bind to free toxins but not to toxins within a cell. Viruses must enter a host cell to replicate. Antibodies can also bind to free virus particles, but they cannot bind to viruses that have already entered a host cell.

Antibodies help defend against pathogens or toxins by binding to antigens. Each antibody can bind to two identical antigens, one at each tip of the Y. When antibodies bind to molecules on the surface of a virus, they prevent the virus from entering a host cell. Antibodies can also bind to toxins produced by pathogens, forming a coat around them and preventing them from damaging host tissues. The antibody coating attracts phagocytic cells to the area and makes the pathogens or toxins more easily engulfed and removed from the body by those phagocytes. An antibody can bind to two antigens, causing them to clump together and precipitate, enhancing both phagocytosis and inflammation. Certain antibodies activate the complement system, which then pokes holes through the membrane of the target cell and causes it to burst.

There are five classes of antibodies, each with a special role to play in protecting against invaders. Antibodies are also called **immunoglobulins** (Ig), and each class is designated with a letter: IgG, IgM, IgE, IgA, and IgD. As you can see in Table 13.4, in some classes, the antibodies exist as single Y-shaped molecules (monomers), in one class they exist as two attached molecules

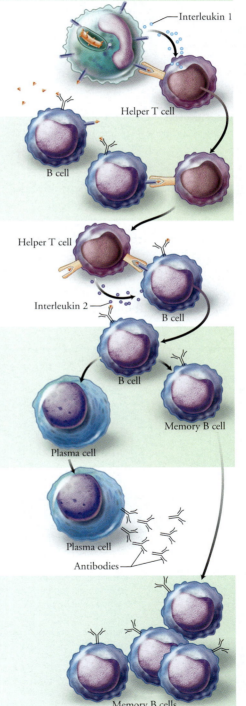

1. THREAT
• An invader enters the body

2. DETECTION
• A macrophage encounters, engulfs, and digests the invader (e.g., a bacterium)

• The macrophage places a piece of the invader (antigen) on its surface with the "self" (MHC) marker

3. ALERT
• The macrophage presents the antigen to a helper T cell
• The macrophage secretes interleukin 1, which activates the helper T cell

• A virgin, or memory B, cell is sensitized by binding to the antigen

• The B cell sensitized by an antigen binds to a helper T cell that has been activated by a macrophage

4. ALARM
• The helper T cell secretes interleukin 2, which stimulates the B cell to begin dividing
• Two populations of B cells are formed: memory cells and plasma cells

5. BUILDING SPECIFIC DEFENSE
• The B cell divides and forms plasma cells and memory cells

6. DEFENSE
• Plasma cells secrete antibodies specific for that antigen
• Antibodies circulate in the blood, where they neutralize and agglutinate the target antigen and activate the complement system

7. CONTINUED SURVEILLANCE
• Memory B cells remain and mount a quick response if the invader is encountered again

FIGURE **13.13**
Antibody-mediated immune response

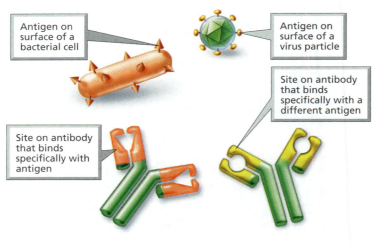

Antigen on surface of a bacterial cell

Antigen on surface of a virus particle

Site on antibody that binds specifically with a different antigen

Site on antibody that binds specifically with antigen

FIGURE **13.14**

An antibody is a Y-shaped protein designed to recognize a specific antigen. The recognition of a specific antigen occurs because of the shape of the tips of the Y in the antibody molecule.

(dimers), and in one class they exist as five attached molecules (pentamers) radiating outward like the spokes of a wheel.

Cytotoxic T cells mount a cell-mediated immune response to destroy antigen-bearing cells

Cytotoxic T cells[4] are the effector T cells responsible for the cell-mediated immune response. Each cell is programmed to recognize the same type of antigen that is bound to MHC (self)

markers on the surface of a cellular pathogen, an infected or cancerous body cell, or on cells of a tissue or organ transplant. A cytotoxic T cell becomes activated to destroy a target cell when two events occur simultaneously, as shown in Figure 13.15. First, the cytotoxic T cell must recognize and bind to the combination of a self marker plus an antigen on the surface of an antigen-presenting cell (either a macrophage or an infected body cell). Second, the cytotoxic T cell must receive additional stimulation from interleukin 2 released from a helper T cell. When sufficiently activated, the cytotoxic T cell divides, producing memory cells and effector cytotoxic T cells. An effector cytotoxic T cell releases chemicals, called **perforins**, that cause holes to form in the target cell membrane. The holes are large enough to allow some of the cell's contents to leave the cell. The cell then disintegrates. The cytotoxic T cell then detaches from the target cell and seeks another cell having the same type of antigen.

Cytotoxic T cells and the natural killer cells we described earlier use the same methods to destroy their targets. The difference between the attacks of these two types of cell is specificity. Cytotoxic T cells destroy cells bearing a specific antigen. Natural killer cells attack any cell that is not recognized as self.

stop and think

Rejection of an organ transplant occurs when the recipient's immune system attacks and destroys the cells of the transplanted organ. Why would this attack occur? Which branch of the immune system would be most involved?

TABLE **13.4**

CLASSES OF ANTIBODIES				
CLASS	**STRUCTURE**	**LOCATION**	**CHARACTERISTICS**	**PROTECTIVE FUNCTIONS**
IgG	Monomer	Blood, lymph, and the intestines	Most abundant of all antibodies in body; involved in primary and secondary immune responses; can pass through placenta from mother to fetus and provides passive immune protection to fetus and newborn	Enhances phagocytosis; neutralizes toxins; triggers complement system
IgA	Dimer or monomer	Present in tears, saliva, and mucus as well in secretions of gastrointestinal system and excretory systems; present in breast milk	Levels decrease during stress, raising susceptibility to infection	Prevents pathogens from attaching to epithelial cells of surface lining
IgM	Pentamer	Attached to B cell where it acts as a receptor for antigens; free in blood and lymph	First Ig class released by plasma cell during primary response	Powerful agglutinating agent (10 antigen binding sites); activates complement
IgD	Monomer	Surface of many B cells; blood and lymph	Life span of about 3 days	Thought to be involved in recognition of antigen and in activating B cells
IgE	Monomer	Secreted by plasma cells in skin, mucous membranes of gastrointestinal and respiratory systems	Become bound to surface of mast cells and basophils	Involved in allergic reactions by triggering release of histamine and other chemicals from mast cells or basophils

[4]Other sources may refer to these as cytolytic cells, T8 cells, or CD8 cells.

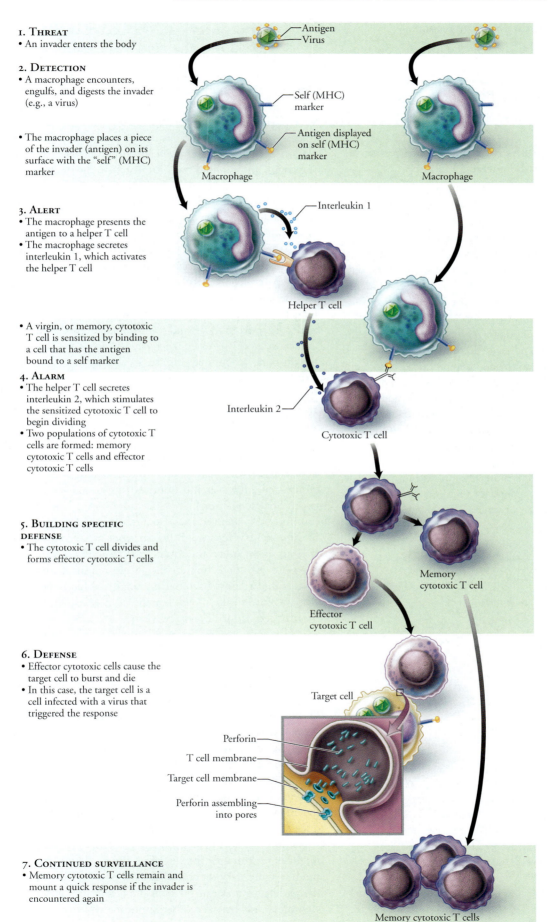

1. THREAT
• An invader enters the body

2. DETECTION
• A macrophage encounters, engulfs, and digests the invader (e.g., a virus)

• The macrophage places a piece of the invader (antigen) on its surface with the "self" (MHC) marker

Antigen
Virus

Self (MHC) marker

Antigen displayed on self (MHC) marker

Macrophage

Macrophage

3. ALERT
• The macrophage presents the antigen to a helper T cell
• The macrophage secretes interleukin 1, which activates the helper T cell

Interleukin 1

Helper T cell

• A virgin, or memory, cytotoxic T cell is sensitized by binding to a cell that has the antigen bound to a self marker

4. ALARM
• The helper T cell secretes interleukin 2, which stimulates the sensitized cytotoxic T cell to begin dividing
• Two populations of cytotoxic T cells are formed: memory cytotoxic T cells and effector cytotoxic T cells

Interleukin 2

Cytotoxic T cell

5. BUILDING SPECIFIC DEFENSE
• The cytotoxic T cell divides and forms effector cytotoxic T cells

Memory cytotoxic T cell

Effector cytotoxic T cell

6. DEFENSE
• Effector cytotoxic cells cause the target cell to burst and die
• In this case, the target cell is a cell infected with a virus that triggered the response

Target cell

Perforin
T cell membrane
Target cell membrane
Perforin assembling into pores

7. CONTINUED SURVEILLANCE
• Memory cytotoxic T cells remain and mount a quick response if the invader is encountered again

Memory cytotoxic T cells

FIGURE **13.15**
Cell-mediated immune response

▌Suppressor T cells turn off the immune response

As the immune system begins to conquer the invading organism and the level of antigens declines, another type of T cell, the **suppressor T cell**, releases chemicals that dampen the activity of both B cells and T cells. Suppressor T cells turn off the immune response when the antigen no longer poses a threat.

▌Immunological memory allows for a more rapid response on subsequent exposure to the antigen

The first time an antigen enters the body, only a few lymphocytes can recognize it. Those lymphocytes must be located and stimulated to divide to produce an army of lymphocytes ready to eliminate that particular antigen. As a result, the **primary response**, the one that occurs during the body's first encounter with a particular antigen, is relatively slow. During the primary response, a lapse of several days occurs before the antibody concentration begins to rise. The concentration does not peak until 1 to 2 weeks after the initial exposure to the antigen (Figure 13.16).

Following subsequent exposure to the antigen, the **secondary response** is strong and swift. Recall that when virgin B cells and T cells were stimulated to divide, they produced both effector cells that actively defended against the invader and memory cells. These memory B cells and T cells live for years or even decades. Thus, the number of lymphocytes programmed to respond to that particular antigen is greater than it was before the first exposure. When the antigen is encountered again, each of those memory cells divides and produces effector cells and memory cells specific for that antigen. As a result, the number of effector cells rises quickly, and within 2 or 3 days reaches a higher peak during the secondary response. This process is the reason we do not get chickenpox twice.

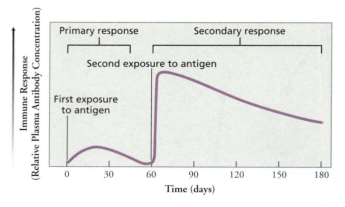

FIGURE **13.16**

The primary and secondary immune responses. During the primary response, which occurs after the first exposure to an antigen, a delay of several days occurs before the concentration of circulating antibodies begins to increase. It takes 1 to 2 weeks for the antibody concentration to peak because the few lymphocytes programmed to recognize that particular antigen must be located and activated. (The T cells show a similar pattern of response.) The secondary response following a subsequent exposure to an antigen is swifter and stronger than the primary response. The long-lived memory cells produced during the primary response cause this difference. The memory cells create a larger pool of lymphocytes programmed to respond to that particular antigen.

Immunity can be Active or Passive

In **active immunity** the body actively participates by producing memory B cells and T cells following exposure to an antigen. Active immunity happens naturally whenever a person gets an infection. Fortunately, active immunity can also develop through **vaccination**, a procedure that introduces a harmless form of the disease-causing microbe into the body to stimulate immune responses against that antigen. In some vaccines, those for whooping cough and typhoid fever for instance, the microbe is inactivated before the vaccine is prepared. These inactivated vaccines usually consist of the cell walls of bacteria, the toxins of bacteria that have been made ineffective, or the protein coats of viruses. Other vaccines must be made from live organisms to be effective. In these cases, the microbes are first weakened so that they can no longer cause disease. Some vaccines, including the one against smallpox, are prepared from microbes that cause related but milder diseases.

Because memory cells are produced, active immunity is relatively long lived. The first dose of a vaccine causes the primary immune response, and antibodies and some memory cells are generated. In certain cases, especially when inactivated antigens are used in the vaccine, the immune system may "forget" its encounter with the antigen after a time. A booster is administered periodically to be sure the immune system does not forget. The booster results in a secondary immune response and enough memory cells to provide for a quick response should a potent form of that pathogen ever be encountered.

Vaccinations have saved millions of lives. In fact, they have been so effective in preventing diseases such as whooping cough and tetanus that many people think that those diseases have been eliminated. However, most of the diseases that vaccines prevent still exist. So vaccinations are still important. Children should be immunized (given vaccines) on a recommended schedule, as shown in Table 13.5. Hepatitis B vaccinations are recommended for adolescents and for certain adults, including those who are at risk of exposure to blood. Some colleges are requiring that freshman be immunized against bacterial meningitis. Elderly people and those who are chronically ill will benefit from yearly flu shots and pneumonia vaccinations every few years.

Although adverse reactions to vaccines are possible, the risk of being harmed by a vaccine is much lower than that associated with the diseases they prevent. For example, a severe paralytic illness called Guillain-Barré syndrome may develop in one or two people out of a million persons receiving the influenza (flu) vaccine. But the flu kills tens of thousands of people each year. In most cases, a reaction to a vaccine is mild: warmth, redness, and tenderness at the site of injection.

Passive immunity results when a person receives antibodies that were produced by another person or animal. For instance, antibodies produced by a pregnant woman can cross the placenta and give the growing fetus some immunity. These maternal antibodies remain in the infant's body for as long 3 months, after which the infant can produce its own antibodies. Antibodies in breast milk also provide passive immunity to nursing infants, especially against pathogens that might enter through the intestinal lining. The mother's antibodies can be a temporary yet critical blanket of protection. Indeed, most of the pathogens that can threaten the health of a newborn have been seen previously by the mother's immune system.

TABLE **13.5**

RECOMMENDED IMMUNIZATION SCHEDULE IN 2003–2004	
VACCINE	**AGE**
Hepatitis B #1	Birth–2 months
Hepatitis B #2	1–4 months
Hepatitis B #3	6–18 months
DPT (Diphtheria, Tetanus, Pertussis)	2, 4, & 6 months
Booster	15–18 months
Booster	4–6 years
DT (Diphtheria Tetanus)	11–18 years
Haemophilus influenzae type b	2, 4, & 6 months
Booster	12–15 months
Polio	2 & 4 months
	6–18 months
	4–6 years
MMR (Measles, Mumps, Rubella)	12–15 months
Booster	4–6 years
Varicella (chickenpox)	12–18 months
PCV (pneumococcal conjugated vaccine)	2–15 months

Passive immunity is also possible in adulthood when antibodies produced in another person or animal are injected into another person. In this case, passive immunity is a good news-bad news situation. The good news is that the effects are immediate. For this reason, gamma globulin (a preparation of antibodies) is used to help people who have been exposed to diseases such as hepatitis B or who are already infected with the microbes that cause tetanus, measles, or diphtheria. Gamma globulin is often given to travelers before they visit a country where viral hepatitis is common. The bad news is that the protection is short lived. The borrowed antibodies circulate for 3 to 5 weeks before being destroyed in the recipient's body. Because the recipient's immune system was not stimulated to produce memory cells, protection disappears with the antibodies.

stop and think

The viruses that cause influenza (the flu) mutate rapidly, so the antigens in the protein coat continually change. Why does this characteristic make it difficult to develop a vaccine against the flu that will be effective for several consecutive years?

what would you do?

Some parents have increasing concern about vaccine safety. Indeed, some parents refuse to have their children vaccinated. Most schools require proof of immunization against certain diseases for admission. Whose rights do you think should prevail—the parents' or the school's? If you were (or are) a parent, would you have your child vaccinated?

Monoclonal Antibodies Are Used in Research, Clinical Diagnosis, and Disease Treatment

Suppose you wanted to determine whether a particular antigen was present in a solution, tissue, or even somewhere in the body. An antibody specific for that antigen would be just the tool you would need. Because of such specificity, any such antibody would go directly to that target. If a label (a radioactive tag or a molecule that fluoresced, for instance) were attached to the antibody, the antibody could reveal the location of the antigen. We see, then, that it is sometimes desirable to have a supply of identical antibodies that will react with a specific antigen. Such groups of identical antibodies that bind to one specific antigen are called **monoclonal antibodies**.

Monoclonal antibodies have many uses. Home pregnancy tests contain monoclonal antibodies produced to react with a hormone (human chorionic gonadotropin; see Chapter 18) produced by membranes associated with the developing embryo. A monoclonal-antibody drug that will detect the early stages of inhalation anthrax infection is now being tested. Monoclonal antibodies have also proved useful in screening for certain diseases, including Legionnaire's disease, hepatitis, certain sexually transmitted diseases, and certain cancers, including those of the lung and prostate. Some monoclonal antibodies are used in cancer treatment. The radioactive material or chemical treatment is attached to a monoclonal antibody that targets the tumor cells but has little effect on other cells.

The Immune System Can Cause Problems

The immune system protects us against myriad threats from agents not recognized as belonging in the body. However, sometimes the defenses are misguided. In autoimmune disease, the body's own cells are attacked. Allergies result when the immune system protects us against substances that are not harmful. Tissue rejection following organ transplant is also caused by the immune system (see the Health Issue essay, *Rejection of Organ Transplants*).

Autoimmune disorders occur when the immune system attacks the body's own cells

Autoimmune disorders occur when the immune system fails to distinguish between self and nonself and attacks the tissues or organs of the body. Thus, if the immune system can be called the body's military defense, then autoimmune disease is the equivalent of "friendly fire."

As we have seen, during their development, lymphocytes are programmed to attack a specific foreign antigen while still tolerating self antigens. Lymphocytes that do not learn to make this distinction are usually destroyed. Unfortunately, some lymphocytes that are primed to attack self antigens escape destruction. These cells are like time bombs ready to attack the body's own cells at the first provocation. For example, if a virus or bacterium activates these renegade lymphocytes, they may direct their attack against healthy body cells as well as the invading organism.

Rejection of Organ Transplants

Each year, tens of thousands of people receive a gift of life in the form of a kidney, heart, lungs, liver, or pancreas. Although these transplants seem commonplace today, they have been performed for only about 30 years.

As we have seen, the effector T cells of the immune system attack and kill cells that lack self markers. When transplanted tissue is killed by the host's immune system, we say that the transplant has been rejected.

We see, then, that the success of the transplant depends on the similarity between the host and transplanted tissues. Thus, it is clear that the most successful transplants would involve tissue taken from one part of a person's body and transplanted to another part. In cases of severe burns, for example, healthy skin can replace badly burned areas of skin. Because identical twins are genetically identical, their cells have the same self markers, and organs can be transplanted from one twin to another with little fear of tissue rejection. But few of us have an identical twin.

The next best source for tissue for a transplant, and the most common source, is a person whose cell surface markers closely match those of the host. Usually the transplanted tissue comes from a person who has recently died. The donor is usually brain-dead, but his or her heart is kept beating by life-support equipment. Some organs—primarily kidneys—can be harvested from someone who has died and whose heart has stopped beating. This discovery could increase the supply of organs available for transplant. In some cases, living people can donate organs; one of two healthy kidneys can be donated to a needy recipient, as can sections of liver.

Regardless of the improving odds for successful transplants, the waiting list of patients in need of an organ from a suitable donor has outpaced the supply. Some researchers believe that, in the future, organs from nonhuman animals may fill the gap between supply and demand for organs. So far, attempts to transplant animal organs into people have failed. The biggest obstacle is hyperacute rejection. Within

minutes to hours after transplant, the animal organ dies because its blood supply is choked off by the human immune system.

The animal most likely to supply these organs is surprising—the pig. It turns out that pigs, which are easily bred, have a physiology similar to ours. Humans can receive small pig parts, such as heart valves, if their immune systems are suppressed after the transplant, but whole organs are rejected. That situation may change, however, through genetic engineering (Chapter 21). Scientists first altered the genes of pigs to make the pig cell membrane proteins more similar to those of humans. Now researchers have found a way to disable another gene in pigs that causes hyperacute rejection. Furthermore, the genetically altered pigs have been cloned, showing that a supply of genetically identical piglets is possible.

Other dangers may remain, even if the rejection problem is solved. Animals carry infectious agents that are harmless to their hosts but that could become deadly if introduced to humans in transplanted tissue.

Autoimmune disorders are often classified as organ-specific or non-organ-specific (Table 13.6). As the name implies, organ-specific autoimmune disorders are directed against a single organ. The thyroid gland, for example, is attacked in Hashimoto's thyroiditis. T cells that have gone awry usually cause organ-specific autoimmune disorders. In contrast, non-organ-specific autoimmune disorders tend to have effects throughout the body. In systemic lupus erythematosis, for instance, connective tissue is attacked. Because connective tissue is found throughout the body, almost any organ can be affected. There may be skin lesions or rashes, especially a butterfly-shaped rash centered on the nose and spreading to both cheeks. It may affect the heart (pericarditis), joints (arthritis), kidneys (nephritis), or nervous system (seizures). Antibodies generally cause non-organ-specific autoimmune disorders (Figure 13.17).

A number of autoimmune disorders occur because portions of disease-causing organisms resemble antigens found on normal body cells. If the immune system mistakes the body's antigens for the foreign antigens, it may attack them. For instance, the body's attack on certain streptococcal bacteria that cause a sore throat may result in the production of antibodies that target the streptococcal bacteria as well as similar molecules that are found in the valves of the heart and joints. The result is an autoimmune disorder known as rheumatic fever.

Treatment of autoimmune disorders is usually two pronged. First, deficiencies caused by the disorder are corrected. In diabetes

mellitus type 1, insulin-producing cells in the pancreas are destroyed by the immune system (discussed in Chapter 6). Treatment for diabetes would, therefore, include replacement of insulin. Second, immune system activity is suppressed with drugs.

Allergies are immune responses to harmless substances

An **allergy** is an overreaction by the immune system to an antigen, in this case called an **allergen**. The immune response in an allergy is considered an overreaction because the allergen usually is not harmful to the body (Table 13.7). The most common allergy is hay fever, which, by the way, is not caused by hay and does not cause a fever. Hay fever is more correctly known as allergic rhinitis (*rhino*, a nose; *-itis*, inflammation of). The symptoms of hay fever—sneezing and nasal congestion—occur when an allergen is inhaled, triggering an immune response in the respiratory system. Mucous membranes of the eyes may also respond, causing red, watery eyes. Common causes of hay fever include pollen, mold spores, animal dander, and the feces of dust mites, creatures that are found all over your house (Figure 13.18). The same allergens, however, can cause asthma. During an asthma attack, the small airways in the lung (bronchioles) constrict and make breathing difficult. In food allergies, the immune response occurs in the digestive system and may cause nausea, vomiting, abdominal cramps, and diarrhea. Foods can also cause hives, a skin condition in which patches of skin temporarily become red and swollen.

TABLE **13.6**

AUTOIMMUNE DISORDERS		
AUTOIMMUNE DISORDER	**TARGET OF IMMUNE SYSTEM ATTACK**	**EFFECT**
Organ-specific		
Hashimoto's thyroiditis	Thyroid gland	Decreased production of thyroid hormone
Pernicious anemia	Cells in stomach lining that produce a chemical needed to absorb vitamin B_{12}, which is needed for red blood cell production	Decreased production of red blood cells
Addison's disease	Adrenal glands	Adrenal failure
Diabetes mellitus type 1	Insulin-producing cells in the pancreas	Elevated blood sugar
Graves' disease	Thyroid gland	Increased rate of chemical reactions in body
Multiple sclerosis	Myelin sheath of nerve cells	Short-circuiting of neural impulses resulting in sensory and/or motor defects
Ankylosing spondylitis	Joints between vertebrae	Spine bent and fused, inflammation
Myasthenia gravis	Connections between nerve and muscle	Muscle weakness
Glomerulonephritis	Kidney	Kidney failure
Encephalitis	Brain	Impaired brain functions, headaches, irritability, double vision, impaired speech
Non-organ-specific		
Lupus erythematosus	Connective tissue	Butterfly-shaped rash on face, skin lesions, joint pain
Rheumatoid arthritis	Collagen fibers of joints	Joint pain
Dermatomyositis	Skin inflammation	Body rash
Rheumatic fever	Heart valves, joints	Joint pain, kidney failure

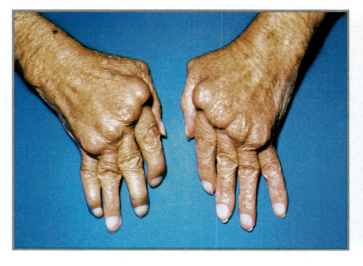

FIGURE **13.17**

Rheumatoid arthritis is an autoimmune disease, a condition caused when the body's immune cells turn the attack against the body's own cells.

TABLE **13.7**

COMMON ALLERGIES			
TYPE OF ALLERGIC RESPONSE	**COMMON CAUSES**	**LOCATION OF REACTIVE MAST CELLS**	**SYMPTOMS**
Hay fever (allergic rhinitis)	Pollen, mold spores, animal dander (bits of skin and hair), feces of dust mites	Lining of nasal cavity	Sneezing, nasal congestion
Asthma	Pollen, mold spores, animal dander	Airways of lower respiratory tract	Difficulty breathing
Food allergy	Chicken, eggs, fish, milk, nuts (especially peanuts), shellfish, soybeans, and wheat	Lining of digestive system	Nausea, vomiting, abdominal cramps, and diarrhea
Hives	Foods (especially shellfish, strawberries, chocolate, nuts, and tomatoes); insect bites; certain drugs (especially penicillin and aspirin); chemicals such as food additives, dyes, and cosmetics	Skin	Patches of skin become red and swollen
Anaphylactic shock	Insect stings (especially from bees, wasps, hornets, yellow jackets, fire ants); medicines (especially penicillin and tetracycline); certain foods (especially eggs, seafood, nuts, and grains)	Throughout the body	Widening of blood vessels, causing blood to pool in capillaries and resulting in dizziness, nausea, diarrhea, and unconsciousness; death

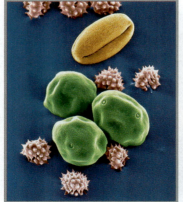

Pollen grains

Dust mite

FIGURE **13.18**
Common causes of allergies are pollen grains and the feces of dust mites, such as the mite shown here.

Anaphylactic shock is an extreme allergic reaction that occurs within minutes after exposure to the substance that a person is allergic to. It can cause pooling of blood in capillaries, which causes dizziness, nausea, and sometimes unconsciousness as well as extreme difficulty in breathing. Anaphylactic shock can cause death, but most people survive. Common triggers of anaphylactic shock include certain foods; medicines, including antibiotics such as penicillin and tetracycline; and insect stings, especially stings from bees, wasps, yellow jackets, and hornets.

An allergy begins when a person is exposed to an allergen and the primary immune response begins (Figure 13.19). Soon, plasma cells churn out a certain type of antibody known as IgE antibodies. These antibodies bind to either basophils or mast cells. Subsequent exposures to that allergen cause an allergic response. This time, the allergen binds to IgE antibodies on the

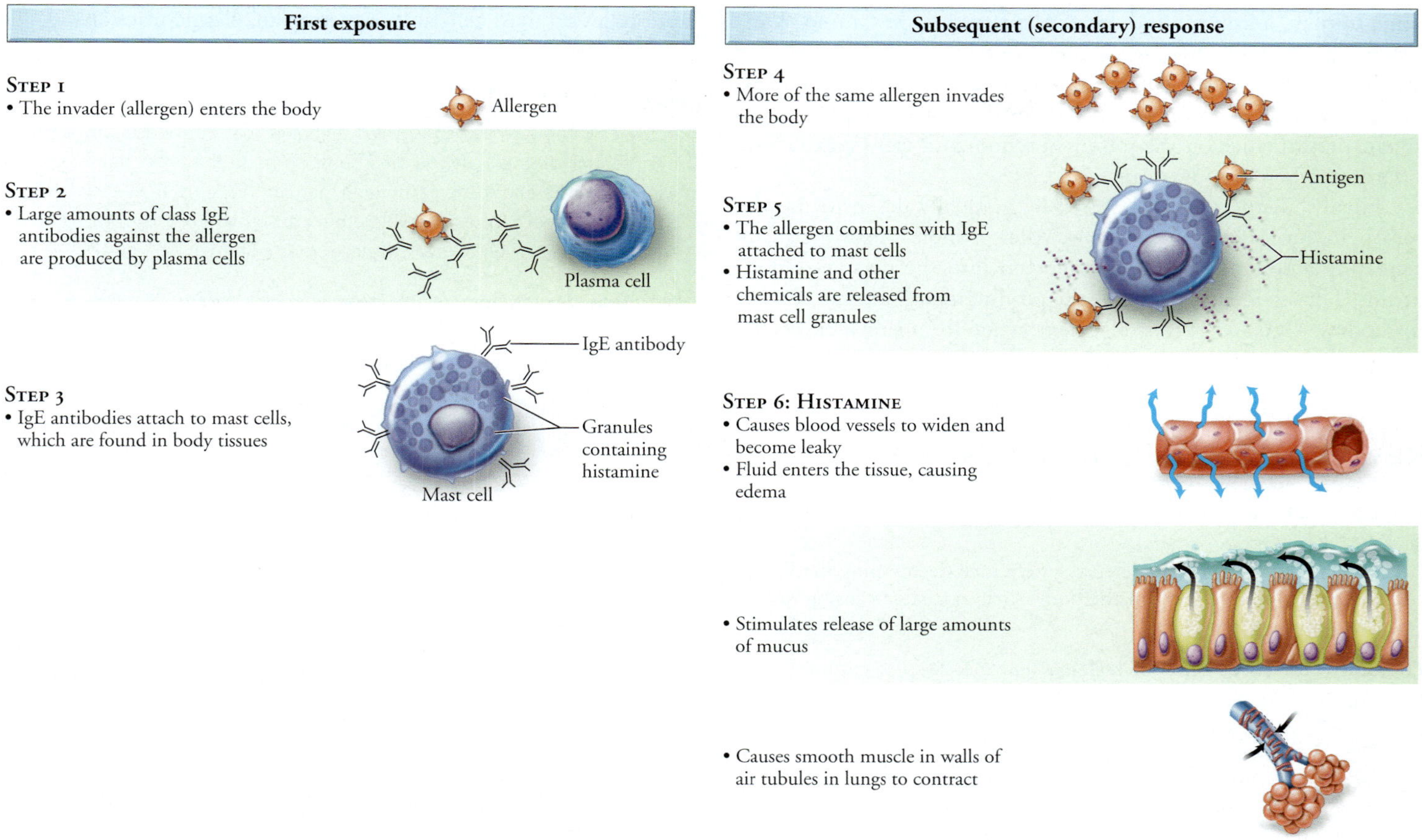

First exposure

STEP 1
• The invader (allergen) enters the body

Allergen

STEP 2
• Large amounts of class IgE antibodies against the allergen are produced by plasma cells

Plasma cell

STEP 3
• IgE antibodies attach to mast cells, which are found in body tissues

IgE antibody

Granules containing histamine

Mast cell

Subsequent (secondary) response

STEP 4
• More of the same allergen invades the body

STEP 5
• The allergen combines with IgE attached to mast cells
• Histamine and other chemicals are released from mast cell granules

Antigen

Histamine

STEP 6: HISTAMINE
• Causes blood vessels to widen and become leaky
• Fluid enters the tissue, causing edema

• Stimulates release of large amounts of mucus

• Causes smooth muscle in walls of air tubules in lungs to contract

FIGURE **13.19**
Steps involved in an allergic reaction

surface of basophils or mast cells. When binding occurs, granules containing histamine release their contents.

Histamine then causes the swelling, redness, and other symptoms of an allergic response. The blood vessels widen, slowing blood flow through them and causing redness. At the same time, the blood vessels become leaky, allowing fluid to flow from the vessel into spaces between tissue cells. The fluid causes swelling of the surrounding tissues. The result in hay fever is a stuffy nose and puffy, red eyelids. Histamine also causes the release of large amounts of mucus. Hence, the nose begins to run. Histamine can also cause smooth muscles of internal organs to contract. In an asthma attack, contraction of smooth muscle in the respiratory airways causes these passageways to constrict and makes breathing difficult. When the smooth muscle of the digestive system contracts, the result can be abdominal cramps and stomach pain. If the allergen spreads

from the area where it entered the body, its effects can be widespread. The result can be anaphylactic shock.

People with allergies often know which substances cause their problems. When the culprits are not known, doctors can identify them using a crude but effective technique. Small amounts of suspected allergens are injected into the skin. If the person is allergic to one of the suspected allergens, a red welt will form at the site of injection.

The simplest way to avoid the miseries of allergies is to avoid exposure to the substances that cause problems. During pollen season, spend as much time as possible indoors, using an air conditioner to filter pollen out of the incoming air. Unfortunately, spores from molds growing in air conditioners and humidifiers are also common triggers of allergies. Some foods, for instance, strawberries, may be easy to avoid. Others, such as peanut oil, can show up in some unlikely foods, including stew, chili, or meat patties.

Certain drugs may also reduce allergy symptoms. As their name implies, antihistamines block the effects of histamine. Antihistamines are most effective if they are taken before the allergic reaction begins. Unfortunately, antihistamines tend to become less effective over time, and most of them cause drowsiness, which can impair performance on the job or in school and can be extremely hazardous when driving a car.

Finally, allergies can be treated by gradually desensitizing the person to the offending allergens. Allergy shots inject gradually increasing amounts of a known allergen into the person's bloodstream. The allergen then causes the production of another class of antibodies—IgG. After this treatment, when the allergen enters the body, IgG antibodies bind to it and prevent it from binding to IgE antibodies on mast cells and triggering an allergic reaction.

exploring further ...

In this chapter, we have considered the ways that our bodies protect us from foreign agents that could cause us harm. In the next chapter, we will learn more about the pathogens that can make us ill. We will also consider the mechanisms by which they cause harm and the ways they are transmitted. Then, we will explore some potential biological weapons.

REVIEWING THE CONCEPTS

The Body's Defense System Targets Pathogens and Cancerous Cells (p. 256)

1. The targets of the body's defense system include anything that is not recognized as belonging in the body, such as disease-causing organisms and cancerous cells.

The Body Has Three Lines of Defense (pp. 256–263)

2. The first line of defense is physical and chemical barriers that prevent entry by pathogens.
3. The skin and mucous membranes provide physical barriers that help keep foreign substances, including pathogens, out of the body. They also produce protective chemicals, such as sweat, oil, tears, and saliva. These are nonspecific defenses that protect against any invader.
4. The second line of defense includes defensive cells and proteins, inflammation, and fever.
5. Protective cells include phagocytes, eosinophils, and natural killer cells. Phagocytes are scavenger cells specialized to engulf and destroy particulate matter, such as pathogens, damaged tissue, and dead cells. Eosinophils kill parasites. Natural killer (NK) cells are lymphocytes that find and kill abnormal cells in the body, primarily cancerous cells and virus-infected cells. NK cells kill by secreting chemicals called perforins that poke holes in the target cell, causing it to burst.
6. Two types of defensive proteins are interferons and complement. Interferons are proteins released by certain cells that are infected by viruses to help curb viral diseases. Complement is a group of at least 20 proteins whose activities enhance other body defense mechanisms.
7. The inflammatory response occurs in response to tissue injury or invasion by foreign microbes. It begins when cells in the injured area release histamine, which increases blood flow by dilating blood vessels to the region and by increasing the permeability of capillaries there. Increased blood flow results in redness and warmth in the region. Fluid leaking from the capillaries causes swelling. Phagocytes and other protective cells are attracted to the area.
8. Fever, an abnormally high body temperature, helps the body fight invading microbes by enhancing several body defense mechanisms and slowing the growth of many pathogens.
9. The third line of defense, the immune system, has specific targets and memory.
10. Substances that trigger immune responses are called antigens.
11. Lymphocytes are white blood cells that are responsible for immune responses. Both B lymphocytes (B cells) and T lymphocytes (T cells) develop in the bone marrow. The B cells mature in the bone marrow, but the T cells mature in the thymus gland.

During maturation, B cells and T cells develop receptors on their surfaces that allow them to recognize one specific antigen. Maturing B cells and T cells learn to recognize the cells that belong in the body.

12. When an antigen is detected, B cells and T cells with receptors specific to that antigen divide repeatedly, forming effector cells that destroy the antigen and memory cells that provide a quick response on subsequent exposure to that antigen.

WEB TUTORIAL 13.1 Inflammatory Response

The Immune System Mounts Antibody-Mediated and Cell-Mediated Responses (pp. 263–270)

13. Macrophages are phagocytic cells that engulf any foreign material or organism they encounter. After engulfing the material, the macrophage places a part of it on its own surface, where it serves as an antigen to alert lymphocytes to the presence of an invader and reveals what the invader looks like. Macrophages also have molecular (MHC) markers on their membranes that identify them as belonging in the body, that is, as "self."
14. A macrophage then presents the antigen to a helper T cell, a cell that serves as the main switch to the entire immune response. The helper T cell must be one that recognizes the antigen on the surface of the macrophage. When this encounter occurs, the macrophage secretes interleukin 1, which activates the helper T cell. The helper T cell, in turn, secretes interleukin 2, which activates the appropriate B cells and T cells (those specific for the antigen that the macrophage engulfed).
15. B cells are responsible for antibody-mediated immune responses, which defend against antigens that are free in body fluids, including bacteria, free virus particles, and toxins. When called into action by a helper T cell, a B cell divides repeatedly, forming two lines of descendant cells: effector cells that transform into plasma cells and memory B cells. Plasma cells secrete Y-shaped proteins called antibodies into the bloodstream. Antibodies bind to the particular antigen and inactivate it or help remove it from the body. Antibodies work by neutralizing the antigen, by agglutinating or precipitating antigens so that phagocytes can more easily engulf them, or by activating complement.
16. Cytotoxic T cells are responsible for cell-mediated immune responses, which are effective against cellular threats, including infected body cells or cancer cells. When a T cell is properly activated, it divides, forming two lines of descendant cells: effector cells, called cytotoxic T cells, and memory T cells. Cytotoxic T cells

secrete perforins that poke holes in the foreign or infected cell, causing it to burst and die.

17. Suppressor T cells dampen the activity of B cells and T cells when antigen levels begin to fall.

18. After the first encounter with a particular antigen, the primary response is initiated, which may take several weeks to be effective against the antigen. However, because of memory cells, a subsequent exposure to the same antigen triggers a quicker response, called a secondary response.

WEB TUTORIAL 13.2 Defense Against Infectious Agents

Immunity Can Be Active or Passive (pp. 270–271)

19. In active immunity, the body actively participates in forming memory cells against a particular antigen. Active immunity may occur when the antigen infects the body, or it may occur through vaccination, a procedure that introduces a harmless form of the antigen into the body. Passive immunity results when a person receives antibodies that were produced by another person or animal. Passive immunity is short lived.

Monoclonal Antibodies Are Used in Research, Clinical Diagnosis, and Disease Treatment (p. 271)

20. Monoclonal antibodies are identical antibodies. They are useful in research and in the diagnosis and treatment of diseases.

The Immune System Can Cause Problems (pp. 271–276)

21. Autoimmune disorders occur when the immune system mistakenly attacks the body's own cells. Organ-specific autoimmune disorders are directed against a single organ and are caused by T cells. Non-organ-specific autoimmune disorders are caused by antibodies and have effects that are widespread throughout the body. Treatments of autoimmune disorders involve correcting the deficiency caused by the disorder and suppressing the immune system.

22. An allergy is a strong immune response against an antigen (called an allergen). An allergy occurs when the allergen causes plasma cells to release large numbers of IgE antibodies. These IgE antibodies bind to mast cells or basophils, causing them to release histamine. Histamine, in turn, causes the redness, swelling, itching, and other symptoms of an allergic response.

KEY TERMS

active immunity   *p. 270*
allergen   *p. 272*
allergy   *p. 272*
antibody   *p. 263*
antibody-mediated immune response   *p. 263*
antigen   *p. 262*
antigen-presenting   cell   (APC)   *p. 264*
autoimmune disorder   *p. 271*

B lymphocyte   *p. 262*
basophil   *p. 260*
cell-mediated   immune   response   *p. 263*
clonal selection   *p. 264*
complement system   *p. 259*
cytotoxic T cell   *p. 268*
helper T cell   *p. 264*
histamine   *p. 260*
immune system   *p. 260*

immunoglobulin   *p. 266*
inflammatory response   *p. 260*
interferon   *p. 258*
lysozyme   *p. 257*
macrophage   *p. 258*
mast cell   *p. 260*
MHC marker   *p. 262*
monoclonal antibody   *p. 271*
natural killer (NK) cell   *p. 258*
passive immunity   *p. 270*

pathogen   *p. 256*
perforin   *p. 268*
phagocyte   *p. 258*
plasma cell   *p. 266*
primary response   *p. 270*
secondary response   *p. 270*
suppressor T cell   *p. 270*
T lymphocyte   *p. 262*
vaccination   *p. 270*

THINKING ABOUT THE CONCEPTS

1. Explain the difference between nonspecific and specific defense mechanisms.   *p. 256*

2. List seven nonspecific defense mechanisms. Explain how each helps protect us against disease.   *pp. 256–260*

3. How does a natural killer cell kill its target cell?   *p. 258*

4. What are interferons? What type of cell produces them? How do they help protect the body?   *pp. 258–259*

5. What are the complement proteins? Explain how they act directly and indirectly to protect the body against disease.   *pp. 259–260*

6. Signs of inflammation include redness, warmth, swelling, and pain. What causes these symptoms? How does inflammation help defend against infection?   *p. 260*

7. What does an antigen-presenting cell do? What is the most common type of antigen-presenting cell? How do other cells recognize the antigen-presenting cell as a "friend"?   *p. 264*

8. What cells are responsible for antibody-mediated immune responses? What are the targets of antibody-mediated immune responses?   *pp. 266–268*

9. Describe an antibody. How do antibodies inactivate or eliminate antigens from the body?   *pp. 266–268*

10. What is responsible for cell-mediated immune responses? What are the targets of cell-mediated immune responses?   *p. 268*

11. How does a natural killer cell differ from a cytotoxic T cell?   *p. 268*

12. Why does a secondary response occur more quickly than the primary response?   *p. 270*

13. Differentiate between active and passive immunity.   *p. 270*

14. What are monoclonal antibodies? What are some medical uses for them?   *p. 271*

15. What is an autoimmune disorder? Give some examples. How are autoimmune disorders treated?   *pp. 271–272*

16. What is an allergy? What causes the symptoms?   *pp. 272–275*

17. What causes a transplanted organ to be rejected? How can the chances of acceptance be increased?   *p. 272*

18. Indicate the *correct* statement.
    a. An antibody is specific to one particular antigen.
    b. Antibodies are held within the cell that produces them.
    c. Antibodies are produced by macrophages.
    d. Antibodies can be effective against viruses that are inside the host cell.

19. An antigen is a
    a. cell that produces antibodies.
    b. receptor on the surface of a lymphocyte that recognizes invaders.
    c. memory cell that causes a quick response to an invader when it is encountered a second time.
    d. large molecule on the surface of an invader that triggers an immune response.

20. Indicate the choice with the *incorrect* pairing of cell type and function.
    a. helper T cell—serves as "main switch" that activates both the cell-mediated immune responses and the antibody-mediated immune responses.
    b. cytotoxic T cells—present antigen to the helper T cell
    c. macrophage—roams the body looking for invaders, which are engulfed and digested when they are found.
    d. suppressor T cells—shut off the immune response when the invader has been removed.

21. When the doctors say they are looking for a suitable donor for a kidney transplant, they are looking for someone
    a. whose tissues have similar "self" markers to those of the recipient.
    b. who lacks antibodies to the recipient's tissues.
    c. who has suppressor T cells that will suppress the immune response against the donor kidney.
    d. lacks macrophages.

22. The piece of the antigen displayed on the surface of a macrophage
    a. stimulates the suppressor T cells to begin dividing.
    b. attracts other invaders to the cell, causing them to accumulate and making it easier to kill the invaders.
    c. informs the other cells in the immune system of the exact nature of the antigen they should be looking for (what the antigen "looks like").
    d. has no function in the immune response.

23. A cell that kills any unrecognized cell in the body and is part of the nonspecific body defenses is the _____.

24. _____ is a chemical released by mast cells and basophils that produces most of the symptoms of an allergy.

25. Antibodies are produced by _____.

26. _____ are important antigen-presenting cells.

27. _____ diseases are those in which the body attacks its own tissues.

APPLYING THE CONCEPTS

1. Barbara was exposed to hepatitis B. Barbara goes to the doctor and asks to be vaccinated against hepatitis B. Instead, the doctor gives her an injection of gamma globulin (a preparation of antibodies). Why wasn't she given the vaccine?

2. More than 100 viruses can cause the common cold. How does this fact explain why you can catch a cold from Ramond immediately after recovering from a cold you caught from Jessica?

3. HIV is a virus that kills helper T cells. HIV is not the direct cause of death in people who have this virus. Instead, people die of diseases caused by organisms that are common in the environment. Explain why HIV-infected persons are susceptible to these diseases.

SPECIAL TOPIC

Infectious Disease

Pathogens Are Disease-Causing Organisms

- Certain bacteria produce toxins that cause disease
- Viruses can damage the host cell as they leave the cell after replication or when incorporated into the cell's chromosomes
- Protozoans cause disease by producing toxins and enzymes
- Fungi often cause disease by secreting enzymes that digest cells
- Parasitic worms cause disease by releasing toxins, feeding off blood, or competing with the host for food
- Prions cause disease by causing normal proteins to become misfolded and form clumps

Disease Is Spread When a Pathogen Enters the Body through Contact, Consumption, or an Animal Vector

Infectious Diseases Remain Cause for Concern

- New diseases are emerging and some old diseases are reappearing
- Epidemiologists track diseases

Biological Organisms and Products May Be Used As Biological Weapons

- Anthrax is caused by a bacterium that forms resistant spores
- Smallpox is caused by a highly contagious virus
- Botulinum toxin is a potent poison

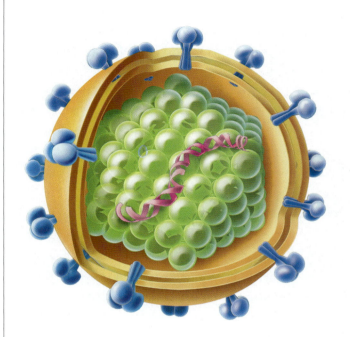

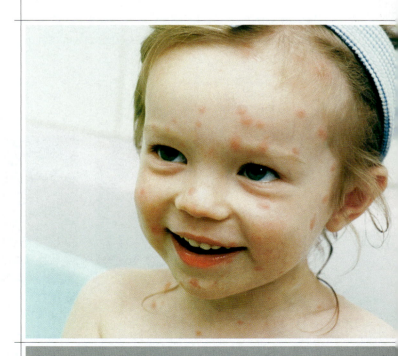

Shortly after breakfast at the dorm cafeteria, while walking to her first class and thinking about nothing in particular, Danielle rubbed her left eye and felt annoyed that the cold February wind was somehow making her tear up. In very little time, her eye began to itch. As the day wore on, she noticed that her left eye was not just itching and tearing; it was also discharging sticky yellow mucus. She went to the rest room. Looking in the mirror, she discovered that there were streaks of red running through the white of her left eye, and her right eye was beginning to show the same signs. After suffering through her next class, she went to the health center, where she was examined after a brief wait. The physician took a sample of the mucus to send to a laboratory for analysis. She examined Danielle's eyes and informed Danielle that she had bacterial conjunctivitis (commonly referred to as pinkeye), an infection she could have picked up from a classmate, roommate, or even the children she babysat a few days earlier. She explained to Danielle that this particular form of conjunctivitis is easily treatable by antibiotic eye drops.

Conjunctivitis is an *infectious disease*; that is, it can be spread from one person to another or from the environment to a person. Although conjunctivitis is highly contagious, it is not usually serious. However, more serious infectious diseases have left and continue to leave their marks on humanity. In the past, the plague, cholera, and diphtheria decimated human populations. Today, tuberculosis is making a comeback. HIV, the human immunodeficiency virus, has caused the deaths of millions of people worldwide. And, we are concerned about the possibility of release of infectious agents by bioterrorists.

Here we will learn more about disease-causing organisms, called *pathogens*, how they cause harm, and how they are transmitted from person to person. ■

Pathogens Are Disease-Causing Organisms

Each variety of pathogen has specific effects on the body. Some pathogens are more familiar than others. We will consider how each type of pathogen causes disease symptoms. However, keep in mind that the immune responses that protect us from pathogens (discussed in Chapter 13) often cause some of the disease symptoms, including fever and body aches.

Virulence is the relative ability of a pathogen to cause disease. The degree of damage done to body cells, and the type of cell damaged are factors that determine a pathogen's virulence.

▮ Certain bacteria produce toxins that cause disease

Bacterial cells differ from the cells that make up our bodies. Recall from Chapter 3 that our bodies are made up of eukaryotic cells that contain a true nucleus and membrane-bound organelles. Bacteria, in contrast, are prokaryotes, which means they lack a nucleus and other membrane-bound organelles (Figure 13a.1). Nearly all bacteria have a semirigid cell wall composed of a strong mesh of peptidoglycan, which consists of sugars and amino acids. The cell wall maintains most types of bacteria in one of three common shapes: a sphere (coccus) that can occur singly, in pairs, or in chains; a rod (bacillus) that usually occurs singly; or a spiral or corkscrew shape (spirilla).

Some bacteria have long, whiplike structures called flagella that allow them to move. Bacteria may also have filaments called pili that help them attach to the cells they are attacking, including those of the human body. Outside the bacterial cell, there is often a capsule that provides protection from white blood cells as well as a means of adhering to a surface.

Bacteria reproduce asexually in a type of cell division called *binary fission* (Figure 13a.2), in which the bacterial genetic material (DNA) is copied, the cell pinches in half, and each new cell contains a complete copy of the original genetic material. Under ideal conditions, certain bacteria can divide every 20 minutes. Thus, if every descendant lived, a single bacterium could result in a massive infection of trillions of bacteria within 24 hours. Their growth rate is significant because the greater the number of bacteria, the greater the potential harm.

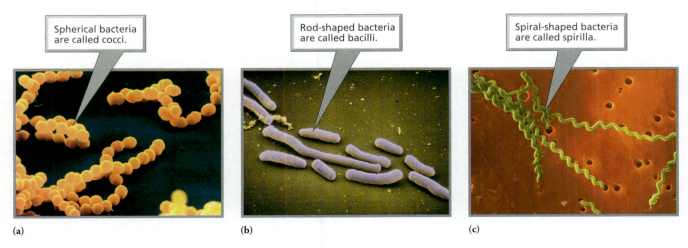

Spherical bacteria are called cocci.

Rod-shaped bacteria are called bacilli.

Spiral-shaped bacteria are called spirilla.

(a)   (b)   (c)

FIGURE **13a.1**

The structure of bacteria. Bacteria are prokaryotic cells. They lack a nucleus and membrane-bound organelles. They have one of three basic shapes: (a) round (coccus), (b) rod-shaped (bacillus), and (c) corkscrew-shaped (spirilla).

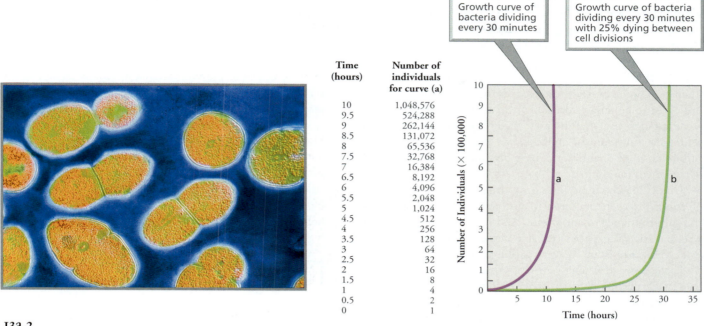

Time (hours)	Number of individuals for curve (a)
10	1,048,576
9.5	524,288
9	262,144
8.5	131,072
8	65,536
7.5	32,768
7	16,384
6.5	8,192
6	4,096
5.5	2,048
5	1,024
4.5	512
4	256
3.5	128
3	64
2.5	32
2	16
1.5	8
1	4
0.5	2
0	1

Growth curve of bacteria dividing every 30 minutes

Growth curve of bacteria dividing every 30 minutes with 25% dying between cell divisions

FIGURE **13a.2**

Bacteria reproduce by binary fission; they copy their genetic information and pinch in half. Bacteria reproduce very rapidly.

BACTERIAL TOXINS

Bacteria can cause disease in a number of ways. Some, such as those that cause tuberculosis or leprosy, directly damage tissue and cause lesions. Most bacteria, however, do their damage by releasing toxins (poisons) into the bloodstream or the surrounding tissues. If the toxins enter the bloodstream, they can be carried throughout the body and disturb body functions.

The disease symptoms depend on which body tissues are affected by the toxin. We find examples of different effects among the toxins produced by bacteria that cause various types of food poisoning. The most common form of food poisoning is caused by *Staphylococcus*. Foods commonly involved with this type of food poisoning include poultry, meat and meat products, and creamy foods such as pudding or salad dressing. The bacteria multiply when food is undercooked or unrefrigerated. The toxins they produce stimulate cells in the immune system to release chemicals that result in inflammation, vomiting, and diarrhea. Eating undercooked contaminated chicken or eggs commonly causes *Salmonella* food poisoning. In this case, the toxin causes changes in the permeability of intestinal cells, leading to diarrhea and vomiting. Contaminated meat, particularly ground meat, is often the cause of *Escherichia coli* (*E. coli*) food poisoning. Several major outbreaks have involved ground beef—hamburgers. In addition to causing vomiting and diarrhea, *E. coli* toxin can cause kidney failure in children and the elderly. The toxin produced by the bacterium (*Clostridium botulinum*) that causes botulism, a type of food poisoning often caused by eating improperly canned food, is one of the most toxic substances known. It interferes with nerve functioning, especially motor nerves that cause muscle contraction. Death occurs because muscle paralysis prevents breathing. If sufficient toxin is consumed, it is almost always fatal.

BENEFICIAL BACTERIA

Not all bacteria are harmful. Some, in fact, are beneficial. They are normal residents in the body that keep potentially harmful microorganisms in check. For example, both group B *Streptococcus* and *E. coli* live harmlessly in most people's intestines. The strep bacteria hold the *E. coli* in check. Although these bacteria are harmless in the intestines, *E. coli* can have devastating, sometimes fatal, effects in the bloodstream. Both can spread to a woman's vagina. If the woman is pregnant, the bacteria can be passed to the newborn during birth and cause serious problems. During the 1990s, many women were given the antibiotic ampicillin during delivery to prevent the spread of strep to the vulnerable newborn. Without the strep bacteria, *E. coli* can flourish. Although strep infections in newborns have declined, *E. coli* infections have risen and are now the most common infection in newborns. Unfortunately, *E. coli* can be more deadly than strep.

ANTIBIOTICS

Fortunately, bacteria can be killed. The human body has its own array of defenses. When the body needs help, we can call on *antibiotics*, chemicals that inhibit the growth of microorganisms. Antibiotics work by reducing the number of bacteria or by slowing the growth rate of the population, allowing time for body defenses to conquer the bacteria. Some antibiotics work by killing bacteria directly by causing them to burst. Others slow bacterial growth by preventing the synthesis of bacterial cell walls. Recall that our body cells lack cell walls (see Chapter 3). Thus, our cells are unaffected by antibiotics that target cell walls. Some antibiotics block protein synthesis by bacteria. These antibiotics prevent bacterial protein synthesis without interfering with cell protein synthesis in human body cells. This selective action is possible because the structure of ribosomes, the organelles on which proteins are synthesized, is slightly different in bacteria and humans. Unfortunately, some bacteria have become resistant to antibiotics, so we are losing our most powerful weapon against bacterial diseases.

ANTIBIOTIC RESISTANCE

When antibiotics were introduced during the 1940s, they were considered to be miracle drugs. For the first time, there was a cure for devastating bacterial diseases, including pneumonia, bacterial meningitis, tuberculosis, and cholera. Today, there are over 160 antibiotics. Indeed, antibiotics have become so commonplace that we take them for granted.

Unfortunately, these life-saving drugs are losing their power. Infections that were once easy to cure can now turn deadly as bacteria gain resistance to antibiotics. And there are now some strains of several bacterial species[1] capable of causing life-threatening illnesses that are resistant to every antibiotic available today.

Bacteria get their ability to resist antibiotics from genes. Some genes enable the bacterium to survive exposure to an antibiotic by preventing the antibiotic from entering the cell, by pumping it out of the cell, or by destroying the antibiotic within the cell. Bacteria can obtain genes for drug resistance in several ways. One way is by inheriting the genes from their ancestors. Genes mutate readily in bacteria, so a mutation for resistance may form spontaneously and be passed to all descendants. What is more alarming, however, is that bacteria can obtain resistance genes from other bacteria. Resistance genes are commonly found on plasmids, small circular pieces of genetic material separate from the bacterial chromosome. Plasmids are duplicated before reproduction and can be transferred to another bacterium during conjugation, which is the bacterial version of sex. In this way, drug resistance can spread through a bacterial population.

It may seem odd, but the use of antibiotics can promote the development of resistant bacteria. When a strain of bacteria is exposed to an antibiotic, the bacteria that are susceptible die. The more resistant bacteria may survive and multiply. If the bacteria are exposed to an antibiotic again, the selection process is repeated. With each exposure to the drug, the resistant bacteria gain a stronger foothold. Making matters worse, antibiotics kill beneficial bacteria along with the harmful ones. Normally, the beneficial bacterial strains help keep the harmful strains in check. Loss of the "good" bacteria can allow the harmful ones to dominate. Furthermore, harmless bacteria that have become antibiotic resistant can pass their resistance genes to disease-causing bacteria in plasmids.

The overuse and misuse of antibiotics are largely to blame for the resistance problem. For example, physicians overuse antibiotics when they prescribe them for illnesses that are viral, such as a cold or flu. However antibiotics have no effect on viruses. Patients misuse antibiotics when they stop taking their medicine as soon as they feel better instead of completing the full course of treatment.

Hospitals use antibiotics heavily, so it should not be surprising that they are breeding grounds for antibiotic-resistant bacteria. The resistant bacteria survive, overgrow susceptible strains, and spread

to other patients. Plasmids with genes for antibiotic resistance can also be transferred from one type of bacteria to another.

Drug-resistant bacteria are selected by overuse in other ways too. Over 40% of the antibiotics used in the United States are given to livestock to promote growth. Farmers also spray crops with antibiotics to control or prevent bacterial infections. Again, plasmids carrying genes for antibiotic resistance can be spread from the bacteria infecting livestock or crops to bacteria capable of infecting humans.

What can you do to slow the spread of drug-resistant bacteria? Use antibiotics responsibly. Do not insist on a prescription for antibiotics against your doctor's advice. Take antibiotics exactly as prescribed, and be sure to complete the treatment. Also, reduce the risk of getting an infection that might require antibiotic treatment by washing your hands frequently, rinsing fruits and vegetables before eating them, and cooking meat thoroughly.

Viruses can damage the host cell as they leave the cell after replication or when incorporated into the cell's chromosomes

Viruses, which are much smaller than bacteria, are responsible for many human illnesses. Some viral diseases, such as the common cold, are usually not very serious. Others, such as yellow fever, can be deadly.

Most biologists do not consider a virus to be a living organism because, on its own, it cannot perform any life processes. For a virus to copy itself, it requires a cell, called the host cell. It exploits the host cell's nutrients and metabolic machinery to make copies of itself that then infect other host cells.

A virus consists of a strand or strands of genetic material, either DNA or RNA, surrounded by a coat of protein, called a capsid (Figure 13a.3). The genetic material carries the instructions for making new viral proteins. These proteins may actually become part of the new viruses; they may serve as enzymes that help carry out biochemical functions important to the virus; or they may be regulatory. Some regulatory proteins trigger the specific viral genes that will be active under a certain set of conditions. Other regulatory proteins may convert the host cell into a virus-producing factory.

Some viruses have an envelope, an outer membranous layer that is studded with glycoproteins. In some viruses, the envelope is actually a bit of plasma membrane from the previous host cell that became wrapped around the virus as it left the host cell. The envelope of certain other viruses, those in the herpes family for instance, is from a previous host cell's nuclear membrane. In any case, the virus produces the glycoproteins on the envelope.

A virus can replicate (make copies of itself) only when its genetic material is inside a host cell. Figure 13a–3 illustrates the steps involved in the multiplication of viruses that infect animal cells:

1. **Attachment.** The virus gains entry by binding to a receptor (a protein or other molecule of a certain configuration) on the host cell surface. Such binding is possible because the viral surface has molecules of a specific shape that fit into the host's receptors. The host cell receptors play a role in normal cell functioning. However, a molecule on the surface of the virus has a similar shape to the chemical that would normally bind to the receptor. Viruses generally attack only certain kinds of cells in certain species

---

[1]Bacteria resistant to all antibiotics available today include some strains of *Staphylococcus aureus* (skin infection, pneumonia), *Mycobacterium tuberculosis* (tuberculosis), *Enterococcus faecalis*, and *Pseudomonas aeruginosa*.

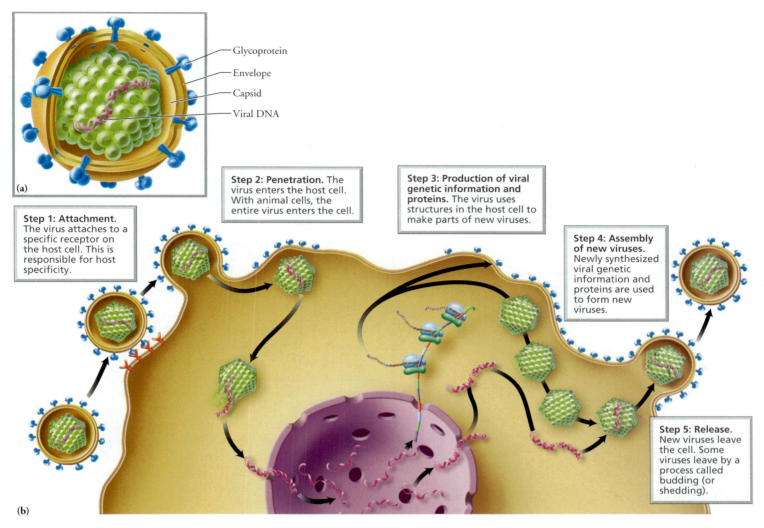

Glycoprotein
Envelope
Capsid
Viral DNA

(a)

Step 2: Penetration. The virus enters the host cell. With animal cells, the entire virus enters the cell.

Step 3: Production of viral genetic information and proteins. The virus uses structures in the host cell to make parts of new viruses.

Step 1: Attachment. The virus attaches to a specific receptor on the host cell. This is responsible for host specificity.

Step 4: Assembly of new viruses. Newly synthesized viral genetic information and proteins are used to form new viruses.

Step 5: Release. New viruses leave the cell. Some viruses leave by a process called budding (or shedding).

(b)

FIGURE **13a.3**

(a) The structure of a typical virus. A protein coat, called a capsid, surrounds a core of genetic information made up of DNA or RNA. Some viruses have an outer membranous layer, called the envelope, from which glycoproteins project. (b) Steps in viral replication.

because a particular virus can infect only cells that bear a receptor that the virus can fit into. Whereas the virus that causes the common cold infects only cells in the respiratory system, the virus that causes hepatitis infects only liver cells.

2. **Penetration.** After a virus has bound to a receptor on an animal cell, the entire virus enters the host cell, often by phagocytosis. Once inside, the virus is stripped of its capsid, leaving only its genetic material.

3. **Production of viral genetic information and proteins.** Viral genes then direct the host cell machinery to make thousands of copies of viral DNA or RNA. Next, viral genes direct the synthesis of viral proteins, including coat proteins and enzymes.

4. **Assembly of new viruses.** Copies of the viral DNA (or RNA) and viral proteins then assemble to form new viruses.

5. **Release.** Enveloped viruses leave the cell in a process called budding, or shedding. The newly formed viruses push through

the host cell's plasma membrane and become wrapped in the membrane, which forms the envelope. Budding need not kill the host cell. Other viruses cause the host cell membrane to rupture, releasing the newly formed viruses and killing the host cell.

Viruses can cause disease in several ways, as summarized in Table 13a.1. Some viruses cause disease when they kill the host's cells or cause the cells to malfunction. The host cell dies or malfunctions when viruses rapidly leave it, and it bursts *(lyses)*. In such cases, the symptoms of the disease will depend on which cells are killed. However, if viruses are shed slowly, the host cell may remain alive and continue to produce new viruses. Slow shedding causes *persistent infections* that can last a long time. Some viruses can produce *latent infections*, in which the viral genes remain in the host cell for a period of time without causing harm to the cell. At any time, however, the virus can begin replicating and cause cell death as new viruses are shed.

TABLE **I3a.I**

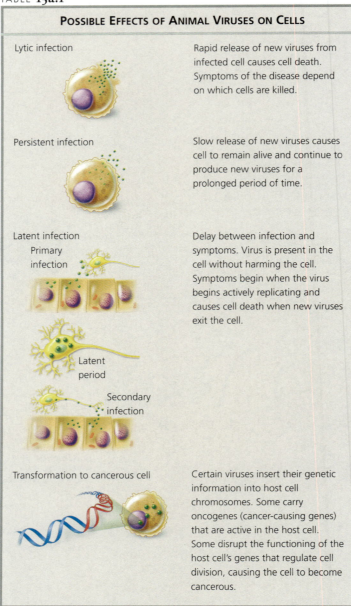

POSSIBLE EFFECTS OF ANIMAL VIRUSES ON CELLS

Lytic infection

Rapid release of new viruses from infected cell causes cell death. Symptoms of the disease depend on which cells are killed.

Persistent infection

Slow release of new viruses causes cell to remain alive and continue to produce new viruses for a prolonged period of time.

Latent infection
Primary infection

Latent period

Secondary infection

Delay between infection and symptoms. Virus is present in the cell without harming the cell. Symptoms begin when the virus begins actively replicating and causes cell death when new viruses exit the cell.

Transformation to cancerous cell

Certain viruses insert their genetic information into host cell chromosomes. Some carry oncogenes (cancer-causing genes) that are active in the host cell. Some disrupt the functioning of the host cell's genes that regulate cell division, causing the cell to become cancerous.

An example of a virus that can act in all of these ways is the herpes simplex virus that causes fever blisters. The viruses are spread by contact and enter the epithelial cells of the mouth, where they actively replicate. Rapid shedding kills the cells, causing fever blisters. Slow shedding may not cause outward signs of infection, but the viruses can still be transmitted. When the blisters are gone, the virus remains in a latent form within nerve cells without causing symptoms. However, stress can activate the virus. It then follows nerves to the skin in the region of the mouth and begins actively replicating, causing new blisters.

Certain viruses can also cause cancer. Some do this when they insert themselves into the host chromosome near a cancer-causing gene and, in doing so, activate that gene. Still other viruses bring cancer-causing genes with them into the host cell.

Unfortunately, viruses are not as easy to destroy as bacteria. A major difficulty is that viruses use host cell structures to carry out their metabolic functions. Thus, most attempts to develop antiviral drugs have failed because the drug was toxic to the host cell. Nonetheless, some antiviral drugs are now available to slow viral growth, and others are being developed. Most of the antiviral drugs available today, including those against the herpes virus and HIV, work by blocking one of the steps necessary for viral replication. As mentioned in Chapter 13, interferons are proteins produced by virus-infected cells that protect neighboring cells from all strains of viruses. Interferons are not as useful as originally hoped, but they have been used for certain viral infections, including hepatitis C and the human papilloma virus that causes genital warts. Because of these difficulties in treating viral infections, the best way to deal with them is through prevention by vaccines (discussed in Chapter 13).

Protozoans cause disease by producing toxins and enzymes

Protozoans are single-celled eukaryotic organisms with a well-defined nucleus. The can cause disease by producing toxins or by releasing enzymes that prevent host cells from functioning normally. Protozoans are responsible for many diseases, including malaria (discussed shortly), sleeping sickness, amebic dysentery, and giardiasis. Giardiasis is a diarrheal disease that can last for weeks. There are frequent outbreaks of giardiasis in the United States, most of them resulting from water supplies contaminated with human or animal feces. Even clear, refreshing lakes and streams in the wilderness can contain *Giardia* (Figure 13a.4). Fortunately, drugs are available to treat protozoan infections.

FIGURE **I3a.4**

Giardia is a protozoan that is commonly found in lakes and streams used as sources of drinking water, even those in remote areas. It causes severe diarrhea that lasts for weeks and can be especially dangerous for children.

Fungi often cause disease by secreting enzymes that digest cells

Like the protozoans, fungi are eukaryotic organisms with a well-defined nucleus in their cells. Some fungi exist as single cells. Others are organized into simple multicellular forms, with not much difference among the cells. There are over 100,000 species of fungi, but less than 0.1% cause human ailments. Fungi obtain food by infiltrating the bodies of other organisms—dead or alive—secreting enzymes to digest the food, and absorbing the resulting nutrients. If the fungus is growing in or on a human, body cells of the human are digested, causing disease symptoms. Some fungi cause serious lung infections, such as histoplasmosis and coccidioidomycosis. Other, less threatening fungal infections occur on the skin and include athlete's foot, ringworm, and vaginitis (discussed in Chapter 17). Most fungal infections can be cured. Fungal cell membranes have a slightly different composition than those of human cells. As a result, the membrane is a point of vulnerability. Some antifungal drugs work by altering the permeability of the fungal cell membrane. Others interfere with membrane synthesis by fungal cells. Fungal infections of the skin, hair, and nails can be combated with a drug that prevents the fungal cells from dividing.

Parasitic worms cause disease by releasing toxins, feeding off blood, or competing with the host for food

The parasitic worms are multicellular animals. They include flukes, tapeworms, and roundworms, such as hookworms and pinworms. They can cause illness by releasing toxins into the bloodstream, feeding off blood, or competing for food with the host. Parasitic worms cause many serious human diseases, including ascariasis, schistosomiasis, and trichinosis.

Ascariasis is caused by a large roundworm, *Ascaris*, that is about the size of an earthworm (Figure 13a.5). People become infected with *Ascaris* when they consume food or drink contaminated with *Ascaris* eggs. The eggs develop into larvae (immature worms) in the person's intestine. The larvae then penetrate the intestinal wall, enter the bloodstream, and travel to the lungs. After developing further, the worms are coughed up and swallowed, returning them to the intestine. Within 2 to 3 months, they mature into male and female worms, which live for about 2 years. During those years, female worms can produce more than 200,000 eggs a day.

As much as 25% of the world population is infected with *Ascaris*, particularly in tropical regions. Up to 50% of the children in some parts of the United States are infected. Many people with ascariasis have no symptoms. However, the worms can cause lung damage and severe malnutrition. When many worms are present, they can block or perforate the intestines, leading to death.

Prions cause disease by causing normal proteins to become misfolded and form clumps

Prions are infectious particles of proteins or, more simply, infectious proteins. They cannot be destroyed by heat, ultraviolet light, or most chemical agents.

The general term for prion diseases is transmissible spongiform encephalopathies (TSEs). They cause degeneration of the brain.

FIGURE **13a.5**

Ascaris is a parasitic roundworm. As adults, these worms live in the small intestine of humans. Mature female worms can grow to over 1 foot long. Adult males are usually 8 to 10 inches long. The worms can cause intestinal blockage and malnutrition.

Prions are responsible for several animal infections, notably mad cow disease, scrapie in sheep, and chronic wasting disease, which is spreading through herds of deer and elk in the United States (Figure 13a.6). Mad cow disease severely damaged the British beef industry in the 1990s. The U.S. and Canadian governments responded quickly when the first cases of mad cow disease appeared in these countries in 2003.

Prions also cause a human neurological disorder called Creutzfeldt-Jakob disease (CJD). Indeed, the prion responsible for mad cow disease is thought to cause one form of CJD. The incubation period for CJD can be months to decades. Symptoms include sensory and psychiatric problems. Once the symptoms begin, death usually occurs within a year.

Prions are misfolded versions of a harmless protein normally found on the surface of nerve cells. The host nerve cell produces a normal version of the protein. However, if a prion is present, it somehow causes the host protein to change its shape to the abnormal form. The misshapen proteins clump together and accumulate in the nerve tissue of the brain. These clumps of prions may damage the plasma membrane or interfere with molecular traffic. Spongelike holes develop in the brain, causing death. Currently, there is no treatment for any disease caused by prions.

How does an animal become infected with prions? In the case of mad cow disease, it appears that the prions eaten in contaminated food are often the source of contamination. The protein supplements fed to cattle to increase their growth and milk production are often the source of contamination. The protein supplements were prepared from the carcasses of animals considered unfit for human consumption, a practice banned in the United States in 1997. If these protein

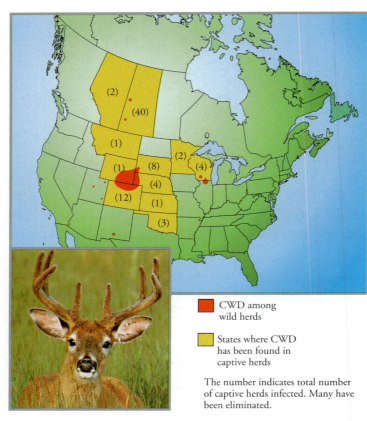

CWD among wild herds

States where CWD has been found in captive herds

The number indicates total number of captive herds infected. Many have been eliminated.

FIGURE **13a.6**

Chronic wasting disease (CWD), which is caused by a prion, is spreading through herds of deer and elk in the United States.

supplements were prepared from infected animals, they would contain prions. The prions pass through the intestinal wall, enter the lymphatic system, and are then transported by nerves to the brain and spinal cord, where they destroy nerve cells. In contrast, chronic wasting disease seems to be able to spread by animal-to-animal contact. This contact may involve touching or ingesting body fluids such as urine or feces from infected animals. For example, deer may consume the prions that cause chronic wasting disease when they chew the bones of dead diseased deer. Scientists also think the prions may remain in the soil or water for years. As a result, healthy animals introduced into a region after diseased animals have been eliminated may become infected. Humans can become infected with prions by eating contaminated material, through tissue transplant, or through contaminated surgical instruments.

Disease Is Spread When a Pathogen Enters the Body Through Contact, Consumption, or an Animal Vector

Obviously, you catch a disease when the pathogen enters your body, but how are diseases spread? The answer to this question varies with the type of pathogen.

One means of transmission is direct contact, as might occur when shaking hands, hugging and kissing, or being sexually inti-

mate. For example, sexually transmitted diseases (STDs) are spread when a susceptible body surface touches an infected body surface (see Chapter 17a). The organisms that cause STDs generally cannot remain alive outside the body for very long, so direct contact is necessary. A few disease-causing organisms, HIV and the bacterium that causes syphilis for instance, can spread across the placenta from a pregnant woman to her growing fetus.

Indirect contact can spread other diseases. Most respiratory infections, including the common cold, are spread by indirect contact (see Chapter 14). When an infected person coughs or sneezes, airborne droplets of moisture full of pathogens are carried through the air (Figure 13a.7). The droplets may be inhaled or land on nearby surfaces. When another person touches an affected surface, the organisms are transmitted. In this way, pathogens may be spread on contaminated inanimate objects, including doorknobs, drinking glasses, and eating utensils.

Certain diseases are transmitted in contaminated food or water. You have read that spoiled food can cause food poisoning. Other diseases transmitted by food or water include hepatitis A, an inflammation of the liver caused by a certain virus. *Legionella*, the bacterium that causes a severe respiratory infection known as Legionnaires' disease, is a common inhabitant of the water in condensers of large air conditioners and cooling towers. The disease-causing bacteria are spread through tiny airborne water droplets. Contaminated water can also cause infection. Coliform bacteria are found in the intestines of humans and are, therefore, an indicator of fecal contamination of water. Their numbers are monitored in drinking and swimming water. To be safe, drinking water should not have any coliform bacteria.

Another means of transmission is by animal vectors that carry pathogens. The most common vector-borne disease in the United States is Lyme disease. It is caused by a bacterium that is transmitted by the deer tick (the vector), which is about the size of the head of a pin (Figure 13a.8). The tick larva picks up the infectious agent when it bites an infected animal. When the tick subsequently bites a human or other mammalian host, the bacteria

FIGURE **13a.7**

Pathogens can be spread through the air in droplets of moisture when an infected person sneezes or coughs.

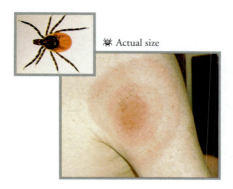

✾ Actual size

FIGURE **13a.8**

A tiny tick, the deer tick, is a vector that transmits the bacterium that causes Lyme disease. One characteristic sign of Lyme disease is a red bull's-eye rash surrounding the tick bite. The rash gradually increases in diameter.

gradually move from the tick's gut to its salivary glands and then to its victim. The incubation period can be as long as 6 to 8 weeks. The early symptoms include a headache, backache, chills, and fever. Often, a rash resembling a bull's eye develops with an intense red center and border. Over a period of weeks, the circle increases in diameter. Weeks to months later, pain, swelling, and arthritis may develop. Cardiovascular and nervous system problems may follow the arthritis.

Mosquitoes transmit the protozoan parasite (*Plasmodium*) that causes malaria. This disease affects about 300 million to 500 million people worldwide and kills 1 million to 3 million people each year. The life cycle of this protozoan involves two hosts: a mosquito and a human. When an infected *Anopheles* mosquito feeds on human blood, the protozoans enter the human's bloodstream (Figure 13a.9). The parasites mature in the liver and then return to

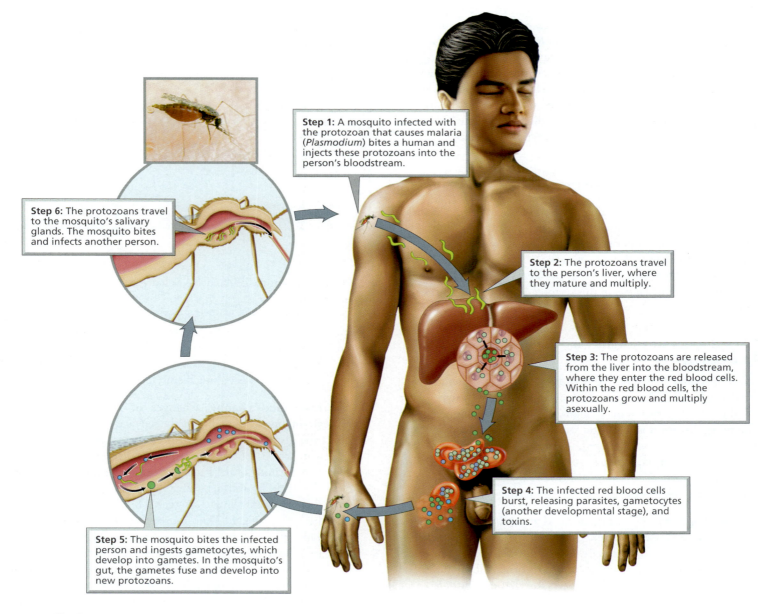

Step 1: A mosquito infected with the protozoan that causes malaria (*Plasmodium*) bites a human and injects these protozoans into the person's bloodstream.

Step 6: The protozoans travel to the mosquito's salivary glands. The mosquito bites and infects another person.

Step 2: The protozoans travel to the person's liver, where they mature and multiply.

Step 3: The protozoans are released from the liver into the bloodstream, where they enter the red blood cells. Within the red blood cells, the protozoans grow and multiply asexually.

Step 4: The infected red blood cells burst, releasing parasites, gametocytes (another developmental stage), and toxins.

Step 5: The mosquito bites the infected person and ingests gametocytes, which develop into gametes. In the mosquito's gut, the gametes fuse and develop into new protozoans.

FIGURE **13a.9**

The life cycle of *Plasmodium*, the protozoan that causes malaria*

the blood, where they infect red blood cells. The protozoans reproduce within the red blood cells, which eventually burst and release more protozoans, which will infect additional red blood cells, and gametocytes (cells that will develop into gametes). Toxins produced by the protozoans are also released when the red blood cells burst, causing chills, fevers, and vomiting. The infection and bursting of red blood cells occur in cycles of 2 to 3 weeks. Several factors related to the loss of red blood cells can cause death. If a mosquito bites an infected person when the protozoans are in the blood, the mosquito will ingest the gametocytes. In the gut of the mosquito, the gametocytes mature into gametes, which will fuse when the protozoan reproduces sexually. The protozoans then migrate to the mosquito's salivary glands. After about a week, the mosquito can infect another person.

Malaria has been known since 1000 BC, so one might think that we would have a cure or means of preventing the deadly disease. We do not. Antimalarial drugs are being developed. Development of vaccines has proved difficult. Another approach to combating malaria is to slow its spread by mosquitoes. Spraying with insecticides to eliminate populations of mosquitoes is the traditional way to control mosquitoes. Scientists have genetically modified mosquitoes with a gene that prevents the development of the protozoan within them. The modified mosquitoes are unable to transmit the protozoan—to mice at least. Scientists still do not know if the same gene will prevent the transmission of malaria in humans.

Infectious Diseases Remain Cause for Concern

A plague is a large-scale outbreak of an infectious disease. The most notorious plagues—the bubonic plague, cholera, diphtheria, and smallpox—are mostly history, but new outbreaks may occur sporadically. However, new diseases and epidemics continue to present problems. We discuss some of these modern-day plagues in this chapter and throughout the text (Table 13a.2).

New diseases are emerging, and some old diseases are reappearing

Since the mid-1970s about 30 new diseases have appeared, causing tens of millions of deaths. We have also witnessed the reemergence of some diseases that were thought to have been conquered. An **emerging disease** is a condition with clinically distinct symptoms whose incidence has increased, particularly over the last two decades. A **reemerging disease** is a disease that has reappeared after a decline in incidence (Figure 13a.10). We will consider two factors that play roles in the emergence and reemergence of disease.

1. **Development of new organisms that can infect humans and drug-resistant organisms.** Mutations (changes in genetic information) occur randomly. Some mutations cause an infectious agent to "jump species" from its original host and infect humans. For example, this process may have been the way the coronavirus that causes severe acute respiratory syndrome (SARS, discussed shortly) gained the ability to infect humans. The virus lived in livestock, and one of the proteins that allows

TABLE **13a.2**

EXAMPLES OF MODERN-DAY PLAGUES

DISEASE	CAUSE	DISCUSSION
Meningitis	Bacterium or virus	Chapter 8
West Nile disease	Virus	Chapter 8
Malaria	Protozoan	Chapter 13a
Lyme disease	Bacterium	Chapter 13a
Transmissible spongiform encephalopathies (TSEs; e.g., mad cow disease, chronic wasting syndrome, Creutzfeldt-Jakob disease)	Prion	Chapter 13a
Hantavirus pulmonary syndrome	Virus	Chapter 13a
Tuberculosis	Bacterium	Chapter 14
Severe acute respiratory syndrome (SARS)	Virus	Chapter 14
Influenza	Virus	Chapter 14
Hepatitis C	Virus	Chapter 15
Chlamydia	Bacterium	Chapter 17a
Gonorrhea	Bacterium	Chapter 17a
Genital herpes	Virus	Chapter 17a
Genital warts	Virus	Chapter 17a
HIV/AIDS	Virus	Chapter 17a

the virus to enter a cell may have changed, allowing the virus to enter a human cell.

Certain bacteria have acquired genes for antibiotic resistance. As a result, some diseases that were once easily cured by antibiotics are now much more difficult to treat. In Chapter 14, we will see that antibiotic resistance is one reason for the reemergence of tuberculosis (a lung disease).

2. **Environmental change.** Changes in local climate—the annual amount of rainfall and the average temperature—can affect the distribution of organisms and change the size of the region in which certain organisms can live. For example, El Niño is a warming of the Pacific Ocean that occurs in cycles of 3 to 7 years. The warmer water causes atmospheric changes that, in turn, cause climate changes. The 1991–1992 El Niño and the 1997–1998 El Niño caused wet and mild winters in the southwestern United States. Scientists think these climate changes were responsible for the outbreak of the hantavirus in the Southwest in 1993 and 1999. The change in climate led to a population explosion of adult rodents. The hantavirus is carried by rodents, especially the deer mouse and the cotton rat. The rodents shed the virus in their urine, feces, and saliva. When fresh urine, droppings, or nest materials are stirred up,

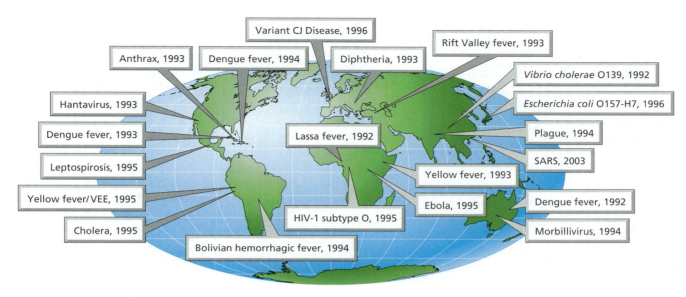

FIGURE **13a.10**
Recent emerging or reemerging diseases

tiny virus-containing droplets get into the air. The virus is transmitted to people when they inhale contaminated air. The virus causes a severe respiratory illness similar to influenza. It can cause respiratory failure and, in some cases, death.

Another important factor causing changes in the environment is the growth of human population along with the development and growth of cities. Swelling numbers of people in cities cause people to move out of the city into surrounding areas. If the surrounding areas were previously undeveloped, the move brings people into contact with animals, including insects, that might carry infectious organisms. The development of suburbs also destroys populations of predators, such as foxes and bobcats. In some regions of New York, the loss of predators led to an increase in the population of tick-carrying mice and an increase in the incidence of Lyme disease.

Infectious disease can also be spread more easily today than in the past. Densely populated cities ease the spread of infectious disease. Air travel allows infectious disease to spread quickly over great distances, as we saw during the outbreak of SARS in 2003 (Figure 13a.11).

▌Epidemiologists track diseases

Epidemiology is the study of the pattern of disease—its occurrence, distribution, and control. Epidemiologists are "disease detectives" who try to determine why a disease is triggered at a particular time and place. The first step in tracking a disease is to verify that there is indeed an outbreak, which is more than the expected number of cases of individuals with similar symptoms in a given area. Next, epidemiologists try to identify the cause of the disease, whether it can be transmitted to other people, and, if it can be, how the disease is transmitted. To identify the cause of the disease, epidemiologists try to isolate the same infectious agent from all people showing symptoms of the condition. In addition, they try to identify factors, including age,

sex, race, personal habits, and geographic location, shared by people with symptoms of the condition. These factors might provide a clue as to whether the condition can be transmitted and how. Information about whether the incidence of the disease changes with seasons can also be important in determining the means of transmission. For example, the incidence of West Nile disease increases in the summer because mosquitoes, which are prevalent during warmer months, transmit the virus responsible for the disease. On the other hand, influenza, which is spread by droplet infection, is more common in the winter, when people are indoors and in closer contact. Intestinal pathogens are usually spread in food or water, so epidemiologists would look for similarities in both the type of food or drink consumed among affected people and the locations in which consumption occurred.

FIGURE **13a.11**
Air travel is one reason that new diseases can spread rapidly, as SARS did in 2003.

As an example of how epidemiologists track a disease, we will consider how the cause of SARS, the first new infectious disease of the millenium, was determined and controlled within a few months during 2003. SARS is an unusual type of pneumonia that involves a high fever and was first reported in China in February 2003. A Chinese physician who cared for patients with pneumonia developed symptoms himself. Nonetheless, he traveled to Hong Kong and stayed in a hotel while visiting relatives. The next day, he was admitted to the hospital and died of respiratory failure a week or so later. Shortly afterward, people displaying symptoms that have come to be described as SARS appeared in Hong Kong, Singapore, Vietnam, Canada, and the United States. By questioning people with symptoms of SARS, epidemiologists determined that the factor that all people had in common was contact with the ill Chinese physician. Some patients were guests in the same hotel, and others were family members. This pattern of occurrence is typical of a disease caused by an infectious agent. Therefore, measures intended to limit the spread of the infectious agent were put in place. These measures included quarantining thousands of people, restricting travel, and even checking the body temperature of people in airports. Although SARS affected more than 8000 people in 30 countries and caused more than 800 deaths in the months that followed, these restrictive measures were eventually successful in limiting the spread of SARS.

Other scientists determined that the cause of SARS was a specific type of virus called a coronavirus. This virus was identified as the cause of SARS because it could be isolated from all SARS patients tested. We now know that SARS is transmitted from person to person in droplets of moisture spread into the air when an infected person coughs. The virus is very stable. It can cause infection after 24 hours on a dry surface.

Biological Organisms and Products May Be Used As Biological Weapons

A biological weapon is an organism or the toxin it produces that is used intentionally to harm humans. Few of us thought about biological weapons until October of 2001, when, after the attack on the World Trade Center, a man in Florida developed anthrax. The bacterium that causes anthrax was found in his workplace and in the nose of a coworker. Several more cases of anthrax were discovered during the next month. Letters containing anthrax spores—resistant resting structures formed by many bacteria—were sent to news media and the U.S. Congress. Fears that terrorists would use biological weapons increased. Heading the list of agents that could be used for bioterrorism are two pathogens, the bacterium that causes anthrax and the virus that causes smallpox.

Anthrax is caused by a bacterium that forms resistant spores

Under adverse conditions, the bacterium that causes anthrax, *Bacillus anthracis*, produces spores, which are a highly resistant, dormant form of the bacterium. Anthrax infection occurs when these spores enter a person's body. Under the more favorable conditions within the body, the toxin-producing bacteria emerge from the spore, become active, and divide.

There are three forms of anthrax. Skin anthrax is the most common but least deadly form of the disease. It is fatal in about 20% of the cases. Lung (inhalation) anthrax is the deadliest form. About 90% of persons with lung anthrax die. A gastrointestinal form is very rare, with fatality ranges from 25% to 60%.

The form of anthrax depends on where the spores enter the body and release their toxins. In skin anthrax, the spores enter the body through minor cuts in the skin. Bacterial toxins then attack surrounding tissue. The affected area of skin swells. This welt progresses to a fluid-filled blister and finally to a black, ulcerous skin sore (Figure 13a.12). Lung anthrax is contracted when airborne spores are inhaled and settle in the lungs. Although the bacteria can emerge from the spores within a week, it may be as long as 60 days before they emerge. After the bacteria emerge, they produce toxins that attack lung tissue, causing bleeding and destruction of lung cells. The initial symptoms of lung anthrax, which usually occur within 7 days of exposure, are similar to those of the flu, including fever and chest pains. As the disease progresses, there may be breathing problems, shock, coma, and death. Gastrointestinal anthrax begins when the spores enter the digestive system when undercooked, contaminated meat is eaten. In this case, the toxins cause nausea and vomiting. They can also cause severe cramps and diarrhea.

Cells of the body defense system carry the anthrax bacteria to the lymph nodes from any of these sites of infection. The bacteria multiply in the lymph nodes, and toxins are carried through the lymphatic system and bloodstream to other organs.

Antibiotics that kill the anthrax bacteria are available, and scientists are working on other means of protection. They are developing drugs that would neutralize the toxin, giving the immune system more time to fight the bacteria. Anthrax can be detected in the environment and in persons who have been exposed. Early detection would surely save lives. There is an anthrax vaccine, but it is currently recommended only for persons considered to have a high risk of exposure to anthrax.

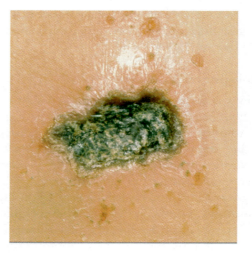

FIGURE **13a.12**
The lesion caused by skin anthrax

❙ Smallpox is caused by a highly contagious virus

It is estimated that 300 million people died from smallpox during the twentieth century alone. The primary weapon in the eradication of smallpox was vaccination. Indeed, an effective program of eradication eliminated smallpox worldwide in 1980. However, the United States and Russia still maintain stocks of smallpox viruses for research. It is feared that terrorists or unreliable foreign governments may also hold hidden stocks. Because vaccination programs were stopped when the disease was conquered, 90% of the world's population is now susceptible to the disease. Smallpox vaccination provides immunity for about 10 years. Therefore, even persons who were vaccinated as children could be affected if bioterrorists released the smallpox virus today.

The smallpox virus is highly contagious, but transmission usually requires direct contact with an infected individual. Only a few virus particles can start an infection. The virus can survive as long as 2 days when airborne. It takes 7 to 17 days for symptoms to appear. The first signs are high fever, fatigue, backache, abdominal pain, and delirium. Next, red spots appear on the skin and eventually develop into painful, pus-filled sores (Figure 13a.13). These blisters crust over within 8 or 9 days, leaving visible scars. One form of smallpox kills roughly 30% of its victims.

❙ Botulinum toxin is a potent poison

The products of certain biological agents, such as the botulism toxin produced by the bacterium *Clostridium botulinum*, are also considered to be potential biological weapons. The botulinum toxin is one of the most poisonous substances known. Scientists believe that a single gram of crystalline toxin could kill 1 million people if it were

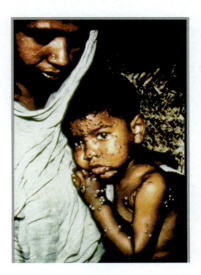

FIGURE **13a.13**
A child with pustules characteristic of smallpox

evenly dispersed and inhaled. Fortunately, technological factors make this efficiency of distribution unlikely. As a weapon, the botulinum toxin could be distributed as an aerosol that would be inhaled or in food that would be consumed. The toxin binds to nerves and prevents muscle contraction, resulting in paralysis and possible respiratory failure. Symptoms begin within 2 hours to several days. A breathing machine and feeding tube may be used to keep an infected person alive. An antitoxin is available that can prevent any free toxin from binding to nerves and causing additional harm.

14

The Respiratory System

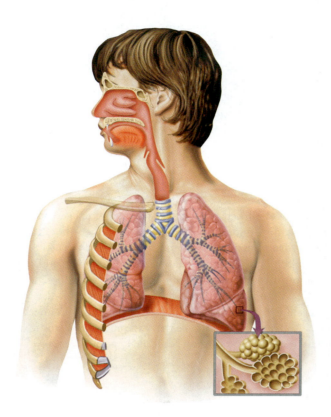

The Respiratory System Allows for the Exchange of Oxygen and Carbon Dioxide Across a Moist Body Surface

- The nose adjusts the quality of incoming air and provides for the sense of smell
- The sinuses lighten the head and adjust air quality
- The pharynx is a passageway for food and air
- The larynx is an adjustable entryway that produces the voice
- The trachea is the windpipe
- The bronchial tree is a system of air tubules that conducts air
- The alveoli of the lungs are surfaces for gas exchange

Pressure Changes within the Lungs Cause Breathing

- Inhalation occurs when the pressure in the lungs decreases
- Exhalation occurs when the pressure in the lungs increases
- The volume of air moved into and out of the lungs is an indication of health

Blood Transports Gases between the Lungs and the Cells

- Most oxygen is carried by hemoglobin
- Most carbon dioxide is transported as bicarbonate ions

Breathing Is Controlled Primarily by Respiratory Centers in the Brain

- Brain centers control the basic breathing pattern
- The changes in depth and rate of breathing are affected by chemoreceptors

Respiratory Disorders Have Many Causes

- The common cold can be caused by many types of viruses
- The flu is caused by three types of virus
- SARS is a serious viral respiratory illness
- Strep throat can have serious consequences
- Bronchitis is an inflammation of the bronchi
- Emphysema is caused by the destruction of alveoli

 Social Issue A Global Health Threat: Tuberculosis

 Health Issue Surviving a Common Cold

 Environmental Issue Air Pollution and Human Health

Collin was feeling awful, and the article he was reading in the morning paper in his hotel room in Singapore was making him feel even worse. "SARS Outbreak," read the headline. The article then listed symptoms, most of which matched the way he felt: aches and pains, coughing, diarrhea, fever, severe congestion in his lungs and nasal passages, sneezing, sore throat, and fatigue. He felt weak just trying to pull enough air into his lungs. "Just what I need on this business trip," he thought, "another obstacle in the way of sealing one of the largest printing deals of my career." What should he do? Call and cancel the meeting? Risk losing the contract? If he *had* contracted SARS, (although he really didn't think he had) wouldn't he be putting everyone in the room in danger of contagion if he participated in this meeting? He wasn't thinking clearly, so he called his wife in Liverpool for advice. "What in the devil am I supposed to do?" he asked. "This is the most important negotiation of my career." After 20 minutes of discussion, Collin decided to cancel the meeting with the printing company. Further, he planned to go directly to the nearest hospital emergency room to get tested for SARS. Once at the hospital, he was immediately placed in a quarantined ward during the testing period but was released within a few days after he learned that he had a bad case of the flu and nothing more. Within 10 days, he not only felt much better but he also had successfully negotiated and sealed the printing deal.

In this chapter, we discuss why the respiratory system is so vital to life that certain respiratory diseases can be life threatening. We will first follow the course of air as it moves to the lungs and study the mechanics of breathing. Next, we will consider the transport of oxygen and carbon dioxide between the lungs and the cells and examine the control of respiration. Finally, we will discuss several disorders of the respiratory system. ■

The Respiratory System Allows for the Exchange of Oxygen and Carbon Dioxide Across a Moist Body Surface

Without oxygen, we would die within a few minutes. Oxygen plays an essential role in extracting energy from food molecules (see Chapter 3). We store that extracted energy in a molecule called ATP (adenosine triphosphate). The stored energy can later be released to do the work of the cell. Indeed, a cell needs ATP to remain alive. Our cells can make a little ATP without oxygen but not enough to supply the energy needs of the body. Cells can make 18 times more ATP if oxygen is present. The same chemical reactions that require oxygen for the production of ATP produce carbon dioxide as a by-product. In solution, carbon dioxide forms carbonic acid, which can be harmful to cells.

The function of the **respiratory system** is to provide the body with essential oxygen and dispose of carbon dioxide, an exchange that also regulates the acidity of body fluids. Four processes play a part in respiration (Figure 14.1).

- **Breathing (ventilating):** moving oxygen-rich air into the lungs and carbon dioxide–laden air away from the lungs.

- **External respiration:** the exchange of oxygen and carbon dioxide between the lungs and the blood. Oxygen moves from the lungs into the blood, and carbon dioxide moves from the blood into the lungs.

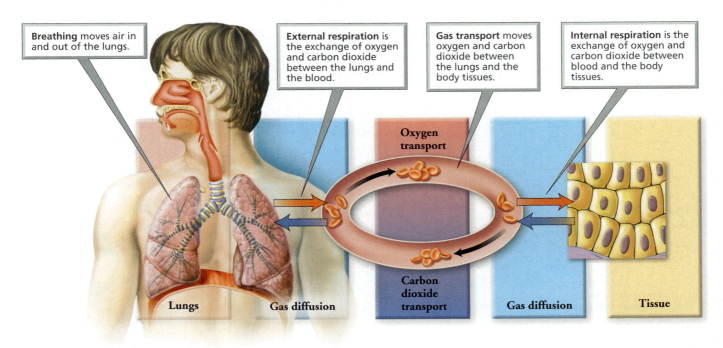

Breathing moves air in and out of the lungs.

External respiration is the exchange of oxygen and carbon dioxide between the lungs and the blood.

Gas transport moves oxygen and carbon dioxide between the lungs and the body tissues.

Internal respiration is the exchange of oxygen and carbon dioxide between blood and the body tissues.

Oxygen transport

Carbon dioxide transport

Lungs   Gas diffusion   Gas diffusion   Tissue

FIGURE **14.1**
An overview of respiration

- **Gas transport:** transport of oxygen from the lungs to the cells and of carbon dioxide from the cells to the lungs.

- **Internal respiration:** the exchange of oxygen and carbon dioxide between the blood and the cells. Oxygen moves from the blood to the cells, and carbon dioxide produced by the cells moves into the blood.

One way to see how humans accomplish the tasks of gaining oxygen and losing carbon dioxide is to follow the path of air from the nose to the lungs. The structures the air passes along the way are identified and described in Figure 14.2 and in Table 14.1. The pathway of air is summarized in Figure 14.3. The respiratory system is generally divided into upper and lower regions. The nose (nasal cavities) and pharynx make up the *upper respiratory system*. The *lower respiratory system* consists of the larynx, trachea, bronchi, bronchioles, and lungs.

The nose adjusts the quality of incoming air and provides for the sense of smell

However large someone's nose might seem from the outside, the inside is not as roomy as you might imagine. One reason is that a thin partition of cartilage and bone called the **nasal septum** divides the inside of the nose into two **nasal cavities**. In addition, much of the space within the nasal cavities is taken up by three convoluted, shelflike bones. These bones increase the surface area of the nasal cavities and divide each cavity into three narrow passageways through which the air flows. Moist mucous membrane covers the entire inner surface of the nasal cavities.

We all know what a nose looks like, but what does a nose do? Your nose has three important functions: (1) filtration and cleansing, (2) conditioning the air, and (3) olfaction (smell).

FILTRATION AND CLEANSING

The nose helps clear particles from the air that moves through its passages. Considering that each of us inhales about 150,000 bacteria along with a great deal of pollen and dust each day, filtering the air is an important job. Most pollen and dust particles are removed from the air before they reach the lungs.

The nose and air tubules clean inhaled air in a variety of ways. Hairs inside the nose filter out the largest particles. In addition, certain cells in the membrane lining the surface of the nasal cavities and air tubules produce mucus, a sticky substance that catches dust particles. Cilia, tiny projections extending from the membranous lining, then sweep the mucus, trapped

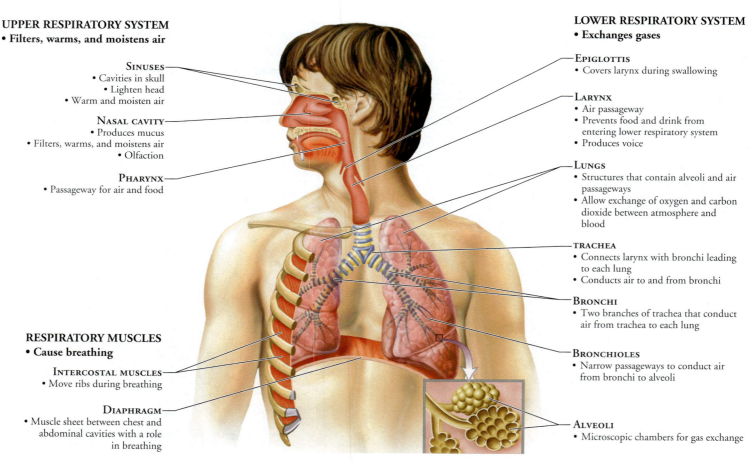

UPPER RESPIRATORY SYSTEM
- **Filters, warms, and moistens air**

SINUSES
- Cavities in skull
- Lighten head
- Warm and moisten air

NASAL CAVITY
- Produces mucus
- Filters, warms, and moistens air
- Olfaction

PHARYNX
- Passageway for air and food

RESPIRATORY MUSCLES
- **Cause breathing**

INTERCOSTAL MUSCLES
- Move ribs during breathing

DIAPHRAGM
- Muscle sheet between chest and abdominal cavities with a role in breathing

LOWER RESPIRATORY SYSTEM
- **Exchanges gases**

EPIGLOTTIS
- Covers larynx during swallowing

LARYNX
- Air passageway
- Prevents food and drink from entering lower respiratory system
- Produces voice

LUNGS
- Structures that contain alveoli and air passageways
- Allow exchange of oxygen and carbon dioxide between atmosphere and blood

TRACHEA
- Connects larynx with bronchi leading to each lung
- Conducts air to and from bronchi

BRONCHI
- Two branches of trachea that conduct air from trachea to each lung

BRONCHIOLES
- Narrow passageways to conduct air from bronchi to alveoli

ALVEOLI
- Microscopic chambers for gas exchange

FIGURE **14.2**

The respiratory system

TABLE **14.1**

STRUCTURES OF THE RESPIRATORY SYSTEM		
STRUCTURE	DESCRIPTION	FUNCTION
Upper respiratory system		
Nasal cavity	Cavity within the nose, divided into right and left halves by nasal septum; has three shelflike bones	Filters and conditions (moistens and warms incoming air); olfaction (sense of smell)
Pharynx (throat)	Chamber connecting nasal cavities to esophagus and larynx	Common passageway for air, food, and drink
Lower respiratory system		
Larynx	Cartilaginous boxlike structure between the pharynx and trachea that contains the vocal cords and the glottis	Allows air but not other materials passage to the lower respiratory system; source of the voice
Epiglottis	Flap of tissue reinforced with cartilage	Covers the glottis during swallowing
Trachea	Tube reinforced with C-shaped rings of cartilage that leads from the larynx to the bronchi	The main airway; conducts air from larynx to bronchi
Bronchi (primary)	Two large branches of the trachea reinforced with cartilage	Conduct air from trachea to each lung
Bronchioles	Narrow passageways leading from bronchi to alveoli	Conduct air to alveoli; adjust airflow in lungs
Lungs	Two lobed, elastic structures within the thoracic (chest) cavity containing surfaces for gas exchange	Exchange oxygen and carbon dioxide between blood and air
Alveoli	Microscopic sacs within lungs, bordered by extensive capillary network	Provide immense, internal surface area for gas exchange

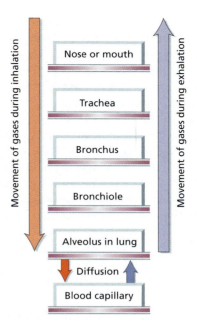

FIGURE **14.3**

The path of air during inhalation and exhalation

OLFACTION

Our sense of smell is due to the olfactory receptors located on the mucous membranes high in the nasal cavities behind the nose. The sense of smell is discussed in more detail in Chapter 9.

stop and think

Very cold temperatures can slow the action of the cilia in the nasal cavities. Explain why the loss of ciliary action can cause a runny nose on a very cold day.

■ The sinuses lighten the head and adjust air quality

Connected to the nasal cavities are large air-filled spaces in the bones of the face, called the **sinuses**. The sinuses make the head lighter and help warm and moisten the air we breathe. In addition,

dirt particles, and bacteria toward the throat, where they can be swallowed and destroyed by digestive enzymes (Figure 14.4). Particles that do not become trapped in the nasal cavities or the air tubules are deposited in the lungs.

CONDITIONING THE AIR

The nose also warms and moistens the inhaled air before it reaches the delicate lung tissues. Warming the air before it reaches the lungs is extremely important in cold climates because frigid air can kill the delicate cells of the lung. Moistening the inhaled air is also essential because oxygen cannot cross dry membranes. Mucus helps moisten the incoming air so that lung surfaces do not dry out.

Tears drain from the eyes through a canal that connects to the nasal cavity. When we cry, tear production increases, causing a runny nose.

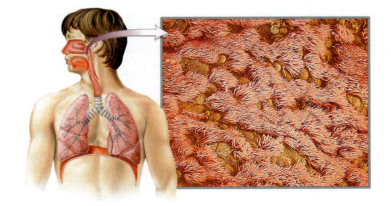

FIGURE **14.4**

The respiratory passageways are lined with clumps of short hairlike structures called cilia interspersed between mucus-secreting cells. The cilia are pink in this color-enhanced electron micrograph.

the sinuses are part of the resonating chamber that affects the quality of the voice. When you have a cold, the mucous membranes of the sinuses swell and produce excess fluid. The swelling and excess fluid results in a change in the sound of your voice.

Because the air spaces of the sinuses are continuous with those of the nasal cavities, any excess mucus and fluids drain from the sinuses into the nasal cavities. However, when the mucous membranes of the sinuses become inflamed, as they do in **sinusitis** (-*itis*, inflammation of), the swelling can block the connection between the nasal cavities and the sinuses, making it difficult for the sinuses to drain the mucous fluid they produce. The pressure caused by the accumulation of fluids in the sinuses causes pain over one or both eyes or in the cheeks or jaws; this condition is usually called a sinus headache. Decongestant nasal sprays reduce the swelling in the tubes connecting the sinuses with the nasal cavity, allowing the sinuses to drain more easily. Sinusitis may be caused by the virus responsible for a cold or by a subsequent bacterial infection.

The pharynx is a passageway for food and air

The **pharynx**, commonly called the throat, is the space behind the nose and mouth. It is a passageway for air, food, and drink. Small passageways, called the **Eustachian tubes**, connect the upper region of the pharynx with the middle ear. These passages help equalize the air pressure in the middle ear with that of the pharynx. Unfortunately, the Eustachian tubes may also allow the organisms causing a sore throat to spread to the ears, resulting in an ear infection. Middle ear infections are more common in children than in adults because the Eustachian tubes become tilted downward during growth. In children, the tubes are nearly horizontal, making it easier for infectious organisms to travel through them than through the tilted, curved tubes of adults.

The larynx is an adjustable entryway that produces the voice

After moving through the pharynx, the air next passes through the **larynx**, which is commonly called the voice box or Adam's apple. The larynx is a boxlike structure composed primarily of cartilage (Figure 14.5).

The larynx has two main functions. It is a traffic director for materials passing through the region, allowing air, but not other materials, to enter the lower respiratory system. The larynx is also the source of the voice. Let's consider these two functions in more detail.

1. **An adjustable entrance to the respiratory system.** The larynx provides an opening to the respiratory system that can be adjusted; that is, it can be opened to allow air to pass into the lungs and closed to prevent other matter, such as food, from entering the lungs. Because the esophagus (the tube leading to the stomach) is behind the larynx, material must pass over the opening to the respiratory system to reach the digestive system. If solid material such as food enters the respiratory system, it could lodge in one of the tubes conducting air to the lungs and prevent air flow. Fluid entering the lungs is equally dangerous because it can cover the respiratory surfaces, decreasing the area of gas exchange. Normally, foreign material is prevented from entering the lower respiratory system during swallowing because the larynx rises and

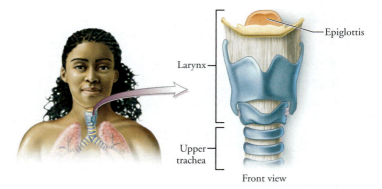

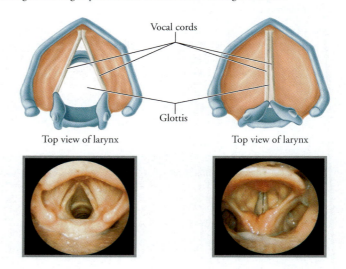

(a) The epiglottis is open during breathing but covers the opening to the larynx during swallowing to prevent food or drink from entering the trachea.

During quiet breathing, the vocal cords are near the sides of the larynx, and the glottis is open.

During speech, the vocal cords are stretched over the glottis and vibrate as air passes through them, producing the voice.

(b) The vocal cords are the folds of connective tissue above the opening of the larynx (the glottis) that produce the voice.

FIGURE **14.5**

The larynx, commonly called the voice box or Adam's apple, is an adjustable entryway to the trachea and the source of the voice.

causes a flap of cartilage called the **epiglottis** to move downward and form a lid over the glottis. In this way, the epiglottis covers the **glottis**, the opening in the larynx through which air passes, and prevents food from entering the lungs. You can feel the movement of the larynx if you put your fingers on your Adam's apple while swallowing. Because of this movement, you cannot breathe and swallow at the same time. (Try it!)

If food or drink accidentally enters the trachea, we usually cough, and the object is expelled. However, if food is lodged in the trachea it may block air flow. The **Heimlich maneuver** can be used to remove the object and restore air flow (Figure 14.6).

2. **Production of voice.** The larynx is the source of the voice. The voice is generated by the vibration of the **vocal cords**, two thick

A person who is choking cannot speak or breathe and needs immediate help.

The **Heimlich maneuver** is a procedure intended to force a large burst of air out of the lungs and dislodge the object blocking air flow.

STEP 1: Stand behind the choking person with arms around the waist.

STEP 2: Make a fist and place the thumb of the fist beneath the victim's rib cage about midway between the navel (belly button) and the breastbone.

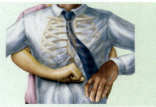

STEP 3: Grasp the fist with your other hand and deliver a rapid "bear hug" up and under the rib cage with the clenched fist. Be careful not to press on the ribs or the breastbone because doing so could cause serious injury.

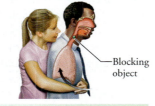

Blocking object

STEP 4: Repeat until the object is dislodged.

FIGURE **14.6**

The Heimlich maneuver can be performed on a choking person who is standing or sitting. If a choking victim is lying on the ground, pushing inward and upward on the upper part of the victim's abdomen will generate the same pressure changes as are produced by the maneuver on a standing victim. If you begin to choke and there is no one to perform the Heimlich maneuver on you, it may be possible to dislodge the food blocking your trachea by throwing your upper abdominal region against a table, chair, or other stationary object.

strands of tissue stretched over the opening of the glottis (Figure 14.5). When you speak, muscles stretch the vocal cords across the air passageway, narrowing the opening of the glottis. Air passing between the stretched vocal cords causes them to vibrate and produce a sound, just as the edges of a balloon vibrate and make noise if you stretch its neck while allowing the air to escape. The vibrations of the vocal cords set up sound waves in the air spaces of the nose, mouth, and pharynx. This resonation is largely responsible for the tonal quality of your voice.

The loudness of the sound is determined by the pressure of the air forced over the vocal cords—the greater the pressure, the louder the sound. So when you shout, you force a great deal of air under intense pressure over the vocal cords, but when you

talk quietly, you gently pass less air between these cords. When you whisper, the vocal cords may not vibrate at all.

The pitch of the voice is altered by changing the tension of the vocal cords. When the cords are stretched, making them thinner and more taut, they vibrate and the pitch of the sound is higher. You can demonstrate the relationship between thickness and pitch for yourself by plucking a rubber band stretched between your thumb and forefinger. The more the rubber band is stretched, the higher the pitch of the twang.

When you suffer from **laryngitis**, an inflammation of the larynx, the vocal cords become swollen and thick. As a result, they cannot vibrate freely, and the voice becomes deeper and huskier. Sometimes, when the vocal cords are particularly inflamed, it is difficult to speak at all because the cords cannot vibrate in that condition.

▮ The trachea is the windpipe

The **trachea**, or windpipe, is a tube that conducts air between the environment and the lungs. It is held open by C-shaped rings of cartilage, giving it the general appearance of a vacuum cleaner hose. You can feel these rings of cartilage in your neck, just below the larynx.

These support rings are necessary in the trachea and its branches to prevent the air tubules from collapsing with the drop in pressure created by the rapid flow of air into the lungs during each breath. Air (or fluid) passing rapidly over a surface causes a lower pressure, or "pull," on that surface. Maybe you have noticed that when you jump in the shower and turn on the water, the shower curtain is drawn in toward you. The curtain moves inward because the moving water lowers air pressure just as the rapid movement of air through the respiratory tubules does. If the trachea were not supported open, lower pressure due to the rapid flow of air would cause it to collapse or flatten.

▮ The bronchial tree is a system of air tubules that conducts air

The trachea divides into two air tubes called primary **bronchi**, with one bronchus (singular) conducting air from the trachea to each lung. The bronchi branch repeatedly within the lung, forming progressively smaller air tubes. The smallest bronchi divide to form yet smaller tubules called **bronchioles**, which finally terminate in *alveoli*, sacs with surfaces specialized for gas exchange (discussed shortly).

The repeated branching of air tubules in the lung is reminiscent of a maple tree in winter. In fact, the resemblance is so close that the system of air tubules is often called the **bronchial tree** (Figure 14.7). All the bronchi are held open by cartilage, just as we saw in the trachea. However, the amount of cartilage decreases with the diameter of the tube. The tiny bronchioles have no cartilage, but their walls do contain smooth muscle, which is controlled by the autonomic nervous system so that air flow can be adjusted to suit metabolic needs (see Chapter 8).

Although the contraction of the muscle in bronchial walls is usually marvelously adjusted to the body's needs, sometimes the bronchial muscles go into spasm, making air flow exceedingly difficult. Such is the case with **asthma**, a chronic condition characterized by recurring attacks of wheezing and difficulty in breathing. The difficult breathing is worsened by persistent inflammation of the airways.

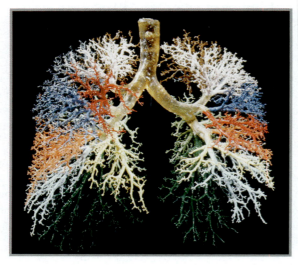

FIGURE **14.7**
A resin cast of the bronchial tree of the lungs. In the body, this branching system of air tubules is hollow and serves as a passageway for the movement of air between the atmosphere and the alveoli, where gas exchange takes place.

An allergy to things such as pollen, dog or cat dander (skin particles), or the feces of tiny mites in household dust trigger many asthma attacks. However, a cold or respiratory infection, certain drugs, inhaling irritating substances, vigorous exercise, and psychological stress can also cause an attack. Some attacks start for no apparent reason. Some inhalants prescribed to treat asthma attacks work by relaxing the bronchial muscles. Other inhalants contain steroids that reduce the inflammation of the air tubules that occurs in asthma.

The alveoli of the lungs are surfaces for gas exchange

Each bronchiole ends with an enlargement called an alveolus (plural, alveoli) or, more commonly, with a grapelike cluster of alveoli. Each **alveolus** is a thin-walled, rounded chamber surrounded by a vast network of capillaries (Figure 14.8). Oxygen diffuses from the alveoli into the blood, where it can be delivered to cells. Carbon dioxide produced by the cells diffuses from the blood into the alveolar air to be exhaled.

Most of the lung tissue is composed of alveoli, making the structure of the lung much more similar to foam rubber than to a balloon, the way the lung is sometimes imagined. The surface area of simple, hollow (balloonlike) lungs the same size as our lungs would be roughly 0.01 m^2 (about 0.2 yd^2). However, each of our lungs contains approximately 300 million alveoli, so the combined alveolar surface area is about 70 to 80 m^2 (about 84 to 96 yd^2), roughly the surface area of a tennis court. In other words, the alveoli increase the surface area of the lung about 8500 times.

For the alveoli to function properly as a surface for gas exchange, they must be kept open. Moist membranes, such as those of the alveolar walls, are attracted to one another because of an attraction between water molecules called surface tension. If the attraction were not prevented by phospholipid molecules called **surfactant**, it would pull the alveolar walls together, collapsing the air chambers.

Surfactant production usually begins during the eighth month of fetal life, so enough surfactant is present to keep the alveoli open when the newborn takes its first breath. Unfortunately, some premature babies have not yet produced a sufficient amount of surfactant to reduce the high surface tension. As a result, their alveoli

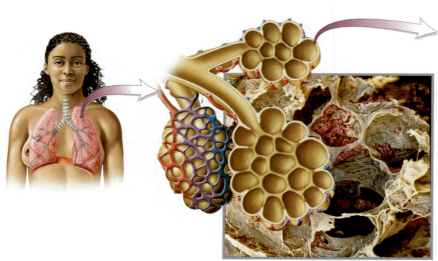
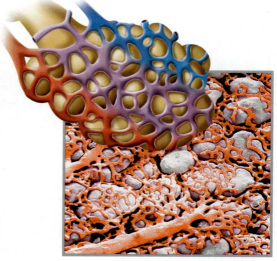

(a) Each alveolus is a cup-shaped chamber. In this section, some of the alveoli have been cut open and you can see into them.

(b) Much of the surface of each alveolus is covered with capillaries. The interface provides a vast surface area for the exchange of gases between the alveoli and the blood.

FIGURE **14.8**
Alveoli in the lungs create a huge surface area where oxygen and carbon dioxide are exchanged between the lungs and the blood. Oxygen diffuses from the alveoli into the blood, and carbon dioxide diffuses from the blood to the alveoli.

collapse after each breath. This condition, called **respiratory distress syndrome (RDS)**, makes each breath difficult for the newborn. Some newborns with RDS die because the task of breathing is too difficult. However, many are saved by using mechanical respirators and artificial surfactant to keep them alive until their lungs mature and produce enough surfactant.

stop and think

Pneumonia is a lung infection that results in an accumulation of fluid and dead white blood cells in the alveoli. Why might this result in lower blood levels of oxygen?

Pressure Changes within the Lungs Cause Breathing

Air moves between the atmosphere and the lungs in response to pressure gradients. Air moves into the lungs when the pressure in the atmosphere is greater than the pressure in the lungs, and it moves out when the pressure in the lungs is greater than the pressure in the atmosphere. The pressure changes in the lungs are created by changes in the volume of the thoracic cavity. Let's consider how the necessary changes in the size of the thoracic cavity are brought about.

■ Inhalation occurs when the pressure in the lungs decreases

Air moves into the lungs when the size of the thoracic cavity increases; this increase causes the pressure in the lungs to drop below atmospheric pressure. The increase is due to the contraction of both the **diaphragm**, a broad sheet of muscle that separates the abdominal and thoracic cavities, and the muscles of the rib cage, called the **external intercostals** (*costa*, rib) (Figure 14.9). This process is called inhalation (or **inspiration**). The external intercostals lie between the ribs so that when the muscles contract they pull the rib cage upward and outward. By placing your hands on your rib cage while you inhale, you can feel the rib cage move up and out. Raising the rib cage increases the size of the thoracic cavity from the front to the back. Meanwhile, the contraction of the diaphragm lengthens the thoracic cavity from top to bottom, as shown in Figure 14.9. The lungs are elastic; that is, they can be stretched and return to their former size. Therefore, an increase in the size of the thoracic cavity increases the volume of the lungs. The increase in lung volume causes the pressure within the lungs to fall below atmospheric pressure, allowing air to move into them.

■ Exhalation occurs when the pressure in the lungs increases

The process of breathing out, called exhalation (or **expiration**), is usually passive. In other words, it occurs when the muscles of the rib cage and the diaphragm relax. The elastic tissues of the lung then recoil; the rib cage falls back to its former lower position; and the diaphragm bulges into the thoracic cavity (Figure 14.9). The pressure within the lungs increases as the volume of the lungs decreases. When the pressure within the lungs exceeds atmospheric pressure, air moves out.

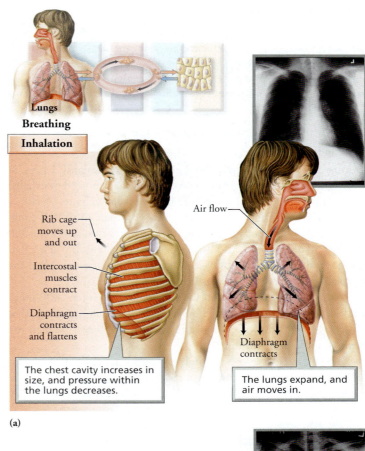

Inhalation

Rib cage moves up and out

Intercostal muscles contract

Diaphragm contracts and flattens

The chest cavity increases in size, and pressure within the lungs decreases.

Air flow

Diaphragm contracts

The lungs expand, and air moves in.

(a)

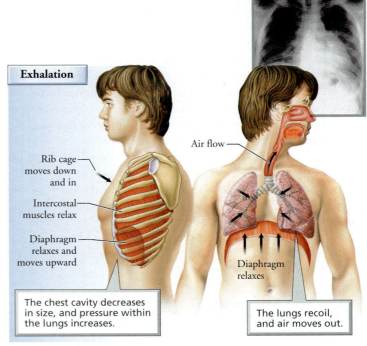

Exhalation

Rib cage moves down and in

Intercostal muscles relax

Diaphragm relaxes and moves upward

The chest cavity decreases in size, and pressure within the lungs increases.

Air flow

Diaphragm relaxes

The lungs recoil, and air moves out.

(b)

FIGURE **14.9**

Changes in the volume of the thoracic cavity bring about inhalation and exhalation. The x-rays show the actual changes in lung volume during inhalation and exhalation.

stop and think

In Victorian times, a woman often wore a corset containing whalebone that formed a band around her waist and lower chest. The corsets would be laced tightly to create a wasplike waistline. These women frequently fainted. What is the most likely cause of their fainting spells?

The volume of air moved into and out of the lungs is an indication of health

During quiet breathing, you move about 500 ml, roughly 1 pint, of air in and out with each breath. The amount of air inhaled or exhaled during a normal breath is called the **tidal volume** (Figure 14.10).

If, after inhaling normally, you were to inhale until you could not take in any more air, you would probably bring another 1900 to 3300 ml of air into your lungs. The additional volume of air that can be brought into the lungs after normal inhalation is called the **inspiratory reserve volume**.

After you have exhaled normally, you can still force 1000 ml of additional air from the lungs. This additional volume of air that can be expelled from the lungs after the tidal volume is called the **expiratory reserve volume**.

The lungs can never be completely emptied, even with the most forceful expiration. The amount of air that remains in the lungs after exhaling as much air as possible, called the **residual volume**, is roughly 1200 ml of air. As we will see later in the chapter, emphysema is a lung condition in which the alveolar walls break down, creating larger air spaces. Thus, the residual volume increases. This residual air is lower in oxygen than is inhaled air, so the person feels short of breath.

If you were to take the deepest breath possible and exhale until you could not force any more air from your lungs, you would measure your **vital capacity**, the maximum amount of air that can be moved into and out of the lungs during forceful breathing. The vital capacity, therefore, equals the sum of the tidal volume, the inspiratory reserve, and the expiratory reserve. Although average values for college-age people are about 4800 ml in males and 3400 ml in women, the values can vary tremendously. In pneumonia, fluid accumulates within the alveoli. This accumulation reduces the person's vital capacity because the fluid takes up space that would normally be occupied by air.

Because some air is always left in the lungs, the vital capacity is not a measure of the total amount of air that the lungs can hold. The **total lung capacity**, the total volume of air contained in the lungs after the deepest possible breath, is calculated by adding the residual volume to the vital capacity. This volume is approximately 6000 ml in men and 4500 ml in women.

Blood Transports Gases between the Lungs and the Cells

We have seen that breathing brings air into the lungs and expels air from the lungs. Recall that three other processes then play roles in delivering the oxygen to the cells and disposing of carbon dioxide from the cells. External respiration occurs in the lungs, where oxygen diffuses into the blood and carbon dioxide diffuses from the blood into the alveoli of the

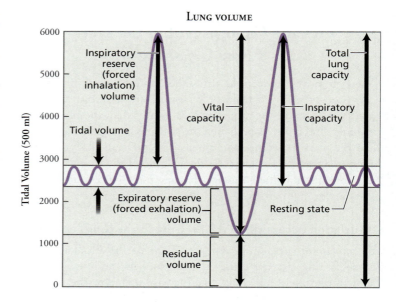

Tidal volume (~500 ml)	Amount of air inhaled or exhaled during an ordinary breath
Inspiratory reserve volume (~1900–3300 ml)	Amount of air that can be inhaled in addition to a normal breath
Expiratory reserve volume (~1000 ml)	Amount of air that can be exhaled in addition to a normal breath
Residual volume (~1100–1200 ml)	Amount of air remaining in the lungs after maximum exhalation
Vital capacity (~3400–4800 ml)	Amount of air that can be inhaled or exhaled in a single breath
Total lung capacity (4500–6000 ml)	Total amount of air in the lungs after maximal inhalation (vital capacity + residual volume)

FIGURE **14.10**

Volumes of air in the lungs

lungs. Blood transports oxygen to the cells and carbon dioxide away from the cells. Internal respiration occurs at the tissues, where oxygen diffuses out of the blood to the tissues and carbon dioxide diffuses from the tissues into the blood (Figure 14.11).

▌Most oxygen is carried by hemoglobin

Oxygen is carried from the alveoli throughout the body by the blood (Figure 14.12). Almost all, about 98.5%, of the oxygen

that reaches the cells is bound to hemoglobin, a protein in the red blood cells. Hemoglobin bound to oxygen is called **oxyhemoglobin** (HbO_2). The remaining 1.5% of the oxygen delivered to the cells is dissolved in the plasma. Whole blood, which has cells as well as plasma, carries 70 times more oxygen than an equal amount of plasma.

Hemoglobin picks up oxygen at the lungs and releases it at the cells. But what determines whether hemoglobin will bind to or release

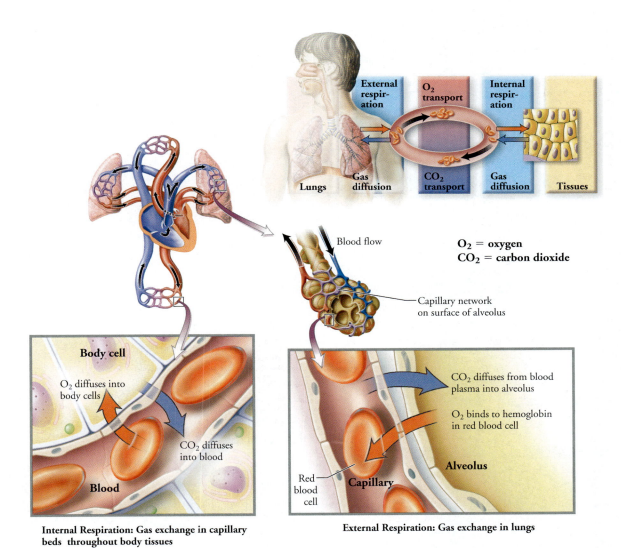

External respir-ation

O_2 transport

Internal respir-ation

Lungs

Gas diffusion

CO_2 transport

Gas diffusion

Tissues

Blood flow

O_2 = oxygen
CO_2 = carbon dioxide

Capillary network on surface of alveolus

Body cell

O_2 diffuses into body cells

CO_2 diffuses into blood

Blood

CO_2 diffuses from blood plasma into alveolus

O_2 binds to hemoglobin in red blood cell

Alveolus

Red blood cell

Capillary

Internal Respiration: Gas exchange in capillary beds throughout body tissues

External Respiration: Gas exchange in lungs

WWW FIGURE 14.11

TUTORIAL 14.3 In the lungs, oxygen diffuses from the alveoli into the blood. Oxygen is carried to the cells in red blood cells. At the cells, oxygen diffuses from the blood to the body cells, which use the oxygen and produce carbon dioxide in the process. Carbon dioxide diffuses into the blood and is carried back to the lungs, where it diffuses from the blood into an alveolus and is exhaled.

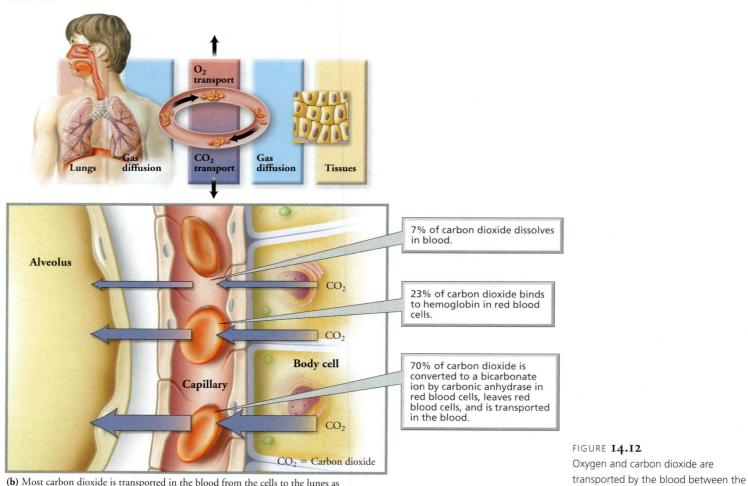

1.5% of oxygen dissolves in blood.

98.5% of oxygen binds to hemoglobin in red blood cells.

Alveolus

O_2

O_2

Capillary

Body cell

O_2 = Oxygen

(a) Most oxygen is carried from the lungs to the cells bound to hemoglobin in red blood cells.

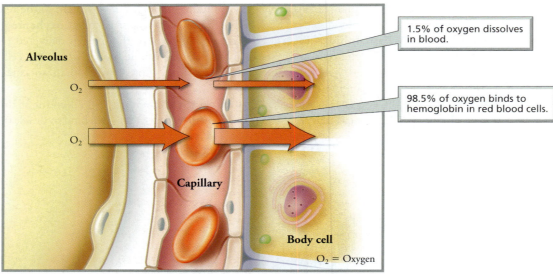

O_2 transport

Lungs Gas diffusion CO_2 transport Gas diffusion Tissues

7% of carbon dioxide dissolves in blood.

23% of carbon dioxide binds to hemoglobin in red blood cells.

70% of carbon dioxide is converted to a bicarbonate ion by carbonic anhydrase in red blood cells, leaves red blood cells, and is transported in the blood.

Alveolus

CO_2

CO_2

Body cell

Capillary

CO_2

CO_2 = Carbon dioxide

(b) Most carbon dioxide is transported in the blood from the cells to the lungs as bicarbonate ions that were formed in red blood cells by carbonic anhydrase.

FIGURE **14.12**

Oxygen and carbon dioxide are transported by the blood between the lungs and the body cells.

oxygen? The concentration (partial pressure[1]) of oxygen is the most important factor in determining whether hemoglobin will bind to

oxygen. Recall from Chapter 3 that substances always diffuse from regions of higher concentration or pressure to regions of lower con-

[1]In a mixture of gases, each gas contributes only part of the total pressure of the whole mixture of gases. The pressure exerted by one of the gases in a mixture, called its partial pressure, is directly related to the concentration of that gas in the mixture.

centration or pressure. In the alveoli of the lungs, where the concentration of oxygen is high, hemoglobin in the red blood cells in nearby capillaries picks up oxygen. The oxygen is then released near the cells, where the oxygen concentration is low.

Interestingly, oxygen delivery is responsive to the needs of the cells. Oxygenated blood passing metabolically inactive cells gives up less oxygen than it does when passing metabolically active cells. When cells are in serious need of oxygen, the amount of oxygen delivered to the cells can be increased more than threefold, even if the rate of blood flow remains constant.

Changes in the conditions around the active cells alter the behavior of hemoglobin and make oxygen delivery responsive to needs of the cells. Cellular metabolism produces carbon dioxide and heat as by-products. When carbon dioxide dissolves in the water of tissue fluid and blood, it forms carbonic acid, and the acidity of the blood rises. Under these conditions of increased temperature and acidity, hemoglobin releases its oxygen load more readily.

▌ Most carbon dioxide is transported as bicarbonate ions

The carbon dioxide produced as the cells use oxygen is removed by the blood. As shown in Figure 14.12, carbon dioxide transport occurs in three fundamental ways:

1. **Dissolved in blood.** Between 7% and 10% of the carbon dioxide is transported dissolved in the blood as molecular carbon dioxide.

2. **Carried by hemoglobin.** Hemoglobin molecules in red blood cells carry slightly more than 20% of the transported carbon dioxide. When carbon dioxide combines with hemoglobin, it forms a compound called **carbaminohemoglobin**.

3. **As a bicarbonate ion.** By far the most important means of transporting carbon dioxide is as a bicarbonate ion dissolved in the plasma. About 70% of the carbon dioxide is transported this way. Carbon dioxide produced by cells diffuses into the blood and into the red blood cells. Carbon dioxide (CO_2) reacts with water (H_2O) in the red blood cells and plasma and forms carbonic acid (H_2CO_3). Carbonic acid quickly dissociates to form hydrogen ions (H^+) and bicarbonate ions (HCO_3^-). At the body cells, the process of bicarbonate ion formation is represented in the following formula:

$$CO_2 + H_2O \xrightarrow{\text{carbonic anhydrase}} H_2CO_3 \rightarrow HCO_3^- + H^+$$

Carbon dioxide　Water　　　　Carbonic acid　　Bicarbonate ion　Hydrogen ion

Although these reactions occur in the plasma as well as in red blood cells, they occur hundreds of times faster in red blood cells. The higher rate of reaction is caused by the enzyme, called **carbonic anhydrase**, that is found within red blood cells but not in the plasma. The hydrogen ions produced by the reaction combine with hemoglobin. In this way, hemoglobin acts as a buffer, and the acidity of the blood changes only slightly as it passes through the tissues. The bicarbonate ions diffuse out of the red blood cells into the plasma and are transported to the lungs. As the negatively charged bicarbonate ions leave the red

blood cells, chloride ions (Cl^-) diffuse in from the plasma, counterbalancing the positive charge that would have been left within the red blood cells.

At the lungs, the process is reversed. When the blood reaches the capillaries of the lungs, carbon dioxide diffuses from the blood into the alveoli because the concentration (partial pressure) of carbon dioxide is comparatively low in the alveoli. Because the concentration of carbon dioxide in the blood is higher than in the alveoli, the chemical reactions we have just described reverse direction. The bicarbonate ions rejoin the hydrogen ions to form carbonic acid. In the presence of carbonic anhydrase within the red blood cells, carbonic acid is converted to carbon dioxide and water. The carbon dioxide then leaves the red blood cells, diffuses into the alveolar air, and is exhaled. The reactions in the lungs are summarized below.

$$HCO_3^- + H^+ \xrightarrow{\text{carbonic anhydrase}} H_2CO_3 \rightarrow CO_2 + H_2O$$

Bicarbonate ion　Hydrogen ion　　Carbonic acid　　Carbon dioxide　Water

Besides providing a means of transporting carbon dioxide, bicarbonate ions are an important part of the body's acid-base buffering system. They help neutralize acids in the blood. If the blood becomes too acidic, the excess hydrogen ions are removed by combination with bicarbonate ions to form carbonic acid. The carbonic acid then forms carbon dioxide and water, which are exhaled.

stop and think

Carbon monoxide binds to hemoglobin much more readily than does oxygen, and it binds in the same site as oxygen. Thus, when carbon monoxide is bound to hemoglobin, oxygen cannot bind. Explain why carbon monoxide poisoning can be fatal.

Breathing Is Controlled Primarily by Respiratory Centers in the Brain

Breathing rate influences the amount of oxygen that can be delivered to cells and the amount of carbon dioxide that can be removed from the body. Neural and chemical controls adjust breathing rate to meet the body's needs.

▌ Brain centers control the basic breathing pattern

As you sit there reading your text, your breathing is probably rather rhythmic, with about 12 to 15 breaths a minute. The basic rhythm is controlled by a **breathing center** located in the medulla of the brain (see Chapter 8). Within the breathing center are an inspiratory area and an expiratory area (Figure 14.13).

During quiet breathing, when you are calm and breathing normally, the inspiratory area shows rhythmic bouts of neural activity, as shown in Figure 14.14. The activity of these neurons increases for about 2 seconds and then ceases for about 3 seconds. While the inspiratory neurons are active, impulses that stimulate contraction

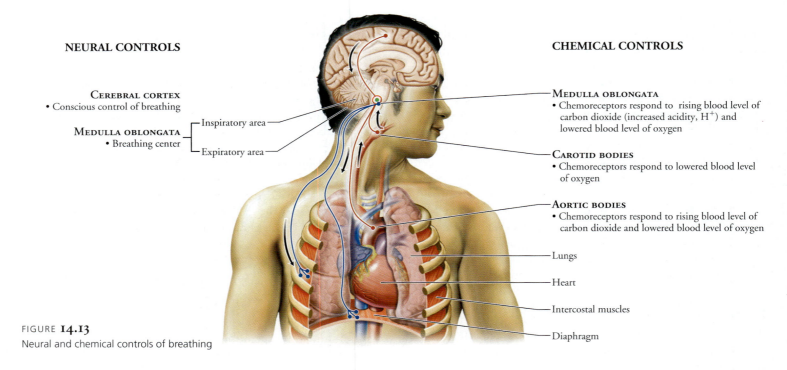

NEURAL CONTROLS

CEREBRAL CORTEX
• Conscious control of breathing

MEDULLA OBLONGATA
• Breathing center

Inspiratory area

Expiratory area

CHEMICAL CONTROLS

MEDULLA OBLONGATA
• Chemoreceptors respond to rising blood level of carbon dioxide (increased acidity, H⁺) and lowered blood level of oxygen

CAROTID BODIES
• Chemoreceptors respond to lowered blood level of oxygen

AORTIC BODIES
• Chemoreceptors respond to rising blood level of carbon dioxide and lowered blood level of oxygen

Lungs

Heart

Intercostal muscles

Diaphragm

FIGURE **14.13**
Neural and chemical controls of breathing

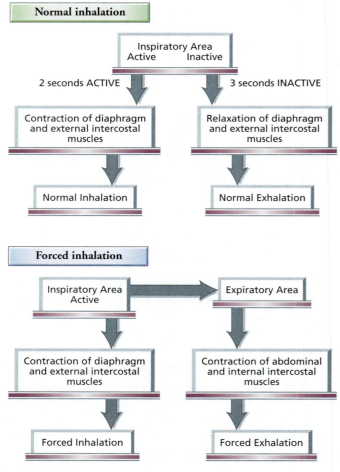

FIGURE **14.14**
The basic breathing pattern is set by the breathing center in the medulla. During quiet breathing, only the inspiratory area plays a role. It is active for 2 seconds, causing inhalation, and then inactive for 3 seconds, causing exhalation. Both the inspiratory and expiratory areas are active during heavy breathing.

are sent to the muscles involved in inhalation (the diaphragm and the external intercostals). As we have seen, contraction of the diaphragm and the external intercostals causes the size of the thoracic cavity to increase, thus moving air into the lungs. When the inspiratory signals cease, the diaphragm and external intercostals relax, and you passively exhale. Then, after the 3-second rest, the inspiratory neurons spontaneously begin to fire again, repeating the cycle.

During heavy breathing, breaths become deeper and faster, increasing the rate of oxygen delivery and carbon dioxide removal. At such times, for example during strenuous activity, exhalation is no longer a passive event. Instead, it is an active process. The inspiratory area is active, as it is in quiet breathing, and inhalation is caused by its impulses contracting both the diaphragm and the external intercostal muscles. During heavy breathing, however, the inspiratory area also sends impulses to the expiratory area. The expiratory area causes the contraction of both the internal intercostals (the inner muscles of the rib cage) and the abdominal muscles. In this way, air is quickly pushed out of the lungs.

Most of the time we breathe without giving it a thought. However, we can voluntarily alter our pattern of breathing through the impulses originating in the cerebral cortex (the "conscious" part of the brain). We control breathing when we speak or sigh, and we can voluntarily pant like a dog. Holding our breath while swimming under water is obviously a good idea, and doing so can sometimes protect us from inhaling smoke or irritating gases.

During forced breathing, as might occur during exercise, stretch receptors in the walls of the bronchi and bronchioles throughout the lungs prevent the overinflation of the lungs. When a deep breath greatly expands the lungs and stretches these receptors, they send impulses over the vagus nerve that inhibit the breathing center, permitting exhalation. As the lungs deflate, the stretch receptors are no longer stimulated.

The changes in depth and rate of breathing are affected by chemoreceptors

The purpose of breathing is to control the blood levels of carbon dioxide and oxygen. So, we will now consider how the levels of these gases control the breathing rate, which in turn, influences the levels of the gases (Figure 14.15).

CARBON DIOXIDE

The most important chemical influencing breathing rate is carbon dioxide. Most of the effect of carbon dioxide on breathing is caused by the hydrogen ions that are formed when carbon dioxide goes into solution and forms carbonic acid:

$$CO_2 + H_2O \rightarrow H_2CO_3 \rightarrow H^+ + HCO_3^-$$

Carbon   Water   Carbonic   Hydrogen   Bicarbonate
dioxide   acid   ion   ion

Chemical control of the breathing rate is based on input from chemoreceptors. Central chemoreceptors are located in a region of the brain called the medulla. Peripheral chemoreceptors are located in the aortic bodies and the carotid bodies, small structures associated with the main blood vessel leaving the heart to the body and the main blood vessels to the head (see Figure 14.13). All of these chemoreceptors respond to changes in carbon dioxide (hydrogen ion) concentration. Those in the aortic bodies and the carotid bodies also respond when the blood level of oxygen is low.

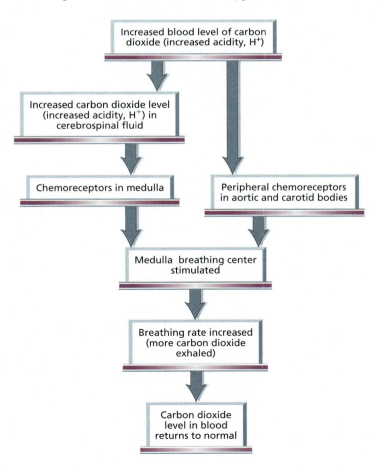

FIGURE **14.15**
The role of carbon dioxide in controlling the breathing rate

The key areas in the chemical control of breathing are in the medulla. The medulla's receptors are near its surface, where they are bathed in cerebrospinal fluid. The brain is protected from many changes in blood chemistry because most substances cannot pass through the specialized capillary walls that form the blood-brain barrier (see Chapter 8). However, carbon dioxide is very soluble and can diffuse into the cerebrospinal fluid, where it raises the hydrogen ion concentration by forming carbonic acid. Because the cerebrospinal fluid lacks the blood's ability to buffer changes in acidity, dissolved carbon dioxide changes the acidity of cerebrospinal fluid more than it does that of blood. When the rising carbon dioxide level (or hydrogen ion concentration) stimulates the chemosensitive areas, breathing rate is increased, causing a decrease in the blood level of carbon dioxide.

OXYGEN

Because oxygen, not carbon dioxide, is essential to survival, it may be somewhat surprising to learn that oxygen does not influence the breathing rate unless its blood level falls dangerously low. The firing of oxygen-sensitive chemoreceptors in the aortic bodies and carotid bodies serves as a warning that the blood oxygen level is at a critical point and initiates a last-minute call to the medulla to increase the breathing rate and raise oxygen levels. If the oxygen level falls much more, the neurons in the inspiratory area die from a lack of oxygen and do not respond well to impulses from the chemoreceptors. As a result, the inspiratory area begins to send fewer impulses to the muscles of inspiration, and the breathing rate decreases and may even cease completely.

stop and think

Divers sometimes take several deep and rapid breaths before diving. Why would this practice allow the diver to remain under water longer? Occasionally, a diver who hyperventilates before diving loses consciousness while under water. Why would hyperventilating cause a diver to pass out?

Respiratory Disorders Have Many Causes

During a restful day, you move more than 86,000 liters of air into and out of your lungs. The inhaled air may contain disease-causing organisms or noxious chemicals and particles. The respiratory system clears most of these materials before they can cause harm. However, some of the disease-causing organisms, including viruses and bacteria, and harmful substances remain and cause problems (see the Social Issue essay, *A Global Threat: Tuberculosis*).

The common cold is caused by many types of viruses

The **common cold** is indeed common. Some 30 million Americans have a cold at this moment. Most people get their first cold before they are a year old and continue to get several colds each year through adulthood. Typically, a cold begins with a runny nose, possibly a sore throat, and sneezing. In the beginning, the nasal discharge is thin and watery, but it becomes thicker as congestion increases. Almost any part of the respiratory system can be affected. Sneezing and a stuffy nose indicate that the infection is in the upper respiratory system. When the pharynx is affected, a sore throat results. The infection may spread to the bronchi, causing a cough, or to the larynx, making your voice hoarse.

SOCIAL ISSUE

A Global Health Threat: Tuberculosis

One of humankind's oldest diseases, *tuberculosis*, or TB, is rapidly becoming one of its deadliest once again. About one-third of the world's population—some 2 billion people—is infected with TB. Fortunately, few of those infected, only about 10%, will actually develop the disease. Because each of those infected individuals (whether or not they display symptoms) will infect 10 to 15 other people, the World Health Organization (WHO) predicts that 1 billion new cases of TB will occur by 2020. Sadly, nearly 3 million people, mostly young adults, die of TB each year. Most of the TB deaths occur in nations whose people cannot afford the cost of medication.

What is tuberculosis, and why has it been so difficult to eradicate? TB is caused by a rod-shaped bacterium, *Mycobacterium tuberculosis*. It is spread when the cough of an infected person sends bacteria-laden droplets into the air, and the bacteria are inhaled into the lungs of an uninfected person. Because the bacteria are inhaled, the lungs are usually the first sites attacked, but the bacteria can spread to any part of the body, especially to the brain, kidneys, or bone.

As a defense against the bacteria, the body forms fibrous connective tissue casings, called tubercles, that encapsulate the bacteria (hence,

the name of the disease). Although the formation of tubercles slows the spread of the disease, it does not actually kill the bacteria. The immune system destroys at least some of the walled-off bacteria and may, in fact, kill them all. When the body's defenses fail, however, pockets of bacteria may persist, undetected, for many years. Later, the disease may progress to the secondary stage as pockets of bacteria become activated again. Furthermore, bacteria may escape from the tubercles and be carried by the bloodstream to other parts of the body. As a result, whenever the victim becomes weak, ill, or poorly nourished, the disease may flare up.

The initial symptoms of tuberculosis, if there are any, are similar to those of the flu. In the secondary stage, the patient usually develops a fever, loses weight, and feels tired. If the infection is in the lungs, as it usually is, it causes a dry cough that eventually produces pus-filled and blood-streaked phlegm. TB can be fatal. Indeed, more people in the United States have been killed by tuberculosis than died in both World Wars.

Treatment for TB usually involves a combination of antibiotics. Because the bacteria grow slowly, antibiotics must be taken for 6 to 12 months. But, for reasons ranging from forgetfulness to poverty, many patients stop taking

their antibiotics when they begin to feel better. Not completing the antibiotic regimen allows the more resistant bacteria that withstood the initial drug onslaught to reproduce, creating strains of bacteria that are difficult or impossible to kill (see Chapter 13A). The resistant bacteria then grow unchecked, and the patient suffers a relapse. When another course of antibiotics is prescribed, the patient may again stop taking it too soon. In this way, resistance to several drugs has developed in tuberculosis bacteria.

The rise of drug-resistant TB bacteria is an important social concern. Recent studies have suggested that drug-resistant bacteria cause 80% of the new cases of TB. The cost of treating multi-drug-resistant TB is staggering; it can cost many tens of thousands of dollars per patient. And, when multi-drug-resistant bacteria are to blame, the fatality rate for TB roughly doubles—to 50%!

Another social concern is that TB is unevenly distributed throughout the world's population. It is more common in developing nations. Because the disease is so contagious, increased crowding in housing among the poor in all countries has contributed further to the increase in tuberculosis. For the same reason, it is prevalent among prison inmates. The homeless and substance abusers are also at increased risk. ✋

As miserable as you feel with a cold, you can take comfort from the fact that your suffering will not last forever (see the Health Issue essay, *Surviving a Common Cold*). A cold is self-limiting, lasting only 1 to 2 weeks. Furthermore, colds are seldom fatal, except occasionally among the very young or very old or in those people already seriously ill with another malady.

Any one of over 200 viruses can cause a cold, a small fact that explains several things about colds. Because there are so many cold-causing viruses, you can get several colds a year, each from a different virus.

Colds are spread when the causative virus is transmitted from an infected person. The viruses are plentiful in nasal secretions. However, transmission of the virus is not usually the result of direct inhalation of infected droplets from a cough or sneeze. It is more likely to be the result of handling an object that is contaminated with the virus. The viruses may remain alive on the skin or an object for several hours, waiting for an unsuspecting person to touch them, thereby contaminating his or her own fingers. Then, when the virus-laden fingers are touched to the mucous membranes of the nose, the transfer is completed. The best ways to prevent a cold are to wash your hands frequently and avoid being around persons who have one.

The flu is caused by three types of virus

Flu is an abbreviation of **influenza**. Viruses cause the flu, but unlike the number of viruses that can cause a cold, there are relatively few viruses responsible for the flu. The viruses that cause the flu in humans are all variants of three major types—A, B, and C. Influenza A is often more serious than B in that it is often accompanied by severe complications and more frequently results in death. Influenza C causes a mild illness with cold symptoms. Although there are only three basic types of influenza virus, there are hundreds of variants.

Symptoms of the flu are similar to those of a cold, but they appear to start suddenly and are more severe. By the time flu symptoms appear, the disease has been incubating for several days. Typically, the flu begins with chills and a high fever, about 103°F (39°C) in adults and perhaps higher in children. Many flu victims experience aches and pains in the muscles, especially in the back. Other common ailments include a headache, sore throat, dry cough, weakness, pain and burning in the eyes, and sensitivity to light. When the flu hits, you usually feel sick enough to go to bed. The flu generally lasts for 7 to 10 days, but it may take an additional week or more before you are completely back on your feet.

Surviving a Common Cold

The only thing more common than the cold is advice on how to treat it. Here we will examine the validity of some frequently suggested treatments for a cold.

1. **Take large doses of vitamin C?** Vitamin C will not prevent a cold unless you are malnourished or under extreme physical stress. Nonetheless, some people who take vitamin C may experience less severe or shorter colds than do people who do not take vitamin C.

2. **Suck on a zinc lozenge?** Some evidence exists that zinc slows viral replication, prevents cold viruses from adhering to nasal membranes, and boosts the immune system. But it is not clear whether zinc lozenges can alleviate cold symptoms. Some people who begin taking zinc within the first day or two of the onset of symptoms and continue taking a lozenge every two or three hours for several days do experience some relief from cold symptoms.

3. **Take echinacea?** Commonly known as the purple coneflower, echinacea has been used for centuries by cold sufferers. Echinacea stimulates white blood cells to fight infection. If it is taken at the first sign of a cold, it *might* lessen the severity of the cold and reduce its duration by a day or two. Echinacea should be taken only when cold symptoms are present, not on a regular basis. Children should not use echinacea.

4. **Take an antibiotic?** Antibiotics are *not* effective against viruses and cannot cure a cold. However, a doctor may prescribe antibiotics to control secondary bacterial infections—such as a middle ear infection, bronchitis, or sinusitis—that may accompany a cold. Unnecessary use of an antibiotic may cause side effects such as diarrhea and can lead to the development of bacterial resistance to the drug. The rise of strains of bacteria that are resistant to antibiotics is likely to be a major threat to public health in the coming years, as discussed in Chapter 13a.

5. **Go to bed?** Bed rest enables the body to muster its resources and fight secondary infections. Staying at home with a cold is also socially responsible, because it helps prevent the spread of the virus. But if you are too busy to spend a few days in bed because of a cold, you probably are not hurting yourself. Bed rest will not cure your cold or shorten its duration.

6. **Have some chicken soup?** Grandmothers have suggested having chicken soup to treat a cold for years, and doctors finally agree that the advice has some merit. You should always consume plenty of fluids when you have a cold. They help loosen secretions in the respiratory tract. Hot fluids, such as chicken soup, are more effective than cold drinks in increasing the flow of nasal mucus. By increasing the flow of nasal secretions, you reduce congestion. As a result, you can breathe more freely, and the amount of time the cold viruses are in contact with the cells lining the respiratory system is reduced.

Although a cold cannot be cured, there are ways to make it more bearable. Some ways to relieve the misery of a cold are as follows.

For Nasal Congestion.

The congestion is caused when the mucous membranes of the nasal cavities become swollen and produce increased amounts of mucus caused by viral infection. The best ways to relieve the congestion are to drink plenty of extra fluids and to inhale moist air from a hot bath, shower, or vaporizer.

By constricting small blood vessels in the nose, decongestants may reduce the accumulation of fluid causing nasal congestion. Unfortunately, if taken orally, decongestants constrict small blood vessels throughout the body; in the process, they raise blood pressure. Other possible side effects of oral decongestants include nervousness, sleeplessness, and dryness of the mouth. Therefore, if you must use a decongestant, it is wise to choose one in nose sprays or drops. Use these at the recommended dosage and only for a few days. Although antihistamines are present in

many cold remedies, there is no evidence that they are effective in reducing nasal congestion caused by colds. Furthermore, antihistamines may cause blurred vision, retention of urine, and dizziness. Drowsiness is a major side effect, so if you do take an antihistamine, avoid driving or other hazardous activities.

For Coughs.

Coughing is a protective reflex controlled by a cough center in the brain. Besides clearing the respiratory tubules of foreign material, coughing loosens and removes phlegm and mucus. Therefore, it is not always a good idea to suppress a cough. If a cough persists for more than a week, you should see a physician.

The safest and cheapest cough remedies are substances such as hard candy or honey, which coat and soothe the throat, and drinking extra fluids, which loosens the mucus and helps relieve some irritation.

Cough medications fall into two general classes: suppressants and expectorants. Cough suppressants work by reducing the activity of the brain's cough center. These are helpful in easing dry, hacking coughs. The most effective nonprescription cough suppressant is dextromethorphan. Codeine also suppresses coughs, but it is available only by prescription. A cough that brings up sputum is performing a useful function and should not be suppressed. Instead, removal of the irritating substances should be assisted with an expectorant, which increases the flow of respiratory tract secretions.

For Fever and Pain Relief.

The most common medicines for relief of fever and pain are aspirin, acetaminophen, and ibuprofen. Children and teenagers should not take aspirin because it is linked to Reye's syndrome, a deadly disease that affects all body organs. Keep in mind that fever is often a good thing. The additional body heat slows the growth and reproduction (or replication) of many disease-causing organisms (discussed in Chapter 13). Unless the fever is very high (above 102°F, 40°C, in an adult), you may want to let it run its course.

The flu is often complicated by secondary infections, which follow the initial disease as other disease-causing organisms take advantage of the body's weakened state. The most common complication is pneumonia, an inflammation of the lungs. Bacteria taking advantage of the weakened state of the body can also cause bronchitis, sinusitis, and ear infections.

One way to prevent the flu is by getting a flu shot—a vaccine made from the strains of viruses that scientists anticipate will cause the next outbreaks of the illness. Flu shots are only about 60% to 70% effective because the viruses they target mutate rapidly, causing new strains to appear. The new strains are not recognized by the body defenses that were programmed by the vaccine. Each flu season brings new strains of flu viruses. The effectiveness of the vaccine lasts only as long as that season's most prevalent strains. As a result, new vaccines must be developed to protect us, and flu shots must be repeated each year.

SARS is a serious viral respiratory illness

Severe acute respiratory syndrome (SARS) is caused by a previously unknown coronavirus (SARS-associated coronavirus, SARS-CoV). SARS was first reported in Asia in February 2003, but it had affected 8098 people in 26 countries by the end of the year (see Chapter 13a).

SARS usually begins with a fever greater than 100.4°F (38°C), a headache, and body aches. Within 2 to 7 days, SARS patients may develop a dry cough. Most patients develop pneumonia, a lung condition in which fluid accumulates in the lungs and causes breathing difficulty. SARS can be fatal.

SARS seems to be spread by close person-to-person contact through droplet infection. When a person with SARS coughs or sneezes, virus-laden droplets are propelled through the air and land on the mucous membranes of the nose, mouth, or eyes of nearby people. SARS can also be spread if a person touches a surface or object contaminated with the virus and then touches a mucous membrane of the nose or mouth. The SARS virus may survive for several days in the environment.

what would you do?

Civet cats, a part of the mongoose family, may have been the source of the SARS virus, which jumped species to infect humans. In China, civet cats were held in outdoor markets and sold to restaurants that prepared them as gourmet delicacies. When a new outbreak of SARS was feared early in 2004, China ordered that traders in wild animal markets kill all animals that might harbor the SARS virus, including civet cats, badgers, and raccoon dogs. An estimated 10,000 civet cats were killed, mostly by drowning, electrocution, or incineration. Do you think it is ethical to kills thousands of animals because they *might* pose a future threat to human health?

Strep throat can have serious consequences

Strep throat, a sore throat that is caused by *Streptococcus* bacteria, is a problem mainly in children 5 to 15 years old. The soreness in the throat is usually accompanied by swollen glands and a fever. The pain may be so mild that a doctor is never consulted.

Ignoring a strep infection can have serious consequences because, if untreated, the *Streptococcus* bacteria can spread to other parts of the body and cause rheumatic fever or kidney problems. The main symptoms of rheumatic fever are swollen, painful joints and a characteristic rash. About 60% of rheumatic fever sufferers develop disease of the heart valves. Another possible consequence of a streptococcal infection is kidney disease (glomerulonephritis). The kidney damage is due to a reaction from the body's own protective mechanisms. The body fights the bacteria by producing antibodies that destroy the bacteria. If these antibodies persist after the bacteria have been killed, they can cause the kidneys to become inflamed. The inflamed kidneys may be unable to filter the blood, and blood may leak into the urine.

Because many viruses can cause sore throats that look like strep infections, the only way to identify the disease is to test for the causative organism. If *Streptococcus* bacteria are found, an antibiotic, usually penicillin, is prescribed to prevent rheumatic fever and kidney disease. So, don't ignore a sore throat.

Bronchitis is an inflammation of the bronchi

Viruses, bacteria, or chemical irritation may cause the mucous membrane of the bronchi to become inflamed, a condition called **bronchitis**. The inflammation results in the production of excess mucus, which triggers a deep cough that produces greenish yellow phlegm.

There are two types of bronchitis: acute and chronic. Acute bronchitis, which often follows a cold, is usually caused by the cold virus itself, but it may be caused by bacteria that take advantage of the body's lowered resistance and invade the trachea and bronchi. An antibiotic will hasten recovery, if the cause is bacterial.

When a cough that brings up phlegm is present for at least 3 months during 2 consecutive years, the condition is called chronic bronchitis, a more serious condition that is usually associated with cigarette smoking or air pollution. (See the Environmental Issue essay, *Air Pollution and Human Health.*) Some people with chronic bronchitis may lack an enzyme that normally protects the air passageways from such irritants. As the disease progresses, it becomes increasingly difficult to breathe. One reason for the labored breathing is that the linings of the air tubules thicken, narrowing the passageway for air. The contraction of muscles in the walls of the bronchioles and the excessive secretion of mucus further obstruct the air tubules.

Chronic bronchitis can cause serious consequences. The degenerative changes in the lining of the air tubules make removal of mucus more difficult. As a result, the patient is more likely to develop lung infections such as pneumonia, which can be fatal, and degenerative changes in the lungs, such as emphysema.

Emphysema is caused by the destruction of alveoli

Emphysema is one of the more common results of smoking, although it can have other causes as well. In **emphysema**, the alveoli break down, causing them to merge, thereby creating fewer and larger alveoli (Figure 14.16). This change has two major effects: a reduction in the surface area available for gas exchange and an increase in the volume of residual, or "dead," air in the lungs. Exhalation, you may recall, is a passive process that depends on the elasticity of lung tissue. In emphysema, the lungs become inelastic, and air becomes trapped in the lungs. As the dead air space increases, adequate ventilation of the lungs requires more forceful inhalation. Forcing the air causes more alveolar walls to rupture, further increasing the dead air space. Lung size gradually increases as the residual volume of air

ENVIRONMENTAL ISSUE

Air Pollution and Human Health

Caution: The air you breathe may be hazardous to your health. It may even kill you—especially if you have heart or respiratory problems. In most major cities, the poor air quality may be obvious at times as a brown haze. However, even in remote national parks, air pollution is often significant enough to reduce visibility. Although the effects of air pollution may be subtle and take a long time to become apparent, they damage the environment as well as human health.

What are the sources of the pollutants in our air? Two major human sources of air pollution are motor vehicle exhaust and industrial emissions. The exhaust pipes on our cars and the smokestacks on factories spew oxides of sulfur, nitrogen, and carbon, a variety of hydrocarbons, and many particulates (such as soot and smoke) into the air. In the atmosphere, sulfur dioxide and nitrogen dioxide dissolve in the moisture and form an aerosol of strong acids—sulfuric acid and nitric acid, respectively. Some of this aerosol may drift upward, forming acidic clouds that may be blown hundreds of miles away by the prevailing winds and then fall as acid raindrops. On the other hand, the acidic aerosol may remain close to its source as a component of the haze created by air pollution. In addition, sunlight can cause hydrocarbons and nitrogen dioxide to react with one another and form a mix of hundreds of substances called photochemical smog. The most harmful component of photochemical smog is ozone, which attacks cells, destroys tissues, irritates the respiratory system, damages plants, and even erodes rubber. Paradoxically, the ozone that is so damaging when it is found at ground level in photochemical smog is the same chemical that is beneficial in the upper layers of the atmosphere,

where it prevents much of the sun's ultraviolet radiation from reaching the earth's surface. Unfortunately, the ozone found in smog does not make its way to the ozone layer of the upper atmosphere because it is converted to oxygen within a few days.

When we breathe polluted air, the respiratory system is, not surprisingly, the first to be affected. As air containing toxic substances fills the lungs, cells lining the airways and within the lungs are injured. Damaged cells release histamine, which causes nearby capillaries to widen and become more permeable to fluid. As a result, fluid leaks from the capillaries and accumulates within the tissues. Even a brief exposure to oxides of sulfur (5 parts per million [ppm] for a few minutes), the oxides of nitrogen (2 ppm for 10 minutes), or ozone leads to fluid accumulation, increased mucus production, and spasms (intense involuntary contractions) of the bronchioles. These effects make air flow more difficult and reduce gas exchange.

People with asthma are usually among those who suffer the most from air pollution. A person with asthma experiences breathing difficulty after inhaling 0.1 ppm nitrogen dioxide for 1 hour. In contrast, a healthy person is not likely to experience the same degree of respiratory distress unless the air inhaled contains 25 times as much nitrogen dioxide (2.5 ppm) for several hours. Ground-level ozone is also a problem for asthmatics.

Long-term irritation of bronchi by pollutants is a cause of chronic bronchitis and emphysema. The process begins as irritation and leads to increased fluid accumulation that, in turn, stimulates mucus production and coughing. The cough and mucus, signs of chronic bronchitis, irritate lungs even more. As

bronchitis continues, the air passageways become narrower, trapping air in the lungs. When the increased pressure accompanying a cough causes the overinflated alveoli to rupture, emphysema begins.

Some pollutants can cause cancer. As the normal mechanisms that cleanse the respiratory system are damaged or overloaded by pollutants, cancer-causing chemicals in the inhaled air are no longer effectively removed. Within the airways these chemicals may then bring about changes in genetic material that can lead to cancer.

With so many documented health consequences of air pollution, you may wonder why the world's great minds have not developed the technology to solve the problem. Indeed, they have. But, unfortunately, it turns out that air pollution is not only a scientific problem. Instead, it is primarily a social, political, and economic problem. We have the technology to prevent air pollution, but prevention costs money. Industry often claims that the cost of installing pollution control devices will have to be covered by an increased price of goods or the loss of jobs. Efforts to improve air quality will require some increased personal cost and inconvenience. How much more are you willing to pay for low-sulfur fuels, for instance? The automobile is a wonderful convenience. Are you willing to take a bus or, better yet, to walk instead of drive to improve your own air quality? Considering that the pollutants in air are often blown hundreds of miles from their source, are you willing to make the same sacrifices to improve a stranger's air quality? Also, who should make the laws that would control pollution when the pollutants are likely to cross state or national borders? Who should enforce those laws? ❦

becomes greater, giving a person with emphysema the characteristic barrel chest. However, gas exchange becomes more difficult. To get an idea of what poor lung ventilation caused by increased dead air space feels like, take a deep breath, then exhale only slightly, and repeat this process several times. Notice how quickly you feel an oxygen shortage if you continue taking very shallow breaths that leave the lungs almost completely filled with air.

Shortness of breath, the main symptom of emphysema, has several causes. Two causes are the decreased surface area for gas exchange and increased dead air space. As the disease progresses, gas exchange becomes even more difficult because the alveolar walls thicken with fibrous connective tissue. The oxygen that does make it to the alveoli has difficulty crossing the connective tissue to enter the blood. Thus, a person with emphysema constantly gasps for air.

(a) Normal alveoli

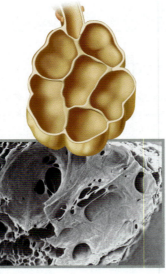

(b) Emphysema causes breakdown of alveolar walls.

exploring further ...

In this chapter, we have seen how the structures of the respiratory system provide an extensive surface for gas exchange between the lungs and the blood, move air into and out of the lungs, and protect the gas exchange surfaces in the lungs. We have also considered some of the disorders of the respiratory system that result from inhaling disease-causing organisms or noxious substances. We cannot help inhaling most potentially harmful substances. However, we can usually control whether we inhale cigarette smoke, a very harmful substance. In the next chapter, we will discuss some of the health effects of smoking cigarettes.

FIGURE **14.16**

A comparison of (a) normal alveoli and (b) alveoli in an individual with emphysema. Notice that, in emphysema, the alveolar walls rupture and there is a decrease in the surface area for gas exchange, an increase in the dead air space, and a thickening of the alveolar walls.

REVIEWING THE CONCEPTS

The Respiratory System Allows for the Exchange of Oxygen and Carbon Dioxide Across a Moist Body Surface (pp. 293–299)

1. The oxygen that we breathe is needed to maximize the number of energy-storing ATP molecules formed from food molecules. Exhaling carbon dioxide, a waste product formed by the same reactions, helps regulate the acid-base balance of body fluids. The role of the respiratory system is to exchange oxygen and carbon dioxide between the air and the blood.

2. During inhalation, the first structure that air usually passes through is the nose, which serves to clean, warm, and moisten the incoming air. Olfactory receptors, located in the nasal cavities, are responsible for our sense of smell. The sinuses are air-filled spaces in the facial bones that also help warm and moisten the air. After leaving the nose, the inhaled air passes through the pharynx, or throat, and then the larynx, or voice box. Reflex movements of the larynx prevent food from entering the airways and lungs. The larynx is the source of the voice. The air passageways include the trachea, which branches to form bronchi, and these branch extensively within the lung to form progressively smaller tubules called bronchioles. The bronchioles terminate at the gas-exchange surfaces in the lungs, the alveoli. The alveoli are thin-walled air sacs surrounded by a capillary network.

WEB TUTORIAL 14.1 Overview of Respiration

WEB TUTORIAL 14.2 Anatomy of the Lung and Alveoli

Pressure Changes within the Lungs Cause Breathing (pp. 299–300)

3. Pressure changes within the lungs caused by changes in the size of the thoracic cavity move air into and out of the lungs. Inspiration occurs when the size of the thoracic cavity increases, causing the pressure in the lungs to decrease below atmospheric pressure. Expiration occurs when the size of the thoracic cavity decreases and pressure in the lungs increases above atmospheric pressure.

Blood Transports Gases between the Lungs and the Cells (pp. 300–303)

4. Oxygen and carbon dioxide are exchanged between the alveolar air

and the capillary blood by diffusion along their concentration (partial pressure) gradients. Oxygen diffuses from the alveoli into the blood, where it binds to hemoglobin within the red blood cells and is delivered to the body cells. A small amount of carbon dioxide is carried to the lungs dissolved in the blood plasma or bound to hemoglobin. Most, however, is transported to the lungs as bicarbonate ions.

WEB TUTORIAL 14.3 Gas Exchange in the Lungs and Tissues

Breathing Is Controlled Primarily by Centers in the Brain (pp. 303–305)

5. The basic rhythm of breathing is controlled by the inspiratory area within the medulla of the brain. The neurons within this center undergo spontaneous bouts of activity. When they are active, messages are sent, causing contraction of the diaphragm and the muscles of the rib cage. As a result, the thoracic cavity increases in size and air is drawn into the lungs. When the inspiratory neurons are inactive, the diaphragm and rib cage muscles relax, and exhalation occurs passively. Also in the medulla is an expiratory area that causes forceful exhalation during heavy breathing.

6. The most powerful stimulant to breathing is an increasing level of carbon dioxide (acting primarily because of the increased number of hydrogen ions formed from carbonic acid). Extremely low levels of oxygen also increase breathing rate.

Respiratory Disorders Have Many Causes (pp. 305–310)

7. Two common respiratory diseases, the common cold and the flu, are caused by viruses. Severe acute respiratory syndrome (SARS) is caused by a coronavirus. SARS usually causes pneumonia and can be fatal. Strep throat, a sore throat caused by *Streptococcus* bacteria, can lead to rheumatic fever (and consequently to disease of the heart valves) or to kidney disease. Acute bronchitis is caused either by bacteria or by a virus. Chronic bronchitis is a persistent irritation of the bronchi. Emphysema is a breakdown of the alveolar walls and thus a reduction in the gas-exchange surfaces. Chronic bronchitis and emphysema are usually caused by smoking or by air pollution.

KEY TERMS

alveoli (sing., alveolus)   *p. 298*
asthma   *p. 297*
breathing center   *p. 303*
bronchi (sing., bronchus)   *p. 297*
bronchial tree   *p. 297*
bronchiole   *p. 294*
bronchitis   *p. 308*
carbaminohemoglobin   *p. 303*
carbonic anhydrase   *p. 303*
common cold   *p. 305*
diaphragm   *p. 299*

emphysema   *p. 308*
epiglottis   *p. 296*
Eustachian tube   *p. 296*
expiration   *p. 299*
expiratory reserve volume   *p. 300*
external intercostals   *p. 299*
glottis   *p. 296*
Heimlich maneuver   *p. 296*
influenza   *p. 306*
inspiration   *p. 299*
inspiratory reserve volume   *p. 300*

laryngitis   *p. 297*
larynx   *p. 296*
nasal cavity   *p. 294*
nasal septum   *p. 294*
oxyhemoglobin   *p. 301*
pharynx   *p. 296*
residual volume   *p. 300*
respiratory distress syndrome (RDS)
  p. 299
respiratory system   *p. 293*
sinus   *p. 295*

sinusitis   *p. 296*
strep throat   *p. 308*
surfactant   *p. 298*
tidal volume   *p. 300*
total lung capacity   *p. 300*
trachea   *p. 297*
vital capacity   *p. 300*
vocal cords   *p. 296*

THINKING ABOUT THE CONCEPTS

1. Why must we breathe oxygen? *p. 293*
2. Trace the path of air from the nose to the cells that use the oxygen. *pp. 294–299*
3. How are most particles and disease-causing organisms removed from the inhaled air before it reaches the lungs? *pp. 294–295*
4. Describe the reflex that normally prevents food from entering the lower respiratory system. *p. 296*
5. How is human speech produced? *pp. 296–297*
6. What is the function of the cartilage rings in the trachea? *p. 297*
7. What is the bronchial tree? *p. 297*
8. How are the pressure changes in the thoracic cavity that are responsible for breathing created? *p. 299*
9. Is tidal volume or the vital capacity a larger volume of air? Explain. *p. 300*
10. How is most oxygen transported to the body cells? *pp. 300–303*
11. How is most carbon dioxide transported from the cells to the lungs? *p. 303*
12. What region of the brain causes the basic breathing rhythm? How does the activity of the brain region differ during quiet breathing and heavy breathing? *pp. 303–304*
13. Explain how blood carbon dioxide levels regulate the breathing rate. *p. 305*
14. What are the causes of the shortness of breath experienced by people with emphysema? *pp. 308–309*
15. Choose the *correct* statement.
    a. During quiet breathing, expiration does not usually involve the contraction of muscles.
    b. Expiration occurs when the diaphragm and the rib muscles contract.
    c. Expiration occurs as the chest (thoracic) cavity enlarges.
    d. The larynx acts like a suction pump to pull air into the lungs.

16. You should be able to hold your breath longer than normal after you hyperventilate (breathe rapidly for a while) because hyperventilating
    a. decreases your blood oxygen levels.
    b. decreases blood carbon dioxide levels.
    c. increases blood oxygen levels.
    d. increases blood carbon dioxide levels.
17. The structure specialized to produce the sound of your voice is the
    a. trachea.
    b. larynx.
    c. bronchiole.
    d. epiglottis.
18. In a healthy person who does not smoke cigarettes, most of the particles that are inhaled into the respiratory system
    a. are trapped in the mucus and moved by cilia to the pharynx (toward the digestive system).
    b. pass through the alveoli into the circulatory system, where they are engulfed by white blood cells.
    c. are caught on the vocal cords.
    d. are trapped in the sinuses.
19. In emphysema
    a. the number of alveoli is reduced.
    b. cartilage rings in the trachea break down.
    c. the diaphragm is paralyzed.
    d. the epiglottis becomes less mobile.
    e. the pharynx is constricted.
20. The _____ is the flap that covers the trachea to prevent food from entering during swallowing.
21. The enzyme in red blood cells that reversibly converts carbonic acid to bicarbonate ions and hydrogen ions is _____.

APPLYING THE CONCEPTS

1. Tatyana is a young woman with iron-deficiency anemia, so her blood does not carry enough oxygen. Would you expect this condition to affect her breathing rate or tidal volume? Why or why not?
2. Cigarette smoke destroys the cilia in the respiratory system. Explain why the loss of these cilia is a reason that cigarette smokers tend to lose more workdays because of illness than do nonsmokers.
3. Rosa has a 4-year-old son, Juan, who threatens to hold his breath until she gives him a candy bar. Should she be worried? Why?

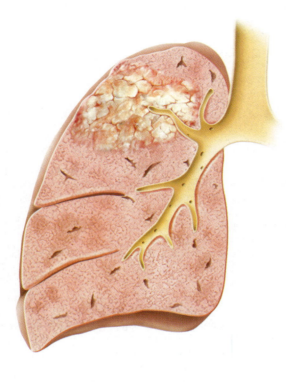

SPECIAL TOPIC

Smoking and Disease

Smoking Is the Leading Cause of Death in the United States

Cigarette Smoke Contains Poisons and Cancer-Causing Substances

Smoking Causes Several Deadly Diseases
- Smoking causes lung disease
- Smoking causes cancer
- Smoking causes heart disease
- Smoking causes other health problems

Smoking Causes Additional Health Risks to Women

Passive Smoking Causes Many Serious Health Problems

No Cigarette Is Safe

The Health Benefits of Quitting Smoking Are Numerous

SURGEON GENERAL'S WARNING: Smoking Causes Lung Cancer, Heart Disease, Emphysema, And May Complicate Pregnancy.

UNDERAGE SALE PROHIBITED

SURGEON GENERAL'S WARNING: Smoking By Pregnant Women May Result in Fetal Injury, Premature Birth, And Low Birth Weight.

Helvi was both hopeful and upset. She had spent the past 2 hours surfing the Web to learn more about emphysema, because she was anxious about her favorite aunt, Aunt Ilta. Helvi received an e-mail earlier in the day from her mother telling her of Ilta's trip to the doctor, the tests, the prognosis, and the new lifestyle that Ilta was going to have to embrace. Helvi couldn't help but feel angry and disappointed that Ilta had not quit smoking 30 years ago when Helvi's mother did. She recalled all of the coughing episodes, shortness of breath, and Ilta's poor health in general for the past several years. Yet Ilta continued to smoke despite Helvi's and other family members' pleas to stop. Now, Ilta had both emphysema and chronic bronchitis. During her Web search of authoritative sites, Helvi learned that one couldn't state with 100% accuracy that Ilta's lung diseases arose from smoking, but she learned that 80% to 90% of patients with lung disease were smokers. Her Web search also yielded some hope that there are treatments to alleviate the symptoms, permitting some short-term relief, but the fact remains that there is no cure. She gathered her strength for the painful call to Aunt Ilta.

In this chapter, we will consider the reasons why cigarette smoking leads to lung disease, heart disease, and cancer. We will see that no cigarette is safe, but much of the harm done to a smoker's body can be repaired after the person quits smoking. ◼

Smoking Is the Leading Cause of Death in the United States

Smoking is the greatest single preventable cause of disease, disability, and death in our society (Figure 14a.1a). In fact, every cigarette pack and cigarette advertisement in the United States must bear a warning from the Surgeon General. Yet, tobacco, which causes bodily harm when used exactly as intended, is legal to sell to anyone at least 18 years old.

Let's put it this way. Each cigarette a person smokes shortens the smoker's life by about 5 to 7 minutes, a little less than the time it takes to smoke the cigarette. Although some smokers' lives may be shortened by less time, on average, smokers shorten their lives by 10 years (Figure 14a.1b). Consequently, each year, more than 3 million lives around the world, more than 400,000 of them American, are snuffed out prematurely because of smoking. Nonetheless, in 2002 more than 46.5 million adults in the United States smoked regularly.

The economic loss because of lost workdays, premature death, and health-care costs caused by cigarette smoke is also staggering. This is one reason nonsmokers want smokers to quit. (Another reason is that sometimes they love them.) According to the 2004 report of the Surgeon General on the health consequences of smoking, smoking-related illnesses cost the United States $157.7 billion a year.

Cigarette Smoke Contains Poisons and Cancer-Causing Substances

The damage caused by cigarettes begins the instant the smoke touches a smoker's lips. Then, the smoke harms every living tissue it touches—mouth, tongue, throat, esophagus, air passageways, lungs, and stomach. When autopsied, even light smokers (those who smoke less than a pack a day) show lung damage. Substances in the smoke are metabolized (broken down) in the liver, but even the breakdown products can injure the bladder, pancreas, and kidneys. Smoking a pipe or cigar is somewhat safer than smoking cigarettes because the pipe or cigar smoke generally is not inhaled. However, the risk of cancer of the lips, mouth, and tongue of a pipe or cigar smoker is still greater than a nonsmoker's. A pipe or cigar smoker's risk of lung cancer or heart disease, although not as great as that of a cigarette smoker, is still above that of a nonsmoker. The effects of chewing smokeless tobacco are similar to those of smoking a pipe.

The average American smoker consumes a pack and a half of cigarettes each day—equal to about 300 puffs a day and over 109,000 puffs a year—year after year. With such exposure, then, we might wonder, what is in the smoke?

The answer is not simple because smoke contains at least 4000 substances. Scientists are still testing the adverse health effects of each of the components. So far, at least 50 have been shown to cause cancer. Also present are some well-known poisons: hydrogen cyanide (the poisonous gas used in gas chambers), carbon monoxide (common in auto exhaust), and cresols (chemicals similar to those used to preserve telephone poles). In addition, tobacco smoke contains significant amounts of radioactive substances (thorium-228, radium-226, and polonium-210). This exposure to radiation may account for as much as 40% of maximum permissible annual exposure to radiation.

The three most dangerous substances in smoke are nicotine, carbon monoxide, and tar. Nicotine is perhaps the most insidious of the three because it creates pleasurable feelings of relaxation. But nicotine is actually a drug that has many harmful effects on the body. More important, nicotine causes smokers to become "hooked" on cigarettes, ensuring continued exposure to the other injurious substances in the smoke. Depending on the brand, a cigarette contains 0.5 to 2.0 mg of nicotine. Each puff delivers about 0.2 mg of the drug to the bloodstream, which reaches the brain within 6 to 7 seconds—twice as fast as injected heroin!

Although nicotine causes a feeling of relaxation, it is actually a stimulant that affects the brain at all levels. When a cigarette delivers nicotine, the heart beats as many as 33 more beats per minute. And nicotine causes the blood vessels to constrict. So, even though the heart is beating faster, it is forcing blood through a less receptive circulatory system. The result is an increase in blood pressure. Recall from Chapter 11 that platelets are cell fragments in the blood that contain chemicals that initiate clotting. Nicotine causes the platelets to become sticky, increasing the likelihood of abnormal clots forming that may lead to heart attacks or strokes.

Nicotine is a powerfully addictive drug, and 95% of smokers are physiologically dependent on it and have withdrawal symptoms when they try to quit. Indeed, some opium addicts have reported

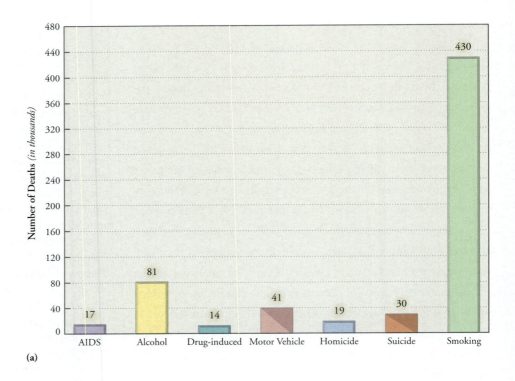

(a)

FIGURE **14a.1**

Cigarette smoking reduces life expectancy.
(a) Smoking causes more than 400,000 deaths each year in the United States, more people each year than the combined number of deaths due to AIDS, alcohol, drugs, car accidents, murder, and suicide. These data are from a report from the Centers for Disease Control and Prevention in 2000. The number of deaths from cigarette smoking have increased since then, but the relative importance of each cause of death remains similar. (b) A study published in 2004 followed nearly 35,000 male British doctors for 50 years. On average, the cigarette smoking cut 10 years off a smoker's life. A quarter of the lifetime smokers died before age 70.

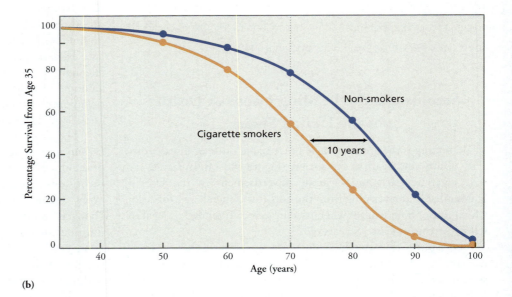

(b)

that it was easier to do without opiates than nicotine. Why is nicotine so addicting? Nerve cells become hyperactive when nicotine is removed. When a smoker tries to quit, this hyperactivity causes withdrawal symptoms: irritability, anxiety, headache, nausea, constipation or diarrhea, craving for tobacco, and insomnia. Most smokers continue to smoke to avoid these strong withdrawal symptoms. Most withdrawal symptoms begin to lessen after a week without nicotine, but some may continue for weeks or even months. Certain symptoms, such as drowsiness, difficulty concentrating, and craving for a cigarette, seem to worsen about 2 weeks after quitting. Consequently, many smokers return to their habit

and continue exposing themselves to the harmful substances in smoke, including carbon monoxide and tar.

The amount of carbon monoxide in cigarette smoke is 1600 ppm (parts per million), which greatly exceeds the 10 ppm considered dangerous in industry. Furthermore, carbon monoxide lingers in the bloodstream for up to 6 hours after smoking a cigarette. You may recall from Chapter 11 that carbon monoxide is a poison that prevents red blood cells from transporting oxygen. In fact, carbon monoxide reduces the oxygen-carrying capacity of the blood by about 12%, reducing oxygen delivery to every part of the body, including the brain and heart. The diminished oxygen supply to the brain can impair judgment, vision, and attentiveness to sounds. For these reasons, smoking can be hazardous for drivers. Furthermore, whereas nicotine makes the heart beat faster, carbon monoxide makes it more difficult to deliver the oxygen the heart needs in order to beat faster, thus straining the heart. In addition, the reduced oxygen supply to muscles can hinder athletic performance—explaining why most serious athletes do not smoke.

Tar is a collection of thousands of substances in the smoke that settles out as a brown sticky substance when the smoke cools within the body. A pack-a-day smoker coats his or her respiratory system with about 50 mg of tar each day. Besides the cancer-causing chemicals in tar, other chemicals destroy the elasticity of the lung.

Smoking Causes Several Deadly Diseases

Most of the health problems associated with cigarette smoking are caused by increases in the risk of three diseases, all of which can kill: lung disease, cancer, and heart disease (Figure 14a.2). We will now examine these three diseases. But keep in mind that the health risks of smoking are far greater than just these diseases.

▍Smoking causes lung disease

Because the objective of smoking is to bring smoke into the lungs, obviously we can look for some of its most damaging effects there. In fact, even teenage smokers show some damage to airways and lungs, but because teenagers are young and strong, the damage may be apparent only in breathing tests or during athletic activities. Young smokers are more likely to be short-winded than their nonsmoking friends. If they continue to smoke, the effects become much more apparent. By age 60, most smokers have significant changes in their airways and lungs.

The damage to the respiratory system of smokers is gradual and progressive. The injury begins as the smoke hampers the actions of two of the lungs' cleansing mechanisms—cilia and macrophages. Recall from Chapter 14 that the cells lining the airways are covered with hairlike cilia. Even the first few puffs from a cigarette slow down the movement of the cilia, making them less effective in sweeping debris from the air passageways. Smoking an entire cigarette prevents the cilia from moving for an hour or longer. With continued smoking, the nicotine and sulfur dioxide in the smoke paralyze the cilia, and the cyanide destroys the ciliated cells (Figure 14a.3).

Cigarette smoke causes a smoker's lungs to be chronically inflamed. For this reason, many macrophages, wandering cells that engulf foreign debris, enter the lungs in a vain attempt to clean the lung surfaces. But, just as smoke paralyzes the cilia, it also paralyzes the macrophages, further hampering the cleansing efforts. As the cilia and the macrophages become less effective, greater quantities of tar and disease-causing organisms remain within the respiratory

The cilia on healthy cells help cleanse the airways of debris.

(a)

Cigarette smoke destroys the cilia in airways.

(b)

FIGURE **14a.3**
Changes in the ciliated linings of the air passageways accompanying smoking. (a) The cilia on the cells lining the airways of a healthy nonsmoker cleanse the airways of debris. (b) Cigarette smoke first paralyzes and then destroys the cilia. As a result, hazardous materials can accumulate on the surfaces of the air passageways.

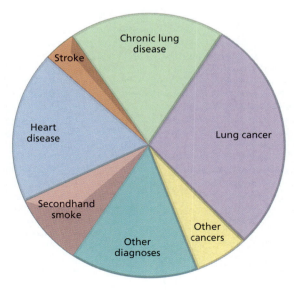

FIGURE **14a.2**
Smoking-related causes of death. Most of these deaths are due to lung disease, diseases of the heart or blood vessels, and cancer.

system. As a result, cigarette smokers are sick in bed and lose more workdays each year than nonsmokers do.

At the same time that the cilia and macrophages are being slowed, the smoke stimulates the mucus-secreting cells in the linings of the respiratory passageways. As a result, the smaller airways become plugged with mucus, making breathing more difficult. At this point, if not before, "smoker's cough" begins. Coughing is a protective reflex, and initially the smoker coughs simply because smoke irritates air passageways. However, as smoking continues and the cilia become increasingly less able to remove mucus and debris, the only way to remove the material from the passageways is to cough. The cough is generally worse in the morning as the body attempts to clear away the mucus that accumulated during the night.

Gradually, the inflammation and congestion within the lungs, along with the constant irritation from smoke, lead to chronic bronchitis, a disease characterized by a persistent deep cough that brings up mucus. The air passageways become narrow because of the thickening of their linings caused by repeated infection, the accumulation of mucus, and the contraction of the smooth muscle in their walls. Air flow becomes difficult, resulting in breathlessness and wheeziness. Bronchial infections now become more common because the air passageways are not cleared of disease-causing organisms. Bacterial infections may be treated with antibiotics, bringing slight temporary relief, but chronic bronchitis will continue as long as smoke irritates the airways.

Emphysema is often the next stage in the progressive damage to the lungs. Emphysema is a lung disease in which the walls of alveoli, which are the surfaces for gas exchange, are destroyed. The elasticity of the airways and alveolar walls is lost as tar causes body defense cells to secrete enzymes that break down the walls of alveoli and destroy the elasticity of lung tissues. The increase in pressure that accompanies a cough is no longer absorbed by flexible walls of airways and alveoli. The pressure is, therefore, directed at the delicate alveolar walls, which, like soap bubbles, break.

With more and more alveoli destroyed, the surface area for gas exchange is reduced, so less oxygen is delivered to the body. Furthermore, as the alveolar walls break down, the alveoli become larger, allowing more air to be trapped in the lungs. It becomes gradually more difficult to exhale. As a result, cyanide, formaldehyde, and carcinogens from the smoke remain in the lungs, killing even more cells.

As the alveoli are damaged, the small blood vessels to and from the alveoli rupture. If the blood vessels sustain enough damage, stress on the right ventricle of the heart increases because it must push the same amount of blood to lungs through fewer vessels.

Smoking causes cancer

Here we consider the link between smoking and cancer (Figure 14a.4). Cancers can develop because the smoke contains about 50 carcinogens, or cancer-causing chemicals, that cause cells to lose control over cell division. Some of the carcinogens change the structure of the genetic material, DNA. Others, such as formaldehyde, cause enzyme changes that allow cells to become cancerous. Still other components of the smoke work as co-carcinogens, chemicals that enhance the action of other carcinogens or promote cell division and, therefore, tumor growth once the cancer has begun.

Smoking is the major single cause of lung cancer, and it causes other cancers as well (Table 14a.1). In fact, smoking is responsible for 30% of all cancer deaths. Sadly, cancer-causing chemicals in tobacco damage every tissue they touch.

FIGURE **14a.4**

Tobacco use and lung cancer. Lung cancer usually takes about 20 years to develop. Notice that the number of deaths from lung cancer increases and decreases with tobacco use with about a 20-year delay. The lung cancer death rate of females is lower than that of men because there are fewer female smokers than there are male smokers. SOURCE: http:www.cancer.org/downloads/PRO/cancer%20statistics%202 02004.ppt. Death rates: U.S. Mortality Public Use Tapes, 1960–2000, U.S. Mortality Volumes, 1930–1959, National Center for Health Statistics, Centers for Disease Control and Prevention, 2002. Cigarette consumption: U.S. Department of Agriculture, 1900–2000.

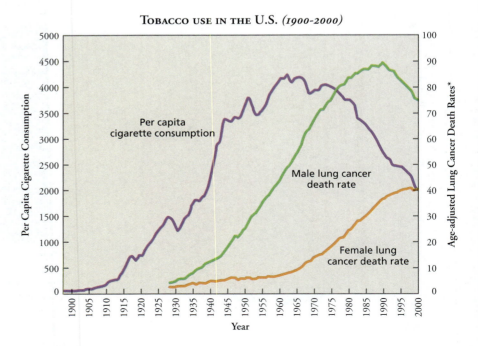

TOBACCO USE IN THE U.S. *(1900-2000)*

*Age-adjusted to 2000 U.S. standard population.

TABLE **14a.1**

TYPES OF INCREASED CANCER RISK DUE TO SMOKING	
TYPE	**INCREASED RISK**
Mouth and lips	4 times
Larynx	5 times in light smokers (less than 1 pack per day)
	20–30 times in heavy smokers (more than 1 pack per day)
Esophagus	2–9 times
Kidney and bladder	2–10 times
Pancreas	2–5 times (especially if alcohol is also consumed)

It is sobering to realize that between 85% and 90% of all cases of lung cancer are caused by smoking and are, therefore, preventable. Unfortunately, nearly 90% of individuals diagnosed as having lung cancer die within 5 years. Lung cancer usually has no symptoms until it is quite advanced. Therefore, it is not usually detected in time for a cure (Figure 14a.5).

Admittedly, not every smoker gets lung cancer, but smokers are 15 to 25 times more likely to get it than are lifetime nonsmokers. The likelihood of a smoker getting lung cancer depends on several factors, such as the number of cigarettes smoked, the number of years of smoking, the age at which smoking began, how deeply the smoke is inhaled, and the amount of tar and nicotine in the brand of cigarette smoked. There are also individual differences in genetic and biological makeup that influence cancer risk.

The progression to lung cancer is marked by changes in the cells of the airway linings of smokers (Figure 14a.6). In a non-smoker, the lining of the air passageways has a basement membrane underlying basal cells and a single layer of ciliated columnar cells. In a smoker, one of the first signs of damage is an increase in the number of layers of basal cells. Next, the ciliated columnar cells die and disappear. The nuclei of the basal cells then begin to change as

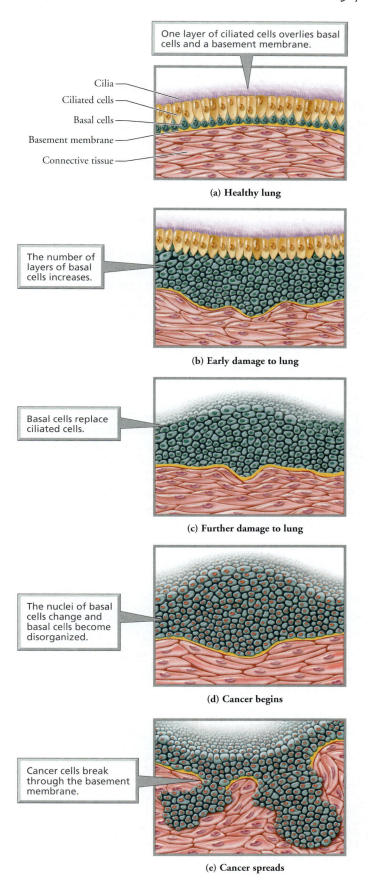

One layer of ciliated cells overlies basal cells and a basement membrane.

Cilia
Ciliated cells
Basal cells
Basement membrane
Connective tissue

(a) Healthy lung

The number of layers of basal cells increases.

(b) Early damage to lung

Basal cells replace ciliated cells.

(c) Further damage to lung

The nuclei of basal cells change and basal cells become disorganized.

(d) Cancer begins

Cancer cells break through the basement membrane.

(e) Cancer spreads

FIGURE **14a.6**

Lung cancer developing in the lining of a bronchus, an air passageway in the lung

FIGURE **14a.5**

Lung cancer. The tumor is the light-colored solid mass shown in the upper region of the lung.

mutations accumulate, and the cells become disorganized. This is the beginning of cancer. When cancer cells break through the basement membrane, they can spread to other parts of the lung and to the rest of the body, a process called metastasis (see Chapter 21a).

Smoking causes heart disease

When we think about the hazards of tobacco smoke, lung cancer generally leaps to mind. But the increased risk of cardiovascular (heart and blood vessel) disease is even more significant. Each year, cardiovascular disease kills many more people than does lung cancer, and smokers have a twofold to threefold increase in the risk of heart disease (Figure 14a.7). The American Heart Association estimates that about 25% of all fatal heart attacks are caused by cigarette smoke. This translates to roughly 200,000 heart attacks a year in the United States that could have been prevented by not smoking.

Smoking stresses both the heart and blood vessels in many ways. Some of the effects are immediate and direct. For instance, nicotine makes the heart beat faster at the same time that carbon monoxide reduces oxygen delivery to the heart. Nicotine also constricts blood vessels, raising blood pressure.

A less immediate but no less important way that smoking leads to cardiovascular disease is by increasing atherosclerosis, a condition in which lipid deposits, primarily composed of cholesterol, form in the walls of blood vessels, restricting the flow of blood (see Chapter 12). Smoking influences atherosclerosis in two ways. One is by decreasing the levels of protective cholesterol-transport particles, called HDLs, that carry cholesterol to the liver, perhaps even removing cholesterol from cells, so that it can be eliminated from the body. With fewer HDLs, more cholesterol begins to clog the arteries. A second way that smoking promotes cholesterol deposits is by raising blood pressure. The elevated blood pressure causes rapid, turbulent blood flow that damages the walls of the arteries, making

them more susceptible to cholesterol deposit. The cholesterol deposits cause inflammation, which leads to atherosclerosis. The narrowing of blood vessels caused by atherosclerosis results in starved tissue downstream and an increase in blood pressure. When these deposits form in the arteries that supply blood to the heart, as they often do, the blood supply to the heart may be reduced or shut down completely, causing heart cells to die. The death of heart cells is called a heart attack. A stroke is caused when a vessel to the head is blocked so that brain cells are damaged.

Smoking also increases the chances of forming blood clots in the vessels. The clots form for several reasons. Nicotine stimulates an increase in the number of platelets and makes them stickier. The levels of fibrinogen, a blood protein important in clotting, increase as well. Either condition may promote clotting. These clots may break loose from the site where they form and travel through the bloodstream until they lodge in a small vessel where they block the blood flow. Thus, these clots may result in a heart attack or stroke.

Smoking can also harm the heart itself. It can initiate coronary artery spasms, sudden violent contractions of the smooth muscle in the arteries supplying the heart, that may lead to heart attack.

Smoking causes other health problems

Smoking causes wrinkles. The reduction in the amount of oxygen reaching tissues impairs the body's ability to make collagen, a main supportive protein in certain connective tissues, including skin and bones. One result is wrinkles. The extent of wrinkling increases with cigarette consumption and the duration of the smoking habit. Heavy smokers are 3.5 times more likely than nonsmokers to show crow's feet, wrinkles around the eyes. In smokers, impaired collagen production also slows down the healing of wounds in bones and skin.

Smoking can dim your vision. In the United States, smoking is responsible for 20% of all cases of cataracts, a clouding of the lens of the eye that leads to blindness if not corrected surgically. Unfortunately, ex-smokers continue to have a higher risk of cataracts. So, unlike many of the other effects of smoking, quitting may not solve this problem.

Urinary incontinence (lack of urinary control) is more common among smokers. The hacking smoker's cough weakens the muscle that normally holds urine in the bladder. Nicotine also causes the muscles of the bladder to contract, allowing urine to leak.

Smoking Causes Additional Health Risks to Women

More than a quarter of women of child-bearing age smoke, and almost half of them have tried to quit. Unfortunately, the hazards of smoking may be greater for women than for men. For instance, some studies suggest that women smokers have a greater chance of getting lung cancer than do male smokers. Although the reasons for this difference are not clear, genes may play a role. Women are more than three times more likely than men to carry a certain genetic mutation, the *K-ras* mutation. This mutation helps the lung tumor to grow, particularly in response to the female hormone estrogen. Another gene that promotes lung cancer growth

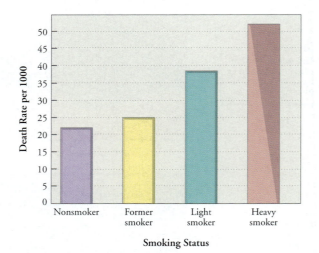

FIGURE **14a.7**

Death rate due to heart disease among nonsmokers and smokers. Notice that the death rate from heart disease increases with the number of cigarettes smoked per day. People who smoke over a pack per day have more than twice the risk of death due to heart disease than do people who have never smoked. In any case, a smoker who successfully quits is much less likely to die of heart disease than if the smoking habit continues.

(gastrin-releasing peptide receptor, or GRPR) is activated by nicotine. This gene is located on the X chromosome; because women have two X chromosomes and men have only one, women have twice as many copies of the gene than men do, making women more susceptible to the cancer-causing effects of smoking.

Furthermore, some risks are unique to women. Women who smoke are three times more likely than nonsmokers to develop cervical cancer. In addition, women who smoke reach menopause 2 to 3 years earlier than nonsmokers. Female smokers also have higher rates of osteoporosis—a condition more common in women than men, in which bones become less dense and, therefore, weaker.

Women of reproductive age may be particularly at risk because the combination of smoking and oral contraceptives can be deadly; both elevate blood pressure and increase the likelihood of abnormal clot formation. The death rate of women smokers who were also taking birth control pills is three times that of nonsmokers who were on the pill. The increased death rate is due to a much higher incidence of strokes, heart attacks, and blood clots in the legs.

Smoking causes many adverse effects on a woman's reproductive ability. A woman smoker who would like to become pregnant will have much better luck doing so if she quits smoking. Specifically, women who smoke more than a pack a day are half as fertile as women who do not smoke. Furthermore, if a woman smoker does become pregnant and continues to smoke, she is twice as likely to miscarry than a nonsmoker. Smoking during pregnancy increases the risk of complications such as premature separation of the placenta from the uterus, which shuts off the oxygen supply to the fetus, and a poorly positioned placenta that may partially or completely block the birth canal. (Incidentally, exposure to someone else's cigarette smoke can also affect a fetus if its nonsmoking mother is exposed to the smoke while she is pregnant.)

In a very real sense, tobacco smoke suffocates the growing fetus because nicotine constricts its lifeline, the blood vessels of the umbilical cord, and because carbon monoxide reduces the amount of oxygen carried by the blood. Smoking two packs a day blocks approximately 40% of the oxygen supply to the fetus. As a result, the fetus's heart rate and blood pressure increase, and the acid-base balance of its blood is shifted.

It is estimated that smoking leads to 5000 preventable fetal deaths annually that occur during the 8 weeks before or 7 days after birth. The incidence of stillbirth among pregnant women who smoke is 7.8% compared with 4.1% among nonsmoking women. Furthermore, a newborn of a woman who smoked during pregnancy has a one-third higher risk of dying soon after birth. Among the reasons for the increased infant mortality are premature birth and low birth weight. Newborns of women who smoke weigh an average of 200 grams (about $\frac{1}{2}$ lb) less than those born to nonsmoking women.

Passive Smoking Causes Many Serious Health Problems

It is not true that people who smoke are hurting only themselves. After all, their smoke pollutes the air of others who do not smoke. This dangerous pollution takes the form of exhaled smoke and the sidestream smoke that comes from the burning end of the cigarette. As a result, two-thirds of a cigarette's smoke actually enters the environment. And, because it is not filtered through tobacco, this smoke contains higher levels of many hazardous materials than the smoke that is inhaled through a cigarette. For example, the sidestream smoke contains more cadmium, which is related to high blood pressure, chronic bronchitis, and emphysema. It also contains twice the tar and nicotine of inhaled smoke and five times as much carbon monoxide. Because the concentration of cancer-causing nitrosamines in sidestream smoke is 50 times greater than in inhaled smoke, after an hour in a smoke-filled room, a nonsmoker may inhale an amount of nitrosamines equal to smoking 15 filter cigarettes.

When people are smoking during a typical campus party, the number of particulates in the air of a room is 40 times greater than the U.S. standard for air quality. After 30 minutes in a smoky room, a nonsmoker's heartbeat and blood pressure begin to increase. The carbon monoxide level in the blood begins to rise and is enough to hamper one's ability to distinguish time intervals or to distinguish the relative brightness of two lights. (This impairment could be of critical significance if the lights to be distinguished are the headlights of oncoming cars.) After a single hour in a very smoky room, a nonsmoker's blood levels of carbon monoxide and nicotine match those of someone who smoked a cigarette. After leaving the room, a person's blood contains carbon monoxide for 3 to 4 hours.

Ironically, because we generally spend the most time with people we care about the most, the loved ones of smokers are often hurt the most. Exposure to smoke most often occurs at home. In other cases, secondhand smoke is encountered in the workplace.

We have seen that the major health risks of smokers are increased risk of lung disease, cancer, and heart disease. Long-term exposure to secondhand tobacco smoke jeopardizes the health of nonsmokers in the same ways.

No Cigarette Is Safe

There is no such thing as a safe cigarette. True, filters do reduce some risks by trapping some tar, nicotine, carbon monoxide, and other poisonous gases. As a result, smokers who smoke cigarettes with filters have a slightly lower risk of lung cancer than those who smoke cigarettes without filters. However, the filters do not trap everything, and these smokers still have a risk of developing lung cancer that is 6.5 times that of a nonsmoker.

Because the tars and nicotine cause many of the harmful effects of cigarette smoke, the risks associated with low-tar, low-nicotine cigarettes might be expected to be lower than those associated with other brands. However, this is not always the case. Because of the craving for nicotine, smoking habits often change so that the blood level of nicotine remains constant regardless of the type of cigarette smoked. For example, there is evidence that smokers who switch to low-tar, low-nicotine cigarettes inhale more deeply or puff more often. They may even smoke more cigarettes.

It is easier to smoke a cigarette with low tar and low nicotine. Such cigarettes have made it easier for a whole new group—young girls and women—to become hooked. Females are more sensitive than males to unpleasant side effects of nicotine, but now they can smoke because they tolerate the low levels of nicotine more easily than the levels in regular cigarettes.

The Health Benefits of Quitting Smoking Are Numerous

Three of four smokers say they want to quit, and 60% say they have tried. On any one serious try, chances are only one in five of successfully quitting. The chances of success go up to three in five with repeated attempts. Despite the difficulty of quitting, more than 30 million Americans have quit.

The health benefits of quitting are enormous, and the major payoff is a longer life for you, your friends, your colleagues, and your loved ones. A study published in 2004[1] followed almost 35,000 British male doctors for 50 years. On average, lifelong smokers in this study cut about 10 years off their lives. Those who stopped smoking at age 50 lost only 5 years. Smokers who quit by age 30 had the same average life expectancy as nonsmokers did.

Much of the damage caused by smoke is reversible once you quit. Blood pressure and heart rate begin to decrease toward normal levels. The risk of heart attack begins to drop a year after quitting. After 5 years, the risk of heart attack for someone who smoked a pack a day is the same as for someone who never smoked. The risk of lung cancer also drops, but it never falls as low as that for people who never smoked (Figure 14a.8).

If you are a smoker who would like to break the habit, here are some suggestions from the Centers for Disease Control and Prevention:

1. **Prepare.** Set a date to quit. Throw away all cigarettes and ashtrays in your home, car, and workplace. Do not allow other people to smoke inside your home. If this is not the first time you are trying to quit, review the actions that helped your efforts and those that did not help.

2. **Do not have even a puff of any kind of cigarette.** Low tar, low nicotine cigarettes do not help you quit.

3. **Enlist support and encouragement from your family, friends and health care provider.**

4. **Make a list of the reasons you want to quit.** Do you want to gain control of your life? Do you want better health? Do you want to protect your family and friends from the smoke from your cigarette? Is there a special person who makes you want a longer life?

5. **Learn new skills and behaviors.** Distract yourself from smoking by walking, talking, or staying busy with a task. Change your routine to avoid behaviors or places that you associate with smoking. Find ways to reduce your stress; you could try taking a hot bath, exercising, or reading a book.

6. **Recognize that quitting will be difficult.** You are addicted to nicotine. There are medications to help you quit. Most of these medications provide nicotine in a form other than smoking: gum, patch, inhaler, or nasal spray. Buprion SR is an antidepressant drug available by prescription that can help if you feel depressed after quitting.

7. **Be prepared to face difficult situations.** Try to avoid drinking alcohol because drinking lowers the likelihood of success. Try to avoid places where people will be smoking because the smoke will increase your cravings. Be prepared to gain weight. Many people gain weight when they quit smoking, but the gain is usually less than 10 pounds. Do not let weight gain discourage your attempt to quit smoking.

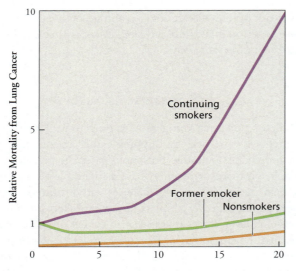

FIGURE **I4a.8**

A comparison of lung cancer deaths among lifetime nonsmokers, ex-smokers, and continuing smokers. Quitting dramatically reduces a smoker's risk of death due to lung cancer.

[1]Doll, R., R. Peto, et al. (2004). "Mortality in relation to smoking: 50 years' observations on male British doctors." *British Medical Journal* 328: 1519-153.

The Digestive System

The Digestive System Consists of a Long Tube That Runs through the Body along with Accessory Glands

The Digestive System Is Divided into Specialized Compartments for Food Processing

- The mouth begins mechanical digestion and the chemical digestion of starch
- The pharynx is an area shared by the digestive and respiratory systems
- The esophagus conducts food from the pharynx to the stomach
- The stomach stores and liquefies food and begins protein digestion
- The small intestine is the primary site of digestion and absorption
- The pancreas, liver, and gallbladder are accessory organs that aid the processes of digestion and absorption within the small intestine
- The large intestine absorbs water and other useful substances

Control of Digestive Activities Involves Nerves and Hormones

HEALTH ISSUE Heartburn and Peptic Ulcers—Those Burning Sensations

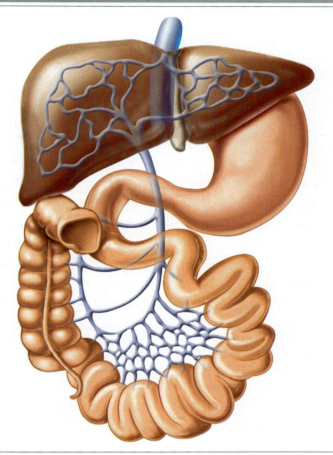

Balamani, an expert computer programmer, was stuck. The programming task itself was simple. The problem was that part of the animation seemed all too real. The pharmaceutical company had asked for an instructional animation that would show how the gastrointestinal (GI) tract worked, along with examples of specific malfunctions or diseases of the GI tract for which they had products. Balamani was amazed at how complex the system was, how the simple act of swallowing was like the orchestration of a three ring circus, how the system had to manage the flow of food and air down the same passage without getting food into the lungs, and then how food was processed. Incredible, he thought. Of immediate concern to him, however, was a malfunction along the GI tract described in the storyboards. It was called gastroesophageal reflux. The symptoms matched his own. He often experienced heartburn, occasional backing up of stomach content, when he could taste acidic fluid, and once in a while his stomach hurt. Antacids didn't seem to help. He imagined that the symptoms were due to the drastic changes in his diet since he had moved from India to the United States and that they would go away after a while. The storyboards, however, explained how "acid reflux" was caused by a problem at the base of the esophagus where it empties into the stomach. After a visit to his doctor, Balamani had all the information he needed to deal with his condition without taking one of the drugs promoted in his animation: elevate the head of his bed to allow gravity to work for his esophagus (keeping fluids down), cut back on fatty foods, cut out coffee and orange juice, reduce the size of his meals while increasing the frequency, and by all means pay fewer visits to fast food restaurants. Within a few weeks, Balamani was feeling much better, and about the same time he finalized the animation.

Although Balamani was having a problem with reflux, we will learn in this chapter that food usually travels in one direction through the digestive system. We will also see how the digestive system breaks down food molecules and makes them available for use as a source of energy or as raw material for the growth and repair of cells. Food is processed in specialized compartments along a long tube, the **gastrointestinal (GI) tract**. Processing of food involves digestion and absorption. Undigested materials are eliminated from the tube at the other end of the body. ■

The Digestive System Consists of a Long Tube That Runs through the Body along with Accessory Glands

There is some truth to the saying "You are what you eat." Rest assured, however, that no matter how many hamburgers you eat, you will never become one. Instead, the hamburger becomes *you*. The transformation is possible largely because of the activities of the digestive system. Like an assembly line in reverse, the digestive system takes the food we eat and breaks the complex organic molecules into their chemical subunits. The subunits are molecules small enough to be absorbed into the bloodstream and delivered to body cells. Food molecules ultimately meet one of two fates; they may be used to provide energy for daily activities, or they may provide materials for growth and repair of the body. Imagine that you just ate a hamburger on a bun. The starch in the bun may fuel a jump for joy; the protein in the beef may be used to build muscle; and the fat may become sheaths that insulate nerve fibers. Food is needed to provide energy and to provide raw material for maintenance, repair, and growth. Without the digestive system, however, food would be useless to us because it could never reach the cells.

The **digestive system** consists of a long, hollow tube, called the gastrointestinal (GI) tract, into which various accessory glands release their secretions (Figure 15.1). The hollow area of the tube that food and fluids travel through is called the *lumen*. Along most of its length, the walls of the gastrointestinal tract have four basic layers, as shown in Figure 15.2.

- **Mucosa**. The innermost layer is the moist, mucus-secreting layer called the mucosa. The mucus helps lubricate the tube, allowing food to slide through easily. Mucus also helps protect the lining cells from rough materials in food and from digestive enzymes. In some regions of the digestive system, cells in the mucosa also secrete digestive enzymes. The mucosa in some digestive organs is highly folded, which increases the surface area for absorption.

- **Submucosa**. The next layer, the submucosa, consists of connective tissue containing blood vessels, lymph vessels and nerves. The blood supply maintains the cells of the digestive system and, in some regions, picks up and transports the products of digestion. The nerves are important in coordinating the contractions of the next layer.

- **Muscularis**. The next layer, the muscularis, is responsible for movement of materials along the GI tract and for mixing materials with digestive secretions. In most sections of the tube, the muscularis is a double layer of muscle. (The stomach, as you will read later, is an exception; it has three layers of muscle.) The muscles of the inner layer circle the tube, causing a constriction when the muscles contract. The muscles in the outer layer run lengthwise, causing shortening when they contract. The following analogy may help you understand the effects of muscle contraction. Imagine a partially inflated, long balloon. Squeezing the balloon in its center extends the ends. This extension is the equivalent of the contraction of circular muscles. Compressing the ends of the balloon causes its center to bulge outward. This compression is the equivalent of the longitudinal muscle contracting. The muscle layers churn the food until it is liquefied, mix the resulting liquid with enzymes, and propel the food along the digestive system in a process called *peristalsis*, which you will read about later in the chapter.

- **Serosa**. The GI tract is wrapped in a thin layer of connective tissue called the serosa. It secretes a fluid that reduces friction with contacting surfaces.

ORGANS

ACCESSORY STRUCTURES

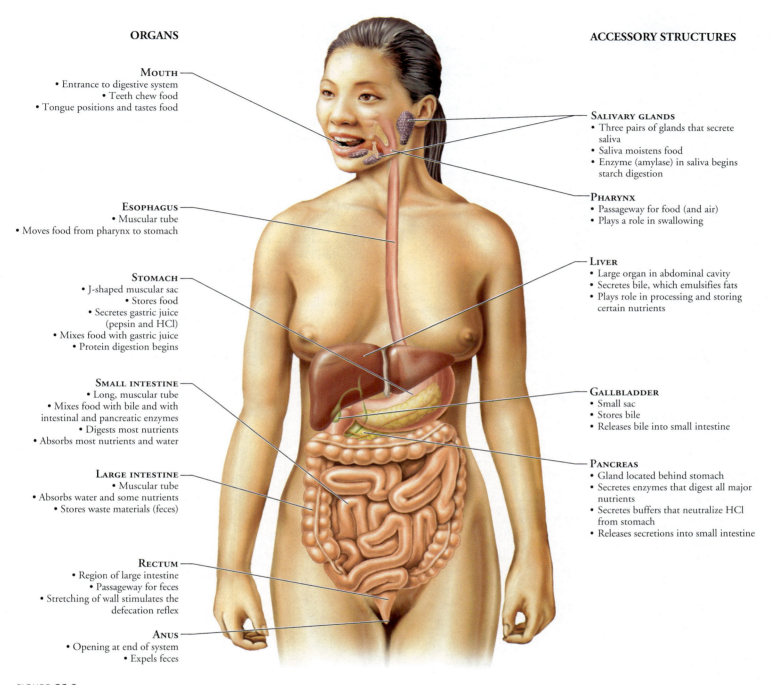

MOUTH
- Entrance to digestive system
- Teeth chew food
- Tongue positions and tastes food

ESOPHAGUS
- Muscular tube
- Moves food from pharynx to stomach

STOMACH
- J-shaped muscular sac
- Stores food
- Secretes gastric juice (pepsin and HCl)
- Mixes food with gastric juice
- Protein digestion begins

SMALL INTESTINE
- Long, muscular tube
- Mixes food with bile and with intestinal and pancreatic enzymes
- Digests most nutrients
- Absorbs most nutrients and water

LARGE INTESTINE
- Muscular tube
- Absorbs water and some nutrients
- Stores waste materials (feces)

RECTUM
- Region of large intestine
- Passageway for feces
- Stretching of wall stimulates the defecation reflex

ANUS
- Opening at end of system
- Expels feces

SALIVARY GLANDS
- Three pairs of glands that secrete saliva
- Saliva moistens food
- Enzyme (amylase) in saliva begins starch digestion

PHARYNX
- Passageway for food (and air)
- Plays a role in swallowing

LIVER
- Large organ in abdominal cavity
- Secretes bile, which emulsifies fats
- Plays role in processing and storing certain nutrients

GALLBLADDER
- Small sac
- Stores bile
- Releases bile into small intestine

PANCREAS
- Gland located behind stomach
- Secretes enzymes that digest all major nutrients
- Secretes buffers that neutralize HCl from stomach
- Releases secretions into small intestine

FIGURE **15.1**

The digestive system consists of a long tube, called the gastrointestinal tract, into which accessory glands release their secretions.

Regions of the GI tract are specialized to process food in particular ways. Part of that processing involves *mechanical digestion*, which is physically breaking food into smaller pieces, and part involves *chemical digestion*, which is breaking chemical bonds so that complex molecules are broken into smaller subunits. Chemical digestion produces molecules that can be absorbed into the bloodstream and used by the cells. We will trace the path food travels along the GI tract to see how it is processed and absorbed.

The Digestive System Is Divided into Specialized Compartments for Food Processing

A s food moves along the gastrointestinal tract, it passes through the mouth, pharynx, esophagus, stomach, small intestine, and large intestine. The salivary glands, liver, and pancreas

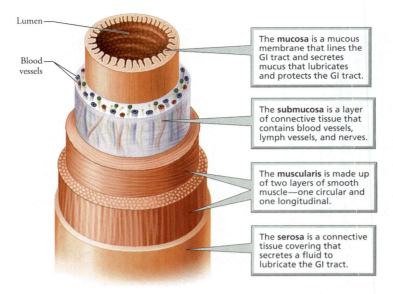

Lumen

Blood vessels

The **mucosa** is a mucous membrane that lines the GI tract and secretes mucus that lubricates and protects the GI tract.

The **submucosa** is a layer of connective tissue that contains blood vessels, lymph vessels, and nerves.

The **muscularis** is made up of two layers of smooth muscle—one circular and one longitudinal.

The **serosa** is a connective tissue covering that secretes a fluid to lubricate the GI tract.

FIGURE **15.2**

Along most of its length, the wall of the digestive system has four basic layers: the mucosa, the submucosa, the muscularis, and the serosa.

add secretions along the way. Most nutrients are absorbed from the small intestine. Additional water is absorbed in the large intestine. Undigested and indigestible materials pass out the anus. Table 15.1 identifies the structures of the digestive system and describes their roles in both mechanical and chemical digestion. You can refer to the table as you read about each of the digestive structures below.

The mouth begins mechanical digestion and the chemical digestion of starch

The entryway to the digestive tract and the first stop on food's journey through the digestive tract is the *mouth*, also called the oral cavity. The roof of the mouth is called the *palate*. The region of the palate closest to the front of the face, the hard palate, is reinforced with bone. Farther to the back of the mouth is the soft palate, which consists of only muscle and prevents food from entering the nose during swallowing. The mouth serves several functions: (1) Mechanical and, to some extent, chemical digestion begin; (2) food quality is monitored; and (3) food is moistened and manipulated so that it can be swallowed. Next, we will consider the roles that the teeth, salivary glands, and tongue play in performing these functions.

TABLE **15.1**

STRUCTURES OF THE DIGESTIVE SYSTEM			
STRUCTURE	DESCRIPTION/FUNCTIONS	MECHANICAL DIGESTION	CHEMICAL DIGESTION
Gastrointestinal tract			
Mouth	Receives food; contains teeth and tongue; tongue manipulates food and monitors quality	Teeth tear and crush food into smaller pieces	Digestion of carbohydrates begins
Pharynx	Area that both food and air pass through	None	None
Esophagus	Tube that transports food from mouth to stomach	None	None
Stomach	J-shaped muscular sac for food storage	Churning of stomach mixes food with gastric juice, creating liquid chyme	Protein digestion begins
Small intestine	Long tube where digestion is completed and nutrients are absorbed	Segmental contractions mix food with intestinal enzymes, pancreatic enzymes, and bile	Carbohydrate, protein, and fat digestion completed
Large intestine	Final tubular region of GI tract; absorbs water and ions; houses bacteria; forms and expels feces	None	Some digestion is carried out by bacteria
Anus	Terminal outlet of digestive tract	None	None
Accessory structures			
Salivary glands (sublingual, submandibular, parotid)	Secrete saliva, a liquid that moistens food and contains an enzyme (amylase) for digesting carbohydrates	None	Saliva contains enzymes that begin carbohydrate digestion
Pancreas	Digestive secretions include bicarbonate ions that neutralize acidic chyme and enzymes that digest carbohydrates, proteins, fats, and nucleic acids	None	Pancreatic enzymes assist in digestion of carbohydrates, proteins, fats, and nucleic acids
Liver	Digestive function is to produce bile, a liquid that emulsifies fats, making chemical digestion easier and facilitating absorption	Bile emulsifies fats	Bile facilitates digestion and absorption of fats
Gallbladder	Stores bile and releases it into small intestine	None	None

TEETH AND MECHANICAL DIGESTION

As we chew, our teeth prepare food for swallowing. The sharp, chisel-like incisors in the front of the mouth (see Figure 15.3a) slice the food as we bite into it. At the same time, the pointed canines to the sides of the incisors tear the food. Foods, such as fruits and vegetables, are then ground, crushed, and pulverized by the premolars and molars, which lie along the sides of the mouth. Thus, teeth mechanically break food into smaller fragments, making it easier to swallow.

Teeth are alive. In the center of each tooth is the pulp, which contains the tooth's life support systems—blood vessels that nourish the tooth and nerves that sense heat, cold, pressure, and pain (Figure 15.3b). Surrounding the pulp is a hard, bonelike substance, called dentin. The crown of the tooth (the part visible above the gum line) is covered with enamel, a nonliving material that is hardened with calcium salts. The root of the tooth (the part below the gum line) is covered with a calcified, yet living and sensitive connective tissue called cementum. The gums of young people adhere tightly to the enamel, but the gums recede with age, exposing a region of the cementum. A receding gum line, therefore, increases the sensitivity of the tooth to temperature, explaining why older people sometimes experience pain when eating food that is very hot or very cold. The roots of the teeth fit into sockets in the jawbone. Blood vessels and nerves reach the pulp through a tiny tunnel through the root called the root canal.

Tooth decay is caused by acid produced by bacteria living in the mouth. When you eat, food particles become trapped between the teeth and in the regions where the teeth meet the gums. Bacteria in the mouth are nourished by the sugar in these food particles. As bacteria break down the sugar, acid is produced. The acid dissolves the calcium and phosphate that make the enamel hard, causing a cavity to form. *Plaque*, an invisible film of bacteria, mucus, and food particles, promotes tooth decay because it holds acid produced by bacteria against the enamel (Figure 15.3c,d). Daily brushing and flossing helps remove plaque, reducing the chance of tooth decay.

After the enamel has been penetrated, bacteria can invade the softer dentin beneath. If a dentist does not fill the resulting cavity, the bacteria can infect the pulp (Figure 15.4).

Gum disease, which affects two of three middle-aged people in the United States, is a major cause of tooth loss in adults. *Gingivitis* (*gingiv*, the gums; *itis*, inflammation of), an early stage of gum disease, occurs when plaque that has formed along the gum line causes the gums to become inflamed and swollen. Although gingivitis is not painful at this point, the swollen gums can bleed, and they do not fit as tightly around the teeth as they once did. The resulting pocket that forms between the tooth and the gum traps additional plaque. The bacteria in the plaque can then destroy the bone and soft tissues around the tooth, causing inflammation of tissues around the tooth and of the gums; this condition is called *periodontitis* (*peri*, around; *dont*, teeth; *itis*, inflammation of). As the tooth's bony socket and the tissues that hold the tooth in place are eroded, the tooth becomes loose. Indeed, periodontitis is the most common cause of tooth loss in adults.

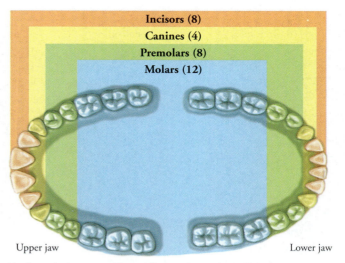

(a) The teeth slice, tear, and grind food until it can be swallowed.

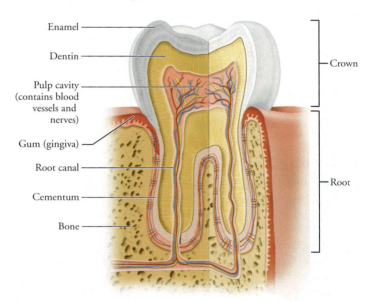

(b) The structure of the human tooth is suited for its function of breaking food into smaller pieces.

(c) Plaque is a layer of bacteria, mucus, and food debris that forms on the surface of teeth, shown here as a yellow coating over the grey enamel of the tooth.

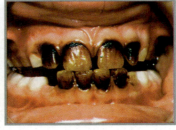

(d) The plaque on these teeth is revealed by a harmless dye. The plaque is the darkly colored material on the teeth, along the gum line, and between the teeth.

FIGURE **15.3**
Adult human teeth

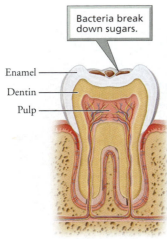

Enamel —
Dentin —
Pulp —

Bacteria break down sugars.

STEP 1
Tooth decay is caused by acid produced when bacteria living on the tooth surface break down sugars in food particles adhered to teeth.

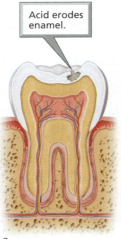

Acid erodes enamel.

STEP 2
The acid erodes the tooth's enamel, causing a cavity to form.

Blood vessels widen, pressing on nerves in the pulp.

STEP 3
The body responds by widening blood vessels in the pulp, to increase delivery of white blood cells to fight the infection. When widened blood vessels press on nerves within the pulp, a toothache results.

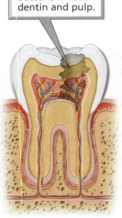

Bacteria invade dentin and pulp.

STEP 4
Bacteria can then infect the softer dentin beneath the enamel and later the pulp at the heart of the tooth.

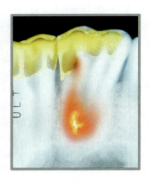

A dental cavity can be seen as a orange area in this x-ray of teeth.

FIGURE **15.4**
The development of a dental cavity

what would you do?

For the past 50 years, some communities in regions where the water is not naturally fluoridated have added low doses of fluoride to the public water supply to reduce tooth decay. Nearly 98% of Americans have, or have had, tooth decay. Over 6 million teeth are removed each year. Proponents of fluoridation point out that the incidence of tooth decay has declined after the addition of fluoride to drinking water. Opponents of the practice argue that (1) fluoridation of the water supply is a form of forced medication; (2) the dosage cannot be adequately controlled because it will vary with a person's body weight and the amount of water consumed; and (3) excessive fluoride intake causes teeth to become brown and mottled. If you had to vote on fluoridation of your public water supply, what additional information would you want to vote intelligently? How would you vote? Explain your reasons.

SALIVARY GLANDS AND CHEMICAL DIGESTION

Three pairs of salivary glands—the sublingual (below the tongue), submandibular (below the jaw), and parotid—release their secretions, which are collectively called **saliva**, into the mouth (Figure 15.5). The salivary glands produce as much as 1 liter (2 pints) of saliva each day. As we chew, food is mixed with saliva. Water in saliva moistens food, and mucus binds food particles together.

Saliva also contains an enzyme, called **salivary amylase**, that begins to chemically digest starches into shorter chains of sugars. You will notice the result of salivary amylase activity if you chew a piece of bread for several minutes: The bread will begin to taste sweet. Try it.

THE TONGUE: TASTE AND FOOD MANIPULATION

The *tongue* is a large skeletal muscle studded with taste buds. Our ability to control the position and movement of the tongue is critical to both speech and the manipulation of food within the mouth. Once food is dissolved in saliva, the chemicals can stimulate receptors in taste buds located primarily on the tongue. Information from the taste buds, along with input from the olfactory receptors in the nose, helps us to monitor the quality of food. For instance, spoiled or poisonous food usually tastes bad, so we can spit it out before

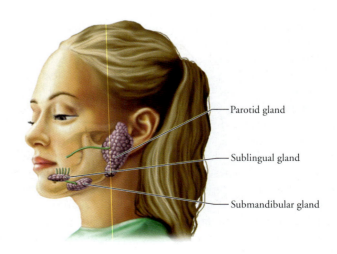

— Parotid gland

— Sublingual gland

— Submandibular gland

FIGURE **15.5**
Three pairs of salivary glands release their secretions into the mouth. These secretions, collectively called saliva, make food easier to swallow, dissolve substances so they can be tasted, and begin the chemical digestion of starch.

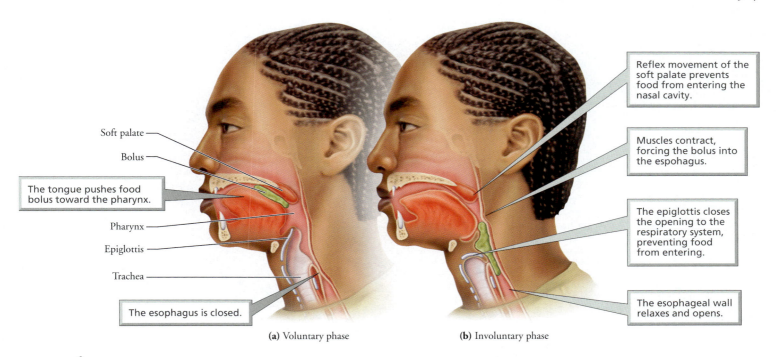

Soft palate

Bolus

The tongue pushes food bolus toward the pharynx.

Pharynx

Epiglottis

Trachea

The esophagus is closed.

Reflex movement of the soft palate prevents food from entering the nasal cavity.

Muscles contract, forcing the bolus into the espohagus.

The epiglottis closes the opening to the respiratory system, preventing food from entering.

The esophageal wall relaxes and opens.

(a) Voluntary phase   **(b)** Involuntary phase

FIGURE **15.6**
Swallowing consists of (a) voluntary and (b) involuntary phases.

swallowing. The tongue manipulates food so that it is crushed and ground by teeth, mixed with saliva, and shaped into a small, soft mass, called a *bolus*, that is easily swallowed. The tongue also initiates swallowing by pushing the bolus to the back of the mouth.

The pharynx is an area shared by the digestive and respiratory systems

When we swallow, food is pushed from the mouth, through the pharynx, and along the **esophagus**, the muscular tube that connects the pharynx to the stomach. The **pharynx** is the area shared by the respiratory and digestive systems and is commonly called the throat. Swallowing consists of a voluntary component followed by an involuntary one. When a person begins to swallow, the tongue pushes the bolus of softened and moistened food into the pharynx (Figure 15.6). Once food is in the pharynx, it is too late to change one's mind about swallowing. Sensory receptors in the wall of the pharynx detect the presence of food and stimulate the involuntary swallowing reflex. Reflex movements of the soft palate prevent food from entering the nasal cavity or respiratory pathways. Muscle contraction pushes the larynx (the voice box, commonly called the Adam's apple) upward. The movement of the larynx causes a cartilaginous flap called the *epiglottis* to move, covering the opening to the airways of the respiratory system (the glottis). The movement of the epiglottis prevents food from entering the airways. Instead, food is pushed into the esophagus.

The esophagus conducts food from the pharynx to the stomach

Recall that the esophagus is a muscular tube that conducts food from the pharynx to the stomach. Food is moved along both the

esophagus and the entire digestive tract by rhythmic waves of muscle contraction called **peristalsis** (Figure 15.7). Gravity is not essential in moving food along the digestive tract. You, therefore, can swallow while standing on your head or while floating in the weightless conditions of outer space.

The stomach stores and liquefies food and begins protein digestion

The **stomach** is a muscular sac that is well designed to carry out its three important functions: (1) storage of food, (2) liquefaction of food, and (3) the initial chemical digestion of proteins (Figure 15.8).

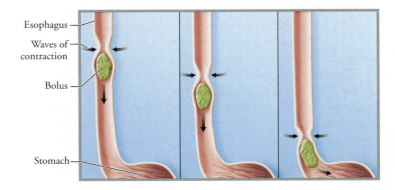

Esophagus

Waves of contraction

Bolus

Stomach

FIGURE **15.7**
Peristalsis is a wave of muscle contraction that pushes food along the esophagus and the entire GI tract. When circular muscles contract, the tube is narrowed and food is pushed forward.

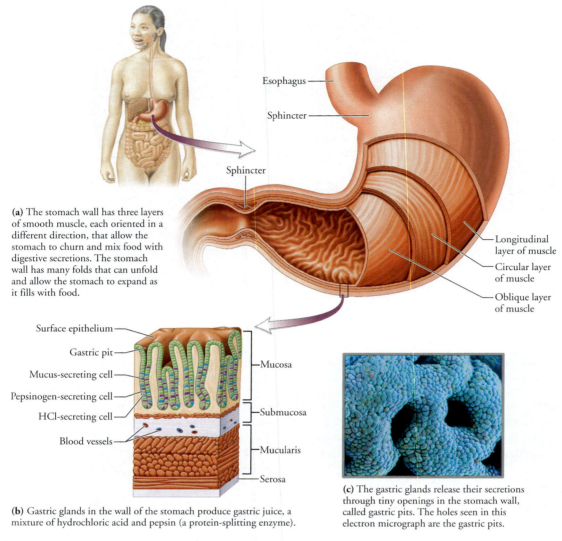

(a) The stomach wall has three layers of smooth muscle, each oriented in a different direction, that allow the stomach to churn and mix food with digestive secretions. The stomach wall has many folds that can unfold and allow the stomach to expand as it fills with food.

Esophagus

Sphincter

Sphincter

Longitudinal layer of muscle

Circular layer of muscle

Oblique layer of muscle

Surface epithelium

Gastric pit

Mucus-secreting cell

Pepsinogen-secreting cell

HCl-secreting cell

Blood vessels

Mucosa

Submucosa

Mucularis

Serosa

(b) Gastric glands in the wall of the stomach produce gastric juice, a mixture of hydrochloric acid and pepsin (a protein-splitting enzyme).

(c) The gastric glands release their secretions through tiny openings in the stomach wall, called gastric pits. The holes seen in this electron micrograph are the gastric pits.

FIGURE **15.8**
The structure of the stomach is well suited to its functions of churning food and digestive secretions, storing food, and beginning protein digestion.

STORAGE OF FOOD

Like any good storage compartment, the stomach is expandable and has adjustable openings that can close to seal the contents within or can open to fill or empty the compartment. When empty, the stomach is a small J-shaped sac, about the size of a sausage, that can hold only about 50 ml (a quarter of a cup). The wall of the empty stomach has folds that can unfold, allowing the stomach to expand as it fills. The smooth muscle of the stomach wall can stretch to accommodate large meals. When fully expanded, the stomach can hold several liters of food. Stretch receptors in the stomach wall signal the appetite center in the hypothalamus of the brain when the stomach is full. The appetite center then helps to quell the urge to eat.

Bands of circular muscle called sphincters guard the opening at each end of the stomach. Contraction of a sphincter closes an opening, and relaxation of a sphincter allows material to pass through.

LIQUEFACTION OF FOOD

Food is generally stored and processed within the stomach for 3 to 5 hours. Mechanical digestion occurs during this time as the food is churned and mixed with secretions produced by the glands of the stomach until it is a soupy mixture called **chyme**. The stomach wall has three layers of smooth muscle (shown in Figure 15.8), each oriented in a different direction. The coordinated contractions of these layers twist, knead, and compress the stomach contents, physically breaking food into smaller pieces.

CHEMICAL DIGESTION OF PROTEIN

In an adult, chemical digestion in the stomach is limited to the initial breakdown of proteins. The lining of the stomach has millions of gastric pits, within which are **gastric glands** that contain several types of secretory cells. Certain secretory cells produce hydrochloric acid (HCl), which kills most of the bacteria swallowed with food or drink.

Hydrochloric acid breaks down the connective tissue of meat and activates pepsinogen (the inactive form of **pepsin**, a protein-digesting enzyme). Other cells in the gastric glands secrete pepsinogen. While still within the gastric gland, HCl breaks off a portion of the pepsinogen molecule, forming pepsin. The mixture of pepsin and HCl, which is called *gastric juice*, is released into the stomach, where the pepsin begins the chemical digestion of protein in food. The gastric glands produce about 3 liters (6 pints) of gastric juice each day. Still other cells within the gastric glands secrete mucus, which helps protect the stomach wall from the action of gastric juice. Although not related to the digestive function of the stomach, a very important material secreted by the gastric glands is *intrinsic factor*, a protein necessary for the absorption of vitamin B_{12} from the small intestine.

The stomach wall is composed of the same materials that gastric juice attacks, so you might wonder why the stomach does not digest itself. One reason is mucus, as mentioned. Mucus forms a thick protective coat that prevents gastric juice from reaching the cells of the stomach wall. The alkalinity of mucus helps to neutralize the HCl. In addition, pepsin is produced in an inactive form that cannot digest the cells that produce it. Neural and hormonal reflexes regulate the production of gastric juice so that little is released unless food is present. Food absorbs and dilutes gastric juice. Finally, if the stomach lining is damaged, it is quickly repaired. Indeed, the high rate of cell division of the cells of the stomach lining replaces a half million cells every minute. As a result, you have a new stomach lining every 3 days!

Very little absorption of food materials occurs in the stomach because food simply has not been broken down into molecules small enough to be absorbed. Notable exceptions are alcohol and aspirin. The absorption of alcohol from the stomach is the reason its effects can be felt so quickly, especially if there is no food present to dilute it. The absorption of aspirin can cause bleeding of the wall of the stomach, which is the reason aspirin should be avoided by people who have stomach ulcers (discussed in the Health Issue essay, *Heartburn and Peptic Ulcers—Those Burning Sensations*).

When the stomach has adequately prepared the food, the contractions of peristalsis squirt chyme into the small intestine. Through neural and hormonal reflexes, several factors work together to ensure that the small intestine is not overloaded with more chyme than it can process. On average, the stomach is emptied within 4 hours of finishing a meal, but there can be considerable variation. Large meals leave the stomach more quickly than do small meals, and liquids such as soup leave more quickly than do solids. Foods rich in carbohydrates pass through the stomach the most quickly, proteins next, then fats.

The small intestine is the primary site of digestion and absorption

The next region of the digestive tract, the **small intestine**, has two major functions: chemical digestion and absorption. The small intestine is called "small" because its diameter is only about 2.5 cm (1 in.) compared with its length, about 6 m (20 ft). As food moves along this twisted tube, it passes through three specialized regions: the duodenum, the jejunum, and the ileum. The **duodenum**, first region of the small intestine, receives chyme from the stomach and digestive juices from the pancreas and liver. Most chemical digestion and absorption occur in the jejunum and the ileum.

The walls of the duodenum are protected from the acidic chyme in several ways. The acid is neutralized by bicarbonate ions in the secretions of the pancreas that empty into the region. In addition, the duodenum secretes mucus, which forms a protective blanket preventing the acid from reaching the cells. Cells that are killed by the acid despite these protective mechanisms are rapidly replaced.

CHEMICAL DIGESTION WITHIN THE SMALL INTESTINE

Within the small intestine, a battery of enzymes completes the chemical digestion of virtually all the carbohydrates, proteins, fats, and nucleic acids in food. Although both the small intestine and the pancreas contribute enzymes, most of the digestion that occurs in the small intestine is due to pancreatic enzymes (Table 15.2).

TABLE **15.2**

MAJOR DIGESTIVE ENZYMES			
SITE OF PRODUCTION	**ENZYME**	**SITE OF ACTION**	**SUBSTRATE**
Salivary glands	Salivary amylase	Mouth	Polysaccharides to shorter molecules
Stomach	Pepsin	Stomach	Proteins into protein fragments (polypeptides)
Pancreas	Trypsin	Small intestine	Proteins and polypeptides to smaller fragments
	Chymotrypsin	Small intestine	Proteins and polypeptides to smaller fragments
	Amylase	Small intestine	Polysaccharides to disaccharides
	Carboxypeptidase	Small intestine	Polypeptides to amino acids
	Lipase	Small intestine	Triglycerides (fats) to fatty acids and glycerol
	Nucleases (deoxyribonuclease and ribonuclease)	Small intestine	DNA or RNA to nucleotides
Small intestine	Maltase	Small intestine	Maltose to glucose units
	Sucrase	Small intestine	Sucrose to glucose and fructose
	Lactase	Small intestine	Lactose to glucose and galactose
	Aminopeptidase	Small intestine	Peptides to amino acids

Fats present a special digestive problem. Fats are insoluble in water, as you have observed when oil quickly separates from the vinegar in your salad dressing. Thus, tiny fat droplets tend to coalesce into large globules. A problem arises because lipase, the enzyme that chemically breaks down fats, is soluble in water and not in fats. As a result, lipase can work only at the surface of a fat globule. Large fat globules have less combined surface area than do smaller droplets.

Bile, a mixture of water, ions, cholesterol, bile pigments, and bile salts, plays an important role in the digestion of fats. Bile is produced by the liver, is stored in the gallbladder, and acts in the small intestine. Bile salts emulsify fats; that is, they keep fats separated into small droplets. This separation exposes a large combined surface area to lipase, facilitating fat digestion and absorption.

We will have more to say about the pancreas, liver, and gallbladder shortly.

STRUCTURE OF THE SMALL INTESTINE

Each day, as much as 10 liters (approximately 2.5 gallons) of food and liquid enters the small intestine, but only 0.5 to 1 liters (approximately 2 to 4 cups) reaches the large intestine. We see, then, that the small intestine is the primary site of absorption in the digestive system.

The small intestine is able to absorb materials so effectively because it is long and has a vast surface area due to several structural specializations (Figure 15.9). First, the entire lining of the small intestine is pleated, like an accordion, into circular folds. These circular folds increase the surface area for absorption and cause chyme to flow through the small intestine in a spiral pattern. The spiral flow helps mix the chyme with digestive enzymes and increases contact with the absorptive surfaces. Covering the entire lining surface are tiny projections, called **villi** (singular, villus). The villi give the lining a velvety appearance. Like the pile on a bath towel, the villi increase the absorptive surface. Indeed, these 1-mm projections increase the surface area of the small intestine tenfold. Also, thousands of microscopic projections, called *microvilli*, cover the surface of each villus, increasing the surface area of the small intestine another 20 times. The microvilli form a fuzzy border, known as the *brush border*, on the surface of absorptive epithelial cells. The circular folds, villi, and microvilli create a surface area of 300 to 600 m^2—greater than the size of a tennis court!

FIGURE **15.9**

The small intestine is specialized for the absorption of nutrients by structural modifications that increase its surface area. (a) Its wall contains accordion-like pleats called circular folds. (b) This electron micrograph shows the intestinal villi. (c) The lining has numerous fingerlike projections called villi. (d) The surface of each villus is bristled with thousands of microscopic projections of cell membranes called microvilli. In the center of each villus, a network of blood capillaries, which carries away absorbed products of protein and carbohydrate digestion as well as ions and water, surrounds a lacteal (a small vessel of the lymphatic system) that carries away the absorbed products of fat digestion. (e) An electron micrograph of the microvilli that cover each villus.

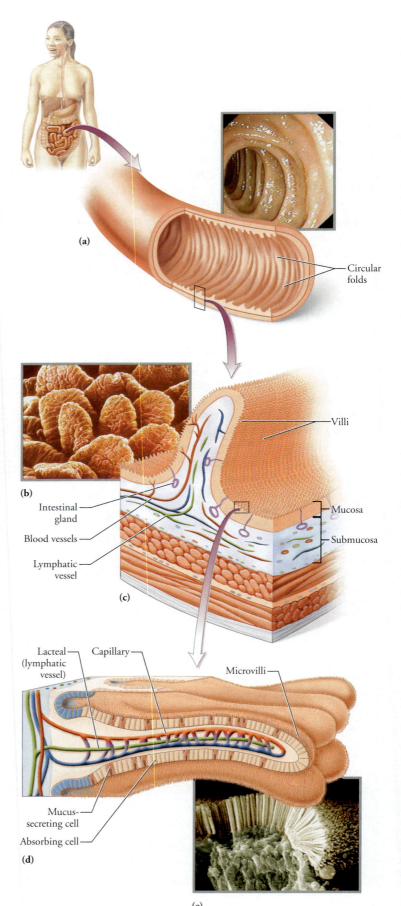

The core of each villus is penetrated by a network of capillaries and a lacteal, which is a lymphatic vessel. Thus, as substances are absorbed from the small intestine, they must cross only two cell layers: the epithelial cells of the villi and the wall of either the capillary or the lacteal. Most materials enter the epithelial cells by active transport, facilitated diffusion, or simple diffusion (Figure 15.10; see Chapter 3). Monosaccharides, amino acids, water, ions, vitamins, and minerals then diffuse across the capillary wall into the bloodstream and are delivered to body cells. The products of fat digestion quickly combine with bile salts, creating particles called micelles. When a micelle contacts an epithelial cell of a villus, the products of fat digestion easily diffuse into the cell. Within an epithelial cell, glycerol and fatty acids are reassembled into triglycerides, mixed with cholesterol and phospholipids, and coated with special proteins, creating a complex known as a **chylomicron**. The protein coating makes the fat soluble in water, allowing it to be transported throughout the body. The chylomicrons leave the epithelial cell by exocytosis (see Chapter 3). Chylomicrons are too large to pass

through capillary walls. However, they easily enter the more porous lacteal and enter the lymphatic system, which carries them to the bloodstream.

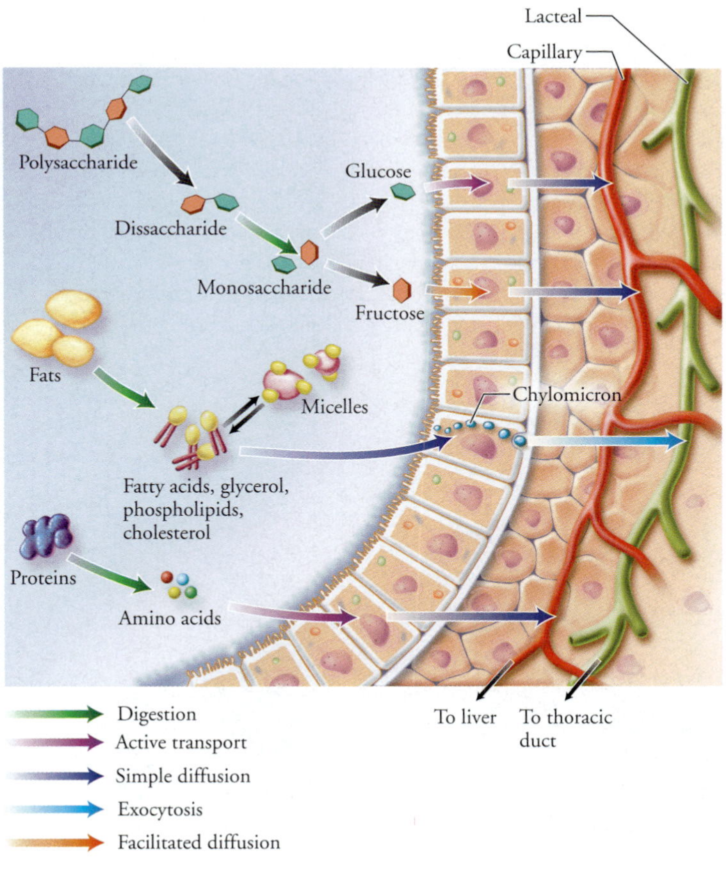

The pancreas, liver, and gallbladder are accessory organs that aid the processes of digestion and absorption within the small intestine

The pancreas, liver, and gallbladder (see Figure 15.1) are not part of the GI tract. However, they play vital roles in digestion by releasing their secretions into the small intestine.

THE PANCREAS

The **pancreas** is an accessory organ that lies behind the stomach, extending toward the left from the small intestine. In addition to enzymes, pancreatic juice contains water and ions, including bicarbonate ions that are important in neutralizing the acid in chyme. Neutralization is essential for optimal enzyme activity in the small intestine. Pancreatic juice drains into the pancreatic duct, which fuses with the common bile duct from the liver just before entering the duodenum of the small intestine (Figure 15.11).

Together, the pancreatic enzymes and intestinal enzymes break nutrients into their component building blocks: proteins to amino acids, carbohydrates to monosaccharides, and triglycerides (a type of lipid) to fatty acids and glycerol.

THE LIVER

The nutrient-laden blood from the capillaries in the villi travels through the hepatic portal vein to the **liver**, the largest internal organ in the body, which has a variety of metabolic and regulatory

FIGURE **15.10**

The small intestine is the primary site for chemical digestion and absorption. The digestive products, such as monosaccharides, amino acids, fatty acids, and glycerol, enter absorptive epithelial cells of villi by active transport, facilitated diffusion, or diffusion. Monosaccharides and amino acids, along with water, ions, and vitamins, then enter the capillaries within the villus and are carried to body cells by the bloodstream.

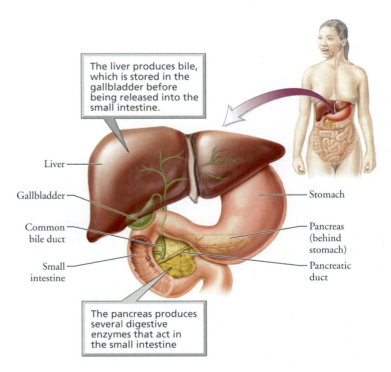

The liver produces bile, which is stored in the gallbladder before being released into the small intestine.

The pancreas produces several digestive enzymes that act in the small intestine

FIGURE **15.11**

The pancreas, liver, and gallbladder are accessory organs of the digestive system.

Heartburn and Peptic Ulcers—Those Burning Sensations

Heartburn is a burning sensation behind the breastbone that occurs when acidic gastric juice backs up into the esophagus. At least 30% of all Americans suffer from heartburn on a monthly basis, and 10% on a daily basis. Heartburn occurs when the pressure of the stomach contents overwhelms the sphincter at the lower end of the esophagus. In some people, this sphincter is weak and is ineffective in keeping stomach contents out of the esophagus. A person with a normal esophageal sphincter will experience heartburn when the stomach contents exert a greater pressure than usual against the sphincter, as might occur after a large meal, during pregnancy, when lying down, or when constipated. Acidic foods, such as tomatoes and citrus fruits, and spicy, fatty, or caffeine-containing products can aggravate the problem by causing the sphincter to relax and open. Tight clothing also worsens heartburn.

There are several treatments for heartburn. Antacids will provide temporary relief. Unfortunately, long-term use of certain antacids can have undesirable side effects. The fizzing bicarbonate types are high in sodium and should be avoided by persons with high blood pressure. Others are high in calcium, which might promote the formation of kidney stones and also stimulate the production of stomach acid. Newer medications, such as Tagamet®, Zantac®, Pepcid®, Axid®, Prilosec®, and Nexium®, decrease the production of stomach acid.

People who have chronic heartburn—at least one attack per week—have an increased risk of developing cancer of the esophagus. The risk increases with the frequency of attacks and the number of years of experiencing heartburn. Those who have one attack per week have a risk eight times higher than normal. Esophageal cancer is particularly aggressive and has become more common in recent years. It is detected using a test called endoscopy, in which a thin, lighted tube is snaked down the throat.

In nearly 13% of Americans, the mechanisms that protect the stomach and duodenum from their acidic contents fail at some point in their lives, allowing the acidic gastric juice to erode the lining of some region in the gastrointestinal tract. The resulting sore resembles a canker sore of the mouth and is called a peptic ulcer (Figure 15.A). Although a peptic ulcer may form in the esophagus or the stomach, the most common site is the duodenum of the small intestine. An ulcer is usually between 10 and 25 mm (0.33 and 1 in.) in diameter, and ulcers may occur singly or in multiple locations.

The symptoms of an ulcer are variable. A common symptom is abdominal pain, which can be quite severe. Vomiting, loss of appetite, bloating, indigestion, and heartburn are other common symptoms. However, some people with ulcers, especially those who are taking nonsteroidal anti-inflammatory drugs (NSAIDs), have no pain. Unfortunately, the degree of pain is a poor indicator of the severity of ulceration. Often, people with no symptoms of ulcers are unaware of the problem until serious complications develop. Gastric juice can erode the lining of the GI tract until it bleeds. In some cases, the ulcer can eat a hole completely through the gut wall (a perforated ulcer). Recurrent ulcers can cause scar tissue to form, which may narrow or block the lower end of the stomach or duodenum.

Although acidic gastric juice is the direct cause of peptic ulcers, factors that interfere with the mechanisms that normally protect the lining of the gastrointestinal tract from the acid are considered to be the real causes. NSAIDs, which include aspirin, ibuprofen, and naproxen, can cause ulcers because they slow the production of chemicals called prostaglandins, which normally help protect the lining of the GI tract from damage by acid.

The leading cause of peptic ulcers, however, is thought to be infection with the bacterium *Helicobacter pylori*. More than 80% of persons with ulcers in the stomach or duodenum are infected with *H. pylori*. These corkscrew-shaped bacteria live in the layer of mucus that protects the lining of the GI tract. Here, partially protected from gastric juice, the bacteria attract body defense cells, called macrophages and neutrophils, that cause inflammation leading to ulcer formation. Toxic chemicals produced by the bacteria also cause ulcers. *H. pylori* causes an infection that may last for years. It affects more than a billion people throughout the world and approximately 50% of the people in the United States who are over 60 years of age. For some reason, however, only about 10% to 15% of those who are infected actually develop peptic ulcers.

Besides ulcers, an *H. pylori* infection is a risk factor for stomach cancer. It is hoped that it will soon be possible to vaccinate children against this bacterium, thereby preventing both peptic ulcers and stomach cancer. ♟♙

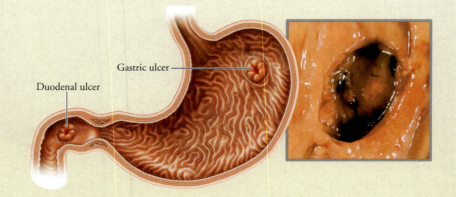

Gastric ulcer

Duodenal ulcer

FIGURE 15.A

A peptic ulcer is a raw area that forms when gastric juice erodes the lining of the esophagus, stomach, or, most commonly, the duodenum. The ulcer shown in this stomach wall is bleeding. The most common symptom is abdominal pain that occurs when the stomach is empty.

roles (Figure 15.12). We have already seen that its primary role in digestion is the production of bile. One of the liver's other roles is to control the glucose level of the blood, removing excess glucose and storing it as glycogen or breaking down glycogen to raise blood glucose levels. Thus, the liver keeps the glucose levels of the blood at a steady level. The liver also packages lipids with protein-carrier molecules to form lipoproteins that transport lipids in the blood. After the composition of blood is adjusted by the liver, the blood is returned to the general circulation through the hepatic veins.

In addition, the liver removes poisonous substances, including lead, mercury, and pesticides, from the blood and, in some cases, breaks them down into less harmful chemicals. Also, the liver converts the breakdown products of amino acids into urea, which can then be excreted by the kidney. These are but a few of the approximately 500 functions of the liver.

With so many vital functions, it should not be surprising that diseases of the liver can be serious and life threatening. *Cirrhosis* is a condition in which the liver becomes fatty and liver cells are eventually replaced by scar tissue. The liver gradually deteriorates. Since cirrhosis is often caused by prolonged, excessive alcohol use, it is discussed in Chapter 8a.

Hepatitis is inflammation of the liver. It is most commonly caused by one of six viruses, designated as A, B, C, D, E, and G. Hepatitis viruses attack and destroy liver cells. Healthy liver cells remove bilirubin, a yellowish pigment produced by the breakdown of red blood cells, from the bloodstream and use it to make bile. Liver cells injured by hepatitis viruses stop filtering bilirubin from the blood. The accumulating bilirubin is deposited in the skin and the whites of the eyes, giving them a yellowish tint. This condition, called jaundice, is characteristic of diseases that damage the liver, including hepatitis.

Although all the hepatitis viruses attack the liver, they may have different means of transmission and present different symptoms (Table 15.3). Hepatitis A, which is spread through sewage-contaminated food or water, and hepatitis B, which is spread primarily through sexual contact, have been known for decades. Vaccines exist to prevent these forms of hepatitis. A person who has been exposed to either hepatitis A or B viruses also can be given injections of antibodies that will prevent the likelihood of contracting the disease.

Currently, about 4 million people in the United States have hepatitis C, and most of them have no idea that they are infected. For years, the disease has spread silently because it has no outward warning signs or very mild symptoms—vague fatigue or flu-like muscle and joint pain.

Hepatitis C is spread primarily through contaminated blood. Hundreds of thousands of intravenous drug users have been infected by sharing contaminated needles. Hepatitis C can also be spread through contaminated needles used in body piercing or tattooing.

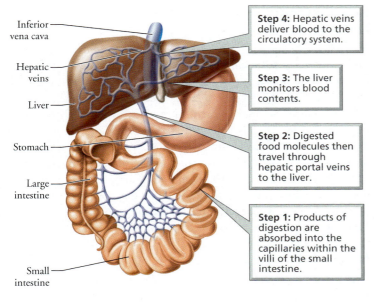

FIGURE **15.12**

The hepatic portal system is the circulation in which blood flows from the capillaries of the villi through the hepatic portal vein to the liver, which monitors blood content and processes nutrients before they are delivered to the bloodstream. A portal system transports blood from one capillary bed to another.

TABLE **15.3**

FORMS OF VIRAL HEPATITIS		
HEPATITIS VIRUS	**MEANS OF TRANSMISSION**	**SYMPTOMS**
A	Sewage-contaminated water or food	Acute infection: fatigue, jaundice, fever, abdominal pain, loss of appetite
		Recovery within several months
B	Sexual contact, contact with contaminated blood, mother to newborn at birth	Acute infection: fatigue, jaundice, fever, abdominal pain, loss of appetite, joint pain
		Chronic, with possible liver failure and death
		Can lead to liver cancer
C	Contaminated blood or body fluids	Acute infection: fatigue, jaundice, dark urine, abdominal pain, nausea
		Chronic, with possible liver failure and death
		Can lead to liver cancer
D	Sexual contact, contact with contaminated blood, mother to newborn at birth	Causes symptoms only in persons already infected with hepatitis B
		Infection with B and D causes progressive and severe liver disease; can be acute or chronic
E	Sewage-contaminated food or water	Acute infection: fatigue and jaundice
		Potentially lethal in pregnant women
G	Contaminated blood? (appears in same populations as hepatitis C)	Recently described; little known

Before there was a way to test for the virus, many people became infected when they received transfusions with contaminated blood.

About 85% of the people with hepatitis C develop a chronic infection. This means that the virus remains in the liver and slowly attacks it. Most people with hepatitis C never develop liver disease. However, cirrhosis occurs in 10% to 20% of those infected, and liver cancer develops in 1% of them.

what would you do?

Egypt suffers one of the world's highest hepatitis C infection rates. About 15% to 25% of the population is infected. The high infection rate is the result of a government-sponsored campaign to vaccinate people against another disease, one that is caused by a parasite. The same needle was used to vaccinate many people. Treatment for hepatitis C is expensive and sometimes has debilitating side effects. The government does not have enough money to pay for treatment for all those infected. Can you think of sources of funds to treat hepatitis C in Egypt? Should this be considered a global concern? What can you do to help?

THE GALLBLADDER

After its production by the liver, bile is stored in a muscular pear-sized sac, called the **gallbladder**. During storage, bile is modified and concentrated. When chyme is present in the small intestine, a hormone causes the gallbladder to contract, squirting its bile through the common bile duct into the duodenum of the small intestine.

Bile is rich in cholesterol, and, sometimes, if the balance of dissolved substances in bile becomes upset, a tiny particle precipitates out of solution. As cholesterol and other substances build up around the particle, a *gallstone* forms (Figure 15.13). Many people form more than one gallstone.

About 25 million people in the United States have gallstones, and roughly a million new cases develop annually. One-third to one-half of these people do not experience symptoms. A problem can develop, however, if the gallstones block the flow of bile from the gallbladder. If a gallstone prevents the gallbladder from emptying after a meal, as it normally would, the pressure within the gallbladder builds, causing intense pain in the right side or center of the upper abdomen. If a gallstone becomes lodged in the common bile duct, bile may build up behind the stone and cause jaundice.

Although there are some nonsurgical treatments for gallstones, these are not always successful. For instance, there are medicines that can dissolve cholesterol gallstones. The stones do not always dissolve completely, however. Even when the stones do completely dissolve, they form again within 5 years in half of the patients. Lithotripsy, which is treatment with shock waves delivered through water, is most successful in patients who have only one stone. Unfortunately, it is unusual to form only one gallstone. For these reasons, surgical removal of the gallbladder remains the most common treatment. If the gallbladder is removed, bile drains directly from the liver into the small intestine. Thus, people who have had their gallbladder removed can still digest fat. However, fatty foods should still be avoided because there is a slow trickle of bile instead of a good store of it.

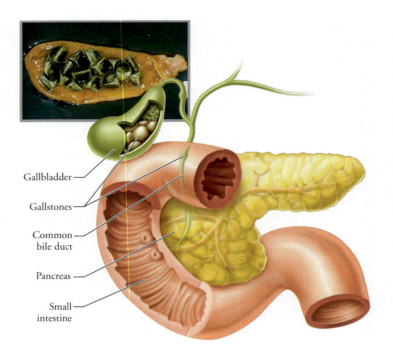

FIGURE **15.13**
Gallstones are composed primarily of cholesterol that has precipitated from bile during storage in the gallbladder. A gallstone can intermittently or continuously block the ducts that drain bile into the small intestine. The photograph shows the gallbladder and several gallstones.

Gallbladder
Gallstones
Common bile duct
Pancreas
Small intestine

The large intestine absorbs water and other useful substances

Now that we have discussed the accessory organs that assist digestion and absorption in the small intestine, we will continue to follow the movement of material through the gastrointestinal tract. Materials that have not been absorbed in the small intestine move into the final segment of the gastrointestinal tract, the **large intestine**. The principal functions of the large intestine are to absorb most of the water remaining in the indigestible food residue (thereby adjusting the consistency of the waste material called feces), to store the feces, and to eliminate them from the body. The large intestine is home to many types of bacteria, some of which produce vitamins that may be absorbed for use by the body.

stop and think

In a gastric bypass, the esophagus is detached from the stomach and reattached farther along the GI tract. Which part of the GI tract would you want to bypass to decrease food absorption?

The diameter of the large intestine is more than two and a half times that of the small intestine, but it is shorter. It measures 1.5 m (about 5 ft) in length.

The large intestine has four regions: the cecum, colon, rectum, and anal canal, as shown in Figure 15.14. The cecum is a pouch that hangs below the junction of the small and large intestines. Extending from the cecum is another slender, wormlike pouch, called the **appendix**. The appendix has no digestive function. (Some scientists believe the appendix plays a role in the immune system, which protects the body against disease.)

Each year, about 1 of 500 people develops *appendicitis*, inflammation of the appendix. Appendicitis is usually caused by an infection that develops in the appendix after it becomes blocked by a piece of hardened stool, food, or a tumor. At first, appendicitis is usually experienced as vague bloating, indigestion, and a mild pain in the region of the navel (bellybutton). As the condition worsens, the pain becomes more severe and is localized in the region of the appendix, the lower right abdomen. The pain is typically accompanied by fever, nausea, and vomiting. Diagnosis of appendicitis is often difficult because symptoms can be variable, and other conditions, such as kidney stones, urinary infections, and, in women, ovarian cysts, have similar symptoms. The difficulty in diagnosis is unfortunate because an untreated infection in the appendix usually causes the appendix to rupture, allowing its contents to spill into the abdominal cavity. Spillage from the appendix usually leads to infection and inflammation throughout the abdomen (a condition called peritonitis), which is potentially fatal.

The largest region of the large intestine, the **colon**, is composed of the ascending colon on the right side of the abdomen, the transverse colon across the top of the abdominal cavity, and the descending colon on the left side (see Figure 15.14). Although much of the water that was originally in chyme was absorbed in the small intestine, the material entering the colon is still quite liquid. As it passes through the

colon, 90% of the remaining water and sodium and potassium ions are absorbed. The remaining material, known as feces, consists primarily of undigested food, sloughed-off epithelial cells, water, and millions of bacteria. The brown color of feces comes from bile pigments.

The bacteria, which account for nearly a third of the dry weight of feces, are not normally disease causing and, in fact, are beneficial. Intestinal bacteria produce several vitamins that we are unable to produce on our own, including vitamin K and some of the B vitamins. Some of the vitamins are then absorbed from the colon for our own use. Indeed, there are roughly 50 species of bacteria, including the well-known *Escherichia coli* (*E. coli,*) living in the healthy colon. The bacteria are nourished by undigested food and material that we are unable to digest, including certain components of plant cells. When the intestinal bacteria use the undigested food for their own nutrition, their metabolic processes liberate gas that sometimes has a foul odor. Although most of the gas produced is absorbed through the intestinal walls, the remaining gas can produce some embarrassing moments when released as flatus.

stop and think

Certain foods, beans for instance, are notorious for producing intestinal gas. Beans contain large amounts of certain short-chain carbohydrates that our bodies are unable to digest. Explain how the nutritional content of beans is related to flatulence.

People who are lactose intolerant lack the enzyme lactase, which is normally produced by the small intestine, where it also acts. Lactase breaks down lactose, the primary sugar in milk, into its component monosaccharides. Without lactase, then, lactose moves into the colon, where it provides a nutritional bonanza for the bacteria living there. As a result, when people who are lactose intolerant consume milk products, the bacteria in the colon ferment the lactose and produce the gases carbon dioxide and methane that, in turn, produce bloating and abdominal discomfort. Although lactose intolerance is common in adults, it is not dangerous. Problems can usually be avoided by swallowing capsules or tablets of lactase or by modifying the diet to avoid dairy products.

Periodic peristaltic contractions move material through the large intestine. These contractions are slower than those in the small intestine. Consequently, it generally takes about 18 to 24 hours for material to pass through the colon but only 3 to 10 hours to pass through the small intestine. Eventually, the feces are pushed into the **rectum**, stretching the rectal wall and initiating the *defecation reflex*. Nerve impulses from the stretch receptors in the rectal wall travel to the spinal cord, which sends back motor impulses that stimulate muscles in the rectal wall to contract and propel the feces into the **anal canal**. Two rings of muscles, called sphincters, must relax to allow *defecation*, the expulsion of feces. The internal sphincter relaxes as part of the defecation reflex. The external sphincter is under voluntary control, allowing us to decide whether the situation is appropriate for defecation. If it is, conscious contraction of abdominal muscles can increase abdominal pressure and help expel the feces.

The water absorption that occurs in the colon adjusts the consistency of feces. When material passes through the colon too rap-

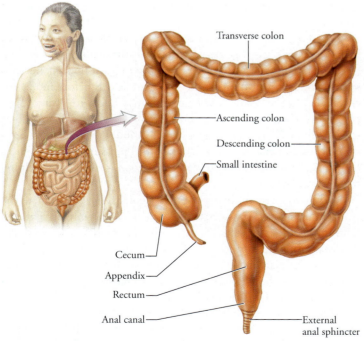

Transverse colon

Ascending colon

Descending colon

Small intestine

Cecum

Appendix

Rectum

Anal canal

External anal sphincter

FIGURE **15.14**

The large intestine consists of the cecum, colon, rectum, and anal canal. It absorbs water from undigested material, forming the feces, and houses bacteria.

idly, as might occur when colon contractions are stimulated by toxins produced by microorganisms or by excess food or alcoholic drink, too little water is absorbed. As a result, the feces are very liquid. This condition, which results in frequent loose stools, is called *diarrhea*. Diarrhea can be dangerous, especially in an infant or young child, because it can lead to dehydration. It is one of the major causes of death worldwide.

On the other hand, if material passes through the colon too slowly, too much water is absorbed, resulting in infrequent, hard stools, a condition called *constipation*. People who are constipated may find it necessary to strain during bowel movements. Straining increases pressure within veins in the rectum and anus, causing them to stretch and enlarge. The resulting varicose veins are hemorrhoids. The wall of the large intestine also experiences great pressure during a strained bowel movement. Then, like the inner tube of an old tire, the weaker spots in the intestinal wall can begin to bulge outward, forming small outpouchings called diverticula (singular, diverticulum) (Figure 15.15). Diverticula are very common in people older than age 50. When diverticula do not cause problems or symptoms, the condition is called diverticulosis. But if the diverticula become infected with bacteria and inflamed, the condition is then called diverticulitis, which can cause abrupt, cramping abdominal pain, a change in bowel habits, fever, and rectal bleeding.

stop and think

Stimulant laxatives stimulate peristalsis in the large intestine but do not affect the small intestine. Some people with eating disorders use laxatives to speed the movement of food through the digestive system, thinking that the calories will not be absorbed. After purging in this way, the person may weigh less on the bathroom scale. What accounts for the weight loss? Could laxative use help a person lose body fat?

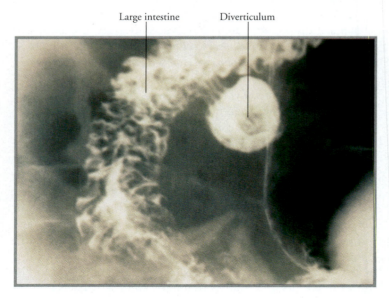

Large intestine   Diverticulum

FIGURE **15.15**

A diverticulum is a small pouch that forms in the wall of the large intestine that is usually caused by repeated straining during bowel movements. A high-fiber diet results in softer, bulkier stools that are easier to pass. Thus, fiber makes it less likely that diverticula will form.

Cancers of the colon or rectum are common and can be deadly. Colorectal cancer is the second leading cause of cancer deaths. Early detection and treatment cut the risk of death dramatically. Colorectal cancer begins with a polyp, a small noncancerous growth. Although polyps are most common in men and women over the age of 50, they can form in younger people. Most polyps stop growing, but a small percentage continue to grow. As they do, genetic mutations accumulate—mutations that can transform a cell into a cancerous tumor. Polyps often take 10 years to grow and turn cancerous, allowing plenty of time for detection. Polyps can bleed, so a sign of colorectal cancer is blood in the stool. This blood is not usually visible and must be detected with a diagnostic test. However, because the polyps may not bleed, screening methods that allow a direct view of the wall of the rectum and colon are more effective. A long, flexible fiber optic tube is threaded through the first third of the colon in a sigmoidoscopy or through the entire colon in a colonoscopy. If a polyp is detected, it can be removed and biopsied to determine whether it is cancerous (see Chapter 21a).

Control of Digestive Activities Involves Nerves and Hormones

As we have seen, food material moves along the gastrointestinal tract, stopping for specific treatment along the way. For digestion to occur, enzymes must be present in the right place at the right time. However, your body is composed of many of the same materials found in food, so it is best to release enzymes only when food is present. Both nerves and hormones play a role in orchestrating the release of digestive secretions, timing the release of each to the presence of food at each stop.

Food spends little time in the mouth, so, to be effective, saliva must be released quickly. Because nervous stimulation is faster than hormonal stimulation, it is not surprising to learn that the nervous system controls salivation. Some saliva is released before food even enters the mouth. Indeed, your mouth may begin to "water" simply at the *thought* of food—and certainly begins at the sight or smell of food. The major trigger for salivation, however, is the presence of food in the mouth—its flavor and pressure. Salivary juices continue to flow for some time after the food is swallowed, helping to rinse out the mouth.

While food is still being chewed, neural reflexes stimulate the stomach lining to begin secreting gastric juice and mucus. The distention of the stomach, along with the presence of partially digested proteins, stimulates cells in the stomach lining to release the hormone **gastrin**. Gastrin enters the bloodstream and circulates throughout the body and back to the stomach, where it increases the production of gastric juice.

The presence of acidic chyme in the small intestine is the most important stimulus for the release of enzymes from both the small intestine and the pancreas, as well as bile from the gallbladder. Local nerve reflexes that are triggered by acid chyme are the most important factor regulating motility and the release of intestinal secretions. Acid chyme also causes the small intestine to release several hormones that, in turn, are responsible for the release of digestive enzymes and bile. For instance, one hormone, **vasoactive intestinal peptide (VIP)**, is released from the small intestine into the bloodstream and is carried back to the small intestine, where it causes the release of intestinal juices. At the same time, the small intestine releases a second hormone, **secretin**, which stimulates the release of sodium bicarbonate from the pancreas into the small intestine, where it helps neutralize the acidity of chyme. A third hormone from the small intestine is **cholecystokinin**, which causes the pancreas to release its digestive enzymes and the gallbladder to contract and release bile. The neural and hormonal controls on the activities of the digestive system are summarized in Tables 15.4 and 15.5.

exploring further ...

In this chapter, we have seen how the digestive system breaks down food molecules to use as an energy source for metabolic activities or to provide building blocks for growth and repair of the body. In the next chapter, we will learn about your body's use of nutrients and consider ways in which the composition of your diet influences health.

TABLE **15.4**

NEURAL CONTROLS ON DIGESTIVE ACTIVITY	
STIMULUS	**EFFECT**
Sight of food, thought of food, presence of food in mouth	Release of saliva from salivary glands
Chewing food	Release of gastric juice (enzymes from stomach and HCl) and mucus from cells of stomach lining
Presence of acidic chyme in small intestine	Release of enzymes from small intestine into the small intestine; increased motility in small intestine

TABLE **15.5**

HORMONAL CONTROLS ON DIGESTIVE ACTIVITY				
HORMONE	**STIMULUS**	**ORIGIN**	**TARGET**	**EFFECTS**
Gastrin	Distention of stomach by food; presence of partially digested proteins in stomach	Stomach	Stomach	Release of gastric juice (enzymes from stomach and HCl)
Vasoactive intestinal peptide (VIP)	Presence of acidic chyme in small intestine	Small intestine	Small intestine	Release of enzymes from small intestine
Secretin	Presence of acidic chyme in small intestine	Small intestine	Pancreas	Release of sodium bicarbonate into small intestine to neutralize acidic chyme
Cholecystokinin	Arrival of chyme containing lipids	Small intestine	Pancreas	Release of enzymes from pancreas
			Gall bladder	Contraction of gallbladder and release of bile

REVIEWING THE CONCEPTS

The Digestive System Consists of a Long Tube That Runs through the Body along with Accessory Glands (pp. 322–323)

1. The digestive system consists of a long tube called the gastrointestinal tract (mouth, pharynx, esophagus, stomach, small intestine, and large intestine) and several accessory organs (salivary glands, pancreas, liver, and gallbladder).

The Digestive System Is Divided into Specialized Compartments for Food Processing (pp. 323–336)

2. The mouth serves several functions. Teeth tear and grind food, making it easier to swallow. The salivary glands produce salivary amylase, which is released into the mouth, where it begins the chemical break-

down of starches. Taste buds help monitor the quality of food. Finally, the tongue manipulates food so that it can be swallowed.

3. The pharynx is a space shared by the digestive and respiratory systems.

4. The esophagus, a tube that leads to the stomach, conducts food from the pharynx to the stomach. When we swallow, food is pushed from the mouth, and waves of muscle contraction called peristalsis push the food along the esophagus.

5. The stomach stores food, liquefies it by mixing it with gastric juice, begins the chemical digestion of proteins, and regulates the release of material into the small intestine. Gastric juice consists of hydrochloric acid (HCl) and pepsin, a protein-splitting enzyme. Pepsin is produced in an inactive form, called pepsinogen, that is activated by HCl.

6. The small intestine is the primary site of digestion and absorption. Enzymes produced by the small intestine and by the pancreas work here to chemically digest carbohydrates, proteins, and fats into their component subunits. Bile, which is produced by the liver and stored in the gallbladder, emulsifies fat (breaks it into tiny droplets) in the small intestine, thereby increasing the combined surface area of fat droplets. Bile's action makes fat digestion by water-soluble lipase faster and more complete.

7. The small intestine's surface area for absorption is increased by circular folds in its lining, fingerlike projections called villi, and microscopic projections covering the villi, called microvilli.

8. Products of digestion are absorbed into the epithelial cells of villi by active transport, facilitated diffusion, or simple diffusion. Most materials, including monosaccharides, amino acids, water, and ions, then enter the capillary blood network in the center of each villus. However, fatty acids and glycerol are resynthesized into triglycerides, combined with cholesterol, and covered in protein, forming droplets called chylomicrons. The chylomicrons enter a lymphatic vessel called a lacteal in the core of each villus. They are then delivered to the bloodstream by way of lymphatic vessels.

9. The pancreas, liver, and gallbladder are accessory organs that aid in digestion and absorption within the small intestine. The pancreas secretes enzymes to digest most nutrients. The liver produces bile, which is stored in the gallbladder. Bile emulsifies fats, making fat digestion easier.

10. The large intestine consists of the cecum, colon, rectum, and anal canal. The large intestine absorbs water, ions, and vitamins. It is home to millions of beneficial bacteria that live on undigested material that has passed from the small intestine. The bacteria produce several vitamins, some of which we then absorb for our own use.

11. Material left in the large intestine after passing through the colon is called feces. Feces consist of undigested or indigestible material, bacteria, sloughed-off cells, and water.

WEB TUTORIAL 15.1 Physical and Chemical Digestion

Control of Digestive Activities Involves Nerves and Hormones (pp. 336–337)

12. Neural and hormonal mechanisms regulate the release of digestive secretions. Neural reflexes trigger the release of saliva, initiate the secretion of some gastric juice, and are the most important factors regulating the release of intestinal secretions. When stimulated by the presence of protein-containing food in the stomach, cells of the stomach lining release the hormone gastrin into the bloodstream. Gastrin is the most important factor controlling the secretion of gastric juice. The small intestine releases several hormones: vasoactive intestinal peptide (VIP), which triggers the release of intestinal juices; secretin, which causes the release of sodium bicarbonate from the pancreas; and cholecystokinin, which causes the release of pancreatic digestive enzymes and the contraction of the gallbladder, leading to the release of bile.

KEY TERMS

anal canal *p. 335*	duodenum *p. 329*	pancreas *p. 331*	small intestine *p. 329*
appendix *p. 335*	esophagus *p. 327*	pepsin *p. 329*	stomach *p. 327*
bile *p. 330*	gallbladder *p. 334*	peristalsis *p. 327*	vasoactive intestinal peptide (VIP)
cholecystokinin *p. 337*	gastric gland *p. 328*	pharynx *p. 327*	*p. 337*
chylomicron *p. 331*	gastrin *p. 336*	rectum *p. 335*	villi (sing., villus) *p. 330*
chyme *p. 328*	gastrointestinal (GI) tract *p. 322*	saliva *p. 326*	
colon *p. 335*	large intestine *p. 334*	salivary amylase *p. 326*	
digestive system *p. 322*	liver *p. 331*	secretin *p. 337*	

THINKING ABOUT THE CONCEPTS

1. List the structures of the gastrointestinal tract in the order in which food passes through them. *p. 323–324*

2. Describe how food is processed in the mouth. What are the functions of the teeth and tongue? *pp. 324–327*

3. Describe the structure of a tooth. What causes tooth decay? *pp. 325–326*

4. What are the functions of the stomach? What are the digestive functions of gastric juice? *pp. 327–329*

5. Why are so few substances absorbed from the stomach? *p. 329*

6. Describe the structural features that increase the surface area for absorption in the small intestine. *pp. 330*

7. Which structures produce the digestive enzymes that act in the small intestine? *pp. 329–330*

8. How does bile assist the digestion and absorption of fats? *p. 330*

9. Where are carbohydrates digested? Proteins? Fats? *pp. 326, 327-328, 329–330*

10. What are the functions of the large intestine? *pp. 334–336*

11. Describe the neural and/or hormonal mechanisms that regulate the release of digestive juices from each structure involved in digestion. *pp. 336–337*

12. Fingerlike projections in the wall of the small intestine are called _____ and they _____
    a. villi; increase the surface area for absorption.
    b. peristalsis; help mix the food with digestive enzymes.
    c. microvilli; secrete bile.
    d. gastric pits; secrete digestive enzymes.

13. It is possible to swallow while standing on one's hands because
    a. valves in the digestive system keep food from moving backward.
    b. peristalsis pushes food along the digestive tract in the right direction.
    c. bacteria clog the digestive tube and prevents food from moving in the wrong direction.
    d. of a wave of muscle contraction called emulsification.

14. The most important function of the stomach is
    a. absorption of nutrients.
    b. chemical digestion.
    c. mucus secretion.
    d. storage of food.

15. You go out with your friends to celebrate your birthday and share a sausage pizza. The digestion of the oil *begins* in the
    a. mouth.
    b. esophagus.
    c. stomach.
    d. small intestine.

16. _____ is a liquid mixture of food and gastric juices found in the stomach.
17. The _____ is the organ that produces bile.
18. Saliva contains an enzyme that begins the chemical digestion of _____.
19. _____ is a hormone produced by the small intestine that causes the gallbladder to contract and release bile.

APPLYING THE CONCEPTS

1. Barbara is a 20-year-old woman with cystic fibrosis, a genetic disease in which abnormally thick mucus is produced. The thick mucus sometimes blocks the pancreatic duct that allows pancreatic juice to enter the small intestine. Explain why cystic fibrosis has caused Barbara to be malnourished.

2. Dehydration occurs when too much body fluid is lost. It can be a serious, even fatal, condition. Doctors worry about dehydration if a person, especially an infant, has severe diarrhea lasting several days. Why?

3. A 35-year-old man sporting a tattoo that says "Janet" visits the dermatologist to have the tattoo removed. When asked why he wants the tattoo removed, he replies that Janet was his first wife. He is getting remarried, and the new wife does not want to see "Janet" on her husband's bicep. Noticing that the patient's skin has a yellowish tint, the dermatologist strongly recommends that the patient visit a physician. Why? What is the dermatologist concerned about?

4. You spend the day with your Aunt Sally on her fifty-fifth birthday. She is overweight, so she has a salad for lunch and has no problems. However, she indulges in fried chicken and French fries for dinner. A few hours later, she is rushed to the emergency room with a severe pain on the right side of her abdomen. She reports that she has had bouts of pain over the last month or so. What do you think is Aunt Sally's problem? Why isn't the pain continuous?

Nutrition and Weight Control

The Food Guide Pyramid Helps Us Plan a Healthy Diet

Nutrients Provide Energy or Have a Structural or Functional Role in the Body

- Lipids include fats, oils, and cholesterol
- Carbohydrates include sugars, starches, and dietary fiber
- Proteins are chains of amino acids
- Vitamins are needed in small amounts to promote and regulate the body's chemical reactions
- Minerals play structural and functional roles in the body
- Water is essential and needed in large amounts

Body Energy Balance Depends on Calories Gained in Food and Calories Used

Obesity Is Body Weight 20 Percent or More above the Body Weight Standard

Dietary Guidelines for Americans Promote Health

Successful Weight-Loss Programs Usually Involve Reducing Calorie Intake, Increasing Calorie Use, and Changing Behavior

Anorexia Nervosa and Bulimia Are Eating Disorders That Create Calorie Deficits

Cassandra was excited about spending the weekend with her former college roommate, Dana. She hadn't seen Dana for over a year and looked forward to the kind of intimate long talks and companionship she could share with no one else. Secretly, Cassandra had to admit, she also couldn't wait to show off her newly slimmed figure. Dana responded to the honk of her horn and immediately rushed to the car door, pulled it open and gave Cassandra a long, tight hug. "It's so good to see you!" she exclaimed, followed immediately by, "My, my, haven't we lost some weight! You look great."

Cassandra couldn't help but blush with pride. "It's nothing—just one of those low-carbohydrate diets." An hour later, after a bit of catching up and questions about her diet, Cassandra admitted that although she had lost more than 20 pounds, she wasn't feeling well. She was frequently constipated, and she had developed intense cravings for fruit and any kind of bread product. Dana, who had taken a few nutrition courses in college, told her that, although she was no expert, it seemed to her that it was time to switch to a balanced diet. She pointed out that the lack of fiber was probably causing her constipation. She also pointed out that this next phase of her diet could be even more difficult if she wanted to keep weight off because she would have to focus on nutritional balance while restricting carbohydrates. Cassandra, nonetheless, was both hopeful and determined and told Dana that she was going to make an appointment with a nutritionist when she returned home. They toasted her commitment, shared the fun and perils of their work lives over dinner, and settled deeply into the comfort of their trusted friendship for the next three days.

In this chapter, we will consider carbohydrates as well as the other nutrients—proteins, fats, vitamins, minerals, and water. We will see how the body uses these nutrients and examine ways to combine the foods we eat to promote health and fitness. ■

The Food Guide Pyramid Helps Us Plan a Healthy Diet

Planning a healthy diet requires more than choosing foods to avoid deficiencies of particular nutrients. Choosing the right balance of foods can help improve health and reduce the risk of serious chronic diseases, such as heart disease, cancer, and diabetes.

The Food Guide Pyramid, released by the U.S. Department of Agriculture (USDA) can help us plan a well-balanced diet even if we do not understand the reasons for choosing certain foods (Figure 15a.1). The pyramid translates nutritional knowledge into recommended eating patterns. The pyramid shape indicates the relative contribution each type of food should make to the diet. Take a moment now to consider how the food you ate today fits with the eating pattern recommended in this pyramid.

FIGURE **15a.1**
The food guide pyramid helps us plan a well-balanced diet without being an expert nutritionist.

The USDA revises the Dietary Guidelines and the Food Guide Pyramid every 5 years. New guidelines and a new pyramid should be available in 2005. It is highly likely that there will be some changes in the recommendations. After we learn about the nutrients, we will consider the nature of some likely changes in the USDA's dietary recommendations.

Nutrients Provide Energy or Have a Structural or Functional Role in the Body

We regularly make decisions about what to eat. When we make conscious choices about what we eat, a number of factors are usually involved: personal preferences, family and peer pressure, our ethnic and religious background, media advertising, and how hungry we are at the time. The choices we make can influence how healthy we will be in the future. A high-fat diet, for instance, promotes certain cancers as well as diseases of the heart and blood vessels. On the other hand, a high-fiber diet is good for the heart and the intestines.

What does the body do with the food you eat? Food provides fuel, building blocks, metabolic regulatory molecules, and water.

- Fuel is needed for all cellular activities. (Energy is measured in a unit called a calorie, which is the amount of energy needed to raise the temperature of 1 gram of water 1° Celsius. In discussions of biochemical reactions, the energy exchanges are usually reported in kilocalories. A kilocalorie is 1000 calories of energy. In popular usage, the *kilo* is dropped. We will follow this tradition and refer to kilocalories simply as calories.[1])
- Building blocks are needed for cell division, maintenance, and repair.
- Metabolic regulators, such as vitamins, are needed to coordinate life's processes. Water is necessary for the proper cellular environment and for certain cellular reactions.

As we saw in Chapter 15, the digestive system breaks complex molecules of carbohydrates, proteins, fats, and nucleic acids into their component subunits. The process of digestion provides building blocks that can be used to build molecules in the body. Most cells of the body, especially those of the liver, are able to convert one type of molecule into certain others. Your body, however, cannot synthesize some amino and fatty acids, at least not in quantities sufficient to meet body needs. These substances are called essential amino acids and essential fatty acids. *Essential* means that these substances must be included in the diet.

A nutrient is a substance in food that provides energy or plays a structural or functional role to promote normal growth, maintenance, or repair. Table 15a.1 identifies the components in food and summarizes their sources and functions in the body. Three nutrients—fats (lipids), carbohydrates, and proteins—can provide energy. Although proteins can provide energy, they are usually used to build structures in the cell or for regulatory molecules, such as enzymes or certain hormones. Vitamins, minerals, water, and fiber do not provide energy, but they are essential to cellular functioning, as we will see.

[1]In technical writing, kilocalories are written as Calories.

Lipids include fats, oils, and cholesterol

To weight-conscious Americans, *fat* is a dirty word. It is especially nasty because you do not have to eat fat to get fat. When you regularly consume more calories than are used for energy, your body makes its own fat.

Lipid is the more technical name for what we have been calling fat. You may recall from Chapter 2 that there are several types of lipids. In fact, 95% of the lipids found in food are triglycerides—the neutral fats that we commonly think of when we hear the term *fat*. A triglyceride is a molecule made from three fatty acids (hence, *tri*) attached to a molecule of glycerol (hence, *glyceride*). The fatty acids in the triglyceride give the molecule its characteristics.

An important way in which fatty acids can differ is in their degree of saturation, which refers to the number of places in the fatty acid where a carbon could hold additional hydrogen atoms. Recall from Chapter 2 that a saturated fatty acid contains all the hydrogen it can hold. A polyunsaturated fatty acid could hold four or more additional hydrogens, and a monounsaturated fatty acid could hold two more hydrogens. In general, *saturated fats* are solid at room temperature, and they usually come from animal sources. In contrast, *unsaturated fats* are liquid at room temperature. They are oils and usually come from plant sources.

FUNCTIONS OF LIPIDS

Although cultural attitudes regarding body fat have varied throughout history, the biological need for some fat in the diet has never changed. Fat is an excellent storage form of energy. It holds 9 calories of energy per gram, compared with the 4 calories per gram stored in carbohydrate or protein. Fat is usually found between muscle cells, including those of the heart, to provide a ready source of energy. In addition, fat is a poor conductor of heat, so the layer of fat stored just beneath the skin provides some insulation against extreme heat or cold. Fat deposits also cushion and protect many vital organs, including the kidneys and the eyeballs. Certain lipids are essential components of all cell membranes; some are used in the construction of myelin sheaths that insulate nerve fibers. Lipids are also needed for the absorption of the fat-soluble vitamins A, D, E, and K (discussed below). These vitamins are often absorbed from the intestines, along with the products of fat digestion. Lipids also can carry these vitamins in the bloodstream to the cells that use them. Oils keep skin soft and prevent dryness. Certain other lipids provide the raw materials for the construction of chemicals used in communication among cells. Another lipid, cholesterol, is the structural basis of the steroid hormones, including the sex hormones.

DIETARY FAT, OBESITY, AND ATHEROSCLEROSIS

Unfortunately for most of us, dietary fat quickly becomes too much of a good thing. Although the daily need for dietary fat is a mere tablespoon, the average American adult consumes 6 to 8 tablespoons of fat (78 to 196 g a day), which can quickly add inches to the waistline! By itself, obesity is associated with health problems such as high blood pressure and increased risk of diabetes. The high-fat diet that causes obesity is also related to certain cancers, including cancer of the colon, prostate gland, lungs, and perhaps the breast.

TABLE **15a.1**

COMPONENTS OF FOOD			
NUTRIENT	**GOOD SOURCES**	**FUNCTIONS**	**RECOMMENDED QUANTITY**[*]
Energy-containing nutrients			
Fats (lipids)	Milk, cheese, meat, vegetable oils, nuts	Provides 9 calories per gram; component of cell membranes; component of nerve sheaths; insulates body; forms protective cushions around vital body organs	20%–35% of daily calories
Carbohydrates	Cereal, bread, pasta, vegetables, fruits, sweets	Provides 4 calories per gram; primary fuel for all cells	45%–65% of daily calories (not more than 25% from added sugar)
Protein and amino acids	Meat, poultry, fish, legumes, nuts	Provides 4 calories per gram; important component of all cells; structural proteins including muscle fibers; regulatory proteins including enzymes and certain hormones	10%–35% of daily calories
Other nutrients			
Vitamins	Many vegetables, fruits, whole grain, meats, dairy products	Most function as co-enzymes that allow the cellular reactions of the body to take place fast enough to support life	Depends on the specific vitamin
Minerals	Many vegetables, fruits, whole grains, meats, seeds, nuts	Structural roles, including hardness of bones and teeth; functional roles including oxygen transport in blood; electrolyte balance; proper nerve and muscle function	Depends on the specific mineral
Water	Nearly every food and all beverages	Solvent; transport of materials; medium for all and participant in some chemical reactions; lubricant; protective cushion; regulation of body temperature	The equivalent of 8 8-ounce glasses
Fiber (not absorbed from intestines)	Whole grains, vegetables, fruits	Soluble fiber good for health of heart and blood vessels; insoluble fiber promotes intestinal health	Men 50 and younger: 38 grams/day Women 50 and younger: 25 grams/day Men over 50: 30 grams/day Women over 50: 21 grams/day

[*]The Food and Nutrition Board, the National Academy of Sciences, and the Institute of Medicine made these recommendations in 2002.

The clearest health risk of a high-fat diet is that of developing atherosclerosis, a condition in which fatty deposits form in the walls of blood vessels. The deposits promote an inflammatory response in the artery wall, thereby increasing the risk of heart attack and stroke (see Chapter 12). The risk of atherosclerosis increases with the blood level of cholesterol (Table 15a.2). In general, blood cholesterol levels under 200 mg/dl[2] are recommended for adults. The bad news is that the average blood cholesterol level for a middle-aged adult in the United States is 215 mg/dl. The good news is that people who lower their blood cholesterol levels can slow, or even reverse, atherosclerosis and, therefore, their risk of heart attack.

Total blood cholesterol, however, does not paint a complete picture of a person's risk of atherosclerosis. Before cholesterol can be transported in the blood or lymph, it is combined with protein and triglycerides to form a lipoprotein, which makes it soluble in water. Low-density lipoproteins (LDLs) are considered to be a *bad* form of cholesterol because, in addition to bringing cholesterol to the cells that need it to sustain life, they deposit cholesterol in artery walls. In contrast, high-density lipoproteins (HDLs) help the body eliminate cholesterol. HDLs are considered to be a *good* form of cholesterol and to be protective against heart disease.

TABLE **15a.2**

BLOOD CHOLESTEROL LEVELS AND THE RISK OF CARDIOVASCULAR DISEASE			
	DESIRABLE	**BORDERLINE HIGH**	**HIGH**
Total cholesterol/ 100 ml of blood	<200 mg/dl	200–239 mg/dl	≥240 mg/dl
LDL cholesterol/ 100 ml blood	100–129 mg/dl	130–159 mg/dl	≥160 mg/dl
Ratio of total cholesterol to HDL cholesterol	<4:1 mg/dl		

Because of the roles they play in transporting cholesterol, the ratio of HDLs to LDLs is considered more important than the total blood cholesterol level. The ratio of total cholesterol to HDLs should not be greater than 4:1. HDL levels higher than 60 mg/dl are considered to reduce the risk of heart disease.

Blood cholesterol comes from one of two sources: the diet or the liver. Of the two, cholesterol production by the liver is more signifi-

[2]dl = deciliter (100 milliliters).

cant. Indeed, *most* of the cholesterol in blood comes from the liver and not from the food we eat. Nonetheless, diet is still important, and, surprisingly, saturated fat in the diet raises blood levels of cholesterol more than dietary cholesterol does. Why? The body handles cholesterol differently depending on a person's genetic makeup and the amount of cholesterol consumed in a meal. As a result, there is not a clear relationship between diet and blood cholesterol.

The types of fat in the diet can influence blood cholesterol level and, more importantly, the ratio of LDLs to HDLs (Figure 15a.2). Saturated fats and *trans* fats are bad for the heart. Saturated fat consistently boosts blood levels of harmful LDL cholesterol both by directly stimulating the liver to step up its production of LDLs and by slowing the rate at which LDLs are cleared from the bloodstream. *Trans* fatty acids may behave like saturated fatty acids and raise LDL levels. *Trans* fatty acids are formed when hydrogens are added to unsaturated fats (oils) to stabilize them or to solidify them, as when margarine is formed from vegetable oil. In contrast, monounsaturated fats and polyunsaturated fats, such as omega-3 fatty acids and omega-6 fatty acids,[3] lower total blood cholesterol and LDLs. Monounsaturated fats are found in olive, canola, and peanut oils and in nuts. Omega-3 fatty acids are found in the oils of certain fish, such as Atlantic mackerel, lake trout, herring, tuna, and salmon. The most important omega-6 fatty acid is linoleic acid, an essential fatty acid found in corn and safflower oils.

RECOMMENDED AMOUNT OF DIETARY FATS

It is recommended that 20% to 35% of calories in the diet come from fats. The goal is to balance the types of fats in the diet to lower blood cholesterol and, therefore, reduce the risk of atherosclerosis. The bad fats that raise LDL levels and lower HDL levels—the saturated fats and *trans* fats—should be minimized. Monounsaturated and polyunsaturated fats lower LDL levels and raise HDL levels and should represent most of the fat calories in the diet.

For most of us, achieving this balance of fats would require a drastic reduction in the amount of saturated fat in the diet. Foods high in saturated fat include meat (especially red meat), butter, cheese, whole milk, and other dairy products, as shown in Figure 15a.3.

Carbohydrates include sugars, starches, and dietary fiber

Carbohydrates in our diets include sugars, starches, and roughage (dietary fiber). The basic unit of a carbohydrate is a monosaccharide. Sugars are *simple carbohydrates* and are generally monosaccharides or disaccharides (two monosaccharides linked together). Sugars taste sweet and are naturally present in whole foods such as fruit and milk. In whole foods, sugars are packaged with other nutrients, including vitamins and minerals. However, refined sugars have been separated from their plant sources and contain few other nutrients, so they are considered to be empty calories. Refined sugars are found in candies, cookies, cakes, and pies. *Complex carbohydrates,* such as starches and fiber, are usually packed with other nutrients. Starches are polysaccharides—long, sometimes branched chains of hundreds or thousands of linked molecules of the monosaccharide glucose. Plants store energy in starches. Common sources of starches include wheat, rice, oats, corn, potatoes, and legumes. Fiber is the indigestible parts of edible plants.

FUNCTIONS OF CARBOHYDRATES

The most important function of carbohydrates is to provide fuel for the body. Of the carbohydrates, the monosaccharide glucose is

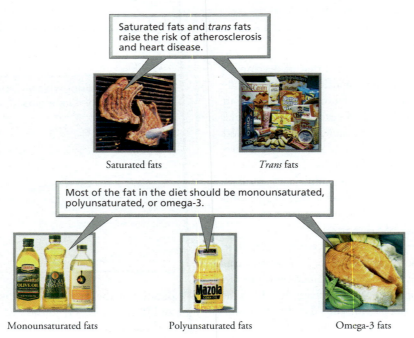

Saturated fats and *trans* fats raise the risk of atherosclerosis and heart disease.

Saturated fats   *Trans* fats

Most of the fat in the diet should be monounsaturated, polyunsaturated, or omega-3.

Monounsaturated fats   Polyunsaturated fats   Omega-3 fats

FIGURE **15a.2**

Types of dietary fats. The total fat content of the diet should be moderate. The types of fats in the diet can influence the risk of heart attack and stroke.

---

[3]The number in omega-3 and omega-6 refers to the location of the first double bond in the carbon chain of the fatty acid.

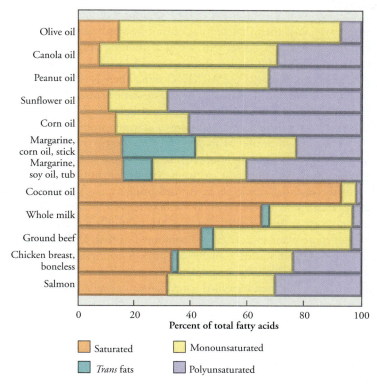

Percent of total fatty acids

Olive oil
Canola oil
Peanut oil
Sunflower oil
Corn oil
Margarine, corn oil, stick
Margarine, soy oil, tub
Coconut oil
Whole milk
Ground beef
Chicken breast, boneless
Salmon

0   20   40   60   80   100

■ Saturated   ■ Monounsaturated
■ *Trans* fats   ■ Polyunsaturated

FIGURE **15a.3**
Foods contain a mixture of saturated, monounsaturated, and polyunsaturated fats. Most Americans would be well advised to reduce their total fat intake, especially the saturated fats. Saturated fats increase blood levels of cholesterol.

the fuel that cells use most of the time. Indeed, even a temporary shortage of blood glucose can impair brain function and kill nerve cells. One gram of carbohydrate yields 4 calories—less than half of the energy provided by one gram of fat.

DIETARY FIBER

Fiber is found in all plants that are eaten for food, including fruits, vegetables, dried beans, and whole grains. Dietary fiber is a carbohydrate in the plant food that humans cannot digest into component monosaccharides. Because only monosaccharides can be absorbed from the small intestine, dietary fiber is passed along to the large intestine. Some fiber is digested by the bacteria living in the large intestine, and the remaining fiber gives bulk to feces.

Although fiber cannot be digested or absorbed, it is still an important part of a healthful diet. Water-soluble fiber is good for the heart and blood vessels. It lowers LDLs and total cholesterol but does not lower the beneficial HDL cholesterol levels.

Intestinal disorders such as constipation, hemorrhoids, and diverticulosis improve when the amount of fiber in the diet is increased (discussed in Chapter 15). Soluble fiber can absorb an amazing amount of water, thereby softening stools and making them easier to pass.

RECOMMENDED AMOUNTS OF CARBOHYDRATES

Nutritionists recommend that 45% to 65% of the calories in your diet come from carbohydrates. Nutritionists often classify digestible

carbohydrates as simple carbohydrates (sugars) or complex carbohydrates (starches and fiber). Simple carbohydrates, especially refined sugars, in the diet should be limited. Most of the carbohydrate calories in the diet should come from complex carbohydrates (Figure 15a.4). Unlike sugars, starches are generally packaged with other nutrients, including vitamins and minerals, and with dietary fiber. Nutritionists generally agree that the emphasis should be on whole grains, vegetables, and fruits.

Some nutritionists also rank the healthfulness of a carbohydrate by how it affects blood sugar. Recall that the digestive system breaks down all carbohydrates (except fiber) to simple sugars, primarily glucose, which is absorbed into the bloodstream. The glycemic load is a measure of how quickly a serving of food is converted to blood sugar and how high the spike in blood sugar is. Sugar and starchy foods such as white bread, potatoes, and white rice have a high glycemic load; they cause the blood sugar level to rise high and quickly. Foods with a low glycemic load, including whole fruit, whole grain foods, brown rice, and barley, cause a more modest and gradual increase in blood sugar. The glycemic load of your diet can affect your risk of heart disease and diabetes (see Chapter 10). A healthful diet contains more foods with low glycemic loads than foods with high glycemic loads.

The average American diet does not contain an adequate amount of fiber. Figure 15a.5 shows some good sources of dietary fiber.

Sugar consumption in the United States averages 125 pounds per person per year. Although it seems unbelievable, it is possible to consume this much sugar because so much is hidden in processed foods, even ones that do not taste sweet. Even people who faithfully read food labels may not be aware of the sugar content in processed food because simple sugars have so many different names, including honey, brown sugar, dextrose, maltose, molasses, fructose, levulose, and corn syrup. Foods that taste sweet often contain whopping amounts of sugar. A single glazed donut, for example, contains 6 teaspoons (24 g) of sugar. And who would expect that 1 tablespoon of ketchup contains 1 teaspoon (4 g) of sugar?

Simple carbohydrates are empty calories, because they provide only energy.

Complex carbohydrates provide energy along with other nutrients.

FIGURE **15a.4**
Types of carbohydrates. Simple carbohydrates, especially refined sugars, should be minimized in the diet. Complex carbohydrates should represent most of the carbohydrate in the diet.

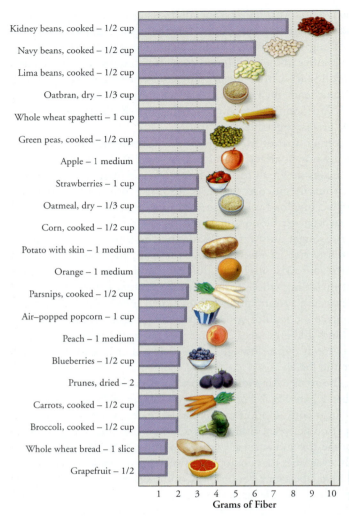

FIGURE **15a.5**
Good sources of dietary fiber include fruits, vegetables, dried beans, and whole grains.

Proteins are chains of amino acids

Protein seems to have a privileged status in our culture. When we think of protein, images of vitality, strength, muscle mass, and energy often come to mind. Proteins were named from the Greek word *proteios*, which means "of prime importance." Before we consider why proteins are so important, we should become familiar with what they are.

A protein consists of one or more chains of amino acids. Human proteins contain 20 different kinds of amino acids. The protein in the food you eat is digested into its component amino acids and then absorbed into the bloodstream and delivered to the cells, creating a pool of amino acids. Your cells then draw the amino acids needed to build the proteins in your body from those available in the pool. The body is able to synthesize 11 of the amino acids from nitrogen and molecules derived from carbohydrates, fats, or other amino acids. Your body cannot synthesize the 9 remaining amino acids. These *essential amino acids* must be supplied by the diet.

When your body makes a protein, the amino acids are put together in a specific order, depending on which protein is being

made. If a particular amino acid is needed for a certain protein but is not available, then that protein cannot be synthesized. Consider this analogy: A sign maker with a bag containing only one copy of the letter *R* and 100 copies of the other 25 letters in the alphabet could make only one NO PARKING sign. There is simply no way to substitute for the limiting letter, *R*. The same principle applies to protein synthesis; if a needed amino acid is lacking, protein production stops or the body breaks down existing proteins to get it.

The pool of amino acids available for protein synthesis must always contain sufficient amounts of all the essential amino acids. The *complete proteins* that you eat are those that contain ample amounts of all of the essential amino acids. Animal proteins are generally complete. Plant proteins are generally *incomplete proteins* and are low in one or more of the essential amino acids. Incomplete proteins from two or more plant sources can be combined so that, after digestion, the pool of amino acids available for protein synthesis will contain ample amounts of all the essential amino acids (Figure 15a.6). Combinations such as these are called *complementary proteins*; together they supply enough of all the essential amino acids (Figure 15a.7). Thus, a vegetarian must make careful food choices to be sure to consume complementary proteins.

FUNCTIONS OF PROTEIN

Protein is an essential structural component of every cell in your body. Indeed, if the water were removed from your body, half of your remaining body weight would be protein. Protein forms the framework of bones and teeth. It forms the contractile part of muscle and helps red blood cells carry oxygen. Dietary protein provides the raw materials to replace or repair cells that are stressed by everyday wear and tear and to form new tissues as you grow.

Proteins regulate body processes. Certain proteins are enzymes that help the chemical reactions of the body take place at body temperature and at rates fast enough to support life. Other proteins are hormones, which are chemical messengers that enable cells in distant parts of the body to communicate with one another.

Proteins known as antibodies help the body defend itself against foreign invaders, such as bacteria. Also, proteins carried in the blood help maintain the water balance in the body.

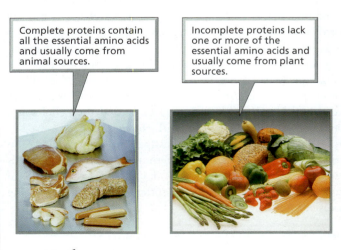

Complete proteins contain all the essential amino acids and usually come from animal sources.

Incomplete proteins lack one or more of the essential amino acids and usually come from plant sources.

FIGURE **15a.6**
Types of proteins

• Hummus (chickpeas and sesame seeds)
• Tofu and cashew stir-fry

• Trail mix (roasted soybeans and nuts)
• Tahini (sesame seeds) and peanut sauce

Nuts and seeds

Legumes

• Rice and beans
• Black-eyed peas and corn bread
• Bean burrito in corn tortilla

• Peanut butter on bread
• Rice and tofu
• Rice and lentils

Grains

FIGURE **15a.7**

Complementary proteins are combinations of two incomplete proteins that supply all the essential amino acids.

Proteins can also be used for energy. When 1 gram of protein is oxidized, it yields 4 calories. However, protein is not the preferred fuel for the body. Protein is used for energy only if the supply of carbohydrates and fat is insufficient or if dietary protein exceeds the body's needs.

RECOMMENDED AMOUNT OF PROTEIN

Nutritionists suggest that protein calories represent 10% to 35% of the calories in the diet. Most North Americans have no trouble meeting this goal. Your specific protein requirement can be determined by multiplying your body weight in kilograms[4] by 0.8. Thus, a person weighing 68 kg (about 150 lb) would require 55 grams of protein a day. In North America, however, the typical person consumes about 100 grams of protein a day—nearly twice the daily requirement.

Although the choice of protein foods must ensure an adequate supply of all the essential amino acids, the protein need not come from animal sources. Indeed, the tendency to make meat the focal point of every meal is the main reason that the typical diet in the United States contains excess fat. Animal protein usually comes with a lot of fat, whereas plant protein usually brings along carbohydrates (Figure 15a.8). Thus, eating a variety of plant proteins will supply all the essential amino acids and will help reduce the percentage of fat calories in the diet as well as boost the percentage of calories from complex carbohydrates. When choosing among animal sources of protein, reduce the amount of red meat in the diet in favor of chicken (with the fatty skin removed) or fish.

Vitamins are needed in small amounts to promote and regulate the body's chemical reactions

A vitamin (*vita*, life giving) is an organic (carbon-containing) compound that, although essential for health and growth, is needed only in minute quantities—milligrams or micrograms. All the vitamins you need in a day would fill only an eighth of a teaspoon. Vitamins are needed in only tiny amounts because they are not destroyed during use. Most function as coenzymes, which are nonprotein molecules that are necessary for certain enzymes to function. Enzymes and coenzymes are continuously recycled and, therefore, can be used repeatedly by the body.

There are two categories of vitamins: the water-soluble vitamins, which dissolve in water, and the fat-soluble vitamins, which are stored in fat. Of the 13 vitamins known to be needed by humans, 9 are water soluble (C and the various B vitamins) and 4 are fat soluble (A, D, E, and K). Table 15a.3 lists the vitamins, their functions, good sources of them, and the problems associated with deficiencies or excesses.

Except for vitamin D, our cells cannot make vitamins, so we must obtain them in our food. A varied, balanced diet is the best way to ensure an adequate supply of all vitamins. No one

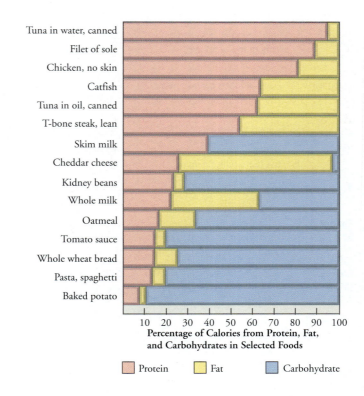

FIGURE **15a.8**

The percentage of calories from protein, fat, and carbohydrate in certain plant and animal foods. Animal foods contain a greater percentage of fats than do plant foods.

[4]Weight in kilograms is determined by dividing weight in pounds by 2.2.

TABLE 15a.3

		VITAMINS		
VITAMIN	GOOD SOURCES	FUNCTION	EFFECTS OF DEFICIENCY	EFFECTS OF EXCESS
Fat-soluble vitamins				
A	Liver, egg yolk, fat-containing and fortified dairy products; formed from carotene (found in deep yellow and deep green leafy vegetables)	Components of rhodopsin, the eye pigment responsible for black-and-white vision; maintains epithelia; cell differentiation	Night-blindness; dry, scaly skin; dry hair; skin sores; increased respiratory, urogenital, and digestive infections; xerophthalmia (the leading cause of preventable blindness worldwide); most common vitamin deficiency in world	Drowsiness; headache; dry, coarse, scaly skin; hair loss; itching; brittle nails; abdominal and bone pain
D	Fortified milk, fish liver oil, egg yolk; formed in skin when exposed to ultraviolet light	Increases absorption of calcium; enhances bone growth and calcification	Bone deformities in children, rickets, bone softening in adults	Calcium deposits in soft tissues, kidney damage, vomiting, diarrhea, weight loss
E	Whole grains, dark green vegetables, vegetable oils, nuts, seeds	May inhibit effects of free radicals; helps maintain cell membranes; prevents oxidation of vitamins A and C in gut	Rare; possible anemia and nerve damage	Muscle weakness, fatigue, nausea
K	Primary source from bacteria in large intestine; leafy green vegetables, cabbage, cauliflower	Important in forming proteins involved in blood clotting	Easy bruising, abnormal blood clotting, severe bleeding	Liver damage and anemia
Water-soluble vitamins				
C (ascorbic acid)	Citrus fruits, cantaloupe, strawberries, tomatoes, broccoli, cabbage, green pepper	Collagen synthesis; may inhibit free radicals; improves iron absorption	Scurvy, poor wound healing, impaired immunity	Diarrhea, kidney stones; may alter results of certain diagnostic lab tests
Thiamin (B_1)	Pork, legumes, whole grains, leafy green vegetables	Coenzyme in energy metabolism; nerve function	Water retention in tissues, nerve changes leading to poor coordination, heart failure, beriberi	None known
Riboflavin (B_2)	Dairy products such as milk; whole grains, meat, liver, egg whites, leafy green vegetables	Coenzyme used in energy metabolism	Skin lesions	None known
Niacin (B_3)	Nuts, green leafy vegetables, potatoes; can be formed from tryptophan found in meats	Coenzyme used in energy metabolism	Contributes to pellagra (damage to skin, gut, nervous system)	Flushing of skin on face, neck, and hands; possible liver damage
B_6	Meat, poultry, fish, spinach, potatoes, tomatoes	Coenzyme used in amino acid metabolism	Nervous, skin, and muscular disorders; anemia	Numbness in feet, poor coordination
Pantothenic acid	Widely distributed in foods, animal products, and whole grains	Coenzyme in energy metabolism	Fatigue, numbness and tingling of hands and feet, headaches, nausea	Diarrhea, water retention
Folic acid (folate)	Dark green vegetables, orange juice, nuts, legumes, grain products	Coenzyme in nucleic acid and amino acid metabolism	Anemia (megaloblastic and pernicious), gastrointestinal disturbances, nervous system damage, inflamed tongue, neural tube defects	High doses mask vitamin B_{12} deficiency
B_{12}	Poultry, fish, red meat, dairy products except butter	Coenzyme in nucleic acid metabolism	Anemia (megaloblastic and pernicious), impaired nerve function	None known
Biotin	Legumes, egg yolk; widely distributed in foods; bacteria of large intestine	Coenzyme used in energy metabolism	Scaly skin (dermatitis), sore tongue, anemia	None known

food contains every vitamin, but most contain some. Vitamins are often more easily available for absorption when the foods containing them are cooked. Cooked carrots, for instance, are a better source of vitamin A than are raw carrots. However, water-soluble vitamins are just that, so these vitamins are likely to be lost if the vegetables containing them are cooked by boiling them in water. Steaming vegetables, then, is a better way to preserve their vitamin content.

It is important to ensure that you obtain adequate daily supplies of certain vitamins. Folic acid, one of the B vitamins that is particularly abundant in dark leafy greens, plays a role in preventing birth defects, such as spina bifida, that involve the brain and spinal cord. It now seems that folic acid, along with vitamins B_6 and B_{12}, may also help prevent heart disease. It is thought that five daily servings of fruits and vegetables would provide enough of these vitamins to protect the heart. Furthermore, it has been suggested that vitamins that act as antioxidants, including vitamin C, vitamin E, and beta-carotene, may slow the aging process and protect against cancer, atherosclerosis, and macular degeneration (the leading cause of irreversible blindness in people over age 65). Spinach, collard greens, and carrots are good sources of antioxidant vitamins.

Are vitamin supplements too much of a good thing? It depends. For one thing, the components of your everyday diet will influence how beneficial vitamin supplements will be. Even though you may know that eating a well-balanced diet, including vegetables, is the best way to obtain all the necessary nutrients, the pressures of time or force of habit may keep you from eating well. In that case, a daily multivitamin supplement is a good idea. Other factors influencing whether vitamin supplements will be beneficial or harmful are the amount and type of vitamin consumed. Excess water-soluble vitamins are usually excreted in the urine. In contrast, excess fat-soluble vitamins are stored in fat and can accumulate in the body, causing serious problems if they reach high levels.

▍Minerals play structural and functional roles in the body

A mineral is an inorganic substance that is essential to a wide range of life processes. We need fairly large, although not megadose, amounts of seven minerals: calcium, phosphorus, potassium, sulfur, sodium, chloride, and magnesium. In addition, we need trace amounts of about a dozen others (Table 15a.4).

We obtain minerals from the foods we eat. As with vitamins, we must be careful how we cook foods to avoid losing minerals. Many minerals are water soluble and can be lost during food preparation.

Sodium, a component of table salt, is essential to health, but most Americans consume too much salt. High salt intake causes high blood pressure in some people. The Dietary Recommendations for Americans advises that salt intake should not exceed 2400 mg (slightly more than 1 teaspoon) a day. Prepared foods are especially high in salt, which is added to preserve food and to enhance the taste. Salt is found in nearly every food, including canned vegetables, cheese, bread, and processed meats. To reduce your salt intake, use salt sparingly when cooking fresh food and read the labels on prepared food.

▍Water is essential and needed in large amounts

Water is perhaps the most essential nutrient. We can live without food for about 8 weeks but without water for only about 3 days. Although you look solid, within your body is an internal sea. Even bone is about one-fourth water! A newborn is, in fact, 85% water. The water content gradually decreases with age to between 55% to 65% in an adult female and between 65% to 75% in an adult male. Females contain less water than males do because their bodies contain a greater proportion of fat, which does not hold water as well as muscle.

Water is so common—found in nearly every food and beverage we consume—that we rarely think about its importance. It is an excellent solvent. Thus, in the blood and lymph, it transports materials, including nutrients, metabolic wastes, and hormones. Water provides a medium in which chemical reactions can take place, and it participates in many of those reactions. One example is hydrolysis (*hydro,* water; *lysis,* splitting) reactions, in which chemical bonds are broken by the addition of water. For instance, the food molecules we eat are split into their component subunits during digestion by hydrolysis. Water is also a lubricant. It keeps your joints from creaking and the soft tissues of your body from sticking together. Water forms a protective cushion within the eyes and around the brain and spinal cord. During pregnancy, water in the amniotic fluid protects the fetus. In addition, water plays an important role in the regulation of body temperature. Perspiration, for instance, cools the body.

It is recommended that we consume eight 8-ounce glasses (2 quarts) of water a day. Water in fruits and vegetables makes up about half of that requirement for the average adult. Although the rest of the requirement does not have to be plain water, it should not come from carbonated sweet drinks, caffeinated beverages, or alcoholic beverages. Carbonation interferes with water absorption, and sugar adds empty calories. Caffeine and alcohol increase water loss in urine.

Body Energy Balance Depends on Calories Gained in Food and Calories Used

As we have seen, our bodies obtain energy from the carbohydrates, proteins, and fats that we eat. Carbohydrates and proteins provide 4 calories per gram, and fats provide 9 calories per gram.

Although energy can be neither created nor destroyed, it can be changed from one form to another. These energy conversions take place in the biochemical reactions that occur in the body. They are the reason that the energy in a potato you eat can fuel muscle contraction, allowing you to run to catch the bus. There is a dynamic balance between the energy the body takes in (in food) and its expenditure of energy. Food energy that is not used for the body's various activities is stored as fat or glycogen, explaining why you gain weight when you consume more calories than you use.

The expenditure of energy is just as important as the input of energy. The body requires energy for maintenance of basic body functions, physical activity, and processing the food that is eaten. The part of the need for body energy that goes to maintenance is called the *basal metabolic rate (BMR)*; it is the minimum energy needed to keep an awake, resting body alive, and it generally rep-

TABLE 15a.4

MINERALS

MINERAL	GOOD SOURCES	FUNCTION	EFFECTS OF DEFICIENCY	EFFECTS OF EXCESS
Major minerals				
Calcium	Milk, cheese, dark green vegetables, legumes	Hardness of bones, tooth formation, blood clotting, nerve and muscle action	Stunted growth, loss of bone mass, osteoporosis, convulsions	Impaired absorption of other minerals, kidney stones
Phosphorus	Milk, cheese, red meat, poultry, whole grains	Bone and tooth formation; component of nucleic acids, ATP, and phospholipids; acid-base balance	Weakness, demineralized bone	Impaired absorption of some minerals
Magnesium	Whole grains, green leafy vegetables, milk, dairy products, nuts, legumes	Component of enzymes	Muscle cramps, neurologic disturbances	Neurologic disturbances
Potassium	Available in many foods including meats, fruits, vegetables, and whole grains	Body water balance, nerve function, muscle function, role in protein synthesis	Muscle weakness	Muscle weakness, paralysis, heart failure
Sulfur	Protein-containing foods including meat, legumes, milk, and eggs	Component of body proteins	None known	None known
Sodium	Table salt	Body water balance, nerve function	Muscle cramps, reduced appetite	High blood pressure in susceptible people
Chloride	Table salt, processed foods	Formation of hydrochloric acid in stomach, role in acid-base balance	Muscle cramps, reduced appetite, poor growth	High blood pressure in susceptible people
Trace minerals				
Iron	Meat, liver, shellfish, egg yolk, whole grains, green leafy vegetables, nuts, dried fruit	Component of hemoglobin, myoglobin, and cytochrome (transport chain enzyme)	Iron-deficiency anemia, weakness, impaired immune function	Liver damage, heart failure, shock
Iodine	Marine fish and shellfish, iodized salt, dairy products	Thyroid hormone function	Enlarged thyroid	Enlarged thyroid
Fluoride	Drinking water, tea, seafood	Bone and tooth maintenance	Tooth decay	Digestive upsets, mottling of teeth, deformed skeleton
Copper	Nuts, legumes, seafood, drinking water	Synthesis of melanin, hemoglobin, and transport chain components; collagen synthesis; immune function	Rare; anemia, changes in blood vessels	Nausea, liver damage
Zinc	Seafood, whole grains, legumes, nuts, meats	Component of digestive enzymes; required for normal growth, wound healing, and sperm production	Difficulty in walking, slurred speech, scaly skin, impaired immune function	Nausea, vomiting, diarrhea, impaired immune function
Manganese	Nuts, legumes, whole grains, leafy green vegetables	Role in synthesis of fatty acids, cholesterol, urea, and hemoglobin; normal neural function	None known	Nerve damage

resents between 60% and 75% of the body's energy needs. A male usually has a higher metabolic rate than does a female because a male's body has more muscle and less fat than a female's. Muscles use more energy than fat does. So, while a man and a woman of equal size sit on the couch and watch television together, he burns 10% to 20% more calories than she does. As you age, muscle mass and metabolic rate both decline. Together, these factors reduce caloric needs. If other adjustments in lifestyle are not made to compensate, these metabolic changes can add extra pounds each year after age 35.

The second most important use of energy is physical activity (Table 15a.5). Exercise is an excellent way to burn calories. It not only boosts your needs during the activity but also speeds up metabolic rate for a while afterward.

TABLE **15a.5**

APPROXIMATE NUMBER OF CALORIES BURNED PER HOUR BY VARIOUS ACTIVITIES			
ACTIVITY	100-LB PERSON	150-LB PERSON	200-LB PERSON
Bicycling, 6 mph	160	240	312
Bicycling, 12 mph	270	410	534
Jogging, 5.5 mph	440	660	962
Jogging, 10 mph	850	1280	1664
Jumping rope	500	750	1000
Swimming, 25 yd/min	185	275	358
Swimming, 50 yd/min	325	500	650
Walking, 2 mph	160	240	312
Walking, 4.5 mph	295	440	572
Tennis (singles)	265	400	535

Source: American Heart Association

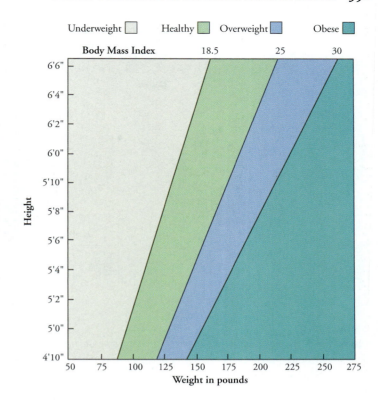

FIGURE **15a.9**

The Body Mass Index (BMI) evaluates body weight relative to height. A BMI over 30 is usually considered to be unhealthy and a sign of obesity.

Obesity Is Body Weight 20 Percent or More above the Body Weight Standard

Although *overweight* and *obese* are both terms used to describe people who have excess body weight, they do not have exactly the same meaning. An obese person is overweight because of excess fat. An athletic person who weighs more than the desirable weight listed on height-weight tables because of well-developed muscle is not obese.

The Body Mass Index (BMI) evaluates your weight in relation to your height (Figure 15a.9). A BMI greater than 30 is generally considered unhealthy and an indication of obesity. However, a person in the healthy weight range might have a lot of fat and little muscle. Conversely, a very muscular person may have a BMI above 30 and not be obese. The Centers for Disease Control and Prevention (CDC) estimates that 64% of Americans are overweight and that almost 31% are obese. The number of obese people has been steadily rising since 1985 (Figure 15a.10).

Although most dieters are motivated to slim down for "cosmetic" reasons, more important reasons are the health risks associated with obesity. For instance, obesity leads to disease of the heart and blood vessels. Indeed, individuals who are just slightly overweight still have an increased risk of having a heart attack. Obesity also raises total cholesterol levels in the blood while lowering levels of the beneficial HDL cholesterol. In addition, it increases the risk of high blood pressure, which can lead to death from heart attack, stroke, or kidney disease. Obesity has harmful effects besides those on the heart and blood vessels. It can induce diabetes, which results in elevated blood glucose levels; obesity is also a major cause of gallstones and can worsen degenerative joint diseases. Worldwide, about 750 million people are overweight, and 300 million are

obese. In the United States, 400,000 people die of weight-related illnesses each year. In 2004, the American Medical Association reported that the combination of sedentary lifestyle and poor eating habits is now the second leading preventable cause of death, surpassed only by tobacco use. If current trends continue, obesity will become the leading avoidable cause of death by 2005. Furthermore, experts predict that by 2024, obesity-linked disease in the United States will cancel health strides from medical technology and disease-fighting measures, such as vaccinations.

Dietary Guidelines for Americans Promote Health

The USDA and the U.S. Department of Human Health and Human Services published the Dietary Guidelines for Americans 2000 to serve as a set of broad-based recommendations to promote health and fitness and to reduce the frequency of chronic diseases (Figure 15a.11). These guidelines, which tie together many of the ideas discussed in this chapter, are organized into three tiers. The first tier, Aim For Fitness, recommends that we aim for a healthy weight and participate in daily physical activity. The second tier, Build A Healthy Base, provides guidelines on choosing a variety of foods and handling them safely. The third tier, Choose Sensibly, recommends limiting certain types of foods.

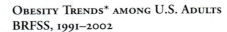

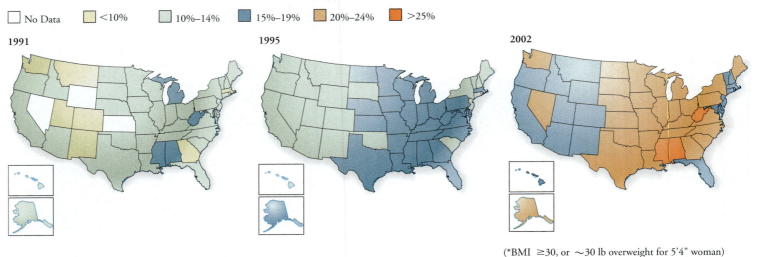

(*BMI ≥30, or ∼30 lb overweight for 5'4" woman)

FIGURE **15a.10**

Obesity trends among adults in the United States. Obesity is defined as a high amount of body fat relative to total body weight. There has been an alarming increase in the number of obese people over the last decade. (Centers for Disease Control and Prevention)

NUTRITION AND YOUR HEALTH
DIETARY GUIDELINES FOR AMERICANS

AIM FOR FITNESS...
▲ Aim for a healthy weight
▲ Be physically active each day

BUILD A HEALTHY BASE...
■ Let the pyramid guide your food choices
■ Choose a variety of grains daily, especially whole grains
■ Choose a variety of fruits and vegetables daily
■ Keep food safe to eat

CHOOSE SENSIBLY...
● Choose a diet that is low in saturated fat and cholesterol and moderate in total fat
● Choose beverages and foods to moderate your intake of sugars
● Choose and prepare foods with less salt
● If you drink alcoholic beverages, do so in moderation

...FOR GOOD HEALTH

FIGURE **15a.11**

The Dietary Guidelines for Americans promote healthy diets and fitness. (Dietary Guidelines for Americans 2000, USDA)

The USDA is mandated to revise its dietary recommendations every 5 years. A new Food Guide Pyramid is expected in 2005. Some nutritionists predict that the revised dietary guidelines and Food Guide Pyramid will recognize the differences among types of lipids and types of carbohydrates. We have seen that when it comes to promoting health, not all lipids are bad, and not all carbohydrates (even complex carbohydrates) are good. Saturated fats and *trans* fats are bad for the heart because they raise blood levels of cholesterol, which increases the risk of atherosclerosis. On the other hand, polyunsaturated fats, omega-3 fats, and omega-6 fats lower blood levels of cholesterol and protect the heart and blood vessels. In the past, most complex carbohydrates have been promoted as healthful. We now know that some carbohydrates—those with a high glycemic load—rapidly raise blood glucose levels, which elevates the level of the hormone insulin and may adversely affect long-term health. Other complex carbohydrates such as vegetables and whole grains are high in fiber and rich in vitamins and minerals; these carbohydrates also have a low glycemic load. So we might find some shifting of the place of oils and of carbohydrates with a high glycemic load in the revised Food Guide Pyramid.

Successful Weight–Loss Programs Usually Involve Reducing Calorie Intake, Increasing Calorie Use, and Changing Behavior

When weight-loss programs are successful, they generally have three components: (1) a reduction in the number of calories consumed, while keeping the balance of calories in line with the recommended guidelines; (2) an increase in energy expenditure; and (3) behavior modification. Gradual changes in eating habits are most likely to lead to permanent lifestyle changes.

To determine the number of calories needed each day to maintain a desirable weight, you must take activity level and age into account (Figure 15a.12). A pound of fat contains approximately 3500 calories, so to lose 1 pound a week, a person should reduce calorie consumption by 500 calories a day (500 calories × 7 days = 3500 calories) or increase calorie use by 500 calories a day or any equivalent combination. The lowest daily caloric intake recommended for a female is 1200 calories, and for a male, 1500 calories, unless they are in a medically supervised program.

LOW CALORIE, LOW FAT DIETS

The easiest way to reduce calorie intake is to cut back on fatty foods, especially those that contain saturated fat. Recall that fat contains more than twice as many calories as an equivalent weight of carbohydrate or protein. Fat contains more calories, and it is easier for the body to store fat as body fat than as protein or carbohydrate. Indeed, most of us do end up wearing the fat we eat.

Another healthful diet tip is to increase the amount of fiber in the diet. High-fiber foods, such as fruits and vegetables, tend to be low in calories and fat but also high in vitamins and minerals. Because the fiber is bulky, these foods also are filling.

LOW CARBOHYDRATE DIETS

The Atkin's diet and other low carbohydrate diets have helped many people lose weight. In the Atkin's diet, the severe restriction of carbohydrates in the induction phase forces the body to burn fat for energy instead of carbohydrates. The rationale for continuing on a low carbohydrate diet is that carbohydrates are digested to glucose, which is absorbed into the blood. When blood glucose levels rise, the pancreas releases the hormone insulin, which allows cells to take in glucose. Blood glucose levels plummet as glucose is taken into cells, causing the brain to send out hunger signals.

Insulin also stimulates fat storage. So proponents of these diets believe that eating carbohydrates stimulates appetite and promotes an increase in body fat.

Factors other than regulating insulin levels help people on a low carbohydrate diet to lose weight. One reason might be that a low carbohydrate diet is also usually a low calorie diet. Snack foods, desserts, and soda are high in calories. A second reason might be that a *moderate* amount of fat makes food more satisfying and filling, which makes it easier to eat less and consume fewer calories.

An important criticism of the Atkin's diet is that it doesn't distinguish among sources of protein, allowing dieters to indulge heavily in animal sources of protein. Animal sources of protein (meat) are usually high in saturated fats, which nearly all nutrition experts agree are bad.

THE YO-YO EFFECT

Approximately 60% to 90% of dieters who lose weight will later regain all the weight they lost (Figure 15a.13). Often, the weight is regained because the weight loss was achieved by drastically cutting back calories, which can be unhealthy and is very difficult to sustain for long periods. As old eating habits return, so do the pounds. When the determination to shed some pounds returns, the diet begins again. This process is commonly known as "the yo-yo effect."

What causes the yo-yo effect? When there is a severe restriction in calories, as occurs in the typical crash diet, the body adopts a calorie-sparing defense by reducing the resting metabolic rate by as much as 45%. This response conserves energy and evolved long ago to help our ancestors survive during times of food scarcity. Today, it makes weight loss during successive diets progressively more difficult. Furthermore, with each diet, a person usually loses both lean and fat tissue, especially if increased exercise is not part of the weight-loss program. When the weight is regained, most of it is fat. Fat is less metabolically active than lean muscle tissue, so the person's metabolic rate drops even lower. Thus, repeated crash dieting may be a "no-lose situation."

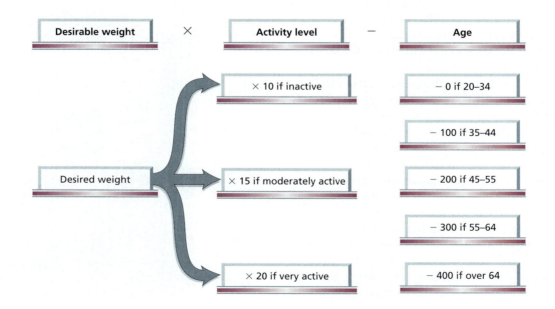

FIGURE **15a.12**

The amount of calories needed to maintain a desirable weight is influenced by one's activity level and age.

(a) 1986   (b) 1988   (c) 2004

FIGURE **15a.13**
Typical dieters change their energy balance temporarily, and, when old habits are resumed, the weight is regained.
The cycle is then repeated. The "yo-yo effect" makes weight loss more difficult with successive diets. The popular
talk show host Oprah Winfrey has been vocal about her struggle with weight regain.

Anorexia Nervosa and Bulimia Are Eating Disorders That Create Calorie Deficits

Obesity can be considered an eating disorder associated with overeating. The obese and overweight would be wise to follow a nutritionally sound weight-loss program.

But dieting can, indeed, go too far. Most people with the eating disorders anorexia nervosa and bulimia begin by dieting. Anorexia nervosa is a deliberate self-starvation. A person whose body weight is 85% or less than expected for his or her height and weight is considered to be anorexic. In contrast, bulimia involves binge eating a large quantity of food followed by purging by self-induced vomiting, enemas, laxatives, or diuretics.

The change in eating habits associated with these eating disorders is thought to be the result of psychological, social, and physiological factors. Both disorders are associated with a preoccupation with body size and shape. Anxious depression also seems to play a role. In addition, anorexics often have a low tolerance for change, explaining why the problem often begins when the person leaves home for college.

The behavior patterns involved in anorexia and bulimia differ, but both result in a severe deficit in calories. An anorexic eats very little food and, therefore, consumes few calories. But, anorexia, which means lack of appetite, is misnamed. Although anorexics deny hunger, they often refuse to eat because of an intense fear of becoming fat—*not* from lack of appetite. Indeed, anorexics are often preoccupied with food and may develop strange rituals about eating. For instance, food may be measured into extremely small amounts or cut into tiny pieces. Excessive exercise is also typical in anorexia. No matter how much weight is lost, however, it can never be enough, because an anorexic has a distorted body image. The person perceives the body as fat even when emaciated (Figure 15a.14).

In contrast, a bulimic eats a huge amount of food but then eliminates it from the body. During a bulimic binge, which may last as long as 8 hours, as many as 20,000 calories may be consumed. Several shorter binges may occur in a single day. Each binge is followed by attempts to purge the body of the calories, usually by self-induced vomiting or by laxatives. Some bulimics take as many as 200 laxatives a week, expecting to rid the body of unwanted calories. It is estimated, however, that laxatives can eliminate only 10% of the calories consumed.

FIGURE **15a.14**
Anorexia is a form of self-starvation. No matter how emaciated the person becomes, the individual still perceives the body as being fat.

Eating disorders have many negative effects on a person's body. One major side effect of anorexia is a severe decrease in bone health. Although the excessive exercise practiced by most anorexics may have a slight strengthening effect on bones, many other factors associated with anorexia work to weaken bones. Amenorrhea (cessation of menstruation), malnutrition, and low body weight, particularly low body fat, are just several potential factors contributing to poor bone health. There is a loss of bone mass, particularly from vertebrae and the long bones of the arms and legs. This makes the bones fragile and increases the risk of fractures, particularly hip fractures.

A bulimic's practice of purging by either self-induced vomiting or overuse of laxatives brings with it special problems. Purging can disturb the balance of electrolytes, especially sodium and potassium; this disturbance, in turn, can lead to muscle cramping, an irregular heartbeat, and cardiac arrest. Laxatives and vomiting also cause dehydration, leading to neurological problems and kidney damage. When vomiting is frequently self-induced, it can cause serious problems. Glands in the face and neck may become swollen. Repeated bathing in acidic stomach contents can cause tooth decay, chronic heartburn, and sores in the mouth or lips. In addition, the acts involved in inducing vomiting can tear the throat or esophagus.

Without treatment, up to 20% of people with serious eating disorders die. Even with treatment, 2% to 3% die. Why? The effects of anorexia are not so different from those of starvation. In the early phase of the illness, an anorexic typically chooses a diet that is low in energy-dense foods but rather high in proteins and other essential nutrients. Dietary protein, combined with the high activity levels characteristic of a person with anorexia nervosa, has a nitrogen-sparing effect. As a result, the initial weight loss is almost entirely due to loss of fat tissue. However, when fat reserves are exhausted and refusal of food becomes more severe, the body begins to break down its own proteins to use as an energy source. The primary sources of these proteins are skeletal and heart muscle. At the same time, water loss is accelerated, especially from the fluids *within* the body cells. The water loss then leads to disturbances in metabolism and electrolytes.

Heart problems are the most common cause of death in anorexics. Starvation, dehydration, and electrolyte disturbances cause the heartbeat to slow (bradycardia) and blood pressure to fall (hypotension). The shrinkage of the heart muscle, which occurs as the patients lose muscle mass as well as body fat, can cause heart murmurs. The decrease in heart size can lead to abnormal blood flow through the heart. In addition, the person can develop congestive heart failure if fluids are replaced too quickly during treatment.

A potential cause of death associated with both eating disorders is hypoglycemia, an abnormally low blood glucose level. Because the brain depends entirely on glucose for its metabolism, hypoglycemia can cause unconsciousness and death.

16

The Urinary System

Organs from Several Body Systems Eliminate Waste

The Kidneys, Ureters, Bladder, and Urethra Make Up the Urinary System

- The kidneys form urine and maintain homeostasis
- Nephrons are the functional units of the kidneys
- The kidneys help maintain acid-base balance
- The kidneys help conserve water
- Hormones influence kidney function
- The kidneys help produce red blood cells and activate vitamin D
- Dialysis and transplant surgery help when kidneys fail

Urination Has Involuntary and Voluntary Components

Bacteria Can Enter the Urethra and Cause Urinary Tract Infections

HEALTH ISSUE Kidney Stones: An Ancient Problem with New (and Shocking) Solutions

HEALTH ISSUE Urinalysis: What Your Urine Says about You

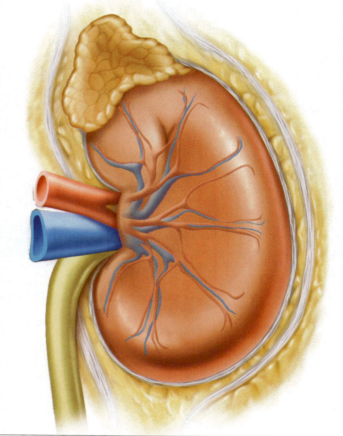

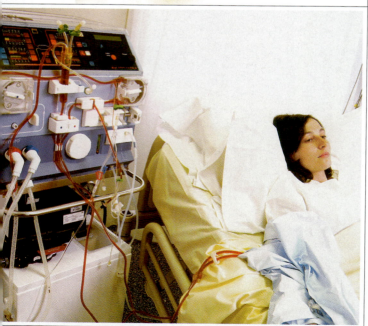

Amanda was feeling depressed. She was 20 years old and doing well in college, but the rest of her life was on hold as she waited for a kidney donor. Her best friend was taking the extended European vacation they had both planned for this summer while she remained tethered by her need for a dialysis machine. With Amanda in mind, do you carry an organ donor card indicating your wish to donate organs or tissues at the time of your death (Figure 16.1)? Have you discussed your decision with your family members? Consider that as of 4:54 P.M. on March 15, 2003, the number of candidates on the waiting list for organs in the United States totaled 80,719. Although the total number of candidates is updated minute by minute, the vast majority of these candidates—53,883—were waiting for kidneys. Why did kidneys top the list? What important roles do kidneys play in our bodies? What medical options—transplantation and others—are available when such critical organs fail? Could your kidney save someone else's life? Could someone else's kidney save you?

In this chapter we will examine kidney structure and consider the critical roles kidneys play in removing wastes from the body and in maintaining homeostasis. We then describe disorders of the kidneys and ways, such as kidney transplantation, to maintain kidney function. We also consider what the characteristics of our urine can say about us and why medical personnel so frequently ask us to provide a urine sample. We finish the chapter with a discussion of urination and urinary tract infections. ◾

Organs from Several Body Systems Eliminate Waste

Kidneys are one of several organs that eliminate wastes—wastes that would severely upset homeostasis if they accumulated in our bodies. Consider that our cells, like tiny laboratories, are sites for the multitude of chemical reactions that constitute metabolism. Just like the day-to-day activities of laboratories, the cells' ongoing synthesis and breakdown of molecules produce wastes. Examples of metabolic wastes include carbon dioxide, water, heat, salts, and nitrogen-containing molecules such as ammonia, urea, and uric acid. Our bodies must diligently dispose of all of them.

Wastes and excess essential ions—such as hydrogen (H^+), sodium (Na^+), potassium (K^+), and chloride (Cl^-)—are eliminated from our bodies through the actions of several organs (Figure 16.2). Lungs and skin eliminate heat, water, and carbon dioxide (as the gas CO_2 from the lungs and as bicarbonate ions, HCO_3^-, from the skin). Skin also excretes salts and urea. Organs of our gastrointestinal tract eliminate solid wastes and many of the other metabolic wastes. Our focus in this chapter is the kidneys, which are the organs of the urinary system responsible for the formation of urine. We will describe their role in excreting nitrogen-containing wastes, as well as water, carbon dioxide (as HCO_3^-), inorganic salts, and hydrogen ions.

FIGURE **16.1**

An organ donor card available through http://www.organdonor.gov. In addition to carrying this card, it is important to let family members know of your decision to donate your organs or tissues. Your intent to donate also can be indicated on your driver's license.

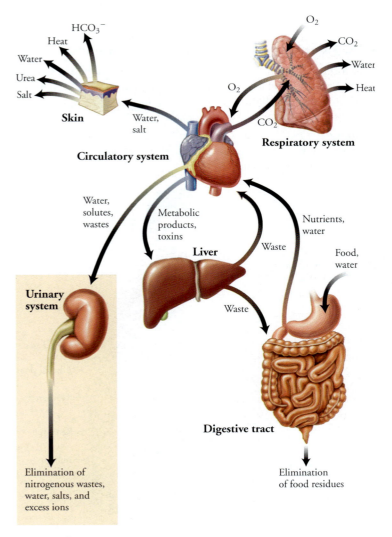

FIGURE **16.2**

Organs from several systems are involved in the removal of wastes from the body. The kidneys, organs of the urinary system, play a major role in the excretion of nitrogenous wastes, water, salts, and excess ions.

The Kidneys, Ureters, Bladder, and Urethra Make Up the Urinary System

The **urinary system** consists of two kidneys, two ureters, one urinary bladder, and one urethra (Figure 16.3). The main function of this system is to regulate the volume, pressure, and composition of the blood. The **kidneys** are the organs of the urinary system that accomplish this task by regulating the amount of water and dissolved substances that are added to or removed from the blood. Wastes and excess materials removed from the blood form **urine**, the yellowish fluid produced by each kidney.

Urine from the kidneys travels down tubelike organs called **ureters** to the urinary bladder. The **urinary bladder** is a muscular organ that temporarily stores urine until it is excreted from the body. Urine leaves the body through the **urethra**, a muscular tube that transports urine from the urinary bladder to the outside of the body through an external opening. The female urethra transports only urine. The male urethra, however, carries urine and reproductive fluids (but not simultaneously). The reproductive function of the male urethra is discussed in Chapter 17. Table 16.1 summarizes the components of the urinary system and their functions.

The kidneys form urine and maintain homeostasis

Our kidneys are reddish in color and shaped like kidney beans (see Figure 16.3). Each kidney is about the size of a fist. The kidneys are located just above the waist between the back wall of the abdominal cavity and the parietal peritoneum (the membrane that lines the abdominal cavity). The slightly indented, or concave, border of each kidney faces the midline of the body. Perched on top of each kidney is an adrenal gland.

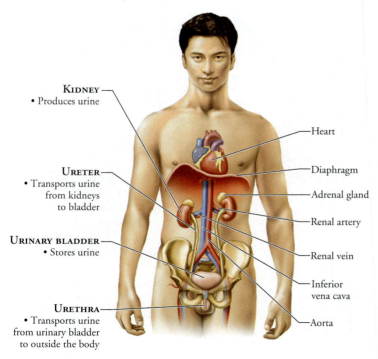

KIDNEY
• Produces urine

URETER
• Transports urine from kidneys to bladder

URINARY BLADDER
• Stores urine

URETHRA
• Transports urine from urinary bladder to outside the body

Heart

Diaphragm

Adrenal gland

Renal artery

Renal vein

Inferior vena cava

Aorta

FIGURE **16.3**
Organs of the urinary system and major blood vessels. Digestive organs are not shown.

TABLE **16.1**

COMPONENTS OF THE URINARY SYSTEM AND THEIR FUNCTIONS	
COMPONENT	**FUNCTION**
Kidneys	Filter wastes and excess material from the blood
	Help regulate blood pressure and pH
	Maintain fluid balance by regulating the volume and composition of blood and urine
	Release erythropoietin, which stimulates production of red blood cells
	Transform vitamin D into its active form
Ureters	Transport urine from kidneys to urinary bladder
Urinary bladder	Stores urine
	Contracts and expels urine into urethra
Urethra	Transports urine from urinary bladder to outside the body
	In males, transports semen to outside the body

The kidneys are covered and supported by several layers of connective tissue (Figure 16.4). The outermost layer anchors each kidney and its adrenal gland to the abdominal wall and surrounding tissues. Beneath this layer is a protective cushion of fat. Sometimes, in very thin people, one or the other of these layers is not substantial enough, and a kidney will slip from its normal position. This painful condition is known as floating kidney. The condition is dangerous because the ureter of the displaced kidney may kink, preventing normal flow of urine from the kidney down the ureter to the bladder. The backup of urine and increased pressure can damage the kidney. The innermost layer covering the kidneys is a transparent layer of fibrous connective tissue. This layer protects the kidneys from trauma and infection. The numerous protective barriers and cushions surrounding our kidneys highlight the importance of these organs to our daily existence.

STRUCTURE OF THE KIDNEYS

The ureter leaves the kidney at a notch in the concave border, as shown in Figure 16.4. This notch is also the area where blood vessels enter and exit the kidney. The renal arteries branch off the aorta and carry blood to the kidneys. The renal veins carry filtered blood away from the kidneys to the inferior vena cava, which transports the blood to the heart.

Each kidney has three regions: an outer region, the **renal cortex**; an inner region, the **renal medulla**; and an inner chamber, the **renal pelvis** (see Figure 16.4). The renal cortex begins at the outer border of the kidney, and portions of it, called renal columns, extend between cone-shaped structures of the renal medulla. These cone-shaped structures of the medulla are called renal pyramids. The narrow end of each renal pyramid joins a cuplike extension of the renal pelvis. As we will soon see, urine produced by the kidneys eventually drains into the renal pelvis and out the ureter to the urinary bladder.

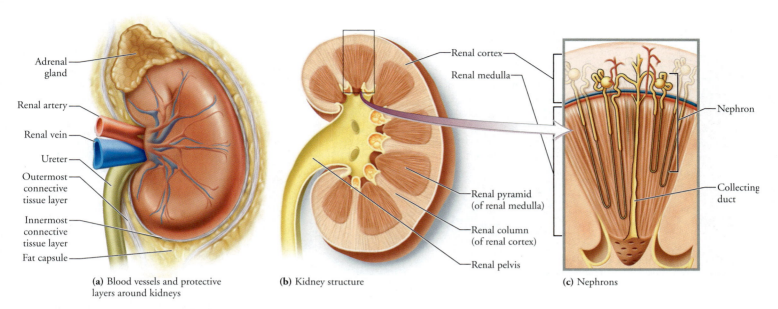

(a) Blood vessels and protective layers around kidneys

(b) Kidney structure

(c) Nephrons

FIGURE **16.4**

Structure of the kidney. These views of the left kidney show (a) the blood vessels and three protective outer layers and (b) the basic internal structure of the kidney. (c) The location of nephrons within these regions is also shown.

▮ Nephrons are the functional units of the kidneys

Nephrons are the functional units of the kidneys and are responsible for the formation of urine. Each kidney contains 1 million to 2 million nephrons.

STRUCTURE OF NEPHRONS

A nephron has two basic parts: the renal corpuscle and the renal tubule, as shown in Figure 16.5. The **renal corpuscle** is the portion of the nephron where fluid is filtered. It consists of a tuft of capillaries, the **glomerulus**, and a surrounding cuplike structure, **Bowman's (glomerular) capsule**. Bowman's capsule has two layers, as would the end of the finger of a glove if it were pushed inward. The space between the two layers is called **Bowman's space**. Blood enters a glomerulus by an afferent (incoming) arteriole. Once within the glomerular capillaries, water and many solutes move from the plasma into Bowman's space and then into the renal tubule. Blood leaves the glomerulus by an efferent (outgoing) arteriole. The **renal tubule** is the site of reabsorption and secretion by the nephron. This tubule has three sections: the **proximal convoluted tubule**, the **loop of Henle**, and the **distal convoluted tubule**. The loop of Henle resembles a hairpin turn, having a descending limb and an ascending limb. The distal convoluted tubules of several nephrons empty into a single collecting duct that eventually drains into the renal pelvis. The renal pelvis is connected to the ureter. Urine exits the kidney by way of the ureter and moves into the urinary bladder.

About 80% of the nephrons in our kidneys have short loops of Henle and are confined almost entirely to the renal cortex. The remaining 20% have long loops of Henle that extend from the cortex into the renal medulla. Once in the medulla, the loops of these nephrons turn abruptly upward, back into the cortex, where they lead into distal convoluted tubules. As we will see, the nephrons whose loops extend into the medulla play an important role in water conservation.

FUNCTIONS OF NEPHRONS

The kidneys are key to maintaining homeostasis. They filter wastes and excess materials from the blood. The kidneys also assist the respiratory system in the regulation of blood pH. Finally, the kidneys maintain fluid balance by regulating the volume and composition of blood and urine.

To understand the functions of the kidneys, we must examine the work of nephrons. Nephrons perform three functions: (1) glomerular filtration, (2) tubular reabsorption, and (3) tubular secretion. In some ways, ridding the blood of wastes is analogous to cleaning a bedroom closet. First, you would remove most of the small items—the valuable along with the unwanted. This action is analogous to glomerular filtration, which removes all materials small enough to fit through the pores of the kidney's filter. (The filter consists of the lining of the glomerular capillaries and the inner lining of Bowman's capsule. Small molecules and other substances pass through the filter into Bowman's space; see below) . The next step in cleaning the closet would be to put back "the good stuff," the items worth saving. Returning valuable materials to the closet is analogous to tubular reabsorption, which returns useful materials to the blood. The final step in cleaning the closet might be to selectively remove any worthless items. This last step is analogous to tubular secretion, in which wastes and excess materials are removed from the blood and added to the filtrate that will eventually leave the body as urine. Here are the three steps in more detail.

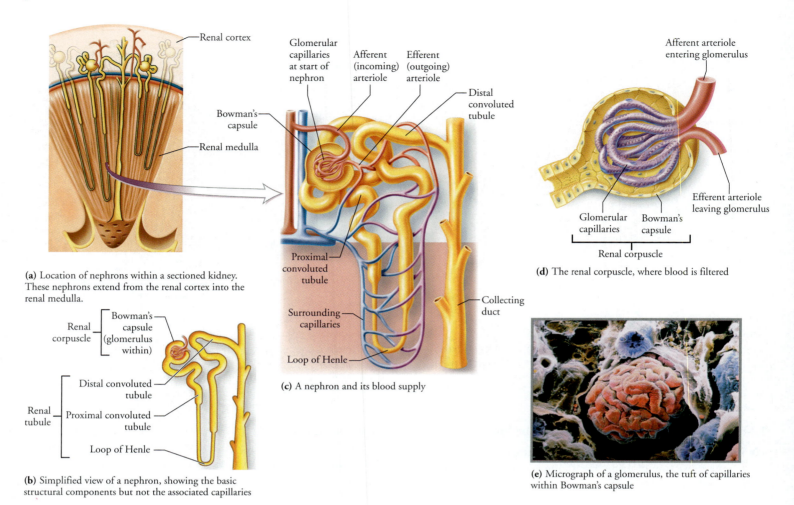

(a) Location of nephrons within a sectioned kidney. These nephrons extend from the renal cortex into the renal medulla.

(b) Simplified view of a nephron, showing the basic structural components but not the associated capillaries

(c) A nephron and its blood supply

(d) The renal corpuscle, where blood is filtered

(e) Micrograph of a glomerulus, the tuft of capillaries within Bowman's capsule

FIGURE **16.5**
Structure of the nephron

1. **Glomerular filtration** occurs as blood pressure forces water and dissolved substances from the blood in the glomerulus to the inside of Bowman's capsule. (The blood pressure is high in the glomerulus because the diameter of the arteriole delivering blood to it is wider than the diameter of the arteriole leaving it.) The filter consists of two cell layers: the lining of the glomerular capillaries and the inner lining of Bowman's capsule. In between the two cell layers is a basement membrane (Figure 16.6). Large molecules and blood cells cannot pass through the pores of the filter. However, water, ions, and smaller molecules pass through the filter into Bowman's space. These substances are known collectively as glomerular filtrate. The concentrations of these molecules in the filtrate are approximately the same as in the plasma. The composition of the filtrate is then adjusted through reabsorption and secretion to form urine.

   Several things can change the rate of filtration by the glomerulus. An increase in the diameter of afferent (incoming) arterioles produces higher pressure in the glomerular capillaries. Higher pressure results in higher filtration rates. A reduction in the diameter of efferent (outgoing) arterioles also produces higher pressure in glomerular capillaries and higher filtration rates. Finally, increases in blood pressure can produce higher filtration rates.

2. **Tubular reabsorption** is the process that removes useful materials from the filtrate and returns them to the blood. This process occurs in the renal tubule, primarily in the proximal convoluted tubule. The structure of the proximal convoluted tubule makes it an ideal location for reabsorption (Figure 16.7). The epithelial cells that make up a proximal convoluted tubule have numerous microvilli (projections of the plasma membrane) that reach into the lumen (passageway) of the tubule. These microvilli dramatically increase the surface area for the reabsorption of materials. Remarkably, as the glomerular filtrate passes through the renal tubule, about 99% of it is returned to capillaries. Thus, only about 1% of the glomerular filtrate is eventually excreted as urine. Put another way, of the approximately 180 liters (48 gallons) of filtrate that enter Bowman's space each day, between 178 and 179 liters (47 gallons) are returned to the blood by reabsorption. The remaining 1 to 2 liters (approximately 1 gallon) are excreted as urine. Reabsorption returns water, essential ions, and nutrients to the blood. For example, about 99% of water molecules and sodium ions are reabsorbed. Almost 100% of glucose molecules and amino acids are reabsorbed. Imagine how much water and food we would have to consume if we did not have reabsorption to offset the losses from glomerular filtration! Some

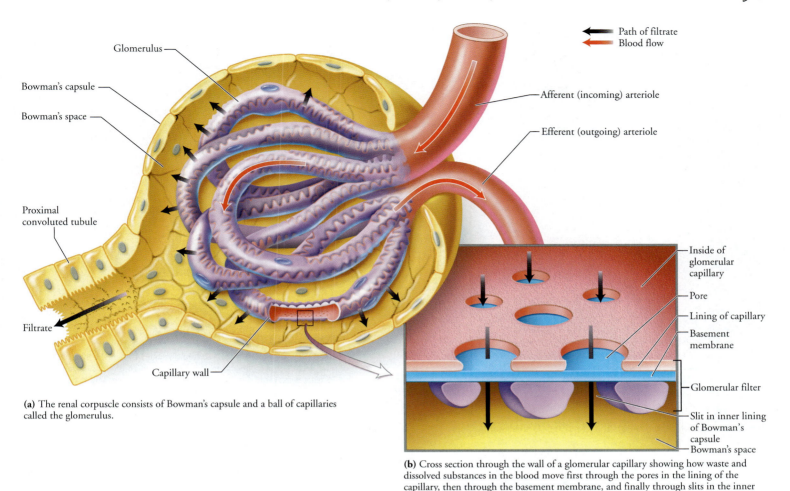

Glomerulus

Bowman's capsule

Bowman's space

Proximal
convoluted tubule

Filtrate

Capillary wall

Path of filtrate
Blood flow

Afferent (incoming) arteriole

Efferent (outgoing) arteriole

Inside of
glomerular
capillary

Pore

Lining of capillary

Basement
membrane

Glomerular filter

Slit in inner lining
of Bowman's
capsule

Bowman's space

(a) The renal corpuscle consists of Bowman's capsule and a ball of capillaries called the glomerulus.

(b) Cross section through the wall of a glomerular capillary showing how waste and dissolved substances in the blood move first through the pores in the lining of the capillary, then through the basement membrane, and finally through slits in the inner lining of Bowman's capsule into Bowman's space.

FIGURE **16.6**
The renal corpuscle is the site of glomerular filtration.

wastes are not reabsorbed at all; others, such as urea, are partially reabsorbed. We will see that antidiuretic hormone (ADH), released by the posterior pituitary gland, regulates the amount of water reabsorbed in parts of the renal tubule.

3. **Tubular secretion** removes wastes and excess ions that escaped glomerular filtration from the blood. Secretion also removes drugs such as penicillin. These wastes are added to the filtered fluid that will become urine. For example, hydrogen ions (H^+), potassium ions (K^+), and ammonium ions (NH_4^+) in the blood are actively transported into the renal tubule, where they become part of the filtrate to be excreted. Tubular secretion occurs primarily along the proximal and distal convoluted tubules. The regions of the nephron and their roles in filtration, reabsorption, and secretion are shown in Figure 16.8 and summarized in Table 16.2.

By the end of filtration, tubular reabsorption, and tubular secretion, blood leaving the kidneys contains most of the water, nutrients, and essential ions that it contained upon entering the kidneys. Wastes and excess materials have been removed, and the blood has been cleansed. This purified blood within capillaries moves into small veins that eventually form the renal vein. Urine contains all of

the materials that were not reabsorbed from the filtrate. Urine within the distal convoluted tubules of nephrons empties into collecting ducts. Here, more water may be reabsorbed, further concentrating

TABLE **16.2**

REGIONS OF THE NEPHRON AND THEIR ROLES IN FILTRATION, REABSORPTION, AND SECRETION	
REGION OF NEPHRON	**ROLE**
Renal corpuscle (Bowman's capsule and glomerulus)	Filters the blood, removing water, glucose, amino acids, ions, nitrogenous wastes, and other small molecules
Proximal convoluted tubule	Reabsorbs water, glucose, amino acids, some urea, Na^+, Cl^-, and HCO_3^- Secretes drugs, H^+, NH_4^+
Loop of Henle	Reabsorbs water, Na^+, Cl^-, and K^+
Distal convoluted tubule	Reabsorbs water, Na^+, Cl^-, and HCO_3^- Secretes drugs, H^+, K^+, NH_4^+

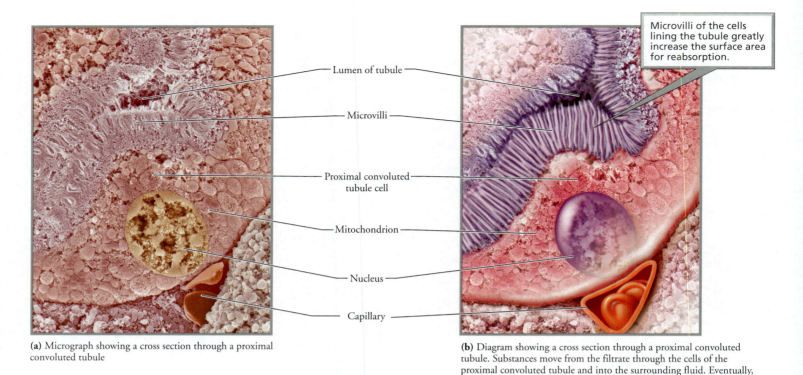

Microvilli of the cells lining the tubule greatly increase the surface area for reabsorption.

Lumen of tubule

Microvilli

Proximal convoluted tubule cell

Mitochondrion

Nucleus

Capillary

(a) Micrograph showing a cross section through a proximal convoluted tubule

(b) Diagram showing a cross section through a proximal convoluted tubule. Substances move from the filtrate through the cells of the proximal convoluted tubule and into the surrounding fluid. Eventually, the substances move into the capillaries nearby.

FIGURE **16.7**
The proximal convoluted tubule is the site of tubular reabsorption.

the urine. From the collecting ducts, urine moves into the renal pelvis and leaves each kidney through a ureter. Urine travels down the ureters to the urinary bladder, which stores urine until it is eliminated from the body through the urethra.

▮ The kidneys help maintain acid-base balance

In addition to removing wastes, the kidneys help regulate the pH of blood. Recall from Chapters 2 and 14 that blood pH must be regulated precisely for proper functioning of the body. This precise regulation is achieved through the actions of the kidneys, respiration, and buffer systems. Buffer systems pick up hydrogen ions (H^+) when such ions are in excess and release hydrogen ions when they are in short supply. In previous discussions, we described the importance of carbonic acid as a buffer in the blood. When carbon dioxide (CO_2) is added to water, it forms carbonic acid (H_2CO_3), which dissociates into hydrogen ions (H^+) and bicarbonate ions (HCO_3^-), as shown in the formula below:

$$CO_2 \; + \; H_2O \; \leftrightarrow \; H_2CO_3 \; \leftrightarrow \; H^+ \; + \; HCO_3^-$$

| Carbon dioxide | Water | Carbonic acid | Hydrogen ion | Bicarbonate ion |

The buffering action of carbonic acid results from the fact that when levels of hydrogen ions (H^+) increase in the blood, the H^+ combine with bicarbonate ions (HCO_3^-) and are removed from solution. This removal of H^+ prevents dramatic changes in pH. Alternatively, when levels of H^+ decrease, carbonic acid (H_2CO_3) dissociates and adds H^+ to solution, which also prevents dramatic changes in pH. The role of the kidneys in helping maintain pH is twofold. First, through

reabsorption of bicarbonate ions, the kidneys aid in restoring the carbonic acid buffer system by resupplying bicarbonate to the blood. Second, through secretion of hydrogen ions, the kidneys remove excess hydrogen ions from the blood.

▮ The kidneys help conserve water

Our kidneys enable us to conserve water through the production of concentrated urine. This task is performed by the 20% of our nephrons with long loops of Henle that dip into the renal medulla. The mechanism of urine concentration is based on an increasing concentration of solutes in the interstitial fluid (the fluid that fills the spaces between cells) from the cortex to the medulla of the kidneys. The main solutes of interest are sodium chloride (NaCl) and urea. The loop of Henle is responsible for maintaining the gradient of sodium chloride in the interstitial fluid. The collecting ducts are sites where urea diffuses out of the filtrate into the interstitial fluid of the medulla. To understand the mechanism of urine concentration, we must trace the path of filtrate as it flows through the renal tubule. We will focus on what happens in the loop of Henle and collecting ducts (Figure 16.9).

The solute concentration of filtrate passing from Bowman's capsule to the proximal tubule is about the same as that of blood. As the filtrate moves through the proximal convoluted tubule, large amounts of water *and* salt are reabsorbed. This reabsorption produces dramatic reductions in the volume of filtrate but little change in its solute concentration. Next, the filtrate enters the descending limb of the loop of Henle; this is where major changes in solute concentration begin (see Figure 16.9). This path takes it from the cortex to the medulla. Along the descending limb, water leaves the filtrate by osmosis (see Chapter 3). The departure of water creates an increase

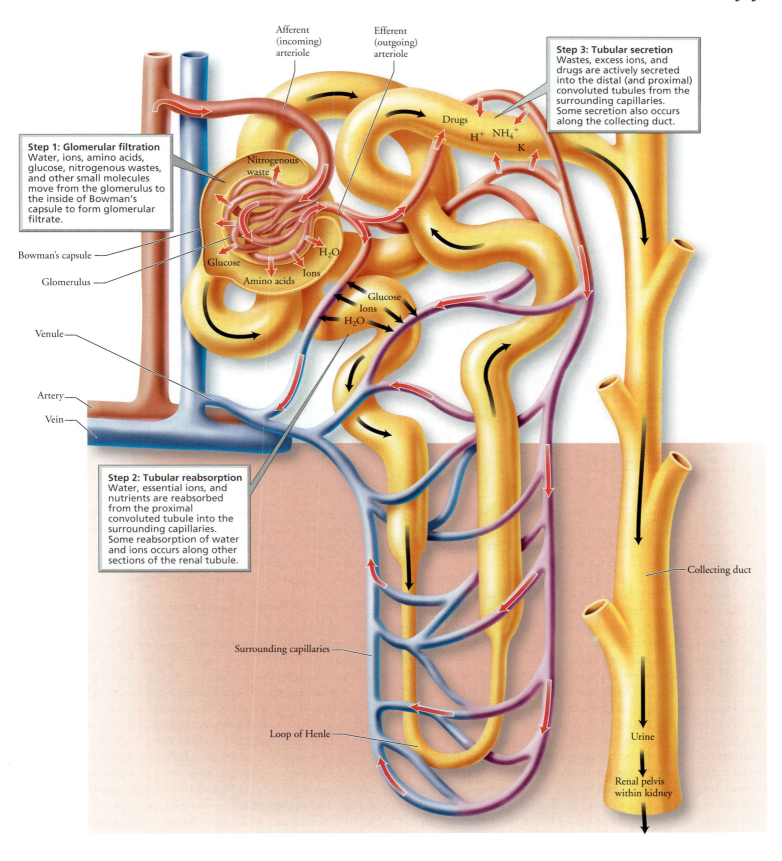

Afferent (incoming) arteriole

Efferent (outgoing) arteriole

Step 3: Tubular secretion
Wastes, excess ions, and drugs are actively secreted into the distal (and proximal) convoluted tubules from the surrounding capillaries. Some secretion also occurs along the collecting duct.

Drugs

H^+ NH_4^+ K

Step 1: Glomerular filtration
Water, ions, amino acids, glucose, nitrogenous wastes, and other small molecules move from the glomerulus to the inside of Bowman's capsule to form glomerular filtrate.

Nitrogenous waste

Bowman's capsule

Glomerulus

Glucose H_2O

Amino acids   Ions

Glucose
Ions
H_2O

Venule

Step 2: Tubular reabsorption
Water, essential ions, and nutrients are reabsorbed from the proximal convoluted tubule into the surrounding capillaries. Some reabsorption of water and ions occurs along other sections of the renal tubule.

Artery

Vein

Collecting duct

Surrounding capillaries

Loop of Henle

Urine

Renal pelvis within kidney

WWW   TUTORIAL 16.2   FIGURE **16.8**
Summary of glomerular filtration, tubular reabsorption, and tubular secretion along the nephron

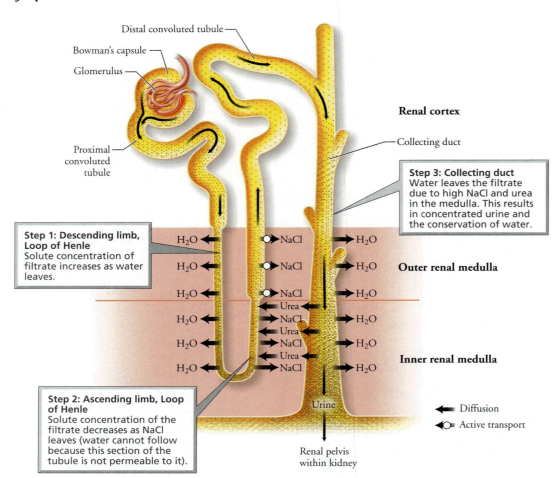

Distal convoluted tubule

Bowman's capsule

Glomerulus

Proximal convoluted tubule

Renal cortex

Collecting duct

Step 3: Collecting duct
Water leaves the filtrate due to high NaCl and urea in the medulla. This results in concentrated urine and the conservation of water.

Step 1: Descending limb, Loop of Henle
Solute concentration of filtrate increases as water leaves.

H_2O $NaCl$ H_2O

H_2O $NaCl$ H_2O

Outer renal medulla

H_2O $NaCl$ H_2O
$Urea$

H_2O $NaCl$ H_2O
$Urea$

H_2O $NaCl$ H_2O
$Urea$

Inner renal medulla

H_2O $NaCl$ H_2O

Step 2: Ascending limb, Loop of Henle
Solute concentration of the filtrate decreases as NaCl leaves (water cannot follow because this section of the tubule is not permeable to it).

Urine

◀── Diffusion
◀─○ Active transport

Renal pelvis within kidney

FIGURE **16.9**
Some nephrons have loops of Henle that extend into the medulla. These nephrons are responsible for water conservation. The steps by which these nephrons concentrate urine and conserve water are shown here. Stippling within the nephron indicates solute concentration.

in the concentration of solutes, including sodium chloride (NaCl), within the filtrate. The concentration of salt in the filtrate peaks at the curve of the loop of Henle, setting the stage for the next step in the process of urine concentration. As the filtrate moves up the ascending limb of the loop of Henle, large amounts of sodium chloride move out of the filtrate into the interstitial fluid of the medulla. Water remains in the filtrate because the ascending limb is not permeable to it, and water cannot pass out of the ascending limb to the interstitial fluid. When the filtrate reaches the distal convoluted tubule in the cortex, it is quite dilute. In fact, the filtrate is hypotonic to body fluids because it has a lesser solute concentration than body fluids (see Chapter 2). The filtrate then moves into a collecting duct and begins its descent, once again, toward the medulla. This pathway is one of increasing salt concentration in the interstitial fluid. Collecting ducts are permeable to water but not to salt. Thus, as the filtrate encounters increasing concentrations of salt in the fluid of the inner medulla, water leaves the filtrate by osmosis. With the departure of large amounts of water, urea is now concentrated in the filtrate. Some of the urea moves into the interstitial fluid of the medulla from lower portions of the collecting duct. The remaining urea is excreted. This leakage of urea contributes to the high solute concentration of the inner medulla and thus aids in concentrating the filtrate. At its most concentrated, urine is hypertonic to blood and interstitial fluid in all parts of the body except the inner medulla, where it is isotonic. (Recall from Chapter 2 that *hypertonic* means having a greater solute concentration than another fluid. *Isotonic*

means having the same solute concentration as another fluid.) Together, then, the loop of Henle and collecting ducts maintain solute gradients in the interstitial fluid of the kidneys. It is the maintenance of these gradients that makes possible the concentration of urine and conservation of water by the kidneys.

Hormones influence kidney function

Our body's maintenance of salt and water levels near the optimum values is essential to health. As we have seen, maintaining these levels is an important job of the kidneys. This job can be challenging because our activities change levels of salt and water in our bodies. For example, on a hot day or after exercise, we may lose body water through perspiration. Eating a tub of popcorn at the movies boosts our salt intake. The kidneys must deal with these events and adjust the concentration of solutes in the urine to keep water and salt levels in our body relatively constant.

Three hormones—antidiuretic hormone, aldosterone, and atrial natriuretic peptide—play important roles in adjusting kidney function to meet the body's needs. **Antidiuretic hormone (ADH)** is manufactured by the hypothalamus and then travels to the posterior pituitary for storage and release (see Chapter 10). Antidiuretic hormone regulates the amount of water reabsorbed by the distal convoluted tubules and collecting ducts of nephrons. The hypothalamus responds to changes in the concentration of water in the blood by increasing or decreasing secretion of ADH. Decreases in the concentration of water in the blood stimulate increased secretion of ADH, as shown in Figure 16.10. Higher levels of ADH in the bloodstream increase the perme-

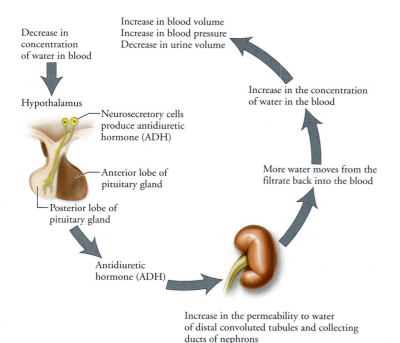

Decrease in concentration of water in blood

Increase in blood volume
Increase in blood pressure
Decrease in urine volume

Hypothalamus

Neurosecretory cells produce antidiuretic hormone (ADH)

Anterior lobe of pituitary gland

Posterior lobe of pituitary gland

Antidiuretic hormone (ADH)

Increase in the concentration of water in the blood

More water moves from the filtrate back into the blood

Increase in the permeability to water of distal convoluted tubules and collecting ducts of nephrons

FIGURE **16.10**

The pathway by which antidiuretic hormone (ADH) influences kidney function and blood volume and blood pressure

ability to water of the distal convoluted tubules and collecting ducts of nephrons, resulting in more water being reabsorbed from the filtrate. The movement of increased amounts of water from the filtrate back into the blood results in increased blood volume and pressure and production of small amounts of concentrated urine. Just the

opposite occurs when the concentration of water in the blood increases. Increases in water concentration inhibit the release of ADH, thereby reducing water reabsorption from the filtrate. Reduced water reabsorption causes reduced blood volume and pressure and production of large amounts of dilute urine. Alcohol, too, inhibits the secretion of ADH, causing reduced water reabsorption by the kidneys and production of large amounts of dilute urine. Thus, it makes little sense after exercising to try to quench your thirst and restore body fluids by drinking an alcoholic beverage. Substances such as alcohol that promote urine production are called **diuretics**. **Diabetes insipidus** is a disease characterized by excretion of large amounts of dilute urine. This condition is caused by a deficiency of ADH.

Aldosterone is a steroid hormone released by the adrenal glands (specifically the adrenal cortex; see Chapter 10). Aldosterone increases reabsorption of sodium by the distal convoluted tubules and collecting ducts. This process is important because water follows sodium. As more sodium is transported out of the nephron into the capillaries, increased amounts of water follow, resulting in increased blood volume and pressure and production of small amounts of concentrated urine. Caffeine, a familiar chemical found in coffee, tea, and many soft drinks, has just the opposite effect. Caffeine decreases reabsorption of sodium, thus decreasing the amount of water moving out of tubules back into the bloodstream. Ingestion of caffeine causes decreased blood volume and pressure and large amounts of dilute urine. Because caffeine, like alcohol, increases urine output, it too is considered a diuretic.

What stimulates the release of aldosterone? Ultimately, it is the blood pressure in the afferent (incoming) arteriole carrying blood to the glomerulus. Our focus will be on a region of the nephron called the **juxtaglomerular apparatus**, a group of cells located where the distal convoluted tubule contacts the afferent arteriole (Figure 16.11).

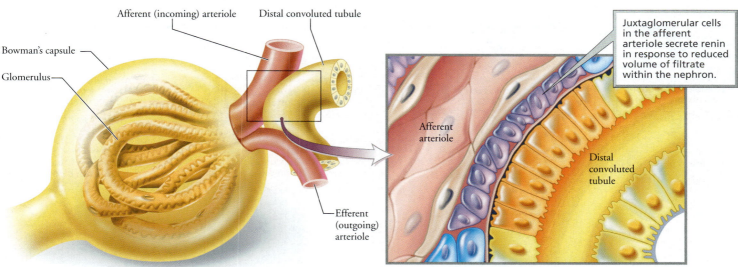

Afferent (incoming) arteriole

Distal convoluted tubule

Bowman's capsule

Glomerulus

Efferent (outgoing) arteriole

Juxtaglomerular cells in the afferent arteriole secrete renin in response to reduced volume of filtrate within the nephron.

Afferent arteriole

Distal convoluted tubule

(a) The juxtaglomerular apparatus (within the square) is a section of the nephron where the distal convoluted tubule contacts the afferent arteriole. The nearby renal corpuscle is shown in ghosted view to reveal its components, Bowman's capsule and the glomerulus.

(b) Close-up view of the juxtaglomerular apparatus

FIGURE **16.11**

The juxtaglomerular apparatus

The juxtaglomerular apparatus monitors blood pressure in the afferent arteriole. Look at the central pathway in Figure 16.12 as we describe each step in the process that leads to aldosterone release. When the blood pressure in the afferent arteriole drops, so does the glomerular filtration rate. This drop in filtration rate, in turn, reduces the volume of filtrate within nephrons. In response, cells within the juxtaglomerular apparatus release the enzyme renin. **Renin** converts angiotensinogen, a protein produced by the liver, into another protein, **angiotensin I**. Angiotensin I is then converted by yet another enzyme into angiotensin II. **Angiotensin II** is the active form of the protein that stimulates the adrenal gland to release aldosterone. Aldosterone increases reabsorption of sodium and water by the distal convoluted tubules and collecting ducts of nephrons, resulting in increased blood volume and pressure. These changes increase the filtration rate within the glomerulus, resulting in an increased volume of filtrate within the nephron.

Besides prompting release of aldosterone, angiotensin II influences blood pressure and volume in three other ways, as shown in the side pathways in Figure 16.12. First, angiotensin II constricts arterioles throughout the circulatory system, causing increases in blood pressure. Second, angiotensin II stimulates the thirst center in the hypothalamus to increase our intake of water. Finally, angiotensin II prompts the posterior lobe of the pituitary gland to release antidiuretic hormone.

stop and think

Estrogens are female sex hormones that are chemically similar to aldosterone. As a result of this similarity, estrogens have similar effects on the distal convoluted tubules and collecting ducts of the kidneys. How might this similarity explain the water retention experienced by many women as their estrogen levels rise during the menstrual cycle?

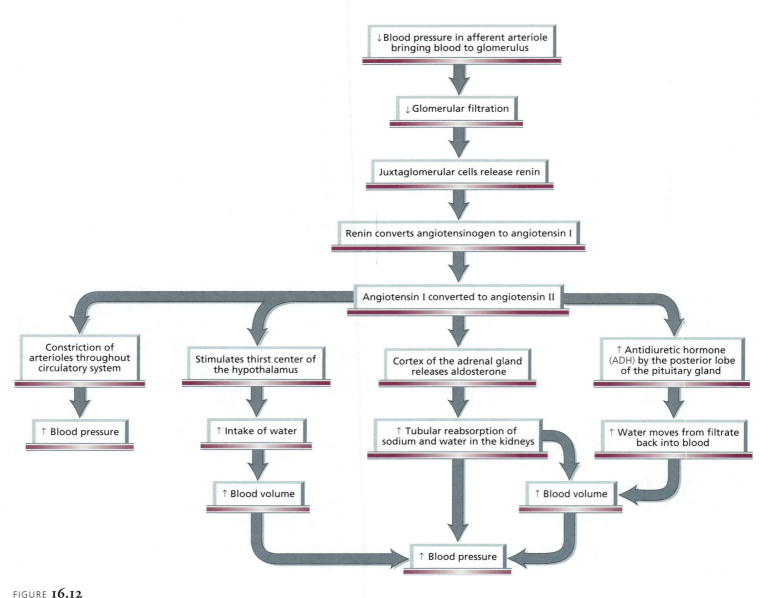

FIGURE **16.12**

The renin-angiotensin system and the many routes by which it influences blood pressure

A final hormone that influences kidney function is **atrial natriuretic peptide (ANP)**. Cells in the right atrium of the heart release this hormone. Release of ANP occurs in response to stretching of the heart caused by increased blood volume and pressure. Atrial natriuretic peptide decreases water and solute reabsorption by the kidneys either by increasing the permeability of the glomerular filter or by dilating afferent arterioles. Regardless of its precise site of action, ANP causes declines in blood volume and pressure and the production of large amounts of urine. Researchers hope that ANP can be used to treat high blood pressure and edema (the accumulation of fluid in tissues). Finally, ANP also appears to influence kidney function indirectly by inhibiting secretion of ADH, aldosterone, and renin.

The kidneys help produce red blood cells and activate vitamin D

The kidneys have two additional functions that are important to homeostasis but are not directly related to the urinary system. First, the kidneys release **erythropoietin,** a hormone that travels to the red bone marrow, where it stimulates production of red blood cells. The second function of the kidneys unrelated to the urinary system involves vitamin D. Vitamin D is a substance provided by certain foods in our diet or produced by the skin in response to sunlight. The kidneys transform vitamin D into its active form, calcitriol, which promotes absorption and the use of calcium and phosphorus by the body.

Dialysis and transplant surgery help when kidneys fail

Renal failure is a decrease or complete cessation of glomerular filtration. In other words, the kidneys stop working. Such failure can be acute or chronic. **Acute renal failure** refers to an abrupt, complete or nearly complete, cessation of kidney function. This condition typically develops over a few hours or days and is characterized by little output of urine. Causes of acute renal failure include nephrons damaged by severe inflammation, certain drugs, and poison. Acute renal failure also may be caused by low blood volume due to profuse bleeding or obstruction of urine flow by kidney stones (see the Health Issue essay, *Kidney Stones: An Ancient Problem with New (and Shocking) Solutions*). **Chronic renal failure**, on the other hand, is a progressive and often irreversible decline in the rate of glomerular filtration. This decline in function occurs over a period of months or years. Kidney disease may destroy nephrons and cause progressive declines in filtration rate. Nephrons lost to kidney disease cannot be replaced. Polycystic disease, for example, is an inherited and progressive condition in which fluid-filled cysts and tiny holes form throughout kidney tissue. Few symptoms are apparent at the beginning of chronic renal failure because the remaining nephrons enlarge and take over for those that have been destroyed. With time, however, the loss of nephrons becomes so severe that symptoms of decreased glomerular filtration rate appear. For example, increased levels of nitrogenous wastes are found in the blood. By end-stage renal failure, about 90% of the nephrons have been lost, making necessary a kidney transplant or the use of an artificial kidney machine.

Renal failure, whether acute or chronic, has many consequences. Among the consequences are (1) acidosis, a decrease in blood pH caused by the inability of the kidneys to excrete hydrogen ions; (2) anemia, low numbers of red blood cells caused by the failure of damaged kidneys to release erythropoietin; (3) edema, the buildup of fluid in the tissues because of water and salt retention; (4) hypertension, an increase in blood pressure caused by failure of the renin-angiotensin system and salt and water retention; and (5) accumulation of nitrogenous wastes in the blood. In short, failure of the kidneys severely disrupts homeostasis. Untreated kidney failure will lead to death within a few days. The cause of death often is cardiac arrest induced by abnormal contractions of heart muscle. Such contractions result from high levels of potassium ions (K^+) in the blood and tissue fluids.

DIALYSIS

A common way of coping with failure or severe impairment of the kidneys is **hemodialysis**, the use of artificial devices to cleanse the blood. This process frequently involves the use of an artificial kidney machine (Figure 16.13). A tube is inserted into an artery of the patient's arm. Blood flows into the tube and into the kidney machine, where it is filtered and then returned to the patient's body. Within the machine, the blood flows through tubing made of a selectively permeable membrane. The tubing is surrounded by a dialysis solution called the dialysate. The selectively permeable membrane permits wastes and excess small molecules to move from the blood into the dialysate. The membrane prevents the passage into the dialysate of blood cells and most proteins, which are too large. Nutrients, which move into the blood, are sometimes provided in the dialysate. The composition of the dialysate is precisely controlled to maintain proper concentration gradients between the solution and the blood. Once the blood has completed the circuit through the tubing, it is returned, free of wastes, to a vein in the arm. Patients requiring hemodialysis typically undergo the procedure about three times a week. Each session lasts several hours.

In past years, people went to a hospital for hemodialysis. Today, some patients dialyze themselves at home. Several things are necessary for home dialysis to work. First, the patient must be highly motivated to learn and follow the precise routine. Second, the patient must have a small room that can be devoted to the kidney machine and its associated paraphernalia. Finally, the patient must have someone to help with setting up and coming off the machine. Another advance in hemodialysis is the miniature portable kidney. This small artificial kidney is available for persons who wish to be away from home or the hospital for 2 to 3 weeks. The miniature portable kidney makes vacations possible.

Another option for removing wastes from the blood is **continuous ambulatory peritoneal dialysis (CAPD)**. In this procedure, the peritoneum, one of the body's own selectively permeable membranes, is used as the dialyzing membrane (Figure 16.14). The peritoneum lines the abdominal cavity and covers the internal organs. Dialyzing fluid held in a plastic container suspended over the patient flows down a tube inserted into the abdomen. As the fluid bathes the peritoneum, wastes move from the blood vessels that line the abdomen across the peritoneum and into the solution.

HEALTH ISSUE

Kidney Stones: An Ancient Problem with New (and Shocking) Solutions

Ask anyone who has experienced passing a good-sized kidney stone—one with a diameter of at least 5 mm (0.2 in.)—and the words most often used to describe the event are "excruciatingly painful." Kidney stones, by the way, are not strictly a modern ailment. They have been found, for example, in Egyptian mummies, and we can only assume that their passing caused as much distress in ancient times as today. What are these pebble-sized "hell raisers" within the urinary tract? What makes their passing so memorable? How can they be treated? How can they be prevented?

Kidney stones, also called renal calculi, are small, hard crystals. These crystals form when substances such as calcium or uric acid precipitate out of urine because of higher than normal concentrations. For example, increased levels of calcium in the urine may result from increased levels of calcium in the blood. Excessive dietary intake of calcium, the destruction of bone by disease, or overactive parathyroid glands may cause the high levels of blood calcium. (Parathyroid glands produce parathormone, a hormone that increases calcium in the blood; see Chapter

10.) Kidney stones also are associated with some rare metabolic disorders. More generally, higher-than-normal concentrations of stone-forming substances in the urine result from dehydration and the production of small amounts of concentrated urine. Urine normally has substances that prevent the formation of crystals. These substances may not function in people with a history of kidney stones.

Kidney stones often begin as just a speck in the renal pelvis. The stones grow over a period of years as more and more material is deposited on the outside of the original particle (Figure 16.A). Some stones may attain diameters of 25 mm (1 in.). Before reaching anything near that size, however, many stones are flushed out of the kidneys and down the ureters to the bladder, and from there they are expelled with urine out the urethra. These tiny stones rarely cause problems, and people may not even be aware of their passing. Somewhat larger stones may cause considerable pain during their journey through the urinary tract. The sharp edges of large stones may gouge into the walls of the ureters and sometimes the urethra. The pain is

called renal colic, and it comes in waves. Each wave may cause the person to double over. Renal colic may be localized in the back or pelvic area, often on one side, depending on which ureter the stone is moving through. Nausea, vomiting, and chills accompany the pain, and blood may appear in the urine. Even more serious problems arise when large stones become lodged in the kidneys or ureters, blocking the flow of urine. The high internal pressure that results from the obstruction can seriously damage the nephrons and impair kidney function.

Kidney stones usually are diagnosed with x-rays of the kidneys, ureters, and bladder. Blood and urine tests also are performed to determine what might be causing the problem. In the past, stones lodged in the kidneys or ureters were removed through major abdominal surgery, and recovery took several weeks. Today, doctors use extra-corporeal shock wave lithotripsy (ESWL), a relatively painless technique in which high-energy shock waves are created outside the body and then directed at the stones within. During the procedure, the patient lies in a water bath or on a soft cushion. The shock waves pulverize the

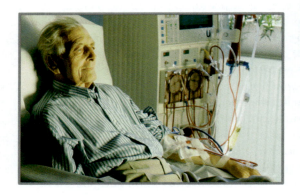

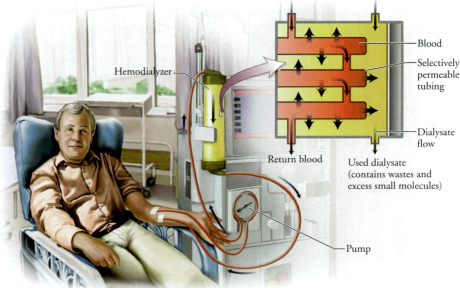

FIGURE **16.13**

An artificial kidney machine can be used to cleanse the blood when the kidneys fail.

stones into tiny pieces that can be passed painlessly in the urine. Because this new technique does not involve making an incision, recovery is more rapid, and the procedure is less costly than surgery. Occasionally, particularly large stones cannot be broken apart by lithotripsy, and surgical removal is needed. Certain stones, particularly those formed of uric acid, can be dissolved with drugs.

How common are kidney stones, and who gets them? Kidney stones seem to run in families and are more common among people living in hot climates than among people in cool climates. People living in the tropics often do not consume enough fluid to counteract the large amounts of water lost through breathing and sweating. This water loss results in the production of small amounts of urine with high concentrations of stone-forming substances. In addition, men seem more susceptible to kidney stones than do women, and stones tend to develop in people between 20 and 40 years of age. Overall, about 10% of people in the United States and Canada will have a kidney stone at some point in their lives.

What can be done to prevent the formation of kidney stones? A simple solution is to drink large quantities of water, about 2.4 to 2.8 liters (5 to 6 pints) a day. Drinking this amount of liquid helps keep the urine dilute and flushes any existing stones from the urinary tract. Depending on the type of stone, dietary changes may be recommended. Finally, people with a history of developing kidney stones usually can take specific medications to prevent formation of new stones once the old ones have been removed. 👫

Kidney stones

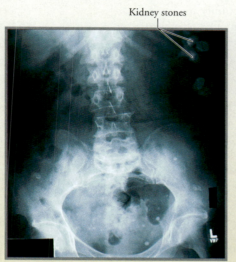

(a) An x-ray showing several kidney stones in the left kidney

(b) Example of a kidney stone that was removed through surgery

FIGURE **16.A**
Kidney stones

Dialysate bag

Fresh dialyzing fluid

Fresh dialyzing fluid flows into the abdominal cavity, where wastes in nearby blood vessels move across the peritoneum and into the dialyzing fluid.

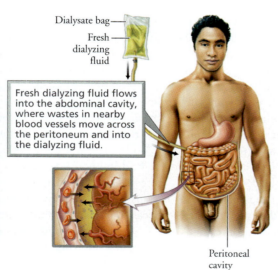

Dialyzing fluid containing wastes is drained from the abdominal cavity and discarded.

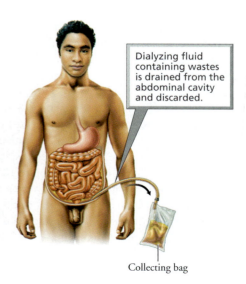

Peritoneal cavity

Collecting bag

FIGURE **16.14**
Continuous ambulatory peritoneal dialysis uses the peritoneum (the membrane that lines the abdominal cavity and covers the organs within) as the dialyzing membrane.

The solution is then returned to the plastic container and discarded. Typically, fluid is passed into the abdomen, left for a few hours, and then removed and replaced with new fluid. CAPD involves about three or four fluid changes each day. However, the patient is free to move around between fluid changes while dialysis proceeds internally. This procedure can be done at home or in the hospital.

Continuous ambulatory peritoneal dialysis is easier to learn and less costly than hemodialysis. This treatment also is gentler in its waste removal. However, CAPD requires several daily changes of dialysis fluid. Each change requires hooking up a new container of dialysate. Thus, with CAPD there is ample opportunity for bacteria to move down the tube and into the abdomen. These bacteria may cause peritonitis (inflammation of the peritoneum). Even the most meticulous patients suffer at least one episode of peritonitis a year. Each episode requires hospitalization and treatment with an antibiotic. Eventually, the peritoneum may become so scarred from bouts of peritonitis that CAPD is no longer possible.

KIDNEY TRANSPLANTATION

The ultimate hope for many people whose kidneys fail is to receive a healthy kidney from another person. The kidney was one of the first organs to be successfully transplanted. Availability of dialysis was critical to the success of kidney transplants. Dialysis keeps people alive until a suitable donor organ can be found. Indeed, progress in liver and heart transplantation has been slower because of the difficulty of providing temporary mechanical replacements for those organs.

What makes a suitable donor organ? The main obstacle to successful acceptance of a transplanted organ is the rejection of foreign tissue by the patient's own immune system (see Chapter 13). The most suitable kidney would come from a patient's identical twin. It is not surprising, then, that the first successful transplant occurred in 1955 between identical twins. A kidney from a close relative, such as a parent or sibling, would be the next best choice. Removal of a healthy kidney is a safe operation, and the donor who gave up a kidney can get along fine with the remaining kidney. About 90% of kidneys donated by close relatives are still functioning 2 years after transplantation. Most donated kidneys come from individuals unrelated to the patient who agreed before their death to donate their organs. Such individuals usually died suddenly in an accident. About 75% percent of kidneys donated from unrelated individuals matched as closely as possible to the patient's tissue and blood type are still functioning 2 years after transplantation. Finally, even if a transplant fails after a few years, successful second and third transplants may extend a patient's life.

In most situations, the kidneys of the patient needing a transplant are not removed. The donor kidney is simply transplanted into a protected area within the pelvis (Figure 16.15). The ureter of the donated kidney is attached to the recipient's bladder, and blood vessels of the donated kidney are attached to the recipient's vessels. The recipient's own kidneys would be removed if they were infected or if they were causing problems such as high blood pressure.

The high success rate of kidney transplants is linked to the use of drugs that suppress the recipient's immune system. Typically, transplant recipients must take two or more of these medications for the rest of their lives. The ability of these drugs to inhibit rejection of transplanted organs has revolutionized organ transplantation. Nevertheless, there are serious concerns about the side effects

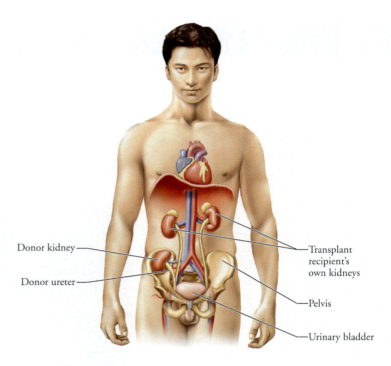

Donor kidney

Donor ureter

Transplant recipient's own kidneys

Pelvis

Urinary bladder

FIGURE **16.15**
In most transplant situations, the donor organ is located in a safe region within the pelvis, and the recipient's own kidneys are left in place. The ureter from the transplanted kidney is attached to the recipient's bladder. Blood vessels from the transplanted kidney are attached to the recipient's vessels.

and long-term use of drugs that suppress the immune system. The side effects include weight gain, acne, growth of facial hair, and increased susceptibility to infection. Over the long term, these drugs may increase the risk of cancer, heart disease, and liver damage. Recently, a team of researchers developed a way to perform kidney transplants without the recipients' needing to permanently remain on immune-suppressing drugs. The suppressing drugs were given to recipients after transplant surgery. Next, the recipients received several radiation treatments, which were directed at reducing the number of cells in their immune system capable of attacking the new organ. Finally, the recipients received blood stem cells from the person who donated the transplanted kidney. These new stem cells moved to the recipient's bone marrow and generated new blood cells and immune cells. In essence, each recipient now had a hybrid immune system, a mix of his or her own cells and those of the donor. The hybrid immune systems did not recognize the transplanted kidneys as foreign, and recipients were gradually weaned off the immune-suppressing medications.

Despite major advances in kidney transplantation, there is still room for improvement. For example, donor kidneys can be kept alive and healthy for only about 24 to 48 hours. This short window of opportunity necessitates rapid location of a recipient, shipment of the donor kidney, and transplantation (Figure 16.16). Perhaps most important, kidneys available for transplantation are always in short supply. Many patients wait for months or years for a new kidney.

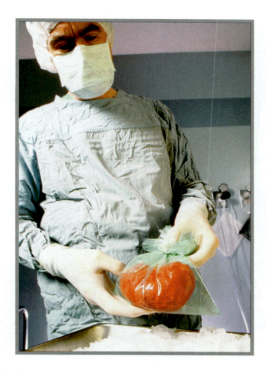

FIGURE **16.16**
Before transplantation, donor kidneys must be kept in a cool salt solution under sterile conditions. Even under such conditions, kidneys will deteriorate after about 1 or 2 days.

what would you do?

The constant shortage of donor organs has led some people to suggest a change in the laws regarding organ donation. At present, a person must give permission to use his or her organs after death. If a potential donor is incapable of giving such permission, then medical personnel may consult close relatives. In recent years, individuals concerned about severe shortages of organs have suggested changing the laws so that a person would have to provide written proof that they did *not* want their organs made available for transplantation after death. In other words, without anything in writing, the law would presume that the person had no objection to organ donation. Many healthy people probably have no objections to organ donation but simply do not think about death and the future use of their organs. Thus, it is likely that such a change in law would go a long way to solving organ shortages. Nevertheless, there are some concerns. Would such a system promote an image of organ transplantation as body snatching? Would relatives claim that doctors and nurses let their loved one die because someone else needed his or her organs? If you were given the responsibility of deciding whether to change the laws regarding organ donation, what would you do?

Urination Has Involuntary and Voluntary Components

Urination is the process by which the urinary bladder is emptied. This process involves both involuntary and voluntary actions and is summarized in Figure 16.17. Recall that the kidneys produce urine around the clock. The urine trickles down the ureters to the urinary bladder, where it is temporarily stored. The walls of the ureters contain smooth muscle whose contractions, called peristaltic contractions, help move urine to the bladder (gravity also plays a role). Each ureter has a slitlike opening into the bladder. This shape of opening helps prevent the backflow of urine from the bladder up into the ureters. When at least 200 ml (0.42 pints) of urine has accumulated, stretch receptors in the wall of the bladder send impulses along sensory neurons to the lower part of the spinal cord. From the spinal cord, impulses are sent along motor neurons back to the bladder, where they cause a band of smooth muscle in the wall of the bladder, called the detrusor muscle, to contract. The impulses also cause the **internal urethral sphincter**, a thickening of smooth muscle located at the junction of the bladder and urethra, to relax. These combined actions push stored urine into the urethra. Upon arrival of sensory impulses in the lower spinal cord, information travels up to the brain and a few moments later initiates the desire to void. At about this time, we feel the need to urinate. If the time for urination is appropriate, then the brain permits voluntary relaxation of the **external urethral sphincter**, a small band of skeletal muscle farther down the urethra. With relaxation of the external sphincter, urine exits the body. The external sphincter usually is kept in a constant state of contraction by impulses from the brain and spinal cord. Thus, it is only when such impulses are inhibited that the muscle relaxes, letting urine flow. If, however, we deem the time to be inappropriate for urination, the brain does not permit the external sphincter to relax. Under these conditions, we can, within reason, store the urine until a better time. Urine will exit the body later, when we allow the external sphincter to relax. Any urine left in the bladder is pushed out by further contractions of the detrusor muscle at the time of voiding. Then the muscle relaxes and allows the bladder to fill once again with urine. Urination is a reflex, a relatively rapid response to a stimulus that is mediated by the nervous system. Nevertheless, urination can be started and stopped voluntarily because of the control exerted by the brain over the external urethral sphincter.

Not everyone can control his or her external urethral sphincter. Lack of voluntary control over urination is called **urinary incontinence**. Incontinence is the norm for infants and children younger than 2 or 3 years of age because nervous connections to the external urethral sphincter are incompletely developed. The very young, then, do not have conscious control over urination. Infants and young children void whenever their bladder fills with enough urine to activate its stretch receptors. Toilet training, a phase often dreaded by parents, occurs when toddlers learn to bring urination under conscious control (Figure 16.18). This step is made possible by the development of complete neural connections to the external sphincter. Damage to the external sphincter in adults may cause incontinence. In men, surgery on the prostate gland may damage the external sphincter. The prostate gland surrounds the male urethra just below the urinary bladder and contributes substances to semen (see Chapter 17). Incontinence also may occur when bladder muscles contract before the bladder is full. Such bladders often are described as being overactive or spastic. Strokes, multiple sclerosis, and spinal cord injuries also can cause incontinence. These conditions may affect nerves supplying the bladder or those carrying information regarding fullness of the bladder to the brain. Infection of the urinary system (see below) can cause incontinence in any age group.

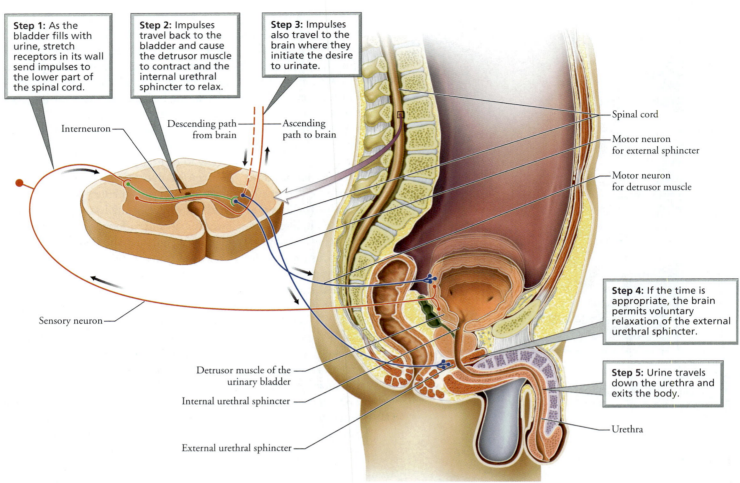

Step 1: As the bladder fills with urine, stretch receptors in its wall send impulses to the lower part of the spinal cord.

Step 2: Impulses travel back to the bladder and cause the detrusor muscle to contract and the internal urethral sphincter to relax.

Step 3: Impulses also travel to the brain where they initiate the desire to urinate.

Interneuron

Descending path from brain

Ascending path to brain

Spinal cord

Motor neuron for external sphincter

Motor neuron for detrusor muscle

Sensory neuron

Step 4: If the time is appropriate, the brain permits voluntary relaxation of the external urethral sphincter.

Step 5: Urine travels down the urethra and exits the body.

Detrusor muscle of the urinary bladder

Internal urethral sphincter

External urethral sphincter

Urethra

FIGURE **16.17**
The steps involved in urination

Mild incontinence, especially of the form called stress incontinence, is common in adults. **Stress incontinence** is characterized by the escape of small amounts of urine when sudden increases in abdominal pressure force urine past the external sphincter. Laughing, sneezing, or coughing may cause these sudden increases in pressure. Stress incontinence is common in women, particularly after childbirth. Giving birth may stretch or damage the external sphincter, making it less effective in controlling the flow of urine. Such damage can be minimized or repaired by doing exercises designed to strengthen the external sphincter and muscles along the floor of the pelvic cavity.

Urinary retention is the failure to expel, completely or normally, urine from the bladder. This condition may result from a lack of the sensation to urinate, as might occur temporarily after general anesthesia. Sometimes, urinary retention results from contraction or obstruction of the urethra. Enlargement of the prostate gland in men may obstruct the urethra. Immediate treatment for retention usually involves the use of a urinary catheter, a tube used by medical personnel to drain urine from the bladder.

Bacteria Can Enter the Urethra and Cause Urinary Tract Infections

Healthy urinary systems contain no microorganisms. Nevertheless, microorganisms that make their way into the urinary system can thrive there and cause **urinary tract infections (UTIs)**. Some bacteria arrive at the kidneys by way of the bloodstream. Most bacteria, however, enter the urinary system

FIGURE **16.18**
Conscious control over urination is usually acquired by the age of 3, when neural connections to the external urethral sphincter are fully developed.

by moving up the urethra from outside the body. Bacteria called *Escherichia coli* normally inhabit the colon and often cause UTIs. Sometimes, microorganisms called *Chlamydia* and *Mycoplasma* cause UTIs. In contrast to *E. coli*, these microorganisms are sexually transmitted (see Chapter 17a).

The urethras of males and females differ in length, as shown in Figure 16.19. The urethra in females is 3 to 4 cm (1.5 in.) long and lies in front of the vagina. The urethra in males is about 20 cm (8 in.) long and opens to the outside at the tip of the penis. Urinary tract infections are more common in females than in males. The shorter urethras of females may make them more susceptible. In females, the bacteria need to travel only a short distance from the external urethral orifice (opening) through the urethra to the bladder. In addition, the external urethral orifice is closer to the anus in females than it is in males. Thus, improper (from back to front) wiping after defecation can easily carry fecal bacteria to the female urethra. Bacteria also may enter the urethra during sexual intercourse. Women are advised to urinate soon after sex to flush bacteria from the lower urinary tract. In fact, to avoid UTIs, everyone is advised to drink plenty of water each day and to urinate frequently throughout the day. Holding urine can allow bacteria to gain a foothold in the urinary tract. Other tips to prevent UTIs include avoiding wet or tight clothing and nylon underwear. Such clothing allows the opening of the urethra to stay moist and creates an environment where bacteria can thrive. Showers are recommended over baths. Bubble baths, in particular, may irritate the urethra and increase susceptibility to UTIs. Finally, some physicians recommend drinking plenty of cranberry juice. This juice acidifies the urine and inhibits the growth of some bacteria.

Bacteria may infect only the urethra and cause inflammation there. This type of infection is called **urethritis**. Sometimes, however, the bacteria travel farther into the urinary system. These bacteria may cause **cystitis**, infection of the urinary bladder. Bacteria also can be introduced into the bladder by a catheter. Although the process of inserting a catheter does pose a risk of infection, use of these devices can be lifesaving. Sometimes bacteria do not stop at the bladder and move through the ureters and into the kidneys. An infection of the kidneys is called **pyelonephritis**.

Symptoms of urinary tract infections include fever, blood in the urine, painful and frequent urination, and bed-wetting in young children. In addition, there may be lower abdominal pain (if the bladder is involved) or back pain (if the kidneys are involved). Physicians usually diagnose such infections by having the urine checked for bacteria and blood cells. People who have recurrent UTIs may wish to purchase nonprescription dipsticks to check for infection. The dipstick detects nitrite, a compound made by the bacteria. When held in the stream of the first morning urine, these dipsticks can detect about 90% of UTIs. A change in the color of the stick indicates infection.

Urinary tract infections are treated with antibiotics. Infections of the lower urinary tract (the urethra and bladder) should be treated immediately and the prescribed antibiotics taken for their full term. These steps are critical to prevent the spread of infection to the kidneys, where very serious damage can occur. Many disorders of the urinary system can be diagnosed by a detailed examination of the urine in a medical laboratory. See the Health Issue essay, *Urinalysis: What Your Urine Says about You.*

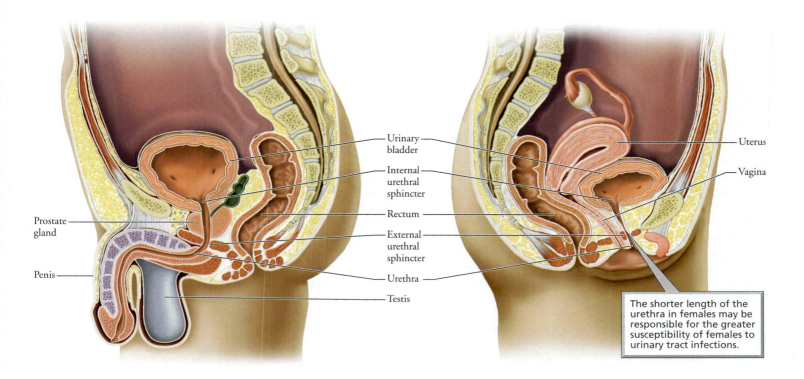

The shorter length of the urethra in females may be responsible for the greater susceptibility of females to urinary tract infections.

FIGURE **16.19**

The urinary bladder and urethra in a male (left) and a female (right)

HEALTH ISSUE

Urinalysis: What Your Urine Says about You

It is often said that much can be learned about people by examining their handwriting, meeting their parents, or looking into their eyes. Probably none of those things, however, can reveal as much about someone as can urinalysis, an analysis of the person's urine. Ancient Greek and Roman physicians knew this to be true, and they routinely studied the urine of their patients when diagnosing ailments. Today, urine is still under intense scrutiny. Indeed, most of us have been handed a small cup by medical personnel and asked to provide a urine sample. Such samples are usually studied for their microorganism content and their physical and chemical properties. In this essay, we consider what a urine sample reveals about us.

Upon exiting the body, healthy urine contains no microorganisms. The presence of bacteria in a properly collected urine sample usually signals infection of the urinary system. Bacteria found in a sample may be cultured to determine their identity. This information can then help in diagnosing the particular infection. Some bacteria may enter a urine sample during improper collection. For example, if the perineal area (in males, the area between the anus and the scrotum, and in females, the area between the anus and the vulva) has not been cleaned and the cup or urine is allowed to contact the area, bacteria may be introduced. So it is important to be careful when collecting a urine sample. Urine also may be screened for fungi or protozoans that cause inflammation within the urinary tract.

The following physical characteristics are routinely checked in urine samples: color, turbidity, pH, and specific gravity. The color of urine comes from urochrome, a yellow pigment produced as a waste product by the liver during the breakdown of hemoglobin in red blood cells. The urochrome travels in the bloodstream from the liver to the kidneys, where it is filtered from the blood and excreted with urine. The color of urine ranges from straw to yellow to amber. The color varies somewhat according to diet. Beets and

blackberries, for example, lend urine a red color, and asparagus causes a green tinge. The color of urine also varies with concentration. More concentrated urine, such as that collected first thing in the morning, is darker than more dilute urine. An abnormal color of urine—particularly red, when followed by microscopic confirmation of the presence of red blood cells—can indicate kidney stones and trauma to urinary organs. Because menstrual blood can contaminate urine samples, women should always inform their doctor if they are menstruating at the time of urine collection. Freshly voided urine is usually transparent. Cloudy, or turbid, urine may indicate a urinary tract infection, particularly if white blood cells are detected. Healthy urine has a pH of about 6, although considerable variation may occur in response to diet. Vegetarian diets produce alkaline urine, and high-protein diets produce acidic urine. Values ranging from 4.6 to 8.0 are not cause for alarm. An alkaline pH also is associated with some bacterial infections. Urine color and turbidity are usually determined without specialized equipment. The pH of urine is checked by recording color changes in test strips dipped in the urine sample.

The specific gravity of a substance is the ratio of its weight to the weight of an equal volume of distilled water. Commonly known as density, specific gravity is used as a measure of the concentration of solutes in urine; the higher the specific gravity, the higher the concentration of solutes. Thus, specific gravity provides a measure of the concentrating ability of the nephrons within our kidneys. When urine becomes highly concentrated for long periods, substances such as calcium and uric acid may precipitate out and form kidney stones. Specific gravity is measured with an instrument called a urinometer, or densitometer. Samples are sometimes spun in a centrifuge to separate residue from fluid. The residue may be examined under a microscope.

In the past, it was not unusual for physicians to taste urine to assess its sweetness or

saltiness. Today, quantitative chemical tests are run to assess the levels of specific chemical constituents of urine. We know, for example, that water is the major constituent of urine, making up about 95% of its total volume. The remaining 5% is made of solutes from outside sources (such as drugs) or the metabolic activities of our cells. Tests run on urine specimens highlight abnormal constituents of urine (Table 16.A) or normal constituents present in abnormal amounts (Table 16.B).

Although daily urine volume varies considerably, most people expel between 1000 and 2000 ml (about 1 to 2 quarts) of urine each day. Factors such as diet, temperature, and blood volume and pressure influence urine volume. For example, high blood volume and pressure cause decreased reabsorption of water in the renal tubules of nephrons, resulting in increased urine output. Thus, when used in conjunction with other tests, an analysis of urine volume may indicate high blood pressure. Typically, volume is not measured at the time of providing a routine urine sample. It is only when a problem with the urinary tract is suspected that people are asked to monitor their urine output over a 24-hour period. Such monitoring may reveal polyuria, excessive urine production (well over 2 liters or 2.1 quarts per day), or oliguria, inadequate urine production (less than 500 ml, or 0.53 quarts per day). Failure to produce any urine is an extreme health crisis called anuria.

The kidneys are our body's filtering system, and, as such, their product, urine, contains substances that originate from the work of almost all of our organs. This is the reason that a detailed analysis of the constituents of urine tells us how the organs of the urinary tract are functioning and about the general health of our other organ systems. In other words, a complete urinalysis involving measures of volume, microorganisms, and the physical and chemical properties of urine provides a check on our basic health.

TABLE **16.A**

SOME ABNORMAL CONSTITUENTS OF URINE AND THEIR CLINICAL IMPLICATIONS	
ABNORMAL CONSTITUENT	**CLINICAL IMPLICATIONS**
Albumin (plasma protein)	Increased permeability of filtration membranes of glomerulus and Bowman's capsule caused by high blood pressure, injury, or kidney disease
Bile pigments	Liver disease (cirrhosis, hepatitis)
	Obstruction of bile ducts
Glucose	Diabetes mellitus
Red blood cells	Inflammation of urinary tract from disease or infection
	Kidney stones
	Tumors
	Trauma
White blood cells	Infection of urinary tract

TABLE **16.B**

SOME NORMAL CONSTITUENTS OF URINE AND THEIR CLINICAL IMPLICATIONS WHEN PRESENT IN ABNORMAL AMOUNTS	
CONSTITUENT	**CLINICAL IMPLICATIONS**
Calcium (Ca^{2+})	Values depend on dietary intake
	Increases signal overactive parathyroid glands or cancer of breast or lungs
	Decreases signal vitamin D deficiency or underactive parathyroid glands
Chloride (Cl^-)	Values depend on dietary salt intake
	Increases signal Addison's disease (undersecretion of glucocorticoids by adrenal cortex), dehydration, or starvation
	Decreases signal diarrhea or emphysema
Creatinine	Increases signal infection
	Decreases signal muscular atrophy, anemia, or kidney disease
Potassium (K^+)	Increases signal chronic renal failure, dehydration, starvation, or Cushing's syndrome (oversecretion of glucocorticoids by adrenal cortex)
	Decreases signal diarrhea or underactive adrenal cortex
Sodium (Na^+)	Values depend on dietary salt intake
	Increases signal dehydration, starvation, or low blood pH from diabetes
	Decreases signal diarrhea, acute renal failure, emphysema, or Cushing's syndrome
Urea	Increases signal high protein intake
	Decreases signal impaired kidney function
Uric acid	Increases signal gout (excessive uric acid in blood causes crystals to form in joints, soft tissues, and kidneys), leukemia (cancer of white blood cells), or liver disease
	Decreases signal kidney disease

REVIEWING THE CONCEPTS

Organs from Several Body Systems Eliminate Waste (p. 357)

1. Several organs from different body systems remove waste from our bodies. Lungs and skin eliminate carbon dioxide (in gaseous form from the lungs and as bicarbonate ions from the skin), heat, and water. The skin also excretes salt and urea. Organs of the digestive system eliminate solid waste, and those of the urinary system excrete nitrogen-containing wastes, water, carbon dioxide (as bicarbonate ions), inorganic salts, and hydrogen ions.

The Kidneys, Ureters, Bladder, and Urethra Make Up the Urinary System (pp. 358–371)

2. Two kidneys, two ureters, the urinary bladder, and the urethra make up the urinary system. The main function of this system is to regulate the volume, pressure, and composition of the blood. The kidneys filter excess materials and wastes from the blood and form urine. Urine travels down the ureters to the urinary bladder, where it is stored until being excreted from the body through the urethra. The male urethra transports reproductive fluid in addition to urine (but not at the same time). The female urethra transports only urine.

3. The kidneys are located just above the waist against the back wall of the abdominal cavity. Each kidney has an outer portion (the renal cortex), an inner region (the renal medulla), and an inner chamber (the renal pelvis).

4. Properly functioning kidneys are critical to maintaining homeostasis. These organs filter wastes and excess materials from the blood, help regulate blood pH, and maintain fluid balance by regulating the volume and composition of blood and urine.

5. Nephrons are tiny tubules responsible for the formation of urine. There are two basic parts to a nephron. The first is the renal corpuscle, which consists of a tuft of capillaries, the glomerulus, and a surrounding cup-like structure called Bowman's capsule. The second part of a nephron is the renal tubule. The renal tubule has three sections: the proximal convoluted tubule, the loop of Henle, and the distal convoluted tubule. The distal convoluted tubules of several nephrons empty into a single collecting duct. Urine in collecting ducts moves into a large chamber within the kidneys called the renal pelvis. From the renal pelvis, urine leaves the kidneys by traveling down the ureters to the urinary bladder.

6. Glomerular filtration is the process by which only certain substances are allowed to pass out of the blood and into the nephron. High pressure within glomerular capillaries forces water, ions, and small molecules in the blood across a filter into Bowman's space within the nephron. Large molecules and formed elements of the blood normally cannot cross the filter.

7. Once in Bowman's space, the filtrate—consisting of water, ions, and small molecules filtered from the blood—moves into the renal tubule of the nephron. As the filtrate passes through the renal tubule, about 99% of it is returned to the blood in a process known as tubular reabsorption. Almost all the water, ions, and nutrients in the filtrate are returned to the blood. Wastes are either partially reabsorbed or not reabsorbed at all. Most reabsorption occurs in the proximal convoluted tubule.

8. During tubular secretion, drugs and any wastes or excess essential ions that may have escaped glomerular filtration are removed from the blood of capillaries surrounding nephrons and added to the filtered fluid that will become urine. This process occurs primarily along the proximal and distal convoluted tubules and collecting ducts.

9. The kidneys work with the lungs and buffer systems to regulate pH. Specifically, tubular secretion plays a role in regulating blood pH by removing excess hydrogen ions from the blood.

10. Most nephrons have short loops of Henle that are restricted almost entirely to the renal cortex. About 20% of our nephrons have long loops of Henle that extend into the renal medulla. The latter nephrons are responsible for the water-conserving ability of the kidneys because they maintain gradients in the solute concentration of the interstitial fluid within the kidneys. Maintenance of the interstitial gradient enables large amounts of water to move out of collecting ducts by osmosis and leads to the production of concentrated urine.

11. Several hormones influence kidney function: antidiuretic hormone (ADH), aldosterone, atrial natriuretic peptide (ANP), and the renin-angiotensin system.

12. Antidiuretic hormone is produced by the hypothalamus and released by the posterior lobe of the pituitary gland. ADH increases water reabsorption in the final portions of the distal convoluted tubule and collecting ducts of nephrons.

13. Aldosterone is a steroid hormone secreted by the adrenal cortex. Aldosterone increases sodium reabsorption by the distal convoluted tubules and collecting ducts. Because water follows sodium, aldosterone indirectly increases water reabsorption by nephrons.

14. Atrial natriuretic peptide (ANP) is a hormone released by cells of the right atrium of the heart. ANP causes decreased water and solute reabsorption by the kidneys.

15. The renin-angiotensin system is a series of steps by which the protein renin is converted to angiotensin II. Angiotensin II stimulates the adrenal cortex to release aldosterone and the posterior pituitary to release ADH. Angiotensin II also stimulates the thirst center in the hypothalamus to increase water intake. Finally, angiotensin II constricts arterioles and increases blood pressure.

16. The kidneys release the hormone erythropoietin, which is necessary for red blood cell production. The kidneys also convert vitamin D into its active form that promotes the body's absorption of calcium and phosphorus.

17. Renal failure is a decrease or complete cessation of glomerular filtration. Such failure may be acute and occur over a period of a few hours or days. Alternatively, renal failure may be chronic and develop over a period of months or years. Failure of the kidneys severely disrupts homeostasis. Conditions such as acidosis, anemia, edema, hypertension, and a toxic buildup of nitrogenous wastes in the blood may result when the kidneys stop working.

18. Treatments for renal failure include hemodialysis (using artificial devices to cleanse the blood), continuous ambulatory peritoneal dialysis (using the patient's own peritoneum as the dialyzing membrane), and kidney transplantation.

WEB TUTORIAL 16.1 Urinary System Anatomy
WEB TUTORIAL 16.2 Nephron Function

Urination Has Involuntary and Voluntary Components (pp. 371–372)

19. Urination is the process by which the urinary bladder is emptied. This emptying entails involuntary and voluntary actions. Moderate filling of the bladder stimulates stretch receptors in the wall of the bladder, and impulses are sent along sensory neurons to the lower spinal cord. From the spinal cord, impulses move along motor neurons back to the bladder, where they cause the muscular wall of the bladder to contract. The impulses also cause the internal urethral sphincter, a thickening of smooth muscle at the junction of the bladder and urethra, to relax. These actions push stored urine into the urethra. If the time for urination is appropriate, the brain then permits voluntary relaxation of the external urethral sphincter, a band of skeletal muscle farther down the urethra. Once this sphincter relaxes, urine exits the body.

Bacteria Can Enter the Urethra and Cause Urinary Tract Infections (pp. 372–373)

20. Urinary tract infections (UTIs) are more common in females than in males, possibly because bacteria can more readily move from outside the body up the shorter urethra of females to the bladder. Infections may occur in the urethra, bladder, and kidneys, with infections of the kidneys being the most serious.

KEY TERMS

acute renal failure *p. 367*
aldosterone *p. 365*
angiotensin I *p. 366*
angiotensin II *p. 366*
antidiuretic hormone (ADH) *p. 364*
atrial natriuretic peptide (ANP) *p. 367*
Bowman's capsule *p. 359*
Bowman's space *p. 359*
chronic renal failure *p. 367*
continuous ambulatory peritoneal dialysis (CAPD) *p. 367*
cystitis *p. 373*

diabetes insipidus *p. 365*
distal convoluted tubule *p. 359*
diuretic *p. 365*
erythropoietin *p. 367*
external urethral sphincter *p. 371*
glomerular filtration *p. 360*
glomerulus *p. 359*
hemodialysis *p. 367*
internal urethral sphincter *p. 371*
juxtaglomerular apparatus *p. 365*
kidney *p. 358*
loop of Henle *p. 359*

nephron *p. 359*
proximal convoluted tubule *p. 359*
pyelonephritis *p. 373*
renal corpuscle *p. 359*
renal cortex *p. 358*
renal failure *p. 367*
renal medulla *p. 358*
renal pelvis *p. 358*
renal tubule *p. 359*
renin *p. 366*
stress incontinence *p. 372*
tubular reabsorption *p. 360*

tubular secretion *p. 361*
ureter *p. 358*
urethra *p. 358*
urethritis *p. 373*
urinary bladder *p. 358*
urinary incontinence *p. 371*
urinary retention *p. 372*
urinary system *p. 358*
urinary tract infection (UTI) *p. 372*
urination *p. 371*
urine *p. 358*

THINKING ABOUT THE CONCEPTS

1. List the components of the urinary system and their functions. *p. 358*
2. Describe the ways in which kidneys maintain homeostasis. *p. 358*
3. Describe the structure of a nephron. *p. 359*
4. Explain glomerular filtration, tubular reabsorption, and tubular secretion by nephrons. Where in the nephron does each process occur? *p. 360*
5. How do nephrons contribute to the regulation of blood pH? *p. 362*
6. How do the kidneys conserve water? *p. 362*
7. Explain how each of the following influences kidney function: antidiuretic hormone (ADH), aldosterone, atrial natriuretic peptide (ANP), and the renin-angiotensin system. *p. 364*
8. Describe acute and chronic renal failure. What treatments are available for persons with severely impaired kidneys? *p. 367*
9. Describe the process of urination. What is urinary retention? *p. 371*
10. How do the urethras of males and females differ? What are the clinical implications of these differences? *p. 373*
11. Glomerular filtration
    a. occurs at the renal corpuscle.
    b. returns useful substances to the blood.
    c. occurs along the renal tubule.
    d. is the last step of three in cleansing the blood.
12. Which of the following organs does *not* play a role in eliminating metabolic wastes?
    a. skin
    b. lungs
    c. gallbladder
    d. kidneys

13. Diabetes insipidus is caused by a deficiency in
    a. aldosterone.
    b. antidiuretic hormone.
    c. erythropoietin.
    d. atrial natriuretic peptide.
14. Which of the following is *not* performed by the kidneys?
    a. regulation of pH
    b. conservation of water
    c. transformation of vitamin D into active form
    d. contraction to initiate urination
15. Tubular reabsorption
    a. allows only certain substances to pass from the blood into the nephron.
    b. returns useful substances from the filtrate to the blood.
    c. removes wastes and excess essential ions that escaped filtration from the blood.
    d. occurs at the renal corpuscle.
16. _____ are the functional units of the kidneys.
17. _____ is the use of artificial devices to cleanse the blood.
18. The _____ urethral sphincter is made of _____ muscle and is involuntary. The _____ urethral sphincter is made of _____ muscle and allows voluntary control over urination.
19. _____ is an enzyme released by cells of the juxtaglomerular apparatus of nephrons. This enzyme initiates a series of steps that ultimately stimulates the adrenal gland to release aldosterone.
20. The _____ transport urine from the kidneys to the bladder.

APPLYING THE CONCEPTS

1. Rachel has noticed that since the birth of her daughter, a small amount of urine escapes whenever she sneezes. What condition might Rachel have? What explains her condition? What can she do to improve it?
2. Miguel received a urinary catheter after surgery. A few days later, he had a fever, pain in his lower abdomen, and painful urination. What do you think Miguel might be suffering from? How would you diagnose and treat his condition?

3. Sean has just finished an exhausting game of basketball on a very hot day. He thinks that a cold beer might be just the thing to quench his thirst and restore body fluids. What do you think, and why?
4. Anne has polycystic disease, an inherited and progressive condition that is destroying her kidneys. What will be the health consequences of her chronic renal failure? What are her options for restoring kidney function?

17

Reproductive Systems

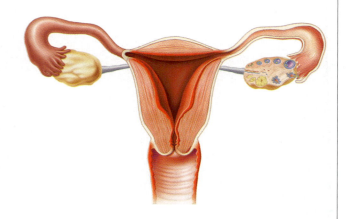

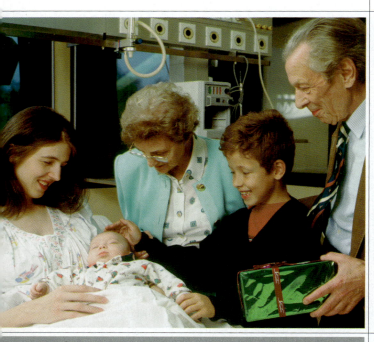

Tim and Andrea were awestruck by the beauty of their newborn baby girl, Paula, who was asleep on Andrea's chest. Each felt like the proudest, happiest parent on earth. Grandparents surrounded the hospital bed and pointed out to each other all of Paula's features and who they came from. Later that evening, Tim thought back in amazement to the time 18 months earlier when they were ready to give up on ever having a child. They had tried for 2 years to get Andrea pregnant, trying everything from doctors' advice to old wives' tales. Nothing worked. Finally, they went to a highly recommended fertility clinic for advice. After many tests, examinations, and thorough questioning about medical history, Dr. Perry told them that Tim was likely to be the source of the problem. He explained that Tim had "poor–performing sperm—slow swimmers and too few of them." He went on to explain that these problems with Tim's sperm were probably caused by varicoceles, enlargements of the veins that drain the testes. The varicoceles create a slightly warmer environment for sperm, and the higher temperature can reduce sperm count and cause the development of deformed sperm. Tim had surgery to correct the problem, and in less than a year Andrea was pregnant.

In this chapter, we will gain an understanding of why Tim and Andrea had difficulty conceiving a child. We will look at the structures of the reproductive systems and how they function. We will also consider the sex hormones and the ways in which they regulate reproductive processes. ■

The Gonads Produce Gametes and Sex Hormones

In both the male and the female, the **gonads** (testes or ovaries) are the most important structures in the reproductive system. The gonads serve two important functions: (1) they produce the **gametes** (eggs and sperm—the cells that will fuse and develop into a new individual), and (2) they produce the sex hormones. In the male, the gonads are the **testes**, and the gametes they produce are the **sperm**. The testes produce the male sex hormone **testosterone**. The **ovaries** are the female gonads. The gametes they produce are the **eggs**. The female sex hormones they produce are **estrogen** and **progesterone**.

A Male's Reproductive Role Differs from the Role of a Female

A male's reproductive role differs from a female's role. A male and a female each make the same genetic contribution (DNA packaged into chromosomes) to the next generation—one copy of each chromosome. A male's reproductive strategy is to produce millions of sperm, deliver them to the female reproductive system, and hope that one sperm reaches an egg. In contrast, a female usually produces only one egg approximately once a month. If sperm delivery is appropriately timed with egg production, the sperm and egg may fuse in a process called **fertilization**. The cell created by fertilization, called a **zygote**, will develop into the new individual (see Chapter 18). A male's role in the reproductive process ends with delivery of sperm, but a female's role continues after egg production; she nourishes and protects the offspring until birth. The egg is packed with nutrients to provide for early development, and the uterus serves as a nourishing, protective environment for the developing offspring.

Sperm and eggs are specialized for their reproductive roles by a reduction in the number of chromosomes. Each of your body cells contains 23 pairs of chromosomes. One member of each pair came from your father's sperm; the other member came from your mother's egg. Each chromosome contains a portion of the instructions for making and maintaining your body. It is important, therefore, that one member of each pair (one complete set of instructions) be present in a gamete. However, it is equally important that *only* one member of each pair of chromosomes be in a gamete so that, when the egg and sperm nuclei fuse during fertilization, the zygote does not have more than two copies of each chromosome.

The development of gametes involves two types of cell division. The first is *mitosis,* a process in which a cell replicates its chromosomes and divides into two daughter cells, each with two complete sets of chromosomes. The alterations in genetic content occur during the second type of cell division, called *meiosis,* which begins with a cell that has two copies of each chromosome (called a *diploid* cell) and ends with up to four cells with only one copy of each chromosome (called *haploid* cells). Meiosis involves two rounds of cell division. (The details of mitosis and meiosis are described in Chapter 19.)

The Male Reproductive System Delivers Sperm to the Egg

The male reproductive system consists of the testes, a system of ducts through which the sperm travel, the penis, and various accessory glands. Accessory glands produce secretions that help protect and nourish the sperm as well as providing a transport medium that aids the delivery of sperm to the outside of a male's body. The structure of the male reproductive system is shown in Figure 17.1 and summarized in Table 17.1.

▌ The testes produce sperm and male hormones

The male reproductive system has two testes (singular, testis). Each is located externally in a sac of skin called the **scrotum**. The temperature within the scrotum is several degrees cooler than within the abdominal cavity. The lower temperature is important in the production of healthy sperm.

Reflexes in the scrotum help keep the temperature within the testes fairly stable. Cooling of the scrotum, such as might occur when a male jumps into frigid water, triggers contraction of a muscle that pulls the testes closer to the warmth of the body. In a hot shower, however, the muscle relaxes, and the testes hang low, away from the heat of the body. The skin of the scrotum is also amply supplied with sweat glands that help cool the testes.

SPERM PRODUCTION

Beginning at puberty, which usually occurs during the teenage years, a healthy male produces over 100 million sperm each day for the rest of his life. The microscopic sites of sperm production are the **seminiferous tubules**, which make up 80% of the mass of each testis (Figure 17.2). The length of the seminiferous tubules is enormous. In fact, if the seminiferous tubules from one testis were placed end to end, they would stretch about a half mile. We will consider the development of sperm in more detail later in this chapter.

HORMONE PRODUCTION

The **interstitial cells** are located between the seminiferous tubules of the testis. They produce the male steroid sex hormones, collec-

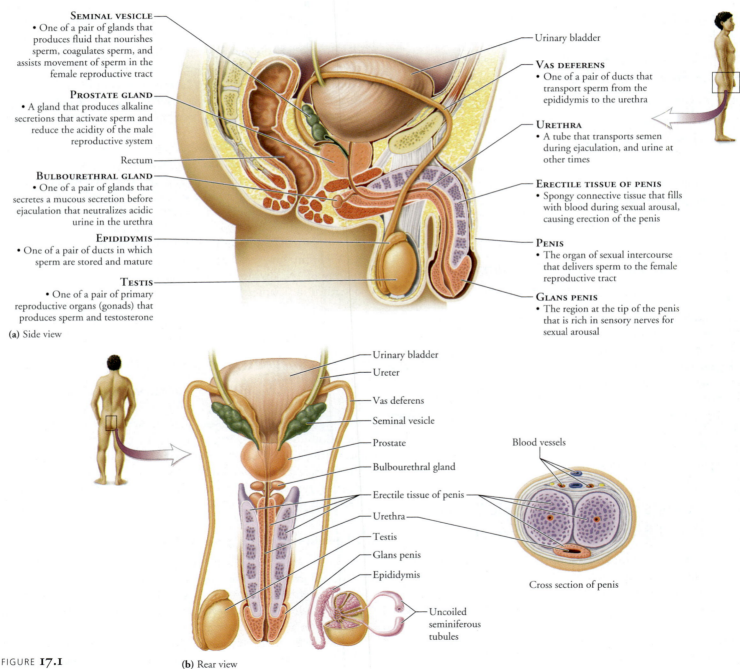

SEMINAL VESICLE
• One of a pair of glands that produces fluid that nourishes sperm, coagulates sperm, and assists movement of sperm in the female reproductive tract

PROSTATE GLAND
• A gland that produces alkaline secretions that activate sperm and reduce the acidity of the male reproductive system

Rectum

BULBOURETHRAL GLAND
• One of a pair of glands that secretes a mucous secretion before ejaculation that neutralizes acidic urine in the urethra

EPIDIDYMIS
• One of a pair of ducts in which sperm are stored and mature

TESTIS
• One of a pair of primary reproductive organs (gonads) that produces sperm and testosterone

(a) Side view

Urinary bladder

VAS DEFERENS
• One of a pair of ducts that transport sperm from the epididymis to the urethra

URETHRA
• A tube that transports semen during ejaculation, and urine at other times

ERECTILE TISSUE OF PENIS
• Spongy connective tissue that fills with blood during sexual arousal, causing erection of the penis

PENIS
• The organ of sexual intercourse that delivers sperm to the female reproductive tract

GLANS PENIS
• The region at the tip of the penis that is rich in sensory nerves for sexual arousal

Urinary bladder
Ureter
Vas deferens
Seminal vesicle
Prostate
Bulbourethral gland
Erectile tissue of penis
Urethra
Testis
Glans penis
Epididymis
Uncoiled seminiferous tubules

Blood vessels

Cross section of penis

FIGURE **17.1**
The male reproductive system

(b) Rear view

TABLE **17.1**

THE MALE REPRODUCTIVE SYSTEM	
STRUCTURE	**FUNCTION**
Testes	Produce sperm and testosterone
Epididymis	Location of sperm storage and maturation
Vas deferens	Conducts sperm from epididymis to urethra
Urethra	Tube through which sperm or urine leaves the body
Prostate gland	Produces secretions that make sperm mobile and that counteract the acidity of the female reproductive tract
Seminal vesicles	Produce secretions that make up most of the volume of semen
Bulbourethral glands	Produce secretions just before ejaculation; may lubricate; may rinse urine from urethra
Penis	Delivers sperm to female reproductive tract

tively called *androgens*. The most important androgen is *testosterone*, which is needed for sperm production and the maintenance of male reproductive structures.

The duct system stores and transports sperm

Sperm produced in the seminiferous tubules next enter a highly coiled tubule called the **epididymis**, where they mature and are stored (see Figure 17.1). Sperm that enter the epididymis look as if they are mature, but they cannot yet function as mature sperm. During their stay in the epididymis, the sperm become capable of fertilizing an egg and of moving on their own, although they do not yet do so.

Each **vas deferens** is a tube that conducts sperm from the epididymis to the urethra. Some sperm may be stored in the part of the vas deferens closest to the epididymis. When a man reaches sexual climax, rhythmic waves of muscle contraction propel the sperm along the vas deferens.

The **urethra** conducts urine from the urinary bladder or sperm from the vas deferens out of the body through the *penis*, the male organ of sexual intercourse and urination. Sperm and urine do not pass through the urethra at the same time. A circular muscle contracts and pinches off the connection to the urinary bladder during sexual excitement.

The accessory glands produce most of the volume of semen

Semen, the fluid released through the urethra at sexual climax, is composed of sperm and the secretions of the accessory glands: the prostate gland, the paired seminal vesicles, and paired bulbourethral glands. Interestingly, very little of the volume of semen is made up of sperm.

About the size of a walnut, the **prostate gland** surrounds the urethra, just beneath the urinary bladder. The secretions of the prostate gland make up between 13% and 33% of the volume of semen. Prostate secretions are slightly alkaline and serve both to activate the sperm, making them fully motile, and to counteract the acidity of the female reproductive tract.

The prostate often begins enlarging when a man reaches middle age. The enlarged prostate may squeeze the urethra and restrict urine flow, making urination difficult. This benign enlargement of the prostate gland that accompanies aging is not related to prostate cancer, which is a growing concern for men. See the Health Issue essay, *Testicular and Prostate Cancers*.

The secretions of the **seminal vesicles** make up roughly 60% of the volume of semen. The secretion contains citric acid, fructose,

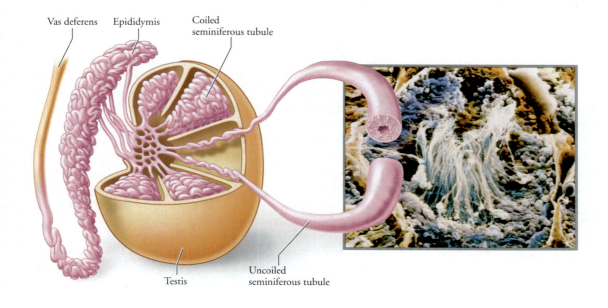

Vas deferens   Epididymis   Coiled seminiferous tubule

Testis

Uncoiled seminiferous tubule

FIGURE **17.2**
The internal structure of the testis and epididymis

HEALTH ISSUE

Testicular and Prostate Cancers

Testicular cancer affects only 1% of all men. It is, however, the most common form of cancer among men between the ages of 15 and 35 years. Nearly three-quarters of all the men who develop testicular cancer are younger than age 50. Testicular cancer affects nearly 7000 American men each year, and the incidence has been increasing worldwide over the past few decades. This cancer is 30 to 50 times more likely in men whose testes did not descend into the scrotum or descended after 6 years of age. Because this cancer does not usually cause pain, it is important for every man to examine his testes each month to feel for a lump or a change in consistency. If a lump is felt, it is not necessarily cancerous because swellings in the epididymis are common. However, a physician should be consulted with the detection of any lump because the cure rate for tumors caught in the early stages is nearly 100%. Treatment usually involves removal of the diseased testis, followed with radiation and chemotherapy. The healthy testis is usually left in place, so the man is still potent and fertile.

Prostate cancer affects 1 of every 11 American males and kills nearly 30,000 of them every year. The incidence of prostate cancer is rising. Prostate cancer is now the second highest cause of cancer deaths among men older than age 55. Part of the explanation for the increase is that men are living longer, and prostate cancer usually affects older men.

There are two ways of detecting prostate cancer: a rectal exam and a blood test. Most physicians recommend that both tests be used in the diagnosis. During a rectal exam for prostate cancer, a physician inserts a gloved finger into the rectum and, through the wall of the rectum, feels the prostate, which is located nearby (see Figure 17.1). A prostate gland that has merely enlarged with age feels diffuse and soft, but a cancerous prostate is firm or may contain a hard lump. One-third of the tumors detected by a rectal exam are already too large to be removed surgically. The blood test for prostate cancer measures the amount of a protein called prostate specific antigen (PSA). Because this protein is produced only by the prostate gland, the amount of PSA in the blood reflects the size of the prostate. As a tumor within the prostate grows, the blood level of PSA usually rises.

When prostate cancer is detected while it is still contained in the prostate, the chances of survival for at least 5 years are 95%. If the cancer has spread, the odds drop to a 50% chance of living another 2 or 3 years.

Surgical removal of the prostate is usually the first step in treating the cancer. In the past, this procedure often damaged nerves in the area, causing erectile dysfunction and incontinence. Recently, however, the technique has improved greatly, and those side effects can be prevented in most patients. Radiation therapy often follows surgery.

A more recent treatment is the implantation of radioactive pellets in the prostate. The implants may be as effective as surgery and radiation in treating early prostate cancer. Inserting the pellets allows patients to resume activity the day after implantation.

Although the frequency of surgery for prostate cancer that remains contained within the gland has increased dramatically in the past few years, it is not clear that treating prostate cancer results in a longer life. Most cases of prostate cancer grow very slowly. Thus, if a man is older than 70 at the time of diagnosis, he could easily die of some other cause before his prostate cancer kills him. Furthermore, the operation is certainly not risk free, especially for older men.

With prostate cancer on the rise, men might wonder how they can reduce their risk. One way might be to reduce the amount of red meat in the diet. A fat found in meat and butter (alpha-linolenic acid) is associated with prostate cancer. Dietary fat increases the rate at which prostate cancer grows. Thus, dietary fat could cause a tumor that might otherwise have grown too slowly to cause problems or become life threatening. Another way to reduce the risk of prostate cancer might be to move to a hot climate. It is thought that a vitamin produced as a result of exposure to sunshine may be protective because men who live in hot climates have a lower risk of prostate cancer than do men who live in cold climates.

---

amino acids, and prostaglandins.[1] Fructose is a sugar that provides energy for the sperm's long journey to the egg. Some of the amino acids coagulate the semen. Coagulation helps keep the sperm within the vagina and also protects the sperm within the clump from the acidic environment of the vagina. The prostaglandins serve to cut the viscosity of cervical mucus and to cause uterine contractions that assist the movement of sperm.

The **bulbourethral glands** release a clear, slippery liquid immediately before ejaculation. This fluid may rinse the slightly acidic urine remnants from the urethra before the sperm pass through.

▌ The penis transfers sperm to the female

The **penis** is a cylindrical organ whose role in reproduction is to deliver the sperm to the female reproductive system. The tip of the penis is enlarged, forming a smooth rounded head known as the *glans penis,* which has many sensory nerve endings and is important in sexual arousal. When a male is born, a cuff of skin called the *foreskin* covers the glans penis. The foreskin can be pulled back to expose the glans penis. The surgical removal of the foreskin is called circumcision.

The transfer of sperm to the female reproductive system usually requires that the penis be erect. An erection involves increases in the length, width, and firmness of the penis, due to changes in the blood supply to the organ. Within the penis are three columns of spongy *erectile tissue,* which is a loose network of connective tissue with many empty spaces (see Figure 17.1). During sexual arousal, the arterioles that pipe blood into the spongy tissue dilate (widen), and the spongy tissue fills with blood, causing the penis to become larger. At the same time, the spongy tissue squeezes shut the veins that drain the blood from the penis. As a result, the blood

---

[1]Prostaglandins are chemicals secreted by one cell that alter the activity of other cells.

flows into the penis faster than it can leave, causing the spongy tissue to fill with blood and press against a connective tissue casing. This makes the penis firm (like a water balloon), larger and erect.

Erectile dysfunction (ED, also called impotence) is a male's inability to achieve or maintain an erection. Thus, it generally leaves a man unable to have sexual intercourse. Affecting approximately 30 million men in the United States alone, ED is hardly an uncommon problem. Indeed, it is not unusual for a man to experience erectile dysfunction at some point in his life. There are a number of psychological causes for erectile dysfunction, including worry, stress, a quarrel with the partner, and depression. However, there are also many physical causes. Nerve damage, such as that which often accompanies chronic alcoholism and sometimes diabetes, might be responsible. Because an erection depends on adequate blood supply, fatty deposits in the arteries (atherosclerosis) serving the penis can also cause erectile dysfunction. Medications, especially certain drugs used to treat high blood pressure, antihistamines, antinausea and antiseizure drugs, antidepressants, sedatives, and tranquilizers, may cause the problem. Cigarette smoking, excessive alcohol consumption, or marijuana use can also cause impotency.

The first step in treating impotency is to eliminate the cause of the problem, whenever possible. Should impotency continue, Viagra®, Levitra®, and Cialis® are drugs for men with erectile dysfunction. They can help a man achieve and maintain an erection when he is sexually aroused. They work by prolonging the effect of a chemical (nitric oxide) that is released when the man becomes sexually aroused and causes the widening of the arterioles in the penis. As the arterioles widen, blood flow increases, and an erection results.

stop and think

Explain why Viagra®, Levitra®, or Cialis® cannot cause an erection in a man who is not sexually aroused.

Sperm development involves changes in the number of chromosomes and in the structure and function of cells

The sequence of events within the seminiferous tubules that gives rise to sperm is called **spermatogenesis**. This process involves a reduction in the number of chromosomes to one member of each pair and changes in the shape and functioning of the cells to make the sperm efficient chromosome delivery vehicles (Figure 17.3).

The process of spermatogenesis begins in the outermost layer of each seminiferous tubule, where undifferentiated diploid cells called *spermatogonia* (singular, spermatogonium) develop. Each spermatogonium divides by mitosis to produce two new diploid spermatogonia. One spermatogonium remains at the periphery of the tubule and divides again to give rise to new spermatogonia. Therefore, there is a constant supply of cells that can develop into sperm. The other spermatogonium pushes deeper into the wall of the tubule, where it will enlarge to form a *primary spermatocyte,* which is also a diploid cell. Primary spermatocytes undergo the two divisions of meiosis forming *secondary spermatocytes,* which mature into *spermatids.*

Although the spermatids are haploid and have the correct set of chromosomes to join with the egg during fertilization, numerous structural changes must still occur to create cells capable of swim-

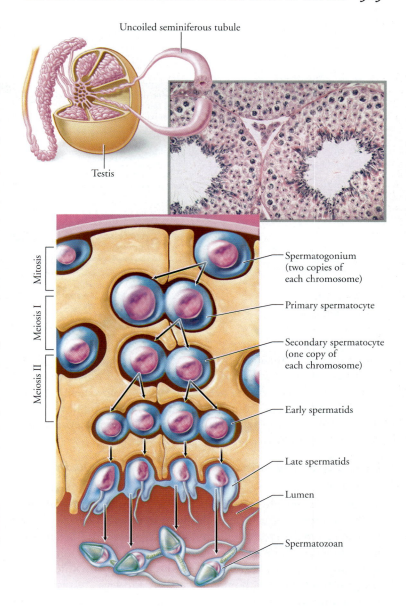

FIGURE **17.3**
The stages of spermatogenesis in the wall of a seminiferous tubule. As the cells that will become sperm develop, they are pushed from the outer wall of the tubule to the central canal.

ming to the egg and fertilizing it. These changes convert spermatids to streamlined *spermatozoa,* or sperm, equipped to deliver the father's genetic contribution to the next generation.

The mature sperm cell has three distinct regions: the head, the midpiece, and the tail (Figure 17.4). The head of the sperm is a flattened oval that contains little else besides the densely packed chromosomes. Recall that the sperm is haploid and contains only one set, or 23, chromosomes. Positioned like a ski cap on the head of the sperm is the **acrosome**, a membranous sac containing enzymes. A few hours after the sperm have been deposited in the female reproductive system, the membranes of the acrosomes break down. The enzymes then spill out and digest through the layers of cells surrounding the egg, assisting fertilization. Within

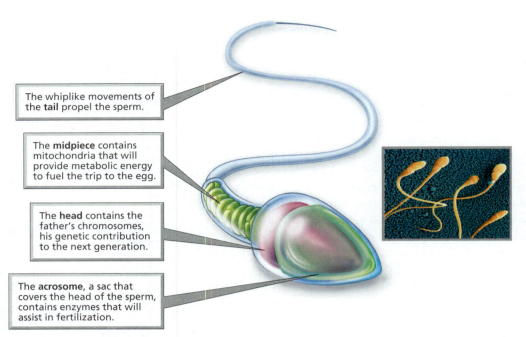

The **whiplike** movements of
the **tail** propel the sperm.

The **midpiece** contains
mitochondria that will
provide metabolic energy
to fuel the trip to the egg.

The **head** contains the
father's chromosomes,
his genetic contribution
to the next generation.

The **acrosome**, a sac that
covers the head of the sperm,
contains enzymes that will
assist in fertilization.

FIGURE **17.4**
The structure of a mature sperm
(spermatozoan)

the midpiece, mitochondria are arranged in a spiral. The mitochondria are the powerhouses of the cell that will provide energy
in the form of ATP that is needed to fuel the movements of the
tail. The tail contains contractile filaments. The whiplike movements of the tail propel the sperm during its long journey through
the female reproductive system.

The interplay of hormones controls male reproductive processes

Testosterone, which is secreted by the interstitial cells of the
testes, is important for sperm production as well as many other
male characteristics. At puberty, testosterone turns boys into
men (literally). The male reproductive organs, including the
testes and the penis, become larger. Testosterone also stimulates
the growth of the long bones of the arms and legs. Testosterone
is, therefore, responsible for the growth spurt that accompanies
puberty. Eventually, testosterone causes the closure of the
growth regions of these bones and the young man stops growing taller. In addition, testosterone is responsible for the development and maintenance of the male secondary sex characteristics, which are features associated with "masculinity" but are not
directly related to reproductive functioning. For example, the
growth of muscles and the skeleton tends to result in wide
shoulders and narrow hips. Patterns of hair growth begin to
change. Pubic hair develops, as does hair under the arms. A
beard appears, perhaps accompanied by hair on the chest.
Meanwhile, the voice box enlarges and the vocal cords thicken,
causing a deepening of the male voice. Testosterone also stimulates the activity of oil and sweat glands. The secretions of these
glands nourish bacteria living on the skin, the result of which
can be acne and body odor. And, importantly, testosterone is
responsible for sex drive.

We can see, then, that testosterone is important, but what
keeps its levels steady? Its production is regulated by a negative
feedback loop (see Chapter 10) involving hormones from the
hypothalamus, the anterior pituitary gland, and the testes. The
hormones involved in the regulation of male reproductive
processes are summarized in Figure 17.5 and Table 17.2. Testosterone levels increase when a region of the brain called the hypothalamus releases **gonadotropin-releasing hormone**, or **GnRH**.
This hormone stimulates the anterior pituitary gland, a pea-sized
endocrine gland that lies at the underside of the brain, to secrete
luteinizing hormone (LH). In turn, LH stimulates the production of testosterone by the interstitial cells of the testis. (For this
reason, LH is sometimes called *interstitial cell–stimulating
hormone*, or *ICSH*.) The rising testosterone level inhibits the
release of GnRH from the hypothalamus. As a result, the amount
of LH produced by the anterior pituitary drops. This decrease in
the LH level lowers the amount of testosterone secretion, removing the inhibition on the hypothalamus. In this way, testosterone
levels remain constant.

Sperm production is regulated by another negative feedback
loop. Besides secreting LH, the anterior pituitary gland produces
follicle-stimulating hormone, or **FSH**. Follicle-stimulating hormone stimulates sperm production by making the cells that will
become sperm more sensitive to the stimulatory effects of testosterone. FSH works by causing certain cells within the seminiferous
tubules to secrete a protein that binds and concentrates testosterone.

The seminiferous tubules produce a hormone called
inhibin in addition to sperm. Inhibin production increases with
sperm count, so it serves as an indicator of sperm numbers. As
its name implies, inhibin serves to inhibit the production of
FSH from the anterior pituitary. It also may inhibit the hypothalamic secretion of GnRH. Rising levels of inhibin cause a
decline in testosterone level and sperm production. As the
sperm count falls, so does the level of inhibin. Released from
inhibition, the anterior pituitary then increases production of
FSH, and the hypothalamus produces GnRH, which increases
sperm production again.

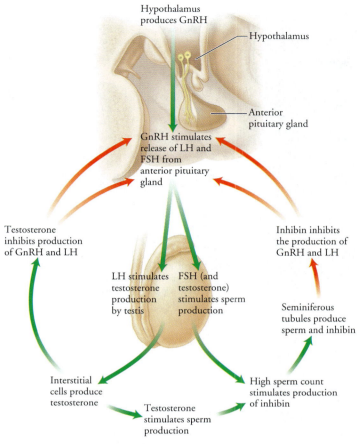

Hypothalamus produces GnRH

Hypothalamus

Anterior pituitary gland

GnRH stimulates release of LH and FSH from anterior pituitary gland

Testosterone inhibits production of GnRH and LH

Inhibin inhibits the production of GnRH and LH

LH stimulates testosterone production by testis

FSH (and testosterone) stimulates sperm production

Seminiferous tubules produce sperm and inhibin

Interstitial cells produce testosterone

Testosterone stimulates sperm production

High sperm count stimulates production of inhibin

FIGURE **17.5**

The feedback relationships among the hypothalamus, anterior pituitary, and the testes control the production of both sperm and testosterone.

TABLE **17.2**

HORMONES IMPORTANT IN THE REGULATION OF MALE REPRODUCTIVE PROCESSES		
HORMONE	**SOURCE**	**EFFECTS**
Testosterone	Interstitial cells in testes	Sperm production; development and maintenance of male reproductive structures, male secondary sex characteristics; sex drive
Gonadotropin-releasing hormone (GnRH)	Hypothalamus (in brain)	Stimulates the anterior pituitary gland to release LH
Luteinizing hormone (LH)	Anterior pituitary gland (in brain)	Stimulates interstitial cells of testis to produce testosterone
Follicle-stimulating hormone (FSH)	Anterior pituitary gland (in brain)	Enhances sperm formation
Inhibin	Seminiferous tubules in testes	Inhibits FSH secretion by anterior pituitary gland, causing a decrease in sperm production and testosterone production

stop and think

Anabolic steroids are taken by some male athletes to build muscles. A side effect of steroid abuse is a reduction in testis size. Considering that the anabolic steroids mimic the effect of testosterone, how can the shrinking testis size be explained?

The Female Reproductive System Produces the Eggs and Nurtures the Embryo and Fetus

The female reproductive system consists of the ovaries, the oviducts, the uterus, and the vagina. Its structure is shown in Figure 17.6, and its functions are summarized in Table 17.3.

The ovaries produce eggs and female hormones

Each of a woman's two ovaries is almond-shaped and measures 5 cm by 2.5 cm (2 in. by 1 in.). The ovaries have two important functions: (1) to produce eggs in a process called **oogenesis** and (2) to produce the female hormones estrogen and progesterone. The production of eggs and hormones is intimately related and occurs in cycles, as we will see.

The oviducts transport the immature egg and zygote

Two **oviducts**, also known as the fallopian tubes or uterine tubes, extend from the uterus. They transport the immature egg, called an *oocyte,* from the ovary to the uterus. The end of each oviduct nearest the ovary is open and funnel shaped. This end has many ciliated, fingerlike projections that drape over the ovary, but they rarely directly contact it. When released from the ovary during ovulation, the oocyte is first cast into the abdominal cavity. If it does not enter the oviduct, the oocyte will be lost there. However, about the time of ovulation, the projections from the tubes begin to wave. The currents they create and those caused by the cilia lining the tubes help to draw the oocyte into the oviduct. Inside, the oviduct is a tunnel about twice the thickness of a human hair. The cilia lining the tubes beat, creating a current toward the uterus. If fertilization occurs, it usually takes place in the oviduct, near the ovary. The resulting zygote begins development into an embryo in the oviduct. The beating cilia and the rhythmic muscular contractions of the oviduct sweep the egg or the early embryo along the oviduct toward the uterus, a distance of about 10 cm (4 in.).

The uterus supports the growth of the developing embryo

The **uterus** is a hollow organ shaped like an inverted pear that receives and nourishes the developing baby (the embryo and later the fetus[2]). During pregnancy, the uterus expands to about 60 times its original size as the fetus grows. After childbirth, the uterus never quite returns to its prepregnancy size.

The wall of the uterus has main two layers: a muscular layer and a lining called the **endometrium** (*endo-,* within; *metr-,* uterus; *-ium,*

[2]After 8 weeks, the embryo is called a fetus.

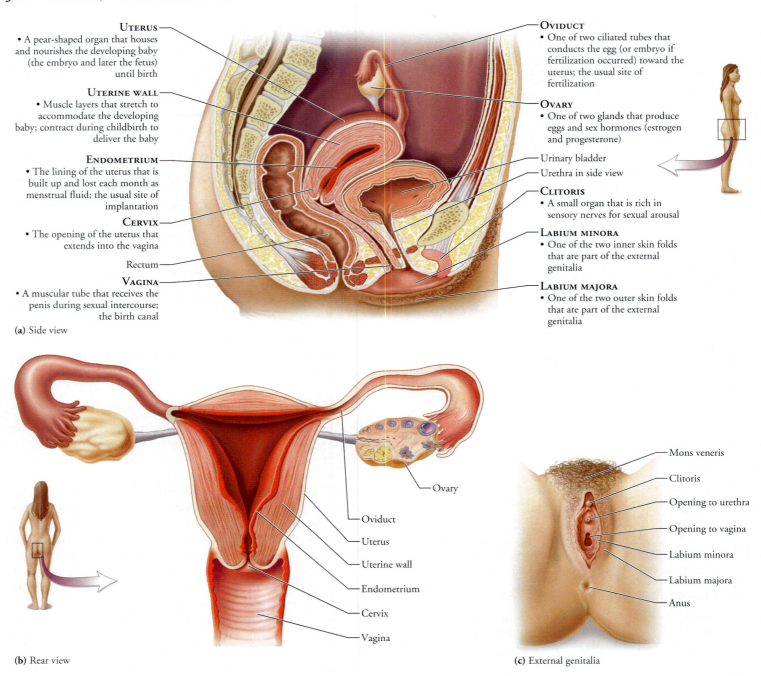

UTERUS
• A pear-shaped organ that houses and nourishes the developing baby (the embryo and later the fetus) until birth

UTERINE WALL
• Muscle layers that stretch to accommodate the developing baby; contract during childbirth to deliver the baby

ENDOMETRIUM
• The lining of the uterus that is built up and lost each month as menstrual fluid; the usual site of implantation

CERVIX
• The opening of the uterus that extends into the vagina

Rectum

VAGINA
• A muscular tube that receives the penis during sexual intercourse; the birth canal

(a) Side view

OVIDUCT
• One of two ciliated tubes that conducts the egg (or embryo if fertilization occurred) toward the uterus; the usual site of fertilization

OVARY
• One of two glands that produce eggs and sex hormones (estrogen and progesterone)

Urinary bladder
Urethra in side view

CLITORIS
• A small organ that is rich in sensory nerves for sexual arousal

LABIUM MINORA
• One of the two inner skin folds that are part of the external genitalia

LABIUM MAJORA
• One of the two outer skin folds that are part of the external genitalia

Ovary
Oviduct
Uterus
Uterine wall
Endometrium
Cervix
Vagina

(b) Rear view

Mons veneris
Clitoris
Opening to urethra
Opening to vagina
Labium minora
Labium majora
Anus

(c) External genitalia

FIGURE **17.6**
The female reproductive system

region). The smooth muscle of the uterine wall contracts rhythmically in waves during childbirth and forces the infant out. The thickness of the endometrium varies during cycles of approximately one month. At the beginning of each cycle, the endometrium gradually becomes thicker and more vascular (more heavily penetrated by blood vessels). If an embryo is formed, it implants (embeds) in the endometrium, where it will remain for the duration of the pregnancy. If an embryo does not form, the endometrium that built up during that cycle will be lost as menstrual flow.

The narrow neck of the uterus, which projects into the vagina, is called the **cervix**. The **vagina** is a muscular tube that opens to the

outside of the body. The vagina receives the penis during sexual intercourse. Sperm that are deposited in the vagina can enter the uterus through an opening in the cervix and then swim to the oviduct to meet the egg. At birth, the infant is pushed through the cervix and then the vagina on its way to greet the world.

ECTOPIC PREGNANCY

If an embryo implants in an area other than the uterus, the condition is described as an ectopic (*ect-*, outside) pregnancy. The most common type of ectopic pregnancy is a tubal pregnancy, in which the embryo implants in an oviduct. A tubal pregnancy must be sur-

TABLE **17.3**

THE FEMALE REPRODUCTIVE SYSTEM	
STRUCTURE	**FUNCTION**
Ovary	Produces eggs and the hormones estrogen and progesterone
Oviducts	Transport ovulated egg (or embryo if fertilization occurred) from ovary to uterus; the usual site of fertilization
Uterus	Receives and nourishes embryo
Vagina	Receives penis during intercourse; serves as birth canal
Clitoris	Contributes to sexual arousal
Breasts	Produce milk

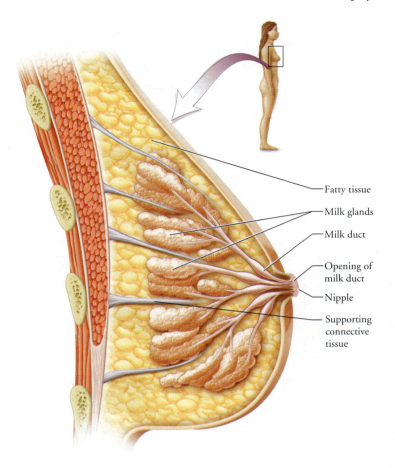

Fatty tissue

Milk glands

Milk duct

Opening of milk duct

Nipple

Supporting connective tissue

FIGURE **17.7**
Breast structure

gically terminated because it places the life of the mother in danger. If the embryo is permitted to continue growing, it will eventually rupture the oviduct, and the rupture can cause the mother to bleed to death internally.

▌ External genitalia lie outside the vagina

The female reproductive structures that lie outside the vagina are collectively known as the external genitalia, or the *vulva*. Two hair-covered folds of skin surround the vaginal opening. The *labia majora* ("big lips") enclose two thinner skin folds, the *labia minora* ("little lips"). The anterior portions of the labia minora form a hood over the **clitoris**. The clitoris develops from the same embryological structure that, in a male, would form the tip of the penis. Like the penis, the clitoris has many nerve endings sensitive to touch. During tactile stimulation, the erectile tissue in the clitoris becomes swollen with blood and contributes to a woman's sexual arousal.

▌ The breasts produce milk to nourish the baby

The *breasts,* or mammary glands, are present in both sexes, but they produce milk to nourish a newborn only in females. Inside the breast are 15 to 25 groups of milk-secreting glands. A milk duct drains each group through the nipple. Interspersed around the glands and ducts is fibrous connective tissue that supports the breast (Figure 17.7).

The monthly changes in ovarian hormones estrogen and progesterone levels prepare the breast for the possibility of pregnancy. Early in each cycle, estrogen stimulates the growth of the glands, ducts, and fibrous tissue of the breast. Progesterone then triggers the initial steps of the milk-secreting process. (Milk production does not begin for 2 to 3 days after the newborn begins suckling; see Chapter 18.) At the same time, blood flow to the breasts increases. Together, these changes may cause feelings of fullness and tenderness that peak just before a woman's period begins. If pregnancy does not occur, the secretions and cells are usually reabsorbed. The swelling and tenderness then subside. In some women, the monthly buildup of the breast is too extensive or else the reabsorption process is insufficient. As a result, pockets of cell debris and trapped secretions form lumps in the breast

that may be painful, a condition called *fibrocystic breast disease.* These lumps may also make it more difficult to detect a cancerous mass (see the Health Issue essay, *Breast Cancer*).

▌ The ovarian cycle involves egg production and ovulation

The changes in the ovary that produce the egg occur in cycles of about one month and are called the **ovarian cycle**. Although the ovaries do not produce eggs until a female reaches puberty, the preparations for egg production begin before she is born. Recall that the egg is a haploid cell formed by meiosis, the type of cell division that forms gametes. Early in fetal development, cells that may eventually develop into eggs migrate from the yolk sac (a structure that forms from embryonic membranes) to the ovaries, where they later develop into diploid cells called *oogonia*. The oogonia divide by mitosis, and their numbers swell to several million. In the third month of fetal development, the oogonia enlarge and begin to store nutrients; some begin meiosis by making a copy of each of their chromosomes, and the copies of each chromosome remain attached to one another. The cells that prepare for meiosis are called *primary oocytes* (immature eggs). A single layer of flattened cells, called follicle cells (granulosa cells), surrounds each primary oocyte. The entire structure is called a *primary follicle.*

Breast Cancer

One in nine American women will develop breast cancer at some point in her life. Breast cancer kills more than 40,000 women each year in the United States, making it the second biggest cancer killer in women after lung cancer.

Breast cancer usually begins with the abnormal growth of the cells lining the milk ducts of the breast, but it sometimes begins in the milk glands themselves. Some breast tumors grow quite large without spreading. Other, more life-threatening types of breast cancer aggressively invade surrounding tissues. Typically, cancerous cells begin to spread when the tumor is about 20 mm (about 3/4 in.) in diameter. At this point, they break through the membranes of the ducts or glands where they initially formed and move into the connective tissue of the breast. They may then move into the lymphatic vessels or blood vessels permeating the breast or into both; the vessels can often transport the cells throughout the body.

Detecting Breast Cancer

Early detection is a woman's best defense against breast cancer. A monthly breast self-exam (BSE) is helpful in detecting a lump early (Figure 17.A). If a woman begins breast self-exam in early adulthood, she becomes familiar with the consistency of her breast tissue. Later in life, it is easy to notice changes that might be telltale signs of breast cancer.

Mammograms, x-rays of breast tissue, are also helpful in early detection of breast cancer, because they can detect a tumor too small to be felt as a lump.

Early detection can make the difference between life and death. A tumor large enough to be felt contains a billion or more cells—a few of which may have already spread from the tumor to other tissues of the body. After cancer cells spread, the woman's chance of survival decreases dramatically. An added benefit of mammograms is that they can detect tumors that are small enough to be removed by a type of surgery, called lumpectomy, that removes the lump but spares the breast. At later stages of breast cancer, the entire breast may have to be removed, a type of surgery called mastectomy.

Risk Factors

Many things increase a woman's risk of breast cancer that she cannot do anything about. One of these is who her parents are. The genes she inherits from her parents can influence her risk of developing breast cancer. At least two genes increase a woman's risk of breast cancer. Although only 5% to 10% of all breast cancers are related to these genes, the unlucky women who inherit either of them have an 85% chance of developing breast cancer at some point in their life.

There is a common thread—exposure to estrogen—that runs through the tapestry of risk factors associated with breast cancer. During each menstrual cycle, estrogen stimulates breast cells to begin dividing in preparation for milk production, in case the egg is fertilized. Excessive estrogen exposure, therefore, may push cell division to a rate characteristic of cancer.

One factor that influences estrogen levels is the number of times a woman ovulates during her lifetime, because estrogen is produced by both the maturing ovarian follicle and the corpus luteum. The number of times a woman ovulates is, in turn, affected by factors such as the following.

1. Age when menstruation begins. Ovulation usually occurs in each menstrual cycle. Thus, the younger a woman is when menstruation begins, the more opportunities there are for ovulation and the greater her exposure to estrogen.

2. Menopause after age 55. The later menopause occurs in life, the more menstrual cycles a woman is likely to experience. Estrogen levels are low and ovulation ceases after Menopause.

3. Childlessness and late age at first pregnancy. Ovulation does not occur during pregnancy. Thus, pregnancy gives the ovaries a rest. Furthermore, the hormonal patterns of pregnancy appear to transform breast tissue in a way that protects against cancer. As a result, delaying pregnancy until after age 30 or remaining childless increases a woman's risk of developing breast cancer later in life.

Long before birth, all of a woman's potential eggs have formed. Although she may have formed as many as 2 million primary follicles, only about 700,000 remain when she is born. The eggs remain in this immature state until she reaches puberty, usually at 10 to 14 years of age, when her reproductive years begin. By the time of puberty, a female's lifetime supply of potential eggs has dwindled to between 200,000 and 400,000.

Beginning at puberty, a group of about a dozen primary follicles will continue development (Figure 17.8) each month. The follicle cells begin dividing, forming layers of cells and secreting a fluid that contains estrogen. Soon, one of these primary follicles will become dominant and will continue development. The others will degenerate. The ovarian cycle of egg development involves the following steps:

- **Follicle maturation.** The follicle cells continue dividing, and fluid begins to accumulate between them. As the fluid accu-

mulates, the wall of follicle cells splits into two layers. The inner layer of follicle cells directly surrounds the primary oocyte. The outer layer forms a balloonlike sphere enclosing the fluid and the oocyte. This structure grows rapidly.

- **Formation of a mature follicle.** Within about 10 to 14 days after its development began, the follicle assumes its mature form, called a Graafian follicle. The primary oocyte then completes the first meiotic division, which it prepared for years earlier. When the primary oocyte divides, it forms two cells of unequal size: a large cell that contains most of the cytoplasm and nutrients, called a *secondary oocyte,* and a tiny cell, called the first *polar body.* The polar body is essentially a garbage bag for one set of chromosomes. It has very few cellular constituents and plays no further role in reproduction. The secondary oocyte is a much larger cell that is packed

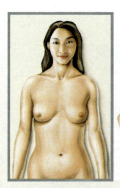

Step 1: Stand in front of the mirror and look at each breast to see if there is a lump, a depression, a difference in texture, or any other change in appearance.

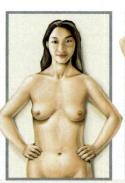

Step 2: Get to know how your breasts look. Be especially alert for any changes in the nipples' appearance.

Step 3: Raise both arms and check for any swelling or dimpling in the skin of your breasts.

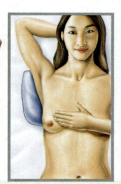

Step 4: Lie down with a pillow under your right shoulder and put your right arm behind your head. Perform a manual breast examination. With the nipple as the center, divide your breast into imaginary quadrants.

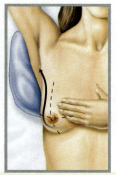

Step 5: With the pads of the fingers of the left hand, make firm circular movements over each quadrant, feeling for unusual lumps or areas of tenderness. When you reach the upper, outer quadrant of your breast, continue toward your armpit. Press down in all directions.

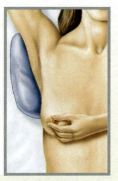

Step 6: Feel your nipple for any change in size and shape. Squeeze your nipple to see if there is any discharge. Repeat from step 4 on the other breast.

FIGURE **17.A**

A monthly breast self-exam can help a woman find lumps in her breast before they have spread to surrounding tissue. It also makes a woman familiar with the consistency of her breast tissue, and therefore able to detect changes in consistency that might indicate cancer.

4. Breast-feeding. Women who breast-feed their infants have a 20% lower risk of developing breast cancer before they reach menopause than do women who bottle-feed their infants. Nursing may guard against breast cancer by blocking ovulation or by causing physiological changes that leave the breast tissue more resistant to cancer-producing chemicals in the environment.

5. Exercise. Even moderate exercise can suppress ovulation in adolescents and women in their twenties (but, unfortunately, exercise is not as likely to supress ovulation in older women).

Other factors may also influence estrogen levels. For instance, obese women have higher estrogen levels than do thin ones because estrogen is produced in fat cells, as well as by the ovaries. Indeed, obesity, especially if the fat is carried above the waist, translates to a three-fold elevated risk of breast cancer. A weight gain as small as 10 pounds when a woman is in her twenties also increases her risk of breast cancer later in life.

with nutrients that will nourish the embryo until it reaches the uterus.

- **Ovulation.** About 12 hours after the secondary oocyte has formed, the mature follicle pops, like a blister, releasing the oocyte mass, which is about the size of the head of a pin. The release of the oocyte from the ovary is called **ovulation**. If a sperm penetrates the secondary oocyte, meiosis is completed. The replicate chromosomes are separated. Division of the cytoplasm is again unequal. One set of chromosomes goes into a small cell, called the second polar body, and the other set into a large cell, the mature **ovum**, or egg. If fertilization does not occur, the egg will not complete meiosis.

- **Formation of the corpus luteum.** The cells that formed the outer sphere of the mature Graafian follicle remain in the ovary, and luteinizing hormone (LH) transforms these cells into an endocrine structure called the **corpus luteum** (which means yellow body). The corpus luteum secretes both estrogen and progesterone. Unless pregnancy occurs, the corpus luteum degenerates. If pregnancy occurs, the corpus luteum will be maintained by a hormone from the embryo called human chorionic gonadotropin (HCG), as we will see later in this chapter.

The interplay of hormones coordinates the ovarian and menstrual cycles

A woman's fertility is cyclic. At approximately monthly intervals, an egg matures and is released from an ovary. Simultaneously, the uterus is readied to receive and nurture the young embryo. These events must be coordinated. If fertilization does not occur, the uterine provisions are discarded as menstrual flow. The ovaries and uterus will prepare again during the next cycle. The events in the ovary, known

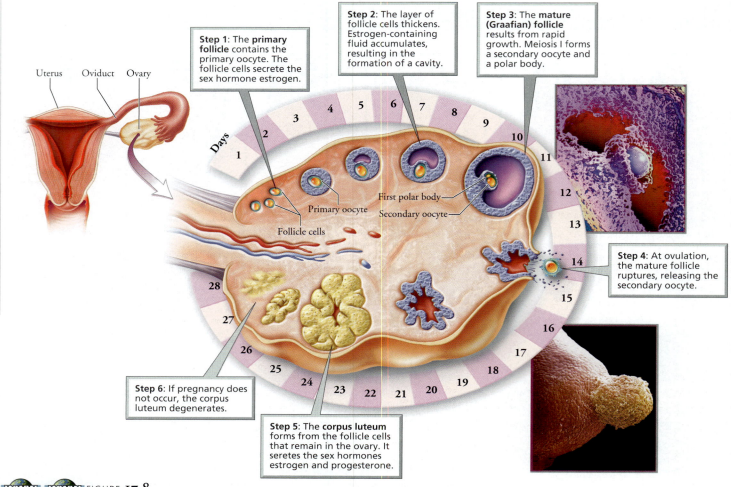

Step 1: The **primary follicle** contains the primary oocyte. The follicle cells secrete the sex hormone estrogen.

Step 2: The layer of follicle cells thickens. Estrogen-containing fluid accumulates, resulting in the formation of a cavity.

Step 3: The mature (Graafian) follicle results from rapid growth. Meiosis I forms a secondary oocyte and a polar body.

Step 4: At ovulation, the mature follicle ruptures, releasing the secondary oocyte.

Step 5: The **corpus luteum** forms from the follicle cells that remain in the ovary. It seretes the sex hormones estrogen and progesterone.

Step 6: If pregnancy does not occur, the corpus luteum degenerates.

Uterus Oviduct Ovary

Days

Primary oocyte

Follicle cells

First polar body

Secondary oocyte

FIGURE **17.8**

TUTORIAL 17.2 TUTORIAL 17.3

The ovarian cycle. A follicle does not move around the ovary during its development, as depicted here. The steps indicate the sequence of events that occurs in a single place in the ovary each month as they might occur in a cycle of 28 days.

as the ovarian cycle, must be closely coordinated with those in the uterus, known as the **uterine cycle** (or the *menstrual cycle*).

A female's fertility is coordinated by the interplay of hormones (Table 17.4). Events in both the uterus and the ovary are coordinated by interactions between hormones from the anterior pituitary gland and from the ovary (Figure 17.9). The anterior pituitary gland produces follicle-stimulating hormone (FSH) and luteinizing hormone (LH). In the female, these hormones cause the ovary to release its hormones, estrogen and progesterone. As in the male, the release of FSH and LH is regulated by a hormone from the hypothalamus called gonadotropin-releasing hormone (GnRH). Progesterone and estrogen (except at a very high level) exert negative feedback on the anterior pituitary, which causes the levels of FSH and LH to decline. As a result, the levels of estrogen and progesterone decline, reducing the inhibition on the anterior pituitary gland. And so, the hormones cycle.

MENSTRUATION   Traditionally, the first day of menstrual flow is considered day 1 of the uterine cycle because it is the point that is most easily noted (Table 17.5). The bleeding usually lasts an average of 5 days. Between 25 and 65 ml (4 to 6 tablespoons) of blood is lost. At this time, the ovarian hormones, estrogen and proges-

TABLE **17.4**

HORMONES INVOLVED IN THE REGULATION OF FEMALE REPRODUCTIVE PROCESSES		
HORMONE	**SOURCE**	**EFFECTS**
Estrogen	Ovaries (follicle cells and corpus luteum)	Maturation of the egg; development and maintenance of female reproductive structures, secondary sex characteristics; thickens endometrium of uterus in preparation for implantation of embryo; cell division in breast tissue
Progesterone	Ovaries (corpus luteum)	Further prepares uterus for implantation of embryo; maintains endometrium
Follicle-stimulating hormone (FSH)	Anterior pituitary gland (in brain)	Stimulates development of a follicle in the ovary
Luteinizing hormone (LH)	Anterior pituitary gland (in brain)	Triggers ovulation; causes formation of the corpus luteum

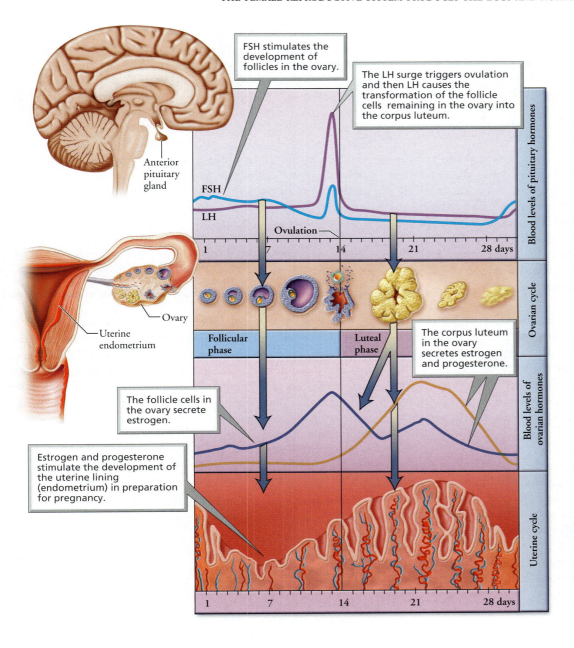

FSH stimulates the development of follicles in the ovary.

The LH surge triggers ovulation and then LH causes the transformation of the follicle cells remaining in the ovary into the corpus luteum.

Anterior pituitary gland

FSH

LH

Ovulation

Blood levels of pituitary hormones

1   7   14   21   28 days

Ovary

Uterine endometrium

Follicular phase

Luteal phase

Ovarian cycle

The corpus luteum in the ovary secretes estrogen and progesterone.

The follicle cells in the ovary secrete estrogen.

Estrogen and progesterone stimulate the development of the uterine lining (endometrium) in preparation for pregnancy.

Blood levels of ovarian hormones

Uterine cycle

1   7   14   21   28 days

WWW **TUTORIAL 17.4** FIGURE **17.9**
The ovarian and uterine cycles are coordinated by the interplay of hormones from the anterior pituitary gland and the ovary. The timing of events shown is for a 28-day cycle.

TABLE **17.5**

OVARIAN AND UTERINE CYCLES			
OVARIAN CYCLE		**UTERINE CYCLE**	
Approximate timing in 28-day cycle	Events	Approximate timing in 28-day cycle	Events
Days 1–13	Follicle develops, caused by FSH	Day 1	Onset of menstrual flow (breakdown of endometrium)
	Follicle cells produce estrogen	Day 6	Endometrium begins to get thicker
Day 14	Ovulation is triggered by LH surge		
Days 15–21	Corpus luteum forms and secretes estrogen and progesterone	Days 15–23	Endometrium is further prepared for implantation of the embryo by estrogen and progesterone
Days 22–28	Corpus luteum degenerates, causing estrogen and progesterone level to decline	Days 24–28	Endometrium begins to degenerate owing to declining maintenance by progesterone

terone, are at their lowest levels, allowing the anterior pituitary gland to produce its hormones, especially FSH. In turn, FSH causes a number of egg follicles in the ovary to develop and produce estrogen. (Hence its name, follicle-stimulating hormone.) Thus, even as the uterus loses the endometrial lining it prepared during the previous cycle in which no egg was fertilized, the egg that will be released in the next cycle is developing.

ENDOMETRIUM THICKENS As the egg follicle develops and the number of follicle cells increases, estrogen levels rise. Estrogen causes the cells in the endometrial lining of the uterus to divide. These cells store nutrients to nourish a future embryo during its early stages of development. In addition, estrogen inhibits the release of FSH through a negative feedback cycle (see Chapter 10). When the egg and follicle are nearly mature, the estrogen level rises rapidly and causes a sudden and spectacular release of LH (and FSH) from the anterior pituitary gland.[3]

OVULATION AND FORMATION OF CORPUS LUTEUM (IN OVARY) The LH surge causes several important events. It causes the egg to undergo its first meiotic division. Next, LH triggers ovulation, and the egg bursts out of the ovary to begin its journey along the oviduct. Continued secretion of LH then transforms the remaining follicle cells into the corpus luteum. The corpus luteum continues the estrogen secretion begun by the follicle cells and, importantly, also secretes progesterone.

ENDOMETRIUM FURTHER PREPARED FOR IMPLANTATION Together estrogen and progesterone make the endometrial lining of the uterus a hospitable place for the embryo. The blood supply to the endometrium increases. Uterine glands develop that secrete a mucous material that can nourish the young embryo when it arrives in the uterus.

The rising estrogen and progesterone levels inhibit pituitary secretion of FSH and LH. As FSH levels decline, the development of new follicles is inhibited.

CORPUS LUTEUM DEGENERATES If fertilization does not occur, the corpus luteum will degenerate within about 2 weeks (14 days plus or minus 2 days, regardless of the length of the menstrual cycle). The degeneration of the corpus luteum results in falling levels of estrogen and progesterone.

MENSTRUATION BEGINS AGAIN Progesterone is essential to the maintenance of the endometrium. As progesterone levels drop because of the degeneration of the corpus luteum, the blood vessels nourishing the endometrial cells collapse. The cells of the endometrium then die and are sloughed off, along with mucus and blood, as menstrual flow. No longer inhibited by estrogen and progesterone, FSH and LH levels begin to climb. Thus, the cycle begins again.

[3]Above a critical level, estrogen exerts a positive feedback effect on the brain and anterior pituitary gland.

what would you do?

In 2004, the International Union of Pure and Applied Chemistry (IUPAC) issued a report indicating that certain pollutants are causing changes in the gender of certain animals, including alligators, polar bears, seals, and some fish. Some of these pollutants, including the pesticide DDT and bisphenol A (a breakdown product of plastics), mimic the effects of estrogen. Others, such as certain PCBs (polychlorinated biphenyls) that were once used in coolants and plastics, block the effects of natural estrogens. Some pollutants bind to receptors for androgens and can feminize a male by blocking the action of natural androgens. Although effects on humans are not yet certain, they could involve an increase in the rates of breast and testicular cancer, lower sperm count, and smaller penis size. IUPAC recommends research of these gender-bending pollutants and global monitoring of their levels in the environment. Implementing these recommendations would be costly. As a taxpayer, are you willing to pay for these precautions? Would you vote for a political candidate who supported their implementation?

If the egg is fertilized, however, a hormone from the embryo, called **human chorionic gonadotropin (HCG)** maintains the corpus luteum. If pregnancy occurs, HCG will prevent the degeneration of the corpus luteum, keeping estrogen and progesterone levels high enough to prevent endometrial shedding. HCG is detectable in the mother's blood within 7 to 9 days after fertilization and in her urine less than 2 weeks after fertilization. Between the second and third month of development the placenta has developed sufficiently to take over the production of estrogen and progesterone during the rest of the pregnancy. The corpus luteum then degenerates.

Menopause ends a woman's reproductive ability

A woman's fertility usually peaks when she is in her twenties and then gradually declines. By the time she reaches 45 to 55 years of age, few of the potential eggs prepared before birth remain in the ovary. Those that do remain become increasingly less responsive to follicle-stimulating hormone (FSH) and luteinizing hormone (LH), the hormones that cause egg development and ovulation. The ovaries, therefore, gradually stop producing eggs, and the levels of estrogen and progesterone fall. During this time, menstrual cycles become increasingly irregular. Eventually, ovulation and menstruation stop completely, an event called **menopause**.

The drop in estrogen levels has a number of physiological effects that can range in severity from annoying to life threatening. At one end of the scale is the loss of a layer of fat that was formerly promoted by estrogen. This loss results in a reduction in breast size and the appearance of wrinkles. Estrogen also plays an important role in regulating a woman's body thermostat. So, as estrogen levels fall, many women experience hot flashes, waves of warmth spreading upward from the trunk to the face. During a hot flash, the skin may redden, and sweating may occur. Although a single hot flash lasts only a few minutes, hot flashes may occur periodically throughout the day and night. Indeed, "night sweats" caused by hot flashes may awaken a woman several times during the night. The absence of estrogen can cause vaginal dryness that can make sexual intercourse painful. Without estrogen, the male hormones produced by the adrenal gland predominate and can cause facial hair to grow.

More serious problems associated with the estrogen deficit that accompanies menopause include an increased risk of diseases of the heart and blood vessels and weak bones. Estrogen offers some protection against atherosclerosis, a condition in which fatty deposits clog the arteries. Thus, after menopause, this protective effect on the heart is lost. Estrogen is also important in the body's ability to absorb calcium from the digestive system and to deposit it in bone. Without calcium, bone becomes weak and porous, a condition known as osteoporosis (discussed in Chapter 5).

Possible Problems with the Female Reproductive System Vary in the Severity of Health Consequences

Premenstrual syndrome (PMS) is a collection of symptoms that appear 7 to 10 days before a woman's period begins. These symptoms include depression, irritability, fatigue, and headaches. Indeed, 20% to 40% of all menstruating women feel that PMS significantly interferes with their daily living.

Some researchers suggest that a progesterone deficiency is to blame for the symptoms of PMS. Progesterone, for example, has a calming effect on the brain. It also decreases fluid retention. As progesterone levels plummet in the days just before a woman's period begins, the nervous system may be stimulated and fluids may be retained. Fluid retention causes an uncomfortable, bloated feeling.

Treatments for PMS are varied. For women whose symptoms are severe, drugs that elevate the levels of serotonin—a chemical in the brain that some nerve cells use for communication—are administered. Some women with milder symptoms find relief from changes in diet. Caffeine, alcohol, fat, and sodium should be avoided. On the other hand, foods high in calcium, potassium, manganese, and magnesium should be increased. Aerobic exercise for at least half an hour stimulates the release of enkephalins and endorphins, which may provide some relief.

Prostaglandins, chemicals used in communication between cells in many parts of the body, are the primary cause of *menstrual cramps*. Endometrial cells produce prostaglandins that, among other things, cause the smooth muscle cells of the uterus to contract, causing cramps. High levels of prostaglandins can cause sustained contractions, called muscle spasms, of the uterus. These muscle spasms may cut down blood flow and, therefore, the oxygen supply to the uterine muscles, resulting in pain.

Endometriosis is a condition in which tissue from the lining of the uterus is found outside the uterine cavity—commonly in the oviducts, on the ovaries, or on the outside surface of the uterus, the bladder, or the rectum. We see, then, that endometrial tissue can move out the open ends of the oviducts to the abdominal cavity. Endometrial tissue, wherever it is, grows and breaks down with the hormonal changes that occur during each menstrual cycle. These cyclic changes can cause extreme pain associated with the menstrual cycle. Women with endometriosis often have difficulty becoming pregnant.

Vaginitis is an inflammation of the vagina that is most commonly caused by one of three types of organisms: yeast (*Candida albicans*), a protozoan (*Trichomonas vaginalis*), or a bacterium (*Gardnerella vaginalis*). Whereas a yeast infection is generally not sexually transmitted, infections with *Gardnerella* or *Trichomonas* are usually transmitted through sexual intimacy with an infected person. Although vaginal discharge is always the primary sign of vaginitis, the nature of the discharge varies with the causative organism. The discharge accompanying a yeast infection, for instance, is white and curdy, like cottage cheese. It causes intense itching and sometimes burning and itching of the outer lips of the vagina. An infection with *Trichomonas* causes a watery, foamy, greenish or yellowish discharge, vaginal itching, and pain. In contrast, the discharge caused by *Gardnerella* is gray or white and has a fishy odor. It does not usually cause itching. It is important to identify the cause of a vaginal infection because each has a different treatment.

Pelvic inflammatory disease (PID), which affects more than a million women each year in the United States alone, is a catch-all term used to refer to an infection of the pelvic organs. The symptoms of PID—abdominal pain or tenderness, lower-back pain, pain during intercourse, abnormal vaginal bleeding or discharge, fever, or chills—can come on gradually or suddenly and vary in intensity from mild to severe.

Pelvic inflammatory disease can be caused by a number of organisms, but the most common culprits are the sexually transmitted bacteria that cause chlamydia and gonorrhea (discussed in Chapter 17a). The bacteria ascend from the vagina through the cervix to the uterus. There, they may infect the endometrium and then spread to the wall of the uterus and to the oviducts. Near the ovary, the oviduct opens to the abdominal cavity, allowing the infection to spread into the abdominal cavity. The infection can be curbed by treatment with antibiotics, which kill the bacteria.

If PID is not treated quickly, its consequences can be long lasting and severe: reduced fertility and an increased risk of ectopic pregnancy. As the infection spreads through the oviducts, the body attempts to defend itself by secreting fibrous material to wall off the area. The resulting scar tissue can block the tubes. If both tubes are completely blocked, the woman is sterile (unable to have children), because the sperm cannot reach the egg. If a tube is partially blocked, the woman may have difficulty conceiving a child. If pregnancy does occur, the scarring also increases the odds that the pregnancy will be ectopic. An egg is many times larger than sperm. Thus, sperm may be able to swim past the scar tissue and fertilize the egg. The resulting embryo, however, being too big to pass the scar tissue and reach the uterus, may implant in the oviduct.

The Human Sexual Response Involves Four Stages

In both men and women, sexual arousal and sexual intercourse involve two basic physiological changes: certain tissues fill with blood (vasocongestion) and certain muscles show sustained or rhythmic contractions (myotonia). However, men and women differ in which tissues fill with blood and which muscles contract. The sequence of events that accompany sexual arousal and intercourse is called the sexual response cycle. Let's consider the four stages of the human response cycle in men and in women.

1. **Excitement.** During the excitement phase of the sexual response cycle, sexual arousal increases. This stage occurs during foreplay, the activities that precede sexual intercourse, and prepares the penis and vagina for intercourse. In a male, blood fills the spongy tissue of the penis, which becomes erect, as described earlier. In a female, blood flow to the breasts, nipple, labia, and clitoris causes these structures to swell. Partly due to an increase in blood flow, the vaginal walls seep fluid that serves as a lubricant, which makes the insertion of the penis easier. In both men and women, muscle contraction can cause the erection of nipples and tension in the arms and legs. Breathing rate and heart rate increase.

2. **Plateau.** During the plateau stage, sexual arousal is maintained at a high level. The changes that began during the first phase will continue through the plateau stage until orgasm. In men, increased blood flow causes the testes to enlarge, and muscle contraction pulls them closer to the body. In women, increased blood flow causes the outer third of the vagina to enlarge and the clitoris to retract under the clitoral hood. In both men and women, breathing rate and heart rate continue to increase. Muscle spasms in the hands, feet, and face may begin.

3. **Orgasm** The peak of the sexual response cycle, orgasm, is brief, but usually intensely pleasurable. In both sexes, orgasm is characterized by muscle contractions of the reproductive structures. In men, ejaculation occurs during orgasm. Ejaculation takes place in two stages. In the first stage, contractions of glands and ducts in the reproductive system force sperm and the secretions of the accessory glands to the urethra. During the second stage of ejaculation, the urethra contracts and semen is forcefully expelled from the penis. Orgasm in women involves rhythmic muscle contractions in the vagina and uterus. Some women experience orgasm rarely or not at all, and others can experience orgasm repeatedly. The absence of orgasm does not affect a woman's ability to reproduce.

4. **Resolution.** During the resolution stage, the body slowly returns to its normal level of functioning. Men usually experience an interval of time during the resolution stage when it is it impossible to achieve another erection or orgasm. Resolution usually takes longer in women than it does in men.

Birth Control Is the Prevention of Pregnancy

Sexual activity is actually a double-edged sword: One edge is the possibility of pregnancy, and the other is the possibility of getting a sexually transmitted disease (STD). The way in which couples dodge one of the edges of the sword influences the chances of being cut by the other one. For this reason, as we discuss various means of contraception, we will also consider whether they reduce the risk of spreading sexually transmitted infections.

There are many means of preventing pregnancy available. Different methods interrupt the reproductive process at different points, have varying degrees of effectiveness, and have varying degrees of protection against STDs. Different types of birth control are summarized in Table 17.6.

▌Abstinence involves refraining from intercourse

Abstinence, that is, not to have intercourse at all, is the most reliable way to avoid both pregnancy and the spread of STDs. However, abstinence does not always seem to be the most popular means of contraception.

▌Sterilization involves cutting and sealing gamete transport tubes

Well before the end of their reproductive life spans, most people have had all the children they want. Other than abstinence, *sterilization* is the most effective way to ensure that pregnancy does not occur. However, unlike abstinence, sterilization offers no protection against STDs.

Sterilization in men involves an operation called a *vasectomy* that cuts the vas deferens on each side and prevents sperm from leaving the man's body (Figure 17.10a). The procedure, which usually lasts about 20 minutes, can be performed in a physician's office under local anesthesia. The physician simply makes small openings in the scrotum through which each vas deferens can be pulled and cut; a small segment is removed and at least one end is sealed shut. Because sperm make up only about 1% of the semen, the volume of semen the man ejaculates is not noticeably reduced. His interest in sex is not lessened because testosterone, which is responsible for the sex drive, is released from the interstitial cells of the testis and carried around the body in the blood.

Female sterilization, called *tubal ligation,* involves blocking the oviducts to prevent the egg and sperm from meeting (Figure 17.10b). Commonly, the tubes are cut and the ends seared shut or mechanically blocked with clips or rings. Tubal ligation is frequently done using a procedure called a laparoscopy, in which two small incisions are made in the abdominal cavity. Laparoscopy is generally performed in a hospital under general anesthesia. Because the abdominal cavity is opened, the risk of infection is greater after tubal ligation than it is after a vasectomy. A woman continues to menstruate after a tubal ligation, because the ovarian hormones continue to be produced in a cyclic manner.

Sterilization should be considered permanent even though it is sometimes possible to reverse the procedure. At best, the reversal procedure is expensive and requires that the surgeon have special training in microsurgical techniques. Even then, success is not guaranteed.

▌Hormonal contraception interferes with regulation of reproductive processes

Hormonal contraception is currently available only to females. There are two basic types: the methods that combine estrogen and progesterone and the progesterone-only means of contraception.

TABLE 17.6

METHODS OF BIRTH CONTROL AVAILABLE IN THE UNITED STATES						
METHOD	PROCEDURE	HOW IT WORKS	PERCENTAGE FAILURE (TYPICAL USE)	PERCENTAGE FAILURE (PERFECT USE)	RISKS	PROTECTION FROM STDs
Abstinence	Abstain from sexual activity	Sperm never contacts egg	0	0	None	Yes
Sterilization						
Vasectomy	Cut and seal each vas deferens	No sperm in semen	<1	<1	No risks presently known	None
Tubal ligation	Oviducts blocked	Sperm cannot reach egg	<1	<1	Infection from surgery	None
Hormonal Methods *Combination estrogen and progesterone*						
Oral (the pill)	Hormone pill taken daily	Prevents egg development and release	3	0.1	Problems with heart or blood vessels; blood clot formation; stroke	None
Injection (Lunelle®)	Estrogen and progesterone injection every month	Same as pill	Better than pill	0.1	Assumed same as pill	None
Vaginal ring (NuvaRing®)	Ring inserted in vagina by woman for 3 weeks and removed for 1 week	Same as pill	Better than pill	1–2	Assumed same as pill	None
Skin patch (Ortho Evra®)	Patch adheres to skin of abdomen or buttocks	Same as pill	1.24	0.99	Assumed same as pill	None
Progesterone only						
Minipill	Hormone pill taken daily	Thickens cervical mucus; endometrium not properly prepared; sperm movement impaired; egg does not mature	13.2	1.1	Ovarian cysts	None
Injection (Depro-Provera®)	Progesterone injection every 3 months	Same as minipill	0.3	0.3	Possible link to osteoporosis	None
Intrauterine device (IUD)	Small plastic device inserted into uterus by physician	Interferes with both fertilization and implantation	<2	<2	Increased risk of pelvic inflammatory disease following insertion	None
Barrier Methods						
Diaphragm	Inserted into vagina before intercourse	Covers cervix and prevents sperm from entering	18	9	No risks presently known	Some for woman
Cervical cap	Inserted into vagina before intercourse	Same as diaphragm (partly by suction)	18	9	No risks presently known	Some for woman
Male condom	Fits over penis before intercourse	Prevents sperm/penis from contacting vagina	12	3	No risks presently known	Latex, excellent; "skin," poor
Female condom	Held in vagina by flexible rim rings	Prevents sperm/penis from contacting vagina	21	5	No risks presently known	Very good
Spermicides	Inserted into vagina before intercourse	Kill sperm for 1 hour after application	21	6	No risks presently known	Increased for women

TABLE **17.6** (*Continued*)

METHODS OF BIRTH CONTROL AVAILABLE IN THE UNITED STATES						
METHOD	PROCEDURE	HOW IT WORKS	PERCENTAGE FAILURE (TYPICAL USE)	PERCENTAGE FAILURE (PERFECT USE)	RISKS	PROTECTION FROM STDs
Fertility Awareness						
	Abstain from sex on days that eggs and sperm may meet	Sperm never come into contact with egg	20	2–9	None	None
Emergency contraception (morning-after pill)						
Combined estrogen and progesterone (Preven®)	First dose taken within 72 hours of unprotected intercourse; second dose 12 hours later	Inhibit or delay ovulation; prevent fertilization; thicken the cervical mucus; alter the endometrium		Reduces likelihood of pregnancy by 75%	Nausea; menstrual cycle disturbance	None
Progesterone-only (Plan B®)	Same as combination	Same as combination		Same as combination	Same as combination	None

THE COMBINATION OF ESTROGEN AND PROGESTERONE

Several forms of birth control combine synthetic forms of estrogen and progesterone. Each of these methods works by mimicking the effects of natural hormones that would ordinarily be produced by the ovaries. Among these effects is the suppression of follicle-stimulating hormone (FSH) and luteinizing hormone (LH) release from the anterior pituitary gland. Without these pituitary hormones, the egg does not mature and is not released from the ovary.

The oldest and most familiar means of combined hormonal birth control is "the pill." A hormone-containing birth control pill is taken daily for 3 weeks, followed by a week of daily pills without hormones.

Over the past few years, additional methods of combined hormonal contraception have been developed. There is a vaginal ring (NuvaRing®) that slowly releases the hormones. A woman inserts the ring so that it encircles the cervix. It is worn for 3 weeks and removed for the fourth. In addition, there is a skin patch (Ortho Evra®) that is worn, usually on the abdomen or buttocks, for a week and then replaced. Patches are worn for 3 weeks, followed by a week without a patch. When a woman uses the birth control pill, the vaginal ring, or the patch, she menstruates during the week without hormones. There is now an injection (Lunelle®) that is given once a month. These newer methods of combined hormone contraception have approximately the same effectiveness and the same risks as the combined birth control pills. Women who have difficulty remembering to take a pill each day may prefer these newer options.

When most people speak about "the pill" they are referring to the combination birth control pill, so named because it contains synthetic forms of both estrogen and progesterone. For most healthy, nonsmoking women younger than 35 years of age, the pill is an extremely safe means of contraception. Nonetheless, a small number of pill users do die each year from a pill complication, so it should be noted as a possibility. The risk increases with age and with cigarette smoking. The pill and cigarette smoking can both increase blood pressure and the risk of abnormal blood clot formation. Together, they have a greater effect than does either alone. Problems with the circulatory system are the most important of the serious (sometimes fatal) complications of pill use. Most pill-associated deaths are caused by heart attack or stroke, which occurs when the blood supply to the heart or a region of the brain is blocked. The pill also increases the risk of abnormal blood clot formation.

Pill users have a greater risk of getting certain STDs from an infected partner than do women who use other forms of birth control or no birth control at all. One way the pill increases the likelihood of transmission of these diseases is by making the vaginal environment more alkaline, a condition that favors the growth of bacteria that cause gonorrhea and chlamydia. Another reason that women on the pill have a greater risk of getting STDs is that it makes the user's cervix more vulnerable to disease-causing organisms. However, because the pill causes cervical mucus to become thicker, it decreases the chances that infections can spread to the uterus and cause pelvic inflammatory disease (PID).

PROGESTERONE-ONLY CONTRACEPTION

Three types of progesterone-only contraception currently exist: an oral contraceptive called the minipill, injections, and subcutaneous implants. As with the combined birth control pill, a woman must take the minipill faithfully every day. However, the minipill is generally less effective than the combined pill. Progesterone injections (Depo-Provera®) given every 3 months, on the

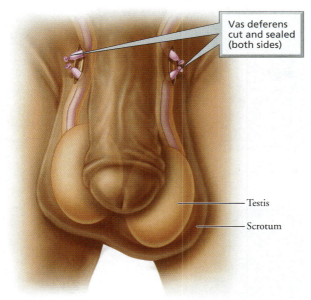

Vas deferens cut and sealed (both sides)

Testis

Scrotum

(a) Vasectomy

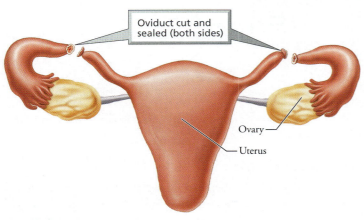

Oviduct cut and sealed (both sides)

Ovary

Uterus

(b) Tubal ligation

FIGURE **17.10**

Sterilization procedures prevent fertilization by cutting and sealing the tubes through which the gametes would travel and meet.

other hand, are an extremely effective means of contraception. Progesterone-containing implants (Implanon® and Jadelle®) are slender, flexible silicone capsules about the size of a matchstick. Implants inserted under the skin in a woman's upper arm are barely visible and provide extremely effective protection for several years as the progesterone slowly diffuses out of the capsules. Progesterone-containing implants are not currently available in the United States.

There are several mechanisms through which progesterone-only contraceptives may prevent pregnancy. All types may prevent ovulation, but they vary in their ability to do so. They all cause thickening of cervical mucus, which makes it more difficult for the sperm to swim through it to reach the egg. In addition, without estrogen, the endometrium is not prepared properly for the implantation of an embryo should fertilization occur.

Another important drawback is that, as with all hormonal contraceptives, these provide no protection against sexually transmitted diseases.

what would you do?

In 2004, a male subcutaneous implant containing progesterone began tests in 14 medical centers in Europe. The implants last a year and reduce sperm levels enough to make a man infertile. A male is given testosterone injections every 3 months to maintain normal levels of the male hormone. It is expected that sperm production will resume within 3 to 6 months after the implant is removed. Researchers believe that these implants will be a realistic alternative means of contraception for couples in a long-term relationship who do not want to use the pill, condoms, or a vasectomy. The implants do not protect against STDs. If you were a male in a long-term relationship, would you volunteer for this study? If you were the female in this relationship, would you want your partner to participate in the study? What criteria would you use to decide?

Intrauterine devices prevent the union of sperm and egg and/or implantation

An *intrauterine device* (IUD), is a small device that is inserted into the uterus by a physician to prevent pregnancy (Figure 17.11a). It can be left in place for several years and should be removed by a physician. An IUD is highly effective in preventing pregnancy. It can interfere with both fertilization and implantation. But, if the embryo does implant, the IUD usually physically dislodges it. However, IUDs offer no protection against the spread of sexually transmitted diseases.

A risk associated with the use of an IUD is pelvic inflammatory disease. The risk is greatest in the weeks following IUD insertion, because disease-causing organisms, usually sexually transmitted organisms, can enter the uterus from the vagina at this time. The risk of pelvic infection due to an IUD is low, less than 1%. However, if it does occur, it can lead to sterility or ectopic pregnancy.

Barrier methods of contraception prevent the union of sperm and egg

Barrier methods of contraception include the diaphragm, cervical cap, contraceptive sponge, and male and female condoms. They work, as their name suggests, by creating a barrier between the egg and sperm.

A *diaphragm* is a dome-shaped soft rubber cup on a flexible ring. It is inserted into the vagina before intercourse so that it covers the cervix, preventing sperm from entering (Figure 17.11b). A *cervical cap* is smaller than a diaphragm and fits snugly over the cervix. Before insertion, spermicidal cream or jelly should be added to the inner surface of the diaphragm or cervical cap. Because a diaphragm covers the cervix, it offers limited protection to the woman against important STDs, including chlamydia and gonorrhea. By protecting the cervix from the virus that causes genital warts, it also lessens the risk of cervical cancer. A cervical cap offers no protection against STDs.

Unlike a diaphragm or cervical cap, a contraceptive sponge or condom can be purchased without a prescription. A *contraceptive sponge* is a small sponge that contains a spermicide. One side of the

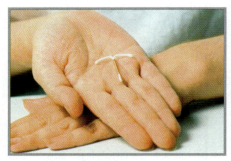

(a) Intrauterine device (IUD)

(b) Diaphragm and spermicidal cream or jelly

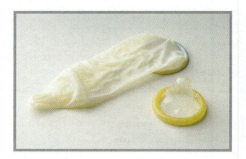

(c) Male latex condom

(d) Female latex condom

FIGURE **17.11**
Selected methods of birth control

sponge has an indentation into which the cervix fits. The other side has a strap that makes removal easier. The sponge must be thoroughly moistened before use to activate the spermicide. It is then inserted and offers protection against pregnancy for the next 24 hours. The sponge is not currently available in the United States but is available in Canada.

The *male condom* is a thin sheath of latex, polyurethane, or natural membranes ("skin") that is rolled onto an erect penis, where it fits like a glove (Figure 17.11c). The sperm are Trapped within the condom and cannot enter the vagina. The effectiveness of condoms in preventing pregnancy depends largely on how consistently they are used.

Other than complete abstinence from sexual activity, the latex condom is the best means of preventing the spread of STDs available today. The skin condoms may offer greater sensitivity, but they do allow some microorganisms, particularly viruses, to pass. Latex has no pores, so microorganisms cannot pass through it. It is highly recommended that people who might be exposed to STDs use a condom for disease protection, even if they are already using another means of birth control to prevent pregnancy. Keep in mind, however, that a condom can prevent disease transmission only between the body surfaces it separates. Diseases can still be spread if contact occurs between other, unprotected body surfaces, such as the external genitalia or thighs.

The *female condom* is a loose sac of polyurethane, a type of clear plastic that resembles that of a food-storage bag. At each end of the sac is a flexible ring that helps to hold the device in place (Figure 17.11d). The female condom does reduce the risk of spreading sexually transmitted infections. Female condoms offer

women who cannot count on their male partner to use condoms a means to protect themselves.

stop and think

A latex condom provides an impenetrable barrier to disease-causing organisms. Yet, studies indicate that women who use a diaphragm or vaginal sponge routinely have lower rates of STDs than do women who rely on their male partners to use a condom. How can you explain this seemingly illogical difference?

Spermicidal preparations kill sperm

Spermicidal preparations consist of a sperm-killing chemical (nonoxynol-9) in some carrier, such as foam, cream, jelly, film, or tablet. When used without any other means of contraception, foams are the most effective form of spermicide to use, because they are effective immediately and disperse more evenly than the others. The sperm-killing effect lasts for about 1 hour after the spermicide has been activated. Laboratory tests have shown that nonoxynol-9 kills the organisms responsible for many STDs. However, this chemical also damages the cells lining the vagina, and this damage could increase a woman's susceptibility to STDs.

Fertility awareness is the avoidance of intercourse when fertilization could occur

Fertility awareness, which also goes by the names of *natural family planning* or the *rhythm method,* is a way to reduce the risk of pregnancy by avoiding intercourse on all days on which sperm and egg might meet. This sounds easier than it is. Sperm can live in the female reproductive tract for 2 to 5 days, but the lower value is most likely. An egg lives only 12 to 24 hours after ovulation. That means that there are only 4 days in each cycle during which fertilization might occur. But, which 4 days? Therein lies the problem. It is difficult enough to pinpoint when ovulation occurs, much less predicting it several days in advance. The situation is not entirely hopeless, however. In almost all women, regardless of their cycle length, ovulation occurs 14 (plus or minus 2) days *before* the onset of flow in the next cycle. Thus, if the woman has a stable cycle length, she can count back 14 days from the date she expects her next period and add 2 days on either side to determine the most likely days of ovulation. Intercourse should be avoided for 3 days before the earliest date on which ovulation might occur, because sperm may remain alive that long. It should also be avoided for 1 day after the latest date on which ovulation is expected, because the egg survives for 24 hours (Figure 17.12). Irregularities in cycle length further complicate the problem, of course.

▌ Morning-after pills are emergency contraception

The so-called morning-after pill is a means of emergency contraception that can actually be used during the *first few days* after unprotected intercourse. One type (Preven®) combines estrogen and progesterone. The second type (Plan B®) contains only progesterone. With either type, the first dose must be taken within 72 hours of unprotected intercourse and the second 12 hours later. Although we do not fully understand the precise mechanisms of

FIGURE **17.12**
Couples using the rhythm method reduce the risk of pregnancy by avoiding intercourse at times that viable sperm and egg might meet.

action, they could inhibit or delay ovulation, prevent fertilization, thicken the cervical mucus, and alter the endometrium, making it an inhospitable place for implantation of the young embryo.

REVIEWING THE CONCEPTS

The Gonads Produce Gametes and Sex Hormones (p. 379)

1. The gonads (testes and ovaries) are the reproductive structures that produce the gametes (sperm and eggs), as well as sex hormones. The male sex hormone is testosterone. The female sex hormones are estrogen and progesterone. A sperm and egg fuse at fertilization, producing a cell called a zygote that will develop into a new individual.

A Male's Reproductive Role Differs from the Role of a Female (p. 379)

2. A male produces a large number of sperm and delivers them to the female reproductive system. A female usually produces only one nutrient-filled egg during each approximately monthly interval. The female also nourishes and protects the developing baby (the embryo and later the fetus) in her uterus.

The Male Reproductive System Delivers Sperm to the Egg (pp. 379–385)

3. The male reproductive system is composed of the testes, a series of ducts (the epididymis, vas deferens, and urethra), accessory glands (the prostate, seminal vesicles, and bulbourethral glands), and the penis.
4. The testes are located outside the body cavity in a sac called the scrotum, which helps regulate the temperature of the testes to ensure proper sperm development.
5. Within the seminiferous tubules of the testis, the production of gametes, called spermatogenesis, is a continuous process. Sperm are haploid gametes produced by meiosis.

6. After leaving the seminiferous tubules, sperm are stored and mature in the epididymis. At ejaculation, sperm travel through the vas deferens and the urethra to leave the body.
7. Semen is the fluid released when a man ejaculates. Most of the semen consists of the secretions of the accessory glands.
8. The penis transfers sperm to the female. Erectile dysfunction is the inability to achieve an erection. There are several options for treatment.
9. Each sperm has three distinct regions. The head of the sperm contains the male's genetic contribution to the next generation. The midpiece is packed with mitochondria, which produce ATP to power movement. The whiplike tail propels the sperm to the egg.
10. The male hormone testosterone is produced by the interstitial cells, which are located within the testes between the seminiferous tubules.
11. Male reproductive processes are regulated by an interplay of hormones from the anterior pituitary gland in the brain (LH and FSH), from the hypothalamus (GnRH), and from the testis (testosterone and inhibin).

WEB TUTORIAL 17.1 Spermatogenesis

The Female Reproductive System Produces the Eggs and Nurtures the Embryo and Fetus (pp. 385–393)

12. The female reproductive system consists of the ovaries, oviducts, uterus, vagina, and external genitalia.
13. The ovaries produce eggs and female hormones, estrogen and progesterone.

14. The oviducts transport the immature egg, zygote, then the early embryo to the uterus. Fertilization usually occurs in the oviduct.

15. The uterus supports the growth of the developing baby (embryo then the fetus). The embryo implants in the lining of the uterus, the endometrium, which provides nourishment during development. The muscular wall of the uterus allows the uterus to stretch as the fetus grows, and contractions of the muscle force the baby out of the uterus during childbirth. The cervix is the opening of the uterus.

16. The breasts produce milk to nourish the baby.

17. The ovarian cycle involves egg production and ovulation. A female is born with a finite number of primary follicles. A primary follicle is a primary oocyte (an immature egg) surrounded by a single layer of follicle cells. The primary oocytes will remain in this state until puberty, when (usually) one each month will continue development and be ovulated as a secondary oocyte.

18. Hormones regulate the ovarian cycle (which prepares an egg for fertilization) and the uterine, or menstrual cycle (which prepares the endometrium of the uterus for implantation of the embryo). Day 1 of the menstrual cycle is the first day of menstrual bleeding. At this time, the endometrium is being shed because the levels of estrogen and, in particular, progesterone are low. At the same time, follicle-stimulating hormone (FSH), secreted by the anterior pituitary gland, is stimulating the development of primary follicles. Follicle cells secrete estrogen. Thus, as the follicle develops, estrogen levels climb. Estrogen causes the endometrium to become thicker and more vascular. Ovulation is the release of the egg from the ovary. Luteinizing hormone (LH) triggers ovulation. After ovulation, LH transforms the follicle cells remaining in the ovary after ovulation into a temporary endocrine structure called the corpus luteum. The corpus luteum secretes estrogen and progesterone. These hormones ready the endometrium for implantation. If the egg is not fertilized, the corpus luteum degenerates, causing the levels of estrogen and progesterone to fall. This decline leads to the onset of menstruation. However, if the egg is fertilized, the young embryo produces human chorionic gonadotropin (HCG), which maintains the corpus luteum. HCG is the hormone that pregnancy tests detect.

19. At menopause, which usually occurs between the ages of 45 and 55, menstruation and ovulation stop. As a result, the levels of estrogen and progesterone drop. Menopause has many psychological and physiological effects on a woman's body.

Possible Problems with the Female Reproductive System Vary in the Severity of Health Consequences (p. 393)

20. Premenstrual syndrome (PMS), a collection of symptoms that occur several days before a woman's period begins, may be caused by low levels of progesterone or of the brain chemical serotonin. Menstrual cramps are caused by prostaglandins, which cause muscle contraction. Endometriosis is a condition in which endometrial tissue is found outside the uterus. Vaginitis is an inflammation of the vagina caused by yeast, a protozoan, or a bacterium. Pelvic inflammatory disease is a bacterial infection of the pelvic organs that can increase a woman's risk of becoming infertile or of having an ectopic pregnancy.

The Human Sexual Response Involves Four Stages (pp. 393–394)

21. The human sexual response is the sequence of events that occur during sexual intercourse. This cycle involves four phases: excitement (increased arousal), plateau (continued arousal), orgasm (climax), and resolution (return to a normal level of functioning).

Birth Control Is the Prevention of Pregnancy (pp. 394–399)

22. Abstinence (refraining from intercourse) and sterilization (vasectomy or tubal ligation) are the most effective ways to prevent pregnancy. Hormonal contraception (combined estrogen plus progesterone or progesterone-only methods) interferes with the regulation of reproductive processes. An intrauterine device prevents the union of sperm and egg and/or implantation. Barrier methods of contraception (the diaphragm, cervical cap, contraceptive sponge, and male or female condom) prevent the union of sperm and egg. Spermicidal preparations kill sperm. Natural family planning (the rhythm method) involves abstinence at times when fertilization could occur. Morning-after pills are emergency contraception that can reduce the risk of an unwanted pregnancy resulting from unprotected intercourse.

KEY TERMS

acrosome *p. 383*
bulbourethral gland *p. 382*
cervix *p. 386*
clitoris *p. 387*
corpus luteum *p. 389*
egg *p. 379*
endometrium *p. 385*
epididymis *p. 381*
estrogen *p. 379*
fertilization *p. 379*
follicle-stimulating hormone (FSH) *p. 384*

gamete *p. 379*
gonad *p. 379*
gonadotropin-releasing hormone (GnRH) *p. 384*
human chorionic gonadotropin (HCG) *p. 392*
inhibin *p. 384*
interstitial cell *p. 380*
luteinizing hormone (LH) *p. 384*
menopause *p. 392*
oogenesis *p. 385*
ovarian cycle *p. 387*

ovary *p. 379*
oviduct *p. 385*
ovulation *p. 389*
ovum *p. 389*
penis *p. 382*
progesterone *p. 379*
prostate gland *p. 381*
scrotum *p. 379*
semen *p. 381*
seminal vesicle *p. 381*
seminiferous tubule *p. 380*
sperm *p. 379*

spermatogenesis *p. 383*
testes *p. 379*
testosterone *p. 379*
urethra *p. 381*
uterine cycle *p. 390*
uterus *p. 385*
vagina *p. 386*
vas deferens *p. 381*
zygote *p. 379*

THINKING ABOUT THE CONCEPTS

1. Name the male and female gonads. What are the functions of these organs?   *p. 379*
2. How is the temperature maintained in the testes? Why is temperature control important?   *pp. 379–380*
3. Trace the path of sperm from their site of production to their release from the body, naming each tube the sperm pass through.   *p. 381*
4. Name the male accessory glands and give their functions.   *pp. 381–382*
5. What is the function of the penis? Describe the process by which the penis becomes erect.   *pp. 382-383*
6. Name and describe the functions of the three regions of a sperm cell.   *pp. 383–384*
7. List the hormones from the hypothalamus, the anterior pituitary gland, and the testes that are important in the control of sperm production. Explain the interactions among these hormones.   *pp. 384-385*
8. What are the two main layers to the wall of the uterus?   *p. 385*
9. List the major structures of the female reproductive system and give their functions.   *pp. 385-386*
10. What is an ectopic pregnancy? Why is it dangerous to the mother's health?   *pp. 386-387*
11. Describe the structure of breasts. What is their function?   *p. 387*
12. Describe the ovarian cycle. Include in your description primary oocytes, primary follicles, mature Graafian follicles, and the corpus luteum.   *pp. 387-389*
13. Describe the interplay among hormones from the anterior pituitary and from the ovaries that is responsible for the menstrual cycle.   *pp. 389-392*
14. What is menopause? Why does menopause lower estrogen levels? What are some effects of lowered estrogen levels?   *pp. 392-393*
15. What are the stages of the human sexual response? *pp. 393-394*
16. Describe a vasectomy and a tubal ligation.   *p. 394*
17. How does the combination birth control pill reduce the chances of pregnancy?   *p. 394*
18. What health risks are associated with use of the pill?   *p. 396*
19. How do progesterone-only means of contraception reduce the risk of pregnancy?   *pp. 396-397*
20. What is an IUD? How does it work to prevent pregnancy?   *p. 397*

21. How do the diaphragm, male condom, and female condom prevent pregnancy?   *pp. 395-396*
22. The interstitial cells
    a. are found in the seminal vesicles.
    b. produce a secretion that makes up most of the volume of semen.
    c. secrete testosterone.
    d. store sperm.
23. Choose the *incorrect* statement about semen.
    a. It helps lubricate passageways in the male reproductive system to make it easier for sperm to travel through them.
    b. Sperm cells make up most of the volume of semen.
    c. It helps reduce the acidity of the female reproductive system, thereby increasing sperm survival.
    d. It contains nourishment for the sperm.
24. After sperm are released in the female reproductive system, they can swim to the egg and fertilize it. Which is the correct order of structures through which the *sperm* will pass?
    a. vagina, cervix, body of uterus, oviduct
    b. ovary, cervix, body of uterus, oviduct
    c. endometrium, cervix, oviduct, ovary
    d. ovary, oviduct, endometrium, cervix
25. Which is the *correct pairing* of a structure with its function?
    a. endometrium, the usual site of fertilization
    b. corpus luteum, production of estrogen and progesterone
    c. epididymis, production of testosterone
    d. seminal vesicles, store sperm while they mature
26. The controversial "abortion pill" RU486 works by preventing progesterone from acting. This effect would cause an abortion because progesterone is needed to
    a. trigger ovulation.
    b. cause the formation of the corpus luteum.
    c. maintain the endometrium.
    d. increase the levels of LH.
27. Sperm are stored and mature in the _____.
28. Fertilization usually occurs in the _____.

APPLYING THE CONCEPTS

1. Katie is a 25-year-old woman who had pelvic inflammatory disease 2 years ago. She was treated with antibiotics and cured. She and her husband have been trying unsuccessfully to conceive a child for the last year. What is the most likely reason that she is having difficulty conceiving a child?
2. You are a health care provider in a family planning clinic. Your first client is a woman who is 22 years old and in excellent health. She is about to marry her high school sweetheart. She and her fiancé have never dated anyone else. They want a very effective means of contraception because they would like to wait until he finishes graduate school to start a family. She does not smoke. Which means of birth control would you recommend for this client? Why?

3. Endometriosis, a condition in which endometrial tissue implants on pelvic organs, causes pain in the pelvic area. Why would the pain be greatest around the time of menstruation?
4. Doctors in fertility clinics sometimes use FSH as a fertility drug to treat women who are having difficulty conceiving a child. Women who use FSH as a fertility drug are more likely than women who conceive naturally to have several babies at once. Why would FSH increase the incidence of multiple births?

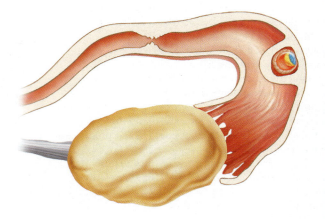

Sexually Transmitted Diseases and AIDS

STDs Are Extremely Common and Can Have Long-Lasting Effects

STDs Caused by Bacteria Can Be Cured with Antibiotics
- Chlamydia can cause pain during urination, PID, or no symptoms
- Gonorrhea can cause pain during urination, PID, or no symptoms
- Syphilis can progress through three stages when untreated

STDs Caused by Viruses Can Be Treated But Not Cured
- Genital herpes can cause painful, fluid-filled blisters
- Genital warts can lead to cervical, penile, or anal cancer

An HIV Infection Progresses to AIDS
- HIV/AIDS is a global epidemic
- HIV consists of RNA and enzymes encased in a protein coat
- HIV enters the cell, rewrites its RNA as DNA, inserts DNA into the host chromosome, and replicates
- Most HIV is transmitted through sexual contact, intravenous drug use, or from a pregnant woman to her fetus
- Sites of HIV infection include the immune system and the brain
- An HIV infection progresses through several stages as helper T cell numbers decline
- Treatments for HIV infection are designed to block specific steps in HIV's replication cycle

Kathie's physical was several months away, and, at 19, she felt she was in perfect health. Nonetheless, she had recently experienced both an urge to urinate frequently and pain during urination, so she made an appointment with her physician. Dr. Spradley took a urine sample and gave it to a technician for testing. In the meantime, Dr. Spradley asked Kathie many questions, including whether she was sexually active. Kathie replied that she was not and asked her why information about her sexual history was important. Dr. Spradley replied that her symptoms were those of a urinary tract infection and a common cause is a bacterium called *Chlamydia trachomatis*. In Kathie's case, she could rule out *Chlamydia* as a cause of the infection because this bacterium is transmitted through sexual contact. She explained that infections caused by *Chlamydia* could make it difficult to become pregnant in the future. Dr. Spradley left to get the results of Kathie's urine test. "Sure enough," she said. "You do have a urinary tract infection. Another bacterium, called *E. coli*, is the most likely cause for your infection because you are not sexually active. We will culture your urine to be sure. I'll prescribe an antibiotic. You should feel better in a day or two, but be sure to continue taking the antibiotic until the entire prescription is used."

In this chapter, we will consider sexually transmitted diseases (STDs), diseases that are transmitted by sexual contact. We will consider three STDs that are caused by bacteria: chlamydia, gonorrhea, and syphilis. Then we will consider three STDs that are caused by viruses: genital herpes, genital warts, and HIV/AIDS. We discuss HIV in this chapter because sexual contact is the most common means of transmission, but there are other ways of transmitting HIV. ■

STDs Are Extremely Common and Can Have Long-Lasting Effects

Sexually transmitted diseases (STDs) have an impressive impact on humans, in terms of their direct effects on the victims and the cost to society in general. Thirteen million infections occur each year in the United States. Two-thirds of these infections affect people younger than age 25. The cost of treating these infections is in the millions of dollars.

We should also mention another trait of STDs. In an age when it seems that political correctness is all-important, STDs are sexist. They affect women more severely than men, causing sterility, ectopic pregnancy (a pregnancy in which the embryo begins development outside the uterus), and cervical cancer.

One of the reasons that STDs are rampant is that many people are unaware they are infected. STDs often lack symptoms. Indeed, the prevalence of asymptomatic cases has led to use of the term sexually transmitted *infection* (STI) rather than sexually transmitted *disease*. Also, the symptoms of some STDs, such as syphilis, disappear without treatment, leading the person to believe that he or she is cured. However, this is not so, as we will see.

STDs Caused by Bacteria Can be Cured with Antibiotics

We will consider three STDs that are caused by bacteria: chlamydia, gonorrhea, and syphilis. Characteristics of these diseases (their symptoms, diagnosis, treatment, and effects) are summarized in Table 17a.1.

Chlamydia can cause pain during urination, PID, or no symptoms

Chlamydia is the most frequently reported infectious disease in the United States, with nearly 835,000 cases reported to the Centers for Disease Control and Prevention (CDC) in 2002. Experts estimate that there are actually 3 million to 10 million new cases each year. Chlamydial infections are rapidly becoming epidemic because they are highly contagious and because they may not cause noticeable symptoms that would prompt the infected person to seek treatment (Figure 17a.1). In fact, approximately 75% of infected women and 50% of infected men have no symptoms. Many people learn that they have chlamydia only because a responsible partner diagnosed with this infection has informed them. If symptoms do develop, they may take weeks or months to appear. Even without outward signs of infection, the disease can be passed to others.

Chlamydia is caused by a very small bacterium (*Chlamydia trachomatis*) that cannot grow outside a human cell. It is completely dependent on another cell for energy, because it cannot produce its own ATP molecules.

Symptoms of chlamydia depend largely on the site of the infection. The bacteria can infect the cells of mucous membranes, including those of the urethra, vagina, cervix, oviducts, throat, anus, and eyes. Worldwide, the major cause of blindness is infections by *Chlamydia trachomatis*.

The most common symptoms of chlamydia are those of urinary tract infection. Inflammation of the urethra causes a burning sensation during urination, itching or burning around the opening of the urethra, and a clear or white discharge from the urethra. The burning pain experienced while urinating generally spurs the person to promptly seek medical attention. In males, the most common site of chlamydial infection is the urethra, because this is the mucous membrane most likely to be exposed to the bacteria during sexual intimacy.

Chlamydial infections in women, however, are more likely to affect the pelvic organs than the urethra, because the vagina and cervix are their primary sites of contact during sexual intimacy. The general term for an infection of the pelvic organs is *pelvic inflammatory disease* (PID; see Chapter 17). Symptoms of PID include vaginal discharge, bleeding between periods (which is actually from the cervix), and pain during intercourse. Unfortunately, pelvic infections are less likely to produce symptoms than are urinary tract infections, and the untreated infection may permanently damage a female's reproductive system. Therefore, it is important that a man who is diagnosed with chlamydia inform everyone with whom he has been sexually intimate.

TABLE 17a.1

SEXUALLY TRANSMITTED DISEASES			
DISEASE	SYMPTOMS	DIAGNOSIS AND TREATMENT	EFFECTS
STDs caused by bacteria			
Chlamydia	First symptoms occur 7–21 days after contact Up to 75% of women and 50% of men show no symptoms *Women:* Vaginal discharge Vaginal bleeding between periods Pain during urination and intercourse Abdominal pain accompanied by fever and nausea *Men:* Urethral discharge Pain during urination	*Diagnosis:* Urine test for chlamydial DNA *Treatment:* Antibiotics	Long-term reproductive consequences such as sterility Infection can pass to infant during childbirth Can cause rupture of the protective membrane surrounding the fetus
Gonorrhea	First symptoms occur 2–21 days after contact About 30%–40% of men and women show no symptoms *Women:* Vaginal discharge Pain during urination and bowel movement Cramps and pain in lower abdomen More pain than usual during menstruation *Men:* Thick yellow or white discharge from penis Inflammation of the urethra Pain during urination and bowel movements	*Diagnosis:* Examination of penile discharge or cervical secretions Urine test for DNA of the bacterium that causes gonorrhea Cell culture *Treatment:* Antibiotics	Can cause long-term reproductive consequences such as sterility Infection can pass to infant during childbirth Can cause heart trouble, arthritis, and blindness
Syphilis	*Stage 1:* Occurs 2–8 weeks after contact. Chancre forms at site of contact. Lymph nodes in groin area swell *Stage 2:* Occurs 6 weeks to 6 months after contact Reddish brown rash appears anywhere on the body Flulike symptoms present Ulcers or warty growth may appear Patches of hair may be lost *Stage 3:* Lesions appear on skin and internal organs May affect nervous system Blindness Brain damage	*Diagnosis:* Identification of the bacterium from a chancre Blood test to detect antibodies to the bacterium that causes syphilis *Treatment:* Large doses of antibiotics over a prolonged period of time	Infection can pass to fetus during pregnancy Can cause heart disease, brain damage, blindness, and death

Chlamydia can have long-term reproductive consequences because it often causes scar tissue to form in the tubes through which gametes travel. If the vas deferens or oviducts are completely blocked by scar tissue, the man or woman is sterile. Sterility is more likely to occur in women than in men, because women are less likely to have symptoms that prompt treatment. In a woman, partial blockage of the oviduct may allow the tiny sperm to reach the egg and fertilize it, but the blockage will not allow the much larger embryo to move past the scar tissue to reach the uterus (Figure 17a.2). The embryo may then implant in the oviduct, resulting in an ectopic pregnancy that places the woman's life in danger.

TABLE **17a.1** (*Continued*)

SEXUALLY TRANSMITTED DISEASES			
DISEASE	**SYMPTOMS**	**DIAGNOSIS AND TREATMENT**	**EFFECTS**
STDs caused by viruses			
Genital herpes	First symptoms appear 2–20 days after contact Some people have no symptoms Flulike symptoms present Small, painful blisters that can leave painful ulcers appear Blisters go away but the virus remains Symptoms recur periodically	*Diagnosis:* Examination of blisters Laboratory tests on the fluid from the sore to detect the presence of the virus Blood test for antibodies *Treatment:* Antiviral drugs can ease symptoms	Cannot be cured Reoccurrences of blisters Infection can pass to fetus, causing miscarriage or stillbirth Can cause brain damage in newborns
Genital warts	First symptoms appear 1–6 months after exposure Small warts appear on sex organs May cause itching, burning, irritation, discharge, bleeding	*Diagnosis:* Appearance of growth In women, Pap test may help *Treatment:* For removal: freezing, burning, laser surgery	Formation of additional warts Closely associated with cervical and penile cancer Infection can pass to infant during childbirth

Chlamydial infections during pregnancy place the fetus at risk. Specifically, chlamydia can cause the protective membranes around the fetus to rupture early, and the rupture can kill the fetus. In addition, newborns can be infected as they pass through an infected cervix or vagina at birth. If untreated, the child may remain infected for 2 or more years. Chlamydial infections picked up during birth can affect the mucous membrane of the eye (the conjunctiva), the throat, vagina, rectum, or lungs and can be fatal to an infant.

Diagnosis of chlamydia can now be done in a physician's office with quick, accurate tests, including a urine test that detects the DNA of *Chlamydia*. Considering the suffering chlamydia can cause, it is almost ironic that it can be so easily cured with antibiotics.

Gonorrhea can cause pain during urination, PID, or no symptoms

One of the oldest known sexually transmitted diseases, gonorrhea, is caused by the bacterium *Neisseria gonorrhoeae*. Gonorrhea, like

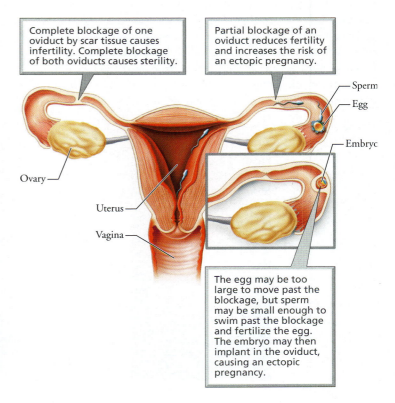

FIGURE **17a.2**
Chlamydia and gonorrhea can cause pelvic inflammatory disease, which can cause scar tissue to form in the oviducts.

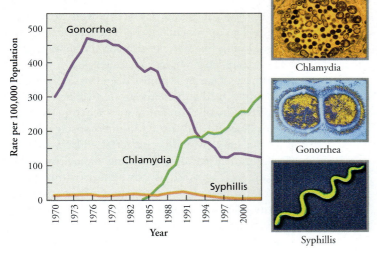

FIGURE **17a.1**
The relative rates of chlamydia, gonorrhea, and syphilis from 1984 to 2002. The bacteria that cause these STDs are shown at the right.

chlamydia, primarily infects the mucous membranes of the genital or urinary tract, throat, or anus. The bacteria that cause gonorrhea cannot survive long if exposed to air, so gonorrhea bacteria are generally transferred by direct contact of an infected mucous membrane and an uninfected one during sexual intimacy. A woman who has unprotected intercourse with a man who has gonorrhea has a 50% chance of becoming infected. If she is taking birth control pills, however, it is almost certain she will get it. The pill creates an alkaline environment in which the bacteria thrive. On the other hand, a man who has intercourse with a woman who has gonorrhea has only about a 22% chance of becoming infected. The difference in transfer rates is most likely because of differences in the surface area of the uninfected membrane that is exposed during intercourse—the small opening of the urethra in a male versus the entire vaginal lining and cervix in a female. Gonorrhea can also spread to the lining of the anus during anal intercourse or to the throat during oral sex. Furthermore, it can infect the eyes if the eye is touched after the infected area is touched. In the case of infants, eye infection can occur as the baby passes through an infected vagina or cervix during the birth process.

The symptoms of gonorrhea are similar to those described for chlamydia. In males, symptoms of urethritis (inflammation of the urethra) usually begin within 2 to 20 days of exposure. Typically, there is a discharge from the urethra that is thin and watery at first. Several days to a week later the discharge becomes thicker and turns to yellow or white (Figure 17a.3). Urination can produce a painful, burning sensation. In a manner similar to chlamydia, gonorrhea can spread to the vas deferens, epididymis, prostate, or bladder. In women, the urethra may become infected, causing a discharge and burning sensation. However, it is more likely to infect the vagina and cervix and eventually result in PID. In the long run, gonorrhea can cause sterility in either sex and increase the risk of ectopic pregnancy in women. Rarely, gonorrhea can cause arthritis or fatal heart and liver infections.

Keep in mind that gonorrhea does not always produce symptoms. About 30% to 40% of infected men and 60% to 90% of

infected women have no symptoms and are, therefore, unaware they have the disease. Nonetheless, they are contagious.

Gonorrhea can be diagnosed in several ways. A urine test can detect the DNA in *Neisseria* to confirm gonorrhea. Gonorrhea can also be diagnosed by observing the bacterium in a smear of cells taken from the infected area or grown in a laboratory.

At one time, gonorrhea could be easily cured with antibiotics, such as penicillin or ceftriaxone. Unfortunately, because of the misuse of antibiotics, some strains of the gonococcal bacteria are now drug resistant. Therefore, it is very important to be retested for gonorrhea when the antibiotic regimen has been completed to be sure the infection is cured. Immunity to gonorrhea does not develop, so you can become infected with each new exposure. Also, if not all partners are treated and cured, the disease can "Ping-Pong" among them.

Syphilis can progress through three stages when untreated

Syphilis rates have shown a steady decline since 1990. The CDC hopes to eliminate syphilis from the United States. Until then, however, we should not forget syphilis, or its frequency might rise again.

Syphilis is caused by a corkscrew-shaped bacterium (a spirochete) called *Treponema pallidum*. As terrible as these bacteria are, they are extremely delicate. These bacteria cannot survive drying or even minor temperature changes. Thus, it is not possible to get syphilis from toilet seats, wet towels, or drinking glasses. You become infected only by direct contact with an infected sexual partner. The bacteria can invade any mucous membrane or enter through a break in the skin. Unfortunately, the bacteria can also cross the placenta and infect the growing fetus if a pregnant woman is infected.

If untreated, syphilis progresses through three stages. The first stage is characterized by a painless bump, called a chancre, that forms at the site of contact, usually within 2 to 8 weeks of the initial contact (Figure 17a.4). Because syphilis is sexually transmitted, the chancre usually appears on the genitals. The chancre can, however, form anywhere on the body, because minute scratches on the skin will allow the bacteria to enter. The chancre appears as a hard, reddish brown bump with raised edges that make it resemble a crater. It can be small or as large as a dime. Unfortunately, a chancre is not always noticed, because it can form in places that are difficult to see and because it may be mistaken for something else. The chancre lasts for one to a few weeks. During this time, it ulcerates, becomes crusty, and disappears. The disease, however, has not disappeared.

During the primary stage of syphilis, diagnosis is made by identifying the bacterium in the discharge from a chancre. Treatment at this stage usually involves an antibiotic such as penicillin.

If syphilis is not detected and treated in the first stage, it can progress to the second stage, which is characterized by a reddish brown rash covering the entire body, including the palms of the hands and soles of the feet. The rash usually appears within a few weeks to a few months after the disappearance of the chancre. The person with the rash may confuse it with an ordinary skin rash, such as might be caused by an allergy or German measles. The rash does not hurt or itch. Each bump eventually breaks open, oozes fluid, and becomes crusty. The ooze contains millions of bacteria. Therefore, the secondary stage is the most contagious stage of syphilis.

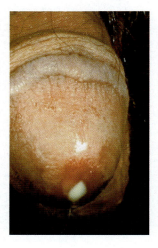

FIGURE **17a.3**

A yellowish white discharge from the urethra caused by gonorrhea

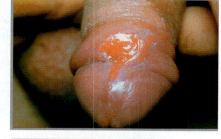

Stage 1: The first stage of syphilis is characterized by a chancre, a hard, painless, crater-shaped bump at the place in the body where the bacteria entered, usually the genitals.

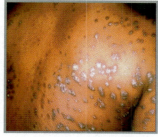

Stage 2: A reddish-brown rash covers the entire body, including the palms of the hands and the soles of the feet.

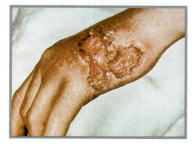

Stage 3: Lesions, called gummas, shown here on the hand, are characteristic of the third stage of syphilis. These lesions can also form on the blood vessels and the bones.

FIGURE **17a.4**
Untreated syphilis goes through three stages.

Secondary syphilis usually has indications besides the rash, and these symptoms may produce the first pain associated with the disease. At first, there may be flulike symptoms, including a slight temperature, chills, a sore throat, and gastrointestinal upset. In addition, warty growths may appear in the genital area. Patches of hair may fall out from the head, eyebrows, and eyelashes.

Diagnosis at this stage of syphilis is done by a blood test to detect antibodies for *Treponema* bacteria. Syphilis *can* be cured in the second stage with antibiotics such as penicillin. But, as the disease progresses, it becomes increasingly difficult to treat.

Again, the symptoms go away whether or not they are treated. However, in some people, the symptoms recur periodically. In others, all outward signs of syphilis are gone for years.

The third stage brings the drama to a grisly conclusion. At this stage, lesions called gummas may appear on the skin or certain internal organs. Gummas often form on the aorta, the major artery that delivers blood from the heart to the rest of the body. As a result, the artery wall may be weakened and may burst, causing the person to bleed to death internally. The bacteria that cause syphilis can also infect the nervous system, damaging the brain and spinal cord. As a result, the person may have difficulty walking, become paralyzed, or become insane. The disease may also cause blindness by affecting the optic nerve, the iris, and other parts of the eye.

In its third stage, syphilis becomes more difficult to diagnose and very difficult to treat. General blood tests may be negative, but specialized blood tests are usually positive. Treatment requires massive doses of antibiotics over a prolonged period. The damage that has already been done to the body cannot be repaired.

STDs Caused by Viruses Can Be Treated But Not Cured

Unlike STDs that are caused by bacteria, viral STDs cannot be cured with antibiotics. One can treat the symptoms, but one can never be certain that the virus has been eliminated. Therefore, it is always important to take precautions not to pass the virus on to others. Characteristics of genital herpes and genital warts, both caused by viruses, are summarized in Table 17a.1.

▌Genital herpes can cause painful, fluid-filled blisters

Genital herpes is caused by herpes simplex viruses (HSV). There are actually two types of HSV that cause similar sores, but they tend to be active in different parts of the body. Type 1 (HSV-1) is most commonly found above the waist, where it causes fever blisters or cold sores. Type 2 (HSV-2) is more likely to be found below the waist, where it causes genital herpes on the genitals, buttocks, or thighs. As a result of oral-genital sex with an infected person, however, HSV-1 can cause sores on the genitals, and HSV-2 can cause cold sores (Figure 17a.5).

(a) Oral herpes: Cold sores (also called fever blisters) are most often caused by HSV-1.

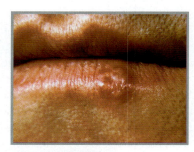

(b) Genital herpes, shown here on the external genitalia of a female, is usually caused by HSV-2.

(c) Genital herpes, shown here on the penis of a male, is usually caused by HSV-2.

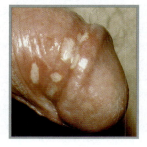

FIGURE **17a.5**
Herpes simplex viruses cause blisters to form at the site of infection.

Herpes simplex virus is quite contagious and can be spread by direct contact with viruses that are being shed or in the fluid on the blisters. Mucous membranes are most susceptible. Skin is a good barrier, unless there is a cut, abrasion, burn, acne, eczema, or other break in the skin.

The first hints of genital herpes infection begin about 2 to 20 days (an average of 6 days) after the initial contact. The initial bout may be severe, often accompanied by fever, aching muscles, and swollen glands in the groin. Soon, blisters appear, accompanied by local swelling, itching, and possibly burning, especially if the blisters get wet during urination. The blisters form at the site of contact and last about 2 days before they ulcerate, leaving small, painful sores.

When the symptoms subside, the virus is not gone from the body. During dormant periods, the virus retreats to ganglia (clusters of nerve cells) near the spinal cord. Then, at times of stress (emotional or physical), the virus may be reactivated and blisters may re-form.

Unfortunately, most people with herpes apparently never develop symptoms, or the symptoms were so mild that they went unnoticed. Eighty percent of people whose blood contains antibodies to HSV-2 (indicating that the virus is in their bodies) do not recall any previous signs of genital herpes. Nonetheless, these asymptomatic individuals can unwittingly transmit the virus. In fact, *most* cases of genital herpes are transmitted by partners who never had symptoms or whose symptoms had been atypical and undiagnosed!

Genital herpes is most contagious when active sores are present, so sexual contact should be avoided during an outbreak. Unprotected contact should also be avoided as soon as there is a hint of a recurrence. Genital herpes is sometimes spread when there are no symptoms.

A herpes infection can cause problems during pregnancy. Most women with herpes have successful pregnancies and normal deliveries. In some cases, however, the infection spreads to the fetus as it is growing in the uterus and can cause miscarriage or stillbirth. HSV can also be transmitted to the newborn during vaginal delivery, especially if active sores are present. In the newborn, a herpes infection can cause local infections in the eyes, skin, or mucous membranes or full-blown infections involving the central nervous system. Full-blown infections are often fatal. Even survivors may have neurological problems. To avoid exposing the baby to HSV, physicians generally recommend that the baby be delivered by Cesarean section (surgical removal from the uterus) if the mother has active sores at the time of delivery.

Clinicians diagnose herpes by examining the sores or by testing the fluid from sores for the presence of the virus. It is also possible to identify the DNA of the herpes virus in the material wiped off when the genital area is gently swabbed. Blood tests can also detect antibodies to the herpes viruses. However, blood tests may not be useful during an initial attack because the antibodies may not show up for several months. Newer tests can distinguish HSV-2 from HSV-1. This distinction is helpful because most people have had a cold sore and will test positive for HSV-1. However, HSV-2 is usually the cause of genital herpes.

Although there is no cure for HSV, there are three antiviral drugs—Zovirax®, Famvir®, Valtrex®—that help ease symptoms during the initial outbreak and reduce the frequency of recurrences. Unfortunately, strains of HSV that are resistant to antiviral drugs are beginning to crop up.

New therapies that are intended to stimulate immune responses against HSV-2 are now being tested. One is a vaccine. Although the vaccine's effectiveness is limited to a select group of women, there is hope that it could reduce genital herpes in the general population if girls were vaccinated before they are exposed to herpes viruses. In addition, a gel is available that boosts the immune response. When applied to the affected area, this gel lengthens the amount of time between recurrences.

Genital warts can lead to cervical, penile, or anal cancer

Genital warts may be caused by any of several human papilloma viruses (HPV). These are not the same viruses that cause warts on the hands and feet. Although chlamydia holds the title of being the most common of all STDs in the United States, genital warts is the most common of the viral STDs. It is estimated that 3 million cases are diagnosed each year.

Several factors contribute to the prevalence of genital warts. HPVs are slow-growing viruses. Thus, there may be a long delay, an average of 2 to 3 months (although longer time periods are possible), between exposure to the virus and the formation of a wart. From the time of infection, months before warts actually appear, the virus can be spread, even though the person has no idea that he or she has been infected. Warts can form in a visible location (Figure 17a.6), but they often form in locations where they are not likely to be discovered—the vagina, cervix, or anus. Furthermore, warts are highly contagious. About 80% of people who have regular contact with a partner with genital warts will also develop warts.

Diagnosis of warts is usually on the basis of their appearance. In women, a Pap test, a test that looks for precancerous cells on the cervix, is also helpful. A Pap test is a painless test in which the cervix is gently swabbed, and the cells are studied under a microscope to look for abnormalities.

Treatments for genital warts are intended to kill the cells that contain the virus. Methods for removing genital warts include: (1) freez-

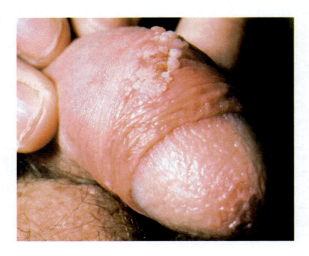

FIGURE **17a.6**
Genital warts on the penis

ing (cold cautery), (2) burning with an electrical instrument (hot cautery), (3) laser (high-intensity light), (4) surgery, and (5) podophyllin (a chemical that is painted onto the warts and washed off after the prescribed time period, before it burns the skin). Creams are also available that are applied by the patient to shrink genital warts. The creams work by boosting the body's defense mechanisms against the virus. These treatments may destroy visible warts, but HPV may remain in nearby normal-looking tissue, which can cause new warts to form weeks or months after old ones have been destroyed.

HPV is closely linked to both cervical cancer in women and penile cancer in men. It can also cause anal cancer. Indeed, HPV can be isolated in 90% of women with cervical cancer. For this reason, any woman who has ever had genital warts should have Pap tests at least once a year. HPV is also thought to be responsible for one-third of all cases of penile cancers in the United States.

An HIV Infection Progresses to AIDS

Acquired *i*mmune *d*eficiency *s*yndrome, or AIDS, has already left its mark on many aspects of society, including medicine, science, law, economics, and education. The name of the condition is apt because, first, it is not inherited, but rather it is acquired. Second, the symptoms are those associated with a damaged immune system. Finally, a syndrome is a set of symptoms that tend to occur together, and people with AIDS experience a devastating set of symptoms.

We now know that a virus, the human immunodeficiency virus (HIV), causes AIDS. However, as we will learn shortly, AIDS is just the final stage in an HIV infection, during which the immune system—the body's defense system—is slowly devastated. One of the primary targets of HIV is helper T cells, which serve as the main switch for the entire immune response (see Chapter 13). Once HIV enters a helper T cell, the T cell stops functioning well, although this cessation of function is not immediately apparent. HIV will, in the end, kill the infected helper T cell. The infection and eventual death of helper T cells cripple the immune system, giving disease-causing organisms that routinely surround us the opportunity to cause infection. These are the "opportunistic infections" that characterize AIDS and, in time, cause death.

▮ HIV/AIDS is a global epidemic

The World Health Organization (WHO) estimates that at the end of 2003, 34 million to 46 million people globally were infected with HIV. In North America, 790,000 to 1.2 million persons were living with HIV/AIDS. With more than 28 million persons infected, Africa is currently the area hardest hit by HIV.

▮ HIV consists of RNA and enzymes encased ▮ in a protein coat

The genetic material in HIV is RNA, rather than DNA. The RNA, along with several virus-specified enzymes, is encased in a protein coat. The RNA and protein coat constitute the core of the virus (Figure 17a.7).

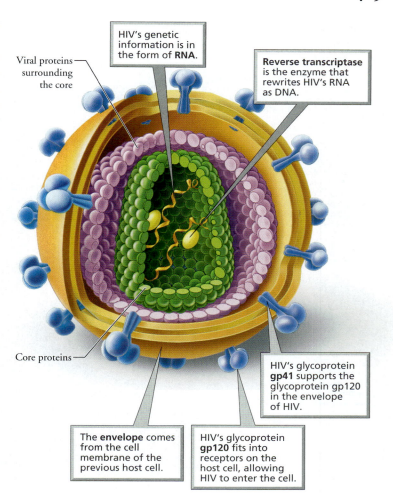

HIV's genetic information is in the form of **RNA**.

Viral proteins surrounding the core

Reverse transcriptase is the enzyme that rewrites HIV's RNA as DNA.

Core proteins

HIV's glycoprotein **gp41** supports the glycoprotein gp120 in the envelope of HIV.

The **envelope** comes from the cell membrane of the previous host cell.

HIV's glycoprotein **gp120** fits into receptors on the host cell, allowing HIV to enter the cell.

FIGURE **17a.7**
The structure of HIV (human immunodeficiency virus)

Surrounding the protein coat is an outer covering, or envelope, consisting of glycoprotein units embedded in a lipid membrane. The lipid membrane is actually a piece of plasma membrane stolen from the previous host cell and altered for use by the virus. Each glycoprotein unit has two parts: a spherical region, called gp120, and a sticklike region, called gp41, which projects into the lipid layer. The proteins of the viral envelope are responsible for binding the virus to the host cell.

▮ HIV enters the cell, rewrites its RNA as DNA, inserts ▮ DNA into the host chromosome, and replicates

HIV binds to an uninfected cell when the spherical glycoprotein in its outer envelope (gp120) fits into a receptor on the host cell surface. The host cell receptor is a surface protein called CD4. Helper T cells and macrophages are the predominant cell types with these CD4 receptors. Thus, these cells, which are important in body defense mechanisms, are the most common targets of HIV. The usual function of CD4 receptors is in exchanging information between cells of the immune system. Nonetheless, as bad luck would have it, the viral surface protein fits into the CD4 receptors like a key in a lock. When gp120 is properly docked in the CD4 receptor, another HIV protein

FIGURE **17a.8**
The life cycle of HIV

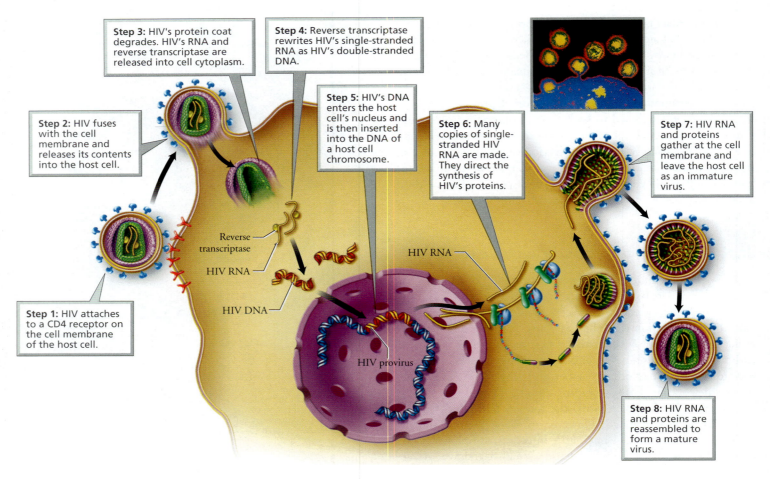

Step 2: HIV fuses with the cell membrane and releases its contents into the host cell.

Step 3: HIV's protein coat degrades. HIV's RNA and reverse transcriptase are released into cell cytoplasm.

Step 4: Reverse transcriptase rewrites HIV's single-stranded RNA as HIV's double-stranded DNA.

Step 5: HIV's DNA enters the host cell's nucleus and is then inserted into the DNA of a host cell chromosome.

Step 6: Many copies of single-stranded HIV RNA are made. They direct the synthesis of HIV's proteins.

Step 7: HIV RNA and proteins gather at the cell membrane and leave the host cell as an immature virus.

Step 1: HIV attaches to a CD4 receptor on the cell membrane of the host cell.

Reverse transcriptase

HIV RNA

HIV DNA

HIV RNA

HIV provirus

Step 8: HIV RNA and proteins are reassembled to form a mature virus.

can fit into a co-receptor on the surface of the T cell. Then the contents of the virus enter the host cell (Figure 17a.8).

Once within the host cell, the RNA of HIV undergoes a process called reverse transcription. During this process, the viral RNA is rewritten as double-stranded DNA. Because going from RNA to DNA is the reverse of the usual genetic information transfer, viruses that work this way are known as retroviruses (*retro-,* backward). The backward writing of genetic information from RNA to DNA is performed by an enzyme called reverse transcriptase, which is inserted into the host cell along with viral RNA.

The newly formed viral DNA, which contains all the instructions necessary for producing thousands of new viruses, is then spliced into the host DNA. Once HIV is integrated into the host cell chromosome, it is called the HIV provirus. After it has been incorporated into the host cell DNA, that cell treats the HIV genes as it would its own. Each time the cell reproduces, the viral DNA is copied along with its own. When HIV is transmitted to another person, it is usually in the form of a latent HIV provirus.

The HIV genes can reside in a helper T cell chromosome for years, until the cell is activated to respond to some foreign antigen, such as another kind of virus, a fungus, or a parasite. At this point, the virus begins making copies of itself instead of allowing the cell to fight the invader. The viral genome is activated and turns the cell

into a virus factory. Some of the newly produced HIV RNA will become genetic material for new viruses, and some will be used to produce viral proteins. The viral components gather at the cell membrane and bud off from the host cell. The HIV RNA and proteins self-assemble into new, mature viruses. The mature viruses can move through the bloodstream to infect new cells (Figure 17a.9).

Most HIV is transmitted through sexual contact, intravenous drug use, or from a pregnant woman to her fetus

HIV is found in bodily fluids—blood, semen, vaginal secretions, breast milk, saliva, tears, urine, cerebrospinal fluid, and amniotic fluid. However, HIV is not easily transmitted. Only the first four of the bodily fluids listed contain HIV concentrations high enough to cause infection in another person.

Today, the major modes of transmission are sexual contact without a latex condom and contact with contaminated blood through intravenous (IV) drug use, but HIV can also be transmitted from a pregnant woman to her fetus. HIV *cannot* be transmitted by casual contact.

- **Unprotected sexual activity.** HIV can enter the body through any break in the skin or a mucous membrane. Therefore, the virus could be transmitted through unprotected sexual activity of any

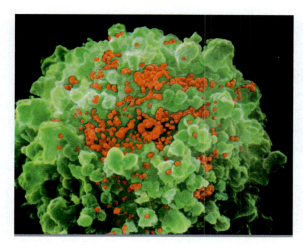

FIGURE **17a.9**
HIV particles (red) on the surface of a lymphocyte (green)

kind—that is, activity in which skin or mucous membranes of the vagina, vulva, penis, mouth, or anus of the partners is not separated by a barrier that the virus cannot penetrate. Sores or lesions caused by other sexual infections, including herpes and syphilis, provide an easy route of entry for HIV and, therefore, increase the risk of transmission. Vaginal intercourse is the most common way that HIV is transmitted in much of the world. During vaginal intercourse, HIV can enter a woman's body through tears in the vagina or by being directly absorbed through the mucous membranes of the vagina or cervix. HIV can enter a male's body through the mucous membrane of the urethra (the opening at the tip of the penis) or through breaks in the skin of the penis. Anal intercourse is particularly risky for the receptive person, because HIV in the infected semen can directly infect cells in the rectum and anus and because macrophages roaming the mucous membranes of the anus may be directly infected. Furthermore, the lining of the rectum is rich in blood vessels and tears more easily than the lining of the vagina. Unprotected oral sex (mouth to penis or mouth to vagina) can also transmit HIV. In the mouth, HIV can enter through open cuts or possibly by absorption through the mucous membranes lining the mouth.

- **Intravenous drug use.** Contact with infected blood is another way that HIV can be transmitted. Since 1985, all blood donated to blood banks is tested for HIV; this testing greatly reduced the risk of getting HIV from a blood transfusion. However, intravenous drug users who share needles *do* have an increased risk of HIV infection. When injecting a drug, the user first draws blood into the syringe to be sure the needle is in a vein. The blood then mixes with the drug. If the syringe is then used by another person, that blood, which may be infected, could be injected directly into the bloodstream along with the drug.

- **Infected mother to child before or during birth.** During pregnancy, maternal and fetal blood supplies are brought close together (see Chapter 18), allowing nutrients and oxygen to be exchanged, but they are kept apart by membranes in the placenta. During the last months of fetal development, however, small tears may occur in the placenta and may allow the infec-

tion to spread to the bloodstream of the fetus. Bleeding during the birth process may also cause infection in the newborn. Treatment with the drug AZT and delivery by Cesarean section can reduce the risk of transmission. HIV can also be passed to the infant in breast milk.

Sites of HIV infection include the immune system and the brain

HIV can infect any cell that has a CD4 receptor. The most important of these cells is the helper T cell, which turns on the entire immune response (see Chapter 13). The hallmark of an HIV infection is the progressive decline in helper T cell numbers. This loss is devastating because it leaves the body increasingly defenseless against other infections.

Macrophages are the other major cell type that becomes infected with HIV. Once inside a macrophage, HIV directs the production of a protein that causes the macrophage to secrete chemicals that attract T cells and activate them. HIV can then infect the activated T cell. An infected macrophage may also carry the viruses within it to different organs, especially the brain, lungs, and bone marrow.

HIV can also infect the brain, killing nerve cells. When the brain is infected, the symptoms can include forgetfulness, impaired speech, inability to concentrate, depression, seizures, and personality changes. Roughly 60% of people with AIDS have signs of dementia.

An HIV infection progresses through several stages as helper T cell numbers decline

An HIV infection usually progresses through a series of stages: the initial infection, an asymptomatic stage, initial disease symptoms, early immune failure, AIDS, and death. The progress of an HIV infection is usually monitored by following both the decline in the number of helper T cells and the increase in viral load (the number of HIV free in the blood). The changes in T cell number and viral load are summarized in Figure 17a.10. The symptoms associated with each stage of an HIV infection are summarized in Table 17a.2.

INITIAL INFECTION

During the initial stages of infection, the virus actively replicates, and the circulating level of HIV rises. The body's immune system produces antibodies against the virus in an attempt to eliminate it. Antibodies can usually be detected within 8 weeks or so, but it may be 6 months or perhaps even years before they can be detected. An HIV test looks for the presence of antibodies to HIV in the blood. If antibodies are found, the person is said to be HIV-positive (HIV$^+$).

Many people have no symptoms when they first become infected; others experience some mild disease symptoms during the initial infection. The typical symptoms at the time of the initial infection fall into two categories. The first, and most common, type of symptoms are flulike: swollen lymph glands throughout the body (lymphadenopathy), sore throat, fever, and perhaps a skin rash. These symptoms are typical of many viral infections and are, therefore, not helpful in diagnosing an HIV infection. The second type of symptoms includes those of brain infection (encephalopathy), in which there may be brain swelling or inflam-

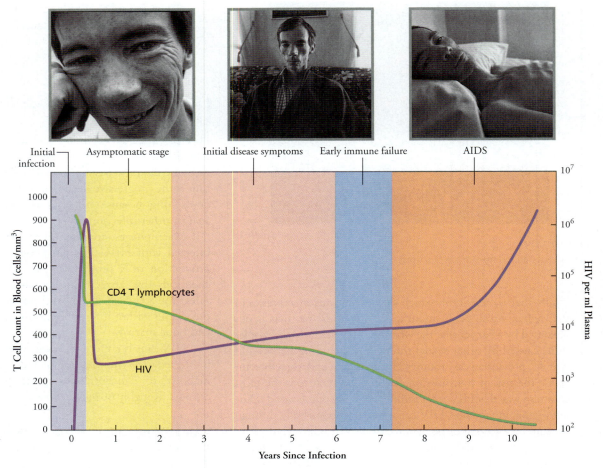

FIGURE **17a.10**

An HIV infection can be monitored by following the decline in the number of T cells and the rise in viral load (number of viruses in the bloodstream).

mation, especially of the outer protective membranes (the meninges). The brain inflammation may result from an immune response to infection or from effects of molecules from infected cells that affect brain cells. Whatever the cause, the results include headaches, fever, and difficulty concentrating, remembering, or solving problems.

ASYMPTOMATIC STAGE

Weeks to months after the initial stage of infection, the person usually feels well again, often for several years. This is the period of asymptomatic infection. During this stage, the immune system mounts a defense strong enough to control, but not conquer, the infection. The rate of T cell production increases. For a while, T cell production matches destruction.

During the asymptomatic stage, HIV is far from idle. It is "hiding out" in certain lymphatic organs, including the spleen, tonsils, and adenoids, but most importantly in the lymph nodes, where it replicates. Consequently, the spaces between cells in the lymph nodes become packed with virus particles. Most helper T cells reside in the lymph nodes, and many of them have the DNA form of HIV genes incorporated into their own chromosomes. Because helper T cells frequently circulate through the lymph nodes, this is an excellent place to encounter healthy T cells. Thus, millions of T cells become infected as they move through the lymph nodes on their way to other parts of the body.

INITIAL DISEASE SYMPTOMS

Eventually, the immune system begins to falter as the virus gains the upper hand. As helper T cell numbers gradually drop, symptoms set in again, signaling the beginning of the end. The initial disease symptoms fall into three classes.

- **Wasting syndrome** involves an otherwise unexplained loss in body weight of more than 10% of the total body weight, often accompanied by diarrhea. The weight loss is similar to that of a cancer patient. (This wasting is the source of the common name for AIDS in parts of Africa—slim disease.) Also part of the wasting syndrome are fevers, which usually occur at night and cause periods of excessive perspiration, called night sweats.

- **Lymphadenopathy** describes the swelling of the lymph nodes in the neck, armpits, and groin. These structures may remain swollen.

- **Neurological symptoms** may appear either because HIV has spread to the brain or because other organisms have caused a brain infection. Among the neurological symptoms are dementia, weakness, and paralysis caused by spinal cord damage. Other symptoms include pain, burning, or a tingling sensation, usually in hands or feet, caused by peripheral nerve damage.

EARLY IMMUNE FAILURE

As T cell numbers continue their gradual decline, the body becomes increasingly vulnerable to infection. Early signs of immune failure include the following.

TABLE **17a.2**

		THE STAGES OF AN HIV INFECTION			
STAGE	**HIV**	**IMMUNE SYSTEM**	**T CELL COUNT/mm³**	**SYMPTOMS**	
Initial infection	Enters body and replicates	Fights back; produces antibodies	~1000–800	• None • Flulike • Neurological	
Asymptomatic	Replicating in lymph nodes	Vigorously produces helper T cells	~800–600	• None • Considered HIV positive when antibodies against HIV detected in blood	
Initial disease symptoms	• Replicating • Viral load gradually increasing	Production of helper T cells cannot keep pace with destruction of helper T cells	~600–400	• Wasting syndrome • Lymphadenopathy • Neurological	
Early immune failure	• Replicating • Viral load gradually increasing	Helper T cell number continues to decline	~400–200	• Thrush (and vaginal yeast infections) • Shingles • Hairy leukoplakia • Herpes simplex recurrences	
AIDS	• Replicating • Viral load rapidly increasing	Helper T cell number continues to decline	200–below		

- **Thrush** is a yeast (fungus) infection in the mouth caused by *Candida*, which is normally held in check by other microbes and the immune system. Thrush causes white, sore, furry-feeling areas to form in the mouth. Although antifungal drugs can be used to control thrush, it is difficult to eliminate completely. The yeast can spread down the esophagus, causing a burning sensation when the person eats.

- **Shingles** is a painful rash caused by the virus that causes chickenpox (*Varicella zoster*). The virus can remain dormant in nerve cells long after a person recovers from chickenpox, and it may become active again when the immune system is compromised.

- **Hairy leukoplakia** is an overgrowth of the papillae (the bumps that give the tongue a rough texture) that causes white plaques to form on the surface of the tongue. The overgrowth, which resembles cancer cells, seems to result from infection with Epstein-Barr virus, the same virus that causes mononucleosis.

AIDS

AIDS (sometimes called frank, or full-blown, AIDS) is the final stage of an HIV infection. The time from HIV infection to diagnosis of AIDS can be 10 or more years. By this time, the body has fought long and hard but is beginning to lose the battle. The rate of progression varies with the number of other organisms the person is exposed to and the health of the immune system at the start. When the immune system becomes this crippled, the body becomes vulnerable to certain characteristic fungal, protozoan, bacterial, and viral infections, as well as cancers.

A diagnosis of AIDS is made when an HIV$^+$ person develops one of the following conditions: (1) a helper T cell count below 200/mm³

of blood; (2) 1 of 26 opportunistic infections, the most common of which are *Pneumocystis carinii* pneumonia and Kaposi's sarcoma, a cancer of connective tissue that affects primarily the skin; (3) a loss of more than 10% of body weight (wasting syndrome); or (4) dementia (mental incapacity such as forgetfulness or inability to concentrate).

One of the most common of the fungal infections is *Pneumocystis carinii*, which infects the lungs and is a leading cause of death in people with AIDS. It causes a dry cough and progressive shortness of breath.

Protozoans, single-celled organisms, can also cause serious illness when HIV weakens the immune system. For instance, one of these protozoans (*Cryptosporidium*) can infect the lining of the intestinal tract and cause long and severe bouts of diarrhea. Another protozoan (*Toxoplasma gondii*) is an intracellular parasite that can infect many organs. When the brain becomes infected, symptoms such as convulsions, disorientation, and dementia develop.

Important among the bacterial infections that commonly plague people with AIDS is tuberculosis. People with AIDS usually get an atypical form of tuberculosis that may infect both the lungs and the bone marrow.

Cytomegalovirus (CMV), a member of the herpes family, is a common virus that infects many people in childhood, but it can be reactivated in AIDS patients. For healthy people, infection with CMV does not cause major problems. Healthy children may have no symptoms at all, and healthy adults typically get mononucleosis-like symptoms. When CMV infection recurs in people with AIDS, however, it can cause blindness, an adrenal hormone imbalance, or pneumonia.

Kaposi's sarcoma and lymphomas are cancers more common in people with AIDS than in the general population. In Kaposi's sarcoma, tumors form in connective tissue, especially in blood vessels, and appear as pink, purple, or brown spots on the skin (Figure 17a.11). The spots are caused by the cancer's means of obtaining nourishment. The tumor cells cause a network of blood vessels to form that tap into surrounding blood vessels. In addition, they cause neighboring blood vessels to leak and bathe the tumor cells in nourishing blood. The tumors spread and eventually may affect all linings of the body.

A virus in the herpes family, HHV8, is thought to be the most probable cause of Kaposi's sarcoma. This virus is almost always present in lesions caused by Kaposi's sarcoma but is rare in other lesions. HHV8 is found in higher concentrations in saliva than in genital secretions. As a result, some researchers think that deep kissing can transmit HHV8.

Treatments for HIV infection are designed to block specific steps in HIV's replication cycle

The cause of death in a person with AIDS is usually one of the opportunistic infections. Thus, treating opportunistic infections and, in some cases, preventing their onset can help improve the quality of life and extend the length of life for people with AIDS.

However, AIDS is just the final stage in an HIV infection, so it is important to slow the progress of the infection. The current strategy is to slow the rate at which HIV can make new copies of itself. Currently, there are three classes of drugs that slow viral replication: reverse transcriptase inhibitors, protease inhibitors, and fusion inhibitors. Each class of drugs prevents HIV from replicating by blocking a different step in HIV's life cycle.

The original class of antiviral drugs used to slow the progression of an HIV infection includes those that act on reverse transcriptase (AZT, ddI, ddC). Recall that reverse transcriptase is the enzyme necessary for rewriting the RNA of HIV as DNA that can then be inserted into the DNA of the host cell chromosome.

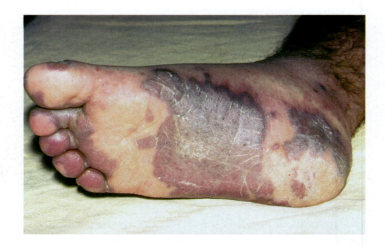

FIGURE **17a.11**
Kaposi's sarcoma is a cancer common among persons living with AIDS in which the tumor cells obtain nourishment by forming a network of blood vessels to tap into nearby blood vessels. The blood vessels then leak blood that bathes the tumor cells. This leakage results in the formation of purple, pink, or brown spots on the skin.

A second class of antiviral drugs is the protease inhibitors. Recall that after the HIV genetic information has been inserted into the DNA of the host cell, it can begin producing proteins needed to form new copies of HIV. The proteins initially produced are too big to be used, however. These proteins are cut to the proper size by an enzyme called a protease. Protease inhibitors block the action of this enzyme, preventing the preparation of the proteins for the assembly of new viruses.

The newest class of antivial drugs is the fusion inhibitors. These block the proteins on the surface of HIV (gp120, gp41) that bind to the CD4 receptor on host cells, thereby blocking HIV from entering the host cell. The first of these drugs was approved in 2003.

Highly active antiretroviral therapy (HAART), also known as a drug cocktail, consists of combinations of drugs. Typically, it contains two inhibitors of reverse transcriptase and a protease inhibitor. The idea behind this drug combination is that a protease inhibitor will slow the rate of formation of new viruses so that the development of resistance to the other drugs will be minimized. Combination drug treatments are prolonging the lives of HIV$^+$ people.

HAART can reduce the viral load to undetectable levels, but the treatment is not a cure. Drugs that slow viral replication cannot eliminate the virus in its latent state as a provirus integrated into the host cell chromosome. Thus, viruses in the latent state serve as a reservoir for HIV during drug treatment. If treatment is stopped, HIV can leave the latent state and begin actively replicating again.

A variety of other approaches to combating an HIV infection are under active investigation. Clearly, though, the best way to fight the devastation caused by HIV is to prevent it from establishing the initial infection. Theoretically, this could be accomplished by the development of a vaccine similar to those for other viral infections such as small pox and measles. Recall from Chapter 13 that vaccines work by causing a person's body to produce antibodies against the organisms that cause a disease without actually causing the disease. Unfortunately, there are several characteristics of HIV that have thwarted efforts to develop a vaccine:

1. The mutation rate in HIV is very high, especially for the proteins in its outer coat, the proteins that an antibody would recognize. Indeed, it has been estimated that each time reverse transcriptase rewrites the RNA of HIV as DNA there is at least one mutation. Furthermore, a person infected with HIV probably produces about a billion new virus particles a day. Thus, the vaccine could stimulate the production of antibodies against one protein, but it would fail to recognize the mutant protein.

2. There are two types of HIV and many strains of each type. Antibodies would have to recognize the exact strain entering the body.

3. HIV establishes a latent state of infection when its genetic material has become incorporated into the host DNA. Antibodies cannot attack the virus while it is hidden within the host cell.

4. HIV can pass directly from an infected cell to an uninfected one, going from one cell interior to another. Antibodies can attack only viruses outside the cell, they cannot prevent the spread of the virus by this means of transmission.

Development and Aging

Human Life Has Two Main Periods of Development

The Prenatal Period Begins at Fertilization and Ends at Birth

- Fertilization, cleavage, and implantation characterize the pre-embryonic period
- Tissues and organs form during the embryonic period
- Rapid growth characterizes the fetal period

Birth Is the Transition from Prenatal to Postnatal Development

- True labor consists of three stages
- The placenta may send the signal to initiate labor
- Drugs and breathing techniques can relieve pain during childbirth

Environmental Disruptions During the Embryonic Period Cause Major Birth Defects

The Mother's Mammary Glands Produce Milk

The Postnatal Period Begins with Birth and Continues into Old Age

- Several factors cause aging
- Medical advances and a healthy lifestyle can help achieve a high-quality old age

HEALTH ISSUE   Making Babies, But Not the Old-Fashioned Way

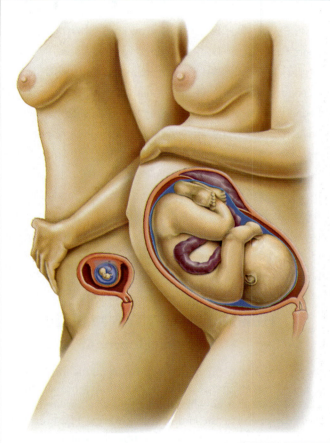

Carl looked into the mirror and studied his face, looking carefully at his eyes, skin, and hair. He had just celebrated his forty-first birthday, and he recalled telling everyone at the party that he had more energy now than when he was 21. And it seemed true: He played in two hockey leagues and jogged three times a week. He also lifted weights at least twice a week. He maintained a healthy diet and had lately taken an interest in yoga, but he had no idea how this would fit into his schedule. As a successful trial lawyer, he felt he had to keep up his appearance and his energy. Treating his body well fueled his self-esteem. Yet the face looking back at him showed wrinkles around his eyes, and a hairline that had receded significantly over the years. He felt a pang of anxiety as he began to think about advertisements he had seen for hair restoration, hair replacement, and antiwrinkle creams. He even wondered about surgical procedures. "Enough of that," he thought, "I'm just beginning to show the effects of age." Like millions of other men and women who have begun to face the prospect of aging, Carl would like to postpone the inevitable, to recapture the freshness of youth, the vivacity that is usual in the young. In this chapter, we will describe the major milestones of human development. We will focus on the extremes—when life begins and when it ends. ■

Human Life Has Two Main Periods of Development

A newborn enters the world outside his mother's womb and rests (Figure 18.1). How did this tiny human develop? Consider that he—like you—began as a fertilized egg, no bigger than the period at the end of this sentence. The miracle of human development begins with **fertilization**, the union of a sperm and an egg. Early development occurs within the female reproductive system (see Chapter 17). Birth occurs about 266 days after fertilization,

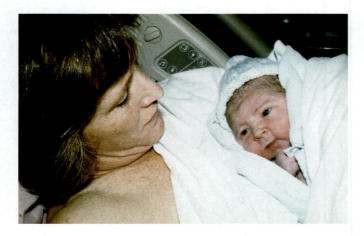

FIGURE **18.1**
A newborn and his mother. Birth is the transition from prenatal to postnatal development.

marking the transition to development outside the mother's body. The period of development before birth is the **prenatal period**. The period after birth is the **postnatal period**. The postnatal period includes infancy, childhood, puberty, and adolescence and culminates in adulthood. Relative to prenatal development, postnatal development is a lengthy process, taking from 20 to 25 years to complete. Bodily changes do not cease once adulthood is reached. Indeed, our bodies continue to change as part of the aging process.

The Prenatal Period Begins at Fertilization and Ends at Birth

We begin our discussion of human development at the moment it all begins—when sperm meets egg. In this section, we consider the major events of prenatal development. We consider what normally happens and what can happen when things go wrong.

From the standpoint of human development, the prenatal period is divided into three periods: (1) the **pre-embryonic period**, from fertilization through the second week (during which the developing human is called a **pre-embryo**); (2) the **embryonic period**, from week 3 through week 8 (during which the developing human is called an **embryo**); and (3) the **fetal period**, from week 9 until birth (during which the developing human is called a **fetus**). We will consider the developmental events that define each stage.

Fertilization, cleavage, and implantation characterize the pre-embryonic period

The pre-embryonic period begins with fertilization, the union between the nucleus of an egg and the nucleus of a sperm. Fertilization takes about 24 hours from start to finish and usually occurs in a widened portion of the oviduct, not far from the ovary, as shown in Figure 18.2. Fertilization is a beautifully orchestrated performance by egg and sperm, aided by the actions of the oviducts and uterus.

FERTILIZATION

Let's begin our discussion of the pre-embryonic period with an overview of how egg and sperm meet. At ovulation, the ovary releases the egg, technically a secondary oocyte (see Chapter 17). The egg is swept into the oviduct by fingerlike projections called fimbriae (singular, fimbria; see Figure 18.2). Cilia and waves of peristalsis move the oocyte along the oviduct toward the uterus. The oocyte slowly drifts along and then, not far into its journey, it pauses in the widened portion of the oviduct.

The trip made by sperm is a considerably more competitive and precarious venture than the movements of the typically lone egg. During sexual intercourse, from 200 million to 600 million sperm are deposited in the vagina and on the cervix, the neck of the uterus that extends into the vagina. Many sperm are trapped at the boundary of the vagina and cervix. Indeed, fewer than 1% of deposited sperm actually make it into the uterus. Sperm that escape entrapment use their whiplike tails to swim into the uterus and eventually the oviduct. Along the way, the sperm are aided by small uterine contractions, stimulated by chemicals (prostaglandins) in the semen. Of

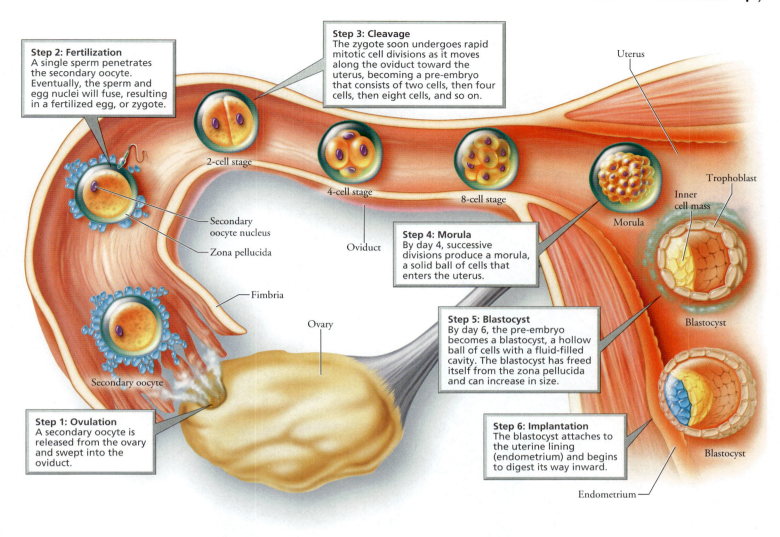

Step 2: Fertilization
A single sperm penetrates the secondary oocyte. Eventually, the sperm and egg nuclei will fuse, resulting in a fertilized egg, or zygote.

Step 3: Cleavage
The zygote soon undergoes rapid mitotic cell divisions as it moves along the oviduct toward the uterus, becoming a pre-embryo that consists of two cells, then four cells, then eight cells, and so on.

Step 4: Morula
By day 4, successive divisions produce a morula, a solid ball of cells that enters the uterus.

Step 5: Blastocyst
By day 6, the pre-embryo becomes a blastocyst, a hollow ball of cells with a fluid-filled cavity. The blastocyst has freed itself from the zona pellucida and can increase in size.

Step 1: Ovulation
A secondary oocyte is released from the ovary and swept into the oviduct.

Step 6: Implantation
The blastocyst attaches to the uterine lining (endometrium) and begins to digest its way inward.

Uterus

2-cell stage

4-cell stage

8-cell stage

Morula

Trophoblast

Inner cell mass

Blastocyst

Secondary oocyte nucleus

Zona pellucida

Oviduct

Fimbria

Ovary

Secondary oocyte

Blastocyst

Endometrium

FIGURE **18.2**
Early stages in the reproductive process

the vast number of sperm originally deposited in the female reproductive tract, only about 200 sperm reach the site of fertilization in the widened portion of the oviduct. Some sperm reach the site after only a few minutes and others after almost an hour.

Timing is everything when it comes to fertilization. Typically, an egg cannot be fertilized more than 24 hours after its release from the ovary. Most eggs are fertilized within 12 hours. In addition, the vast majority of sperm survive no more than 24 hours in the female reproductive tract.

Two critical processes must precede fertilization (Figure 18.3). First, secretions from the uterus or oviducts must alter the surface of the acrosome. Recall from Chapter 17 that the acrosome is the enzyme-containing cap on the head of a sperm. This process usually requires about 6 to 7 hours to complete. It appears to involve the removal of cholesterol and possibly other substances from the surface of the sperm. These changes destabilize the sperm's plasma membrane and seem to be a prerequisite for fertilization.

The second process occurs when sperm with unstable membranes contact the corona radiata, a layer of cells surrounding the

secondary oocyte (Figure 18.3). Once such contact occurs, perforations develop in the plasma membrane and outer acrosomal membrane of the sperm. Enzymes spill out of the acrosome and disrupt the attachments between cells of the corona radiata. This disruption enables sperm to pass through the corona radiata to the layer below. The zona pellucida is a thick noncellular layer that immediately surrounds the secondary oocyte. Enzymes released from the acrosome also break apart the zona pellucida, creating a pathway by which the sperm reach the oocyte.

Several sperm may arrive at the secondary oocyte at about the same time and enter the zona pellucida. However, usually only one sperm crosses this layer and fertilizes the oocyte. Two mechanisms prevent **polyspermy,** an abnormal condition in which more than one sperm enters an egg. First, when the sperm contacts the plasma membrane of the oocyte, it triggers a rapid depolarization (change in electrical state) of the membrane, which prevents other sperm from crossing the membrane and entering the oocyte. Second, enzymes released by granules near the plasma membrane of the oocyte cause the zona pellucida to harden and prevent passage of

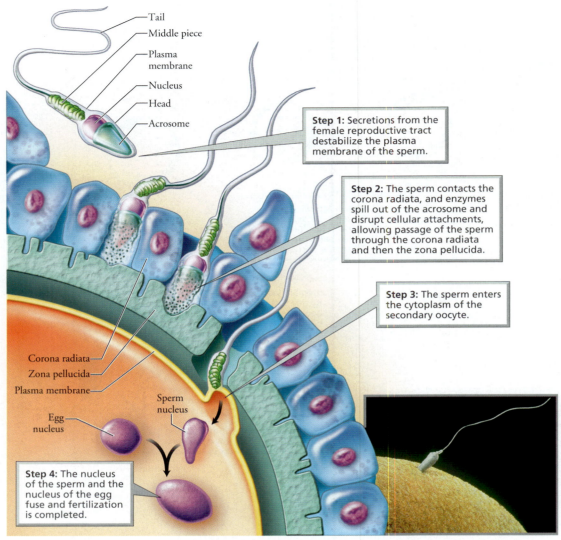

Tail
Middle piece
Plasma membrane
Nucleus
Head
Acrosome

Step 1: Secretions from the female reproductive tract destabilize the plasma membrane of the sperm.

Step 2: The sperm contacts the corona radiata, and enzymes spill out of the acrosome and disrupt cellular attachments, allowing passage of the sperm through the corona radiata and then the zona pellucida.

Step 3: The sperm enters the cytoplasm of the secondary oocyte.

Corona radiata
Zona pellucida
Plasma membrane

Egg nucleus

Sperm nucleus

Step 4: The nucleus of the sperm and the nucleus of the egg fuse and fertilization is completed.

(a) Diagram showing the processes that occur before and during fertilization

(b) Micrograph of a sperm contacting a secondary oocyte

FIGURE **18.3**
Fertilization

other sperm. These blocks to polyspermy ensure equal genetic contributions from each parent and prevent abnormal numbers of chromosomes in the resulting embryo. Too many chromosomes, as would result if two or more sperm fertilized an egg, impair cell division and cause the death of the embryo.

After one sperm penetrates the zona pellucida, the plasma membranes of the sperm and secondary oocyte fuse, and the oocyte undergoes its second meiotic division and is now technically considered an ovum. (Meiosis, discussed in Chapters 17 and 19, involves two divisions and results in the formation of gametes—eggs and sperm—with 23 chromosomes. Sperm complete meiosis while in the testes, whereas oocytes do not complete meiosis until fertilization.) The nucleus of the ovum and the nucleus of the sperm fuse (Figure 18.3). The fertilized ovum is called a **zygote**. Just visible to the unaided eye, the zygote contains genetic material from both the mother (23 chromosomes) and father (23 chromosomes), and thus has 46 chromosomes.

CLEAVAGE

About one day after fertilization, the zygote undergoes **cleavage**, a rapid series of mitotic cell divisions. During cleavage, the single-celled zygote becomes a pre-embryo, first consisting of two cells, then four cells, then eight cells, and so on. Cleavage usually occurs as the pre-embryo moves along the oviduct toward the uterus (see Figure 18.2). The cleaving pre-embryo becomes a solid ball of cells by about day 3. By day 4, the pre-embryo is called a **morula**, a solid ball of 12 or more cells produced by successive divisions of the zygote. The name reflects its resemblance to the fruit of the mulberry tree. These early cell divisions do not result in an overall increase in size. Increases in overall size are prevented by the tight-fitting zona pellucida that still covers the morula. Instead, the cells within the ball become progressively smaller as divisions occur. The morula is about the same size as the fertilized ovum.

Sometimes during an early stage of cleavage, the mass of cells splits into two, and two embryos are formed. **Identical twins**, also called monozygotic twins ("from one zygote"), develop in this way. Such twins have identical genetic material and always are the same gender. Sometimes the splitting of the pre-embryo is incomplete and **conjoined twins** result. Such twins may be surgically separated after birth once doctors have evaluated which structures are shared and the likelihood of successful separation. **Fraternal twins** occur when two secondary oocytes are released from the ovaries and fertilized by different sperm. Such twins also are called dizygotic twins ("from two zygotes"). Fraternal twins may or may not be the same gender and are as genetically similar as any siblings. Identical twins (separate and conjoined) and fraternal twins are shown in Figure 18.4.

stop and think

Given what you know about the events necessary to produce identical and fraternal twins, how might triplets—two of which are identical and one fraternal—form?

The morula enters the uterus about 5 days after fertilization. Spaces soon begin to appear between the cells at the center of the ball. Fluid from the uterine cavity accumulates within the spaces, converting the morula into a hollow sphere of cells. By day 6, the pre-embryo is a hollow ball of cells with a fluid-filled cavity; at this stage it is called a **blastocyst.** As shown in Figure 18.2, the blastocyst is made up of two parts: an inner cell mass and a trophoblast. The **inner cell mass** is a group of cells that will become the embryo proper and some of the extraembryonic membranes that will extend from or surround the embryo (see below). The **trophoblast** is a thin layer of cells that will give rise to the extraembryonic membrane that is the embryo's contribution to the placenta. The **placenta** is the organ that delivers oxygen and nutrients to the embryo and carries carbon dioxide and other wastes away. Before the blastocyst becomes implanted in the uterine wall and the placenta develops, the blastocyst floats freely in the uterus. Increases in overall size of the blastocyst eventually become possible when it frees itself from the zona pellucida. The zona pellucida then degenerates.

IMPLANTATION

The blastocyst attaches to the **endometrium** (the lining of the uterus) about 6 days after fertilization and begins to digest its way inward. **Implantation** is the process by which the blastocyst becomes embedded in the endometrium (Figure 18.5). Implantation normally occurs high up on the back wall of the uterus. Sometimes, however, a blastocyst implants outside the uterus, and an **ectopic pregnancy** results. The vast majority of implantations outside the uterus occur in the oviducts. Most ectopic pregnancies occur because of impaired passage of the dividing pre-embryo through the oviduct. Factors that hinder passage of the pre-embryo include structural abnormalities and scar tissue from surgery or pelvic inflammatory disease (see Chapters 17 and 17a). The embryo is removed because rupture of the oviduct and hemorrhage can be fatal to the mother.

Successful early development and implantation require precise conditions. Indeed, an estimated one-third to one-half of all zygotes fail to become blastocysts and to implant. These zygotes or early pre-embryos are either reabsorbed by cells of the endometrium or expelled from the uterus in an early spontaneous abortion (such an end to a pregnancy is described as spontaneous because it was not medically induced). Causes of early spontaneous abortion include chromosomal abnormalities in the zygote and an inhospitable uterine environment for implantation. The latter might result from the presence of an intrauterine device (IUD) or inadequate production of the hormones estrogen and progesterone by the corpus luteum. (Recall from Chapter 17 that an IUD is a small plastic birth control device that is inserted into the uterus, and the corpus luteum is a glandular structure that forms from the ovarian follicle after ovulation.) Many women who spontaneously abort in the first few weeks after fertilization are unaware of their pregnancy. They simply notice a delay in their menstrual period and an unusually heavy flow when the period begins.

(a) Identical twins result when a single fertilized egg splits in two very early in development.

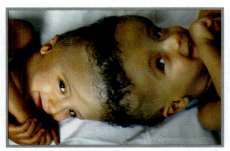

(b) Rarely, the splitting is incomplete and conjoined twins result.

(c) Fraternal twins result from the fertilization of two oocytes by different sperm.

FIGURE **18.4**
Twins

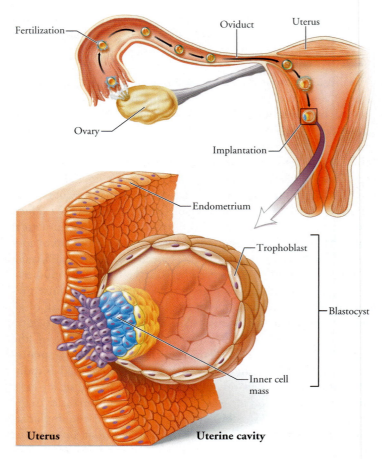

FIGURE **18.5**

Implantation. About 6 days after fertilization, the blastocyst attaches to the endometrium of the uterus and begins to digest its way inward.

When conditions are right, the blastocyst completes implantation by the end of the second week of the pre-embryonic period. During this time, cells within the blastocyst produce **human chorionic gonadotropin (HCG),** a hormone that enters the bloodstream of the mother and is excreted in her urine. The presence of this hormone in the mother's urine forms the basis for many pregnancy tests. Enough HCG is produced by the end of the second week to be detected by the pregnancy test and yield a positive result. Thus, a pregnancy can be detected as early as 2 weeks from the time of fertilization. Its usefulness in pregnancy tests aside, the physiological function of HCG is to maintain the corpus luteum and to stimulate it to continue producing the hormone progesterone. Progesterone is essential for maintenance of the endometrium, the very lining in which the embryo is implanting.

Infertility is the inability to conceive (become pregnant) or to cause conception (in the case of males). Implantation is a major hurdle in the series of steps leading to a successful pregnancy. We describe the many options available to infertile couples in the Health Issue essay, *Making Babies, But Not the Old-Fashioned Way.*

EXTRAEMBRYONIC MEMBRANES

Toward the end of the pre-embryonic period, four membranes—the amnion, yolk sac, allantois, and chorion—begin to form around the pre-embryo (Figure 18.6). Formation of these membranes extends into the embryonic period, when the developing human is called an embryo. The membranes lie outside the embryo and are called **extraembryonic membranes**. These membranes protect and nourish the embryo and later the fetus. The **amnion** surrounds the entire embryo, enclosing it in a fluid-filled space called the amniotic cavity. Amniotic fluid forms a protective cushion around the embryo that later can be examined as part of prenatal testing in a procedure known as amniocentesis (see Chapter 20). The **yolk sac** is the primary source of nourishment for embryos in many species of vertebrates. In embryonic humans, however, it remains quite small and does not provide nourishment. Human embryos receive nutrients from the placenta. The yolk sac in humans is a site of early blood cell formation. It also contains **primordial germ cells** that migrate to the gonads (testes or ovaries), where they differentiate into immature cells that will eventually become sperm or oocytes. The **allantois** is a small membrane whose blood vessels become part of the **umbilical cord**, the ropelike connection between the embryo and the placenta. The umbilical cord consists of blood vessels and supporting connective tissue. Finally, the **chorion** is the outermost membrane. It develops largely from the trophoblast and becomes the embryo's major contribution to the placenta. The amnion, yolk sac, and allantois develop from the inner cell mass.

THE PLACENTA

We mentioned that developing humans rely on the placenta, rather than the yolk sac, for nourishment. Indeed, the placenta orchestrates all interactions between the mother and the fetus. The placenta often is called the afterbirth because it is expelled from the uterus after the

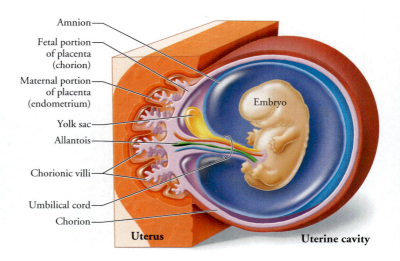

FIGURE **18.6**

Extraembryonic membranes. The amnion, yolk sac, chorion, and allantois form during the second to third week after fertilization.

Making Babies, But Not the Old-Fashioned Way

Some couples desperately want a child, but they cannot conceive one. The inability to become pregnant or to cause a pregnancy is called infertility. Couples are considered infertile if conception does not occur after 1 year of unprotected sexual intercourse. Infertility is not uncommon, striking about 1 in 6 American couples. In some of these couples, the woman is infertile. Hormonal imbalances may disrupt ovulation or implantation, or scarring from disease may block the oviducts and prevent passage of gametes. Some women have had their uterus removed because of cervical cancer. In women older than 40, aging eggs may be the problem. In other couples, it is the man who is infertile. Male infertility often is caused by production of few or sluggish sperm or sperm with structural abnormalities. Such problems with sperm may make fertilization unlikely, if not impossible. Also, in a strange twist of fate, otherwise compatible couples may be incompatible at the cellular level. Women, for example, are sometimes allergic to their partner's sperm and produce antibodies that kill the sperm before they are able to fertilize an egg. Finally, in many of the most frustrating cases the cause of infertility is unknown. Can anything be done to help those who are infertile? For roughly half of the couples seeking help the answer is yes, although the road to reproduction may be long, expensive, and filled with emotional ups and downs. The procedures to treat infertility are collectively called assisted reproductive techniques (ARTs).

Some cases of female infertility can be treated by administering hormones. For example, hormones that trigger ovulation may be given to women whose ovaries fail to properly release eggs. Women who tend to miscarry (spontaneously abort) may be given progesterone to enhance the receptivity of their uterine lining. However, many conditions—including scarred oviducts and a low sperm count—do not respond to hormone therapy. What can be done in these cases?

Artificial insemination is one option available to couples whose infertility is caused by a low sperm count. In this procedure, sperm that have been donated and stored at a sperm bank are deposited (usually with a syringe) in the woman's cervix or vagina at about the time of ovulation. Sperm from the male member of the couple may

be concentrated and then used. Alternatively, couples may use semen from an anonymous donor. (Artificial insemination with donor sperm also is frequently used by lesbian couples seeking to have children.) Another possibility for couples in which the man has few sperm, or sperm that lack the strength or enzymes necessary to penetrate an egg, is intracytoplasmic sperm injection (ICSI). In this procedure, a tiny needle is used to inject a single sperm into an egg (Figure 18.A).

Some cases of infertility can be treated with in vitro fertilization (IVF). This technique was made famous in England in 1978 with the birth of Louise Joy Brown, the first "test-tube" baby. Literally meaning "fertilization in glass," IVF involves placing eggs and sperm together in a glass dish in the laboratory. First, the woman is treated with hormones, such as follicle-stimulating hormone (FSH) and luteinizing hormone (LH), to trigger superovulation. Superovulation is the ovulation of several oocytes, rather than the typical lone oocyte. An abundance of eggs is the key to many infertility treatments. Large numbers of eggs can be fertilized to produce many embryos. Having several embryos available allows selection of the healthiest ones for immediate use and freezing of other healthy embryos for future use. Next, the oocytes are removed from the woman's ovaries with a suction device and placed into a dish that contains a sample of the man's sperm. If fertilization occurs, the zygotes are transferred to a solution that will support further development. A few days later, when the embryos are at an early blastocyst stage, one or more of them is transferred to the woman's uterus. The uterus has been primed with estrogen and progesterone, and it is hoped that the embryos will implant and complete development there. Typically, three or four embryos are transferred. Transfer of fewer embryos is associated with a lower probability of pregnancy. Transfer of more than four embryos does not increase the chances for pregnancy and brings with it risks associated with multiple fetuses. Women with blocked oviducts may conceive with IVF because the technique bypasses the oviducts altogether.

Failure of embryos to implant is, in part, responsible for the relatively low success rate of IVF. (Implantation is also a major hurdle for fertile couples, who lose about one in three pregnancies at this time.) The problem with IVF seems to be

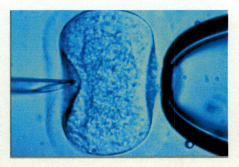

FIGURE **18.A**

Intracytoplasmic sperm injection. An egg, held in place with a blunt-nosed pipette (right), is injected with a single sperm.

the rather violent squirt of the embryo into a reproductive tract that is already traumatized from hormone treatments and retrieval of eggs. Two new procedures avoid the embryo's problematic arrival in the uterus. In gamete intrafallopian transfer (GIFT), eggs and sperm are collected from a couple and then inserted into the woman's oviduct, where fertilization may occur. Then, any resulting embryos drift naturally (and gently) into the uterus. The second procedure is zygote intrafallopian transfer (ZIFT). In this procedure, eggs and sperm are collected and brought together in a dish in the laboratory. If fertilization occurs, the resulting zygotes are inserted into the woman's oviducts, where they can travel on their own to the uterus. Normal healthy oviducts are needed for GIFT and ZIFT to work.

A couple in which the woman lacks a uterus can hire a surrogate mother to gestate their baby. The egg and sperm may come from the couple or from other individuals. One other option for such couples may be on the horizon. In March 2002, doctors in Saudi Arabia announced the first uterine transplant. The doctors transplanted a uterus from a 46-year-old woman undergoing a hysterectomy into a 26-year-old woman. (The recipient's uterus had previously been removed because of excessive bleeding following a Cesarean section.) The transplanted uterus had to be removed after 99 days because of blood clotting. Nevertheless, Saudi doctors viewed the transplant as a success and were hopeful that future transplants would have even better outcomes. However, many problems remain. Of special concern is rejection of the transplanted organ

continued ⟶

and possibly a fetus, should the recipient of a uterus become pregnant.

At the beginning of this essay we stated that about half of couples who seek treatment for infertility are eventually rewarded with the birth of a child. Success rates of ARTs range from about 20% to 28% live births per egg retrieval. Although costs of the techniques vary somewhat, each attempt typically costs between $8000 and $12,000, and multiple attempts often are needed. Many health insurance plans do not cover these expenses because the procedures are considered experimental or elective.

What about the couples who do not conceive with ARTs? Is there hope that some day they might also conceive? The answer is a decided yes. Advances in reproductive research and technology are rapidly providing potential treatments for infertility. For example, scientists are currently working on techniques with mice in which spermatogonia, the stem cells deep within the testes that produce sperm, are transferred from a fertile animal to a sterile one. The once-sterile mouse can then sire offspring. Some day it might be possible to remove stem cells from a man with a low sperm count, encourage the cells to multiply

under laboratory conditions, and then return them—now in much greater numbers—to their rightful owner, who might then father children.

What about research with eggs? Although sperm and embryos are routinely frozen, eggs quickly become inviable when manipulated outside the body and are thus difficult to store. However, new techniques for freezing eggs are being developed. These techniques might one day permit a young woman intent on pursuing a career to freeze her eggs. Years later, when her career is well established, she might remove her eggs from the deep freeze and have them fertilized. The ability to successfully freeze eggs would allow women who postponed childbearing to avoid the problem of trying to conceive with old eggs. Research also is under way on ovarian tissue transplantation as a way to produce functioning ovaries. Tissue from ovaries rich in potential eggs could be removed from a young female, frozen, and then placed back in her body years later. It might also be possible, though more difficult, to transplant ovarian tissue between females. Other research focuses on using eggs from fetuses. Fetal ovaries are a rich source of eggs, containing about 2 million as compared

with the estimated 400 eggs ovulated by most women over the course of their reproductive years. Scientists are investigating the possibility of transplanting into infertile women tissue from the ovaries of fetuses made available from elective or spontaneous abortions. Fetal eggs might one day be stored at an egg bank, where they could be used by infertile women. The babies produced from such eggs would be in the truly unusual position of having a biological mother who had been aborted. Indeed, the aborted fetus with her wealth of eggs could potentially become the biological mother of many children.

Obviously, current and future reproductive technologies raise many ethical questions. Among them, do people who donate sperm or eggs have a claim to their biological offspring? What happens to viable embryos that are created in the laboratory but turn out to be "extra"? Do such embryos have a right to life? And who gets the embryos when couples divorce or die? Finally, is it in our long-term interest to help obviously defective sperm fertilize eggs or to have women old enough to be grandmothers bearing children? We need to think carefully about these issues. ♀♂

---

baby is born. The placenta forms from the chorion of the embryo and a portion of the endometrium of the mother (specifically, the endometrium in the area where implantation occurred; Figure 18.7).

A major function of the placenta is to allow oxygen and nutrients to diffuse from maternal blood into embryonic blood. Wastes such as carbon dioxide and urea diffuse from embryonic blood into maternal blood. The placenta also produces hormones such as human chorionic gonadotropin (HCG), estrogen, and progesterone. Placental hormones, as mentioned earlier, are essential for the continued maintenance of pregnancy. Nevertheless, their high levels may be responsible for **morning sickness**, the nausea and vomiting experienced by some women early in pregnancy. Morning sickness is not restricted to the morning and may be caused by high levels of HCG from the placenta.

Formation of the placenta begins shortly after implantation. During implantation, cells of the outer layer of the trophoblast rapidly divide and invade the endometrium. Cavities form that fill with blood from maternal capillaries severed by the invading cells. Soon, the inner layer of the trophoblast invades the endometrium, and fingerlike processes of the chorion, called **chorionic villi,** grow into the endometrium (refer again to Figure 18.7). These chorionic villi grow, divide, and continue their invasion of maternal tissue, causing ever larger cavities and pools of maternal blood to form. The placenta is fully developed by the third month of pregnancy. At this time, it weighs about 680 g (1.5 lb).

Chorionic villi contain blood vessels connected to the developing embryo. Oxygen and nutrients within the pools of maternal blood diffuse through the capillaries of the villi into the umbilical vein and travel to the embryo. Wastes leave the embryo by the umbilical arteries, move into capillaries within the chorionic villi, and diffuse into maternal blood. Thus, the chorionic villi provide exchange surfaces for diffusion of nutrients, oxygen, and wastes. Some chorionic villi also anchor the embryonic portion of the placenta to maternal tissue. Maternal and fetal blood vessels are close together within the placenta. Under normal circumstances, however, there is no direct mixing of maternal and fetal blood. All exchanges occur across capillary walls. Nevertheless, this intimate connection between mother and fetus causes the fetus to share in maternal nutrition, habits, and lifestyle. Substances that cross the placenta include drugs, alcohol, caffeine, toxins in cigarette smoke, and HIV (human immunodeficiency virus, the virus that causes AIDS).

Usually, the placenta forms in upper portions of the uterus. **Placenta previa** is a condition in which the placenta forms in the lower half of the uterus and covers the cervix, the neck of the uterus that projects into the vagina. Placenta previa may cause premature birth or maternal hemorrhage and usually makes vaginal delivery impossible. About 75% of women with diagnosed placenta previa have their baby delivered by Cesarean section before labor begins. **Cesarean section**, often shortened to C-section, is the procedure by which the fetus and placenta are removed surgically from the

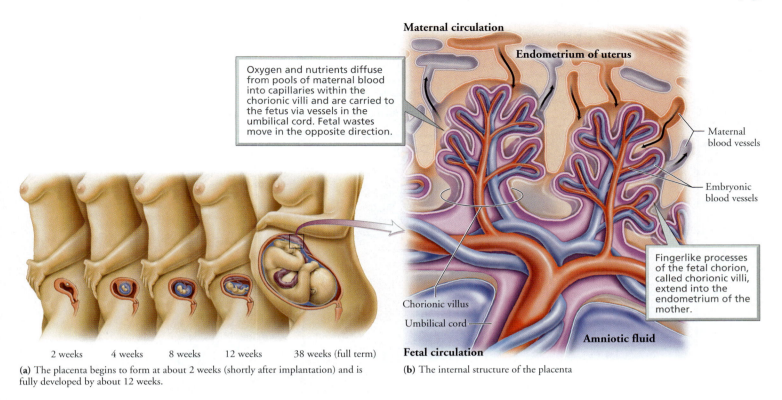

Maternal circulation

Endometrium of uterus

Oxygen and nutrients diffuse from pools of maternal blood into capillaries within the chorionic villi and are carried to the fetus via vessels in the umbilical cord. Fetal wastes move in the opposite direction.

Maternal blood vessels

Embryonic blood vessels

Fingerlike processes of the fetal chorion, called chorionic villi, extend into the endometrium of the mother.

Chorionic villus

Umbilical cord

Amniotic fluid

Fetal circulation

2 weeks 4 weeks 8 weeks 12 weeks 38 weeks (full term)

(a) The placenta begins to form at about 2 weeks (shortly after implantation) and is fully developed by about 12 weeks.

(b) The internal structure of the placenta

FIGURE **18.7**

The placenta is formed from the chorion of the embryo and the endometrium of the mother.

uterus through an incision in the abdominal wall and uterus. The symptoms and signs of placenta previa include painless, bright-red bleeding late in pregnancy as the placenta pulls away from the expanding lower uterus. Suspected cases of placenta previa are confirmed with ultrasound. Physicians usually recommend bed rest in the hope that reduced maternal activity will aid in prolonging the pregnancy until the baby can survive outside the mother's body. In the past, placenta previa posed a serious health risk. Today, however, the vast majority of mothers with diagnosed placenta previa survive delivery without problems, as do their babies.

Tissues and organs form during the embryonic period

The embryonic period extends from the third to the eighth week. It is a time of great change, characterized by the formation of three distinct germ layers from which all tissues and organs develop. Two major milestones occur during this period: the development of the central nervous system and the development of the reproductive organs. By the end of the embryonic period, all organs have formed, and the embryo has a distinctly human appearance.

GASTRULATION

Morphogenesis is the development of body form that begins during the third week after fertilization. Just before morphogenesis, during implantation, the inner cell mass moved away from the surface of the blastocyst and the amniotic cavity formed. The amniotic cavity is filled with amniotic fluid and is lined by the amnion, one of the four extraembryonic membranes. The inner cell mass becomes a flattened platelike structure called the **embryonic disk** (Figure 18.8). The embryonic disk will become the embryo proper. At this time, cells within the embryonic disk differentiate and migrate. These cells form three **primary germ layers** known as ectoderm, mesoderm, and endoderm. The cell movements that establish the primary germ layers are called **gastrulation,** and the embryo during this period is called a **gastrula** (Figure 18.8).

All tissues and organs develop from the primary germ layers. **Ectoderm** forms the nervous system and the outer layer of skin and its derivatives such as hair, nails, oil glands, sweat glands and mammary glands. **Mesoderm** gives rise to muscle, bone, connective tissue, and organs such as the heart, kidneys, ovaries, and testes. **Endoderm** forms some organs and glands (for example, the pancreas, liver, thyroid gland, and parathyroid glands) and the lining of the urinary, respiratory, and digestive tracts. Gastrulation, thus, initiates the processes by which the developing embryo's body takes shape and becomes organized.

As gastrulation proceeds, a flexible rod of mesoderm tissue called the **notochord** develops where the vertebral column will form. The notochord defines the long axis of the embryo and gives the embryo some rigidity. The notochord also prompts overlying ectoderm to begin formation of the central nervous system, as discussed below. Vertebrae eventually form around the notochord, and the notochord degenerates. The pulpy, elastic material in the center of intervertebral disks (pads that help cushion the bones of the vertebral column) is all that remains of our notochord.

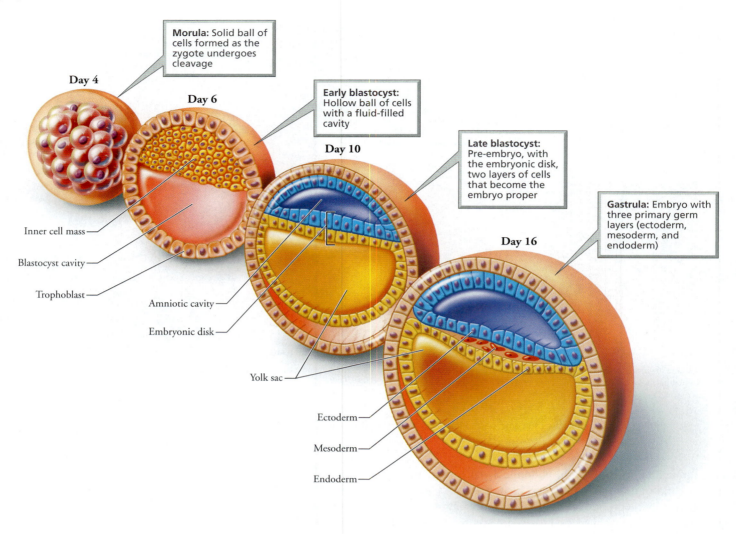

Day 4

Day 6

Morula: Solid ball of cells formed as the zygote undergoes cleavage

Early blastocyst: Hollow ball of cells with a fluid-filled cavity

Day 10

Late blastocyst: Pre-embryo, with the embryonic disk, two layers of cells that become the embryo proper

Day 16

Gastrula: Embryo with three primary germ layers (ectoderm, mesoderm, and endoderm)

Inner cell mass

Blastocyst cavity

Trophoblast

Amniotic cavity

Embryonic disk

Yolk sac

Ectoderm

Mesoderm

Endoderm

FIGURE **18.8**

Early stages of development in cross section

NEURULATION

A major milestone in embryonic development is the formation of the central nervous system (brain and spinal cord) from ectoderm, as shown in Figure 18.9. The developing notochord induces overlying ectoderm to thicken. This thickened region of ectoderm above the notochord is called the neural plate. The neural plate folds inward, forming a groove that extends the length of the embryo, along its back surface. The raised sides of the groove, known as neural folds, grow upward and eventually meet and fuse to form the **neural tube**, a fluid-filled tube that will become the central nervous system. This series of events is called **neurulation**, and the embryo during this period is called a **neurula**. The anterior portion of the neural tube develops into the brain, and the posterior portion forms the spinal cord. Alongside the neural tube, mesoderm cells organize into blocks called somites. **Somites** eventually form skeletal muscles of the neck and trunk, connective tissues, and vertebrae.

Failure of the neural tube to develop and close properly results in neural tube defects. **Spina bifida** ("split spine") is a type of neural tube defect in which part of the spinal cord develops abnormally as does the

adjacent area of the spine. An infant with spina bifida often has a cyst on his or her back within which lies an exposed portion of the spinal cord. Such infants face the constant threat of infection of the nervous system and also may experience areas of skin without sensation and paralysis of some muscles. The severity of spina bifida varies greatly, and some cases can be improved through surgery. **Anencephaly** is a neural tube defect that involves incomplete development of the brain and results in stillbirth or death shortly after birth.

Genetic and environmental factors appear to be involved in neural tube defects. Maternal nutrition, an environmental factor, has an important impact on the incidence of neural tube defects. Specifically, taking multivitamins containing folic acid (one of the B vitamins) before conception appears to reduce the incidence of spina bifida, a defect that now strikes about 2 of every 1000 newborns in the United States. If parents have a child with spina bifida, their chances of having another child with the birth defect are 1 in 35. However, these chances can be reduced by more than 70% if folic acid is added to the mother's diet. We will have more to say about defects present at birth later in the chapter.

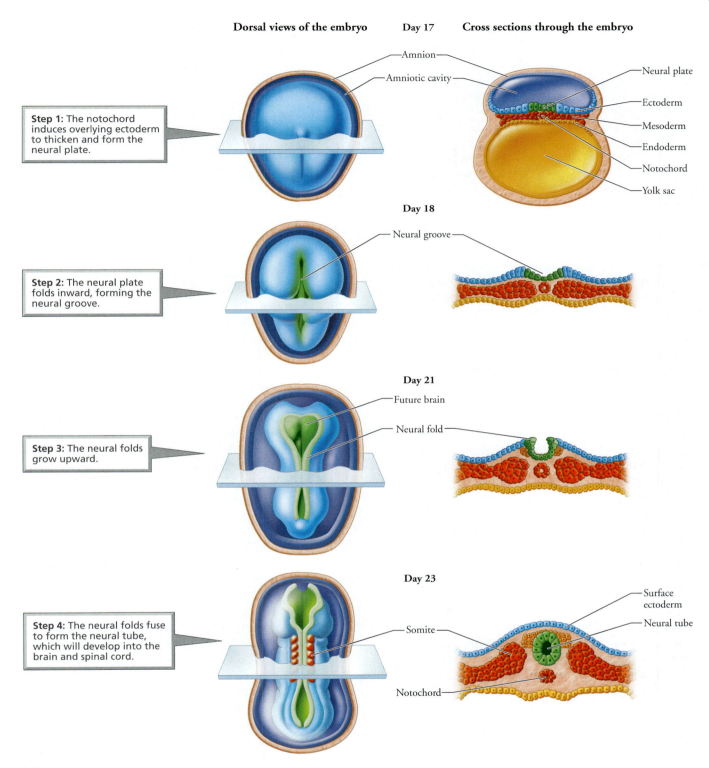

Dorsal views of the embryo   **Day 17**   **Cross sections through the embryo**

Amnion
Amniotic cavity
Neural plate
Ectoderm
Mesoderm
Endoderm
Notochord
Yolk sac

Step 1: The notochord induces overlying ectoderm to thicken and form the neural plate.

Day 18

Neural groove

Step 2: The neural plate folds inward, forming the neural groove.

Day 21

Future brain
Neural fold

Step 3: The neural folds grow upward.

Day 23

Surface ectoderm
Neural tube
Somite
Notochord

Step 4: The neural folds fuse to form the neural tube, which will develop into the brain and spinal cord.

FIGURE **18.9**
Formation of the central nervous system from ectoderm

GENDER DETERMINATION

Among the 23 chromosomes provided by the mother's egg is an X chromosome. Among the 23 chromosomes provided by the father's sperm is an X or a Y chromosome. The X and Y chromosomes are called sex chromosomes, and they determine the gender of human embryos. More specifically, the gender of an embryo is determined at fertilization by the type of sperm that fertilizes the egg. If an X-bearing sperm fertilizes the egg, then the zygote is XX and will normally develop as a female. If a Y-bearing sperm fertilizes the egg, then the zygote is XY and will normally develop as a male.

what would you do?

Methods are currently available to separate X-bearing sperm from Y-bearing sperm. The sorting is possible because X chromosomes have almost 3% more DNA than Y chromosomes. During the procedure, sperm are treated with a fluorescent dye that binds to DNA. X-bearing and Y-bearing sperm glow differently because of their different amounts of DNA. They can be sorted by an automated machine. Such "sperm sorting" makes possible the selection of a child's gender. With this technology, couples could decide whether to use X-bearing sperm to produce a female baby or Y-bearing sperm to produce a male baby. Some couples want to select the gender of their child to avoid X-linked genetic diseases (see Chapter 20); others want to balance their families with respect to numbers of sons and daughters. For whatever reason, once the choice of gender has been made, the man donates a semen sample, and the sperm are sorted according to the type of chromosome (X or Y) they contain. Then, the chosen sperm are deposited within the woman's vagina or cervix at a time when pregnancy is likely to occur. (The process of depositing semen within the female reproductive tract is called artificial insemination.) Should parents be allowed to choose the gender of their children? What implications would gender selection have for families and for society? If you were given the responsibility of deciding whether such technology should continue to be made available to parents, what would you do?

DEVELOPMENT OF SEX ORGANS

Male and female embryos have the same external anatomy for the first few weeks, and their internal reproductive organs have not differentiated. The embryos at this time are described as being "sexually indifferent." This situation changes, however, about 6 weeks after fertilization when a region of the Y chromosome—called SRY for sex-determining region of the Y chromosome—initiates development of testes in XY embryos. The testes soon begin producing testosterone, the sex hormone that directs development of male reproductive organs. Female embryos lack the Y chromosome, and, in its absence, ovaries develop. The development of female reproductive structures is not influenced by hormones and will occur even if ovaries are absent. Thus, it is the absence of the Y chromosome and testosterone that leads to female development. We will focus on development of the external genitalia, but keep in mind that internal reproductive organs also are undergoing differentiation.

Both male and female embryos of about 6 weeks of age have an elevated region of tissue between their legs called the genital tubercle. Within this tubercle is a shallow depression called the urogenital groove. The urogenital groove is surrounded by urogenital folds, which, in turn, are surrounded by labioscrotal swellings. In the male embryo, testosterone produced by the newly differentiated testes causes the urogenital groove to elongate and completely close. The genital tubercle then becomes the glans penis; the urogenital folds develop into the shaft of the penis; and the labioscrotal swellings become the scrotum (into which the testes will later descend). In the female, the urogenital groove also elongates, but unlike in the male, it remains open. The genital tubercle becomes the clitoris in females, and the urogenital folds develop into the labia minora. The labioscrotal swellings of females become the labia majora. Figure 18.10 summarizes differentiation of the external male and female genitalia.

Rapid growth characterizes the fetal period

The fetal period extends from the ninth week after fertilization until birth. Although tissues and organs continue to differentiate during this period, the most notable changes concern growth. Growth during the fetal period is extremely rapid, reflected in phenomenal increases in weight (from 8 to 3400 g, about 0.3 to 120 oz) and substantial increases in length (crown-to-rump length changes from 50 to 360 mm, about 2 to 14 in.), when the pregnancy continues for the full 38 weeks. Also known as "sitting height," crown-rump length is a measurement of the distance from the head to the rump. One striking change is that the rate of growth of the head slows relative to the rate of growth of other regions of the body, altering the way the body is proportioned. At the start of the fetal period, the head is about half the crown-rump length of a fetus (Figure 18.11). In contrast, shortly before birth, the head makes up only about one-third of the crown-rump length. Change in the relative rates of growth of various parts of the body is called **allometric growth**. Such growth continues after birth and helps to shape developing humans and other organisms.

The fetal circulatory system differs from circulation after birth because several organs—the lungs, kidneys, and liver, to name a few—do not perform their postnatal functions in the fetus. As a result, most blood is shunted past these organs through temporary vessels or openings. Recall that the fetus receives its oxygen and nutrients from maternal blood and gives up its carbon dioxide and wastes to maternal blood at the placenta. We will now trace the path of blood flow from the placenta to the fetus and then back to the placenta. We will note the circulatory bypasses unique to life before birth. We end with a description of the changes in circulation that occur at birth when the organs of the newborn begin to function. This discussion is summarized in Figure 18.12.

The umbilical vein (within the umbilical cord) carries blood rich in oxygen and nutrients from the placenta to the fetus (Figure 18.12a). Some of this blood flows to the fetal liver, which produces red blood cells but does not yet function in digestion. Most of the blood from the placenta, however, bypasses the liver by means of a temporary vessel called the **ductus venosus** and enters the inferior vena cava (see Chapter 12), where it mixes with blood low in oxygen. This mixed blood then enters the right atrium of the heart. Some of the blood in the right atrium passes to the right ventricle and on to the lungs, as it does in circulation after birth. Most of the blood, however, moves into the left atrium through a small hole in the wall between the atria, called the **foramen ovale**. Like the liver, fetal lungs are not yet functional and need only a small amount of blood for growth and removal of wastes from their cells. The blood in the left atrium moves into the left ventricle and out to the body of the fetus. Another shunt, called the **ductus arteriosus**, connects the pulmonary trunk to the aorta and also functions to divert blood away from the lungs. Thus, blood flowing out of the right ventricle into the pulmonary trunk is shunted away from the lungs and into the aorta, where it travels to all areas of the fetus. Fetal blood must return to the pla-

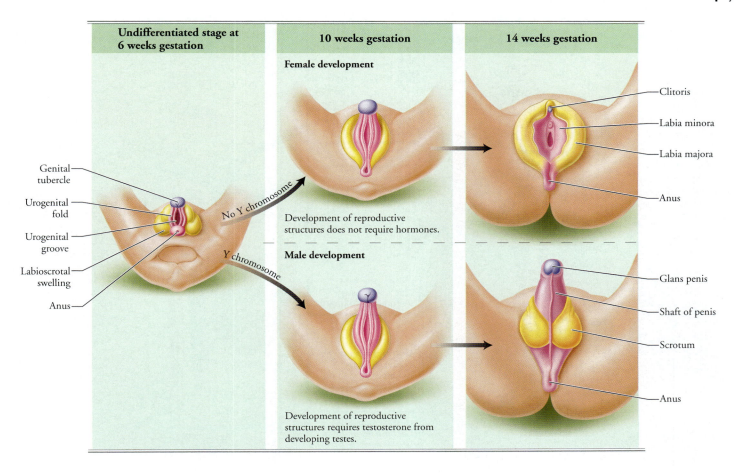

FIGURE **18.10**
Development of external genitalia

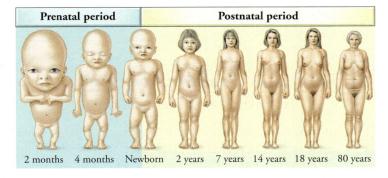

FIGURE **18.11**
Allometric growth. Changes in body proportions occur throughout prenatal and postnatal development. For comparison, all stages are drawn to the same total height.

centa to rid itself of carbon dioxide and wastes and to pick up oxygen and nutrients. It does so by flowing through umbilical arteries that branch off large arteries in the legs. The two umbilical arteries run through the umbilical cord to the placenta.

Another glance at Figure 18.12a reveals that much of the blood traveling through the fetus is moderate to low in oxygen. Indeed, the umbilical vein is the only fetal vessel that carries fully oxygenated blood.

At birth, when the lungs, liver, and other organs begin their postnatal functions, fetal circulation converts to the postnatal pattern (Figure 18.12b). Blood flow to the placenta ceases when the umbilical cord is tied and cut off. The scar left by the cord becomes the baby's navel, or belly button. Within the infant, the umbilical arteries, umbilical vein, ductus venosus, and ductus arteriosus constrict, shrivel, and form ligaments. The foramen ovale normally closes shortly after birth, leaving a small depression in the wall between the two atria. With the closure of the foramen ovale, blood from the right atrium passes to the right ventricle and to the lungs, which are now functional in respiration. In some newborns, the foramen ovale fails to close. Known as **blue babies**, these infants have a bluish appearance because much of their blood still bypasses the lungs and is low in oxygen. Fortunately, this defect can be corrected with surgery.

The major events that occur during prenatal development are summarized by developmental period in Table 18.1. Several stages of human development are shown in Figure 18.13.

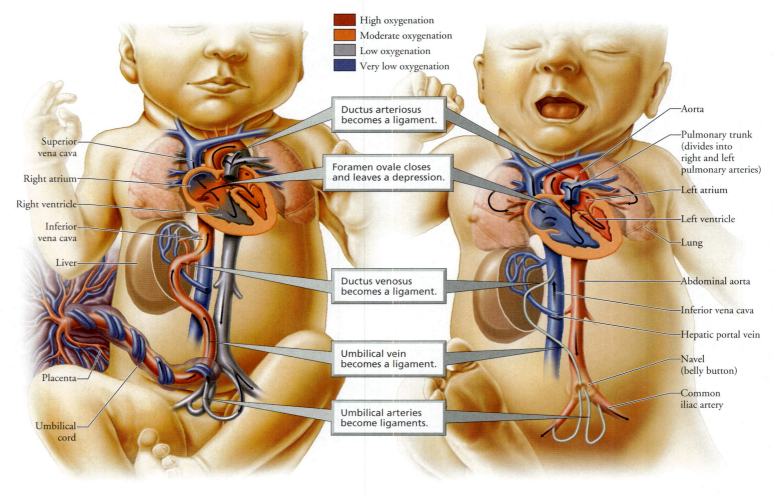

High oxygenation
Moderate oxygenation
Low oxygenation
Very low oxygenation

Ductus arteriosus becomes a ligament.

Foramen ovale closes and leaves a depression.

Ductus venosus becomes a ligament.

Umbilical vein becomes a ligament.

Umbilical arteries become ligaments.

Superior vena cava
Right atrium
Right ventricle
Inferior vena cava
Liver
Placenta
Umbilical cord

Aorta
Pulmonary trunk (divides into right and left pulmonary arteries)
Left atrium
Left ventricle
Lung
Abdominal aorta
Inferior vena cava
Hepatic portal vein
Navel (belly button)
Common iliac artery

(a) Fetal circulation is characterized by a connection to the placenta (the umbilical cord) and several bypasses around organs that do not yet perform their postnatal functions.

(b) At birth, the umbilical cord is tied off and cut, leaving the navel (belly button). The bypasses close, allowing more blood to reach the now functional organs of the newborn.

FIGURE **18.12**
Fetal circulation and changes at birth

TABLE **18.1**

MAJOR EVENTS THAT OCCUR DURING PRENATAL DEVELOPMENT	
PERIOD	**MAJOR EVENTS**
Pre-embryonic (fertilization–week 2)	Fertilization
	Cleavage of the zygote
	Formation and implantation of the blastocyst
	Beginning of formation of extraembryonic membranes and placenta
Embryonic (week 3–week 8)	Gastrulation
	Formation of tissues, organs, and organ systems
Fetal (week 9–birth)	Continued differentiation and growth of tissues and organs
	Increase in crown-rump length
	Increase in weight

Birth Is the Transition from Prenatal to Postnatal Development

Birth, also called parturition, usually occurs about 38 weeks after fertilization. The process by which the fetus is expelled from the uterus and moved through the vagina to the outside world is called **labor**, an appropriate term given the phenomenal effort often required. (In fact, *labor* in Latin means "toil" or "suffering.") Uterine contractions during true labor occur at regular intervals, are usually painful, and intensify with walking. Recall from Chapter 10 that the hormone oxytocin released from the posterior pituitary gland causes contractions of the smooth muscle of the uterus. These contractions, in turn, stimulate further release of oxytocin in a positive feedback loop. As true labor proceeds, contractions become more intense, and the interval between them decreases. In contrast, false labor is characterized by irregular contractions that fail to intensify or change with walking. Expectant mothers who arrive at the hospital in false labor are sent home to await the arrival of true labor.

(a) A zygote

(b) A blastocyst implanting into the wall of the uterus

(c) An embryo about 7 weeks old

(d) A fetus about 4 months old

(e) A fetus about 5 months old

FIGURE **18.13**
The developing human

▌ True labor consists of three stages

True labor can be divided into three stages (Figure 18.14). The first stage, known as the **dilation stage**, begins with the onset of regular contractions and ends when the cervix has fully dilated to 10 cm (4 in.). This stage usually lasts about 6 to 7 hours, although it can be shorter in women who previously have given birth. During the dilation stage, the amniotic sac usually ruptures, releasing amniotic fluid. Known commonly as "breaking the water," rupture of the amniotic sac is done deliberately by a doctor or midwife if it does not happen spontaneously. The second stage, the **expulsion stage**, begins with full dilation of the cervix and ends with delivery of the baby. This stage may last one or more hours. Some physicians make an incision to enlarge the vaginal opening, just before passage of the baby's head. This procedure, an episiotomy, facilitates delivery and avoids ragged tearing. Once delivery of the baby is complete, the umbilical cord is clamped and cut. The newborn takes his or her first breath, and the conversion of fetal circulation to the postnatal pattern begins. Labor,

however, is not over. The third (and final) stage of true labor is the **placental stage**, which begins with delivery of the newborn and ends when the placenta is expelled from the mother's body. The placenta is typically delivered about 15 minutes after the baby. Continuing and powerful uterine contractions aid in expulsion of the placenta and constriction of maternal blood vessels torn during delivery. Any vaginal incision or tearing is sutured once labor has ended.

Most babies are born head first, facing the vertebral column of their mother. Some babies, however, are born "buttocks first" and their delivery is called **breech birth**. Breech births are associated with difficult labors and umbilical cord accidents such as compression or looping of the cord around the baby's neck. Thus, many physicians attempt to turn breech babies into the headfirst position before delivery. Turning is done by a trained physician applying his or her hands to the expectant mother's abdomen and gently guiding the fetus into the head-down position. The baby may be delivered by Cesarean section if turning is not successful.

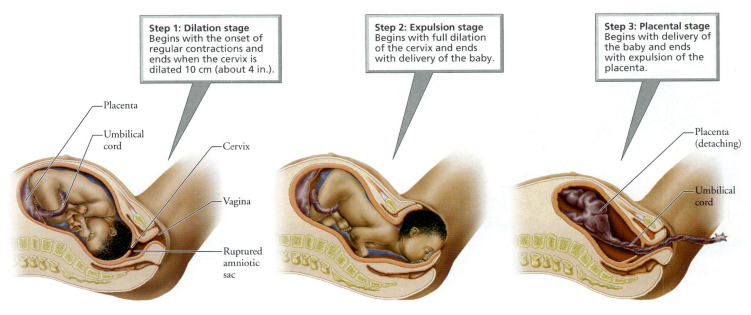

Step 1: Dilation stage
Begins with the onset of regular contractions and ends when the cervix is dilated 10 cm (about 4 in.).

Step 2: Expulsion stage
Begins with full dilation of the cervix and ends with delivery of the baby.

Step 3: Placental stage
Begins with delivery of the baby and ends with expulsion of the placenta.

Placenta

Umbilical cord

Cervix

Vagina

Ruptured amniotic sac

Placenta (detaching)

Umbilical cord

FIGURE **18.14**
The stages of true labor

stop and think

Compared with humans, our closest relatives—the apes—have a fairly easy time giving birth. A female ape, for example, has a relatively large pelvis, and the small head of her infant passes easily through a roomy birth canal (the channel formed by the cervix, vagina, and vulva). Infant apes are born face up, usually with little fuss. Birth in humans is decidedly more difficult and risky. With its large head, the typical human fetus may spend hours making its way through a narrow birth canal. Furthermore, because the canal changes dimensions along its length, the fetus enters it facing sideways and then must turn to enter the world face down. What evolutionary trends among humans might have produced the tight squeeze of human birth?

what would you do?

Babies born before 25 weeks gestation typically have a low chance of survival, even with excellent medical care. In addition, they have a very high probability of severe disabilities, if they do survive. Should efforts be made to save all such infants? Or should criteria be used to decide which infants doctors should try to save and which they should let die? If you were given the responsibility of developing such criteria, what would you suggest? More to the point, if your infant were born before 25 weeks, what would you ask the doctor to do?

Babies born 38 weeks after fertilization are called **full-term infants**. These babies weigh, on average, 3400 g (7.5 lb). Some babies are born 1 or more weeks late. Because the placenta begins to deteriorate with time, labor is typically induced if a pregnancy continues for 41 or more weeks. Babies born before 37 weeks of gestation are called **premature infants** (*gestation* refers to the time a pre-embryo, embryo, or fetus is carried in the female reproductive tract). About 12% of babies born in the United States are born prematurely. Their chances for survival increase with time spent in the female reproductive tract. Infants born at 22 weeks of age usually are not revived because major disabilities are almost inevitable. Such babies weigh about 630 g (1.4 lb). Amazingly, infants born at 25 weeks gestation and weighing about 1 kg (2.2 lb) have a 50% to 80% chance of surviving, provided they receive intensive care. Intensive medical care is needed because the organ systems of these infants have not matured sufficiently to take over the functions normally performed by the mother's body. Care for premature infants often includes numerous blood transfusions, tube feeding, and maintenance on a respirator (Figure 18.15). Even with excellent intensive care, health problems may persist well beyond infancy. For example, individuals born prematurely and classified as very low-birth-weight babies may have lower IQ scores in adulthood than individuals born at normal birth weight.

FIGURE **18.15**

Many premature infants require intensive care because their organs are not yet functional.

▌The placenta may send the signal to initiate labor

In humans, the signal to initiate labor appears to come from the placenta in the form of **corticotropin-releasing hormone (CRH)**. Levels of CRH begin to rise in fetal and maternal bloodstreams beginning about 12 weeks into gestation. Sharp increases in CRH occur during the last few weeks of gestation. This placental hormone stimulates the anterior pituitary gland of the fetus to release adrenocorticotropic hormone (ACTH), which, in turn, prompts the fetal adrenal glands to release a hormone that is converted by the placenta into estrogen. As a result of this conversion, levels of estrogen increase relative to progesterone, causing a cascade of events that initiates labor. These events include coordinated uterine contractions, dilation of the cervix, and release of oxytocin. Some recent studies support CRH as the initiating signal. In these studies, the level of CRH between weeks 16 and 20 of gestation predicted the timing of birth. Women with the highest levels of CRH were likely to deliver prematurely. Those with the lowest levels were likely to deliver late. Prevention of premature deliveries and the consequent health risks to infants is a major goal for understanding the events that initiate labor.

▌Drugs and breathing techniques can relieve pain during childbirth

Most women experience pain during childbirth. The level of pain varies greatly and seems to depend on factors such as degree of fear and tension, the physical surroundings, and the support received. Typically, the more prepared and relaxed a woman is, the less pain she feels. Reduced stress is one reason health care workers encourage expectant mothers and their partners to enroll in childbirth classes. These classes provide information on the stages of labor and medical procedures associated with delivery. Childbirth classes also teach breathing and relaxation techniques.

Drugs may be administered to reduce the pain of childbirth. Analgesics—substances that relieve pain—may be administered intramuscularly or intravenously. Regional anesthetics, substances that produce loss of sensation in specific areas of the body, also are used. For example, in epidural anesthesia, an anesthetic is injected in the lower back into the space between the dura mater (the outer membrane covering the spinal cord) and the vertebral canal. This injection blocks pain and motor impulses to the abdomen. Because motor impulses are blocked, epidural anesthesia makes pushing out the baby impossible. Epidural procedures are performed fairly early during labor and not toward the end, when pushing by the mother is critical for successful vaginal delivery. Epidural anesthesia is associated with slowing the fetal heartbeat. Thus, its use necessitates continuous fetal monitoring. General anesthesia removes sensation of pain and produces unconsciousness. Although once a popular method for relieving pain during childbirth, general anesthesia today is used almost exclusively for emergency Cesarean sections, when it is preferred over regional anesthetics because of the speed with which it takes effect.

Many women rely on breathing or relaxation techniques rather than medications to relieve the pain of childbirth. Childbirth without the use of pain-killing drugs often is called **natural childbirth**. Whether to use medication during childbirth is a decision that is best made by weighing the costs and benefits to both mother and baby. Also, because it is impossible to predict the precise circumstances of labor and delivery, all expectant mothers should keep an open mind.

Environmental Disruptions During the Embryonic Period Cause Major Birth Defects

The development of a fertilized egg into a baby is an incredibly wondrous yet complex process. In view of its complexity, it is not difficult to imagine that things might go wrong along the way. Indeed, the potential for mistakes seems enormous. Thus, we should not be surprised to learn that not every zygote develops into a baby, and not all babies are born healthy. Developmental defects present at birth are called **birth defects**. Such defects may involve structure, function, behavior, or metabolism. The study of birth defects is called teratology (*teratos*, monster). Sadly, the name of the field reflects the term used in earlier times to describe severely deformed infants.

The causes of birth defects may be genetic or environmental. Genetic causes include mutant genes or changes in the number or structure of chromosomes. Environmental causes include drugs, chemicals, radiation, deficiencies in maternal nutrition, and certain viruses such as herpes simplex and rubella (the cause of German measles). Environmental agents that disrupt development are called **teratogens**. Birth defects resulting from genetic causes will be discussed in Chapter 19. Those associated with drugs were considered in Chapter 8a. Here we focus on *when* in development disruptive agents have their greatest impact.

Teratogens have their greatest effects on the developing embryo during periods of rapid differentiation. At such times, cells and tissues acquire specific functions and form organs. Recall that these events largely occur during the embryonic period (week 3 through week 8). Exposure to disruptive agents during the embryonic period can cause major birth defects (Figure 18.16). Such defects involve major structural abnormalities and occur in 2% to 3% of newborns. Spina bifida is an example of a major birth defect. Disruptive agents produce more minor defects during the fetal period, when most organs are growing rather than differentiating. Minor birth defects occur in about 4% to 5% of newborns.

We have said that all organs and organ systems develop primarily during the embryonic period. Nevertheless, different organs develop at somewhat different times and rates. Such variation results in each organ system having a critical period in which it is most susceptible to disruption by teratogens. The central nervous system (CNS) has a critical period extending from week 3 through week 16. During this time, the developing nervous system is highly sensitive, and teratogens can cause major birth defects. Exposure to

teratogens after week 16 tends to induce relatively minor defects. The CNS has the longest critical period. It extends beyond the embryonic period and into the fetal period. This long critical period may explain why the brain is the organ in which defects in structure and function are most commonly found. The critical period for development of the upper and lower limbs is relatively short, extending from about week 4 to week 6. It is during this short window of time that teratogens cause major abnormalities such as tiny limbs or complete absence of limbs.

Another look at Figure 18.16 reveals that exposure to harmful agents during the first 2 weeks after fertilization is not known to cause birth defects. Development during this period (the preembryonic period) centers on formation of structures outside the embryo, such as the extraembryonic membranes described earlier. However, the developing pre-embryo is not immune to environmental influences at this time. Teratogens in the pre-embryonic period may interfere with cleavage or implantation and cause a spontaneous abortion. Chromosomal abnormalities also account for a large proportion of spontaneous abortions at this time.

what would you do?

During the late 1950s and early 1960s, the drug thalidomide was widely used in Europe as a sedative and as relief for morning sickness. Previous animal testing had indicated that thalidomide was safe; although such testing reportedly did not involve pregnant animals. The drug turned out to be a potent teratogen, causing defects of the limbs such as complete absence of limbs or abnormally small limbs that lacked elbows and knees. Thalidomide also caused abnormalities of the heart (some of which caused death shortly after birth), ears, and urinary and digestive systems. About 12,000 infants whose mothers had taken thalidomide were affected. In later years, thalidomide was found to be an effective treatment for conditions such as leprosy and cancer of the bone marrow. Thalidomide also appears to relieve some of the conditions associated with HIV infection, such as the wasting syndrome and ulcers of the mouth. Today, thalidomide is available by prescription in the United States and is sold over the counter in South America, where "thalidomide babies" are still being born. Should a drug known to cause severe birth defects be made available for treatment of seriously ill adults? Is there any way, other than a complete ban of its use, to ensure that pregnant women do not mistakenly take thalidomide? If you were responsible for developing a policy regarding the use of thalidomide, what would you do?

The Mother's Mammary Glands Produce Milk

Lactation is the production and ejection of milk from the mammary glands. Recall from Chapter 10 that the hormone prolactin from the anterior pituitary gland promotes milk production. Recall also that oxytocin from the posterior pituitary gland stimulates milk ejection, or letdown. The structure of the breast was described in Chapter 17.

Prolactin levels increase during pregnancy and reach their peak at birth. Milk production does not begin during pregnancy because high levels of estrogen and progesterone inhibit the actions of prolactin. Prolactin can initiate milk production only when levels of

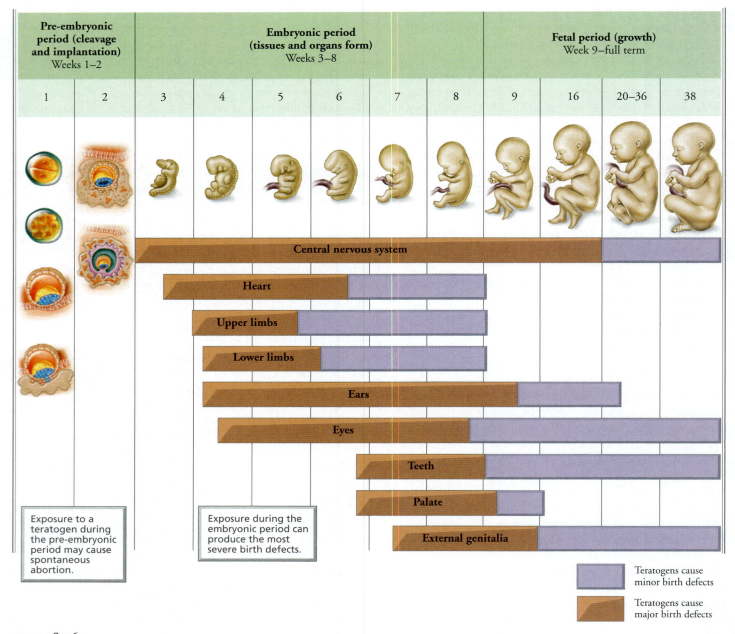

FIGURE 18.16
Critical periods in development

estrogen and progesterone decline at birth. Milk usually is available about 3 days later. The breasts produce **colostrum**, a cloudy yellowish fluid of different composition than milk, in the interval after birth, when milk is not yet available. Milk production is maintained after birth, when suckling by the infant stimulates the release of prolactin. Once suckling ends, so too does milk production.

Which is best—bottle or breast? Most people today believe that breast-feeding is best for infants. Advocates of breast-feeding point out that compared with commercial formulas based on cow's milk, breast milk is more digestible and much less likely to cause an allergic reaction, constipation, or obesity in infants. In addition, maternal milk and colostrum contain antibodies and special pro-

teins that boost the immune system. This boost makes breast-fed babies less susceptible than bottle-fed babies to illness during the first few months of life. Finally, obtaining milk from a breast requires more effort than sucking on a bottle. Health professionals think this greater effort promotes optimum development of the infant's jaws, teeth, and facial muscles.

Breast-feeding also is beneficial to the mother. Nursing helps the mother's uterus return more rapidly to its prepregnant size. Recall that oxytocin stimulates uterine contractions and milk letdown. When oxytocin is released in response to an infant's sucking, it helps the uterus shrink. Breast-feeding also is convenient and economical. Bottles and formula do not have to be purchased or carried while traveling. Perhaps most important, breast-feeding establishes prolonged periods of skin-to-skin contact between mother and infant.

Despite its benefits, breast-feeding is not for everyone. Some women simply prefer bottle-feeding. Others are warned against breast-feeding because the medications they take pass into breast milk. Finally, mothers who are HIV positive should not breast-feed because the virus can be transmitted in breast milk.

Although we might think of breast-feeding as "natural" and "easy," it can be difficult during the first few weeks. At this time, mothers experience sore nipples, and the feeding relationship is just being established. Indeed, many health care professionals urge new mothers to give breast-feeding a fair trial. They suggest allowing 6 or more weeks to establish a successful breast-feeding relationship.

The Postnatal Period Begins with Birth and Continues into Old Age

The period of growth and development after birth is the postnatal period. The postnatal period includes the following stages: infancy (from birth to 12 months), childhood (from 13 months to 12 or 13 years), puberty (from 12 to 15 years in girls and from 13 to 16 years in boys), adolescence (from puberty to 17 years), and adulthood (generally reached by at least 21 years of age). The start of adulthood is imprecise because of the many factors—physical, behavioral, societal—that influence how we define an adult. Growth and formation of bone usually are completed by age 25. Developmental changes occur quite slowly after this time. Aging is a process that begins only a few years after growth has ceased. You might wonder why we discuss aging in a textbook geared toward college students. Our reasoning is simple; you can do a lot right now to ensure a high quality of life during old age (Figure 18.17).

Aging is the normal and progressive alteration in the structure and function of our bodies. Observable characteristics of aging—graying and thinning hair, wrinkles and sagging skin, reduced mus-

FIGURE **18.17**

Wise and informed lifestyle decisions made when young can lead to a healthy old age.

cle mass, and stooped posture—are familiar to us. However, we also age on the inside. For example, bones weaken and blood vessels stiffen. Nephrons, the functional units of the kidneys, decline in number and efficiency, ultimately challenging efforts by the kidneys to maintain the balance of fluids in our bodies. Aging also strikes our nervous system and sense organs, as evidenced by our deteriorating ability to remember, see, hear, taste, and smell. Table 18.2 summarizes some of the changes to organ systems that typically occur with old age. What causes these changes? Are such changes inevitable? Is there anything we can do to delay or reverse the aging process?

▌Several factors cause aging

Scientists do not agree about what causes our bodies to age. Some, with a "whole body" view, suggest that aging results from changes in critical body systems. For example, aging might be prompted by a decline in function of the immune system or by changing levels of certain hormones. Recall from Chapter 10 that the production of some hormones, such as growth hormone in both sexes and estrogen in females, declines as we age. Perhaps such declines induce the changes in bodily structure and function that characterize aging.

Other scientists seek explanations at the cellular and molecular levels. Ongoing cell division is necessary to replace cells that die. Such replacement is essential for most tissues and organs to continue to function effectively. Nevertheless, cell division appears to slow in aging animals. Supporting the idea that cessation of cell division is genetically programmed is the observation that cells grown in culture in the laboratory do not divide indefinitely. In fact, such cells divide only a certain number of times. The number of divisions seems to be correlated with the age of the individual that donated the cells (be the donor a roundworm, mouse, or human) and the life span of the particular species. Telomeres, protective pieces of DNA at the tips of chromosomes, may help cells keep track of how many times they have divided (see Chapter 21a). A tiny piece of each telomere is sliced off each time the DNA is copied before cell division. After a certain number of cell divisions, the telomere is gone, and the chromosome is no longer protected from the slicing. Chromosomes that sustain such damage can no longer participate in cell division—and the cell dies. Cells grown in culture demonstrate this phenomenon, yet a clear link to human longevity has not been established.

Other researchers postulate that highly reactive molecules known as free radicals disrupt cell processes and lead to aging. Free radicals are by-products of normal cellular activities. They have an unpaired electron and readily combine with and damage DNA, proteins, and lipids. Mitochondrial DNA seems particularly susceptible to such damage. Recall that mitochondria play a critical role in processing energy for cells (discussed in Chapter 3). Thus, damage to mitochondria threatens the supply of energy needed to sustain cellular activities. Aging also is associated with a decline in the ability of cells to repair damaged DNA. An inability to "fix" DNA might lead to an accumulation of gene mutations and ultimately to a decline in cell function. In addition, scientists are focusing on **Werner's syndrome**, a rare inherited disease characterized by premature aging. People with Werner's syndrome develop the appearance and ailments of the elderly while still in their twenties.

TABLE **18.2**

CHANGES IN BODY SYSTEMS COMMON IN OLD AGE	
ORGAN SYSTEM	**SOME CHANGES THAT OCCUR WITH AGING**
Integumentary	Wrinkles appear as skin becomes thinner and less elastic.
	Sweat glands decrease in number, making regulation of body temperature more challenging.
	Hair thins owing to death of hair follicles and turns gray as pigment-producing cells die.
Skeletal	Bones become lighter and more brittle, especially in women after menopause.
	About 7.6 cm (3 in.) of height are lost as the intervertebral disks deteriorate and the vertebrae move closer together.
	Joints become stiff and painful because of decreased production of synovial fluid.
Muscular	Muscle mass decreases owing to loss of muscle cells and decreased size of remaining muscle cells.
Nervous	Brain mass decreases.
	Movements and reflexes slow as conduction velocity of nerve fibers decreases and release of neurotransmitters slows.
	Hearing becomes less acute as hair cells in the inner ear stiffen.
	Ability of the eye to focus declines as the lens of the eye stiffens.
	Smell and taste become less acute.
Endocrine	In women, production of estrogen and progesterone decreases with menopause.
	In men, production of testosterone decreases.
	In both sexes, production of growth hormone decreases.
Circulatory	Cardiac output decreases as walls of the heart stiffen.
	Blood pressure rises as arteries become less elastic and clogged by fatty deposits.
Respiratory	Lung capacity decreases as alveoli break down and lung tissue becomes less elastic.
Digestive	Basal metabolic rate declines.
	Ability of the liver to detoxify substances declines.
Urinary	In both sexes, kidney mass declines, as does the rate of filtration of the blood by nephrons.
	Particularly in women, the external urethral sphincter weakens, causing incontinence.
	In men, the prostate gland enlarges, causing painful and frequent urination.
Reproductive	In men, fewer viable sperm are produced.
	In women, ovulation and menstruation cease at menopause.

They typically die in their late forties. An understanding of this condition may provide information on the normal aging process.

Finally, some scientists have suggested that glucose—our main source of fuel—is the culprit. Glucose changes proteins such as collagen by causing cross-linkages to form between their molecules. This shackling of protein molecules may cause the stiffening of connective tissue and heart muscle associated with aging.

Is there a primary cause of aging? Most scientists believe that aging is not caused by a single factor but rather by several processes that interact to ensure our eventual deterioration and demise.

Medical advances and a healthy lifestyle can help achieve a high-quality old age

The news today is full of stories about antiaging products and regimens. For example, human growth hormone (HGH) is billed as being able to reverse many of the effects of aging. (Recall from Chapter 10 that growth hormone is secreted by the anterior pituitary gland.) Administration of HGH seems to increase lean body mass and the density of some bones. It also appears to reduce thinning of the skin and the amount of adipose (fat) tissue. However, the administration of HGH has side effects such as fluid accumulation, increased blood pressure, and diabetes. In addition, some animal studies have suggested that administering growth hormone might actually shorten life.

Aging is a normal biological process that, at present, cannot be slowed, stopped, or reversed. However, it is possible to stave off the disease and disability often associated with aging. The keys to a healthy old age lie in medical advances and lifestyle. Physicians are now able to treat many conditions of old age. For example, worn-out hip or knee joints can be replaced with artificial joints. The clouding of vision caused by cataracts is now treated readily by replacing the old lens of the eye with an artificial lens. New drugs may break the cross-links that glucose creates in proteins and thereby restore elasticity to our arteries. In addition, research is under way to determine if antioxidants (substances that inhibit the formation of free radicals) are effective in delaying certain aspects of aging. Potential antioxidants include the hormone melatonin and vitamins E and C. At present, however, the evidence does not justify the use of antioxidants as antiaging products.

Today, the maximum documented life span for humans is 122 years, a record recently established by Madame Jeanne Calment of France. Most of us will likely never reach her age. In fact, life expectancy for babies born in the United States today is about 77 years. Nevertheless, we can do a lot to promote an enjoyable and illness-free old age. The human life span appears to be determined by genes, environment, and lifestyle. Lifestyle is the factor over which we have the most personal control. Some components of a healthy lifestyle include proper nutrition, plenty of exercise and sleep, no smoking, and routine medical checkups. Indeed, the lifestyle choices we make when young can delay some aspects of aging. For example, we can delay wrinkling and aging of the skin by avoiding excessive exposure to the sun. Although it is never too late to change lifestyle habits, the earlier healthy habits are followed the better.

REVIEWING THE CONCEPTS

Human Life has Two Main Periods of Development (p. 416)

1. The period of development before birth is the prenatal period. The period of development after birth is the postnatal period. Stages in postnatal development include infancy, childhood, puberty, adolescence, and adulthood.

The Prenatal Period Begins at Fertilization and Ends at Birth (pp. 416–427)

2. The prenatal period can be subdivided into the pre-embryonic period, embryonic period, and fetal period. The pre-embryonic period runs from fertilization through the second week and is characterized by formation and implantation of the blastocyst. The extraembryonic membranes and placenta also begin to form at this time. The embryonic period extends from week 3 through week 8 and is characterized by gastrulation and the formation of organs and organ systems. The fetal period runs from the ninth week until birth and is a period of intense growth.

3. Fertilization is the union of an egg and a sperm to form a single cell called the zygote. Before fertilization can occur, sperm must first undergo two processes. During the first, secretions from the female reproductive tract destabilize the plasma membrane of the sperm. During the second, enzymes spill out of the cap on the head of the sperm and digest a pathway through the layers surrounding the secondary oocyte. A rapid depolarization of the oocyte's plasma membrane and changes in the zona pellucida ensure that only one sperm enters and fertilizes it. These mechanisms block polyspermy, an abnormal condition in which more than one sperm enters an oocyte.

4. Cleavage is a rapid series of mitotic cell divisions that transforms the zygote into a morula (a solid ball of cells) and then a blastocyst (a hollow ball of cells with a fluid-filled cavity). Within the blastocyst are the inner cell mass, which will become the embryo and some of the extraembryonic membranes, and the trophoblast, which will form part of the placenta.

5. The placenta is the organ that delivers nutrients and oxygen to the embryo (and later the fetus) and carries wastes away. The placenta also produces hormones such as estrogen, progesterone, human chorionic gonadotropin (HCG), and corticotropin-releasing hormone (CRH). Toward the end of the first week, the blastocyst begins implantation, the process by which the embryo becomes embedded in the endometrium of the uterus.

6. Extraembryonic membranes—the amnion, yolk sac, chorion, and allantois—lie outside the embryo and begin to form during the second or third week after fertilization. The amnion surrounds a fluid-filled cavity (the amniotic cavity) that protects the embryo from physical trauma. The yolk sac is a site of early blood cell formation and contains primordial germ cells that migrate to the gonads, where they form cells that mature into gametes. The chorion is the embryo's contribution to the placenta. The blood vessels of the allantois become part of the umbilical cord, the ropelike connection between the embryo and placenta.

7. The placenta orchestrates all interactions between mother and fetus. It forms from the chorion of the embryo and the endometrium of the mother. Cells of the embryonic trophoblast invade the endometrium, creating pools of maternal blood. Then, chorionic villi, fingerlike processes of the embryo's chorion, grow into the endometrium and contact the pools of blood. Oxygen and nutrients within the pools of maternal blood diffuse into the capillaries within the chorionic villi and move into the umbilical vein and on to the embryo. Wastes leave the embryo by the umbilical arteries, move into capillaries within the chorionic villi, and diffuse into maternal blood. Although maternal and fetal blood vessels are close together within the placenta, there is no direct mixing of blood.

8. Gastrulation includes the cell movements by which the primary germ layers—ectoderm, mesoderm, and endoderm—are established. Ectoderm forms the nervous system and the top layer of skin and its derivatives (hair, nails, oil glands, sweat glands, and mammary glands). Mesoderm forms muscle, bone, connective tissue, and organs such as the heart, ovaries, and testes. Endoderm forms organs such as the liver, some endocrine glands, and the lining of the urinary, respiratory, and digestive tracts.

9. Neurulation is the series of events leading to formation of the neural tube from ectoderm. The anterior portion of the neural tube forms the brain, and the posterior portion forms the spinal cord. Somites, blocks of mesoderm organized alongside the neural tube, eventually form vertebrae, connective tissue, and skeletal muscles of the neck and trunk.

10. The gender of a human embryo is determined at fertilization by the type of sperm (X-bearing or Y-bearing) that fertilizes the egg (which is X-bearing). An XX zygote will develop into a female, and an XY zygote into a male. Early embryos are sexually indifferent, having undifferentiated internal sex organs and external genitalia. About 6 weeks after fertilization in XY embryos, a region of the Y chromosome initiates development of testes. The testes soon begin production of testosterone, which directs the development of male internal and external reproductive structures. Female embryos lack a Y chromosome, and, in its absence, female internal and external reproductive structures develop.

11. Growth during the fetal period is extremely rapid. The rate of growth of the head slows relative to that of other parts of the body, producing changes in the way the body is proportioned. Such change in the relative rates of growth of different parts of the body is called allometric growth.

12. Fetal circulation differs from circulation after birth in that most blood is shunted through temporary vessels or openings past organs such as the lungs and liver, which do not yet perform their postnatal functions. The shunts of fetal circulation include the ductus venosus (a vessel that diverts most blood away from the liver and into the inferior vena cava), the foramen ovale (an opening between the two atria that diverts blood away from the lungs), and the ductus arteriosus (a vessel that diverts blood away from the lungs by connecting the pulmonary trunk to the aorta). At birth, when organs begin their postnatal functions, the bypasses of fetal circulation begin to close.

WEB TUTORIAL 18.1 Overview of Fetal Development

WEB TUTORIAL 18.2 Early Development

Birth is the Transition from Prenatal to Postnatal Development (pp. 428–431)

13. Parturition (birth) usually occurs about 38 weeks after fertilization. Birth marks the transition from the prenatal period of development to the postnatal period.

14. Labor is the process by which the fetus is expelled from the uterus and moved through the vagina into the outside world. Labor occurs in three stages: (1) the dilation stage, from the onset of uterine contractions to full dilation of the cervix; (2) the expulsion stage, from full dilation of the cervix to delivery of the baby; and (3) the placental stage, from delivery of the baby to expulsion of the placenta.

15. A hormonal signal from the placenta appears to cause a cascade of events that ultimately initiates labor.

16. The pain of childbirth can be relieved through analgesics (painkillers), anesthetics (substances that produce loss of sensation), and breathing and relaxation techniques.

Environmental Disruptions During the Embryonic Period Cause Major Birth Defects (p. 431)

17. Birth defects are developmental defects present at birth. Such defects may be caused by genetic factors (mutant genes or changes in the number or structure of chromosomes) or teratogens (environmental agents such as drugs or radiation). Teratogens have their greatest effects on the developing embryo during periods of rapid differentiation when organs and organ systems are forming. Thus, exposure to teratogens during the embryonic period may cause major birth defects. Exposure during the fetal period, when most organs are growing rather than differentiating, produces more minor defects.

The Mother's Mammary Glands Produce Milk (pp. 431–433)

18. Lactation is the production and ejection of milk from the mammary glands. At birth, when levels of estrogen and progesterone drop, prolactin initiates milk production. For the first few days after birth, the breasts produce colostrum, a cloudy fluid of different composition than milk. Milk becomes available about 3 days after birth. Suckling stimulates the release of prolactin, and thus milk production is maintained as long as the infant suckles.

The Postnatal Period Begins With Birth and Continues into Old Age (pp. 433-434)

19. Aging is the progressive decline in the structure and function of the body. Changes occur in all organ systems. For example, the skin, lungs, and vessels of the circulatory system become less elastic, and the mass of muscle and bone decreases. Production of some key hormones declines, and movements and reflexes slow owing to less rapid conduction of nerve impulses. Many of our senses become less acute. Men produce fewer viable sperm, and women pass through menopause, when ovulation and menstruation cease. Our digestive and urinary systems also change as we age.

20. Possible causes of aging include genetically programmed cessation of cell division; damage to DNA, protein, and lipids caused by free radicals; the inability of cells to repair damaged DNA; alteration of proteins by glucose such that cross-linkages form and produce a stiffening in tissues; and declines in the functioning of key organ systems.

21. Although aging is a normal process, it is often associated with disease and disability. We can stave off disease and disability through medical advances and a healthy lifestyle.

KEY TERMS

aging   *p. 433*
allantois   *p. 420*
allometric growth   *p. 426*
amnion   *p. 420*
anencephaly   *p. 424*
birth defect   *p. 431*
blastocyst   *p. 419*
blue baby   *p. 427*
breech birth   *p. 429*
Cesarean section   *p. 422*
chorion   *p. 420*
chorionic villi   *p. 422*
cleavage   *p. 418*
colostrum   *p. 432*
conjoined twins   *p. 419*
corticotropin-releasing hormone (CRH)   *p. 430*
dilation stage   *p. 429*

ductus arteriosus   *p. 426*
ductus venosus   *p. 426*
ectoderm   *p. 423*
ectopic pregnancy   *p. 419*
embryo   *p. 416*
embryonic disk   *p. 423*
embryonic period   *p. 416*
endoderm   *p. 423*
endometrium   *p. 419*
expulsion stage   *p. 429*
extraembryonic membranes   *p. 420*
fertilization   *p. 416*
fetal period   *p. 416*
fetus   *p. 416*
foramen ovale   *p. 426*
fraternal twins   *p. 419*
full-term infant   *p. 430*
gastrula   *p. 423*
gastrulation   *p. 423*

human chorionic gonadotropin (HCG)   *p. 420*
identical twins   *p. 419*
implantation   *p. 419*
infertility   *p. 420*
inner cell mass   *p. 419*
labor   *p. 428*
lactation   *p. 431*
mesoderm   *p. 423*
morning sickness   *p. 422*
morphogenesis   *p. 423*
morula   *p. 418*
natural childbirth   *p. 431*
neural tube   *p. 424*
neurula   *p. 424*
neurulation   *p. 424*
notochord   *p. 423*
placenta   *p. 419*
placenta previa   *p. 422*

placental stage   *p. 429*
polyspermy   *p. 417*
postnatal period   *p. 416*
pre-embryo   *p. 416*
pre-embryonic period   *p. 416*
premature infant   *p. 430*
prenatal period   *p. 416*
primary germ layers   *p. 423*
primordial germ cells   *p. 420*
somite   *p. 424*
spina bifida   *p. 424*
teratogen   *p. 431*
trophoblast   *p. 419*
umbilical cord   *p. 420*
Werner's syndrome   *p. 433*
yolk sac   *p. 420*
zygote   *p. 418*

THINKING ABOUT THE CONCEPTS

1. List the stages of prenatal development. What ages and major developmental milestones are associated with each stage? *pp. 416, 428*
2. What effect do secretions from the female reproductive tract have on sperm? *p. 417*
3. What is polyspermy? How is it prevented? *p. 417*
4. Describe implantation. Where does it usually occur? What happens if it occurs elsewhere? *p. 419*
5. What are the functions of the four extraembryonic membranes? *p. 420*
6. What is the placenta, and how does it form? Explain how nutrients, oxygen, and wastes are exchanged between fetal and maternal blood. *p. 420*
7. What is gastrulation? What tissues and organs are formed from each of the three primary germ layers? *p. 423*
8. How does the central nervous system form? *p. 424*
9. How is the gender of a human embryo determined? *p. 425*

10. What are teratogens? Explain the relationship between the effects of teratogens and embryonic age. *p. 431*
11. How does fetal circulation differ from circulation after birth? *p. 426*
12. What are the three stages of labor? What initiates labor? *pp. 428-429*
13. Why is breast-feeding considered better than bottle-feeding for most infants and mothers? *p. 432*
14. What are some possible causes of aging? What can you do today to improve the quality of your old age? *p. 433*
15. Fertilization
    a. typically occurs in the uterus.
    b. often involves more than one sperm uniting with an egg.
    c. is the process by which the zygote becomes embedded in the uterine lining.
    d. typically occurs in an oviduct.

16. The yolk sac
    a. is a primary source of nourishment for human embryos.
    b. is the embryo's major contribution to the placenta.
    c. contains primordial germ cells.
    d. encloses the embryo in a fluid-filled sac.
17. Which of the following does *not* characterize the placenta?
    a. produces estrogen, progesterone, and human chorionic gonadotropin
    b. is the site where fetal and maternal blood directly mix
    c. appears to send the signal to initiate labor in humans
    d. is expelled after birth
18. Neurulation
    a. is the process by which the neural tube forms.
    b. is the process by which the notochord forms.
    c. occurs just before gastrulation.
    d. involves endoderm.
19. The fetal period
    a. is the time when tissues and organs form.
    b. is characterized by rapid growth.
    c. runs from week 2 to week 8.
    d. is when exposure to teratogens is most likely to produce major birth defects.
20. Which of the following statements regarding childbirth is *true*?
    a. Most babies are born feet first.
    b. Delivery of the baby follows delivery of the placenta.
    c. Full-term babies are those born at 34 weeks gestation.
    d. Infants born prematurely at very low-birth-weight require intensive neonatal care.
21. Within the blastocyst, the _____ becomes the embryo proper and some extraembryonic membranes, and the _____ becomes part of the placenta.
22. Soon after fertilization, the zygote undergoes _____, a series of rapid mitotic divisions without an increase in overall size.
23. The placenta forms from the _____ of the embryo and the _____ of the mother.
24. _____ is the process by which primary germ layers form.
25. The small hole between the right and left atria of the fetus is called the _____.
26. The hormone _____ stimulates milk production, and the hormone _____ stimulates milk letdown.
27. _____ are highly reactive molecules produced by normal cellular metabolism that damage DNA and may lead to aging.

APPLYING THE CONCEPTS

1. Tanya had pelvic inflammatory disease 2 years ago. Now she is a few weeks pregnant and is experiencing severe abdominal pain on the right side. What condition might Tanya have? What is typically done for women with this condition?
2. Juan was described as a "blue baby" at birth and underwent corrective surgery. What did the surgeons correct and why?
3. Why might exposure to thalidomide during weeks 4 to 6 of gestation produce more severe birth defects than a similar level of exposure toward the end of gestation?
4. Dora is expecting her baby in about a month. She is deciding whether to breast-feed or bottle-feed. You are her physician. First describe how milk is produced and ejected from the mammary glands. Then, present the advantages and disadvantages of both methods of feeding.

19

Chromosomes and Cell Division

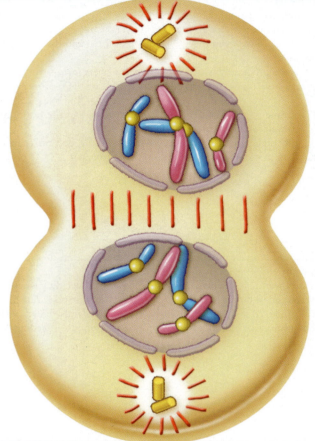

All Humans Have a Life Cycle

Chromosomes Consist of DNA and Protein

Our Cells Divide in a Characteristic Cyclic Pattern
- Interphase is a period of growth and preparation for cell division
- Division of body cells entails division of the nucleus and the cytoplasm
- Mitosis has four stages
- Cytokinesis occurs toward the end of mitosis

Meiosis Forms Haploid Gametes
- Meiosis keeps the chromosome number constant through generations and increases genetic variability in the population
- Meiosis involves two cell divisions
- Crossing over and independent assortment cause genetic recombination during meiosis
- Failure of chromosomes to separate during meiosis creates cells with extra or missing chromosomes

SOCIAL ISSUE   Cloning

As her little brother walked across the stage to receive his high school diploma, Lynn had never been prouder. When he stopped the graduation procession, looked around, held up his diploma, and waved at her with that huge infectious smile of his, people in the auditorium cheered—cheered for her little brother. Everyone in school knew Chris. He wore the mascot suit at home games and encouraged everyone to cheer for the teams. He designed the Picasso-like background scenery for the junior play and helped 10 other students paint the huge murals and anchor them into place. It was the most original Iowa set *The Music Man* had ever had.

Chris hadn't always had it so easy. He was born with an extra chromosome 21. He had endured several heart surgeries as a young child and spent countless hours with speech therapists. People had not always been kind to Chris either, and sometimes his feelings were hurt. Lynn nevertheless envied the people who would meet Chris for the first time next fall when he went to community college. He was going to study art and music, but Lynn knew that his most precious gift was teaching. He would teach other students to value and appreciate their own unique gifts. Chris has Down syndrome.

In this chapter we will learn why it is important that each cell has the correct number of chromosomes. We will begin by considering the physical basis of heredity: the chromosomes. Then we will consider how the chromosomes are parceled out during two types of cell division: mitosis and meiosis. ■

All Humans Have a Life Cycle

We begin life as a single cell called a **zygote**, formed by the union of an egg and a sperm. By adulthood, our bodies consist of trillions of cells. What happened in the intervening years? How did we go from a single cell to the multitude of cells that make up the tissues of a fully functional adult? To put it simply, cell division happened—over and over again. Even as an adult, many of your cells continue to divide for growth and repair of body tissues. With very few exceptions, each of those cells carries the same genetic information as its ancestors. The type of cell division that results in identical body cells is called *mitosis*.

In Chapter 17 you learned that males and females produce specialized reproductive cells called gametes (eggs or sperm). *Meiosis* is a special type of cell division that gives rise to gametes. In females, meiosis occurs in the ovaries and produces eggs. In males, meiosis occurs in the testes and produces sperm. Meiosis is important because the gametes end up with half the amount of genetic information (half the number of chromosomes) in the original cell. When an egg and sperm unite (fertilization), the chromosome number is restored to that of the original cell. As a result, the number of chromosomes in body cells remains constant between generations.

The process of cell division of body cells by mitosis and the formation of gametes by meiosis are summarized in the diagram of the human life cycle in Figure 19.1. You will learn more about both mitosis and meiosis later in this chapter.

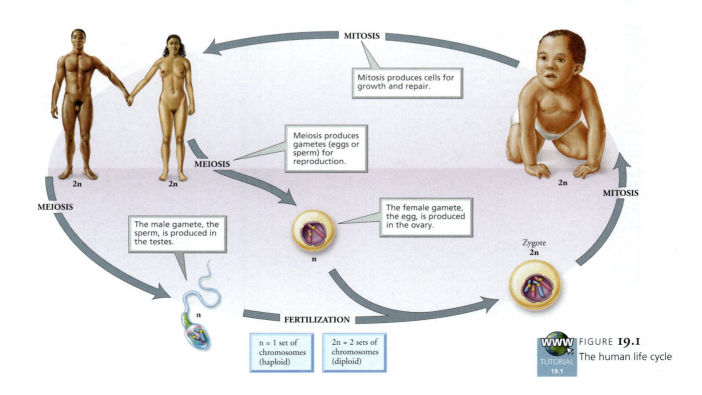

MITOSIS

Mitosis produces cells for growth and repair.

Meiosis produces gametes (eggs or sperm) for reproduction.

MEIOSIS

MEIOSIS

The male gamete, the sperm, is produced in the testes.

The female gamete, the egg, is produced in the ovary.

Zygote
2n

MITOSIS

2n

2n

2n

n

n

FERTILIZATION

n = 1 set of chromosomes (haploid)

2n = 2 sets of chromosomes (diploid)

WWW FIGURE **19.1**
The human life cycle

TUTORIAL 19.1

Chromosomes Consist of DNA and Protein

A chromosome is a combination of a DNA molecule that contains the genetic information of a cell and specialized proteins called histones. Chromosomes are found in the cell nucleus. The information contained in the DNA molecule directs the development and maintenance of the body. The proteins combined with the DNA are for support and control of gene expression. A **gene** is a specific segment of the chromosome's DNA that directs the synthesis of a protein, which in turn plays a structural or functional role within the cell. In this way, a gene determines the expression of a particular characteristic, or trait. Each chromosome in a human cell contains a specific assortment of genes, as shown in Figure 19.2. Like beads on a string, genes are arranged in a fixed sequence along the length of specific chromosomes. The point on a chromosome where a particular gene is found is called its **locus** (plural, loci).

In the human body, **somatic cells**, that is, all cells except for eggs or sperm, have 46 chromosomes. Those chromosomes actually consist of two sets of 23 chromosomes—one set of chromosomes came from the egg and the other from the sperm. Thus, each cell contains two chromosomes with genes for the same traits, called **homologous pairs** of chromosomes, or homologues for short. Any cell with two sets of chromosomes is described as being **diploid** (written as 2n). In diploid cells, then, genes occur in pairs, and the members of each pair are located at the same locus on homologous chromosomes.

Keep in mind, however, that the forms of the genes may differ. For example, one form of a gene for eye color may direct the formation of blue eyes and another brown eyes. The alternative forms of a gene are called **alleles**. Because homologous chromosomes may possess different alleles for the same traits, and there are many genes on one chromosome, the homologues are usually not identical.

One pair of chromosomes of the 23 pairs is the **sex chromosomes** that are involved in determining whether a person is male or female. There are two types of sex chromosomes, X and Y. A person who is XX is genetically female, and one who is XY is genetically male. The other 22 pairs of chromosomes are called the **autosomes**. The autosomes determine the expression of most of a person's inherited characteristics.

Our Cells Divide in a Characteristic Cyclic Pattern

In **mitosis,** one nucleus divides into two daughter nuclei with the same number and kinds of chromosomes (see the Social Issue essay, *Cloning*). But mitosis is only one phase during the life of a dividing cell. The entire sequence of events that a cell goes through from its origin in the division of a parent cell through its own division into two cells is called the **cell cycle** (Figure 19.3). The cell cycle consists of two major phases: interphase and cell division.

Interphase is a period of growth and preparation for cell division

Interphase is the period of the cell cycle between cell divisions. Interphase usually accounts for most of the time that elapses during a cell cycle. Depending on the type of cell, an entire cell cycle might take about 16 to 24 hours to complete, and only 1 to 2 hours are spent in division. Interphase is not a "resting period" as once thought. Instead, interphase is a time when the cell carries out its

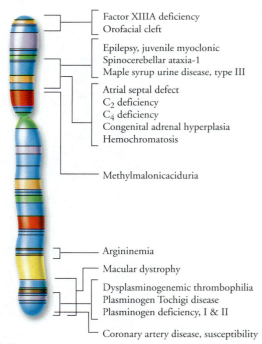

FIGURE **19.2**

A map of a human chromosome. Each chromosome contains a specific assortment of genes in a specific location along the chromosome. Genes are often identified through the disease caused when they do not function normally.

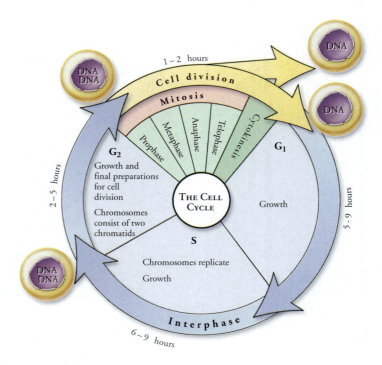

FIGURE **19.3**
The cell cycle

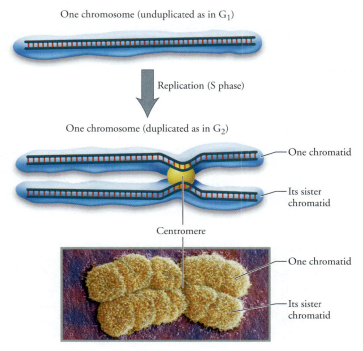

One chromosome (unduplicated as in G₁)

Replication (S phase)

One chromosome (duplicated as in G₂)

One chromatid

Its sister chromatid

Centromere

One chromatid

Its sister chromatid

FIGURE **19.4**

Changes in chromosome structure because of DNA replication during interphase

functions and grows. If the cell is going to divide, interphase is a time of intense preparation for cell division. It is a time when the DNA and organelles are duplicated. Such preparations ensure that when the cell divides, each of its resulting cells, called daughter cells, will receive the essentials for survival.

Interphase consists of three parts: G_1 (first "gap"), S (DNA synthesis), and G_2 (second "gap"). All three parts of interphase are times of cell growth, characterized by the production of organelles and the synthesis of proteins and other macromolecules. There are, however, some events specific to certain parts of interphase.

- G1: Time during interphase before DNA synthesis begins
- S: Time of interphase during which DNA is synthesized
- G2: Time during interphase after DNA is synthesized and before mitosis begins

The details of DNA synthesis (replication) are described in Chapter 21. Our discussion here introduces some basic terminology regarding the cell cycle.

During G_1 at the start of interphase, chromosomes consist of a DNA molecule and proteins. When DNA is replicated during the S phase, the copies of the chromosome remain attached, as shown in Figure 19.4. The two copies, each an exact replicate of the original chromosome, remain attached to one another at a region called the **centromere**. As long as the replicate copies remain attached, each copy is called a **chromatid.** The two chromatids that are banded together are genetically identical and are called *sister chromatids.*

Division of body cells entails division of the nucleus and the cytoplasm

Body cells divide continually in the developing embryo and fetus. Such division also plays an important role in the growth and repair of body tissues in children. In the adult, some cells, such as nerve cells, have been arrested in interphase for some time, having completely lost their ability to divide. Other adult cells, such as liver cells, stop dividing but retain the ability to undergo cell division should the need for tissue repair and replacement arise. Still other cells actively divide throughout life. Skin cells, for example, continue to divide in adults. The ongoing cell division in skin cells serves to replace the enormous numbers of cells worn off each day.

The division of body cells consists of two processes that overlap in their timing. The first process, division of the nucleus, is called *mitosis.* The second process is **cytokinesis,** which is the division of the cytoplasm that occurs toward the end of mitosis (Figure 19.5).

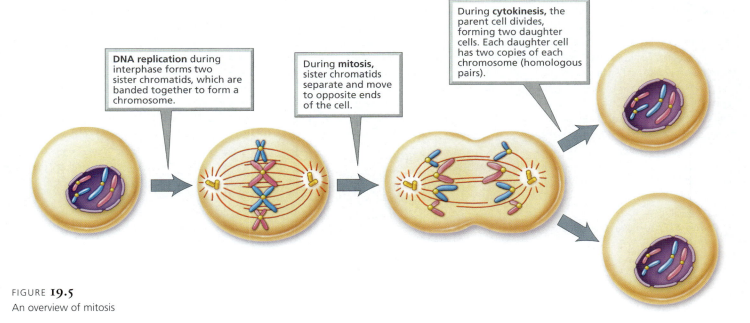

DNA replication during interphase forms two sister chromatids, which are banded together to form a chromosome.

During **mitosis,** sister chromatids separate and move to opposite ends of the cell.

During **cytokinesis,** the parent cell divides, forming two daughter cells. Each daughter cell has two copies of each chromosome (homologous pairs).

FIGURE **19.5**

An overview of mitosis

SOCIAL ISSUE

Cloning

What do the Dalai Lama and a lamb named Dolly have in common? They prompt us to look at the way we live our lives with a critical eye. Born on July 5, 1996, Dolly was the first sheep—indeed, the first mammal—to be cloned *from an adult*. (A clone is a genetically identical copy or copies of a cell or individual.) The goal of Dr. Ian Wilmut, the Scotsman whose work produced Dolly, was to develop techniques that would eventually lead to the production of animals that could be used as factories to manufacture proteins, such as hormones that would be beneficial to humans. In other words, cloning could make genetic engineering more efficient (see Chapter 21). With the ability to clone, a scientist would need to engineer an animal with a particular desired trait only once. Then, when the animal was old enough for scientists to be sure it had the desired trait, the animal could be duplicated exactly, producing multiple copies. Cloning from an adult instead of from an embryo is advantageous because you can be certain about the traits it possesses. In other words, "What you see is what you get."

Dr. Wilmut created Dolly by taking cells from the mammary gland of a pregnant ewe—cells already specialized to produce milk. (Dolly was named after Dolly Parton, a country singer who is perhaps known as much for her mammary cells as she is for her voice.) The nucleus was sucked out of an egg cell from another ewe. The enucleated egg cell and the mammary cell were then fused, and the new cell was activated to begin cell division. After about 6 days, when a naturally created embryo would implant in the uterus, the cloned embryos were implanted into the uterus of a host ewe. In the experiment that created Dolly, there were 277 artificially created "zygotes," 29 of which began to develop. Only one lamb, Dolly, was born. Dolly died in February 2003 at the age of 6 years—about half the usual life span for sheep. She was euthanized because she developed a progressive lung disease—a tumor caused by a virus.

By mid-2004, scientists had successfully cloned a variety of mammals, including mice, pigs, goats, cattle, rabbits, a cat ("cc" for CopyCat), a mule, wildcats, and deer. Attempts to clone a dog, monkey, a rare species of wild ox, and a rare Javan banteng (a cattle-like animal native to Asian jungles) failed or the offspring did not survive long. Most people are happy to see the cloning of a thoroughbred race horse, a nearly extinct species, a goat genetically engineered to produce drugs, or pigs with heart valves that will not be rejected if transplanted into humans.

However, discussions about cloning often become heated when it turns to human cloning. A distinction is often drawn between reproductive cloning and therapeutic cloning. Reproductive cloning produces a new individual with a known genetic makeup. The cloning of nonhuman animals previously described is reproductive cloning.

Experts agree that we are still far from being able to produce a cloned human baby.[1] Reproductive cloning is an extremely inefficient process. It requires hundreds of cloned embryos to get one born mammal. When cloned embryos do continue to develop, they grow faster and larger than normal. Indeed, a cloned banteng calf had to be euthanized because it was nearly twice the usual size. There is also a question about whether cloned mammals age prematurely. For example, there are indications that Dolly was genetically older than her birth age. Telomeres are protective structures on chromosomes that get shorter each time a cell divides (see Chapter 18). The length of Dolly's telomeres suggested that she was 6 years old at birth. In addition, Dolly had arthritis, a painful joint condition that is usually found only in older animals.

The idea of therapeutic cloning is to create a clone of "replacement cells" with the same genetic makeup as the patient; these replacement cells would be used in medical therapies. They would be stem cells, unspecialized cells that retain the ability to develop into any cell type. More precisely, the cells would be *embryonic* stem cells created using a technique similar to the one that created Dolly. A cell would be taken from the patient and the nucleus put into an enucleated donor egg cell, which would then begin development into an embryo (Figure 19.A). (If the patient's disease were genetic, a normal form of the gene would be substituted for the defective gene before the nucleus was put in the enucleated egg. Although it is possible to replace specific genes in mice, it has not yet been done in humans.) Stem cells could then be removed from the young embryo and coaxed to develop into the type of cell type needed for treatment by growth factors. The stem cells could then be transplanted into the patient to replace faulty cells. As examples, neurons might be transplanted to cure Parkinson's disease or to repair spinal cord damage, and insulin-producing cells might be transplanted to cure diabetes. Because the embryonic stem cells would have the same genetic makeup as the patient, they would not be attacked as foreign by the patient's body defense responses. Alternatively, instead of using genetic engineering to repair the faulty gene in each individual, genetic engineering might be used to create cloned adult cells with the healthy form of the gene in question but with other genetic alterations that would prevent the cloned cells from being rejected.

How far have we progressed in therapeutic cloning? Early in 2004, South Korean scientists created a cloned human embryo and extracted stem cells. The embryo was created by transferring the nucleus from an adult cell into an enucleated egg cell. The next step would be to develop the techniques to coax the stem cells to develop into specific cell types to be used for therapeutic purposes.

If it does become possible to clone humans someday, what might this technology mean to

---

[1] In late 2002 and early 2003, a sect called the Raelians and the scientific group they work with (Clonaid) claimed to have produced several cloned human babies, but they have not provided any proof or permitted DNA tests to verify these claims.

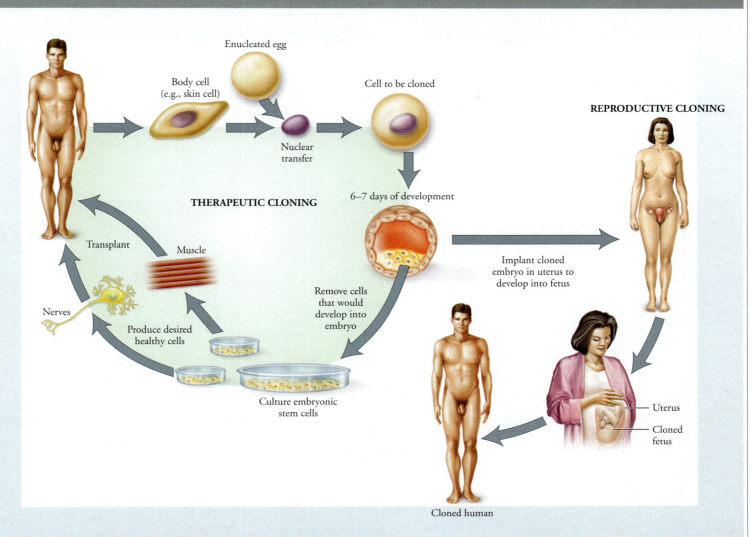

FIGURE **19.A** Methods for human therapeutic and reproductive cloning

you and me? How would it affect *our* lives? It would not give us the power to create an exact copy of a specific person. Environment *and* genes determine personality characteristics. Although cloning could exactly replicate the genetic makeup of a specific person, there would be no way to replicate all the past experiences in the life of the person being cloned, so the clone would be a different person than the original one. Thus, we could not replace a loved one who has died or decide that we should live forever. We could, however, produce an identical twin of a person.

Cloning could offer a solution to those medical problems that could be cured by a transplant—say bone marrow or a kidney—by providing an organ that has a matching tissue type. A family with a child in desperate need of a bone marrow transplant to cure leukemia might be glad to raise a second, identical child to save the life of the first.

The use of embryonic stem cells for therapeutic purposes has always been controversial. In August 2001, President George W. Bush announced that federal funding for stem cell research could be used only for existing stem cell lines. Since then, some scientists have left the United States to do stem cell research in other countries. Other U.S. scientists continue research on embryonic stem cells in privately funded research institutions. The debate regarding the use of embryonic stem cells was reopened when Ronald Reagan Jr., the son of the former President, spoke in favor of therapeutic cloning at the Democratic National Convention in July 2004.

Where do you stand on this issue? If human cloning does become feasible, should it be done? For what reasons? Who should decide? These issues will be debated into the next century. What do you think? ✋

Mitosis has four stages

For convenience, mitosis is usually divided into four stages: prophase, metaphase, anaphase, and telophase. The major events of each stage are depicted in Figure 19.6 and are described below.

PROPHASE

Mitosis begins with **prophase,** a time when changes occur both in the nucleus and in the cytoplasm. In the nucleus, the chromosomes are packaged for easy separation. During interphase, the chromosomes are long, thin threads, which are often called *chromatin.* They twist around one another like tangled strands of yarn. In this state, DNA can be synthesized (replicated) and genes can be active. However, during prophase the chromosomes begin to condense as the DNA wraps around histones. The DNA then becomes looped and twisted, forming a compact structure (Figure 19.7). When DNA is in this condensed state, it cannot be replicated, and gene activity is shut down. These thicker and shorter chromosomes become visible when stained and viewed under a light microscope. In this condensed state, the chromatids are easier to separate without breaking. About this time, the nucleolus disappears, and the nuclear membrane also begins to break down.

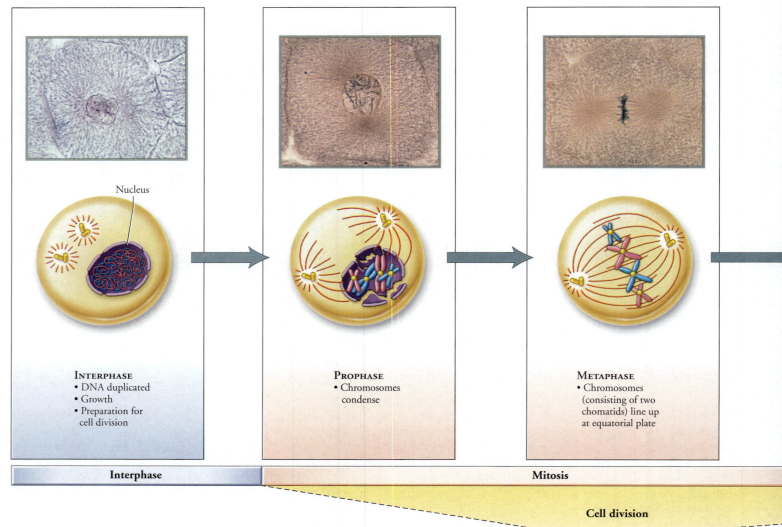

INTERPHASE
- DNA duplicated
- Growth
- Preparation for cell division

PROPHASE
- Chromosomes condense

METAPHASE
- Chromosomes (consisting of two chomatids) line up at equatorial plate

Interphase

Mitosis

Cell division

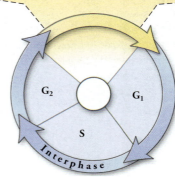

FIGURE **19.6**

The stages of cell division (mitosis and cytokinesis) depicted in light micrographs and schematic drawings

Outside the nucleus in the cytoplasm, the mitotic spindle forms. The mitotic spindle is made of microtubules associated with the centrioles (discussed in Chapter 3). During prophase, the centrioles, duplicated during interphase, move away from each other toward opposite ends of the cell.

METAPHASE

During the next stage of mitosis, **metaphase,** the chromosomes, guided by the fibers of the mitotic spindle, form a line at the center of the cell. The actual location, called the **equatorial plate,** is equidistant between the two poles of the mitotic spindle. As a result of this align-

ment, each daughter cell receives one chromatid from each chromosome when the chromosomes split at the centromere. Thus, each daughter cell receives a complete set of the parent cell's chromosomes.

ANAPHASE

Anaphase begins when the chromatids of each chromosome begin to separate, splitting at the centromere. Now separate entities, the chromatids are considered chromosomes in their own right. The chromosomes move toward opposite poles of the cell. By the end of anaphase, equivalent collections of chromosomes are located at the two poles of the cell.

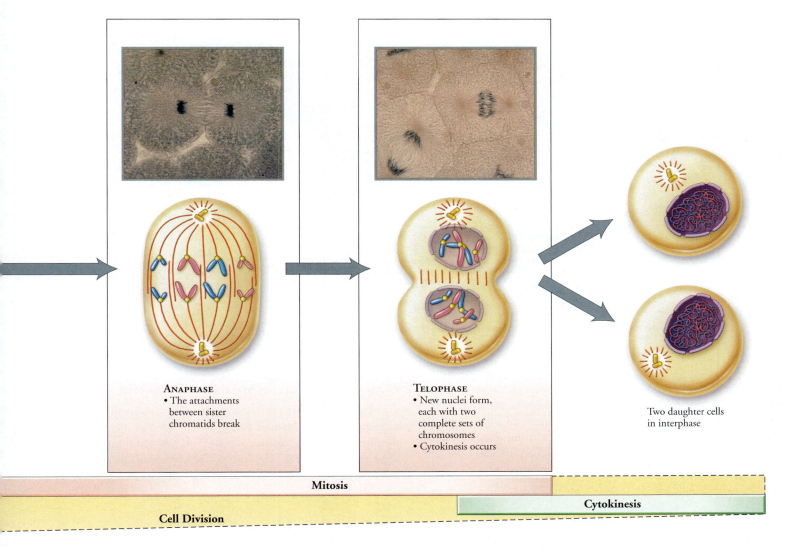

ANAPHASE
• The attachments between sister chromatids break

TELOPHASE
• New nuclei form, each with two complete sets of chromosomes
• Cytokinesis occurs

Two daughter cells in interphase

Mitosis

Cytokinesis

Cell Division

TELOPHASE

During **telophase,** a nuclear envelope forms around each group of chromosomes at each pole. The mitotic spindle disassembles, and nucleoli reappear. The chromosomes also become more threadlike in appearance.

stop and think

Cancer cells have lost control over cell division. One type of drug used in cancer chemotherapy inhibits the formation of spindle fibers. Why is this an effective anticancer treatment?

KARYOTYPES As we have seen, one of the major features of cell division is the shortening and thickening of the chromosomes. Visible with a light microscope, the chromosomes can be used for diagnostic purposes, such as when potential parents want to check their own chromosomal makeup for defects. One often-used method involves taking a blood sample. The white blood cells are separated from the sample and then grown in a laboratory with nutrients and other substances required by the cells. This culture of white blood cells is then treated with colchicine, a drug that destroys the mitotic spindle. Destruction of the spindle prevents separation of the chromosomes, halting cell division at metaphase. The cells in the culture are fixed,

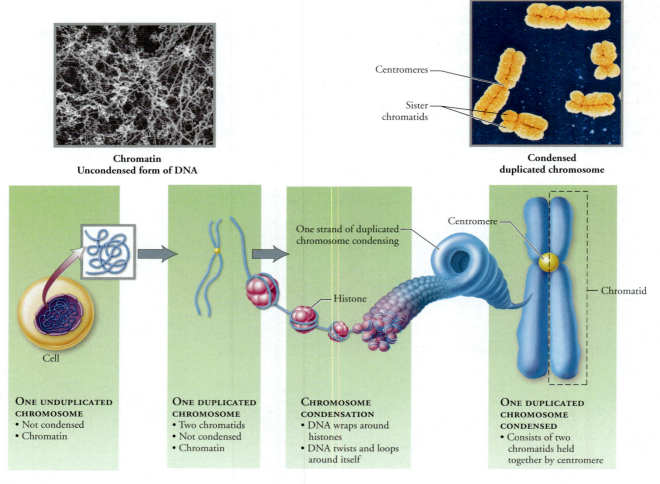

Chromatin
Uncondensed form of DNA

Condensed
duplicated chromosome

Centromeres

Sister
chromatids

One strand of duplicated
chromosome condensing

Centromere

Histone

Chromatid

Cell

ONE UNDUPLICATED CHROMOSOME
• Not condensed
• Chromatin

ONE DUPLICATED CHROMOSOME
• Two chromatids
• Not condensed
• Chromatin

CHROMOSOME CONDENSATION
• DNA wraps around histones
• DNA twists and loops around itself

ONE DUPLICATED CHROMOSOME CONDENSED
• Consists of two chromatids held together by centromere

FIGURE **19.7**
During prophase, duplicated chromosomes condense from a threadlike form called chromatin to compact packages of DNA.

stained, and photographed. The images of the chromosomes can then be arranged on the basis of physical characteristics of the chromosomes such as location of the centromere and overall length of the chromosomes. This arrangement of chromosomes is called a **karyotype** (Figure 19.8). Karyotypes can be checked for irregularities in number or structure of chromosomes.

▌Cytokinesis occurs toward the end of mitosis

Cytokinesis—division of the cytoplasm—begins sometime during late anaphase or early telophase. At this time, a band of microfilaments in the area that was the equatorial plate contracts and forms a furrow, as shown in Figure 19.9. The furrow deepens, eventually pinching the cell in two. The phases of the cell cycle are summarized in Table 19.1.

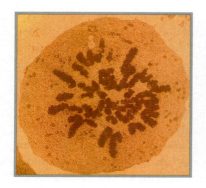

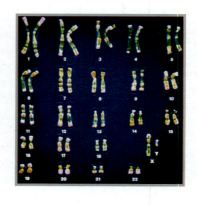

FIGURE **19.8**
Chromosomes in dividing cells can be examined for defects in number or structure. (a) Photomicrograph of metaphase chromosomes from a human white blood cell. (b) A karyotype constructed by arranging the chromosomes from photographs like that in (a). The chromosomes have been arranged on the basis of location of centromere and size.

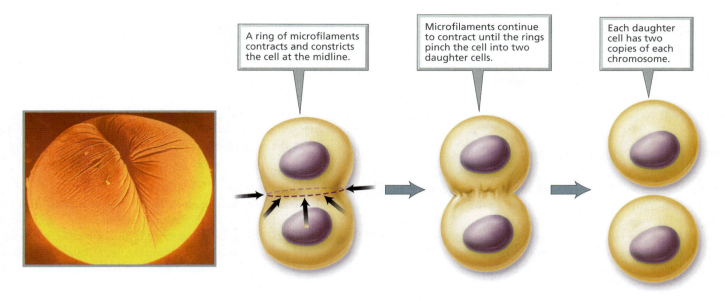

A ring of microfilaments contracts and constricts the cell at the midline.

Microfilaments continue to contract until the rings pinch the cell into two daughter cells.

Each daughter cell has two copies of each chromosome.

FIGURE **19.9**

Cytokinesis is the division of the cytoplasm to form two daughter cells.

TUTORIAL 19.3

Meiosis Forms Haploid Gametes

We have seen that the somatic cells contain two copies of every chromosome, one from the father and one from the mother. (Recall that a cell with two sets of chromosomes is described as being diploid, 2n.) The gametes—eggs or sperm—differ from somatic cells in that they are **haploid** (written as n), meaning that they have only one copy of each chromosome. As you read earlier in the chapter, gametes are produced by **meiosis**, a type of cell division that is actually two divisions that result in four haploid daughter cells. When a sperm fertilizes an egg, a new cell, the zygote, is created. Because the egg and sperm both contribute a set of chromosomes to the zygote, it is diploid. After trillions of mitotic cell divisions, the zygote eventually develops into a new individual.

Meiosis keeps the chromosome number constant through generations and increases genetic variability in the population

Meiosis serves two important functions:

- Meiosis keeps the chromosome number constant through generations.
- Meiosis increases genetic variability in the population.

Meiosis keeps the chromosome number constant through generations because it creates haploid gametes with only one copy of each chromsome. If gametes were produced by mitosis, they would be diploid; each sperm and egg would contain 46 chromosomes instead of 23. Then, when a sperm containing 46 chromosomes fertilized an egg with 46 chromosomes, the zygote would have 92 chromosomes. The zygote of the next generation would have 184

TABLE **19.1**

PHASES OF THE CELL CYCLE	
PHASE	**MAJOR EVENTS**
Interphase	Production of proteins and other macromolecules and organelles; duplication of centriole and DNA
Cell division	
Mitosis	
Prophase	Chromosomes condense; nucleoli disappear; nuclear envelope breaks down; centrioles divide and move to opposite ends of the cell; mitotic spindles form and attach to chromosomes
Metaphase	Chromosomes line up at the equatorial plate
Anaphase	Centromere splits; chromatids begin to separate, each becoming a chromosome
Telophase	Chromosomes reach opposite poles; nuclear envelopes form; mitotic spindle disassembles; nucleoli reappear; chromosomes uncoil
Cytokinesis	
	Cytoplasm divides, forming two daughter cells

chromosomes, having been formed by an egg and sperm each containing 92 chromosomes. The next generation would have 368 chromosomes in each cell, and the next one 736—and so on. You can see how quickly the chromosome number would become unwieldy and, more important, alter the amount of genetic information in each cell. As we will see later in this chapter, even one extra copy of a single chromosome usually causes an embryo to die.

Meiosis also increases genetic variability in the population. Later in this chapter we will consider the mechanisms by which it accomplishes this increase. Genetic variability is important because it provides the raw material through which natural selection can act, leading to changes in populations that are called evolution. The relationship between genetic variability and evolution is discussed in Chapter 22.

Meiosis involves two cell divisions

First, let's consider how meiosis keeps the chromosome number constant. Meiosis and mitosis begin the same way. Both are preceded by the same event—the replication of chromosomes. Unlike mitosis, however, meiosis involves *two* divisions. In the first division, the chromosome number is reduced, because the members of each pair of chromosomes are separated into two cells, each bearing one complete set of chromosomes. In the second division, the replicated copies of the single set of chromosomes are separated. We see, then, that meiosis begins with one diploid cell and, two divisions later, produces four haploid cells. The orderly movements of chromosomes during meiosis ensure that each haploid gamete produced contains one copy of each chromosome. The stages in meiosis are summarized in Figure 19.10. Although not shown in the summary figure, each of the

two meiotic divisions has four stages similar to those in mitosis: prophase, metaphase, anaphase, and telophase.

Changes in the shape and functioning of the four haploid cells result in functional gametes. In males, one diploid cell results in four functional sperm. In contrast, in females, meiotic divisions of one diploid cell result in only one functional egg and up to three nonfunctional polar bodies. As a result, the egg contains most of the nutrients found in the original diploid cell, and these nutrients will nourish the early embryo (Figure 19.11).

MEIOSIS I

The first meiotic division—meiosis I—is called **reduction division** because it produces two cells, each with 23 chromosomes, as shown in Figure 19.12. However, the daughter cells do not contain a random assortment of any 23 chromosomes. Instead, each daughter cell contains *one complete set of chromosomes* (one member of each homologous pair), with each chromosome consisting of two chromatids.

During reduction division, it is important that each daughter cell receive a complete set of chromosomes. It would not do if one of the daughter cells had two copies of chromosome 3 and no copy of chromosome 6. Although there would still be 23 chromosomes present, part of the instructions for the structure and function of the body (chromosome 6) would be missing.

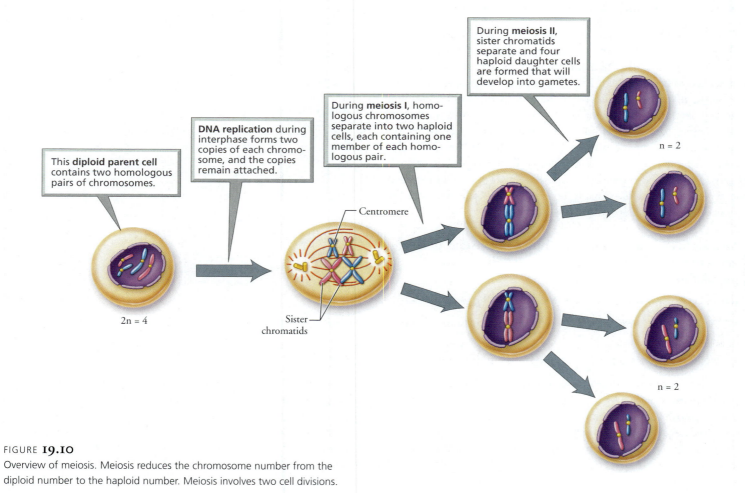

FIGURE **19.10**

Overview of meiosis. Meiosis reduces the chromosome number from the diploid number to the haploid number. Meiosis involves two cell divisions.

The separation of chromosomes into complete sets occurs reliably during meiosis I because during prophase I members of homologous pairs line up next to one another, a phenomenon called **synapsis** ("bringing together"). For example, chromosome 1 that was originally from your father would line up with chromosome 1 from your mother. Paternal chromosome 2 would pair with maternal chromosome 2, and so on. During metaphase I, matched homologous pairs become positioned at the midline of the cell, the so-called equatorial plate, and attach to spindle fibers. The pairing of homologous chromosomes helps ensure that the daughter cells will receive one copy of each homologous pair. Consider the following analogy. Pairing your socks before putting them in a drawer ensures that you will be more likely to put on a matching pair than if you randomly pulled out two socks.

Next, during anaphase I, the members of each homologous pair of chromosomes separate, and each homologue moves to opposite ends of the cell. During telophase I, cytokinesis begins, resulting in two daughter cells, each with one member of each chromosome pair. Each chromosome still consists of two chromatids. Telophase I is followed by interkinesis, a brief interphase-like period. *Interkinesis* differs from mitotic interphase in that there is no replication of DNA during interkinesis.

MEIOSIS II

During the second meiotic division—meiosis II—each chromosome lines up in the center of the cell independently (as occurs in mitosis), and the sister chromatids (attached replicates) making up each chromosome separate. Separation of the sister chromatids occurs in both daughter cells produced in meiosis I,

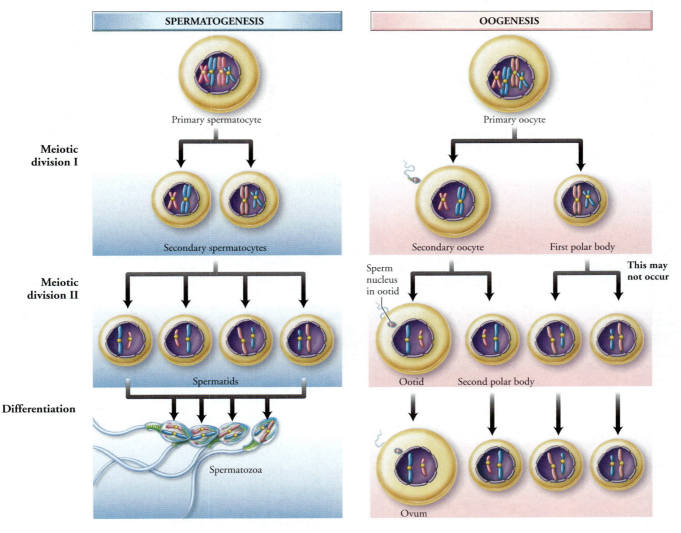

FIGURE **19.11**

Comparison of spermatogenesis and oogenesis. Meiosis results in haploid cells that differentiate into mature gametes. Spermatogenesis produces four sperm cells that are specialized to transport the male's genetic information to the egg. Oogenesis produces three polar bodies and one ovum that is packed with nutrients to nourish the early embryo.

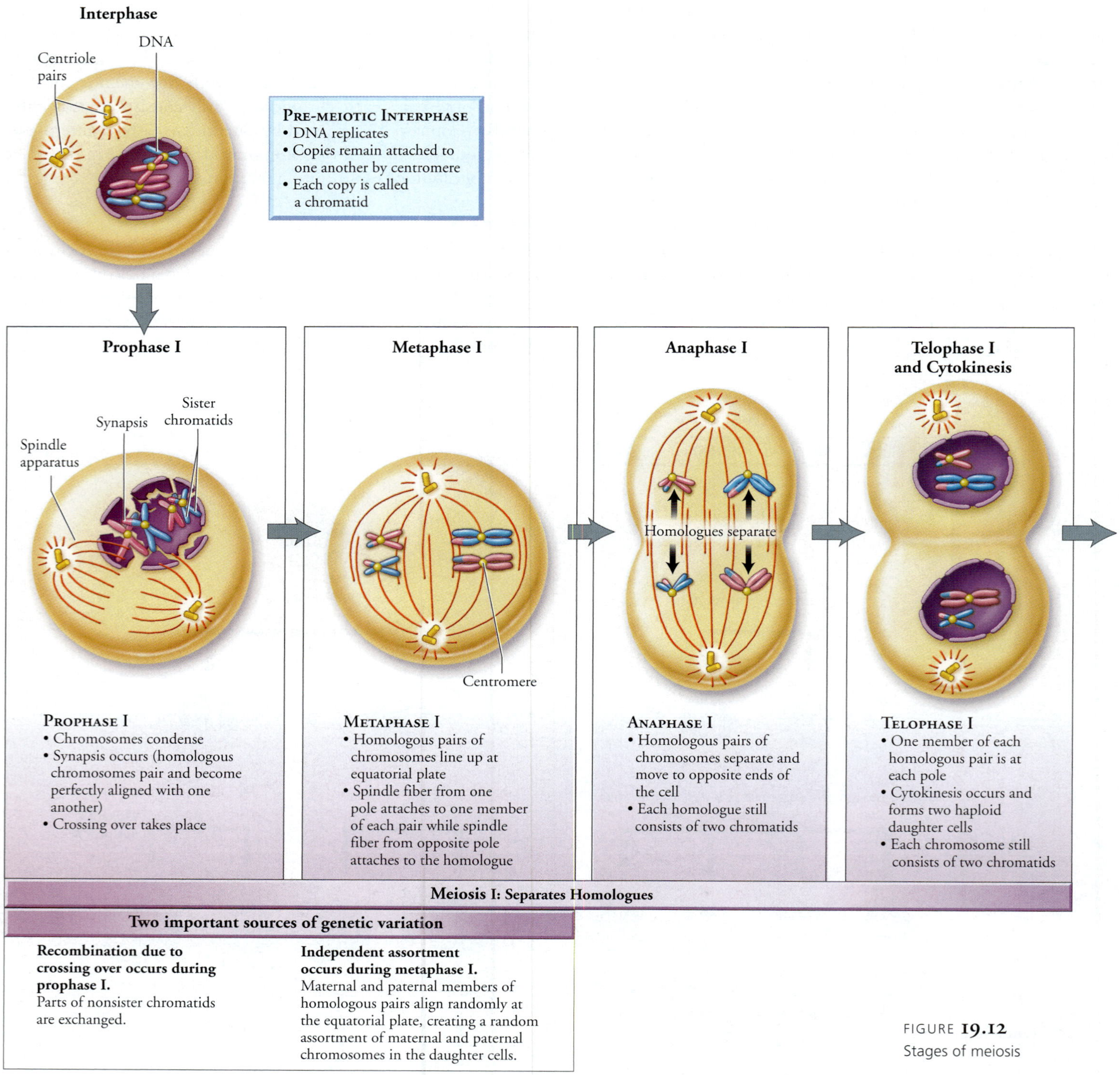

Interphase

Centriole pairs

DNA

PRE-MEIOTIC INTERPHASE
• DNA replicates
• Copies remain attached to one another by centromere
• Each copy is called a chromatid

Prophase I	**Metaphase I**	**Anaphase I**	**Telophase I and Cytokinesis**
Spindle apparatus Synapsis Sister chromatids		Homologues separate	
	Centromere		
PROPHASE I • Chromosomes condense • Synapsis occurs (homologous chromosomes pair and become perfectly aligned with one another) • Crossing over takes place	**METAPHASE I** • Homologous pairs of chromosomes line up at equatorial plate • Spindle fiber from one pole attaches to one member of each pair while spindle fiber from opposite pole attaches to the homologue	**ANAPHASE I** • Homologous pairs of chromosomes separate and move to opposite ends of the cell • Each homologue still consists of two chromatids	**TELOPHASE I** • One member of each homologous pair is at each pole • Cytokinesis occurs and forms two haploid daughter cells • Each chromosome still consists of two chromatids

Meiosis I: Separates Homologues

Two important sources of genetic variation

Recombination due to crossing over occurs during prophase I. Parts of nonsister chromatids are exchanged.	**Independent assortment occurs during metaphase I.** Maternal and paternal members of homologous pairs align randomly at the equatorial plate, creating a random assortment of maternal and paternal chromosomes in the daughter cells.

FIGURE **19.12**
Stages of meiosis

resulting in four cells, each containing one copy of each chromosome. The events of meiosis II are similar to those of mitosis, except that there are only 23 chromosomes lining up independently in meiosis II compared with the 46 chromosomes aligning independently in mitosis. Table 19.2 and Figure 19.13 compare mitosis and meiosis.

stop and think

If you were examining dividing cells under a microscope, how could you determine whether a particular cell was in metaphase of mitosis or metaphase I of meiosis?

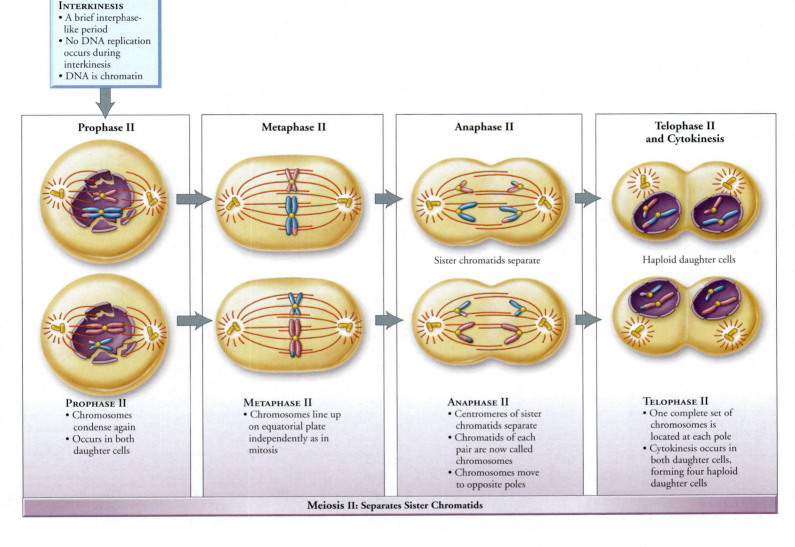

INTERKINESIS
- A brief interphase-like period
- No DNA replication occurs during interkinesis
- DNA is chromatin

Prophase II

PROPHASE II
- Chromosomes condense again
- Occurs in both daughter cells

Metaphase II

METAPHASE II
- Chromosomes line up on equatorial plate independently as in mitosis

Anaphase II

Sister chromatids separate

ANAPHASE II
- Centromeres of sister chromatids separate
- Chromatids of each pair are now called chromosomes
- Chromosomes move to opposite poles

Telophase II and Cytokinesis

Haploid daughter cells

TELOPHASE II
- One complete set of chromosomes is located at each pole
- Cytokinesis occurs in both daughter cells, forming four haploid daughter cells

Meiosis II: Separates Sister Chromatids

Crossing over and independent assortment cause genetic recombination during meiosis

At the moment of fertilization, when the nuclei of an egg and a sperm fuse, a new, *unique* individual is formed. Although certain family characteristics may be passed along, each child bears its own assortment of genetic characteristics (Figure 19.14). Only identical twins are exactly alike genetically, because they are formed from a single zygote.

Genetic diversity arises largely because of the shuffling of maternal and paternal forms of genes during meiosis. One way this mixing occurs is through a process called **crossing over**, in which corresponding pieces of chromatids of maternal and paternal homologues (nonsister chromatids) are exchanged during synapsis when the homologues are aligned side by side

TABLE **19.2**

MITOSIS AND MEIOSIS COMPARED	
MITOSIS	**MEIOSIS**
Involves one cell division	Involves two cell divisions
Produces two diploid cells	Produces four haploid cells
Occurs in most somatic cells	Occurs only in ovary and testes during the formation of gametes (egg and sperm)
Results in growth and repair	Results in gamete (egg and sperm) production
No exchange of genetic material	Parts of chromosomes are exchanged in crossing over
Daughter cells are genetically similar	Daughter cells are genetically dissimilar

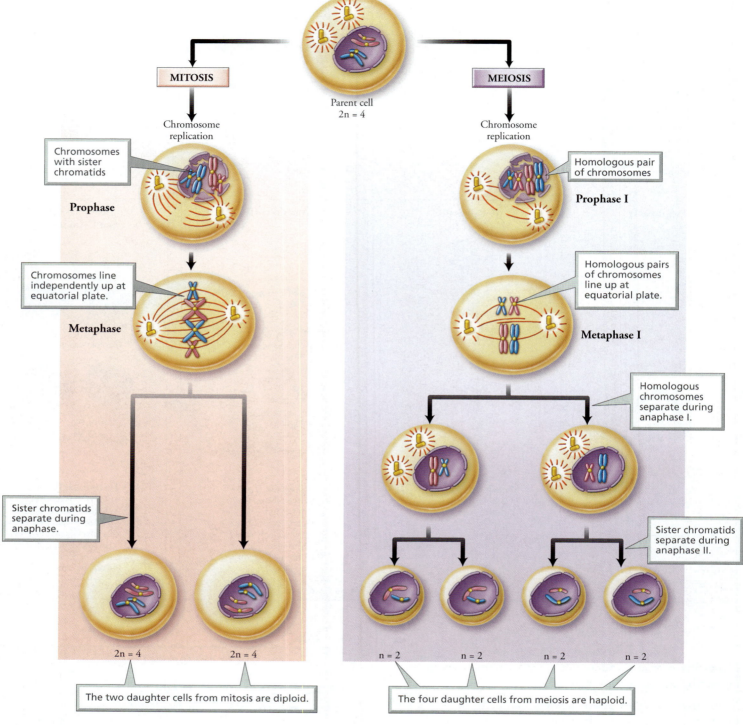

FIGURE **19.13**
A comparison of meiosis and mitosis

(Figure 19.15). After crossing over, the resulting chromatids have a mixture of DNA from the two parents. Because the homologues align gene by gene during synapsis, the exchanged segments contain genetic information for the same traits. However, because the genes of the mother and those of the father may direct different expressions of the trait, attached or unat-

tached earlobes for instance, the chromatids that result from crossing over have a new, novel combination of genes. Thus, crossing over increases the genetic variability of gametes.

Independent assortment is a second way that meiosis provides for the shuffling of genes between generations (Figure 19.16). Recall that the homologous pairs of chromo-

FIGURE **19.14**
Each child inherits a unique combination of maternal and paternal genetic characteristics because of the shuffling of chromosomes that occurs during meiosis. This photograph shows Mary and Eric Goodenough with their four sons: Derick, Stephen, David, and John.

Failure of chromosomes to separate during meiosis creates cells with extra or missing chromosomes

Most of the time, meiosis is a precise process that results in the chromosomes being distributed evenly to gametes. Nonetheless, meiosis is not foolproof, and occasionally a mistake is made. A pair of chromosomes or sister chromatids may adhere so tightly to one another that they do not separate during anaphase. As a result, both go to the same daughter cell, and the other daughter cell has none of this type of chromosome (Figure 19.17). The failure of homologous chromosomes to separate during meiosis I or sister chromatids to separate during meiosis II is called **nondisjunction**.

What happens if nondisjunction creates a gamete with an extra or a missing chromosome and that gamete is then united with a normal gamete during fertilization? The excess or deficit of chromosomes also occurs in the resulting zygote. For instance, if the abnormal gamete has an extra chromosome, the resulting zygote will have three copies of one type of chromosome and two copies of the rest. This condition, in which there are three representatives of one chromosome, is called **trisomy**. If, on the other hand, a gamete that is missing a representative of one type of chromosome joins with a normal gamete during fertilization, the resulting zygote will have only one copy, rather than the normal two copies, of one type of chromosome. The condition in which there is only one representative of a particular chromosome in a cell is called **monosomy**. **Aneuploidy** is a general term that describes gametes or cells with too many or too few chromosomes compared with the normal number. The imbalance of chromosomes usually causes abnormalities in development. Most of the time, the resulting malformations are severe enough to cause the death of the fetus, which will result in a miscarriage. Indeed, in about 70% of miscarriages, the fetus has an abnormal number of chromosomes.

When a fetus inherits an abnormal number of certain chromosomes, for instance chromosome 21 or the sex chromosomes, the resulting condition is usually not fatal. The upset in chromosome balance does, however, cause a specific syndrome. (A syndrome is a group of symptoms that generally occurs together.)

somes line up at the equatorial plate during metaphase I. However, the orientation of the members of the pair relative to the poles of the cell is random. Thus, like the odds that a flipped coin will come up heads, there is a fifty-fifty chance that a given daughter cell will receive the maternal chromosome. Each of the 23 pairs of chromosomes orients independently during metaphase I. The alignments of all 23 pairs will determine the assortments of maternal and paternal chromosomes in the daughter cells. The general formula for determining the possible combinations of maternal and paternal chromosomes in gametes due to independent assortment is 2^n, where n is the haploid number. In humans, the haploid number is 23, so each person can produce 2^{23}, approximately 8 million, different gametes solely due to independent assortment.

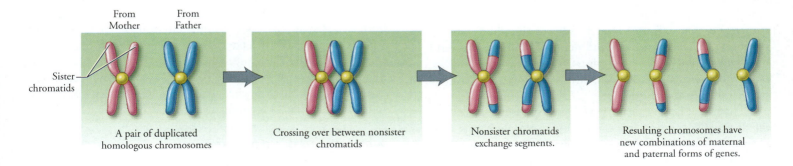

FIGURE **19.15**
Crossing over. During synapsis, when the homologous chromosomes of the mother and the father are closely aligned, corresponding segments of nonsister chromatids are exchanged. Each of the resulting chromatids has a mixture of maternal and paternal genetic information.

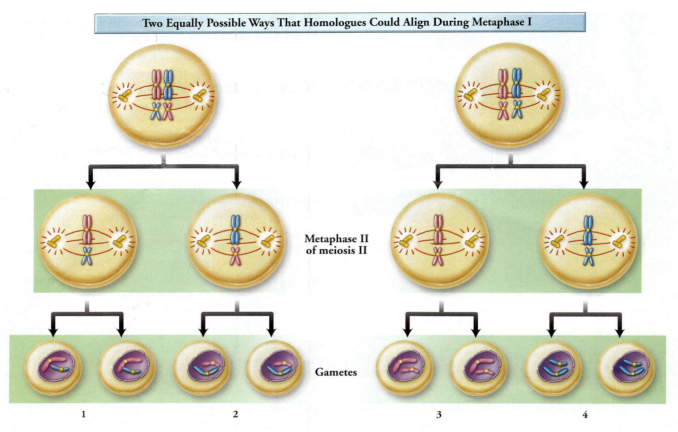

| Two Equally Possible Ways That Homologues Could Align During Metaphase I |

Metaphase II
of meiosis II

Gametes

1   2   3   4

FIGURE **19.16**

Independent assortment. The alignment of homologous maternal and paternal chromosomes relative to the poles of the cell is random. The members of each homologous pair orient independently of the other pairs. Notice that with only two homologous pairs, there are four possible combinations of chromosomes in the resulting gametes.

DOWN SYNDROME

One in every 800 to 1000 infants born will have three copies of chromosome 21 (trisomy 21), a condition known as **Down syndrome**. The symptoms of Down syndrome include moderate to severe mental retardation, short stature or shortened body parts due to poor skeletal growth, and characteristic facial features (Figure 19.18). Individuals with Down syndrome typically have a flattened nose, a forward-protruding tongue that forces the mouth open, upward-slanting eyes, and a fold of skin at the inner corner of each eye. Approximately 50% of all infants with Down syndrome have heart defects, and many of them die as a result of this defect. Blockage in the digestive system, especially in the esophagus or small intestine, is also common and may require surgery shortly after birth.

The risk of having a baby with Down syndrome increases with the age of the mother. Indeed, a 30-year-old woman is twice as likely to give birth to a child with Down syndrome as is a 20-year-old woman. After age 30, the risk rises dramatically, so at age 45 a mother is 45 times as likely to give birth to a Down syndrome infant as is a 20-year-old woman.

The increased incidence of Down syndrome babies among older mothers has several possible explanations. Potential eggs form and meiosis begins before a woman is even born. Although homologous chromosomes in potential eggs pair in prophase I before a woman is born, they do not separate until shortly before that egg is ovulated. The probability that the chromosomes will remain stuck together during anaphase may increase with the length of time they have been paired. Exposure to environmental factors, including ionizing radiation and viruses, that are known to increase the rate of nondisjunction may also play a role. The older a woman is, the longer her potential eggs have been exposed to these environmental factors. However, some studies suggest that the chances that a zygote will form with trisomy 21 may be the same regardless of the mother's age. The incidence of Down syndrome babies may differ with maternal age because an older mother is more likely to carry an affected fetus to term than is a younger mother.

what would you do?

It is possible to determine whether a 2-month-old fetus has Down syndrome. If you were a 30-year-old woman who became pregnant (or the father of the child), would you want to know whether the child would have Down syndrome? What criteria would you use to decide?

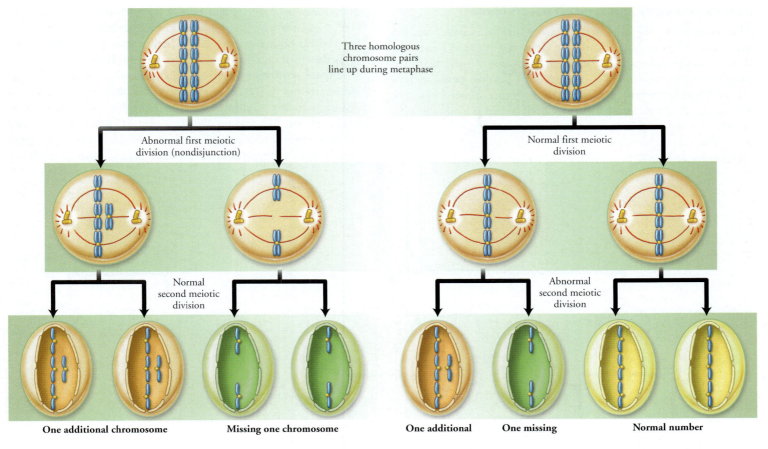

Three homologous chromosome pairs line up during metaphase

Abnormal first meiotic division (nondisjunction)

Normal first meiotic division

Normal second meiotic division

Abnormal second meiotic division

One additional chromosome **Missing one chromosome** **One additional** **One missing** **Normal number**

FIGURE **19.17**

Nondisjunction is a mistake that occurs during cell division in which homologous chromosomes or sister chromatids fail to separate during anaphase. One of the resulting daughter cells will have an extra copy of one chromosome, and the other will be missing that type of chromosome.

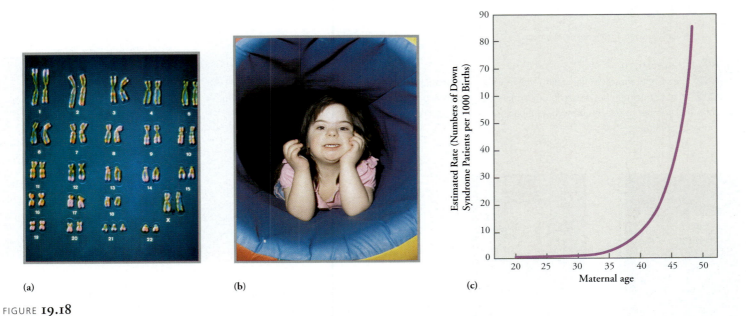

(a)   (b)   (c)

FIGURE **19.18**

Three copies of chromosome 21 (trisomy 21), as seen here (a), cause Down syndrome. (b) A person with Down syndrome is moderately to severely mentally retarded and has a characteristic appearance, including a flattened nose, a tongue that protrudes forward, upward-slanting eyes, and a fold of skin at the inner corner of each eye. (c) The incidence of Down syndrome increases with the age of the mother, especially after age 35.

TURNER SYNDROME AND KLINEFELTER SYNDROME

Like autosomes, sex chromosomes may fail to separate during anaphase. This error can occur during either egg or sperm formation. A male is chromosomally XY, so when the X and Y separate during anaphase, equal numbers of X-bearing and Y-bearing sperm are produced. However, if nondisjunction occurs during sperm formation, half of the resulting sperm will carry both X and Y chromosomes, whereas the other resulting sperm will not contain any sex chromosome. A female is chromosomally XX, so each of the eggs she produces should contain a single X chromosome. When nondisjunction occurs,

however, an egg may contain two X chromosomes or none at all. When a gamete with an abnormal number of sex chromosomes is joined with a normal gamete during fertilization, the resulting zygote has an abnormal number of sex chromosomes (Figure 19.19).

Turner syndrome occurs in individuals who have only a single X chromosome (XO). Approximately 1 in 2000 female infants is born with Turner syndrome, but this represents only a small percentage of the XO zygotes that are formed. Most of these XO zygotes are lost as miscarriages. Such an individual has the external appearance of a female (Figure 19.20a). Before puberty, she looks

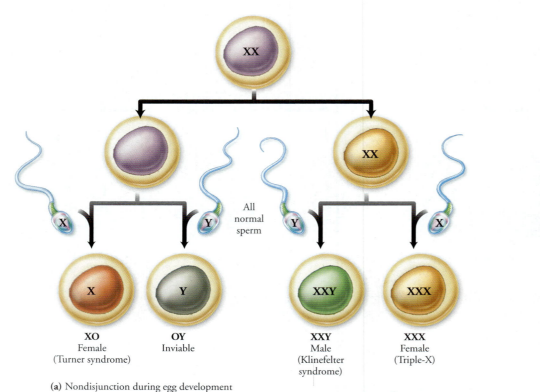

(a) Nondisjunction during egg development

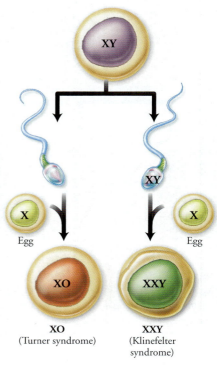

(b) Nondisjunction during sperm development

FIGURE **19.19**

The sex chromosomes may fail to separate during formation of either eggs or sperm. When a gamete with an abnormal number of sex chromosomes joins a normal gamete in fertilization, the resulting zygote has an abnormal number of sex chromosomes. Imbalances among sex chromosomes upset normal development of reproductive structures.

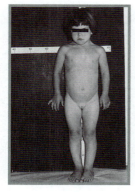

(a)   (b)

FIGURE **19.20**

Missing or extra sex chromosomes cause Turner syndrome and Klinefelter syndrome. (a) A female with Turner syndrome has one rather than the normal two X chromosomes. The signs of Turner syndrome are a skin fold at the neck, short stature, a wide chest, and underdeveloped breasts. Because the ovaries are underdeveloped, she is usually infertile. (b) A male with Klinefelter syndrome is XXY. Klinefelter syndrome is characterized by small testes and sometimes breast enlargement.

like a normal, XX female. The only hint of Turner syndrome may be a thick fold of skin on the neck. As she ages, however, she generally is noticeably shorter than her peers. Her chest is wide, and her breasts underdeveloped. In 90% of the women with Turner syndrome, the ovaries are also poorly developed, leading to infertility.

Injections of growth hormone allow girls with Turner syndrome to reach a normal height. Orally administered sex hormones at an appropriate age will promote breast development. Although the ovaries of a woman with Turner syndrome cannot produce eggs, pregnancy may be possible through in vitro fertilization (see Chapter 18), in which a fertilized egg from a donor is implanted in her uterus.

Klinefelter syndrome is observed in males who are XXY. Although the extra X chromosome can be inherited as a result of nondisjunction during either egg or sperm formation, it is twice as likely to come from the egg. As with Down syndrome, increased maternal age may increase the risk slightly.

Klinefelter syndrome is fairly common. Approximately 1 in 500 to 1 in 1000 of all newborn males is XXY. However, not all XXY males display the symptoms of having an extra X chromosome. In fact, many of them live their lives without ever suspecting that they are XXY. Signs that a male has Klinefelter syndrome do not usually show up until puberty. During the teenage years, the testes of an XY male gradually increase in size. In contrast, the testes of an XXY male remain small and do not produce an adequate amount of the male sex hormone, testosterone. As a result of the testosterone insufficiency, he may grow 5 to 11.4 cm (2 to 4.5 in.) taller than average. Secondary sex characteristics, such as facial and body hair, fail to develop fully. His arms and legs may be exceptionally long, and his breasts may develop slightly (Figure 19.20b). His penis is usually of normal size, but his testes do not produce sperm, so men with Klinefelter syndrome are usually sterile.

Testosterone injections, especially if they begin during adolescence, bring about changes that make a male with Klinefelter syndrome more similar to his XY peers—more developed muscles, deeper voice, more facial and body hair, and testicular development. These treatments may also ease the emotional and psychological problems experienced by many males with Klinefelter syndrome. Sterility, however, remains irreversible.

Nondisjunction can also result in a female with three X chromosomes (XXX, triple X syndrome) or a male with two Y chromosomes (XYY, Jacob syndrome). Women with triple X syndrome (XXX) may have menstrual disturbances and reach menopause earlier in life than do women with two X chromosomes. Males with two Y chromosomes (XYY) are often taller than normal, and some have a slightly lower intelligence than do XY males.

REVIEWING THE CONCEPTS

All Humans Have a Life Cycle (p. 439)

1. The human life cycle requires two types of cell division—mitosis and meiosis. Mitosis creates cells that are exact copies of the original cell. Mitosis is used for growth and repair. Meiosis creates cells with half the number of chromosomes as were in the original cell. Gamete production requires meiosis.

WEB TUTORIAL 19.1 Cell Division in Humans

Chromosomes Consist of DNA and Protein (p. 440)

2. A chromosome contains DNA and proteins called histones. A gene is a segment of DNA that codes for a protein that plays a structural or functional role in the cell. Genes are arranged along a chromosome in a specific order. Each of the 23 different kinds of chromosomes in human cells contains a specific sequence of genes.

3. Somatic cells (all cells except for eggs and sperm) are diploid; that is, they contain two sets of chromosomes, one from each parent. Homologous chromosomes carry genes for the same traits. In humans, the diploid number of chromosomes is 46, two sets of 23 homologous pairs. One pair of chromosomes, the sex chromosomes, determines gender. Males are XY, and females are XX. The other 22 pairs of chromosomes are called autosomes. Eggs and sperm are haploid and contain only one set of chromosomes.

Our Cells Divide in a Characteristic Cyclic Pattern (pp. 440–446)

4. The cell cycle consists of two major phases: interphase and cell division. Interphase is the period between cell divisions.

5. During interphase, the cell carries out the activities it is specialized to perform. DNA and organelles are replicated in preparation for the cell to divide and produce two identical daughter cells. Somatic cell division consists of mitosis (division of the nucleus) and cytokinesis (division of the cytoplasm).

6. In mitosis, the original cell first replicates its genetic material and then distributes it equally between its two daughter cells. There are four stages of mitosis: prophase, metaphase, anaphase, and telophase.

7. Cytokinesis, division of the cytoplasm, begins sometime during late anaphase or early telophase. A band of microfilaments at the midline of the cell contracts and forms a furrow. The furrow deepens and eventually pinches the cell in two.

WEB TUTORIAL 19.2 Cell Cycle and Mitosis

Meiosis Forms Haploid Gametes (pp. 447–457)

8. Meiosis, a special type of cell division that occurs in the ovaries or testes, begins with a diploid cell and produces four haploid cells that will become gametes (egg or sperm).

9. Meiosis is important because it halves the number of chromosomes in gametes, thereby keeping the chromosome number constant between generations. When a sperm fertilizes an egg, a diploid cell, called a zygote, is created. After many mitotic divisions, the zygote will develop into a new individual.

10. Before meiosis begins, the chromosomes are replicated, and the copies remain attached to one another by centromeres. The attached replicated copies are called sister chromatids. There are two cell divisions in meiosis. During the first meiotic division (meiosis I), members of homologous pairs are separated. Thus, the daughter cells contain only one member of each homologous pair (although each chromosome still consists of two chromatids). During the second meiotic division (meiosis II), the sister chromatids are separated.

11. Genetic recombination during meiosis results in variation among offspring from the same two parents. One cause of genetic recombination is crossing over, in which corresponding segments of DNA are exchanged between maternal and paternal homologues, creating new combinations of alleles in the resulting chromatids. Crossing over occurs during synapsis.

12. A second cause of genetic recombination is the independent assortment of maternal and paternal homologues into daughter cells during meiosis I. The orientation of the members of the pair relative to the poles of the cell determines whether a daughter cell will receive the maternal or the paternal chromosome. Each pair aligns independently of the others.

13. Nondisjunction is the failure of homologous chromosomes or sister chromatids to separate during cell division. It results in an abnormal number of chromosomes in the resulting cells. Nondisjunction of chromosome 21 can result in Down syndrome. Nondisjunction of the sex chromosomes can cause Turner syndrome (XO) or Klinefelter syndrome (XXY).

WEB TUTORIAL 19.3 Meiosis

WEB TUTORIAL 19.4 Determining Sex

KEY TERMS

allele *p. 440*
anaphase *p. 445*
aneuploidy *p. 453*
autosome *p. 440*
cell cycle *p. 440*
centromere *p. 441*
chromatid *p. 441*
chromosome *p. 440*
crossing over *p. 451*

cytokinesis *p. 441*
diploid *p. 440*
Down syndrome *p. 454*
equatorial plate *p. 445*
gene *p. 440*
haploid *p. 447*
homologous pair *p. 440*
independent assortment *p. 452*
interphase *p. 440*

karyotype *p. 446*
Klinefelter syndrome *p. 457*
locus *p. 440*
meiosis *p. 447*
metaphase *p. 445*
mitosis *p. 440*
monosomy *p. 453*
nondisjunction *p. 453*
prophase *p. 444*

reduction division *p. 448*
sex chromosome *p. 440*
somatic cell *p. 440*
synapsis *p. 449*
telophase *p. 445*
trisomy *p. 453*
Turner syndrome *p. 456*
zygote *p. 439*

THINKING ABOUT THE CONCEPTS

1. Explain the relationship between genes and a chromosome. *p. 440*
2. Define mitosis and cytokinesis. *p. 440-441*
3. Why is meiosis important? *pp. 447–448*
4. Describe the alignment of chromosomes at the equatorial plate during meiosis I and meiosis II. Explain the importance of these alignments in creating haploid gametes from diploid cells. *pp. 448–451*
5. Explain how crossing over and independent assortment result in genetic recombination that causes variability among offspring from the same two parents. *pp. 451–453*
6. Define nondisjunction. Explain how nondisjunction can result in abnormal numbers of chromosomes in a person. *p. 453*
7. What causes Down syndrome? What are the usual characteristics of the condition? *pp. 454–455*
8. Differentiate between Turner syndrome and Klinefelter syndrome by describing the cause of the conditions and the characteristics of each. *pp. 456–457*
9. The process of mitosis results in
    a. two haploid cells.
    b. two diploid cells.
    c. four haploid cells.
    d. four diploid cells.

10. DNA is synthesized (replicated) during
    a. interphase.
    b. prophase.
    c. metaphase.
    d. anaphase.
11. Crossing over occurs during which stage of meiosis?
    a. prophase I
    b. metaphase I
    c. prophase II
    d. metaphase II
12. During meiosis, the processes of _____ and _____ increase genetic diversity.
13. _____ chromosomes carry genes for the same traits.
14. _____ is the pairing of chromosomes during meiosis.
15. The stage of *mitosis* during which sister chromatids separate is _____.
16. The stage of *meiosis* during which sister chromatids separate is _____.

APPLYING THE CONCEPTS

1. Cathy is a 35-year-old woman who just finished graduate school. She and her husband, Bob, would like to have a family. When she became pregnant, the doctor performed an amniocentesis. This allowed the doctor to create a karyotype of the fetus's genes and, in so doing, to discover that the fetus had three copies of chromosome 21. How could this condition occur? How will it affect the child?

2. A cell biologist is studying the cell cycle. She is growing the cells in culture, and they are actively dividing mitotically. One particular cell has half the DNA as most of the other cells. Which stage of mitosis is this cell in? How do you know?

20

The Principles of Inheritance

Principles of Inheritance Help Us Predict How Simple Traits Are Passed to the Next Generation

- During gamete formation alleles segregate and assort independently
- Mendelian genetics considers how genes are transmitted from parents to offspring
- Pedigrees help us to determine genotype
- A dominant allele often produces a protein that the recessive allele does not
- Certain genes have multiple alleles in a population
- Most traits are controlled by many genes
- Genes on the same chromosome are usually inherited together
- Sex-linked genes are located on the sex chromosomes
- Sex-influenced genes are autosomal genes whose expression is influenced by hormones

Breaks in Chromosomes Change Chromosomal Structure and Function

Certain Genetic Disorders Can Be Detected before Birth by Amniocentesis or Chorionic Villi Sampling

 **SOCIAL ISSUE**   Gene Testing

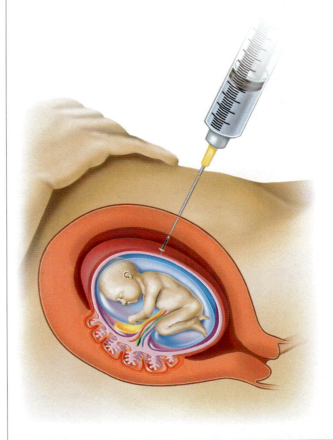

Steve's brother, Dan, is a risk-taker. Two years ago, he hiked through the Andes in Chile with only a backpack and a knife. Four years ago, he learned to scuba dive while on an Adventure Cruise in Alaska. Recently, he took up sky diving.

Dan hasn't always been like this. Six years ago, he was finishing law school and looking forward to a career in politics. Then their father was diagnosed with Huntington's disease, a disease caused by the degeneration of brain cells. This devastating disease usually appears in middle age with symptoms of clumsiness and forgetfulness. Now Dad is still able to work, but his handwriting has become uncontrollable and it is difficult for him to sit for long periods without twitching.

The bad news for Dan and Steve is that Huntington's disease is genetic. There is a 50% chance that they also carry the gene. Because Huntington's disease is caused by a dominant gene, if they do carry the gene, they will surely develop the disease in middle age. After Dad's diagnosis, Dan had decided to have his genes analyzed to see if he carried the gene that affected their father. Unfortunately, he did. When Dan finished school, he started to live the life of a risk-taker—what did he have to lose, he thought? Steve, on the other hand, was happily married to Jessica. He elected not to have his genes tested but to continue to live the nearly perfect life he and Jessica had always enjoyed. But now Jessica was pregnant. Jessica wanted to have the child, but not if it would be born with the gene for Huntington's disease. She insisted that the baby's genes be tested. Fortunately for Steve, Jessica, and the baby, the test was negative for the Huntington's disease gene.

Most of us do not need to worry about whether we will develop Huntington's disease. However, as you flip through the pages of the family album, you may notice distinctive traits scattered through the generations. Physical characteristics may be the most obvious way in which heredity has shaped you, but they are a very small part of your genetic legacy. The genes you received at the moment of conception influence all the biochemical reactions taking place inside your cells, your susceptibility to disease, certain behavior patterns, and even your life span. Although environment influences the expression of genes, your genes provide the basic outline for your possibilities and limitations.

In this chapter, we will consider the genetic foundation that has been so important in shaping who you are. We will learn how traits are passed through generations and how to predict the appearance of traits in future children. ■

Principles of Inheritance Help Us Predict How Simple Traits Are Passed to the Next Generation

An understanding of meiosis helps us answer important questions about inheritance. Why is your brother the only sibling with Mom's widow's peak (a hairline that comes to a point on the forehead) and freckles? How can you have blue eyes when both your parents are brown eyed? Will you be bald at 40, like Dad? Let's take a second look, then, to see how chromosomes, meiosis, and heredity are related.

Before beginning, you may want to review some of the terms that were introduced in Chapter 19 and are summarized in Figure 20.1. As you may recall, somatic (body) cells have two copies of every chromosome. Thus, human cells have 23 pairs of chromosomes. One member of each pair was inherited from the female parent, and the other from the male parent. The chromosomes that carry genes for the same traits are a **homologous pair of chromosomes**, or homologues. Chromosomes are made of DNA and protein. Certain segments of the DNA of each chromosome function as genes. A **gene** directs the synthesis of a specific polypeptide (protein) that can play either a structural or a functional role in the cell.[1] In this way, the gene-determined protein can influence whether a certain **trait**, or characteristic, will develop. For instance, the formation of your brother's widow's peak was directed by a protein coded for by a gene that he inherited from Mom.

There are different forms of genes, called **alleles**. Alleles produce different versions of the trait they determine. One of the sev-

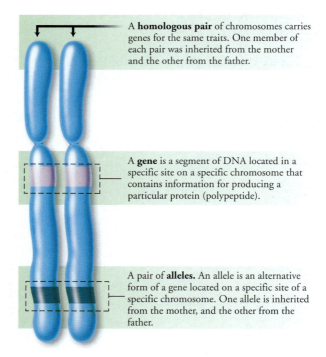

A **homologous pair** of chromosomes carries genes for the same traits. One member of each pair was inherited from the mother and the other from the father.

A **gene** is a segment of DNA located in a specific site on a specific chromosome that contains information for producing a particular protein (polypeptide).

A pair of **alleles.** An allele is an alternative form of a gene located on a specific site of a specific chromosome. One allele is inherited from the mother, and the other from the father.

FIGURE **20.1**
Important terms in genetics

[1]This is a simple definition of a gene. In some genes, the polypeptide produced is only part of a functional protein. Furthermore, as we will see in Chapter 21, some genes have regulatory regions of DNA within their boundaries. Also, some genes code for RNA molecules that are needed for the production of the polypeptide but are not part of it.

eral genes that determines eye color, for instance, dictates whether the brown pigment melanin will be deposited in the iris. When this allele is present, eye color will range from green to dark brown, depending on the amount of melanin present. The other allele for eye color does not lead to melanin deposit. Generally, there are two alleles for each trait in body cells, one allele on each homologous chromosome. If neither homologue bears an allele for melanin deposition, eye color will be blue.

Individuals with two copies of the same allele of a gene are said to be **homozygous** (*homo*, same; *zygo*, joined together) for that trait. Those with different alleles of a given gene are said to be **heterozygous** (*hetero*, different; *zygo*, joined together). When the effects of a certain allele can be detected, regardless of whether an alternative allele is also present, the allele is described as **dominant**. Dimples and freckles are human traits dictated by dominant alleles. The allele whose effects are masked in the heterozygous condition is described as **recessive**. Because of this masking, only homozygous recessive alleles are expressed. Conditions such as cystic fibrosis, in which excessive mucus production impairs lung and pancreatic function, and albinism, in which the pigment melanin is missing in the hair, skin, and eyes, result from recessive alleles. By convention, the dominant allele is designated with a capital letter, and the recessive allele with a lowercase letter—*A* and *a*, for example.

We observe the dominant form of the trait whether the individual is homozygous dominant (*AA*) or heterozygous (*Aa*) for that trait. Thus, we cannot always tell which alleles are present. However, it is often useful to distinguish between an individual's genetic makeup and an individual's appearance. **Genotype** refers to the precise alleles that are present: that is, whether the individual is homozygous or heterozygous for different gene pairs. **Phenotype**, on the other hand, refers to the observable physical traits of an individual. Figure 20.2 shows the genotype and phenotype of several inherited traits in humans. Table 20.1 summarizes terms commonly used in genetics as they apply to the inheritance of freckles, which is a dominant trait.

During gamete formation alleles segregate and assort independently

Recall from Chapter 19 that the members of each homologous pair of chromosomes segregate during meiosis I, each homologue going to a different daughter cell. Thus, an egg or a sperm has only one member of each homologous pair. Consider what this fact means for inheritance. Because the alleles for each gene segregate during gamete formation, half the gametes bear one allele, and half bear the other. This principle is known as the **law of segregation**.

Furthermore, each pair of homologous chromosomes lines up at the equatorial plate during meiosis I independently of the other pairs. The orientation of the paternal and maternal homologues relative to the poles of the cell is entirely random. What this fact means for inheritance is that each pair of alleles (on different chromosomes) segregates into gametes independently. This principle is the **law of independent assortment**.

Mendelian genetics considers how genes are transmitted from parents to offspring

During the nineteenth century, Gregor Mendel, a monk who grew up in a region of Austria that is now part of the Czech Republic,

Freckles: *FF* or *Ff*   No freckles: *ff*

Widow's peak: *WW* or *Ww*   Straight hairline: *ww*

Attached earlobes: *EE* or *Ee*   Unattached earlobes: *ee*

tt   *Tt*   *TT* or *Tt*

Tongue rolling

FIGURE **20.2**

Genotypes and phenotypes of selected inherited human traits

worked out much of what we know today about the laws of heredity by performing specific crosses of pea plants. Although Mendel knew nothing about chromosomes, his ideas about the inheritance

TABLE **20.1**

COMMON TERMS IN GENETICS		
GENOTYPE: **THE ALLELES THAT** **ARE PRESENT**	**DESCRIPTION**	**PHENOTYPE:** **THE OBSERVABLE TRAIT**
FF	Homozygous dominant: • Two dominant alleles present. • Dominant phenotype expressed.	Freckles
Ff	Heterozygous: • Different alleles present. • Dominant phenotype expressed.	Freckles
ff	Homozygous recessive • Two recessive alleles present. • Recessive phenotype expressed.	No freckles

of traits are consistent with what we now know about the movement of chromosomes during meiosis. Mendel's results are used today to predict the outcome of hereditary crosses.

Mendel began by studying **monohybrid crosses**, which are crosses that consider the inheritance of a single trait from individuals differing in the expression of that trait. Consider, for example, the inheritance of freckles, a characteristic determined by a dominant allele. Suppose a freckled female who is homozygous dominant (*FF*) mated with a homozygous recessive male with no freckles (*ff*) (Figure 20.3). We know that alleles segregate during meiosis, but, because both parents are homozygous, each can produce only one type of gamete (as far as freckles are concerned). The female produces gametes (eggs) with the dominant allele (*F*), and the male produces gametes (sperm) with the recessive allele (*f*).

A **Punnett square** is a useful tool for determining the probable outcome of genetic crosses. In a Punnett square, columns are set up and labeled to represent each of the possible gametes of one parent. Remember, the alleles segregate during meiosis. In this case, then, there would be two columns to represent the gametes of the male without freckles, each labeled with a recessive allele, *f*. In a similar manner, rows are established and labeled to represent all the possible gametes formed by the other parent, in this case, the woman with freckles. There would be two rows in this Punnett square, each labeled *F*. Each square is then filled in by combining

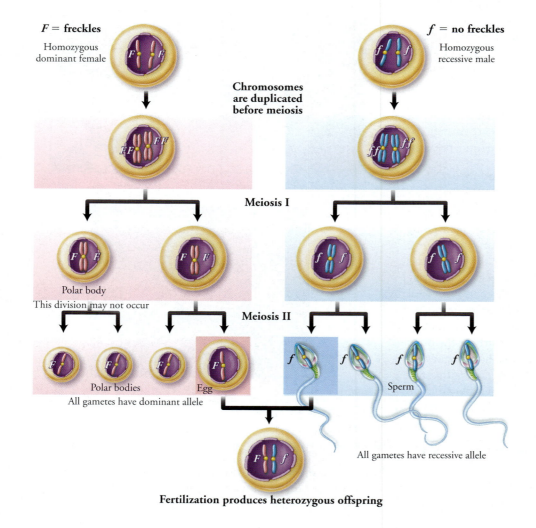

FIGURE **20.3**

Gamete formation by a female who is homozygous dominant for freckles (*FF*) and a male who is homozygous recessive for no freckles (*ff*) and the heterozygous (*Ff*) individual resulting from the union of these gametes.

the labels on the corresponding rows and columns. These squares represent the possible offspring of these matings. All the children would have freckles and would be heterozygous for the trait (*Ff*). Figure 20.4 shows the Punnett square for the first generation.

Now consider why it is possible for parents who are both heterozygous for the freckle trait (*Ff*) to have a child without freckles. The *F* and *f* alleles segregate during meiosis. So half the gametes bear the *F* allele, and half bear the *f* allele (Figure 20.5a). There would be two rows and two columns in the Punnett square, labeled with *F* and *f* (Figure 20.5b). Combining the labels on rows and columns, we see that the probable genotypes of the children would be homozygous dominant (*FF*), heterozygous (*Ff*), and homozygous recessive (*ff*, for a genotypic ratio of 1 *FF*: 2 *Ff*: 1 *ff*. The ratio of phenotypes would be 3 freckled children (*FF* and *Ff*):1 child without freckles (*ff*). This ratio translates to a 3 in 4, or 75%, chance that a child born to this couple will have freckles. But there is a 1 in 4, or 25%, chance that the child will not have freckles.

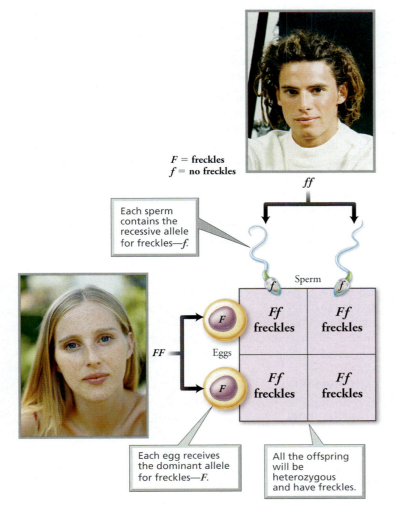

FIGURE **20.4**

This Punnett square illustrates the probable offspring from a cross between a homozygous dominant female with freckles (*FF*) and a homozygous recessive male without freckles (*ff*). Each column is labeled with the possible gametes the male could produce. Each row is labeled with possible gametes the female could produce. Combining the labels on the corresponding rows and columns yields the genotype of possible offspring.

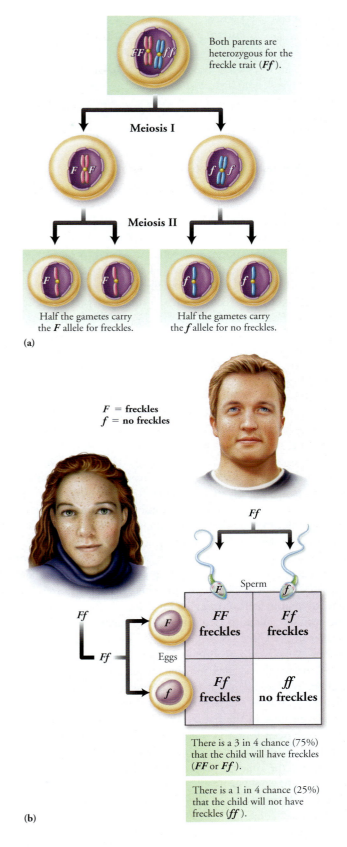

FIGURE **20.5**

(a) Gamete formation by a person who is heterozygous for the freckle trait (*Ff*). (b) A Punnett square showing the probable outcome of a mating between two people who are heterozygous for the freckle trait (*Ff*).

stop and think

At the beginning of the section on inheritance, a question was raised about why only one sibling (your hypothetical brother) inherited your mother's freckles. Implicit in the way the question was phrased was that there is at least one other sibling (you), who, like your father, does not have freckles. Explain why the mother in this example must be heterozygous for the freckle trait.

▮ Pedigrees help us to determine genotype

It is sometimes important to determine the genotype of a human with a dominant phenotype for a particular trait. We can often deduce the unknown genotype by looking at the expression of the trait in the ancestors or descendants of the person in question. A chart showing the genetic connections among individuals in a family is called a **pedigree**. Family or medical records are used to fill in the pattern of expression of the trait in question for as many family members as possible. Pedigrees are useful, not just in determining an unknown genotype but also in predicting the chances that one's offspring will display the trait.

Genetic disorders are often determined by recessive alleles, so knowledge of whether the parents carry the allele will help predict whether their child could be born with the condition. For example, cystic fibrosis (CF), a disorder in which abnormally thick mucus is produced, is controlled by a recessive allele. Approximately 1 in every 2000 infants in the United States is born with CF. It is a leading cause of childhood death. The average age of survival is 24 years. The three primary signs of CF are salty sweat, digestive problems, and respiratory problems. Many children with CF suffer from malnutrition because mucus clogs the pancreatic ducts, preventing pancreatic digestive enzymes from reaching the small intestine, where they function. Thick mucus also plugs the respiratory passageways, making breathing difficult and increasing susceptibility to lung infections such as pneumonia (Figure 20.6).

Because CF is inherited as a recessive allele, children with CF are usually born to normal, healthy parents who had no idea they carried the trait. A **carrier** is someone who displays the dominant phenotype but is heterozygous for a trait and can, therefore, pass the recessive allele to descendants. Approximately 1 in 22 Caucasians in the United States is a carrier for CF. (CF is less common among Asians and rare among African Americans.)

Consider the pedigree showing the inheritance of CF in eight generations of descendants of five ancestral couples (shown in blue in Figure 20.7). It tells us that all the parents in the seventh generation (shown in pink in the figure) were heterozygous carriers of the CF allele because at least one child from each marriage was homozygous for the trait and was, therefore, born with CF (shown in black in the figure). Of the children born to two carriers, we would predict that three-fourths would be unaffected and that one-fourth would have CF. In this pedigree, we see at least some children with CF in 8 families. Of the 33 children in these 8 families, 14 had CF, a greater number than would be expected. On the basis of the expected ratios from a monohybrid cross, we would predict that only 8 children, 25% of all the children of these carriers, would be affected. The actual number is biased toward affected offspring because it is so difficult to identify any heterozygous parents who, by chance, did not have an affected child.

FIGURE **20.6**

Cystic fibrosis, controlled by a recessive allele, is a condition in which abnormally thick mucus is produced, causing serious digestive and respiratory problems. This child with cystic fibrosis is using a therapeutic toy to enhance air flow. When she exhales with enough force, the ribbons wave.

▮ A dominant allele often produces a protein that the recessive allele does not

You may wonder what makes an allele dominant or recessive. In many cases, the dominant allele produces a normal, functional protein, but the recessive allele does not produce that protein or it produces the protein in an altered form that does not function properly. Consider the inheritance of **albinism**, the inability to produce the brown pigment melanin that normally gives color to the eyes, hair, and skin (Figure 20.8). Because of the lack of melanin, an albino has pale skin and white hair. As a child, an albino has pink eyes, but the eye color darkens to blue in an adult. Because there is no melanin in the skin to protect against sunlight's ultraviolet rays, an albino is very vulnerable to sunburn and skin cancer.

The ability to produce melanin depends on the enzyme tyrosinase. The dominant allele, which results in normal skin pigmentation, produces a functional form of tyrosinase. A single copy of the dominant allele can produce all of this enzyme that is needed. The recessive allele that causes albinism produces a nonfunctional form of tyrosinase, and melanin cannot be formed.

CODOMINANCE

The situations we have described so far involve **complete dominance**: In a heterozygote, the dominant allele produces a functional protein, and the protein's effects are apparent. But the recessive allele produces a less functional protein or none at all, and its effects are not apparent. Complete dominance is not always the case, however. In some situations, both alleles for certain traits produce functional proteins. In this case, which is described as **codominance**, the effects of *both* alleles are separately apparent in a heterozygote.

Consider the inheritance of type AB blood as an example of codominance. As we will see shortly, two alleles, I^A and I^B, specify the production of a different polysaccharide on the surface of red blood cells. In persons with type AB blood both alleles are expressed, and their red blood cells have both A and B polysaccharides on their surfaces.

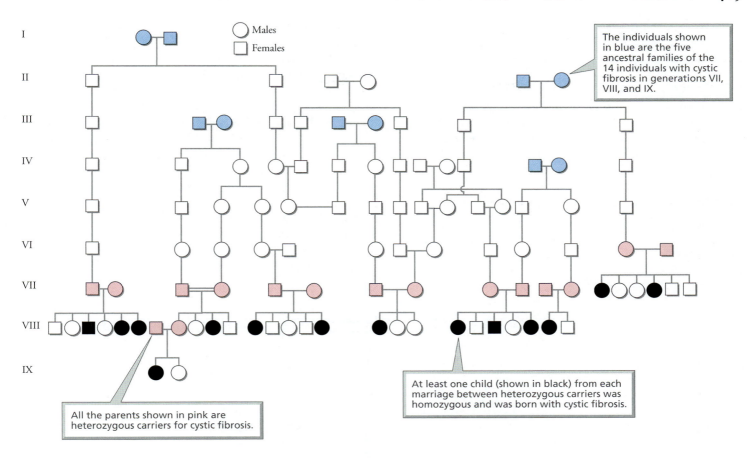

The individuals shown in blue are the five ancestral families of the 14 individuals with cystic fibrosis in generations VII, VIII, and IX.

All the parents shown in pink are heterozygous carriers for cystic fibrosis.

At least one child (shown in black) from each marriage between heterozygous carriers was homozygous and was born with cystic fibrosis.

FIGURE **20.7**

The inheritance of cystic fibrosis is shown in this pedigree. A pedigree is constructed so that each generation occupies a different horizontal line, numbered from top to bottom with the most ancestral at the top. Males are indicated as squares, and females as circles. A horizontal line connects the partners in a marriage. An affected individual is indicated with a solid black symbol.

FIGURE **20.8**

An albino lacks the brown pigment melanin in the skin, hair, and irises of the eyes.

INCOMPLETE DOMINANCE

In **incomplete dominance**, the expression of the trait in a heterozygous individual is somewhere in between the expression of the trait in a homozygous dominant individual and the expression of the trait in a homozygous recessive individual. Sickle-cell hemoglobin provides an example of incomplete dominance. Recall from Chapter 11 that hemoglobin is the pigment in red blood cells that carries oxygen. A red blood cell filled with normal hemoglobin (Hb^A) is a biconcave disk. The allele for sickling hemoglobin (Hb^S) produces an abnormal form of hemoglobin that is less efficient in binding oxygen. In the homozygous sickling condition $(Hb^S Hb^S)$, called sickle-cell anemia, the red blood cells contain only the abnormal form of hemoglobin. When the oxygen content of the blood drops below a certain level, as might occur during excessive exercise or respiratory difficulty, the red blood cells become sickle shaped and tend to clump together. The clumped cells can clog capillaries and break open, causing great pain. Vital organs may be damaged by lack of oxygen. People who are homozygous for sickle-cell anemia usually die at a young age.

The sickle-cell allele shows incomplete dominance, so people who are heterozygous $(Hb^S Hb^S)$ have the *sickle-cell trait*. They are generally healthy, but sickling and clumping of red cells may occur if there is a prolonged drop in the oxygen content of the blood, as might occur when traveling at high elevations. (Sickle-cell anemia is discussed in more detail in Chapter 11. The maintenance of the sickle cell trait in a population is discussed in Chapter 22.)

PLEIOTROPY

In addition to providing an example of incomplete dominance, sickle-cell anemia is an example of **pleiotropy**: one gene having many effects. As you can see in Figure 20.9, the sickling of red blood cells caused by the abnormal hemoglobin has effects throughout the body.

▌Certain genes have multiple alleles in a population

Many genes have more than two alleles. When three or more forms of a given gene exist, they are referred to as **multiple alleles**. Keep in mind, however, that one individual has only two alleles for a given gene, even if multiple alleles exist in the population. These alleles segregate independently during meiosis in the same manner as alleles for any other gene.

The ABO blood groups (discussed in Chapter 11) provide an example of multiple alleles. Blood type is determined by the presence of certain polysaccharides (sugars) on the surface of red blood cells. Type A blood has the A polysaccharide; type B has the B polysaccharide; type AB has both A and B polysaccharides; and type O has neither. The synthesis of each of these polysaccharides is directed by a specific enzyme. The enzyme, in turn, is specified by an allele of the gene.

The gene controlling ABO blood groups has three alleles, I^A, I^B, and I^O. Alleles I^A and I^B specify the A and B polysaccharides, respectively. When both these alleles are present, both polysaccharides are produced. I^A and I^B are, therefore, codominant. I^O is recessive to both I^A and I^B. The relationship between these alleles and the resulting blood type is shown in Table 20.2.

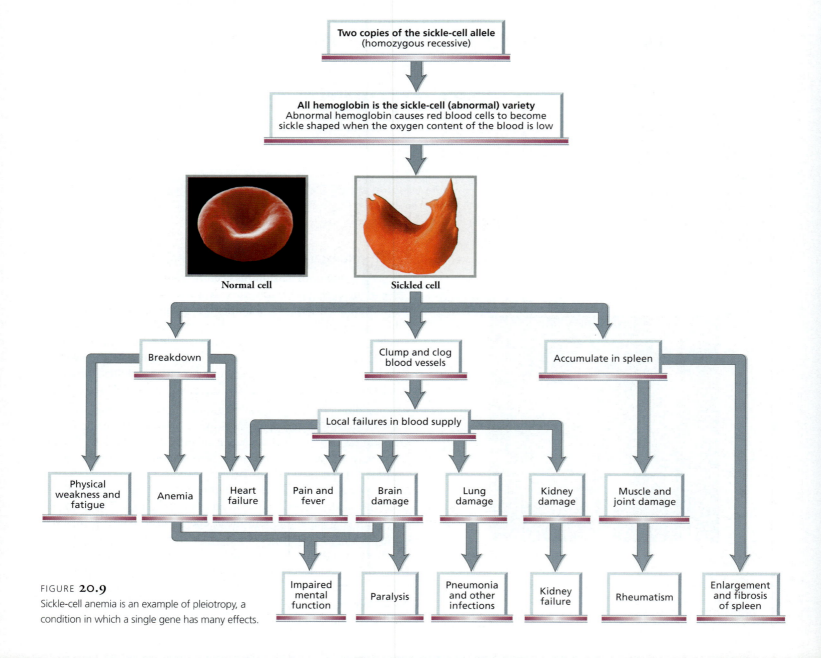

FIGURE **20.9**

Sickle-cell anemia is an example of pleiotropy, a condition in which a single gene has many effects.

TABLE **20.2**

THE RELATIONSHIP BETWEEN GENOTYPE AND ABO BLOOD GROUPS	
GENOTYPE	BLOOD TYPE
$I^A I^A$, $I^A I^O$	A
$I^B I^B$, $I^B I^O$	B
$I^A I^B$	AB
$I^O I^O$	O

stop and think

A man who has blood type AB is accused of fathering a child with blood type O. The mother has blood type B. Is it possible for the accused man to be the father of this child? (*Hint*: What are the possible genotypes of the three people involved? Given each possible genotype of the parents, what gametes could each produce? Use Punnett squares to determine whether any combination of possible parental genotypes could produce a child with the same genotype as this child.)

▌Most traits are controlled by many genes

So far, we have discussed traits that occur with distinct classes of possible phenotypes. In most cases in which a single gene controls a trait, the trait is present or it is not, although the environment may modify its expression. In the case of multiple alleles, there may be several distinct classes, as in A, B, AB, and O blood types.

The expression of most traits is much more variable, however. Indeed, many traits, including height, skin color, and eye color, vary almost continuously from one extreme to the other. Environment can play a role in smoothing out variations. For instance, diet and disease influence adult height, and exposure to sunlight darkens skin color. But, all environmental factors being equal, there is still considerable variation in the expression of certain traits. Such variation results from **polygenic inheritance**, that is, the involvement of two or more genes in the determination of the trait. The more genes involved, the smoother the gradations and the greater the extremes of expression.

Although human height is probably controlled by more than three genes, we will simplify things to see the variation in expression possible when as few as three genes—*A, B,* and *C*—are involved in determining a trait. Assume that the dominant alleles (*A, B, C*) of each gene add height, and the recessive alleles (*a, b, c*) do not. How tall would one expect the children to be if both of the parents were of medium height and heterozygous for all three genes? As you can see in Figure 20.10, there would be seven genetic height classes, ranging from very short to very tall. The probability of the children having either extreme of stature is slim, 1/64. It is most likely (a 20/64 chance), however, that the children will be of medium height, like their parents.

Skin color is also determined by several genes. The allele for albinism prevents melanin production. So if a person is homozygous recessive for this allele, no melanin can be deposited in the

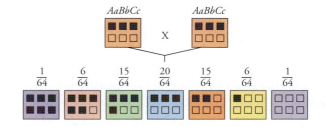

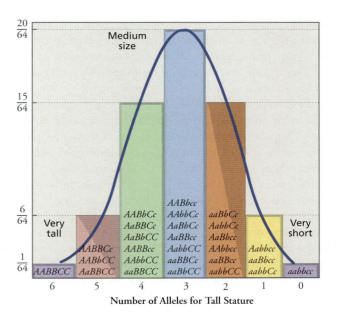

(a)

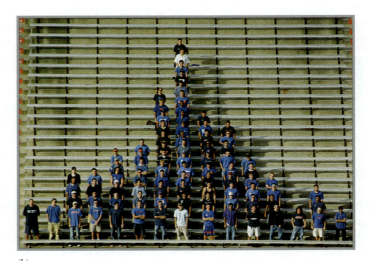

FIGURE **20.10**

Human height varies in a continuous manner. (a) One reason is that height is determined by more than one gene (polygenic inheritance). This figure shows the distribution of alleles for tallness in children of two parents of medium height, assuming that three genes are involved in the determination of height. The top line shows the parental genotypes, and the second line indicates the possible genotypes of the offspring. Alleles for tallness are indicated with dark squares. (b) Students organized according to height.

skin. In addition, there are probably at least four other genes involved in determining the amount of melanin deposited in the skin. Two alleles of four genes would create nine classes of skin color, ranging from pale to dark.

Genes on the same chromosome are usually inherited together

Scientists estimate that there are about 30,000 to 40,000 human genes packaged onto the 23 kinds of chromosomes. Thus, each chromosome bears a great number of genes. Genes on the same chromosome tend to be inherited together because an entire chromosome moves into a gamete as a unit. Genes that tend to be inherited together are described as being **linked**. We see, then, that linked genes do not *usually* assort independently.

Usually is emphasized here because there is a mechanism that can unlink genes on the same chromosome: crossing over. You may recall from Chapter 19 that crossing over occurs during prophase I of meiosis when the homologous chromosomes pair and align themselves so that the genes for the same traits are next to one another. Bridges form between nonsister chromatids, and segments of DNA are exchanged. The result is an exact exchange of alleles. After crossing over occurs, one resulting chromatid can have a different assortment of alleles than its sister chromatid. For instance, consider the possible gametes that could be formed in a person who is heterozygous with linked genes *A, B, C,* and *D* on one chromosome and *a, b, c, d* on the other. If no crossing over occurred, two types of gametes would be formed, *ABCD* and *abcd*. If crossing over occurred between genes *B* and *C*, however, four types of gametes would be formed: *ABCD, abcd, ABcd,* and *abCD*. Notice that the alleles were recombined, creating two new kinds of gametes. We know that crossing over has occurred when we see a recombination of alleles in the offspring.

Sex-linked genes are located on the sex chromosomes

Recall from Chapter 19 that one pair of chromosomes is called the sex chromosomes. There are two kinds of sex chromosomes, X and Y. They are not truly homologous because the Y chromosome is much smaller than the X chromosome, and they do not carry all of the same genes.[2] The Y chromosome carries very few genes, but it is more important than the X chromosome in determining gender. If a particular gene on the Y chromosome is present, an embryo will develop as a male. In the absence of that gene, an embryo will develop as a female. In contrast, most of the genes on an X chromosome have nothing to do with sex determination. For instance, genes for certain blood-clotting factors and for the pigments in cones (the photoreceptors responsible for color vision) are found on the X chromosome but not on the Y chromosome. Furthermore, the X chromosome has about as many genes as an autosome, but the Y chromosome has relatively few. Thus, most genes on the

X chromosome have no corresponding alleles on the Y chromosome and are known as **X-linked genes**.

Because most X-linked genes have no homologous allele on the Y chromosome, they have a different pattern of inheritance than do autosomes. A male is XY and therefore will express virtually all of the alleles on his single X chromosome, even those that are recessive. A female, on the other hand, is XX, so she does not always express recessive alleles. As a result, the recessive phenotype is much more common in males than in females (Figure 20.11). Furthermore, a son cannot inherit an X-linked recessive allele from his father. To be male, a child must have inherited his father's Y chromosome, not his X chromosome. Consequently, a son can inherit an X-linked recessive allele only from his mother. A daughter, however, can inherit an X-linked recessive allele from either parent. She must be homozygous for the recessive allele to show the recessive phenotype. If she is heterozygous for the trait, she will have a normal phenotype but be a carrier for that trait.

Among the disorders caused by X-linked recessive alleles are red-green color blindness, two forms of hemophilia, and Duchenne muscular dystrophy. Red-green color blindness, the inability to distinguish red and green, was discussed in Chapter 9. Hemophilia (discussed in Chapter 11) is a bleeding disorder caused by a lack of certain blood-clotting factors: Hemophilia A is a lack of blood-clotting factor VIII, and hemophilia B is a lack of clotting factor IX.

Duchenne muscular dystrophy is an X-linked recessive condition in which there is progressive muscle weakness because the muscle cells break down and are gradually lost. The responsible

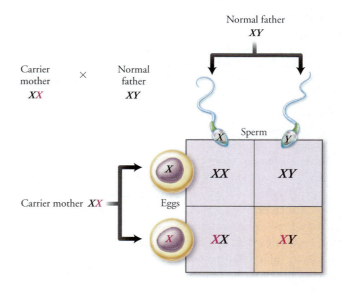

FIGURE **20.11**

Genes that are X linked have a different pattern of inheritance than do genes on autosomes, as seen in this cross between a carrier mother and a father who is normal for the trait. The recessive allele is indicated in red.

---

[2]X and Y are considered to be a homologous pair because they each have a small region at one end that carries some of the same genes. During meiosis, the tiny homologous region on X and Y will pair in synapsis. As a result, they segregate into gametes in the same manner that autosomes segregate.

allele produces a defect in an important protein (dystrophin) that supports the plasma membrane of muscle cells. This disorder affects only boys, with rare exceptions. The first muscles affected are those of the shoulders, hips, thighs, and calves. Therefore, early symptoms include difficulty climbing stairs or rising to an upright position when bending over, falling easily, and a waddling gait. Difficulty in walking, which starts when the boy is between 1 and 3 years old, gradually worsens until the boy is between 8 and 11 years old, when walking becomes impossible. With time, the disease spreads to all muscles. Death usually occurs by age 20, because of heart or respiratory failure.

stop and think

Explain why Duchenne muscular dystrophy is inherited from one's mother but is usually expressed only in sons. (*Hint*: Consider its mode of inheritance.)

Sex-influenced genes are autosomal genes whose expression is influenced by hormones

The expression of certain autosomal genes, those that are not located on the sex chromosomes, is powerfully influenced by the presence of sex hormones, so their expression differs in males and females. These traits are described as **sex-influenced** traits.

Male pattern baldness, premature hair loss on the top of the head but not on the sides, is an example of a sex-influenced trait. Male pattern baldness is much more common in men than in women because its expression depends on both the presence of the allele for baldness and the presence of testosterone, the male sex hormone. The allele for baldness, then, acts as a dominant allele in males because of their high level of testosterone and as a recessive allele in females because females have a much lower testosterone level. A male will develop pattern baldness whether he is homozygous or heterozygous for the trait. However, only women who are homozygous for the trait will develop pattern baldness. When a woman does develop pattern baldness, it usually appears later in life than it does in a man. The allele is expressed in women because the adrenal glands produce a small amount of male hormone. After menopause, when the supply of estrogen declines, adrenal male hormones may cause the expression of the baldness gene. However, balding in women may be merely thinning of hair.

stop and think

Male pattern baldness was passed from father to son through at least four generations of the Adams family. John Adams (1735–1826), the second U.S. President, passed it to his son, John Quincy Adams (1767–1848), the sixth U.S. President. He, in turn, passed the gene to his son, Charles Frances Adams (1807–1886), a diplomat, who passed it to his son, Henry Adams (1838–1918), a historian. Explain why father-to-son transmission of the trait rules out X-linked inheritance.

Breaks in Chromosomes Change Chromosomal Structure and Function

Chromosomes can break, allowing alterations in structure. Breakage can be caused by certain chemicals, radiation, or viruses. It also occurs as an essential part of crossing over. Although breakage does not happen often, chromosomes can be misaligned when crossing over occurs. Then, when the pieces reattach, one chromatid will have lost a segment, and the other will have gained a segment.

The loss of a piece of chromosome is called a **deletion**. The most common type of deletion occurs when the tip of a chromosome breaks off and is not included in a daughter cell during cell division. Deletion of more than a few genes on an autosome is usually lethal, and the loss of even small regions causes disorders.

In humans, the most common deletion, the loss of a small region near the tip of chromosome 5, causes cri-du-chat syndrome (meaning "cry of the cat"). An infant with this syndrome has a high-pitched cry that sounds like a kitten meowing. The unusual sound of the cry is caused by an improperly developed larynx (voice box). Infants with this syndrome have a round face; wide-set, downward-sloping eyes with a fold of skin at the corner of each eye; and misshapen ears (Figure 20.12). Although the condition is not usually fatal, it does cause severe mental retardation.

An added piece of chromosome is called a **duplication**. The effects of a duplication depend on its size and position. In general, however, a small duplication is less harmful than a deletion of com-

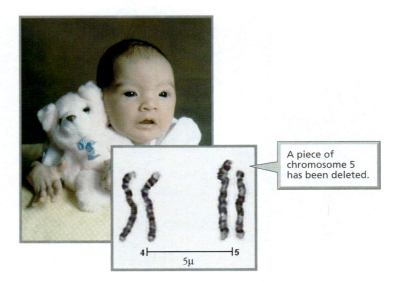

A piece of chromosome 5 has been deleted.

FIGURE **20.12**

Occurring in 1 out of every 50,000 live births, cri-du-chat syndrome is the most common deletion found in humans. It is caused by the loss of a small region near the tip of chromosome 5. The name of the syndrome, which means "cry of the cat," is given because of the sound of the cry of affected babies. Infants have a characteristically round face and wide-set eyes.

parable size. There is a small region of chromosome 9 that can be duplicated, resulting in cells containing three copies of this segment. The result is mental retardation, accompanied by facial characteristics that may include a bulbous nose, wide-set squinting eyes, and a lopsided grin.

Genetic disorders also occur when certain sequences of three subunits of DNA (nucleotides) are duplicated multiple times. The fragile X syndrome provides an example. The syndrome is so named because the long sequence of repeats makes the X chromosome fragile and easily broken. Besides making the chromosome fragile, the repeated subunits can shut down the activity of the entire chromosome. Fragile X syndrome is the most common form of inherited mental retardation, affecting roughly 1 in 1250 males and 1 in 2500 females. It is not known, however, exactly how fragile X syndrome causes the retardation. Other characteristics may include attention deficit, hyperactivity, large ears, long face, and flat feet (Figure 20.13).

Certain Genetic Disorders Can Be Detected Before Birth by Amniocentesis or Chorionic Villi Sampling

More than 4000 disorders have their roots in our genes. Tests are now available to look for predisposition to many of these genetic disorders. Some tests can even confirm the presence of a suspected disease-related allele in a particular person. Knowledge about the presence or absence of a faulty gene can be very helpful to couples who are planning a family. Normal, healthy parents can carry recessive alleles for disorders such as cystic fibrosis and Tay-Sachs disease (a disorder of lipid metabolism that causes death, usually between the ages of 1 and 5; discussed in Chapter 3). Pedigree analysis can help prospective parents determine whether they *might* be carriers of recessive alleles, but it cannot answer the question with certainty. However, this information would allow the parents to weigh the risks of passing a lethal allele to their children or to know that their child would not be affected. You can read more about the pros and cons of genetic testing in the Social Issue essay, *Gene Testing*.

Prenatal testing is usually recommended when a defective gene runs in the family or when the mother is older than 35 (because age increases the risk of problems due to nondisjunction; see Chapter 19). There are two available procedures for diagnosing genetic problems in the fetus: **amniocentesis** and **chorionic villi sampling (CVS)** (Figure 20.14). Although it is possible to look for more than 100 disorders with these procedures, tests are run only for those disorders that are common and those that are of particular concern in that pregnancy. Reassuring results from either form of prenatal testing cannot guarantee a healthy baby, despite the high degree of accuracy with which they can detect genetic disorders.

In amniocentesis, a needle is inserted through the lower abdomen into the uterus and a small amount of amniotic fluid, 10 to 20 ml (about 2 to 4 tsp), is withdrawn. Because it is important that the needle not injure the fetus, umbilical cord, or placenta, ultrasound is used to find the safest spot for insertion. Floating in the amniotic fluid are living cells that have sloughed

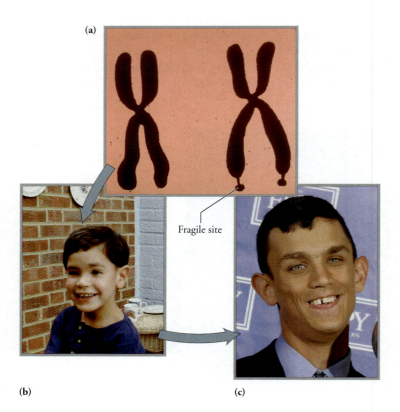

(a)

(b)   (c)

Fragile site

FIGURE **20.13**

Fragile X syndrome. (a) Duplication of a region on the X chromosome makes the chromosome fragile and easily broken. (b) A child with fragile X syndrome appears normal. (c) Characteristics of an adult with fragile X syndrome include a long face and large ears. In addition, fragile X syndrome causes mental retardation.

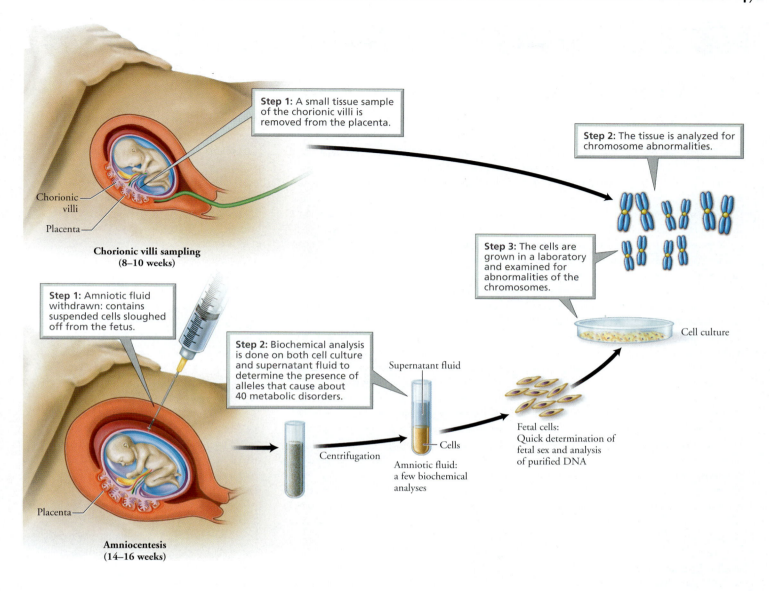

Step 1: A small tissue sample of the chorionic villi is removed from the placenta.

Step 2: The tissue is analyzed for chromosome abnormalities.

Step 3: The cells are grown in a laboratory and examined for abnormalities of the chromosomes.

Chorionic villi

Placenta

**Chorionic villi sampling
(8–10 weeks)**

Step 1: Amniotic fluid withdrawn: contains suspended cells sloughed off from the fetus.

Step 2: Biochemical analysis is done on both cell culture and supernatant fluid to determine the presence of alleles that cause about 40 metabolic disorders.

Supernatant fluid

Cell culture

Centrifugation

Cells

Amniotic fluid: a few biochemical analyses

Fetal cells: Quick determination of fetal sex and analysis of purified DNA

Placenta

**Amniocentesis
(14–16 weeks)**

FIGURE **20.14**

Amniocentesis and chorionic villi sampling (CVS) are procedures available for prenatal genetic testing.

off the fetus. These cells are grown in the laboratory for a week or two and then are examined for abnormalities in the number of chromosomes and the presence of certain alleles that are likely to cause specific diseases. Biochemical tests are also done on the fluid to look for certain chemicals that are indicative of problems. For example, a high level of alpha-fetoprotein, a substance produced by the fetus, suggests a problem with the development of the central nervous system (a neural tube defect). Amniocentesis is generally safe for both the mother and the fetus, but there

is a small risk of triggering a miscarriage or of the needle injuring the mother and causing infection or bleeding.

Amniocentesis is usually done between 14 and 18 weeks after the woman's last menstrual period, when there is enough amniotic fluid, about 250 ml (about 1 cup), to minimize the risk of injuring the fetus. (A few medical centers are now able to perform amniocentesis at 12 weeks of pregnancy.)

Chorionic villi sampling (CVS) involves taking a small piece of chorionic villi, which are small, fingerlike projections of the part of

SOCIAL ISSUE

Gene Testing

Genetic screening involves the testing of people who have no symptoms to determine whether they carry genes that will influence their chances of developing certain genetic diseases. It is a technology whose time has come. Like a snowball rolling downhill, the practice is gaining momentum. However, genetic screening raises many ethical issues.

Among the advantages of gene testing is that knowing one has a treatable or preventable condition enables a person to take steps to reduce the risk of developing the disorder. It can also save suffering in future generations when the disorder is caused by a recessive allele that can remain hidden for generations or by a dominant allele that is not expressed until late in life. Consider, for instance, Tay-Sachs disease. This disorder causes the death of children, usually by the age of 5. The infant appears healthy at birth, but at about 6 months old it gradually stops smiling, crawling, or turning over. Eventually, the child becomes blind, paralyzed, and unaware of its surroundings. Tay-Sachs disease is especially prevalent in descendants from Jewish communities from eastern Europe. As a result of voluntary screening programs, the number of children born with Tay-Sachs disease has decreased by 10-fold in many communities.

Genetic testing also has a dark side. The psychological consequences of test results can be devastating. Many genetic disorders cannot be prevented or treated. One must wonder, then, do you really want to know *now* what will cause your death? Huntington's disease is caused by a dominant allele that provides no hints of its existence until relatively late in life, usually past child-bearing years. About 60% of the people with Huntington's disease are diagnosed between the ages of 35 and 50. The gene causes degeneration of the brain, leading to muscle spasms, personality disorders, and death, usually within 10 to 15 years. Because Huntington's disease is caused by a dominant allele, a bearer has a 50% chance of passing it to children. Thus, a person whose parent died of Huntington's disease could be relieved if a gene test did not detect the allele. But, it is equally likely that the allele *will* show up. Many persons at risk for Huntington's disease prefer to live without the knowledge of their possible fate.

A gene test confirming the risk of a serious disease can cause anxiety and depression in carriers and feelings of guilt in siblings who do not carry the allele. Much of the carrier's worry may be unnecessary, because predictive gene tests deal in probabilities, not in certainties. A mutation in the *BRCA-1* gene on chromosome 17 increases a woman's risk of developing breast cancer and ovarian cancer. Carriers have an 85% chance of developing breast cancer by age 65. A woman who learns she is a carrier must make medical decisions that could influence the length of her life. One option would be to have frequent mammograms in the hope of discovering breast cancer, if it should develop, early enough for treatment. Of course, there is a chance that cancer would not be detected before it spread, making effective treatment difficult. A second option would be to have a mastectomy quickly. However, the cancer may have already spread. Also, there is a 15% chance that she would not develop breast cancer.

There is also the threat that the results of gene tests will not remain private information, but instead be used by employers as well as by life and health insurers. As an employer, if you had information about the genetic makeup of prospective employees, would you choose to invest time and money in training a person who carried an allele that increased the risk of cancer, heart disease, Alzheimer's disease, or alcoholism? As an insurer, would you knowingly cover a carrier?

The results of gene testing can have both positive and negative consequences for those being tested and for their families: Who should decide whether screening should be done, for which genes, on whom, and in which communities? A flippant answer might be, "There oughta be a law!" Should we leave ethical issues to judges and legislators? Should moral matters be decided by society or clergy? Or should they be personal decisions?

If gene testing is done, who should be told the results? If the affected person is an infant, should the parents always be told the results, even if the condition is poorly understood? How do we balance helping such children with the possibility of stigmatizing them?

We live in a world of limited resources. Thus, as soon as it is decided who should be tested, we must decide who should pay the bill. Both testing and treatment are expensive. Should testing be done only when treatment or preventive measures are available? How much say should the agent that pays for the procedure have in who is tested and who receives medical treatment?

There are no easy answers. It is time for each of us to think about the issues raised by these and similar questions. ✍

---

the placenta called the chorion (see Chapter 18). Cells of the chorion have the same genetic composition as those of the fetus. Guided by ultrasound, a small tube is inserted through the vagina and the cervix to where the villi are located. Gentle suction is then used to remove a small tissue sample, which can be analyzed for genetic abnormalities.

There are pros and cons to CVS. Advantages are that it can be performed 6 to 8 weeks earlier in the pregnancy than amniocentesis, and the results are available within a few days. More than 95% of the high-risk women who opt for prenatal tests receive good news, and an early diagnosis saves weeks of worry over the health of the fetus. Also, if there is a genetic problem in the fetus and the couple wishes to terminate the pregnancy, the procedure can be performed earlier in the pregnancy, when it is safer for the mother. When abortion is not chosen, early diagnosis allows more time to plan the safest time, location, and method of delivery. A disadvantage of CVS is that it has a slightly greater risk of triggering miscarriage than does amniocentesis.

If detected early enough, certain birth defects can be prevented. Congenital adrenal hyperplasia (an overgrowth of the adrenal glands), for instance, will cause a female fetus to develop abnormal genitalia, unless she is treated with hormones from week 10 to week 16 of gestation. Early diagnosis with CVS can tell a physician whether hormone treatment is needed.

Newborns are routinely screened for phenylketonuria (PKU), an inherited metabolic disorder. This simple blood test can prevent mental retardation. People with PKU have inherited an allele responsible for a defective enzyme that prevents them from converting phenylalanine (an amino acid in food) to tyrosine. As a result, they have too much phenylalanine in their bodies and too little tyrosine, an imbalance that somehow causes brain damage. Although nothing can be done to correct the enzyme, brain damage can be prevented with a strict diet that excludes most phenylalanine. Nearly all proteins contain phenylalanine, so it is often necessary to substitute a specially prepared mixture of amino acids for most protein-containing foods, such as meat. It is also necessary to avoid foods containing the artificial sweetener NutraSweet®, which contains aspartame. Aspartame consists of the amino acids phenylalanine and aspartic acid. When aspartame is digested, phenylalanine is separated from aspartic acid and can reach dangerous levels.

Many predictive genetic tests are now available or are being developed. These tests identify people who are at risk of getting a disease, *before* symptoms appear. The procedure is simple and can usually be done with a small blood sample. When steps can be taken to prevent the disease, predictive gene tests can be lifesaving. For instance, colon cancer will develop in nearly everyone who has the alleles for familial adenomatous polyposis, a condition in which thousands of benign polyps grow in the intestine. If the colon is routinely inspected for polyps, and they are removed, cancer can be prevented.

Other predictive gene tests look for alleles that might predispose one to a disorder. A protein that transports cholesterol in the blood, called *ApoE*, comes in three forms, each specified by a different allele. Having two alleles for one of these, *ApoE-2*, causes catastrophically high blood cholesterol levels, which can lead to heart attack and stroke. Knowing that a person had this genetic makeup, a physician could prescribe medication to lower blood cholesterol.

what would you do?

Having two copies of another *ApoE* allele, *ApoE-4*, increases a person's risk of heart disease by 30% to 50%. It also nearly guarantees that Alzheimer's disease will develop by 80 years of age. Alzheimer's disease is an untreatable condition in which brain tissue degenerates (see Chapter 7). A person is gradually robbed of memories, of the ability to function normally in society, and eventually of life itself. Suppose that a gene test that is performed out of concern for a person's risk of heart disease reveals that two copies of *ApoE-4* are carried. In your opinion, is a physician with this knowledge morally obligated to tell the patient about the inevitability of Alzheimer's disease, or should factors in the patient's life be considered in making the decision of whether to tell? If your genes were tested and found to have two copies of *ApoE-4* would you want to be told? Why or why not?

REVIEWING THE CONCEPTS

Principles of Inheritance Help Us to Predict How Simple Traits Are Passed to the Next Generation (pp. 460–469)

1. Different forms of a gene are called alleles. An individual who has two of the same alleles is said to be homozygous. An individual with two different alleles for a gene is said to be heterozygous. The allele that is expressed in the heterozygous condition is described as being dominant. The allele that is masked in a heterozygote is described as recessive.

2. The genotype is the precise alleles present in an individual. The phenotype refers to the observable traits.

3. The law of segregation states that alleles for each gene separate during gamete formation so that half the gametes receive one allele and the other half receive the other allele.

4. The law of independent assortment states that each pair of alleles that are located on separate chromosomes separate into gametes independently of other pairs of alleles.

5. A monohybrid cross is one in which only one trait is considered. In a monohybrid cross, the individuals who mate exhibit contrasting forms of the trait being studied. If a homozygous dominant individual (*AA*) mated with a homozygous recessive (*aa*), the genotypes of the offspring would be heterozygous (*Aa*), and the phenotypes would be dominant. If heterozygous individuals (*Aa*) mate, the probable genotypes of children will be 1 in 4 (25%) homozygous dominant (*AA*):2 in 4 (50%) heterozygous dominant (*Aa*):1 in 4 (25%) homozygous recessive (*aa*). Their phenotypes will be 3 dominant:1 recessive.

6. Pedigrees, which are constructed to show the genetic relationships among individuals in an extended family, are often useful in determining the unknown genotypes of humans showing dominant phenotypes.

7. A dominant allele often produces a functional protein. A recessive allele often produces a nonfunctional protein or no protein at all.

8. In complete dominance, the dominant allele completely masks the recessive allele. However, when two alleles are codominant, both are separately apparent in the phenotype. In incomplete dominance, the expression of the trait in a heterozygous individual is somewhere in between the expression of the trait in a homozygous dominant individual and the expression of the trait in a homozygous recessive individual.

9. When there are three or more alleles for a particular gene in a population, they are called multiple alleles. ABO blood groups are determined by three alleles: I^A, I^B, and I^O. Blood types are determined by the presence of certain polysaccharides on the surface of red blood cells.

10. Many traits, including height, skin pigmentation, and eye color, are determined by more than one gene (polygenic inheritance). Such traits are variable in their manner of expression. When many genes are involved in determining a trait, the variation in expression can be continuous.

11. Linked genes are those that are inherited together because they are physically on the same chromosome, unless crossing over occurs between them.

12. An X-linked gene is one located on the X chromosome that has no corresponding allele on the Y chromosome. A recessive X-linked allele will always be expressed in a male, but it will be expressed only in the homozygous condition in a female. Red-green color blindness, hemophilia, and Duchenne muscular dystrophy are disorders caused by X-linked recessive alleles.

13. The expression of a sex-influenced trait depends on both the presence of the allele and the presence of sex hormones. Therefore, the expression of the allele depends on one's sex. For example, the expression of the allele for male pattern baldness requires the presence of testosterone.

WEB TUTORIAL 20.1 Mendelian Inheritance

WEB TUTORIAL 20.2 Incomplete Dominance

WEB TUTORIAL 20.3 Sex Linked Traits

Breaks in Chromosomes Change Chromosomal Structure and Function (pp. 469–470)

1. The loss of a piece of a chromosome is called a deletion. The gain of a piece of chromosome is called a duplication. Either chromosome abnormality can cause a genetic disorder.

Certain Genetic Disorders Can Be Detected before Birth by Amniocentesis or Chorionic Villi Sampling (pp. 470–473)

1. Tests such as amniocentesis and chorionic villi sampling (CVS) are available to determine whether someone is likely to develop a genetic disease. In amniocentesis, a sample of amniotic fluid is taken. The fetal cells in the fluid are grown in the laboratory and then analyzed for genetic problems. Tests on amniotic fluid can reveal certain developmental problems, such as neural tube defects. CVS involves sampling cells from the chorion of the placenta and analyzing them for genetic disorders.

WEB TUTORIAL 20.4 Some Human Genetic Disorders

KEY TERMS

albinism *p. 464*	dominant *p. 461*	incomplete dominance *p. 465*	phenotype *p. 461*
allele *p. 460*	duplication *p. 469*	law of independent assortment *p. 461*	pleiotropy *p. 466*
amniocentesis *p. 470*	gene *p. 460*		polygenic inheritance *p. 467*
carrier *p. 464*	genotype *p. 461*	law of segregation *p. 461*	Punnett square *p. 462*
chorionic villi sampling *p. 470*	heterozygous *p. 461*	linked *p. 468*	recessive *p. 461*
codominance *p. 464*	homologous pair of chromosomes *p. 460*	monohybrid cross *p. 462*	sex-influenced *p. 469*
complete dominance *p. 464*		multiple alleles *p. 466*	trait *p. 460*
deletion *p. 469*	homozygous *p. 461*	pedigree *p. 464*	X-linked gene *p. 468*

THINKING ABOUT THE CONCEPTS

1. State the law of segregation. Explain how this law is related to events that take place during meiosis I. *p. 461*
2. State the law of independent assortment. Explain how this law is related to events that take place during meiosis I. *p. 461*
3. Define a monohybrid cross. *p. 462*
4. What is the ratio of genotypes and phenotypes in the offspring resulting from a monohybrid cross between a homozygous dominant individual and a homozygous recessive one? *p. 462-463*
5. What is a pedigree? What can a family pedigree reveal about the inheritance of a trait? *pp. 464, 465*
6. Using an example, explain what is meant by codominance. *p. 464*
7. Differentiate between multiple alleles and polygenic inheritance. *p. 466-467*
8. What are linked genes? Why are they usually inherited together? What can cause such genes to become unlinked? *p. 468*
9. Explain why the pattern of inheritance for recessive X-linked genes is different from the pattern for recessive autosomal alleles. *p. 468*
10. What two procedures are used for prenatal testing? How do they differ? *pp. 470-472*
11. Which of the following crosses could produce offspring with the recessive phenotype?

    a. *AA × aa*

    b. *Aa × Aa*

    c. *AA × Aa*

    d. *AA × AA*

12. All of the following crosses will have a 50% probability of producing 50% of the offspring with a heterozygous genotype *except*

    a. *AA × aa*

    b. *Aa × Aa*

    c. *AA × Aa*

    d. *Aa × aa*

13. A trait controlled by many genes is described as being _____.
14. The _____ of an individual refers to the appearance of the trait in a person, and the _____ refers to the precise alleles that are present.
15. The genotype of a person with two copies of the same allele is _____, the genotype of a person with two different alleles for a trait is _____.
16. Genes for different traits that are located on the same chromosome are described as being _____.

APPLYING THE CONCEPTS

1. Klaus has red-green color blindness. His wife, Helen, is homozygous for normal color vision. (Normal color vision is dominant to red-green color blindness.) What is the probability of their having a color-blind daughter? What is the probability of their having a color-blind son?
2. George and Sue have an infant, Sammy, who is lethargic, is vomiting, and has liver disease. Sammy is diagnosed with galactosemia, an autosomal recessive disorder in which the affected individuals are unable to metabolize the milk sugar galactose. Sammy is placed on a diet free

of lactose and galactose and slowly recovers. George and Sue would like to have a second child. What are the chances that the second child will have galactosemia? (*Hint*: What are the genotypes of George and Sue?)
3. The parents of first cousins are siblings. Explain why the offspring of first cousins are more likely to have harmful recessive traits than are offspring of unrelated individuals?

DNA and Biotechnology

21

DNA Is a Double Helix Consisting of Two Strings of Nucleotides

During Replication of DNA Each Original Strand Serves As a Template for a New Strand

DNA Codes for RNA, Which Codes for a Protein

- Transcription is RNA synthesis
- Translation is protein synthesis

Point Mutations Result from Nucleotide Substitution, Insertion, or Deletion

Gene Activity Can Be Turned On or Off

- Coiling and uncoiling of chromosomes regulate gene activity at the chromosome level
- Certain genes regulate the activity of other genes
- Chemical signals regulate gene activity

Genetic Engineering Is the Manipulation of DNA for Human Purposes

- Recombinant DNA is made of DNA from different sources
- Genetic engineering produces proteins of interest or transgenic organisms
- Gene therapy replaces faulty genes with functional genes
- Faulty information in mRNA may someday be correctable
- Gene silencing may someday adjust gene activity
- DNA fingerprinting can identify individuals

The Purpose of the Human Genome Project Is to Identify the Complete DNA Sequence of the Human Genome

 **SOCIAL ISSUE**   Stem Cells—The Body's Repair Kit

HEALTH ISSUE   Genetically Modified Food

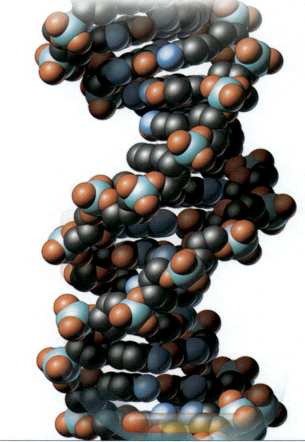

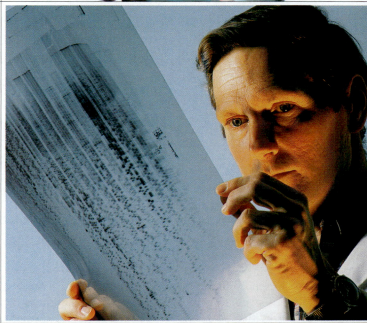

"Can DNA evidence really be used to solve crimes that occurred decades ago?" asked Sarah, a student in the high school class that Detective Jim Young was speaking to.

"Yes, it's possible," Det. Young explained. "In the lab, DNA can be extracted from archived biological samples such as blood on clothing, saliva on a cigarette butt, or semen from a swab in a rape kit. Using methods that mimic natural DNA replication, the sample can be multiplied and then characterized. The tests focus on several sites in the DNA that are highly variable among individuals. If a suspect's DNA matches those sites, it is almost certain that he or she was the source of the original biological sample."

"But if a crime was committed years ago, who is considered a suspect?" asked Sean.

"Good question," Det. Young said. "Each case is unique, but one approach is to compare DNA evidence to a database of DNA samples collected from convicted criminals. A number of "cold" cases have been solved because the guilty party was later convicted of a different crime and became a part of the database."

Another hand went up. "But I thought that DNA evidence was used to get people out of jail, not to put them in jail," Erin commented.

Det. Young explained, "Those are the cases we hear about, and there have been over 140 innocent people released from jail after serving years for crimes they didn't commit. Because DNA evidence can help convict or to clear a person, DNA evidence can be the most powerful tool in my arsenal against crime."

In this chapter we will learn that DNA fingerprints are possible because the DNA of each person (except for identical twins) is unique. We will become familiar with the structure and function of **DNA (deoxyribonucleic acid)** to understand how this molecule can be the basis of our genetic inheritance and legacy as well as the basis of the evolutionary process that has provided the diversity of life on the earth. We will begin to understand that the importance of DNA on a personal level is that it directs the synthesis of specific polypeptides (proteins) that play structural or functional roles in our bodies. Then, we will consider the technology that our understanding of DNA has already made available and what such technology may hold for the future. ■

DNA Is a Double Helix Consisting of Two Strings of Nucleotides

DNA is sometimes referred to as the thread of life—and a very slender thread it is. If the DNA from a single cell were unwound, it would be a mere 50-trillionths of an inch in diameter. With all the DNA strands in a single cell fastened together end to end, the thread would stretch over 5 feet in length. DNA might also be considered the thread that ties all life together, because the DNA of organisms ranging from bacteria to humans is built from the same kinds of subunits. The order of these subunits encodes the information needed to make the proteins that build and maintain life.

DNA is a double-stranded molecule resembling a ladder that is gently twisted to form a spiral, called a double helix, as shown in Figure 21.1. Each side of the ladder and half of each rung are made from a string of repeating subunits called nucleotides. A **nucleotide** is composed of one sugar (deoxyribose), one phosphate, and one nitrogenous base. DNA contains four types of nitrogenous bases: adenine (A), guanine (G) thymine (T), and cytosine (C). The sides of the ladder are composed of alternating sugars and phosphates; the rungs consist of paired bases. Following the rules of **complementary base pairing**, adenine pairs *only* with thymine (an A-T pair), and cytosine pairs *only* with guanine (a C-G pair). Each base pair is held together by weak hydrogen bonds. The pairing of complementary bases is specific because of the shapes of the bases and the number of hydrogen bonds that can form between them. Recall from Chapter 2 that a molecule formed by the joining of nucleotides is called a nucleic acid. Thus, DNA is a nucleic acid.

Because of the specificity of base pairing, the bases on one strand of DNA are *always* complementary to the bases on the other strand. Thus, the order of bases on one strand determines the sequence of bases on the other strand. For instance, if the sequence of bases on one strand were CATATGAG, what would the complementary sequence be? Remember, cytosine (C) always pairs with guanine (G), and adenine (A) always pairs with thymine (T). As a result, the complementary sequence on the opposite strand would be GTATACTC.

The DNA within each human cell has an astounding 3 billion base pairs. Although the pairing of adenine with thymine and cytosine with guanine is specific, the sequence of bases along the length of the DNA molecule can vary in myriad ways. As we will see, genetic information is encoded in the exact sequence of bases.

During Replication of DNA Each Original Strand Serves As a Template for a New Strand

For DNA to be the basis of inheritance, its genetic instructions must be passed from one generation to the next. Moreover, for DNA to direct the activities of each cell, its instructions must be present in every cell. These requirements dictate that DNA be copied before both mitotic and meiotic cell division (see Chapter 19). It is important that the copies be exact. The key to the precision of the copying process, or **replication**, is in the complementarity of the bases.

DNA replication begins when an enzyme breaks the hydrogen bonds that hold together the two nucleotide strands of the double helix, thereby "unzipping" and unwinding the strands. As a result, the nitrogenous bases on the separated regions of each strand are exposed. Bases of nucleotides floating free in the nucleus then attach to complementary bases on the open DNA strands. Enzymes called **DNA polymerases** link the new nucleotides together to form a new strand. In this way, two DNA strands form, each identical to the original one. As each of the new double-stranded DNA molecules forms, it twists and forms a double helix (Figure 21.2).

Parental molecule of DNA

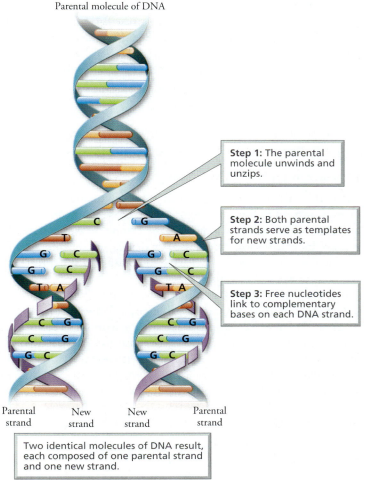

Step 1: The parental molecule unwinds and unzips.

Step 2: Both parental strands serve as templates for new strands.

Step 3: Free nucleotides link to complementary bases on each DNA strand.

| Parental strand | New strand | New strand | Parental strand |

Two identical molecules of DNA result, each composed of one parental strand and one new strand.

FIGURE **21.2**

DNA replication is called semiconservative because the new DNA consists of one "old" strand and one "new" strand.

Each strand of the original DNA molecule serves as a template for the formation of a new strand. This process is called **semiconservative replication** because, in each of the new double-stranded DNA molecules, one original strand is saved (conserved) and the other strand is new. Look at Figure 21.2. Notice the original (parental) strand of nucleotides in each new molecule of DNA. Complementary base pairing creates two new DNA molecules that are identical to the parent molecule.

DNA Codes for RNA, Which Codes for a Protein

We have seen how the process of replication ensures that genetic information is passed accurately from a parent cell to daughter cells and from generation to generation. The next obvious question is, "How does DNA issue commands that direct cellular activities?" The answer is that DNA directs the synthesis of another nucleic acid, **ribonucleic acid**, or **RNA** (described shortly). RNA, in turn, directs the synthesis of a polypeptide or protein. The protein can be a structural part of the cell or an enzyme that speeds up certain chemical reactions within the cell.

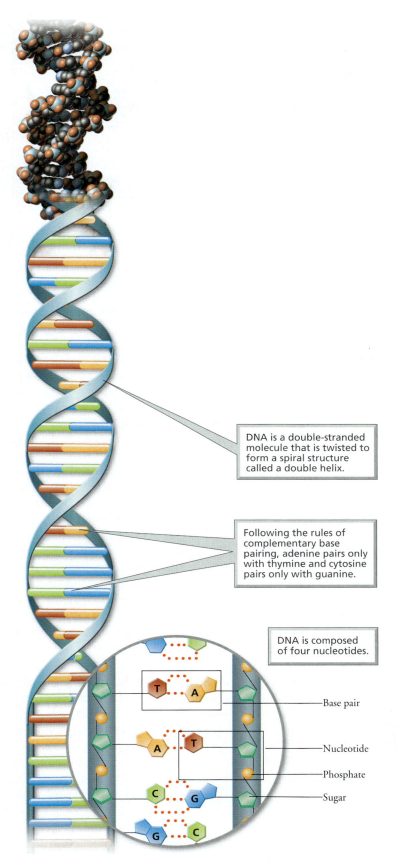

DNA is a double-stranded molecule that is twisted to form a spiral structure called a double helix.

Following the rules of complementary base pairing, adenine pairs only with thymine and cytosine pairs only with guanine.

DNA is composed of four nucleotides.

Base pair

Nucleotide

Phosphate

Sugar

FIGURE **21.1**

DNA is a double-stranded molecule that is twisted to form a spiral structure called a double helix.

Recall from Chapter 20 that a **gene** is a segment of DNA that contains the instructions for producing a specific protein (polypeptide).[1] We say that the gene is *expressed* when the protein it codes for is produced. The protein whose synthesis is directed by a gene is the molecular basis of the inherited trait; it determines the phenotype.

Gene expression:

$$DNA \rightarrow RNA \rightarrow Protein$$

To gain a fuller appreciation for how DNA works, we will consider each step in slightly more detail.

▌ Transcription is RNA synthesis

Just as the CEO of a major company issues commands from headquarters instead of from the factory floor, DNA issues instructions from the cell nucleus and not from the cytoplasm where the cell's work is done. RNA is the intermediary that carries the information encoded in DNA from the nucleus to the cytoplasm and directs the synthesis of the specified protein.

Like DNA, RNA is composed of nucleotides linked together, but there are some important differences between DNA and RNA, as shown in Table 21.1. First, the nucleotides of RNA contain the sugar ribose, instead of deoxyribose, as in DNA. Second, in RNA the nucleotide uracil (U) pairs with adenine, whereas in DNA thymine pairs with adenine. Third, most RNA is single-stranded. Recall that DNA is a double-stranded molecule.

The first step in converting the DNA message to a protein is copying the message as RNA, a process called **transcription**.

$$DNA \xrightarrow{\text{transcription}} RNA \rightarrow Protein$$

Three types of RNA are produced in cells. Each plays a different role in protein synthesis (Table 21.2). **Messenger RNA**

TABLE **21.1**

COMPARISON OF **DNA** AND **RNA**		
	DNA	**RNA**
Similarities	Are nucleic acids	
	Are composed of linked nucleotides	
	Have a sugar-phosphate backbone	
	Have four types of bases	
Differences	Is a double-stranded molecule	Is a single-stranded molecule
	Has the sugar deoxyribose	Has the sugar ribose
	Contains the bases adenine, guanine, cytosine, and thymine	Contains the bases adenine, guanine, cytosine, and uracil (instead of thymine)
	Functions primarily in the nucleus	Functions primarily in the cytoplasm

TABLE **21.2**

FUNCTIONS OF **RNA** MOLECULES	
MOLECULE	**FUNCTIONS**
Messenger RNA (mRNA)	Carries DNA's information in the sequence of its bases (codons) from the nucleus to the cytoplasm
Transfer RNA (tRNA)	Binds to a specific amino acid and transports it to the appropriate region of mRNA
Ribosomal RNA (rRNA)	Combines with protein to form ribosomes (structures on which polypeptides are synthesized)

(mRNA) carries DNA's instructions for synthesizing a particular protein from the nucleus to the cytoplasm. The order of bases in mRNA specifies the sequence of amino acids in the resulting protein, as we will see. **Transfer RNA (tRNA)** binds to a specific amino acid and transports it to the appropriate region of mRNA. **Ribosomal RNA (rRNA)** combines with proteins to form the ribosomes, which are structures on which protein synthesis occurs.

Transcription begins with the unwinding and unzipping of the specific region of DNA to be copied, actions that are directed by an enzyme. The DNA message is determined by the order of bases in the unzipped region of the DNA molecule. One strand of the DNA molecule serves as the template. RNA nucleotides that are floating free in the nucleus pair with their complementary bases on the template—cytosine with guanine and uracil with adenine (Figure 21.3). The signal to start transcription is given by a specific sequence of bases on DNA, called the **promoter**. An enzyme called **RNA polymerase** binds with the promoter on DNA and then moves along the DNA strand, aligning the appropriate RNA nucleotides and linking them together. Another sequence of bases serves as a stop signal. When RNA polymerase reaches this stop signal, transcription ceases, and the newly formed strand of RNA, called the *RNA transcript*, is released from the DNA.

Messenger RNA is usually modified before it leaves the nucleus (Figure 21.4). Most of the DNA between a promoter and the stop signal includes regions that do not contain codes that will be translated into protein. These unexpressed regions of DNA are called **introns**, for intervening sequences. Introns are snipped out of the newly formed mRNA strand by enzymes before the strand leaves the nucleus. The remaining segments, called **exons** for expressed sequences, actually direct the synthesis of a protein.

▌ Translation is protein synthesis

The newly formed mRNA carries the genetic message in DNA from the nucleus to the cytoplasm, where it is translated into protein. Just as we might translate a message written in Spanish into

---

[1] We will develop this concept as the chapter proceeds. Some genes code for a polypeptide that is only part of a functional protein. A gene can also code for RNA that forms part of the ribosomes, and there are genes that regulate the activity of other genes.

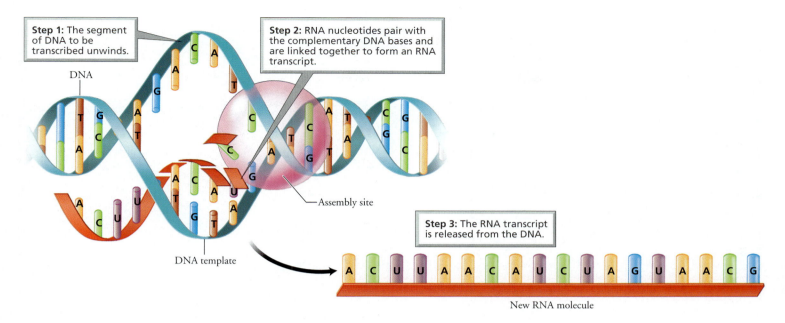

Step 1: The segment of DNA to be transcribed unwinds.

Step 2: RNA nucleotides pair with the complementary DNA bases and are linked together to form an RNA transcript.

DNA

Assembly site

DNA template

Step 3: The RNA transcript is released from the DNA.

A C U U A A C A U C U A G U A A C G

New RNA molecule

FIGURE **21.3**

Transcription is the process of producing RNA from a DNA template.

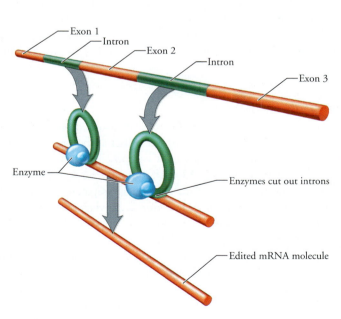

Exon 1
Intron
Exon 2
Intron
Exon 3
Enzyme
Enzymes cut out introns
Edited mRNA molecule

FIGURE **21.4**

Messenger RNA is modified before it leaves the nucleus. Noncoding regions of mRNA called introns are snipped out of the molecule. Segments of mRNA that code for protein are then spliced together.

English, **translation** converts the nucleotide language of mRNA into the amino acid language of a protein.

$$\text{DNA} \xrightarrow{\text{transcription}} \text{RNA} \xrightarrow{\text{transcription}} \text{Protein}$$

Before examining the process of translation, we should become more familiar with the language of mRNA.

THE GENETIC CODE

The **genetic code** is used to convert the linear sequence of bases in DNA to the sequence of amino acids in proteins. The "words" in the code, called **codons**, are sequences of three bases on mRNA that specify 1 of the 20 amino acids or the beginning or end of the protein chain. All the codons of the genetic code are shown in Table 21.3. For instance, the codon UUC on mRNA specifies the amino acid phenylalanine. (The complementary sequence on DNA would be AAG.)

stop and think

Look at the strand of mRNA in Figure 21.3. Notice that the codon at the end of the mRNA strand is AGC. Which amino acid does this specify? (Use Table 21.3.)

The four bases in RNA (A, U, C, and G) could form 64 combinations of three-base sequences. The number of possible codons, therefore, exceeds the number of amino acids. So, we find that several codons may code for the same amino acid. The codon AUG can either serve as a start signal to initiate translation or can specify the amino acid methionine, depending on where it occurs in the mRNA molecule. In addition, three codons (UAA, UAG, and UGA) are stop codons that specify

TABLE **21.3**

THE GENETIC CODE

First base	Second base				Third base
	U	**C**	**A**	**G**	
U	UUU ⎤ Phenylalanine UUC ⎦ UUA ⎤ Leucine UUG ⎦	UCU ⎤ UCC UCA Serine UCG ⎦	UAU ⎤ Tyrosine UAC ⎦ UAA Stop UAG Stop	UGC ⎤ Cysteine UGC ⎦ UGA Stop UGG Tryptophan	U C A G
C	CUU ⎤ CUC CUA Leucine CUG ⎦	CCU ⎤ CCC CCA Proline CCG ⎦	CAU ⎤ Histidine CAC ⎦ CAA ⎤ Glutamine CAG ⎦	CGU ⎤ CGC CGA Arginine CGG ⎦	U C A G
A	AUU ⎤ AUC Isoleucine AUA ⎦ AUG Methionine Start	ACU ⎤ ACC ACA Threonine ACG ⎦	AAU ⎤ Asparagine AAC ⎦ AAA ⎤ Lysine AAG ⎦	AGU ⎤ Serine AGC ⎦ AGA ⎤ Arginine AGG ⎦	U C A G
G	GUU ⎤ GUC GUA Valine GUG ⎦	GCU ⎤ GCC GCA Alanine GCG ⎦	GAU ⎤ Asparagine GAC ⎦ GAA ⎤ Glutamic Acid GAG ⎦	GGU ⎤ GGC GGA Glycine GGG ⎦	U C A G

the end of a protein. If we think of the codons as genetic words, then a stop codon functions as the period at the end of the sentence.

TRANSFER RNA

A language interpreter translates a message from one language to another. Transfer RNA (tRNA) serves as an interpreter that converts the genetic message carried by mRNA into the language of protein, which is a particular sequence of amino acids. To accomplish this conversion a tRNA molecule must be able to recognize both the codon on mRNA and the amino acid that the codon specifies (Figure 21.5).

There are many kinds of tRNA—at least one for each of the 20 amino acids. Each type of tRNA molecule binds to a particular type of amino acid. Enzymes ensure the specificity of this binding. The tRNA then ferries the amino acid to the correct location along a strand of mRNA.

How does the tRNA know the correct location? The location is determined by a sequence of three nucleotides on the tRNA called the **anticodon**. In a sense, the anticodon "reads" the language of mRNA by binding to a codon on the mRNA molecule following the complementary base-pairing rules. When the tRNA's anticodon binds to the mRNA's codon, a specific amino acid is brought to the growing polypeptide

chain. For example, a tRNA molecule with the anticodon AAA binds to the amino acid phenylalanine and ferries it to the mRNA molecule, where the codon UUU is presented for translation. Phenylalanine will then be added to the growing amino acid chain.

RIBOSOMES

Ribosomes function as the workbenches on which proteins are built from amino acids. Recall from Chapter 3 that a ribosome consists of two subunits (small and large), each composed of ribosomal RNA (rRNA) and protein (Figure 21.6). The subunits form in the nucleus and are shipped to the cytoplasm. They remain separate except during protein synthesis. As you can see in Figure 21.6, when the two subunits fit together to form a functional ribosome, a groove for mRNA is formed. The ribosome has two binding sites for tRNA molecules. It also contains an enzyme that will promote the formation of a peptide bond between the amino acids that are attached to the tRNA molecules in the binding sites.

The role of the ribosome in protein synthesis is to bring the tRNA bearing an amino acid close enough to the mRNA to interact. The smaller subunit binds to mRNA. Two binding sites position tRNA molecules so that bonds can be formed between their amino acids. Enzymes in the ribosome then link the new amino acid to the growing amino acid chain.

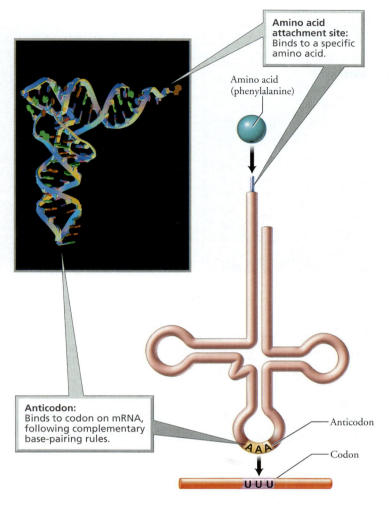

Amino acid attachment site: Binds to a specific amino acid.

Amino acid (phenylalanine)

Anticodon: Binds to codon on mRNA, following complementary base-pairing rules.

Anticodon

Codon

FIGURE **21.5**

A tRNA molecule is a short strand of RNA that twists and folds on itself. The job of tRNA is to ferry a specific amino acid to the ribosome and insert it in the appropriate position in the growing peptide chain.

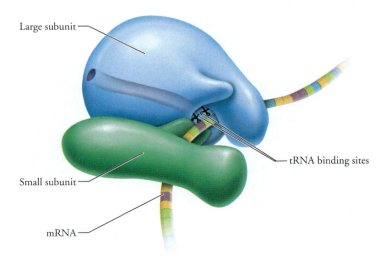

Large subunit

tRNA binding sites

Small subunit

mRNA

PROTEIN SYNTHESIS

Newly formed mRNA and ribosomal subunits move from the nucleus to the cytoplasm, where translation (protein synthesis) occurs. Translation can be divided into three stages: initiation, elongation, and termination (Figure 21.7).

1. **During initiation, the major players in protein synthesis (mRNA, tRNA, and ribosomes) come together.** First, the small ribosomal subunit attaches to the mRNA strand at the start codon, which is usually AUG. The tRNA with the complementary anticodon, UAC, bearing the amino acid methionine quickly pairs with the AUG codon. The larger ribosomal subunit then joins the smaller one to form a functional, intact ribosome. When the ribosomal subunits come together, mRNA is positioned in a groove between the two subunits. Initiation is important because it determines the point from which all codons will be read during protein synthesis and, therefore, the order and identity of amino acids in the protein.

2. **Elongation of the protein occurs as additional amino acids are added to the chain.** With the start codon positioned in one binding site, the next codon is aligned in the other binding site. The tRNA bearing an anticodon that will pair with the exposed codon slips into place at the binding site, and the amino acid it bears binds to the previous amino acid with the assistance of enzymes. The first tRNA then leaves the ribosome. The ribosome moves along the mRNA molecule, positioning the next codon in the open site. The appropriate tRNA slips into the open site, and its amino acid binds to the previous one. The tRNA in the first binding site leaves the ribosome. Then the mRNA and ribosome move relative to one another, and the next codon is aligned in the slot. The addition of each amino acid takes about 60 milliseconds.

   Numerous ribosomes may glide along an mRNA strand at the same time, each producing its own copy of the protein directed by that mRNA (Figure 21.8). As soon as one ribosome moves past the start codon, another ribosome can attach. A cluster of ribosomes simultaneously translating the same mRNA strand is called a **polysome**.

3. **Termination occurs when a stop codon—UAA, UAG, or UGA—moves into the ribosome.** There are no tRNA anticodons that pair with the stop codons, so protein synthesis is terminated. Certain proteins, called release factors, cause the newly synthesized protein, the mRNA strand, and the ribosomal sub-

FIGURE **21.6**

A ribosome consists of two subunits of different sizes. When the two subunits fit together to form a functional ribosome, a groove for mRNA is formed. The ribosome has two binding sites for tRNA molecules. It also contains an enzyme that will promote the formation of a peptide bond between the amino acids that are attached to the tRNAs in the binding sites.

Initiation: Translation Begins

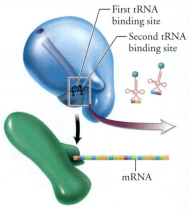

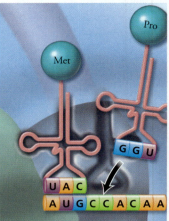

STEP 1
- The small ribosomal subunit joins to mRNA at the start codon, AUG

STEP 2
- A tRNA with a complementary anticodon pairs with the start codon
- Ribosomal subunits join to form a functional ribosome

Elongation: Amino Acids are Added to the Growing Chain One at a Time

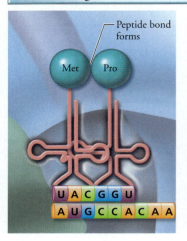

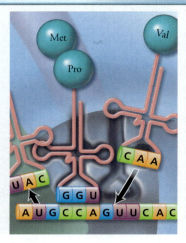

STEP 3
- A tRNA with the appropriate anticodon pairs with the next codon on mRNA
- Enzymes link the amino acids

STEP 4
- The first tRNA leaves the ribosomes
- The ribosome moves along the mRNA, exposing the next codon
- Enzymes link the amino acids

FIGURE **21.7**
Translation occurs in three steps: initiation, elongation, and termination.

units to separate from one another. The steps in gene expression are summarized in Figure 21.9.

stop and think

Streptomycin is an antibiotic, a drug taken to slow the growth of bacteria and allow body defense mechanisms more time to destroy the bacteria causing an infection. Streptomycin works by binding to the ribosome and preventing an accurate reading of mRNA. Why would this process slow bacterial growth?

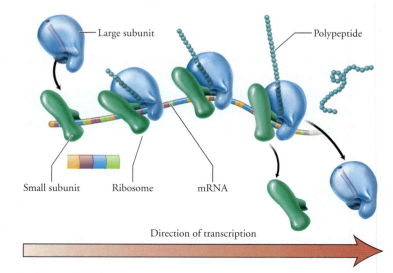

FIGURE **21.8**
Several ribosomes can read an mRNA molecule at the same time.

Point Mutations Result from Nucleotide Substitution, Insertion, or Deletion

DNA is remarkably stable, and the processes of replication, transcription, and translation generally occur with incredible precision. However, sometimes DNA is altered, and this alteration can change its message. Changes in DNA are called **mutations**. One type of mutation occurs when whole sections of chromosomes become rearranged, duplicated, or deleted, as discussed in Chapter 20. Now that we are familiar with the chemical structure of DNA and how it directs the synthesis of proteins, we can consider another type of mutation—point mutation. A **point mutation** involves changes in one or a few nucleotides in DNA. Although a point mutation can occur in any cell, it can be passed to one's child only when it is present in a cell that will become an egg or a sperm. A mutation that occurs in a body cell can affect the functioning of that cell, sometimes with disastrous effects, but it cannot be transmitted to offspring.

One way in which a point mutation can occur is through the replacement of one nucleotide pair by another pair of nucleotides. During DNA replication, bases may accidentally pair incorrectly. For example, adenine might mistakenly pair with cytosine instead of thymine. Repair enzymes normally replace the incorrect base with the correct one. However, sometimes the enzymes recognize that the bases are incorrectly paired but mistakenly replace the original base on the old strand. The result is a complementary base pair consisting of the wrong nucleotides (Figure 21.10).

Base-pair substitutions can have different consequences depending on its impact on protein synthesis. The effect on the resulting protein depends on the nature and location of the change in DNA.

Termination: The Newly Synthesized Protein is Released

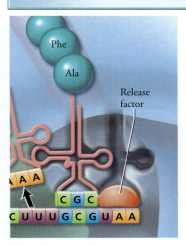

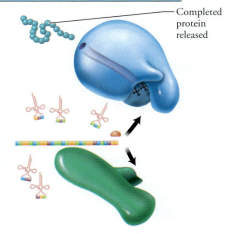

Completed protein released

Release factor

(Many shifts later)

STEP 5
- The tRNA in the first binding site leaves the ribosome
- The ribosome moves along the mRNA, exposing the next codon
- Enzymes link the amino acids
- The process is repeated many times

STEP 6
- The stop codon moves into the ribosome

STEP 7
- Release factors cause the release of the newly formed protein and the separation of the ribosomal subunits and the mRNA

Transcription: DNA to RNA
- DNA is transcribed to form mRNA.
- DNA's message is the sequence of bases.
- The sequence of bases in mRNA is complementary to that of DNA.
- The sequence of three base units in mRNA, called codons, specifies the order of amino acids in the resulting protein.

mRNA Processing
- Introns are cut out, and exons are spliced together.

The edited mRNA leaves the nucleus and carries DNA's information in its codons to a ribosome.

Transcription: RNA to Protein
- Ribosome subunits, mRNA, and tRNA assemble.
- mRNA is moved through the ribosome one codon at a time.
- Amino acid-bearing tRNA with appropriate anticodons pairs with codons in the ribosome.
- Amino acids are added to the growing polypeptide one at a time.
- Translation is terminated when a stop codon is encountered.

The protein is released, and the ribosome disassembles.

FIGURE **21.9**

A summary of the steps involved in gene expression

- **No change in protein structure.** Because of the redundancy of the genetic code, base substitutions do not always change the genetic message. For example, the third base pair in a DNA sequence that would be transcribed to the mRNA codon UCU could be altered so that the new codon read UCC. This substitution would not change the genetic message because the codons UCU and UCC both specify the amino acid serine.

- **Protein function remains the same.** Other substitutions may change the genetic message so that the wrong amino acid is added to the protein chain, but the change may not affect the functioning of that protein. Recall from Chapter 2 that the sequence of amino acids of a protein determines how the amino acid chain bends and folds into a particular shape. Recall also that the shape of a protein is important to its function and that some regions of a protein are more important to its function than are other regions. Some base substitutions may cause a change in a protein that does not affect either its shape or its function.

- **Protein function is altered.** Many substitutions do have drastic effects on the functioning of the resulting protein. A single base-pair substitution in the DNA coding for hemoglobin causes sickle-cell anemia, a condition in which the red blood cells become distorted when the oxygen content of the blood is low. Whereas the normal hemoglobin DNA reads CTC, the DNA that results in sickle-cell hemoglobin reads CAC. As a result, valine is incorporated into the protein instead of glutamic acid. This small change makes a big difference in the way that the resulting hemoglobin functions.

- **Protein function is destroyed.** If a point mutation produces a stop codon, protein synthesis stops immediately. The resulting protein will be incomplete and nonfunctional.

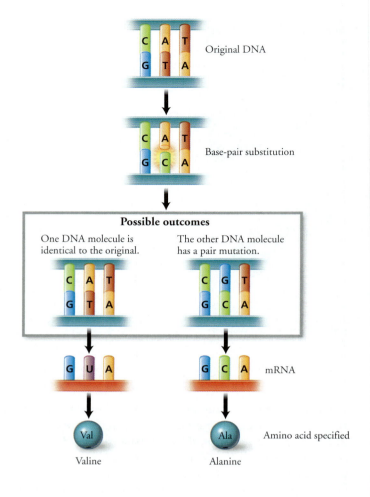

FIGURE **21.10**

A base-pair substitution is a point mutation in which one base is incorrectly paired with another. This may change the amino acid specified by the mRNA and alter the structure of the protein.

Gene Activity Can Be Turned On or Off

At the moment of your conception, you received one set of chromosomes from your father and one set from your mother. The resulting cell then began a remarkable series of cell divisions—some of which continue in many types of body cells to this day. With each cell division, the genetic information was faithfully replicated, and exact copies were parceled into the daughter cells. Thus, every cell in your body, except eggs or sperm, contains a complete set of identical genetic instructions for making every structure and performing every function in your body (see the Social Issue essay, *Stem Cells—The Body's Repair Kit*).

How, then, can liver, bone, blood, muscle, and nerve cells look and act differently from one another? The answer is deceptively simple: Only certain genes are active in a certain type of cell; most genes are turned off. Indeed, as cells become specialized for specific jobs, the timing of the activity of specific genes is critical. The active genes produce specific proteins, which determine the structure and function of that particular cell. For example, genes for producing the hormone insulin are present in every cell, but those genes are active only in certain cells of the pancreas.

As is often the case in biology, the answer to one question prompts other questions. What controls gene activity? How are genes turned on or off? The answers to those questions are a bit more complex because gene activity is controlled in several ways.

Coiling and uncoiling of chromosomes regulate gene activity at the chromosome level

At the chromosome level, gene activity is affected by an intricate system of folding and coiling of the DNA. When the DNA is tightly coiled, or condensed, the genes are not expressed. When a particular protein is needed in a cell, the region of the chromosome containing the necessary gene unwinds, allowing transcription to take place. Presumably, the uncoiling allows enzymes needed for transcription to reach the DNA in that region of the chromosome. Other regions of the chromosome remain tightly coiled and are not expressed.

Another type of point mutation, one that is caused by the insertion or deletion of a single nucleotide, generally has more serious effects than do those caused by substitutions. Recall that the mRNA is translated in units of three nucleotides (codons). *All* the triplet codons that follow the insertion or deletion are likely to change. Thus, point mutations due to the insertion or deletion of a nucleotide change the resulting protein. A sentence consisting of three-letter words ("codons") illustrates the result. Consider how deleting a single letter in the sentence "The big fat dog ran" renders the sentence nonsensical:

Original: THE BIG FAT DOG RAN

After deletion of the E in THE: THB IGF ATD OGR AN

Certain genes regulate the activity of other genes

Gene activity is also regulated by "master genes" that turn on a battery of other genes; the turning on of those genes, in turn, leads to the production of proteins that cause the development of a particular cell type. For instance, a master gene in an immature muscle cell produces a protein called myoD1. MyoD1 then turns on other genes that produce the muscle-specific proteins needed to transform the immature cell into a muscle cell.

If master genes are likened to the ignition switch in a car, then other regions of DNA, called enhancers, are similar to the accelerator. Enhancers are segments of DNA that increase the *rate* of transcription of certain genes and, therefore, the amount of a specific protein that is produced.

Chemical signals regulate gene activity

Chemical signals can also regulate gene activity. You may recall from Chapter 10 that one of the ways that certain hormones bring about their effects is by turning on specific genes. Steroid hormones, for instance, bind to receptors within a target cell. The hormone-receptor complex then finds its way to the nuclear chromatin and turns on specific genes. This complex, for example, turns on the genes in cells that produce facial hair—explaining why your father may have a beard, but your mother probably does not even though she has the necessary genes to grow one. In this case, the male hormone testosterone binds to a receptor and turns on hair-producing genes. Facial cells of both men and women have the necessary testosterone receptors. However, women usually do not produce enough testosterone to activate the hair-producing genes, so bearded women are rare.

stop and think

Considering the mechanism by which testosterone turns on hair-producing genes in certain cells on the face, explain why female athletes who inject themselves with testosterone to stimulate muscle development can develop facial hair.

In addition to hormones that turn on specific genes, regulatory proteins can activate a gene. An activator protein, for instance, can bind to DNA somewhere near the promoter for a gene. The activator protein then helps to position and activate RNA polymerase so that transcription begins.

Genetic Engineering Is the Manipulation of DNA for Human Purposes

The manipulation of genetic material for human purposes, the practice of **genetic engineering**, began almost as soon as we began to understand the language of DNA. Genetic engineering has allowed us to produce pharmaceuticals and hormones, improve diagnosis and treatment of human diseases, increase food production from plants and animals, and gain insight into the growth processes of cells. Genetic engineering is a subset of a broader endeavor, **biotechnology**, which involves making a living cell perform a task considered useful by humans in a controlled way.

Recombinant DNA is made of DNA from different sources

The basic idea behind genetic engineering is to put the "gene of interest," the one that produces the protein or trait that is considered useful, into another piece of DNA to create **recombinant DNA**, which is DNA created from two sources. The recombinant DNA then carries the gene of interest into another rapidly multiplying cell that quickly produces many copies of the gene of interest. Then the harvest—large amounts of the gene product or many copies of the gene of interest—is reaped. Let's consider the procedure in slightly more detail, one step at a time.

1. **The gene of interest is sliced out of one organism and spliced into vector DNA.** The *source DNA* which contains the gene of interest, and the *vector DNA*, which receives the transferred genes, are both cut at specific places by a **restriction enzyme**. Bacteria produce restriction enzymes as a defense against invasion by viruses. Viruses cause infection by inserting their own DNA into the host cell and taking over its genetic machinery. Restriction enzymes evolved to "chop up" viral DNA in bacteria and render it harmless. There are many different restriction enzymes. Each makes a staggered cut between specific base pairs, leaving several unpaired bases on the cut end of each DNA strand. The region of unpaired bases is called a "sticky end" because of its tendency to pair with single-stranded regions of complementary base sequences on the ends of other DNA molecules that were cut with the same restriction enzyme (Figure 21.11).

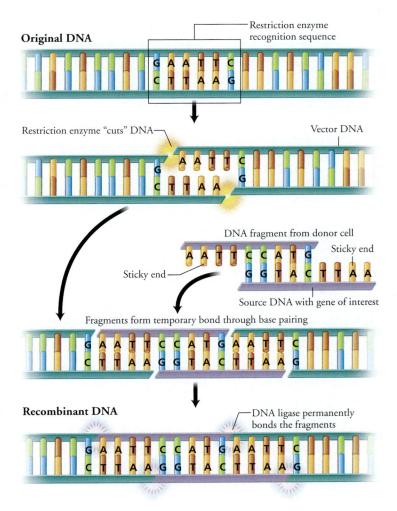

FIGURE **21.11**

DNA from different sources can be spliced together using a restriction enzyme to make cuts in the DNA. A restriction enzyme makes a staggered cut at specific sequences of DNA, leaving a region of unpaired bases on each cut end. The region of single-stranded DNA at the cut end is called a sticky end, because it tends to pair with the complementary sticky end of any other piece of DNA that has been cut with the same restriction enzyme, even if the pieces of DNA came from different sources.

SOCIAL ISSUE

Stem Cells—The Body's Repair Kit

The possibilities are mind boggling. Imagine growing new heart cells to repair damage from a heart attack. Think about curing Parkinson's disease by restoring the dopamine-producing neurons that are lacking in such patients or generating insulin-producing pancreatic islet cells to cure diabetes. Can we find a cure for multiple sclerosis, Alzheimer's disease, Lou Gehrig's disease, and spinal cord injuries? How? These cures may someday be possible by using the body's own repair kit, stem cells—cells that retain the ability to develop into many types of cells—to produce the needed type of cell.

Most of the trillions of cells in your body have become specialized to perform a particular job. Muscle cells are specialized to contract, and neurons are specialized to conduct nerve impulses. Recall that all cells in your body have the same genetic makeup. Cells become specialized for different functions as genes are turned on or off in different cells during development.

Stem cells, however, remain relatively unspecialized. They divide continually, giving rise either to more cells like themselves or to specialized cells of various types. A rich source of stem cells is the early embryo, because those embryonic cells will ultimately produce all the cells of the new individual. Adults also have stem cells, but not as many. Sprinkled in the bone marrow, brain, skin, liver, and other organs throughout the body, adult stem cells wait, ready to generate new cells for repair or replacement of old cells in the body. Diseased or dying cells provide biological cues that summon stem cells and prompt them to differenti-ate into the needed type of cell.

Adult stem cells do not seem to be as versatile as those from an embryo. Whereas embryonic stem cells can transform into any cell type in the body, adult stem cells usually form cells of a particular lineage. Adult stem cells in the bone marrow form blood cells. Those in the skin form epithelial cells. However, as we learn more about the signals that direct differentiation, we may be able to coax adult stem cells to form many more types of cells than they normally do in the body. The results of early studies attempting to direct the differentiation of adult stem cells have not been clear. Some scientists conclude that true differentiation occurred; others argue that stem cells simply fused with adult cells.

As a cell becomes specialized during development, its course is directed by biological cues in its immediate environment, including growth factors, surface proteins, pressure, salts, and touch. Neighboring cells usually provide these cues.

Although the technique is still in its infancy, the idea is to harvest stem cells and then provide the developmental cues that will direct them to differentiate into the type of cell needed to repair damage or to replace diseased or nonfunctional cells. The same stem cells may be coaxed into forming different types of cells depending on the environment in which they are placed. In some cases, the proper environment is created in a laboratory culture dish. In other cases, stem cells may be transplanted into the needy organ. The neighboring cells would then provide the environmental triggers that transform the stem cells into the necessary cell type. Almost all of the research to date has been done on nonhuman animals. Only a few pioneer studies have been done using humans. However, scientists are hopeful that the day when this technique will be used to improve the quality of human life will soon be here.

Embryonic stem cells are fairly easy to harvest (Figure 21.A), but their use has generated ethical and political turmoil. The cells come from embryos that were created for reproductive purposes but were not used. Only a few days old, the embryos left in fertility clinics are destined for destruction. Removing the stem cells destroys the embryo, preventing it from becoming a fetus.

In May 2003, scientists discovered a "master gene" in mouse and human embryonic stem cells that allows the cells that retain their potential to form any cell the body might need. This master gene has been named *nanog*, in reference to the mythological Celtic land of Tir Nan Og, where the residents remain forever young. *Nanog* is a regulatory gene, and its protein turns other genes on or off. The pattern of gene activity caused by the nanog protein is usually seen in 4- to 5-day-old human embryos. At this time, human cells still have the ability to develop into any type of cell. The human *nanog* gene was inserted into a mouse embryo and subjected to the type of laboratory conditions that usually cause the cells to mature into one type of tissue. The *nanog* gene prevented the cell from specializing. In the future, scientists hope to turn on the *nanog* gene in adult human cells, causing them to behave as embryonic stem cells and thereby eliminating the need for embryonic stem cells. But first, scientists must identify the signals that normally regulate

---

The sticky ends are the secret to splicing the gene of interest and the host DNA. The sticky ends of DNA from different sources will be complementary and stick together as long as they have been cut with the same restriction enzyme. The initial attachment between sticky ends is temporary, but they can be pasted together permanently by another enzyme, DNA ligase. The resulting recombinant DNA contains DNA from two sources.

2. **The recombinant DNA is transferred to a new host cell by the vector.** Biological carriers that ferry the recombinant DNA to the host cell are called **vectors**. Bacterial **plasmids**[2] are small, circular pieces of self-replicating DNA separate from the bacterial chromosome, and they are common vectors used to carry the gene of interest into bacteria. The source DNA and plasmid (vector) DNA are subjected to the same restriction enzyme. Fragments of source DNA, some of which will contain the gene of interest, will be incorporated into plasmids when their sticky ends join. Under the right conditions, some of the bacterial cells will take up the recombined plasmids.

3. **Cloning generates many copies of the gene of interest.** Next, the bacteria are grown in a laboratory under conditions that will select for bacteria that contain recombinant DNA. A plasmid of recombinant DNA can replicate within the bacterium, creating multiple copies. Also, each time a bacterium containing recom-

[2]Plasmids seem to have evolved as a means of moving genes among bacteria. A plasmid can replicate itself and pass, with its genes, into another bacterium.

nanog, and identifying those signals requires working with human embryonic stem cells.

Where do you stand on this thorny issue? Does the hope of lifesaving treatments justify the use of a human embryo? The embryos will still be destroyed even if they are not used for research. Should harvesting the stem cells from embryos be considered equivalent to salvaging organs from the dead for transplant to others whose lives may be saved? Or should we avoid using embryonic stem cells altogether and shift our focus entirely to research using adult stem cells? ✋

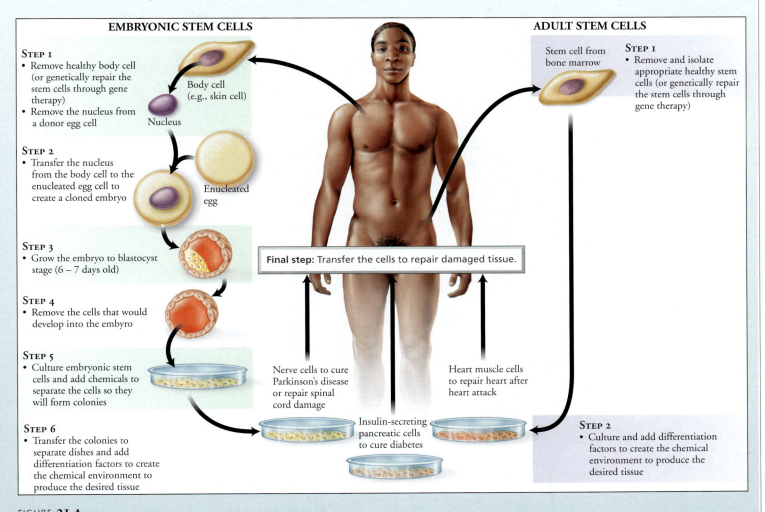

EMBRYONIC STEM CELLS

STEP 1
- Remove healthy body cell (or genetically repair the stem cells through gene therapy)
- Remove the nucleus from a donor egg cell

Body cell (e.g., skin cell)

Nucleus

STEP 2
- Transfer the nucleus from the body cell to the enucleated egg cell to create a cloned embryo

Enucleated egg

STEP 3
- Grow the embryo to blastocyst stage (6 – 7 days old)

STEP 4
- Remove the cells that would develop into the embyro

STEP 5
- Culture embryonic stem cells and add chemicals to separate the cells so they will form colonies

STEP 6
- Transfer the colonies to separate dishes and add differentiation factors to create the chemical environment to produce the desired tissue

Final step: Transfer the cells to repair damaged tissue.

Nerve cells to cure Parkinson's disease or repair spinal cord damage

Insulin-secreting pancreatic cells to cure diabetes

Heart muscle cells to repair heart after heart attack

ADULT STEM CELLS

Stem cell from bone marrow

STEP 1
- Remove and isolate appropriate healthy stem cells (or genetically repair the stem cells through gene therapy)

STEP 2
- Culture and add differentiation factors to create the chemical environment to produce the desired tissue

FIGURE **21.A**
The proposed procedures for using adult stem cells and embryonic stem cells

binant DNA divides, copies of the recombinant DNA go to each daughter cell. Each cell divides many times, forming a colony. Thus, each colony is a **clone**—a group of identical organisms all descended from a single ancestor. In this case, all members of a clone carry the same recombinant DNA plasmid. Bacteria reproduce rapidly, so cloning can produce a billion copies of a particular desirable gene in only 10 hours.

Although the basic idea is usually the same, there may be variations on this theme. For instance, the gene of interest is sometimes combined with viral DNA. The viruses are then used as vectors to insert the recombinant DNA into a host cell. Hosts other than bacteria, including yeast or animal cells, can also be used.

4. **The recombinant with the gene of interest is identified and isolated from the mixtures of recombinants.** A restriction enzyme cuts the source DNA into a mixture of many fragments, only some of which contain the gene of interest. Thus, thousands of recombinant plasmids are created, each containing a segment of the source DNA. Each recombinant plasmid is introduced into a bacterial cell, and the cells are grown into a colony. Different colonies contain different recombinant plasmids. The bacteria containing the gene of interest must be identified and isolated.

5. **The gene is further amplified through bacterial cloning or by using polymerase chain reaction.** After the colony containing the gene of interest has been identified, it is helpful to amplify

(duplicate) the gene. Gene amplification can be accomplished using one of two techniques: bacterial cloning or polymerase chain reaction. Bacteria containing the plasmid with the gene of interest can be grown in huge numbers by cloning. Each time a bacterium divides, another copy of the gene of interest is produced. Later the plasmids can be separated from the bacteria, a process that partially purifies the gene of interest. Then the plasmids can be taken up by other bacteria that will be used to perform a service deemed useful by humans, such as cleaning up oil spills, or by bacteria that will transfer the plasmid to plants. For other purposes, the plasmid can be injected into animal eggs. This procedure is summarized in Figure 21.12.

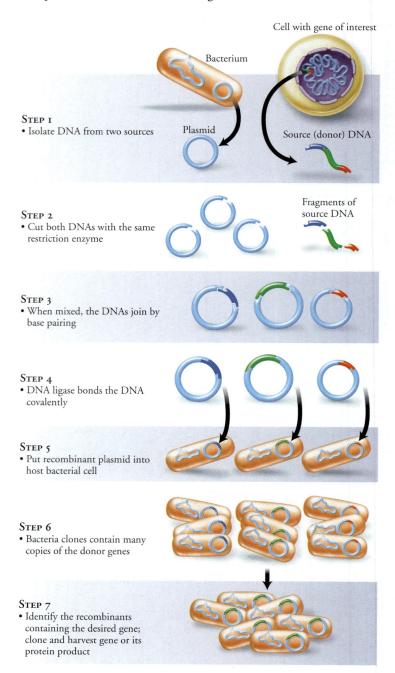

Cell with gene of interest

Bacterium

STEP 1
• Isolate DNA from two sources

Plasmid

Source (donor) DNA

STEP 2
• Cut both DNAs with the same restriction enzyme

Fragments of source DNA

STEP 3
• When mixed, the DNAs join by base pairing

STEP 4
• DNA ligase bonds the DNA covalently

STEP 5
• Put recombinant plasmid into host bacterial cell

STEP 6
• Bacteria clones contain many copies of the donor genes

STEP 7
• Identify the recombinants containing the desired gene; clone and harvest gene or its protein product

FIGURE **21.12**

An overview of genetic engineering using plasmids

The second technique for increasing the amount of specific DNA is a **polymerase chain reaction (PCR)** (Figure 21.13). In the PCR, the DNA of interest is unzipped, forming single strands. This unzipping is accomplished by heating the DNA. The single strands of DNA will serve as templates. These templates are mixed with primers (special short pieces of nucleic acid), one complementary primer to each strand. The primers serve as start tags for DNA replication. Nucleotides and the enzyme DNA polymerase, which promotes DNA replication, are also added. The mixture is then cooled to allow base pairing. A complementary strand forms for each single strand. This procedure is then repeated many times. Each time this is done, the number of copies of the DNA of interest is doubled. In this way, billions of copies of the DNA of interest can be produced in a short time.

what would you do?

Genetic engineering involves altering an organism's genes—adding new genes and traits to microbes, plants, or even animals. Do you think we have the right to "play God" and alter life forms in this way? The U.S. Supreme Court has approved patenting of genetically engineered organisms, first microbes and later mammals. If you were asked to decide whether it is ethical to patent a new life form, how would you respond?

Genetic engineering produces proteins of interest or transgenic organisms

Genetic engineering has been used in two general ways.

• **Genetic engineering provides a way to produce large quantities of a particular gene product.** The useful gene is simply transferred to another cell, usually a bacterium or a yeast cell, that can be grown easily in large quantities. The cells are then manipulated so that the gene is expressed. The protein is then harvested. For example, genetically engineered bacteria have produced large quantities of human growth hormone (Figure 21.14a). Treatment with growth hormone allows children with an underactive pituitary gland to grow to nearly normal height.

• **Genetic engineering allows a gene for a trait considered useful by humans to be taken from one species and transferred to the cells of another species.** These transgenic organisms, organisms that contain genes from another species, then show the desired trait. For example, scientists have endowed salmon with a gene from an eel-like fish. This gene causes the salmon to produce growth hormone year-round (Figure 21.14b). As a result, the salmon grow several times larger than usual.[3]

THE ENVIRONMENT

Genetic engineering also has environmental applications. In sewage treatment, genetically engineered microbes lessen the amount of phosphate and nitrate discharged into waterways. Phosphate and nitrate can cause excessive growth of aquatic plants, which could choke waterways and dams, and algae, which can produce chemicals that are poisonous to fish and livestock. Microorganisms are also being genetically engineered to modify or destroy chemical

[3]In mid-2004, the Food and Drug Administration was considering approval for genetically modified salmon to be marketed.

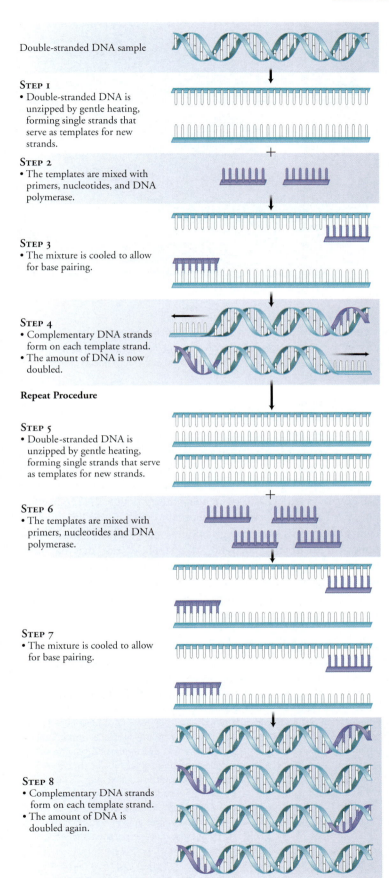

Double-stranded DNA sample

STEP 1
• Double-stranded DNA is unzipped by gentle heating, forming single strands that serve as templates for new strands.

STEP 2
• The templates are mixed with primers, nucleotides, and DNA polymerase.

STEP 3
• The mixture is cooled to allow for base pairing.

STEP 4
• Complementary DNA strands form on each template strand.
• The amount of DNA is now doubled.

Repeat Procedure

STEP 5
• Double-stranded DNA is unzipped by gentle heating, forming single strands that serve as templates for new strands.

STEP 6
• The templates are mixed with primers, nucleotides and DNA polymerase.

STEP 7
• The mixture is cooled to allow for base pairing.

STEP 8
• Complementary DNA strands form on each template strand.
• The amount of DNA is doubled again.

The procedure is repeated many times, doubling the amount of DNA with each round.

wastes or contaminants so that they are no longer harmful to the environment. For instance, oil-eating microbes that can withstand the high salt concentrations and low temperatures of the oceans have proven useful in cleaning up after marine oil spills.

LIVESTOCK

Genetic engineering has also been used on livestock. Vaccines have been created to protect piglets against a form of dysentery called scours, sheep against foot rot and measles, and chickens against bursal disease (a viral disease affecting the bursa[4] and liver that is often fatal). Genetically engineered bacteria produce bovine somatotropin (BST), a hormone naturally produced by a cow's pituitary gland that enhances milk production. Injections of BST can boost milk production by nearly 25%.

Transgenic animals have been created in a petri dish by injecting a fertilized egg with the gene of interest. In this way, extra copies of the gene for growth hormone have been introduced into cows and pigs, creating fast-growing animals that yield leaner meat.

PHARMACEUTICALS

Genes have been put into a variety of cells, ranging from microbes to mammals, to produce proteins for treating allergies, cancer, heart attacks, blood disorders, autoimmune disease (see Chapter 13), and infections. Among these products are hormones, including insulin for treating diabetes and human growth hormone.

Genetically engineered bacteria have been used to create vaccines. You may recall from Chapter 13 that a vaccine typically uses an inactivated bacterium or virus to stimulate the body's immune response to that type of organism. The idea is that the body will learn to recognize proteins on the surface of the infectious organism and mount defenses against any organism bearing those proteins. Because the organism used in the vaccine was rendered harmless, the vaccine cannot trigger an infection. Vaccines that are produced in the traditional way work well, but the production of vaccines is time consuming and expensive. In a genetically engineered vaccine, the gene for the surface protein of the infectious organism is put into bacteria. The bacteria then produce large quantities of that protein, which can be used as a vaccine. The vaccine cannot cause infection because only the surface protein is used.

Plants have also been used to produce therapeutic proteins. Engineering bananas to produce an altered form of the hepatitis B virus has produced a vaccine against hepatitis B, a liver disease. Someday, you may eat a banana instead of receiving an injection to be vaccinated against hepatitis B. Plants are also being engineered to produce "plantibodies," antibodies produced by plants. For example, soybeans are being cultivated that contain human antibodies to the herpes simplex virus that causes genital herpes. A human gene for an antibody that binds to tumor cells has been transplanted into corn. The antibodies can then deliver radioisotopes to cancer cells, selectively killing them.

---

[4]The bursa of Fabricus is an organ that defends against disease in chickens.

FIGURE **21.13**

The polymerase chain reaction (PCR) rapidly produces a multitude of copies of a single gene or any specific DNA. PCR amplifies DNA more quickly than does bacterial cloning. It has many uses besides genetic engineering, including DNA fingerprinting.

(a)

(b)

FIGURE **21.14**

Genetic engineering is used to produce large quantities of a desired protein or to create an organism with a desired trait. (a) This girl has an underactive pituitary gland. The undersecretion of growth hormone would have caused her to be very short, even as an adult. However, growth hormone from genetically engineered bacteria has helped her grow to almost a normal height. (b) This genetically modified salmon has a gene from an eel-like fish that allows it to produce growth hormone year-round, causing it to grow to several times its normal size.

Pharming is a word that comes from the combination of the words farming and pharmaceuticals. In gene pharming, transgenic animals are created that produce a protein with medicinal value in their milk, eggs, or blood. The protein is then collected and purified for use as a pharmaceutical. When the pharm animal is a mammal, the gene is usually inserted into the chromosome at a site that regulates milk production so that it is expressed only in mammary glands. The desired protein is then collected in milk (Figure 21.15). For example, the gene for the protein alpha-1-antitrysin (AAT) has been inserted into sheep that then secrete AAT in their milk. People with an inherited, potentially fatal form of emphysema (a lung disease) take AAT as a drug. It is also being tested as a drug to prevent lung damage in people with cystic fibrosis.

Researchers are currently working on a vaccine carried in goat's milk against malaria, a disease that affects 300 million to 500 million people a year. If the researchers are successful, eight goats could produce enough vaccine to innoculate 20 million people.

PROTEIN PRODUCTS

Genetic engineering is also used to produce useful proteins. For example, spider silk is five times stronger than steel and still lightweight. Attempts have been made to farm spiders for their silk, but spiders are too aggressive to live close together. It is hoped that when the goats that were genetically engineered to possess the gene for spider silk mate, they will create a herd of goats that will secrete spider silk in their milk (Figure 21.16). The spider silk proteins could then be spun into a fine thread called BioSteel®, which could be used for bullet-proof clothing, thinner surgical thread, and stronger yet lighter racing cars and aircraft.

AGRICULTURE

We experience some of the results of genetic engineering at our dinner tables (see the Health Issue essay, *Genetically Modified Food*). The most common traits that have been genetically engineered into crops are either resistance to pests or resistance to herbicides. Scientists also developed two virus-resistant strains of papaya and distributed them to papaya growers in Hawaii, saving the industry (Figure 21.17). In addition, rice has been genetically engineered to resist disease-causing bacteria. Other plants have been genetically engineered to be more nutritious. For example, golden rice is a strain of rice that has been genetically engineered to produce high levels of beta-carotene, which is in short supply in certain parts of the world. Other crops have also been created that grow faster, produce greater yields, and have longer shelf lives.

what would you do?

Some ecologists are concerned about the harm to the environment or humans that could occur if genetically modified plants or animals escaped from the places where they were bred. The U.S. Department of Agriculture (USDA) has approved tests of plants that are genetically altered to produce pharmaceuticals. In Hawaii, where most of these tests have been conducted, environmental groups and consumer groups filed a lawsuit in 2003 asking the judge to order the USDA to better regulate these tests. The suit was filed because ecologists are concerned that tradewinds might carry pollen from some of these biopharm plants out of the test area and the altered genes might spread to native plants and animals. This spread could threaten Hawaii's fragile biodiversity and harm humans. Ecologists also warn that the escape of the gene-altered fast-growing salmon could decimate the population of wild salmon. Do you think that the potential benefits of genetically engineered plants and animals outweigh the harm that might occur if the genetically altered organisms escaped? Would you object if a biopharm were developed near your home? What are your reasons?

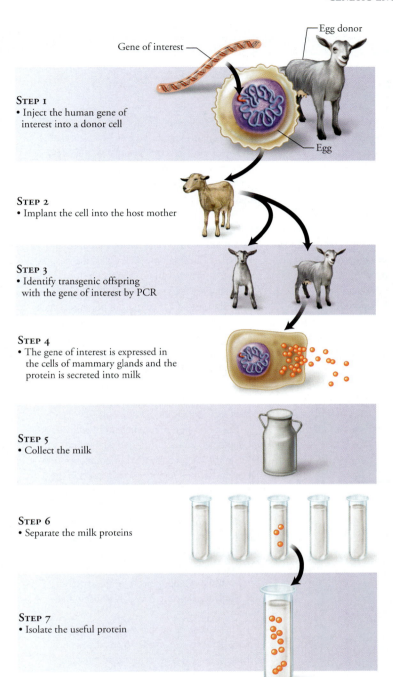

Gene of interest

Egg donor

STEP 1
• Inject the human gene of interest into a donor cell

Egg

STEP 2
• Implant the cell into the host mother

STEP 3
• Identify transgenic offspring with the gene of interest by PCR

STEP 4
• The gene of interest is expressed in the cells of mammary glands and the protein is secreted into milk

STEP 5
• Collect the milk

STEP 6
• Separate the milk proteins

STEP 7
• Isolate the useful protein

FIGURE **21.15**
The procedure for producing a transgenic animal that will produce a useful protein in its milk

FIGURE **21.16**
This transgenic goat has the gene for making spider silk, one of the strongest substances known. The spider silk protein can be extracted from the goat's milk and spun into threads that can be used for products in which strength and light weight are important qualities.

FIGURE **21.17**
The papaya industry in Hawaii was rescued by the introduction of genetically engineered papaya plants that are resistant to a particular virus.

▐ Gene therapy replaces faulty genes with functional genes

The problems associated with many genetic diseases result because a faulty gene fails to produce its normal protein product. The goal of **gene therapy** is to cure genetic diseases by putting healthy, functional genes into the body cells that were affected by the faulty gene, the so-called target cells. The healthy gene then produces the needed protein. We will consider some specific examples shortly.

METHODS OF DELIVERING A HEALTHY GENE

One way a healthy gene is transferred to an affected cell is by using viruses. Viruses generally attack only one type of cell. For instance, an adenovirus, which causes the common cold, typically attacks cells of the respiratory system. A virus consists of little more than genetic material, commonly DNA, surrounded by a protein coat (see Chapter 13a). Once inside the cell, the viral DNA uses the cell's metabolic machinery to produce its

Genetically Modified Food

From dinner tables to diplomatic circles, people are now discussing genetically modified (GM) food.

This current interest may seem somewhat surprising because people in the United States have been eating GM food since the mid-1990s. Today, more than 60% of the food in U.S. supermarkets contains genetically modified soybeans, corn, or canola. Worldwide, 130 million acres in 13 countries are planted with GM crops. Yet many people vehemently object to GM Food.

Why is something as common as GM food still controversial? The concerns about GM food generally fall into one of three categories: health concerns, social concerns, and environmental concerns (Table 21.A). Let's explore what is known about concerns in each of those categories.

Health Concerns In July 2004, a panel of the National Academy of Sciences (NAS) issued a report saying that genetically engineered crops do not pose health risks that cannot also be caused by crops created by conventional breeding. However, because genetic engineering could produce unintended harmful changes in food, the NAS panel recommends scrutiny of GM foods before they can be marketed. Currently, the U.S. Department of Agriculture (USDA), the Food and Drug Administration (FDA), and the Environmental Protection Agency (EPA) regulate genetically modified foods. The NAS panel concluded that the GM foods already on the market are safe.

A common safety concern is that GM foods may contain allergens (substances that cause allergies). After a protein is produced, the cell modifies it in various ways. The protein may be modified in the genetically modified plant differently than it would be in an unmodified cell. The likelihood is reduced by rigorous testing. Most known allergens share certain properties. They are proteins, relatively small molecules, and resistant to heat, acid, and digestion in the stomach. The FDA considers any protein produced by a GM plant that has any of those properties or that is structurally similar to a known allergen to be a potential allergen and require that the protein undergo additional allergy testing.

Bacterial resistance to antibiotics is likely to be a major threat to public health in this decade (see Chapter 13a). When bacteria are resistant to an antibiotic, the drug will not kill the bacteria. As a result, we can no longer cure the human disease caused by that bacterium. Genes for resistance to an antibiotic are put into GM crops as markers to identify them as genetically modified. Plant seedlings thought to be genetically modified are grown in the laboratory in the presence of an antibiotic. Only those seedlings with the gene for resistance will survive. Because of the way they were engineered, the surviving plants also contain the "useful" gene.

Concerns exist that genes for resistance to antibiotics could be transferred to bacteria, making the bacteria resistant to antibiotics. The receiving bacteria might be those living in the human digestive system or bacteria ingested with food. It is not known whether genes can be transferred from a plant to a bacterium. However, it is known that bacteria can easily and quickly transfer genes for antibiotic resistance to one another. Thus, a harmless bacterium in the gut could transfer the gene for antibiotic resistance to a disease-causing bacterium.

The transfer of antibiotic-resistance genes from GM plants to bacteria could have serious consequences. As a result, antibiotic-resistance marker genes are being phased out in favor of other marker genes, such as a green fluorescent protein. Scientists also have developed a way to

TABLE **21.A**

CONCERNS ABOUT GENETICALLY MODIFIED FOOD	
Health concerns	Is GM food safe for human consumption? Could GM crops increase bacterial resistance to antibiotics?
Environmental concerns	What effects will GM crops have on the level of use of pesticides? Can GM crops harm other organisms? Will GM crops become "superweeds"?
Social concerns	Can GM foods reduce world hunger? Who is responsible if GM foods prove to be harmful to humans or the environment?

own proteins. So, if the healthy gene were spliced into the DNA of a virus that had been rendered harmless, the virus would theoretically deliver the healthy gene to the host cell and ensure that the desired gene product was produced (Figure 21.18a). Another type of virus used in gene therapy is a retrovirus, a virus whose genetic information is in RNA, rather than DNA. Once inside the target cell, a retrovirus rewrites its genetic information as double-stranded DNA and inserts the viral DNA into a chromosome of the target cell.

Other methods exist for delivering genes into cells for gene therapy. The gene can be enclosed in a capsule of fatty material called a liposome. Liposomes can fuse with cell membranes and deliver the genetic material to the inside of the cell. Some scientists attach molecules that will bind to special cell receptors to the liposome or directly to the DNA. Once bound to these receptors, the liposome or DNA is engulfed by the cell membrane and enters the cell. Another experimental method of delivery of DNA is to inject it directly into the cell.

inactivate the antibiotic-resistance gene if it is transferred to bacteria.

Environmental Concerns Proponents of GM foods argue that herbicide-resistant and pesticide-resistant crops reduce the need for spraying with herbicides and pesticides. But have pest-resistant corn or cotton or herbicide-resistant soybeans reduced the amount of chemical spraying? The answer depends on the crop. Pest-resistant cotton has substantially reduced the use of pesticides, but pest-resistant corn probably has not. Farmers who grow herbicide-resistant crops still spray with herbicides, but they change the type of herbicide they use to a type that is less harmful to animals.

However, engineered crops containing insecticides could have some undesirable effects on insects. Genetically engineering insecticides into plants could hasten the development of insect resistance to that insecticide, making the insecticide ineffective—not just for the genetically modified crop but for all crops.

Another concern is that genetically modified organisms could harm other organisms. One example is that pollen from pest-resistant corn has been shown to harm monarch butterfly caterpillars. Fortunately, monarch caterpillars rarely encounter enough pollen to be harmed, and most of the pest-resistant corn grown in the United States today does not produce pollen that is harmful to monarch butterflies. A second example of a genetically modified organism that has the potential to harm other organisms is the salmon that produce more growth hormone and grow several times faster than their wild relatives do. These salmon are grown on fish farms. If the FDA grants approval, the gene-altered salmon could dramatically cut costs for fish farmers and consumers. However, when the genetically modified salmon are grown in tanks with ordinary salmon and food is scarce, the genetically modified salmon eat most of the food *and* some of their ordinary companions. What would happen if the genetically modified salmon escaped from their pens on the fish farm? They might mate with the wild salmon and create less healthy offspring or outcompete wild salmon for food, which could eventually cause the extinction of the wild salmon. Escape is possible. During the past few years, hundreds of thousands of fish have escaped from fish farms when floating pens were ripped apart by storms or sea lions. To minimize the risk that genetically modified salmon could destroy the population of wild salmon, scientists plan to breed the fish inland, sterilize the offspring, and ship only sterile fish to coastal pens. The sterilization procedure is effective in small batches of fish, but it is not known whether the procedure is completely effective in large batches of fish.

Critics of GM foods are also concerned that crops that are genetically engineered to resist herbicides could become "superweeds" that could not be controlled with existing chemicals. In Canada, canola crops in several areas have been genetically modified, each for resistance to a different herbicide. Because pollen from canola plants can spread about 8 kilometers (5 miles), neighboring crops have cross-pollinated, producing seeds that are resistant to many herbicides. Could seeds left behind after a harvest grow and begin to invade other areas? Some scientists worry that they could. Others point out that there are more kinds of herbicides than there are resistance genes, so the modified plants could still be easily controlled.

Social Concerns Proponents of GM food suggest that GM food can assist in the battle against world hunger. We have seen that genetic engineering can produce crops that resist pests and disease. It can also produce crops with greater yields and crops that will grow on land that will not support conventional farming because of drought, depleted soil, or excess salt, aluminum, or iron.

Foods can also be genetically modified to contain higher amounts of specific nutrients. One example is "golden rice," which was genetically modified to contain high levels of beta-carotene, a precursor for vitamin A. More than 100 million children worldwide suffer from vitamin A deficiency, and 500,000 of them go blind every year because of the deficiency. Although golden rice cannot supply a complete recommended daily dose of vitamin A, the amount it contains could be helpful to a person whose diet is very low in vitamin A.

Critics of GM food argue that there is enough food to feed all the hungry people in the world. The problem is distributing food to the people who need it.

Finally, if any of the fears of those who oppose GM foods are realized, who should be held accountable? Farmers buying GM seeds from Monsanto, one of the largest producers of GM foods, must sign an agreement that Monsanto is not liable for any damages that might result from the use of their product. At the same time, Monsanto insists that there are no dangers. Indeed, there are farmers in Canada who get premium prices for their canola because they keep their crops free of pesticides and genetic modifications. However, some of their canola fields have now tested positive for genes that have been introduced into crops hundreds of miles away. The value of their crops has dropped precipitously. Many of these farmers have joined in a suit against the companies that produce genetically modified canola. The companies have argued that the individual farmers who buy their genetically modified seeds are responsible for any damage done to their neighbors.

Do you consider GM food to be a blessing or a danger to the world? What are your reasons? 👫

EXAMPLES OF CONDITIONS SUITABLE FOR GENE THERAPY

More than 4000 human diseases result from a defect in a single gene. Although the Food and Drug Administration (FDA) has not yet approved any gene therapy treatment, there are hundreds of clinical trials on gene therapy for many conditions.

SCID The first condition to be treated with gene therapy was a disorder referred to as severe combined immunodeficiency disease (SCID). The immune system of children with SCID is devastated, leaving them vulnerable to every germ. The cause of the problem is a mutant gene that prevents the production of an enzyme called adenosine deaminase (ADA). Without ADA, white blood cells never mature and die while still developing in the bone marrow. The first gene therapy trial began in 1990 when white blood cells of 4-year-old Ashanthi DeSilva, a child with SCID, were genetically engineered to contain the ADA gene and then returned to her tiny body. Her own gene-altered white blood cells began producing ADA, and her body defense mechanisms were strengthened. Ashanthi's life began to change. She was not

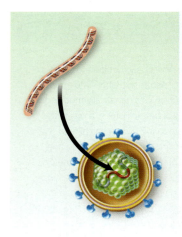

STEP 1
Incorporate a healthy form of the gene into the virus.

(a)

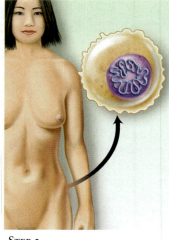

STEP 2
Remove bone marrow stem cells from the patient.

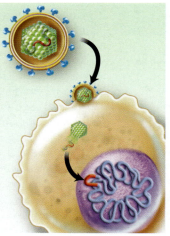

STEP 3
Infect the patient's stem cells with the virus that is carrying the healthy form of the gene.

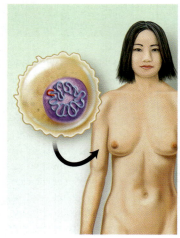

STEP 4
Return the genetically engineered stem cells to the patient. The gene is expressed to produce the needed protein.

(b)

FIGURE **21.18**

In gene therapy a healthy gene is given to a patient who has a genetic disease caused by a faulty gene. (a) The procedure for gene therapy using a virus. (b) Rhys Evans is the first person to be cured of severe combined immunodeficiency disease (SCID) by gene therapy. SCID is caused by a mutant gene that fails to produce an enzyme needed for the maturation of white blood cells for body defense. His immune system was strengthened, and he can now go out into public places without fearing contact with people who might carry germs. He can play with other children.

ill as often as she had been before. She could play with other children. However, the life span of white blood cells is measured in weeks, and when the number of gene-altered cells declined, new gene-altered cells must be infused. Ashanthi is now a teenager with a reasonably healthy immune system. However, she still needs repeated treatments.

French scientists believe they have cured 10 children with SCID using gene therapy (Figure 21.18b). It is still too soon to know for certain whether all 10 children will require ADA injections in the future. Two children in the French studies developed leukemia, as we will discuss shortly.

CYSTIC FIBROSIS   The genetic disease cystic fibrosis is caused by the lack of a functional gene that codes for a protein that transports chloride ions across membranes. The lack of this protein disrupts the flow of water and salt into and out of cells, causing thick, sticky mucus to

form in the lungs and airways. Researchers hope that restoring a functional copy of the gene that codes for the transport protein will end the damage to lung cells. In a 2003 study on gene therapy for cystic fibrosis, the gene was inserted into cells lining the nasal cavity. In most patients, the treated cells showed some correction. The next trials will focus on inserting the gene into lung tissue.

CANCER   Although still in its infancy and considered highly experimental, gene therapy is being tested as treatment for several forms of cancer, including cancers of the brain, lung, ovary, kidney, and skin (melanoma, an especially deadly form of skin cancer). Many forms of cancer involve defects in genes that control cell division (see Chapter 21a). Gene therapy for cancer involves several strategies, including inserting (1) functional genes to replace the faulty ones that cause unrestrained cell division, (2) genes to stimulate the immune response against cancer cells, and (3) genes into tumor cells that will kill only the cancerous cells while leaving the cells of surrounding tissue unharmed. We will consider these techniques in more detail in Chapter 21a, after we are more familiar with the genetic basis of cancer.

RISKS OF GENE THERAPY Unfortunately, gene therapy using viruses has resulted in several deaths. An 18-year-old boy died while participating in a gene therapy trial in 1999. In 2003, two toddlers in France developed leukemia following gene therapy using retroviruses to treat severe combined immunodeficiency disease (SCID), the rare inherited disease discussed earlier.

what would you do?

Without treatment, children with SCID die at a young age. The gene therapy treatment for SCID that caused leukemia in two French boys does appear to cure this deadly disorder. If you had a child with SCID, would you want him or her to have this gene therapy treatment? Why or why not? What factors would you consider in reaching your decision?

▌Faulty information in mRNA may someday be correctable

The kind of gene therapy that uses a functional gene to replace a defective one can be helpful if adding a functional protein is enough to alleviate the problem. A newer approach, which has not yet been tested on humans, is to leave the faulty gene in the DNA but to correct the mRNA message before it is translated to protein.

Earlier in this chapter, we saw that DNA often contains regions called introns that are not translated into protein. The introns are transcribed to mRNA but are snipped out of the RNA by enzymes. These enzymes recognize a specific sequence of bases on mRNA, snip the RNA wherever those sequences are found, and then splice the cut ends together. Scientists are working to design enzymes that will snip the errors out of mRNA produced by a faulty gene. The technique has the potential to cure sickle-cell anemia, cystic fibrosis, and some cancers.

▌Gene silencing may someday adjust gene activity

Gene silencing, also called RNA interference, is used to adjust the level of activity of a specific gene or to prevent the protein product of a faulty gene from being produced. Cells use short pieces of double-stranded RNA called short, intervening RNAs, or siRNAs, to cut mRNA with a specific sequence. If an siRNA is designed to match mRNA from a faulty gene, then the abnormal protein will not be produced. This technique may someday be used to treat genetic diseases caused by a dominant allele, such as Huntington's disease (see Chapter 19). The technique would silence the faulty allele and leave the normal allele to produce its protein. It also could be used to silence viral genes, preventing viral infection, or to silence overactive genes that cause cancer (see Chapter 21a).

▌DNA fingerprinting can identify individuals

DNA fingerprints, like the more conventional ones on fingers, can help identify individuals in a large population. DNA fingerprints are possible because there are many regions of DNA that do not code for known traits. Some of these regions are composed of small, specific sequences of DNA that are repeated many times. The number of times these sequences are repeated varies considerably from person to person, from as few as four to several hundred repeats. Because the length of segments containing repeating sequences varies throughout the population, these segments can be used to help identify a specific person.

The first step in preparing a DNA fingerprint is to extract DNA from a tissue sample. The type of tissue does not matter, so sources that are readily available or that are not too painful to remove, including blood, semen, skin, and hair follicles, are commonly used.

If only a minute tissue sample is available, a spot of blood from a crime scene for instance, the amount of DNA can be greatly increased using PCR, as described earlier. The DNA can then be cut into pieces of varying length by using a restriction enzyme. The result is a mixture of millions of tiny DNA fragments of many different sizes.

If we were to sort and arrange these DNA fragments according to size, the pattern that would result from the DNA from one person would always be identical. *But*, if we were to compare the cut-up DNA from different people, the pattern would always be different because the number and size of the fragments are determined by the sequence of bases along the DNA strand. The pattern of DNA fragments that have been cut by a restriction enzyme and sorted by size is called a **DNA fingerprint**.

DNA fingerprinting has many applications, but the most familiar is probably in crime investigations. In general, the DNA fingerprint is created from a sample of tissue such as blood or hair follicles that was collected at the crime scene. A fingerprint can be produced from tissue left at the scene years before. That DNA fingerprint is then compared with the DNA fingerprints of various suspects. A match of DNA fingerprints reveals the culprit with a high degree of certainty (Figure 21.19).

Of course, the degree of certainty of a match between DNA fingerprints depends on how carefully the analysis was done. DNA fingerprints should look like a bar code, but sometimes the bars run together, lessening the certainty of a match. Because DNA fingerprints are being used as evidence in an increasing number of court cases each year, it is important that national standards be set to ensure the reliability of these molecular witnesses.

It is generally easier to declare with certainty that a DNA sample is *not* a match to a person than it is to be sure that it is a match. Many convicts, including more than 110 on death row, have been found innocent through DNA testing since 1990.

The Purpose of the Human Genome Project Is to Identify the Complete DNA Sequence of the Human Genome

A genome is an organism's complete set of DNA, including all of its genes. The Human Genome Project, completed in 2003, was a worldwide research effort to reach a common goal: determining the location of genes along the 23 pairs of chromosomes and sequencing (determining the order of) the estimated 3 billion base pairs that make up those chromosomes. The DNA of model organisms, including the mouse, the fruit fly, a roundworm, and yeast, are also being mapped with the hope of gaining some

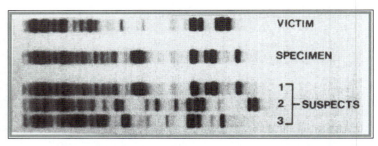

FIGURE **21.19**

The pattern of banding in a DNA fingerprint is determined by the sequence of bases in a person's DNA and is, therefore, unique to each person. A match of DNA fingerprints between the crime scene and a suspect can reveal the culprit with a high degree of certainty. Which one of the suspects' DNA fingerprint matches the specimen found at the crime scene?

insight into basic biology—the organization of genes within the genome, gene regulation, and molecular evolution.

We have learned many things from this monumental effort.

- The human genome consists of 30,000 to 40,000 genes, not 100,000 as originally thought. One reason for the smaller number of actual genes is that many gene families have related or redundant functions. Thus, fewer genes are needed to carry out body functions. A second reason is that many genes are now known to code for more than one protein.

- Genes for more than 1400 genetic diseases have been identified and mapped to specific locations on specific chromosomes.

- More than 95% of human DNA does not code for protein. Some of this DNA consists of regulatory regions that determine when, where, and how much of certain proteins are produced.

- Humans share many genes with other organisms. For example, we share half of our genes with the fruit fly and 90% with the mouse. The genetic similarities are evidence of our common evolutionary past. The genes we share with other organisms are likely to be important in determining body plan as well as development and aging.

It is hoped that this information will give us a greater ability to diagnose, understand, and eventually treat many of the 4000 diseases

of humans that have a genetic basis. Researchers have already identified and cloned the genes responsible for many genetic diseases, including Duchenne muscular dystrophy, retinoblastoma, cystic fibrosis, and neurofibromatosis. These isolated genes can now be used to test for their presence in people. As we saw in Chapter 20, gene tests can be used to identify people who are carriers for certain genetic diseases such as cystic fibrosis, allowing families to make choices based on known probabilities of bearing an affected child, and they can be used for prenatal diagnosis and for diagnosis before symptoms of the disease begin. It is further hoped that, after a disease-related gene has been identified, scientists can learn more about the protein it codes for and this information can suggest ways to correct the problem. We now know that there are single-nucleotide polymorphisms (SNPs, or "snips")—differences in one nucleotide in a DNA sequence that vary among individuals—that influence how we respond to stress and disease. As we learn more about SNPs, we may be able to develop treatments tailored to the genetic makeup of each individual. Also, finding a disease-related gene may open the door for gene therapy.

what would you do?

Some people worry that once we know the location of every gene and have perfected the techniques of gene therapy, we will no longer limit gene manipulation to repairing faulty genes and begin to modify genes to enhance human qualities. We could begin to design our babies by choosing "good" genes. What do you think? Where should the line be drawn? Who should draw that line? Who should decide which genes are "good"?

exploring further ...

In Chapter 19, we learned about the cell cycle. We learned about genes, their inheritance, and their regulation in Chapters 20 and 21. We also considered how mutations affect gene functions. In Chapter 21a, we will use what we have learned to understand cancer—a family of diseases caused by a loss of control over cell division caused by mutations in genes that regulate the cell cycle.

REVIEWING THE CONCEPTS

DNA Is a Double Helix Consisting of Two Strings of Nucleotides (p. 476)

1. DNA consists of two strands of linked nucleotides, twisted together to form a double helix. Each nucleotide consists of a phosphate, a sugar called deoxyribose, and one of four nitrogenous bases: adenine, thymine, cytosine, or guanine. Alternating sugar and phosphate molecules are linked, forming the two sides of the molecule. Two bases are paired in the interior of the double helix, like rungs on a ladder.

2. According to the rules of complementary base pairing, adenine binds only with thymine, and cytosine binds only with guanine.

WEB TUTORIAL 21.1 DNA Structure

During Replication of DNA Each Original Strand Serves As a Template for a New Strand (pp. 476–477)

3. DNA replication is semiconservative. Each new double-stranded DNA molecule consists of one old and one new strand. The enzyme DNA polymerase "unzips" the two strands of the parent molecule, allowing

each strand to serve as a template for the formation of a new strand. Complementary base pairing ensures the accuracy of replication.

WEB TUTORIAL 21.2 The Genetic Code

DNA Codes for RNA, Which Codes for a Protein (pp. 477–482)

4. Genetic information is transcribed from DNA to RNA and then translated into a protein.

5. RNA is assembled on a DNA template following base-pairing rules during a process called transcription. RNA differs from DNA in that the sugar ribose replaces deoxyribose, and the base uracil replaces thymine. Most RNA is single-stranded.

6. Messenger RNA (mRNA) carries the DNA genetic message to the cytoplasm, where it is translated into protein. The words in the genetic code are sequences of three RNA nucleotides; each triplet is called a codon. Each of the 64 codons specifies a particular amino acid or indicates the point where translation should start or stop.

7. Transfer RNA (tRNA) interprets the genetic code. At one end of the tRNA molecule is a sequence of three nucleotides called the anticodon that pairs with a codon on mRNA following base-pairing rules. The tail of the tRNA binds to a specific amino acid.

8. Each of the two subunits of a ribosome consists of ribosomal RNA (rRNA) and protein. A ribosome brings tRNA and mRNA together, allowing protein synthesis.

9. Translation begins when the two ribosomal subunits and an mRNA assemble, with the mRNA sitting in a groove between the two subunits of a ribosome, attached to the ribosome at the start codon. During translation, the ribosome slides along the mRNA molecule, reading one codon at a time. Molecules of tRNA ferry amino acids to the mRNA and insert them into a growing protein chain. Translation stops when a stop codon is encountered. The protein chain then separates from the ribosome.

WEB TUTORIAL 21.3 Manufacturing Proteins

Point Mutations Result from Nucleotide Substitution, Insertion, or Deletion (pp. 482–484)

10. A point mutation is a change in the nucleotide sequence of a gene. When one nucleotide is mistakenly substituted for another, the function of the resulting protein may or may not affect the function of the protein. The insertion or deletion of a nucleotide always changes the resulting protein.

WEB TUTORIAL 21.4 Nature of Mutations

Gene Activity Can Be Turned On or Off (pp. 484–485)

11. Gene activity is regulated at several levels. Most DNA is usually folded and coiled. For a gene to be active, that region of DNA must be uncoiled. Gene activity can be affected by other segments of DNA. Master genes can turn on other genes. Other regions of DNA, called enhancers, can increase the amount of RNA produced. Chemical signals such as regulatory proteins or hormones can also affect gene activity.

Genetic Engineering Is the Manipulation of DNA for Human Purposes (pp. 485–495)

12. Genetic engineering is the manipulation of genetic material to achieve an end desired by humans. It can be used to produce large quantities of a particular gene product or to transfer a desirable genetic trait from one species to another species or to another member of the same species.

13. Genetic engineering requires a restriction enzyme to cut the source DNA, which contains the gene of interest, and the host DNA at specific places, creating sticky ends composed of unpaired complementary bases. A vector is then needed to transfer the recombinant DNA to a host cell. Common vectors include bacterial plasmids and viruses. The host cell is often a type of cell that reproduces rapidly, often a bacterium or yeast cell. With each host cell division, copies of the gene of interest enter each daughter cell. Introducing a gene from one species into a different one results in the creation of transgenic plants or animals.

14. Genetic engineering has had many applications in plant and animal agriculture, the environment, and medicine.

15. Gene therapy involves introducing a healthy form of a gene into body cells to correct problems caused by a defective one.

15. Scientists are working to create enzymes that will remove the errors located on mRNA resulting from a faulty gene.

17. Scientists are working to create small pieces of double-stranded RNA that can be used to prevent a faulty gene from producing its protein.

18. Each individual has a unique set of DNA restriction fragments, called a DNA fingerprint.

WEB TUTORIAL 21.5 Restriction Enzymes and Plasmids

WEB TUTORIAL 21.6 Polymerase Chain Reaction (PCR)

The Purpose of the Human Genome Project Is to Identify the Complete DNA Sequence of the Human Genome (pp. 495–496)

19. Scientists have determined the sequence of nucleotides of every human chromosome. It consists of 30,000 to 40,000 genes. The information may be helpful in treating genetic diseases.

WEB TUTORIAL 21.7 Sequencing DNA

KEY TERMS

anticodon   *p. 480*
biotechnology   *p. 485*
clone   *p. 487*
codon   *p. 479*
complementary base pairing   *p. 476*
deoxyribonucleic acid (DNA)   *p. 476*
DNA fingerprint   *p. 495*
DNA polymerase   *p. 476*
exon   *p. 478*

gene   *p. 478*
gene therapy   *p. 491*
genetic code   *p. 479*
genetic engineering   *p. 485*
genome   *p. 495*
intron   *p. 478*
messenger RNA (mRNA)   *p. 478*
mutation   *p. 482*
nucleotide   *p. 476*

plasmid   *p. 486*
point mutation   *p. 482*
polymerase chain reaction (PCR)   *p. 488*
polysome   *p. 481*
promoter   *p. 478*
recombinant DNA   *p. 485*
replication   *p. 476*
restriction enzyme   *p. 485*

ribonucleic acid (RNA)   *p. 477*
ribosomal RNA (rRNA)   *p. 478*
RNA polymerase   *p. 478*
semiconservative replication   *p. 477*
transcription   *p. 478*
transfer RNA (tRNA)   *p. 478*
translation   *p. 479*
vector   *p. 486*

THINKING ABOUT THE CONCEPTS

1. Describe the structure of DNA.   *p. 476*
2. Explain why complementary base pairing is crucial to exact replication of DNA.   *pp. 476–477*
3. Why is DNA replication described as semiconservative?   *p. 477*
4. Explain the roles of transcription and translation in converting the DNA message to a protein.   *pp. 477–482*
5. In what ways does RNA differ from DNA?   *pp. 476, 478*
6. What roles do mRNA, tRNA, and rRNA play in the synthesis of protein?   *p. 478*
7. Define codon. What role do codons play in protein synthesis?   *p. 479*
8. What type of molecule has an anticodon? What is the role of an anticodon?   *p. 480*

9. Describe the events that occur during the initiation of protein synthesis, the elongation of the protein chain, and the termination of synthesis. *pp. 481–482*
10. What is a point mutation? *p. 482*
11. How is gene activity regulated? *pp. 484-485*
12. Define genetic engineering. Explain the roles of restriction enzymes and a vector in genetic engineering. *p. 485*
13. Describe some of the ways in which genetic engineering has been used in farming and medicine. *pp. 488–489*
14. What is gene therapy? *pp. 491–495*
15. What is a DNA fingerprint? *p. 495*
16. The complementary base for thymine is
    a. adenine.
    b. cytosine.
    c. guanine.
    d. uracil.
17. Although the amount of any particular base in DNA will vary among individuals, the amount of guanine will always equal the amount of
    a. thymine.
    b. adenine.
    c. cytosine.
    d. uracil.

18. A codon is located on
    a. DNA.
    b. mRNA.
    c. tRNA.
    d. rRNA.
19. The process of transcription involves the production of
    a. a polypeptide chain.
    b. mRNA complementary to a template strand of DNA.
    c. tRNA complementary to mRNA.
    d. rRNA.
20. The anticodon is located on a molecule of _____.
21. In RNA, the nucleotide _____ binds with adenine.
22. In genetic engineering, the staggered cuts in DNA that allow genes to be spliced together are made by _____.

APPLYING THE CONCEPTS

Use the genetic code in Table 21.3 (p. 480) to answer questions 1 and 2.

1. What would be the amino acid sequence in the polypeptide resulting from a strand of mRNA with the following base sequence?
   AUG ACA UAU GAG ACG ACU
2. Below are base sequences in four mRNA strands: one normal and three with mutations. (Keep in mind that translation begins with a start codon and ends with a stop codon.)
   Which of the mutations is likely to have the most severe effects? Why? Which would have the least severe effects? Why?

   Normal mRNA: AUG ACA UAU GAG ACG ACU
   Mutation 1:   AUG ACC UAC GAA ACG ACC
   Mutation 2:   AUG ACU UAA GAG ACG ACA
   Mutation 3:   AUG ACG UAU GAG ACG ACG

3. You are a geneticist asked to testify at a murder trial. The DNA fingerprints shown below are those of a bloodstain at the murder scene (not the victim's) and those of seven suspects (1-7). Which of the suspects' blood matches the bloodstain from the crime scene?

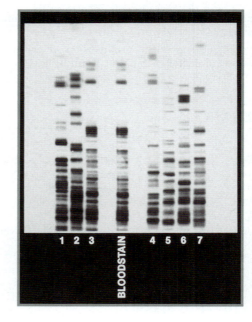

Cancer

Cancer Is Uncontrolled Cell Division

- Tumor development progresses through stages

Cancer Begins with a Single Cell That Escapes Normal Control Mechanisms

- Cancer cells lose restraints on cell division
- Cancer cells do not self-destruct when their DNA is damaged
- Cancer cells divide indefinitely
- Cancer cells attract a blood supply
- Cancer cells do not adhere to neighboring cells
- Body defense cells destroy cancer cells

Viruses, Certain Chemicals, and Radiation Can Cause Cancer

- Certain viruses can disrupt genetic control of cell division
- Some chemicals can cause mutations
- Radiation can cause mutations

Certain Lifestyle Habits Reduce the Risk of Cancer

There Are Several Ways to Diagnose Cancer

Surgery, Radiation, and Chemotherapy Are Conventional Ways to Treat Cancer

- Surgery is used to remove tumors
- Radiation therapy is used to kill localized cancer cells
- Chemotherapy is used to kill cancer cells throughout the body
- Immunotherapy boosts the immune responses against cancer cells
- Inhibition of blood vessel formation may slow the spread of cancer cells
- Gene therapy may someday help fight cancer in several ways

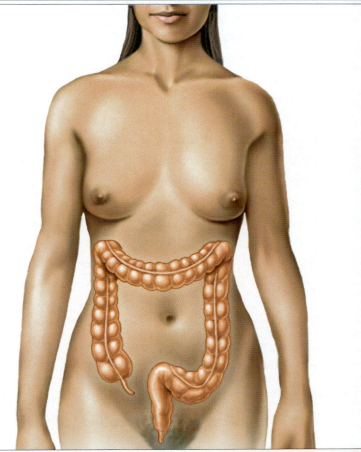

When Margaret turned 50, her doctor insisted that she schedule a screening sigmoidoscopy, a procedure that uses a short, flexible lighted tube to look inside the first part of the colon (large intestine). He explained that the test was important to make sure that Margaret had no signs of colon cancer. It would also help him monitor any future changes in the lining of her colon that may signal precancerous conditions. Margaret reluctantly agreed.

During the sigmoidoscopy, Dr. Greer found a small growth, or polyp. He removed the polyp and sent it to a laboratory for testing. He explained to Margaret that her risk of colon cancer was very low. He later called and explained that the polyp was not cancerous, but she should schedule another sigmoidoscopy in 5 years. Margaret realized she had been foolish to be reluctant to have the procedure and was thankful that her doctor had insisted on it. Now she encourages all of her friends to have a sigmoidoscopy when she sends them a birthday card on their fiftieth birthday.

Cancer, the "Big C," is perhaps the disease we dread the most—and with good reason. Cancer touches the lives of nearly everyone. One of every three people in the United States will develop cancer at some point in life, and the other two are likely to have a friend or relative with cancer (Figure 21a.1).

In this chapter, we will consider how cancer cells escape the normal controls over cell division. We will then learn about some causes of cancer and how we can reduce our risk of developing cancer. Finally, we will identify some means of diagnosing and treating cancer. ■

Cancer Is Uncontrolled Cell Division

Although we tend to think of cancer as one disease, it is, in fact, a family of more than 200 diseases, usually named for the organ in which they arise (Table 21a.1). However, all forms of cancer share one characteristic—uncontrolled cell division. Cancer cells behave like aliens taking over our bodies, but they are actually traitorous body cells.

▌Tumor development progresses through stages

An abnormal growth of cells can form a mass of tissue called a tumor, or a neoplasm, meaning new growth. However, not all tumors are cancerous. In fact, tumors can be either benign or malignant. Surrounded by a capsule of connective tissue, a benign tumor is an abnormal mass of tissue that usually remains at the site where it forms. Its cells do not invade surrounding tissue or spread to distant locations. In most cases, a benign tumor does not threaten health, because it can be removed completely by surgery. Benign tumors can be harmful when they press on nearby tissues enough to interfere with functioning of the tissues. In some locations, however, such as in the brain, a benign tumor may be inoperable and can be life threatening. Only malignant tumors, those that can invade surrounding tissue and spread to multiple locations throughout the

body, are properly called cancerous. The spread of cancer cells from one part of the body to another is called *metastasis*.

Cells on their way to becoming cancerous are accumulating genetic damage. As a result, precancerous cells typically look different from normal cells. *Dysplasia* is the term used to describe the changes in shape, nuclei, and organization within tissues of precancerous cells. Their ragged edges give precancerous cells an abnormal shape. Their nuclei become unusually large and atypically shaped and may contain increased amounts of DNA. We can see the differences by comparing the chromosomes of a normal cell with those from a cancer cell. Notice in Figure 21a.2 that the cancer cell has extra copies of some chromosomes, is missing parts of some chromosomes, and has extra parts to other chromosomes. As a group, precancerous cells form a disorganized clump and, significantly, have an unusually high percentage of cells in the process of dividing.

Eventually, the tumor will reach a critical mass consisting of about a million cells. Although the tumor is still only a millimeter or two in diameter (smaller than a BB), the cells in the interior cannot get a sufficient supply of nutrients, and their own waste is poisoning them. This tiny mass is now called carcinoma *in situ*, which literally means "cancer in place."

Unless the tumor is removed, at some point, perhaps years later, some of its cells will start to secrete chemicals that cause blood vessels to invade the tumor. This process marks an ominous point of transition, because the tumor cells now have supply lines bringing in nutrients to support continued growth and to carry away waste. Of equal importance, the tumor cells have an escape route; they can enter the blood (or lymphatic) vessels and travel to numerous, distant locations throughout the body. Like cancerous seeds, the cancer cells that spread, or metastasize, can begin to grow into tumors in their new locations (Figure 21a.3).

As long as a tumor stays in place, it can grow quite large and still be removed by a surgeon (depending on the location). However, once cancer cells leave the original tumor, they usually spread to so many locations that a surgeon's scalpel is no longer an effective weapon. At this point, chemotherapy or radiation (discussed in greater detail later in the chapter) is generally used to kill the cancer cells wherever they are hiding. The original tumor is rarely a cause of death. Instead, the tumors that form in distant sites in the body are responsible for 90% of the deaths of people with cancer.

Once the cancer cells have separated from the original tumor, they usually enter the circulatory or lymphatic system, which carries the renegade cells to distant sites. Thus, circulatory pathways in the body often explain the patterns of metastasis. Cancer cells escaping from tumors in most parts of the body, including the skin, encounter the next capillary bed in the lungs. Thus, many cancers spread to the lungs. However, blood leaving the intestine travels directly to the liver, so colon cancer typically spreads to the liver.

How does cancer cause death? The simplest answer is that cancer kills by interfering with the ability of body cells to function normally. For instance, cancer cells are greedy. They deprive normal cells of nutrients, thereby weakening them, sometimes to the point of death. Cancer cells can also prevent other cells from performing their usual functions. In addition, tumors can block blood vessels or air passageways in the respiratory system or press on vital nerve pathways in the brain.

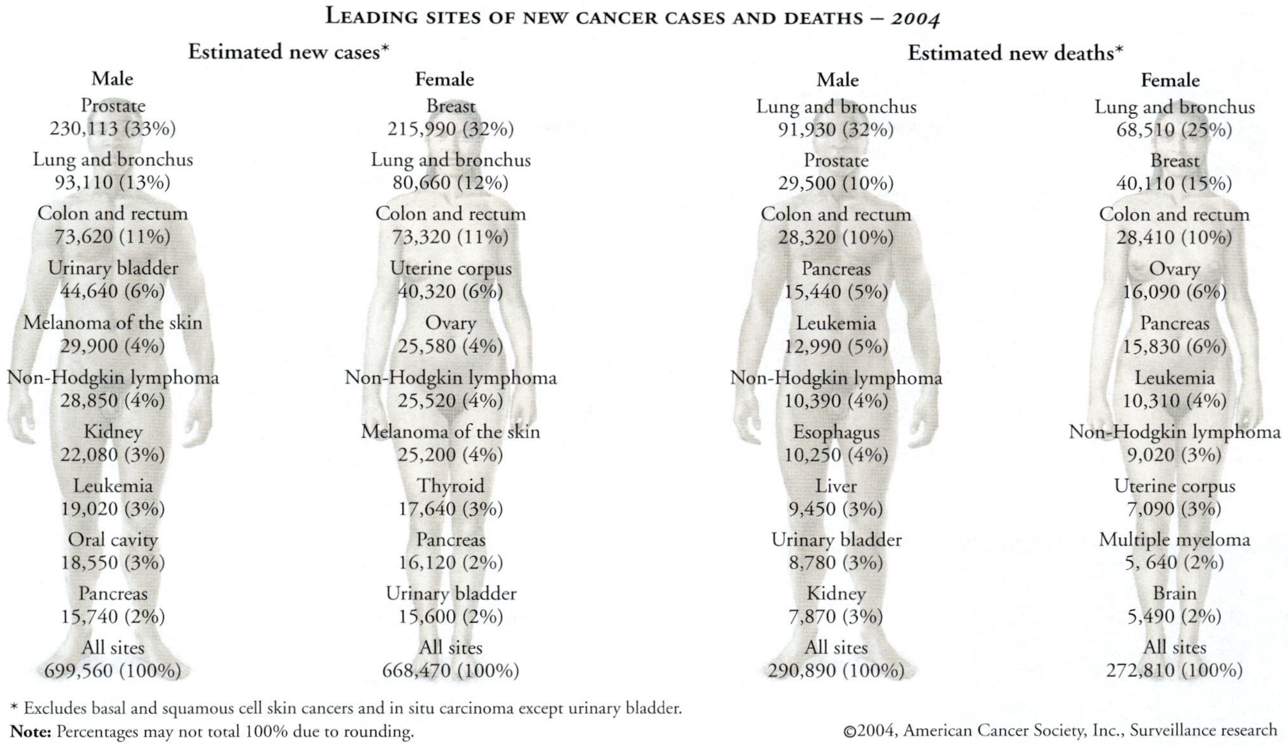

LEADING SITES OF NEW CANCER CASES AND DEATHS — 2004

Estimated new cases*

Male

Prostate
230,113 (33%)

Lung and bronchus
93,110 (13%)

Colon and rectum
73,620 (11%)

Urinary bladder
44,640 (6%)

Melanoma of the skin
29,900 (4%)

Non-Hodgkin lymphoma
28,850 (4%)

Kidney
22,080 (3%)

Leukemia
19,020 (3%)

Oral cavity
18,550 (3%)

Pancreas
15,740 (2%)

All sites
699,560 (100%)

Female

Breast
215,990 (32%)

Lung and bronchus
80,660 (12%)

Colon and rectum
73,320 (11%)

Uterine corpus
40,320 (6%)

Ovary
25,580 (4%)

Non-Hodgkin lymphoma
25,520 (4%)

Melanoma of the skin
25,200 (4%)

Thyroid
17,640 (3%)

Pancreas
16,120 (2%)

Urinary bladder
15,600 (2%)

All sites
668,470 (100%)

Estimated new deaths*

Male

Lung and bronchus
91,930 (32%)

Prostate
29,500 (10%)

Colon and rectum
28,320 (10%)

Pancreas
15,440 (5%)

Leukemia
12,990 (5%)

Non-Hodgkin lymphoma
10,390 (4%)

Esophagus
10,250 (4%)

Liver
9,450 (3%)

Urinary bladder
8,780 (3%)

Kidney
7,870 (3%)

All sites
290,890 (100%)

Female

Lung and bronchus
68,510 (25%)

Breast
40,110 (15%)

Colon and rectum
28,410 (10%)

Ovary
16,090 (6%)

Pancreas
15,830 (6%)

Leukemia
10,310 (4%)

Non-Hodgkin lymphoma
9,020 (3%)

Uterine corpus
7,090 (3%)

Multiple myeloma
5, 640 (2%)

Brain
5,490 (2%)

All sites
272,810 (100%)

* Excludes basal and squamous cell skin cancers and in situ carcinoma except urinary bladder.
Note: Percentages may not total 100% due to rounding.

©2004, American Cancer Society, Inc., Surveillance research

FIGURE **21a.1**

The American Cancer Society's 2004 estimates for the leading sites of new cancer cases and deaths

TABLE **21a.1**

CANCER NOMENCLATURE BY TISSUE TYPE	
TYPE OF CANCER	**TISSUE TYPE**
Carcinomas	Cancers of the epithelial tissues that infiltrate and spread (metastasize); include skin cancers, such as squamous cell, basal cell, and melanoma
Leukemias	Cancers of the bone marrow stem cells that produce the white blood cells
Sarcomas	Cancers of muscle, bone, cartilage, and connective tissues; may involve connective and muscle tissues of the bladder, kidneys, liver, lungs, parotid gland, and spleen
Lymphomas	Cancers of the lymph tissues such as lymph nodes; include Hodgkin lymphoma, non-Hodgkin lymphoma, and lymphosarcoma
Adenocarcinomas	Cancers of glandular epithelia, including liver, salivary glands, and breast

Cancer Begins with a Single Cell That Escapes Normal Control Mechanisms

The 30 trillion to 50 trillion cells in the human body generally work together, much as any organized society does. There are "rules," or controls, that tell a cell when and how often to divide, to self-destruct, and to stay in place. However, cancer cells are outlaws. They evade the many controls that would normally maintain order in the body. As a result, cancer cells can become a threat to health. We will consider the normal systems of checks and balances that regulate healthy cells and see how cancer cells are able to get around these safeguards to grow and spread out of control.

Healthy cells have a system of damage control. Recall from Chapter 19 that the cell cycle is the life cycle of the cell. During the cell cycle, there are checkpoints at which the cell determines whether conditions are right before moving on to the next stage. During the cell cycle, if a healthy cell detects damage, such as a mutated gene, it stops the cell cycle, assesses the damage, and begins repair (Figure 21a.4). If the damage is successfully repaired, the cell returns to the cell cycle. If the damage is too severe to repair, a program of cell death is initiated, as will be discussed shortly. However, if the damage repair is unsuccessful or incomplete, genetic damage accumulates and can lead to cancer.

Most of us are familiar with this type of decision making with car damage rather than cell damage. When something goes wrong with your car, you take it to a repair shop, where someone will tell you the extent of damage. On the basis of that information, you will either ask a mechanic to fix the car or have it towed to the junkyard. But suppose, as sometimes happens, you ask the mechanic to repair the problem, but it does not get fixed correctly or completely. Suppose, for example, your brakes were not functioning properly and the mechanic fixed only the brake pads but ignored damage to the drums or rotors. With faulty brakes, you could drive the car, but you might be unable to stop at a critical moment.

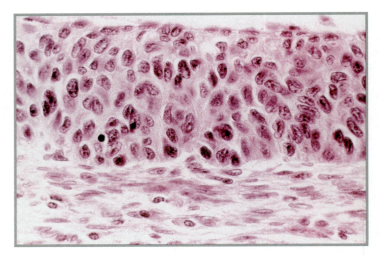

(a)

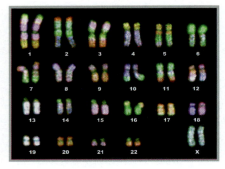

(b)

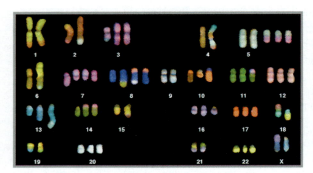

(c)

FIGURE **21a.2**
Cancer cells have an abnormal appearance. (a) In dysplasia, precancerous cells have large irregularly shaped nuclei that contain increased amounts of DNA. The chromosomes of a (b) normal cell and (c) a cancerous cell. These arrangements of chromosomes based on physical characteristics are called karyotypes. Notice that in the cancer karyotype there are extra chromosomes and chromosomes with extra pieces.

In the cell, tumor-suppressor genes function in damage control. Some tumor-suppressor genes detect damaged DNA.[1] If damage is detected, other tumor-suppressor genes serve as "managers of the cell's repair shop." These genes assess the damage and coordinate the activities of other genes that serve as the "mechanics" that

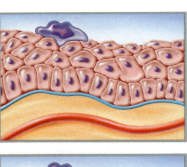

INITIAL TUMOR CELL
• One cell acquires mutations, causing loss of control of cell division

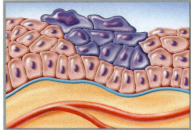

CELL DIVIDES MORE RAPIDLY THAN OTHERS

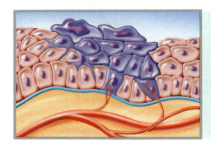

CARCINOMA IN SITU
• Tumor remains at its site of origin

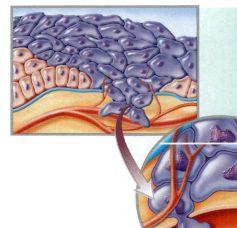

MALIGNANT TUMOR
CANCER CELLS:
• Attract a blood supply
• Gain the ability to leave the other cells
• Spread to distant sites (metastasize)

FIGURE **21a.3**
The progression of cancer from the initial cancer cell to a malignant tumor

repair the damage. If the damage is too extensive, the manager activates still other genes that cause cell death. A particularly important tumor-suppressor gene is *p53*. We will consider some of the activities of *p53* as we discuss the relationship between genes and cancer.

▌ Cancer cells lose restraints on cell division

When genes that regulate cell division are mutated, they usually do not function properly, and the cell involved loses control over cell division. Cancer, we will see, is fundamentally a genetic disease.

---

[1]Recall that genes act by producing proteins. It is really the proteins produced by genes that are acting.

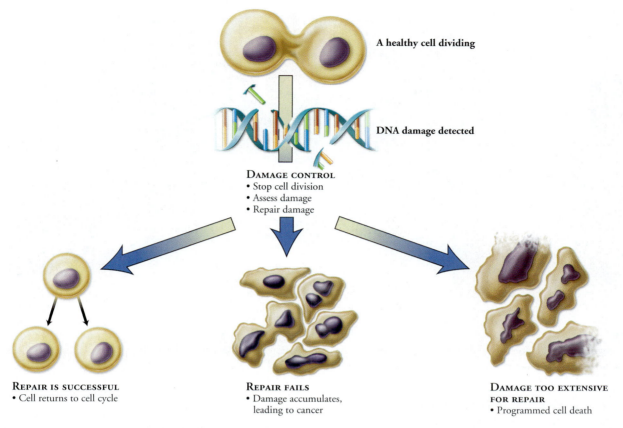

A healthy cell dividing

DNA damage detected

DAMAGE CONTROL
• Stop cell division
• Assess damage
• Repair damage

REPAIR IS SUCCESSFUL
• Cell returns to cell cycle

REPAIR FAILS
• Damage accumulates, leading to cancer

DAMAGE TOO EXTENSIVE FOR REPAIR
• Programmed cell death

FIGURE **21a.4**
Steps in controlling DNA damage during the cell cycle

MUTATIONS AND CANCER

Two types of genes usually regulate cell division: proto-oncogenes and tumor-suppressor genes. Proto-oncogenes stimulate cell division, usually by producing or affecting the function of growth factors. In contrast, tumor-suppressor genes slow or stop cell division. Thus, tumor-suppressor genes function like brakes on cell division. The combined activities of these two types of genes allow the body to control cell division. In this way, the body can grow and develop normally, defective cells can be repaired, and dead cells can be replaced.

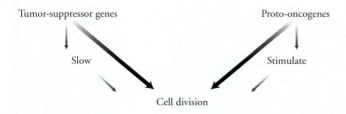

When mutations change the instructions of these genes, the normal system of checks and balances that regulates cell division goes awry, and the disruption can result in the unrestrained cell division that characterizes cancer. A mutation in a tumor-suppressor gene can promote cancer by taking the brakes off cell division. The tumor-suppressor gene *p53* regulates another gene whose job is to produce a protein that keeps cells in a nondividing state. When *p53* mutates,

cell division is no longer curbed. Mutant *p53* seems to be an important factor in more than half of all cancers. Mutation in another tumor-suppressor gene, called *RB*, causes retinoblastoma, which is a rare form of childhood eye cancer. The normal product of *RB* turns off a proto-oncogene that stimulates cell division. When *RB* mutates, the activity of the proto-oncogene is no longer inhibited, and the rate of cell division increases. Two tumor-suppressor genes that play a role in breast cancer—*BRCA1* and *BRCA2*—initiate DNA repair. When these two genes mutate, damaged DNA is not repaired.

A mutation in a proto-oncogene can destroy the regulation of cell division. The mutated gene is called an oncogene, and it speeds the rate of cell division. An oncogene does to the rate of cell division what a stuck accelerator would do to the speed of a car. The proteins produced by many proto-oncogenes are growth factors or receptors for growth factors. When the gene mutates and becomes an oncogene, it often causes too much of the protein to be produced or makes the protein more active than usual. The product of the *ras* gene normally signals the presence of a growth factor, stimulating cell division. The *ras* oncogene protein is hyperactive and stimulates cell division even in the absence of growth factors. The *ras* oncogenes are thought to be important in most pancreatic and colon cancers, as well as some lung cancers. Other oncogenes play a role in leukemia and many of the most deadly forms of breast and ovarian cancer.

Damage must occur in *at least* two genes before cancer occurs. For instance, colon cells must accumulate damage in at least one oncogene and three tumor-suppressor genes before they become cancerous (Figure 21a.5). The number of damaged genes needed to produce can-

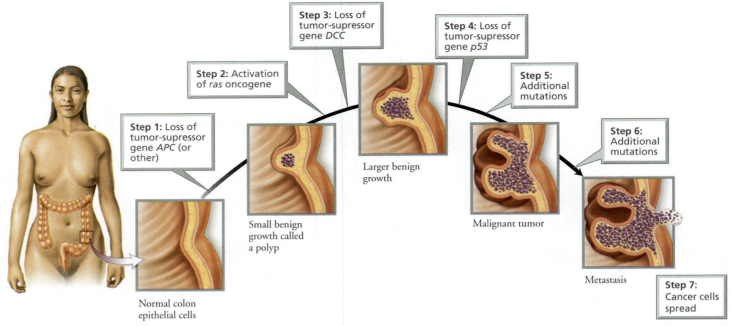

Step 1: Loss of tumor-supressor gene *APC* (or other)

Step 2: Activation of *ras* oncogene

Step 3: Loss of tumor-supressor gene *DCC*

Step 4: Loss of tumor-supressor gene *p53*

Step 5: Additional mutations

Step 6: Additional mutations

Step 7: Cancer cells spread

Normal colon epithelial cells

Small benign growth called a polyp

Larger benign growth

Malignant tumor

Metastasis

FIGURE **21a.5**

Multiple mutations must occur before a cell becomes cancerous. At least one oncogene and three tumor-suppressor genes must mutate in colon cells before they become cancerous.

cer explains why it is possible to inherit a predisposition to a certain form of cancer. A person who inherits only one mutant gene may be predisposed to cancer, but a second event, a mutation in at least one other gene, is required before uncontrolled cell division is unleashed.

LOSS OF CONTACT INHIBITION

Most cells are in contact with appropriate neighboring cells and are anchored in place, forming tissues and organs. Normal cells have signaling systems that inform them about contact with other cells. When normal cells contact a neighbor, they stop dividing. This phenomenon is called contact inhibition. But cancer cells do not show contact inhibition. Instead, they continue to divide and form a tumor.

Cancer cells do not self-destruct when their DNA is damaged

When the genes that regulate cell division are faulty, backup systems normally swing into play to protect the body from the renegade cell. One such system, programmed cell death, occurs when cells activate a genetic suicide program in response to a biochemical or physiological signal. Activation of the so-called death genes prompts cells to manufacture proteins that they then use to kill themselves. Often, the condemned cells go through a predictable series of physical changes called apoptosis. During apoptosis, the outer membrane of the condemned cell boils up and pinches off to produce bulges called blebs (Figure 21a.6).

The defective DNA in cancer cells does not trigger genetic self-destruction. Although not the only way that cancer cells evade this safeguard, a faulty tumor-suppressor gene *p53* is often at least partly responsible. Besides producing a protein that inhibits cell division, *p53* normally prevents the replication of damaged DNA and, if damage is detected, *p53* halts cell division until the DNA can be mended. If the damage is beyond repair, *p53* triggers the events that will lead to programmed cell death.

In a cancer cell, however, a faulty *p53* gene fails to initiate the events leading to cellular self-destruction, so the cells are free to proliferate uncontrollably. Tumors containing cells with damaged *p53* grow aggressively and spread easily and quickly to new locations in the body. Mutations in *p53* also make it difficult to kill the cancer cells by radiation or chemotherapy, because these techniques are intended to damage the DNA of the cancer cell to trigger programmed self-destruction. However, the cancer cell is simply unable to self-destruct in response to DNA damage.

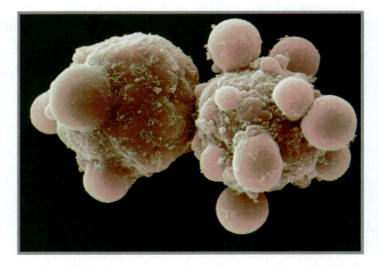

FIGURE **21a.6**

Programmed cell death is a backup system that protects the body from a cell in which the genes regulating cell division have been damaged. Defective DNA, for instance an oncogene or a faulty tumor-suppressor gene, normally triggers a genetic suicide program that causes the cell to self-destruct, as these cells are doing. As the cell goes through programmed cell death, its plasma membrane forms bulges called blebs. Cancer cells are able to evade this protective mechanism.

Cancer cells divide indefinitely

Healthy cells have yet another safeguard against unrestrained cell division—a mechanism that limits the number of times a cell can divide during its lifetime. When grown in the laboratory, for instance, human cells divide only about 50 or 60 times before entering a nondividing state called senescence. Like sand running through an hourglass, no matter how quickly or slowly the events occur, there is a predetermined end.

How does a cell count the number of times it has divided? The answer might lie in telomeres—pieces of DNA at the tips of chromosomes that protect the ends of the chromosomes like the plastic pieces on the ends of shoelaces—or in telomerase, the enzyme that constructs the telomeres (Figure 21a.7). Soon after an embryo is fully developed, most cells stop producing telomerase, so the construction of additional telomeres stops as well. Each time DNA is copied in preparation for cell division, a tiny piece of every telomere is shaved off, shortening the chromosome slightly. When the telomeres are completely gone, the chromosome tip is no longer protected, and parts of important genes will be sliced off each time DNA is copied. Telomeres, then, may be the cell's way of limiting the number of times division can occur. When the telomeres are gone, time is up for that cell. Thus, telomere length may serve both as a gauge of a cell's age and as an indicator of how long that line of cells will continue to divide.

It is currently suspected that the "fountain of youth" that bestows immortality on cancer cells, allowing them to divide continuously, may be their unceasing production of telomerase. This enzyme reconstructs the telomeres after each cell division, stabilizing telomere length and protecting the important genes at chromosome tips. Although most types of mature human cells can no longer produce telomerase, the genes for telomerase apparently become turned on in most cancer cells. Telomerase is present in nearly 90% of biopsies of human tumors.

Cancer cells attract a blood supply

We have seen that cancer cells have escaped the normal cellular controls on cell division and are unable to issue the orders that would lead to their own death. Instead, they multiply and form a tumor.

As mentioned earlier, when a tumor reaches a critical size, about a million cells, its growth will stop unless it can attract a blood supply that will deliver the nutrients needed to support its growth and remove waste. Cancer cells release special growth factors that cause capillaries to invade the tumor (Figure 21a.8). These tiny blood vessels become the lifeline of the tumor, removing wastes and delivering fresh nutrients and additional growth factors that will spur tumor growth. They also serve as pathways for the cancer cells to leave the tumor and spread to other sites in the body.

In a healthy person, blood vessel formation is uncommon; it is an event usually reserved for repairing cuts or other wounds. Unwanted blood vessels that invade tissues can cause damage. For instance, when blood vessels invade the eye's light-sensitive retina, blindness can result. When vessels invade joints, they can cause arthritis. To prevent such damage, cells produce a protein that prevents blood vessels from spreading into tissues. The gene that normally produces this protein is by now familiar—*p53*. Mutations in *p53* can block the production of the protein that prevents the attraction of blood vessels, so blood vessels invade the tumor.

Cancer cells do not adhere to neighboring cells

With blood vessels in place, the cancer cells can begin to spread. Unlike most cells, cancer cells have the ability to travel about the body—providing yet another example of their ability to escape cellular control mechanisms. You may recall from Chapter 3 that normal cells are "glued" in place by special molecules on their surfaces called cellular adhesion molecules (CAMs). When most normal cells become "unglued" from other cells, they stop dividing, and their genetic program for self-destruction is activated.

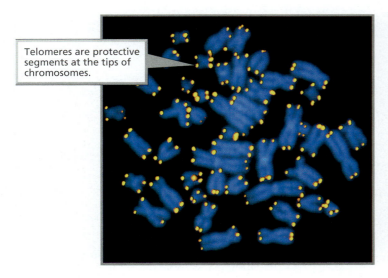

Telomeres are protective segments at the tips of chromosomes.

FIGURE **21a.7**
Telomeres, shown here in yellow, protect the tips of chromosomes, like the small plastic pieces at the ends of shoelaces. Telomeres may serve as molecular counting mechanisms that limit the number of times a cell can divide. Cancer cells retain the ability to construct new telomeres to replace the bits that are shaved off.

FIGURE **21a.8**
Tumor cells release growth factors that cause blood vessels to penetrate the tumor mass. These vessels bring nutrients and additional growth factors and remove wastes. They also provide a pathway for cancer cells to leave the tumor and spread to other locations in the body. Here, blood vessels have formed in a tumor on the spinal cord.

Cancer cells must become "unglued" from other cells to travel through the body. One way cancer cells break loose is by secreting enzymes that break down the glue that holds them and their neighbors in place. In this way, their anchors are broken, and mechanical barriers, such as basement membranes, that would prevent metastasis are breached. But how can unanchored cancer cells continue dividing and evade self-destruction? Their oncogenes send a false message to the nucleus saying that the cell is properly attached (Table 21a.2).

Body defense cells destroy cancer cells

Despite all the safeguards that prevent cells from becoming cancerous, cancer cells develop in our bodies every day. Fortunately, certain body defense cells—natural killer cells and cytotoxic T cells—(see Chapter 13) usually kill those cancer cells. The processes that lead to cancer also produce new and slightly different proteins on cancer cell membranes. The defense cells recognize these proteins as "nonself" and destroy the cancer cells. But sometimes cancer cells evade destruction. Some types of cancer cells actively inhibit the

defense cells, and some simply multiply so quickly that the defense cells cannot destroy them all. Then the tumor can grow and spread.

Viruses, Certain Chemicals, and Radiation Can Cause Cancer

We have seen that many of the "tactics" a cancer cell uses to evade normal cellular safeguards are possible because of changes in genes. Those genetic changes are usually brought about by viruses or by mutations caused by certain chemicals or radiation.

Certain viruses can disrupt genetic control of cell division

Some human cancers are caused by viruses (Table 21a.3). It is estimated that viruses cause about 5% of the cancers in the United States. You may wonder how viruses can alter the genetic makeup of a cell, leading to cancer. The viruses that cause cancer insert copies of

TABLE **21a.2**

CONTROL MECHANISMS AND METHODS AND PREVENTION OF CANCER CELL EVASION		
MECHANISM THAT PROTECTS CELLS FROM CANCER	**METHOD OF EVASION USED BY CANCER CELLS**	**TREATMENT AIMED AT PREVENTING EVASION TACTICS**
Genetic controls on cell division		
Proto-oncogenes stimulate cell division through effects on growth factors	Oncogenes accelerate the rate of cell division	Gene therapy
Tumor-suppressor genes inhibit cell division	Mutations in tumor-suppressor genes take the "brakes" off cell division	Gene therapy
Programmed cell death		
A genetic program that initiates events that lead to the death of the cell when damaged DNA is detected or another signal is received	Mutations in tumor-suppressor genes: Mutant gene *p53* no longer triggers cell death when damaged DNA is detected	Gene therapy
Limitations on the number of times a cell can divide		
Telomeres protect the ends of chromosomes, but a fraction of each is shaved off each time the DNA is copied; when the telomeres are gone, DNA containing important genes is damaged when the DNA is copied, and the cell can no longer divide	Genes to produce telomerase, the enzyme that reconstructs telomeres, are turned on in cancer cells so telomere length is stabilized	Drugs to block telomerase activity or turn off the genes that produce it
Controls that prevent the formation of new blood vessels		
These controls are normally in effect except in a few instances such as wound healing	Cancer cells produce growth factors that attract new blood vessels and proteins that counter the normal proteins that inhibit blood vessel formation	Drugs to block blood vessel formation
Controls that keep normal cells in place		
Cellular adhesion molecules (CAMs) hold cells in place; unanchored cells stop dividing and self-destruct	Cancer cell oncogenes send a false message to the nucleus that the cell is properly anchored	Gene therapy to insert the gene that produces CAMs into tumor cells; using drugs to block certain CAMs on cancer cells so that they cannot grip blood vessel walls as they metastasize

TABLE **21a.3**

SOME VIRUSES LINKED TO HUMAN CANCERS	
VIRUS	**TYPES OF CANCER**
Human papilloma viruses (HPVs)	Cervical, penile, and other anogenital cancers in men and women
Hepatitis B and C	Liver cancer
Epstein–Barr virus	B cell lymphomas, especially Burkitt's lymphoma; nasopharyngeal carcinoma
Human T cell leukemia virus (HTLV–1)	Adult T cell leukemia
Cytomegalovirus (CMV)	Lymphomas and leukemias
Herpes viruses (HHV8)	Kaposi's sarcoma

their genes into the DNA of the host cell. Oncogenes are found among the genes of a few viruses. When the viral oncogene is inserted into the host cell DNA, it behaves as a host oncogene would, and the cell is one step closer to becoming cancerous. A viral oncogene is partly the reason that the human papilloma virus that causes genital warts can cause cervical and penile cancer. In other cases, the viral DNA may be inserted into the host DNA in a location that disrupts the functioning of one of the host genes. The viral DNA could, for example, be inserted into a regulatory gene that controls a proto-oncogene, breaking the switch that turns the gene off. Still other viruses may interfere with the function of the immune system, lessening its ability to find and destroy cancer cells as they arise.

Some chemicals can cause mutations

A carcinogen is an environmental agent that fosters the development of cancer. Some chemicals, especially certain organic chemicals, cause cancer by causing mutations. As we saw in Chapter 21, a change in as little as one nucleotide in a DNA sequence can alter a gene's message. Thus, even a small alteration in DNA can wreak havoc with a cell's regulatory mechanisms and lead to cancer.

Chemical carcinogens are around us most of the time—in our air, food, water, and physical surroundings. We can choose to avoid contact with some of them. For instance, tobacco smoke contains a host of chemical carcinogens. Among the carcinogens in tobacco smoke is one that specifically mutates the tumor-suppressor gene *p53* and another that specifically mutates one of the *ras* oncogenes. Tobacco smoke is responsible for 30% of all cancer deaths and may contribute to as many as 60% of all cancer deaths in the United States. Excessive alcohol consumption is another cancer risk factor that can be avoided. Other chemical carcinogens are more difficult to avoid. These include benzene, formaldehyde, hydrocarbons, certain pesticides, and chemicals in some dyes and preservatives.

Other chemicals contribute to the development of cancer by stimulating cell division rather than by causing mutations. If a cancerous cell has already formed, this stimulation will speed up the rate at which the cancer progresses. Certain hormones can promote cancer in this way. For example, the female hormone estrogen stimulates cell division in the tissues of the breast and in the lining of the uterus (the endometrium). Sustained, high levels of estrogen are linked with

breast cancer.[2] Also, the incidence of endometrial cancer is elevated in postmenopausal women whose hormone replacement therapy involves estrogen alone. Estrogen does not seem to promote endometrial cancer when taken in combination with progesterone. However, since 2002 several studies have shown that estrogen taken in combination with progesterone slightly increases a woman's risk of developing breast cancer. Furthermore, breast tumors in women taking combined hormone pills were larger and more likely to have spread than were tumors in women not taking hormone pills.

Radiation can cause mutations

Radiation can also cause cancer by causing mutations in DNA. It is impossible to avoid exposure to radiation from natural sources, such as the ultraviolet light from the sun, cosmic rays, radon, and uranium. However, we can take reasonable precautions to minimize our risks. For example, sunlight's ultraviolet rays cause skin cancer. Although we probably would not choose to spend our lives entirely indoors to reduce the risk of skin cancer, it would be wise to avoid sunbathing, as well as tanning lamps and tanning parlors. It would also be a good idea to use a sunscreen whenever exposure to the sun is unavoidable.

Certain Lifestyle Habits Reduce the Risk of Cancer

The good news is that some lifestyle changes can greatly decrease your risk of developing cancer (Table 21a.4). Tobacco use and unhealthy diet are responsible for two-thirds of all cancer deaths in the United States. Tobacco smoke is the leading carcinogen, and it is obvious how to modify that risk. So here we will focus on diet as a way to reduce your cancer risk.

TABLE **21a.4**

TIPS FOR REDUCING YOUR CANCER RISK
1. Do not use tobacco. If you do, quit. Avoid exposure to secondhand smoke.
2. Reduce the amount of saturated fat in your diet, especially the fat from red meat.
3. Minimize your consumption of salt-cured, pickled, and smoked foods.
4. Eat at least five servings of fruits and vegetables every day.
5. Avoid excessive alcohol intake. If you consume alcohol, one or two drinks a day should be the maximum.
6. Watch your caloric intake, and keep your body weight proper for your height.
7. Avoid excessive exposure to sunlight. Wear protective clothing. Use sunscreen.
8. Avoid unnecessary medical x-rays.
9. Have the appropriate screening exams on a regular basis. Women should have PAP tests and mammograms. Men should have prostate tests. All adults should have tests for colorectal cancer.

[2]The link between estrogen and breast cancer is discussed in Chapter 17 and in the Heath Issue essay, *Is It Hot in Here, or Is It Me? Hormone Replacement Therapy and Menopause* in Chapter 10.

A few simple food choices may reduce your risk of developing cancer. The best rules are to eat a well-balanced diet and to eat all foods in moderation. For instance, a high-fat diet is linked to colon and breast cancers. Most people in the United States consume far too much fat. Thus, it is wise to reduce fat intake, especially saturated fat, which comes from animal sources such as red meat. Consuming large quantities of smoked, salt-cured, and nitrite-cured foods, including ham, bologna, and salami, increases the risk of cancers of the esophagus and stomach.

A diet rich in fruits and vegetables can reduce your cancer risk because they are high in fiber, which can dilute the contents of the intestines, bind to carcinogens, and reduce the amount of time the carcinogens spend in the intestine by speeding passage of intestinal contents. The colorful vegetables are usually high in antioxidant vitamins, and these vitamins may play a role in protecting against cancer. As their name implies, antioxidants interfere with oxidation, a process that can result in the formation of molecules called free radicals that can damage DNA and lead to cancer. The three major antioxidants are beta-carotene, vitamin E, and vitamin C. The first two are common in red, yellow, and orange fruits and vegetables, and the last abounds in citrus fruits, among other sources.

There Are Several Ways to Diagnose Cancer

Early detection is critical to cancer survival because successful treatment is much more likely before the cancer has spread. You know your own body better than anyone else does. For this reason, it is wise for everyone to be aware of cancer's seven warning signs, the first letters of which spell the word CAUTION (Table 21a.5). There are also many routine tests that can detect cancer in people without symptoms. You can perform some of the tests on yourself; others require a visit to a medical professional (Table 21a.6).

When cancer is suspected, there are many noninvasive techniques to confirm (or contradict) the suspicion by examining blood, urine, feces, or other bodily secretions. For example, DNA analysis can identify gene mutations associated with lung cancer in sputum, bladder cancer in urine, and colon cancer in feces. There are other telltale signs of cancer that can be detected by other tests. For instance, the enzyme telomerase is produced by cancer cells but rarely by normal ones. A test for telomerase appears to be helpful in diagnosing certain cancers, but this test is still in an experimental stage.

Other tests look for tumor markers, which are chemicals produced either by the cells of the tumor or by body cells in response

TABLE **21a.5**

CANCER'S SEVEN WARNING SIGNS
Change in bowel or bladder habit or function
A sore that does not heal
Unusual bleeding or bloody discharge
Thickening or lump in breast or elsewhere
Indigestion or difficulty swallowing
Obvious change in wart or mole
Nagging cough or hoarseness
Source: *American Cancer Society.*

TABLE **21a.6**

RECOMMENDED CANCER SCREENING TESTS
Guidelines suggested by the American Cancer Society for the early detection of cancer in people without symptoms, age 20 to 40
Cancer-related checkup every 3 years
Should include the procedures listed below plus health counseling (such as tips on quitting cigarette smoking) and examinations for cancers of the thyroid, testes, prostate, mouth, ovaries, skin, and lymph nodes. Some people are at higher than normal risk for certain cancers and may need to have tests more frequently.

Breast	• Exam by doctor every 3 years • Self-exam every month • One baseline breast x-ray ages 35–40 Higher risk for breast cancer: Personal or family history of breast cancer; never had children; had first child after 30
Uterus	• Pelvic exam every 3 years
Cervix	• Yearly PAP test beginning at age 18 or when sexual activity begins Higher risk for cervical cancer: Early age at first intercourse; multiple sex partners

Guidelines suggested by the American Cancer Society for the early detection of cancer in people without symptoms, age 40 and over
Cancer-related checkup every year
Should include the procedures listed below plus health counseling (such as tips on quitting cigarette smoking) and examinations for cancers of the thyroid, testes, prostate, mouth, ovaries, skin, and lymph nodes. Some people are at higher than normal risk for certain cancers and may need to have tests more frequently.

Breast	• Exam by doctor every year • Self-exam every month • Breast x-ray every year after 40 Higher risk for breast cancer: Personal or family history of breast cancer; never had children; had first child after 30
Uterus	• Pelvic exam every year
Cervix	• Yearly PAP test Higher risk for cervical cancer: Early age at first intercourse; multiple sex partners
Endometrium	• Endometrial tissue sample at menopause if at risk Higher risk for endometrial cancer: Infertility, obesity, failure of ovulation, abnormal uterine bleeding, estrogen therapy
Prostate	• Yearly prostate-specific antigen (PSA) blood test and digital rectal exam after age 50
Colon and rectum	• Fecal occult blood test every year after age 50 • Flexible sigmoidoscopy beginning at age 50 and every 5 years thereafter Higher risk for colorectal cancer: Personal or family history of colon or rectal cancer; personal or family history of polyps in the colon or rectum; ulcerative colitis

to a tumor. Prostate cells produce prostate specific antigen (PSA). Thus, elevated blood PSA levels suggest the presence of prostate cancer. Although PSA is the only tumor marker that is useful in the original diagnosis of a cancer, other tumor markers may reveal whether the cancer has spread or returned. Blood levels of TA–90 can help determine whether melanoma (a type of skin cancer) has spread. The tumor marker CA 125 can identify ovarian cancer. CA–15–3 indicates a recurrence of breast cancer, and CEA indicates a recurrence of colon cancer.

Many imaging techniques allow physicians to look inside the body and identify tumors. These include x-rays, computerized tomography (CT) scan, magnetic resonance imaging (MRI), and ultrasound. A biopsy involves removing a small piece of tissue suspected to be cancerous. A biopsy is often done using a needle instead of surgery. In either case, cells are then examined under a microscope to see whether they have the characteristic appearance of cancer cells.

FIGURE **21a.9**
Radiation therapy is often used to treat cancer that is localized to one area.

Surgery, Radiation, and Chemotherapy Are Conventional Ways to Treat Cancer

The conventional cancer treatments—surgery, radiation therapy, and chemotherapy—are still the mainstays of cancer treatment. But many new treatments hold the promise of a brighter future.

Surgery is used to remove tumors

When the tumor is accessible and can be removed without damaging vital surrounding tissue, surgery is usually performed to try to eradicate the cancer or remove as much as possible. If every cancer cell is removed, as can be done with early tumors (carcinoma *in situ*), a complete cure is possible. However, if cancer cells have begun to invade surrounding tissue or have spread to distant locations, surgery cannot "cure" the cancer. If the cancer has spread, surrounding tissue and perhaps even nearby lymph nodes may also be removed. Unfortunately, more than half of all tumors have already metastasized by the time of diagnosis, so further treatment is necessary.

Radiation therapy is used to kill localized cancer cells

If the cancer has spread from the initial site but is still localized, surgery is usually followed by radiation therapy (Figure 21a.9). In some cases, such as cancer of the larynx (voice box), localized tumors that are difficult to remove surgically without damaging surrounding tissue may be treated with radiation alone.

As we have seen, radiation damages DNA, and extensive DNA damage triggers programmed cell death. The greatest damage is done to cells that are rapidly dividing. The intended targets of radiation, cancer cells, are actively dividing but so are several types of body cells, called renewal tissues, that normally continue dividing throughout life. These tissues include cells of the reproductive system, cells that replace layers of skin or the lining of the stomach, cells that give rise to blood cells, or cells that give rise to hair. Unfortunately, radiation cannot distinguish between cancer cells and renewal tissue, so good cells are sacrificed to kill the harmful ones. The destruction of renewal tissues leads to the side effects of radiation such as temporary sterility, nausea, anemia, and hair loss.

Chemotherapy is used to kill cancer cells throughout the body

When it is thought that the cancer may have spread by the time of diagnosis, chemotherapy is often used. The drugs used generally reach all parts of the body and kill all rapidly dividing cells, just as radiation does. Some of the drugs used in chemotherapy block DNA synthesis, others damage DNA, and a few others prevent cell division by interfering with other cellular processes. The side effects are similar to those that accompany radiation therapy.

As we have seen, radiation and chemotherapy damage DNA in rapidly dividing cells in the hopes that the cell will commit suicide. However, *p53*, the gene that detects DNA damage and initiates programmed cell death, is mutant in more than half of all cancers. As a result, even though the treatment successfully damages DNA in cancer cells, the cells do not self-destruct and treatment fails.

Immunotherapy boosts the immune responses against cancer cells

Cytotoxic T cells of the immune system continually search the body for abnormal cells, such as cancer cells, and kill those they find. The goal of immunotherapy, then, is to boost the patient's immune system so that it becomes more effective in destroying cancer cells. One form of immunotherapy involves the administration of factors normally secreted by lymphocytes, including interleukin-2 (which stimulates lymphocytes that attack cancer cells), interferons (which stimulate the immune system and also have direct effects on the tumor cells), and tumor necrosis factor (which has direct effects on cancer cells).

Vaccines that stimulate T cells to attack and kill the cancer cells are being tested in clinical trials with promising results. Unlike most vaccines, these vaccines cannot *prevent* disease (cancer); they can only treat it. Cancer vaccines contain dead cancer cells, parts of cells, or proteins from cancer cells. A vaccine causes the immune system to fight other cells, in this case the patient's cancer cells, with the same characteristics. The vaccine strategy is now being tested against melanoma (a deadly form of skin cancer) and leukemia, as well as cancers of the kidney, prostate, colon, and lung.

Inhibition of blood vessel formation may slow the spread of cancer cells

In recent years, there has been a major change in the way we think about cancer. Gone are the days when the only aim was killing the cancer cells. As we have learned more about the molecular biology of cancer, new ways of slowing its progression or dealing with it as a genetic disease have been developed.

The formation of blood vessels is a critical step in the life of a tumor, because they bring nourishment and provide a pathway for cell migration. Researchers are working on ways to cut off these lifelines and starve the tumor. Many such drugs have been developed, and some are nearing the end of clinical trials. For example, taking a drug that blocks blood vessel formation along with chemotherapy boosts the survival rate of patients with colorectal cancer.

Gene therapy may someday help fight cancer in several ways

There are currently over 50 clinical trials using gene therapy to treat cancer but no means of gene therapy has been approved by the Food and Drug Administration (FDA) for cancer treatment.

Nearly all these studies are in very early stages. There are several strategies. One strategy has been to insert normal tumor-suppressor genes into the cancerous cells. For instance, you may recall that gene *p53* normally triggers programmed cell death when DNA damage is detected, but this gene is often faulty in cancer cells. It is hoped that inserting a healthy form of *p53* into cancer cells will lead to their death, causing the tumor to shrink.

Another strategy is to insert a piece of DNA into a cancer cell that will prevent an oncogene from exerting its effects. This so-called antisense DNA is a piece of DNA that is complementary to the mRNA produced by the oncogene. The antisense DNA would be expected to bind to the mRNA produced by the oncogene and prevent it from being translated into protein, thereby preventing the effects of the oncogene. Currently, this is being tried as a means of treating leukemia.

One of the most promising approaches using gene therapy, inserting a gene that makes tumor cells sensitive to a drug that will kill them, is now being evaluated as treatment for brain, ovarian, and prostate cancers. In this case, a viral gene for the enzyme thymidine kinase is inserted into the tumor cells. When the gene is expressed, the resulting enzyme makes the cell sensitive to a drug called ganciclovir. The drug is activated in only those cells that produce the enzyme coded for by the inserted gene. Thus, only the tumor cells are killed.

Evolution: Basic Principles and Our Heritage

Life Evolved on the Earth about 3.8 Billion Years Ago
- Inorganic molecules formed small organic molecules
- Small organic molecules joined to form larger molecules
- Genetic material originated
- Organic molecules aggregated into droplets

The Scale of Evolutionary Change May Be Small or Large
- Microevolution occurs at the genetic level
- Macroevolutionary change occurs above the species level

Evidence of Evolution Comes from Diverse Sources
- The fossil record provides evidence of evolution
- Geographic distributions reflect evolutionary history
- Comparative anatomy and embryology reveal common descent
- Comparative molecular biology also reveals evolutionary relationships

Human Roots Trace Back to the First Primates
- Primates have distinct characteristics
- Discussions of human origins often evoke controversy
- Misconceptions distort the picture of human evolution
- Walking on two feet was a critical step in hominid evolution

SOCIAL ISSUE   Conducting Research on Our Relatives

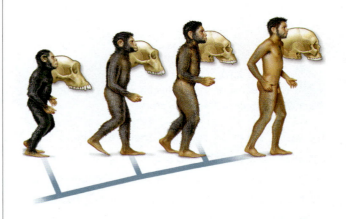

WWW TUTORIAL 22.1 FIGURE **22.1**
Charles Darwin shortly after his return to England from his voyage on the H.M.S. *Beagle*

The year is 1831. The country is England. A young man, 22 years of age, walks to the ship he is to board and upon which he will live for the next several years. On this day, he has no thoughts about revolutionizing the pervasive view of the earth and its inhabitants and forever changing the course of scientific thinking. Charles Darwin, an avid naturalist and somewhat disenchanted student of medicine and theology, left Great Britain in 1831 to spend 5 years aboard the H.M.S. *Beagle* (Figure 22.1). The goal of the voyage was to chart the coastline of South America—a goal that allowed Darwin to examine, describe, and collect plants and animals from the South American coast and nearby islands. In 1836, Darwin returned to Great Britain, where he began to mull over all that he had seen and read. Twenty-three years later, in 1859, his book *On the Origin of Species* was published. In his book, Darwin proposed that the incredible diversity of life on the earth was the product of evolution. Broadly defined, **evolution** is descent with modification from a common ancestor. It is the process by which life forms on the earth have changed from their earliest beginnings to today.

How did life arise and evolve on the earth? How does evolution work? How has evolution shaped species, including our own? What were our ancestors like? We address those questions in this chapter. ■

Life Evolved on the Earth about 3.8 Billion Years Ago

The earth is estimated to be 4.5 billion years old. Evidence from physical and chemical changes in the earth's crust and atmosphere suggests that life has existed on the earth for about 3.8 billion years. The environment of the early earth was very different from that of today and would have been an extremely hostile place for most organisms (Figure 22.2). The earth's crust was hot and volcanic. There were intense lightning and ultraviolet radiation, and there was almost no gaseous oxygen (O_2) in the atmosphere. The early atmosphere likely consisted of the following gases: hydrogen (H_2), nitrogen (N_2), carbon dioxide (CO_2), methane (CH_4), and carbon monoxide (CO). How could life have evolved under such conditions? In the sections that follow, we present a plausible sequence of events that many scientists believe explains the origin of life. The sequence, known as **chemical evolution**, suggests that life evolved from chemicals slowly increasing in complexity over a period of perhaps 300 million years.

Inorganic molecules formed small organic molecules

During the 1920s, two scientists who were studying evolution independently suggested that conditions of the early earth favored the synthesis of small organic molecules from inorganic molecules. The scientists were A. I. Oparin of Russia and J. B. S. Haldane of Great Britain. They hypothesized that the low-oxygen atmosphere of the primitive earth encouraged the joining of simple molecules to form complex molecules. The low-oxygen atmosphere was important because oxygen attacks chemical bonds. Oparin and Haldane reasoned that the energy required for the joining of simple molecules could have come from the lightning and intense ultraviolet (UV) radiation striking the primitive earth. (Scientists assume that UV radiation was more intense during those times than it is now because young suns emit more UV radiation than do old suns, and the early earth lacked an ozone layer to shield it from much of the UV radiation.)

Stanley Miller and Harold Urey of the University of Chicago tested the Oparin-Haldane hypothesis in 1953. In their laboratory, these scientists re-created conditions presumed to be similar to those of the early earth (Figure 22.3). Miller and Urey discharged electric sparks (meant to simulate lightning) through an atmos-

The atmosphere of the early earth contained little gaseous oxygen (O_2) and probably consisted of gases such as hydrogen (H_2), methane (CH_4), carbon dioxide (CO_2), and carbon monoxide (CO).

FIGURE **22.2**
Representation of the early earth

phere that contained some of the gases thought to be present in the early atmosphere. The scientists generated a variety of small organic molecules, including amino acids. Their results supported the Oparin-Haldane hypothesis that organic molecules could be synthesized from inorganic ones.

▌Small organic molecules joined to form larger molecules

Scientists hypothesize that these small organic molecules accumulated in the early oceans. Over long periods, these molecules formed a complex mixture. Somehow, perhaps by being washed onto clay or hot sand or lava, smaller molecules joined to form larger molecules. In this way, for example, amino acids may have joined to form proteins. Some scientists suggest that deep-sea vents were important locations for the synthesis of small organic molecules and their joining to form larger molecules.

▌Genetic material originated

Some time later, genetic material originated. The origin of genetic material made possible the transfer of information from one generation to the next. Cells today store their genetic information as DNA. Many researchers cite evidence that RNA was the first genetic material and that DNA later evolved from an RNA template.

(a) Using this apparatus, Stanley Miller (shown) and Harold Urey demonstrated that simple organic molecules could be synthesized from inorganic ones in the low-oxygen environment of the primitive earth.

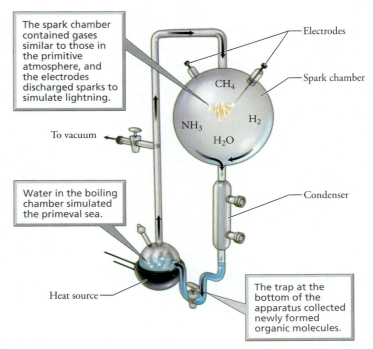

The spark chamber contained gases similar to those in the primitive atmosphere, and the electrodes discharged sparks to simulate lightning.

Electrodes

Spark chamber

CH_4

H_2

NH_3

H_2O

To vacuum

Water in the boiling chamber simulated the primeval sea.

Condenser

Heat source

The trap at the bottom of the apparatus collected newly formed organic molecules.

(b) Diagram of the apparatus

FIGURE **22.3**
Testing the Oparin-Haldane hypothesis

One constant feature throughout the evolution of life on the earth has been the presence of self-replicating molecules of nucleic acids— RNA and DNA. The stability of the genetic code for billions of years of the earth's history remains one of the most powerful insights of our time. This remarkably constant and simple set of genetic codes applies to all life and suggests a single common origin for life.

Organic molecules aggregated into droplets

Scientists theorize that the newly formed organic molecules, including genetic material, aggregated into droplet-like structures. The early droplets displayed some of the same properties as living cells, such as the ability to maintain an internal environment different from surrounding conditions.

Fossil evidence indicates that the earliest cells were prokaryotes. Recall from Chapter 3 that eukaryotic cells, such as those in your body, differ from prokaryotic cells in many ways. Eukaryotic cells have an extensive system of internal membranes, a cytoskeleton, and membrane-bound organelles such as the nucleus.

So the question arises, "How did more complex cells arise from the earliest prokaryotic cells?" It is possible that some of the organelles within eukaryotic cells appeared when other, smaller organisms became incorporated into the cells. Mitochondria, for example, may be descendants of once free-living bacteria. These bacteria either invaded or were engulfed by the ancient cells and formed a mutually beneficial relationship with them. Evidence to support this possibility includes the similarity of the genetic material of mitochondria and free-living bacteria. In addition, mitochondria are about the size of bacteria, have bacteria-like ribosomes, and have inner membranes similar to the plasma membranes of bacteria. This hypothesis is called the **endosymbiont hypothesis**. It is generally accepted by the biological community to explain the origin of some features of eukaryotic cells, including mitochondria. At present, scientists are not certain whether endosymbiosis was involved in the origin of other eukaryotic features such as the cytoskeleton or membrane-bound nucleus. An alternative possibility is that infolding of the plasma membrane of some ancient prokaryote produced some of the organelles (endoplasmic reticulum, Golgi apparatus) found in eukaryotic cells of today.

Prokaryotic cells were first to evolve on the ancient earth. Fossils of prokaryotic cells have been dated at 3.5 billion years old, and scientists estimate that the first prokaryotic cells probably arose around 3.8 billion years ago. Eukaryotic cells evolved some 2 billion years later. Multicellularity evolved in eukaryotes, which eventually led to organisms such as plants, fungi, and animals. Figure 22.4 summarizes the possible steps leading to the origin of life on the earth.

The Scale of Evolutionary Change May Be Small or Large

Evolution occurs on two levels. One level is small (microevolution), and the other is grand (macroevolution). **Microevolution** involves changes in populations at the genetic level. **Macroevolution**, on the other hand, involves phenomena on a larger scale, such as mass extinctions and the evolution of mammals. Macroevolutionary change might result from long-term changes in the

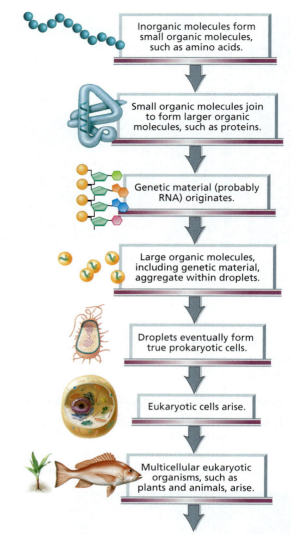

FIGURE **22.4**
Possible steps in the origin of life on the earth

climate or position of the continents. We begin our coverage with microevolution. We first examine genetic variation within populations and then the causes of microevolution. We later expand to the larger scale phenomena of macroevolution.

Microevolution occurs at the genetic level

Look around. You will see variation in almost any population. Consider your classmates. They do not all look alike, and, unless you are an identical twin, you probably do not precisely resemble your brothers or sisters. Before we look at what makes individuals in a population different, let's define some basic terminology. A **gene pool** includes all of the alleles of all of the genes of all individuals in a population. (Recall from Chapter 20 that a **gene** is a segment of DNA on a chromosome that directs the synthesis of a specific protein, whereas **alleles** are different forms of a gene.) A **population** is a group of individuals of the same species living in a particular area. Now let's take a closer look at what makes individuals in a population different and examine some of the ways that variation can appear in populations.

Sexual reproduction and mutation produce variation in populations. Sexual reproduction shuffles alleles already present in the population. As discussed in Chapter 20, the gametes (eggs or sperm) of any one individual show substantial genetic variation due to crossing over and independent assortment. Crossing over creates new combinations of alleles. Independent assortment mixes maternal and paternal chromosomes in gametes. Also, the combination of gametes at fertilization is a chance event. Of the millions of sperm produced by a male, only one fertilizes the egg. This union produces a new individual with new combinations of alleles.

The only source of new characteristics in a population is **mutation,** a change in the DNA of genes. Mutations occur at a low rate in any set of genes. For example, the average mutation rate for genes in humans that lead to visible phenotypic (observable physical or physiological) effects is estimated at one mutation per million gametes per generation. Because mutations occur at such a low rate, their contribution to genetic diversity in large populations is quite small. Mutations can appear spontaneously from mistakes in DNA replication, or they can be caused by outside sources such as radiation or chemical agents.

Recall that microevolution involves change at the genetic level. More specifically, microevolution involves changes in the frequency of certain alleles relative to others within a gene pool. Some of the processes that produce those changes are genetic drift, gene flow, natural selection, and mutation.

GENETIC DRIFT

Genetic drift occurs when allele frequencies within a population change randomly because of chance alone. This process is usually negligible in large populations. However, in small populations (fewer than about 100 individuals), chance events can cause allele frequencies to drift randomly from one generation to the next. Genetic drift often is associated with the bottleneck effect and the founder effect.

THE BOTTLENECK EFFECT Sometimes, dramatic reductions in population size occur because of disasters such as fires and earthquakes that kill large numbers of individuals at random. Consider a population that experiences a flood in which most members die. The genetic makeup of the survivors may not be representative of the original population. This change in the gene pool is the **bottleneck effect**. The effect gets its name because the population

experiences a dramatic decrease in size much as the size of a bottle decreases at the neck. Certain alleles may be more or less common in the survivors than in the original population simply by chance alone. In fact, some alleles may have been completely lost, resulting in reduced overall genetic variability among survivors. One example of the bottleneck effect concerns the northern elephant seal, a large marine mammal found along the west coast of North America (Figure 22.5). By the late 1800s, the population of northern elephant seals was reduced through hunting to about 20 individuals. Once protected, the population increased and currently stands at about 30,000 individuals. Although robust in numbers, the population lacks genetic variability. Indeed, scientists found no genetic variation in the 24 gene loci (singular, locus) examined from members of the elephant seal population. (Recall from Chapter 20 that a locus is the point on a chromosome where a particular gene can be found.) Clearly, bottleneck effects can be a serious problem for species threatened by extinction (the death of all species members).

THE FOUNDER EFFECT Genetic drift also occurs when a few individuals leave their population and establish themselves in a new, somewhat isolated place. By chance alone, the genetic makeup of the colonizing individuals is probably not representative of the population they left. Genetic drift in new, small colonies is called the **founder effect**. This effect, which has been demonstrated in human populations, typically results in a relatively high frequency of inherited disorders.

The relatively high incidence of Ellis-van Creveld syndrome among the Amish in Lancaster County, Pennsylvania, has been linked to the founder effect. This inherited condition is associated with dwarfism and extra fingers or toes (Figure 22.6). Its incidence in the general population of the United States is about 1 in 60,000 live births. Its incidence in the Old Order Amish is a striking 5 in 1000 live births (or 300 in 60,000). About 200 Amish moved to Pennsylvania from Switzerland between 1722 and 1770. Either a Mr. Samuel King or his

FIGURE **22.5**
The population of northern elephant seals provides an example of the bottleneck effect, which occurs when a population experiences a dramatic reduction in size that results in alleles from the original population being less common or completely lost in survivors.

FIGURE **22.6**
An Amish woman holding her child with Ellis-van Creveld syndrome, a disorder characterized by dwarfism and extra digits. The founder effect (genetic drift in new, small colonies) may explain the relatively high incidence of this genetic disorder among the Amish in Lancaster County, Pennsylvania.

wife, who emigrated to Pennsylvania in 1744, probably carried the recessive allele for Ellis-van Creveld syndrome. The couple and their descendants were apparently quite prolific, and the frequency of the deleterious allele drifted higher in subsequent generations.

GENE FLOW

Another cause of microevolution is **gene flow**, which occurs when individuals move into and out of populations. As individuals come and go, they carry with them their unique sets of genes. Gene flow occurs if these individuals successfully interbreed (mate and produce offspring) with the resident population.

The question of gene flow is important to the formation of new species. A **species** is a population or group of populations whose members are capable of successful interbreeding. Such interbreeding must occur under natural conditions and produce fertile offspring. Consider a population that becomes geographically isolated from other populations when a new river forms. The isolated population may begin to take a separate evolutionary route as distinctly different sets of allele frequencies and mutations accumulate. Indeed, the isolated population may become so different that a new species is formed. This process is called **speciation**.

stop and think

We have said that the criterion of interbreeding can be used to define a species. According to this definition, members of populations that can interbreed in nature and produce fertile offspring belong to the same species. Can you think of any problems with this definition of a species? Can you imagine instances in which this definition would not work? How else might a species be defined?

NATURAL SELECTION

In his book *On the Origin of Species* (1859), Charles Darwin argued that species were not specially created, unchanging forms. He suggested that modern species are descendants of ancestral species. In other words, present-day species evolved from past species. Darwin also proposed that evolution occurred by the process of natural selection. His ideas can be briefly summarized as follows:

1. Individual variation exists within a species. Some of this variation is inherited.

2. Some individuals live longer and have more offspring than do others because their particular inherited characteristics make them better suited to their local environment; this is the process of **natural selection**.

3. Evolutionary change occurs as the traits of individuals that survive and reproduce become more common in the population. Traits of less successful individuals become less common.

Given Darwin's ideas, an individual's evolutionary success can be measured by fitness (sometimes called Darwinian fitness). **Fitness** is the average number of reproductively viable offspring left by an individual. Individuals with greater fitness have more of their genes represented in future generations. To succeed in evolution, one must reproduce (Figure 22.7). Indeed, you could live more than 100 years, but

FIGURE **22.7**
Fitness is the number of offspring left by an individual. These parents clearly have high fitness.

your individual fitness would be zero if you did not reproduce. Some of the diseases or conditions discussed in earlier chapters are associated with zero fitness because they cause death before reproduction (such as Tay-Sachs disease) or sterility (such as Turner's syndrome).

One result of natural selection is that populations become better suited to their particular environment by a process called **adaptation**. If the environment changes, individuals who had become so finely attuned to the initial environment would lose their advantage. If the environmental change were great enough, individuals with different alleles may be selected by nature to leave more offspring. If the environment once again stabilizes, the population reestablishes itself around a new set of allele frequencies that better meet the new conditions.

Natural selection often proceeds slowly and is difficult to observe and to measure. Nevertheless, in some cases we have been able to see natural selection in action. One example concerns the rise of antibiotic-resistant bacteria. Consider what happens when you take an antibiotic to treat a bacterial infection. Many bacteria are killed by the treatment, but some survive. Those that survive have genes that confer resistance to the antibiotic. The survivors reproduce and pass on the trait for resistance to future generations of bacteria. Over time, antibiotic resistance is more common in bacterial populations than before. In other words, natural selection has caused evolutionary change. Failure to take the full course of antibiotics also allows more resistant bacteria to survive, setting the stage for the evolution of bacterial resistance. Strains of antibiotic-resistant bacteria have been identified for many diseases. The evolution of drug-resistant strains may prove to be one of the greatest threats to humankind (Chapter 13a).

Natural selection does not lead to perfect organisms. Sometimes genetic variation is limited, and the variation available does not include the ideal traits. Natural selection can act only on available variation. Historical constraints also limit natural selection. Natural selection modifies existing structures rather than starting from scratch. Finally, organisms have to do many different things—such as escaping from predators and finding food, shelter, and mates—and it simply is not possible to be perfect at all things simultaneously. Indeed, adaptations often are compromises between the many competing demands faced by organisms.

stop and think

Some people have suggested that natural selection no longer operates within the human species because modern medicine levels the playing field. Medical technology, they argue, helps "fit" and "unfit" individuals alike to survive and reproduce. Others believe that natural selection continues to operate in modern society, suggesting that some present-day conditions act as selective agents. Air pollution, for example, might select for individuals with genetic resistance to respiratory disease. Which argument do you favor and why?

MUTATION

Mutations, you will recall, are rare changes in the DNA of genes. Mutations produce new alleles that, when transmitted in gametes, cause an immediate change in the gene pool. Essentially the new (mutant) allele is substituted for another allele. If the frequency of the mutant allele increases in a population, it is not because mutations are suddenly occurring with greater frequency. Instead, possession of the mutant allele might confer some advantage to enable individuals with the mutant allele to produce more offspring than those without it. In other words, the increased frequency of the mutant allele relative to others results from natural selection.

▌ Macroevolutionary change occurs above the species level

Large-scale evolutionary change is macroevolution. Whereas microevolution involves changes in the frequencies of alleles within populations, macroevolution involves major changes in groups of species. First, we examine how species are named. Then we consider how their evolutionary histories can be interpreted and depicted.

SCIENTIFIC NAMES

Systematic biology (also known as systematics) deals with the naming, classification, and evolutionary relationships of organisms. A stable system of names is essential for communicating information about organisms. The Swedish naturalist Carl Linnaeus developed our naming system more than 200 years ago. The center of Linnaeus's naming scheme is the **Latin binomial**. This two-part name consists of the genus name followed by the species name. For example, humans are genus *Homo,* species *sapiens*. The genus and species names are italicized, and the first letter of the genus name is always capitalized. The species name is traditionally given in all lowercase letters. Sometimes, the genus name is abbreviated, such as *H. sapiens*.

Linnaeus also developed a system for grouping species into increasingly broad categories, or **taxa** (singular, taxon): species, genus, family, order, class, phylum, and kingdom. Similar species were placed in the same genus; similar genera (the plural of *genus*) were placed in the same family; similar families in the same order, and so on. Figure 22.8 depicts such a classification for humans.

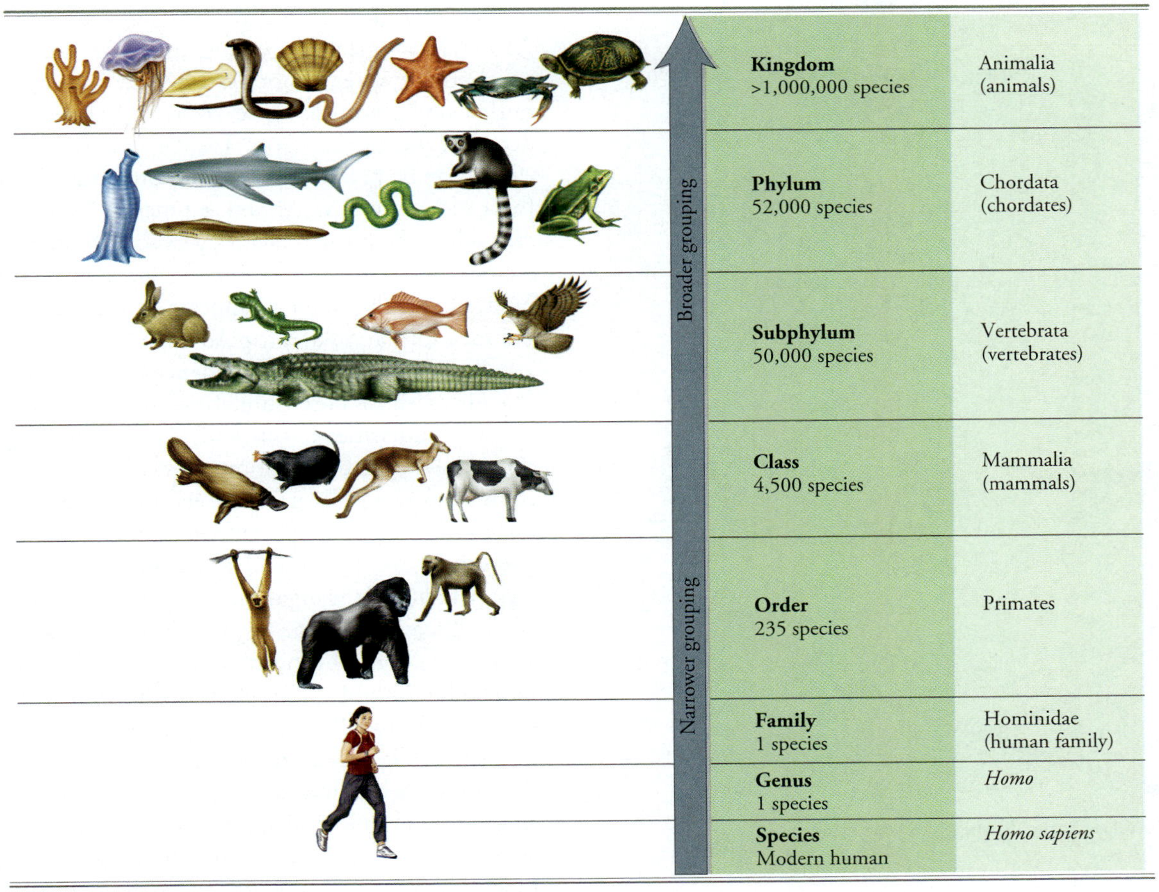

Broader grouping / Narrower grouping		
Kingdom >1,000,000 species	Animalia (animals)	
Phylum 52,000 species	Chordata (chordates)	
Subphylum 50,000 species	Vertebrata (vertebrates)	
Class 4,500 species	Mammalia (mammals)	
Order 235 species	Primates	
Family 1 species	Hominidae (human family)	
Genus 1 species	*Homo*	
Species Modern human	*Homo sapiens*	

FIGURE **22.8**

Linnaeus's scheme for classifying life. Any organism, including a human being, can be classified using a hierarchy of increasingly general categories. Modern humans are called *Homo sapiens*. We are the only living species in our genus (*Homo*) and in our family (Hominidae).

Now that we know how organisms are named and classified, let's consider how we can study their evolutionary history.

PHYLOGENETIC TREES

Phylogenetic trees describe the history of life. They depict hypotheses about evolutionary relationships among species or higher taxa and can illustrate in simple graphic form concepts that are difficult to express in words.

A major goal of systematic biology is the development of character matrices for constructing phylogenetic trees. Such matrices often consist of columns of taxa and rows of "characters." The characters score the presence or absence of particular features in the taxa of interest. For example, consider a simple three-taxon matrix with a shark (fish), frog (amphibian), and human (mammal). As character data, we might score for the presence or absence of two paired appendages (fins or limbs), individual digits, and hair (Figure 22.9a). We can array these features on a phylogenetic tree as shown in Figure 22.9b. It is easy to see from this example how phylogenetic trees can be generated from character data and how character data can be well illustrated on phylogenetic trees. Sometimes, more than one phylogenetic tree is

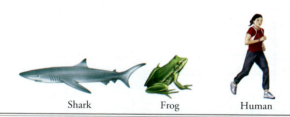

	Shark	Frog	Human
Two paired appendages	Yes	Yes	Yes
Digits	No	Yes	Yes
Hair	No	No	Yes

(a) A character matrix is used to construct a phylogenetic tree.

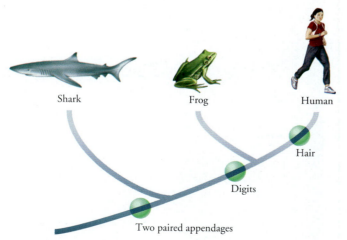

(b) A sample tree developed from the above character matrix

FIGURE **22.9**

A phylogenetic tree depicts hypotheses about evolutionary relationships among organisms.

generated for the taxa of interest. Under these circumstances, scientists prefer the simplest tree, the one requiring the fewest evolutionary events (having the fewest branches).

Evidence of Evolution Comes from Diverse Sources

We know that evolution has occurred because the physical evidence of evolution surrounds us. Such evidence comes from a variety of sources, including the fossil record, biogeography, the comparison of anatomical and embryological structures, and molecular biology.

The fossil record provides evidence of evolution

The earth is littered with silent monuments to past organisms (Figure 22.10). We find, for example, tiny spiders preserved in resin that dripped as sticky sap from some ancient tree. Also, we find mineralized bones and teeth, hardened remains that tell us much about the ancestry of today's vertebrates (animals with a backbone). These preserved remnants and impressions of past organisms are **fossils**. Most fossils occur in sedimentary rocks, such as limestone, sandstone, shale, and chalk. These rocks form when accumulated minerals and organic particles are cemented together in still water.

Fossilization is the process by which fossils form (Figure 22.11). In a typical case, an organism dies and settles to the bottom of a body of water. If not destroyed by scavengers, the organism is buried under accumulating layers of sediment. Soft parts usually decay or are carried away. Hard parts such as bones, teeth, and shells may be preserved if they become impregnated with minerals from surrounding water and sediments. As new layers of sediment are added, the older (lower) layers solidify under the pressure generated by overlying sediments. Eventually, when the sediments are uplifted and the water disappears, wind may erode the surface of the rock formation and expose the fossil.

Fossils provide strong evidence of evolution. Fossils of extinct organisms show both similarities to and differences from living species. Similarities to other fossil and modern species are used to assess the degree of evolutionary relationships. Often, fossils reveal combinations of features not seen in any living forms. Such combinations help us understand how major new adaptations arose. Sometimes we are lucky enough to find transitional forms that closely link ancient organisms to modern species. Consider a fossil whale discovered in the last few years that had hind limbs and an aquatic lifestyle. This transitional form links older terrestrial forms that had hind limbs to modern whales, which are aquatic and have no hind limbs. The relative ages of fossils can be determined because fossils found in deeper layers of rock are typically older than those found in layers closer to the surface. In this way, we can study the chronological appearance of lineages. For example, fishes were the first fossil vertebrates to appear in deep (old) layers of rock. Fossils closer to the surface are those of amphibians, then reptiles, then mammals, and finally birds. This chronological sequence for the appearance of the major groups of vertebrates has been sup-

(a) A spider preserved in amber.

(b) An ancient act of predation preserved in sedimentary rocks.

FIGURE **22.10**
A sampling of past life in fossils

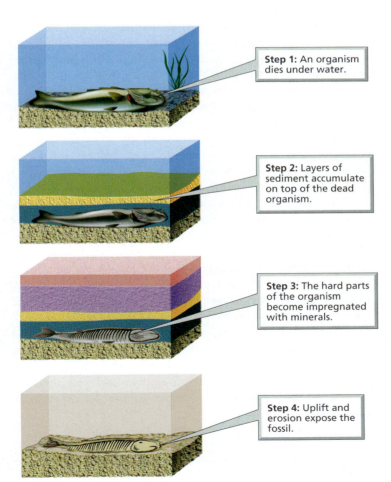

Step 1: An organism dies under water.

Step 2: Layers of sediment accumulate on top of the dead organism.

Step 3: The hard parts of the organism become impregnated with minerals.

Step 4: Uplift and erosion expose the fossil.

FIGURE **22.11**
A typical sequence for fossilization

ported by other lines of evidence, some of which we consider later in this section. Other large-scale evolutionary events revealed by the fossil record include changes in diversity over time and changes in faunas (animal life) and floras (plant life) as continents moved.

Although the fossil record tells us much about past life, it has limitations. First, fossils are rare. When most animals or plants die, their remains are eaten by predators or scavengers or are broken down by microorganisms, chemicals, or mechanical processes such as friction. Even if a fossil should form, the chances are small that it will be exposed by erosion or other forces and not be destroyed by those same forces before it is discovered. Second, the fossil record represents a biased sampling of past life. Aquatic plants and animals have a much higher probability of burial under water than do terrestrial organisms. Thus, preservation of aquatic organisms is more likely. Large animals with a hard skeleton are far more likely to be preserved than are small animals with soft parts. Organisms from large, enduring populations are more likely to be represented in the fossil record than are those from small, short-lived populations. Despite these limitations, fossils document that life on the earth has not always been the same as it is today. The simple fact of these changes is potent evidence of evolution.

Geographic distributions reflect evolutionary history

Biogeography is the study of the geographic distribution of organisms. Geographic distributions often reflect evolutionary history and relationships because related species are more likely to be found in the same geographic area than are unrelated species. A careful comparison of the animals in a given place with those occurring elsewhere can yield clues about the relationship of the groups. If animals have been separated, biogeography can tell us how long ago the separation occurred.

For example, today, we find many species of marsupials—mammal species, such as opossums and kangaroos, in which the female has a pouch—in Australia but only a few in North and

South America. Australia is an island, remote from other major continental landmasses. The presence of so many species of marsupials there suggests that they arose from distant ancestors whose descendants were not replaced by animals arriving from other places. New distributions of organisms occur by two basic mechanisms. In the first mechanism, the organisms disperse to new areas. In the second mechanism, the areas occupied by the organisms move or are subdivided. The evolutionary history of the Australian marsupials involves both dispersal and the movement of continents. Fossil evidence suggests that marsupials evolved in North America and that some dispersed southward into South America, then to Antarctica, and later to Australia, to which Antarctica was attached at the time (Figure 22.12). As Australia and other landmasses slowly shifted to form the modern continental arrangement, the ancestors of today's marsupials were carried away from their place of origin to evolve in isolation in Australia. As the continental landmasses separated and shifted, they carried populations of other living things with them as well. It should come as no surprise, then, that trees, freshwater fishes, and even birds are quite different in Australia than in other parts of the world. Australia's unique flora and fauna reflect its long history of isolation and continental movement.

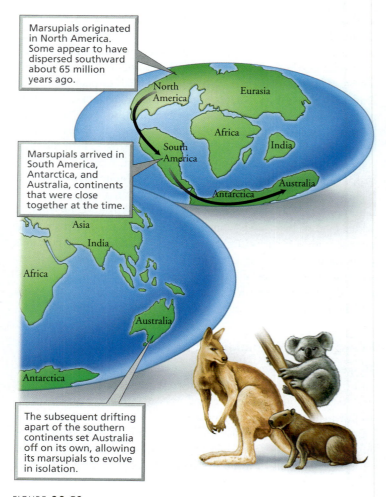

Marsupials originated in North America. Some appear to have dispersed southward about 65 million years ago.

Marsupials arrived in South America, Antarctica, and Australia, continents that were close together at the time.

The subsequent drifting apart of the southern continents set Australia off on its own, allowing its marsupials to evolve in isolation.

FIGURE **22.12**

The story of marsupials and Australia involves both dispersal and movement of continents.

Dispersal played a critical role in human evolution. Scientists studying our DNA suggest that there were several major waves of dispersal out of Africa to other parts of the world. Australia, it turns out, was probably colonized by humans arriving by boat more than 60,000 years ago. Our ancestors, as you will see, were a restless bunch.

Comparative anatomy and embryology reveal common descent

Comparative anatomy, as its name suggests, is the comparison of the anatomies (physical structures) of different species. Comparative anatomists often look for common, or shared, traits among different species. These shared traits have long been considered a measure of relatedness. Put simply, species with more shared traits are considered more closely related than are species without shared traits. For example, the vast majority of vertebrates alive today are united by the presence of jaws and paired fins or limbs. As another example, many very different vertebrates share similar structures, such as the bones of forelimbs (Figure 22.13). These similarities suggest that the animals share a common ancestor. Structures that are similar and that probably arose from a common ancestry are called **homologous structures**. The forelimbs that support bird wings and bat wings are homologous structures. Sometimes, however, similarities are not inherited from a common ancestor. For example, the wings of birds and those of insects both permit flight, but they are made from entirely different structures. Whereas bird wings consist of forelimbs, the wings of insects are not true appendages; they are extensions of the cuticle (exoskeleton). Thus, the wings of birds and insects do not reflect common ancestry. Instead, birds and insects independently evolved wings because of similar ecological roles and selection pressures, in a process known as **convergent evolution**. Structures that are similar because of convergent evolution are called **analogous structures**.

Homologous structures arise from the same kind of embryonic tissue. Hence, comparative embryology, the comparative study of development, also can be a useful tool for studying evolution. Common embryological origins can be considered evidence of common descent. For example, 4-week-old human embryos closely resemble embryos of other vertebrates, including fish. Indeed, human embryos at 4 weeks gestation come complete with a tail and gill pouches, as shown in Figure 22.14. As development proceeds, the gill pouches of fish develop into gills. The gill pouches of humans develop into other structures such as the Eustachian tubes connecting the middle ear and throat. Nevertheless, the fact that human, fish, and all other vertebrate embryos look very similar at early stages of development indicates common descent from an ancient ancestor.

Comparative molecular biology also reveals evolutionary relationships

Just as visible traits such as forelimbs can be compared, so can the molecules that are the basic building blocks of life. We have already mentioned that a common genetic code is potent evidence that all

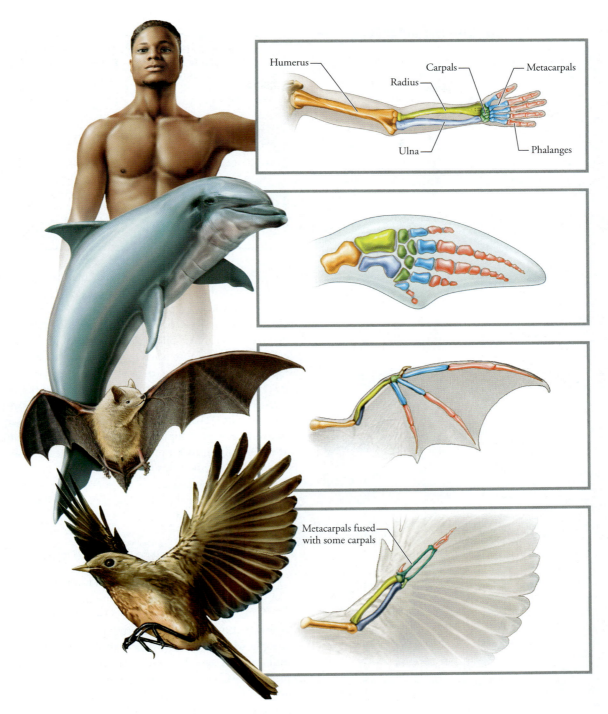

FIGURE **22.13**
Homologous structures. The similarity of the bones of the forelimbs of humans, dolphins, bats, and birds suggests that these organisms share a common ancestry.

life is related. What can differences tell us? For example, scientists can compare the sequences of amino acids in proteins or the nucleotide sequences in DNA. Such comparisons of DNA sequences tell us that only about 2.5% of our nucleotide sequences differ from those of chimpanzees (Table 22.1). Expressed another way, we share about 98% of our genetic material with chimpanzees. This information indicates that humans and chimpanzees diverged fairly recently from a common ancestor. As can anatomical characters, molecular characters can be used to develop phylogenetic trees. In fact, there is hope that the use of both molecular characters and anatomical characters will provide new insights into the evolution of life.

Development →

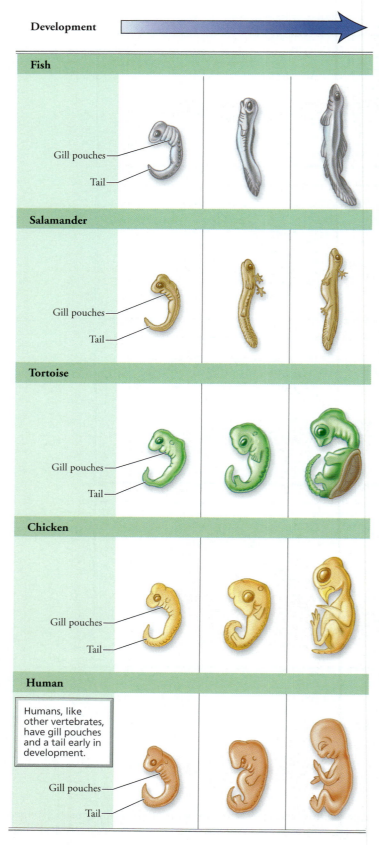

FIGURE **22.14**

Resemblance early in development indicates common descent. The embryos in the different groups have been drawn to the same approximate size to permit comparison.

TABLE **22.1**

DIFFERENCES IN DNA SEQUENCES INDICATING PHYLOGENETIC DISTANCES BETWEEN PAIRS OF PRIMATE SPECIES	
SPECIES PAIRS	**PERCENTAGE DIFFERENCE IN NUCLEOTIDE SEQUENCES**
Human–chimpanzee	2.5
Human–gibbon	5.1
Human–Old World monkey	9.0
Human–New World monkey	15.8
Human–lemur (prosimian)	42.0

Data from Stebbins, G. L. 1982. *Darwin to DNA, Molecules to Humanity.* San Francisco: W. H. Freeman.

The study of comparative molecular biology suggests the existence of a **molecular clock**, a constant rate of divergence of macromolecules (such as proteins) from one another. The molecular clock hypothesis is based on the notion that single nucleotide changes in DNA, called point mutations, and their resultant changes in the amino acids of proteins occur with steady, clocklike regularity. If this is true, then we can compare molecular sequences to estimate the ages of separation between species. The more differences in sequence, the more time that has elapsed since the common ancestor. For example, a comparison of such sequences tells us that humans and chimpanzees diverged from a common ancestor about 6 million to 8 million years ago. (See the Social Issue essay, *Conducting Research on Our Relatives.*)

We have learned about microevolution, macroevolution, and the evidence for evolution. Now let's look at our own past and see how humans evolved.

Human Roots Trace Back to the First Primates

We begin our discussion of human evolution with the primates, an order of mammals that includes humans, apes, monkeys, and related forms (such as lemurs). Scientists do not agree when the first primates evolved. Paleontologists (people who study fossils) believe that primates evolved about 65 million years ago. This estimate takes into account that the oldest primate fossils are about 55 million years old, and organisms always evolve earlier than when they first appear in the fossil record. In contrast, molecular biologists have suggested that primates evolved about 90 million years ago. Their estimate is based on comparisons of DNA sequences. A third group, composed of mathematicians and biologists, developed a computer model to examine the origin of primates. Their model estimates that primates evolved 85 million years ago, a value in line with estimates from the molecular data.

stop and think

As we look through the stages of our ancestry, we will be reviewing the species that led to us. At some point the line *became* us. Before getting into the details of that ancestry, what qualities would you now say are necessary for a species to qualify as human?

Conducting Research on Our Relatives

Chimpanzees look and behave somewhat differently than we do. Nevertheless, it is hard to look into their eyes and not see something of ourselves. Chimpanzees are our closest living relatives, sharing about 98% of our genes. Is it ethical to maintain chimpanzees in laboratory cages and use them for invasive scientific research that might benefit us? What about other primate species to which we are more distantly related?

An estimated 57,000 primates were used in research in the United States during the year 2000. Despite the controversy and cost of using primates, they often are preferred as subjects in certain experiments precisely because they are so similar to humans. For example, human and nonhuman primates have brains with similar organization, develop comparable plaques in their arteries, and experience many of the same changes in anatomy, physiology, and behavior with age. Human and nonhuman primates also have similar immune systems. In some cases, Nobel Prize–winning research has resulted from the contributions of research with nonhuman primates, including development of vaccines for yellow fever (1951) and polio (1954) and illuminating how visual information is processed in the brain (1981). Research with nonhuman primates has also led to significant advances in our understanding of Alzheimer's disease, AIDS, and SARS (severe acute respiratory syndrome).

The care and use of primates (and other animals) in research are regulated by federal agencies such as the Public Health Service (PHS) and the U.S. Department of Agriculture (USDA). Animal research also is regulated at the local level. Each college, university, or research center has an Institutional Animal Care and Use Committee whose members include veterinarians, teachers, researchers, and members from the public. In addition to federal and local oversight, scientists and animal care personnel are making substantial efforts to provide appropriate housing for captive nonhuman primates and to consider their psychological well-being. Even with such efforts, however, controversy and questions remain. Should we ban the use of nonhuman primates in medical research? And if we do, will progress in fighting diseases such as AIDS and Alzheimer's be slowed?

▌Primates have distinct characteristics

Despite debate over when primates evolved, most scientists agree that the first primates probably arose from an insect-eating mammal that lived in trees. The ancestor to primates might have looked something like a modern tree shrew (Figure 22.15). Primates share many characteristics that reflect an arboreal lifestyle specialized for the visual hunting and manual capture of insects. Primates have flexible, rotating shoulder joints and exceptionally mobile digits with sensitive pads on their ends. Flattened nails replace claws. In many primate species, the big toe is separated from the other toes, and thumbs are opposable to other fingers. Thus, primates have grasping feet *and* hands, features that help in the pursuit and capture of insects along branches. A complex visual system and a large brain relative to body size make possible the well-developed depth perception, hand-eye coordination, and neuromuscular control needed by arboreal insectivores. Also, most primates give birth to only one infant at a time and provide extensive parental care—characteristics that may reflect the difficulty of carrying and rearing infants in trees. Humans are the most terrestrial of primates, yet we retain many of the basic characteristics that evolved in the trees. The characters that distinguish primates from other mammals are summarized in Table 22.2.

Modern primates are divided into two groups, or suborders: prosimians and anthropoids. The **prosimians** ("premonkeys") include the lemurs, lorises, pottos, and tarsiers (Figure 22.16). These modern

FIGURE **22.15**
A tree shrew. These modern animals resemble the arboreal, insect-eating mammals from which primates probably evolved over 65 million years ago.

TABLE **22.2**

CHARACTERISTICS OF PRIMATES
Flexible shoulder joints
Grasping forelimbs; some species with fully opposable thumbs
Grasping feet, with big toe separated from other toes (except humans)
Sensitive digits with flattened nails instead of claws
Forward-facing eyes
Small litter size
Complex social behavior, including extensive parental care
Reliance on learned behavior

(a) Lemur   (b) Loris   (c) Potto   (d) Tarsier

FIGURE **22.16**
Modern prosimians retain ancestral primate features.

species have been grouped together because they retain ancestral primate features such as small body size, nocturnal habits, and fairly generalized diets (omnivorous or insectivorous); they also have a relatively small brain when compared with anthropoids. The **anthropoids** include monkeys, apes, and humans. Monkeys are further divided into two groups: New World and Old World (Figure 22.17).

Within the anthropoids are four genera of modern apes: gibbons *(Hylobates)*, orangutans *(Pongo)*, gorillas *(Gorilla)*, and chimpanzees and bonobos *(Pan)*. An example from each genus is shown in Figure 22.18. The final modern member of the anthropoids is *Homo sapiens*—humans. Apes and humans are termed **hominoids**. The term **hominid** is used for members of the human family, such as those in the genera *Australopithecus* and *Homo* (discussed later in this chapter).

(a) New World monkeys, such as the spider monkey shown here, are arboreal, have a prehensile tail, and have nostrils that open to the side.

(b) Old World monkeys, such as the baboons shown here, lack a prehensile tail, and have nostrils that open downward.

FIGURE **22.17**
New World and Old World monkeys

Recall that phylogenetic trees depict hypothesized relationships among organisms; Figure 22.19 shows such a tree for primates. Let's focus on the hominoids, a group that diverged from the Old World monkeys at least 40 million years ago. As described earlier, molecular evidence suggests that the lines leading to modern humans and chimpanzees diverged about 6 million to 8 million years ago. Molecular data further indicate that beyond chimpanzees, gorillas are our next closest living relatives, followed by orangutans and then gibbons.

Let's now consider the origin and evolution of humans. We begin by describing the problems associated with studying human evolution and dispel some common misconceptions about our evolutionary history. Then we discuss some of the hominid species. Keep in mind that as new hominid fossils are found and genetic studies conducted, the dates for the origin of many species will likely be pushed back.

▌ Discussions of human origins often evoke controversy

Paleoanthropology is the scientific study of human origins and evolution. This field combines the disciplines of paleontology (the study of fossils) and anthropology (the study of humans). Many debates about human evolution stem from past practices of paleoanthropologists. For example, past paleoanthropologists tended to give each new fossil discovery a new name and to consider it an example of a separate line. More recent studies indicate that many of these "finds" are probably of the same lineage. The confusion was compounded when some paleoanthropologists extrapolated from sketchy information, such as a partial skull or a few teeth. Another problem stems from variation in a population. Two specimens that differ slightly in some traits may be interpreted as different lines when this interpretation is not warranted. Then, there is the problem of the overall paucity of fossils. There just have not been enough hominid remains found to reconstruct a solid, unequivocal history of our species. Finally, many arguments undoubtedly arise because of certain misconceptions, as we see in the next section.

(a) Gibbon (*Hylobates*) **(b)** Orangutan (*Pongo*) **(c)** Gorilla (*Gorilla*) **(d)** Chimpanzee (shown) and bonobo (*Pan*)

FIGURE **22.18**
Apes. Modern apes are organized into four genera. A representative species of each genus is shown.

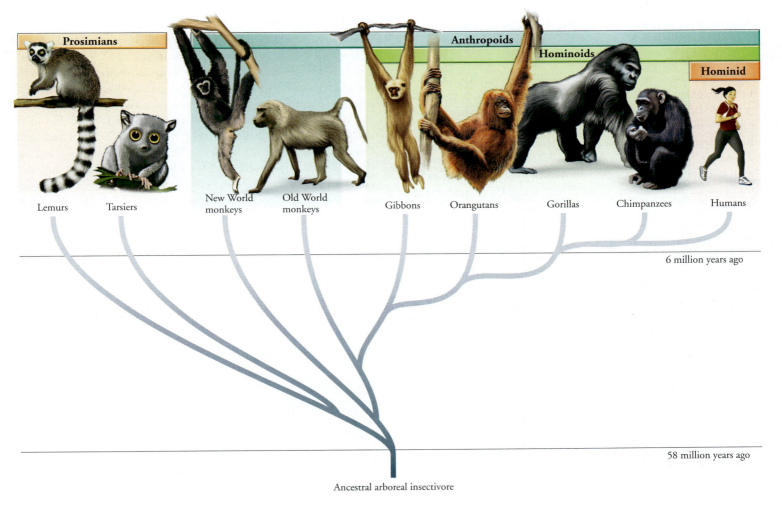

FIGURE **22.19**
Summary of primate evolution. (Figures are not drawn to scale.)

Misconceptions distort the picture of human evolution

One misconception is that we descended from chimpanzees or any of the other modern apes. Humans and chimpanzees represent separate branches of the anthropoid tree, branches that diverged at least 6 million to 8 million years ago. Thus, the common ancestor of humans and chimpanzees was different from any modern species of ape.

Another misconception is that modern humans evolved in an orderly, stepwise fashion. We often see such a stepwise progression illustrated, sometimes humorously, and its appeal lies in its simplicity. However, as is so often the case, simplicity is deceiving. The path to modern humans has been fraught with "unsuccessful phenotypes" leading to dead end after dead end. In fact, the path looks more like a family "bush" than an orderly progression to modern humans.

A final misconception is that, over the course of human evolution, the various bones and organ systems evolved together at the same rate. They did not. There is no reason to believe that the human brain evolved at the same rate as, say, the appendix or the foot. Instead, various traits evolved at their own rate, a phenomenon known as **mosaic evolution**.

Walking on two feet was a critical step in hominid evolution

Several evolutionary trends are apparent within hominids. **Bipedalism** (walking on two feet) evolved early and probably set the stage for the evolution of other characteristics, such as increases in brain size. Cultural trends, such as tool use and language, are linked to increases in brain size. Once the hands were freed from the requirements of locomotion, they could be used for tasks such as making tools. Evidence that bipedalism preceded increases in brain size and cultural trends comes from fossilized hominid footprints found in Tanzania, Africa. These footprints, dated at about 3.6 million years old, were apparently made by two adults and a child (Figure 22.20). The footprints clearly predate the oldest stone tools, from 2.6 million years ago.

Other changes associated with upright posture include the S-shaped curvature of the vertebral column ("the lumbar curve"); modifications to the bones and muscles of the pelvis, legs, and feet; and positioning of the skull on top of the vertebral column. The faces of hominids also changed (see Figure 22.24). For example, the forehead changed from sloping to vertical, and sites of muscle attachment, such as the brow ridges and crests on the skull, became smaller. The jaws became shorter, and the nose and chin more prominent. The overall size difference between males and females decreased. Males of our early ancestors appear to have been 1.5 times the size of females. Modern human males weigh about 1.2 times as much as females. A difference in appearance between males and females is called **sexual dimorphism**. Table 22.3 summarizes major trends in hominid evolution.

When considering our ancestors, we will focus on the hominids that we know the most about—those in the genus *Australopithecus* and the genus *Homo*. We will also mention some recent finds of apparently older hominids.

FIGURE **22.20**

These hominid footprints from Laetoli, Tanzania, predate the oldest known tools and thus provide evidence that bipedalism preceded increases in brain size and cultural trends such as tool making. The larger prints were made by two individuals, one following in the footsteps of the other. The smaller prints to the side may have been made by a child.

AUSTRALOPITHICINES

The first hominid remains discovered were classified in the genus *Australopithecus*, meaning "southern ape." The seven recognized species of *Australopithecus* are sometimes collectively termed australopithecines. The most spectacular australopithecine fossil found to date is that of a young adult female of the species *Australopithecus afarensis* (because she was found in the Afar region of Ethiopia). She was named Lucy because the Beatles' song "Lucy in the Sky with Diamonds" was playing the night Donald Johanson and his coworkers celebrated her discovery. When more than 60 pieces of Lucy's bones were arranged, scientists estimated that, at death, she was about 1 m (3 ft) tall and weighed about 30 kg (66 lb) (Figure 22.21). Her bones were dated to be 3.2 million years old. As more remains were found, it became apparent that the males of Lucy's species were somewhat taller (about 1.5 m, or about 5 ft) and heavier (about 45 kg, or 99 lb). The brain of *A. afarensis* was similar in size to that of modern chimpanzees or gorillas, about 380 to 450 cc (23 to 28 in.^{3}). Other remains of *A. afarensis* were found in Tanzania and dated at between 3.6 million and 3.8 million

TABLE **22.3**

MAJOR TRENDS IN HOMINID EVOLUTION
• Bipedalism
• Increasing brain size and associated cultural trends such as tool use, language, and behavioral complexity
• Shortening of the jaw and flattening of the face
• Reduced difference in size between males and females

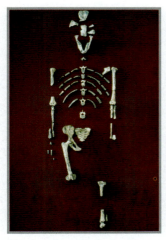 

(a) Lucy is a remarkably complete skeleton of a young female *Australopithecus afarensis*, a hominid in existence over 3 million years ago.

(b) Reconstruction of Lucy on display at the St. Louis Zoo

FIGURE **22.21**
"Lucy"

years old. Although many aspects of the anatomy of *A. afarensis* suggest adaptations for living in trees, the remains also indicate bipedalism. In *A. afarensis*, we see an example of mosaic evolution—bipedalism evolving before substantial increases in brain size.

stop and think

Brain size and shape can be estimated by making an endocast, a model of the inside of the skull. Such reconstructions are possible as long as some parts of the skull are discovered and reassembled. Endocasts reveal general brain anatomy and size, allowing scientists to draw tentative conclusions about the cognitive capabilities of the skull's owner. Endocasts do not, however, capture the folds or convolutions on the surface of a brain. Do you think endocasts and brain volumes make reliable indicators of intelligence? Why or why not?

Since the mid-1990s, researchers have found hominid remains older than those of *Australopithecus afarensis*. Some remains (17 pieces, mostly teeth and a few bones) from Ethiopia have been assigned to the species *Ardipithecus ramidus*. The remains were dated at 4.4 million years old. Some debate exists about whether *A. ramidus* displayed bipedalism, although most researchers think it did. It is also unclear whether *A. ramidus* is the ancestor of *Australopithecus* or a separate line that became extinct. Even older remains were discovered in Chad, Africa. The remains are believed to be those of a hominid and have been dated at nearly 7 million years old. This date is close in time to the hypothesized 6 million- to 8 million-year-old split between chimpanzees and humans. The presumed hominid was named *Sahelanthropus tchadensis*, a reference to the Sahel region in Chad where the nearly complete skull was found. Initial evaluation of the remains suggests that the species walked upright.

About 3 million years ago, when *Australopithecus afarensis* had been in existence for nearly 1 million years, several new hominid species appeared in the fossil record. Scientists believe that *Australopithecus africanus*, one of these new species, was a hunting-and-gathering omnivore. *Autralopithecus africanus*, like *A. afarensis*, was a gracile, or slender, australopithecine. Three more "robust" australopithecines (*A. aethiopicus*, *A. robustus*, and *A. boisei*) also appeared and are thought to have been savanna-dwelling vegetarians. The robust species had a massive skull, heavy facial bones, a pronounced brow, and huge teeth. Whatever they ate, it required a lot of chewing. It is unclear whether Lucy's species, *A. afarensis*, gave rise to these other australopithecines or simply lived at the same time as they did. The robust australopithecines appear to have been evolutionary dead ends. The gracile species may have lead to ancestors of our genus, the genus *Homo*.

HOMO HABILIS

Homo habilis ("handy man"), the first member of the modern genus of humans, appeared in the fossil record about 2.5 million years ago. The remains of *H. habilis* are highly variable, causing some researchers to question their taxonomic status. Some researchers believe the remains are varied enough to represent more than one species. *Homo habilis* differed from *Australopithecus afarensis* primarily in brain size. The cranial capacity of *H. habilis* has been estimated at between 500 and 650 cc (30 and 40 in.^{3}). Some scientists theorize that *H. habilis* was the first hominid to use stone tools. Simple stone tools, dating from 2.5 million to 2.7 million years ago have been found in Africa. Whether these tools were used by *H. habilis* or one of the species of robust australopithecines is unclear. It also is possible that *H. habilis* was capable of rudimentary speech. Casts of one brain made from reassembled skull fragments indicate a bulge in Broca's area, the area of the brain important in speech (see Chapter 8).

HOMO ERECTUS

A new hominid, *Homo erectus* ("upright man"), appeared in the fossil record about 1.8 million years ago. *Homo erectus* may have originated in East Africa and coexisted there for several thousand years with some of the robust australopithecines. Nevertheless, *H. erectus* was a wanderer, believed by many to be the first hominid to migrate out of Africa. Bones of *H. erectus* have been found in Africa, India, China, Indonesia, and Europe. *Homo erectus* differed from earlier hominids in being larger (up to 1.85 m, or 6 ft tall, and weighing at least 65 kg, or 143 lb) and less sexually dimorphic. Later specimens of *H. erectus* had a brain volume as large as 1200 cc (73 in.^{3}); this volume is close to that of modern humans (1300 cc, or 79 in.^{3}). Evidence indicates that *H. erectus* used more sophisticated tools and weapons than had other hominids. They may have used fire. Most researchers believe that *H. erectus* is the closest species to modern humans. *Homo erectus* disappeared between 200,000 and 300,000 years ago.

HOMO SAPIENS

We now come to our own species, *Homo sapiens* ("thinking man"), which evolved around the same time that *H. erectus* disappeared. *Homo sapiens* differs from *H. erectus* in having a slightly larger brain, a more robust skull, larger teeth, and a less projecting jaw. *Homo sapiens* differs from *H. erectus* only in a matter of measurements—not in the presence or absence of specific characters. There seem to have been two lines of descendants of *H. erectus*.

Neanderthals evolved from *H. erectus* living in Europe. Neanderthals had distinct features apparently adapted for life in a cold climate. Some Neanderthals lived in caves ("the cave men"), and others made temporary camps (Figure 22.22a). Neanderthal burial sites have been discovered, making them the first people known to have buried their dead. Also, the discovery of 50,000-year-old remains of sick, injured, and elderly individuals suggests that Neanderthals cared for the less fortunate among them.

Interestingly, Neanderthals had a larger brain case than does *Homo sapiens*. However, this feature may not have been correlated with intelligence but rather with Neanderthals' more massive body. They had larger bones, suggesting heavier musculature, and rather short legs. They also had a thick brow ridge, large nose, broad face, and well-developed incisors and canines. Some anthropologists consider Neanderthals to be a subspecies of *H. sapiens*, calling them *H. sapiens neanderthalensis* (modern humans are known as *H. sapiens sapiens*). Most, however, assign Neanderthals species status and call them *Homo neanderthalensis*.

Neanderthals vanished some 30,000 years ago for still mysterious reasons. Some scientists suggest they were outcompeted or simply killed outright by a form of *H. sapiens sapiens* called Cro-Magnons (Figure 22.22b). Other scientists suggest that interbreeding between anatomically modern humans and Neanderthals might have resulted in the loss of the Neanderthal phenotype. The extent and type of interactions between Neanderthals and Cro-Magnons are still debated. It is clear, however, that the Cro-Magnons had superior tools and weapons, including bows and arrows, knives, and stone-tipped spears. It is generally believed that Cro-Magnons were quite similar to modern humans in appearance and lived in small to moderate-sized groups. They also were accomplished hunters and artists.

what would you do?

Human remains can tell scientists much about the diets, diseases, lifestyles, and genetic relationships of our ancestors, and such information can help piece together our evolutionary past. Sometimes, however, keeping human remains for scientific study conflicts with the wishes of descendants of the deceased individual, who wish to have the remains returned to them for reburial. In the United States, for example, many museums and universities are developing policies regarding the treatment and disposition of Native American and Native Hawaiian remains. How should human remains be treated, and, in the end, who should get them? If you were director of a museum and were asked to develop a policy for the treatment of human remains, what would you do?

Did modern humans evolve several times in different regions? The idea that they did is known as the **multiregional hypothesis**. It suggests that modern humans evolved independently in Europe, Asia, Africa, and Australia from distinctive local populations of *Homo erectus*. The alternative idea is known as the **Out of Africa hypothesis**. It suggests a single origin for all *H. sapiens*. According to the Out of Africa hypothesis, anatomically modern humans evolved from *H. erectus* in Africa and only later migrated to Europe, Asia, and Australia, where they replaced the diverse descendants of *H. erectus*. The two hypotheses are depicted in Figure 22.23.

The Out of Africa hypothesis for the origin of modern humans has wide support. Nevertheless, it may need modification in light of recent evidence. The results from a new genetic study in which DNA sequences were compared from populations around the world suggest two things. First, the results suggest two waves of migration out of Africa—not just one. In addition to the migration of 100,000 years ago postulated by the Out of Africa hypothesis, the new data

(a) The Neanderthals had an appearance apparently adapted for life in a cold climate. Some anthropologists consider Neantherthals to be a subspecies of our own species; others consider them a species separate from our own.

(b) The Cro-Magnons were quite similar to us in appearance, and some anthropologists believe they were responsible for the disappearance of the Neantherthals.

FIGURE **22.22**
Representatives of the genus *Homo*, after *H. erectus*

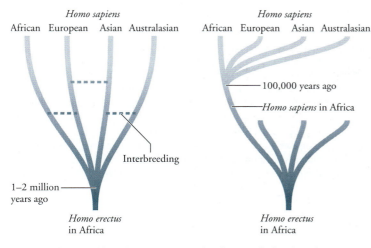

(a) The multiregional hypothesis suggests that modern humans evolved independently in several areas of the world from distinctive populations of *H. erectus.*

(b) The Out of Africa hypothesis suggests that modern humans evolved from *H. erectus* in Africa and later migrated to other parts of the world, replacing other descendants of *H. erectus.*

FIGURE **22.23**
Two hypotheses for the origin of modern humans

TABLE **22.4**

SOME MILESTONES IN HUMAN EVOLUTION		
HOMINID	**YEARS**	**MILESTONE**
Sahelanthropus tchadensis	6–7 million years ago	Bipedalism?
Homo habilis	2.4–1.6 million years ago	Tool use, speech
Homo erectus	1.8 million–300,000 years ago	Fire, migration
*Homo neanderthalensis**	165,000–30,000 years ago	Buried their dead

*Some scientists consider Neanderthals to be a subspecies of *Homo sapiens* and call them *Homo sapiens neanderthalensis.*

indicate a migration some time between 420,000 and 840,000 years ago. Second, the new results suggest that, on each of those migrations, the more modern people from Africa did not completely replace the locals they encountered but instead interbred with them. If these conclusions are correct, then some of us today may have genes from the local people—folks such as the Neanderthals and other descendants of *H. erectus*—that our ancestors met.

A possible scenario for human evolution is shown in Figure 22.24. Major milestones in human evolution are summarized in Table 22.4.

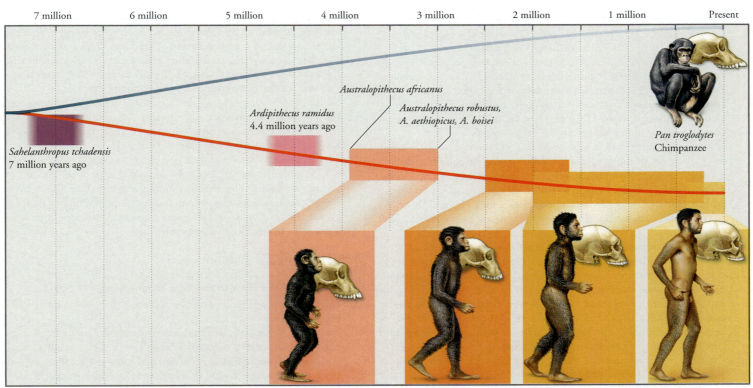

FIGURE **22.24**
One hypothesis of hominid evolution

REVIEWING THE CONCEPTS

Life Evolved on the Earth about 3.8 Billion Years Ago (pp. 512–514)

1. The earth is estimated to be 4.5 billion years old. Physical and chemical changes in the earth's crust suggest that life originated about 3.8 billion years ago. A plausible sequence of events known as chemical evolution suggests that life arose from chemicals that grew increasingly complex over a long period of time.

2. A. I. Oparin and J. B. S. Haldane suggested that conditions of the early earth favored the synthesis of small organic molecules from inorganic molecules. Laboratory re-creations of the conditions of the primitive earth have yielded small organic molecules, including amino acids. Over time, it is believed, small organic molecules accumulated in the early oceans and joined to form larger molecules. Next, genetic material arose, which allowed the transfer of information from one generation to the next. The original genetic material may have been RNA, not DNA. Eventually, the large organic molecules, including genetic material, aggregated into droplets. These droplets displayed some of the properties of life and were the precursors to cells.

3. The earliest cells were prokaryotic. According to the endosymbiont hypothesis, more complex cells formed as smaller organisms were incorporated into the cells, forming organelles such as mitochondria. Other organelles may have formed by infolding of the plasma membrane. Multicellularity evolved, leading to the appearance on the earth of organisms such as plants and animals.

WEB TUTORIAL 22.1 Principles of Evolutions

The Scale of Evolutionary Change May Be Small or Large (pp. 514–518)

4. Microevolution involves change at the genetic level. Macroevolution is large-scale evolution involving change among groups of species.

5. Populations are groups of individuals of the same species that live in a particular area. A gene pool is a collection of all of the alleles of all of the genes of all of the individuals in a population.

6. Sexual reproduction and mutation produce variation in populations. With respect to sexual reproduction, sources of variation include crossing over and independent assortment during meiosis. Also, fertilization produces a zygote with new combinations of alleles. Mutations, rare changes in the DNA of genes, are the only source of new characteristics within populations.

7. Microevolution involves change in the frequency of certain alleles relative to others in the gene pool. Causes of microevolution include genetic drift, gene flow, natural selection, and mutation.

8. Genetic drift is random changes in allele frequencies due to chance within a small population. Populations small enough for genetic drift to occur often are associated with two circumstances: the bottleneck effect and the founder effect.

9. A species is a population or group of populations whose members are capable of interbreeding to produce fertile offspring under natural conditions. Sometimes, populations of a species become geographically isolated and accumulate distinct sets of allele frequencies and mutations. Such populations may become so different from the original population that successful interbreeding between the original and new populations is no longer possible; in such a case, speciation (the formation of a new species) has occurred.

10. Charles Darwin argued that species living today are descendants of ancestral species. He also proposed that evolution could occur by natural selection. According to Darwin, evolutionary change occurs as the traits of successful individuals (those that survive and reproduce) become more common in a population, and traits of less successful individuals become less common.

11. Mutation produces new alleles that, when transmitted in sperm or eggs, cause an immediate change in the gene pool.

12. Systematic biology deals with the naming, classification, and evolutionary relationships of organisms. Organisms are named according to a scheme developed by Carl Linnaeus in which each organism is given a two-part name consisting of the genus name followed by the species name. In addition, organisms may be classified using a hierarchy of increasingly more general categories (taxa): species, genus, family, order, class, phylum, and kingdom.

13. A phylogenetic tree is a schematic diagram depicting evolutionary relationships among species or higher taxa. Such trees represent hypotheses and are constructed from data matrices.

WEB TUTORIAL 22.2 Mutations: Agents of Change

WEB TUTORIAL 22.3 Natural Selection

Evidence of Evolution Comes from Diverse Sources (pp. 518–522)

14. Evidence of evolution comes from a variety of sources, including the fossil record, biogeography, the comparison of anatomical and embryological structures, and molecular biology.

15. Fossils are the preserved remnants and impressions of past organisms. The fossil record documents that life on the earth has not always been the same as it is today and thus provides evidence of evolution.

16. Biogeography is the study of the geographic distribution of organisms. New distributions of organisms occur either when the organisms disperse to a new location or when the areas they occupy move or are subdivided. In general, related species are more likely to be found in the same geographic area than are unrelated species.

17. Comparative anatomy also provides evidence of evolution. In general, species with more shared traits are considered more likely to be related. For example, the bones of the forelimbs are common to many different vertebrates, and this similarity in structure suggests that the vertebrates share a common ancestry. Structures that have arisen from common ancestry are called homologous structures and usually arise from the same embryonic tissue. Thus, common embryological origins can also be viewed as evidence of common descent.

18. Molecules that are the basic building blocks of life can be compared for evidence of evolution. For example, scientists compare the sequences of amino acids in proteins or the nucleotide sequences in DNA of different species to gauge relatedness.

19. The sequences of amino acids in proteins or the nucleotide sequences in DNA of different species can be compared to estimate the time of divergence from a most recent common ancestor. This idea is based on the presumed existence of a molecular clock, the notion that single nucleotide changes in DNA and amino acid changes in proteins occur with steady, clocklike regularity.

WEB TUTORIAL 22.4 Evolution

Human Roots Trace Back to the First Primates (pp. 522–529)

20. Humans are primates, an order of mammals that also includes monkeys and apes.

21. Primates have several characteristics that set them apart from other mammals. Primates have forward-facing eyes, flexible shoulder joints, and grasping forelimbs and feet. Flattened nails—rather than claws—cover their sensitive digits. In addition, primates provide extensive parental care to a small number of offspring.

22. There are two suborders of modern primates: the prosimians (lemurs, lorises, pottos, and tarsiers) and the anthropoids (monkeys, apes, and humans). Monkeys include New World and Old World groups, and apes include gibbons, orangutans, gorillas, and chimpanzees. Modern humans, *Homo sapiens*, are the final members of the anthropoid sub-

order. The term *hominoid* refers to apes and humans, and the term *hominid* refers to members of the human family, such as species within the genera *Australopithecus* and *Homo*.

23. Paleoanthropology is the study of human origins and evolution. Intense argument about human evolution stems, in part, from intraspecific variation, the paucity of fossils, and the tendency of some anthropologists to give each fossil a new name and to consider it a new line.

24. The following three common misconceptions about human evolution stir controversy: (1) Humans descended from chimpanzees (in fact, humans and chimps represent separate branches that diverged from a common ancestor); (2) Human evolution occurred in an orderly progression from ancient to modern forms (in actuality, there have been several times when two or more species of hominids lived at the same time, and some of those lines were evolutionary dead ends); and (3) Traits of humans evolved at the same rate (instead, evidence indicates that traits of humans evolved at different rates, a phenomenon known as mosaic evolution).

25. Bipedalism probably evolved early on in hominids. Walking on two feet set the stage for the evolution of other characteristics such as an increase in brain size, which, in turn, was associated with cultural trends such as tool use and language.

26. The oldest hominid remains are likely those of *Sahelanthropus tchadensis*, dated to be nearly 7 million years old. The next hominid remains are those of *Ardipithecus ramidus*, about 4.4 million years old. Excellent specimens have been found of *Australopithecus afarensis* ("Lucy," for example), a species about 3.8 million years old. About 3 million years ago, several new hominid species appeared (*Australopithecus africanus* and three more robust species of *Australopithecus*). It is unclear whether *A. afarensis* gave rise to these new forms or simply coexisted with them.

27. About 2.5 million years ago, *Homo habilis* remains appear in the fossil record in Africa. About 1.8 million years ago, *H. erectus* remains appear in the fossil record. Fossils of *H. erectus* are not restricted to Africa. About 200,000 years ago, *H. erectus* disappeared, and *H. sapiens* emerged.

28. The multiregional hypothesis suggests that *Homo sapiens* evolved independently in different regions from distinctive populations of *H. erectus*. The Out of Africa hypothesis suggests that modern humans evolved from *H. erectus* in Africa and then migrated to other regions where they replaced the diverse descendants of *H. erectus*.

KEY TERMS

adaptation *p. 516*	evolution *p. 512*	hominoids *p. 524*	Out of Africa hypothesis *p. 528*
allele *p. 514*	fitness *p. 516*	homologous structure *p. 520*	paleoanthropology *p. 524*
analogous structure *p. 520*	fossil *p. 518*	Latin binomial *p. 517*	phylogenetic trees *p. 518*
anthropoids *p. 524*	fossilization *p. 518*	macroevolution *p. 514*	population *p. 514*
biogeography *p. 519*	founder effect *p. 515*	microevolution *p. 514*	prosimians *p. 523*
bipedalism *p. 526*	gene *p. 514*	molecular clock *p. 522*	sexual dimorphism *p. 526*
bottleneck effect *p. 515*	gene flow *p. 516*	mosaic evolution *p. 526*	speciation *p. 516*
chemical evolution *p. 512*	gene pool *p. 514*	multiregional hypothesis *p. 528*	species *p. 516*
convergent evolution *p. 520*	genetic drift *p. 515*	mutation *p. 515*	taxon *p. 517*
endosymbiont hypothesis *p. 514*	hominid *p. 524*	natural selection *p. 516*	

THINKING ABOUT THE CONCEPTS

1. How might life have evolved from inorganic molecules to complex cells? *pp. 512–514*
2. Distinguish between microevolution and macroevolution. *p. 514*
3. What are four sources of variation within populations? *pp. 514–515*
4. How does genetic drift lead to microevolution? *pp. 515–516*
5. Define speciation and relate it to gene flow. *p. 516*
6. Define natural selection. How can variation within populations be maintained in the face of natural selection? *p. 516*
7. Describe the binomial system by which organisms are named and the hierarchical system by which they are classified. *p. 517*
8. What is a phylogenetic tree? *p. 518*
9. What is a fossil? Describe the process of fossilization and relate it to limitations of the fossil record. *p. 518-519*
10. How do new distributions of organisms arise? *pp. 519–520*
11. Distinguish between homologous and analogous structures. *p. 520*
12. Describe how comparative embryology provides evidence of evolution. *p. 520*
13. What is a molecular clock? *p. 522*
14. What characters set primates apart from other mammals? *p. 523*
15. Why do discussions of human evolution inspire controversy? *p. 524*
16. What are three popular misconceptions about human evolution? *p. 526*
17. Describe the major species of hominids and trends in hominid evolution. *pp. 526–529*

18. Which of the following does *not* produce variation in populations?
    a. bottleneck effect
    b. mutation
    c. crossing over
    d. independent assortment
19. Which of the following occurs when fertile individuals move into and out of populations?
    a. genetic drift
    b. speciation
    c. mutation
    d. gene flow
20. Which of the following was *absent* from the early earth?
    a. lightning
    b. volcanoes
    c. oxygen
    d. ultraviolet radiation
21. The earliest cells
    a. were prokaryotic.
    b. were eukaryotic.
    c. arrived from outer space.
    d. were part of multicellular organisms.

22. Which of the following primates is our closest living relative?
    a. gorilla
    b. orangutan
    c. chimpanzee
    d. gibbon
23. Which of the following would promote fossilization?
    a. terrestrial existence
    b. member of a small population
    c. soft body parts
    d. aquatic existence
24. Which hominid was probably the first to migrate out of Africa?
    a. *Australopithecus afarensis*
    b. *Homo sapiens*
    c. *Homo erectus*
    d. *Australopithecus africanus*
25. The major events of human evolution occurred in
    a. Africa.
    b. South America.
    c. Europe.
    d. North America.

26. Which of the following hominid characters evolved the earliest?
    a. bipedalism
    b. large brain
    c. language
    d. use of fire
27. _____ involves change in the frequency of certain alleles relative to others in the gene pool. In contrast, _____ involves evolutionary phenomena such as mass extinction brought about by climate change.
28. _____ is differential survival and reproduction.
29. _____ is when different traits evolve at different rates.
30. The _____ model of human evolution suggests that modern humans evolved independently in several places from local populations of *Homo erectus*.

APPLYING THE CONCEPTS

1. A friend of yours believes that all organisms were specially created. You believe in evolution and want to present your case to your friend. What would you say?
2. Can life arise on the earth from inanimate material today? Why or why not?
3. You are a biologist exploring the dense tropical rain forests of Brazil. You come upon an unfamiliar medium-sized mammal moving above you in the trees. How would you determine whether this animal is a primate? How might you identify its closest living relatives?

4. Tay-Sachs disease has zero fitness because it causes death before the individual reaches reproductive age. How does a trait with zero fitness persist in a population?

Ecology, the Environment, and Us

23

The Earth Is a Closed Ecosystem with Energy As the Only Input

The Biosphere Is the Part of the Earth Where Life Exists

Ecological Succession Is the Sequence of Change in Species over Time

Energy Flows through Ecosystems from Producers to Consumers
- Food chains and food webs depict feeding relationships
- Energy is lost as it is transferred through trophic levels
- Ecological pyramids depict energy or biomass at each trophic level
- Ecological pyramids have health and environmental ramifications

Chemicals Cycle through Ecosystems
- Water cycles between the atmosphere and land
- Carbon cycles between the environment and living bodies
- Nitrogen cycles through several nitrogenous compounds
- Phosphorus cycles between rocks and living organisms

Humans Can Upset Biogeochemical Cycles
- Humans sometimes cause shortage or pollution of water
- Atmospheric levels of carbon dioxide affect global warming
- Disruptions to the nitrogen and phosphorus cycles can cause eutrophication

 **ENVIRONMENTAL ISSUE** Global Warming

John stopped the car next to the river. He heard that the eagles had started to arrive in Illinois a couple of days ago. Bald eagles were once plentiful throughout North America. Three feet tall, with a majestic white head and impressive 7-foot wing span, they could be seen nesting and fishing along most major waterways. Unfortunately, human settlers killed many eagles, and as the human population grew, there were fewer places where eagles could nest. Moreover, when natural waters became polluted with chemicals such as mercury, or when pesticides were sprayed on crops, these substances eventually found their way into fish. Every time an eagle ate a contaminated fish, it was at risk. On July 4, 1976, America's national bird, the bald eagle, was officially listed as an Endangered Species. But now the eagles were coming back. John was eager to share these birds with his 3-year-old son, Jason. "Look. There's a huge eagle up on the bluff!"

"Where?" asked Jason, followed by "Wow! Look at the size of that thing! He is as tall as I am!"

In this chapter, we will see that humans are but one of a host of species that share our small planet. We have many of the same opportunities as do other species, and we face many of the same threats. If we are ever to understand the world around us and our place in it, we must have some knowledge of its physical characteristics and of the other species with which we share it. Here, then, let's take a look at this place we call home, keeping in mind that many other creatures also call it home. Let's pay particular attention to just how we and they, together, can influence the well-being of each other and see what lessons nature might have for our kind.

We will begin this chapter with an overview of the part of the earth where life exists and then break that down into smaller, more manageable units. We will look at how the planet's physical traits influence life and how living things influence each other as well. Then we will consider how energy passes from one level to another in living systems. The great lesson we should learn from all this is the interdependence of living things and, in turn, their dependence on the nonliving world.

Our focus here is ecology, which comes from the Greek words *oikos*, meaning home, and *logos*, meaning to study. Thus, ecology is the study of our home, the earthly environment, including both its living (biotic) components and nonliving (abiotic) components. More precisely, **ecology** is the study of the interactions among organisms and between organisms and the environment. The key word is *interaction*. Ecologists are the scientists who study these interactions. ■

FIGURE **23.1**

A view of the earth from space shows us that the earth is isolated. Because there is no regular input of materials, important elements must be cycled from organism to organism and between the living and nonliving components of the earth. The only input to the system is energy from the sun, which sustains nearly all life on the earth.

The Earth Is a Closed Ecosystem with Energy As the Only Input

We can make two important observations as we view the earth from a distance (Figure 23.1). First, we see that the earth is isolated. With respect to materials, this observation is accurate. Aside from an occasional meteorite, bits of debris from space, and perhaps sprinklings of cosmic rain from celestial "snowballs" that disintegrate high above the earth's surface, there is no source of new materials. Many of the materials that came together to form our planet some 4.5 billion years ago have cycled from organism to organism and between the living and nonliving components of our world. A carbon atom in your body may once have been part of a dinosaur or of Aristotle. Second, the light reflected from the earth's surface reminds us that the earth receives one very important contribution—energy in the form of sunlight. As we will see, that energy is captured by green plants and transferred from organism to organism, sustaining nearly all life on the earth.

The Biosphere Is the Part of the Earth Where Life Exists

The part of the earth where life exists is the **biosphere**. The biosphere consists of many **ecosystems**, each consisting of the organisms in a specific geographic area and their physical environment. All the living species in an ecosystem that can potentially interact form a **community**. A **population** is all the individuals of the *same species* that can potentially interact. Thus, populations of different species form a community.

On a still smaller scale, we can consider an individual organism and describe it according to its niche (sometimes called *ecological niche*). The **niche** is the organism's role in the ecosystem, including all physical, chemical, and biological factors that the organism requires to live, remain healthy, and reproduce. Such factors include the nature of the organism's food and how it obtains food, its predators, its specific needs for shelter, and its interactions

This thin mat of lichen is helping break down the bare rock, beginning soil formation.

After soil has begun to accumulate, plant species appear that often include shrubs and dwarf trees.

Trees later become the dominant plant form.

FIGURE **23.2**

Primary succession is the sequence of changes in the species composition of a community that begins where no life previously existed.

with other species. The niche also includes the physical conditions, such as temperature, light, water, and oxygen, required for survival. The organism's **habitat**—the place (physical area) where it lives—is also a part of its niche. You may want to think of an organism's habitat as its address and its niche as its profession.

Ecological Succession Is the Sequence of Change in Species over Time

Ecosystems change over long time periods, as does everything else on the earth. The sequence of changes in the species making up a community is called **ecological succession**. There are two types of ecological succession: primary and secondary.

Primary succession occurs where no community previously existed (Figure 23.2). Such places may be on rocky outcroppings, where lava has covered everything, or where an island has appeared. When primary succession begins, no soil exists.

The first living things to invade such an area are called pioneer species. Among the most prominent of these are lichens, which are actually two species—a fungus and a photosynthetic organism, usually an alga. The fungus provides the attachment to the barren surface and retains water; the alga provides the food. The lichens secrete acid, which helps break down the rock, beginning soil formation. In time, the lichens die and their remains mix with the rock particles, furthering soil formation in cracks and crevices.

Soil building is a slow process, but once the "dirt" is in place, things happen quickly. Plants begin to appear. Their roots force themselves into every crevice, and their leaves fall and decompose, accelerating the pace of soil building and new plant colonization. In some areas, trees eventually become the dominant plant form.

Each species that enters the area changes the environmental conditions slightly and changes the resources available for others, favoring some species and not others. The community that eventually forms and remains if no disturbances occur is called the **climax community** (Figure 23.3). The nature of a climax community is determined by many factors, including temperature, rainfall, nutrient availability, and exposure to sun and wind.

When an existing community becomes cleared, either by natural means or by human activity, it undergoes a sequence of changes in species composition known as **secondary succession**. Secondary succession occurs where soil is already in place. Such areas include old deserted fields and farms that have been cleared, planted by humans, and then abandoned, or in areas damaged by fire, flood, wind, or grazing (Figure 23.4). If the community remains undisturbed, the initial invaders, such as grasses, weeds, and shrubs, are gradually outcompeted as other plants move in from the surrounding community. In some

FIGURE **23.3**

Selected climax communities of the earth. Environmental conditions, particularly the temperature and the availability of water, affect the distribution of organisms. Each species is adapted to certain conditions.

Temperate deciduous forest. These forests receive 75–125 cm (30–50 in.) of rainfall per year. Summers are hot, and winters are cold. Trees lose their leaves in the winter, helping to avoid water loss when it is too cold to photosynthsize. Insects, mice, squirrels, and many species of birds are common in these forests.

Temperate grasslands. These grasslands receive 25–75 cm (10–30 in.) of rainfall per year. Long dry periods and fire are important factors in maintaining grasslands. Grazing animals such as antelope and burrowing animals such as prairie dogs are common.

Desert. Lack of water defines the desert community. Deserts receive less than 25 cm (10 in.) of rain each year. Most deserts are hot, but some are cold. Both plants and animals must be able to conserve water. Many desert plants are succulents with leaves that retain and store water. Animals may tend to avoid the sun by foraging at night.

Taiga. The taiga is composed of evergreen forests with highly variable rainfall between 50–100 cm (20–40 in.) per year. Winters are long and cold, and summers are short. The needles on evergreen trees help save water by providing little surface through which water can leave. Animals such as the grizzly bear, moose, wolf, and snowshoe hare are found in the taiga.

Tropical rain forest. Tropical rain forests may receive 200–1000 cm (80–400 in.) of rain each year. It is hot throughout the year. Tropical rain forests have a tremendous diversity of life.

Immediately after a fire there are no visible signs of life. Dead trees stand as ghostly reminders of the forest that had existed, and gray ash covers the forest floor.

The following spring young plants, such as fireweed and heartleaf arnica, appear and begin the stages of secondary succession.

FIGURE **23.4**

Secondary succession following the 1988 fires in Yellowstone National Park

areas, pines move in next, but these may eventually resign their dominance to hardwood trees, such as oak. The area finally "heals," and the community that forms often resembles the one that existed before the disturbance.

Energy Flows through Ecosystems from Producers to Consumers

Virtually all the energy that propels life on this small planet comes from the sun. Only a small fraction of the sun's energy that reaches the earth's surface worldwide is captured by living organisms. Nonetheless, the life that abounds on the earth owes its existence to that captured energy, which is shifted, shuffled, channeled, and scattered through various systems from one level to the next.

Food chains and food webs depict feeding relationships

The flow of energy through the living world begins when light from the sun is absorbed by photosynthetic organisms, such as plants, algae, and cyanobacteria.[1] Photosynthesis essentially captures light energy and transforms it into the chemical energy of the sugar glucose, which is manufactured from carbon dioxide and water. The amount of light energy that is converted to chemical energy in the bonds of organic molecules during a given period is called gross **primary productivity**. The photosynthesizers themselves use some of the energy stored in these glucose molecules to fuel their own metabolic activities. Any remaining energy, called the net primary productivity, can be used for growth and repro-

duction. Once the photosynthesizer uses the energy in glucose to make new organic molecules, these molecules may then become food for an animal.

We see, then, that the photosynthesizers, called the **producers**, form the first **trophic level** (*trophic*, feeding). All other animals are **consumers** and use the energy that producers store (Figure 23.5). Consumers are grouped on the basis of their food source.

- **Herbivores**, or **primary consumers**, eat plants.

- **Carnivores**, or **secondary consumers**, feed on herbivores. Some carnivores eat other carnivores, forming still higher trophic levels—tertiary and quaternary consumers.

- **Omnivores** eat both plants and animals.

- **Decomposers**, such as bacteria, fungi, and worms consume dead organic material for energy and release inorganic material that can then be used by producers.

At one time, the feeding relationships responsible for the flow of energy through ecological systems was described as a **food chain**—a linear sequence in which A eats B, which eats C, which eats D, and so on. However, now we know that such "chains" are too simplistic because many organisms eat at several trophic levels. To illustrate, consider the number of trophic levels on which humans feed. Whereas eating chicken makes us secondary consumers, eating tuna (a predatory fish) makes us tertiary consumers (or quarternary consumers if the tuna had eaten a predatory fish). But we also can be considered primary consumers because we eat vegetables (Figure 23.6). These more realistic patterns of interconnected food chains give rise to what are called **food webs**. Figure 23.7 describes part of a food web in a community where land meets water.

[1]Cyanobacteria are a group of photosynthetic bacteria that usually live in water.

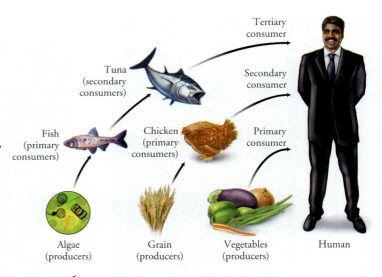

FIGURE **23.6**
Humans eat at several trophic levels.

FIGURE **23.5**
Trophic levels

▌Energy is lost as it is transferred through trophic levels

Energy is lost as it is transferred from one trophic level to the next. Although the efficiency of transfer can vary greatly among organisms, on average only 10% of the energy available at one trophic level is transferred to the next higher level. As a result, ecosystems rarely have more than four or five trophic levels.

As we consider why energy transfer between trophic levels is so inefficient, keep in mind that only the energy that is converted to **biomass** (the dry weight of body mass), is available to the next higher trophic level. Let's consider some of the energy losses that might occur between trophic levels (Figure 23.8). An animal must obtain its food, usually by grazing or hunting, activities that use energy. Furthermore, not all the food available at a given trophic level is captured and consumed. Some of the food eaten cannot be digested and is lost as feces. The energy in the indigestible material is unavailable to the next higher trophic level. Roughly two-thirds of the energy in the food that is digested is used by that animal as a source of energy for cellular respiration. The remaining energy can be converted to biomass and will be available to the next higher

trophic level. As these losses become multiplied at successively higher trophic levels, a point is reached where there simply is not enough food or energy to feed another level.

▌Ecological pyramids depict energy or biomass at each trophic level

An **ecological pyramid** is a diagram that shows the relationships among trophic levels. The **pyramid of energy** in Figure 23.9 shows the amount of energy available at each trophic level. Because only about 10% of energy is transferred from one trophic level to the next, the tiers representing successively higher trophic levels are smaller, forming a pyramid.

Pyramids of biomass describe the number of individuals at each trophic level multiplied by their dry weight. In many ecosystems, the number of individuals decreases at each higher trophic level. Both the loss of energy between trophic levels and the tendency for predators to be larger than their prey cause the declining numbers. We see this in a grassland in which there are far more grass plants than there are herbivores, such as zebras and wildebeests, grazing on them and far fewer predators than herbivores. However, in some ecosystems, certain forests for instance, pyramids of numbers can be partly inverted because the primary producers may be large trees that provide food for tremendous numbers of small insects.

▌Ecological pyramids have health and environmental ramifications

We have seen that materials such as nutrients are passed from one feeding level to the next. As a result, nondegradable substances will accumulate in higher levels of the food chain. When these substances are toxic, the accumulation can have harmful effects on top predators, including humans. We have also seen that energy is lost each time it is transferred to another organism. For this reason, more humans could be nourished on a vegetarian diet than on a diet containing meat.

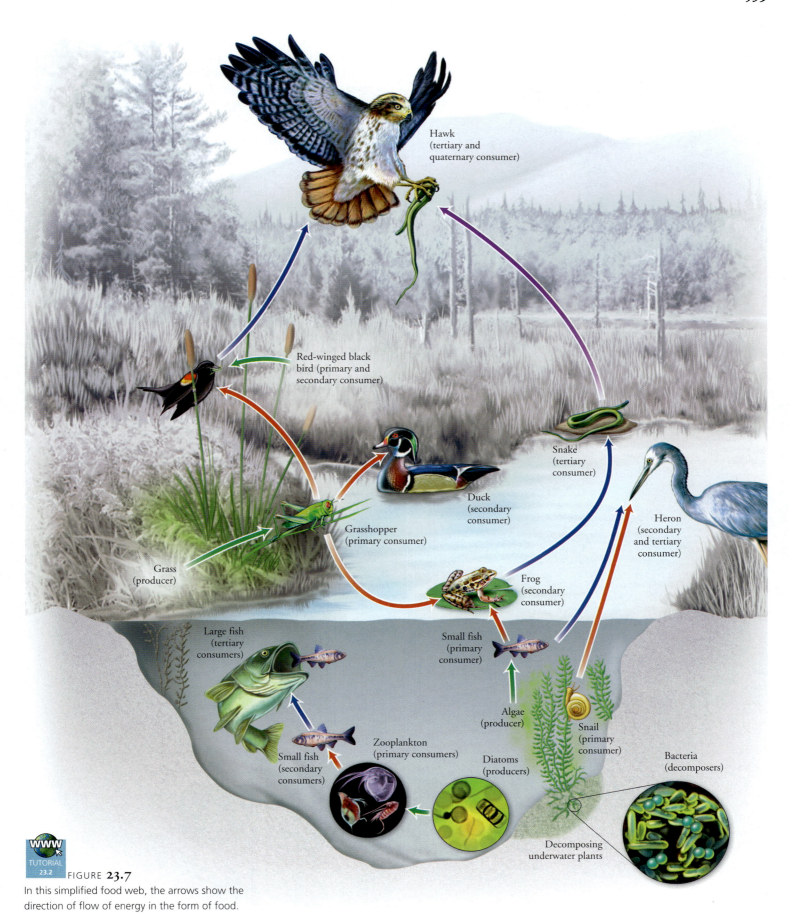

FIGURE 23.7

In this simplified food web, the arrows show the
direction of flow of energy in the form of food.

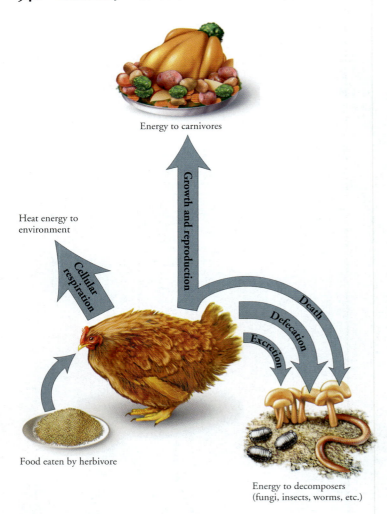

FIGURE **23.8**

As energy flows through a food web, only a small amount of it is stored as body mass and becomes available to the next higher trophic level. Some of the food energy is simply not captured; some is undigested and is lost in fecal waste; and some is used for cellular respiration. Only the remaining energy becomes available to the next higher trophic level.

BIOLOGICAL MAGNIFICATION

Chemicals that are essential to life—carbon, hydrogen, oxygen, nitrogen, and phosphorus—are passed from one trophic level to the next, being continuously recycled from organism to organism. Organic molecules are broken down and then either metabolized for energy or put together to form the biomass of another individual.

However, certain molecules—including chlorinated hydrocarbon pesticides such as DDT, heavy metals such as mercury, and radioactive isotopes—are broken down or excreted very slowly. Molecules such as these tend to stay in the body, often stored in fatty tissues such as liver, kidneys, and the fat around the intestines. As more of these types of molecules are consumed, they accumulate and become concentrated in the animal's body.

The concentration of these substances in the bodies of consumers becomes "magnified" at each higher trophic level. This magnification results from the fact that, on average, only 10% of the energy available at one trophic level transfers to the next. To see the picture more clearly, let's consider a simplified example. Assume that a nondegradable substance such as mercury pollutes the water. Phytoplankton (photosynthetic protists that are abundant in aquatic environments) will absorb the mercury, and mercury levels in the phytoplankton will be in some dilute concentration that we

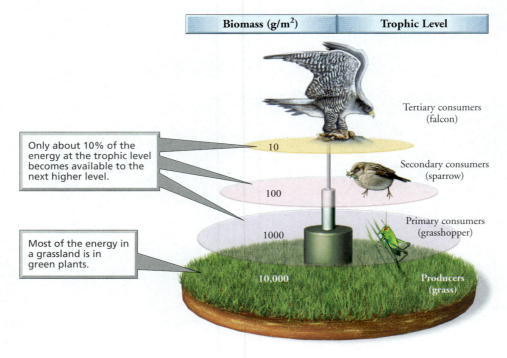

Only about 10% of the energy at the trophic level becomes available to the next higher level.

Most of the energy in a grassland is in green plants.

Biomass (g/m²)	Trophic Level
10	Tertiary consumers (falcon)
100	Secondary consumers (sparrow)
1000	Primary consumers (grasshopper)
10,000	Producers (grass)

FIGURE **23.9**

An ecological pyramid. Note that each level contains less energy and less biomass than the level below it.

will represent as 1. An herbivore such as a zooplankton may eat phytoplankton to stay alive. The mercury is concentrated in the zooplankton's tissues, so the mercury level in zooplankton becomes about 10 times greater than in phytoplankton.. The zooplankton (nonphotosynthetic protists) may be eaten by a small fish. Once in the fish's body, the mercury stays there and accumulates. When these mercury–containing fish are eaten by a tuna, the mercury passes to the tuna's body and accumulates. Because many small fish must be consumed to keep a tuna alive, the mercury concentration in the tuna's body might be 1000 times greater than that in water. Indeed, we might expect the substance to be 10 times more concentrated at each successive level on the trophic scale. This tendency of a nondegradable chemical to become more concentrated in organisms as it passes along the food chain is known as **biological magnification** (Figure 23.10).

If humans are wise, we will learn an important lesson from observations of the effects of biological magnification. We are often top carnivores, and we continue to pollute our environment with nondegradable, potentially harmful substances. During the 1950s in Minamata, Japan, methyl mercury, a heavy metal, was released into Minamata Bay as industrial waste. The mercury accumulated in higher levels of the food chain. Many people living in the region developed mercury poisoning because they ate fish contaminated with mercury. Mercury is a toxin that affects the nervous system. It can cause muscle tremors, personality disorders, and birth defects (see Chapter 7).

Mercury contamination is still a problem today. It enters water through volcanic activity, coal combustion, and medical waste. Bac-

teria convert mercury to a particularly toxic form, methyl mercury. The mercury then works its way through food webs and accumulates in fish, particularly long-lived predators, such as tuna, swordfish, or shark. In 2004, the Food and Drug Administration (FDA) revised its recommendations on fish consumption for pregnant women, breast-feeding women, women of child-bearing age, and children. The FDA now recommends that people in these groups stop eating shark, marlin, and swordfish. These people should also limit their intake of albacore (white) tuna to one 6-oz. serving a week.

Polychlorinated biphenyls (PCBs) are chemicals that may cause cancer and nervous system damage. PCBs have been called environmental estrogens because they mimic the effects of the female sex hormone estrogen or enhance estrogen's effects. We know that PCBs have had feminizing effects on certain male animals. There is currently a debate among scientists about whether PCBs reduce human fertility, affect the rates of cancer in certain reproductive structures (breasts, ovaries, prostate gland, and testes), or reduce human sperm count. Similar to mercury, PCBs are non-degradable and accumulate in the food web. PCBs can be found in chicken, beef, and dairy products. In addition, some farm-raised Atlantic salmon have PCB levels high enough to trigger health warnings from the Environmental Protection Agency. Wild salmon have lower levels of PCBs.

what would you do?

DDT is a pesticide that was widely used in the 1950s. During the 1960s and 1970s, DDT was blamed for the alarming decline in certain species of birds.

The United States banned the use of DDT during the 1970s, and many other countries followed suit. Malaria is a disease that infects 300 million people a year and kills 200 million people a year. Malaria is transmitted by a mosquito (see Chapter 13a). DDT is an affordable and effective way to kill the mosquitoes that transmit malaria. Neither the United States nor the World Health Organization will fund the use of DDT to control malaria-causing mosquitoes. Some of the areas that are hit the hardest by malaria—Africa, Asia, and Latin America—are too poor to afford other pesticides, which cost four to six times more than DDT. What do you think should be done to curb the transmission of malaria? Do you think the use of DDT should be permitted? What are your reasons?

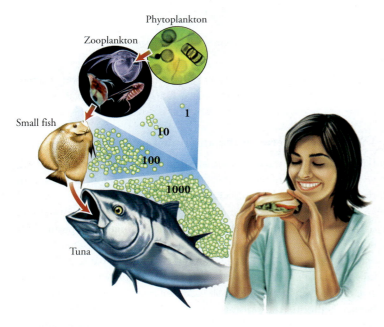

FIGURE **23.10**

Substances, such as mercury, that are broken down or excreted slowly tend to accumulate in the body. The concentration becomes magnified at each successive trophic level, because the energy losses at each level require that a consumer eat many individuals at a lower trophic level to stay alive. The mercury concentration in the tuna eaten by a human at the end of this food chain is about 1000 times greater than the mercury concentration in the phytoplankton.

WORLD HUNGER

The world's human population has grown at an alarming rate. The pyramid of energy suggests a way to feed the growing population more efficiently—eat lower on the food chain. Because only about 10% of the energy available on one trophic level becomes available to the next one, we see that:

10,000 calories of corn energy can produce → 1000 calories of beef energy can produce → 100 calories of human energy

However, if humans began to eat one level lower on the food chain about 10 times more energy would be available to them.

10,000 calories of grain (corn, wheat, rice, etc.) can produce → 1000 calories of human energy

We see, then, that more people could be fed and less land would have to be cultivated if we adopted a largely, or exclusively,

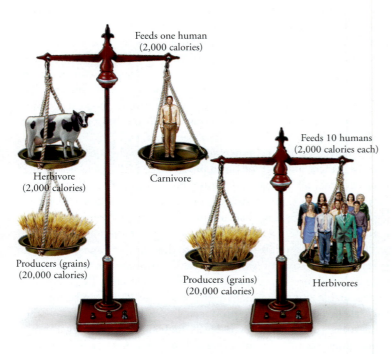

FIGURE **23.11**
Energy pyramids may hold an important lesson for humans. Because only approximately 10% of the energy available at one trophic level is available at the next higher level, a greater number of people could be fed if they ate a vegetarian diet rather than a diet containing meat.

vegetarian diet (Figure 23.11). This fact is one explanation why people in densely populated regions of the world, including China and India, are primarily vegetarians. It seems likely that, as the human population continues to expand, meat will become even more of a luxury throughout the world than it is today.

stop and think

A traditional diet of Inuits (Eskimos), one of the native people of the coastal areas of North America, involves a relatively long food chain:

Diatoms (producers) → Tiny marine animals → Fish → Seals → Inuits

Explain how the length of this food chain may be one factor contributing to the small population size of Inuit groups. Further, seal blubber contains high levels of DDT and PCBs. Explain why the levels of DDT and PCBs are four to seven times higher in the breast milk of Inuit women who eat blubber than in the breast milk of women in Quebec who eat lower on the food chain, on fish for example.

Chemicals Cycle Through Ecosystems

Resources on the earth are limited. Life on the earth is demanding. Many of the earth's reserves would be quickly depleted if it were not for nature's cycling. Materials move through a series of transfers, from living to nonliving systems and back again (Figure 23.12). Let's look at some of the more important of these **biogeochemical cycles**, the recurring pathways of materials between living and nonliving systems.

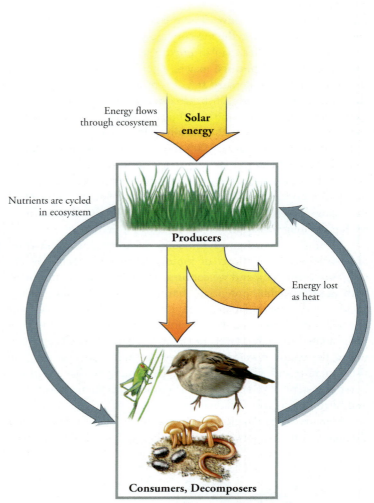

FIGURE **23.12**
Through biogeochemical cycles, matter cycles among living organisms and the physical environment. This cycling of matter is in contrast to the path of energy, which flows through the ecosystem in one direction.

❚ Water cycles between the atmosphere and land

Each drop of rain teaches us something about water recycling. Indeed, water, in the form of rain, snow, sleet, or hail, continuously cycles from the atmosphere to the land, where it collects in ponds, lakes, or oceans and moves back into the atmosphere as it evaporates. Most of the rain or snow that falls to the earth returns to the sea at some point. This cycle provides us with a renewable source of drinking water. Because water is so critical to life, large amounts temporarily pause in the bodies of living things. In living cells, water helps regulate temperature and acts as a solvent for many biological reactions. The very oxygen we breathe is produced from water through the reactions of photosynthesis. Water also cycles back to the environment from living things as plants return 99% of the water they absorb to the atmosphere in the process of transpiration (the evaporation of water from the leaves and stems of plants). The water cycle is shown in Figure 23.13.

❚ Carbon cycles between the environment and living bodies

If you have ever seen a dead opossum beside the road, you know something about carbon cycling. Put simply, in the carbon cycle,

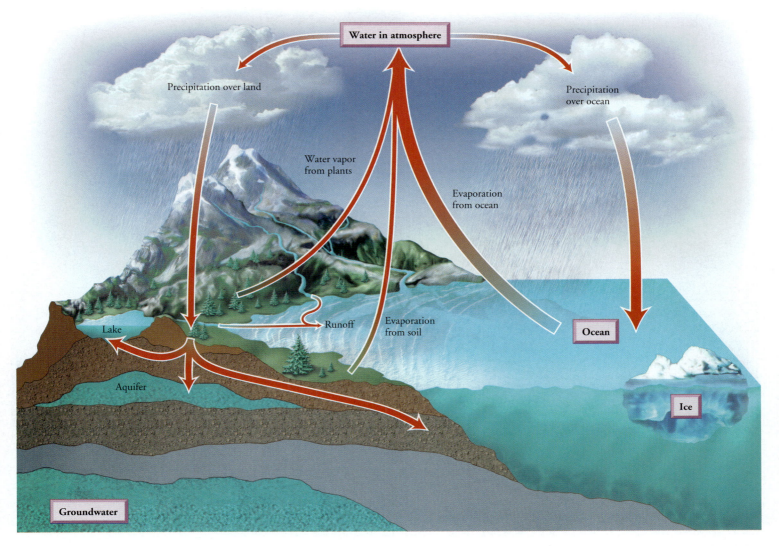

FIGURE **23.13**

The water cycle. Water, which is essential to life, cycles from the atmosphere to the land as precipitation and then back to the atmosphere as it evaporates from oceans and other bodies of water where it has collected. Water also returns to the atmosphere in the water vapor lost from the leaves of plants. This cycling provides us with a renewable source of drinking water.

carbon moves from the environment, into the bodies of living things, and back to the environment (Figure 23.14). Life and the carbon cycle are intimately related. Carbon is essential to organisms because it is a part of molecules such as proteins, carbohydrates, fats, and nucleic acids. Also, certain processes of living organisms—photosynthesis and respiration—are critical to carbon cycling.

Carbon moves from the environment primarily during photosynthesis, as plants, algae, and cyanobacteria use carbon dioxide (CO_2) to produce sugars and other organic molecules. When the photosynthesizers are eaten by herbivores, the organic molecules serve as a carbon source for the herbivores. The herbivores, in turn, use the carbon to produce their own organic molecules, which then serve as a carbon source for carnivores. When these organisms die, as did the

opossum beside the road, the organic molecules in their bodies will serve as a carbon source for decomposers. However, while alive, all organisms cycle carbon back to the environment through the reactions of cellular respiration, which breaks down organic molecules to CO_2.

In some cases, there may be a significant delay before carbon cycles back into the environment. For example, carbon may remain tied up in the wood of some trees for hundreds of years. Most of the carbon that has left the carbon cycle is thought to be in limestone, a type of sedimentary rock that formed from the shells of marine organisms that sank to the bottom of the ocean floor and were covered by sediments. Also, fossil fuels, so named because they formed from the remains of organisms that lived millions of years ago, are the vast stores of products of photosynthesis.

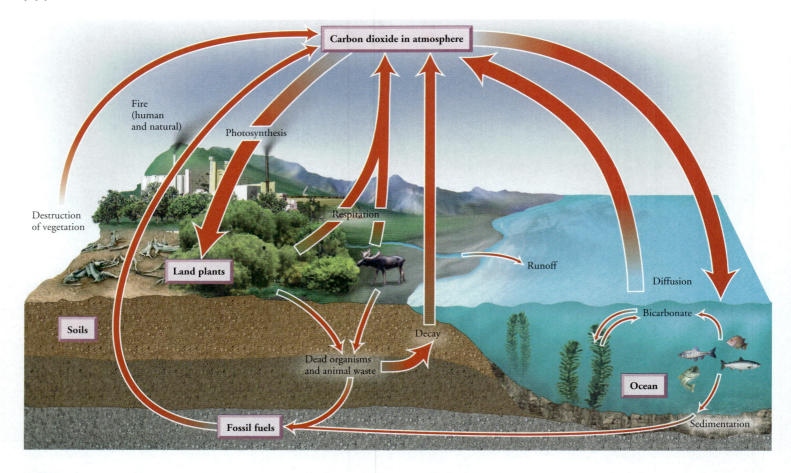

FIGURE **23.14**

The carbon cycle. Carbon cycles between the environment and living organisms. Carbon dioxide (CO_2) is removed from the environment as producers use it to synthesize organic molecules in photosynthesis. The carbon in those organic molecules then moves through the food web, serving as a carbon source for herbivores, carnivores, and decomposers. Carbon is returned to the atmosphere as CO_2 when organisms use the organic compounds in cellular respiration.

Three processes are involved in returning carbon from long-term storage to the environment: respiration, erosion, and combustion. The trees will eventually die, and the natural process of decomposition will make the carbon available for new organisms, which will respire and release CO_2 to the atmosphere. The carbon in limestone is recycled through erosion. Millions of years after it forms, sedimentary rock containing limestone can be lifted to the earth's surface, where it is eroded by chemical and physical weathering. Carbon is returned to the environment and is available to cycle through the food web once again. Combustion, or burning, returns the carbon in fossil fuels to the environment. Today, fossil fuels such as coal, oil, and natural gas are being burned, and the carbon they contain is being returned to the atmosphere as CO_2.

Nitrogen cycles through several nitrogenous compounds

Nitrogen is a principal constituent of a number of critical molecules including proteins and nucleic acids. Unfortunately, nitrogen is often in short supply to living systems, so its cycling can be critical (Figure 23.15).

The largest reservoir of nitrogen is the atmosphere, which is about 79% nitrogen gas (N_2). However, nitrogen gas cannot directly interact with life. So, as you sit there, you are bathed in nitrogen gas without interacting with it. Before it can be used, nitrogen gas must be converted to a form that living organisms can use—ammonium (NH_4^+). The process of converting nitrogen gas to ammonium is called **nitrogen fixation**. This process is done by nitrogen-fixing bacteria, many of which live in nodules on the roots of leguminous plants such as peas and alfalfa. Ammonium is then converted to nitrite (NO_2^-) and then to nitrate (NO_3^-) by nitrifying bacteria living in the soil in a process called **nitrification**.

The nitrogen in ammonia or nitrates is assimilated in the proteins or nucleic acids of living organisms. The ammonia or nitrates are first absorbed by plants, and the nitrogen is used to form plant proteins and nucleic acids. The nitrogen then passes through the food web and is incorporated into the nitrogen-containing compounds of animals.

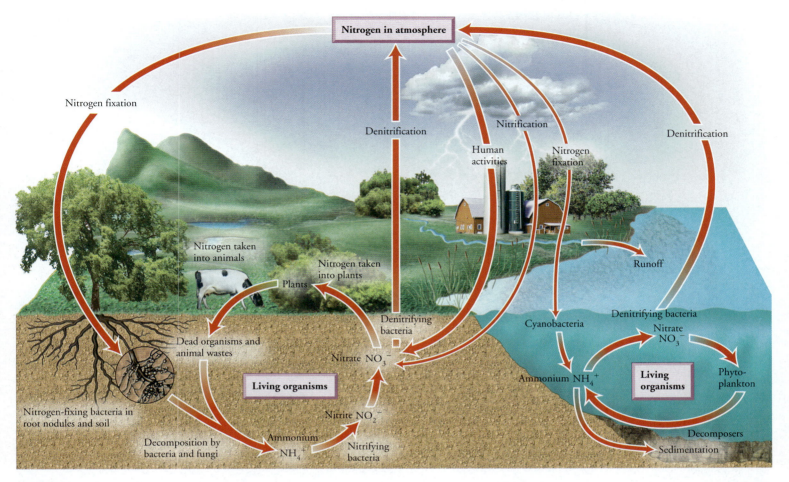

FIGURE **23.15**

The nitrogen cycle. Atmospheric nitrogen can be converted to ammonia by nitrogen-fixing bacteria, which then convert the ammonia to nitrate, the main form of nitrogen absorbed by plants. Plants use nitrate to produce proteins and nucleic acids. Animals eat the plants and use the plant's nitrogen-containing chemicals to produce their own proteins and nucleic acids. When plants and animals die, their nitrogen-containing molecules are converted to ammonium by bacteria. Denitrifying bacteria return nitrogen to the atmosphere.

When plants and animals die, decomposers such as bacteria break down the waste products and dead bodies of plants and animals, producing ammonium (NH_4^+). Much of the ammonium is converted to nitrate by nitrifying bacteria. The process of denitrification, in which the nitrates that are not assimilated into living organisms are converted to nitrogen gas by denitrifying bacteria, balances the nitrogen cycle. Denitrifying bacteria are found in wet soil, swamps, and estuaries.

Phosphorus cycles between rocks and living organisms

Phosphorus is an important component of many biological molecules, including the genetic material DNA, energy transfer molecules such as ATP, and phospholipids found in membranes. Phosphorus is also essential in vertebrate bones and teeth.

Unlike the water, carbon, and nitrogen cycles, the phosphorus cycle does not have an atmospheric component (Figure 23.16). Instead, the reservoir for phosphorus is sedimentary rock, where it is found in the form of phosphate. The cycle begins when erosion caused by rainfall or runoff from streams dissolves the phosphates in the rocks. The dissolved phosphate is readily absorbed by producers and incorporated into their biological molecules. When animals eat the plants, the phosphates are passed through food webs. Decomposers return the phosphates to the soil or water, where they become available to plants and animals once again. Much of the phosphate is lost to the sea. Although some of this phosphate may cycle through marine food webs, most of it becomes bound to sediment. The phosphates in sediment become unavailable unless geological forces bring the sediment to the surface.

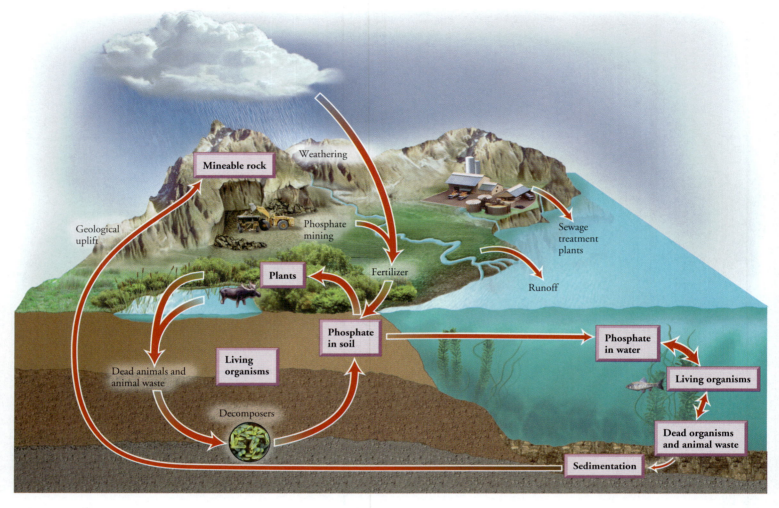

FIGURE **23.16**

The phosphorus cycle. Phosphates dissolve from rocks in rainwater. Phosphates are absorbed by producers and are passed to other organisms in food webs. Decomposers return phosphates to the soil. Some phosphates are carried to the oceans and form marine sediments.

Humans Can Upset Biogeochemical Cycles

Throughout most of history, biogeochemical cycles have worked just as we have described. However, human activity, particularly since the time of the Industrial Revolution, which began around the mid-1800s, has disrupted the normal pattern of cycling. Some of these disturbances result from new technologies, but most are caused by the demands of an ever-increasing human population.

Humans sometimes cause shortage or pollution of water

The world's water supply is threatened by overuse by humans. Both water shortages and water pollution can result from human activities.

WATER SHORTAGE

The earth has been called the Water Planet because we have so much of it. Still, relatively little is available for human use because most of the earth's water is in the oceans. You might think that at least the fresh water would be available, but much of it is tied up in ice or clouds, flowing in underground rivers, or otherwise inaccessible for immediate human use.

Several problems are associated with water. In some areas, there simply is not enough water. Indeed, the World Health Organization estimates that the lack of clean water for drinking, washing, and sewage disposal can be blamed for 80% of the disease in developing countries. Although water shortages in other parts of the world are more severe, there are regions in the United States, particularly in the agricultural regions of the South and West, where water shortages are being experienced. The water tables beneath the highly populated northeastern United States are also dwindling dangerously.

In some places, water shortages are caused by drought. A drought has affected the western United States for nearly a decade. In some regions, farming has been severely affected. For example, in 2003 and 2004 farmers who were receiving water from the Salt River Project in Arizona had their water allocations cut by one-third. Drought also increases the risk of wildfires that kill wildlife and humans and destroy property.

In most regions the underlying cause is too many people drawing from a limited water supply. North Americans use an exceptionally large quantity of water. In the United States, each person uses about 7500 liters (1950 gallons) of water a day, whereas a person in a less developed country uses less than 75 liters (19.5 gallons) of water a day.

You might respond to these numbers defensively and claim that *you* certainly do not use that much water. In a sense you would be right. Personal use accounts for only a small percentage of total water use. Most of the water is used in agriculture. In an effort to feed a growing human population, irrigation has become increasingly important as a way to make arid areas fruitful. The rest of the water is used primarily in industry for steam generation or the cooling of power plants.

Irrigation has pros and cons. On one hand, irrigation makes it possible to grow crops in areas that would otherwise be barren, allowing us to feed many people. On the other hand, in the long run, irrigation can make land unfit for agriculture. Irrigation water contains dissolved minerals. Whereas runoff from natural rainfall would carry salts away, irrigation water soaks into the soil. Then, when the water evaporates from the soil, the salts are left behind. The resulting accumulation of salts in the soil is called **salinization**. Worldwide, salinization destroys the fertility of 5000 km^2 (1500 mi^2) of irrigated land each year.

There are two main sources of water: surface water, such as rivers and lakes, and groundwater, which is water found under the earth's surface in porous layers of rock, such as sand or gravel. When too much surface water is used in an area, ecosystems are affected. As much as 30% of a river's flow can usually be removed without affecting the natural ecosystem. However, in some parts of the southwestern United States, as much as 70% of the surface water has been removed. Such overdrawing of surface water can disrupt ecosystems. For example, in 1941 much of the surface water that feeds Mono Lake in eastern California was diverted to supply water to Los Angeles. By 1982, Mono Lake's water level had dropped 13.72 meters (45 ft), and the lake's saltiness had nearly doubled. These changes drastically affected the plants and animals living in or around the lake. In 1994, Los Angeles was ordered to reduce the amount of water diverted from the lakes and streams feeding Mono Lake. At the same time, Los Angeles began a program promoting water conservation and recycling. As a result of this program, Los Angeles' water use in 2004 is the same as it was in the 1970s. Following the reduction in water diversion, Lake Mono's water level has risen approximately 3 meters (10 ft).

stop and think

Conservationists sometimes buy wetlands, hoping to preserve habitat of endangered wildlife. Despite the protection by conservationists, these wetlands sometimes dry up. Explain why it would also be important to buy the rights to the water that feeds the wetland.

Human use of groundwater is depleting the aquifers (porous layers of underground rock where groundwater is found). Water is

added to aquifers by rainfall or melting snow. In some areas, however, water is being removed from aquifers faster than it can be replenished. The Ogallala Aquifer underlies eight states in the Great Plains region of the United States. Water pumped up from the Ogallala Aquifer is often used for irrigation and this irrigation has made this region some of the most productive farmland in the country. However, the water table in this aquifer is dropping by 6 ft a year.

What steps can be taken to reduce water shortages?

1. **Reduce water use.** Each of us could help by individual efforts. However, because most water is used in agriculture, the biggest benefit would come from improved irrigation methods. Between 1950 and 1980, for example, Israel reduced its water wastage from 83% to a mere 5%, primarily by changing from spray irrigation to drip irrigation.

2. **Raise the price of water.** The cost of water in the United States is substantially below its cost in European countries. History has shown us that the consumer's interest in conserving water is directly related to its cost. Clearly, economic and legal incentives for conserving water are needed.

what would you do?

Because of water shortages in Colorado, some farmers are giving up their farms. In 2004, some farmers in the Arkansas Valley sold their water rights to Aurora, a suburb of Denver. About half of Aurora's water is used to irrigate the landscape around the suburban homes. With these sales, the valley lost 2,600 acres of farmland. Neighbors worry that the loss of farmland will harm the local economy and are lobbying the state legislature to prevent people holding water rights from selling those rights to buyers outside the local area. The farmers holding water rights, their neighbors, and the people in Aurora have conflicting interests. What do you think would be an equitable way to reduce the conflict over water use?

WATER POLLUTION

Another problem is that much of the water we have is polluted, often by our own activities. Factories, refineries, and waste treatment plants sometimes release fluids of varying quality directly into our water supplies. We have already discussed the effects of mercury and PCBs in water. Water can also become polluted from indirect sources. Contaminants in the atmosphere, such as gaseous emissions from automobiles and factories, can also pollute water. Runoff of fertilizer or pesticides also pollutes water. We will consider some of the effects of fertilizer runoff later in this chapter.

In a recent study of 80 American cities, 250 synthetic chemicals were found in the water supply. We have no idea what effects many of these chemicals have on life. Relatively few have been tested for any links to cancer or birth defects.

Atmospheric levels of carbon dioxide affect global warming

The level of atmospheric carbon dioxide (CO_2) is increasing (Figure 23.17). Two human activities—burning fossil fuels and deforestation—are important causes of the increase in atmospheric

ENVIRONMENTAL ISSUE

Global Warming

Why is there such hoopla about global warming? Why should anyone really care whether the atmosphere becomes 1° to 2°C (1.8° to 3.6°F) warmer? Let's consider some of the possible effects of global warming.

1. **Rising sea level.** An increase of just a few degrees in the overall temperature of the earth could cause glaciers and polar ice caps to begin melting. At the same time, the ocean water would expand as it warmed. The rising sea level would have the greatest effect on coastal countries such as Bangladesh, Egypt, Vietnam, and Mozam-

bique. Seawater could then cover a significant amount of land in these densely populated areas. For example, according to the 2001 report of the United Nations Intergovernmental Panel on Climate Change (IPCC), flooding in Bangladesh could increase by 40% in this century. If the global temperature increased by 6°C (10.8°F), the worst-case scenario of the IPCC, more than half of Bangladesh would be submerged for most of the year. In the United States, major cities such as New York, Miami, Jacksonville, Boston, San Francisco, and Los Angeles could be largely under water.

2. **Changing weather patterns.** Global warming would cause changes in both temperature and rainfall patterns. You may recall that the distribution of climax communities, each of which has characteristic plant and animal life, is primarily determined by temperature and rainfall. If global warming occurs, some species will thrive, and others may become extinct. There will be a shift in the agricultural regions of the world, because there will be a change in the pattern of rainfall and temperature. Food production in certain areas, including the central plains of the United States and

CO_2. Fossil fuels, such as coal, oil, and natural gas, were formed from deposits of dead plants and animals that were buried in sediments and escaped decomposition between 345 million and 280 million years ago. High temperatures and pressure over millions of years converted the deposits to fossil fuels. Burning fossil fuels returns carbon to the environment in the form of CO_2. Defor-

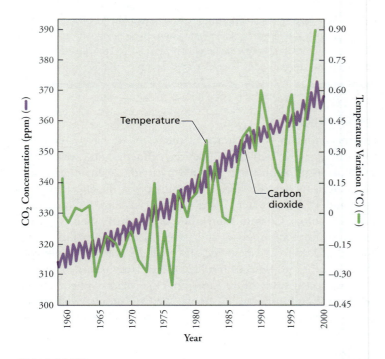

FIGURE **23.17**
The concentration of carbon dioxide (CO_2) has been increasing slowly for many years. Carbon dioxide is being added to the atmosphere through the burning of fossil fuels and by a reduction in the removal of CO_2 from the atmosphere by deforestation. Global temperatures have also risen during this time period.

estation, the removal of a forest without adequate replanting, as is occurring in areas of the Amazon rain forest and the Pacific Northwest of the United States, increases atmospheric CO_2 in two ways. First, it *reduces the removal* of CO_2 from the atmosphere through the trees' photosynthesis. Second, the trees are often burned after cutting, and burning *adds* CO_2 to the atmosphere. About three-quarters of the CO_2 added to the atmosphere comes from cars and factories and the remaining quarter largely from deforestation.

The rise in atmospheric CO_2 raises concerns because CO_2 is one of the "greenhouse" gases that play a role in warming our atmosphere. Other greenhouse gases include methane, nitrous oxide, chlorofluorocarbons (CFCs), and ozone in the lower atmosphere. These gases are accumulating in the atmosphere, largely because of human activity. Like the glass on a greenhouse, these gases allow the sunlight to pass through to the earth's surface, where it is absorbed and radiated back to the atmosphere as heat (long-wave infrared radiation). Both greenhouse glass and greenhouse gases reflect the infrared radiation, thus trapping the heat (Figure 23.18). Because of this effect, there is a concern that the increase in atmospheric CO_2 is leading to a rise in temperatures throughout the world—global warming.

Is the earth getting warmer? A 2001 report from the United Nations Intergovernmental Panel on Climate Change (IPCC) states that the global average surface temperature has increased by 0.6°C since the late 1950s. In addition, snow and ice cover have diminished about 10%. During the last century, sea level has risen about 10 to 20 cm (about 4 to 8 in.), and the temperature of the oceans has increased. The 1990s was the hottest decade since records began in 1860. The three hottest years on record, in order, are 2002, 1998, and 1999. Although the rate and extent of the rise are not clear, scientists believe that global temperatures will continue to rise with the level of greenhouse gases. The IPCC projects a rise of 1.4° to 5.8°C (2.2° to 10°F)

Canada, could drop. Areas that now produce enough food for export may not be able to produce enough for their own citizens. On the other hand, other regions, parts of India and Russia, for instance, might receive increased rainfall that would benefit agriculture. As the food-producing regions of the world shift with changing weather patterns, so will the distribution of "haves" and "have-nots." How will we, as a world community, deal with the shifts in economic and political power brought on by global warming?

3. **Human health.** The IPCC predicts that global warming will have several negative effects on human health. The change in climate is likely to increase the number of deaths due to heat waves, flooding, and droughts leading to starvation. Climate change could also increase air pollution. There is also concern that mosquitoes and ticks, as well as other vectors of disease, would thrive in a warmer climate. Thus, there could be an increase in malaria or West Nile virus encephalitis (inflammation of the brain), which are transmitted by mosquitoes, and Lyme disease, which is transmitted by deer ticks.

What can individuals do? An important step, which would have other benefits besides slowing global warming, would be to reduce our production of greenhouse gases. We will soon run out of fossil fuels anyway, probably within your lifetime. Therefore, it would be wise to walk, ride a bicycle, or take public transportation instead of driving. When driving is necessary, use car pools whenever possible. When you purchase appliances, choose those that are energy efficient.

The good news is that *you* can make a difference. Whenever possible walk; do not drive or ride. Drive a fuel-efficient vehicle. Wear clothing that will minimize the use of heating or cooling systems in your home. In short, conserve energy. The environment will benefit; the exercise will improve your health; and you will save money. 🌲

by the year 2100. Even a rise in temperature at the low end of the prediction could have serious consequences (see the Environmental Issue essay, *Global Warming*). The range of the prediction for temperature increase is large because of uncertainties in the models concerning the effects of many interacting factors, including cloud cover due to evaporation, cooling effects of atmospheric particulates such as sulfates, possible increased primary productivity due to the increase in CO_2, and the ability of the ocean to absorb CO_2.

Some skeptics do not believe that rising atmospheric CO_2 is increasing global temperature. Some point out that, although ground temperatures may have risen in the last decades, large areas of the upper atmosphere have actually cooled. There is also some evidence that the amount of energy that the sun is producing is increasing, and this factor alone might increase global temperatures. Others suggest that domesticated animals, such as cattle, sheep, and goats, are to blame. Bacteria in the guts of these animals produce methane, a greenhouse gas whose heat-trapping ability is 20 times greater than that of carbon dioxide. The collective belches and flatulence of all the cattle on the earth accounts for 20% of the methane gas released into the atmosphere.

stop and think

Scientists are working on a diet for cattle and sheep that is low in nitrogen and will reduce the amount of methane produced by these domesticated animals. If they are successful, how might this diet affect global warming?

Disruptions to the nitrogen and phosphorus cycles can cause eutrophication

Human activity is also disturbing the nitrogen and phosphorus cycles. Inadequate supplies of nitrogen and phosphorus can slow the growth of plants. For this reason, fertilizers contain both nitrogen (as ammonium) and phosphates. The demand for fertilizer has increased with the demand for food to feed a growing human population. Humans artificially "fix" nitrogen using energy from fossil fuels such as coal or oil. A problem arises, however, because the natural denitrification processes cannot keep pace with the rate of industrial nitrogen fixation and the amount of nitrogen that is added to the land in commercial fertilizers. The result is an excess of fixed nitrogen.

Nitrogen and phosphate from fertilizer can wash into nearby streams, rivers, ponds, or lakes in runoff. Fertilizer runoff is one cause of **eutrophication**, the enrichment of water in a lake or pond by nutrients. The nutrient boost can lead to

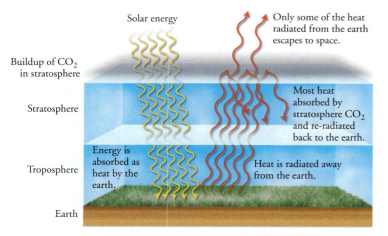

FIGURE **23.18**

Carbon dioxide and other greenhouse gases promote global warming by trapping heat (infrared radiation) in the atmosphere, in much the same way the glass on a greenhouse traps heat. Thus, the global warming that may be occurring because these gases are accumulating in the atmosphere is known as the greenhouse effect.

the explosive growth of photosynthetic organisms, such as algae, and in shallow areas, weeds. The problem is that, when these organisms die, organic material accumulates at the bottom of the lake. Decomposers then deplete the water of dissolved oxygen. Gradually, fish, such as pike, sturgeon, and whitefish, that require higher oxygen concentration in the water, are replaced by others, such as catfish and carp, that can tolerate lower levels of dissolved oxygen.

REVIEWING THE CONCEPTS

The Earth Is a Closed Ecosystem with Energy As the Only Input (p. 534)

1. Ecology is the study of the interactions among organisms and between organisms and their environment.
2. Materials on the earth are recycled among living organisms and the environment. The only input to the earth is energy from the sun.

The Biosphere Is the Part of the Earth Where Life Exists (pp. 534–536)

3. The biosphere consists of all the ecosystems on the earth. An ecosystem consists of a community of organisms and their physical environment. A population consists of all the organisms of the same species in a community. An organism's niche is the role it plays in the community.

Ecological Succession Is the Sequence of Change in Species over Time (pp. 536-537)

4. Ecological succession is the sequence of changes in communities over long time periods. Primary succession occurs where no community previously existed. The first invaders are called pioneer species. A climax community eventually forms and remains as long as no disturbances occur. Secondary succession describes the sequences of changes that occur when an existing community is disturbed by human or natural means.

Energy Flows through Ecosystems from Producers to Consumers (pp. 537–542)

5. Most of the energy in living systems comes from the sun when that energy is absorbed and stored in the molecules of photosynthetic organisms, called producers. The energy that producers have stored in their molecules enters the animal world through plant-eating herbivores (primary consumers), which may be eaten by carnivores (secondary consumers) or by animals that eat other carnivores (tertiary consumers). Decomposers consume dead organic material, releasing inorganic material that can be used by producers. The position of an organism in these feeding relationships is referred to as a trophic level.
6. The feeding relationships that allow for the one-way flow of energy through the ecosystem and for the cycling of materials among organisms are called food chains or food webs. A food chain describes the simple linear sequence of who eats whom. Many food chains in an ecosystem interconnect to form a complicated pattern called a food web.
7. Pyramids of energy show the loss of energy from one trophic level to another. On average, only 10% of the energy available at one trophic level is available at the next higher level. Ecosystems generally have only four or five trophic levels.
8. Ecological pyramids depict the amount of energy or biomass at each trophic level.
9. Biological magnification—the tendency for a nondegradable substance to become more concentrated in organisms as it passes through food webs—is a consequence of the energy loss between trophic levels. Thus, top carnivores are most likely to be poisoned by nondegradable toxic substances in the environment.
10. Because energy is lost with each transfer in the trophic scale, one way to feed more people would be for humans to adopt a largely vegetarian diet.

WEB TUTORIAL 23.1 Energy Flow in an Ecosystem

WEB TUTORIAL 23.2 Food Webs

Chemicals Cycle Thorugh Ecosystems (pp. 542–545)

11. In biogeochemical cycles, materials cycle between organisms and the environment.
12. The water cycle is the pathway of water as it falls as precipitation; collects in ponds, lakes, and seas; and returns to the atmosphere through evaporation or transpiration.
13. The carbon cycle is the worldwide circulation of carbon from the abiotic environment (carbon dioxide in air) to the biotic environment (carbon in organic molecules of living organisms) and back to the air. Carbon enters living systems when photosynthetic organisms incorporate carbon dioxide into organic materials. Carbon dioxide is formed again when the organic molecules are used by the living organism for cellular respiration.
14. The nitrogen cycle is the worldwide circulation of nitrogen from nonliving to living systems and back again. Atmospheric nitrogen (N_2) cannot enter living systems. Nitrogen-fixing bacteria living in nodules on the roots of leguminous plants convert N_2 to ammonium (NH_4^+), which is then converted to nitrites (NO_2^-) and then to nitrates (NO_3^-) by nitrifying bacteria. The ammonia and nitrates are then available to plants to use in their proteins and nucleic acids. The nitrogen is then transferred to organisms that consume the plants. Nitrates that are not assimilated into living organisms can be converted to nitrogen gas by denitrifying bacteria.
15. Phosphorus in the form of phosphates is washed from sedimentary rock by rainfall. The dissolved phosphates are used by producers to produce important biological molecules, including DNA and ATP. When animals eat producers or other animals, phosphates are passed along the food webs. Decomposers release phosphates into the soil or water from dead organisms.

WEB TUTORIAL 23.3 Water Cycle

WEB TUTORIAL 23.4 Carbon Cycle

WEB TUTORIAL 23.5 Nitrogen Cycle

Humans Can Upset Biogeochemical Cycles (pp. 546–550)

16. The water cycle can be disturbed as humans cause shortages through overuse of water supplies. Most water use is in agriculture, primarily for irrigation. Irrigation can cause salts to accumulate in the soil (salinization), which can make the land unusable for agriculture.
17. Humans have disturbed the carbon cycle through activities that have increased carbon dioxide levels in the atmosphere: burning fossil fuels, which directly adds carbon dioxide, and deforestation, which decreases the removal of carbon dioxide from the atmosphere. The rise of atmospheric carbon dioxide is a concern because it is a greenhouse gas (along with methane, nitrous oxide, CFCs, and ozone). These gases trap heat in the earth's atmosphere and could cause global warming.
18. Humans have disrupted the nitrogen and phosphorus cycles through the industrial fixation of nitrogen to produce fertilizer and the addition of phosphates to fertilizer. Some of this fixed nitrogen and phosphates washes into nearby waterways. Fertilizer runoff is one cause of eutrophication.

WEB TUTORIAL 23.6 Greenhouse Effect and Seasons

KEY TERMS

biogeochemical cycle *p. 542*	ecological pyramid *p. 538*	niche *p. 535*	pyramid of biomass *p. 538*
biological magnification *p. 541*	ecological succession *p. 536*	nitrification *p. 544*	pyramid of energy *p. 538*
biomass *p. 538*	ecology *p. 534*	nitrogen fixation *p. 544*	salinization *p. 547*
biosphere *p. 534*	ecosystem *p. 534*	omnivore *p. 537*	secondary consumer *p. 537*
carnivore *p. 537*	eutrophication *p. 549*	population *p. 535*	secondary succession *p. 536*
climax community *p. 536*	food chain *p. 537*	primary consumer *p. 537*	trophic level *p. 537*
community *p. 535*	food web *p. 537*	primary productivity *p. 537*	
consumer *p. 537*	habitat *p. 536*	primary succession *p. 536*	
decomposer *p. 537*	herbivore *p. 537*	producer *p. 537*	

THINKING ABOUT THE CONCEPTS

1. What is ecological succession? How does primary succession differ from secondary succession? *p. 536*
2. Explain how energy flows through an ecosystem. *p. 537*
3. Define the following terms: producer, primary consumer, secondary consumer, and decomposer. Explain the role each plays in cycling nutrients through an ecosystem. *p. 535*
4. Explain why the feeding relationships in a community are more realistically portrayed as a food web than as a food chain. *pp. 537–538*
5. Explain the reasons for the inefficiency of energy transfer from one trophic level to the next higher one. Why does this loss of energy limit the number of possible trophic levels? *p. 538*
6. What does a pyramid of energy mean? *p. 538*
7. What is meant by a pyramid of biomass? *p. 538*
8. Define biological magnification. Explain how biological magnification is a consequence of the energy loss between trophic levels. Why should humans be concerned about biological magnification? *pp. 540–541*
9. Explain why more people could be fed on a vegetarian diet than on a diet in which meat provides most of the protein calories. *pp. 541–542*
10. Describe the water cycle, the carbon cycle, the nitrogen cycle, and the phosphorus cycle. *pp. 542–546*
11. Explain some of the ways that humans are disturbing the water cycle. *pp. 546–547*
12. Which human activities are primarily responsible for the rising level of atmospheric carbon dioxide? *pp. 547–548*
13. What is meant by the greenhouse effect? Explain why there is concern that the rising level of atmospheric carbon dioxide could lead to global warming. *pp. 548–549*
14. Which of the following is *not* constantly recycled in the biosphere?
    a. energy
    b. water
    c. carbon
    d. nitrogen
15. The greenhouse effect is due to
    a. phosphorus.
    b. nitrogen.
    c. carbon dioxide.
    d. carbon monoxide.
16. Secondary succession is most likely to be found
    a. on rocky outcroppings.
    b. in areas destroyed by fire.
    c. where an island has appeared.
    d. in areas where estuaries have filled in to form land.
17. An organism's role in a community is its _____.
18. An animal that eats an herbivore is called a _____.

APPLYING THE CONCEPTS

1. In a particular grassland ecosystem energy flows in the following path:

    Grass → Crickets → Frogs → Herons

    On the assumption that the efficiency of energy transfer is typical, if there were 100,000 calories of grass, how many calories will the herons receive?

2. Explain why the mercury levels in a lake may be low enough that the water is safe to drink yet fish from the same lake may be poisonous.

24

Human Population Dynamics

Population Dynamics Describe How Population Size Changes
- Rates of addition and subtraction determine population growth rate
- The size of the initial population influences how quickly population size increases
- The age structure of a population influences future growth
- Immigration and emigration affect population size

The Human Population Cannot Continue Growing Exponentially

Environmental Factors Regulate Population Size

The Growing Human Population Has Caused Several Problems
- The human population may be reaching the earth's carrying capacity
- Human activities cause pollution
- Human activities deplete the earth's resources
- Human activities have reduced biodiversity

Our Future Depends on the Decisions We Make Today

SOCIAL ISSUE   Maintaining Our Remaining Biodiversity

Brad and Krista walked toward the boat, ready for their parasailing adventure. "Isn't there a lot of wind today?" Krista asked Dale, the driver.

Sensing she was nervous, Dale said "No, this is a great day to go up."

Not satisfied she asked, "How long have you been working with the parasailing adventure, Dale?"

"About three years now," he explained, "ever since the fishing gave out around here, and I had to give it up. The big trawlers have dangerously overexploited the waters around here. I guess it's just as well because we would have had to stop fishing anyway. Some company dumped a load of electronic equipment—mostly cell phones—off the coast a while back, and now they've found lead and some other metals in the shellfish catches."

"Doesn't that stuff just get diluted?" Krista asked.

Dale sighed again. "Well, that's what most people think," he said. "The truth is that there are over six billion people in this world. Every time one of them throws something on the beach, or dumps something into a river, or flushes something down their sink, there's a chance it will wind up in the ocean. I hope we all realize soon that the ocean isn't an infinite resource." The trio grew silent.

As they reached the boat, Krista said, "I love the ocean. I didn't realize that it could be depleted of fish or polluted. It seems so huge! I thought nothing could harm it."

Dale replied, "Overfishing by trawlers and pollution have already cost many of us our traditional livelihood. For this part of the ocean to become a reliable fishery again, many countries will have to work together to make conservation a top priority, and they have to place heavy fines on polluters."

In this chapter, we will consider overfishing and other problems that stem from the growth of the human population. We will start by reviewing general principles regarding the changes in population size. Then we will consider how the rapid growth of the human population has led to pollution, a depletion of the earth's resources, and a reduction in biodiversity. Finally, we will recognize the choices we make today will affect the quality of life on earth in the future. ∎

Population Dynamics Describe How Population Size Changes

There is no way to discuss human populations without arousing concern. The concern stems from two kinds of information. First, the increase in our numbers is startling, and we just do not know how long it can continue. It took until 1804 for the human population to reach 1 billion, but since then we have been adding a billion people to our numbers at a much faster rate (Figure 24.1). In mid-2004, the world population was approximately 6.4 billion people. It is expected that by 2014 our numbers will reach more than 7.4 billion people.

Population dynamics describe how populations change in size. The human population, we can say, consists of all individuals alive at any point in time. This enormous population can be broken down into smaller groups. Thus, we can speak of the population of the United States or Colombia or China. We can break it down further and consider, say, urban American populations versus rural ones.

The basic principle of population dynamics is simple: Population size changes when individuals are added and removed at different rates. A population grows when more individuals are added than are subtracted.

Rates of addition and subtraction determine population growth rate

A population's growth rate is determined by the rates at which individuals are added to and subtracted from the population. Individuals are added to a population by births and removed from the population by deaths. Births and deaths are usually expressed as rates. Usually, we indicate the rate as the number of births or deaths per 1000 persons per year.

Expressing a population's growth as a rate keeps the change in population size in perspective. For example, we would consider the addition of 1000 individuals to a population differently if it took place overnight than if it occurred during a year. Furthermore, the addition of 1000 individuals would be more important if the initial population size were 100 than if it were 100,000.

In 2003, the *birth rate* of the world's population was 22 births per 1000 individuals per year, and the *death rate* was 9 deaths per 1000 individuals per year. Thus the growth rate of the world's population was:

$$\frac{\text{Births - Deaths}}{1000} = \frac{22 - 9}{1000} = \frac{13}{1000} = \frac{1.3}{100} = 1.3\%$$

The age at which a female has her first offspring has a dramatic impact on the birth rate of a population. The younger women are when they begin to reproduce, the faster the population grows. In fact, the age when reproduction begins is the most important factor influencing a female's overall reproductive potential. Indeed, the increasing proportion of women who are postponing childbirth until their later reproductive years has been helping to slow the rate of population growth of the United States. The birth rate in the United States is declining for women of all ages. However, the largest drop in birth rate has occurred among women in their twenties. One reason may be that increasing numbers of women want to establish careers before beginning motherhood. Regardless of the reasons, this trend has helped slow the growth of the U.S. population. Indeed, world birth rate has declined steadily over the last 20 years.

Despite the steady decline in birth rate, the human population has continued to grow because the drop in death rate has been greater than the drop in birth rate. The discovery of antibiotics was certainly instrumental in the decline of death rates. However, even more important were improvements in sanitation, hygiene, and nutrition.

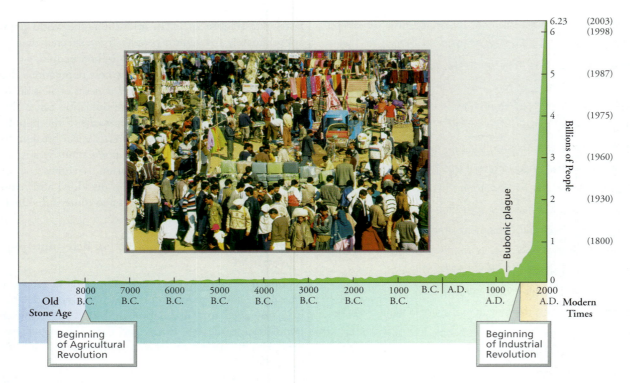

FIGURE **24.1**

The human population has grown steadily throughout history. It has skyrocketed since the Industrial Revolution.

The size of the initial population influences how quickly population size increases

Sometimes, when deciding how many new schools to build, for instance, we want to predict how many new individuals will be added to the population. To predict the *number* of births that will occur during a year, birth rate is not enough information; we must also know the number of individuals in the starting population. Because added individuals also reproduce, population growth occurs in much the same way as compound interest accumulates in the bank; the more you start with, the more you end up with. At *the same growth rate*, the larger the size of the starting population, the greater the number of individuals produced.

The age structure of a population influences future growth

When we want to predict future growth, a population's **age structure**—the relative number of individuals of each age—can be helpful. The age structure of the population is important because only individuals within a certain age range reproduce. Among humans, for instance, toddlers and octogenarians are counted as members of the population, but they do not reproduce. The ages are often grouped into prereproductive, reproductive, and postreproductive categories. Generally, only those individuals of reproductive age add to the size of the population. However, the prereproductive group (those who are currently too young to reproduce) usually get older, and when they enter the reproductive class, they have children, adding additional members to the population. Even if the birth rate remains the same, the overall size of the population will grow more rapidly as the size of the reproductive class increases.

stop and think

Health care has improved greatly in the last century and has increased population growth. Why would measures that decrease child mortality (death) contribute more to population growth than measures that increase life expectancy?

We see, then, that the relative size of the base of the age structure of a population determines how quickly numbers will be added to the population. Whereas a wide base to any age structure reflects a growing population, a narrow base is characteristic of a population that is getting smaller (Figure 24.2). A large base to any such pyramid, such as in India, spells trouble in terms of future demands on the country's commodities. Moderately stable nations, such as the United States, show a more consistent age distribution with some fluctuation, but with a small prereproductive population. The prereproductive base of the age structure of Canada is smaller than the reproductive base. Canada's population will probably decline in the future.

Immigration and emigration affect population size

If we consider the human population of the world, the effects of immigration and emigration on population size are eliminated. However, immigration and emigration can have dramatic effects on the population size of countries, cities, and regions. Each year, 1 million legal immigrants and 500,000 illegal immigrants enter the United States. Also, immigration is constantly changing the face of America. In 1950, 75% of minorities in the United States were African Americans. However, by 2003, Hispanics were the largest minority in the United States. These shifts in populations can have social, political, and economic effects.

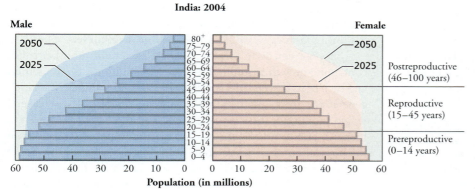

India: 2004

SOURCE: U.S. Census Bureau, International Data Base.

Expanding

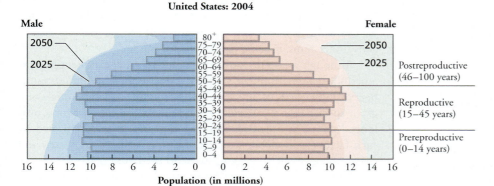

United States: 2004

SOURCE: U.S. Census Bureau, International Data Base.

Moderately stable

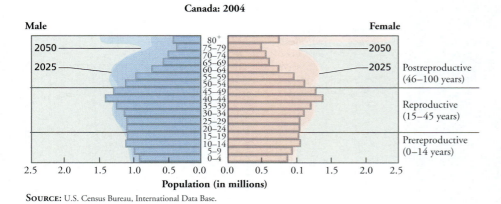

Canada: 2004

SOURCE: U.S. Census Bureau, International Data Base.

Declining

FIGURE **24.2**

Age structures of India, the United States, and Canada. The age structures shown are estimates for 2004, 2025, and 2050.

what would you do?

The United States is currently struggling with several issues related to immigration and the use of public funds. Should an illegal immigrant be able to obtain a driver's license? Should our schools permit bilingual education? Should a student who is an illegal immigrant be eligible for financial aid to attend college? Should certain illegal immigrants be given the opportunity to become legal immigrants? If so, under what circumstances should privileges be given? What do you think? What are the reasons for your opinion?

The growth rate in some parts of the world is greater than in other parts. The growth rate of the world population has been slowing steadily during the past decade. However, it is not *reversing*. (Another way of looking at the situation is to compare it with a bus heading toward a cliff at 60 miles an hour. The bus may slow to 50 miles an hour just before it goes over the edge, but it does not change direction by slowing down a little. It will still go over the cliff.)

Although a growth rate of a few percent a year may not seem startling, its importance becomes clearer when it is expressed as **doubling time**—the number of years it will take for a population to double at that rate of growth. Think about your own family. How many living descendants do your grandparents have (Figure 24.3)?

stop and think

A pair of cockroaches is placed on a small island. Assume that the doubling time for a population of cockroaches is 1 month. After 20 years, the island contains half of the total number of roaches it could possibly hold. How much longer will it take for the island to become completely filled with roaches? (*Hint:* What is the doubling time for the population?)

To get an idea of how important doubling time is, consider this page. If it is $\frac{1}{250}$ of an inch thick and were doubled exponentially only eight times it would be an inch thick. At 12 more doublings, it would be as thick as a football field is long. At 42 times, it would reach the moon, and at only 50 times it would reach the sun. Obviously, knowing the doubling time of a population can yield some interesting predictions.

The Human Population Cannot Continue Growing Exponentially

Let's return to an idea we introduced earlier in this chapter—the importance of population size to the rate at which new individuals are added to a population. Recall that *when the rate of growth remains constant*, the number of individuals in a population increases more rapidly the larger the size of that population.

This unrestricted growth at a constant rate is called **exponential growth**, and when the size of a population growing in this fashion is shown graphically, we see a **J-shaped growth curve**.

Suppose you were to place a single bacterium on a petri dish containing growth medium. Given ideal conditions, the bacterium and each of its descendants could divide every 30 minutes. With each generation, their numbers are effectively doubled. Two bacteria would produce 4, these would leave 8, and those 8 would leave 16, then 32, and so on (Figure 24.4). Note that numbers are added to the population at an accelerating pace even though the growth rate remains constant, because the size of the population is increasing.

Exponential growth occurs in ideal environments, where there are plenty of resources and adequate waste removal. In the real world, however, these conditions rarely exist for long. What happens when the food supply begins to become depleted, when space runs out, or when wastes begin to accumulate? The answer to this question is of more than theoretical interest to humans. Notice in Figure 24.1 that the human growth curve is J-shaped. Let's see what we can learn from the changes in growth patterns of other organisms under more realistic conditions.

Although a population may be able to grow exponentially for a period, the environment simply cannot support unlimited numbers of individuals of any species. The number of individuals of a given species that a particular environment can support for a prolonged period is called its **carrying capacity**. It is determined by such factors as availability of resources, including food, water, and space; ability to clean away wastes; and predation pressure.

If there are environmental limits placed on population growth, we may wonder what might happen if a population overshoots the carry-

FIGURE **24.3**

Gladys Davis is surrounded by her six great-grandchildren. She had four children, two of whom had children of their own, producing three children. They, in turn, produced these six great-grandchildren. By mid-2004, two of the children shown here had started their own families, producing six children. Although the size of each of these families was small, there are currently 15 living descendants of Jack and Gladys Davis. Thus, although the sizes of individual families are small, this extended family has grown sevenfold. How much has your family grown in the last few generations?

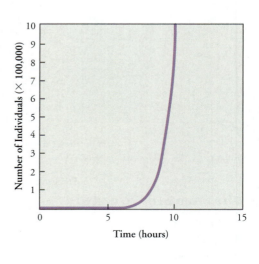

Time (hours)	Number of Individuals for Curve
10	1,048,576
9$^1/_2$	524,288
9	262,144
8$^1/_2$	131,072
8	65,536
7$^1/_2$	32,768
7	16,384
6$^1/_2$	8,192
6	4,096
5$^1/_2$	2,048
5	1,024
4$^1/_2$	512
4	256
3$^1/_2$	128
3	64
2$^1/_2$	32
2	16
1$^1/_2$	8
1	4
$^1/_2$	2
0	1

FIGURE **24.4**

A J-shaped growth curve describes unrestricted growth at a constant rate. This curve shows the number of bacteria in a colony, assuming that the bacteria divide every 30 minutes and all descendants survive.

ing capacity of the environment. One possibility is that environmental resources may become critically depleted and the population may decline precipitously. More commonly, however, population growth begins rapidly, but it becomes increasingly slower as the population size approaches the carrying capacity of its environment because of environmental pressures, such as limited resources or accumulating wastes. Eventually, growth levels off and the population size fluctuates slightly around the carrying capacity—a pattern that forms an **S-shaped growth curve** (Figure 24.5). This pattern is called **logistic growth**.

Environmental Factors Regulate Population Size

Now let's look at two ways populations can be reduced. We will place these ways in the context of what we have just learned.

Density-independent regulating factors are events that bring about death that are not related to the density of individuals in a population (the number of individuals in a given area of habitat). These factors are typically natural disasters, such as floods, mudslides, earthquakes, hurricanes, and fires. For example, in Haiti and the Dominican Republic floods killed an estimated 1300 people within a few weeks in mid-2004. Such events are all density independent because adding or removing people from the population would not alter the event itself.

Density-dependent regulating factors are events that have a greater impact on the population as conditions become more crowded. For example, starvation may be a density-dependent effect. If only so much food is available, an increasing population would result in there being less food to go around until, finally, some individuals would begin to starve (Figure 24.6). The greater the density, the greater the effects of starvation on the population. Because disease-causing organisms and parasites spread more easily in crowded conditions than in uncrowded conditions, these too are density-dependent factors that regulate population sizes.

The Growing Human Population Has Caused Several Problems

Whatever your personal predictions for the future of humankind, one thing should be clear: The world is becoming an increasingly crowded place. You may recall from Chapter 23 that there is no input of new resources on the earth. Therefore, we run the risk of using up our resources and making our world unlivable because of our waste. In fact, the size of the human population contributes to most of the problems we face today.

The human population may be reaching the earth's carrying capacity

It is difficult to estimate the earth's carrying capacity for humans. Indeed, the earth's ability to support people is influenced both by natural constraints, such as limited resource availability, and by human choices on a variety of issues, including economics, politics, technology, and values. Thus, the earth's carrying capacity for humans is uncertain and constantly changing. For instance, technological advances in agriculture and pollution control have raised the carrying capacity. At the same time, however, we have lowered the carrying capacity by consuming the earth's resources faster than they can be replaced.

Estimates of the earth's carrying capacity vary widely, ranging from 5 billion to 20 billion people. Those who accept the lower values point to resource depletion and pollution as evidence that we

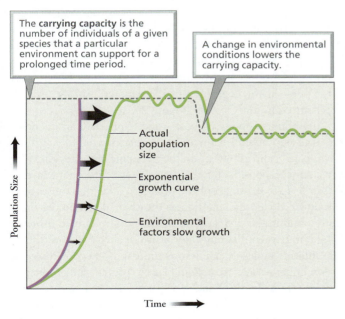

The **carrying capacity** is the number of individuals of a given species that a particular environment can support for a prolonged time period.

A change in environmental conditions lowers the carrying capacity.

Population Size

Actual population size

Exponential growth curve

Environmental factors slow growth

Time

FIGURE **24.5**

A logistic growth pattern produces an S-shaped growth curve. Population growth is often restrained by environmental factors, including availability of food, water, and space, and the ability to prevent wastes from accumulating. Environmental factors such as these determine the environment's carrying capacity. Eventually, growth levels off and the population size fluctuates around the carrying capacity. If a change in the environment lowers the carrying capacity, the population will stabilize at a smaller size.

FIGURE **24.6**

Food shortages may reflect density-dependent population controls because they have a greater impact as population density increases.

have already exceeded the earth's carrying capacity with a world population of more than 6.4 billion.

World hunger is one consequence of human population growth. Most of us know what hungry means, but not hunger. Hungry is the feeling we get when lunch is late; hunger is a condition in which lack of food causes physiological changes within our body. Hunger can be brought on by two conditions: *undernourishment*, when not enough food is eaten, and *malnourishment*, when the diet is not balanced and the right foods are not eaten.

Hunger particularly affects children, pregnant women, and nursing mothers. Children are notably susceptible to two very dangerous nutritional diseases. Marasmus is caused by a diet low in proteins and calories. The result is wrinkled skin, a startling appearance of old age, and thin limbs caused by wasting muscles. Kwashiorkor also results from a diet low in proteins and calories. It often appears when a child is weaned from breast milk. One of the many symptoms of the disease is fluid retention (edema), which is typically seen as a pronounced swelling of the abdomen. It isn't clear why a diet low in proteins and calories progresses to marasmus in some people and into kwashiorkor in other people.

Chronic undernourishment (starvation) is associated with irreversible changes in growth and brain development. If children receive less than 70% of the required daily calories, their growth and activity fall below normal levels, the number of brain cells decreases, and brain chemistry is altered. Thus, undernourishment can lower intelligence.

One reason for hunger is that food is unequally distributed throughout the world. North Africa, for example, produces very little food, because of a combination of drought, poor farming practices, and warfare. On the other hand, both food abundance and scarcity may exist in the same country. For example, southern Brazil produces a great deal of food, but the northern part of the country does not, so the people living in the north may suffer from malnutrition. The United States is an important exporter (our population has not yet caught up with our food production), but even here we find hunger in certain areas.

Agricultural production has increased largely because of the so-called *Green Revolution*—the development of high-yield varieties of crops and the use of modern cultivation methods, including the use of farm machinery, fertilizers, pesticides, and irrigation. The Green Revolution has boosted crop production in many countries. For example, Indonesia imported more rice than any other country in the past, but now it has enough rice to feed its people, as well as some to export.

Nonetheless, the Green Revolution has a big price tag: high energy costs and environmental damage. Although crop yields may be four times greater than with more traditional methods, modern farming practices use as much as a hundred times more energy and mineral resources. It takes energy to produce fertilizers. In addition, fossil fuels are needed to provide power for tractors and combines and to install and operate irrigation systems.

Overfishing is lowering the carrying capacity of the earth. Fishing has long provided humans with an abundant food supply, but now many fish populations are being depleted to such levels that certain kinds of fish can no longer be caught in certain areas. Recent examples include Atlantic cod in New England and salmon in the northwestern United States. The problem is that we have not given the fish enough time to replace their numbers at their reduced population size. As numbers of fish have dwindled, the response of humans has been to fish harder, adding more boats and new techniques, including electronic searches. Some species, the Atlantic bluefin tuna for instance, may already be so overfished that they will become extinct.

Human activities cause pollution

We discussed water pollution in Chapter 23, the health consequences of air pollution in Chapter 14, and acid rain caused by air pollution in Chapter 2. Here we will focus on another consequence of air pollution: the destruction of the ozone layer.

Ozone, composed of three atoms of oxygen (O_3), can be a good thing or a bad thing. If ozone is close to the earth, say anywhere in the first few miles above the surface, it is generally a bad thing. In fact, ozone is the primary component of photochemical smog, the noxious gas produced largely by sunlight interacting with air pollutants. Ozone irritates the eyes, skin, lungs, nose, and throat. (The effects of photochemical smog on the respiratory system are discussed in the Environmental Issue essay, in Chapter 14.) Ozone can also damage forests and crops and dissolve rubber.

However, naturally produced ozone is an essential part of the stratosphere, a layer of the lower atmosphere that encircles the earth about 10 to 45 kilometers (6 to 28 miles) above its surface. Without this layer of ozone, most terrestrial forms of life could not exist on the earth. The ozone layer acts as a shield against ultraviolet (UV) radiation from the sun. Ultraviolet radiation causes cataracts, aging of the skin, sunburn, and snow blindness, and it is the primary cause of skin cancer that kills 12,000 Americans each year. In addition, UV radiation inhibits the immune system and can interfere with plant growth.

The ozone in the atmosphere reaches levels of only about 1 molecule in 100,000. If all the ozone were compressed into one layer, its depth would be equal to only the diameter of a pencil lead. Still, ozone is able to shield the earth from excessive UV radiation.

In 1985, British scientists reported a sharp drop in the concentration of ozone over the Antarctic. This "hole" appears on a yearly cycle—at the beginning of the Antarctic spring (early September through mid-October; Figure 24.7).

Chlorofluorocarbons (CFCs)—once important in cooling systems, as an aerosol propellant, in the manufacture of plastic foam such as Styrofoam®, and as solvents in the electronic industry—are the primary culprit responsible for the destruction of the ozone layer. The CFCs drift up to the stratosphere, where UV radiation causes them to break down to chlorine, fluorine, and carbon. Then, under the conditions found in the stratosphere, chlorine can react with ozone, converting it to oxygen. Because chlorine is not altered in this reaction, a single chlorine molecule can destroy thousands of ozone molecules.

A number of laws are now in place to prevent the production or release of CFCs. In 1990 and 1992, industrialized countries signed agreements to reduce the production of CFCs. These countries have now phased out CFC production. It is expected that developing countries will phase out CFC production by 2010. Substitutes for CFCs have been found for aerosol cans, refrigerators, and air conditioners. There is evidence that the reduction in CFC production is paying off. In 1996, two years after CFC use peaked, chlorine levels in the lower atmosphere were slightly lower than before. By 1999, chlorine levels in the stratosphere were reduced. The ozone hole did shrink during 2001 and 2002, causing some scientists to suggest that

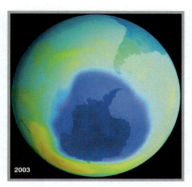

A satellite image of the hole in the ozone layer over Antarctica. The ozone hole is shown in blue.

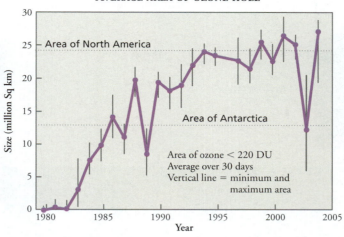

AVERAGE AREA OF OZONE HOLE

Area of North America

Area of Antarctica

Area of ozone < 220 DU
Average over 30 days
Vertical line = minimum and maximum area

The size of the ozone hole is now about the size of North America. Although the size of the hole shrank in 2002, it increased again in 2003.

The ozone layer absorbs ultraviolet (UV) radiation. A thinner ozone layer allows more UV radiation to reach the earth, increasing the cancer risk.

FIGURE **24.7**
The hole in the ozone layer

the ozone hole will get smaller by 2010 and may be eliminated by 2050 if international goals are met. However, it is still too soon to determine whether their prediction will be correct, because the ozone hole expanded nearly to its record size in 2003. In September of 2003, the ozone hole was 10.8 million square miles—an area three times the size of the United States, excluding Alaska.

▌Human activities deplete the earth's resources

An intact ecosystem usually retains soil and remains fertile. However, overuse and misuse of land can destroy its fertility.

SOIL EROSION

As the population grows, land is cultivated, grazed, and stripped of vegetation faster than it can recover. Wind and rain then carry away the topsoil, and the productivity of farmland declines. In this way, overfarming and overgrazing transform marginal farmlands to deserts, a process called **desertification** (Figure 24.8).

DEFORESTATION

You often do not know what you have until it is gone. This may be true for forests. Some of us live in areas where trees are so common that we take them for granted. However, trees, especially when they are grouped together to form forests, are an essential part of the global ecosystem. As we learned in Chapter 23, forests play an important role in water, carbon, and nitrogen cycles. Tree roots also reduce erosion by holding soil in place. If trees are cut down, rainwater runs off the land, carrying away soil and causing floods. In addition, trees are part of local ecosystems that include an incredible variety of life. They also influence local and global climate, including temperature and rainfall. And, in hot weather, their shade is pleasant.

Why, then, are forests disappearing? The cause is simple: The human population is expanding. People need space in which to live. Wood is needed for homes and furniture. Livestock cannot graze in forests. So trees are cut. Most are not replanted.

Deforestation is the removal of trees from an area without replacement. Deforestation is taking place in many regions of the world (Figure 24.9), including the United States. Before colonization, forests covered most of the eastern seacoast of the United States. Cities have now replaced many of those forests. In the Pacific Northwest, 90% to 95% of the forests have been cut.

Nonetheless, tropical forests are falling the fastest. Why? The first and most important reason that land is cleared is so that native families can feed themselves. A second reason is commercial logging. The third reason is cattle ranching. Cattle can graze on the land for about 6 to 10 years after trees have been cleared from a tropical forest before shrubs take over and make the area unsuitable for rangeland. Foreign companies own most of these cattle ranches, and the beef is often exported to fast-food restaurant chains. Thus, the deci-

FIGURE **24.8**
Desertification is the process by which the overuse of farmland or, in this case rangeland, converts the region to desert.

Deforestation is removing trees from an area and not replacing them.

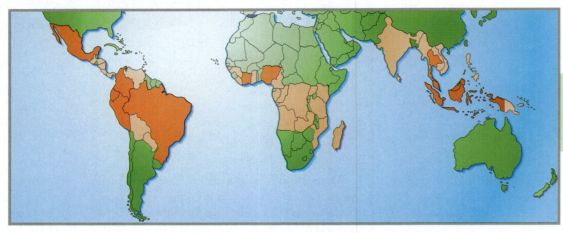

The countries in which the most deforestation is occurring are shown in red and orange.

Rate of deforestation:
Red = 2,000–14,280 km^2 per year
Orange = 100–1900 km^2 per year

FIGURE **24.9**
Deforestation damages soil quality.

sions we make about what to eat for lunch may have an indirect influence on the rate of tropical deforestation. Less important reasons include mining and the development of hydroelectric dams.

Whatever the reason for the tropical deforestation, soil fertility declines rapidly when these forests are cut. When it does, native people who depend on the soil for their living become even poorer.

▮ Human activities have reduced biodiversity

Biodiversity can be described as species richness. In specific terms, **biodiversity** is the number and variety of all living things in a given area. Recently, scientists have raised an alarm worldwide because there are indications that globally, and especially in certain critical areas, biodiversity is decreasing. We are losing species, perhaps as many as 100 species each day, in large part because of human activity.

Mass extinctions, the loss of many species from the earth, have occurred previously. This time, though, many other species are being forced to compete with us as we change the environment to suit ourselves, rendering it unsuitable for them.

How many species are there? The answer is that no one knows. Many scientists believe, however, that there are 8 million to 30 million species now in existence, with about 90% of them living on land. However, we are aware of only about 1.7 million of these species, and only 3% of those have been studied.

Where are these species? Most are in the tropics. Although tropical rain forests cover only 7% of the earth's surface, they are home to between 50% and 80% of the earth's species. It is estimated that 7 million to 8 million species live in the tropics (see the

Social Issue essay, *Maintaining Our Remaining Biodiversity*). Nonetheless, most of the *known* species come from the temperate regions, because most scientists live in the temperate regions, and they tend to investigate their own surroundings.

Habitat destruction is largely responsible for the loss of biodiversity. The primary reason habitats are destroyed is to create living space and economic development for an expanding human population. It has been estimated that most of the original tropical rain forest will be destroyed by 2040. This means that in the next 35 years, we could lose a quarter of all the species on the earth. By the end of the next century, perhaps two-thirds of the species living today will be gone. Evolutionary biologists tell us that, currently, we are losing species at the rate of 1000 to 10,000 times the average rate for the last 65 million years, since the extinction of the dinosaurs.

The rain forests are not the only type of habitat being destroyed. Lumbering and acid rain are destroying northern forests. Also, marine ecosystems are being destroyed by pollution, overfishing, and coastal development. For example, a report released in 2003 indicates that populations of all but 2 of 17 shark species in the North Atlantic and the Gulf of Mexico plunged by more than half between 1986 and 2000. Hammerhead sharks were particularly hard hit. Their populations declined by 89%. Commercial fishing is largely to blame for the reduction in the size of shark populations. People eat shark meat and use the fins to make soup. Some sharks, hammerheads for example, prey on herring and squid, which are often used as bait by people fishing for tuna or swordfish. Thus, catching sharks is a routine part of fishing for other species. Because sharks are attracted by

Maintaining Our Remaining Biodiversity

In the face of powerful evidence that we are undergoing a marked reduction in the earth's biodiversity, there is some guarded optimism that we may be able to reverse the trend. Let's consider some of the measures that are being suggested.

First, both developed (richer) and developing (poorer) countries must carefully census their life forms. They need to increase their reference libraries and their taxonomic inventories of the species they harbor. The need is especially acute in tropical countries. Unfortunately, these are often the least likely to be able to afford such programs, so the effort must be worldwide, with developed countries subsidizing the research in less affluent areas.

Second, we must link economic development of impoverished regions, particularly the tropics, to conservation. Lending agencies must stipulate that certain areas must be set aside

and left undeveloped. The decisions regarding which locations should be spared development should be made with the advice of ecologists so that the effects on the biodiversity of the region would be minimized. (One cannot help wondering what effect such rules would have had on the development of the United States.)

Third, educating the people whose decisions would most immediately affect biodiversity should be a priority. For example, recent studies have shown that native people are able to make more money through sustainable use of the forest (as in gathering nuts and fruits) than would have been possible if the forests had been cleared for agriculture or ranching. Many people in developing countries are also learning that scientists worldwide are interested in the medicines and healing knowledge they have, so in a sense, the educational opportunities are reciprocal.

Fourth, biological research worldwide must emphasize multiple land-use planning. In other words, biologists must begin to work more closely with zoning and land-use personnel to maximize the use of areas being cleared or cultivated. It is suggested that crops be rotated to maintain healthy soil, which will increase biodiversity. It is also suggested that single-species forests planted by paper companies, for example, might be better replaced by a more diverse forest that could harbor more species of animals. In this latter case, of course, the situation becomes more complicated than it seems because, as paper companies point out, why should they be asked to reduce their profits to increase biodiversity? This is one of the big problems with any ecological issue. We may know what is good for the environment, but it may not always be good, or appear to be good, for all concerned. ✸

---

the bait, they may be accidentally caught by people fishing for tuna or swordfish. In the United States, there are federal regulations that restrict shark fishing. However, shark fishing is not regulated in Europe. If shark numbers continue to fall, we may not be able to save them from extinction because sharks grow slowly, mature late in life, have few young at a time, and females do not reproduce every year.

We might wonder, however, why we should care about the loss of species, especially those we did not even know existed. Two reasons immediately come to mind. The first reason is that biodiversity preserves genetic diversity. That genetic diversity, encoded in the chromosomes of the vast array of species, is useful for cross-breeding. By breeding strains that possess different desirable traits, the traits of two different strains can be joined. For example, a new kind of corn was found on a Mexican hillside. (The farmer had to be persuaded not to plow it under to plant his usual corn crop.) The newly discovered corn was a perennial and proved to be resistant to a disease called corn smut, a fungal disease that causes the kernels to turn black and become oddly shaped. Plans were made to cross the perennial corn with common forms of agricultural corn, and these sorts of experiments continue.

Interestingly, the usefulness of genetic diversity as a reservoir of genetic stores has increased dramatically with modern technology. With our recently acquired abilities in genetic engineering, specific genes can be isolated and moved into other kinds of organisms. Thus, the maintenance of genetic stores becomes even more important as we use genetic engineering to maximize our crop yields or otherwise create new forms of organisms that might be useful to society (see Chapter 21).

The second benefit to maintaining biodiversity lies in developing new kinds of medicines. About 25% of all known drugs come from plants. The rosy periwinkle found in Madagascar is a source of two anticancer drugs, vincristine and vinblastine, and the Mexican yam was once a source of oral contraceptives. Most of the plants with medicinal value have been found in the tropics, and the tropics are being rapidly destroyed (discussed in Chapter 1).

Our Future Depends on the Decisions We Make Today

In many ways, our grandmothers were right—older ideas *were* better, at least in some ways. Today's economy is wasteful. We seem to have an idea that the world is here for humans alone.

In Chapter 23 and this chapter, we have seen that humans are part, a *small* part, of a large ecosystem. We can no longer think that we are the pinnacle of all life forms. We must realize that we are subject to the same principles that govern all other life, indeed, all of the earth. Instead of trying to conquer nature, we must work with natural laws. Above all, we must recognize that resources on the earth are limited and must be shared with all living organisms.

This dramatic turnabout in thinking will not be easy. It will require changes in the way governments and businesses operate and the ways in which individuals think. However, as was said earlier, humans are, by nature, problem solvers. We now have the opportunity to intelligently approach one of the greatest problems to ever confront humankind.

Jane Goodall, famed animal behavior researcher and founder of the Jane Goodall Institute, a center for environmental studies, is optimistic about our future. In her words, "My reasons for hope are fourfold: (1) the human brain; (2) the resilience of nature; (3) the energy and enthusiasm that is found or can be kindled among young people worldwide; and (4) the indomitable human spirit."[1] We have to be optimistic that a solution is possible and work together to achieve it.

[1]Jane Goodall (with Phillip Berman), *Reasons for Hope. A Spiritual Journey* (New York: Warner Books, 1999), 233.

REVIEWING THE CONCEPTS

Population Dynamics Describe How Population Size Changes (pp. 553–556)

1. Population dynamics describes how populations change in size. Population size changes when the number of individuals added differs from the number of individuals removed. Individuals are added through births and immigration. Individuals are removed through death and emigration.

2. The change in population is usually expressed as a rate: the number of births or deaths for a set number of individuals during a set time period (usually 1000 individuals per year). If we rule out immigration and emigration, we can say the growth rate of a population equals the birth rate minus the death rate.

3. The age at which females have their first offspring is the most important factor in determining their reproductive potential and, therefore, has a great influence on the rate at which a population grows.

4. The discovery of antibiotics and improvements in sanitation, hygiene, and nutrition have decreased the death rate.

5. Initial population size is important in determining how quickly new individuals will be added. At the same *rate* of increase, the number of individuals in the population increases faster, the larger its initial size of the population.

6. The age structure of a population is important in determining future changes in population size. The age structure of a population is the number of individuals in each age category: prereproductive, reproductive, and postreproductive. A growing population has a large base of prereproductive individuals. A stable population has approximately equal numbers of individuals in each age category. A population that is getting smaller has a large proportion of individuals who are past reproductive age.

7. Immigration (movement into a population) and emigration (movement out of a population) can influence local population size.

8. The doubling time for a population is the length of time it will take for the population to double in size at that rate of growth. In general, less developed countries have faster doubling times than do more developed countries.

WEB TUTORIAL 24.1 Population Growth

WEB TUTORIAL 24.2 Age Structure and Population Growth

The Human Population Cannot Continue Growing Exponentially (pp. 556–557)

9. A J-shaped growth curve results when a population grows exponentially and is not limited by environmental factors.

10. The carrying capacity of the environment is the number of individuals (of a particular species) it can support over a long period of time. Carrying capacity is determined by the amount of available resources. If a population overshoots its carrying capacity and depletes its resources, a population crash, in which the numbers drop rapidly, can follow.

11. Commonly, as a population grows and its size approaches the carrying capacity of the environment, the growth rate slows and the population size levels off, fluctuating around the carrying capacity. This so-called logistic growth produces an S-shaped growth curve.

Environmental Factors Regulate Population Size (p. 557)

12. Population size can be regulated by density-independent regulating factors, events such as natural disasters whose effects are not influenced by the size of the population, and by density-dependent regulating factors, which include factors that have a greater impact as the population becomes more crowded. Density-dependent factors include food availability and disease.

The Growing Human Population Has Caused Several Problems (pp. 557–561)

13. Humans have raised the carrying capacity of the earth through technological advances, but they have also decreased it through resource depletion and pollution. The size of the human population contributes to many of the problems we face today.

14. Hunger can be caused by undernourishment (starvation) or by malnourishment (a diet that does not contain enough of the right kinds of nutrients). Hunger has its greatest effects on children, pregnant women, and nursing mothers.

15. The Green Revolution has increased agricultural production through the use of high-yield varieties of crops and modern cultivation methods. The Green Revolution also has high energy costs, depletes natural resources, and causes environmental damage. Many fishing areas are becoming depleted because of overfishing.

16. Air pollution in the form of chlorofluorocarbons (CFCs) is destroying the ozone layer in the lower atmosphere (the stratosphere). The ozone layer traps ultraviolet (UV) light, protecting life on the earth from the harmful effects of UV radiation.

17. Overgrazing and overfarming are converting marginal farmlands to deserts, a process called desertification.

18. Deforestation—removing trees from an area without replacing them—is a growing problem throughout the world, especially in the tropical rain forests.

19. Biodiversity, the number and variety of living things, is being reduced dramatically, largely because of human activity. Most of the loss is occurring in the tropics and most is due to habitat destruction. Two practical reasons for concern over the loss of biodiversity are that these species could have genes that would someday prove useful or that they could be found to produce chemicals with medicinal qualities.

Our Future Depends on the Decisions We Make Today (pp. 561–562)

20. Humans are a small part of a large ecosystem. We must find ways to work within natural laws to preserve ecosystem earth and its biodiversity.

KEY TERMS

age structure   *p. 554*
biodiversity   *p. 560*
carrying capacity   *p. 556*
deforestation   *p. 559*

density-dependent regulating
  factors   *p. 557*
density-independent regulating
  factors   *p. 557*

desertification   *p. 559*
doubling time   *p. 556*
exponential growth   *p. 556*
J-shaped growth curve   *p. 556*

logistic growth   *p. 557*
population dynamics   *p. 553*
S-shaped growth curve   *p. 557*

THINKING ABOUT THE CONCEPTS

1. How is the growth rate of a population calculated?   *p. 553*
2. Explain why population growth is usually expressed as a rate.   *p. 553*
3. Explain why the size of the initial population influences how quickly it grows even when the growth rate remains constant.   *p. 554*
4. What is meant by the age structure of a population? How is a population's age structure related to predictions of its future growth?   *p. 554*
5. What is meant by the doubling time of a population? Why is this an important measure?   *pp. 555–556*
6. Under what conditions is a population likely to have a J-shaped growth curve?   *p. 556*
7. What is the carrying capacity of the environment? How is it related to the S-shaped growth curve?   *pp. 556–557*
8. Differentiate between density-independent and density-dependent regulating factors. Give examples of each.   *p. 557*
9. Describe some of the health consequences of hunger.   *p. 558*
10. What is the Green Revolution? What are its benefits and drawbacks?   *p. 558*
11. What determines whether ozone is helpful or harmful? What is causing the destruction of the ozone layer?   *p. 558*
12. What causes desertification?   *p. 559*
13. Define deforestation. Where is it occurring most rapidly? What are the primary reasons for deforestation?   *pp. 559, 560*

14. Define biodiversity. Where is it greatest? Why should we be concerned about the loss of biodiversity?   *pp. 560–562*
15. Which of the following is an example of a density-dependent regulating factor?
    a. earthquake
    b. starvation
    c. flood
    d. drought
16. If a population reaches the carrying capacity of the environment
    a. exponential growth will occur.
    b. the population will crash.
    c. food and other resources will increase.
    d. the population size will fluctuate around this level.
17. The ozone layer of the atmosphere absorbs
    a. carbon dioxide.
    b. chlorine.
    c. nitrogen dioxide.
    d. ultraviolet radiation.
18. A population that is increasing at a constant rate is experiencing _____ growth.
19. The number of individuals an environment can support is the _____.

APPLYING THE CONCEPTS

1. The populations of Kenya and Italy (shown on the right) have the same number of people, and the growth rate of the two populations is the same. Assuming that the growth rate stays the same, will the populations have the same number of people in 50 years? Explain.
2. Around 1993, four species of Asian carp escaped from fish farms in Mississippi and Arkansas, and by 1997 they had spread from Arkansas to Louisiana. By 2003, the number of Asian carp in Louisiana had increased dramatically. Asian carp are very large fish that eat plankton, which is also consumed by many other freshwater fish.
   a. Explain why the Asian carp population was able to increase extremely rapidly in a new environment (Louisiana).
   b. Explain why scientists are concerned that Asian carp will reduce the population size of other freshwater fish, even though the carp do not eat these fish.

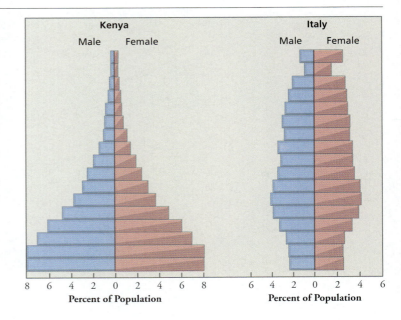

Glossary

Accommodation A change in the shape of the lens of the eye brought about by contraction of the smooth muscle of the ciliary body that changes the degree to which light rays are bent so that an image can be focused on the retina.

Acetylcholine A neurotransmitter found in both the central nervous system and the peripheral nervous system. It is the neurotransmitter released at neuromuscular junctions and causes muscle contraction.

Acetylcholinesterase An enzyme that breaks the neurotransmitter acetylcholine into its inactive component parts, acetate and choline. Acetylcholinesterase stops the action of acetylcholine at a synapse.

Acid Any substance that increases the concentration of hydrogen ions in solution.

Acinar cells Exocrine cells of the pancreas that secrete digestive enzymes into ducts that empty into the small intestine.

Acromegaly A condition characterized by thickened bones and enlarged soft tissues caused by overproduction of growth hormone in adulthood.

Acrosome A membranous sac at the tip of a sperm cell that contains enzymes that facilitate sperm penetration into the egg during fertilization.

Actin The contractile protein that makes up the major portion of the thin filaments in muscle cells. An actin (thin) filament is composed of actin, troponin, and tropomyosin. In muscle cells, contraction occurs when actin interacts with another protein called myosin.

Actin filaments The thin filaments in muscle cells composed primarily of the protein actin and essential to muscle contraction. In addition to actin, thin filaments contain two proteins important in the regulation of muscle contraction: tropomyosin and troponin.

Action potential A nerve impulse. An electrochemical signal conducted along an axon. A wave of depolarization caused by the inward flow of sodium ions followed by repolarization caused by the outward flow of potassium ions.

Active immunity Immune resistance in which the body actively participates by producing memory B cells and T cells after exposure to an antigen, either naturally or through vaccination.

Active site A specific location on an enzyme where the substrate binds.

Active transport The movement of molecules across the plasma membrane, usually against a concentration gradient (from a region of lower concentration to one of higher concentration) with the aid of a carrier protein and energy (usually in the form of adenosine triphosphate, or ATP) supplied by the cell.

Acute renal failure An abrupt, complete or nearly complete, cessation of kidney function.

Adaptation The process by which populations become better attuned to their particular environments as a result of natural selection.

Adaptive trait A characteristic (structure, function, or behavior) of an organism that makes an individual better able to survive and reproduce in its natural environment. Adaptive traits arise through natural selection.

Addison's disease An autoimmune disorder characterized by fatigue, loss of appetite, low blood pressure, and increased skin pigmentation resulting from undersecretion of glucocorticoids and aldosterone.

Adenosine triphosphate (ATP) A nucleotide that consists of the sugar ribose, the base adenine, and three phosphate groups. ATP is the energy currency of all living cells.

Adipose tissue A type of loose connective tissue that contains cells specialized for storing fat.

Adolescence The stage in postnatal development that begins with puberty and ends by about age 17. It is a period of rapid physical and sexual maturation during which the ability to reproduce is achieved.

Adrenal cortex The outer region of the adrenal gland that secretes glucocorticoids, mineralocorticoids, and gonadocorticoids.

Adrenal glands The body's two adrenal glands are located on top of the kidneys. The outer region of each adrenal gland secretes glucocorticoids, mineralocorticoids, and gonadocorticoids, and the inner region secretes epinephrine and norepinephrine.

Adrenaline See epinephrine.

Adrenal medulla The inner region of the adrenal gland that secretes epinephrine and norepinephrine.

Adrenocorticotropic hormone (ACTH) The anterior pituitary hormone that controls the synthesis and secretion of glucocorticoid hormones from the cortex of the adrenal glands.

Adulthood The stage in postnatal development that is generally reached somewhere between 18 and 21 years of age. Growth and formation of bone are usually completed by age 25, and thereafter developmental changes occur quite slowly.

Afferent (sensory) neuron A nerve cell specialized to conduct nerve impulses from the sensory receptors *toward* the central nervous system.

Age structure Of a population, the number of males and females of each age in a population. The ages are often grouped into prereproductive, reproductive, and postreproductive categories. Generally, only individuals of reproductive age add to the size of the population.

Agglutinate To clump together.

Aging The normal and progressive alteration in the structure and function of the body. Aging is possibly caused by declines in critical body systems, disruption of cell processes by free radicals, slowing or cessation of cell division, and decline in the ability to repair damaged DNA.

Agranulocytes The white blood cells without granules in their cytoplasm, including monocytes and lymphocytes.

AIDS Acquired immune deficiency syndrome. A diagnosis of AIDS is made when an HIV-positive person develops one of the following conditions: (1) a helper T cell count below $200/mm^3$ of blood; (2) one of 26 opportunistic infections, the most common of which are *Pneumocystis carinii* pneumonia and Kaposi's sarcoma, a cancer of connective tissue that affects primarily the skin; (3) a loss of more than 10% of body weight (wasting syndrome); or (4) dementia (mental incapacity such as forgetfulness or inability to concentrate).

Albinism A genetic inability to produce the brown pigment melanin that normally gives color to the eyes, hair, and skin.

Alcohol fermentation Process by which glucose is broken down by yeast when oxygen becomes scarce in the environment.

Aldosterone A hormone (the primary mineralocorticoid) released by the adrenal cortex that stimulates the reabsorption of sodium within kidney nephrons.

Allantois The extraembryonic membrane that serves as a site of blood cell formation and later becomes part of the umbilical cord, the ropelike connection between the embryo and the placenta.

Allele An alternative form of a gene. One of two or more slightly different versions of a gene that code for different forms of the same trait.

Allergen An antigen that stimulates an allergic response.

Allergy A strong immune response to an antigen (an allergen) that is not usually harmful to the body.

Allometric growth The change in the relative rates of growth of various parts of the body. Such growth helps shape developing humans and other organisms.

Alveolus (plural, alveoli) A thin-walled rounded chamber. In the lungs, the alveoli are the surfaces for gas exchange. They form clusters at the end of each bronchiole that are surrounded by a vast network of capillaries. The alveoli greatly increase the surface area for gas exchange.

Amino acid The building blocks of proteins consisting of a central carbon atom bound to a hydrogen atom, an amino group (NH_2), a carboxyl group (COOH), and a side chain designated by the letter R. There are twenty amino acids important to human life; some can be synthesized by our bodies (nonessential amino acids), whereas others cannot be synthesized and must be obtained from the foods we eat (essential amino acids).

Amniocentesis A method of prenatal testing for genetic problems in a fetus in which amniotic fluid is withdrawn through a needle so that the fluid can be tested biochemically and the cells can be cultured and examined for genetic abnormalities.

Amnion The extraembryonic membrane that encloses the embryo in a fluid-filled space called the amniotic cavity. Amniotic fluid forms a protective cushion around the embryo that later can be examined as part of prenatal testing in a procedure known as amniocentesis.

Amoeboid movement Cell movement along a surface caused by extensions of the cell called pseudopodia. Microfilaments are involved in this form of movement.

Ampulla A wider region in a canal or duct. In the inner ear, the ampulla is found at the base of a semicircular canal in the inner ear.

Anabolic steroids Synthetic hormones that mimic testosterone and stimulate the body to build muscle and increase strength. Steroid abuse can have many dangerous side effects.

Anabolism The building (synthetic) chemical reactions within living cells, as when cells build complicated molecules from simple ones. Compare with catabolism.

Anal canal The canal between the rectum and the anus. Feces pass through the anal canal.

Analgesic A substance, such as Demerol, that relieves pain.

Analogous structure A structure of one organism that is similar to that of another organism because of convergent evolution and not because the organisms arose from a common ancestry. Compare with homologous structures.

Anaphase In mitosis, the phase when the chromatids of each chromosome begin to separate, splitting at the centromere. Now separate entities, the chromatids are considered chromosomes, and they move toward opposite poles of the cell.

Anaphylactic shock An extreme allergic reaction that occurs within minutes after exposure to a substance to which a person is allergic and that can cause pooling of blood in capillaries, which causes dizziness, nausea, and sometimes unconsciousness as well as extreme difficulty in breathing. Anaphylactic shock can cause death.

Androgen A steroid sex hormone secreted by the testes in males and produced in small quantities by the adrenal cortex in both sexes.

Anemia A condition in which the blood's ability to carry oxygen is reduced. It can result from too little hemoglobin, too few red blood cells, or both.

Anencephaly A neural tube defect that involves incomplete formation of the brain and results in stillbirth or death shortly after birth.

Anesthesia The drug-induced loss of the sensation of pain. It may be general or regional.

Aneuploidy A general term describing the condition whereby gametes or cells contain too many or too few chromosomes compared with the normal number.

Aneurysm A blood-filled sac in the wall of an artery caused by a weak area in the artery wall.

Angina pectoris Choking or strangling chest pain, usually experienced in the center of the chest or slightly to the left, that is caused by a temporary insufficiency of blood flow to the heart. It begins during physical exertion or emotional stress, when the demands on the heart are increased and the blood flow to the heart muscle can no longer meet the needs.

Angioplasty A procedure that widens the channel of an artery obstructed because of atherosclerosis. It involves inflating a tough, plastic balloon inside the artery.

Angiotensin I Renin converts angiotensinogen into this protein.

Angiotensin II A protein that stimulates the adrenal gland to release aldosterone.

Anorexia nervosa An eating disorder characterized by deliberate self-starvation, a distorted body image, and low body weight.

Antagonistic pairs Muscles arranged in pairs so that the actions of the members of the pair are opposite to one another. This arrangement is characteristic of most skeletal muscles.

Anthropoids The suborder of modern primates that includes monkeys, apes, and humans.

Antibody A Y-shaped protein produced by plasma cells during an immune response that recognizes and binds to a specific antigen because of the shape of the molecule. Antibodies defend against invaders in a variety of ways, including neutralization, agglutination and precipitation, or activation of the complement system.

Antibody-mediated immune responses Immune system responses conducted by B cells that produce antibodies and that defend primarily against enemies that are free in body fluids, including toxins or extracellular pathogens, such as bacteria or free viruses.

Anticodon A three-base sequence on transfer RNA (tRNA) that binds to the complementary base pairs of a codon on the mRNA.

Antidiuretic hormone (ADH) A hormone manufactured by the hypothalamus but stored in and released from the posterior pituitary. It regulates the amount of water reabsorbed by the distal convoluted tubules and collecting ducts of nephrons. ADH causes water retention at the kidneys and elevates blood pressure. It is also called vasopressin.

Antigen A substance that is recognized as foreign by the immune system. Antigens trigger an immune response.

Antigen-presenting cell (APC) A cell that presents an antigen to a helper T cell, initiating an immune response toward that antigen. The most important type of antigen-presenting cell is a macrophage.

Aorta The body's main artery that conducts blood from the left ventricle toward the cells of the body. The aorta arches over the top of the heart and gives rise to the smaller arteries that feed the capillary beds of the body tissues.

Apoptosis A series of predictable physical changes in a cell that is undergoing programmed cell death. Apoptosis is sometimes used as a synonym for programmed cell death.

Appendicular skeleton The part of the skeleton comprising the pectoral girdle (shoulders), the pelvic girdle (pelvis), and the limbs (arms and legs).

Appendix A slender closed pouch that extends from the large intestine near the juncture with the small intestine.

Aqueous humor The fluid within the anterior chamber of the eye. It supplies nutrients and oxygen to the cornea and lens and carries away their metabolic wastes.

Arachnoid The middle layer of the meninges (the connective tissue layers that protect the central nervous system).

Areolar connective tissue A type of loose connective tissue composed of cells in a gelatinous matrix. It serves as a universal packing material between organs and anchors skin to underlying tissues and organs.

Arrector pili The tiny, smooth muscles attached to the hair follicles in the dermis.

Arteriole A small blood vessel located between an artery and a capillary. Arterioles serve to regulate blood flow through capillary beds to various regions of the body. They also regulate blood pressure. Arterioles are barely visible to the naked eye.

Artery A large-diameter muscular tube (blood vessel) that transports blood away from the heart toward the cells of body tissues. Arteries conduct blood low in oxygen to the lungs and blood high in oxygen to the body tissues. Arteries typically have thick muscular and elastic walls that dampen the blood pressure pulsations caused by heart contractions.

Arthritis An inflammation of a joint.

Artificial insemination   A treatment for infertility in which sperm are deposited in the woman's cervix or vagina at about the time of ovulation.

Association neuron   An interneuron. These neurons are located within the central nervous system between sensory and motor neurons and serve to integrate information.

Asthma   A condition marked by spasms of the muscles of bronchioles, making air flow difficult. It is often triggered by allergy.

Astigmatism   Irregularities in the curvature of the cornea or lens that cause distortions of a visual image because the irregularities cause light rays to converge unevenly.

Atherectomy   A technique for scraping lipid deposits associated with atherosclerosis away from an artery wall.

Atherosclerosis   A narrowing of the arteries caused by thickening of the arterial walls and a buildup of lipid (primarily cholesterol) deposits. Atherosclerosis reduces blood flow through the vessel, choking off the vital supply of oxygen and nutrients to the tissues served by that vessel.

Atom   A unit of matter that cannot be further broken down by chemical means; it is composed of subatomic particles, which include protons (positively charged particles), neutrons (with no charge), and electrons (with negative charges).

Atomic number   The number of protons in the nucleus of an atom.

Atria   (singular, atrium) The upper chambers of the heart that receive blood from the veins and pump it to the ventricles.

Atrial fibrillation   Rapid, ineffective contractions of the atria of the heart.

Atrial natriuretic peptide (ANP)   The hormone released by cells in the right atrium of the heart in response to stretching of the heart caused by increased blood volume and pressure. ANP decreases water and solute reabsorption by the kidneys, resulting in the production of large amounts of urine.

Atrioventricular bundle   A tract of specialized cardiac muscle cells that runs along the wall between the ventricles of the heart and conducts an electrical impulse that originated in the sinoatrial (SA) node and was conducted to the AV node to the ventricles. The bundle forks into right and left branches and then divides into many other specialized cardiac muscle cells, called Purkinje fibers, that penetrate the walls of the ventricles.

Atrioventricular (AV) node   A region of specialized cardiac muscle cells located in the partition between the two atria. It receives an electrical signal that spreads through the atrial walls from the sinoatrial node and relays the stimulus to the ventricles by means of a bundle of specialized muscle fibers, called the atrioventricular bundle, that runs along the wall between the ventricles.

Atrioventricular (AV) valves   Heart valves located between the atria and the ventricles that keep blood flowing in only one direction, from the atria to the ventricles. The right AV valve consists of three flaps of tissue and is also called the tricuspid valve. The left AV valve consists of two flaps of tissue and is also called the bicuspid or the mitral valve.

Autoimmune disorder   An immune response misdirected against the body's own tissues.

Autonomic nervous system   The part of the peripheral nervous system that governs the involuntary, unconscious activities that maintain a relatively stable internal environment. The autonomic nervous system has two branches: the sympathetic and the parasympathetic.

Autosomes   The 22 pairs of chromosomes (excluding the pair of sex chromosomes) that determine the expression of most of the inherited characteristics of a person.

Autotroph   An organism that makes its own food (organic compounds) from inorganic substances. The autotrophs include photoautotrophs, which use the energy of light, and chemoautotrophs, which use the energy in chemicals.

Axial skeleton   The part of the skeleton comprising the skull, the vertebral column, the breastbone (sternum), and the rib cage.

Axon   A long extension from the cell body of a neuron that carries an electrochemical message away from the cell body toward another neuron or effector (muscle or gland). The tips of the axon release a chemical called a neurotransmitter that can affect the activity of the receiving cell. Typically, there is one long axon on a neuron.

Axon terminal   The tip of a branch of an axon that releases a chemical (neurotransmitter) that alters the activity of the target cell. A synaptic knob.

B cell   See B lymphocyte.

B lymphocyte   B cell. A type of white blood cell important in antibody-mediated immune responses that can transform into a plasma cell and produce antibodies.

Balanced polymorphism   A phenomenon in which natural selection maintains two or more alleles for a trait in a population from one generation to the next. It occurs when the environment changes frequently or when the heterozygous condition is favored over either homozygous condition.

Ball-and-socket joint   A joint that allows motion in all directions, such as the shoulder and hip joints.

Barr body   A structure formed by a condensed, inactivated X chromosome in the body cells of female mammals.

Basal body   The structure that anchors the microtubules of a cilium or flagellum to a cell. It contains nine triplets of microtubules arranged in a ring.

Basal cell carcinoma   The most common type of skin cancer, occurring in the rapidly dividing cells of the basal layer of the epidermis.

Basal ganglia   A region of the cerebrum consisting of collections of cell bodies located deep within it called the white matter. They are important in coordinating movement and may also play a role in cognition and memory of learned skills.

Basal metabolic rate (BMR)   A measure of the minimum energy required to keep an awake, resting body alive. It generally represents between 60% and 75% of the body's energy needs.

Base   Any substance that reduces the concentration of hydrogen ions in solution.

Basement membrane   A noncellular layer beneath epithelial tissue that binds the epithelial cells to underlying connective tissue. It helps epithelial tissue resist stretching and forms a boundary.

Basophil   A white blood cell that releases histamine, a chemical that both attracts other white blood cells to the site and causes blood vessels to widen during an inflammatory response.

Benign tumor   An abnormal mass of tissue that usually remains at the site where it forms.

Bicuspid valve   A heart valve located between the left atrium and ventricle. It is also called the mitral valve or the left atrioventricular (AV) valve.

Bile   A mixture of water, ions, cholesterol, bile pigments, and bile salts that emulsifies fat (keeps fat as small globules), facilitating digestion. Bile is produced by the liver, stored in the gallbladder, and acts in the small intestine.

Bilirubin   A red pigment produced from the breakdown of the heme portion of hemoglobin by liver cells. It is excreted by the liver in bile.

Binary fission   A type of asexual reproduction in which the genetic information is replicated and then a cell, a bacterium for example, or organism divides into two equal parts.

Biodiversity   The number and variety of all living things in a given area. It includes genetic diversity, species diversity, and ecological diversity.

Biofeedback   The use of artificial signals to provide feedback about unconscious visceral and motor activity, particularly that associated with stress.

Biogeochemical cycle   The recurring process by which materials (for example, carbon, water, nitrogen, and phosphorus) cycle between living and nonliving systems and back again.

Biogeography   The study of the geographic distribution of organisms. New distributions of organisms occur when organisms move to new areas (dispersal) and when areas occupied by the organisms move or are subdivided.

Biological magnification   The tendency of a nondegradable chemical to become more concentrated in organisms as it passes along the food chain.

Biomass   In ecosystems, the dry weight of the body mass of a group of organisms in a particular habitat.

Biopsy   The removal and examination, usually microscopic, of a piece of tissue to diagnose a disease, usually cancer.

Biosphere   The part of the earth in which life is found. It encompasses all of the earth's living organisms.

Biotechnology The industrial or commercial use or alteration of living organisms, cells, or molecules to achieve specific useful goals.

Bipedalism Walking on two feet. The trait that evolved early in hominid evolution and set the stage for the evolution of other characteristics, such as increases in brain size.

Bipolar neuron A neuron that has only two processes. The axon and the dendrite extend from opposite sides of the cell body. Bipolar neurons are receptor cells found only in some of the special sensory organs, such as in the retina of the eye and in the olfactory membrane of the nose.

Birth defects Developmental defects present at birth. Such defects involve structure, function, behavior, or metabolism and may or may not be hereditary.

Birth rate The number of births per a specified number of individuals in the population during a specific time.

Bladder A muscular saclike organ that receives urine from the two ureters and temporarily stores it until release into the urethra.

Blastocyst The stage of development consisting of a hollow ball of cells. It contains the inner cell mass, a group of cells that will become the embryo, and the trophoblast, a thin layer of cells that will give rise to part of the placenta.

Blind spot The region of the retina where the optic nerve leaves the eye and on which there are no photoreceptors. Objects focused on the blind spot cannot be seen.

Blood Connective tissue that consists of cells and platelets suspended in plasma, a liquid matrix.

Blood-brain barrier A mechanism that protects the central nervous system by selecting the substances permitted to enter the cerebrospinal fluid from the blood. The barrier results from the relative impermeability of the capillaries in the brain and spinal cord.

Blood pressure The force exerted by the blood against the walls of the blood vessels. It is caused by the contraction of the ventricles and is influenced by vasoconstriction.

Blue babies Newborns whose foramen ovale, the fetal opening between the right and left atria of the heart, fails to close. As a result, much of their blood still bypasses the lungs and is low in oxygen. The condition can be corrected with surgery.

BMR See basal metabolic rate.

Bolus A small, soft ball of food mixed with saliva.

Bone marrow The soft material filling the cavities in bones. Yellow bone marrow serves as a fat-storage site. Red bone marrow is the site where blood cells are produced.

Bone Strong connective tissue with specialized cells in a hard matrix composed of collagen fibers and mineral salts.

Bone remodeling The ongoing process of bone deposition and absorption in response to hormonal and mechanical factors.

Bottleneck effect The genetic drift associated with dramatic, unselective reductions in population size so that by chance alone the genetic makeup of survivors is not representative of the original population.

Bowman's (glomerular) capsule A cuplike structure surrounding the glomerulus of a nephron.

Bowman's space The space between the two layers of Bowman's capsule. During filtration, water and many solutes move from the plasma into Bowman's space and then into the renal tubule.

Brain The organ composed of neurons and glial cells that receives sensory input and integrates, stores, and retrieves information.

Brain waves The patterns recorded in an EEG (electroencephalogram) that reflect the electrical activity of the brain and are correlated with the person's state of alertness.

Breast The front of the chest, especially either of the two protuberant glandular organs (mammary glands) that in human females and other female mammals produce milk to nourish newborns.

Breathing center A region in the medulla of the brain that controls the basic breathing rhythm.

Breech birth Delivery in which the baby is born buttocks first rather than head first. It is associated with difficult labors and umbilical cord accidents.

Bronchi (singular, bronchus) The respiratory passageways between the trachea and the bronchioles that conduct air into the lungs.

Bronchial tree The term given to the air tubules in the respiratory system because their repeated branching resembles the branches of a tree.

Bronchioles A series of small tubules branching from the smallest bronchi inside each lung.

Bronchitis Inflammation of the mucous membranes of the bronchi, causing excess mucus and a deep cough.

Brush border A fuzzy border of microvilli on the surface of absorptive epithelial cells of the small intestine.

Buffer A substance that prevents dramatic changes in pH by removing excess hydrogen ions from solution when concentrations increase and adding hydrogen ions when concentrations decrease.

Bulbourethral glands Cowper's glands. Male accessory reproductive glands that release a clear slippery liquid immediately before ejaculation.

Bulimia An eating disorder characterized by binge eating followed by purging by means of self-induced vomiting, enemas, laxatives, or diuretics.

Bursa (plural, bursae) A flattened sac containing a thin film of synovial fluid that surrounds and cushions certain synovial joints. Bursae are common in locations where ligaments, muscles, skin, or tendons rub against bone.

Bursitis Inflammation of a bursa (a sac in a synovial joint that acts as a cushion). Bursitis causes fluid to build up within the bursa, resulting in intense pain that becomes worse when the joint is moved and cannot be relieved by resting in any position.

Calcitonin (CT) A hormone secreted by the thyroid gland when blood calcium levels are high. It stimulates the removal of calcium from the blood and inhibits the breakdown of bone.

Callus A mass of repair tissue formed by collagen fibers secreted from fibroblasts or woven bone that forms around and links the ends of a broken bone.

Capillary A microscopic blood vessel between arterioles and venules with walls only one cell layer thick. It is the site where the exchange of materials between the blood and the tissues occurs.

Capillary bed A network of true capillaries servicing a particular area. Precapillary sphincters regulate blood flow through the capillary bed. When the sphincters relax, blood fills the capillary bed and materials can be exchanged between the blood and the tissues. When the sphincters contract, blood flows directly from an arteriole to a venule, bypassing the capillaries.

Carbaminohemoglobin The compound formed when carbon dioxide binds to hemoglobin.

Carbohydrate An organic molecule that provides fuel for the human body. Carbohydrates, which we know as sugars and starches, can be classified by size into the monosaccharides, disaccharides, and polysaccharides.

Carbonic anhydrase An enzyme in the red blood cells that catalyzes the conversion of unbound carbon dioxide to carbonic acid.

Carcinogen A substance that causes cancer.

Carcinoma in situ A tumor that has not spread; "cancer in place."

Cardiac cycle The events associated with the flow of blood through the heart during a single heartbeat. It consists of systole (contraction) and diastole (relaxation) of the atria and then of the ventricles of the heart.

Cardiac muscle A contractile tissue that makes up the bulk of the walls of the heart. Cardiac muscle cells are cylindrical and have branching interconnections between cells. Cardiac muscle cells are striped (striated) and have a single nucleus. Contraction of cardiac muscle is involuntary.

Cardiovascular Pertaining to the heart and blood vessels.

Cardiovascular system The organ system composed of the heart and blood vessels. The cardiovascular system distributes blood, delivers nutrients, and removes wastes.

Carnivore An animal that obtains energy by eating other animals. A secondary consumer.

Carpal tunnel syndrome A condition of the wrist and hand whose symptoms may include numbness or tingling in the affected hand, along with pain in the wrist, hand, and fingers that is caused by repeated motion in the hand or wrist, causing the tendons to become inflamed and press against the nerve.

Carrier An individual who displays the dominant phenotype but is heterozygous for a trait and can therefore pass the recessive allele to descendants.

Carrying capacity The number of individuals of a given species that a particular environment can support for a prolonged time period. The carrying capacity of the environment is determined by such factors as availability of resources, including food, water, and space; ability to clean away wastes; and predation pressure.

Cartilage A type of specialized connective tissue with a firm gelatinous matrix containing protein fibers for strength. The cartilage cells (chondrocytes) lie in small spaces (lacunae) within the matrix.

Catabolism Chemical reactions within living cells that break down complex molecules into simpler ones, releasing energy from chemical bonds. Compare with anabolism.

Cataract A cloudy or opaque lens. Cataracts reduce visual acuity and may be caused by glucose accumulation associated with Type I diabetes, excessive exposure to sunlight, and exposure to cigarette smoke.

CD4 cell See helper T cell.

Cell adhesion molecule (CAM) A molecule that pokes through the plasma membranes of most cells and helps hold cells together to form tissues and organs.

Cell body The part of a neuron that contains the organelles and nucleus needed to maintain the cells.

Cell cycle The entire sequence of events that a cell goes through from its origin in the division of a parent cell through its own division into two daughter cells. The cell cycle consists of two major phases: interphase and cell division.

Cell-mediated immune responses Immune system responses conducted by T cells that protect against cellular threats, including body cells that have become infected with viruses or other pathogens and cancer cells.

Cellular respiration The oxygen-requiring pathway by which glucose is broken down by cells to yield carbon dioxide, water, and energy.

Cellulose A structural polysaccharide found in the cell walls of plants. Humans lack the enzymes necessary to digest cellulose, and thus it passes unchanged through our digestive tract. Although cellulose has no value as a nutrient, it is an important form of dietary fiber known to facilitate the passage of feces through the large intestines.

Cementum A calcified but sensitive part of a tooth that covers the root.

Central nervous system (CNS) The brain and the spinal cord.

Centriole A structure, found in pairs, within a centrosome. Each centriole is composed of nine sets of triplet microtubules arranged in a ring.

Centromere The region of a replicated chromosome at which sister chromatids are held together until they separate during cell division.

Centrosome The region near the nucleus that contains centrioles. It forms the mitotic spindle during prophase.

Cerebellum A region of the brain important in sensory-motor coordination. It is largely responsible for posture and smooth body movements.

Cerebral cortex The extensive area of gray matter covering the surfaces of the cerebrum. It is often referred to as the conscious part of the brain. The cerebral cortex has sensory, motor, and association areas.

Cerebral white matter A region of the cerebrum beneath the cortex consisting primarily of myelinated axons that are grouped into tracts that allow various regions of the brain to communicate with one another.

Cerebrospinal fluid The fluid bathing the internal and external surfaces of the central nervous system. It serves as a shock absorber, supports the brain, nourishes the brain, delivers chemical messengers, and removes waste products.

Cerebrovascular accident See stroke.

Cerebrum The largest and most prominent part of the brain, composed of the cerebral hemispheres. It is responsible for thinking, sensory perception, originating most conscious motor activity, personality, and memory.

Cervical cap A barrier means of contraception consisting of a small rubber dome that fits snugly over the cervix and is held in place partly by suction. It prevents sperm from reaching the egg.

Cervix The narrow neck of the uterus that projects into the vagina whose opening provides a passageway for materials moving between the vagina and the body of the uterus.

Cesarean section A procedure by which the baby and placenta are removed from the uterus through an incision in the abdominal wall and uterus. The term is often shortened to C-section.

Chancre A painless bump that forms during the first stage of syphilis at the site of contact, usually within 2 to 8 weeks of the initial contact.

Chemical digestion A part of the digestive process that involves breaking chemical bonds so that complex molecules are broken into their component subunits. Chemical digestion produces molecules that can be absorbed into the bloodstream and used by the cells.

Chemical evolution The sequence of events by which life evolved from chemicals slowly increasing in complexity over perhaps 300 million years.

Chemoreceptor A sensory receptor specialized to respond to chemicals. We describe the input from the chemoreceptors of the mouth as taste (gustation) and from those of the nose as smell (olfaction). Other chemoreceptors monitor levels of chemicals in body fluids, such as carbon dioxide, oxygen, and glucose.

Childhood The stage in postnatal development that runs from about 13 months to 12 or 13 years of age. Growth during this period is initially rapid and then slows. Toward the end of childhood there is a dramatic increase in growth known as a growth spurt.

Chitin A structural polysaccharide found in the exoskeletons (hard outer coverings) of animals such as insects, spiders, and crustaceans.

Chlamydia A genus of bacteria. In this text, it is an infection (usually sexually transmitted) caused by *Chlamydia trachomatus*, commonly causing urethritis and pelvic inflammatory disease.

Cholecystokinin A hormone secreted by the small intestine that stimulates the pancreas to release its digestive enzymes and the gallbladder to contract and release bile.

Chordae tendineae Strings of connective tissue that anchor the atrioventricular valves to the wall of the heart, preventing the backflow of blood.

Chorion The extraembryonic membrane that becomes the embryo's major contribution to the placenta.

Chorionic villi Fingerlike projections of the chorion that grow into the uterine lining of the mother during formation of the placenta and become part of the placenta.

Chorionic villi sampling (CVS) A procedure for screening for genetic defects of a fetus by removing a piece of chorionic villi and examining the cells for genetic abnormalities.

Choroid The pigmented middle layer of the eyeball that contains blood vessels.

Chromatid One of the two identical replicates of a duplicated chromosome. The two chromatids that make up a chromosome are held together by a centromere and are referred to as sister chromatids. During cell division, the two strands separate and each becomes a chromosome in one of the two daughter cells.

Chromatin DNA and associated proteins in a dispersed, rather than condensed, state.

Chromosomal mutation A change in DNA in which a section of a chromosome becomes rearranged, duplicated, or deleted.

Chromosome DNA (which contains the genetic information of a cell) and specialized proteins, primarily histones.

Chronic renal failure A progressive and often irreversible decline in the rate of glomerular filtration.

Chylomicron A particle formed when proteins coat the surface of the products of lipid digestion, making lipids soluble in water and allowing them to be transported throughout the body.

Chyme The semifluid, creamy mass created during digestion once the food has been churned and mixed with the gastric juices of the stomach.

Cilia Extensions of the plasma membrane found on some cells, such as those lining the respiratory tract, that move in a back-and-forth motion. They are usually shorter and much more numerous than flagella but have the same 9 + 2 arrangement of microtubules at their core.

Ciliary body A portion of the middle coat of the eyeball near the lens that consists of smooth muscle and ligaments. Contractions of the smooth muscle of the ciliary body change the shape of the lens, which then focuses images on the retina.

Circulatory system An organ system composed of the cardiovascular system (heart and blood vessels).

Circumcision The surgical removal of the foreskin of the penis, usually performed when the male is an infant.

Cirrhosis A chronic disease of the liver in which the liver becomes fatty and the liver cells are gradually replaced with scar tissue.

Citric acid cycle The cyclic series of chemical reactions that follows the transition reaction and yields two molecules of adenosine triphosphate (one from each acetyl CoA that enters the cycle) and several molecules of nicotine adenine dinucleotide (NADH) and flavin adenine dinucleotide ($FADH_2$), carriers of high-energy electrons that enter the electron transport chain. This phase of cellular respiration occurs inside the mitochondrion and is sometimes called the Kreb's cycle.

Cleavage A rapid series of mitotic cell divisions in which the zygote first divides into two cells, and then four cells, and then eight cells, and so on. Cleavage usually begins about one day after fertilization as the zygote moves along the oviduct toward the uterus.

Climax community The relatively stable community that eventually forms at the end of ecological succession and remains if no disturbances occur.

Clitoris A small, erectile body in the female that plays a role in sexual stimulation and is developed from the same embryological structure that develops into the penis in a male.

Clonal selection The hypothesis that, by binding to a receptor on a lymphocyte surface, an antigen selectively activates only those lymphocytes able to recognize that antigen and programs that lymphocyte to divide, forming an army of cells specialized to attack the stimulating antigen.

Clone A population of identical cells descended from a single ancestor.

Cochlea The snail-shaped portion of the inner ear that contains the actual organ of hearing (the organ of Corti).

Codominance The condition in which the effects of both alleles are separately expressed in a heterozygote.

Codon A three-base sequence on messenger RNA (mRNA) that specifies one of the 20 common amino acids or the beginning or end of the protein chain.

Coenzyme An organic molecule such as a vitamin that functions as a cofactor and helps enzymes convert substrate to product.

Cofactor A nonprotein substance such as zinc, iron, and vitamins that helps enzymes convert substrate to product. It may permanently reside at the active site of the enzyme or may bind to the active site at the same time as the substrate.

Collagen fibers Strong insoluble protein fibers common in many types of connective tissues.

Colon The division of the large intestine composed of the ascending colon, the transverse colon, and the descending colon.

Colostrum A cloudy yellowish fluid produced by the breasts in the interval after birth when milk is not yet available. Its composition is different from that of milk.

Columnar epithelium A type of epithelial tissue composed of tall, rectangular cells that are specialized for secretion and absorption.

Coma An unconscious state caused by trauma to neurons in regions of the brain responsible for stimulating the cerebrum, particularly those in the reticular activating system or thalamus. Coma can be caused by mechanical shock, such as might be caused by a blow to the head, tumors, infections, drug overdose (from barbiturates, alcohol, opiates, or aspirin), or failure of the liver or kidney.

Combination birth control pill A means of hormonal contraception that consists of a series of pills that contain synthetic forms of estrogen and progesterone. The hormones in the pills mimic the effects of natural hormones ordinarily produced by the ovaries and inhibit FSH and LH secretion by the anterior pituitary gland and, therefore, prevent the development of an egg and ovulation.

Common cold An upper respiratory infection caused by one of the adenoviruses.

Community An assortment of organisms of various species interacting in a defined habitat.

Compact bone Very dense, hard bone, containing internal spaces of microscopic size and narrow channels that contain blood vessels and nerves. It makes up the shafts of long bones and the outer surfaces of all bones.

Complementary base pairing The process by which specific bases are matched: adenine with thymine (in DNA) or with uracil (in RNA) and cytosine with guanine. Each base pair is held together by weak hydrogen bonds.

Complementary proteins A combination of foods each containing incomplete proteins that provide ample amounts of all essential amino acids when combined.

Complement system A group of about 20 proteins that enhances the body's defense mechanisms. The complement system destroys cellular pathogens by creating holes in the plasma membrane, making the cell leaky, enhancing phagocytosis, and stimulating inflammation.

Complete dominance In genetic inheritance, the dominant allele in a heterozygote completely masks the effect of the recessive allele. Complete dominance often occurs because the dominant allele produces a functional protein and the recessive allele produces a less functional protein or none at all.

Complete protein A protein that contains ample amounts of all the essential amino acids. Animal sources of protein are generally complete proteins.

Compound A molecule that contains two or more different elements.

Computed tomography (CT scanning) A method of visualizing body structures, including the brain, using an x-ray source that moves in an arc around the body part to be imaged, thereby providing different views of the structure.

Concentration gradient A difference in the number of molecules or ions between two adjacent regions. Molecules or ions tend to move away from an area where they are more concentrated to an area where they are less concentrated. Each type of molecule or ion moves in response to its own concentration gradient.

Condom, female A barrier means of contraception used by a female that consists of a loose sac of polyurethane held in place by flexible rings (one at each end). It is used to prevent sperm from entering the female tract.

Condom, male A barrier means of contraception consisting of a thin sheath of latex or animal intestines that is rolled onto an erect penis, where it prevents sperm from entering the vagina. A latex condom also helps prevent the spread of sexually transmitted diseases.

Cone One of the three types of photoreceptors responsible for color vision.

Confounding variable In a controlled experiment, it is a second variable between the control group and the experimental group.

Conjoined twins Individuals that develop from a single fertilized ovum that fails to completely split in two at an early stage of cleavage. Such twins have identical genetic material and thus are always the same gender. They may be surgically separated after birth.

Connective tissue Tissue that binds together and supports other tissues of the body. All connective tissues contain cells in an extracellular matrix, which consists of protein fibers and a noncellular ground substance.

Constipation Infrequent, difficult bowel movements of hard feces. Constipation occurs when feces move through the large intestine too slowly and too much water is reabsorbed.

Consumers In ecosystems, an organism that obtains energy and raw materials by eating the tissues of other organisms.

Contact inhibition The phenomenon whereby cells placed in a dish in the presence of growth factors stop dividing once they have formed a monolayer. This may result from competition among the cells for growth factors and nutrients. Cancer cells do not display density-dependent contact inhibition and continue to divide, piling up on one another until nutrients run out.

Continuous ambulatory peritoneal dialysis (CAPD) A method of hemodialysis whereby the peritoneum, one of the body's own selectively permeable membranes, is used as the dialyzing membrane. It is an alternative to the artificial kidney machine during kidney failure.

Contraceptive sponge A barrier means of contraception consisting of a sponge containing spermicide.

Controlled experiment An experiment in which the subjects are divided into two groups, usually called the control group and the experimental group. Ideally, the groups differ in only the factor(s) of interest.

Convergence In vision, the process by which the eyes are directed toward the midline of the body as an object moves closer. Convergence is necessary to keep the image focused on the fovea of the retina.

Convergent evolution The process by which two species become more alike because they have similar ecological roles and selection pressures. For example, birds and insects independently evolved wings.

Core temperature The temperature in body structures below the skin and subcutaneous layers.

Cornea A clear, transparent dome located in the front and center of the eye that both provides the window through which light enters the eye and helps bend light rays so that they focus on the retina.

Coronary artery bypass A technique for bypassing a blocked coronary blood vessel to restore blood flow to the heart muscle. Typically, a segment of a leg vein is removed and grafted so that it provides a shunt between the aorta and a coronary artery past the point of obstruction.

Coronary artery disease (CAD) A condition in which fatty deposits associated with atherosclerosis form on the inside of coronary arteries, obstructing the flow of blood. It is the underlying cause of the vast majority of heart attacks.

Coronary circulation The system of blood vessels that services the tissues of the heart itself.

Coronary sinus A vessel that returns deoxygenated blood collected from the heart muscle to the right atrium of the heart. The coronary sinus is formed from the merger of cardiac veins.

Corpus callosum A band of myelinated axons (white matter) that connects the two cerebral hemispheres so they can communicate with one another.

Corpus luteum A structure in the ovary that forms from the follicle cells remaining in the ovary after ovulation. The corpus luteum functions as an endocrine structure that secretes estrogen and progesterone.

Corticotropin-releasing hormone (CRH) A hormone released from the placenta that seems to initiate labor.

Covalent bond A chemical bond formed when outer shell electrons are shared between atoms.

Cowper's glands See bulbourethral glands.

Cranial nerves Twelve pairs of nerves that arise from the brain and service the structures of the head and certain body parts such as the heart and diaphragm. Cranial nerves can be sensory, motor, or mixed.

Cranium The skeletal portion of the skull that forms the cranial case. It is formed from eight (sometimes more) flattened bones including the frontal bone, two parietal bones, the occipital bone, two temporal bones, the sphenoid bone, and the ethmoid bone.

Creatine phosphate A compound stored in muscle tissue that serves as an alternative energy source for muscle contraction.

Cretinism A condition characterized by dwarfism, mental retardation, and slowed sexual development. It is caused by undersecretion of thyroid hormone during fetal development or infancy.

Cristae Infoldings of the inner membrane of a mitochondrion.

Cross-bridges Myosin heads. Club-shaped ends of a myosin molecule that bind to actin filaments and can swivel, causing actin filaments to slide past the myosin filaments, which causes muscle contraction.

Crossing over The breaking and rejoining of nonsister chromatids of homologous pairs of chromosomes during meiosis (specifically at prophase I when homologous chromosomes pair up side by side). Crossing over results in the exchange of corresponding segments of chromatids and increases genetic variability in populations.

Cross-tolerance The development of tolerance for a drug that is not used, caused by the development of tolerance to another, usually similar, drug.

Crown The part of a tooth that is visible above the gum line. It is covered with enamel, a nonliving material that is hardened with calcium salts.

CT scanning See computed tomography.

Cuboidal epithelium A type of epithelial tissue composed of cube-shaped cells that are specialized for secretion and absorption.

Culture Social influences that produce an integrated pattern of knowledge, belief, and behavior.

Cupula A pliable gelatinous mass covering the hair cells within the ampulla of the semicircular canals of the inner ear and whose movement bends hair cells, triggering the generation of nerve impulses that are interpreted by the brain as movement of the head.

Cushing's syndrome A condition characterized by accumulation of fluid in the face and redistribution of body fat caused by prolonged exposure to cortisol.

Cutaneous membrane The skin. It is thick, relatively waterproof, and dry.

Cystitis Inflammation of the urinary bladder caused by bacteria.

Cytokinesis The division of the cytoplasm and organelles into two daughter cells during cell division. Cytokinesis usually occurs during telophase.

Cytomegalovirus Any of a group of highly host-specific herpes viruses that produce an infection characterized by unique large cells with nuclear inclusions. Cytomegalovirus infects many people in childhood but can be reactivated in people with an HIV infection.

Cytoplasm The part of a cell that includes the aqueous fluid within the cell and all the organelles with the exception of the nucleus.

Cytoskeleton A complex network of protein filaments within the cytoplasm that gives the cell its shape, anchors organelles in place, and functions in the movement of entire cells or certain organelles or vesicles within cells. The cytoskeleton includes microtubules, microfilaments, and intermediate filaments.

Cytotoxic T cell A type of T lymphocyte that directly attacks infected body cells and tumor cells by releasing chemicals called perforins that cause the target cells to burst.

Darwinian fitness See fitness.

Decomposer An organism that obtains energy by consuming the remains or wastes of other organisms. Decomposers release inorganic materials that can then be used by producers. Bacteria and fungi are important decomposers.

Deductive reasoning A logical progression of thought proceeding from the general to the specific. It involves making specific deductions based on a larger generalization or premise. The statement is usually in the form of an "if … then" premise.

Deforestation Removing trees from an area without replacing them.

Dehydration synthesis The process by which polymers are formed. Monomers are linked together through the removal of a water molecule.

Deletion Pertaining to chromosomes, the loss of a nucleotide or segment of a chromosome.

Denaturation The process by which changes in the environment of a protein, such as increased heat or changes in pH, cause it to unravel and lose its three-dimensional shape. Change in the shape of a protein results in loss of function.

Dendrite A process of a neuron specialized to pick up messages and transmit them toward the cell body. There are typically many short branching dendrites on a neuron.

Dense connective tissue Connective tissue that contains many tightly woven fibers and is found in ligaments, tendons, and the dermis.

Density-dependent regulating factor One of many factors that have a greater impact on the population size as conditions become more crowded. Such factors include disease and starvation.

Density-independent regulating factor One of many factors that regulate population size by causing deaths that are not related to the density of individuals in a population. Such factors include natural disasters.

Dentin A hard, bonelike substance that forms the main substance of teeth. It is covered by enamel on the crown and by cementum on the root.

Deoxyribonucleic acid (DNA) The molecular basis of genetic inheritance in all cells and some viruses. A category of nucleic acids that usually consists of a double helix of two nucleotide strands. The sequence of nucleotides carries the instructions for assembling proteins.

Depolarization A change in the difference in electrical charge across a membrane that moves it from a negative value toward 0 mV. During a nerve impulse (action potential), depolarization is caused by the inward flow of positively charged sodium ions.

Depo-Provera® A means of hormonal contraception that consists of an injection of progesterone every 3 months.

Dermis The layer of the skin that lies just below the epidermis and is composed of connective tissue. The dermis contains blood vessels, oil glands, sensory structures, and nerve endings. The dermis does not wear away.

Desertification The process by which overfarming and overgrazing transform marginal farmlands and rangelands to deserts.

Desmosome A type of junction between cells that anchors adjacent cells together.

Detrital food web Energy flow begins with detritus (organic material from the remains of dead organisms) that is eaten by a primary consumer.

Detritivore An organism that obtains energy by consuming the remains or wastes of other organisms. Detritivores release inorganic materials that can then be used by producers.

Detrusor muscle A layer of smooth muscle within the walls of the urinary bladder. It plays a role in urination.

Diabetes insipidus A condition characterized by excessive urine production caused by inadequate ADH production.

Diabetes mellitus A condition characterized by excessive urine production in which large amounts of glucose are lost as a result of insulin deficiency.

Diaphragm A broad sheet of muscle that separates the abdominal and thoracic cavities. When the diaphragm contracts, inhalation occurs.

Diaphragm, contraceptive A barrier means of contraception that consists of a dome-shaped soft rubber cup on a flexible ring. A diaphragm is used in conjunction with a spermicide to prevent sperm from reaching the egg.

Diarrhea Abnormally frequent, loose bowel movements. Diarrhea occurs when feces pass through the large intestine too quickly and too little water is reabsorbed. Diarrhea sometimes leads to dehydration.

Diastole Relaxation of the heart. Atrial diastole is the relaxation of the atria. Ventricular diastole is the relaxation of the ventricles.

Diastolic pressure The lowest blood pressure in an artery during the relaxation of the heart. In a typical, healthy adult, the diastolic pressure is about 80 mm Hg.

Digestive system The organ system that digests and absorbs food. The digestive system includes the mouth, esophagus, stomach, small intestine, and large intestine. Associated structures include the teeth, tongue, salivary glands, liver, gallbladder, and pancreas.

Dilation stage The first stage of true labor. It begins with the onset of contractions and ends when the cervix has fully dilated to 10 cm (4 in.).

Diploid The condition of having two sets of chromosomes in each cell. Somatic (body) cells are diploid.

Disaccharide A molecule formed when two monosaccharides covalently bond to each other through dehydration synthesis. It is known as a double sugar.

Distal convoluted tubule The section of the renal tubule where reabsorption and secretion occur.

Diuretic A substance that promotes urine production. Alcohol is a diuretic.

DNA See deoxyribonucleic acid.

DNA fingerprint The pattern of DNA fragments that have been cut by a restriction enzyme and sorted by size. Each person has a characteristic, individual DNA fingerprint.

DNA library A large collection of cloned recombinant DNA fragments containing the entire genome of an organism.

DNA ligase An enzyme that catalyzes the formation of bonds between the sugar and the phosphate molecules that form the sides of the DNA ladder during replication, repair, or the creation of recombinant DNA.

DNA polymerase Any one of the enzymes that catalyze the synthesis of DNA from free nucleotides using one strand of DNA as a template.

Dominant allele The allele that is fully expressed in the phenotype of an individual who is heterozygous for that gene. The dominant allele usually produces a functional protein, whereas the recessive allele does not.

Dopamine A neurotransmitter in the central nervous system thought to be involved in regulating emotions and in the brain pathways that control complex movements.

Dorsal nerve root The portion of a spinal nerve that arises from the back (posterior) side of the spinal cord and contains axons of sensory neurons. It joins with the ventral nerve root to form a single spinal nerve, which passes through the opening between the vertebrae.

Doubling time The number of years required for a population to double in size at a given, constant growth rate.

Down syndrome A collection of characteristics that tend to occur when an individual has three copies of chromosome 21. It is also known as trisomy 21.

Drug cocktail A combination of drugs used to treat people who are HIV1 positive. The combination usually includes a drug that blocks reverse transcription and a protease inhibitor.

Ductus arteriosus A small vessel in the fetus that connects the pulmonary artery to the aorta. It diverts blood away from the lungs.

Ductus venosus A small vessel in the fetus through which most blood from the placenta flows, bypassing the liver.

Duodenum The first region of the small intestine. The duodenum receives chyme from the stomach and digestive juices from the pancreas and liver.

Duplication Pertaining to chromosomes, the duplication of a region of a chromosome that often results from fusion of a fragment from a homologous chromosome.

Dura mater The tough, leathery outer layer of the meninges that protects the central nervous system. Around the brain, the dura mater has two layers that are separated by a fluid-filled space containing blood vessels.

Dysplasia The changes in shape, nuclei, and organization of adult cells. It is typical of precancerous cells.

Eardrum See tympanic membrane.

Ecological pyramid A diagram in which blocks represent each tropic (feeding) level.

Ecological succession The sequence of changes in the species making up a community over time.

Ecology The study of the interactions among organisms and between organisms and their environment.

Ecosystem All the organisms living in a certain area that can potentially interact, together with their physical environment.

Ectoderm The primary germ layer that forms the nervous system, epidermis, and epidermal derivatives such as hair, nails, and mammary glands.

Ectopic pregnancy A pregnancy in which the embryo (blastocyst) implants and begins development in a location other than the uterus, most commonly in the oviducts (a tubal pregnancy).

Edema Swelling caused by the accumulation of interstitial fluid.

EEG See electroencephalogram.

Effector cells Lymphocytes that are responsible for the attack on cells or substances not recognized as belonging in the body.

Efferent (motor) neuron A neuron specialized to carry information *away from* the central nervous system to an effector, either a muscle or a gland.

Egg A mature female gamete. An ovum. The egg contains the mother's genetic contribution to the next generation and nutrients.

Elastic cartilage The most flexible type of cartilage because of an abundance of wavy elastic fibers in its matrix.

Elastic fibers Coiled proteins found in connective tissues that allow the connective tissue to be stretched and recoil. Elastic fibers are common in tissues that require elasticity.

Electrocardiogram (ECG or EKG) A graphical recording of the electrical activities of the heart.

Electroencephalogram (EEG) A graphical record of the electrical activity of the brain.

Electron transport chain A series of carrier proteins embedded in the inner membrane of the mitochondrion that receives electrons from the molecules of nicotine adenine dinucleotide (NADH) and flavin adenine dinucleotide ($FADH_2$) produced by glycolysis and the citric acid cycle. During the transfer of electrons from one molecule to the next, energy is released and this energy is then used to make adenosine triphosphate (ATP). Oxygen is the final electron acceptor in the chain.

Element Any substance that cannot be broken down into simpler substances by ordinary chemical means.

Embolus A blood clot that drifts through the circulatory system and can lodge in a small blood vessel and block blood flow.

Embryo The developing human from week 3 through week 8 of gestation (the embryonic period).

Embryonic disk The flattened platelike structure that will become the embryo proper. It develops from the inner cell mass.

Embryonic period The period of prenatal development that extends from week 3 through week 8 of gestation. It is when tissues and organs form.

Emigration The departure of individuals from a population for some other area.

Emphysema A condition in which the alveolar walls break down, thicken, and form larger air spaces, making gas exchange difficult. This change results in less surface area for gas exchange and an increase in the volume of residual air in the lungs.

Enamel A nonliving material that is hardened with calcium salts and covers the crown of a tooth.

Encephalitis An inflammation of the meninges around the brain.

Endocardium A thin layer that lines the cavities of the heart.

Endocrine gland A gland that lacks ducts and releases its products (hormones) into the fluid just outside the cells.

Endocytosis The process by which large molecules and single-celled organisms such as bacteria enter cells. It occurs when a region of the plasma membrane surrounds the substance to be ingested, then pinches off from the rest of the membrane, enclosing the substance in a saclike structure called a vesicle that is released into the cell. Two types of endocytosis are phagocytosis ("cell eating") and pinocytosis ("cell drinking").

Endoderm The primary germ layer that forms some organs and glands (for example, the pancreas, liver, thyroid gland, and parathyroid glands) and the epithelial lining of the urinary, respiratory, and gastrointestinal tracts.

Endometriosis A painful condition in which tissue from the lining of the uterus (the endometrium) is found outside the uterine cavity.

Endometrium The inner layer of the uterus consisting of connective tissue, glands, and blood vessels. The endometrium thickens and develops with each menstrual cycle and is then lost as menstrual flow. It is the site of embryo implantation during pregnancy.

Endoplasmic reticulum The network of internal membranes within eukaryotic cells. Whereas rough endoplasmic reticulum (RER) has ribosomes attached to its surface, smooth endoplasmic reticulum (SER) lacks ribosomes and functions in the production of phospholipids for incorporation into cell membranes.

Endorphin A chemical released by nerve cells that binds to the so-called opiate receptors on the pain-transmitting neurons and quells the pain. The term is short for *endo*genous mor*phine*-like substance.

Endosymbiont hypothesis The hypothesis that organelles such as mitochondria were once free-living prokaryotic organisms that either invaded or were engulfed by primitive eukaryotic cells with which they established a symbiotic (mutually beneficial) relationship.

Endothelium The lining of the heart, blood vessels, and lymphatic vessels. It is composed of flattened, tight-fitting cells. The endothelium forms a smooth surface that minimizes friction and allows the blood or lymph to flow over it easily.

Enkephalin A chemical released by nerve cells that binds to the so-called opiate receptors on the pain-transmitting neurons and quells the pain.

Enzyme A substance (usually a protein, but sometimes an RNA molecule) that speeds up chemical reactions without being consumed in the process.

Enzyme-substrate complex Complex formed when a substrate binds to an enzyme at the active site.

Eosinophil The type of white blood cell important in the body's defense against parasitic worms. It releases chemicals that help counteract certain inflammatory chemicals released during an allergic response.

Epidermis The outermost layer of the skin, composed of epithelial cells.

Epididymis A long tube coiled on the surface of each testis that serves as the site of sperm cell maturation and storage.

Epiglottis A part of the larynx that forms a movable lid of cartilage covering the opening into the trachea (the glottis).

Epinephrine Adrenaline. A hormone secreted by the adrenal medulla, along with norepinephrine, in response to stress. They initiate the physiological "fight-or-flight" reaction.

Epiphyseal plate A plate of cartilage that separates the head of the bone from the shaft, permitting the bone to grow. In late adolescence, the epiphyseal plate is replaced by bone, and growth stops. The epiphyseal plate is commonly called the growth plate.

Episiotomy An incision made to enlarge the vaginal opening, just before passage of the baby's head at the end of the second stage of labor.

Epithelial tissue One of the four primary tissue types. The tissue that covers body surfaces, lines body cavities and organs, and forms glands.

Equatorial plate A plane at the midline of a cell where chromosomes line up during mitosis or meiosis.

Erythrocyte A red blood cell. A nucleus-free biconcave cell in the blood that is specialized for transporting oxygen to cells and assists in transporting carbon dioxide away from cells.

Erythropoietin A hormone released by the kidneys when the oxygen content of the blood declines that stimulates red blood cell production.

Esophagus A muscular tube that conducts food from the pharynx to the stomach using peristalsis.

Essential amino acid Any of the eight amino acids that the body cannot synthesize and, therefore, must be supplied in the diet.

Estrogen A steroid female sex hormone produced by the follicle cells and the corpus luteum in the ovary. Estrogen helps oocytes mature, stimulates cell division in the endometrium and the breast with each uterine cycle, and maintains secondary sex characteristics. The adrenal cortex also secretes estrogen.

Eukaryotic cell A cell with a nucleus and extensive internal membranes that divide it into many compartments and enclose organelles. Eukaryotes include cells in plants, animals, and all other organisms except bacteria and archaea.

Eustachian tubes Small tubes that join the upper region of the pharynx (throat) with the middle ear. They help to equalize the air pressure between the middle ear and the atmosphere.

Eutrophication The enrichment of water in a lake or pond by nutrients. Eutrophication is often caused by nitrogen or phosphate that washes into bodies of water.

Evolution Descent with modification from a common ancestor. It is the process by which life forms on the earth have changed over time.

Excitatory synapse A synapse in which the response of the receptors for that neurotransmitter on the postsynaptic membrane increases the likelihood that an action potential will be generated in the postsynaptic neuron. The postsynaptic cell is excited because it becomes less negative than usual (slightly depolarized), usually because of the inflow of sodium ions.

Exhalation Breathing out (expiration) involves the movement of air out of the respiratory system into the atmosphere.

Exhaustion phase The final phase of the general adaptation syndrome that occurs in response to extreme and prolonged stress. During the exhaustion phase, the heart, blood vessels, and adrenal glands begin to fail after the taxing metabolic and endocrine demands of the resistance phase.

Exocrine glands Glands that secrete their product through ducts onto body surfaces, into the spaces within organs, or into a body cavity. Examples include the salivary glands of the mouth and the oil and sweat glands of the skin.

Exocytosis The process by which large molecules leave cells. It occurs when products packaged by cells in membrane-bound vesicles move toward the plasma membrane. Upon reaching the plasma membrane, the membrane of the vesicle fuses with it, spilling its contents outside the cell.

Exon The nucleotide sequences of a newly synthesized messenger RNA (mRNA) that are spliced together to form the mature mRNA that is ultimately translated into protein.

Exophthalmos A condition characterized by protruding eyes that is caused by the accumulation of interstitial fluid due to oversecretion of thyroid hormone.

Expiration The process by which air is moved out of the respiratory system into the atmosphere. It is also called exhalation.

Expiratory reserve volume The additional volume of air that can be forcefully expelled from the lungs after normal exhalation.

Exponential growth Unrestricted growth at a constant rate over a period of time. When exponential growth is plotted, it produces a characteristic J-shaped curve.

Expulsion stage The second stage of true labor. It begins with full dilation of the cervix and ends with delivery of the baby.

External auditory canal The canal leading from the pinna of the ear to the eardrum (tympanic membrane).

External intercostals The external layer of muscles of the rib cage. When the external intercostals contract, the rib cage is lifted and inhalation occurs.

External respiration The exchange of oxygen and carbon dioxide between the lungs and the blood.

External urethral sphincter A sphincter made of skeletal muscle that surrounds the urethra. This voluntary sphincter helps stop the flow of urine down the urethra when we wish to postpone urination.

Exteroceptor A sensory receptor that is located near the surface of the body and that responds to changes in the environment.

Extracellular fluid The watery solution outside cells. It is also called interstitial fluid.

Extraembryonic membranes Membranes that lie outside the embryo, where they protect and nourish the embryo and later the fetus. They include the amnion, yolk sac, chorion, and allantois.

Facial bones The bones that form the face. They are composed of 14 bones that support several sensory structures and serve as attachments for most muscles of the face.

Facilitated diffusion The movement of a substance from a region of higher concentration to a region of lower concentration with the aid of a membrane protein that either transports the substance from one side of the membrane to the other or forms a channel through which it can move.

Farsightedness A condition in which distant objects are seen more clearly than near ones. Farsightedness occurs either because the eyeball is too short or the lens is too thin, causing the image to be focused behind the retina.

Fascicle A bundle of skeletal muscle fibers (cells) that forms a part of a muscle. Each fascicle is wrapped in its own connective tissue sheath.

Fast-twitch cells Muscle fibers that contract rapidly and powerfully, with little endurance. They have few mitochondria and large glycogen reserves. They depend on anaerobic pathways to generate adenosine triphosphate (ATP) during muscle contraction.

Fatigue A state in which a muscle is physiologically unable to contract despite continued stimulation. Muscle fatigue results from a relative deficit of adenosine triphosphate (ATP).

Fat-soluble vitamin A vitamin that does not dissolve in water and is stored in fat. This group includes vitamins A, D, E, and K.

Feces Waste material discharged from the large intestine during defecation. Feces consist primarily of undigested food, sloughed-off epithelial cells, water, and millions of bacteria.

Fermentation A pathway by which cells can harvest energy in the absence of oxygen. It nets only 2 molecules of ATP as compared with the approximately 36 molecules produced by cellular respiration.

Fertilization The union between an egg (technically a secondary oocyte) and a sperm. It takes about 24 hours from start to finish and usually occurs in a widened portion of the oviduct, not far from the ovary.

Fetal alcohol syndrome A group of characteristics in children born to mothers who consumed alcohol during pregnancy. It can include certain facial features including an eye fold near the bridge of the nose, mental retardation, and slow growth.

Fetal period The period of prenatal development that extends from week 9 of gestation until birth. It is when rapid growth occurs.

Fetus The developing human from week 9 of gestation until birth (the fetal period).

Fever An abnormally elevated body temperature. Fever helps the body fight disease-causing invaders in a number of ways.

Fibrillation Rapid, ineffective contraction of the heart.

Fibrin A protein formed from fibrinogen by thrombin that forms a web that traps blood cells, forming a blood clot.

Fibrinogen A plasma protein produced by the liver that is important in blood clotting. It is converted to fibrin by thrombin.

Fibroblasts Cells in connective tissue that secrete the protein fibers that are found in the matrix of the connective tissue. Fibroblasts also secrete collagen fibers for the repair of body tissues.

Fibrocartilage Cartilage with a matrix containing many collagen fibers. Fibrocartilage is found around the edges of joints and the intervertebral disks.

Fibrous joints Joints that are held together by connective tissue and lack a joint cavity. Most fibrous joints do not permit movement.

Fight-or-flight response The body's reaction to stress or threatening situations by the stimulation of the sympathetic division of the autonomic nervous system.

Fitness The average number of reproductively viable offspring left by an individual. It is sometimes called Darwinian fitness.

Flagellum A whiplike appendage of cells that moves in an undulating manner. It is composed of an extension of the plasma membrane containing microtubules in a 9 + 2 array. It is found on some cells, such as sperm cells.

Floating ribs The last two ribs that do not attach directly to the sternum (breastbone).

Fluid mosaic A term used to describe the structure of the plasma membrane. Proteins interspersed throughout the lipid molecules give the membrane its mosaic quality. The ability of some proteins to move sideways gives the membrane its fluid quality.

Follicle-stimulating hormone (FSH) A hormone secreted by the anterior pituitary gland that in females stimulates the development of the follicles in the ovaries, resulting in the development of ova (eggs) and the production of estrogen, and in males stimulates sperm production.

Fontanels The "soft spots" in the skull of a newborn. The membranous areas in the skull of a newborn that hold the skull bones together before and shortly after birth. The fontanels are gradually replaced by bone.

Food chain The successive series of organisms through which energy (in the form of food) flows in an ecosystem. Each organism in the series eats or decomposes the preceding one. It begins with the photosynthesizers and flows to herbivores and then to carnivores.

Food web The interconnection of all the feeding relationships (food chains) in an ecosystem.

Foramen magnum The opening at the base of the skull (in the occipital bone) through which the spinal cord passes.

Foramen ovale In the fetus, a small hole in the wall between the right atrium and left atrium of the heart that allows most blood to bypass the lungs.

Formed elements Cells or cell fragments found in the blood. They include platelets, white blood cells, and red blood cells.

Fossilization The process by which fossils form. Typically an organism dies in water and is buried under accumulating layers of sediments. Hard parts become impregnated with minerals from surrounding water and sediments. Eventually the fossil may be exposed when sediments are uplifted and wind erodes the rock formation.

Fossils The preserved remnants or impressions of past organisms. Most fossils occur in sedimentary rocks.

Founder effect Genetic drift associated with the colonization of a new place by a few individuals so that by chance alone the genetic makeup of the colonizing individuals is not representative of the population they left.

Fovea A small region on the retina that contains a high density of cones but no rods. Objects are focused on the fovea for sharp vision.

Frameshift mutation A mutation that occurs when the number of nucleotides inserted or deleted is not a multiple of three, causing a change in the codon sequence on the mRNA molecule as well as a change in the resulting protein.

Fraternal twins Individuals that develop when two oocytes are released from the ovaries and fertilized by different sperm. Such twins may or may not be the same gender and are as similar genetically as any siblings. They are also called dizygotic twins.

Free radicals Molecular fragments that contain an unpaired electron.

FSH See follicle-stimulating hormone.

Full-term infant Baby born 38 weeks after fertilization.

Gallbladder A muscular pea-sized sac that stores, modifies, and then concentrates bile. Bile is released from the gallbladder into the small intestine.

Gamete A reproductive cell (sperm or egg) that contains only one copy of each chromosome. A sperm and egg fuse at fertilization, producing a zygote.

Gamete intrafallopian transfer (GIFT) A procedure in which eggs and sperm are collected from a couple and then inserted into the woman's oviduct, where fertilization may occur. If fertilization occurs, resulting embryos drift naturally into the uterus.

Ganglion (plural, ganglia) A collection of nerve cell bodies outside the central nervous system.

Gap junction A type of junction between cells that links the cytoplasm of adjacent cells through small holes, allowing physical and electrical continuity between cells.

Gastric glands Any one of several glands in the stomach mucosa that contribute to the gastric juice (hydrochloric acid and pepsin).

Gastric juice The mixture of hydrochloric acid (HCl) and pepsin released into the stomach.

Gastrin A hormone released from the stomach lining in response to the presence of partially digested proteins. Gastrin triggers the production of gastric juice by the stomach.

Gastrointestinal (GI) tract A long tubular system specialized for the processing and absorption of food that begins at the mouth and continues to the esophagus, stomach, intestines, and anus. Several accessory glands empty their secretions into the GI tract to assist digestion.

Gastrula The embryo during gastrulation, when cells move to establish primary germ layers.

Gastrulation Cell movements that establish the primary germ layers of the embryo. The embryo during this period is called a gastrula.

Gated ion channel An ion channel that is opened to allow ions to pass through or closed to prevent passage by changes in the shape of a protein that functions as a gate.

Gene A segment of DNA on a chromosome that directs the synthesis of a specific polypeptide that will play a structural or functional role in the cell. Some genes have regulatory regions of DNA within their boundaries. Also, some genes code for RNA molecules that are needed for the production of the polypeptide but are not part of it.

Gene flow Movement of alleles between populations as a result of the movement of individuals. It is a cause of microevolution.

Gene pool All of the alleles of all of the genes of all individuals in a population.

General adaptation syndrome (GAS) A series of physiological adjustments made by our bodies in response to prolonged, extreme stress.

General senses The sensations that arise from receptors in the skin, muscles, joints, bones, and internal organs and include touch, pressure, vibration, temperature, a sense of body and limb position, and pain.

Gene therapy Treating a genetic disease by inserting healthy functional genes into the body cells that are affected by the faulty gene.

Genetic code The base triplets in DNA that specify the amino acids that go into proteins or that function as start or stop signals in protein synthesis. It is used to convert the linear sequence of bases in DNA to the sequence of amino acids in proteins.

Genetic drift The random change in allele frequencies within a population due to chance alone. It is a cause of microevolution that is usually negligible in large populations.

Genetic engineering The manipulation of genetic material for human practical purposes.

Genital herpes A sexually transmitted disease caused by the herpes simplex virus (HSV) that is usually characterized by painful blisters on the genitals. It is usually caused by HSV-2 but can be caused by HSV-1.

Genital warts Warts that form in the genital area caused by the human papilloma viruses (HPV). These viruses also cause cervical cancer and penile cancer.

Genome The complete set of DNA of an organism, including all of its genes.

Genotype The genetic makeup of an individual. It refers to precise alleles that are present.

Giantism A condition characterized by rapid growth and unusual height caused by abnormally high levels of growth hormone in childhood.

Gingivitis An inflammation of the gums.

Glandular epithelium A type of epithelial tissue that is a component of endocrine and exocrine glands and is specialized to produce secretions.

Glaucoma A condition in which the pressure within the anterior chamber of the eye increases as a result of the buildup of aqueous humor. It can cause blindness.

Glial cells Nonexcitable cells in the nervous system that are specialized to support, protect, and insulate neurons. Also called neuroglial cells.

Global warming A long-term increase in atmospheric temperatures caused by a buildup of carbon dioxide (CO_2) and other greenhouse gases in the atmosphere. Greenhouse gases trap heat in the atmosphere.

Glomerular filtration The process by which water and dissolved substances move from the blood in the glomerulus to the inside of Bowman's capsule.

Glomerulus A tuft of capillaries within the renal corpuscle of a nephron.

Glottis The opening to the airways of the respiratory system from the pharynx into the larynx.

Glucagon The hormone secreted by the pancreas that elevates glucose levels in the blood.

Glucocorticoids Hormones secreted by the adrenal cortex that affect glucose homeostasis, thereby influencing metabolism and resistance to stress.

Gluconeogenesis The conversion of noncarbohydrate molecules to glucose.

Glycolysis The splitting of glucose, a six-carbon sugar, into two three-carbon molecules called pyruvate. Glycolysis takes place in the cytoplasm of a cell and is the starting point for cellular respiration and fermentation.

GnRH See gonadotropin-releasing hormone.

Goiter, simple An enlarged thyroid gland caused by iodine deficiency.

Golgi complex An organelle consisting of flattened membranous disks that functions in protein processing and packaging.

Golgi tendon organs The highly branched nerve fibers located in the tendons that sense the degree of muscle tension.

Gonad An ovary in a female or a testis in a male. The gonads produce gametes (eggs or sperm) and sex hormones.

Gonadocorticoids The male and female sex hormones, androgens and estrogens, secreted by the adrenal cortex.

Gonadotropin-releasing hormone (GnRH) A hormone produced by the hypothalamus that causes the secretion of luteinizing hormone and follicle-stimulating hormone from the anterior pituitary gland.

Gonorrhea A sexually transmitted disease caused by the bacterium *Neisseria gonorrhoeae* that commonly causes urethritis and pelvic inflammatory disease. It may not cause symptoms, especially in women.

Graafian follicle A mature follicle in the ovary.

Graded potential A temporary local change in the membrane potential that varies directly with the strength of the stimulus.

Granulocytes White blood cells with granules in their cytoplasm, including neutrophils, eosinophils, and basophils.

Graves' disease An autoimmune disorder caused by undersecretion of thyroid hormone. It is characterized by increased heart and metabolic rates, sweating, nervousness, and exophthalmos.

Gray matter Regions of the central nervous system that contain neuron cell bodies and unmyelinated axons. These regions are gray because they lack myelin. Gray matter is important in neural integration.

Greenhouse effect An increase in atmospheric carbon dioxide (CO_2) that could lead to a rise in temperatures throughout the world. Greenhouse gases trap heat in the atmosphere.

Growth factor A type of signaling molecule that stimulates growth by stimulating cell division in target cells.

Growth hormone (GH) An anterior pituitary hormone with the primary function of stimulating growth through increases in protein synthesis, cell size, and rates of cell division. GH stimulates growth in general, especially bone growth.

Gumma A large sore that forms during the third stage of syphilis.

Habitat The natural environment or place where an organism, population, or species lives.

Hair cells A type of mechanoreceptor that generates nerve impulses when bent or tilted. Hair cells in the inner ear are responsible for hearing and equilibrium.

Hair root plexus The nerve endings that surround the hair follicle and are sensitive to touch.

Haploid The condition of having one set of chromosomes, as in eggs and sperm.

HCG See human chorionic gonadotropin.

Heart A muscular pump that keeps blood flowing through an animal's body. The human heart has four chambers: two atria and two ventricles.

Heart attack The death of heart muscle cells caused by an insufficient blood supply; a myocardial infarction.

Heartburn A burning sensation behind the breastbone that occurs when acidic gastric juice backs up into the esophagus.

Heart murmur Heart sounds other than "lub dup" that are created by turbulent blood flow. Heart murmurs can indicate a heart problem, such as a malfunctioning valve.

Heimlich maneuver A procedure intended to force a large burst of air out of the lungs to dislodge material lodged in the trachea.

Helper T cell The kind of T lymphocyte that serves as the main switch for the entire immune response by presenting the antigen to B cells and by secreting chemicals that stimulate other cells of the immune system. It is also known as a T4 cell or a CD4 cell, after the receptors on its surface.

Hematocrit The percentage of red blood cells in blood by volume. It is a measure of the oxygen-carrying ability of the blood.

Hemodialysis The use of artificial devices, such as the artificial kidney machine, to cleanse the blood during kidney failure.

Hemoglobin The oxygen-binding pigment in red blood cells. It consists of four subunits, each made up of an iron-containing heme group and a protein chain.

Hemolytic disease of the newborn A condition in which the red blood cells of an Rh-positive fetus or newborn are destroyed by anti-Rh antibodies previously produced in the bloodstream of an Rh-negative mother.

Hemophilia An inherited blood disorder in which there is insufficient production of blood-clotting factors. Hemophilia results in excessive bleeding in joints, deep tissues, and elsewhere. Hemophilia usually occurs in males.

Hepatitis An inflammation of the liver. Hepatitis is usually caused by one of six viruses.

Herbivore An animal that eats primary producers (green plants or algae). A primary consumer.

Heterotroph An organism that cannot make its own food from inorganic substances and instead consumes other organisms or decaying material.

Heterozygote advantage The phenomenon in which the heterozygous condition is favored over either homozygous condition. It maintains genetic variation within a population in the face of natural selection.

Heterozygous The condition of having two different alleles for a particular gene.

Hinge joint A joint that permits motion in only one plane, such as the knee joint or the elbow.

Hippocampus The part of the limbic system of the brain that plays a role in converting short-term memory into long-term memory.

Histamine A substance released by basophils and mast cells during an inflammatory response that causes blood vessels to widen (dilate) and become more permeable.

Homeostasis The ability of living things to maintain a relatively constant internal environment in all levels of body organization.

Hominid A member of the human family, such as members of the genera *Australopithecus* and *Homo*.

Hominoids Apes and humans.

Homologous chromosomes A pair of chromosomes that bear genes for the same traits. One member of each pair came from each parent. Homologous chromosomes are the same size and shape and line up with one another during meiosis I.

Homologous structures Structures that have arisen from a common ancestry. Compare with analogous structure.

Homozygous The condition of having two identical alleles for a particular gene.

Hormonal implants A means of hormonal contraception consisting of silicon rods containing progesterone that are implanted under the skin in the upper arm and prevent pregnancy for up to 5 years.

Hormone A chemical messenger released by cells of the endocrine system that travels through the circulatory system to affect receptive target cells.

Host (vector) DNA The DNA that is recombined with pieces of DNA from another source (that might contain a desirable gene) in the formation of recombinant DNA.

HSV-1 Herpes simplex virus 1. HSV-1 usually infects the upper half of the body and causes cold sores (fever blisters), but it can cause genital herpes if contact is made with the genital area.

HSV-2 Herpes simplex virus 2. HSV-2 causes genital herpes, but it can cause cold sores if contact is made with the mouth.

Human chorionic gonadotropin (HCG) A hormone produced by the cells of the early embryo (blastocyst) and the placenta that maintains the corpus luteum for approximately the first 3 months of pregnancy. HCG enters the bloodstream of the mother and is excreted in her urine. HCG forms the basis for many pregnancy tests.

Human papilloma virus (HPV) One of the group of viruses that commonly cause genital warts.

Hyaline cartilage A type of cartilage with a gel-like matrix that provides flexibility and support. It is found at the end of long bones as well as parts of the nose, ribs, larynx, and trachea.

Hydrocephalus A condition resulting from the excessive production or inadequate drainage of cerebrospinal fluid.

Hydrogen bond A weak chemical bond formed between a partially positively charged hydrogen atom in a molecule and a partially negatively charged atom in another molecule or in another region of the same molecule.

Hydrolysis The process by which polymers are broken apart by the addition of water.

Hyperglycemia An elevated blood glucose level.

Hypertension High blood pressure. A high upper (systolic) value usually suggests that the person's arteries have become hardened, a condition known as arteriosclerosis, and are no longer able to dampen the high pressure of each heartbeat. The lower (diastolic) value is generally considered more important because it indicates the pressure when the heart is relaxing.

Hyperthermia Abnormally elevated body temperature.

Hypertonic solution A solution with a higher concentration of solutes than plasma, for example.

Hypodermis The layer of loose connective tissue below the epidermis and dermis that anchors the skin to underlying tissues and organs.

Hypoglycemia Depressed levels of blood glucose often resulting from excess insulin.

Hypophysis See pituitary gland.

Hypothalamus A small brain region located below the thalamus that is essential to maintaining a stable environment within the body. The hypothalamus influences blood pressure, heart rate, digestive activity, breathing rate, and many other vital physiological processes. It acts as the body's "thermostat"; regulates food intake, hunger, and thirst; coordinates the activities of the nervous system and the endocrine (hormonal) system; is part of the circuitry for emotions; and functions as a master biological clock.

Hypothermia Abnormally low body temperature.

Hypothesis   A testable explanation for a specific set of observations that serves as the basis for experimentation.

Hypotonic solution   A solution with a lower concentration of solutes than plasma, for example.

Identical twins   Individuals that develop from a single fertilized ovum that splits in two at an early stage of cleavage. Such twins have identical genetic material and thus are always the same gender. They are also called monozygotic twins.

Immigration   Movement of new individuals from other populations into an area.

Immune response   The body's response to specific targets not recognized as belonging in the body.

Immune system   The system of the body directly involved with body defenses against specific targets—pathogens, cells, or chemicals not recognized as belonging in the body.

Immunoglobulin (Ig)   Any of the five classes of proteins that constitute the antibodies.

Implantation   The process by which a blastocyst (pre-embryo) becomes embedded in the lining of the uterus. It normally occurs high up on the back wall of the uterus.

Impotence   The inability to achieve or maintain an erection long enough for sexual intercourse.

Incomplete dominance   In genetic inheritance, expression of the trait in a heterozygous individual is somewhere in between expression of the trait in a homozygous dominant individual and homozygous recessive individual.

Incomplete proteins   Proteins that are deficient in one or more of the essential amino acids. Plant sources of protein are generally incomplete proteins.

Incontinence   A condition characterized by the escape of small amounts of urine when sudden increases in abdominal pressure, perhaps caused by laughing, sneezing, or coughing, force urine past the external sphincter. This condition is common in women, particularly after childbirth, an event that may stretch or damage the external sphincter, making it less effective in controlling the flow of urine.

Incus   The middle of three small bones in the middle ear that transmit information about sound from the eardrum to the inner ear. The incus is also known as the anvil.

Independent assortment   Of chromosomes, the process by which homologous chromosomes and the alleles they carry segregate randomly during meiosis, creating mixes of maternal and paternal chromosomes in gametes. This is an important source of genetic variation in populations.

Independent assortment, law of   In genetic inheritance, a principle that states that the alleles of unlinked genes (those that are located on different chromosomes) are randomly distributed to gametes.

Inductive reasoning   A logical progression of thought proceeding from the specific to the general. It involves the accumulation of facts through observation until the sheer weight of the evidence forces some general statement about the phenomenon. A conclusion is reached on the basis of a number of observations.

Infancy   The stage in postnatal development that roughly corresponds to the first year of life. It is a time of rapid growth when total body length usually increases by one-half and weight may triple.

Infertility   The inability to conceive (become pregnant) or to cause conception (in the case of males).

Inflammatory response   A nonspecific body response to injury or invasion by foreign organisms. It is characterized by redness, swelling, heat, and pain.

Influenza   The flu. Influenza is a viral infection caused by any of the variants of influenza A or influenza B viruses.

Informed consent   A document that a person must sign before participating in an experiment that lists all possible harmful consequences that might result from participation.

Inguinal canal   A passage through the abdominal wall through which the testes pass in their descent to the scrotum and that contains the testicular artery and vein and the vas deferens.

Inhalation   Breathing in. Inhalation (inspiration) involves the movement of air into the respiratory system.

Inhibin   A hormone produced in the testes that increases with sperm count and inhibits follicle-stimulating hormone secretion and, therefore, inhibits sperm production.

Inhibiting hormone   A hormone that inhibits secretion of another hormone.

Inhibitory synapse   A synapse in which the response of the receptors for that neurotransmitter on the postsynaptic membrane decreases the likelihood that an action potential will be generated in the postsynaptic neuron. The postsynaptic cell is inhibited because its resting potential becomes more negative than usual.

Inner cell mass   A group of cells within the blastocyst that will become the embryo proper and some extraembryonic membranes.

Inner ear   A series of passageways in the temporal bone that houses the organs for hearing (cochlea) and the sense of equilibrium (vestibular apparatus).

Insertion   The end of the muscle that is attached to the bone that moves when the muscle contracts.

Insoluble fiber   A fiber that does not easily dissolve in water. These fibers include cellulose, hemicellulose, and lignin.

Inspiration   Inhalation. The movement of air into the respiratory system.

Inspiratory reserve volume   The volume of air that can be forcefully brought into the lungs after normal inhalation.

Insulin   The hormone secreted by the pancreas that reduces glucose levels in the blood.

Insulin shock   A condition that results from severely depressed glucose levels in which brain cells fail to function properly, causing convulsions and unconsciousness.

Integral proteins   Proteins embedded in the plasma membrane, either completely or incompletely spanning the bilayer.

Integumentary system   The skin.

Intercalated disks   Thickenings of the plasma membranes of cardiac muscle cells that strengthen cardiac tissue and promote rapid conduction of impulses throughout the heart.

Interferon   A type of defensive protein produced by T lymphocytes that slows the spread of viruses already in the body by interfering with viral replication. Interferons also attract macrophages and natural killer cells, which kill the viral-infected cell.

Interleukin 1   A chemical secreted by a macrophage that activates helper T cells in an immune response.

Interleukin 2   A chemical released by a helper T cell that activates both B cells and T cells.

Intermediate filament   A component of the cytoskeleton made from fibrous proteins. It functions in the maintenance of cell shape and anchoring organelles such as the nucleus.

Internal intercostals   The internal layer of muscles between the ribs. Contraction of the internal intercostals lowers the rib cage and helps push air out during heavy breathing.

Internal respiration   Movement of oxygen from the blood to the tissues, and movement of carbon dioxide from the tissues to the blood.

Internal urethral sphincter   A thickening of smooth muscle at the junction of the bladder and the urethra. The action of this sphincter is involuntary and keeps urine from flowing into the urethra while the bladder is filling.

Interneuron   An association neuron. Neurons located within the central nervous system between sensory and motor neurons that serve to integrate information.

Interoceptor   A sensory receptor located inside the body, where it monitors conditions. Interoceptors play a vital role in maintaining homeostasis. They are an important part of the feedback loops that regulate blood pressure, blood chemistry, and breathing rate. Interoceptors may also cause us to feel pain, hunger, or thirst, thereby prompting us to take appropriate action.

Interphase   The period between cell divisions when the DNA, cytoplasm, and organelles are duplicated and the cell grows in size. Interphase is also the time in the cell's life cycle when the cell carries out its functions in the body.

Interstitial cells Cells located between the seminiferous tubules in the testes that produce the male steroid sex hormones, collectively called androgens.

Intervertebral disks Pads of cartilage that help cushion the bones of the vertebral column.

Intracytoplasmic sperm injection (ICSI) A procedure in which a tiny needle is used to inject a single sperm cell into an egg. It is an option for treating infertility when the man has few sperm or sperm that lack the strength or enzymes necessary to penetrate the egg.

Intrauterine device (IUD) A means of contraception consisting of a small plastic device that either is wrapped with copper wire or contains progesterone. It is inserted into the uterus to prevent pregnancy.

Intrinsic factor A protein secreted by the stomach necessary for the absorption of vitamin B_{12} from the small intestine.

Intron A portion of a newly formed mRNA that is cut out of the mature mRNA molecule and is not expressed (used to form a protein).

In vitro fertilization A procedure in which eggs and sperm are placed together in a dish in the laboratory. If fertilization occurs, early-stage embryos are then transferred to the woman's uterus, where it is hoped they will implant and complete development. It is a common treatment for infertility resulting from blocked oviducts.

Ion An atom or group of atoms that carries an electric charge resulting from the loss or gain of electrons.

Ion channel A protein-lined pore or channel through a plasma membrane through which one type or a few types of ions can pass. Nerve cell ion channels are important in the generation and propagation of nerve impulses.

Ionic bond A chemical bond that results from the mutual attraction of oppositely charged ions.

Iris The colored portion of the eye. The iris regulates the amount of light that enters the eye.

Iron-deficiency anemia A reduction in the ability of the blood to carry oxygen due to an insufficient amount of iron in the diet, an inability to absorb iron from the digestive system, or blood loss.

Ischemia A temporary reduction in blood supply caused by obstructed blood flow. It causes reversible damage to heart muscle.

Islets of Langerhans See pancreatic islets.

Isotonic solution A solution with the same concentration of solutes as plasma, for example.

Isotopes Atoms that have the same number of protons but different numbers of neutrons.

Jaundice A condition in which the skin develops a yellow tone caused by the buildup of bilirubin in the blood and its deposition in certain tissues such as the skin. It is an indication that the liver is not handling bilirubin adequately.

Joint A point of contact between two bones; an articulation.

J-shaped growth curve A graphical representation of the growth of a population growing exponentially (unrestrained growth at a constant rate).

Junctional complexes Membrane specializations that attach adjacent cells to each other to form a contiguous sheet. There are three kinds of junctions between cells: tight junctions, desmosomes, and gap junctions.

Juxtaglomerular apparatus The region of the kidney nephron where the distal convoluted tubule contacts the afferent arteriole bringing blood into the glomerulus. Cells in this area secrete renin, an enzyme that triggers events eventually leading to increased reabsorption of sodium and water by the distal convoluted tubules and collecting ducts of nephrons.

Kaposi's sarcoma A cancer that forms tumors in connective tissue and manifests as pink or purple marks on the skin. It is common in people with a suppressed immune system, such as people living with HIV/AIDS, and is thought to be associated with a new virus in the herpes family, HHV8.

Karyotype The arrangement of chromosomes based on physical characteristics such as length and location of the centromere. Karyotypes can be checked for defects in number or structure of chromosomes.

Keratinocytes Cells of the epidermis that undergo keratinization, the process in which keratin gradually replaces the contents of maturing cells.

Ketoacidosis A lowering of blood pH resulting from the accumulation of breakdown by-products of fats.

Kidneys Reddish, kidney bean–shaped organs responsible for filtering wastes and excess materials from the blood, assisting the respiratory system in the regulation of blood pH, and maintaining fluid balance by regulating the volume and composition of blood and urine.

Kidney stones Small, hard crystals formed in the urinary tract when substances such as calcium (the most common constituent), uric acid, or magnesium ammonium phosphate precipitate from urine as a result of higher than normal concentrations. They are also called renal calculi.

Klinefelter syndrome A genetic condition resulting from nondisjunction of the sex chromosomes in which a person inherits an extra X chromosome that results in an XXY genotype. The person has a male appearance.

Labia majora One of two elongated skin folds lateral to the labia minora. They are part of the female external genitalia.

Labia minora One of two small skin folds on either side of the vagina and interior to the labia majora. They are part of the external genitalia of a female.

Labor The process by which the fetus is expelled from the uterus and moved through the vagina and into the outside world. During true labor, uterine contractions occur at regular intervals, are often painful, and intensify with walking. Labor is usually divided into the dilation stage, expulsion stage, and placental stage. False labor is characterized by irregular contractions that fail to intensify or change with walking.

Lactation The production and ejection of milk from the mammary glands. The hormone prolactin from the anterior pituitary gland promotes milk production, and the hormone oxytocin released from the posterior pituitary gland makes milk available to the suckling infant by stimulating milk ejection, or let-down.

Lactic acid fermentation The process by which glucose is broken down by muscle cells when oxygen is low during strenuous exercise.

Lacuna (plural, lacunae) A tiny cavity. It contains osteocytes (bone cells) in the matrix of bone and cartilage cells in the matrix of cartilage.

Lanugo Soft, fine hair that covers the fetus beginning about the third or fourth month after conception.

Large intestine The final segment of the gastrointestinal tract, consisting of the cecum, colon, rectum, and anal canal. The large intestine helps absorb water, forms feces, and plays a role in defecation.

Laryngitis An inflammation of the larynx in which the vocal cords become swollen and can no longer vibrate and produce sound.

Larynx The voice box or Adam's apple. A boxlike cartilaginous structure between the pharynx and the trachea held together by muscles and elastic tissue.

Latent period Pertaining to muscle contraction, the interval between the reception of the stimulus and the beginning of muscle contraction.

Latin binomial The two-part scientific name that consists of the genus name followed by the species name.

Leukemia A cancer of the blood-forming organs that causes white blood cell numbers to increase. The white blood cells are abnormal and do not effectively defend the body against infectious agents.

Leukocytes White blood cells. They are cells of the blood including neutrophils, eosinophils, basophils, monocytes, B lymphocytes, and T lymphocytes. Leukocytes are involved in body defense mechanisms and the removal of wastes, toxins, or damaged, abnormal cells.

LH See luteinizing hormone.

Life expectancy The average number of years a newborn is expected to live.

Ligament A strong band of connective tissue that holds the bones together, supports the joint, and directs the movement of the bones.

Limbic system A collective term for several structures involved in emotions and memory.

Linkage The tendency for a group of genes located on the same chromosome to be inherited together.

Lipid A compound, such as a triglyceride, phospholipid, or steroid, that does not dissolve in water.

Liver A large organ that functions mainly in the production of plasma proteins, the excretion of bile, the storage of energy reserves, the detoxification of poisons, and the interconversion of nutrients.

Locus The point on a chromosome where a particular gene is found.

Logistic growth A pattern in which population growth begins rapidly but levels off as population size reaches the carrying capacity.

Longitudinal fissure A deep groove that separates the cerebrum into two hemispheres.

Long-term memory Memory that stores a large amount of information for hours, days, or years.

Loop of Henle A section of the renal tubule that resembles a hairpin turn. Reabsorption occurs here.

Loose connective tissue Connective tissue, such as areolar and adipose tissue, that contains many cells in which the fibers of the matrix are fewer in number and more loosely woven than those found in dense connective tissue.

Lower esophageal sphincter A ring of muscle at the juncture of the esophagus and the stomach that controls the flow of materials between the esophagus and the stomach. It relaxes to allow food into the stomach and contracts to prevent too much food from moving back into the esophagus.

Lumen The hollow cavity or channel of a tubule through which transported material flows.

Luteinizing hormone (LH) A hormone secreted by the anterior pituitary gland that in females stimulates ovulation and the formation of the corpus luteum (which produces estrogen and progesterone) and prepares the mammary glands for milk production. In males, it stimulates testosterone production by the interstitial cells within the testes.

Lymph The fluid within the vessels of the lymphatic system. It is derived from the fluid that bathes the cells of the body (interstitial fluid).

Lymphatic system A body system consisting of lymph, lymphatic vessels, and lymphatic tissue and organs. The lymphatic system helps return interstitial fluid to the blood, transports the products of fat digestion from the digestive system to the blood, and assists in body defenses.

Lymphatic vessels The vessels through which lymph flows. A network of vessels that drains interstitial fluid and returns it to the blood supply and transports the products of fat digestion from the digestive system to the blood supply.

Lymph nodes Small nodular organs found along lymph vessels that filter lymph. The lymph nodes contain macrophages and lymphocytes, cells that play an essential role in the body's defense system.

Lymphocyte A type of white blood cell important in nonspecific and specific (immune) body defenses. The lymphocytes include B lymphocytes (B cells) that transform into plasma cells and produce antibodies and T lymphocytes (T cells) that are important in defense against foreign or infected cells.

Lymphoid organs Various organs that belong to the lymphatic system, including the tonsils, spleen, thymus, and Peyer's patches.

Lymphoid tissue The type of tissue that predominates in the lymphoid organs except the thymus. The organs in which lymphoid tissue is found are the lymph nodes, tonsils, and spleen. Lymphoid tissue is an important component of the immune system.

Lymphoma Cancer of the lymphoid tissues. In people with AIDS, it commonly affects the B cells.

Lysosomal storage diseases Disorders such as Tay-Sachs disease caused by the absence of lysosomal enzymes. In these disorders, molecules that would normally be degraded by the missing enzymes accumulate in the lysosomes and interfere with cell functioning.

Lysosome An organelle that serves as the principal site of digestion within the cell.

Lysozyme An enzyme present in tears, saliva, and certain other body fluids that kills bacteria by disrupting their cell walls.

Macroevolution Large-scale evolutionary changes such as those that might result from long-term changes in the climate or position of the continents. Examples include mass extinctions and the evolution of mammals.

Macromolecule A giant molecule of life such as a nucleic acid, protein, or polysaccharide. A macromolecule is formed by the joining together of smaller molecules.

Macrophage A large phagocytic cell derived from a monocyte that lives in loose connective tissue and engulfs anything detected as foreign.

Magnetic resonance imaging (MRI) A means of visualizing a region of the body, including the brain, in which the picture results from differences in the way the hydrogen nuclei in the water molecules within the tissues vibrate in response to a magnetic field created around the area to be pictured.

Malignant tumor A cancerous tumor. An abnormal mass of tissue that can invade surrounding tissue and spread to multiple locations throughout the body.

Malleus The first of three small bones in the middle ear that transmit information about sound from the eardrum to the inner ear. The malleus is also known as the hammer.

Malnourishment A form of hunger that occurs when the diet is not balanced and the right foods are not eaten.

Mammary glands The milk-producing glands in the breasts.

Mass The measure of how much matter is in an object.

Mast cells Small, mobile connective tissue cells often found near blood vessels. In response to injury, mast cells release histamine, which dilates blood vessels and increases blood flow to an area, and heparin, which prevents blood clotting.

Matter Anything that takes up space and has mass. Matter is made up of atoms.

Mechanical digestion A part of the digestive process that involves physically breaking food into smaller pieces.

Mechanoreceptor A sensory receptor that is specialized to respond to distortions in the receptor itself or in nearby cells. Mechanoreceptors are responsible for the sensations we describe as touch, pressure, hearing, and equilibrium.

Medulla oblongata The part of the brain stem containing reflex centers for some of life's most vital physiological functions: the pace of the basic breathing rhythm, the force and rate of heart contraction, and blood pressure. It connects the spinal cord to the rest of the brain.

Medullary cavity The cavity in the shaft of long bones that is filled with yellow marrow.

Medullary rhythmicity center The region of the brain stem controlling the basic rhythm of breathing.

Meiosis A type of cell division that occurs in the gonads and gives rise to gametes. As a result of two divisions (meiosis I and meiosis II), haploid gametes are produced from diploid germ cells.

Meissner's corpuscles Encapsulated nerve cell endings that are common on the hairless, sensitive areas of the skin, such as the lips, nipples, and fingertips, and tell us exactly where we have been touched.

Melanin A pigment produced by the melanocytes of the skin. There are two forms of melanin: black-to-brown and yellow-to-red.

Melanocytes Spider-shaped cells located at the base of the epidermis that manufacture and store melanin, a pigment involved in skin color and absorption of ultraviolet radiation.

Melanocyte-stimulating hormone (MSH) The anterior pituitary hormone that stimulates melanin production.

Melanoma The least common and most dangerous form of skin cancer. It arises in the melanocytes, the pigment-producing cells of the skin.

Melatonin A hormone secreted by the pineal gland that may influence daily rhythms, fertility, and aging.

Membranous epithelium A type of epithelial tissue that forms linings and coverings. Depending on its location, it may be specialized to protect, secrete, and absorb.

Memory cell A lymphocyte (B cell or T cell) of the immune system that forms in response to an antigen and that circulates for a long period of time; such cells are able to mount a quick immune response to a subsequent exposure to the same antigen.

Meninges Three protective connective tissue membranes that surround the central nervous system: the dura mater, the pia mater, and the arachnoid.

Meningitis An inflammation of the meninges (protective coverings of the brain and spinal cord).

Menopause The end of a female's reproductive potential when ovulation and menstruation cease.

Menstrual cycle   The sequence of events that occurs on an approximately 28-day cycle in the uterine lining (endometrium) that involves the thickening of and increased blood supply to the endometrium and the loss of the endometrium as menstrual flow. It is also called the uterine cycle.

Merkel cells   Cells of the epidermis found in association with sensory neurons where the epidermis meets the dermis.

Merkel disk   The Merkel cell–neuron combination that functions as a sensory receptor for light touch, providing information about objects contacting the skin. It is found on both the hairy and the hairless parts of the skin.

Mesoderm   The primary germ layer that gives rise to muscle, bone, connective tissue, and organs such as the heart, kidneys, ovaries, and testes.

Messenger RNA (mRNA)   A type of RNA synthesized from and complementary to a region of DNA that attaches to ribosomes in the cytoplasm and specifies the amino acid order in the protein. It carries the DNA instructions for synthesizing a particular protein.

Metabolism   The sum of all chemical reactions within the living cells.

Metaphase   In mitosis, the phase when the chromosomes, guided by the fibers of the mitotic spindle, form a line at the center of the cell. As a result of this alignment, when the chromosomes split at the centromere, each daughter cell receives one chromatid from each chromosome and thus a complete set of the parent cell's chromosomes.

Metastasize   To spread from one part of the body to another part not directly connected to the first part. Cancerous tumors metastasize and form new tumors in distant parts of the body.

MHC markers   Molecules on the surface of body cells that label the cell as "self."

Microevolution   Changes in populations at the genetic level. The causes include genetic drift, gene flow, natural selection, and mutation.

Microfilament   A component of the cytoskeleton made from the globular protein actin. Microfilaments form contractile units in muscle cells and are responsible for amoeboid movement.

Microtubule   A component of the cytoskeleton made from the globular protein tubulin. Microtubules are responsible for the movement of cilia and flagella and serve as tracks for the movement of organelles and vesicles.

Microvilli   Microscopic cytoplasm-filled extensions of the cell membrane that serve to increase the absorptive surface area of the cell.

Middle ear   An air-filled space in the temporal bone that includes the tympanic membrane (eardrum) and three small bones (the hammer, anvil, and stapes). It serves as an amplifier of sound pressure waves.

Mineralocorticoids   Hormones secreted by the adrenal cortex that affect mineral homeostasis and water balance.

Minipill   A birth control pill that contains only progesterone.

Mitochondrion (plural, mitochondria)   An organelle within which most of cellular respiration occurs in a eukaryotic cell. Cellular respiration is the process by which oxygen and an organic fuel such as glucose are consumed and energy is released and used to form ATP.

Mitosis   A type of cell division occurring in somatic cells in which two identical cells, called daughter cells, are generated from a single cell. The original cell first replicates its genetic material and then distributes a complete set of genetic information to each of its daughter cells. Mitosis is usually divided into prophase, metaphase, anaphase, and telophase.

Mitral valve   A heart valve located between the left atrium and left ventricle that prevents the backflow of blood from the ventricle to the atrium. It is also called the bicuspid valve or the left atrioventricular (AV) valve.

Molecular clock   The idea that there is a constant rate of divergence of macromolecules (such as proteins) from one another. It is based on the notion that single nucleotide changes in DNA (point mutations) and the amino acid changes in proteins that can be produced by some point mutations occur with steady, clocklike regularity. It permits comparison of molecular sequences to estimate the time of separation between species (the more differences in sequence, the more time that has elapsed since the common ancestor).

Molecule   A chemical structure composed of atoms held together by covalent bonds.

Monoclonal antibodies   Defensive proteins specific for a particular antigen secreted by a clone of genetically identical cells descended from a single cell.

Monocyte   The largest white blood cell. Monocytes are active in fighting chronic infections, viruses, and intracellular bacterial infections. A monocyte can transform into a phagocytic macrophage.

Monohybrid cross   A genetic cross that considers the inheritance of a single trait from individuals differing in the expression of that trait.

Monomer   A small molecule that joins with identical molecules to form a polymer.

Mononucleosis   A viral disease caused by the Epstein-Barr virus that results in an elevated level of monocytes in the blood. It causes fatigue and swollen glands, and there is no available treatment.

Monosaccharide   The smallest molecular unit of a carbohydrate. Monosaccharides are known as simple sugars.

Monosomy   A condition in which there is only one representative of a chromosome instead of two representatives.

Mons veneris   A round fleshy prominence over the pubic bone in a female. Part of the female external genitalia.

Morning sickness   The nausea and vomiting experienced by some women early in pregnancy. It is not restricted to the morning and may be caused, in part, by high levels of the hormone human chorionic gonadotropin (HCG).

Morphogenesis   The development of body form that begins during the third week after fertilization.

Morula   A solid ball of 12 or more cells produced by successive divisions of the zygote. The name reflects its resemblance to the fruit of the mulberry tree.

Mosaic evolution   The phenomenon whereby various traits evolve at their own rates.

Motor (efferent) neuron   A neuron specialized to carry information *away from* the central nervous system to an effector, either a muscle or a gland.

Motor unit   A motor neuron and all the muscle fibers (cells) it stimulates.

MRI   See magnetic resonance imaging.

mRNA   See messenger RNA.

Mucosa   The innermost layer of the gastrointestinal tract. It secretes mucus that helps lubricate the tube, allowing food to slide through easily.

Mucous membranes   Sheets of epithelial tissue that line many passageways in the body that open to the exterior. Mucous membranes are specialized to secrete and absorb.

Mucus   A sticky secretion that serves to lubricate body parts and trap particles of dirt and other secretions. It also helps protect the stomach from the action of gastric juice.

Multiple alleles   Three or more alleles of a particular gene existing in a population. The alleles governing the ABO blood types provide an example.

Multiple sclerosis   An autoimmune disease in which the body's own defense mechanisms attack myelin sheaths in the nervous system. As a patch of myelin is repaired, a hardened region called a sclerosis forms.

Multipolar neuron   A neuron that has at least three processes, including an axon and a minimum of two dendrites. Most motor neurons and association neurons are multipolar.

Multiregional hypothesis   The idea that modern humans evolved independently in several different areas from distinctive local populations of *Homo erectus*. Compare with Out of Africa hypothesis.

Muscle   Tissue composed of muscle cells that contract when stimulated and passively lengthen to the resting state. There are three types of muscle tissue: skeletal, smooth, and cardiac.

Muscle fiber   A muscle cell.

Muscle spindles   Specialized muscle fibers with sensory nerve cell endings wrapped around them that report to the brain whenever a muscle is stretched.

Muscle twitch   Contraction of a muscle in response to a single stimulus.

Muscularis   The muscular layers of the gastrointestinal tract. These layers help move food through the gastrointestinal system and mix food with digestive secretions.

Mutation A change in the base sequence of the DNA of a gene. A mutation may occur spontaneously or be caused by outside sources, such as radiation or chemical agents. Mutations are the only source of new characteristics in a population.

Myelin sheath An insulating layer around axons that carry nerve impulses over relatively long distances that is composed of multiple wrappings of the plasma membrane of certain glial cells. Outside the brain and spinal cord, Schwann cells form the myelin sheath. The myelin sheath greatly increases the speed at which impulses travel and assists in the repair of damaged axons. The Schwann cells that form the myelin sheath are separated from one another by short regions of exposed axon called nodes of Ranvier.

Myocardial infarction A heart attack. A condition in which a part of the heart muscle dies because of an insufficient blood supply.

Myocardium Cardiac muscle tissue that makes up the bulk of the heart. The contractility of the myocardium is responsible for the heart's pumping action.

Myofibril A rodlike bundle of contractile proteins (myofilaments) found in skeletal and cardiac muscle cells essential to muscle contraction.

Myofilament A contractile protein within muscle cells. There are two types: myosin (thick) filaments and actin (thin) filaments.

Myoglobin An oxygen-binding pigment in muscle fibers.

Myometrium The smooth muscle layer in the wall of the uterus.

Myosin filaments The thick filaments in muscle cells composed of the protein myosin and essential to muscle contraction. A myosin molecule is shaped like a golf club with two heads.

Myosin heads Club-shaped ends of a myosin molecule that bind to actin filaments and can swivel, causing actin filaments to slide past the myosin filaments, which causes muscle contraction. They are also called cross-bridges.

Myxedema A condition characterized by swelling of the facial tissues because of the accumulation of interstitial fluids caused by undersecretion of thyroid hormone.

Nasal cavities Two chambers in the nose, separated by the septum.

Nasal conchae The three convoluted bones within each nasal cavity that increase surface area and direct airflow.

Nasal septum A thin partition of cartilage and bone that divides the inside of the nose into two nasal cavities.

Natural childbirth Childbirth without the use of pain-killing drugs.

Natural killer (NK) cells A type of cell in the immune system, probably lymphocytes, that roam the body in search of abnormal cells and quickly kill them.

Natural selection The process by which some individuals live longer and produce more offspring than other individuals because their particular inherited characteristics make them better suited to their local environment.

Nearsightedness Myopia. Nearsightedness is a visual condition in which nearby objects can be seen more clearly than distant objects. Nearsightedness occurs because either the eyeball is elongated or the lens is too thick, causing the image to be focused in front of the retina.

Negative feedback mechanism The homeostatic mechanism in which the outcome of a process feeds back on the system, shutting down the process.

Nephrons Functional units of the kidneys responsible for the formation of urine. These microscopic tubules number 1 million to 2 million per kidney and perform filtration (only certain substances are allowed to pass out of the blood and into the nephron), reabsorption (some useful substances are returned from the nephron to the blood), and secretion (the nephron directly removes wastes and excess materials in the blood and adds them to the filtered fluid that becomes urine).

Nerve A bundle of parallel axons, dendrites, or both from many neurons. A nerve is usually covered with tough connective tissue.

Nervous tissue Tissue consisting of two types of cells, neurons and neuroglia, that make up the brain, spinal cord, and nerves.

Neural tube The embryonic structure that gives rise to the brain and spinal cord.

Neuroglial cells Cells of the nervous system that support, insulate, and protect nerve cells; also called glial cells.

Neuromuscular junction The area of contact between the terminal end of a motor neuron and the sarcolemma of a skeletal muscle fiber. When an action potential reaches the terminal end of the motor neuron, acetylcholine is released, triggering events that can lead to muscle contraction.

Neurons Nerve cells involved in intercellular communication. A neuron consists of a cell body, dendrites, and an axon. Neurons are excitable cells in the nervous system specialized to generate and transmit electrochemical signals called action potentials or nerve impulses.

Neurotransmitter A chemical released from the axon tip of a neuron that affects the activity of another cell (usually a nerve, muscle, or gland cell) by altering the electrical potential difference across the membrane of the receiving cell.

Neurula The embryo during neurulation, formation of the brain and spinal cord from ectoderm.

Neurulation A series of events during embryonic development when the central nervous system (brain and spinal cord) forms from ectoderm. During this period, the embryo is called a neurula.

Neutrophil A phagocytic white blood cell important in defense against bacteria and removal of cellular debris. Most abundant of white blood cells.

New World monkeys A group of monkeys whose members are arboreal, have nostrils that open to the side, and have a grasping (prehensile) tail. They occur in South and Central America and include spider monkeys, capuchins, and squirrel monkeys.

Niche The role of a species in an ecosystem. The niche includes the habitat, food, nest sites, and so on. It describes how a member of a particular species uses materials in the environment and how it interacts with other organisms.

Nitrification The conversion of ammonia to nitrate (NO_3^-) by nitrifying bacteria living in the soil.

Nitrogen fixation The process of converting nitrogen gas to ammonium (a nitrogen-containing molecule that can be used by living organisms). The process of nitrogen fixation is carried out by nitrogen-fixing bacteria living in nodules on the roots of leguminous plants such as peas and alfalfa.

Node of Ranvier A region of exposed axon between Schwann cells forming a myelin sheath. In myelinated nerves, the impulse jumps from one node of Ranvier to the next, greatly increasing the speed of conduction. This type of transmission is called saltatory conduction.

Nondisjunction Failure of the members of a pair of homologous chromosomes or the sister chromatids to separate during mitosis or meiosis. Nondisjunction results in cells with an abnormal number of chromosomes.

Norepinephrine Noradrenaline. A hormone secreted by the adrenal medulla, along with epinephrine, in response to stress. They initiate the physiological "fight-or-flight" reaction. Norepinephrine is also a neurotransmitter found in both the central and peripheral nervous systems. In the central nervous system, it is important in the regulation of mood, in the pleasure system of the brain, arousal, and dreaming sleep. Norepinephrine is thought to produce an energizing "good" feeling. It is also thought to be essential in hunger, thirst, and sex drive.

Notochord The flexible rod of tissue that develops during gastrulation and signals where the vertebral column will form. The notochord defines the axis of the embryo, gives the embryo some rigidity, and prompts overlying ectoderm to begin formation of the central nervous system. During development, vertebrae form around the notochord. The notochord eventually degenerates, existing only as the pulpy, elastic material in the center of intervertebral disks.

Nuclear envelope The double membrane that surrounds the nucleus.

Nuclear pore An opening in the nuclear envelope that permits communication between the nucleus and the cytoplasm.

Nucleolus A specialized region within the nucleus that forms and disassembles during the course of the cell cycle. It plays a role in the generation of ribosomes, organelles involved in protein synthesis.

Nucleoplasm The chromatin and the aqueous environment within the nucleus.

Nucleotide　A subunit of DNA composed of one five-carbon sugar (either ribose or deoxyribose), one phosphate group, and one of five nitrogenous bases. Nucleotides are the building blocks of nucleic acids (DNA and RNA).

Nucleus　The command center of the cell containing almost all the genetic information.

Oil glands　Glands associated with the hair follicles that produce sebum. They are also called sebaceous glands.

Old World monkeys　A group of monkeys whose members lack a prehensile tail, often having a short tail or none at all, and have nostrils that open downward. Some Old World monkeys are arboreal and some are terrestrial. Examples include baboons and rhesus monkeys.

Omnivore　An organism that feeds on a variety of food types, such as plants and animals.

Oocyte　A cell whose meiotic divisions will produce an ovum and up to three polar bodies.

Oogenesis　The production of ova (eggs), including meiosis and maturation.

Oogonium　(plural, oogonia) A germ cell in an ovary that divides, giving rise to oocytes.

Opsin　One of several proteins that can be bound to retinal to form the visual pigments in rods and cones.

Optic nerve　One of two nerves, one from each eye, responsible for bringing processed electrochemical messages from the retina to the brain for interpretation.

Organ　A structure composed of two or more different tissues with a specific function.

Organelle　A specialized membrane-bound compartment within a eukaryotic cell.

Organ of Corti　The portion of the cochlea in the inner ear that contains receptor cells that sense vibrations caused by sound. It is most directly responsible for the sense of hearing.

Organ system　A group of organs with a common function.

Origin　In reference to a muscle, the end of the muscle that is attached to the bone that remains relatively stationary during a movement.

Osmosis　A special case of diffusion in which water moves across the plasma membrane or any other selectively permeable membrane from a region of lower concentration of solute to a region of higher concentration of solute.

Osteoarthritis　An inflammation in a joint that is caused by degeneration of the surfaces of the joint caused by wear and tear.

Osteoblast　A bone-forming cell.

Osteoclast　A large cell responsible for the breakdown and absorption of bone.

Osteocytes　Mature bone cells found in lacunae that are arranged in concentric rings around the central canal. Cytoplasmic projections from osteocytes extend tiny channels that connect with other osteocytes.

Osteon　The structural unit of compact bone that appears as a series of concentric circles of lacunae. The lacunae contain bone cells around a central canal containing blood vessels and nerves.

Osteoporosis　A decrease in bone density that occurs when the destruction of bone outpaces the formation of new bone, causing bone to become thin, brittle, and susceptible to fracture.

Otoliths　Granules of calcium carbonate embedded in gelatinous material in the utricle and saccule of the ear. Otoliths cause the gelatin to slide over and bend sensory hair cells when the head is moved. The bending generates nerve impulses that are sent to the brain for interpretation as the position of the head.

Out of Africa hypothesis　The idea that modern humans evolved from *Homo erectus* in Africa and later migrated to Europe, Asia, and Australia. It suggests a single origin for *Homo sapiens*. Compare with multiregional hypothesis.

Outer ear　The external appendage on the outside of the head (pinna) and the canal (the external auditory canal) that extends to the eardrum. It functions as the receiver for sound vibrations.

Oval window　A membrane-covered opening in the inner ear (cochlea) through which vibrations from the stirrup (stapes) are transmitted to fluid within the cochlea.

Ovarian cycle　The sequence of events in the ovary that lead to ovulation. The cycle is approximately 28 days long and is closely coordinated with the menstrual cycle.

Ovaries　The female gonads. The female reproductive organs that produce the ova (eggs) and the hormones estrogen and progesterone.

Oviduct　One of two tubes that conduct the ova from the ovary to the uterus in the female reproductive system. It is also called a uterine tube or a fallopian tube.

Ovulation　The release of the secondary oocyte from the ovary.

Ovum　A mature egg; a large haploid cell that is the female gamete.

Oxidation　The loss of electrons from one atom or molecule to another.

Oxygen debt　The amount of oxygen required after exercise to oxidize the lactic acid formed during exercise.

Oxyhemoglobin　Hemoglobin bound to oxygen.

Oxytocin (OT)　The hormone released at the posterior pituitary that stimulates uterine contractions and milk ejection.

Pacemaker　See sinoatrial (SA) node.

Pacinian corpuscle　A large encapsulated nerve cell ending that is located deep within the skin and near body organs that responds when pressure is first applied. It is important in sensing vibration.

Pain receptor　A sensory receptor that is specialized to detect the physical or chemical damage to tissues that we sense as pain. Pain receptors are sometimes classified as chemoreceptors, because they often respond to chemicals liberated by damaged tissue, and occasionally as mechanoreceptors, because they are stimulated by physical changes, such as swelling, in the damaged tissue.

Palate　The roof of the mouth. The front region of the palate, the hard palate, is reinforced with bone. The tongue pushes against the hard palate while mixing food with saliva. The soft palate is farther to the back of the mouth and consists of only muscle. The soft palate prevents food from entering the nose during swallowing.

Paleoanthropology　The scientific study of human origins and evolution.

Pancreas　An accessory organ behind the stomach that secretes digestive enzymes, bicarbonate ions to neutralize the acid in chyme, and hormones that regulate blood sugar.

Pancreatic islets　Small clusters of endocrine cells in the pancreas; also called islets of Langerhans.

Parasympathetic nervous system　The branch of the autonomic nervous system that is active during restful conditions. Its effects generally oppose those of the sympathetic nervous system. The parasympathetic nervous system adjusts bodily functions so that energy is conserved during nonstressful times.

Parathormone　See parathyroid hormone.

Parathyroid glands　Four small, round masses at the back of the thyroid gland that secrete parathyroid hormone (parathormone).

Parathyroid hormone (PTH)　A hormone released from the parathyroid glands that regulates blood calcium levels by stimulating osteoclasts to break down bone. PTH, also called parathormone, is secreted when blood calcium levels are too low.

Parental generation　The parents—the individuals in the earliest generation under consideration.

Parkinson's disease　A progressive disorder that results from the death of dopamine-producing neurons that lie in the heart of the brain's movement control center, the substantia nigra. Parkinson's disease is characterized by slowed movements, tremors, and muscle rigidity.

Parturition　Birth, which usually occurs about 38 weeks after fertilization.

Passive immunity　Temporary immune resistance that develops when a person receives antibodies that were produced by another person or animal.

Pathogen　A disease-causing organism.

Pectoral girdle　The bones that connect the arms to the rib cage. The pectoral girdle is composed of the shoulder blades (scapulae) and the collarbones (clavicles).

Pedigree　A chart showing the genetic connections among individuals in an extended family that is often used to trace the expression of a particular trait in that family.

Pelvic girdle   The bones that connect the legs to the vertebral column. The pelvic girdle is composed of the paired hipbones.

Pelvic inflammatory disease (PID)   A general term for any bacterial infection of a woman's pelvic organs. PID is usually caused by sexually transmitted bacteria, especially those that cause chlamydia and gonorrhea.

Penis   The cylindrical external reproductive organ of a male through which most of the urethra extends and that serves to deliver sperm into the female tract during sexual intercourse.

Pepsin   A protein-splitting enzyme initially secreted in the stomach in the inactive form of pepsinogen that is activated into pepsin by hydrochloric acid.

Peptic ulcer   A local defect in the surface of the stomach or small intestine characterized by dead tissue and inflammation.

Peptide   A chain containing only a few amino acids.

Perforin   A type of protein released by a natural killer cell that creates numerous pores (holes) in the target cell, making it leaky. Fluid is then drawn into the leaky cell because of the high salt concentration within, and the cell bursts.

Pericardium   The fibrous sac enclosing the heart that holds the heart in the center of the chest without hampering its movements.

Perichondrium   The layer of dense connective tissue surrounding cartilage that contains blood vessels, which supply the cartilage with nutrients.

Periodontitis   An inflammation of the gums and the tissues around the teeth.

Periosteum   The membranous covering that nourishes bone.

Peripheral nervous system   The part of the nervous system outside the brain and spinal cord. It keeps the central nervous system in continuous contact with almost every part of the body. It is composed of nerves and ganglia. The two branches are the somatic and the autonomic nervous systems.

Peripheral proteins   Proteins attached to the inner or outer surface of the plasma membrane.

Peristalsis   Rhythmic waves of muscular contraction and relaxation in the walls of hollow tubular organs, such as the digestive organs, that serve to push contents through the tubes.

Peroxisome   An enzyme-containing organelle within which chemical reactions occur that generate hydrogen peroxide. Such reactions have diverse functions ranging from the breakdown of fats to the detoxification of poisons and alcohol.

PET scan   See positron emission tomography.

Phagocytes   Scavenger cells specialized to engulf and destroy particulate matter, such as pathogens, damaged tissue, or dead cells.

Phagocytosis   The process by which cells such as white blood cells ingest foreign cells or substances by surrounding the foreign material with cell membrane. It is a type of endocytosis.

Pharynx   The space shared by the respiratory and digestive systems that is commonly called the throat. The pharynx is a passageway for air, food, and liquid.

Phenotype   The observable physical and physiological traits of an individual. Phenotype results from the inherited alleles and their interactions with the environment.

Phospholipid   An important component of cell membranes. It has a nonpolar "water-fearing" tail (made up of fatty acids) and a polar "water-loving" head (containing an R group, glycerol, and phosphate).

Photoreceptor   A sensory receptor specialized to detect changes in light intensity. Photoreceptors are responsible for the sensation we describe as vision.

pH scale   A scale for measuring the concentration of hydrogen ions. The scale ranges from 0 to 14, with a pH of 7 being neutral, a pH of less than 7 being acidic, and a pH of greater than 7, basic.

Phylogenetic trees   Generalized descriptions of the history of life. They depict hypotheses about evolutionary relationships among species or higher taxa.

Physical dependence   A condition in which continued use of a drug is needed to maintain normal cell functioning.

Pia mater   The innermost layer of the meninges (the connective tissue layers that protect the central nervous system).

Piloerection   Contraction of the arrector pili muscles causing hairs to stand on end and form a layer of insulation.

Pineal gland   The gland that produces the hormone melatonin and is located at the center of the brain.

Pinealocyte   A cell of the pineal gland that secretes melatonin.

Pinna   The visible part of the ear on each side of the head that gathers sound and channels it to the external auditory canal.

Pinocytosis   A type of endocytosis in which droplets of fluid and the dissolved substances therein are engulfed by cells.

Pituitary dwarfism   A condition caused by insufficient growth hormone during childhood.

Pituitary gland   The endocrine organ connected to the hypothalamus by a short stalk. It consists of the anterior and posterior lobes and is also called the hypophysis.

Pituitary portal system   The system in which a capillary bed in the hypothalamus connects to veins that lead into a capillary bed in the anterior lobe of the pituitary gland. It allows hormones of the hypothalamus to control the secretion of hormones from the anterior pituitary gland.

Placebo   In a controlled experiment to test the effectiveness of a drug, it is a substance that appears to be identical to a drug but has no known effect on the condition for which it is taken.

Placenta   The organ that delivers oxygen and nutrients to the embryo and carries carbon dioxide and wastes away from it. The placenta is also called the afterbirth.

Placental stage   The third (and final) stage of true labor. It begins with delivery of the baby and ends when the placenta detaches from the wall of the uterus and is expelled from the mother's body.

Placenta previa   The condition in which the placenta forms in the lower half of the uterus, entirely or partially covering the cervix. It may cause premature birth or maternal hemorrhage and usually makes vaginal delivery impossible.

Plaque   A bumpy layer consisting of smooth muscle cells filled with lipid material, especially cholesterol, that bulges into the channel of an artery and reduces blood flow. Another type of plaque is a buildup of food material and bacteria on teeth that leads to tooth decay.

Plasma   A straw-colored liquid that makes up about 55% of blood. It serves as the medium for transporting materials within the blood. Plasma consists of water (91%–93%) with substances dissolved in it (7%–9%).

Plasma cell   The effector cell, produced from a B lymphocyte, that secretes antibodies.

Plasma membrane   The thin outer boundary of a cell that controls the movement of substances into and out of the cell.

Plasma proteins   Proteins dissolved in plasma, including albumins, which are important in water balance between cells and the blood; globulins, which are important in transporting various substances in the blood; and antibodies, which are important in the immune response.

Plasmid   A small, circular piece of self-replicating DNA that is separate from the chromosome and found in bacteria. Plasmids are often used as vectors in recombinant DNA research.

Plasmin   An enzyme that breaks down fibrin and dissolves blood clots. Plasmin is formed from plasminogen.

Plasminogen   A plasma protein. It is the inactive precursor of plasmin.

Platelet   A cell fragment of a megakaryocyte that releases substances necessary for blood clotting. It is formed in the red bone marrow from a precursor cell called a megakaryocyte. It is sometimes called a thrombocyte.

Platelet plug   A mass of platelets clinging to the protein fiber collagen at a damaged blood vessel to prevent blood loss.

Pleiotropy   One gene having many effects.

PMS   See premenstrual syndrome.

Point mutation   A mutation that involves changes in one or a few nucleotides in DNA.

Polar body   Any of three small nonfunctional cells produced during the meiotic divisions of an oocyte. The divisions also produce a mature ovum (egg).

Polygenic inheritance   Inheritance in which several independently assorting or loosely linked genes determine the expression of a trait.

Polymer   A large molecule formed by the joining together of many smaller molecules of the same general type (monomers).

Polymerase chain reaction (PCR)   A technique used to amplify (increase) the quantity of DNA in vitro using primers, DNA polymerase, and nucleotides.

Polypeptide A chain containing 10 or more amino acids.

Polysaccharide A complex carbohydrate formed when large numbers of monosaccharides (most commonly glucose) join together to form a long chain through dehydration synthesis. Most polysaccharides store energy or provide structure.

Polysome A cluster of ribosomes simultaneously translating the same messenger RNA (mRNA) strand.

Polyspermy The abnormal condition in which more than one sperm enters an oocyte. It is usually prevented by a rapid depolarization of the plasma membrane of the oocyte as well as by the release of enzymes by the oocyte that make the zona pellucida impermeable to other sperm.

Pons A part of the brain that connects upper and lower levels of the brain.

Population A group of potentially interacting individuals of the same species living in a distinct geographic area.

Population dynamics Changes in population size over time.

Portal system A system whereby a capillary bed drains to veins that drain to another capillary bed.

Positive feedback mechanism The mechanism by which the outcome of a process feeds back on the system, further stimulating the process.

Positron emission tomography (PET) A method that can be used to measure the activity of various brain regions. The person being scanned is injected with a radioactively labeled nutrient, usually glucose, that is tracked as it flows through the brain. The radioisotope emits positively charged particles, called positrons. When the positrons collide with electrons in the body, gamma rays are released. The gamma rays can be detected and recorded by PET receptors. Computers then use the information to construct a PET scan that shows where the radioisotope is being used in the brain.

Postnatal period The period of development after birth. It includes the stages of infancy, childhood, puberty, adolescence, and adulthood.

Postsynaptic neuron The neuron located after the synapse. The receiving neuron in a synapse. The membrane of the postsynaptic neuron has receptors specific for certain neurotransmitters.

Precapillary sphincter A ringlike muscle that acts as a valve that opens and closes a capillary bed. Contraction of the precapillary sphincter squeezes the capillary shut and directs blood through a thoroughfare channel to the venule. Relaxation of the precapillary sphincter allows blood to flow through the capillary bed.

Pre-embryo The developing human from fertilization through the second week of gestation (the pre-embryonic period).

Pre-embryonic period The period during prenatal development that extends from fertilization through the second week. Cleavage and implantation follow fertilization.

Prefrontal cortex An association area of the cerebral cortex that is important in decision making.

Premature infant A baby born before 37 weeks of gestation.

Premenstrual syndrome (PMS) A collection of uncomfortable symptoms, including irritability, stress, and bloating, that appears 7 to 10 days before a woman's menstrual period and is associated with hormonal cycling.

Prenatal period The period of development before birth. It is further subdivided into the pre-embryonic period (from fertilization through the second week), the embryonic period (from the third through the eighth weeks), and the fetal period (from the ninth week until birth).

Presynaptic neuron The neuron located before the synapse. The sending neuron in a synapse. The neuron in a synapse that releases neurotransmitter from its synaptic knobs into the synaptic cleft.

Primary consumer An herbivore. An animal that eats primary producers (green plants or algae).

Primary follicle A spherical structure in the ovary that contains a primary oocyte and is surrounded by follicle cells.

Primary germ layers The layers produced by gastrulation from which all tissues and organs form. They are ectoderm, mesoderm, and endoderm.

Primary motor area A band of the frontal lobe of the cerebral cortex that initiates messages that direct voluntary movements.

Primary productivity In ecosystems, the gross primary productivity is the amount of light energy that is converted to chemical energy in the bonds of organic molecules during a given period. The net primary productivity is the amount of productivity left after the photosynthesizers have used some of the energy stored in organic molecules for their own metabolic activities.

Primary response The immune response that occurs during the body's first encounter with a particular antigen.

Primary somatosensory area A band of the parietal lobe of the cerebral cortex to which information is sent from receptors in the skin regarding touch, temperature, and pain and from receptors in the joints and skeletal muscles.

Primary spermatocyte The original large cell that develops from a spermatogonium during sperm development in the seminiferous tubules. It undergoes meiosis and gives rise to secondary spermatocytes.

Primary structure The precise sequence of amino acids of a protein. This sequence, determined by the genes, dictates a protein's structure and function.

Primary succession The sequence of changes in the species making up a community over time that begins in an area where no community previously existed and ends with a climax community.

Primordial germ cells Cells that migrate from the yolk sac of the developing human to the ovaries or testes, where they differentiate into immature cells that will eventually become sperm or oocytes.

Probe A single-stranded molecule of radioactive DNA or RNA with a sequence of bases complementary to the sequence in the gene of interest that is used to locate the desirable gene.

Prodrome The symptoms that precede recurring outbreaks of a disease such as genital herpes.

Producers In ecosystems, the producers are the organisms that convert energy from the physical environment into chemical energy in the bonds of organic molecules through photosynthesis or chemosynthesis. The producers form the first trophic level.

Progesterone A female sex hormone produced by the corpus luteum in the ovary. Progesterone helps prepare the endometrium (lining) of the uterus for pregnancy and maintains the endometrium.

Programmed cell death A genetically programmed series of events that causes a cell to self-destruct.

Prokaryotic cell A cell that lacks a nucleus and other membrane-enclosed organelles. The prokaryotes include bacteria and archaea.

Prolactin (PRL) An anterior pituitary hormone that stimulates the mammary glands to produce milk.

Promoter A specific region on DNA next to the "start" gene that controls the expression of the gene.

Prophase In mitosis, the phase when the chromosomes begin to thicken and shorten, the nucleolus disappears, the nuclear envelope begins to break down, and the mitotic spindle forms in the cytoplasm.

Prosimians The suborder of modern primates that includes the lemurs, lorises, pottos, and tarsiers.

Prostaglandins The lipid molecules found in and released by the plasma membranes of most cells. They are often called the local hormones because they exert their effects on the secreting cells themselves or on nearby cells.

Prostate gland An accessory reproductive gland in males that surrounds the urethra as it passes from the bladder. Its secretions contribute to semen and serve to activate the sperm and to counteract the acidity of the female reproductive tract.

Proteins The macromolecules composed of amino acids linked by peptide bonds. The functions of proteins include structural support, transport, movement, and regulation of chemical reactions.

Prothrombin A plasma protein synthesized by the liver that is important in blood clotting. It is converted to an active form (thrombin) by thromboplastin that is released from platelets.

Prothrombin activator A blood protein that converts prothrombin to thrombin as part of the blood-clotting process.

Protozoans A group of single-celled organisms with a well-defined eukaryotic nucleus. Protozoans can cause disease by producing toxins or by releasing enzymes that prevent host cells from functioning normally.

Proximal convoluted tubule The section of the renal tubule where reabsorption and secretion occur.

Pseudopodia The temporary extensions of a cell formed by microfilaments and used when feeding and moving. The name means "false foot."

Puberty The stage in postnatal development when secondary sexual characteristics such as pubic and underarm hair develop. This stage usually occurs slightly earlier in girls (from 12 to 15 years of age) than in boys (from 13 to 16 years of age).

Pulmonary arteries Blood vessels that carry blood low in oxygen from the right ventricle to the lungs, where it is oxygenated.

Pulmonary circuit (or circulation) The pathway that transports blood from the right ventricle of the heart to the lungs and back to the left atrium of the heart.

Pulmonary veins Blood vessels that carry oxygenated blood from the lungs to the left atrium of the heart.

Pulp The center of a tooth that contains the tooth's life-support systems.

Pulse The rhythmic expansion of an artery created by the surge of blood pushed along the artery by each contraction of the ventricles of the heart. With each beat of the heart, the wave of expansion begins, moving along the artery at the rate of 6 to 9 meters per second.

Punnett square A diagrammatic method used to determine the probable outcome of a genetic cross. The possible allele combinations in the gametes of one parent are used to label the columns and the possible allele combinations of the other parent are used to label the rows. The alleles of each column and each row are then paired to determine the possible genotypes of the offspring.

Pupil The small hole through the center of the iris through which light passes to enter the eye. The size of the pupil is altered to regulate the amount of light entering the eye.

Purkinje fibers The specialized cardiac muscle cells that deliver an electrical signal from the atrioventricular bundle to the individual heart muscle cells in the ventricles.

Pyelonephritis An infection of the kidneys caused by bacteria.

Pyloric sphincter A ring of muscle between the stomach and small intestine that regulates the emptying of the stomach into the small intestine.

Pyramid of biomass A diagram in which blocks represent the amount of biomass (dry body mass of organisms) available at each trophic (feeding) level.

Pyramid of energy A graphical representation in which blocks represent the decreasing amount of energy available at each trophic (feeding) level.

Pyramid of numbers A graphical representation of the number of individuals at each trophic level.

Pyrogen A fever-producing substance.

Quaternary structure The shape of an aggregate protein. It is determined by the mutually attractive forces between the protein's subunits.

Radioisotopes Isotopes that are unstable and spontaneously decay, emitting radiation in the form of gamma rays and alpha and beta particles.

Receptor A protein molecule located in the cytoplasm and on the plasma membrane of cells that is sensitive to chemical messengers.

Receptor potential An electrochemical message (a change in the degree of polarization of the membrane) generated in a sensory receptor in response to a stimulus. Receptor potentials vary in magnitude with the strength of the stimulus.

Recessive allele The allele whose effects are masked in the heterozygous condition. The recessive allele usually produces a nonfunctional protein.

Recombinant DNA Segments of DNA from two sources that have been combined in vitro and transferred to cells in which their information can be expressed.

Rectum The final section of the gastrointestinal tract. The rectum receives and temporarily stores the feces.

Red blood cell See erythrocyte.

Red marrow Blood cell–forming connective tissue found in the marrow cavity of certain bones.

Reduction The gain of electrons by one atom or molecule from another.

Reduction division The first meiotic division (meiosis I) that produces two cells, each of which contains one member of each homologous pair (23 chromosomes with replicates attached in humans).

Referred pain Pain felt at a site other than the area of origin.

Reflex A simple, stereotyped reaction to a stimulus.

Reflex arc A neural pathway consisting of a sensory receptor, a sensory neuron, usually at least one interneuron, a motor neuron, and an effector.

Refractory period The interval following an action potential during which a neuron cannot be stimulated to generate another action potential.

Relaxin The hormone released from the placenta and the ovaries. It initiates labor and facilitates delivery by dilating the cervix and relaxing the ligaments and cartilage of the pubic bones.

Releasing hormone A hormone that stimulates hormone secretion by another gland.

Renal corpuscle The portion of the nephron where fluid is filtered. It consists of a tuft of capillaries, the glomerulus, and a surrounding cuplike structure called Bowman's (glomerular) capsule.

Renal cortex The outer region of the kidney, containing renal columns.

Renal failure A decrease or complete cessation of glomerular filtration.

Renal medulla The inner region of the kidney. It contains cone-shaped structures called renal pyramids.

Renal pelvis The innermost region of the kidney; the chamber within the kidney.

Renal tubule The site of reabsorption and secretion by the nephron. It consists of the proximal convoluted tubule, the loop of Henle, and the distal convoluted tubule.

Renin An enzyme released by cells of the juxtaglomerular apparatus of nephrons. Renin converts angiotensinogen, a protein produced by the liver and found in the plasma, into another protein, angiotensin I. These actions of renin initiate a series of hormonal events that leads to increased reabsorption of sodium and water by the distal convoluted tubules and collecting ducts of nephrons.

Rennin The gastric enzyme that breaks down milk proteins.

Replication Copying from a template, as occurs during the synthesis of new DNA from preexisting DNA.

Repolarization The return of the membrane potential to approximately its resting value. Repolarization of the nerve cell membrane during an action potential occurs because of the outflow of potassium ions.

Residual volume The amount of air that remains in the lungs after a maximal exhalation.

Resistance phase The phase of the general adaptation syndrome that follows the alarm reaction. It is characterized by anterior pituitary release of adrenocorticotropic hormone (ACTH), thyroid-stimulating hormone (TSH), and growth hormone (GH) and the sustainment of metabolic demands by the body's fat reserves.

Respiratory distress syndrome (RDS) A condition in newborns caused by an insufficient amount of surfactant, causing the alveoli of the lungs to collapse and thereby making breathing difficult.

Respiratory system The organ system that carries out gas exchange. The respiratory system includes the nose, pharynx, larynx, trachea, bronchi, and lungs.

Resting potential The separation of charge across the plasma membrane of a neuron when the neuron is not transmitting an action potential. A neuron's resting potential is the electrical potential difference across its plasma membrane when the neuron is not conducting an impulse. It is primarily caused by the unequal distributions of sodium ions, potassium ions, and large negatively charged proteins on either side of the plasma membrane. The resting potential of a neuron is about -70 mV.

Restriction enzyme An enzyme that recognizes a specific sequence of bases in DNA and cuts the DNA into two strands at that sequence. Restriction enzymes are used to prepare DNA containing "sticky ends" during the creation of recombinant DNA. Their natural function in bacteria is to control the replication of viruses that infect the bacteria.

Reticular activating system (RAS) An extensive network of neurons that runs through the medulla and projects to the cerebral cortex. It filters out unimportant sensory information before it reaches the brain and controls changing levels of consciousness.

Reticular fibers Interconnected strands of collagen in certain connective tissues that branch extensively. Networks of reticular fibers support soft tissues, including the liver and spleen.

Retina The light-sensitive innermost layer of the eye containing numerous photoreceptors (rods and cones).

Retinal The light-absorbing portion of pigment molecules in the photoreceptors. Retinal combines with one of four opsins (proteins) to form the light-absorbing pigments in rods and cones.

Retrovirus Any one of the viruses that contain only RNA and carry out transcription from RNA to DNA (reverse transcription).

Rheumatoid arthritis An inflammation of a joint caused by an autoimmune response. It is marked by inflammation of the synovial membrane and excess synovial fluid accumulation in the joints, causing swelling, pain, and stiffness.

Rhodopsin The light-absorbing pigment in the photoreceptors called rods. Rhodopsin is responsible for black and white vision.

Rhythm method of birth control A method of reducing the risk of pregnancy by avoiding intercourse on all days that might result in sperm and egg meeting.

Ribonucleic acid (RNA) A single-stranded nucleic acid that contains ribose (a five-carbon sugar), phosphate, adenine, uracil, cytosine, or guanine. RNA plays a variety of roles in protein synthesis.

Ribosomal RNA (rRNA) A type of RNA that combines with proteins to form the ribosomes, structures on which protein synthesis occurs. The most abundant form of RNA.

Ribosome The site where protein synthesis begins in a cell. It consists of two subunits, each containing ribosomal RNA and proteins.

Ribozyme An enzyme consisting of RNA that catalyzes reactions during RNA splicing. Ribozymes snip out introns before the protein is formed.

RNA See ribonucleic acid.

RNA polymerase One of the group of enzymes necessary for the synthesis of RNA from a DNA template. It binds with the promoter on DNA that aligns the appropriate RNA nucleotides and links them together.

Rod One of the photoreceptors containing rhodopsin and responsible for black and white vision. Rods are extremely sensitive to light.

Root The part of a tooth that is below the gum line. It is covered with a calcified, yet living and sensitive, connective tissue called cementum.

Root canal A channel through the root of a tooth that contains the blood vessels and nerves.

Rough endoplasmic reticulum (RER) Endoplasmic reticulum that is studded with ribosomes. It produces membrane.

Round window A membrane-covered opening in the cochlea that serves to relieve the pressure caused by the movements of the oval window.

rRNA See ribosomal RNA.

Rugae Folds in the mucosa of the lining of the empty stomach's walls that can unfold, allowing the stomach to expand as it fills.

Salinization An accumulation of salts in soil caused by irrigation over a long period of time that makes the land unusable.

Saliva The secretion from the salivary glands that helps moisten and dissolve food particles in the mouth, facilitating taste and digestion. An enzyme in saliva (salivary amylase) begins the chemical digestion of starch.

Salivary amylase An enzyme in saliva that begins the chemical digestion of starches, breaking them into shorter chains of sugars.

Salivary glands Exocrine glands in the facial region that secrete saliva into the mouth to begin the digestion process.

Saltatory conduction The type of nerve transmission along a myelinated axon in which the nerve impulse jumps from one node of Ranvier to the next. Saltatory conduction greatly increases the speed of nerve conduction.

Sarcomere The smallest contractile unit of a striated or cardiac muscle cell.

Sarcoplasmic reticulum An elaborate form of smooth endoplasmic reticulum found in muscle fibers. The sarcoplasmic reticulum takes up, stores, and releases calcium ions as needed in muscle contraction.

Schizophrenia A mental illness characterized by hallucinations and disordered thoughts and emotions that is caused by excessive activity at dopamine synapses in one part of the brain (the midbrain). As a result, dopamine is no longer in the proper balance with glutamate, a neurotransmitter released by neurons in another brain region (the cerebral cortex).

Schwann cell A type of glial cell in the peripheral nervous system that forms the myelin sheath by wrapping around the axon many times. The myelin sheath insulates axons, increases the speed at which impulses are conducted, and assists in the repair of damaged neurons.

Scientific method A procedure underlying most scientific investigations that involves observation, formulating a hypothesis, making predictions, experimenting to test the predictions, and drawing conclusions. Experimentation usually involves a control group and an experimental group that differ in one or very few factors (variables). New hypotheses may be generated from the results of experimentation.

Sclera The white part of the eye that protects and shapes the eyeball and serves as a site of attachment for muscles that move the eye.

Scrotum A loose fleshy sac containing the testes.

Seasonal affective disorder (SAD) A form of depression associated with winter months when overproduction of melatonin is triggered by short day length.

Sebum An oily substance made of fats, cholesterol, proteins, and salts secreted by the oil glands.

Secondary consumer A carnivore. An animal that obtains energy by eating other animals.

Secondary response The immune response during the body's second or subsequent exposures to a particular antigen. The secondary immune response is much quicker than is the primary response because memory cells specific for the antigen are present.

Secondary spermatocyte A haploid cell formed by meiotic division of a primary spermatocyte during sperm development in the seminiferous tubules.

Secondary structure The bending and folding of the chain of amino acids of a protein to produce shapes such as coils, spirals, and pleated sheets. These shapes form as a result of hydrogen bonding between different parts of the polypeptide chain.

Secondary succession The sequence of changes in the species making up a community over time that takes place after some disturbance destroys the existing life. Soil is already present.

Second messenger A molecule in the cytoplasm of a cell that is activated when a water-soluble hormone binds to a receptor on the surface of the cell. Second messengers influence the activity of enzymes and ultimately the activity of the cell to produce the effect of the hormone.

Secretin A hormone released by the small intestine that inhibits the secretion of gastric juice and stimulates the release of bicarbonate ions from the pancreas and the production of bile in the liver.

Segregation, law of A genetic principle that states that the alleles for each gene separate (segregate) during meiosis and gamete formation, so half of the gametes bear one allele and the other half bear the other allele.

Selectively permeable A characteristic of the plasma membrane because it permits some substances to move across and denies access to others.

Semen The fluid expelled from the penis during male orgasm. Semen consists of sperm and the secretions of the accessory glands.

Semicircular canals Three canals in each ear that are oriented at right angles to one another and contain sensory receptors that precisely monitor any sudden movement of the head. They detect body position and movement.

Semiconservative replication Replication of DNA in which the two strands of a DNA molecule become separated and each serves as a template for a new double-stranded DNA. Each new double-stranded molecule consists of one new strand and one old strand.

Semilunar valves Heart valves located between each ventricle and its connecting artery that prevent the backflow of blood from the artery to the ventricle. Whereas the cusps of the atrioventricular (AV) valves are flaps of connective tissue, those of the semilunar valves are small pockets of tissue attached to the inner wall of their respective arteries.

Seminal vesicles A pair of male accessory reproductive glands located posterior to the urinary bladder. Their secretions contribute to semen and serve to nourish the sperm cells, reduce the acidity in the vagina, and coagulate sperm.

Seminiferous tubules Coiled tubules within the testes where sperm are produced.

Sensory adaptation A gradual decline in the responsiveness of a sensory receptor that results in a decrease in awareness of the stimulus.

Sensory (afferent) neuron A nerve cell specialized to conduct nerve impulses from the sensory receptors *toward* the central nervous system.

Sensory receptors The structures specialized to respond to changes in their environment (stimuli) by generating electrochemical messages that are eventually converted to nerve impulses if the stimulus is strong enough. The nerve impulses are then conducted to the brain, where they are interpreted to build our perception of the world.

Serosa A thin layer of connective tissue that forms the outer layer of the gastrointestinal tract. It secretes a fluid that reduces friction with contacting surfaces.

Serotonin A neurotransmitter in the central nervous system thought to promote a generalized feeling of well-being.

Serous membranes Sheets of epithelial tissue that line the thoracic and abdominal cavities and the organs within them. Serous membranes secrete a fluid that lubricates the organs within these cavities.

Sex chromosomes The X and Y chromosomes. The pair of chromosomes involved in determining gender.

Sex-influenced inheritance An autosomal genetic trait that is expressed differently in males and females, usually because of the presence of sex hormones.

Sexual dimorphism A difference in appearance between males and females within a species.

Short-term memory Memory of new information that lasts for a few seconds or minutes.

Sickle-cell anemia A type of anemia caused by a mutation that results in a change in one amino acid in a globin chain of hemoglobin (the iron-containing protein in red blood cells that transports oxygen). Such a change causes the red blood cell to assume a crescent (sickle) shape when oxygen levels are low. The sickle-shaped cells clog small blood vessels, leading to pain and tissue damage from insufficient oxygen.

Simple diffusion The spontaneous movement of a substance from a region of higher concentration to a region of lower concentration.

Simple epithelial tissue Epithelial tissue with only a single layer of cells.

Simple goiter An enlarged thyroid gland caused by iodine deficiency.

Sinoatrial (SA) node A region of specialized cardiac muscle cells located in the right atrium near the junction of the superior vena cava that sets the pace of the heart rate at about 70 to 80 beats a minute. It is also known as the pacemaker. The SA node sends out an electrical signal that spreads through the muscle cells of the atria, causing them to contract.

Sinuses Large, air-filled spaces in the bones of the face.

Sinusitis Inflammation of the mucous membranes of the sinuses making it difficult for the sinuses to drain their mucous fluid.

Skeletal muscle A contractile tissue. One of three types of muscle in the body. Skeletal muscle cells are cylindrical, have many nuclei, and have stripes (striations). Skeletal muscle provides for conscious, voluntary control over contraction. It attaches to bones and forms the muscles of the body. It is also called striated muscle.

Skeleton A framework of bones and cartilage that functions to support and protect internal organs and to permit body movement.

Sliding filament model A model of the mechanism of muscle contraction in which the myofilaments actin and myosin slide across one another, causing the sarcomere to shorten. When enough sarcomeres shorten, the muscle contracts.

Slow-twitch cells Muscle fibers that are specialized to contract slowly but with incredible endurance when stimulated. They contain an abundant supply of myoglobin and mitochondria and are richly supplied with capillaries. They depend on aerobic pathways to generate adenosine triphosphate (ATP) during muscle contraction.

Small intestine The organ located between the stomach and large intestine responsible for the final digestion and absorption of nutrients.

Smooth endoplasmic reticulum (SER) Endoplasmic reticulum without ribosomes. It produces membrane and detoxifies drugs.

Smooth muscle A contractile tissue characterized by the lack of visible striations and by unconscious control over its contraction. It is found in the walls of blood vessels and airways and in organs such as the stomach, intestines, and bladder.

Sodium-potassium pump A molecular mechanism in a plasma membrane that uses cellular energy in the form of adenosine triphosphate (ATP) to pump ions against their concentration gradients. Typically, each pump ejects three sodium ions from the cell while bringing in two potassium ions.

Soluble fiber A type of dietary fiber that either dissolves or swells in water. It includes the pectins, gums, mucilages, and some hemicelluloses. Soluble fiber has a gummy consistency.

Somatic cells All body cells except for gametes (egg and sperm). Somatic cells contain the diploid number of chromosomes, which in humans is 46.

Somatic nervous system The part of the peripheral nervous system that carries information to and from the central nervous system, resulting in voluntary movement and sensations.

Somites Blocks formed from mesoderm cells of the developing embryo that eventually form skeletal muscles of the neck and trunk, connective tissues, and vertebrae.

Source (donor) DNA DNA containing the "gene of interest" that will be combined with host DNA to form recombinant DNA.

Special senses The sensations of smell, taste, vision, hearing, and the sense of balance or equilibrium.

Speciation The formation of a new species.

Species A population or group of populations whose members are capable of successful interbreeding under natural conditions. Such interbreeding must produce fertile offspring.

Sperm A mature male gamete. A spermatozoon.

Spermatid A haploid cell that is formed by mitotic division of a haploid secondary spermatocyte and that develops into a spermatozoon.

Spermatocyte A cell developed from a spermatogonium during sperm development in the seminiferous tubules.

Spermatogenesis The series of events within the seminiferous tubules that gives rise to physically mature sperm from germ cells. It involves meiosis and maturation.

Spermatogonium (plural, spermatogonia) The undifferentiated male germ cells in the seminiferous tubules that give rise to spermatocytes.

Spermicides A means of contraception that consists of sperm-killing chemicals in some form of a carrier, such as foam, cream, jelly, film, or tablet.

Sphincter A ring of muscle between regions of a system of tubes that controls the flow of materials from one region to another past the sphincter.

Sphygmomanometer A device for measuring blood pressure. A sphygmomanometer consists of an inflatable cuff that wraps around the upper arm attached to a device that can measure the pressure within the cuff.

Spina bifida A birth defect in which the neural tube fails to develop and close properly. The mother's taking vitamins and folic acid before conception appears to reduce the chance of having a baby with spina bifida. Some cases can be improved with surgery.

Spinal cord A tube of neural tissue that is continuous with the medulla at the base of the brain and extends about 45 cm (17 in.) to just below the last rib. It conducts messages between the brain and the rest of the body and serves as a reflex center.

Spinal nerves Thirty-one pairs of nerves that arise from the spinal cord. Each spinal nerve services a specific region of the body. Spinal nerves carry both sensory and motor information.

Spongy bone The bone formed from a latticework of thin struts of bone with marrow-filled areas between the struts. It is found in the ends of long bones and within the breastbone, pelvis, and bones of the skull. Spongy bone is less dense than compact bone and is made of an irregular network of collagen fibers surrounded by a calcium matrix.

Sprain Damage to a ligament (a strap of connective tissue that holds bones together).

Squamous cell carcinoma The second most common form of skin cancer that arises in the newly formed skin cells as they flatten and move toward the skin surface.

Squamous epithelium A type of epithelial tissue composed of flattened cells. It forms linings and coverings.

S-shaped growth curve A graphical representation of logistic growth (growth that levels off as the population size approaches the carrying capacity of the environment).

Stapes The last of three small bones in the middle ear that transmit information about sound from the eardrum to the inner ear. The stapes is also known as the stirrup.

Stem cell A type of cell that can give rise to other types of cells.

Steroid A lipid, such as cholesterol, consisting of four carbon rings with functional groups attached.

Steroid hormones A group of closely related hormones chemically derived from cholesterol and secreted primarily by the ovaries, testes, and adrenal glands.

Stomach A muscular sac that is well designed for the storage of food, the liquefaction of food, and the initial chemical digestion of proteins.

Stratified epithelial tissue Epithelial tissue with several layers of cells.

Strep throat A sore throat that is caused by *Streptococcus* bacteria.

Stress incontinence A mild form of incontinence characterized by the escape of small amounts of urine when sudden increases in abdominal pressure force urine past the external urethral sphincter.

Striated muscle See skeletal muscle.

Stroke A cerebrovascular accident. A condition in which nerve cells die because the blood supply to a region of the brain is shut off, usually because of hemorrhage or atherosclerosis. The extent and location of the mental or physical impairment caused by a stroke depend on the region of the brain involved.

Succession, ecological The sequence of changes in the species making up a community over time.

Summation (of muscle contraction) A phenomenon that results when a muscle is stimulated to contract before it has time to completely relax from a previous contraction. The response to each stimulation builds on the previous one.

Superovulation The ovulation of several oocytes. It is usually prompted by administration of hormones.

Suppressor T cell A type of T lymphocyte that turns off the immune response when the level of antigen falls by releasing chemicals that dampen the activity of both B cells and T cells.

Suprachiasmatic nucleus (SCN) The region of the brain in the hypothalamus where the "master clock" controlling biological rhythms is believed to reside.

Surface area-to-volume ratio The physical relationship that dictates that increases in the volume of a cell occur faster than increases in its surface area. This relationship explains why most cells are small.

Surfactant Phospholipid molecules coating the alveolar surfaces that prevent the alveoli from collapsing.

Survivorship The number of individuals of a group born in the same year that is expected to be alive at any given age. It is one way to represent age-specific mortality.

Sweat glands The sweat glands of the skin. One type of sweat gland is functional throughout life and releases sweat onto the surface of the skin. Another type releases its secretions into hair follicles and becomes functional at puberty.

Sympathetic nervous system The branch of the autonomic nervous system responsible for the "fight-or-flight" responses that occur during stressful or emergency situations. Its effects are generally opposite to those of the parasympathetic nervous system.

Synapse The site of communication between a neuron and another cell, such as another neuron or a muscle cell.

Synapsis The physical association of homologous pairs of chromosomes that occurs during prophase I of meiosis. The term literally means "bringing together."

Synaptic cleft The gap between two cells forming a synapse, for example, two communicating nerve cells.

Synaptic knob A small bulblike swelling of an axon terminal that releases neurotransmitter. An axon terminal.

Synaptic vesicle A tiny membranous sac containing molecules of a neurotransmitter. Synaptic vesicles are located in the synaptic knobs of axon endings, and they release their contents when an action potential reaches the synaptic knob.

Synergistic muscles Two or more muscles that work together to cause movement in the same direction.

Synovial cavity A fluid-filled space surrounding a synovial joint formed by a double-layered capsule. The fluid within the cavity is called synovial fluid.

Synovial fluid A viscous, clear fluid within a synovial cavity that acts as both a shock absorber and a lubricant between the bones.

Synovial joint A freely movable joint. A synovial joint is surrounded by a fluid-filled cavity. This is the most abundant type of joint in the body.

Synovial membranes Membranes that line the cavities of freely movable joints and secrete a fluid that lubricates the joint.

Syphilis A sexually transmitted disease caused by the bacterium *Treponema pallidum*. If untreated, it can progress through three stages and cause death. The first stage is characterized by a painless craterlike bump called a chancre that forms at the site where the bacterium entered the body.

Systematic biology The discipline that deals with the naming, classification, and evolutionary relationships of organisms. It is also called systematics.

Systemic circuit (or circulation) The pathway of blood from the left ventricle of the heart to the cells of the body and back to the right atrium.

Systole Contraction of the heart. Atrial systole is contraction of the atria. Ventricular systole is contraction of the ventricles.

Systolic pressure The highest pressure in an artery during each heartbeat. The higher of the two numbers in a blood pressure reading. In a typical, healthy adult, the systolic pressure is about 120 mm Hg.

T cell See T lymphocyte.

T4 cell A helper T cell. The kind of T lymphocyte that serves as the main switch for the entire immune response by presenting the antigen to B cells and by secreting chemicals that stimulate other cells of the immune system.

T lymphocyte T cell. A type of white blood cell. Some T lymphocytes attack and destroy cells that are not recognized as belonging in the body, such as an infected cell or a cancerous cell.

Tanning The buildup of melanin in the skin in response to ultraviolet (UV) exposure.

Target cell A cell with receptors that recognize and bind a specific hormone.

Taste bud A structure consisting of receptors responsible for the sense of taste surrounded by supporting cells. Taste buds are located primarily on the surface epithelium and certain papillae of the tongue.

Taste hairs Microvilli that project into a pore at the tip of the taste bud that bear the receptors for certain chemicals found in food.

Taxon (plural, taxa) A group under consideration, such as a genus, family, or order.

Tectorial membrane A membrane that forms the roof of the organ of Corti (the actual organ of hearing). It projects over and is in contact with the sensory hair cells. Pressure waves caused by sound cause the sensory cells to push against the tectorial membrane and bend, resulting in nerve impulses that are carried to the brain by the auditory nerve.

Telomerase The enzyme that synthesizes telomeres.

Telomere A piece of DNA at the tips of chromosomes that protects the ends of the chromosomes.

Telophase In mitosis, the phase when nuclear envelopes form around the group of chromosomes at each pole, the mitotic spindle is disassembled, and nucleoli reappear. The chromosomes also become less condensed and more threadlike in appearance.

Temporomandibular joint (TMJ) syndrome A group of symptoms including headaches, toothaches, and earaches caused by physical stress on the mandibular joint.

Tendinitis Inflammation of a tendon caused by excessive stress on the tendon.

Tendon A band of connective tissue that connects muscle to bone.

Teratogens Environmental agents, such as drugs, chemicals, radiation, and certain viruses, that disrupt development.

Terminal hair Thick, strong hair found on the scalp, eyebrows, eyelashes, and after puberty, the underarms and pubic area.

Tertiary structure The three-dimensional shape of proteins formed by hydrogen, ionic, and covalent bonds between different side chains.

Testes The male gonads. The male reproductive organs that produce sperm and the hormone testosterone.

Testosterone A male sex hormone needed for sperm production and the maintenance of male reproductive structures. Testosterone is produced primarily by the interstitial cells of the testes.

Tetanus A smooth, sustained contraction of muscle caused when stimuli are delivered in such rapid succession that there is no time for muscle relaxation.

Thalamus A brain structure located below the cerebral hemispheres that is important in sensory experience, motor activity, stimulation of the cerebral cortex, and memory.

Theory A broad-ranging explanation for some aspect of the universe that is consistent with many observations and experiments.

Thermoreceptor A sensory receptor specialized to detect changes in temperature.

Threshold The degree to which the voltage difference across the plasma membrane of a neuron or other excitable cell must change to trigger an action potential.

Thrombin A plasma protein formed from prothrombin by thromboplastin that is important in blood clotting. It converts fibrinogen to fibrin, which forms a web that traps blood cells and forms the clot.

Thrombocyte See platelet.

Thrombus A stationary blood clot that forms in the blood vessels. A thrombus can block blood flow.

Thymopoietin A hormone produced by the thymus gland involved in the maturation of T lymphocytes.

Thymosin A hormone produced by the thymus gland involved in the maturation of T lymphocytes.

Thymus gland A gland located on the top of the heart that secretes the hormones thymopoietin and thymosin. It decreases in size as we age.

Thyroid gland The shield-shaped structure at the front of the neck that synthesizes and secretes thyroid hormone and calcitonin.

Thyroid hormone (TH) A hormone released by the thyroid gland that regulates blood pressure and the body's metabolic rate and production of heat. It also promotes normal development of several organ systems.

Thyroid-stimulating hormone (TSH) The anterior pituitary hormone that acts on the thyroid gland to stimulate the synthesis and release of thyroid hormones.

Tidal volume The amount of air inhaled or exhaled during a normal breath.

Tight junction A type of junction between cells in which the membranes of neighboring cells are attached, forming a seal to prevent fluid from flowing across the epithelium through the minute spaces between adjacent cells.

Tissue A group of cells that work together to perform a common function.

Tolerance A progressive decrease in the effectiveness of a drug with continued use.

Tongue The large skeletal muscle studded with taste buds that aids in speech and eating.

Total lung capacity The total volume of air contained in the lungs after the deepest possible breath. It is calculated by adding the residual volume to the vital capacity.

Trabecula (plural, trabeculae) A supporting bar or strand of spongy bone that forms an internal strut that braces the bone from within.

Trachea The tube that conducts air into the thoracic cavity toward the lungs. The trachea is reinforced with C-shaped rings of cartilage to prevent it from collapsing during inhalation and exhalation.

Trait A phenotypically expressed characteristic.

Transcription The process by which a complementary single-stranded messenger RNA (mRNA) molecule is formed from a single-stranded DNA template. As a result, the information in DNA is transferred to RNA.

Transfer RNA (tRNA) A type of RNA that binds to a specific amino acid and transports it to the appropriate region of messenger RNA (mRNA). Transfer RNA acts as an interpreter between the nucleic acid language of mRNA and the amino acid language of proteins.

Transgenic organism An organism that contains certain genes from another species that code for a desired trait. It can be created, for example, by injecting foreign DNA into an egg cell or an early embryo.

Transition reaction The phase of cellular respiration that follows glycolysis and involves pyruvate reacting with coenzyme A (CoA) in the matrix of the mitochondrion to form acetyl CoA. The acetyl CoA then enters the citric acid cycle.

Translation Protein synthesis. The process of converting the nucleotide language of messenger RNA (mRNA) into the amino acid language of a protein.

Transverse tubules T tubules. The tiny, cylindrical inpocketings of the muscle fiber's plasma membrane that carry nerve impulses to almost every sarcomere.

Tricuspid valve A heart valve located between the right atrium and right ventricle that prevents the backflow of blood from the ventricle to the atrium. It is also called the right atrioventricular (AV) valve.

Triglycerides The lipids composed of one molecule of glycerol and three fatty acids. They are known as fats when solid and oils when liquid.

Trisomy A condition in which there are three representatives of a chromosome instead of only two representatives.

tRNA See transfer RNA.

Trophic level The feeding level of one or more populations in a food web. The producers form the first trophic level. Herbivores, which eat the producers, form the second trophic level. Carnivores that eat herbivores form the third trophic level. Carnivores that eat other carnivores form the fourth trophic level.

Trophoblast A group of cells within the blastocyst that will give rise to the chorion, the extraembryonic membrane that will become part of the placenta.

Tropic hormone A hormone that influences another endocrine gland.

Tropomyosin A protein on the thin (actin) filaments in muscle cells that works with troponin to prevent actin and myosin from binding in the absence of calcium ions.

Troponin A protein on the thin (actin) filaments in muscle cells that works with tropomyosin to prevent actin and myosin from binding in the absence of calcium ions.

Tubal ligation A sterilization procedure in females in which each oviduct is cut and sealed to prevent sperm from reaching the eggs.

Tubal pregnancy An ectopic pregnancy in which the embryo implants in an oviduct. This is the most common type of ectopic pregnancy.

Tuberculosis (TB) A highly contagious disease caused by a rod-shaped bacterium, *Mycobacterium tuberculosis*. TB is spread by coughing.

Tubular reabsorption The process by which useful materials are removed from the filtrate within the renal tubule and returned to the blood.

Tubular secretion The process by which wastes and excess ions that escaped glomerular filtration are removed from the blood and added to the filtrate within the renal tubule.

Tumor A neoplasm. An abnormal growth of cells. A tumor forms from the new growth of tissue in which cell division is uncontrolled and progressive.

Turner syndrome A genetic condition resulting from nondisjunction of the sex chromosomes in which a person has 22 pairs of autosomes and a single, unmatched X chromosome (XO). The person has a female appearance.

Tympanic membrane The eardrum. A membrane that forms the outer boundary of the middle ear and that vibrates in response to sound waves. Vibrations of the tympanic membrane are transferred through the middle ear by three small bones (hammer, anvil, stirrup) to the inner ear, where hearing occurs when neural messages are generated in response to the pressure waves caused by the vibrations.

Type 1 diabetes mellitus An autoimmune disorder characterized by abnormally high glucose in the blood due to insufficient production of insulin by cells in the pancreas.

Type 2 diabetes mellitus A condition characterized by a decreased sensitivity to insulin caused by decreased numbers of insulin receptors on target cells.

Umbilical cord The ropelike connection between the embryo (and later the fetus) and the placenta. It consists of blood vessels (two umbilical arteries and one umbilical vein) and supporting connective tissue.

Undernourishment Starvation. A form of hunger that occurs when inadequate amounts of food are eaten.

Ureters Tubular organs that carry urine from the kidneys to the urinary bladder.

Urethra The muscular tube that transports urine from the floor of the urinary bladder to the outside of the body. In males, it also conducts sperm from the vas deferens out of the body through the penis.

Urethritis Inflammation of the urethra caused by bacteria.

Urinalysis An analysis of the volume, microorganism content, and physical and chemical properties of urine.

Urinary bladder The muscular organ that temporarily stores urine until it is excreted from the body.

Urinary incontinence Lack of voluntary control over urination. Incontinence is the norm for infants and children younger than 2 or 3 years old, because nervous connections to the external urethral sphincter are incompletely developed. In adults, incontinence may result from damage to the external sphincter (often caused, in men, by surgery on the prostate gland), disease of the urinary bladder, and spinal cord injuries that disrupt the pathways along which travel impulses related to conscious control of urination. In any age group, urinary tract infection can result in incontinence.

Urinary retention The failure to completely or normally expel urine from the bladder. This condition may result from lack of the sensation to urinate, as might occur temporarily after general anesthesia, or from contraction or obstruction of the urethra, a condition caused, in men, by enlargement of the prostate gland. Immediate treatment for retention usually involves use of a urinary catheter to drain urine from the bladder.

Urinary system The system that consists of two kidneys, two ureters, one urinary bladder, and one urethra. Its main function is to regulate the volume, pressure, and composition of the blood.

Urinary tract infection (UTI) An infection caused by bacteria in the urinary system. Most bacteria enter the urinary system by moving up the urethra from outside the body.

Urination The process, involving both involuntary and voluntary actions, by which the urinary bladder is emptied. It is also called voiding or micturition.

Urine The yellowish fluid produced by the kidneys. It contains wastes and excess materials removed from the blood. Urine produced by the kidneys travels down the ureters to the urinary bladder, where it is stored until being excreted from the body through the urethra.

Uterine cycle See menstrual cycle.

Uterus A hollow muscular organ in the female reproductive system in which the embryo implants and develops during pregnancy.

Vaccination A procedure that introduces a harmless form of the disease-causing organism into the body to stimulate immune responses against that antigen.

Vagina A muscular tube in the female reproductive system that extends from the uterus to the vulva and serves to receive the penis during sexual intercourse and as the birth canal.

Vaginitis An inflammation of the vagina.

Variable In a controlled experiment, the factor that differs between the control group and experimental groups.

Varicose veins Veins that have become stretched and distended because blood is prevented from flowing freely and so accumulates, or "pools," in the vein. A common cause of varicose veins is weak valves within the veins.

Vas deferens A tubule that conducts sperm from the epididymis to the urethra.

Vasectomy A sterilization procedure in men in which both of the vas deferens are cut and sealed to prevent sperm from leaving the man's body.

Vasoactive intestinal peptide (VIP) A hormone released by the small intestine into the bloodstream that triggers the small intestine to release intestinal juices.

Vasoconstriction A decrease in the diameter of blood vessels, commonly of the arterioles. Blood flow through the vessel is reduced, and blood pressure rises as a result of vasoconstriction.

Vasodilation An increase in the diameter of blood vessels, commonly of the arterioles. Blood flow through the vessels increases, and blood pressure decreases as a result of vasodilation.

Vasopressin See antidiuretic hormone.

Vector A biological carrier, usually a plasmid or a virus, that ferries the recombinant DNA to the host cell.

Vein A blood vessel formed by the merger of venules that transports blood back toward the heart. Veins have walls that are easily stretched, so they serve as blood reservoirs, holding up to 65% of the body's total blood supply.

Vellus hair Soft, fine hair that persists throughout life covering most of the body surface.

Vena cava One of two large veins that empty oxygen-depleted blood from the body to the right atrium of the heart. The superior vena cava delivers blood from regions above the heart. The inferior vena cava delivers blood from regions below the heart.

Ventilation In respiration, breathing.

Ventral nerve root The portion of a spinal nerve that arises from the front (anterior) side of the spinal cord and contains axons of motor neurons. It joins with the dorsal nerve root to form a single spinal nerve, which passes through the opening between the vertebrae.

Ventricles The two lower chambers of the heart that receive blood from the atria. The ventricles function as the main pumps of the heart. The right ventricle pumps blood to the lungs. The left ventricle pumps blood to body tissues.

Ventricular fibrillation Rapid, ineffective contractions of the ventricles of the heart, which render the ventricles useless as pumps and stop circulation.

Venule A small blood vessel that receives blood from the capillaries. Venules merge into larger vessels called veins. The exchange of materials between the blood and tissues across the walls of a venule is minimal.

Vertebral column The backbone. It is composed of 26 vertebrae (7 cervical, 12 thoracic, 5 lumbar, 1 sacrum, and 1 coccyx) and associated tissues. The spinal cord passes through a central canal within the vertebrae.

Vesicle A membrane-bound sac formed during endocytosis.

Vestibular apparatus A closed fluid-filled maze of chambers and canals within the inner ear that monitors the movement and position of the head and functions in the sense of balance.

Vestibule A space or cavity at the entrance to a canal. In the inner ear, the vestibule is a structure consisting of the utricle and saccule.

Villi (singular, villus) Small fingerlike projections on the small intestine wall that increase surface area for absorption.

Virus A minute infectious agent that consists of a nucleic acid encased in protein. A virus cannot replicate outside a living host cell.

Vital capacity The maximal amount of air that can be moved into and out of the lungs during forceful breathing.

Vitiligo A condition in which melanocytes disappear from areas of the skin, leaving white patches in their wake.

Vitreous humor The jellylike fluid filling the posterior cavity of the eye between the lens and the retina that helps to keep the eyeball from collapsing and holds the thin retina against the wall of the eye.

Vocal cords Folds of tissue in the larynx that vibrate when air passes through them, producing sound.

Water-soluble vitamin A vitamin that dissolves in water. Water-soluble vitamins include vitamin C and the various B vitamins.

Werner's syndrome A rare inherited disease characterized by premature aging.

White blood cells See leukocytes.

White matter Regions of the central nervous system that are white owing to the presence of myelinated nerve fibers. White matter is important in neural communication over distances.

X-linked genes Genes located on the X chromosome. Most X-linked genes have no corresponding allele on the Y chromosome and will be expressed in a male, and in a female if she is homozygous.

Yellow marrow A connective tissue found in the shaft of long bones that stores fat. It forms from red marrow, and, if the need arises, it can convert back to red marrow and form blood cells.

Yolk sac The extraembryonic membrane that is the primary source of nourishment for embryos in many species of vertebrates. In embryonic humans, however, the yolk sac does not provide nourishment and remains quite small (human embryos receive nutrients from the placenta). In humans, the yolk sac is a site of blood cell formation and contains cells, called primordial germ cells, that migrate to the gonads, where they differentiate into immature cells that will eventually become sperm or oocytes.

Zygote The diploid cell resulting from the joining of an egg nucleus and a sperm nucleus. The first cell of a new individual.

Zygote intrafallopian transfer (ZIFT) A procedure in which zygotes created by the union of egg and sperm in a dish in the laboratory are inserted into the woman's oviducts. Zygotes travel on their own from the oviducts to the uterus.

Illustration Credits

Chapter 1 CO.01 Last Refuge Limited; Adrian Warren/www.lastrefuge.co.uk **1.1** ©Gallo Images/CORBIS **1.2a** Lawrence Livermore/Photo Researchers, Inc. **1.2b** Andrew Syred/Photo Researchers, Inc. **1.2c** Larry J. Pierce/Getty Images Inc. **1.2d** Jacques Jangoux/Peter Arnold, Inc. **1.2e** JC Carton/Bruce Coleman Inc. **1.2f** Bob Krist/Corbis/Bettmann **1.2g** ©Kenneth Fink/Photo Researchers, Inc. **1.5** Last Refuge Limited **1.7** Frederick Atwood

Chapter 2 CO.02 Juan Silva Productions/Getty Images Inc.-Image Bank **2.3** Corbis/Bettmann **2.4a,b** Visuals Unlimited **2.5** Orlen N. Johnson **2.7a,b** Richard Megna/Fundamental Photographs **2.7c** Charles D. Winters/Photo Researchers, Inc. **2.7d** Yoav Levy/Phototake NYC **2.12** Robert Daly/Getty Images Inc.-Stone Allstock **2.Ba** Hulton Archive/Getty Images **2.Bb** Richard Megna/Fundamental Photographs **2.Ca** Simon Frasier/Science Photo Library/Photo Researchers, Inc. **2.Cb** F. Harvey Pough **2.13a** Spencer Grant/PhotoEdit **2.13b** Antonio Luiz Hamdan/Getty Images Inc.-Image Bank **2.13c** © Dorling Kindersley Media Library **2.13d** Andrew Syred/Getty Images Inc.-Stone Allstock **2.13e** Rachel Epstein/PhotoEdit **2.13f** Felicia Martinez/PhotoEdit **2.13g** Amy Etra/PhotoEdit **2.17c** Dorling Kindersley Media Library

Chapter 3 CO.03 Getty Images, Inc. **3.1a** Secchi-Lecaque-Roussel-UCLAF/CNRI/Science Photo Library/Photo Researchers, Inc. **3.1b** Visuals Unlimited **3.7a,b,c** © Dr. David Phillips/Visuals Unlimited **3.11b** Secchi-Lecaque-Roussel-UCLAF/CNRI/Science Photo Library/Photo Researchers, Inc. **3.12a** Adrian T. Sumner/Getty Images Inc.-Stone Allstock **3.12b** Brian Eyden/Science Photo Library/Photo Researchers, Inc. **3.13b** Dr. D. Spector/Peter Arnold, Inc. **3.14** R.Bollender-D.Fawcett/Visuals Unlimited **3.15b** Don W. Fawcett/Visuals Unlimited **3.17** Custom Medical Stock Photo, Inc. **3.A** Grant Heilman Photography, Inc **3.B** Visuals Unlimited **3.C** AP Wide World Photos **3.18b** Bill Longcore/Photo Researchers, Inc. **3.19b** David M. Phillips/Visuals Unlimited **3.20.b** © Dr. Gopal Murti/Science Photo Library/Photo Researchers, Inc. **3.21** Photo Researchers, Inc. **3.22** Yorgos Nikas/Getty Images Inc.-Stone Allstock **3.23** David M. Phillips/Visuals Unlimited **3.24** David M. Phillips/Visuals Unlimited **3.25a,b** From Afzelius et al. *Tissue & Cell* Vol. 27, p. 241, 1995 **3.26** Photo Lennart Nilsson/Albert Bonniers Forlag

Chapter 4 CO.04 Steve Mason/Getty Images, Inc.-Photodisc. **4.1a** Peter Arnold, Inc. **4.1b,c,e** Ed Reschke/Peter Arnold, Inc. **4.1d** G. W. Willis, MD/Biological Photo Service **4.1f** Courtesy of Gregory N. Fuller, M.D., Ph.D., Chief, Section of Neuropathology, M. D. Anderson Cancer Center, Houston, Texas. **4.2a** Ed Reschke **4.2b** G. W. Willis/BPS/Getty Images Inc.-Stone Allstock **4.2c** Peter Arnold, Inc. **4.2d** Dorling Kindersley Media Library/Spike Walker (Microworld Services) **4.2e** Custom Medical Stock Photo, Inc. **4.2f** Ed Reschke/Peter Arnold, Inc. **4.3a** © Eric Grave /Phototake **4.3b** Ed Reschke **4.3c** © Carolina Biological Supply/Phototake **4.4** Prof. P. Motta/Department of Anatomy, University "*La Sapienza*," Rome/Science Photo Library/Photo Researchers, Inc. **4.5a,c** Dennis Kunkel/Phototake NYC **4.5b** Visuals Unlimited **4.9** CMCD/Getty Images, Inc.- Photodisc. **4.10** Maury Tannen/The Image Works **4.11a** AP Wide World Photos **4.11b** Robert Van Der Hilst/Getty Images Inc.-Stone Allstock **4.Aa** Phototake NYC **4.Ab** Visuals Unlimited **4.Ac** © Dr. P. Marazzi/Photo Researchers, Inc.

Chapter 5 CO.05 Eric O'Connell/Getty Images, Inc.-Taxi **5.1** Manfred Kage/Peter Arnold, Inc. **5.2** Biophoto Associates/Photo Researchers, Inc. **5.3** Photo Researchers, Inc. **5.Ab** Dr. P. Marazzi/Photo Researchers, Inc. **5.Ac** Yoav Levy/Phototake NYC **5.8** Ralph T. Hutchings **5.13b** Phillip Dowell © Dorling Kindersley **5.16** Lawrence Livermore National Laboratory/Science Photo Library/Photo Researchers, Inc.

Chapter 6 CO.06 Corbis/Bettmann **6.3b** © Eric Grave/Phototake **6.3c** Photo Researchers, Inc. **6.8** Young-Jin Son, MCP Hahnemann University School of Medicine **6.11** G.W. Willis/BPS/Getty Images Inc.-Stone Allstock **6.12** Ron Chapple/Getty Images, Inc. – Taxi **6.A** AP Wide World Photos

Chapter 7 CO.07 David Young-Wolff/PhotoEdit **7.2** © Dr. Dennis Kunkel/Phototake **7.3** C. Raines/Visuals Unlimited **7.6** Dennis Kunkel/Phototake NYC **7.8** E.R. Lewis, T.E. Everhart, Y.Y. Zeevi/Visuals Unlimited **7.9** Corbis/Bettmann **7.10** Ron Sachs/Sygma/Corbis/Bettmann

Chapter 8 CO.08 Tom Grill/Corbis RF **8.3** Martin M. Rotker/Science Source/Photo Researchers, Inc. **8.4** Geoff Tomkinson/Science Photo Library/Photo Researchers, Inc. **8.5** Ralph T. Hutchings **8.6** Courtesy Marcus E. Raichle, M.D., Washington University Medical Center, from research based on S.E. Petersen et al., Positron emission tomographic studies of the cortical anatomy of single-word processing. *Nature* 331:585-589 (1988). **8.9c** Karl E. Deckart/Phototake NYC **8.13** AP Wide World Photos

Chapter 8a CO.08a Stuart McClymont/The Image Bank/Getty Images, Inc. **8a.5** George Steinmetz Photography **8a.6** Fetal Alcohol & Drug Unit (FAS)/Streissguth, A.P., Landesman-Dwyer, S., Martin, J.C., & Smith, D.W. (1980). Teratogenic effects of alcohol in humans and laboratory animals. *Science*, 209, 353-361 **8a.7** Dr. Miles Herkenham/National Institute of Health **8a.8** Greg Smith/AP Wide World Photos

Chapter 9 CO.09 Mark Ludak/The Image Works **9.4** Bruce Rowell/Masterfile Corporation **9.8** Omikron/Photo Researchers, Inc. **9.12** Dr. Goran Bredberg/Science Photo Library/Photo Researchers, Inc. **9.A** Scanning electron micrographs by Dr. Robert S. Preston and Prof. Joseph E. Hawkins, Kresge Hearing Research Institute, University of Michigan Medical School. **9.16d** Caroloina Biological/Visuals Unlimited

Chapter 10 CO.10 Alan Thornton/Getty Images Inc.-Stone Allstock **10.7** Corbis/Bettmann **10.8** ©Dr. William H. Daughaday, University of California/Irvine. *American Journal of Medicine* (20) 1956. With permission of Excerpta Medica Inc. **10.9** Daniel Margulies **10.12b** Ed Reschke/Peter Arnold, Inc. **10.13** John Paul Kay/Peter Arnold, Inc. **10.14** Dr. P. Marazzi/Photo Researchers, Inc. **10.15** Corbis/Bettmann **10.B** Will and Deni McIntyre/Photo Researchers, Inc. **10.18** Sharmyn McGraw **10.19b** Astrid & Hanns-Frieder/Science Photo Library/Photo Researchers, Inc. **10.Ca** Peter Arnold, Inc.

Chapter 11 CO.11 AP Wide World Photos **11.4** Andrew Syred/Getty Images Inc.-Stone Allstock **11.7b** J.C. Revy/Phototake NYC **11.9** Photo Researchers, Inc.

Chapter 12 CO.12 Jim Cummings/Getty Images, Inc. – Taxi **12.2** ©Dr. Richard Kessel & Dr. Randy Kardon/Tissues & Organs/Visuals Unlimited **12.3c** Ed Reschke/Peter Arnold, Inc. **12.5a** T. Kuwabara/Don W. Fawcett/Visuals Unlimited **12.6a** John D. Cunningham/Visuals Unlimited **12.7a** Joseph R. Siebert/Custom Medical Stock Photo, Inc. **12.8.(left)** CNRI/Science Photo Library/Photo Researchers, Inc. **12.8.(right)** Ralph T. Hutchings **12.9a** Will Willner/Biomedical Communications, Wake Forest University School of Medicine **12.9b** Visuals Unlimited **12.11b** Martin Dohrn/Royal College of Surgeons/Science Photo Library/Photo Researchers, Inc. **12.14a** Craig X. Sotres/Pearson Education/PH College **12.16a** Cabisco/Visuals Unlimited **12.16b** Visuals Unlimited **12.16c** Biophoto Associates/Photo Researchers, Inc. **12.19** Photo Lennart Nilsson/Albert Bonniers Forlag AB **12.21** © R. Umesh Chandran, TDR, WHO/Science Photo Library/Photo Researchers, Inc.

Chapter 13 CO.13 Aaron Haupt/Photo Researchers, Inc. **13.1** David M. Phillips/Photo Researchers, Inc. **13.4** Photo Lennart Nilsson/Albert Bonniers Forlag **13.5** Oliver Meckes & Nicole Ottawa/Photo Researchers, Inc. **13.7a,c.** Photo Lennart Nilsson/Albert Bonniers Forlag AB, THE BODY VICTORIOUS **13.7b** Dr. Robert Dourmashkin **13.9** Ken Fisher/Getty Images Inc.-Stone Allstock **13.17a** James Stvenson/Science Photo Library/Photo Researchers, Inc. **13.17b** CNRI Science Photo Library/Photo Researchers, Inc. **13.18a** Oliver Meckes/Photo Researchers, Inc. **13.18b** ©Oliver Meckes/Ottawa/Photo Researchers, Inc.

Chapter 13a CO.13a Eamonn McNulty/Science Photo Library/Photo Researchers, Inc. **13a.1a** Oviver Meckes/Photo Researchers, Inc. **13a.1b** Manfred Kage/Peter Arnold, Inc. **13a.1c** Tina Carvalho/Visuals Unlimited **13a.2a** CNRI/Phototake **13a.4** (top) Terje Rakke/Getty Images Inc.-Image Bank **13a.4.(bottom)** Dr. Fred Hossler/Visuals Unlimited **13a.5** Dr. Daniel H. Connor/Armed Forces Institute of Pathology **13a.6** Stephen Krasemann/Getty Images Inc.-Stone Allstock **13a.7** Grapes/Michaud/Photo Researchers, Inc. **13a.8.(left)** © R. Calentine/Visuals Unlimited **13a.8** (right) James Marshall/The Image Works **13a.9** Martin Dohrn/Science Photo Library/Photo Researchers, Inc. **13a.11** Shizuo Kambayashi/AP Wide World Photos **13a.12** Visuals Unlimited **13a.13** Courtesy of World Health Organization

Chapter 14 CO.14 Paula Bronstein/Allsport Concepts/Getty Images **14.4** Susumu Nishingaga/Science Photo Library/Photo Researchers, Inc. **14.5** (left and right) CNRI/Photo Researchers, Inc. **14.7** Ralph T. Hutchings **14.8a** Oliver Meckes/Photo Researchers, Inc. **14.8b** © Dr. Richard Kessel & Dr. Randy Kardon/Tissues & Organs/Visuals Unlimited **14.9a,b** Peter Arnold, Inc. **14.10** SIU/Peter Arnold, Inc. **14.16a,b** Dr. Andrew Evan, Indiana University Medical Center, Department of Anatomy

Chapter 14a CO.14a James Leynse/Corbis/Bettmann **14a.3a** CNRI/Science Photo Library/Photo Researchers, Inc. **14a.3b** Dr. Andrew Evan, Indiana University Medical Center, Department of Anatomy **14a.5** Photo Researchers, Inc.

Chapter 15 CO.15 Bill Losh/Photographer's Choice/Getty Images, Inc. **15.3c,d** Dr. Tony Brain/Science Photo Library/Photo Researchers, Inc. **15.4** Chris Bjornberg/Photo Researchers, Inc. **15.8** Dr. E. Walker/Science Photo Library/Photo Researchers, Inc. **15.8c** WG/Science Photo Library/Photo Researchers, Inc. **15.9a** ISM/Phototake NYC **15.9b** Photo Lennart Nilsson/Albert Bonniers Forlag **15.9e** Dr. Richard Kessel & Dr. Randy Kardon/Tissues & Organs/Visuals Unlimited **15.A** Dr. E. Walker/Science Photo Library/Photo Researchers, Inc. **15.13** Martin Rotker/Phototake NYC **15.15** Lester V. Bergman/Corbis/Bettmann

Chapter 15a CO.15a Tony Freeman/PhotoEdit **15a.2a** Ian O'Leary © Dorling Kindersley **15a.2b** Charles D. Winters/Photo Researchers, Inc. **15a.2c** Felica Martinez/Photo Edit **15a.2d** Rachel Epstein/PhotoEdit **15a.2e** Tom Main/Getty Images Inc.-Stone Allstock **15a.4**(left) Peggy Greb/USDA/ARS/Agricultural Research Service **15a.4**(right) Tony Craddock/Getty Images Inc.-Stone Allstock **15a.6**(left) Susan C. Bourgoin/PictureArts Corporation

Appendix 1

METRIC-ENGLISH SYSTEM CONVERSIONS

Length

1 inch (in.) = 2.54 centimeters (cm)
1 centimeter (cm) = 0.3937 inch (in.)
1 foot (ft) = 0.3048 meter (m)
1 meter (m) = 3.2808 feet (ft) = 1.0936 yard (yd)
1 mile (mi) = 1.6904 kilometer
1 kilometer (km) = 0.6214 mile (mi)

Area

1 square inch (in.^{2}) = 6.45 square centimeters (cm^2)
1 square centimeter (cm^2) = 0.155 square inch (in.^{2})
1 square foot (ft^2) = 0.0929 square meter (m^2)
1 square meter (m^2) = 10.7639 square feet (ft^2) =
 1.1960 yards (yd^2)
1 square mile (mi^2) = 2.5900 square kilometers (km^2)
1 acre (a) = 0.4047 hectare (ha)
1 hectare (ha) = 2.4710 acres (a) = 10,000 square meters (m^2)

Volume

1 cubic inch (in.^{3}) = 16.39 cubic centimeters (cm^3 or cc)
1 cubic centimeter (cm^3 or cc) = 0.06 cubic inch (in.^{3})
1 cubic foot (ft^3) = 0.028 cubic meter (m^3)
1 cubic meter (m^3) = 35.30 cubic feet (ft^3) =
 1.3079 cubic yards (yd^3)
1 fluid ounce (oz) = 29.6 milliliters (mL) = 0.03 liter (L)
1 milliliter (mL) = 0.03 fluid ounce (oz) =
 1/4 teaspoon (approximate)
1 pint (pt) = 473 milliliters (mL) = 0.47 liter (L)
1 quart (qt) = 946 milliliters (mL) = 0.9463 liter (L)
1 gallon (gal) = 3.79 liters (L)
1 liter (L) = 1.0567 quarts (qt) = 0.26 gallon (gal)

Mass

1 ounce (oz) = 28.3496 grams (g)
1 gram (g) = 0.03527 ounce (oz)
1 pound (lb) = 0.4536 kilogram (kg)
1 kilogram = 2.2046 pounds (lb)
1 ton (tn), U.S. = 0.91 metric ton (t or tonne)
1 metric ton = 1.10 tons (tn), U.S.

Metric Prefixes

Prefix	Abbreviation	Meaning
giga-	G	$10^9 = 1,000,000,000$
mega-	M	$10^6 = 1,000,000$
kilo-	k	$10^3 = 1,000$
hecto-	h	$10^2 = 100$
deka-	da	$10^1 = 10$
		$10^0 = 1$
deci-	d	$10^{-1} = 0.1$
centi-	c	$10^{-2} = 0.01$
milli-	m	$10^{-3} = 0.001$
micro-	μ	$10^{-6} = 0.000001$

Appendix 2

SELECTED ANSWERS TO THINKING ABOUT THE CONCEPTS QUESTIONS

Chapter 1: 10. c, 11. homeostasis, 12. hypothesis, 13. adaptive

Chapter 2: 18. d, 19. b, 20. a, 21. b, 22. a, 23. d, 24. denaturation, 25. nucleotides, 26. primary, 27. Phospholipids, 28. starch; glycogen, 29. ATP, 30. pH, 31. Covalent, 32. Dehydration synthesis. Hydrolysis

Chapter 3: 12. c, 13. a, 14. d, 15. c, 16. b, 17. a, 18. a, 19. c, 20. Cytoplasm, 21. Pinocytosis, 22. nucleolus, 23. Oxygen, 24. lactic acid fermentation

Chapter 4: 15. b, 16. a, 17. d, 18. calcium, 19. homeostasis

Chapter 5: 10. b, 11. c, 12. axial, 13. calcium, 14. osteocytes, 15. sprain

Chapter 6: 12. c, 13. a, 14. actin, myosin, 15. sarcomere

Chapter 7: 14. d, 15. a, 16. c, 17. b, 18. myelin sheath, 19. sodium-potassium pump, 20. dopamine

Chapter 8: 17. a, 18. a, 19. d, 20. hypothalamus, 21. medulla, 22. sympathetic nervous system, 23. cerebrum

Chapter 9: 21. a, 22. b, 23. d, 24. a, 25. b, 26. cones, 27. lens, 28. cochlea, 29. sudden movement of the head, including acceleration and deceleration

Chapter 10: 17. a, 18. d, 19. c, 20. c, 21. a, 22. Lipid-soluble; water soluble, 23. giantism; acromegaly, 24. calcitonin; parathormone, 25. adrenal cortex, 26. insulin; glucagon

Chapter 11: 18. a, 19. c, 20. b, 21. transport oxygen, 22. negative, positive, 23. hemoglobin, 24. fibrin

Chapter 12: 21. a, 22. b, 23. b, 24. b, 25. sinoatrial node, 26. an aneurysm, 27. capillaries

Chapter 13: 18. a, 19. d, 20. b, 21. a, 22. c, 23. natural killer cell, 24. histamine, 25. plasma cells, 26. macrophages, 27. autoimmune

Chapter 14: 15. a, 16. b, 17. b, 18. a, 19. a, 20. epiglottis, 21. carbonic anhydrase

Chapter 15: 12. a, 13. b, 14. d, 15. d, 16. chyme, 17. liver, 18. starch, 19. cholecystokinin

Chapter 16: 11. a, 12. c, 13. b, 14. d, 15. b, 16. Nephrons, 17. Hemodialysis, 18. internal; smooth; external; skeletal, 19. Renin, 20. ureters

Chapter 17: 23. c, 24. b, 25. a, 26. b, 27. c, 28. epididymis, 29. oviduct

Chapter 18: 15. d, 16. c, 17. b, 18. a, 19. b, 20. d, 21. inner cell mass; trophoblast, 22. cleavage, 23. chorion; endometrium, 24. Gastrulation, 25. foramen ovale, 26. prolactin; oxytocin, 27. Free radicals

Chapter 19: 10. b, 11. a, 12. a, 13. crossing over, independent assortment, 14. Homologous, 15. Synapsis, 16. anaphase II

Chapter 20: 11. b, 12. a, 13. polygenic, 14. phenotype, genotype, 15. homozygous, hererozygous, 16. linked

Chapter 21: 16. a, 17. c, 18. b, 19. b, 20. tRNA, 21. uracil, 22. restriction enzyme

Chapter 22: 18. a, 19. d, 20. c, 21. a, 22. c, 23. d, 24. c, 25. a, 26. a, 27. Microevolution; macroevolution, 28. Natural selection, 29. Mosaic evolution, 30. multiregional

Chapter 23: 14. a, 15. c, 16. b, 17. niche, 18. primary consumer, carnivore

Chapter 24: 16. b, 17. d, 18. d, 19. exponential, 20. carrying capacity

HINTS FOR APPLYING THE CONCEPTS QUESTIONS

Chapter 1
1. Design an experiment with a control group and an experimental group.
2. Consider the list of questions on page 12.

Chapter 2
1. What functions do triglycerides perform in your body?
2. Cellulose is an important form of dietary fiber.
3. Consider the following qualities of water: polarity, heat of vaporization, and high heat capacity.

Chapter 3
1. Consider how the plasma membrane functions in cell-cell recognition.
2. What gas is released during the first step of alcohol fermentation?
3. Mitochondria process energy for cells, and thus they occur in large numbers in cells with a high demand for energy.

Chapter 4
1. What are the functions of skin?
2. When you exercise, the body's need for oxygen increases.

Chapter 5
1. What kind of activity builds bone density?
2. Why would a pregnant woman need more calcium than she did before she was pregnant?

Chapter 6
1. What role does acetylcholine play in muscle contraction?
2. What role do calcium ions play in muscle contraction?
3. Are fast twitch or slow twitch fibers darker in color? How do the properties of fast twitch fibers differ from those of slow twitch fibers?

Chapter 7
1. What effect does an inhibitory neurotransmitter have on the postsynaptic neuron?
2. What would happen if a neurotransmitter were not removed from the synapse?
3. What factors cause ions to cross the membrane during an action potential? What role do potassium ions play in the action potential?
4. What role does the myelin sheath play in the conduction of an action potential?

Chapter 8
1. Where in the brain is the language center of most people located? Which side of the brain controls movement on the right side of the body?
2. What are the functions of the spinal cord? Would loss of any of the functions due to injury have varying effects depending on the location of injury?

Chapter 9
1. In what type of vision are distant objects seen more clearly than nearby objects?
2. What part of the ear is responsible for equilibrium?
3. How would loud sounds affect the organ of hearing in the inner ear?

Chapter 10
1. Cortisone is a glucocorticoid.
2. Matt's age can be used to make an educated guess about his type of diabetes.
3. Theresa's symptoms and the timing of their onset suggest melatonin may be involved.

Chapter 11
1. Are the white blood cells that are produced in leukemia functional?
2. What regulates the circulating number of red blood cells?
3. What is the function of red blood cells? How is iron related to the ability of red blood cells to carry out their function?
4. Which antibodies against antigens on red blood cells does Raul have?
5. What could happen if either of Elizabeth's babies had Rh+ blood?
6. What are the functions of platelets?

Chapter 12
1. What would be the result if a blood clot were lodged in a small blood vessel in the heart?
2. What happens in the lymph nodes when you get sick?
3. The pressure that propels blood through the arteries is equal to the pressure against the arterial walls. What force generates this pressure? What is the relationship between high blood pressure and atherosclerosis?

Chapter 13
1. Is a vaccine effective immediately?
2. How specific is the immune response?
3. What role do helper T cells play in the immune response?

Chapter 14
1. What factors regulate breathing rate and tidal volume?
2. What is the function of the cilia in the respiratory tubules?
3. What happens to Juan's blood level of carbon dioxide when he holds his breath? How would this affect his breathing?

Chapter 15
1. What role does the pancreas play in digestion?
2. What is the relationship between diarrhea and water reabsorption?
3. What would cause skin to develop a yellow tone? What might the cause of the yellow skin tone have to do with a tattoo?
4. What substance is released into the small intestine when fatty food enters the small intestine?

Chapter 16
1. Consider the role of the external urethral sphincter in urination.
2. Consider what might have been introduced to Miguel's urinary tract when the catheter was inserted.
3. How does alcohol affect antidiuretic hormone?
4. Consider the functions of the kidneys and the ways to replace these functions.

Chapter 17
1. What are the possible consequences of pelvic inflammatory disease?
2. When making your recommendation for a means of contraception consider its effectiveness in preventing pregnancy, this couple's need for protection against STDs, and the health effects of the means of contraception.

Chapter 18
1. What might pelvic inflammatory disease have done to Tanya's oviduct?
2. How does fetal circulation differ from circulation after birth?
3. Consider the concept of critical periods in development.
4. Consider the hormones involved in milk production and ejection.

Chapter 19
1. What is nondisjunction?
2. Look at Figure 19.3.

Chapter 20
1. Colorblindness is a sex-linked trait.
2. What are the genotypes of George and Sue?
3. Recessive alleles are not expressed.

Chapter 21
1. What is the start codon? What is the stop codon? What amino acids do the other codons code for?
2. Translate each of the mRNA strands. Remember that more than one codon can code for the same amino acid.

3. Are any bands present in a father, but absent in both the mother and the child?

Chapter 22
1. Describe the evidence in support of evolution.
2. Consider chemical evolution.
3. What characteristics distinguish primates from other mammals?
4. Tay-Sachs disease is caused by a recessive allele.

Chapter 23
1. On average, what percentage of the energy available at one trophic level is available to the next level?
2. What happens to mercury as it moves up the food chain?

Chapter 24
1. What will happen to the size of each population as the prereproductive individuals reach reproductive age?
2a. What factors regulate population size?
2b. How might Asian carp affect factors regulating the population size of other species of fish?

Index

Prothrombin, 225f, 226
Proto-oncogenes, 503
Protozoan, 393, 266, 413
    disease and, 284, 287–288
Proximal convoluted tubule, 359, 360f, 362t, 361t
Pseudopodia, 57, 58f
Psychedelic drugs, 162
Psychoactive drugs, 153–163
Puberty, 384, 388
Pulmonary artery, 237f, 239
Pulmonary circuit, 238–239
Pulmonary vein, 237f, 239
Pulp, 325
Pulse, 233
Punnett square, 462–463
Pupil, 168–169
Purkinje fibers, 241
Pyelonephritis, 373
Pyramid of biomass, 538
Pyramid of energy, 538
Pyrogens, 260
Pyruvic acid, muscle contraction and, 112–113

Quaternary structure of protein, 33, 34f

Radiation, cancer and, 507
Radiation therapy, 500, 504, 509
    leukemia and, 221
Radioisotope, 17
Rain forest, 4,
Rapid-eye movement (REM) sleep, 147
Ras gene, 503
Receptor
    in plasma membrane, 43, 189–191
    sensory, 143
Receptor potential, 165, 172
Recessive, 461, 462t, 464
Recombinant DNA technology, hemophilia and, 226
Recombinant DNA, 485–488
Recombination, 450f
Recruitment, of motor units, 110–111
Rectum, 335
Red blood cells, 73, 216t, 218–219
    gas transport and, 301–303
    malaria and, 288
    types, 221–224
Red bone marrow, 90, 91f, 215, 218
Reduction, 59
Reduction division, 448
Reemerging disease, 288–289
Referred pain, 167
Reflex, defecation, 335–336
Reflex arc, 142–143, 144f
Refractory period, 125
Rejection of transplanted tissue, 268, 271, 272
Release factors, 481–482
Releasing hormone, 192
Remodeling of bone, 93, 94
Renal corpuscle, 359, 360f, 361t
Renal cortex, 358, 359f
Renal failure, 367

Renal medulla, 358, 359f
Renal pelvis, 358, 359f
Renal tubule, 359, 360f
Renin, 366
Replication of DNA, 476–477
Replication of viruses, 282–283
Repolarization, 123, 124f
Reproductive cloning, 442–43
Residual volume, 300
Resistance exercise, 113–115
Respiration, 293, 294
Respiratory system, 293–310
Resting neuron, 122, 124f
Resting potential, 122, 124
Restriction enzyme, 485
Reticular activating system, 142
Reticular fibers, 70
Retina, 168t, 169
Retinal, 172
Retinoblastoma, 503
Retrovirus, 410, 492
Reverse transcriptase, 410
Reverse transcriptase inhibitors, 414
Reverse transcription, 410
Reye's syndrome, 307
Rh factor, 222–224
Rheumatic fever, 272, 308
Rheumatoid arthritis, 102, 273f
Rhodopsin 172
Ribonucleic acid, *see* RNA
Ribosomal RNA (rRNA), 478, 480
Ribosome, 49, 480–482
Ribs, 98
Rickets, 90, 91f
Ringworm, 285
RNA (ribonucleic acid), 35, 36f, 477–482
RNA interference, 495
RNA polymerase, 478
RNA synthesis, 478, 479f
RNA transcript, 478
Rod, 168t, 169, 172, 173f
Root canal, 325
Rosy periwinkle, 11
Rough endoplasmic reticulum (RER), 50, 53t
Round window, 176t, 177, 178f
rRNA, *see* ribosomal RNA
Rubella, 179
Ruffini corpuscle, 167

Saccule, 176t, 181–183
Saliva, 184, 326, 336
Salivary glands, 323f, 324f, 326
Salmonella, 281
Salt, 245
Salt, dietary, 349
Saltatory conduction, 121–122
Sarcoma, 501t
Sarcomere, 105–110
Sarcoplasmic reticulm, 109–110
SARS, 288, 308
Saturated fats, 342, 344, 508
Schistosomiasis, 285
Schwann cell, 121–122

Sciatic nerve, 96–97
Sciatica, 97
SCID (Severe combined immunodeficiency disease), 493–494
Scientific method, 6
Sclera, 167, 168t
Scoliosis, 95
Scotopsin, 172
Scrotum, 379–380
Seasonal affective disorder (SAD), 207
Sebum, 85
Second messenger, 191
Secondary immune response, 270
Secondary structure of protein, 31, 34f
Secondhand smoke, 319
Secretin, 337
Sedatives, 162–163
Selectively permeable, 43
Semen, 381–382
Semicircular canals, 176t, 180–181, 182f
Semiconservative replication, 477
Semilunar valve, 237f, 238
Seminal vesicle, 380f, 381–382
Seminiferous tubules, 380, 484
Sensitization, 264
Sensory adaptation, 165
Sensory neuron, 119, 143, 184
Sensory receptors, 165–167, 169,172–174
Serosa, 322, 324f
Serotonin, 128, 129, 142, 162, 393
Serous membrane, 77
Severe acute respiratory syndrome, *see* SARS
Severe combined immunodeficiency disease (SCID), 493–494
Sex chromosomes, 440
Sex–influenced genes, 469
Sexual dimorphism, 526
Sexually transmitted diseases (STDs), 286, 396, 403–414
Shedding of virus, 283
Shingles, 413
Shivering, 86
Short-term memory, 142
Sickle-cell anemia, 31, 220, 465–466, 483
Sickle-cell trait, 463–464
Sigmoidoscopy, 336
Simple carbohydrate, 344, 345
Simple diffusion, 44, 47t, 331
Simple goiter, 198–199
Sinoatrial node, 240
Sinuses, 294f, 295–296
Sinusitis, 296
Skeletal muscle, 73, 105–115
Skeletal system, 90–102
Skeleton, 90, 95–99
Skin, 77–83, 256–257
    cancer, 81, 464, 507, 558
    color, 467–468
Skull, 94, 97, 134
Sleep, 147–148
Sleeping sickness, 284
Sliding filament model, 106–109
Slipped disk, 96–97